# Handbook of Experimental Pharmacology

## *Volume 126*

Springer
*Berlin*
*Heidelberg*
*New York*
*Barcelona*
*Budapest*
*Hong Kong*
*London*
*Milan*
*Paris*
*Santa Clara*
*Singapore*
*Tokyo*

# Platelets and Their Factors

Contributors

K.S. Authi, Y. Banno, E.M. Bevers, A.S. Brown, E. Butt
F. Catella-Lawson, J.-P. Cazenave, J.D. Chang, K.J. Clemetson
P.G. de Groot, Y. Denizot, T.F. Deuel, J.D. Erusalimsky
G.A. FitzGerald, M.M. Frojmovic, M.H. Fukami, C. Gachet
P.V. Halushka, D. Handley, R. Heller, M. Hirata, H. Holmsen
D. Koesling, J.F. Martin, M.L. Martin, E.A. Martinson
L.M. McManus, H. Miszta-Lane, E. Morgenstern, S. Nakashima
S. Narumiya, Y. Nozawa, B. Nürnberg, H. Patscheke, S. Pawate
R.N. Pinckard, J. Polgár, P. Presek, M.W. Radomski
M. Rodrigues, G. Rudnick, A. Ruf, E. Salas, B. Schwippert
W. Siess, H. Sinzinger, J.J. Sixma, D. Tschoepe, F. Ushikubi
G.H. van Zanten, F. von Bruchhausen, U. Walter, J.A. Ware
S.P. Watson, S.T. Weintraub, D.S. Woodard

Editors

F. von Bruchhausen and U. Walter

Springer

Professor Dr. FRANZ VON BRUCHHAUSEN
Institut für Pharmakologie
Freie Universität Berlin
Thielallee 69–73
14195 Berlin
Germany

Professor Dr. med. ULRICH WALTER
Medinzinische Universitätsklinik
Institut für Klinische Biochemie
und Pathobiochemie
Josef-Schneider-Str. 2
97080 Würzburg
Germany

With 89 Figures and 39 Tables

ISBN-13:978-3-642-64488-7 e-ISBN-13:978-3-642-60639-7
DOI:10.1007/978-3-642-60639-7

Library of Congress Cataloging-in-Publication Data. Platelets and their factors/contributors, Authi, K.S. . . . [et al.]: editors, F. von Bruchhausen and U. Walter. p. cm. — (Handbook of experimental pharmacology; v. 126) Includes bibliographical references and index. 1. Blood platelets. 2. Blood platelets — Receptors. 3. Blood platelet disorders. 4. Platelet activating factor. 5. Cellular signal transduction. I. Authi, Kalwant S. II. Bruchhausen, F. von (Franz von), 1929– . III. Walter, U. (Ulrich), 1949– . IV. Series. [DNLM: 1. Blood Platelets. W1 HA51L v. 126 1997/WH 300 P7179 1997] QP905.H3 vol. 126 [QP97] 615′.1 s — dc21 [612.1′17] DNLM/DLC 96-39635



Softcover reprint of the hardcover 1st edition 1997



Cover design: Design & Production GmbH, Heidelberg

Typesetting: Best-set Typesetter Ltd., Hong Kong

SPIN: 10501668 27/3136/SPS – 5 4 3 2 1 0 – Printed on acid-free paper

# Preface

Platelets play a fundamental, life-saving role in hemostasis and blood clotting at sites of vascular injury. Unwanted platelet activation and arterial thombus formation are, however, implicated in the onset of myocardial infarction, stroke, and other cardiovascular diseases. Acceptance that platelets play a major role in the pathogenesis of atherosclerosis including coronary heart disease has revolutionized the pharmacological treatment of cardiovascular diseases, and aspirin is now an essential antiplatelet drug and the golden standard for future developments. Yet the search for better and perhaps safer antiplatelet drugs is one of the most active areas of investigation in both basic and clinical research.

Platelets, especially human platelets, have also emerged as one of the major models for the study of inter- and intracellular signal transduction pathways. Many biochemists, cell biologists, pharmacologists, pathologists, hematologists, and cardiologists find platelets useful for studying processes such as adhesion, inside-out and outside-in signalling through the plasma membrane, channels, calcium homeostasis, protein kinases, the network of intracellular signal transduction cascades, and the release of vasoactive substances.

The aim of the editors has been to compile chapters summarizing the current state-of-the-art information on the biochemistry, cell biology, pharmacology, and physiologic and pathophysiologic roles of human platelets. We hope that this volume represents the major aspects of current platelet research although it is perhaps inevitable that certain areas are covered less thoroughly than others. We would like to acknowledge the excellent help and support of the Springer-Verlag staff, in particular that of Ms. Doris Walker. Finally, we would like to thank our colleagues, the authors of this volume, for their expert contributions and scholarship. Hopefully, this volume will become a valuable source of information for all students and investigators of platelets and their fascinating physiological and pathophysiologial roles.

January 1997

F. VON BRUCHHAUSEN
U. WALTER

# List of Contributors

AUTHI, K.S., Platelet Section, Thrombosis Research Institute, Emanuel Kaye Building, Manresa Road, Chelsea, London SW3 6LR, Great Britain

BANNO, Y., Department of Biochemistry, Gifu University School of Medicine, Tsukasamachi-40, Gifu 500, Japan

BEVERS, E.M., Cardiovascular Research Institute Maastricht, Research School University of Limburg, P.O. Box 616, 6200 MD Maastricht, The Netherlands

BROWN, A.S., Department of Cardiology, King's College Hospital/Denmark Hill, South London SE5 9RS, Great Britain

BUTT, E., Medizinische Universitätsklinik, Institut für Klinische Biochemie und Pathobiochemie, Josef-Schneider-Str. 2, 97080 Würzburg, Germany

CATELLA-LAWSON, F., University of Pennsylvania Health System, School of Medicine, Center for Experimental Therapeutics, Stellar-Chance Laboratories, 422 Curie Boulevard, Philadelphia, PA 19104-6100, USA

CAZENAVE, J.-P., Etablissement de Transfusion Sanguine de Strasbourg, 10 rue Spielmann, B.P. No. 36, 67065 Strasbourg Cedex, France

CHANG, J.D., Harvard-Thorndike Laboratories of the Cardiovascular Division, Beth Israel Hospital, Harvard Medical School, 330 Brookline Avenue, Boston, MA 02215, USA

CLEMETSON, K.J., Theodor Kocher Institute, University of Berne, Freiestr. 1, 3012 Berne, Switzerland

DE GROOT, P.G., Department of Haematology, University Hospital Utrecht, P.O. Box 85500, 3508 GA Utrecht, The Netherlands

DENIZOT, Y., Laboratoire d'Hématologie Expérimentale, Faculté de Médecine, 2 rue Dr. Marcland, 87025 Limoges, France

DEUEL, T.F., Department of Internal Medicine, Jewish Hospital at Washington University, 216 So. Kingshighway Blvd., St. Louis, MO 63110-1092, USA

Erusalimsky, J.D., The Cruciform Project, University College London, 140 Tottenham Court Road, London W1P 9LN, Great Britain

FitzGerald, G.A., University of Pennsylvania Health System, School of Medicine, Center for Experimental Therapeutics, Stellar-Chance Laboratories, 422 Curie Boulevard, Philadelphia, PA 19104-6100, USA

Frojmovic, M.M., Cell Adhesion Laboratory, Physiology Department, McIntyre Medical Building, Montreal, Quebec, Canada

Fukami, M.H., Department of Biochemistry and Molecular Biology, University of Bergen, Årstadveien 19, 5009 Bergen, Norway

Gachet, C., Etablissement de Transfusion Sanguine de Strasbourg, 10 rue Spielmann, B.P. No. 36, 67065 Strasbourg Cedex, France

Halushka, P.V., Department of Cell and Molecular Pharmacology, Division of Clinical Pharmacology, Medical University of South Carolina, 171 Ashley Ave., Charleston, SC 29425, USA

Handley, D., Pharmaceuticals, Sepracor, Inc., 111 Locke Drive, Marlborough, MA 01752, USA

Heller, R., Klinikum der Friedrich-Schiller-Universität Jena, Zentrum für Vaskuläre Biologie und Medizin, Bereich Erfurt, Institut für Biochemie und Molekularbiologie, Nordhäuser Str. 78, 99089 Erfurt, Germany

Hirata, M., Department of Pharmacology, Kyoto University, Faculty of Medicine, Sakyo-ku, Kyoto 606, Japan

Holmsen, H., Department of Biochemistry and Molecular Biology, University of Bergen, Årstadveien 19, 5009 Bergen, Norway

Koesling, D., Institut für Pharmakologie, Freie Universität Berlin, Thielallee 69–73, 14195 Berlin, Germany

Martin, J.F., The Cruciform Project, University College London, 140 Tottenham Court Road, London W1P 9LN, Great Britain

Martin, M.L., Department of Pharmacology, Division of Clinical Pharmacology, Medical University of South Carolina, 171 Ashley Ave., Charleston, SC 29425, USA

Martinson, E.A., Rudolf-Buchheim-Institut für Pharmakologie, Justus-Liebig-Universität Giessen, Frankfurter Str. 107, 35392 Giessen, Germany

McManus, L.M., Department of Pathology, The University of Texas Health Science Center, 7703 Floyd Curl Drive, San Antonio, TX 78284-7750, USA

Miszta-Lane, H., Departments of Pharmacology and Obstetrics, University of Alberta, 950 Medical Research Building, Edmonton, Alberta, Canada T6G 2H7

MORGENSTERN, E., Fachrichtung 3.5: Medizinische Biologie, Fachbereich Theoretische Medizin, Universität des Saarlandes, 66421 Homburg/Saar, Germany

NAKASHIMA, S., Department of Biochemistry, Gifu University School of Medicine, Tsukasamachi-40, Gifu 500, Japan

NARUMIYA, S., Department of Pharmacology, Kyoto University, Faculty of Medicine, Sakyo-ku, Kyoto 606, Japan

NOZAWA, Y., Department of Biochemistry, Gifu University School of Medicine, Tsukasamachi-40, Gifu 500, Japan

NÜRNBERG, B., Institut für Pharmakologie, Freie Universität Berlin, Thielallee 69–73, 14195 Berlin, Germany

PATSCHEKE, H., Institute for Medical Laboratory Diagnostics, Klinikum Karlsruhe, Academic Hospital of the University of Freiburg, P.O. Box 6280, 76042 Karlsruhe, Germany

PAWATE, S., Department of Pharmacology, Division of Clinical Pharmacology, Medical University of South Carolina, 171 Ashley Ave., Charleston, SC 29425, USA

PINCKARD, R.N., Department of Pathology, The University of Texas Health Science Center, 7703 Floyd Curl Drive, San Antonio, TX 78284-7750, USA

POLGÁR, J., Theoder Kocher Institute, University of Berne, Freiestr. 1, 3012 Berne, Switzerland

PRESEK, P., Rudolf-Buchheim-Institut für Pharmakologie, Justus-Liebig-Universität Giessen, Frankfurter Str. 107, 35392 Giessen, Germany

RADOMSKI, M.W., Departments of Pharmacology and Obstetrics, University of Alberta, 950 Medical Research Building, Edmonton, Alberta, Canada T6G 2H7

RODRIGUES, M., Department of Nuclear Medicine, Vienna University Hospital, Währinger Gürtel 18–20, 1090 Vienna, Austria

RUDNICK, G., Department of Pharmacology, Yale University of Medicine, 333 Cedar Street, New Haven, CT 06510-8066, USA

RUF, A., Institute for Medical Laboratory Diagnostics, Klinikum Karlsruhe, Academic Hospital of the University of Freiburg, P.O. Box 6280, 76042 Karlsruhe, Germany

SALAS, E., Departments of Pharmacology and Obstetrics, University of Alberta, 950 Medical Research Building, Edmonton, Alberta, Canada T6G 2H7

SCHWIPPERT, B., Cellular Haemostasis and Clinical Angiology Group, Diabetes Research Institute at the Heinrich Heine University, Auf'm Hennekamp 65, 40225 Düsseldorf, Germany

SIESS, W., Institut für Prophylaxe und Epidemiologie der Kreislaufkrankheiten, Universität München, Pettenkoferstr. 9, 80336 München, Germany

SINZINGER, H., Department of Nuclear Medicine, Vienna University Hospital, Währinger Gürtel 18–20, 1090 Vienna, Austria

SIXMA, J.J., Department of Haematology, University Hospital Utrecht, P.O. Box 85500, 3508 GA Utrecht, The Netherlands

TSCHOEPE, D., Cellular Haemostasis and Clinical Angiology Group, Diabetes Research Institute at the Heinrich Heine University, Auf'm Hennekamp 65, 40225 Düsseldorf, Germany

USHIKUBI, F., Department of Pharmacology, Kyoto University, Faculty of Medicine, Sakyo-ku, Kyoto 606, Japan

VAN ZANTEN, G.H., Department of Haematology, University Hospital Utrecht, P.O. Box 85500, 3508 GA Utrecht, The Netherlands

VON BRUCHHAUSEN, F., Institut für Pharmakologie, Freie Universität Berlin, Thielallee 69/73, 14195 Berlin, Germany

WALTER, U., Medizinische Universitätsklinik, Institut für Klinische Biochemie und Pathobiochemie, Josef-Schneider-Str. 2, 97080 Würzburg, Germany

WARE, J.A., Harvard-Thorndike Laboratories of the Cardiovascular Division, Beth Israel Hospital, Harvard Medical School, 330 Brookline Avenue, Boston, MA 02215, USA

WATSON, S.P., Department of Pharmacology, University of Oxford, Mansfield Road, Oxford OX1 3QT, Great Britain

WEINTRAUB, S.T., Department of Biochemistry, The University of Texas Health Science Center, 7703 Floyd Curl Drive, San Antonio, TX 78284-7760, USA

WOODARD, D.S., Department of Pathology, The University of Texas Health Science Center, 7703 Floyd Curl Drive, San Antonio, TX 78284-7750, USA

# Contents

## Section I: Platelet Development, Morphology and Physiology

CHAPTER 4

**Platelet Aggregation**



**Section II: Platelet Biochemistry, Signal Transduction**

CHAPTER 5

**Platelet Receptors: The Thrombin Receptor**



CHAPTER 6

**Platelet ADP/Purinergic Receptors**

CHAPTER 15

**$Ca^{2+}$ Homeostasis in Human Platelets**

CHAPTER 16

**Regulation of Platelet Function by Nitric Oxide and Other Nitrogen- and Oxygen-Derived Reactive Species**



CHAPTER 17

**Platelet 5-Hydroxytryptamine Transporters**

**Section III: Platelet-Derived Factors**

CHAPTER 18

**Dense Granule Factors**



CHAPTER 19

**Protein Granule Factors**

CHAPTER 20

**Lysosomal Storage**
M.H. FUKAMI and H. HOLMSEN ... 447



CHAPTER 21

**Thromboxane $A_2$ and Other Eicosanoids**
P.V. HALUSHKA, S. PAWATE, and M.L. MARTIN ... 459

## Section IV: Clinical Aspects of Platelets and Their Factors

CHAPTER 25

**Platelets and the Vascular System: Atherosclerosis, Thrombosis, Myocardial Infarction**



CHAPTER 26

**Platelets, Vessel Wall, and the Coagulation System**

CHAPTER 28

**Pathological Aspects of Platelet-Activating Factor (PAF)**
F. VON BRUCHHAUSEN . . . . 645

# Section I
# Platelet Development, Morphology and Physiology

CHAPTER 1

# Megakaryocytopoiesis: The Megakaryocyte/ Platelet Haemostatic Axis

A.S. BROWN, J.D. ERUSALIMSKY, and J.F. MARTIN[1]

## A. Introduction

Megakaryocytes as the precursor of platelets are essential for the maintenance of haemostasis. Platelets are heterogeneous for size and reactivity and large platelets are more haemostatically active than small platelets (THOMPSON et al. 1982). This platelet heterogeneity arises at thrombopoiesis (MARTIN et al. 1983). Several studies suggest that platelets are involved in experimentally produced atherosclerosis (FUSTER et al. 1978; HARKER et al. 1976), and many large scale clinical trials provide convincing evidence for the role of platelets in acute ischaemic syndromes (LEWIS et al. 1983; ISIS-2 1988). Furthermore, the presence of more reactive platelets (i.e. patients with a raised mean platelet volume, MPV) 6 months after a first myocardial infarction (MI) is a powerful predictor of either death or a further ischaemic event (MARTIN et al. 1991). Platelet organelles and enzyme systems are created in the megakaryocyte and changes in platelet volume and reactivity are correlated with and preceded by changes in the megakaryocyte. Therefore, perturbations of megakaryocyte physiology may have important pathological implications for the development of atherosclerosis and thrombosis, which remain a major cause of morbidity and mortality in the Western world. Furthermore, abnormalities of megakaryocyte number and function can lead to haemorrhagic disease.

Megakaryocyte physiology remains incompletely understood partly because of their fragility and partly because of the difficulty in obtaining adequate numbers for analysis; they make up less than 0.05% of all bone marrow cells. Despite this, some understanding of the biology of this unique cell has been achieved, although many fundamental questions remain unanswered. Continuing research is leading to the development of novel therapeutic interventions for altering platelet production and may eventually enable us to manipulate the physiology of platelets such that we can modify the course of vascular disease.

[1] ASB is a British Heart Foundation Junior Research Fellow, JFM is a British Heart Foundation Professor of Cardiovascular Science.

## B. Megakaryocyte Anatomy

### I. Structure

Mature megakaryocytes are large cells, having a mean diameter in humans of approximately 40 $\mu$m. They were first discovered in the bone marrow in 1884. However, it was Howell in 1891 who described these giant cells as megakaryocytes. It was not until 15 years later that Wright (1906) made the all important observation that these cells were the precursors of platelets.

Megakaryocytes develop in the bone marrow from stem cells which eventually give rise to committed megakaryocytic precursors (Fig. 1). In the mature megakaryocyte the nucleus is multilobed. It contains a variable amount of DNA (polyploidy). As with other cells, the nucleus is surrounded by a double nuclear membrane. A varying number of nucleoli are distributed at random throughout the nucleoplasm. The cytoplasm contains organelles which are typical of most other cells. There are ribosomes and mitochondria, as well as rough endoplasmic reticulum. Numerous granules are also found within the cytosol. These are the $\alpha$ granules, the electron-dense granules and lysosomes. The $\alpha$ granules contain fibrinogen, $\beta$-thromboglobulin, thrombospondin, platelet factor 4 (PF4), von Willebrand factor and platelet-derived growth factor (PDGF). The dense granules contain adenosine diphosphate (ADP), calcium and serotonin. A further class of granules distinct from both the $\alpha$ granules and dense bodies can be distinguished by the ultrastructural cytochemical demonstration of aryl sulfatase and acid phosphatase. These lysosomal granules also contain the enzyme cathepsin D.

The demarcation membrane system (DMS) is seen throughout the cytoplasm. This is created by infoldings and convolutions of the copious amounts of plasma membrane produced as a result of endoreduplication, and which otherwise might have formed the membrane of the daughter cell if the megakaryocyte had divided. It has a typical bi-lamellar structure and may form part of the future platelet membrane. Part of the plasma membrane invaginations give rise to the surface-connected canalicular system (SCCS) providing a route for biochemicals into the deeper parts of the cell. There are also microtubules which are distributed at random within the cytoplasm. This contrasts with the orientation of the microtubules in the cytoplasmic extensions of the cell in which case they are orientated parallel to the surface, in a similar fashion to the arrangement found in mature platelets.

### II. Site of Platelet Production

Megakaryocytes are found most abundantly in the bone marrow and it has therefore been generally believed that this is the site of thrombopoiesis. However, the evidence for this is surprisingly poor, being highly subjective and circumstantial. Bone marrow megakaryocytes possess small bud-like processes and long pseudopodia which penetrate the sinusoidal membrane and

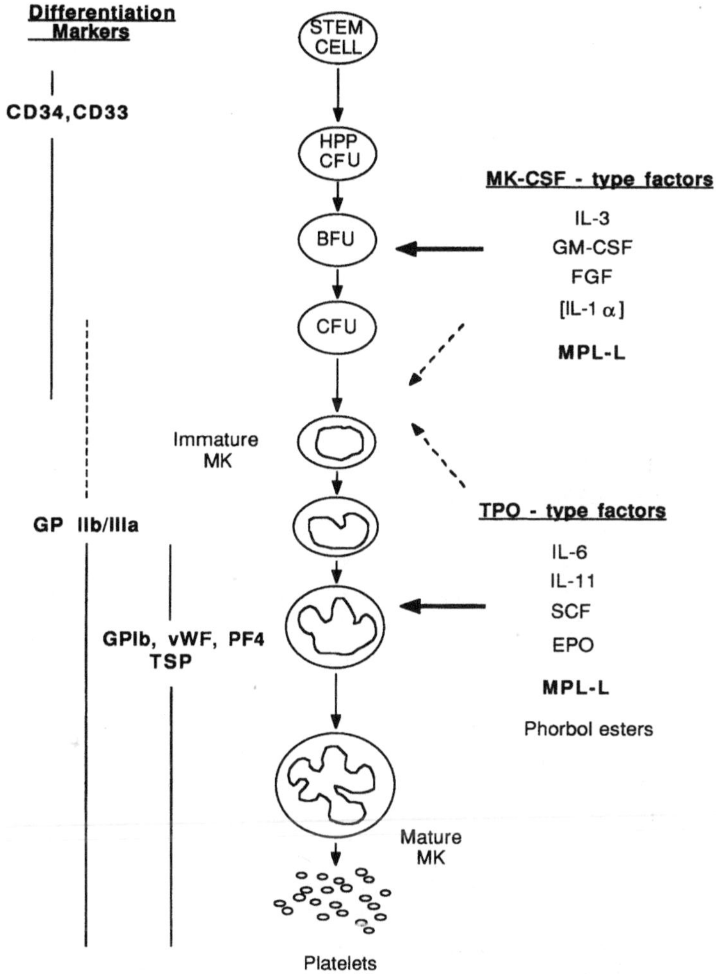


**Fig. 1.** Megakaryocytopoiesis: megakaryocytes develop from stem cells in the bone marrow. During the differentiation process, several classes of megakaryocyte progenitor cells have been functionally defined. These include the high proliferative potential colony-forming unit megakaryocyte (*HPP-CFU*), the burst-forming unit megakaryocyte (*BFU*) and the colony-forming unit megakaryocyte (*CFU*). Megakaryocyte maturation is marked by an increase in size, the development of organelles and acquisition of platelet proteins such as the glycoproteins IIb/IIIa (*GPIIb/IIIa*), GPIb, thrombospondin (*TSP*), von Willebrand factor (*vWF*) and platelet factor 4 (*PF4*). Earlier differentiation markers such as CD34 are now usefully used to isolate pure populations of megakaryocyte precursors. Megakaryocytopoiesis is controlled by a complex network of interacting cytokines including interleukin-3, -1, -6 and -11 (*IL-3*, *IL-1*, *IL-6* and *IL-11*), granulocyte–macrophage colony-stimulating factor (*GM-CSF*), fibroblast growth factor (*FGF*), stem cell factor (*SCF*), erythropoietin (*EPO*), and c-Mpl ligand (*MPL-L*)

project into the blood stream (BEHNKE et al. 1969). However, no one has shown in vivo that these processes actually break off to become platelets. The presence of "naked" megakaryocyte nuclei in the bone marrow has also been considered proof of platelet production there. It has alternatively been argued that too few naked megakaryocytes are present to be consistent with this hypothesis (SLATER et al. 1983), though there have been no data of quantification published. In addition, when thrombopoiesis is accelerated by destroying circulating platelets in animal models, large pieces of megakaryocyte cytoplasm are seen in the circulation, apparently without nuclei (SLATER et al. 1983). Therefore, the presence of naked nuclei in the marrow does not necessarily imply that platelets have been produced there.

A number of investigators have observed megakaryocytes in the venous circulation (MINOT et al. 1922; HUME et al. 1964; BRESLOW et al. 1968). These megakaryocytes pass into the right heart and thence into the pulmonary circulation. At this point, the blood vessels start to become much narrower and the megakaryocytes must either lodge in a vessel, deform and pass through it, or alternatively break up in some way. An increase in the number of circulating and pulmonary megakaryocytes has been reported in several conditions which are sometimes associated with secondary thrombopoiesis, such as inflammation, cancer and haemorrhage (AABO and HANSEN 1978; BRESLOW et al. 1968; SHARNOFF 1959; HUME et al. 1964).

Many workers have observed megakaryocytes in the lungs (SHARNOFF and SCARDINO 1960a; SCOTT et al. 1982; WELLS et al. 1984), their presence being dependent on the venous blood supply to this organ (KAUFMAN et al. 1965). The quantity of megakaryocytes found in the pulmonary vein is 12 times less than that found in the pulmonary artery and nearly all are naked nuclei (KAUFMAN et al. 1965; KALLINKOS-MANIATIS et al. 1969; PEDERSEN 1974; WARHEIT et al. 1980). These data are consistent with the hypothesis that megakaryocytes entering the lungs undergo cytoplasmic fragmentation there. The majority of the resulting naked nuclei are trapped there and become phagocytosed. A minority, however, leave the lungs, having shed all but a thin rim of cytoplasm. Those nuclei that enter the systemic circulation lodge in other organs where they are found in very low numbers. The number of intact megakaryocytes reaching the lungs is great enough to maintain thrombopoiesis, and the increase in platelet count seen across the pulmonary circulation could be attributed to pulmonary platelet production (SHARNOFF and SCARDINO 1960b). However, the hypothesis that the lungs are the site of platelet production remains disputed; furthermore, estimates of the relative contribution of this organ to the overall thrombopoietic capacity of the organism vary widely among different workers who support this theory. Estimates between 7% and 100% of the total platelet production have been calculated by various authors to occur in the lungs (TROWBRIDGE et al. 1984a; KAUFMAN et al. 1965). Thus definitive evidence in support of this hypothesis is still lacking. The development of in vitro culture systems capable of producing platelets

(CHOI et al. 1995) combined with artificial capillary culture systems may help clarify the mechanisms and site(s) of platelet production.

## III. Ploidy

Polyploidy – the result of chromosome reduplication in the absence of chromosome segregation – is a phenomenon widely observed in plants, protozoa and metazoa (for a review see WINKELMANN 1987). The term polyploidy was first introduced by the botanist STRASSBURGER (1910). In humans, *obligate* polyploid tissue is found in the liver parenchyma, heart and megakaryocyte. In addition, *facultative* polyploid cells have also been noted under certain circumstances in the epithelial cells of the seminal vesicles, in the adrenal gland, renal tubular cells, thyroid tissue and the human trophoblast. In these organs, age, stimulation of proliferation and increased metabolic function lead to polyploidisation. In humans it generally follows that those post-mitotic tissues which lose part of their ability for proliferation (partial hepatectomy, partial nephrectomy, heart muscle post MI) compensate the loss by an increase in polyploidisation. In rapidly proliferating tissue such as the skin or intestinal mucosa, polyploidy has not been observed except in tumours. Megakaryocytes are unique in that they are the only mammalian cells in which polyploidy is a sine qua non of maturity. NAGL (1990) has suggested that megakaryocyte ploidy occurs by a unique mechanism.

The development of polyploidisation and its culmination in platelet production enormously amplifies the efficiency of the mammalian haemostatic system. Thus a 32N megakaryocyte generated by four consecutive endomitotic cycles produces over 3000 platelets, whereas the same number of mitotic divisions would produce only 16 nucleated thrombocytes. Megakaryocytes normally exist in a 4N, 8N, 16N or 32N ploidy class, though occasionally cells of 64N and 128N are also seen. The normal modal ploidy of megakaryocytes in the steady state is 16N. As the DNA content increases, there is a parallel increase in nuclear size, number of organelles and megakaryocyte volume (ODELL et al. 1976; CORASH et al. 1987). Ultrastructural studies have shown that megakaryocytes from the larger ploidy classes are characterised by a greater content of intra-cytoplasmic membrane systems compared to those in the lower ploidy classes (PENINGTON et al. 1975, 1976a). As membrane phospholipids are sources of haemostatically active agents, variations in the content of membrane constituents may represent an important factor in the determination of platelet heterogeneity (ERUSALIMSKY and MARTIN 1993). Thus polyploidisation in these cells may fulfil an important haemostatic role by dictating platelet reactivity. Indeed, KRISTIENSEN et al. (1988a) have shown that the bleeding time (an indicator of in vivo platelet reactivity) is inversely related to megakaryocyte nuclear DNA content and size in humans.

Using quantitative in situ hybrididisation techniques, measurement of mRNA expression and ploidy in the same cell have shown that megakaryocytes in the higher ploidy classes contain greater amounts of mRNA for glycoprotein IIb, neuropeptide Y and $\beta$ actin (Hancock et al. 1993). As this relationship was linear, it suggests that polyploidisation does not alter gene expression in steady-state platelet production. It remains to be determined whether this may change with an increase in platelet production rate.

# C. Megakaryocytopoiesis and Thrombopoiesis

## I. Perturbations of the Steady State

The regulation of thrombopoiesis appears to be based on a humoral mechanism which senses changes in the circulating platelet mass. Acute, severe thrombocytopenia, induced by the administration of antiplatelet antibodies causes an increase in size (Odell et al. 1976; Pennington et al. 1970), ploidy (Jackson et al. 1984; Corash et al. 1987) and number and rates of maturation of megakaryocytes (Odell et al. 1969). These changes thus result in accelerated platelet production. Patients with thrombocythaemia also show a shift to higher ploidy megakaryocytes which is associated with an increase in cell size and granularity (Tomer et al. 1989). Conversely, platelet hypertransfusion leads to a decrease in ploidy, size and number of bone marrow megakaryocytes (Harker et al. 1968; Burstein et al. 1981; Odell et al. 1967). Plasma taken from animals or humans with severe thrombocytopenia stimulates thrombopoiesis in rats, and aplastic serum stimulates maturation of human megakaryocytes in vitro (Miyake et al. 1982; Stravena et al. 1987). Furthermore, growth of megakaryocyte cultures is inhibited by incubation with platelets or platelet products (Komatsu et al. 1986), suggesting the presence of a feedback control mechanism.

It has been hypothesised that the regulation of platelet volume and platelet count are under separate hormonal control as changes in these variables may occur independently of each other (Gladwin and Martin 1990a). Mean platelet volume is increased when measured several hours after the induction of immuno-thrombocytopenia, before changes in megakaryocyte ploidy have occurred (Trowbridge et al. 1986; Corash et al. 1987). Thus large platelets can be produced from megakaryocytes without preceding changes in DNA content. The increase in MPV seen in animals following platelet destruction is similar to that observed in humans following cardiopulmonary bypass (Martin et al. 1987). Since changes in platelet volume are determined at thrombopoiesis, these studies would imply that the rapid increase in the size of platelets produced following thrombocytopenia must result from a change in the fragmentation pattern of the megakaryocyte cytoplasm. When a single dose of vincristine was given to rats, thrombocytosis occurred without a preceding thrombocytopenia. This was maximal at day 5, and platelet count returned to normal by day 8. MPV was not changed from control values at any time. Megakaryocyte ploidy was increased from a control modal value of 16N–

32N by day 4 (J.F. MARTIN et al., unpublished results; HARRIS et al. 1984). These experiments indicate that, when increased platelet destruction occurs alone (soon after the induction of immunothrombocytopenia), there is an increase in the MPV, with no change in megakaryocyte DNA content. In contrast, when increased platelet production occurs alone (after administration of vincristine), megakaryocyte ploidy is increased without any alteration in the MPV. However, when platelet production and destruction are both stimulated (in chronic hypoxia and repeated administration of anti-platelet serum), both MPV and megakaryocyte ploidy are increased.

The hormonal regulation and exact mechanisms of platelet production remain incompletely understood. This is in part related to the complexity of megakaryocytopoiesis and in part to the fragility and difficulty in obtaining adequate numbers of megakaryocytes for analysis. Much of the work in attempting to unravel this biological process has been done with megakaryocyte cell cultures, with often conflicting results depending on the type of culture system used and the sort of megakaryocyte progenitor cell cultured.

## II. Megakaryocyte Progenitor Cells

Megakaryocytes develop from a pluripotent haemopoietic stem cell capable of undergoing self-renewal and differentiation into progenitor cells of various haemopoietic lineages. These progenitor cells are functionally defined using a variety of semisolid culture techniques (VAINCHENKER et al. 1979; BRIDDELL et al. 1989, 1990). Several classes of megakaryocyte progenitor cells have been identified so far, including the burst-forming unit–megakaryocyte (BFU-MK) and the colony-forming unit–megakaryocyte (CFU-MK) (HAN et al. 1991). The BFU-MK, the most primitive committed progenitor cell, has greater self-renewal capacity, whereas the CFU-MK are more mature progenitor cells having lower proliferative potential than the BFU MK. Various antigens that appear at different stages of lineage differentiation have been identified on these primitive cells (see Fig. 1). This has allowed for greater accuracy in separating these cells prior to inoculating culture systems. In addition, these markers are now routinely used to characterise megakaryoblastic leukaemias.

## III. Megakaryocyte Growth Factors

A large body of work using in vitro culture techniques has established that for their development megakaryocytes require two classes of trophic factors or activities. One, a megakaryocyte colony-stimulating factor (megakaryocyte-CSF) which affects mainly the proliferation of progenitors. The other, a terminal differentiating activity, acts on more mature cells. Conditioned media from various transformed cell lines such as the myelomonocytic leukaemia cell line as well as human plasma have been used as sources of megakaryocyte-CSF activity (HOFFMAN et al. 1981; KIMURA et al. 1984). Similarly, sources of maturation activities have included conditioned media from macrophages, bone marrow stromal cells and lung cells, as well as plasma from patients with aplastic marrow.

Megakaryocyte precursors require the presence of megakaryocyte-CSFs in order to generate megakaryocyte colonies in vitro. It is now clear that this megakaryocyte-CSF activity is accounted for by several known haemopoietic growth factors including the lymphokine interleukin-3 (IL-3) and granulocyte–macrophage CSF (GM-CSF). IL-3 appears to be more potent than GM-CSF in supporting the growth of megakaryocyte colonies. IL-3 enhances BFU-MK derived colony formation and also increases the number of cells in each CFU-MK derived colony. Indeed, antibodies to IL-3 can completely neutralise all megakaryocyte-CSF activity from mitogen-stimulated spleen cell-conditioned medium, suggesting that IL-3 may be the primary stimulus for megakaryocytopoiesis (WILLIAMS et al. 1985) in this medium. GM-CSF apparently stimulates the growth of megakaryocyte colonies by a mechanism separate from that of IL-3 because the addition of GM-CSF to cultures of murine bone marrow in the presence of optimal levels of IL-3 produces an augmentation in megakaryocyte colony growth (ROBINSON et al. 1987). The ploidy values of megakaryocytes in serum-free cultures supplemented with IL-3 and GM-CSF are mostly 2N and 4N (TERAMURA et al. 1990). In animals, infusions of IL-3 followed by GM-CSF result in an increase in the number of circulating platelets, though IL-3 or GM-CSF alone have no significant influence on platelet numbers which suggests that the effects of these two growth factors are synergistic in vivo as well as in vitro (KRUMWEIH 1988, 1989; ISHIBASHI 1990). These data exemplify the difficulty in extrapolating in vitro work to the in vivo situation and suggest that the true physiological significance of the roles of GM-CSF and IL-3 on thrombopoiesis remains to be clarified.

It would seem that IL-3 and GM-CSF primarily act as megakaryocyte-CSFs, having little effect on megakaryocyte endoreduplication. Indeed, DEBILI et al. (1991) showed that IL-3 inhibits polyploidisation when added to CFU-MK cultures supported by either normal or aplastic plasma. In parallel to its inhibition of ploidy, IL-3 decreased megakaryocyte volume. In this study, the majority of small megakaryocytes obtained with IL-3 contained ultrastructural markers of terminal differentiation such as $\alpha$ granules and demarcation membranes, suggesting that IL-3 inhibits endoreduplication without affecting other maturation parameters. Other workers have shown that megakaryocyte size is increased with IL-3; however, in this case maturing murine megakaryocytes were used rather than primitive precursors (ISHIBASHI et al. 1986). This has led to the hypothesis that IL-3 favours proliferation of megakaryocyte progenitors, possibly by inhibiting a switch from a mitotic to an endomitotic process, but may increase the number of endomitoses in maturing megakaryocyte precursors (DEBILI et al. 1991).

It has been observed that some megakaryocyte trophic activities present in long-term bone marrow cultures and in some macrophage cell lines cannot support megakaryocyte growth alone, but they increase or potentiate the number of detectable megakaryocyte colonies in the presences of megakaryocyte-CSFs. IL-6 and erythropoietin (Epo) appear to have megakaryocyte-potentiating activity though there are conflicting data on their efficacy, particularly with respect to Epo. Initially, partially purified Epo preparations were found to have megakaryocyte-CSF activity but, with the use of increasingly pure preparations, this was not found to be the case (Kolike et al. 1986). Some investigators have found that Epo can interact with other factors to promote megakaryocyte colony formation and can cause an increase in megakaryocyte size and acetylcholinesterase activity in liquid cultures (Ishibashi et al. 1987). Whether these in vitro effects of Epo have any physiological importance is unclear. In this respect, it is noteworthy that the administration of recombinant human Epo to patients with renal failure at doses sufficient to raise the haematocrit do not elevate peripheral platelet or white cell counts. Furthermore, hypoxic mice with raised levels of endogenous Epo do not demonstrate increased platelet production (Evatt et al. 1976).

Interleukin-6 is a pleotrophic cytokine known to act on various cell types (Wong 1988), including megakaryocytes (Asano et al. 1990; Burstein et al. 1992). Several studies have shown that IL-6 has no effect alone but when given with IL-3 leads to an increase in megakaryocyte number (Debili et al. 1993) and ploidy (Teramura and Mizoguchi 1990). In vivo studies in both mice and primates have shown that IL-6 increases platelet number and increases the size and ploidy of bone marrow megakaryocytes (Asano et al. 1990; Ishibashi et al. 1989). The summation of this work has led to its use in clinical trials to assess the efficacy of IL-6 in treating thrombocytopenic patients. In humans, subcutaneous injections of IL-6 led to a significant increase in platelet count but also to an increase in fibrinogen and C-reactive protein levels. There was an increase in the frequency of higher-ploidy megakaryocytes but no significant increase in the number of assayable megakaryocyte progenitor cells. Patients complained of fevers and severe fatigue limiting the use of IL-6 at the doses given in these trials (Weber et al. 1993; Gordon et al. 1995). Further studies using other haemopoietic growth factors and combining lower doses of IL-6 with these agents are under way.

More recently, a host of other non-lineage specific haemopoietic growth factors have been shown to have effects on megakaryocytopoiesis. These include stem cell factor (SCF) or c-kit ligand, IL-11, basic fibroblast growth factor (bFGF), IL-1 and the leukaemia inhibitory factor (LIF). SCF appears to have similar effects to IL-6, being synergistic with IL-3 in stimulating megakaryocyte proliferation. Thus, in combination with IL-3, it increased the absolute number of polyploid cells and induced more cytoplasmic maturation as compared with IL-3 alone (Debili et al. 1993). Furthermore, IL-6 and SCF together induced more complete cytoplasmic maturation and development of demarcation membranes (Debili et al. 1993). IL-11 has similar effects to IL-6, potentiating the effects of IL-3 in promoting both murine (Paul et al. 1990) and human megakaryocyte colony formation (Bruno et al. 1991). In addition, it shifted the megakaryocyte ploidy distribution towards higher values when added in combination to IL-3. This effect was not suppressed by anti-IL-6 antibody (Teramura et al. 1992), suggesting that, like IL-6, it acts directly as a megakaryocyte potentiator.

Basic fibroblast growth factor (bFGF), a multifunctional growth factor produced by bone marrow stromal cells, is known to be a potent modulator of haematopoiesis. Recently, the expression of bFGF by platelets and megakaryocytes has been demonstrated by immunofluorescent staining and enzyme-linked immunoabsorbent assays (Brunner et al. 1993). Bikfalvi et al. (1992) also noted the presence of mRNA for the bFGF receptor in a human leukaemic cell line that expresses megakaryocyte properties, in megakaryocytes themselves and in platelets. This led Han et al. (1992) and Bruno et al. (1993) to explore the hypothesis that bFGF may play a part in regulating megakaryocytopoiesis. They found that the effect of bFGF depended on the cell population assayed. It had no effect on CFU-MK or BFU-MK but exhibited megakaryocyte-CSF activity equivalent to IL-3 when incubated with low-density bone marrow cells. This activity was inhibited with the addition of neutralising antibodies to IL-3 and GM-CSF. The addition of bFGF and IL-3 to cultures containing CFU-MK or BFU-MK increased the size of both colonies compared with IL-3 alone. In addition, bFGF augmented secretion of the cytokines tumour necrosis factor $\alpha$ (TNF$\alpha$) and IL-6 by human megakaryocytes (Avraham et al. 1994). Furthermore, adhesion of human megakaryocytes to bone marrow stromal fibroblasts was enhanced in the presence of bFGF which resulted in significantly increased proliferation of megakaryocytes. These results suggest that the effect of bFGF on various megakaryocyte populations is different and that it may affect megakaryocytopoiesis via modulation of megakaryocyte stromal interactions and via augmentation of cytokine secretion from megakaryocytes.

The macrophage-derived cytokine IL-1$\alpha$ plays a part in the regulation of in vitro haematopoiesis by either stimulating the production of growth factors or interacting with cytokines such as IL-6, IL-3 or GM-CSF. In acute myeloid leukaemia (AML), it induces GM-CSF and TNF production which then stimulates the proliferation of AML blast cells (Dewel et al. 1989). In megakaryocyte progenitor cells, IL-1 induced DNA synthesis which could be inhibited by neutralising anti-TNF and anti-GM-CSF antibodies. Indeed, concentrations of TNF and GM-CSF in the culture medium of normal bone marrow cells after addition of IL-1 are increased (Bot et al. 1990). Other workers have shown that IL-1 alone is incapable of stimulating either CFU-MK or BFU-MK derived colony formation in vitro (Briddell et al. 1989), but it is capable of enhancing the ability of IL-3 to promote BFU-MK-derived colony formation (Briddell et al. 1990). When injected into mice, it suppressed mature erythroid progenitors but increased the number of spleen and marrow immature erythroid, macrophage, granulocyte and megakaryocyte (CFU-MK) progenitor cells (Johnson et al. 1989). IL-1$\alpha$ in combination with IL-3 has also been shown to promote long-term megakaryocytopoiesis in culture systems of up to 14 weeks (Briddell et al. 1992). Multiple animal models and clinical trials have been used to investigate the effects of IL-1, confirming its ability to promote platelet production (CARTI et al. 1992). However, as with other cytokines, its toxicity has been an impediment to its application and will probably limit its clinical utility.

Yet another substance, LIF, has been shown to have effects on megakaryocytopoiesis. When recombinant LIF was injected into mice, they developed a significant increase in the number of megakaryocyte progenitors and mature megakaryocytes followed by elevated platelet counts (METCALF et al. 1990). This finding led to the demonstration that both mature and immature megakaryocytes have receptors for LIF on their surface. When cultured with LIF alone, there was no effect on the survival or proliferation of murine megakaryocytes. However, it did enhance all types of megakaryocyte colony formation and resulted in increased numbers of megakaryocytes when incubated in vitro with IL-3 (METCALF et al. 1991). This culture system used non-purified bone marrow cells, and one possibility is that LIF interacts with or stimulates the production of some other megakaryocyte-stimulating factor. Again, the relevance of these factors in promoting megakaryocyte proliferation in vivo remains to be elucidated.

A further demonstration of the complex network of interactions among factors regulating megakaryocytopoiesis has been elegantly shown by WICKENHAUSER et al. (1995). By means of the reversed haemolytic plaque assay, which is an extremely sensitive method of determining cytokine secretion at a single cell level, the secretory activity of normal human megakaryocytes was studied. Moreover, these investigators also used the reverse transcriptase-polymerase chain reaction to study the expression of IL-6 and IL-6 receptor mRNA in concentrated megakaryocyte preparations. They found that both rhIL-3 and IL-11 exerted a marked effect on IL-6 secretion and that rhIL-3 induced IL-6 expression. Additionally, after stimulation with rhIL-3, a significant enhancement of the secretion of IL-3 and GM-CSF, but not IL-1$\alpha$, could be observed. It is also well recognised that bone marrow stromal cells secrete a variety of cytokines. This, together with the previous data, provides further evidence of the interdependence of the factors modulating megakaryocyte proliferation and maturation and suggests the existence of both autocrine and paracrine mechanisms influencing these processes.

## IV. Thrombopoietin/cMpl Ligand

In the 1950s Epo was discovered to be the regulator of red cell production. This led to the hypothesis that the number of platelets must also be controlled by a humoral regulator (YAMAMOTO 1957). This factor was named thrombopoietin (TPO) KELMEN in 1958. Following the discovery of IL-6 and the description of its effects on megakaryocytopoiesis, much consideration was given to the possibility that it might be a physiological stimulator of thrombopoiesis. However, this is unlikely as IL-6 levels do not correlate with platelet counts (Stravena et al. 1990) and are not raised in mice after the induction of thrombocytopenia (Hill 1990). It is now clear that TPO is a separate entity from IL-6, IL-3 or GM-CSF because the megakaryocyte-CSF activity of aplastic plasma is not neutralised by antibodies to any of these cytokines (Hegyi et al. 1990; Mazur et al. 1988, 1990; Yang et al. 1986). TPO was partially purified from the plasma, serum and urine of thrombocytopenic animals and patients. This semi-purified preparation was found to be free of IL-6, it had a biological activity at much lower concentrations than IL-6 and had different effects on megakaryocytopoiesis to IL-3 and GM-CSF. Unlike IL-6 and many other pleiotrophic cytokines, TPO was found to act specifically on megakaryocytes.

Despite considerable efforts, until recently unequivocal identification of TPO had proven elusive. Progress in this field came with the realisation that the orphan cytokine receptor encoded by the proto-oncogene *c-mpl* was implicated in the regulation of megakaryocytopoiesis. The expression of c-Mpl appears to be restricted to primitive stem cells, megakaryocytes and platelets (Vignon et al. 1992), indicating its down-regulation during the differentiation of all haemopoietic cells except megakaryocytes. Antisense oligonucleotides to *c-mpl* selectively inhibit megakaryocyte colony formation in vitro without affecting the growth of erythroid or granulocyte–macrophage colonies (Methia et al. 1993). Furthermore, the stimulation of human megakaryocytopoiesis by aplastic porcine plasma (APP) is inhibited by the soluble extracellular domain of c-Mpl, which identifies c-Mpl ligand as the source of the thrombopoietic activity present in APP (Wendling et al. 1994). Following this observation, c-Mpl ligand was purified from APP and from dog plasma (de Sauvage et al. 1994; Bartley et al. 1994) and the resulting amino acid sequence used to isolate human c-Mpl ligand-complementary DNA. Concurrently, two other groups isolated cDNA clones encoding murine c-Mpl ligand by expression cloning using c-DNA libraries prepared from c-Mpl ligand-dependent cell lines (Kaushansky et al. 1994; Lok et al. 1994). All megakaryocyte-CSF activity and platelet-elevating activity in thrombocytopenic plasma can be removed by recombinant c-Mpl, indicating that both biological properties can be ascribed to c-Mpl ligand (Wendling et al. 1994; de Sauvage et al. 1994; Lok et al. 1994; Kaushansky et al. 1994). Furthermore, *c-mpl*-deficient mice have been found to have an 85% reduction in the number of platelets and megakaryocytes but to have normal amounts of other haemopoietic cell types. These mice were also found to have a compensatory increase in the levels of circulating TPO (Gurney et al. 1994). Broudy et al. (1995) have shown that recombinant TPO stimulates proliferation of CFU-MK as effectively as IL-3, and that Epo, SCF and IL-11, three cytokines that display minimal megakaryocyte colony-stimulating activity alone, synergise with TPO to promote CFU-MK growth. Taken together, these results show that c-Mpl ligand specifically regulates megakaryocytopoiesis, acting on both the mitotic and endomitotic pathways and in thrombopoiesis, and that this ligand is the elusive TPO. The liver may be an important site of TPO production and the mature human protein appears to have 50% homology with Epo. This structural homology may explain why Epo has some effect on megakaryocytopoiesis.

Several studies have shown that the injection of aplastic plasma and partially purified TPO into animals leads to an increased incorporation of isotopes into the platelets, an increase in platelet number and a small increase in MPV (McDonald et al. 1980, 1987, 1990; Levin et al. 1979, 1982). As yet, the effect of the recombinant c-Mpl on platelet volume has not been evaluated.

Much of the work done to date indicates that TPO is required for the control of platelet production in both animals and humans. Although it effects both the proliferation and maturation of megakaryocytes, TPO appears to be primarily a differentiation factor. Its most impressive feature is its ability to increase platelet counts to previously unobtainable levels and more quickly than other stimulatory cytokines such as IL-6. Furthermore, from a therapeutic point of view, its advantage over the use of other cytokines may reside in its apparent specificity thus limiting many of the unwanted side effects which occur with other multi-functional regulators.

## V. Negative Regulation of Megakaryocytopoiesis

Circumstantial evidence has suggested the existence of a negative feedback control of megakaryocytopoiesis. Murine megakaryocytes produce acetylcholinesterase, and the addition of acetylcholinesterase to megakaryocyte cultures inhibits megakaryocyte progenitor cells (Ebbe et al. 1979). Conversely, the injection of neostigmine (an inhibitor of acetylcholinesterase) into mice enhanced the percentage of small acetylcholinesterase-positive cells (McDonald et al. 1985). Furthermore, other studies have shown that plasma is superior to serum as a supplement for megakaryocyte culture, suggesting platelets contain inhibitors of megakaryocytopoiesis (Vainchenker et al. 1982). The decrease in megakaryocyte size, ploidy and number following platelet hypertransfusion suggests that either platelets suppress TPO production/activity or alternatively that they contain an inhibitor of megakaryocyte development. Several groups have reported that transforming growth factor-$\beta$ (TGF-$\beta$), a cytokine found in abundance in platelet $\alpha$ granules, might be in part responsible for this inhibitory activity. Bruno et al. (1989) found that TGF-$\beta$ was capable of partially inhibiting human megakaryocyte colony formation even in the presence of IL-3 and GM-CSF. This finding suggests that the inhibitory effect occurs at a cellular target common to the action of both GM-CSF and IL-3. Interestingly, Zauli et al. (1992) have shown that platelet lysates had an inhibitory effect on bone marrow CFU-MK. The main factor responsible for the inhibitory activity was TGF-$\beta$ as an anti-TGF-$\beta$ neutralising antibody almost completely reversed the suppressive effect of platelet lysates. TGF-$\beta$ has similar inhibitory effects on other cell types (Mitjavila et al. 1988). PF4, another megakaryocyte/platelet $\alpha$ granule product, also inhibits megakaryocyte colony formation. In contrast to TGF-$\beta$, this factor does not inhibit erythroid or granulocyte/macrophage colony formation (Gewirtz et al. 1989). $\beta$-Thromboglobulin ($\beta$-TG) and connective tissue-activating peptide are also found in platelets and have a structure similar to PF4 (NIEWIAROWSKI et al. 1980; ZUKER et al. 1989; WALZ et al. 1984). They also inhibit human megakaryocyte colony formation by acting on BFU-MK and CFU-MK, though their potency varies somewhat (HAN et al. 1987, 1990a, 1990b). In addition to platelet-derived factors, GANSER et al. (1987) have reported that interferon $\alpha$ is capable of inhibiting megakaryocyte colony formation, suggesting that the negative regulation of megakaryocytopoiesis is also controlled by accessory cell populations.

## D. Signal Transduction Events and Mechanisms of Polyploidisation

### I. Involvement of Protein Kinase C in Megakaryocyte Differentiation

To circumvent the difficulties of obtaining adequate numbers of primitive progenitor cells or mature megakaryocytes, cell lines expressing megakaryocytic features have been used to study the signal transduction mechanisms involved in megakaryocytopoiesis (HOFFMAN 1989). In these cell lines, it has been consistently shown that phorbol esters induce terminal megakaryocytic differentiation. In addition to this effect on cell lines, phorbol esters have been shown to mimic the action of TPO-like activities in human and murine bone marrow cultures. In other systems, the majority of the biological effects of phorbol esters can be attributed to the activation of protein kinase C (PKC) (NISHIZUKA 1986). In agreement with this possibility, a specific PKC inhibitor, the bisindolylmalamide GF109293X, suppresses phorbol ester-induced megakaryocytic differentiation in HEL cells (HONG et al., unpublished results). Surprisingly, this compound stimulates erythroid differentiation, suggesting that PKC may control developmental decisions in these cells and perhaps also in those haemopoietic progenitors which are common to the erythroid and megakaryocytic lineages.

### II. Induction of Tyrosine Phosphorylation by TPO

With the cloning of TPO (c-Mpl ligand), intense interest in the signal transduction pathways engaged by the c-Mpl receptor has recently arisen. Like most receptors for haematopoietic growth factors, c-Mpl is a transmembrane protein lacking intrinsic enzymatic activity in its cytoplasmic domain. Recently it has been shown that TPO induces tyrosine phosphorylation and activation of the Janus protein tyrosine kinase JAK2 in the megakaryocyte cell line MO7e (TORTOLANI et al. 1994). As TPO induces a potent proliferative response in these cells, this finding suggests that JAK2 is involved in TPO-mediated mitogenic signal transduction. Phosphorylation and activation of JAK2 have been previously implicated in Epo-mediated signal transduction and mitogenesis of erythroid cells. Thus JAK2 appears to be involved in the mitogenic effects of both Epo and TPO. It is not clear, however, whether this signal transduction mechanism is also involved in the distinct differentiation responses of these two cytokines. This is particularly intriguing in view of the fact that similar sets of transcription factors appear to be involved in the induction of erythroid and megakaryocytic specific genes (ORKIN et al. 1995).

### III. Cell Cycle Control and Polyploidisation

In most eukaryotic cells, the regular alterations of chromosome reduplication and cell division are controlled by cell cycle checkpoints that prevent progression to the next cell cycle phase unless the preceding phase has been completed (HARTWELL et al. 1989). This led to the hypothesis that cells of the megakaryocyte lineage have lost the checkpoint which in diploid cells would normally prevent the initiation of DNA synthesis when the previous mitosis had not been completed. Suppression of this checkpoint in megakaryocyte cells would allow polyploidisation to occur (ERUSALIMSKY and MARTIN 1993). To test this hypothesis, megakaryocyte-like cell lines were treated with antimicrotubule agents that inhibit spindle formation and therefore the completion of mitosis. When cultured with either colcemid or taxol, cell lines with megakaryocyte features did not proliferate but continued to synthesise DNA becoming polyploid. This effect is not seen in most human non-megakaryocyte cell lines; these cells become arrested in mitosis (KUNG et al. 1990). Thus, in human megakaryocyte cell lines, this effect appears to be inherent in the process of megakaryocyte differentiation, suggesting that the dependency of DNA replication on completion of the previous mitosis is suppressed in the megakaryocytic lineage (VAN DER LOO et al. 1993). In megakaryocyte progenitor cells, inactivation of the surveillance point which then enables the cell to become polyploid may be achieved by differential regulation of gene expression. The timing of the switch from one type of cycle to the other may determine the final ploidy value of the mature megakaryocyte. This timing could be regulated by the concentration of humoral factors that could rise or fall according to the physiological and pathological status of the organism. If loss of the dependency of S phase on mitosis takes place during the early phases of lineage commitment, the transition from a mitotic to an endomitotic cycle can be advanced or delayed to regulate the final ploidy of the megakaryocyte.

## E. Megakaryocytes in Atherosclerosis

Platelets are devoid of DNA and have a negligible capacity for protein synthesis. Although there is some evidence that proteins, like fibrinogen, may be taken up from the circulation (HANDAGAMA and BAINTON 1989), the majority of their protein content is determined by the biosynthetic output of the parental megakaryocyte (Gladwin et al. 1990; Chernoff et al. 1980; Ryo et al. 1980; Benfield-Baker et al. 1982). It is still unclear as to whether it is the degree of megakaryocyte polyploidisation or other factor(s) which dictate the reactivity of the platelet progeny. There is, however, an increasing amount of data suggesting that atherosclerosis and thrombosis are associated with changes in the megakaryocyte.

Megakaryocyte size has been found to be increased 18 days post MI, when compared with controls and in men suffering sudden cardiac death as compared with age-matched victims of road traffic accidents (Trowbridge et al. 1984b). The latter data suggest that these changes occurred prior to the terminal cardiac event. As megakaryocyte size is positively related to megakaryocyte ploidy (Penington et al. 1975), it may be hypothesised that the increase in megakaryocyte ploidy precedes the acute thrombotic event. In humans with coronary atherosclerosis, megakaryocyte ploidy is increased when compared to patients with normal coronary arteries and correlates with serum lipids. It has therefore been hypothesised that hypercholesterolaemia may induce increases in megakaryocyte ploidy and hence the production of hyperfunctioning platelets which could contribute to atherogenesis (Bath et al. 1994).

In some animal models, there are also data suggesting that experimentally produced atherosclerosis is related to changes in the megakaryocyte and platelet physiology. Some studies have shown that, in rabbits and guinea pigs fed a high-cholesterol diet, the platelet half-life is shortened (Wanless et al. 1984) and the induction of atherosclerotic lesions in the aorta is associated with an increase in megakaryocyte size and platelet count (Martin et al. 1985). However, other workers have shown that the development of early atherosclerotic lesions in animals is associated with a decrease in megakaryocyte size and DNA content (Kristensen et al. 1988b). This may reflect either an increased influx of committed stem cells or an increased efflux of large megakaryocytes in response to platelet demand or as a direct effect of cholesterol. These conflicting data illustrate the complexity of the mechanisms regulating thrombopoiesis. Patients with diabetes have a greatly increased risk of developing atherosclerosis. The cause of this is unclear, but presence of larger, more reactive platelets in these patients has been suggested to account for at least part of this risk. We have found that patients with diabetes have a shift towards higher ploidy classes which is more pronounced in those diabetics with atherosclerosis (Brown et al. 1994). Analysis of bone marrow megakaryocytes gives only a "snapshot" of a dynamic situation, but it is plausible that the increased ploidy seen in this diabetic population is related to the increased platelet mass and increased platelet reactivity found in these patients. The absolute relationship between the nature of the circulating platelet and the megakaryocyte that produces them is complex and still to be understood.

## F. Conclusion

The variability of the megakaryocyte response to a wide selection of non-megakaryocyte-specific cytokines demonstrates the complexity of megakaryocytopoiesis. This process is further complicated by cell-to-cell interactions, synergism between growth factors and the autocrine production of cytokines from bone marrow cells themselves. In addition, all these cytokines have overlapping functions acting at all stages of megakaryocyte differentiation. There is a great deal more work required before the physiological role of this network of interacting cytokines is defined and the mechanisms of platelet production are elucidated. With an increased understanding of early but specific epitopes on megakaryocyte progenitors, megakaryocyte isolation and purification techniques continue to improve. The development of flow cytometric techniques now enables the rapid detection of megakaryocytes and their precursors in large enough numbers to provide statistically valid results; furthermore, the use of automated cell sorters may mean that megakaryocytes of different ploidy classes can be separated and cultured. The aforementioned improvements in isolation and culture techniques should hasten our understanding of this elusive field.

There are further fundamental questions which have been raised (Martin 1990) but remain unanswered. Does the megakaryocyte DNA distribution have a fundamental physiological significance? How is the change in bone marrow megakaryocyte DNA content to be interpreted when it provides only a snapshot of a dynamic situation? Why is the modal ploidy usually 16N? What is the relationship between change in megakaryocyte ploidy and gene expression and is there selective control of individual protein production in megakaryocytes? Which genes are involved in the regulation of polyploidisation? Are megakaryocyte size and DNA content separately regulated? Finally, do megakaryocytes of different ploidy classes produce platelets with different reactivity? Despite much research, the understanding of the meaning of megakaryocyte ploidy is limited. The megakaryocyte's uniqueness poses original problems that will require original solutions in biology and medicine.

## References

Aabo K, Hansen KB (1978) Megakaryocyte in pulmonary blood vessels. I. Incidence at autopsy, clinicopathophysiological relations especially to disseminated intravascular coagulation. Acta Pathol Microbiol Scand Sect A 86:285–291

Asano S, Okanao A, Ozawa K et al. (1990) In vivo effects of recombinant human interleukin-6 in primates. Blood 75:1602–1605

Avraham H, Banu N, Scadden DT, Abraham J, Groopman JE (1994) Modulation of megakaryocytopoiesis by human basic fibroblast growth factor. Blood 83:2126–2132

Bartley TD, Bogenberger J, Hunt P, Li Y-S, Lu HS, Martin F, Chang MS, Samal B, Nichol JL, Swift S et al. (1994) Identification and cloning of megakaryocyte growth and development factor that is a ligand for the cytokine receptor Mpl. Cell 77:1117–1124

Bath PMW, Gladwin A-M, Carden N, Martin JF (1994) Megakaryocyte DNA content is increased in patients with coronary artery atherosclerosis. Cardiovascular Res 28:1348–1352

Behnke O (1969) An electron microscope study of the rat megakaryocyte. II. Some aspects of platelet release and microtubules. J Ultrastruct Res 26:111–129

Bentfield-Barker ME, Bainton DF (1982) Identification of primary lysosomes in human megakaryocytes and platelets. Blood 59:472–481

Bikfalvi A, Han ZC, Fuhrumann G (1992) Interaction of fibroblast growth factor (FGF) with megakaryocytopoiesis and demonstration of FGF receptor expression in megakaryocyte and megakaryocytic-like cells. Blood 80:1905–1913

Bot FJ, Schipper P, Broeders L, Delwell R, Kaushansky K, Lowenberg B (1990) Interleukin-1$\alpha$ also induces granulocyte–macrophage colony-stimulating factor in immature normal bone marrow cells. Blood 76:307–311

Breslow A, Kaufman RM, Lawsky AR (1968) The effect of surgery on the concentration of circulating megakaryocytes and platelets. Blood 32:393–401

Briddel RA, Brandt JE, Straeva JE, Srour EF, Hoffman R (1989) Characterisation of the human burst-forming unit – megakaryocyte. Blood 74:145–151

Briddell RA, Hoffman R (1990) Cytokine regulation of the human burst-forming unit – megakaryocyte. Blood 76:516–522

Briddell RA, Brandt JE, Leemhuis TB, Hoffman R (1992) Role of cytokines in sustaining long term human megakaryocytopoiesis in vitro. Blood 79:332–337

Broudy VC, Lin NL, Kaushansky K (1995) Thrombopoietin (c-MPL ligand) acts synergistically with erythropoietin, stem cell factor, and interleukin-11 to enhance murine megakaryocyte colony growth and increases megakaryocyte ploidy in vitro. Blood 85:1719–1726

Brown AS, Hong Y, de Belder A, Jewitt D, Edmonds M, Martin JF, Erusalimsky JD (1994) Diabetics with peripheral vascular disease have an increased mean platelet volume and megakaryocyte ploidy. (Abstract) Br Heart J 71:178

Brunner G, Nguyen H, Gabrilove J, Rifkin DB, Wilson EL (1993) Basic fibroblast growth factor expression in human bone marrow and peripheral blood cells. Blood 81:631–638

Bruno E, Miller M, Hoffman R (1989) Interacting cytokines regulate in vitro human megakaryocytopoiesis. Blood 73:671–677

Bruno E, Briddell RA, Cooper J, Hoffman R (1991) Effects of recombinant interleukin-11 on human megakaryocyte progenitor cells. Exp Hematol 19:378–381

Bruno E, Cooper R, Wilson EL, Gabrilove JL, Hoffman R (1993) Basic fibroblast growth factor promotes the proliferation of human megakaryocyte progenitor cells. Blood 82:430–435

Burstein SA, Adamson JW, Erb SK, Harker LA (1981) Megakaryocytopoiesis in the mouse: response to varying platelet demand. J Cell Physiol 109:333–341

Burstein SA, Downs T, Freise P, Lynam S, Anderson S, Henthorn J, Epstein RR, Sarye K (1992) Thrombocytopoiesis in normal and sublethally irradiated dogs: response to human interleukin-6. Blood 80:420–428

Carti B, Sznol M, Steis R et al. (1992) Thrombopoietic effects of IL-1 alpha in combination with high dose carobplatin. Proc Am Soc Clin Oncol 11:820a

Chernoff A, Levine RF, Goodman DS (1980) Origin of platelet derived growth factor in megakaryocytes in guinea pigs. J Clin Invest 65:926–930

Choi ES, Nichol JL, Hokom MM, Hornkohl AC, Hunt P (1995) Platelets generated in vitro from proplatelet-displaying human megakaryocytes are functional. Blood 85:402–413

Corash L, Chen HY, Levin J et al. (1987) Regulation of thrombopoiesis; effects of degree of thrombocytopenia on megakaryocyte ploidy and platelet volume. Blood 70:177–185

Debili N, Hegyi E, Navarro S, Mouthon M-A, Breton-Gourius J, Vainchenker W (1991) In vitro effects of hemopoietic growth factors on the proliferation, endoreplication and maturation of human megakaryocytes. Blood 77:2326–2338

Debili N, Masse JM, Katz A, Guichard J, Breton-Gouris J, Vainchenker W (1993) Effects of the recombinant hemopoietic growth factors interleukin-3, interleukin-6, stem cell factor, and leukaemia inhibitory factor on the megakaryocytic differentiation of CD34+ cells. Blood 82:84–95

Delwel R, Van Buitenen C, Salem M, Bot F, Gillis S, Kaushansky K, Altrock B, Lowenberg B (1989) Interleukin-1 stimulates proliferation of acute myeloblastic leukaemia cells by induction of endogenous granulocyte–macrophage colony-stimulating factor (GM-CSF) release. Blood 74:586–593

de Sauvage FJ, Hass PE, Spencer SD et al. (1994) Stimulation of megakaryopoiesis and thrombopoiesis by the c-MPL ligand. Nature 369:533–565

Ebbe S, Phallen E (1979) Does autoregulation of megakaryocytopoiesis occur? Blood Cells 5:123–138

Erusalimsky JD, Martin JF (1993) Regulation of megakaryocyte polyploidisation and its implication for coronay artery occlusion. Eur J Clin Invest 23:1–9

Evatt BL, Spivak JL, Levin J (1976) Relationship between thrombopoiesis and erythropoiesis: with studies of the effects of preparations of thrombopoietin and erythropoietin. Blood 48:547–558

Fuster V, Bowie EJW, Lewis JC, Fass DN, Owen CA, Brown AL (1978) Resistance to arteriosclerosis in pigs with von Willebrand's disease. J Clin Invest 61:722–730

Ganser A, Carlo Stella C, Greher J, Volkers B, Hoelzer D (1987) Effect of recombinant interferon alpha and gamma on human bone marrow derived megakaryocyte progenitor cells. Blood 70:1173–1179

Gewirtz AM, Calabretta B, Rucinski B, Niewiarowski S, Xu WY (1989) Inhibition of human megakaryocytopoiesis in vitro by platelet factor 4 and a synthetic COOH-terminal PF4 peptide. J Clin Invest 83:1477–1486

Gladwin A-M, Martin JF (1990a) The control of megakaryocyte ploidy and platelet production: biology and pathology. Int J Cell Cloning 8:291–298

Gladwin AM, Carrier MJ, Beesley JE, Lelchuck R, Hancock V, Martin JF (1990b) Identification of mRNA for PDGF b-chain in human megakaryocytes isolated using a novel immunomagnetic separation method. Br J Haematol 76:333–339

Gordon MS, Nemunaitis J, Hoffman R, Paquette RL, Rosenfield C, Manfreda S, Isaacs R, Nimer SD (1995) A phase 1 trial of recombinant human interleukin-6 in patients with myelodysplastic syndromes and thrombocytopenia. Blood 85:3066–3076

Gurney AL, Carver-Moore K, de Sauvage FJ, Moore MW (1994) Thrombocytopenia in c-mpl deficient mice. Science 265:1445–1447

Hagendama PJ, Bainton DF (1989) Incorporation of a circulating protein into alpha granules of megakaryocytes. Blood Cells 15:59–72

Han ZC, Bellucci S, Tenza D, Caen JP (1990a) Negative regulation of human megakaryocytopoiesis by human platelet factor 4 and beta-thromboglobulin: comparative analysis in bone marrow cultures from normal individuals and patients with essential thrombocythaemia and immune thrombocytopenic purpura. Br J Haematol 74:395–401

Han ZC, Bellucci S, Walz A, Baggiolini M, Caen JP (1990b) Negative regulation of human megakaryocytopoiesis by human platelet factor 4 (PF4)Z and connective tissue activating peptide (CATP-111). Int J Cell Cloning 8:253–259

Han ZC, Sensebe L, Abgrall JF, Briere J (1990c) Platelet factor 4 inhibits human megakaryocytopoiesis in vitro. Blood 75:1234–1239

Han ZC, Bellucci S, Caen JP (1991) Megakaryocytopoiesis: characterisation and regulation in normal and pathologic states. Int J Haematol 54:3–14

Han ZC, Bellucci S, Wan HY, Caen JP (1992) New insights into the regulation of megakaryocytopoiesis by haemopoietic and fibroblastic growth factors and transforming growth factor $\beta 1$. Br J Haematol 81:1–5

Hancock V, Martin JF, Lelchuk R (1993) The relationship between human megakaryocyte nuclear DNA content and gene expression. Br J Haematology 85:692–697
Harker LA (1968) Kinetics of thrombopoiesis. J Clin Invest 47:458–465
Harker LA, Ross R, Slicter SJ, Scott CR (1976) Homocysteine-induced arteriosclerosis – the role of endothelial cell injury and platelet response in its genesis. J Clin Invest 58:731–741
Harris RA, Penington DG (1984) The effects of low dose vincristine on megakaryocyte colony-forming cells and megakaryocyte ploidy. Br J Haematol 57:37–48
Hartwell LH, Weinert TA (1989) Checkpoints: controls that ensure the order of cell cycle events. Science (Wash DC) 246:629–634
Hegyi E, Navarro S, Debili N, Mouthon M-A, Katz A, Breton-Gorius J, Vainchenker W (1990) Regulation of human megakaryocytopoiesis: analysis of proliferation, ploidy and maturation in liquid cultures. Int J Cell Cloning 8:236–244
Hill RJ, Warren K, Levin J (1990) Does interleukin-6 mediate the thrombopoietic response to acute immune thrombocytopenia [Abstract]. Exp Hematol 18:704
Hoffman R, Mazur E, Bruno E et al. (1981) Assay of an activity in the serum of patients with disorders of thrombopoiesis that stimulates formation of megakaryocytic colonies. N Eng J Med 305:533–538
Hoffman R (1989) Regulation of megakaryocytopoiesis. Blood 74:1196–1212
Howell WH (1981) Observations upon the occurrence, structure, and function of the giant cells of the marrow. J Morphol 4:117–129
Hume R, West JT, Malmgren RA, Chu EA (1964) Quantitative observations of circulating megakaryocytes in the blood of patients with cancer. N Engl J Med 270:111–117
Ishibashi T, Burstein SA (1986) Interleukin-3 promotes the differentiation of isolated single megakaryocytes. Blood 67:1512–1514
Ishibashi T, Koziol JA, Burstein SA (1987) Human recombinant erythropoietin promotes differentiation of murine megakaryocytes in vitro. J Clin Invest 79:286–289
Ishibashi T, Kimura H, Shikama Y et al. (1989) Interleukin-6 is a potent thrombopoietic factor in vivo in mice. Blood 74:1241–1244
Ishibashi T, Kimura H, Shikama Y, Ichida T, Kariyone S, Maruyama Y (1990) Effect of recombinant granulocyte–macrophage colony-stimulating factor on murine thrombocytopoiesis in vitro and in vivo. Blood 75:1433–1438
ISIS-2 collaborative group (1988) Randomised trial of intravenous streptokinase, oral aspirin, both, or neither among 17187 cases of suspected acute myocardial infarction: ISIS-2. Lancet ii:349–360
Jackson CW, Brown LK, Somerville BC et al. (1984) Two-colour flow cytometric measurements of DNA distributions of rat megakaryocytes in unfixed, unfractionated marrow cell suspensions. Blood 63:768–778
Johnson CS, Keckler DJ, Topper MI, Braunschweiger PG, Furmanski P (1989) In vivo hemopoietic effects of recombinant interleukin 1$\alpha$ in mice: stimulation of granulocytic, monocytic, megakaryocytic, and early erythroid progenitors, suppression of late-stage erythropoiesis, and reversal of erythroid suppression with erythropoietin. Blood 73:678–683
Kallinilos-Maniatis A (1969) Megakaryocytes and platelets in central venous and arterial blood. Acta Haematol 42:330–335
Kaufman RM, Airo R, Pollack S, Crosby WH (1965) Circulating megakaryocytes and platelet release in the lung. Blood 26:720–731
Kaushansky K, Lok S, Holly RD et al. (1994) Promotion of megakaryocyte progenitor expansion and differentiation by the cMPL ligand thrombopoietin. Nature 369:568–571
Kelemen E, Cserhati I, Tanos B (1958) Demonstration and some properties of human thrombopoietin in thrombocytopenic sera. Acta Haematol 20:350–355
Kimura H, Burstein SA, Thorning D et al. (1984) Human megakaryocytic progenitors (CFU-M) assaying in methylcellulose: physical characteristics and requirements for growth. J Cell Physiol 118:87–96

Kolike K, Shimizu T, Miyake T et al. (1986) Haemopoietic colony stimulation by mouse spleen cells in serum free culture supported by purified erythropoietin and/or interleukin-3. In: Levine RF, Williams N, Levin J, Evatt BL (eds) Megakaryocyte development and function. Alan R Liss, New York, pp 33–49

Komatsu N, Suda T, Sakata Y et al. (1986) Megakaryocytopoiesis in vitro of patients with essential thrombocytopenia: effect of plasma and serum on megakaryocyte colony formation. Br J Haematol 64:241–252

Kristensen SD, Bath PMW, Martin JF (1988a) The bleeding time is inversely related to megakaryocyte nuclear DNA content and size in man. Thromb Haemost 59:357–359

Kristiensen SD, Roberts KM, Lawry J, Martin JF (1988b) Megakaryocyte and vascular changes in rabbits on short term high cholesterol diets. Atherosclerosis 71:121–130

Krumwieh D, Seiler FF (1988) Changes of haematopoiesis in cynomolgus monkeys after application of human colony stimulating factors. Ex Hematol 16:551–554

Krumwieh D, Seiler FR (1989) In vivo effects of recombinant interleukin-3 alone or in various combinations with other cytokines as GM-CSF, G-CSF and EPO in normal cynomolgus monkeys. J Cell Biochem (suppl) 132:51

Kung AL, Sherwood SW, Schimegakaryocytee RT (1990) Cell line-specific differences in the control of cell cycle progression in the absence of mitosis. Proc Natal Acad Sci USA 87:9553–9557

Levin J, Evatt BL (1979) Humoral control of thrombopoiesis. Blood Cells 5:105–121

Levin J, Levin FC, Hull DF et al. (1982) The effects of thrombopoietin on megakaryocyte-CFC, megakaryocytes, and thrombopoiesis: with studies of ploidy and platelet size. Blood 60:989–998

Lewis HD, Davis JW, Archibald DG et al. (1983) Protective effects of aspirin against acute myocardial infarction and death in men with unstable angina. Results of Veteran Administration cooperative study. N Eng J Med 309:396–403

Lok S, Kaushansky K, Holly RD et al. (1994) Cloning and expression of murine thrombopoietin cDNA and stimulation of platelet production in vivo. Nature 369:565–568

MacPherson GG (1971) Development of megakaryocytes in bone marrow of the rat: an analysis by electron microscopy and high resolution autoradiography. Proc R Soc London Ser B 177:265–274

Martin JF, Penington DG (1983) The relationship between the age and density of circulating C-labled platelets in the sub-human primate. Throm Res 30:157–164

Martin JF, Slater DN, Kishk YT, Trowbridge EA (1985) Platelet and megakaryocyte changes in cholesterol-induced experimental atherosclerosis. Arteriosclerosis 5:604–612

Martin JF, Daniels TD, Trowbridge EA (1987) Acute and chronic changes in platelet volume and count after cardiopulmonary bypass induced thrombocytopenia in man. Thromb Haemost 57:55–58

Martin JF (1990) The meaning and control of megakaryocyte polyploidy: questions needing answers. Int J Cell Cloning 8:316

Martin JF, Bath PMW, Burr ML (1991) Influence of platelet size on outcome after myocardial infarction. Lancet 338:1409–1511

Mazur EM, Cohen JL, Newton J, Gesner TG, Mufson RA (1988) Human serum megakaryocyte (megakaryocyte) colony stimulating activity (Meg-CSA) is distinct from interleukin-3 (IL-3), granulocyte macrophage colony-stimulating factor (GM-CSF) and phytohemagglutinin stimulated lymphocyte conditioned medium. Blood 76:290–297

Mazur EM, Cohen JL, Newton J et al. (1990) Human megakaryocyte colony stimulating activity appears to be distinct from interleukin-3, granulocyte–macrophage colony-stimulating factor and lymphocyte-conditioned medium. Blood 76:290–297

McDonald TP (1980) Effect of thrombopoietin on platelet size of mice. Exp Haematol 8:527–532

McDonald TP, Cottrell M, Clift R (1985) Regulation of megakaryocytopoiesis by acetylcholinesterases [Abstract]. Exp Haematol 13:437
McDonald TP (1987) Regulation of megakaryocytopoiesis by thrombopoietin. Ann NY Acad Sci 509:1–24
McDonald TP, Cotterell M, Clift R (1977) Haematological changes and thrombopoietin production in mice after X-irradiation and platelet specific antisera. Exp Hematol 5:291–298
McDonald TP, Jackson CW (1990) Thrombopoietin derived from human embryonic kidney cells stimulates an increase in DNA content of murine megakaryocytes in vivo. Exp Hematol 18:758–763
Metcalf D, Nicolaa NA, Gearing DP (1990) Effects of injected leukaemic inhibitory factor on hemopoietic and other tissues in mice. Blood 76:50–56
Metcalf D, Hilton D, Nicola NA (1991) Leukaemic inhibitory factor can potentiate murine megakaryocye production in vitro. Blood 77:2150–2153
Methia N, Louache F, Vainchenker W, Wendling F (1993) Oligodexoynucleotides antisense to the proto-oncogene c-mpl specifically inhibit in vitro megakaryocytopoiesis. Blood 82:1395–1401
Minot GR (1992) Megakaryocytes in the peripheral circulation. J Exp Med 36:1–8
Mitjavila MT, Vinci G, Villeval JL, Keieffer N et al. (1988) Human platelet alpha granules contain a nonspecific inhibition of megakaryocyte colony formation: its relationship to type $\beta$ transforming growth factor (TGF-beta). J Cell Physiol 134:93–100
Miyake T, Kawakita M, Entomoto K, Murphy MJ (1982) Partial purification and biological properties of thrombopoietin extracted from the urine of aplastic anaemia patients. Stem Cells 2:129–144
Nagl W (1990) Polyploidy in differentiation and evolution. Int J Cell Cloning 8:216–223
Niewiarowski S, Walz DA, James P, Rucinski B, Kueppers F (1980) Identification and separation of secreted platelet proteins by isoelectric focusing. Evidence that low affinity platelet factor 4 is converted to $\beta$-thromboglobulin by limited proteolysis. Blood 55:453–456
Nishizuka Y (1986) Studies and perspectives of protein kinase C. Science 233:305–312
Odell TT, Jackson CW, Reiter RS (1967) Depression of the megakaryocyte platelet system in rats by transfusion of platelets. Acta Haematol 38:34–42
Odell TT Jr, Jackson CW, Friday TJ, Charsha DE (1969) Effects of thrombocytopenia on megakaryocytopoiesis. Br J Haematology 17:91–101
Odell TT, Murphy JR, Jackson CW (1976) Stimulation of megakaryocytopoiesis by acute thrombocytopenia in rats. Blood 48:211–229
Orkin SH (1995) Transcription factors and hemopoietic development. J Bio Chem 270:4955–4958
Paul SR, Bennet F, Calvetti JA, Kelleher K et al. (1990) Molecular cloning of a cDNA encoding interleukin-11, a novel stormal cell-derived lymphopoietic and haemopoietic cytokine. Proct Natl Acad Sci USA 87:7512–7516
Pedersen TN (1974) The pulmonary vessels as a filter for circulating megakaryocytes in rats. Scand J Haematol 13:225–231
Pennington DG, Olsen TE (1970) Megakaryocytes in states of altered platelet production: cell numbers, size and DNA content. Br J Haematol 18:447–463
Pennington DG, Streatfield K (1975) Heterogeneity of megakaryocytes and platelets. Ser Haematol 8:22–48
Pennington DG, Streatfield K, Roxburgh AE (1976a) Megakaryocytes and the heterogeneity of circulating platelets. Br J Haematol 34:639–653
Pennington DG, Lee NYT, Roxburgh AE, McGready JE (1976b) Platelet density and size: the interpretation of heterogeneity. Br J Haematol 34:365–376
Robinson BE, McGrath HE, Quesenberry PJ (1987) Recombinant murine granulocyte-macrophage colony-stimulating factor has megakaryocyte colony-stimulating activity and augments megakaryocyte colony stimulation by interleukin-3. J Clin Invest 79:1648–1652

Ryo R, Proffitt RT, Roger ME, O'Bear R, Deuel TF (1980) Platelet factor 4 antigen in megakaryocytes. Thromb Res 17:645–652
Scott GDB (1982) Circulating megakaryocytes. Histopathology 6:467–475
Sharnoff JG (1959) Increased pulmonary megakaryocytes – probable role in post operative thromboembolism. J Am Med Assoc 169:688–691
Sharnoff JG, Scardino V (1960a) Pulmonary megakaryocytes in human foetuses and premature and full-term infants. Arch Pathol 69:139–141
Sharnoff JG, Scardino V (1960b) Platelet count differences in blood of the rabbit right and left heart ventricles. Nature 187:334–335
Slater DN, Trowbridge EA, Martin JF (1983) The megakaryocyte in thrombocytopenia: a light and electronmicroscope study supporting the theory that all platelets are produced in the lungs. Thromb Res 31:163–176
Sorensen PHB, Mui AL-F, Murthy SC, Krystal G (1989) Interleukin-3, GM-CSF, and TPA induce distinct phosphorylation events in an iterleukin-3 dependent multipotential cell line. Blood 73:406–418
Strassburger E (1910) Chromosomenzahl. Flora 100:398–446
Stravena JE, Yang HH, Hui SL et al. (1987) Effects of megakaryocyte colony stimulating factor on terminal cytoplasmic maturation of human megakaryocytes. Exp Hematol 15:657–663
Straneva JE, Van Besien K, Derigs G et al. (1990) Is interleukin-6 the physiological regulator of thrombopoiesis [Abstract]. Exp Hematol 18:596
Takata K, Singer SJ (1988) Localisation of high concentrations of phosphotyrosine-modified proteins in mouse megakaryocytes. Blood 71:818–821
Teramura M, Mizoguchi H (1990) The effect of cytokines on the ploidy of megakaryocytes. Int J Cell Cloning 8:245–252
Teramura M, Kobayashi S, Hoshino S, Oshimi K, Mizoguchi H (1992) Interleukin-11 enhances human megakaryocytopoiesis in vitro. Blood 79:327–331
Thompson CB, Eaton KA, Princicotta SM, Kushkin CA, Valeri CA (1982) Size dependent platelet subpopulations: relationship of platelet volume to ultrastructure, enzymatic activity and function. Br J Haematol 50:509–519
Tomer A, Friese P, Conklin R, Bales W, Archer L, Harker L, Burstein S (1989) Flow cytometric analysis of megakaryocytes from patients with abnormal platelet counts. Blood 74:594–601
Tortolani J, Johnston JA, Bacon CM, McVicar DW, Shimosaka A, Linnekin D, Longo DL, O'Shea JJ (1995) Trombopoietin induces tyrosine phosphorylation and activation of the Janus kinase, JAK2. Blood 85:3444–3451
Trowbridge EA, Martin JF, Slater DN, Kishk YT, Warren CW, Harley PJ, Woodcock B (1984a) The origin of platelet count and volume. Clin Phys Physiol Meas 5:145–170
Trowbridge EA, Slater DN, Kishk YT, Woodcock BW, Martin JF (1984b) Platelet production in myocardial infarction and sudden cardiac death. Thromb Haemostas 52:167–171
Trowbridge EA, Warren CW, Martin JF (1986) Platelet volume heterogeneity in acute thrombocytopenia. Clin Phys Physiol Meas 7:203–210
Vainchenker W, Brouguet J, Guichard J, Breton-Goris J (1979) Megakaryocyte colony formation from human bone marrow precursors Blood 54:940–945
Vainchenker W, Chapman J, Deschamps JF et al. (1982) Normal human serum contains a factor(s) capable of inhibiting megakaryocyte colony formation. Exp Haematol 10:650–660
Van Der Loo B, Hong Y, Hancock V, Martin JF, Erusalimsky JD (1993) Antimicrotubule agents induce polyploidisation of human leukaemic cell lines with megakaryocytic features. Eur J Clin Invest 23:621–629
Vignon I, Mornon JP, Cocault L, Mitjavila MT, Tamborin P, Gisselbrect S, Souryi M (1992) Molecular cloning and characterization of MPL, the human homolog of the v-mpl oncogene: identification of a member of the haemopoietic growth factor receptor super family. Proc Natl Acad Sci USA 89:5640–5644

Wanless IR (1984) The effect of dietary cholesterol on platelet survival in the rabbit – a study using $^{14}$C-seretonin and $^{51}$C-chromium double labled platelets. Thromb Haemostas 52:85–89
Warheit DB, Barnhart MI (1981) Ultrastructure of circulating and platelet forming megakaryocytes: a combined correlative SEM-TEM and SEM histochemical study. Ann NY Acad Sci 370:30–41
Weber J, Yang JC, Topalian SL, Parkinson DR, Schwartzentruber DS, Ettinghausen SE, Gunn H, Mixon A, Kim H, Cole R, Levin R, Rosenberg SA (1993) Phase-1 trial of subcutaneous interleukin-6 in patients with advanced malignancies. J Clin Oncol 11:499–506
Wells S, Sissons M, Hasleton PS (1984) Quantitation of pulmonary megakaryocytes and fibrin thrombi in patients dying from burns. Histopathology 8:517–527
Wendling F, Maraskovsky E, Debili N et al. (1994) c-MPL ligand is a humoral regulator of megakaryopoiesis. Nature 369:571–574
Wickenhauser C, Lorenzen J, Thiele J, Hillienhof A, Jungheim K, Schmitz B, Hansmann M-L, Fischer R (1995) Secretion of cytokines (interleukins-1$\alpha$, -3, and -6 and granulocyte macrophage colony-stimulating factor) by normal human bone marrow megakaryocytes. Blood 85:685–691
Williams N, Gill K, Yasmeen D, Mclece I (1985) Murine megakaryocyte colony stimulating factor: its relationship to interleukin-3. Leuk Res 9:1487–1491
Winkelmann M, Pfitzer P, Schneider W (1987) Significance of polyploidy in megakaryocytes and other cells in health and tumor disease. Klin Wochenser 65:1115–1131
Wong CG, Clark SC (1988) Multiple actions of interleukin-6 within a cytokine network. Immunol Today 9:137–138
Wright JH (1906) The origin and nature of the blood plates. Boston Med Surg J 154:643–645
Yamamoto S (1957) Mechanism of the development of thrombocytosis due to bleeding. Acta Haematol 20:163–178
Yang HH, Bruno E, Hoffman R (1990) Studies of human megakaryocytopoiesis using an anit-megakaryocytes stimulating factor and lymphocyte-conditioned medium. Blood 76:290–297
Zauli G, Visani G, Catani L, Vianelli N, Gugliotta L, Capitani S (1993) Reduced responsiveness of bone marrow megakaryocyte progenitors to platelet-derived transforming growth factor $\beta$1, produced in normal amount, in patients with essential thrombocythaemia. Br J Haematol 83:14–20
Zuker MB, Katz OR, Thorbecke JT, Milot DC, Holt J (1989) Immunoregulatory activity of peptides related to platelet factor 4. Proc Natl Acad Sci USA 86:7551–7574

CHAPTER 2

# Human Platelet Morphology/Ultrastructure

E. MORGENSTERN

## A. General Morphology: Shape and Properties of Platelets

The mean resting human platelet was specified to be $1 \times 3.1\,\mu m^2$ in dimension, 4–7.6$\mu m^3$ in volume (FROJMOVIC and MILTON 1982) and 10pg in weight (IYENGAR et al. 1979). Carefully prepared platelet-rich plasma (PRP) collected from citrated blood after venipuncture contains discocytes. The number of discocytes with small pseudopodia (echinodiscocytes) does not exceed 10%. Contact activation during centrifugation is the most abundant reason for the presence of activated platelets in PRP. The prefixation of the whole blood with low concentrations of aldehyde before centrifugation is well recommended (STOCKINGER et al. 1969) to preserve the discoid shape of platelets for ultrastructural examination. Proven washing procedures (PATSCHEKE 1981) or the preparation of gel filtered platelets – preconditions for investigations on platelets under physiological concentrations of divalent ions – preserve the resting state of platelets (Figs. 1a, 2a, 3). Additives like adenosine diphosphate (ADP) scavengers (apyrase) or inhibitors of thromboxane $A_2$ are used to prevent platelet activation.

Stimulus-induced activation of platelets leads to a shape change (Sect. D.III.1.) and the formation of echinospherocytes, i.e., spherical cells with pseudopodia (Figs. 1b, 9 and Table 4). An artificial shape change is also induced by EDTA as an anticoagulant (WHITE 1968) or by platelet preparation at low temperatures (WHITE and KRUMWIEDE 1968).

The activated platelets aggregate in the presence of fibrinogen (Fig. 9; Sect. E.I. and Chap. 4). Shape change and aggregation are very fast processes after stimulation with ADP or thrombin (BORN 1970; GEAR et al. 1987). Adhesion of platelets on collagen fibres (Sect. E.I., Table 5 and Chap. 4) induces shape change and strong stimulation (RUF et al. 1991). The platelets spread (Sect. E.I.) while adhering to plane surfaces.

The compilation in this article does not mention human platelets in diseases. Reviews concerning platelet ultrastructure may be found by WHITE (1987a), RAO (1990), and CAEN et al. (1994).

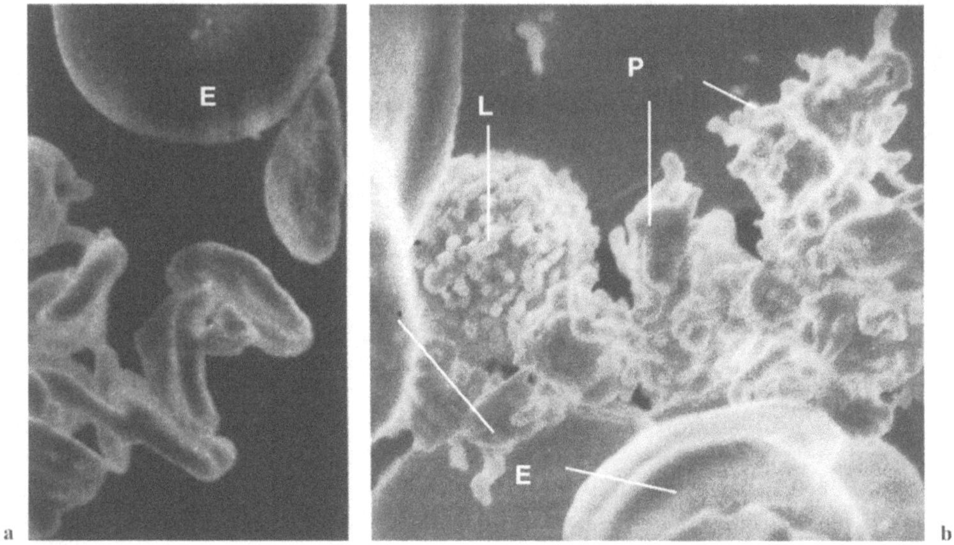


**Fig. 1a,b.** Scanning electron micrographs of discoid resting platelets (**a**) and ADP-stimulated and aggregated platelets (*P* in **b**) which represent echinospherocytes with pseudopodia. In **a** and **b** erythrocytes (*E*) and in **b** a leukocyte (*L*) are shown for comparison. ×14900

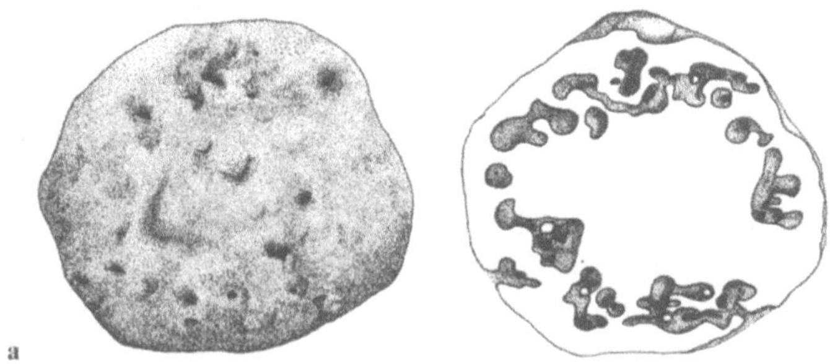

**Fig. 2. a** The surface of a chemically fixed platelet with the openings of the SCS; **b** the arrangement of the SCS within the cytoplasm is seen in this three-dimensional reconstruction. The peripheral localization of the SCS can be observed in many resting platelets. (Werner and Morgenstern 1982)

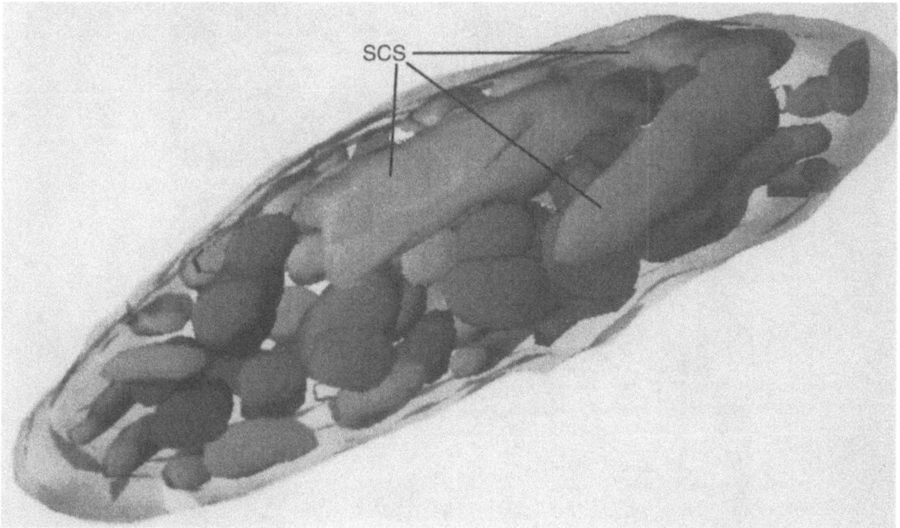


**Fig. 3.** Three-dimensional reconstruction of a cryofixed discoid resting platelet from serial sections. The very dense arrangement of platelet organelles (alpha-granules and mitochondria) and the SCS is evident

## B. Electron-Microscopic Techniques

The micrographs using the transmission electron microscope (TEM) which are demonstrated in this chapter show cryofixed platelets – impact frozen on a liquid nitrogen metal mirror – and after that cryosubstituted with acetone containing osmium tetroxide, if not otherwise mentioned.

Cryofixation (rapid freezing) with a time resolution of <1 ms (Monck and Fernandez 1992; Plattner et al. 1992) enables the direct observation of subcellular dynamics, i.e., fusion events during exocytosis (reviewed in Morgenstern and Edelmann 1989). The investigation of frozen–hydrated preparations within the cryo-TEM is presently the best (but expensive) standard (Fig. 4) (Frederik et al. 1991; Edelmann 1992; O'Toole et al. 1993). Cryofixation followed by cryo-substitution or -drying yields ultrathin sections suited for heavy metal staining, immunolabeling and analysis of water and ion distribution, and the localization of subcellular components (Steinbrecht and Mueller 1987; Morgenstern and Edelmann 1989).

The procedure that is mostly used to investigate platelets by electron microscopy includes chemical fixation followed by dehydration in an organic solvent, embedding in a resin and polymerization. Complex chemical reactions as well as shifts of ions and water take place and thus changes in volume of cellular or subcellular compartments are inevitable. Chemical fixation does not arrest rapid dynamic cell functions, e.g., membrane fusion, in less than 1 s. Detailed descriptions of artifacts caused by chemical fixation are given by

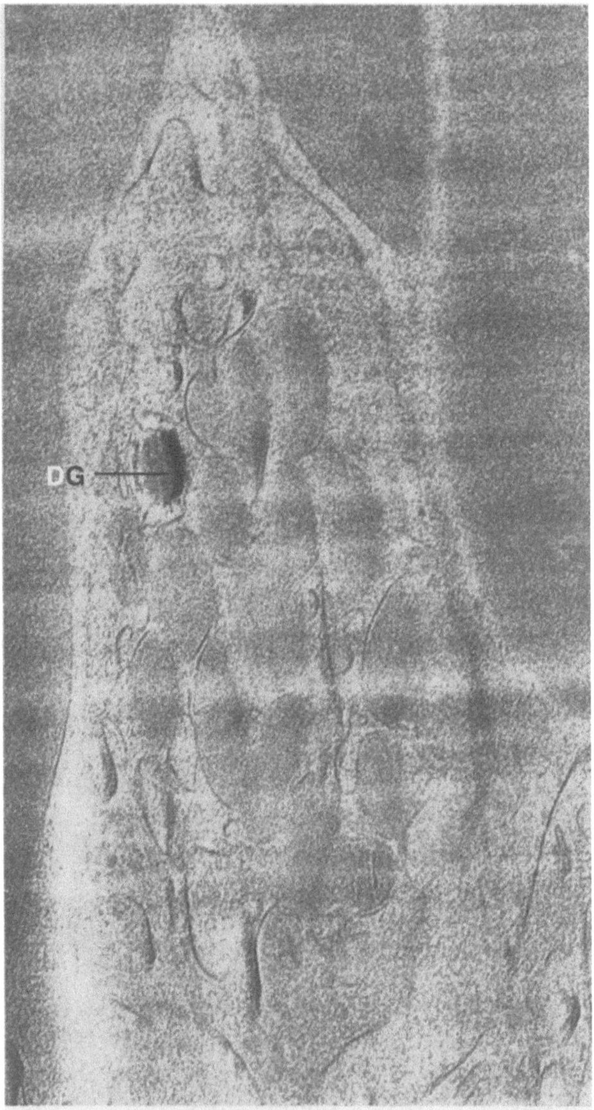


**Fig. 4.** The micrograph from the cryo-TEM shows a platelet in citrated whole blood which was rapidly frozen. The frozen-hydrated section does not need any treatment to recognize organelles and membranes. The dense granule (*DG*) with its electron-dense core is well preserved (cf. Fig. 10a–c). (courtesy of Dr. Edelmann/Homburg) ×50 000

ROBARDS and SLEYTR (1985), KNOLL et al. (1987), MORGENSTERN and EDELMANN (1989), and MORGENSTERN (1991).

The examination of replicas from freeze-fractured (and -etched) cells with the TEM allows the observation of membrane contours as well as the recognition of the inner leaflet of membranes. The distribution of membrane-associated or -integrated macromolecules can be visualized (MORGENSTERN 1980; LUPU et al. 1986). The use of cryo-techniques is also advantageous by application of the mentioned procedures (PLATTNER and KNOLL 1987).

The simple whole-mount preparation of unfixed platelets on coated grids is a valuable method of investigating platelets (WOLPERS and RUSKA 1939) with the TEM. Although the preparation puts up with an activation, it renders possible the quantitative analysis of platelet structures and components as well as the examination of the redistribution of surface bound ligands (MORGENSTERN 1980; FREDERIK et al. 1991; O'TOOLE et al. 1993; ESCOLAR and WHITE 1994).

Scanning electron microscopy (SEM) is widely used for the investigation of functional alterations of the platelet shape and of the interactions of the platelet surface with ligands (WHITE and ESCOLAR 1994; MORGENSTERN et al. 1994) as well as cytoskeletal components after detergent permeabilization (HARTWIG and DESISTO 1991).

Atomic force microscopy was recently introduced, also in the examination of platelets. This technique allows the observation of living platelets at high resolution (>50 nm) under nearly physiological conditions (FRITZ et al. 1994; EPPELL et al. 1995; RADMACHER et al. 1996).

## C. Morphometric Data

Morphometric measurements involve great expense and are therefore not numerous in platelet literature. The published data differ markedly regarding the individual studies (Table 2). The differences may be due to the special preparation of platelets and the fixation and embedding mode as well as the choice of measured section profiles. Nevertheless, the data give important information regarding the distribution of platelet membrane and organelle densities. Tables 1–3 show the values found in measurements on resting platelets in sections.

The change of discocytes to echinospherocytes (see Sect. A.) during activation is accompanied by surface enlargement (Table 4). The sizes of surface ($\mu m^2$) and volume ($\mu m^3$) is 8.02 $\mu m^2$ in resting platelets and 13.0 $\mu m^3$ in

**Table 1.** Mean membrane surface density of unstimulated platelets (calculated on the basis of morphometric measurements by FROJMOVIC and MILTON 1982)

| | Relative membrane surface area (%) | Absolute membrane surface area ($\mu m^2$/cell) |
|---|---|---|
| Plasmalemma | 30 | 18 |
| Surface connected system (SCS) | 24 | 14 |
| Alpha-granules | 24 | 14 |
| Dense granules | 3 | 1 |
| Dense tubular system (DTS) | 9 | 11 |

**Table 2.** Relative volumes of structures in unstimulated platelets

| | a (%) | b (%) | c (%) | d (%) | e (%) | f (%) | g[1] (%) |
|---|---|---|---|---|---|---|---|
| SCS | 15.8 | 6.2 | 20.1 | | 11.8 | | 13.8 |
| Alpha-granules | 15.7 | 9.7 | 13.1 | 17.1 | | 5.5–14.3 | 9.85 |
| Dense granules | 0.3 | 1.2 | | | | | |
| Mitochondria | 2.8 | 1.0 | | | | 1.5–1.7 | 1.98 |
| DTS | 1.4 | | | | | | |

[1] References: a, MORGENSTERN and STARK 1974; MORGENSTERN and KHO 1977; b, STAHL et al. 1978; c, VASKINEL and PETROV 1981; d, BRYON et al. 1978; e, WURZINGER et al. 1987; f, DUYVENE DE WIT and HEYNS 1987; g, ROGER et al. 1992.

**Table 3.** Geometric properties of organelle cross-sections (from ROGER et al. 1992)

| | Alpha-granules ($\mu m^2$) | Mitochondria ($\mu m^2$) |
|---|---|---|
| Area | 0.327 ± 0.146 | 0.319 ± 0.129 |
| Perimeter | 0.640 ± 0.161 | 0.651 ± 0.161 |
| Diameter | 0.209 ± 0.052 | 0.212 ± 0.048 |

**Table 4.** Surface enlargement after platelet shape change (FROJMOVIC and MILTON 1982)

| | Discocyte | Echinospherocyte | Spread platelet |
|---|---|---|---|
| Surface area ($\mu m^2$) | 18 | 30–40 | ~70 |

aggregates of contact-activated platelets (MORGENSTERN and KHO 1977). In the calculations of FROJMOVIC and MILTON (1982), the increase of surface area may be more than twofold after shape change and in spread platelets nearly four times the surface of a discocyte.

The morphometrically measured decrease of membranes of the SCS during shape change (MORGENSTERN and KHO 1977) corresponds with the concept of SCS evagination (Sect. D.III.1.) to provide membrane supplies for this fast process (reviewed in FROJMOVIC and MILTON 1982).

# D. Ultrastructure of Platelets

## I. Cytosol and Cytoskeleton

### 1. Cytosol

The homogenous hyaloplasm contains glycogen particles in a scattered or agglomerated distribution (indicated in Fig. 9). The glycogen content in platelets is found to be comparable with muscle cells (reviewed in MORGENSTERN 1980). Free ribosomes are present in few numbers, preferentially in young cells (MORGENSTERN 1980). This agrees with recent data that synthesis of platelet proteins is finished at the terminal stages of megakaryocyte maturation (SCHICK 1990; DORN et al. 1994).

A marginal bundle of microtubules consisting of about 20 circumferential microtubule coils exists within the discocyte (BEHNKE 1970a; NACHMIAS 1980; WHITE 1987b; XU and AFZELIUS 1988a; for review of earlier work see MORGENSTERN 1980). Conventionally fixed and embedded, the microtubules are often devoid of surrounding cytoplasm whereas they are enclosed by the cytoplasmic matrix in frozen sections or in cryofixed specimens (MORGENSTERN 1980; O'TOOLE et al. 1993). Lying under the plasmalemma (Fig. 5a), the bundle supports the discoid shape and is required for platelet resistance to deformation. Agonist-activated platelets undergo a rearrangement of their microtubules and this results in a change of cell deformability (WHITE et al. 1984; WANG et al. 1992). The microtubules, which are associated with the outer rim of the constricting contractile gel (Figs. 5b and 14a; Sect. D.I.3.), are translocated centripetally from their submembranous site to the cell center (ESCOLAR et al. 1987; BERRY et al. 1989; XU and AFZELIUS 1988b; STARK et al. 1991;

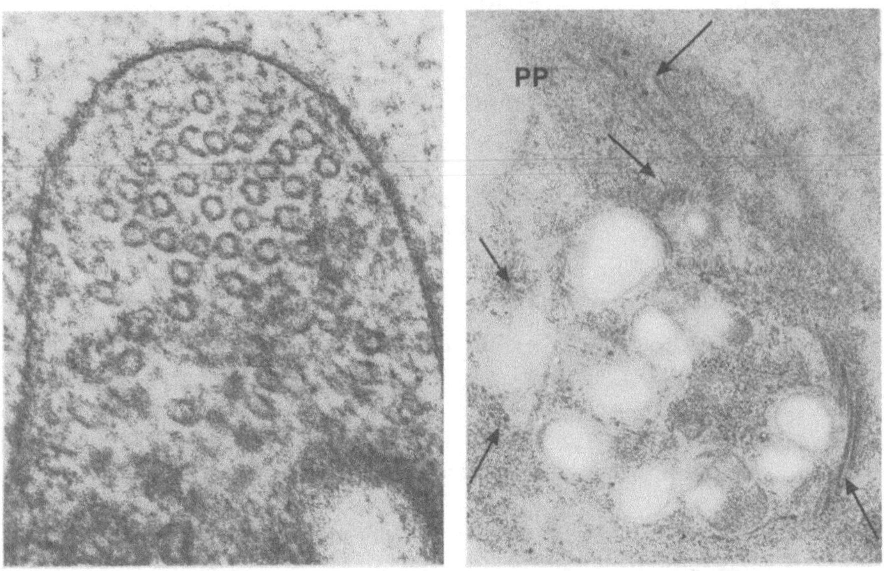


a b

**Fig. 5. a** Cross-sectioned microtubules within the marginal area of a resting platelet. The microtubules are embedded within the hyaoloplasm and sporadic small denser structures (associated proteins?) are seen. **b** Platelet 25 s after thrombin stimulation, which was cryo-substituted in acetone containing 0.25% uranyl acetate to enhance microtubule contrast, is shown. Longitudinally and cross-sectioned microtubules (*arrows*) are situated outside the contractile gel surrounding the centralized mass of organelles (unstained) and radiating into a pseudopodium (*PP*). **a** ×132 000; **b** ×33 800

DIERICHS et al. 1992). Microtubules are also found radiating into the pseudopodia (Fig. 5b).

### 2. Submembranous Cytoskeleton

This membrane cytoskeleton can be visualized with special techniques in the subplasmalemmal region of the discoid (resting) platelet (WHITE 1969; ZUCKER-FRANKLIN 1970; NACHMIAS 1980; BOYLES et al. 1985; NAKATA and HIROKAWA 1987; WHITE 1987b; HARTWIG and DESISTO 1991; HARTWIG 1992). The viscoelastic gel contains actin filaments crosslinked by actin-binding protein that binds to a spectrin-rich submembrane lamina. The actin filaments are connected to the membrane glycoprotein GPIb/IX via actin-binding protein (HARTWIG and DESISTO 1991; LUNA and HITT 1992; FOX 1993; NURDEN et al. 1994).

The submembranous cytoskeleton (SMC) (a) regulates properties of the membrane such as its contours and stability and is, amongst the marginal microtubules (Sect. D.I.1.), responsible for the shape of a resting platelet; (b) mediates the lateral distribution of membrane receptor glycoproteins (GPIb/IX complex, see also Sect. E.II.1. and Chap. 8) as reviewed by FOX (1993); (c) acts as a barrier for exocytosis (see Sect. E.III.1.); (d) its alteration may lead to shedding of so-called microparticles (Sect. E.III.2.) (FOX et al. 1990; FOX 1993), that obviously represent cytoplasmic fragments (MORGENSTERN et al. 1995).

Platelet activation induces a disintegration of the SMC by dissociation of the GPIb-actin-binding protein complex (FOX et al. 1990) or by fragmentation of actin as a result of gelsolin action (HARTWIG et al. 1989a; HARTWIG 1992; LUNA and HITT 1992). The actin-monomer-binding protein profilin that is associated with regions of the plasmalemma devoid of actin filaments reversibly increases after platelet activation (HARTWIG et al. 1989b).

### 3. Contractile Gel

A loose network of long actin filaments exists within the resting platelet (NAKATA and HIROKAWA 1987; FOX and BOYLES 1988). Its actin is connected with the SMC (HARTWIG and DESISTO 1991) or with the cell organelles (NAKATA and HIROKAWA 1987). Agonists effect the formation of filaments from actin and myosin molecules, which are present preponderantly in a non-polymerized state in the resting platelet. This contractile gel comprises the mass of cell organelles, and the constriction of the gel results in organelle centralization as shown in Figs. 9 and 14a–c (NACHMIAS 1980; TANAKA et al. 1986; NAKATA and HIROKAWA 1987; HARTWIG 1992; WHITE 1992c; KOVACSOVICS and HARTWIG 1996). Membrane glycoproteins like the fibrinogen receptor GPIIb/IIIa can associate in focal contacts via linker proteins – apparently talin (BECKERLE et al. 1989) – with the contractile gel and mediate the internalization of ligands (i.e., clot retraction, see Sect. E.II.2. and Chap. 8). Recently it was obseved that GPIb-IX-actin-binding protein complexes, linked to actin filaments, are pulled into the cell center by the contractile gel. Thus, platelets may exert contractile tension not only on ligands of GPIIb/IIIa (fibrinogen, fibrin) but also on vWF bound to GPIb-IX (KOVACSOVICS and HARTWIG 1996).

## II. Plasmalemma and Surface

The plasmalemma shows a non-random distribution of phospholipids that depend on the functional state of platelets (ZWAAL and HEMKER 1982). Agonist-induced redistribution of phospholipids and resulting changes in membrane fluidity are mediated by $Ca^{2+}$ and by an ATP-dependent aminophospholipid translocase (FEIJGE et al. 1990; SMEETS et al. 1994; STUART et al. 1995; WILLIAMSON et al. 1995). Transbilayer movement of phospholipids precedes exocytosis (HEEMSKERK et al. 1993). Cytoskeletal components (actin-binding protein, see Sect. D.I.2.) are found to be involved in movement of phosphatidylserine (COMFURIUS et al. 1985; VERHALLEN et al. 1987; FOX et al. 1990). The reported molecular rearrangements correlate well with morphological observations (FEIJGE et al. 1990). Membrane ATPases are found to be located within the plasmalemma (GONZALEZ-UTOR et al. 1992).

Membrane-integrated glycoproteins (see also Sects. D.IV.3., E.I. and II. and Chaps. 5, 7, and 8) are receptors for physiological platelet agonists (ADP, thromboxane, thrombin), adhesive proteins (fibrinogen, fibronectin, laminin, thrombospondin, vitronectin, vWF) and for fibre-type ligands like collagen (PREISSNER 1991; HYNES 1991; CALVETE 1995; CLEMETSON, 1995). The best-characterized glycoprotein is that of GPIIb/IIIa. The number of GPIIb/IIIa copies on a resting platelet is estimated at 50000 (PIDARD et al. 1983). Intracellular receptor pools from the membranes of the SCS and the alpha-granules can be additionally expressed at the cell surface following platelet activation (WOODS et al. 1986).

The heterodimeric GPIIb/IIIa complex does not bind considerable amounts of soluble fibrinogen on resting platelets. The GPIIb/IIIa binds fibrinogen after platelet activation and aggregation ensues. Similar regulation of receptor function has also been demonstrated for GPIIIa that binds fibronectin (FAULL and GINSBERG 1995). Interestingly, the number of molecules bound to the fibrinogen receptor GPIIb/IIIa on the platelet surface – estimated after ultrastructural examination – corresponds with the biochemical data (cf. MORGENSTERN et al. 1985; CALVETE 1995).

Using certain heavy metal compounds (alcian blue, ruthenium red) during fixation, the platelets show a surface coat (glycocalix) in ultrathin sections (BEHNKE 1968; MORGENSTERN 1980). This coat is caused by surface proteoglycans which are involved in the binding of ligands (collagen, vWF, fibrinogen) (BEHNKE 1987; STEINER 1987) and the interaction of platelets with leukocytes (CERLETTI et al. 1994).

## III. Surface-Connected Membranes

### 1. Surface-Connected (Open Canalicular) System

The openings of the surface-connected system (SCS) into the plasmalemma (BEHNKE 1970b; WHITE 1970) are recognizable with the SEM and at better resolution on replicas obtained after freeze-fracturing/-etching with the TEM (REDDICK and MASON 1973; VAN DEURS and BEHNKE 1980; MORGENSTERN 1980; HOLS et al. 1985). The SCS in the cytoplasm of resting platelets consists of branched channels (Fig. 2b). In human platelets, the dimension of the surface of the SCS and the plasmalemma or the membranes of alpha-granules is quite comparable (Sect. C.; Table 1). Some properties of the SCS membranes resemble those of the plasmalemma: the membrane glycoproteins GPIb, GPIb/IIIa, and others (Sect. D.IV.3.) are localized there (POLLEY et al. 1981; WENCEL-DRAKE et al. 1981; MORGENSTERN et al. 1992). The membrane–GPIIb/IIIa complex is transported from the surface via the SCS into the alpha-granules of resting platelets (Sect. E.II.1.).

During stimulus-induced shape change, the SCS of resting normal human platelets disappears (Figs. 9, 14a–c) (MORGENSTERN et al. 1987; MORGENSTERN and EDELMANN 1989; MORGENSTERN et al. 1995a). Morphometric measurements and findings of comparative studies on normal and "giant" platelets (reviewed in FROJMOWIC and MILTON 1982) or normal and "gray" platelets (MORGENSTERN et al. 1990) have already suggested that the evagination of SCS membranes causes the surface enlargement during platelet shape change (Sect. C.). The evagination of the SCS membranes for surface enlargement during spreading of adherent platelets (WHITE et al. 1990) agrees with this.

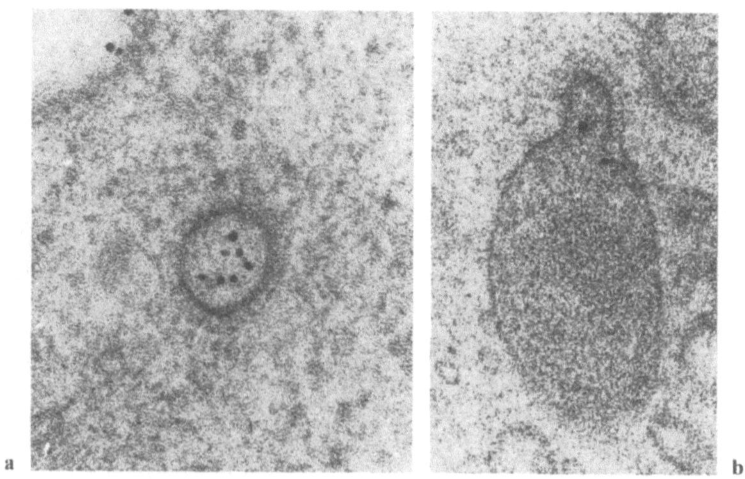

**Fig. 6a,b.** Coated membranes in platelets prepared as in Fig. 10. **a** Cross-sectioned tip of the SCS; **b** coated protrusion of an alpha-granule. (Experiment: A. Ruf; electron microscopy: E. Morgenstern), ×100000

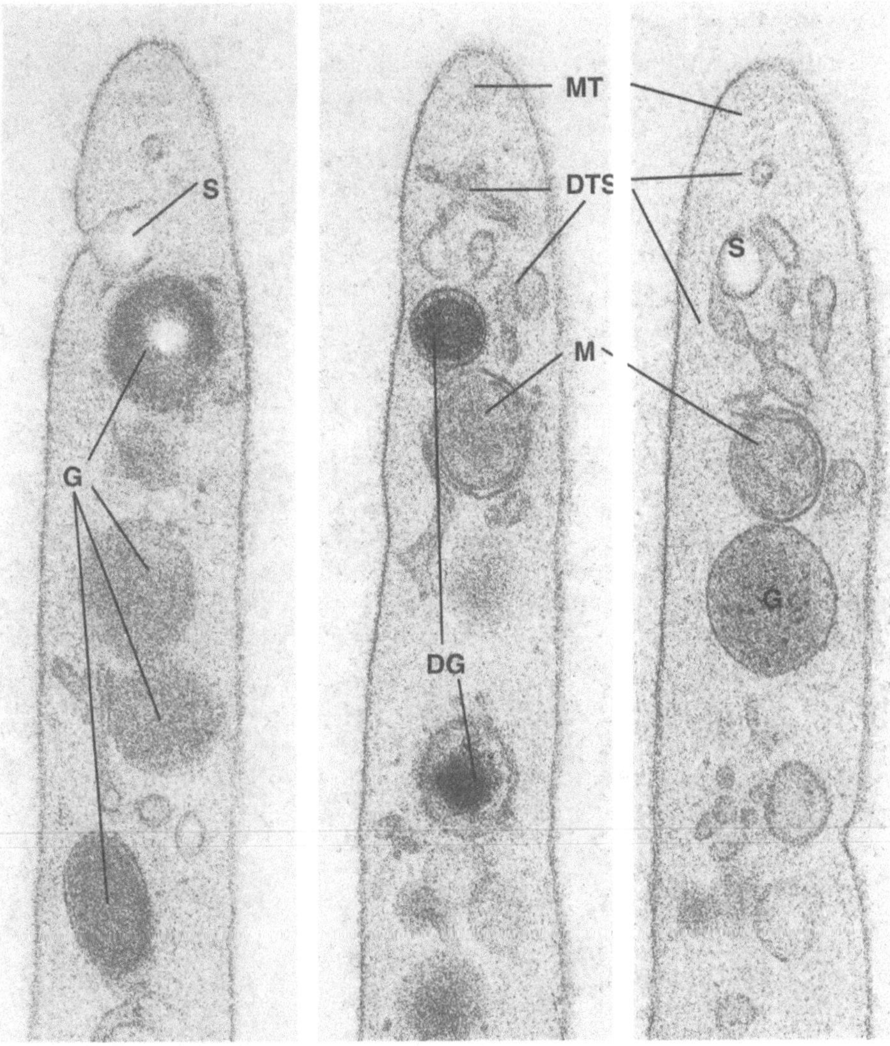


**Fig. 7a–c.** Three sections (30 nm thick) of a series from an unstimulated platelet. The section profiles and the cross-sectioned marginal bundle of the microtubules (*MT*) correspond with the discoid shape of the platelet. Alpha-granules (*G*) and dense granules (*DG* in **b**) are clearly separated from other membranes and from the plasmalemma. The dense core fills up the whole upper organelle but only the center of the organelle seen below in **b**. A mitochondrion (is seen in **b** and **c**). DTS, dense tubular system (**b** and **c**). An opening of the SCS (*S*) is seen in **a**. ×79 000

After full stimulation, a new SCS is formed by internalization of plasmalemmal areas that are connected with receptor-ligand complexes (see Sect. E.II.2.). The role of the SCS in platelet exocytosis is discussed in Sect. E.III.1.

## 2. Coated Membranes

Areas of the plasmalemma, the SCS membranes, or the membranes of alpha-granules are coated with clathrin (Fig. 6). Coated membrane areas may appear as coated pits or, arising from alpha-granules, as coated protrusions (MORGENSTERN 1982). Coated membranes are involved in the transport of cationic ferritin and gold or thorium particles (BEHNKE 1989, 1992). They may also play a role in receptor movement in resting and activated platelets (Sects. E.II.1, 2.). The presence of coated vesicles, sequestered from coated membranes, is quite probable (BEHNKE 1989; MORGENSTERN et al. 1992).

# IV. Cell Organelles

## 1. Mitochondria

Human platelets contain mitochondria (Fig. 7b,c) with a diameter of approximately 200nm (Table 3). The organelles are provided with well-developed cristae and their relative volume comes to 3% of the platelet volume (Sect. C.). This is comparable with the chondriome of smooth muscle cells or monocytes (reviewed in MORGENSTERN 1980). The important role of the mitochondrial contribution to the energy metabolism (reviewed by HOLMSEN 1990) of platelets is well documented by these morphometric data.

## 2. Dense Tubular System

The dense tubular system (DTS) is ubiquitously distributed in platelets. This endomembrane system is found to be situated in the vicinity of the

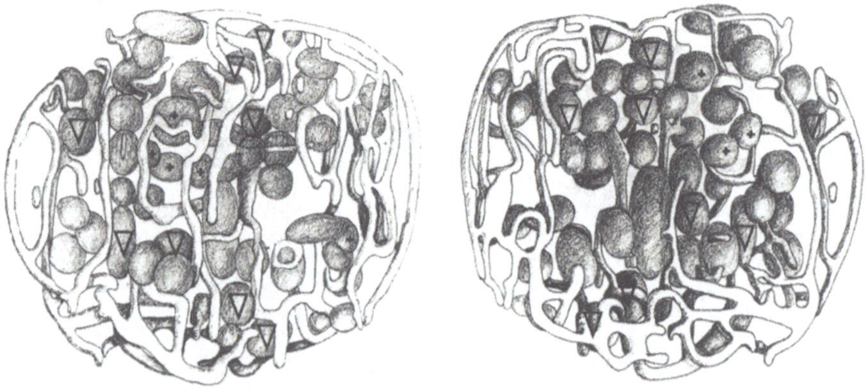


**Fig. 8a,b.** The reconstruction from sections of a resting platelet shows the arrangement of the DTS. The branching tubules are found to be located in the zone of the marginal microtubules forming a circumferential structure. On the other hand, the DTS is situated within the cytoplasm between the cell organelles – alpha-granules, dense granules (*crosses*) and mitochondria (*triangles*). (From WERNER and MORGENSTERN 1982)

microtubules as well as surrounding the organelles (Figs. 7, 8) (Werner and Morgenstern 1980). In many sections a close association of dense tubules and the SCS (Sect. D.III.1.) is observed (White 1972). The dense appearance of the DTS contents after conventional fixation is attributed to lipids, divalent cations, and proteins (Morgenstern 1980; Stenberg and Bainton 1986). Dense tubules can be stained using a diaminobenzidine medium at pH 7.3 (reviewed by Stenberg and Bainton 1986). Regarding their structure and functional properties, they are comparable with the agranular endoplasmic reticulum of eucytes (Nichols et al. 1984). The DTS regulates platelet activation by sequestering or releasing calcium, similarly to the sarcotubules of skeletal muscle. ATPases, the enzymes of the endoperoxide metabolism and adenylate cyclase activity, are localized in the DTS (Cutler et al. 1978; Gonzalez-Utor et al. 1992).

In thrombin- or $Ca^{2+}$ ionophore-activated platelets, a rapidly occurring change from the thin elongated form to a rounded vesicular form is described. Because this change is absent after addition of a protein kinase C activator, it is suggested that the phenomenon is secondary to the increase in cytosolic calcium (Ebbeling et al. 1992).

### 3. Alpha-Granules

The alpha-granules have a relative volume of approximately 15% (see Sect. C.). Within the resting platelet, these spherical organelles are 200 nm (140–400 nm) in diameter (Table 3) and are localized clearly separated from one another (Fig. 7). The dense granular matrix, rich in highly concentrated macromolecules (see Table 34–1 in Niewiarowski and Holt 1987), includes an electron-denser nucleoid. After conventional fixation the matrix sometimes contains small tubular structures of 20-nm diameter (White 1971). On the organelle membrane, coated or uncoated protrusions into the cytoplasm (Fig. 6b) are observable (Morgenstern 1982; Behnke 1989, 1992; Morgenstern et al. 1992). Giant alpha-granules, resulting from fusion of normally sized organelles, are observed under anoxic conditions (White 1992b).

The alpha-granules are storage organelles for a plethora of proteins (Chap. 19) (reviewed regarding ultrastructure in Harrison and Cramer 1993). When released, they take part in platelet reactions during:

1. Platelet aggregation and platelet interaction with other cells or components of the extracellular matrix [fibrinogen, fibronectin, thrombospondin, b-thromboglobulin, vitronectin, von Willebrand Factor and multimerin (reviewed in Hayward and Kelton 1995), platelet factor 4]
2. Clot formation and fibrinolysis [alpha 2-macroglobulin and alpha 2-antiplasmin, factors V and VIII, fibrinogen and other adhesins, plasminogen, PAI-1, tPA-like plasminogen activator (as shown by Wang et al. 1994) and TGF/beta]

3. Inflammation and immune response (chemotactic, permeability or bactericidal factor, PDGF, complement proteins, immunoglobulins) (reviewed in PREISSNER and DEGROOT 1993).

The uptake of proteins originating from blood plasma into the alpha-granules is discussed in Sect. E.II.1.

The granule membrane contains a number of physiologically important receptors including GPIIb/IIIa and P-selectin (reviewed in HARRISON and CRAMER 1993) as well as small amounts of GPIb, GPIX and GPV (BERGER et al. 1996). The integral glycoproteins CD36, a receptor for thrombospondin (BERGER et al. 1993); CD9; and PECAM-1 (CRAMER et al. 1994), a new granule membrane protein 33 (METZELAAR et al. 1992); and the phosphoglycoprotein osteonectin (BRETON-GORIUS et al. 1992) complete the list of proteins that are demonstrated immunocytochemically with the TEM in the granule membrane. Except for P-selectin, the membrane proteins mentioned are also expressed in the plasmalemma and in the membranes of the SCS. As known from P-selectin (STEINBERG et al. 1985), an increase of expression on the plasmalemma of platelets after stimulation and exocytosis (Sect. E.III.1.) is characteristic for these proteins.

#### 4. Dense Granules

Dense granules (dense bodies; see Chap. 18) differ from alpha-granules by their electron-dense contents which may fill the whole or the central matrix of the organelle (Figs. 4, 7b). Special techniques of recognition in the TEM are used. In human platelets, $Ca^{2+}$, inorganic phosphorus, and serotonin are responsible for the characteristic appearance (TRANZER et al. 1966; DAIMON 1992; for review see WHITE 1992a). Moreover, in contrast to alpha-granules, the protein granulophysin is present on the membrane of dense granules (reviewed in NISHIBORI et al. 1993). Also, the kinetic properties of dense granule secretion may differ from those of alpha-granule release (reviewed in MORGENSTERN et al. 1995a). In resting platelets, the dense granules are spherical organelles. A precondition for this finding is to avoid platelet activation (PAYNE 1984; O'TOOLE et al. 1993; MORGENSTERN et al. 1995). The dense cores of the organelles are not stable during sectioning and, instead of the cores, electron-lucent holes arise in many sections, especially after cryofixation (Fig. 13) (DAIMON 1992; MORGENSTERN et al. 1995). As found in ultrathin frozen hydrated sections (Fig. 4), after cryofixation the dense cores are preserved during sectioning. The behavior of dense granules in exocytosis is described in Sect. E.III.

#### 5. Lysosomes and Peroxisomes

Platelets possess primary lysosomes (BENTFELD-BARKER and BAINTON 1982; SIXMA et al. 1985; STENBERG and BAINTON 1986; BEHNKE 1989, 1992). Lysosomes of platelets (see Chap. 20) are small dimensioned (175–250 nm in diameter). The question of whether lysosomes are a particular organelle type (reviewed by STENBERG and BAINTON 1986) was answered by the determination that only in these organelle membranes is a 53-kDa glycoprotein located. It is exposed on the surface only after platelet activation (NIEUWENHUIS et al. 1987; METZELAAR et al. 1990; HAMAMOTO et al. 1994). Secondary lysosomes and residual bodies are observed (BEHNKE 1992) in an immunocytochemical study. There it is suggested that lysosomes take part in autophagic, focal cytoplasmic degradation.

Catalase-reactive peroxisomes can be distinguished from other organelles by the diaminobenzidine reaction at pH 9.7 (BRETON-GORIUS and GUICHARD 1976; STENBERG and BAINTON 1986).

# E. Functional Morphology

## I. Focal Adhesion Contacts

Platelets interact during aggregation, clot formation, and adhesion with adhesive ligands (fibrinogen, fibrin and collagen fibers, or components of the extracellullar matrix) by means of focal contacts. Integrins involved in the platelet reaction bind the ligand either after activation by physiological agonists (GPIIb/IIIa and fibrinogen) or already in the resting state (GPIa/IIa and collagen fibers). Occupation of such integrins induces transmembrane signalling (FISCHER et al. 1990; POLANOWSKA et al. 1993; HAIMOVICH et al. 1996) and as consequences clustering as well as the interactions of patched ligand-receptor complexes with the contractile cytoskeleton (BERTAGNOLLI et al. 1993; HAGMANN 1993; ESCOLAR et al. 1995; for reviews see FOX 1993, 1994b; CLEMETSON 1995). In contrast, a microfilament- and focal contact-associated protein, VASP (vasodilator-stimulated phosphoprotein), is located in focal contacts. This indicates a correlating down-regulating mechanism of platelet activation (REINHARD et al. 1992; HORSTRUP et al. 1994).

The focal contacts between aggregated platelets show bridging filamentous structures (Fig. 9) between the platelets at an average distance of 20nm. Measurements of the width of the contact spaces and immunocytochemical findings prove that these focal contacts are mediated by fibrinogen molecules (MORGENSTERN and REIMERS 1986). Between the aggregated platelets, a tight contact type with the pentalaminar feature of the contacting membranes also exists (SKAER et al. 1979; MORGENSTERN et al. 1985). The function of this tight contact is not yet known.

Platelet-fibrin contacts in a clot are side-to-side contacts and of smaller width than the contacts in an aggregate. The space is spanned by fine filaments at an average distance of 20nm. Interestingly, this distance is similar to the distances found in aggregate contacts (Table 5). The bridging filaments do not run at right angles between the contact structures (MORGENSTERN et al. 1984).

Platelet–collagen type I contacts are narrow (Table 5) and contain bridging structures. No correlation between the bridges and the periodic band

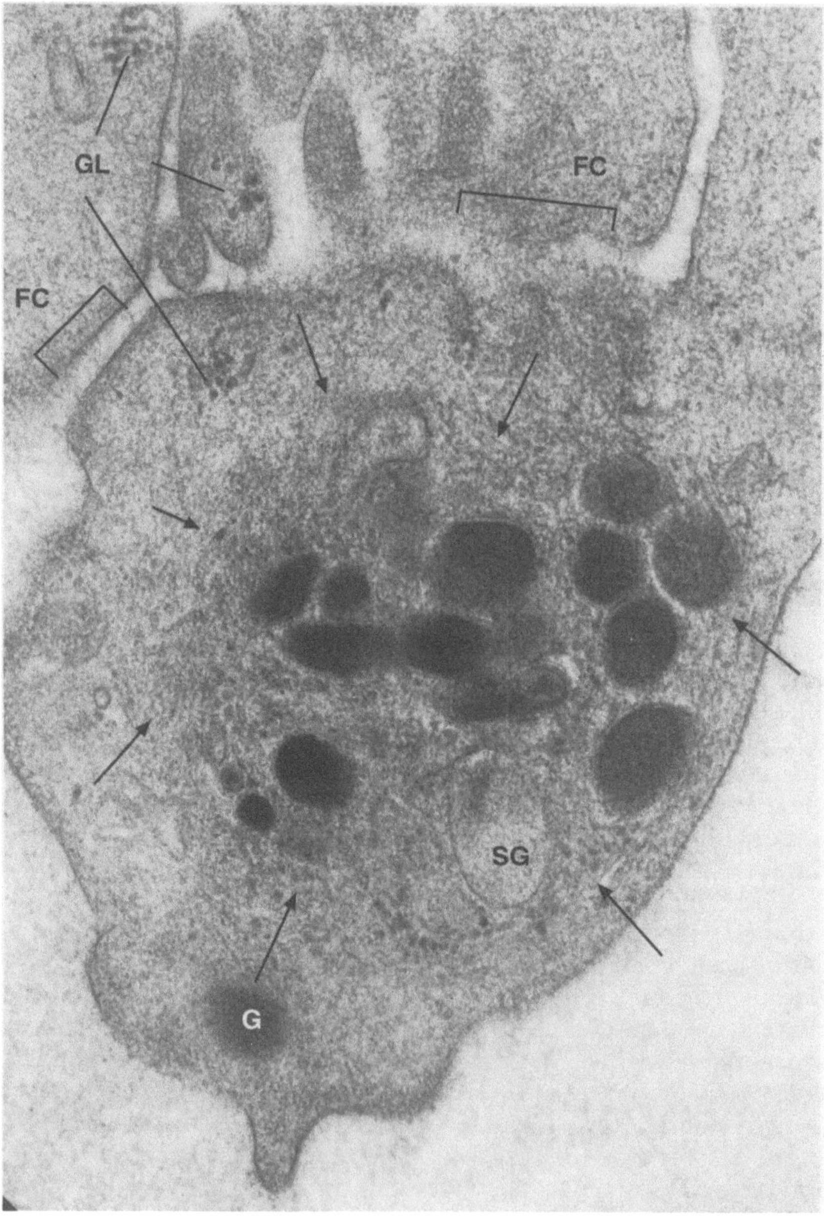


**Fig. 9.** The section of a platelet in an aggregate – induced by ADP in citrated blood plasma – shows focal contacts (*FC*) with the surface of other platelets on the top of the micrograph. In the bordered zones, bridging filamentous structures are seen. The platelet has changed its shape and the SCS is evaginated. The secretory organelles are centralized and surrounded by the contractile gel (*arrows*). An alpha-granule is seen outside the constricting sphere, one granule (*SG*) is swollen and contains a dispersed matrix. *GL*, glycogen. ×61 000

**Table 5.** Properties of platelet focal adhesion contacts

| Type | Width of the space (nm) | Distances of bridging structures (nm) | References |
|---|---|---|---|
| Fibrinogen (platelet contacts in an aggregate) | 40–50 | 10–20<br>~20 | SKAER et al. 1979<br>MORGENSTERN et al. 1985<br>MORGENSTERN and REIMERS 1986 |
| Fibrin | 10–25<br>10–15<br>17–19.5 | ~20 | ERICHSON et al. 1967<br>MORGENSTERN et al. 1984<br>LEWIS et al. 1988 |
| Collagen | 2–3<br>4–10 | | HOVIG et al. 1968<br>RUF et al. 1986 |
| Polysulphone | 13–25 | | MORGENSTERN et al. 1994 |
| Polyurethane | 10–19 | | |

pattern of collagen is recognizable. The presence of fibrillar collagen with repetitive binding sites for its binding to the receptor GPIa/IIa is necessary to induce contact activation of platelets (reviewed in RUGGIERO et al. 1985; RUF et al. 1986, 1991; LÜSCHER and WEBER 1993). The binding mode depends on flow conditions (reviewed in SAELMAN et al. 1994). The collagen–platelet interaction can also be mediated by vWF that binds to GPIb.

Contacts of platelets with artificial surfaces are well investigated. On plane surfaces (see HAGMANN 1993), the platelets spread (BEHNKE and BRAY 1988) and form focal contacts like those in platelets which interact with textured materials. Fibrinogen is the main adhesive protein mediating the platelet–material contact (reviewed in MORGENSTERN et al. 1994).

## II. Internalization and Endocytosis

### 1. Receptor Transport and Membrane Recycling

Molecules from the blood plasma including fibrinogen are transferred into the storage organelles (alpha-granules) of circulating blood platelets (reviewed in HARRISON 1992; HARRISON and CRAMER 1993). The in vitro observation that the fibrinogen receptor GPIIb/IIIa is transported in resting platelets from the surface into the alpha-granules supports the existence of a constitutional, i.e., stimulus-independent, transport (MORGENSTERN et al. 1992). Moreover, the internalization of ligands does not seem to require metabolic energy (ZUCKER-FRANKLIN 1981). The pathway transporting endocytosed material to alpha-granules starts with internalization via coated or uncoated pits that originate from the plasmalemma or the SCS (BEHNKE 1989, 1992; MORGENSTERN et al. 1992). From these studies it is very probable that a sequestration of ligands or tracers in (coated) vesicles precedes the passing into the alpha-granules (Fig.

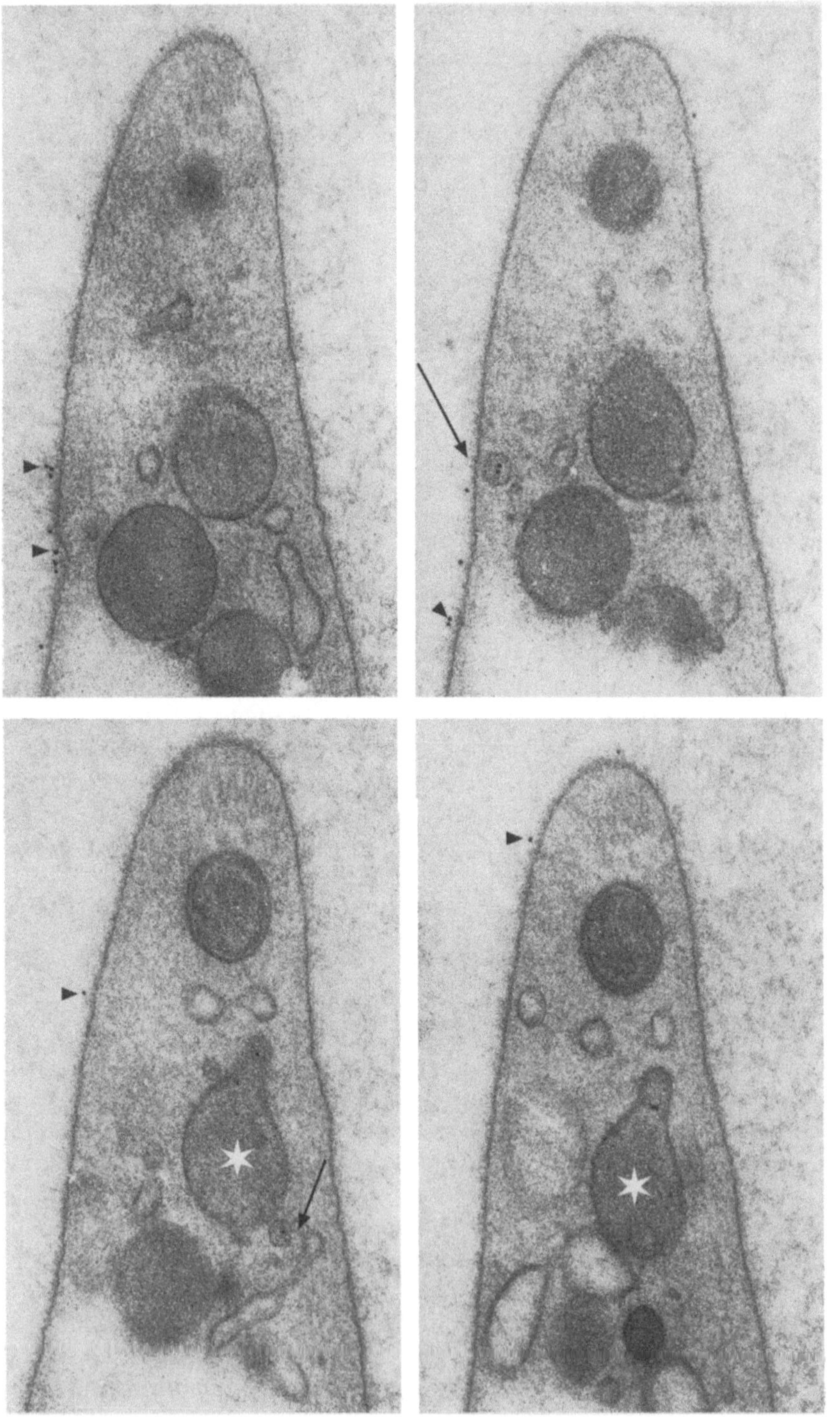

a b c d

10). This pathway seems to be prevalent not only for GPIIb/IIIa and fibrinogen but also for membrane proteins and other adhesive proteins (GRAMER et al. 1994; KLINGER and KLUETER 1995). A degrading, lysosomal pathway of surface-bound ligands via uncoated vesicles is suggested by BEHNKE (1992). The described involved endosomal compartment (multivesicular bodies) also exists after cryofixation.

A recycling of membrane-integrated receptors is indicated (a) in resting platelets by the occurrence of coated protrusions of the alpha-granular membrane and by the presence of gold-labeled membrane receptors in the plasmalemma (Fig. 10) during long incubation periods (MORGENSTERN et al. 1992);

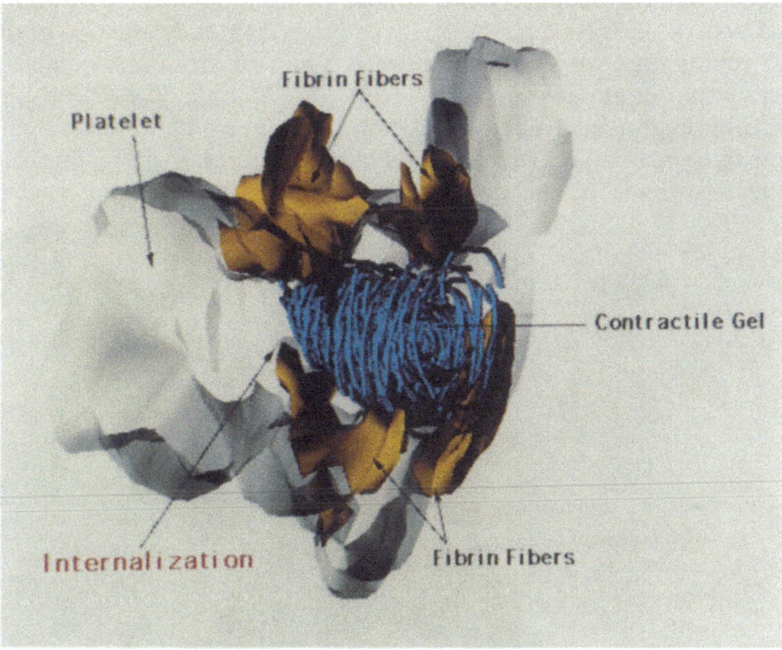


**Fig. 11.** Three-dimensional reconstruction from serial sections of a platelet during thrombin-induced clot formation shows that the platelets retract bound fibrin fibers by internalization with assistance of the contractile gel. After 15 min, the fibrin fibers are found in deep surface invaginations which are associated with the outer rim of the constricting contractile gel

**Fig. 10a–d.** Four consecutive serial sections show endocytotic traffic of the receptor molecule GPIIb/IIIa in a resting platelet. The platelet suspension was incubated for 90 min with the gold-labelled anti-GPIIb/IIIa complex (MORGENSTERN et al. 1992) and afterwards cryofixed. Membrane-bound gold particles are seen on the surface (*arrow heads*) in separated vesicles which are recognizable only in one section respectively (*arrows* in **b** and **c**) and in alpha-granules (*G*). An alpha-granule with gold particles in a protrusion is indicated in **c** and **d** (*star*). (Experiment: A. Ruf; electron microscopy: E. Morgenstern), ×78 000

and (b) in activated platelets by the redistribution of membrane components (ISENBERG et al. 1987; WENCEL-DRAKE 1990; ESCOLAR and WHITE 1991; ESTRY et al. 1991; NURDEN et al. 1994). The glycoprotein lb is not redistributed in resting platelets (MORGENSTERN et al. 1992). However, an activation-dependent redistribution of the glycoprotein IV (CD36) and of the glycoprotein lb-IX can be observed (MICHELSON et al. 1994a,b; WHITE et al. 1994; NURDEN et al. 1994; KOVACSOVICS and HARTWIG 1996).

### 2. Internalization of Ligands by the Contractile Gel

Adhering particular ligands, like cationized metals, latex spheres, or fibrinogen-coated gold, induce internalization into the SCS (reviewed in ESCOLAR and WHITE 1991; BEHNKE 1992). In activated platelets, the tips of the newly formed channels are associated with the constricting contractile gel (MORGENSTERN et al. 1990a).

Fibrillar ligands such as collagen, fibrin, or textured biomaterials adhere to platelets in focal contacts (Sect. E.I.). Platelets retract fibers by internalization with the assistance of the cytoskeleton. The action of the cytoskeleton is

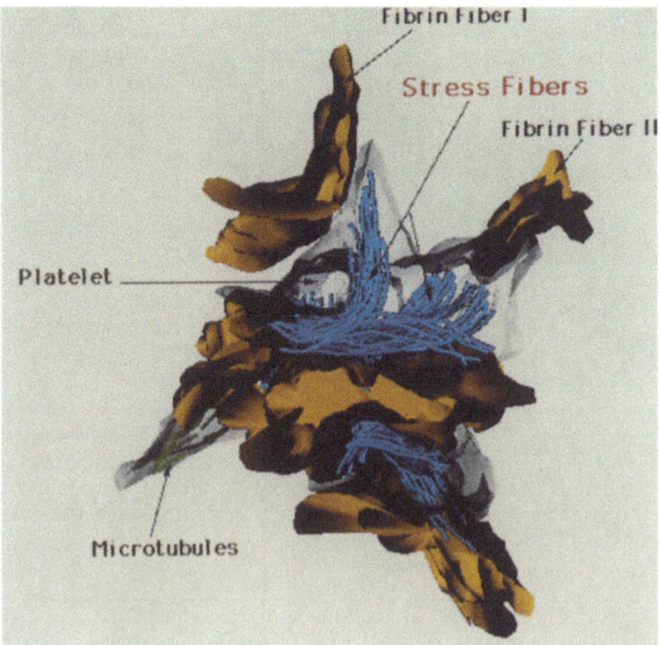


**Fig. 12.** In a three-dimensional reconstruction as shown in Fig. 11, after 30 min, the focal contacts fibrin fibers were situated predominantly on extended surface areas of the platelets and the focal contacts were associated with straightened microfilament bundles ("stress fibers") which insert at a right angle to the plasmalemma – characteristic for isometric contracting platelets under tension

inhibited by cytochalasins (Morgenstern et al. 1990b; Lefebvre et al. 1993). The contractile gel (Sect. D.I.3.) acts as a constricting sphere (Painter et al. 1984) which internalizes surface-bound fibers during the first stages of clot formation (Fig. 11) or during retraction of collagen networks as well as flexible biomaterial fibers (Erichson et al. 1967; Morgenstern et al. 1990; Ruf et al. 1986, 1991; Morgenstern et al. 1994; Ruf and Morgenstern 1995).

In later phases of clot retraction or with inflexible biomaterial fibers, bundles of microfilaments, which connect the contacts with the fibers within the platelet (Fig. 12), are observed in place of the constricted sphere of the contractile cytoskeleton (Cohen et al. 1982; Tanaka et al. 1986; Morgenstern et al. 1994). Myosin filaments are found to be arranged between actin filaments in cable-like bundles comparable to stress fibers of other cells. The formation of stress fiber-like microfilament bundles is induced under the condition of isometric contraction in fiber-adhering platelets (Cohen et al. 1982; Morgenstern et al. 1994).

## III. Exocytosis

### 1. Secretory Pathway

In strongly stimulated blood platelets, the secretory granules (alpha-granules and dense bodies) elongate or form protrusions (Morgensterm et al. 1987; O'Toole et al. 1993) and come in close contact (apposition) with the plasmalemma. The disintegration of the SMC (Sect. D.I.2.) enables the formation of these appositions which are characterized by a pentalaminar membrane aspect. Granules in membrane apposition seem to maintain this position during subsequent – fast running (Sect. A.) – shape change and internal contraction. Apposed granules are moved to the platelet surface (reviewed in Morgenstern and Edelmann 1989). Such a movement was recently verified using the atomic force microscope (Fritz et al. 1994). Both dense granules as well as alpha-granules are able to form initial fusion pores into the plasmalemma (Fig. 13b). After fusion, the organelles swell very fast due to water influx (reviewed in Morgenstern and Edelmann 1989).

The simultaneous commencement of concentric internal contraction (Sect. D.I.3.) drives (movable) platelet organelles to the platelet center and into contact with each other, thus supporting further apposition (Fig. 14). The secretory organelles which are already apposed to the surface membrane (and arrested) induce sequential fusion after pore formation (compound exocytosis, Figs. 13, 15) (Morgenstern et al. 1987; reviewed in Morgenstern and Edelmann 1989). Remnants of the dense granule matrix and remnants from alpha-granules within compound organelles suggest that the secretory pathway is similar for both types of organelles (Morgenstern et al. 1995).

Under strong stimulation and after rapid freezing, no evidence was found for fusion of granule membranes with membranes of the SCS (Hols et al. 1985; Morgenstern et al. 1987, 1995; Morgenstern and Edelmann 1989) as

a b c

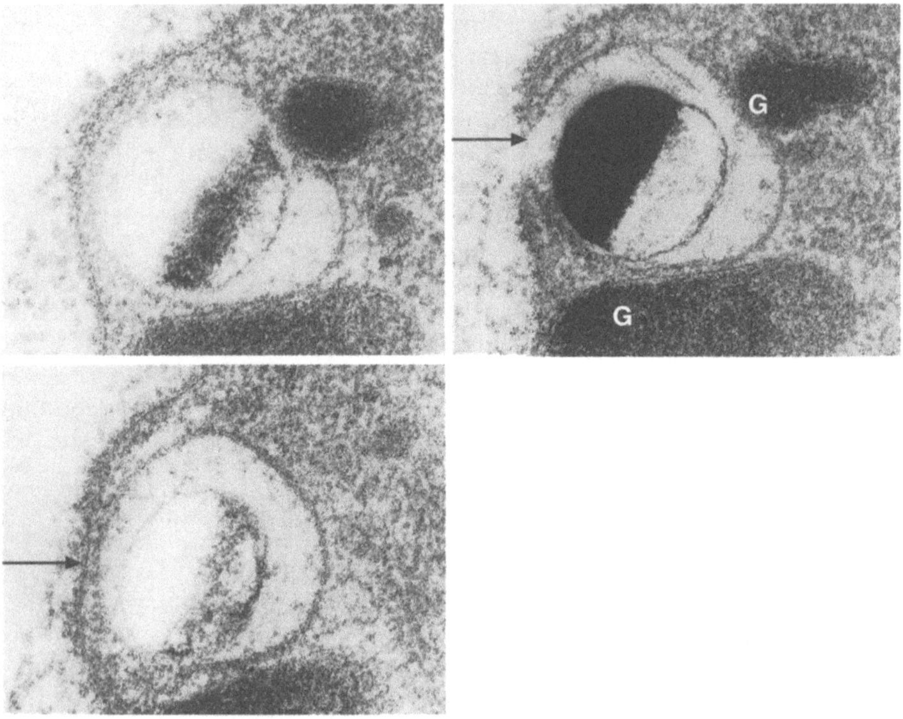


**Fig. 13a–c.** A dense granule in three serial sections of a thrombin-stimulated platelet. The dense granule in **b** shows a small pore with a diameter of 60 nm (*arrow*). A dimpling of the plasmalemma – characteristic for the membrane in the vicinity of a pore – in shown in **c** (*arrow*). The dense granule is seen in close contact (apposition) with two alpha-granules (*G* in **b**). The dense core which is seen in the center of the dense granule has dropped out partially in the consecutive sections shown. The core of the moderately swollen organelle is surrounded by an electron-lucent phase which contains characteristic small electron-dense particles. (From MORGENSTERN et al. 1995) ×100 000

suggested by others (reviewed in WHITE and KRUMWIEDE 1987). The convolute of dilated empty channels in degranulated platelets originates from the membranes of exhausted secretory organelles (MORGENSTERN et al. 1990a).

### 2. Shedding of "Microparticles"

This phenomenon appears only after platelet stimulation. The microparticles presented in the TEM after chemical fixation contain cytoplasmic and membrane constituents (reviewed in FOX 1994a). Recent studies using cryofixation show the origin of such structures from platelets (ZWAAL et al. 1992) and also that membrane fusion gives rise to sequestration (shedding) of blebs, representing cytoplasmic fragments (MORGENSTERN et al. 1995). After incorporation of the pore-forming complement attack complex (reviewed in SOLUM et al. 1994) into the plasmalemma, the sequestration of cytoplasmic fragments is

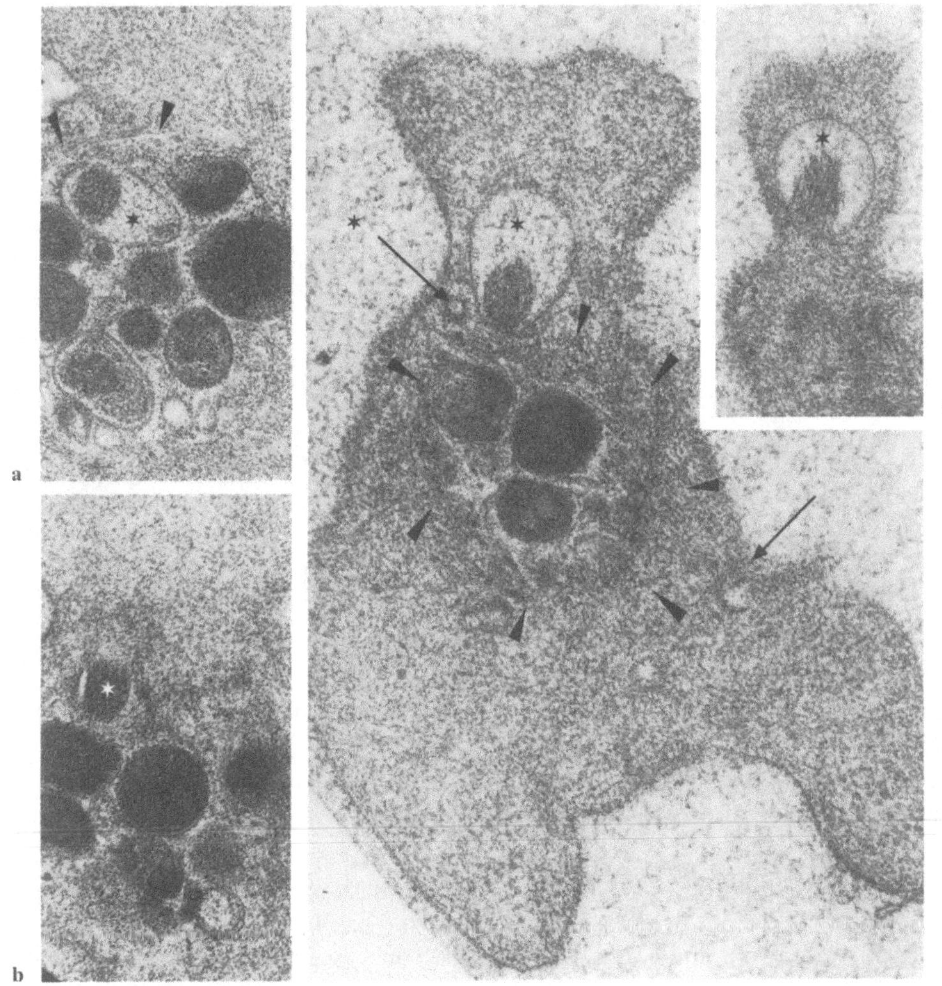


**Fig. 14a–d.** Interactions of the contractile gel with the exocytosis in a collagen-stimulated (40 g/ml citrated plasma, 20 s at 37°C) human blood platelet. In four sections of a series (**a–d**) the mass of the alpha-granules is located within the sphere of the contractile gel and microtubules (*arrowheads*). The surface-connected membrane system is evaginated. The swollen granule (*asterisks*) shows a persisting nucleoid and is situated outside the contractile sphere in (**c–d**), seen to be squeezed by the contractile gel in **b** and located inside the contractile gel (**a**). A pore is not visible in these sections but the swelling and the dispersal of the granular matrix indicate water influx. Note that surface membrane invagination (*arrows* in **c**) takes place simultaneously. This phenomenon is seen within the cytoplasm outside the contractile gel and involves coated membranes (*arrow with asterisk*) ×50 625

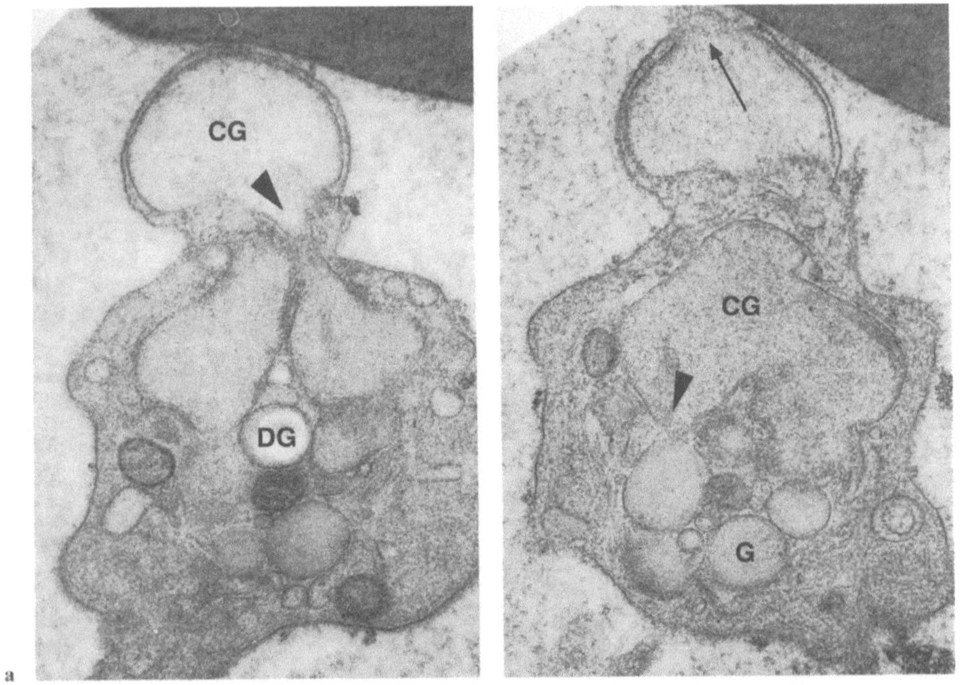


**Fig. 15a,b.** Two sections of a thrombin-stimulated platelet during compound exocytosis after cryo-fixation/-substitution. The compound granules (*CG*) appear swollen and show a fusion pore (*arrow* in **b**). Connections of the compound are indicated with *arrow heads* in **a** and **b**. Several appositions of the compound with secretory organelles (*G* and *DG*) are visible. The dense core of the DG is dropped out. ×40000

induced (Fig. 16). They ensue upon the plasmalemma or upon the tapered filopodia. The finding corresponds with the common suggestions that (a) microparticles are shed by a true budding process when the association between the plasmalemma and the submembranous cytoskeleton is disrupted (Fox et al. 1990; Fox 1993); (b) the fragments express surface-bound procoagulant activity and contain cytoskeletal proteins (reviewed in MORGENSTERN 1991).

# F. Synopsis and Conclusions

The reactions of platelets during hemostasis are enabled by an exceptional changeability of the plasmalemmal surface, the SCS, as well as the cell organelles. Ultrastructural demonstration of the platelet alterations contributes productively in the clarification of platelet function. The resting, discoid platelet is stabilized by the quality of the surface (distribution and state of the receptor molecules), the arrangement of the microtubules as a marginal bundle, the properties of the viscoelastic SMC, the presence of the invaginated

a 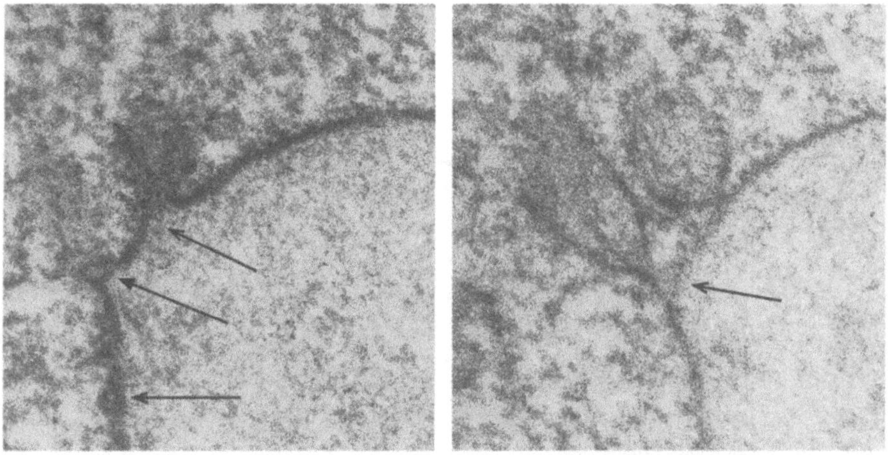 b

**Fig. 16a,b.** Shedding of cytoplasmic fragments ("microparticles") is seen in two consecutive sections of a platelet incubated for 60 s at 37°C after addition of an antibody (FN52) to CD9 – inducing pore forming C5b-9 complex (SOLUM et al. 1994). Originating from the surface, two fragments are demostrated during sequestration (*open arrows*). The *filled arrow* in **a** indicates membrane irregularities within the plasmalemma. (From MORGENSTERN et al. 1995) ×113 000

membranes of the SCS and the separated localization of the secretory organelles within the cytoplasm. A constitutional (?) receptor-mediated transport involving coated membranes claims special interest because such a mechanism may explain the uptake of important molecules from blood plasma into the secretory storage organelles and the recycling of receptors and membrane constituents. The reactions of the stimulated platelet – shape change, formation of focal adhesion contacts, and aggregation – involve the redistribution of membrane phospholipids and receptor molecules, the disintegration of the SMC, and the assembly of a contractile gel. The fast surface enlargement and the formation of pseudopodia during the shape change is rendered possible by the evagination of the SCS. The organization of the contractile gel as a constricting sphere at the outer rim associated with ligand-receptor complexes serves the internalization of surface-bound ligands and the retraction of fibers. The disintegration of the SMC allows the fusion of the secretory organelles with the plasmalemma. Subsequent fusion of secretory organelles (alpha-granules and dense bodies) with each other leads to compound exocytosis. An endoplasmic reticulum (DTS) with regulative missions is present during all functional states within the platelet. The platelets are well equipped with mitochondria and glycogen stores to fulfil their energy-consuming action.

*Acknowledgements*. The author is grateful to Dr. Ludwig Edelmann (Homburg) for helpful advice and for the micrograph in Fig. 4. The author thanks his students Daniel Bastian and Matthias Daub as well as Prof. Dr. Rolf Dierichs (Muenster) for valuable help in three-dimensional reconstruction. The author would also like to thank Anne Vecerdea and Claudia Fischer for excellent technical help.

## References

Beckerle MC, Miller DE, Bertagnolli ME, Locke SJ (1989) Activation-dependent redistribution of the adhesion plaque protein, talin, in intact human platelets. J Cell Biol 109:3333–3346

Behnke O (1968) Electron microscopical observations on the surface coating of human blood platelets. J Ultrastruct Res 24:51–69

Behnke O (1970a) Microtubules in disc shaped blood cells. Int Rev Exp Pathol 9:1–92

Behnke O (1970b) The morphology of blood platelet membrane systems. Ser Haemat 3:3–16

Behnke O (1987) Platelet adhesion to native collagens involves proteoglycans and may be a two-step process. Thromb Haemostas 58:786–789

Behnke O (1989) Coated pits and vesicles transfer plasma components to platelet granules. Thromb Haemost 62:718–722

Behnke O (1992) Degrading and non-degrading pathways in fluid-phase (non-adsorptive) endocytosis in human blood platelets. J Submicrosc Cytol Pathol 24:169–178

Behnke O, Bray D (1988) Surface movements during spreading of blood platelets. Eur J Cell Biol 109:207–216

Bentfeld-Barker ME, Bainton F (1982) Identification of primary lysosomes in human megakaryocytes and platelets. Blood 59:472–481

Berger G, Caen JP, Berndt MC, Cramer EM (1993) Ultrastructural demonstration of CD36 in the alpha-granule membrane of human platelets and megakaryocytes. Blood 82:3034–3044

Berger G, Massé JM, Cramer EM (1996) Alpha-granule membrane mirrors the platelet plasma membrane and contains the glycoproteins Ib, IX and V. Blood 87:1385–1395

Berry S, Dawicki DD, Agarwal KC, Steiner M (1989) The role of microtubules in platelet secretory release. Biochim Biophys Acta 1012:46–56

Bertagnolli ME, Locke SJ, Hensler ME, Bray PF, Beckerle MC (1993) Talin distribution and phosphorylation in thrombin-activated platelets. J Cell Science 106:1189–1199

Boyles J, Fox JEB, Phillips DR, Stenberg DE (1985) Organization of the cytoskeleton in resting, discoid platelets: preservation of actin filaments by a modified fixation that prevents osmium damage. J Cell Biol 101:1463–1472

Breton-Gorius J, Guichard J (1976) Amelioration techniques permettant de reveler la peroxydase plaquettaire. Nouv Rev Fr Hemat 16:381–390

Breton-Gorius J, Clezardin P, Guichard J, Debili N, Malaval L, Vainchenker W, Cramer EM, Delmas PD (1992) Localization of platelet osteonectin at the internal face of the alpha-granule membranes in platelets and megakaryocytes. Blood 79:936–941

Bryon DA, Lagarde M, Dechavanne M (1978) Stereologic evaluation on in vitro thrombin-induced platelet degranulation and contraction. Nouv Rev Fr Hematol 20:173–178

Caen J, Castaldi P, Ruan C (1994) Thrombocytopenias and thrombocytopathies. Rev Invest Clin Suppl:153–162

Calvete JJ (1995) On the structure and function of platelet integrin alpha IIb beta 3, the fibrinogen receptor. Proc Soc Exp Biol Med 208:346–360

Cerletti C, Rajtar G, Marchi E, Degaetano G (1994) Interaction between glycosaminoglycans, platelets, and leukocytes. Sem Thromb Hemostas 20:245–252

Clemetson KJ (1995) Platelet activation: signal transduction via membrane receptors. Thrombos Haemost 74/1:111–116

Cohen I, Gerrard JM, White JG (1982) Ultrastructure of clots during isometric contraction. J Cell Biol 93:775–787
Comfurius P, Bevers EM, Zwaal RFA (1985) The involvement of cytoskeleton in the regulation of transbilayer movement of phospholipids in human blood platelets. Biochim Biophys Acta 815:143–148
Cramer EM, Berger G, Berndt MC (1994) Platelet alpha-granule and plasma membrane share two new components: CD9 and PECAM-1. Blood 84:1722–1730
Cutler L, Rodan G, Feinstein MB (1978) Cytochemical localization of adenylate cyclase, and of calcium ion, magnesium ion-activated ATPases in the dense tubular system of human blood platelets. Biochim Biophys Acta 542:357–371
Daimon T (1992) Freeze-substitution and X-ray microprobe analysis of amine-storage organelles of rat platelets after treatment with reserpine. J Electr Microsc 41:350–356
Deurs van B, Behnke O (1980) Membrane structure of nonactivated and activated human blood platelets as revealed by freeze-fracture: evidence for particle redistribution during platelet contraction. J Cell Biol 87:209–218
Dierichs R, Ahohen-Sann R, Marquardt T (1992) The influence of the microtubule-acting drug, nocoda-zole, and the ATP-depleting system, deoxyglucose-dinitrophenol (DOG/DNP), on the functional morphology of human platelets. Platelets 3:255–263
Dorn GW, Davis MG, D'Angelo DD (1994) Gene expression during phorbol ester-induced differentiation of cultured human megakaryoblastic cells. Am J Physiol 266:C1231–C1239
Duyvene de Wit LJ, Heyns P (1987) Ultrastructural morphometric observations on serial sectioned human blood platelet subpopulations. Eur J Cell Biol 43:408–411
Ebbeling L, Robertson C, McNicol A, Gerrard JM (1992) Rapid ultrastructural changes in the dense tubular system following platelet activation. Blood 80:718–723
Edelmann L (1992) Biological X-ray microanalysis of ions and water: artefacts and future strategies. In: Megias-Megias L, Rodriguez-Garcia MI, Rios A, Aias JM (eds) Electron microscopy 1992. Proc Eur Congr Electron Microsc Vol 1:333–336
Eppell SJ, Simmons SR, Albrecht RM, Marchant RE (1995) Cell-surface receptors and proteins on platelet membranes imaged by scanning force microscopy using immunogold contrast enhancement. Biophys J 68:671–680
Erichson RB, Katz AJ, Cintron JR (1967) Ultrastructural observations on platelet adhesion reactions. I. Platelet-fibrin interaction. Blood 29:385–400
Escolar G, Sauk J, Bravo ML, Krumwiede M, White JG (1987) Immunogold staining of microtubules in resting and activated platelets. Am J Hematol 24:177–188
Escolar G, White JG (1991) The platelet open canalicular system: a final common pathway. Blood-Cells 17:476–485
Escolar G, White JG (1994) Combined use of immunocytochemical techniques and ligand-gold complexes for investigation of platelet membrane responses to surface activation. Microsc Res Tech 28:308–326
Escolar G, Diaz-Ricart M, White JG (1995) Talin does not associate exclusively with alpha(2b)beta(3) integrin in activated human platelets. J Lab Clin Med 125:597–607
Estry DW, Mattson JC, Mahoney GJ, Oesterle JR (1991) A comparison of the fibrinogen receptor distribution on adherent platelets using both soluble fibrinogen and fibrinogen immobilized on gold beads. Eur J Cell Biol 54:196–210
Faull RJ, Ginsberg MH (1995) Dynamic regulation of integrins. Stem Cells Dayt 13:38–46
Feijge MAH, Heemskerk JWM, Hornstra G (1990) Membrane fluidity of non-activated and activated human blood platelets. Biochim Biophys Acta 1025:173–178

Fischer TH, Gatling MN, Lacal JC, White GC (1990) rap1B, a cAMP-dependent protein kinase substrate, associates with the platelet cytoskeleton. J Biol Chem 265:19405–19408
Fox JEB, Boyles JK (1988) Characterization of the platelet membrane cytoskeleton. In: Signal transduction in cytoplasmic organization and cell motility. Alan R Riss Inc, pp 313–324
Fox JE, Austin CD, Boyles JK, Steffen PK (1990) Role of the membrane skeleton in preventing the shedding of procoagulant-rich microvesicles from the platelet plasma membrane. J Cell Biol 111:483–493
Fox JE (1993) Regulation of platelet function by the cytoskeleton. Adv Exp Med Biol 344:175–185
Fox JEB (1994a) Shedding of adhesion receptors from the surface of activated platelets (Review) Blood Coagulat Fibrinolys 5:291–304
Fox JE (1994b) Transmembrane signaling across the platelet integrin glycoprotein IIb–IIa. Ann NY Acad Sci 741:75–87
Frederik PM, Stuart MCA, Bomans PHH, Busing WM, Burger NJ, Verkleij AJ (1991) Perspective and limitations of cryo-electron microscopy. J Microsc 161:253–262
Fritz M, Radmacher M, Gaub HE (1994) Granula motion and membrane spreading during activation of human platelets imaged by atomic force microscopy. Biophys J 66:1328–1334
Frojmovic MM, Milton JG (1982) Human platelet size, shape and related functions in health and disease. Physiol Rev 62:185–261
Gear RL (1984) Rapid platelet morphological changes visualized by scanning-electron microscopy: kinetics derived from a quenched-flow approach. Br J Haematol 56:387–398
Gonzalez-Utor AL, Sanchez-Aguayo I, Hidalgo J (1992) Cytochemical localization of K(+)-dependent P-nitrophenyl phosphatase and adenylate cyclase by using one-step method in human washed platelets. Histochemistry 97:503–507
Hagmann J (1993) Pattern formation and handedness in the cytoskeleton of human platelets. Proc Natl Acad Sci USA 90:3280–3283
Haimovich B, Kaneshiki N, Ji P (1996) Protein kinase C regulates tyrosine phosphorylation of pp125FAK in platelets adherent to fibrinogen. Blood 87:152–161
Hamamoto K, Ohga S, Nomura S, Yasunaga K (1994) Cellular distribution of CD63 antigen in platelets and in three megakaryocytic cell lines. Histochem J 26:367–375
Harrison P, Wilboum B, Debili N, Vainchenker W, Breton-Gorius J, Lawrie AS, Masse JM, Savidge GF, Cramer EM (1989) Uptake of plasma fibrinogen into the alpha granules of human megakaryocytes and platelets. Clin Invest 84:1320–1324
Harrison P (1992) Platelet alpha-granular fibrinogen. Platelets 3:1–10
Harrison P, Cramer EM (1993) Platelet alpha-granules. Blood Rev 7:52–62
Hartwig JH (1992) Mechanisms of actin rearrangements mediating platelet activation. J Cell Biol 118:1421–1442
Hartwig JH, Chambers KA, Stossel TP (1989a) Association of gelsolin with actin filaments and cell membranes of macrophages and platelets. J Cell Biol 108:467–479
Hartwig JH, DeSisto M (1991) The cytoskeleton of the resting human blood platelet: structure of the membrane skeleton and its attachment ot actin filaments. J Cell Biol 112:407–425
Hartwig JH, Chambers KA, Hopcia KL, Kwiatkowski DJ (1989b) Association of profilin with filament-free regions of human leukocyte and platelet membranes and reversible membrane binding during platelet activation. J Cell Biol 109:1571–1579
Hayward CPM, Kelton JG (1995) Multimerein: a multimeric protein stored in platelet alpha-granules. Platelets 6:1–10
Heemskerk JWM, Feijge MAH, Andree HAM, Sage SO (1993) Function of intracellular [Ca++] in exocytosis and transbilayer movement in human platelets surface-

labeled with the fluorescent probe 1-(4-(trimethylammonio)phenyl)-6-phenyl-1.3.5-hexatriene. Biochim Biophys Acta 1147:194–204
Hoak JC (1972) Freeze-etching studies of human platelets. Blood 40:514–522
Holmsen H (1990) Biochemistry and function of platelets. In: Williams WJ, Bentler E, Ersten AJ, Lichtman MA (eds) Hematology. McGraw Hill Publ Comp, New York, pp 1182–1200
Hols H, Sixma JJ, Leunissen-Bijvelt J, Verkleij A (1985) Freeze-fracture studies of human blood platelets activated by thrombin using rapid freezing. Thromb Haemost 54:574–578
Horstrup K, Jablonka B, Hönig-Liedl P, Just M, Kochsiek K, Walter U (1994) Phosphorylation of focal adhesion vasodilator-stimulated phosphoprotein at Ser 157 in intact human platelets correlates with fibrinogen receptor inhibition. Eur J Biochem 225:21–27
Hovig T, Jorgensen L, Packham MA, Mustard JF (1968) Platelet adherence to fibrin and collagen. J Lab Clin Med 71:20–40
Hynes RO (1991) The complexity of platelet adhesion to extracellular matrices. Thromb Haemostas 66:40–43
Isenberg WM, McEver RP, Phillips ER, Shuman MA, Bainton DF (1978) The platelet fibrinogen receptor: an immunogold-surface replica study of agonist-induced ligand binding and receptor clustering. J Cell Biol 104:1655–1663
Iyengar GV, Borberg H, Kasperek K, Kiem J, Siegers M, Feinendegen LE, Gross R (1979) Elemental composition of platelets. I. Sampling and sample preparation of platelets for trace-element analysis. Clin Chem 25:699–704
Klinger MHF, Klueter H (1995) Immunocytochemical colocalization of adhesive proteins with clathrin in human blood platelets: further evidence for coated vesicle-mediated transport of von Willebrand factor, fibrinogen and fibronectin. Cell Tissue Res 279:453–457
Knoll G, Verkleij AJ, Plattner H (1987) Cryofixation of dynamic processes in cells and organelles. In: Steinbrecht RA, Zierold K (eds) Cryotechniques in biological electron microscopy. Springer, Berlin Heidelberg New York, p 259
Kovacsovics TJ, Hartwig JH (1996) Thrombin-induced GPIb-IX centralization on the platelet surface requires actin assembly and myosin II activation. Blood 87:618–629
Lefebvre P, White JG, Krumwiede MD, Cohen I (1993) Role of actin in platelet formation. Eur J Cell Biol 62:194–204
Lewis JC, Johnson C, Ramsomooj P, Hantgan RR (1988) Orientation and specifity of fibrin protofibril binding to ADP-stimulated platelets. Blood 72:1992–2000
Luna EJ, Hitt AL (1992) Cytoskeleton-plasma membrane interactions. Science 258:955–964
Lupu F, Calb M, Scurei C, Simoniescu N (1986) Changes in the organization of membrane lipids during human platelet activation. Lab Invest 54:136–145
Lüscher EF, Weber S (1993) The formation of the haemostatic plug – a special case of platelet aggregation. An experiment and a survey of the literature. Thromb Haemost 70:234–237
Metzelaar MJ, Sixma JJ, Nieuwenhuis HJ (1990) Detection of platelet activation using activation specific monoclonal antibodies. Blood Cells 16:85–96
Metzelaar MJ, Heijnen HF, Sixma JJ, Nieuwenhuis HK (1992) Identification of a 33-Kd protein associated with the alpha-granule membrane (GMP-33) that is expressed on the surface of activated platelets. Blood 79:372–379
Michelson AD, Wencel-Drake JD, Kestin AS, Barnard MR (1994a) Platelet activation results in a redistribution of glycoprotein IV (CD36). Arterioscler Thromb 14: 1193–1201
Michelson AD, Benoit SE, Kroll MH, Li JM, Rohrer MJ, Kestin AS, Barnard MR (1994b) The activation-induced decrease in the platelet surface expression of the glycoprotein Ib–IX complex is reversible. Blood 83:3562–3573
Monck JR, Fernandez JM (1992) The exocytotic fusion pore. J Cell Biol 119:1395-1404

Morgenstern E, Stark G (1975) Morphometric analysis of platelet ultrastructure in normal and experimental conditions. In: Ulutin ON (ed) Platelets. Excerpta Medica, Amsterdam, pp 37–42
Morgenstern E, Kho A (1977) Morphometrische Untersuchungen an Blutplättchen. Veränderungen der Plättchenstruktur bei Pseudopodienbildung und Aggregation. Cytobiologie 15:233–249
Morgenstern E (1980) Ultracytochemistry of human blood platelets. In: Graumann W, Lojda Z, Pearse AGE, Schiebler TH (eds) Progr Histochem Cytochem 12/4 Fischer Stuttgart-New York, pp 1–86
Morgenstern E (1982) Coated membranes in blood platelets. Eur J Cell Biol 26:315–318
Morgenstern E, Korell U, Richter J (1984) Platelets and fibrin strands during clot formation. Thromb Res 33:617–623
Morgenstern E, Edelmann L, Reimers H-J, Miyashita C, Haurand M (1985) Fibrinogen distribution on surfaces and in organelles of ADP stimulated human blood platelets. Eur J Cell Biol 38:292–300
Morgenstern E, Reimers HJ (1986) Ultrastructure of platelet fibrinogen interaction. In: Müller-Berghaus G, Scheefers-Borchel U, Selmayr E, Henschen A (eds) Fibrinogen and its derivates. Amsterdam: Excerpta Medica pp 137–140
Morgenstern E, Neumann K, Patscheke H (1987) The exocytosis of human blood platelets. A fast freezing and freeze-substitution analysis. Eur J Cell Biol 43:273–282
Morgenstern E, Edelmann L (1989) Analysis of dynamic cell processes by rapid freezing and freeze substitution. In: Plattner H (ed) Electron microscopy of subcellular dynamics. CRC Press, Inc Boca Raton, Florida, pp 119–140
Morgenstern E, Patscheke H, Mathieu G (1990) The origin of the membrane convolute in degranulating platelets. A comparative study of normal and "gray" platelets. Blut 60:15–22
Morgenstern E, Ruf A, Patscheke H (1990b) Ultrastructure of the interaction between human platelets and polymerizing fibrin within the first minutes of clot formation. Blood Coagul Fibrinolys 1:543–546
Morgenstern E (1991) Aldehyde fixation causes membrane vesiculation during platelet exocytosis: a freeze-substitution study. Scanning Microsc Suppl 5:109–115
Morgenstern E, Ruf A, Patscheke H (1992) Transport of anti-glycoprotein IIb/IIIa-antibodies into the alpha-granules of unstimulated human blood platelets. Thromb Haemostas 67:121–125
Morgenstern E, Hubertus U, Bastian D (1994) Textured biomaterials as a model for studying formation of focal contacts and rearrangement of the contractile cytoskeleton in platelets. Platelets 5:29–39
Morgenstern E, Bastian D, Dierichs R (1995a) The formation of compound granules from different types of secretory organelles in human platelets (dense granules and alpha-granules). A cryofixation/-substitution study using serial sections. Eur J Cell Biol 68:183–190
Morgenstern E, Holme PA, Solum NO (1995b) Ultrastructural demonstration of membrane vesiculation on platelets after complement-mediated permeabilization using cryofixation technique. Thromb Haemostas 73:1072
Nachmias VT (1980) Cytoskeleton of human platelets at rest and after spreading. J Cell Biol 86:795–802
Nakata T, Hirokawa N (1987) Cytoskeletal reorganization of human platelets after stimulation revealed by quick-freeze deep-etch technique. J Cell Biol 105:1771–1780
Nichols BA, Setzer PY, Bainton DF (1984) Glucose-6-phosphatase as a cytochemical marker of endoplasmic reticulum in human leukocytes and platelets. J Histochem Cytochem 32:165–171
Nieuwenhuis HK, Van Oosterhout JJG, Rozemuller E, Van Iwaarden F, Sixma JJ (1987) Studies with a monoclonal antibody against activated platelets:

evidence that a secreted 53000-molecular weight lysosome-like granule protein is exposed on the surface of activated platelets in the circulation. Blood 70:838–845
Niewiarowski S, Holt JC (1987) Biochemistry and physiology of secreted platelet proteins. In: Colmann RW, Hirsh J, Marder VJ, Salzman MD (eds) Haemostasis and thrombosis. JP Lippincott Comp, Philadelphia, pp 618–630
Nishibori M, Cham B, McNicol A, Shalev A, Jain N, Gerrard JM (1993) The protein CD63 is in platelet dense granules, is deficient in a patient with Hermansky-Pudlak syndrome, and appears identical to granulophysin. J Clin Invest 91:1775–1782
Nurden P, Heilmann E, Paponneau A, Nurden A (1994) Two-way trafficking of membrane glycoproteins in thrombin-activated platelets. Semin Hamatol 31:240–250
O'Toole E, Wray G, Kremer J, Mclntosh JR (1993) High voltage cryomicroscopy of human blood platelets. J Struct Biol 110:55–66
Patscheke H (1981) Shape and functional properties of human platelets washed with acid citrate. Haemostasis 10:14–27
Payne CM (1984) A quantitative ultrastructural evaluation of the cell organelle specificity of the uranaffin reaction in normal human platelets. Am J Clin Pathol 81:62–70
Painter RG, Ginsberg MH (1984) Centripetal myosin redistribution in thrombin-stimulated platelets. Relationship to platelet factor 4 secretion. Exp Cell Res 155:198–212
Pidard D, Montgomery RR, Bennett JS, Kunicki TJ (1983) Interaction of AP-2, a monoclonal antibody specific for the human platelet glycoproteins IIb/IIIa complex, with intact platelets. J Biol Chem 258:12582–12586
Plattner H, Knoll G (1987) Ultrastructural analysis of dynamic cellular processes: a survey of current problems, pitfalls and perspectives. Scanning Microsc 1:1199–1216
Plattner H, Knoll G, Erxleben C (1992) The mechanics of biological membrane fusion. Merger of aspects from electron microscopy and patch-clamp analysis. J Cell Sci 103:613–618
Polanowska-Grabowska R, Geanacopoulos M, Gear AR (1993) Platelet adhesion to collagen via the alpha 2 beta 1 integrin under arterial flow conditions causes rapid tyrosine phosphorylation of pp 125FAK. Biochem J 296:543–547
Polley MJ, Leung LLK, Clark FY, Nachman RL (1981) Thrombin-induced platelet membrane glycoprotein IIb and IIIa complex formation. An electron microscope study. J Exp Med 154:1058–1068
Preissner KT (1991) Structure and biological role of vitronectin. Ann Rev Cell Biol 7:275–310
Preissner KT, de Groot P (1993) Platelet adhesion molecules in natural immunity. In: Sim E (ed) Natural immune system: humoral factors. Oxford Univ Press, pp 281–318
Radmacher M, Fritz M, Kacher CM, Cleveland JP, Hansma PK (1996) Measuring the viscoelastic properties of human platelets with the atomic force microscope. Biophys J 70:556–567
Rao AK (1990) Congenital disorders of platelet function. Hematol Oncol-Clin North Am 4:65–86
Reddick RL, Mason RG (1973) Freeze-etch observations on the plasma membrane and other structures of normal and abnormal platelets. Amer J Path 70:473–482
Reinhard M, Halbrugge M, Scheer U, Wiegand C, Jockusch BM, Walter U (1992) The 46/50 kDa phosphoprotein VASP purified from human platelets is a novel protein associated with actin filaments and focal contacts. EMBO-J 11:2063–2070
Robards AW, Sleytr UB (1985) Freeze-substitution and low temperature embedding. In: Glauert AM (ed) Practical methods in electron microscopy. Vol 10, Low temperature methods in biological electron microscopy. Elsevier, Amsterdam New York Oxford, p 461

Roger M, Huitfeldt HS, Hovig T (1992) Ultrastructural morphometric analysis of human blood platelets exposed to minimal handling procedures. APMIS 100:922–929

Ruf A, Morgenstern E, Janzarik H, Lüscher E (1986) Morphology of the interaction of collagen fibrils with normal human platelets and thrombasthenic platelets. Thrombos Res 44:447–487

Ruf A, Patscheke H, Morgenstern E (1991) Role of internalization in platelet activation by collagen fibers – differential effects of aspirin, cytochalasin D, and prostaglandin $E_1$. Thromb Haemostas 66:708–714

Ruf A, Morgenstern E (1995) Ultrastructural aspects of platelet adhesion on subendothelial structures. Sem Thrombos Hemastas 21:119–122

Ruggiero F, Belleville J, Garrone R, Eloy R (1985) An ultrastructural study of the contact between type I collagen assemblies and the induced human platelet aggregates. J Submicr Cytol 17:11–19

Saelman EUM, Nieuwenhuis HK, Hese KM, de Groot PG, Heijnen HFG, Sage EH, Williams S, McKeown L, Gralnik HR, Sixma JJ (1994) Platelet adhesion to collagen types I–VIII under conditions of stasis and flow is mediated by GPIa/IIa (alpha$_2$beta$_1$-integrin). Blood 85:1244–1250

Schick BP (1990) Synthesis of proteins from [35S]methionine by guinea pig megakaryocytes in vivo and time course of appearance of newly synthesized proteins in platelets. Blood 76:887–891

Sixma JJ, Berg van den A, Hasilik A, Figura von K, Genze HJ (1985) Immuno-electron microscopical demonstration of lysosomes in human blood platelets and megakaryocytes using anti-cathepsin D. Blood 65:1287–1291

Skaer RJ, Emmines JP, Skaer HB (1979) The fine structure of cell contacts in platelet aggregation. J Ultrastruct Res 69:28–42

Smeets EF, Comfurius P, Bevers EM, Zwaal RFA (1994) Calcium-induced transbilayer scrambling of fluorescent phospholipid analogs in platelets and erythrocytes. Biochim Biophys Acta 1195:281–286

Solum NO, Rubach-Dahlberg E, Pedersen TM, Reisberg T, Hogasen K, Funderud S (1994) Complement-mediated permeabilization of platelets by monoclonal antibodies to CD9: inhibition by leupeptin, and effects on the GP Ib-actin-binding protein system. Thromb Res 75:437–452

Stahl K, Themann H, Dame WR (1978) Ultrastructural morphometric investigations on normal human platelets. Haemostasis 7:242–251

Stark F, Golla R, Nachmias VT (1991) Formation and contraction of a microfilamentous shell in saponin-permeabilized platelets. J Cell Biol 903–913

Steinberg PE, McEver RP, Dhuman MA, Jacqes YV, Bainton DF (1985) A platelet alpha-granule membrane protein (GMP-140) is expressed on the membrane after activation. J Cell Biol 101:880–886

Steinbrecht RA, Mueller M (1987) Freeze-substitution and freeze-drying. In: Steinbrecht RA, Zierold K (eds) Cryotechniques in biological electron microscopy. Springer, Berlin Heidelberg New York, p 149

Steiner M (1987) Platelet surface glycosaminoglycans are an effective shield for distinct platelet receptors. Biochim Biophys A 931:286–293

Stenberg PE, Bainton DF (1986) Storage organelles in platelets and megakaryocytes. In: Philipps DR, Shuman MA (eds) Biochemistry of platelets. Acad Press New York, pp 257–294

Stockinger L, Weissel M, Lechner K (1968) Thrombozytenpraeparation und Auswertung. Mikroskopie (Wien) 25:262–277

Stuart MC, Bevers EM, Comfurius P, Zwaal RF, Reutelingsperger CP, Frederik PM (1995) Ultrastructural detection of surface exposed phosphatidylserine on activated blood platelets. Thromb Haemost 74:1145–1151

Tanaka K, Shibata N, Okamoto K, Matsusaka T, Fukuda H, Takagi M, Fujii N (1986) Reorganization of contractile elements in the platelet during clot retraction. J Ultrastr Res 89:98–109

Tranzer JP, DaPrada M, Pletscher A (1966) Ultrastructural localization of 5-hydroxytryptamine in blood platelets. Nature (Lond) 212:1574–1575

Vashkinel VK, Petrov MN (1981) Morphometric analysis of platelet ultra-structure in healthy subjects and in patients with acute leukemia, chronic myeloleukemia and myelofibrosis. Medizina (Moscow) 4:30–36

Verhallen PFJ, Bevers EM, Comfurius P, Zwaal RFA (1987) Correlation between calpain-mediated cytoskeletal degradation and expression of platelet procoagulant activity. A role for the platelet membrane-skeleton in the regulation of membrane lipid asymmetry? Biochim Biophys Acta 903:206–217

Wang DL, Chang YN, Hsu HT, Usami S, Chien S (1992) Prostaglandins and dibutyryl cyclic AMP enhance platelet resistance to deformation. Thromb Res 65:757–768

Wang DL, Pan YT, Wang JJ, Cheng CH, Liu CY (1994) Demonstration of a functionally active tPA-like plasminogen activator in human platelets. Thromb Haemost 71:493–498

Wencel-Drake JD, Plow EF, Kunicki TJ, Woods VL, Keller DM, Ginsberg HH (1982) Localization of internal pools of membrane glycoproteins involved in platelet adhesive responses. Am J Pathol 124:324–334

Wencel-Drake JD (1990) Plasma membrane GPIIb/IIIa. Evidence for a cycling receptor pool. Am J Pathol 136:61–70

Werner G, Morgenstern E (1980) Three-dimensional reconstruction of human blood platelets using serial sections. Eur J Cell Biol 20:276–282

White JG (1968) Effects of ethylene diamine tetraacetic acid (EDTA) on platelet structure. Scand J Haemat 5:241–254

White JG, Krumwiede M (1968) Influence of cytochalasin B on the shape change induced in platelets by cold. Blood 41:241–253

White JG (1969) The submembranous filaments of blood platelets. Am J Pathol 56:267–277

White JG (1970) A search for the platelet secretory pathway using electron dense tracers. Am J Pathol 58:31–49

White JG (1971) Platelet morphology. In: Johnson SA (ed) The circulating platelet. Acad Press, New York London, pp 46–117

White JG (1972) Interactions of membrane systems in blood platelets. Am J Pathol 66:295–305

White JG, Burris SM, Tukey D, Smith C, Clawson CC (1984) Micropipette aspiration of human platelets; influence of microtubules and actin filaments on deformability. Blood 64:210–214

White JG, (1987a) Inherited abnormalities of the platelet membrane and secretory granules. Hum Pathol 18:123–139

White JG (1987b) An overview of platelet structural physiology. Scanning Microsc 1:1677–1700

White JG, Krumwiede M (1987) Further studies of the secretory pathway in thrombin-stimulated human platelets. Blood 69:1196–1203

White JG, Leistikow EL, Escolar G (1990) Platelet membrane responses to surface and suspension activation. Blood Cells 16:43–72

White JG (1992a) The dense bodies of human platelets. In: Meyers KM, Barnes CD (eds) The platelet amine storage granule. CRC Press, Boca Raton Ann Arbor London Tokyo, pp 1–29

White JG (1992b) Ultrastructural changes in stored platelets. Blood Cells 18:461–479

White JG (1992c) Ultrastructural analysis of platelet contractile apparatus. In: Hawiger JJ (ed) Methods in enzymology 215, Part B pp 109–127

White JG, Krumwiede M, Cocking-Johnson D, Escolar G (1994) Influence of combined thrombin stimulation, surface activation, and receptor occupancy on organization of GPIb/IX receptors on human platelets. Br J Haematol 88:137–148

Winokur R, Hartwig JH (1995) Mechanism of shape change in chilled human platelets. Blood 85:1796–1804

Wolpers C, Ruska H (1939) Strukturuntersuchungen zur Blutgerinnung. Klin Wochenschr 18:1077–1081; 1111–1117
Woods VL, Wolff LE, Keller DM (1986) Resting platelets contain a substantial centrally located pool of glycoprotein IIb/IIIa complex which may be accesible to some but not other extracellular proteins. J Biol Chem 261:15242–15251
Wurzinger LJ, Wolf M, Langen H (1987) Vergleichende morphometrische und funktionelle Untersuchungen der Thrombozyten von Mensch und Schaf. Verh Anat Ges 81:781–782
Xu Z, Afzelius A (1988a) The substructure of marginal bundles in human blood platelets. J Ulstrastruct Molecular Struc Res 99:244–253
Xu Z, Afzelius A (1988b) Early changes in the substructure of the marginal bundle in human blood platelets responding to adenosine diphosphate. J Ultrastruct Molecular Struc Res 99:254–260
Zucker-Franklin D (1970) The submembrane fibrils of human blood platelets. J Cell Biol 47:293–299
Zucker-Franklin D (1981) Endocytosis by human platelets: metabolic and freeze fracture studies. J Cell Biol 91:706–715
Zwaal RFA, Hemker HC (1982) Blood cell membranes and haemostasis. Haemostasis 11:12–39
Zwaal RFA, Comfurius P, Bevers EM (1992) Platelet procoagulant activity and microvesicle formation. Its putative role in hemostasis and thrombosis. Biochim Biophys Acta 1180:1–8

CHAPTER 3

# Platelet Adhesion

G.H. van Zanten, P.G. de Groot, and J.J. Sixma

## A. Introduction

Platelet adhesion to the damaged vessel wall is the first step in haemostasis and thrombosis (Sixma and Wester 1977; Baumgartner et al. 1976). Platelet adhesion is followed by spreading and activation, resulting in secretion of the $\alpha$-granula and aggregate formation. The severity of the platelet reaction is dependent on the extent of the vessel wall lesion and on the relative reactivity of the components that have become exposed to the flowing blood. Platelets play an important role in the atherosclerotic narrowing of arteries by promoting thrombus formation, leading to myocardial infarction, stroke and angina, but also to a further development of atherosclerosis (Davies and Thomas 1981; Falk 1985). The adhesion of platelets occurs under flow conditions which require a unique cell adhesion mechanism: platelet adhesion has to occur rapidly and resist shear forces. This has led to the development of a unique set of ligands and receptors. Platelet vessel wall interaction and subsequent platelet–platelet interaction are mediated by glycoproteins on the platelet outer membrane and by adhesive proteins present in plasma, platelet $\alpha$-granules and the vessel wall.

The introduction of in vitro perfusion models allowed the study of platelet adhesion to subendothelium (Baumgartner and Muggli 1976) and to purified components of the vessel wall (Sakariassen et al. 1983) in flowing (anticoagulated) blood under well-controlled conditions. Platelets are transported towards the vessel wall by red blood cells (RBC). The smaller platelets are pushed aside towards the vessel wall due to collisions with the RBC rotating in the shear fields of flow (Wang and Keller 1979; Goldsmith and Turitto 1986; Uijttewaal et al. 1993). RBC play a central role in the adhesion process. Beside haematocrit and red cell size, the deformability of the RBC is an important factor regulating platelet adhesion (Aarts et al. 1983, 1984, 1986; Karino and Goldsmith 1987; Zwaginga et al. 1991). Increasing the rigidity of the red cell membrane, e.g. by increasing the plasma viscosity, results in decreased deformability of the red cell, enhanced platelet transport and thus increased adhesion of platelets (Aarts et al. 1984; van Breugel et al. 1992).

In this chapter, we discuss platelet adhesion to the vessel wall, together with the relevant ligands and receptors involved.

## B. The Vessel Wall

The vessel wall is built up of three layers: the intima, the media and the adventitia (Rodin 1974). The intima consists of endothelium (which is thrombo-resistant) and subendothelium. Endothelial cells separate the circulating platelets from the adhesive proteins in the deeper layers of the vessel wall. The most important products secreted by the endothelium that influence platelet function are prostacyclin and nitric oxide (Moncada et al. 1991; Lüsher 1993; see also Chaps. 16 and 21). Both substances are inhibitors of platelet adhesion (Weiss and Turitto 1979; de Graaf et al. 1992) and platelet aggregation, and are potent vasodilators. Only cells in the proximity of the producing cells are influenced. Locally produced prostacyclin and nitric oxide may reduce platelet deposition on endothelial cell injuries in proximity to the endothelial cells.

The intima is separated from the media by the internal elastic lamina. The adult intima of human arteries is diffusely thickened. The thickness varies between individuals and increases with age, and may correspond to or exceed the thickness of the media. Intimal thickening occurs very early in life and is conceived as a physiological adaptation to mechanical stress, unrelated to atherosclerosis (Stary 1987). The intima is composed of two layers: an inner proteoglycan-rich layer directly beneath the endothelial cells and a basal "musculoelastic layer", so called because of the abundance of longitudinally arranged smooth muscle cells and elastic fibres (Stary et al. 1992). Endothelial cells, smooth muscle cells and scattered macrophages are the principal cellular components of the healthy intima (Gown et al. 1986). The media consists mainly of concentric smooth muscle cells. The adventitia forms a layer of fibrous connective tissue and contains small blood vessels, lymph vessels and nerves. The connective tissue of intima, media and adventitia is composed of collagen, elastin, proteoglycans and non-collagenous glycoproteins. Collagen types I and III are the most prominent collagen types in the arterial wall. Collagen types IV, V, VI (VIII and XIII) are also present, but in smaller amounts (Mayne 1986; Pihlajaniemi and Tamminen 1990). With aging and also in atherosclerosis, the ratio of type I to type III collagen changes in favour of type I collagen (Bihari-Varga 1986; Morton and Barnes 1982; McCullagh 1983; Murata et al. 1986). Fibronectin and laminin are the major non-collagenous glycoproteins present in the extracellular matrix of normal intima. Von Willebrand factor (vWF) is synthesized by endothelial cells and is present in their underlying extracellular matrix, but not in the deeper part of the intima (Gown et al. 1986). Thrombospondin is particularly predominant in proliferating tissues, but is also variably present in vessel walls (Wight et al. 1985). Plasma proteins, such as low-density lipoproteins and fibrinogen, are present in the normal intima in concentrations directly related to the molecular size of the protein and to the plasma concentration (Smith 1990).

20000kDa (WAGNER 1990). After cleaving a propeptide, the mature vWF multimers consist of a subunit of 2050 amino acids (aa). The multimeric structure is essential for proper functioning of vWF as is evidenced by some subtypes of vWD, lacking the high multimers (SADLER 1994b). vWF mediates platelet adhesion by forming a bridge between collagen and other components in the damaged vessel wall and the receptors on the platelets. Platelet adhesion to the extracellular matrix of endothelial cells (ECM) is, to a significant extent, dependent on vWF already present in the ECM (SAKARIASSEN et al. 1979; STEL et al. 1985; TURITTO et al. 1985). vWF binds to two distinct platelet receptors: the GpIb receptor and the GpIIb-IIIa receptor. For the interaction with GpIIb-IIIa, platelets must first be activated. GpIb is already active on nonactivated platelets, but can only interact with vWF when vWF is bound to a surface. In in vitro tests, the antibiotic ristocetin and the snake venom botrocetin can mimic this process (ZUCKER et al. 1977; READ et al. 1978).

The vWF cDNA has been cloned (VERWEIJ et al. 1985; LYNCH et al. 1985; SADLER et al. 1985; BONTHRON et al. 1986) and the protein has been sequenced (TITANI et al. 1986) showing a characteristic composition of repeating homologous domains, the A-D domains. Factor VIII and heparin binding are localized in the first 272 aa of the mature vWF (FOSTER et al. 1987). A second more important heparin-binding site is localized in the A1 domain (aa 565–587) (SOBEL et al. 1992). The GpIb binding site is also located on the A1 domain (FUJIMURA et al. 1986; GIRMA et al. 1986; MOHRI et al. 1989). The arginine–glycine–aspartate (RGD) sequence at aa 1744–1746 is the binding site for GpIIb-IIIa (BERLINER et al. 1988; WEISS et al. 1989; BEACHAM et al. 1992). vWF interacts with collagen types I, III (SANTORO 1983; MORTON et al. 1983) and VI (RAND et al. 1991, 1993; DENIS et al. 1993). The two domains involved in binding are located in the A1 and A3 repeats (ROTH et al. 1986; DENIS et al. 1993). Probably the A3 domain is the major binding site for collagen types I and III since recombinant A1 fragments do not show any binding to collagen (CRUZ et al. 1993), A1-deleted vWF still binds to collagen (SIXMA et al. 1991), and A3-deleted vWF cannot bind to collagen (LANKHOF et al. 1996). The A1 domain probably plays a role in the binding of vWF to collagen type VI (DENIS et al. 1993; HOYLAERTS et al. 1995).

Several studies have indicated that vWF binding to the extracellular matrix of endothelial cells is not to collagen types I and III (DE GROOT et al. 1988; WAGNER et al. 1984). RAND et al. (1991) indicated collagen type VI as the binding site for vWF. vWF does not bind directly to collagen type IV, fibronectin or fibrin when sprayed on coverslips, but adhesion of platelets, particularly at high shear rates, still requires the support of vWF molecules. Localization studies have shown vWF to be localized only between the adhering platelets and the protein surface but not elsewhere on the surface. The presence of platelets is necessary for the interaction between vWF and the adhesive proteins (VAN ZANTEN et al. 1996; BEUMER et al. 1995; ENDENBURG et al. 1995a). Adhesion to isolated vWF under flow conditions needs the presence of the two receptors for vWF, GpIb and GpIIb-IIIa (Savage et al. 1992; Danton et al. 1994; Lankhof et al. 1995). GpIb binding to immobilized vWF supports a slow movement of platelets on the surface (Savage et al. 1996). For vWF-mediated adhesion to collagen type III, only the GpIb receptor is required for adhesion, although the vWF-GpIIb-IIIa interaction is required for optimal platelet spreading and subsequent platelet aggregate formation (Lankhof et al. 1995). Interaction of platelets with vWF via GpIb causes activation of GpIIb-IIIa leading to platelet spreading and firm attachment (Savage et al. 1992). The GpIIb-IIIa–vWF interaction is also involved in platelet spreading and thrombus formation on subendothelium (Sakariassen et al. 1986; Weiss et al. 1989, 1991).

## II. Collagen

Collagen molecules are structural macromolecules of the extracellular matrix of the vessel wall and play a major role in its thrombogenicity (Baumgartner 1977). At present 17 different types of collagen have been identified (Li et al. 1993). In the vessel wall collagen types I, III and IV are the reactive collagens which cause platelet aggregate formation and also platelet adhesion at relative high shear rates; collagen types V, VI and VIII are the less reactive collagens which function at relatively low shear rates (Saelman et al. 1994; Ross et al. 1995). Collagen type V was nonadhesive under flow conditions.

Collagens possess a triple helical structure characterized by the intertwining of three helical polypeptides forming a coiled coil structure. The amino acid sequences of collagenous domains are characterized by the repeating triplet Gly-X-Y. About 30% of the X and Y positions are occupied by proline and hydroxyproline, respectively (for a review about the molecular structures and assemblies of the different collagen types see van der Rest and Garrone 1991). Collagen types I, III, V and XIII belong to the fibrillar collagens. Mixed fibrils consisting of collagen type I and III may exist (Henkel et al. 1982); collagen type V is often associated with these collagens. Collagen type V regulates the fibril diameter of type I collagen fibrils (Birk et al. 1990; Andrikopoulos et al. 1995). The composition and diameter of the collagen fibrils, their association with other collagens, glycoproteins, or proteoglycans (see also Sect. D.V. for proteoglycans) may be of crucial importance for the reactivity towards platelets. Collagen types IV and VIII belong to the sheet-forming collagens. Collagen type IV is the major collagenous constituent of the basement membranes of endothelial cells and around smooth muscle cells. Collagen type VIII is present in Descemet's membrane in the eye and is also expressed by vascular endothelial cells, but the function in the vessel wall is not clear. The structure of collagen type VI is unique within the collagen family. The protein contains a short triple helix with globular domains of unequal size on both ends. Two heterotrimeric molecules aggregate in a head-to-tail orientation to form dimers, then assemble into tetramers which then form beaded filaments. The triple helical region contains 11 RGD sequences, the globular domains in the $\alpha3$-chain of the collagen VI molecule show homology with the A domains of vWF and with the type III repeats of fibronectin (CHU et al. 1988, 1990). Recent data indicate a direct interaction between the RGD sites in collagen type VI and GpIIb-IIIa on the platelet membrane (ROSS et al. 1995). Collagen type VI is present as an independent microfilamentous network between cell surfaces and the main fibrillar network.

Platelet adhesion to collagen types I–VIII under conditions of both stasis and flow is mediated by the $\alpha2\beta1$ integrin (GpIa-IIa receptor on platelets) (SAELMAN et al. 1994). The function of this integrin is dependent on the presence of $Mg^{2+}$ ions (SANTORO 1986; STAATZ et al. 1989). The platelet-collagen interaction can be divided into divalent cation-independent adhesion and $Mg^{2+}$-dependent adhesion (SANTORO 1988). The relationship between the divalent cation-independent mechanism and the $Mg^{2+}$-dependent mechanism is still not known. Under static conditions, the $Mg^{2+}$-dependent mechanism supports a rate and extent of adhesion far greater than that observed by the divalent cation-independent mechanism. $Ca^{2+}$ ions have an inhibitory effect on platelet adhesion (STAATZ et al. 1989).

The effect of divalent cations on platelet adhesion in flow to collagen types I, III and IV was studied with blood in which the plasma concentrations of the calcium and magnesium were varied (VAN ZANTEN et al. 1996). A strong dependence on $Mg^{2+}$ ions was observed at a high shear rate (shear rate $1600\,s^{-1}$). The dependence at a low shear rate of $300\,s^{-1}$ was much smaller, indicating that the GpIa-IIa receptor is the most important receptor for platelet adhesion to collagen at high shear rate. The cation-independent part is dependent on the type of collagen. Platelet adhesion to collagen type IV was almost absent, and adhesion to collagen types I and III was significantly decreased at high shear rate in citrated blood.

The surface antigenicity of GpIa-IIa on platelets from normal donors varies significantly, resulting in a strong variation in platelet reactivity to collagen. A fourfold range of surface antigen correlates with a 20-fold variation in the ability of nonactivated, washed platelets to adhere to type I collagen and a fivefold variation in the adhesion of platelets to collagen type III (KUNICKI et al. 1993).

The GpIa-IIa receptor was first recognized in a patient with a congenital bleeding disorder and a total lack of interaction with collagen (NIEUWENHUIS et al. 1985). Later, two other patients were descried who had a deficiency of GpIa (KEHREL et al. 1988; HANDA et al. 1995). Also a patient with an autoantibody against GpIa has been described with a lack of platelet response to collagen (DECKMYN et al. 1990, 1994).

The platelet receptors for collagen have recently been reviewed (KEHREL 1995). Besides GpIa-IIa, which is probably the major collagen receptor on platelets, GpIV (CD36) (TANDON et al. 1989) and GpVI (MOROI et al. 1989) have been reported to be collagen receptors. GpIV is a 85-kDa transmembrane glycoprotein which binds to collagen and decreases the initial phase of platelet adhesion (Diaz-Ricart et al. 1993). GpIV may be involved in the early phase of the divalent cation-independent mechanism. However, the physiological relevance is questionable since normal Japanese individuals, lacking GpIV on their platelet membranes (Yamamoto et al. 1990), have no bleeding disorder and have normal bleeding times. GpVI is a glycoprotein with a molecular weight of 62kDa. Three different patients have been found with platelets deficient in GpVI and lacking both collagen induced aggregation and adhesion (Moroi et al. 1989; Ryo et al. 1992; Arai et al. 1995). Recent data (Arai et al. 1995) indicate that GpVI is less involved in platelet adhesion to collagen than in aggregation induced by collagen. Deckmyn et al. (1992) reported a patient with a prolonged bleeding time possessing an antibody against an 85–90-kDa platelet glycoprotein which blocked collagen-induced platelet aggregation.

As already described in Sects. C. and D., adhesion to collagen under flow conditions is also dependent on vWF. Recent studies using a deletion mutant of vWF which lacks the A1 domain (ΔA1-vWF) (Sixma et al. 1991) and a RGGS-vWF mutant in which the RGDS sequence is mutated in a RGGS sequence (aspartate to glycine mutation) directly show that the GpIb–vWF interaction is critical for platelet adhesion to collagen. The GpIIbIIIa–vWF interaction plays an important role in subsequent platelet aggregation (Lankhof et al. 1995).

At present, little is known about the sites in the native collagen molecules that are essential for platelet adhesion to collagen under flow conditions. The localization of the reactive sites was explored by using triple-helical cyanogen bromide fragments of the $\alpha$1 chain of collagen type I (Saelman et al. 1993). The CB3, CB7 and CB8 fragments all support (GpIa-IIa-dependent) platelet adhesion. Adhesion to CB3 was the strongest and was accompanied by aggregate formation. The DGEA sequence in the CB3 fragment of collagen type I has been postulated to be involved in the binding to GpIa-IIa (Staatz et al. 1991), but it is not known whether this sequence is also involved in platelet adhesion to collagen under flow conditions.

## III. Fibronectin

Fibronectin (FN) is a ubiquitous glycoprotein composed of two structurally similar subunits of approximately 220–250kDa. It is synthesized by endothelial cells, smooth muscle cells and fibroblasts as a disulfide-bonded dimer. Two forms can be distinguished: soluble plasma FN, which is produced by hepatocytes (Tamkun and Hynes 1983), and insoluble FN present as fibrils in the extracellular matrix of the vessel wall, where it is involved in the assembly of the matrix through its many interactions with other molecules. FN is also present in the $\alpha$-granula of platelets. Each subunit of FN is composed of three types of homologous repeats, 12 type I repeats, two type II repeats and 15 type III repeats, consisting of 45, 60 and 90aa, respectively (Petersen et al. 1983; Schwarzbauer 1990). Tissue FN contains one or two extra type III repeats.

The repeats are organized into functional domains containing binding sites for fibrin, collagen, heparin and cells. The cell-binding domain contains the RGD sequence (repeat III-10). "Synergistic adhesive sites" have been determined on repeats III-7, III-8 and III-9 (Obara et al. 1988; Katayama et al. 1989; Aota et al. 1991).

Plasma FN is required for the adhesion of platelets to monomeric collagen types I and III (Houdijk et al. 1985a). No requirement for plasma FN was observed in platelet adhesion to the fibrillar collagen types I and III, probably because these collagens can release FN from the platelet $\alpha$-granula. FN has also been shown to be implicated in thrombus formation on either subendothelium or collagen (Bastida et al. 1987). The contribution of tissue FN in platelet adhesion to the extracellular matrix of endothelial cells is blocked by antibodies against FN resulting in a 40% decrease in adhesion (Houdijk and Sixma 1985b; Houdijk et al. 1986). Over the years, different receptors for FN have been described. GpIc-IIa, or VLA-5 (Pytela et al. 1985), and GpIIb-IIIa (Ginsberg et al. 1983; Plow et al. 1985a; Parise and Philips 1986) are the two receptors present on the platelet membrane. Both recognize the RGD sequence in FN. VLA-5 binds to immobilized FN and is partially responsible for the platelet–fibronectin interaction (Piotrowicz et al. 1988; Wayner et al. 1988; Beumer et al. 1994). GpIIb-IIIa is only capable of binding to FN after platelets are activated, similarly to the binding of GpIIb-IIIa to vWF. FN sprayed to coverslips supports adhesion at shear rates below $700\,s^{-1}$ and is dependent on GpIIb-IIIa and VLA-5 (Beumer et al. 1994). An antibody against GpIIb-IIIa and an RGD-containing peptide inhibits adhesion almost completely, whereas antibodies against VLA-5 inhibit by 50%. Physiological levels of divalent cations are not required. These results differ with the FN-dependent adhesion to a more complex surface as the ECM, which is not mediated by the RGD site of GpIIb-IIIa (Nievelstein et al. 1988). These data suggest that other mechanisms are involved in FN-dependent adhesion in the extracellular matrix of the vessel wall. Further studies are required to find out what the essential receptor is for FN in flow.

Glycoprotein Ib-vWF also mediates platelet adhesion to FN under flow conditions (Beumer et al. 1995). An antibody to the vWF-binding site on GpIb, absence of the GpIb receptor on platelets in a patient with Bernard-Soulier syndrome and blood from a patient with severe vWD all reduce platelet adhesion to FN to approximately 30% of the control value. The GpIb-mediated adhesion fully depends on vWF, and the presence of platelets is a prerequisite to support vWF binding to the FN surface.

## IV. Fibrin(ogen)

Fibrinogen is a 340-kDa glycoprotein composed of three pairs of disulphide-linked $\alpha$, $\beta$ and $\gamma$ chains (Mosesson 1992). It is synthesized by hepatocytes and is present in plasma, $\alpha$-granula of platelets, and as absorbed protein in the vessel wall. Fibrinogen contains two RGD sequences in the $\alpha$-chain which can interact with GpIIb-IIIa, one near the N terminus (aa 95–97) and one near the C terminus (aa 572–574). The N terminal RGD sequence is only active when platelets are activated (PLOW et al. 1985b; HAVERSTICK et al. 1985). In addition, a dodecapeptide at the C terminus of the $\gamma$-chain (sequence 400–411) also interacts with GpIIb-IIIa. This sequence is specific for fibrinogen (KLOCZEWIAK et al. 1984; LAM et al. 1987; BENNETT et al. 1988). On resting platelets, GpIIb-IIIa only binds to immobilized fibrinogen (KIEFFER et al. 1991). After activation, unoccupied GpIIb-IIIa becomes competent to bind soluble ligands like fibrinogen, vWF, FN, thrombospondin and vitronectin. When in in vitro studies whole blood is perfused over fibrinogen or over fibrin which is formed of purified fibrinogen under static conditions, platelets adhere as a monolayer of extensively spread platelets (HANTGAN et al. 1990, 1992). This adhesion was mediated by GpIIb-IIIa on the platelet membrane. The degree of spreading is determined by the multiple interaction sites on the fibrinogen molecule (SAVAGE and RUGGERI 1991). Adhesion at high shear rates is dependent on vWF interacting with GpIb on the platelet (ENDENBURG et al. 1995a). The RGD sequences in fibrin seem not necessary for platelet adhesion to immobilized fibrin (HANTGAN et al. 1995). Recent data from our laboratory show pronounced aggregate formation on fibrin fibres formed under flow conditions (ENDENBURG et al. 1995b). These data correspond with the thrombi formed on fibrin fibres when native blood was perfused over tissue factor-rich matrices.

## V. Other Proteins

In contrast to the well-established role of collagen and vWF in adhesion to the subendothelium, the role of laminin and thrombospondin in this process is less clear.

### 1. Laminin

Laminin is a ubiquitous adhesive protein in basement membranes in association with collagen type IV, nidogen, and heparan sulphate proteoglycans (YURCHENCO and SCHITTNY 1990). It has a cross-shape structure and a molecular weight of 850 kDa (BECK et al. 1990). It consists of three chains, an A-chain, a B1-chain and a B2-chain. Platelets adhere and spread to purified laminin at shear rates up to $800\,s^{-1}$. The adhesion is completely dependent on VLA-6 on the platelet membrane, as tested by a monoclonal antibody against VLA-6, and the E8 domain on the long arm of the laminin molecule (SONNENBERG et al. 1990; HINDRIKS et al. 1992). Platelet adhesion to purified laminin is strongly dependent on divalent cations, particularly calcium. The antibody against VLA-6 inhibits platelet adhesion to the extracellular matrix of endothelial cells by 40% at shear rates of $300\,s^{-1}$, but has less effect at high shear rates.

## 2. Thrombospondin

Thrombospondin (TSP) is a 450-kDa glycoprotein consisting of a homotrimer linked by disulphide bridges (Frazier 1991). It is present in the $\alpha$-granula of platelets, but not in the plasma. TSP is synthesized and secreted by endothelial cells and smooth muscle cells, in particular by proliferating cells. The TSP molecule contains a heparin domain, two epidermal growth factor domains, 12 calcium-binding domains, an RGD sequence and a C-terminal cell-binding domain. Both adhesive and anti-adhesive properties of TSP have been reported (Tuszynski et al. 1987; Lahav 1988). The capacity of TSP to support platelet adhesion depends on its conformation (Agbanyo et al. 1993). In a $Ca^{2+}$-depleted conformation it is non-adhesive or anti-adhesive. In this conformation it inhibits platelet adhesion to vWF, FN, fibrinogen and laminin. In its normal conformation, platelets adhere to TSP at shear rates up to $1600 s^{-1}$ but, in contrast to FN or laminin, they do not spread. TSP also has a role in platelet aggregation (Leung 1984; Tuszynski et al. 1988). The platelet receptor for TSP is unknown.

## 3. Proteoglycans

Proteoglycans are macromolecules that possess a core protein and one or more covalently attached glycosaminoglycan (GAG) chains. The GAG chains are composed of repeating, unbranched disaccharide backbones consisting of [hexuronic acid-N-acetylhexosamine]$_n$ which can have sulfate ester subunits on either the uronic acid or hexosamin or both. In keratan sulfate the hexuronic acid is replaced by galactose units. Proteoglycans, particularly those substituted with heparan sulfate side chains, have been recognized as major basement membrane components. The most prominent proteoglycan perlecan, which is synthesized by endothelial cells, consists of a 400–500-kDa core protein and three heparan sulfate chains of 40–60kDa (Noonan et al. 1991; Iozzo et al. 1994). Heparan sulfate decreases during atherosclerotic lesion development, dermatan and chondroitin sulfate increase with lesion development (Berenson et al. 1988; Ylä-Herttuala et al. 1986; Robbins et al. 1989).

The small chondroitin/dermatan proteoglycans decorin and biglycan (Ruoslahti and Yamaguchi 1991) are produced by smooth muscle cells and are present in the extracelular matrix of the vessel wall (Riessen et al. 1989). Decorin binds to FN and collagen types I, III and VI (Bidanset et al. 1992). The binding of decorin to the collagen fibrils has an effect on the lateral assembly of triple helical collagen molecules and decreases the fibril diameter (Vogel and Trotter 1987). Biglycan does not show a specific effect on collagen fibril formation in vitro (Hedbom and Heinegard 1989), but competes with decorin for collagen binding (Schönherr et al. 1995). Proteoglycans have not been studied in detail, but preliminary studies have not indicated an adhesive role in platelet adhesion. The modulating effects of decorin and biglycan on collagen fibrils may be of crucial importance for the reactivity towards platelets and will be of interest to study in flow experiments.

## E. Conclusion

In this chapter, we have presented data concerning the role of the different adhesive proteins and their receptors in platelet adhesion. Platelet adhesion is the first step in the response of platelets to vascular injury. During haemostasis and thrombosis in vivo, platelet adhesion always precedes thrombus formation and is probably necessary for platelet activation and subsequent platelet–platelet interaction. This makes it an attractive step to inhibit in the prevention of thrombosis. However, the precise mechanisms which determine whether platelet activation leads to platelet aggregation or not are not known yet. It is conceivable that there exists more than a single mechanism by which platelets adhere to a specific adhesive protein in the vessel wall, become activated and subsequently aggregate. Much remains to be learned about the details in signal transduction, which may lead to new selective approaches to platelet function inhibitors.

*Acknowledgements*. Supported by Grant # 93.112 (G.H. van Zanten) by the Netherlands Heart Foundation.

### References

Aarts PAMM, Bolhuis PA, Sakariassen KS, Heethaar RM, Sixma JJ (1983) Red blood cell size is important for adherence of blood platelets to artery subendothelium. Blood 62:214–217

Aarts PAMM, Heethaar RM, Sixma JJ (1984) Red blood cell deformability influences platelet vessel wall interaction in flowing blood. Blood 64:1228–1233

Aarts PAMM, Banga JD, van Houwelingen HC, Heethaar RM, Sixma JJ (1986) Increased red blood cell deformability due to isoxsuprine administration decreases platelet adherence in a perfusion chamber: a double-blind crossover study in patients with intermittent claudication. Blood 67:1474–1481

Agbanyo FR, Sixma JJ, de Groot PG, Languino LR, Plow EF (1993) Thrombospondin-platelet interactions. Role of divalent cations, wall shear rate and platelet membrane glycoproteins. J Clin Invest 92:288–296

Alevriadou BR, Moake JL, Turner NA, Ruggeri ZM, Folie BJ, Phillips MD, Schreiber AB, Hrinda ME, McIntire LV (1993) Real-time analysis of shear-dependent thrombus formation and its blockade by inhibitors of von Willebrand factor binding to platelets. Blood 81:1263–1276

Andrikopoulos K, Liu X, Keene DR, Jaenisch R, Ramirez F (1995) Targeted mutation in the col5a2 gene reveals a regulatory role for type V collagen during matrix assembly. Nature Genetics 9:31–36

Arai M, Yamamoto N, Moroi M, Akamatsu N, Fukutake K, Tanoue K (1995) Platelet with 10% of the normal amount of glycoprotein VI have an impaired response to collagen that results in a mild bleeding tendency. Br J Haematol 89:124–130

Bastida EB, Escolar G, Ordinas O, Sixma JJ (1987) Fibronectin is required for platelet adhesion and for thrombus formation on subendothelium and collagen surfaces. Blood 70:1437–1442

Baumgartner HR, Muggli R, Tschopp TB, Turitto VT (1976) Platelet adhesion, release and aggregation in flowing blood: effects of surface properties and platelet function. Thromb Haemost 35:124–138

Baumgartner HR (1977) Platelet interaction with collagen fibrils in flowing blood. I. Reaction of human platelets with α-chymotrypsin-digested subendothelium. Thromb Haemostas 37:1–16

Baumgartner HR, Tschopp TB, Meyer D (1980) Shear rate dependent inhibition of platelet adhesion and aggregation on collagenous surfaces by antibodies to human factor VIII/von Willebrand factor. Brit J Haemat 44:127–139

Baumgartner HR, Muggli R (1976) Adhesion and aggregation: morphologic demonstration, quantitation in vivo and in vitro. In: Gordon JL (ed) Platelets in biology and pathology. North Holland, Amsterdam, p 23

Beacham DA, Wise RJ, Turci SM, Handin RI (1992) Selective inactivation of the Arg-Gly-Asp-Ser (RGDS) binding site in von Willebrand factor by site-directed mutagenesis. J Biol Chem 267:3409–3415

Beck K, Hunter I, Engel J (1990) Structure and function of laminin: anatomy of a multidomain glycoprotein. FASEB J 4:148–160

Bennett JS, Shattil SJ, Power JW, Gartner TK (1988) Interaction of fibrinogen with its platelet receptor. Differential efects of α- and γ-chain fibrinogen peptides on the glycoprotein IIb-IIIa complex. J Biol Chem 263:12948–12953

Berenson GS, Radhakrishnamurthy B, Srinivasan SR, Vijayagopal P, Dalferes ERJ (1988) Arterial wall injury and proteoglycan changes in atherosclerosis. Arch Pathol Lab Med 112:1002–1010

Berliner S, Niiya K, Roberts JR, Houghten RA, Ruggeri ZM (1988) Generation and characterization of peptide-specific antibodies that inhibit von Willebrand factor binding to glycoprotein IIb–IIIa without interacting with other adhesive molecules. Selectivity is conferred by Pro1743 and other amino acid residues adjacent to the sequence Arg1744-Gly1745-Asp1746. J Biol Chem 263:7500–7505

Beumer S, IJsseldijk MJW, de Groot PG, Sixma JJ (1994) Platelet adhesion to fibronectin in flow: dependence on surface concentration and shear rate, role of platelet membrane glycoproteins Gp IIb/IIIa and VLA-5, and inhibition by heparin. Blood 84:3724–3733

Beumer S, Heynen HFG, IJsseldijk MJW, Orlando E, de Groot PhG, Sixma JJ (1995) Platelet adhesion to fibronectin in flow: the importance of von Willebrand factor and glycoprotein Ib. Blood 86:3452–3460

Bidanset DJ, Guidry C, Rosenberg LC, Choi HU, Timpl R, Hook M (1992) Binding of the proteoglycan decorin to collagen type VI. J Biol Chem 267:5250–5256

Bihari-Varga M (1986) Collagen, aging, and atherosclerosis. Atherosclerosis Reviews 14:171–181

Birk DE, Fitch JM, Babiarz JP, Doane KJ, Linsenmayer TF (1990) Collagen fibrillogenesis in vitro: interaction of types I and V collagen regulates fibril diameter. J Cell Sci 95:649–657

Bonthron DT, Handin RI, Kaufman RJ, Wasley LC, Orr EC, Mitsock LM, Ewenstein B, Loscalzo J, Ginsburg D, Orkin SH (1986) Structure of pre-pro-von Willebrand factor and its expression in heterologous cells. Nature 324:270–273

Chu M-L, Conway D, Pan T-C, Baldwin C, Mann K, Deutzmann R, Timpl R (1988) Amino acid sequence of the triple-helical domain of human collagen type VI. J Biol Chem 263:18601–18606

Chu M-L, Zhang R-Z, Pan T-C, Stokes D, Conway D, Kuo H-J, Glanville R, Mayer U, Mann K, Deutzmann R, Timpl R (1990) Mosaic structure of globular domains in the human type VI collagen alpha 3 chain: similarity to von Willebrand factor, fibronectin, actin, salivary proteins and aprotinin type protease inhibitors. EMBO J 9:385–393

Coller BS (1994) Inherited disorders of platelet function. In: Bloom AL, Forbes CD, Thomas DP, Tuddenham EGD (eds) Haemostasis and thrombosis, 3rd edn. Churchill Livingstone, Edinburgh, p 721

Cruz MA, Handin RI, Wise RJ (1993) The interaction of von Willebrand factor-A1 domain with platelet glycoprotein Ib/IX. J Biol Chem 268:21238–21245
Danton MC, Zaleski A, Nichols WL, Olson JD (1994) Monoclonal antibodies to platelet glycoproteins Ib and IIb/IIIa inhibit adhesion of platelets to purified solid-phase von Willebrand factor. J Lab Clin Med 124:274–282
Davies MJ, Thomas A (1981) The pathological basis and microanatomy of occlusive thrombus formation in human coronary arteries. Phil Trans R Soc London (Biol) 294:225–229
De Graaf JC, Banga JD, Moncada S, Palmer RMJ, de Groot PhG, Sixma JJ (1992) Nitric oxide functions as an inhibitor of platelet adhesion under flow conditions Circulation 85:2284–2290
De Groot PG, Verweij C, Nawroth PP, De Boer H, Stern DM, Sixma JJ (1987) Interleukin-1 inhibits the synthesis of von Willebrand factor in endothelial cells which results in a decreased reactivity of their matrix towards platelets. Arteriosclerosis 7:605–611
De Groot PG, Ottenhof-Rovers M, Van Mourik JA, Sixma JJ (1988) Evidence that the primary binding site of von Willebrand factor that mediates platelet adhesion to subendothelium is not collagen. J Clin Invest 82:65–73
Deckmyn H, Chew SL, Vermylen J (1990) Lack of platelet response to collagen associated with an autoantibody against glycoprotein Ia: a novel cause of acquired qualitative platelet dysfunction. Thromb Haemostas 64:74–79
Deckmyn H, Van Houtte E, Vermylen J (1992) Disturbed platelet aggregation to collagen associated with an antibody against an 85- to 90-kD platelet glycoprotein in a patient with prolonged bleeding time. Blood 79:1466–1471
Deckmyn H, Zhang J, van Houtte E, Vermylen J (1994) Production and nucleotide sequence of an inhibitory human IgM autoantibody directed against platelet glycoprotein Ia/IIa. Blood 84:1968–1974
Denis C, Baruch D, Kielty CM, Ajzenberg N, Christophe O, Meyer D (1993) Localization of von Willebrand factor binding domains to endothelial extracellular matrix and to type VI collagen. Arterioscler Thromb 13:398–406
Diaz-Ricart M, Tandon NN, Carretero M, Ordinas A, Bastida E, Jamieson GA (1993) Platelets lacking functional CD36 (glycoprotein IV) show reduced adhesion to collagen in flowing whole blood. Blood 82:491–496
Endenburg SC, Hantgan RR, Lindeboom-Blokzijl L, Lankhof H, Jerome WG, Lewis JC, Sixma JJ, de Groot PhG (1995a) On the role of vWF in promoting platelet adhesion to fibrin in flowing blood. Blood 86:4158–4165
Endenburg SC, Lindeboom-Blokzijl L, Sixma JJ, de Groot PhG (1995b) Thrombus formation in flowing whole blood on fibrin formed under flow conditions. Thromb Haemostas 73:1356 (abstract)
Faggiotto A, Ross R (1984) Studies of hypercholesterolemia in the nonhuman primate. II. Fatty streak conversion to fibrous plaque. Arteriosclerosis 4:341–356
Falk E (1985) Unstable angina with fatal outcome: dynamic coronary thrombosis leading to infarction and/or sudden death. Autopsy evidence of recurrent mural thrombosis with peripheral embolization culminating in total vascular occlusion. Circulation 71:699–708
Federici AB, Bader R, Pagani S, Colibretti ML, De Marco L, Mannucci PM (1989) Binding of von Willebrand factor (vWF) to glycoproteins (Gp) Ib and IIb-IIIa complex: affinity is related to multimeric size. Br J Haematol 73:93–99
Foster PA, Fulcher CA, Marti T, Titani K, Zimmerman TS (1987) A major factor VIII binding domain resides within the amino-terminal 272 amino acid residues of von Willebrand factor. J Biol Chem 262:8443–8446
Frazier WA (1991) Trombospondins. Curr Opin Cell Biol 3:792–799
Fressinaud E, Sakariassen KS, Rothschild C, Baumgartner HR, Meyer D (1992) Shear rate-dependent impairment of thrombus growth on collagen in nonanticoagulated

blood from patients with von Willebrand disease and hemophilia A. Blood 80:988–994
Fujikawa K, Thompson AR, Legaz ME, Meyer RG, Davie EW (1973) Isolation and characterization of bovine factor IX (Christmas factor). Biochemistry 12:4938–4945
Fujimura Y, Titani K, Holland LZ, Russell SR, Roberts JR, Elder JH, Ruggeri ZM, Zimmerman TS (1986) Von Willebrand factor. A reduced and alkylated 52/48 kDa fragment beginning at amino acid residue 449 contains the domain interacting with platelet glycoprotein Ib. J Biol Chem 261:381–385
Ginsberg MH, Forsyth J, Lightsey A, Chediak J, Plow EF (1983) Reduced surface expression and binding of fibronectin by thrombin-stimulated thrombasthenic platelets. J Clin Invest 71:619–624
Girma J-P, Kalafatis M, Pietu G, Lavergne J-M, Chopek MW, Edgington TS, Meyer D (1986) Mapping of distinct von Willebrand factor domains interacting with platelet GPIb and GPIIb/IIIa and with collagen using monoclonal antibodies. Blood 67:1356–1366
Goldsmith HL, Turitto VT (1986) Rheological aspects of thrombosis and haemostasis: basic principles and applications. Thromb Haemostas 55:415–435
Gown AM, Tsukada T, Ross R (1986) Human atherosclerosis. II Immunocytochemical analysis of the cellular composition of human atherosclerotic lesions. Am J Pathol 125:191–207
Handa M, Watanabe K, Kawai Y, Kamata T, Koyama T, Nagai H, Ikeda Y (1995) Platelet unresponsivenss to collagen: involvement of glycoprotein Ia-IIa ($\alpha 2\beta 1$ integrin) deficiency associated with a myeloproliferative disorder. Thromb Haemostas 73:521–528
Hantgan RR, Hindriks G, Taylor R, Sixma JJ, De Groot PG (1990) Glycoprotein Ib, von Willebrand factor, and glycoprotein IIb:IIIa are all involved in platelet adhesion to fibrin in flowing whole blood. Blood 76:345–353
Hantgan RR, Endenburg SC, Cavero I, Marguerie G, Uzan A, Sixma JJ, De Groot PG (1992) Inhibition of platelet adhesion to fibrin(ogen) in flowing whole blood by Arg-Gly-Asp and fibrinogen gamma-chain carboxy-terminal peptides. Thromb Haemostasis 68:694–700
Hantgan RR, Endenburg SC, Sixma JJ, De Groot PG (1995) Evidence that fibrin $\alpha$-chain RGDX sequences are not required for platelet adhesion in flowing whole blood. Blood 86:1001–1009
Haverstick DM, Cowan JF, Yamada KM, Santoro SA (1985) Inhibition of platelet adhesion to fibronectin, fibrinogen and von Willebrand factor substrates by a synthetic tetrapeptide derived from the cell-binding domain of fibronectin. Blood 66:946–952
Hedbom E, Heinegård D (1989) The effect of proteoglycan on the morphology of collagen fibrils formed in vitro. J Biol Chem 264:6898–6905
Henkel W, Glanville R (1992) Covalent crosslinking between molecules of type I and III collagen. Eur J Biochem 122:205–213
Hindriks GA, IJsseldijk MJW, Sonnenberg A, Sixma JJ, De Groot PG (1992) Platelet adhesion to laminin: role of Ca2+ and Mg2+ ions, shear rate, and platelet membrane glycoproteins. Blood 79:928–935
Houdijk WPM, Sakariassen KS, Nievelstein PFEM, Sixma JJ (1985a) Role of factor VIII-von Willebrand factor and fibronectin in the interaction of platelets in flowing blood with monomeric and fibrillar collagen types I and III. J Clin Invest 75:531–540
Houdijk WPM, Sixma JJ (1985b) Fibronectin in artery subendothelium is important for platelet adhesion. Blood 65:598–604
Houdijk WPM, De Groot PG, Nievelstein PFEM, Sakariassen KS, Sixma JJ (1986) Subendothelial proteins and platelet adhesion: von Willebrand factor and fibronectin, not thrombospondin, are involved in platelet adhesion to

extracellular matrix of human vascular endothelial cells. Arteriosclerosis 6:24–33
Hoylaerts MF, Nuyts K, Vreys I, Deckmyn H, Vermylen J (1995) Von Willebrand factor binding to endotheleial cell extracellular matrix collagen type VI. Thromb Haemostas 73:1349 (abstract)
Ikeda Y, Handa M, Kawano K, Kamata T, Murata M, Araki Y, Anbo H, Kawai Y, Watanabe K, Itagaki I, Sakai K, Ruggeri ZM (1991) The role of von Willebrand factor and fibrinogen in platelet aggregation under varying shear stress. J Clin Invest 87:1234–1240
Iozzo RV, Cohen IR, Grässel S, Murdoch AD (1994) The biology of perlecan: the multifaceted heparan sulphate proteoglycan of basement membranes and pericellular matrices. Biochem J 302:625–639
Karino T, Goldsmith HL (1987) Rheological factors in thrombosis and haemostasis. In: Bloom AL, Thomas DP (eds) Haemostasis and thrombosis, 2nd edn. Churchill Livingstone, Edinburgh, p 739
Katayama M, Hino F, Odete Y, Goto S, Kimizuka F, Kato I, Titani K, Sekiguchi K (1989) Isolation and characterization of two monoclonal antibodies that recognize remote epitopes on the cell-binding domain of human fibronectin. Exp Cell Res 185:229
Kehrel B, Balleisen L, Kokott R, Mester R, Stenzinger W, Clemetson KJ, van de Loo J (1988) Deficiency of intact thrombospondin and membrane glycoprotein Ia in platelets with defective collagen-induced aggregation and spontaneous loss of disorder. Blood 71:1074–1078
Kehrel B (1995) Platelet receptors for collagens. Platelets 6:11–16
Kieffer N, Fitzgerald LA, Wolf D, Cheresh DA, Phillips DR (1991) Adhesive properties of the $\beta_3$ integrins: comparison of GP IIb-IIIa and the vitronectin receptor individually expressed in human melanoma cells. J Cell Biol 113:451–461
Kloczewiak M, Timmons S, Lukas TJ, Hawiger J (1984) Platelet receptor recognition site on human fibrinogen. Synthesis and structure-function relationships of peptides corresponding to the carboxy-terminal segment of the gamma-chain. Biochemistry 23:1767–1774
Kunicki TJ, Orchekowski R, Annis D, Honda Y (1993) Variability of integrin $\alpha 2\beta 1$ activity on human platelets. Blood 82:2693–2703
Lahav J (1988) Thrombospondin inhibits adhesion of platelets to glass and protein covered substrata. Blood 71:1096–1099
Lam SC-T, Plow EF, Smith MA, Andrieux A, Ryckwaert J-J, Marguerie G, Ginsberg MH (1987) Evidence that arginyl-glycyl-aspartate peptides and fibrinogen $\gamma$-chain peptides share a common binding site on platelets. J Biol Chem 262:947–950
Lankhof H, van Hoeij M, Schiphorst ME, Bracke M, Wu Y-P, Ysseldijk MJW, Vink T, de Groot PG, Sixma JJ (1996) A3 domain of von Willebrand factor with collagen type III. Thromb Haemostas 75:950–958
Lankhof H, Wu YP, Vink T, Schiphorst ME, Zerwes H-G, De Groot PG, Sixma JJ (1995) Role of the GpIb-binding A1-repeat and the RGD-sequence in platelet adhesion to human recombinant von Willebrand factor. Blood 86:1035–1042
Leung LL (1984) Role of thrombospondin in platelet aggregation. J Clin Invest 74:1764–1772
Li K, Tamai K, Tan EML, Uitto J (1993) Cloning of type XVII collagen. J Biol Chem 268:8815–8834
Lüsher TF (1993) Nitric oxide, prostaglandins and endothelins. Ballieres Clin Haematol 6:609–627
Lynch DC, Zimmerman TS, Collins CJ, Brown M, Morin MJ, Ling EH, Livingstone DM (1985) Molecular cloning of cDNA for human von Willebrand factor: authentication by a new method. Cell 41:49–56
Mayne R (1986) Collagenous proteins of blood vessels. Arteriosclerosis 6:585–593
McCullagh KG (1983) Increased type I collagen in human atherosclerotic plaque. Atherosclerosis 46:247–248

Mohri H, Yoshioka A, Zimmerman TS, Ruggeri ZM (1989) Isolation of the von Willebrand factor domain interacting with platelet glycoprotein Ib, heparin, and collagen and characterization of its three distinct functional sites. J Biol Chem 264:17361–17367

Moncada S, Palmer RMJ, Higgs EA (1991) Nitric oxide: physiology, pathophysiology and pharmacology. Pharmacol Rev 43:109–142

Moroi M, Jung SM, Okuma M, Shinmyozu K (1989) A patient with platelets deficient in glycoprotein VI that lack both collagen induced aggregation and adhesion. J Clin Invest 84:1440

Morton LF, Barnes MJ (1982) Collagen polymorphism in the normal and diseased vessel wall. Atherosclerosis 42:41–51

Morton LF, Griffin B, Pepper DS, Barnes MJ (1983) The interaction between collagens and factor VIII/von Willebrand factor: investigation of the structural requirements for interaction. Thromb Res 32:545–556

Mosesson MW (1992) The roles of fibrinogen and fibrin in hemostasis and thrombosis. Semin Hematol 29:177–188

Murata K, Motoyama T, Kotake C (1986) Collagen types in various layers of the human aorta and their changes with the atheroslerotic process. Atherosclerosis 60:251–262

Nieuwenhuis HK, Akkerman JWN, Houdijk WPM, Sixma JJ (1985) Human blood platelets showing no response to collagen fail to express surface glycoprotein Ia. Nature 318:470–472

Nievelstein PFEM, Sixma JJ (1988) Glycoprotein IIB-IIIA and RGD(S) are not important for fibronectin-dependent platelet adhesion under flow conditions. Blood 72:82–88

Noonan DM, Fulle A, Valente P, Cai S, Horigan E, Sasaki M, Yamada Y, Hassell JR (1991) The complete sequence of perlecan, a basement membrane heparan sulfate proteoglycan, reveals extensive similarity with laminin A chain, low density lipoprotein-receptor, and the neural cell adhesion molecule. J Biol Chem 266:22939–22947

Obara M, Kang MS, Yamada KM (1988) Site-directed mutagenesis of the cell-binding domain of human fibronectin: separable, synergistic sites mediate adhesive function. Cell 53:649–657

Parise L, Philips DR (1986) Fibronectin-binding properties of the purified platelet glycoprotein IIb-IIIa complex. J Biol Chem 261:14011–14017

Petersen TE, Thogersen HC, Skorstengaard K, Vibe-Pedersen K, Sahl P, Sottrup-Jensen L, Magnusson S (1983) Partial primary structure of bovine plasma fibronectin: three types of internal homology. Proc Natl Acad Sci USA 80:137–141

Peterson DM, Stathopoulos NA, Giorgio TD, Hellums JD, Moake JL (1987) Shear-induced platelet aggregation requires von Willebrand factor and platelet membrane glycoproteins Ib and IIb-IIIa. Blood 69:625–628

Pihlajaniemi T, Tamminen M (1990) The $\alpha 1$ chain of type XIII collagen consists of three collagenous and four noncollagenous domains, and its primary transcript undergoes complex alternative splicing. J Biol Chem 265:16922–16928

Piotrowicz RS, Orchekowski RP, Nugent DJ, Yamada KY, Kunicki TJ (1988) Glycoprotein Ic-IIa functions as an activation independent fibronectin receptor on human platelets. J Cell Biol 106:1359–1364

Plow EF, McEver RP, Coller BS, Woods VL, Marguerie GA, Ginsberg MH (1985a) Related binding mechanisms for fibrinogen, fibronectin, von Willebrand factor and thrombospondin on thrombin-stimulated platelets. Blood 66:724–727

Plow EF, Pierschbacher MD, Ruoslahti E, Marguerie GA, Ginsberg MH (1985b) The effect of Arg-Gly-Asp-containing peptides on fibrinogen and von Willebrand factor binding to platelets. Proc Natl Acad Sci USA 70:110–115

Pytela R, Pierschbacher MD, Ruoslahti E (1985) Identification and isolation of a 140 kD cell surface glycoprotein with properties expected of a fibronectin receptor. Cell 40:191–198

Rand JH, Sussman II, Gordon RE, Shu SV, Solomon V (1980) Localization of factor VIII-related antigen in human vascular subendothelium. Blood 55:752–756

Rand JH, Patel ND, Schwartz E, Zhou S-L, Potter BJ (1991) 150-kD von Willebrand factor binding protein extracted from human vascular subendothelium is type VI collagen. J Clin Invest 88:253–259

Rand JH, Wu X-X, Potter BJ, Uson RR, Gordon RE (1993) Co-localization of von Willebrand factor and type VI collagen in human vascular subendothelium. Am J Pathol 142:843–850

Read MS, Sherma RW, Brinkhous KM (1978) Venom coagglutin: an activator of platelet aggregation dependent on von Willebrand factor. Proc Natl Acad Sci USA 75:4514–4518

Riessen R, Isner JM, Blessing E, Loushin C, Nikol S, Wight TN (1994) Regional differences in the distribution of the proteoglycans biglycan and decorin in the extracellular matrix of atherosclerotic and restenotic human coronary arteries. Am J Pathol 144:962–974

Robbins RA, Wagner WD, Sawyer LM, Caterson B (1989) Immunolocalization of proteoglycan types in aortas of pigeons with spontaneous or diet-induced atherosclerosis. Am J Pathol 134:615–626

Rodin JAG (1974) Anonymous histology, a text atlas. Oxford University Press, Oxford, p 340

Ross JM, McIntire LV, Moake JL, Rand JH (1995) Platelet adhesion and aggregation on human type VI collagen surfaces under physiological flow conditions. Blood 85:1826–1835

Roth GJ, Titani K, Hoyer LW, Hickey MJ (1986) Localization of binding sites within human von Willebrand factor for monomeric type III collagen. Biochemistry 25:8357–8361

Ruoslahti E, Yamaguchi Y (1991) Proteoglycans as modulators of growth factor activities. Cell 64:867–869

Ryo R, Yoshida A, Sugano W, Yasunaga M, Nakayama K, Saigo K, Adachi M, Yamaguchi N, Okuma M (1992) Deficiency of p62, a putative collagen receptor, in platelets from a patient with defective collagen-induced platelet aggregation. Am J Hematol 39:25–31

Sadler JE, Shelton-Inloes BB, Sorace JM, Harlan JM, Titani K, Davie EW (1985) Cloning and characterization of two cDNAs coding for human von Willebrand factor. Proc Natl Acad Sci USA 82:6394–6398

Sadler JE (1994a) Von Willebrand disease. In: Bloom AL, Forbes CD, Thomas DP, Tuddenham EGD (eds) Haemostasis and thrombosis, 3rd edn. Churchill Livingstone, Edinburgh London Madrid Melbourne New York Tokyo, p 843

Sadler JE (1994b) A revised classification of von Willebrand disease. Thromb Haemostas 71:520–525

Saelman EUM, Horton LF, Barnes MJ, Gralnick HR, Hese KM, Nieuwenhuis HK, De Groot PhG, Sixma JJ (1993) Platelet adhesion to canogen-bromide fragments of collagen $\alpha$1 (I) under flow conditions. Blood 82:3029–3033

Saelman EUM, Nieuwenhuis HK, Hese KM, de Groot PG, Heijnen HFG, Sage EH, Williams S, McKeown L, Gralnick HR, Sixma JJ (1994) Platelet adhesion to collagen types I through VIII under conditions of stasis and flow is mediated by GPIa/IIa ($\alpha_2\beta_1$-integrin). Blood 83:1244–1250

Sakariassen KS, Bolhuis PA, Sixma JJ (1979) Human blood platelet adhesion to artery subendothelium is mediated by factor VIII-von Willebrand factor bound to the subendothelium. Nature 279:635–638

Sakariassen KS, Aarts PAMM, De Groot PG, Houdijk WPM, Sixma JJ (1983) A perfusion chamber developed to investigate platelet interaction in flowing blood with human vessel wall cells, their extracellular matrix and purified components. J Lab Clin Med 102:522–535

Sakariassen KS, Nievelstein PFEM, Coller BS, Sixma JJ (1986) The role of platelet membrane glycoproteins Ib and IIb-IIIa in platelet adherence to human artery subendothelium. Brit J Haematol 63:681–691

Santoro SA (1983) Preferential binding of high molecular weight forms of von Willebrand factor to fibrillar collagen. Biochim Biophys Acta 756:123–126

Santoro SA (1986) Identification of a 160000 dalton platelet membrane protein that mediates the initial divalent cation-dependent adhesion of platelets to collagen. Cell 46:913–920

Santoro SA (1988) Molecular basis of platelet adhesion to collagen. In: Jamieson GA (ed) Platelet membrane receptors: molecular biology, immunology, biochemistry, and pathology. Liss, New York, p 291

Savage B, Ruggeri ZM (1991) Selective recognition of adhesive sites in surface-bound fibrinogen by glycoprotein IIb-IIIa on nonactivated platelets. J Biol Chem 266:11227–11233

Savage B, Shattil SJ, Ruggeri ZM (1992) Modulation of platelet function through adhesion receptors. A dual role for glycoprotein IIb-IIIa (integrin $\alpha$IIb$\beta$3) mediated by fibrinogen and glycoprotein Ib-von Willebrand factor. J Biol Chem 267:11300–11306

Savage B, Saldívar E, Ruggeri ZM (1996) Initiation of platelet adhesion by arrest onto fibrinogen or translocation on von Willebrand Factor. Cell 84:289–297

Schwarzbauer J (1990) The fibronectin gene. In: Sandell LJ, Boyd CD (eds) Extracellular matrix genes. Academic, San Diego, p 195

Schönherr E, Witsch-Prehm P, Harrach B, Robenek H, Rauterberg J, Kresse H (1995) Interaction of biglycan with type I collagen. J Biol Chem 270:2776–2783

Sixma JJ, Wester J (1977) The haemostatic plug. Semin Hemat 14:265–299

Sixma JJ, Sakariassen KS, Stel HV, Houdijk WPM, In der Maur DW, Hamer RJ, De Groot PG, Van Mourik JA (1984) Functional domains on von Willebrand factor. Recognition of discrete fragments by monoclonal antibodies that inhibit interaction of von Willebrand factor with platelets and with collage. J Clin Invest 74:736–744

Sixma JJ, Schiphorst ME, Verweij CL, Pannekoek H (1991) Effect of deletion of the A1 domain of von Willebrand factor on its binding to heparin, collagen and platelets in the presence of ristocetin. Eur J Biochem 196:369–375

Smith EB (1990) Transport, interactions and retention of plasma proteins in the intima: the barrier function of the internal elastic lamina. Eur Heart J 11 Suppl E:72–81

Sobel M, Soler DF, Kermode JC, Harris RB (1992) Localization and characterization of a heparin binding domain peptide of human von Willebrand factor. J Biol Chem 267:8857–8862

Sonnenberg A, Linders KJT, Modderman PW, Damsky CH, Aumailley M, Timpl R (1990) Integrin recognition of different cell-binding fragments of laminin (P1, E3, E8) and evidence that $\alpha 6\beta 1$ but not $\alpha 6\beta 4$ functions as a major receptor for fragment E8. J Cell Biol 110:2145–2155

Staatz WD, Rajpara SM, Wayner EA, Carter WG, Santoro SA (1989) The membrane glycoprotein Ia–IIa (VLA-2) complex mediates the $Mg^{++}$-dependent adhesion of platelets to collagen. J Cell Biol 108:1917–1924

Staatz WD, Fok KF, Zutter MM, Adams SP, Rodriguez BA, Santoro SA (1991) Identification of a tetrapeptide recognition sequence for the $\alpha_2\beta_1$ integrin in collagen. J Biol Chem 266:7363–7367

Stary HC (1987) Macrophages, macrophage foam cells, and eccentric intimal thickening in the coronary arteries of young children. Atherosclerosis 64:91–108

Stary HC, Blankenhorn DH, Chandler AB, Glagov S, Insull W, Richardson M, Rosenfeld ME, Schaffer SA, Schwarz CJ, Wagner WD, Wissler RW (1992) A definition of the intima of human arteries and of its atherosclerotic-prone regions. Circulation 85:391–405

Stel HV, Sakariassen KS, De Groot PG, Van Mourik JA, Sixma JJ (1985) Von Willebrand factor in the vessel wall mediates platelet adherence. Blood 65:85–90

Tamkun JW, Hynes RO (1983) Plasma fibronectin is synthesized and secreted by hepatocytes. J Biol Chem 258:4641–4647

Tandon NN, Kralisz U, Jamieson GA (1989) Identification of glycoprotein IV (CD36) as a pimary receptor for platelet-collagen adhesion. J Biol Chem 264:7576–7583

Titani K, Kumar S, Takio K, Ericsson LH, Wade RD, Ashida K, Walsh KA, Chopek MW, Sadler JE, Fujikawa K (1986) Amino acid sequence of human von Willebrand factor. Biochemistry 25:3171–3184

Turitto VT, Weiss HJ, Zimmerman TS, Sussman II (1985) Factor VIII/von Willebrand factor in subendothelium mediates platelet adhesion Blood. 65:823–831

Tuszynski GP, Rothman V, Murphy A, Siegler K, Smith L, Smith S, Karcewski J, Knudsen KA (1987) Thrombospondin promotes cell-substratum adhesion. Science 236:1570–1573

Tuszynski GP, Rothman VL, Murphy A, Siegler K, Knudsen KA (1988) Thrombospondin promotes platelet aggregation. Blood 72:109–115

Uijttewaal WSJ, Nijhoh EJ, Bronkhorst PJH, den Hartog E, Heethaar RM (1993) Near wall excess of platelets induced by lateral migration of erythrocytes in flowing blood. Am J Physiol 264:H1239–H1244

van Breugel JHFI, De Groot PG, Heethaar RM, Sixma JJ (1992) The role of plasma viscosity in platelet adhesion. Blood 80:953–959

Van Hinsberg VWM, Bertina RM, Van Wijngaarden A, Van Tilburg NH, Emeis JJ, Haverkate F (1985) Activated protein C decreases plasminogen activator-inhibitor activity in endothelial cell-conditioned medium. Blood 65:444–451

Van der Rest M, Garrone R (1991) Collagen family of proteins. FASEB J 5:2814–2823

Van Zanten GH, Saelman EUM, Schut-Hese KM, Wu YP, Slootweg PJ, Nieuwenhuis HK, Sixma JJ (1996) Platelet adhesion to collagen type IV under flow conditions. Blood 88:3862–3871

Verweij CL, de Vries CJM, Distel B, van Zonneveld AJ, Geurts van Kessel A, van Mourik JA, Pannekoek H (1985) Construction of cDNA coding for human von Willebrand factor using antibody probes for colony-screening and mapping of the chromosomal gene. Nucleic Acids Research 13:4699–4717

Vogel KG, Trotter JA (1987) Interaction of a 59-kDa connective tissue matrix protein with collagen I and collagen II. Collagen Relat Res 7:105–114

Wagner DD, Marder VJ (1983) Biosynthesis of von Willebrand protein by human endothelial cells. J Biol Chem 258:2065–2067

Wagner DD, Urban-Pickering M, Marder V (1984) Von Willebrand factor binds to extracellular matrices independently of collagen. Proc Natl Acad Sci USA 81:471–475

Wagner DD (1990) Cell biology of von Willebrand factor. Annu Rev Cell Biol 6:217–246

Wang NH, Keller KH (1979) Solute transport induced by erythrocyte motions in shear flow. Trans Am Soc Artific Org 25:14–18

Wayner EA, Carter WG, Piotrowicz RS, Kunicki TJ (1988) The function of multiple extracellular matrix receptors in mediating cell adhesion to extracellular matrix: preparation of monoclonal antibodies to the fibronectin receptor that specifically inhibit cell adhesion to fibronectin and react with platelet glycoproteins Ic-IIa. J Cell Biol 107:1881–1891

Weiss HJ, Turitto VT (1979) Prostacyclin inhibits platelet adhesion and thrombus formation on subendothelium. Blood 53:244–250

Weiss HJ, Hawiger J, Ruggeri ZM, Turitto VT, Thiagarajan R, Hoffmann T (1989) Fibrinogen independent adhesion and thrombus formation on subendothelium mediated by GPIIb-IIIa complex at high shear rate. J Clin Invest 83:288–297

Weiss HJ, Turitto VT, Baumgartner HR (1991) Further evidence that glycoprotein IIb–IIIa mediates platelet spreading on subendothelium. Thromb Haemostasis 65:202–205

Wight TN, Raugi GJ, Mumby SM, Bornstein P (1985) Light microscopic immunolocation of thrombospondin in human tissues. J Histochem Cytochem 33:295–302

Yamamoto N, Ikeda H, Tandon NN, Herman J, Tomiyama Y, Mitani T, Sekiguchi S, Lipsky R, Kralisz U, Jamieson GA (1990) A platelet membrane glycoprotein (GP) deficiency in healthy blood donors: Nak$^{a-}$ platelets lack detectable GPIV (CD36). Blood 76:1698–1703

Ylä-Herttuala S, Sumuvuori H, Karkola K, Möttönen M, Nikkari T (1986) Glycosaminoglycans in normal and atherosclerotic human coronary arteries. Lab Invest 54:402–407

Yurchenco PD, Schittny JC (1990) Molecular architecture of basement membranes. FASEB J 4:1577–1590

Zucker MB, Kim SJ, McPherson J, Grant RA (1977) Binding of factor VIII to platelets in the presence of ristocetin. Br J Haematol 35:535–549

Zwaginga JJ, IJsseldijk MJW, de Groot PG, Sixma JJ (1991) Treatment of uremic anemia with recombinant erythropoietin also reduces the defects in platelet adhesion and aggregation caused by uremic plasma. Thromb Haemostas 66:638–647

CHAPTER 4

# Platelet Aggregation

A. Ruf, M.M. Frojmovic, and H. Patscheke

## A. Introduction

The formation of a platelet thrombus plays a crucial role in hemostasis and also in pathomechanisms of several arterial disorders, including stroke and myocardial infarction. The initial step in thrombus formation is the adhesion of platelets onto vascular subendothelial connective tissue exposed upon endothelial injury (see Chap. 3). Collagen and von Willebrand factor (vWF) are important constituents of the subendothelial matrix which mediate adhesion and subsequent activation of the platelets. Platelet activation allows interplatelet contact and the formation of platelet aggregates. In this chapter, the molecular mechanisms mediating platelet aggregation and the tests to assess this platelet function in vitro are summarized. In addition, recent progress in the selective inhibition of platelet aggregation as a therapeutical principle is addressed.

## B. Mechanism of Platelet Aggregation

### I. The Glycoprotein IIb-IIIa Complex

It is well established that the platelet membrane glycoproteins (GP) IIb and IIIa are essential for platelet aggregation (Kieffer and Phillips 1990; Nurden 1994). The glycoproteins IIb and IIIa form a heterodimeric complex in the platelet membrane which belongs to the protein superfamily of integrins (Hynes 1987; Ruoslahti 1991). Platelets from patients with thrombasthenia lack the GPIIb–IIIa complex or express it in malfunctional forms and are not able to aggregate (George et al. 1990). The complex constitutes a prosmiscuous receptor for several ligands such as fibrinogen, vWF, fibronectin, vitronectin, and possibly other cytoadhesive proteins (Plow et al. 1985; Haverstick et al. 1985). The complex is exposed on resting platelets in an inactive form and platelet activation is required to transform it into a conformational state which is competent for ligand binding (Sims et al. 1991; Frelinger et al. 1991; Du et al. 1991). Fibrinogen contains distinct amino acid sequences which mediate the binding to GPIIb–IIIa (Hawiger et al. 1982). These include the Arg-Gly-Asp (RGD) sequences in the A$\alpha$-chain and a dodecapeptide located at the carboxy terminal $\gamma$-chain (Fig. 1). Other adhesive

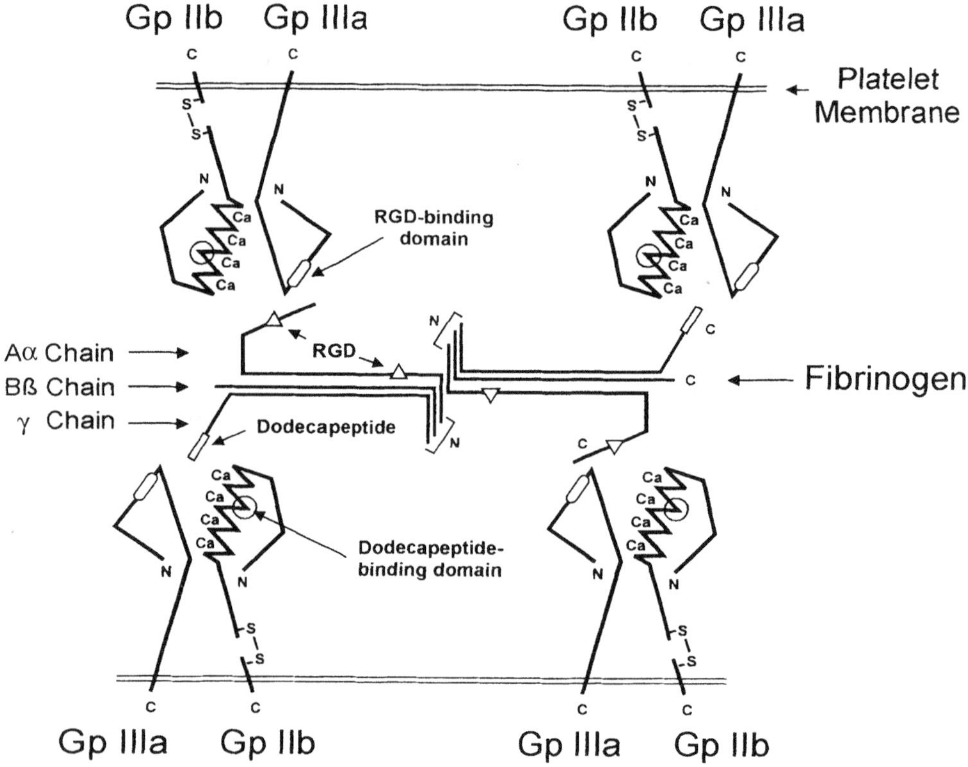


**Fig. 1.** Fibrinogen as a bridging molecule between two adjacent platelets. *Fibrinogen* has six *GpIIb–IIIa* binding sites: four *RGD* sequences on the *Aα-chains*, and two *dodecapeptides*, one at each carboxy terminus of the *γ-chains*. The RGD-peptide has been cross-linked to GPIIIa (residues 109–171), whereas the dodecapeptide cross-links to GPIIIa (residues 244–314). (Adapted from CHARO et al. 1994)

proteins, such as vWF, bind to the activated GPIIb–IIIa via an RGD motif. On the basis of chemical cross-linking experiments, it was concluded that the primary binding site for RGD peptides is located in GPIIIa (residues 109–171), while the dodecapeptide of fibrinogen cross-links essentially to GpIIb at or close to the second putative $Ca^{2+}$-binding domain (PLOW et al. 1992; BENNETT 1996).

The peptide GpIIb alpha 300–312 binds, in addition to fibrinogen, also other adhesive glycoproteins such as vitronectin. Furthermore, this peptide inhibits the adhesion of activated platelets to fibrinogen, fibronectin, vitronectin, and vWF (TAYLOR and GARTNER 1992). Thus this peptide may constitute a common recognition site on GPIIb–IIIa for the dodecapeptide and the RGD-containing ligands. Another binding site for fibrinogen seems to be located at residues 211 through 222 of GPIIIa (CHARO et al. 1991). With respect to most of the published studies, it is reasonable to assume that an activation-dependent conformational change in the GPIIb–IIIa complex leads to the formation of a ligand-binding pocket which is composed of the above peptides.

## II. Adhesive Ligands of the GPIIb-IIIa Complex

It is well established that fibrinogen plays an important role in platelet aggregation under low shear conditions (PEERSCHKE 1985; WEISS et al. 1989; IKEDA et al. 1991; XIA and FROJMOVIC 1994). Human fibrinogen is composed of three pairs of distinct protein chains ($\alpha$, $\beta$, $\gamma$) which are linked by various disulfide bonds and arranged in three main domains: one central E and two distal D domains (DOOLITTLE 1984). The amino acid sequence RGD on the A$\alpha$-chain (residues $\alpha$ 95–97 and $\alpha$ 572–574) and the dodecapeptide HHLGGAKQAGDV on the carboxy terminus of the $\gamma$-chain of fibrinogen (residues $\gamma$ 400–411) are potential recognition sites for the GPIIb–IIIa complex (ANRIEUX et al. 1989). Studies using mutated fibrinogens, synthetic peptides, and antibodies against fibrinogen have highlighted the essential role of the dodecapeptide in intact fibrinogen for platelet aggregation (KLOCZEWIAK et al. 1984, 1989; ANRIEUX et al. 1989; FARREL et al. 1992; KIRSCHBAUM et al. 1992). Recombinant fibrinogens with mutations in RGD sequences supports platelet aggregation as effectively as wild-type plasma fibrinogen. However mutations in the $\gamma$-chain dodecapeptide yield variant fibrinogens which are markedly defective in platelet aggregation (FARRELL et al. 1992).

Fibrinogen is symmetric and possesses two sets of binding domains at each end; thus, it can form a molecular bridge between two adjacent platelets (HAWIGER et al. 1989). Each half of the fibrinogen molecule can potentially bind to two or three receptor sites on the platelet membrane (Fig. 1). The gap between platelets in closely packed aggregates in vivo is 100 to 300 Å wide (SHIRASAWA and CHANDLER 1969; MOON et al. 1990). Since the overall length of the fibrinogen molecule is about 475 Å it is likely that the fibrinogen molecule is oriented such that its long axis is parallel to the contact zone of adhering platelets (Fig. 1).

In contrast to the case of low shear conditions where fibrinogen mediates platelet aggregation, at high shear rates, vWF seems to bind to GpIIb–IIa and thereby mediate platelet aggregation and thrombus formation (FRESSINAUD et al. 1988; WEISS et al. 1989). Platelet aggregation on purified collagen or subendothelium is inhibited by a monoclonal antibody of vWf that selectively blocks vWF binding, but not fibrinogen binding to GPIIb–IIIa (WEISS et al. 1989). Direct evidence that the RGD sequence in vWF is involved was provided by studies in which aggregation was inhibited by monoclonal antibodies that recognize the RGD (residues 1744–1746) sequence in vWF and selectively inhibit vWF binding to GPIIb–IIIa and platelet aggregation at high shear rates (WEISS et al. 1993). Other studies using cone–in–plate viscometers also provided evidence that vWF mediates platelet–platelet interaction at high shear rates whereas fibrinogen does not (IKEDA et al. 1991, 1993; CHOW et al. 1992; RUGGERI 1994). These investigations showed that shear rates >4000 $s^{-1}$ in plasma initiate the binding of multimeric vWF to GPIb, which subsequently induces a transmembrane flux of calcium ions. The increase in intracellular calcium levels leads to platelet activation with an exposure of active GPIIb–IIIa complexes which in turn bind vWF and thereby mediates platelet aggregation. When platelets are extrinsically activated with agonists such as ADP, much lower shear rates (300–1200 $s^{-1}$) will drive platelet aggregation independent of fibrinogen; it is likely that this aggregation is mediated by vWF secreted onto the platelet surface (GOLDSMITH et al. 1994).

### III. Redistribution of the GpIIb-IIIa Complex and Internalization of the Ligands

Stimulation of platelets leads to an increase in the number of GPIIb–IIIa complexes on the platelet surface which may be due to a redistribution from intracellular pools such as the membranes of the dense tubular system and storage organelles (WENCEL-DRAKE et al. 1986; WOODS et al. 1986). The binding of fibrinogen induces a clustering of the GpIIb–IIIa complexes which may be caused by cross-linking of neighboring receptors by the multiple binding domains of the fibrinogen molecule in cis position (ISENBERG et al. 1987). This clustering is followed by an internalization of fibrinogen; the internalized fibrinogen seems to represent the fraction of irreversibly bound fibrinogen (PEERSCHKE 1995; WENZEL-DRAKE et al. 1996). Platelets which have internalized fibrinogen lose their ability to aggregate as measured turbidimetrically (PEERSCHKE 1995; WENZEL-DRAKE et al. 1996). Thus, receptor redistribution and fibrinogen internalization may represent a regulatory mechanism of platelet aggregation. Whether other ligands, such as vWF, can also be internalized is still matter of debate.

## C. Platelet Aggregation Testing

### I. Sample Preparation

Platelet aggregation testing can be performed with various platelet samples such as whole blood, platelet-rich plasma (PRP), and suspensions of washed and gel-filtered platelets (MUSTARD et al. 1972; HUTTON et al. 1974; DAY et al. 1975; PATSCHEKE 1981). The procedures applied during sample preparation may affect the activation state and the functional integrity of the platelets. The factors involved are the blood sampling procedure, the type of anticoagulant used, the pH and temperature during the isolation procedures, and the composition of the platelet suspending media (MUSTARD and PACKHAM 1970; ZUCKER 1972; DAY et al. 1975; KINLOUGH-RATHBONE et al. 1977a; PACKHAM et al. 1978). Artefactual platelet activation is one of the major problems. This seems mainly to be due to the formation of trace amounts of thrombin during blood sampling which although not sufficient to initiate fibrin generation may activate platelets (KAPLAN and OWEN 1981; LEVINE et al. 1981). An artifactual preactivation may enhance the aggregation response induced by added agonists or may decrease it in case of a desensitization (KINLOUGH-RATHBONE et al. 1977b; PACKHAM et al. 1973). Whichever effect occurs is dependent on the time and degree of preactivation and is usually unpredictable.

In addition to the procedure of blood collection, the anticoagulant can also affect the function or the activation state of platelets. Heparin has long been recognized as an inducer of platelet activation (EIKA 1972; ZUCKER 1975). Trisodium citrate depresses the $Ca^{2+}$ concentration, and under these conditions platelets liberate arachidonate when they are brought into close contact with one another (DAY et al. 1975; MUSTARD et al. 1975). Arachidonate is converted to thromboxane $A_2$ ($TXA_2$) which is a potent platelet agonist. Formation of $TXA_2$ mediates the release reaction upon stimulation of citrate-anticoagulated PRP with ADP (MUSTARD et al. 1975). It also accounts for the aggregation of platelets when citrated plasma is subjected to centrifugation during washing procedures. This problem does not occur when blood is collected into acid-citrate-dextrose (ASTER and JANDL 1964) which gives a pH of 6.5 or slightly lower and prevents platelet activation and aggregation upon centrifugation.

Platelet-rich plasma obtained by using hirudin as an anticoagulant exhibits minimal activation of platelets and the arachidonate pathway is not activated when the platelets are aggregated by ADP; this is most likely due to the fact that the ionic calcium concentration in this plasma is not depressed (MUSTARD et al. 1975). EDTA can induce the conformational change of platelets and dissociates the GPIIb–IIIa complex at 37°C (GRANT and ZUCKER 1978; FITZGERALD and PHILLIPS 1985). These platelet alterations may be irreversible if platelets are exposed to EDTA for 10–15 min, depending on the temperature. Thus EDTA should not be used as anticoagulant even if it is diluted during the isolating procedures.

When washing or gel-filtrating procedures are to be applied, the buffer used to maintain the pH of the platelet suspending media should not affect platelet function. In particular TRIS should be avoided since it exerts substantial effects on platelet structure and function (LAGES et al. 1975; ZUCKER et al. 1974). The suspending media for platelets should contain some essential constituents, in particular various ions such as $Ca^{2+}$ and a protein such as albumin (DAY et al. 1975; KINIOUGH-RATHBONE et al. 1977a; PACKHAM et al. 1978). Furthermore, glucose should be added to the media since platelets require a source of metabolic energy (KINIOUGH-RATHBONE et al. 1970, 1972). Some ADP may be released from residual red blood cells during prolonged storage of a suspension of isolated platelets. This may cause the platelets to become refractory to ADP. Addition of enzyme systems such as apyrase or creatine phosphate/creatine phosphokinase to the suspending medium will remove this ADP and maintain platelets sensitive to ADP.

## II. Optical Aggregometry

The turbidimetric determination of platelet aggregation was introduced by Born (1962a,b). The principal of the method is as follows: light shines through a cuvette which contains a platelet suspension and any transmitted light is detected by a photoelectric cell. Stimulation of the platelets leads to shape change and aggregation. Both phenomena affect the light transmittance of a platelet suspension, although it must be emphasized that only shape change and macroaggregation are measurable. It was found that the microaggregation of platelets which parallels the shape change does not contribute to the initial change in light transmittance until >30%–40% of the singlets have aggregated (Milton and Frojmovic 1983; Frojmovic et al. 1989). The disc–to–sphere transformation leads to a decrease and the platelet macroaggregation to an increase in light transmittance.

Many, apparently disparate, theories have been proposed to explain the observed turbidity changes related to platelet activation. The effects are mainly due to the different light scattering properties of resting discoid platelets, activated platelets, and platelet aggregates. Latimer predicted by application of light scattering theories that a change in platelet orientation by flow and of their shape leads to change in light transmittance, which he experimentally verified (Latimer et al. 1977). This accounts for the fact that orientation of discoid platelets perpendicular to the incident light beam by stirring the platelet sample leads to an increase in transmittance. Without stirring, discoid platelets become randomly oriented which causes the light scatter to increase and as consequence the light transmission decreases. A suspension of spheroid platelets does not show a stir-dependent change in transmittance since a change in orientation by flow cannot occur. However, if discoid platelets change their shape upon stimulation, light scatter increases and the transmittance decreases. In addition, pseudopod formation moves dry weight away from the center of mass of the cell which causes total light scattering to be decreased; i.e., formation of pseudopods increases light transmittance. However, the decrease in transmittance due to the disc–to–sphere transformation dominates the increase in transmittance caused by pseudopod formation; thus, the net effect is a decrease in transmittance (Latimer et al. 1977).

Indeed, measurement of the change in shape, from an analysis of the mean platelet axial ratio, within the first 10s upon ADP addition, showed that the slope of the initial decrease in light transmission was in good agreement with the theoretical predictions. However, at later times (>40s), pseudopod formation has an important influence on the change in light transmittance (Milton and Frojmovic 1983). The extent of the optical effect of shape change strongly depends on the optical geometry of the aggregometer (Latimer et al. 1977; Latimer 1975; Frojmovic 1978).

The situation with a suspension of aggregating platelets is even more complex than in suspensions of single plateles. It is a heterogenous system with respect to the size distribution of the particles, i.e., it contains platelets and aggregates of various sizes. Furthermore, it is a dynamic system with a shift in the size distribution towards large aggregates as long as the aggregation process proceeds. Frojmovic has shown that the law of Lambert–Beer is valid for platelet suspensions at a concentration within the physiological range. Furthermore, he proposed that the law of Lambert–Beer may be rewritten as follows to account for particle aggregation (FROJMOVIC 1978):

$$E = L\sum_{n=1}^{\infty} N_n K_n$$

where $L$ is the light path length of the cuvette; $N_n$ is the number of n-tuples per unit volume; $K_n$ is the scattering cross-section for a particular n-tuple, where $n = 1$ for singlets of the same size, $n = 2$ for doublets, $n = 3$ for triplets, etc.

Since the scattering cross-section of an n-tuple relative to that of n singlets decreases with increasing n (CHANG and ROBERTSON 1976), a particular number of single platelets scatter more light than an aggregate consisting of the corresponding platelet number. Thus the light transmission increases as a consequence of platelet aggregation.

The most widespread parameters used for quantitation of platelet aggregation are the maximal net increase in transmission, the slope of the aggregation curve, and the half-time of the aggregation curve (FROJMOVIC 1973). All of these parameters are dependent on several factors such as the optical geometry, wavelength, and light path length of the measuring device as well as shape, size distribution, and concentration of the platelets and the flow conditions in the system (FROJMOVIC 1978). Athough the method has a history of over 30 years there are still major problems regarding standardization between different laboratories. Various types of aggregometer are available, each of which are well defined and stable with respect to the above rheooptical characteristics (MILLS 1969; O'BRIEN 1971; TEN CATE 1972; BORN 1972; SIXMA 1972; FROJMOVIC 1978; LATIMER 1982). However, these characteristics differ significantly between the different aggregometer types and thus, quantitative results obtained with different types are not comparable.

## III. Lumi-aggregometry

Lumi-aggregometers are turbidimeters which are additionally equipped with a photomultiplier rendering them capabable of detecting chemiluminescence (FEINMAN et al. 1977; INGERMAN-WOJENSKI et al. 1983). Thus, these aggregometers allow the determination of ADP release from platelet dense bodies using the Luciferin/luciferase system. Furthermore, with Lumi-aggregometry it is possible to simultaneously measure the activation of platelets and neutrophils or possibly other phagocytosing white blood cells (RUF et al. 1992). With this application the experimental conditions are adjusted in a way that the increase in light transmission predominantly reflects platelet aggregation and the Luminol-enhanced chemiluminescence reflects the oxidative burst of neutrophils.

Another feature of Lumi-aggregometers is that they can be supplied with an impedance module which can assess platelet aggregation electrically. The principle of electrical aggregometry is as follows (CARDINAL and FLOWER 1980). Two electrodes separated by a small gap are located in the cuvette containing a platelet sample. The electrodes become coated with a monolayer of platelets, and if no platelet agonists are added no further interaction with platelets and the electrodes occurs. Thus, in unstimulated platelet samples the conductance between the two electrodes is constant after the initial coating of the electrodes by platelets. When the platelets are stimulated the electrodes become further coated with aggregating platelets which impair the conductance between the two electrodes. Hence, an increase in impedance can be recorded as a consequence of platelet aggregation.

## IV. Determination of Platelet Aggregation by Particle Counting

The concentration of single, free platelets and total particles (platelet doublets, triplets, etc.) decreases as a consequence of platelet aggregation. Thus aggregation can be determined by measuring the concentration of single platelets and/or total particles before and after stimulation of platelet aggregation (LUMLEY and HUMPHREY 1981; GEAR 1982; FROJMOVIC et al. 1989). The aggregation can be quantified as the fraction (PA) of platelets which were incorporated into aggregates:

$$PA = 1 - N_t/N_0$$

where $N_0$ is the concentration of platelets before aggregation and $N_t$, the platelet concentration upon stimulation of aggregation at a time point t. Measurements by electronic particle counters may record platelet aggregates consisting of up to four platelets as "singlets" depending on the platelet size. However, comparisons with microaggregation determined by tedious microscopic analysis of truly singlet platelets have shown comparable sensitivities for the two methods (FROJMOVIC et al. 1989). Both total particle count and "singlets" can also be determined with highly sophisticated flow cytometers, using the technique first described for similar studies of neutrophil aggregation (ROCHON and FROJMOVIC 1993).

## V. Potential and Limitations of Platelet Aggregation Testing

Platelet aggregation testing is still the most often applied functional assay of platelets. It has been used for the characterization of various platelet agonists as well as platelet inhibitors in experimental and clinical research settings. Furthermore, it is an established screening tool for the diagnosis of thrombocytopathies. However, it has to be noted that the use of aggregation testing for assessing the actual in vivo function of platelets is rather limited. This is due to two major problems. Firstly, it is impossible to preserve the in vivo activation state of platelets during sample preparation (see Sect. C.I in this chapter). The second problem is the standardization of the aggregation testing with respect to the different size distributions of platelets from different donors. It is well known that the rate of aggregation is dependent on the number of platelet collisions in a platelet sample. The collision frequency of particles in suspension is dependent not only on the stirring rate and the particle concentration but also on the particle volume (MANLEY and MASON 1952). Large platelets collide more frequently than small ones; thus, large platelets may undergo an enhanced aggregation although no functional differences to small platelets may exist. Thus, if a comparison is to be made and differences in platelet aggregation between two different donors are to be detected, an adjustment for different size distributions is required. To compensate for different size distributions of platelets it was suggested by HOLME and MURPHY (1981) to adjust the extinction of the test and the control platelet sample prior to turbidimetric measurements of aggregation. The rationale for this type of calibration is the fact that the extinction of a platelet sample prior to aggregation reflects the mass and the relative refractive index of the platelets. However, if the extinctions are adjusted, different platelet contentrations between test and control samples have to be accepted.

Among the different methods for aggregation testing turbidimetry is the most widespread application used. There are fundamental differences between the different methods with respect to the material which can be analyzed and the information which can be obtained (Table 1; FROJMOVIC et al. 1989). In contrast to electrical aggregometry and particle counting, turbidimetry can assess platelet aggregation only in suspensions of isolated platelets such as PRP or washed or gel-filtered platelets and not in whole blood. Turbi-

**Table 1.** Characteristics of aggregation testing methods

| | Particle counting | Optical/Lumi-aggregometry |
|---|---|---|
| Advantages | – Applicable to all platelet samples, i.e., whole blood, PRP, washed or gel-filtered platelets<br>– Extremely sensitive for early stages of platelet aggregation, i.e., formation of small aggregates | – Continuous measurements, i.e., kinetic information available<br>– Sensitive for late stages of platelet aggregation, i.e., formation of large aggregates<br>– Simultaneous measurement of other parameters of platelet activation such as shape change or release |
| Disadvantages | – Insensitive for late stages of platelet aggregation, i.e., formation of large aggregates<br>– Discontinuous measurement | – Insensitive for early stages of platelet aggregation, i.e., formation of small aggregates<br>– Applicable only to samples of isolated platelets, i.e., PRP etc., except for Lumi-aggregometers with impedance modules |

dimetric and electrical aggregometry yield kinetic information on the aggregation process. In particular it can easily be seen whether the process is reversible or not. With particle counting this information can only be obtained when samples are taken and measured at different time intervals. Furthermore, counting single platelets is very sensitive for detecting platelet aggregation at early stages of aggregation during which the concentrations of platelets rapidly decreases and only small aggregates are formed; it is completely insensitive at later stages at which mainly the size of the aggregates increases. The opposite is true for optical aggregometry which is very insensitive for the detection of small aggregates composed of few platelets (MILTON and FROJMOVIC 1983). This difference in sensitivity is the reason why, for a given agonist, different concentration–response relationships are obtained when the aggregation is measured by either particle counting or turbidimetry and electrical aggregometry. Therefore, $EC_{50}$ values, i.e., the concentration of agonist producing 50% aggregation, will be much lower with particle counting than with electrical and optical aggregometry. On the other hand, optical and probably also electrical aggregometry is more sensitive to an inhibition of platelet aggregation than particle counting (FROJMOVIC et al. 1983; PEDVIS et al. 1988).

## D. Inhibition of Platelet Aggregation as a Therapeutic Principle

GpIIb–IIIa is an attractive therapeutical target for several reasons. It mediates platelet aggregation induced by all physiological platelet agonists. Thus, in contrast to other antiplatelet drugs such as aspirin or ticlopidine, GPIIb–IIIa antagonists are independent of the platelet agonists. Furthermore, GpIIb–IIIa antagonists do not abolish platelet adhesion and thereby minimize the risk of severe clinical bleedings.

The compounds which currently are in development can be classified as follows: (COOK et al. 1994; COX et al. 1994; COLLER et al. 1995; LEFKOVITS et al. 1995):

1. Antibody fragments, e.g., c7E3 Fab (abciximab, ReoPro; Centocor)
2. Cyclic peptides, e.g., Integrelin (COR Therapeutics)
3. Peptidomimetics, e.g., Ro 44-9883 (Hofman-La Roche); MK-383 (Merck)
4. Orally active compounds, e.g., SC54684 (Searle); Ro 43-8857 (Hofmann-LaRoche); GR144053 (Glaxo); DMP728 (DuPont-Merck)

Disintegrins are not in development due to their antigenicity. Except for the c7E3 Fab, a chimeric recominant Fab version of 7E3 (a murine antibody against GpIIb–IIIa; COLLER 1985) the other agents are based on the RGD sequence. They virtually all retain the basic charge relations of the RGD sequence yielding a positive and a negative charge which are separated approximately by 10–20 Å. The various GpIIb–IIIa inhibitors differ from each other with respect to their receptor specificity, pharmacodynamics, and pharmacokinetics (Cook et al. 1994; Cox et al. 1994; Coller et al. 1995; Lefkovits et al. 1995). The c7E3 Fab (abciximab, ReoPro[R]) was the first compound to be approved for patients undergoing high-risk angioplasty in the US and in several European and Scandinavian countries. All of these agents are currently under further clinical investigation and the results of the Phase-II and Phase-III studies so far are promising. It is likely that GpIIb–IIIa therapy will prove to be beneficial in thrombotic disorders other than angioplasty (Coller et al. 1995; Lefkovits et al. 1995).

# E. Conclusions

Platelet aggregation plays an essential role in thrombosis and hemostasis. The process is mediated by a GpIIb–IIIa heterodimer complex in the platelet membrane which acts as a promiscuous receptor for several ligands such as fibrinogen, vWF, fibronectin, and vitronectin. For the interaction with soluble ligands an activation of the platelets is required which converts the GpIIb–IIIa complex into a conformational state competent for ligand binding. Under low shear conditions fibrinogen is the relevant ligand of GPIIb–IIIa whereas under high shear conditions vWF seems to mediate platelet aggregation. For fibrinogen binding a dodecapeptide on it's $\gamma$-chain is essential whereas the binding of other cytoadhesive proteins such as vWF is mediated by an RGD motif. Since a blockade of GpIIb–IIIa inhibits the induction of platelet aggregation by all physiological agonists, GpIIb–IIIa is a promising target for antithrombotic therapies. Several types of GpIIb–IIIa inhibitors are under clinical development and the results of the clinical studies so far are all promising.

Platelet aggregation testing is one of the major functional assays applied to characterize platelet inhibitors such as GpIIb–IIIa antagonists and various other antiplatelet agents. Furthermore, it is an established screening tool for the diagnosis of thrombocytopathies. Platelet aggregation testing can be performed with several techniques such as optical and electrical aggregometry and particle counting. There are several differences between these methods. Only particle counting and electrical aggregometry can be applied in whole blood studies, whereas for turbidimetric aggregation, samples of isolated platelet are required.

Kinetic information of the aggregation process can easily be obtained with electrical and turbidimetric aggregometry. However, in contrast to particle counting which is a discontinuous measurement, these two methods are only sensitive to late stages of platelet aggregation and are very insensitive to early stages of platelet aggregation.

Finally, it should be noted that aggregation studies have generally been performed with stirred platelet suspensions which have ill-defined flow regimes. The increasing use of microdevices with well-defined laminar flow can be expected to yield important information on shear-dependent platelet aggregation and effects of antithrombotic drugs.

## References


Andrieux A, Hudry-Clergeon G, Ryckewaert JJ (1989) Amino acid sequences in fibrinogen mediating its interaction with its platelet receptor, GP IIb/IIIa. J Biol Chem 264:9258–9265

Aster RH, Jandl JH (1964) Platelet sequestration in man. I. Methods. J Clin Invest 43:843–849

Bennett JS (1996) Structural biology of glycoprotein IIb–IIIa. Trends Cardiovasc Med 6(1):31–37

Born GVR (1962a) Quantitative investigations into the aggregation of blood platelets. J Physiol (Lond) 162:67–72

Born GVR (1962b) Aggregation of blood platelets by adenosine diphosphate and its reversal. Nature 194:927–929

Born GVR (1972) Current ideas on the mechanism of platelet aggregation. Am NY Acad Sci 201:4–11

Cardinal DC, Flower RJ (1980) The electronic aggregometer: a novel device for assessing platelet behavior in blood. J Pharmacol Methods 3:135–158

Chang HN, Robertson CR (1976) Platelet aggregation by laminar shear and brownian motion. Ann Biomed Eng 4:151–157

Charo IF, Nannizzi L, Phillips DR, Hsu MA, Scarborough RM (1991) Inhibition of fibrinogen binding to GPIIb–IIIa by a GPIIIa peptide. J Biol Chem 266:1415–1421

Charo IF, Kieffer N, Phillips DR (1994) Platelet membrane glycoproteins. In: Colman RW, Hirsh J, Marder VJ, Salzman EW (eds) Hemostasis and thrombosis: basic principles and clinical practice, 3rd edn. Lippincott, Philadelphia, p 489

Chow TW, Hellums JD, Moake JL, Kroll MH (1992) Shear stress-induced von Willebrand factor binding to platelet glycoprotein Ib initiates calcium influx associated with aggregation. Blood 80:113–120

Coller BS (1985) A new murine monoclonal antibody reports an activation-dependent change in the conformation and/or microenvironment of the platelet glycoprotein IIb/IIIa complex. J Clin Invest 76:101–108

Coller BS, Anderson K, Weisman HF (1995) New antiplatelet agents: platelet GPIIb/IIIa antagonists. Thromb Haemost 74(1):302–308

Cook NS, Kottirsch G, Zerwes H-G (1994) Platelet glycoprotein IIb/IIIa antagonists. Drugs Future 19:135–159

Cox D, Aoki T, Seki J, Motoyama Y, Yoshida K (1994) The pharmacology of integrins. Med Res Rev 14:195–228

Day HJ, Holmsen H, Zucker MB (1975) Methods for separating platelets from blood and plasma. Thromb Diath Haemorrh 33:648–654

Doolittle RF (1984) Fibrinogen and fibrin. Annu Rev Biochem 53:195–204

Du XP, Plow EF, Frelinger AL III, O'Toole TE, Loftus JC, Ginsberg MH (1991) Ligands "activate" integrin alpha IIb beta 3 (platelet GpIIb–IIIa). Cell 65:409–415

Eika C (1972) The platelet aggregating effect of eight commercial heparins. Scand J Hematol 9:480–489

Farrell DH, Thiagarajan P, Chung DW, Davie EW (1992) Role of fibrinogen alpha and gamma chain sites in platelet aggregation. Proc Natl Acad Sci USA 89(2):10729–10732

Feinman RD, Lubowsky J, Caro IF, Zabinski MP (1977) The lumi-aggregometer: a new instrument for simultaneous measurement of secretion and aggregation. J Lab Clin Med 90:125–136

Fitzgerald LA, Phillips DR (1985) Calcium regulation of the platelet membrane glycoprotein IIb–IIIa complex. J Biol Chem 260:11366–11374

Frelinger AL III, Du XP, Plow EF, Ginsberg MH (1991) Monoclonal antibodies to ligand-occupied conforms of integrin alpha Iib beta 3 (glycoprotein GpIIb–IIIa) alter receptor affinity, specifity, and function. J Biol Chem 266:17106–17112

Fressinaud E, Baruch D, Girma J-P, Sakariassen KS, Baumgartner HR, Meyer D (1988) von Willebrand factor-mediated platelet adhesion to collagen involves

platelet membrane glycoprotein IIb–IIIa as well as glycoprotein Ib. J Lab Clin Med 112:58–67
Frojmovic MM (1973) Quantitative parameterization of the light transmission properties of citrated, platelet-rich plasma as a function of platelet and adenosine diphosphate concentrations and temperature. J Lab Clin Med 82(1):137–153
Frojmovic MM (1978) Rheooptical studies of platelet structure and function. In: Spaet TH (ed) Progress in hemostasis and thrombosis, vol 4. Grund and Stratton, New York, p 279
Frojmovic MM, Milton JG, Duchastel A (1983) Microscopic measurement of platelet aggregation reveal a low ADP-dependent process distinct from turbidimetrically measured aggregation. J Lab Clin Med 101:964–976
Frojmovic MM, Milton JG, Gear AL (1989) Platelet aggregation measured in vitro by microscopic and electronic particle counting. Methods Enzymol 169:134–149
Gear ARL (1982) Rapid reaction of platelets studied by a quenched-flow approach: Aggregation kinetics. J Lab Clin Med 100:866–873
George JN, Nurden AT, Caen JP (1990) Glanzmann's thrombasthenia: the spectrum of clinical disease. Blood 75:1383–1395
Goldsmith HL, Frojmovic MM, Braovac S, McIntosh F, Wong T (1994) Adenosine diphosphate-induced aggregation of human platelets in flow through tubes: III. Shear and extrinsic fibrinogen-dependent effects. Thomb Haemost 71:78–90
Grant RA, Zucker MB (1978) EDTA-induced increase in platelet surface charge associated with the loss of aggregability. Assessment by partition in aqueous two-phase polymer systems an electrophoretic mobility. Blood 52:515
Haverstick DM, Cowan JF, Yamada KM, Santoro SA (1985) Inhibition of platelet binding to fibronectin, fibrinogen and von Willebrand factor substrates by a synthetic tetrapeptide derived from the cell binding domain of fibronectin. Blood 66:439–448
Hawiger J, Timmons S, Kloczewiak M (1982) Gamma and alpha chains of human fibrinogen possess sites reactive with human platelet receptors. Proc Natl Acad Sci USA 79:2068–2074
Hawiger J, Kloczewiak M, Bednarek MA, Timmons S (1989) Platelet receptor recognition domains on the $\alpha$-chain of human fibrinogen: structure–function analysis. Biochemistry 28:2909–2914
Holme H, Murphy S (1981) Influence of platelet count and size on aggregation studies. J Lab Clin Med 97:623–630
Hutton RA, Howard MA, Deykin D et al. (1974) Methods for the separation of platelets from plasma. Thromb Diath Haemorrh 31:119–124
Hynes RO (1987) A family of cell surface receptors. Cell 48:549–554
Ikeda Y, Handa M, Kawano K, Kamata T, Murata M, Araki Y, Anbo H, Kawai K, Watanable I, Itagaki K, Sakai, Ruggeri ZM (1991) The role of von Willebrand factor and fibrinogen in platelet aggregation under varying shear stress. J Clin Invest 87:1234–1240
Ikeda Y, Handa M, Kamata T, Kawano Y, Kawai K, Watanable K, Sakai F, Mayumi F, Itagaki I, Yoshioka A, Ruggeri ZM (1993) Transmembrane calcium influx associated with von Willebrand factor binding to GPIb in the initiation of shear-induced platelet aggregation. Thromb Haemost 69:496–502
Ingerman-Wojenski C, Smith JB, Silver MJ (1983) Evaluation of electrical aggregometry: comparison with optical aggregometry, secretion of ATP, and accumulation of radiolabeled platelets. J Lab Clin Med 101(1):44–51
Isenberg WM, McEver RP, Phillips DR (1987) The platelet fibrinogen receptor: an immunogold-surface replica study of agonist-induced ligand binding and receptor clustering. J Cell Biol 104:1655–1663
Kaplan KL, Owen J (1981) Plasma levels of $\beta$-thromboglobulin and platelet factor 4 as indices of platelet activation in vivo. Blood 57:199–207
Kieffer N, Phillips DR (1990) Platelet membrane glycoproteins: functions in cellular interactions. Annu Rev Cell Biol 6:329–357

Kinlough-Rathbone RL, Packham MA, Mustard JF (1970) The effect of glucose on adenosine diphosphate-induced platelet aggregation. J Lab Clin Med 75:780–788

Kinlough-Rathbone RL, Packham MA, Mustard JF (1972) The effect of glucose on the platelet response to release-inducing stimuli. J Lab Clin Med 80:247–254

Kinlough-Rathbone RL, Mustard JF, Packham MA, Perry DW, Reimers H-J, Cazenave J-P (1977a) Properties of washed human platelets. Thromb Haemost 37:291–298

Kinlough-Rathbone RL, Packham MA, Mustard JF (1977b) Synergism between platelet aggregating agents: the role of the arachidonate pathway. Thromb Res 11:567–574

Kirschbaum NE, Mosesson MW, Amrani DL (1992) Characterization of the gamma chain platelet binding site on fibrinogen fragment D. Blood 79(10):2643–2648

Kloczewiak M, Timmons S, Lukas TL, Hawiger J (1984) Platelet receptor recognition site on human fibrinogen: synthesis and structure–function relationship of peptides corresponding to the carboxy-terminal segment of the gamma chain. Biochemistry 23:1767–1774

Kloczewiak M, Timmons S, Bednarek MA et al. (1989) Platelet receptor recognition domain on the gamma chain of human fibrinogen and its synthetic peptide analogues. Biochem 28:2915–2923

Lages B, Scrutton MC, Holmsen H (1975) Studies on gel-filtered human platelets: isolation and characterization in a medium containig no added $Ca^{2+}$, $Mg^{2+}$, or $K^{+}$. J Lab Clin Med 86:811–819

Latimer P (1975) Transmittance: an index to shape changes of blood platelets. Appl Opt 14:2324–2326

Latimer P (1982) The platelet aggregometer – an anomalous diffraction explanation. Biophys J 37:353–358

Latimer P, Born GVR, Michal F (1977) Application of light-scattering theory to the optical effects associated with the morphology of blood platelets. Arch Bioch Biophys 180:151–159

Lefkovits J, Plow EF, Topol EJ (1995) Platelet glycoprotein IIb/IIIa receptors in cardiovascular medicine. N Engl J Med 332:1553–1559

Levine SP, Suarez AJ, Sorenson RR, Knieriem LK, Ramond NM (1981) The importance of blood collection methods for assessment of platelet activation. Thromb Res 24:433–437

Lumley P, Humphrey PPA (1981) A method for quantitating platelet aggregation and analyzing drug–receptor interactions on platelets in whole blood in vitro. J Pharmacol Methods 6:153–166

Manley RJ, Mason SG (1952) Particle motions in sheared suspensions. II. Collisions of uniform spheres. J Colloid Sci 7:352–359

Mills DCB (1969) Platelet aggregation. In: Bittar EE, Bittar N (eds) The biological basis of medicine, vol 3. Academic, London

Milton JG, Frojmovic MM (1983) Turbidimetric evaluations of platelet activation: Relative contributions of measured shape change, volume and early aggregation. J Pharmacol Methods 9:101–115

Moon DG, Shainoff JR, Gonda SR (1990) Electron microscopy of platelet interactions with heme-octapeptide-labeled fibrinogen. Am J Physiol 259:611–617

Mustard JF, Packham MA (1970) Factors influencing platelet function: adhesion, release and aggregation. Pharmacol Rev 22:97–103

Mustard JF, Perry DW, Ardlie NG et al. (1972) Preparation of suspension of washed platelets. Br J Haematol 22:193–202

Mustard JF, Perry DW, Kinlough-Rathbone RL, Packham MA (1975) Factors responsible for ADP-induced release reaction of human platelets. Am J Physiol 228:1757–1764

Nurden AT (1994) Human platelet membrane glycoproteins. In: Bloom AL, Forbes CD, Thomas DP, Tuddenham EGD (eds) Haemostasis and thrombosis, 3rd edn. Churchill Livingstone, Edinburgh, p 115

O'Brien JR (1971) Factors influencing the optical platelet aggregation test. Acta Med Scand [Suppl] 525:43–52

Packham MA, Guccione MA, Chang P-L, Mustard JF (1973) Platelet aggregation and release: effects of low concentrations of thrombin or collagen. Am J Physiol 225:38–44

Packham MA, Kinlough-Rathbone RL, Mustard JF (1978) Aggregation and agglutination. In: Day HJ, Holmsen H, Zucker MB (eds) Platelet function testing. DHEW Publ no (NIH) 78-1087, US Government Printing Office, Washington DC, p 66

Patscheke H (1981) Shape and functional properties of human platelets washed with acid cirtate. Haemostasis 10:14–18

Pedvis LG, Wong T, Frojmovic MM (1988) Differential inhibition of the platelet activation sequence: shape change, micro- and macroaggregation by a stable prostacylin analogue (Iloprost). Thomb Haemost 59:323–328

Peerschke EI (1985) The platelet fibrinogen receptor. Semin Hematol 22:241–252

Peerschke EI (1995) Regulation of platelet aggregation by post-fibrinogen binding events. Insights provided by dithiothreitol-treated platelets. Thromb Haemost 73(5):862–867

Plow EF, Pierschbacher MD, Ruoslahti E, Marguerie G, Ginsberg MH (1985) The effect of Arg–Gly–Asp-containing peptides on fibrinogen and von Willebrand factor binding to platelets. Proc Natl Acad Sci USA 82:8057–8061

Plow EF, D'Souza SE, Ginsberg MH (1992) Ligand binding to GPIIb–IIIa: a status report. Semin Thromb Hemost 18:324–332

Rochon Y van P, Frojmovic MM (1993) Regulation of human neutrophil aggregation: comparable latent times, activator sensitivities and exponential decay in aggregability for FMLP, PAF and $LTB_4$. Blood 3460–3468

Ruf A, Schlenk RF, Maras A, Morgenstern E, Patscheke H (1992) Contact-induced neutrophil activation by platelets in human cell suspensions and whole blood. Blood 80(5):1238–1246

Ruggeri ZM (1994) New insights into the mechanisms of platelet adhesion and aggregation. Semin Hematol 31:229–239

Ruoslahti E (1991) Integrins. J Clin Invest 87:1–12

Shirasawa K, Chandler AB (1969) Fine structure of the bond between platelets in artificial thrombi and in platelet aggregates induced by adenosine diphosphate. Am J Pathol 57:127–132

Sims PJ, Ginsberg MH, Plow EF, Shattil SJ (1991) Effect of platelet activation on the conformation of the plasma membrane glycoprotein IIb–IIIa complex. J Biol Chem 266:7345–7352

Sixma JJ (1972) Methods for platelet aggregation. In: Mannucci PM, Gorini S (eds) Platelet function and thrombosis. Plenum, New York

Taylor DB, Gartner TK (1992) A peptide corresponding to GPIIb alpha 300–312, a presumptive fibrinogen gamma-chain binding site on the platelet integrin GPIIb/IIIa, inhibits the adhesion of platelets to at least four adhesive ligands. J Biol Chem 267(17):11729–11733

Ten Cate JW (1972) Platelet function testes. Clin Haematol 1:283–297

Weiss HJ, Hoffmann T, Yoshioka A, Ruggeri ZM (1993) Evidence that the $Arg^{1744}Gly^{1745}Asp^{1746}$ sequence in the GPIIb–IIIa-binding domain of von Willebrand factor is involved in platelet adhesion and thrombus formation on subendothelium. J Lab Clin Med 122:324–332

Weiss JH, Hawiger J, Ruggeri ZM, Turitto VT, Thiagarajan P, Hoffmann T (1989) Fibrinogen-independent interaction of platelets with subendothelium mediated by glycoprotein IIb–IIIa complex at high shear rate. J Clin Invest 83:288–297

Wencel-Drake JD, Plow EF, Kunicki TJ, Woods VL, Keller DM, Ginsberg MH (1986) Localization of internal pools of membrane glycoproteins involved in platelet adhesive responses. Am J Pathol 124:324–334

Wencel-Drake JD, Boudignon-Proudhon C, Dieter MG, Criss AB, Parise LV (1996) Internalization of bound fibrinogen modulates platelet aggregation. Blood 87(2):602–612

Woods VL, Wolff LE, Keller DM (1986) Resting platelets contain a substantial centrally located pool of glycoprotein IIb–IIIa complex which may be accessible to some but not other extracellular proteins. J Biol Chem 261:15242–15251
Xia Z, Frojmovic MM (1994) Aggregation efficiency of activated normal or fixed platelets in a simple shear field: effect of shear and fibrinogen occupancy. Biophys J 66:2190–2201.
Zucker MB (1972) Proteolytic inhibitors, contact and other variables in the release reaction of human platelets. Thromb Diath Haemorrh 33:63–68
Zucker MB (1975) Effect of heparin on platelet function. Thromb Diath Haemorrh 33:66–72
Zucker WH, Shermer RW, Mason RG (1974) Ultrastructural comparison of human platelets separated from blood by various means. Am J Pathol 77:255–261

# Section II
# Platelet Biochemistry, Signal Transduction

CHAPTER 5

# Platelet Receptors: The Thrombin Receptor

W. SIESS

## A. Introduction

The serine protease thrombin evokes biological responses from a variety of cells such as platelets, megakaryoblasts, endothelial cells (VU et al. 1991a), monocytes (HOFFMAN and CHURCH 1993; JOSEPH and MACDERMOT 1993), fibroblasts (VOURET-CRAVIARI et al. 1992), mesangial cells (GRANDALIANO et al. 1994), smooth muscle cells (HERBERT et al. 1992; TESFAMARIAM 1994; GLUSA and PAINTZ 1994), even neuronal cells (JALINK and MOOLENAAR 1992) and osteoblasts (JENKINS et al. 1994). All of the effects of thrombin are mediated by receptors on the cell surface. The receptor that mediates most of thrombin's cellular effects has been cloned from human megakaryoblasts (VU et al. 1991a), Chinese hamster lung fibroblasts (RASMUSSEN et al. 1991), and rat smooth muscle cells (ZHONG et al. 1992). The cloned receptor is a member of the seven-transmembrane domain receptor family that couples to G proteins and is activated by thrombin through a unique mechanism (VU et al. 1991a).

Human $\alpha$-thrombin consists of two polypeptide chains of 36 (A-chain) and 259 (B-chain) amino acid residues covalently linked via a disulfide bridge (FENTON and BING 1986; BODE et al. 1989). Thrombin cleaves behind basic amino residues (Arg; Lys). Thrombin's anion binding exosite is important for the binding to macromolecular substrates such as fibrinogen and the cloned thrombin receptor (FENTON et al. 1988). $\gamma$-Thrombin, obtained from $\alpha$-thrombin by controlled trypsin digestion, lacks the anion binding exosite, shows no clotting, yet some amidolytic activity (BREZNIAK et al. 1990), and is able to activate platelets.

Studies on intact platelets have delineated three thrombin binding sites: a high affinity binding site ($K_D = 0.3\,nM$; 50 sites/platelet) which may be located on glycoprotein Ib (GPIb), a moderate-affinity binding site ($K_D = 11\,nM$; 1700 sites/platelet) which might correspond to the G-protein-coupled receptor, and a low-affinity, nonspecific binding site ($K_D = 2.9\,\mu M$; 590000 sites/platelet; HARMON and JAMIESON 1985, 1986, 1988; GRECO and JAMIESON 1991). The G-protein-coupled receptor and GPIb mediate thrombin's actions on platelets. GPV localized on the platelet surface can be hydrolyzed by thrombin, but GPV cleavage does not play a role in platelet activation induced by thrombin (reviewed by SIESS 1989).

# B. The Seven-Transmembrane Domain Receptor

## I. Structure and Activation Mechanism

The cloned human platelet thrombin receptor belongs to the family of the seven-transmembrane domain G-protein-coupled receptors (Vu et al. 1991a). Several studies demonstrate that the platelet thrombin receptor is necessary as well as sufficient for thrombin-induced platelet activation. Thrombin binds to the amino-terminal exodomain of the receptor and interacts with at least two sites (Fig. 1): thrombin's active center (the $S_1$–$S_4$ subsites) interacts with the receptor's cleavage site (LDPR/S), and thrombin's anion-binding exosite binds to the receptor's hirudin-like domain (Liu et al. 1991; Vu et al. 1991a; Hung et al. 1992a). Thrombin then cleaves the receptor C-terminal to Arg-41, exposing a new N terminus that serves as a tethered ligand (Vu et al. 1991a). Peptides containing at least the first five residues of the tethered ligand sequence (SFLLR) can elicit many of the effects of thrombin (see below and Sect. D).

The hirudin-like domain of the receptor (residues 52–60) which is rich in aromatic and acidic residues confers the high affinity for thrombin binding. $\gamma$-Thrombin that has no anion binding exosite requires much higher concentrations to bind and cleave the receptor (Brass et al. 1992; Bouton et al. 1995). Binding of the hirudin-like domain to thrombin causes a conformational change in thrombin's active-site region, thereby altering thrombin's specificity for small chromogenic substrates (Liu et al. 1991). This change may be important for the ability of the receptor to be a suitable thrombin substrate since D-39 in the LDPR sequence of the thrombin receptor is unfavorable to the interaction with thrombin's catalytic site. These results are also supported by crystallographic studies demonstrating conformational changes in the active center of thrombin by the binding of the KYEPF sequence of receptor peptides (Mathews et al. 1994).

The affinity of the thrombin receptor in intact platelets cannot directly be measured because of the presence of other high-affinity binding proteins for thrombin on platelets. Using the residues 28–60 of the thrombin receptor as substrate for thrombin, a $K_M$ of $10\,\mu M$ for substrate hydrolysis has been deter-

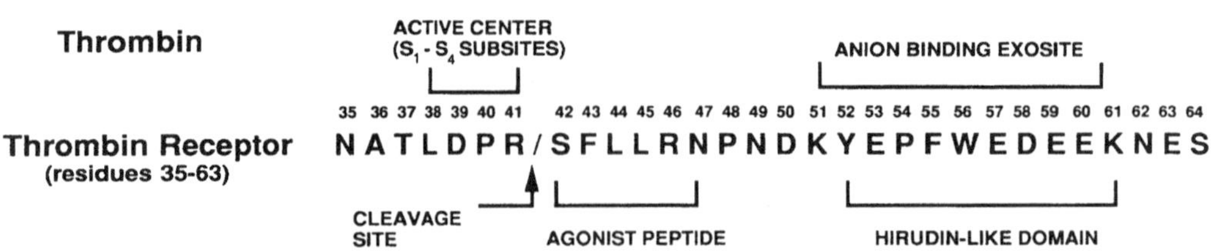


**Fig. 1.** Interaction of thrombin with functionally important regions of the amino-terminal exodomain of the thrombin-receptor. Amino acid residues are indicated by the one-letter code

mined (Vu et al. 1991b). In studies of the interaction of thrombin with a fusion protein of glutathione-S-transferase containing thrombin receptor residues 25–97, the hydrolysis of a tripeptide substrate was inhibited by the fusion protein with a $K_i$ of 0.5 $\mu M$ (Bouton et al. 1995).

Noncleavable thrombin receptor peptides corresponding to residues 38–64 of the thrombin receptor inhibit thrombin-induced platelet aggregation with $IC_{50}$ values of 0.1–0.4 $\mu M$ (Hung et al. 1992b). The thrombin receptor in intact platelets might have a higher affinity than the isolated receptor protein or portions of the receptor. The thrombin receptor in platelets and other cells is heavily glycosylated – the apparent molecular mass on SDS-polyacrylamid gels is 66–70 kDa, the predicted molecular mass only 44 kDa (Brass et al. 1992; Vouret-Craviari et al. 1995). Thus, glycosylation of the receptor might influence its affinity. It has been assumed that the thrombin receptor corresponds to the moderate affinity binding site determined in previous studies of thrombin binding to platelets ($K_D$ = 11 $nM$; 1700 sites per platelet; Greco and Jamieson 1991). Using radiolabelled antibodies against the thrombin receptor, a similar receptor density (1800 molecules per platelet) has been found (Brass et al. 1992).

The tethered ligand peptide (SFLLRN . . .) exposed by cleavage of the thrombin receptor auto-activates the receptor (Vu et al. 1991a). Although the tethered ligand could activate a neighboring thrombin receptor, the intramolecular liganding mechanism is the predominant mode of thrombin receptor activation (Chen et al. 1994). Recent studies in *Xenopus* oocytes showed that the extracellular surface of the thrombin receptor binds the agonist peptide (Gerszten et al. 1994; Bahou et al. 1994). In particular, the second extracellular loop and the residues Glu-83 to Ser-93 on the amino-terminal exodomain upstream of the first transmembrane spanning region were found to be essential for the response to thrombin as well as thrombin receptor activating peptides (TRAPs). The tethered ligand and exogenously added TRAPs most probably bind to these receptor domains, thereby initiating signal transduction events (see Fig. 2).

The phenyl side chain of $F_2$ and the protonated amino group of $S_1$ are essential for the agonist function of the tethered ligand (SFLLRN . . .). The side chains of $L_4$ and $R_5$ are also important (Scarborough et al. 1992; Chao et al. 1992; Coller et al. 1992). Interestingly, plasma aminopeptidase M is able to inactivate TRAPs by cleavage of the N-terminal serine. This enzyme, however, does not inactivate the tethered ligand produced by thrombin's cleavage of the platelet receptor (Coller et al. 1992). Studies comparing TRAPs of various length established that the minimal sequence requirement is four amino acids (SFLL or $TRAP_{1-4}$; Scarborough et al. 1992; Hui et al. 1992; Vassallo et al. 1992; Chao et al. 1992). However, this peptide is less potent ($EC_{50}$ for aggregation of platelet-rich plasma at 300 $\mu M$) than SFLLRN ($TRAP_{1-6}$; $EC_{50}$ = 0.8 or 1.3 $\mu M$). The latter was found to also be more potent than SFLLRNPNDKYEPF ($TRAP_{1-14}$; $EC_{50}$ = 4 or 8 $\mu M$) in two studies (Vassallo et al. 1992; Hui et al. 1992). In contrast, in another study the

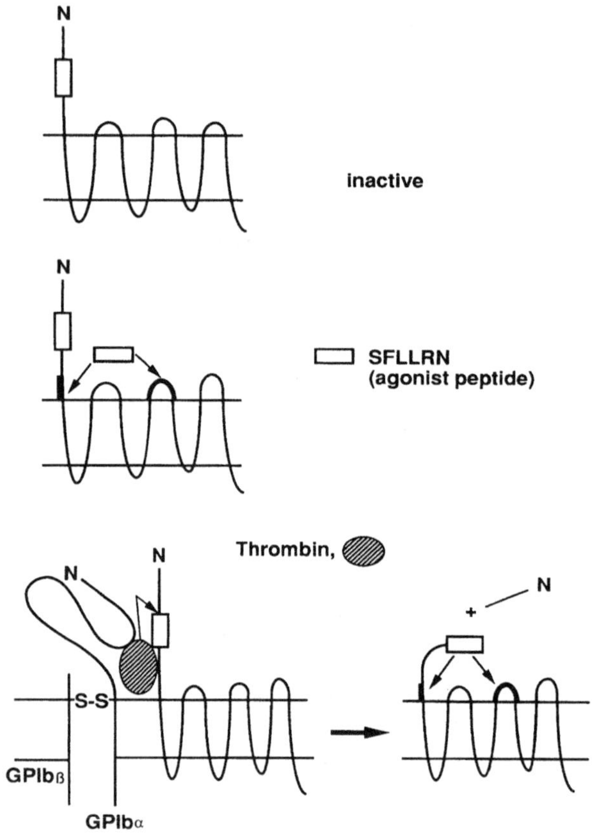


**Fig. 2.** Thrombin receptor activation by SFLLRN or thrombin. *Empty box* indicates sequence of free agonist peptide and tethered ligand. The extracellular receptor domains that are essential for ligand binding are indicated (*middle* and *bottom*). Thrombin binds with its anion-binding exosite to hirudin-like domains of GPIb$\alpha$ and the cloned receptor (see also Fig. 3). The synchronous binding of thrombin to both proteins is hypothetical. Note that in this model GPIb$\alpha$ needs to be bent to bring its thrombin binding site close to the thrombin receptor (*bottom*)

minimal concentration necessary to achieve a full aggregation response in platelet-rich plasma was two- to fivefold higher when shorter peptides ($TRAP_{1-5}$, $TRAP_{1-6}$) were used as compared to $TRAP_{1-14}$ (CHAO et al. 1992).

Thrombin antagonists have been designed based on the thrombin receptor structure (HUNG et al. 1992b). Uncleavable receptor mimicking peptides such as LDPRPFLLRNP ... or FPRPFLLRNP ... were found to inhibit platelet ATP secretion induced by thrombin ($10 nM$) at an $IC_{50}$ of 400 and $100 nM$, respectively. These uncleavable peptides also containing the sequence of the hirudin-like domain bound with high affinity to thrombin (HUNG et al. 1992b). Other peptides such as YFLLRNP were found not only to inhibit platelet activation by thrombin and SFLLRN but also to partially activate platelets (RASMUSSEN et al. 1993). YFLLRNP induced shape change, only minor aggregation, and little or no secretion. Interestingly, the same peptide in endothelial cells was not inhibitory to $Ca^{2+}$-mobilization induced by thrombin and $TRAP_{1-14}$ (KRUSE et al. 1995).

Recombinant thrombin mutants were obtained by changing alanine for serine in the active center at the B-chain position 205. The binding function of both the active site/$S_1$–$S_4$ subsite region and the anion-binding exosite of the recombinant thrombin was preserved. The thrombin mutant inhibited ATP-secretion induced by 1 *nM* thrombin but not by SFLLRN, indicating that the tethered ligand binds to a region of the receptor different from thrombin (HUNG et al. 1992b).

Several groups used antibodies against specific thrombin receptor domains to elucidate structure and function of the thrombin receptor. Monoclonal antibodies produced against Ser-42 to Phe-55 residues of the thrombin receptor (distal to the cleavage site) inhibited the platelet response to $\alpha$-thrombin, $\gamma$-thrombin, and trypsin (BRASS et al. 1992). Therefore, the thrombin receptor is involved in the platelet response not only to thrombin but also to trypsin. Polyclonal antibodies against residues 34–52 of the thrombin receptor (spanning the cleavage site) inhibited responses of platelets and endothelial cells to thrombin, but not to TRAP (BAHOU et al. 1993). In contrast, an antibody against residues 1–160 of the thrombin receptor abrogated both thrombin- and TRAP-induced responses in platelets and endothelial cells, indicating that the tethered ligand exerts its effect within this domain of the receptor (see also above). Since this antibody abolished TRAP-induced responses but did not completely inhibit thrombin-induced $Ca^{2+}$ transients in endothelial cells and platelets, alternative mechanisms or receptors for thrombin were proposed (BAHOU et al. 1993).

## II. Receptor Desensitization and Resensitization

Thrombin responses in platelets as in many other cell types are subject to homologous desensitization. Platelet preincubation with TRAPs desensitized platelets more to subsequent stimulation by TRAPs than to activation by thrombin (SEILER et al. 1991; KINLOUGH-RATHBONE et al. 1993a; WILHELM and SIESS 1993; LAU et al. 1994). Desensitized platelets aggregated fully to high concentrations of $\alpha$- und $\gamma$-thrombin, but less or not at all to TRAP. Some authors suggested that two pathways mediate thrombin-induced platelet activation (SEILER et al. 1991). However, the results can also be explained by incomplete desensitization of the thrombin receptor by TRAP since even maximal concentrations of TRAP only partially activate the thrombin receptor (see Sect. D).

Not much is known about the fate of the activated thrombin receptor in platelets. In a study of HEL cells, which express many platelet proteins and have signal transduction mechanisms similar to platelets (BRASS et al. 1991), desensitization appeared to occur through two mechanisms (BRASS 1992). The first mechanism of desensitization, which was seen only after thrombin, involved receptor cleavage and required protein synthesis for recovery – resensitization to thrombin after cell exposure to thrombin required up to 24h. In platelets where little protein synthesis occurs, no resensitization to thrombin after receptor cleavage is to be expected. The second mechanism observed after HEL cell stimulation with both $TRAP_{1-14}$ and thrombin probably involved receptor phosphorylation by a G-protein-coupled receptor kinase. Indeed, in human-embryonic-kidney 293 cells or Rat1 fibroblasts transfected with human thrombin receptor cDNA, it could be demonstrated that TRAP and thrombin rapidly induced receptor phosphorylation (Ishii et al. 1994; Vouret-Craviari et al. 1995a,b). Potential phosphorylation sites for G-protein receptor kinases are present in the C-terminal portion of the thrombin receptor, and it was observed that these sequences in the C terminus are sufficient to trigger the rapid uncoupling of the receptor from its G protein(s) and downstream effector(s) (Vouret-Craviari et al. 1995a). In *Xenopus* oocytes transfected with the human thrombin receptor, coexpression of a G-protein-coupled receptor kinase (the $\beta$-adrenergic receptor kinase 2) blocked signaling by thrombin receptors (Ishii et al. 1994). Therefore, G-protein-coupled receptor kinases apparently also shut off thrombin receptor signaling in platelets.

Resensitization to TRAP after exposure to thrombin or to thrombin after exposure to TRAP in HEL cells required only 30min and was most likely due to receptor dephosphorylation (Brass 1992). Whether receptor dephosphorylation is a mechanism for receptor resensitization in platelets is unknown. Interestingly, a new mechanism for the resensitization of cell responses to thrombin has been described in transfected Rat1 fibroblasts (Hein et al. 1994) and might also occur in endothelial cells (Woolkalis et al. 1995). In both cell types, an intracellular pool of thrombin receptors has been observed. More than 90% of the thrombin receptor in endothelial cells were found to be associated with a large continuous endosomal network (Horvat and Palade 1995). In Rat1 cells and endothelial cells, the intracellular receptors translocated to the plasma membrane upon activation of cell surface thrombin receptors, thereby providing new uncleaved thrombin receptors on the cell surface (Hein et al. 1994; Woolkalis et al. 1995). Such a mechanism could also explain the sensitization of the thrombin receptor observed after activation of endothelial cells with $TRAP_{1-14}$ (Kruse et al. 1995). It is not known whether platelets contain an intracellular pool of thrombin receptors.

In the megakaryoplastic cell line CHRF-288, receptor activation by thrombin of TRAP caused a rapid (within 10min) and efficient (more than 90%) internalization of thrombin receptors (Brass et al. 1994). Receptor activation, not cleavage, appeared to signal internalization. Once internalized, most of the receptors were routed through endosomes to lysosomes to be ultimately degraded. However, 25%–40% of the internalized receptors were recycled to the cell surface. The recycled uncleaved receptors were no longer desensitized to TRAP or thrombin activation, suggesting that they had been dephosphorylated by endosomal phosphatases. The recycled receptors cleaved after cell treatment with thrombin could be activated by SFLLRN but not by thrombin. Receptor internalization, but not recycling, was also observed in endothelial cells (WOOLKALIS et al. 1995). Whether receptor internalization and recycling occur in platelets is not known. In one study using antibodies against specific regions of the amino-terminal exodomain of the thrombin receptor, it was found that receptor activation by SFLLRN or thrombin did not initiate redistribution or internalization of surface thrombin receptors within 5 min of activation. Such a time period, however, might be too short to observe receptor internalization in platelets (NORTON et al. 1993).

# C. Glycoprotein Ib

Glycoprotein (GP) Ib (165 kDa) is an integral membrane protein spanning the lipid bilayer. It consists of a disulfide-linked $\alpha$-chain (143 kDa) and a $\beta$-chain (22 kDa) and forms a noncovalent hetero-oligomeric complex in the platelet membrane with glycoprotein IX and glycoprotein V (CLEMETSON and LÜSCHER 1988; CLEMETSON 1995).

The extracellular amino-terminal domain of $GPIb_\alpha$ contains high-affinity binding sites for von Willebrand factor and $\alpha$-thrombin: $GPIb_\alpha$ residues within the sequence 251–279 are involved in von Willebrand-factor binding (VICENTE et al. 1990); and $GPIb_\alpha$ residues within the sequence 271–284 participate in $\alpha$-thrombin binding (DE MARCO et al. 1994). Three tyrosine residues at positions 276, 278, and 279 which are sulfated (DONG et al. 1994) are critical for the binding of $\alpha$-thrombin to $GPIb_\alpha$. A single substitution of tyrosine at position 276, 278, or 279 reduced or abolished the binding of $\alpha$-thrombin, but not of von Willebrand factor (MARCHESE et al. 1995).

Binding of $\alpha$-thrombin to GPIb on intact platelets has been demonstrated by cross-linking studies (JANDROT-PERRUS et al. 1988). Thrombin binds to GPIb on intact platelets or to glycocalicin (128 kDa), a proteolytic product of $GPIb_\alpha$, with high affinity (HARMON and JAMIESON 1986). This high-affinity binding is inhibited by monoclonal antibodies against anti-$GPIb_\alpha$ (DE MARCO et al. 1991), by peptides containing the residues 271–284 of $GPIb_\alpha$ (DE MARCO et al. 1994), or by treatment of platelets with proteases (*Serratia marcescens*, elastase, chymotrypsin) that remove GPIb from the platelet surface (HARMON and JAMIESON 1988; WICKI and CLEMETSON 1985). Furthermore, Bernard-Soulier platelets lacking the GPIb/GPIX/GPV complex do not possess this high-affinity binding site (CLEMETSON and LÜSCHER 1988). Since platelets have 28 000 copies of GPIb per cell (GRECO and JAMIESON 1991), only 0.2% (corresponding to 50 sites per cell) of GPIb molecules are involved in the high-affinity binding of thrombin.

The thrombin-binding sequence in $GPIb_\alpha$ has similarities with regions of hirudin and the hirudin-like domain of the thrombin receptor (Fig. 3; DE MARCO et al. 1994). $\alpha$-Thrombin binds to $GPIb_\alpha$, to hirudin, and to the hirudin-like domain with its anion-binding exosite. $\gamma$-Thrombin lacks the anion-binding exosite region and does not bind to GPIb (JANDROT-PERRUS et al.

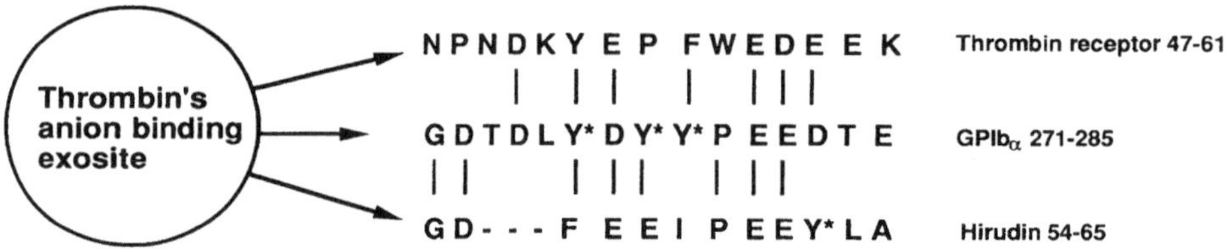


**Fig. 3.** Thrombins anion-binding exosite binds to domains rich in acidic and aromatic residues that show sequence similarities in the case of the thrombin receptor, GPIb$\alpha$, and hirudin. Amino acid residues are indicated by the one-letter code. Identical residues and conserved substitutions are indicated. Sulfated tyrosine residues are indicated by an *asterisk*

1988). Interestingly, the binding of thrombin to the recombinant thrombin receptor is not inhibited by glycocalicin (BOUTON et al. 1995). Therefore, it is possible that both GPIb$_\alpha$ 271–285 and thrombin receptor 47–61 interact with discrete subsites of the large groove that constitutes the anion-binding exosite of thrombin.

The possible function of GPIb in platelet activation by thrombin can be deduced from studies on platelets which lack GPIb or from results obtained after inhibition of GPIb:

(a) Bernard-Soulier platelets are less sensitive to low concentrations of $\alpha$-thrombin. Their maximal aggregation response to high $\alpha$-thrombin concentrations is, however, almost normal (reviewed by SIESS 1989).
(b) Platelets treated with proteases (*Serratia marcescens*, chymotrypsin) show concomitant decreases in GPIb, thrombin responsiveness, and high-affinity thrombin binding – changes similar to Bernard-Soulier platelets (GRECO and JAMIESON 1991). The responses to SFLLRN are not inhibited (KINLOUGH-RATHBONE et al. 1995). However, proteases are also able to cleave the cloned thrombin receptor at a site that results in the loss of the tethered ligand. Indeed, receptor cleavage has been demonstrated for chymotrypsin which N-terminally removed a 7-kDa portion (about 60 amino acid residues) of the receptor (VOURET-CRAVIARI et al. 1995). A similar proteolytic degradation of the receptor has been reported by incubation of platelets with cathepsin G (MOLINO et al. 1995).
(c) Monoclonal antibodies against the thrombin-binding domain of glycoprotein Ib (residues 238–293 of GPIb$_\alpha$) and peptides containing residues 271–284 of GPIb$_\alpha$ inhibited platelet activation by thrombin (DE MARCO et al. 1991, 1994). However, peptides binding to the anion exosite region of thrombin might not only inhibit binding of thrombin to GPIb, but also to the receptor. It is also not clear whether the monoclonal antibodies against GPIb interfere with the binding of thrombin to the cloned receptor.

Once thrombin is bound to the high-affinity binding site of GPIb, this site might come close to the seven-transmembrane domain receptor (see Fig. 2, bottom). GPIb might further facilitate the access of thrombin to the receptor cleavage site (LDPR/S) by providing an additional high-affinity binding site for thrombin, thereby accelerating the rate of receptor cleavage which determines signaling and ultimately the platelet response.

## D. Receptor Signaling and Platelet Responses

All effects that are elicited by thrombin in platelets are also induced by SFLLRN of longer TRAPs ($TRAP_{1-14}$). The equimolar concentrations of TRAPs are about 10000-fold higher than those of thrombin, indicating low affinity of the free agonist peptide for its binding site on the non-cleaved thrombin receptor (see also Fig. 2, middle). The biological effects of TRAPs on platelets include platelet aggregation, secretion of ATP and serotonin from dense granules (Vu et al. 1991a; Kinlough-Rathbone et al. 1993a), secretion of thrombospondin from $\alpha$-granules (Lau et al. 1994), $Ca^{++}$ mobilization, arachidonate release, thromboxane formation, phosphorylation, of cytosolic phospholipase $A_2$ (Lau et al. 1994), production of inositol phosphates and phosphatidic acid (Vassallo et al. 1992), formation of 3-phosphorylated phosphoinositides (Huang et al. 1991), inhibition of adenylate cyclase (Seiler et al. 1991; Vassallo et al. 1992), decrease of the pertussis-toxin-dependent ADP ribosylation of membrane and soluble $G_i$ proteins (Wilhelm and Siess 1993), stimulation of protein phosphorylation by protein kinase C and tyrosine kinases, translocation of integrin $\alpha_{IIb}\beta_3$ (Lau et al. 1994), $pp60^{c\text{-}src}$, $pp125^{FAK}$, rap1B, CDC42Hs, and rac to the cytoskeleton (Dash et al. 1995a,b), and increase of $Na^+/H^+$-exchange (Nieuwland et al. 1994).

In contrast to thrombin, TRAPs are not full agonists for activation and signal transduction in human platelets. This is especially evident when experiments are carried out in the presence of aspirin and ADP scavengers in order to block generation and action of thromboxane $A_2$ and ADP, respectively, which are released from activated platelets (Lau et al. 1994). In platelets from rabbits, dogs, pigs, and hamster, SFLLRN has even lower agonistic activity – it induces only shape change; in rat platelets, it is inactive (Kinlough-Rathbone et al. 1993b; Connolly et al. 1994). The lower intrinsic activity of TRAPs as compared with thrombin in human platelets might be explained by the allosteric Ternary Complex Model for G-protein-linked receptors as outlined by Lau et al. (1994). In this model, agonists (A) stabilize the active receptor (R*) conformation because of a preferentially higher affinity for R*. TRAPs might not be so specific in binding to R* and stabilize the ternary complex AR*G less, leading to the biological response. In the case of receptor activation by thrombin, the proteolytic action of thrombin generates R* which is constitutively active and resembles an active mutant receptor. It is further possible that the binding of thrombin to the receptor induces a conformational change of the receptor that is more favorable to activation by the tethered ligand than to activation of the uncleaved receptor by a free agonist peptide (see also Fig. 2).

In a proteolytic process, the rate of substrate cleavage is proportional to enzyme concentration, while the extent is independent of it. In elegant experiments using Rat1 fibroblasts transfected with a thrombin receptor containing an epitope-tagged activation peptide, the kinetics of thrombin receptor cleavage could be directly measured (ISHII et al. 1993). The rate of thrombin receptor cleavage was proportional to thrombin over the physiologic range, but low thrombin concentrations ultimately cleaved all receptors. Thrombin at 10 $nM$ caused nearly complete cleavage and activation of thrombin receptors within 1 min; Thrombin at 0.3 $nM$ also cleaved and activated all cell receptors, but required over 30 min to do so. Cumulative phosphoinositide hydrolysis in response to thrombin correlated with cumulative receptor cleavage, indicating that each cleaved and activated receptor produces a quantum of phosphotidylinositol hydrolysis and then shuts off. In platelets exposed to thrombin for 1 min, low concentrations of thrombin (0.1 $nM$) induce less aggregation than high concentrations (1 $nM$), which can be explained by a lower rate of receptor cleavage and reduced production of second messenger molecules during that time period. The rate of receptor cleavage is initially high and then decreases over time (ISHII et al. 1993). Thus, the ability of thrombin to cause sustained signals (as opposed to the agonist peptide SFLLRN) can be explained by its ongoing enzymatic action in progressively cleaving further receptors. Hirudin which rapidly inactivates thrombin and can strip it of from its receptor (and possibly from GPIb) quickly terminated receptor cleavage and phosphoinositide breakdown (ISHII et al. 1993). Hirudin added after thrombin stimulation of platelets stopped phosphatidic acid production and arachidonate release and caused accelerated protein dephosphorylation (HOLMSEN et al. 1981; LAU et al. 1994). Leupeptin, a $Ca^{2+}$-protease inhibitor that inhibits thrombin's proteolytic activity produced similar effects when added following thrombin stimulation of platelets (RUGGIERO and LAPETINA 1986).

The initial signaling of the seven-transmembrane domain receptor involves the activation of G proteins. Platelet treatment by both thrombin and $TRAP_{1-14}$ decreased the pertussis-toxin dependent ADP ribosylation of the membrane and soluble $G_{i\alpha}$ (GENNITY and SIESS 1991; WILHELM and SIESS 1993). Activation of $G_i$ protein by thrombin receptor activation leads to inhibition of adenylate cyclase. In addition, $G_i$-protein activation in platelets might – probably through the generation of $\beta\gamma$-subunits – be coupled to stimulation of phospholipase C (BRASS et al. 1986). Receptor activation by thrombin or SFLLRN also leads to stimulation of $G_q$ (BENKA et al. 1995) which activates phospholipase C-$\beta$ and to activation of $G_{12}$ and $G_{13}$ (OFFERMANNS et al. 1994). Thrombin might further activate $G_z$ which is phosphorylated in platelets and HEL cells by protein kinase C (CARLSON et al. 1989; GAGNON et al. 1991). Thus, the thrombin receptor might couple to several G proteins. In addition, thrombin-receptor-mediated G-protein activation in platelets leads to the activation of tyrosine kinases. The early platelet response, i.e., shape change induced by activation of the thrombin receptor with the partial agonist peptide

YFLLRNP, was associated with tyrosine kinase activation which was apparently independent of calcium mobilization, protein kinase C stimulation, and integrin $\alpha_{IIb}\beta_3$ activation (NEGRESCU et al. 1995). There are several tyrosine kinases that could be stimulated by thrombin receptor activation during platelet shape change (see Chap. 13).

It is not known whether thrombin binding to GPIb elicits transmembrane signaling. This might be possible since vWF binding to GPIb induces platelet signaling. vWF that binds to a domain of GPIb near the thrombin-binding region stimulated activation of phospholipase C, protein kinase C, and $Ca^{2+}$ mobilization (KROLL et al. 1991) and produced activation of phosphatidylinositol 3-kinase and $pp60^{c\text{-}src}$ as well as the translocation of these kinases to the cytoskeleton (JACKSON et al. 1994).

# E. Thrombin Receptor Inhibitors

Thrombin receptor inhibitors might not only be useful as antiplatelet agents, but also retard proliferative and inflammatory processes in human vasculature leading to atherosclerosis and restenosis. Increased expression of the thrombin receptor has been found in advanced atherosclerotic lesions (NELKEN et al. 1992). Furthermore, upon balloon catheter vascular injury, thrombin receptor expression in smooth muscle cells occurred very early (WILCOX et al. 1994).

"Mirror image" antagonists of thrombin-induced platelet activation based on thrombin receptor structure have been described. Preincubation of thrombin with uncleavable receptor-mimicking peptides blocked thrombin-induced platelet ATP secretion (HUNG et al. 1992b). Furthermore, the peptide YFLLRNP ($200\,\mu M$) has been shown to antagonize platelet aggregation and $Ca^{2+}$ transients induced by low concentrations of thrombin and SFLLRNP. Interestingly, YFLLRNP also partially activated platelets (RASMUSSEN et al. 1993). In contrast, in human umbilical venous endothelial cells, YFLLRNP failed to prevent $Ca^{2+}$ increases in response to $\alpha$-thrombin or SFLLRNPNDKYEPF (KRUSE et al. 1995). The problem is that most of the described peptides have an affinity to the receptor that is approximately 10000-fold lower than that of thrombin. More potent peptide receptor antagonists are needed but have not yet been described.

Recently, a polyclonal peptide antibody (IgG 9600) produced against the exosite of the thrombin receptor (Lys-51 to Ser-64) has been demonstrated to inhibit experimental arterial thrombosis in the African green monkey. Thrombin-stimulated aggregation and secretion were selectively inhibited by IgG 9600, and other hemostatic parameters (such as fibrin clotting) were not altered. The authors suggested that the thrombin receptor is an attractive antithrombotic target (COOK et al. 1995).

*Note Added in Proof.* A recent study indicates that mouse platelets (unlike mouse fibroblasts) do not contain the cloned thrombin receptor and are activated through a second thrombin receptor [CONNOLLY et al. (1996) Nature 381:516–519].

## References

Bahou WF, Coller BS, Potter CL, Norton KJ, Kutok JL, Goligorsky MS (1993) The thrombin receptor extracellular domain contains sites crucial for peptide ligand-induced activation. J Clin Invest 91:1405–1413

Bahou WF, Kutok JL, Wong A, Potter CL, Coller BS (1994) Identification of a novel thrombin receptor sequence required for activation-dependent responses. Blood 84:4195–4202

Benka ML, Lee M, Wang GR, Buckman SA, Burlacu A, Cole L, DePina A, Dias P, Granger A, Grant B, Hayward-Lester A, Karki S, Mann S, Marcu O, Nussenzweig A, Piepenhagen P, Raje M, Roegiers F, Rybak S, Salic A, Smith-Hall J, Waters J, Yamamoto N, Yanowitz J, Yeow K, Busa WB, Mendelsohn ME (1995) The thrombin receptor in human platelets is coupled to a GTP binding protein of the $G\alpha_q$ family. FEBS Lett 363:49–52

Bode W, Mayr I, Baumann U, Huber R, Stone SR, Hofsteenge J (1989) The refined 1.9 Å crystal structure of human $\alpha$-thrombin: interaction with D-Phe-Pro-Arg chloromethylketone and significance of the Tyr-Pro-Pro-Trp insertion segment. EMBO J 8:3467–3475

Bouton MC, Jandrot-Perrus M, Moog S, Cazenave JP, Guillin MC, Lanza F (1995) Thrombin interaction with a recombinant N-terminal extracellular domain of the thrombin receptor in an acellular system. Biochem J 305:635–641

Brass LF (1992) Homologous desensitization of HEL cell thrombin receptors. Distinguishable roles for proteolysis and phosphorylation. J Biol Chem 267:6044–6050

Brass LF, Laposata M, Banga HS, Rittenhouse SE (1986) Regulation of the phosphoinositide hydrolysis pathway in thrombin-stimulated platelets by a pertussis toxin-sensitive guanine nucleotide-binding protein. Evaluation of its contribution to platelet activation and comparison with the adenylate cyclase inhibitory protein, $G_i$. J Biol Chem 261:16833–16847

Brass LF, Manning DR, Williams AG, Woolkalis MJ, Poncz M (1991) Receptor and G protein-mediated responses to thrombin in HEL cells. J Biol Chem 266:958–965

Brass LF, Vassallo RR, Belmonte E, Ahuja M, Cichowski K, Hoxie JA (1992) Structure and function of the human platelet thrombin receptor. J Biol Chem 267:13795–13798

Brass LF, Pizarro S, Ahuja M, Belmonte E, Blanchard N, Stadel JM, Hoxie JA (1994) Changes in the structure and function of the human thrombin receptor during receptor activation, internalization, and recycling. J Biol Chem 269:2943–2952

Brezniak DV, Brower MS, Witting JI, Walz DA, Fenton JW II (1990) Human $\alpha$- to $\zeta$-thrombin cleavage occurs with neutrophil cathepsin G or chymotrypsin while fibrinogen clotting activity is retained. Biochemistry 29:3536–3542

Carlson KE, Brass LF, Manning DR (1989) Thrombin and phorbol esters cause the selective phosphorylation of a G protein other than $G_i$ in human platelets. J Biol Chem 264:13298–13305

Chao BH, Kalkunte S, Maraganore JM, Stone SR (1992) Essential groups in synthetic agonist peptides for activation of the platelet thrombin receptor. Biochemistry 31:6175–6178

Chen J, Ishii M, Wang L, Ishii K, Coughlin SR (1994) Thrombin receptor activation. Confirmation of the intramolecular tethered liganding hypothesis and discovery of an alternative intermolecular liganding mode. J Biol Chem 269:16041–16045

Clemetson KJ (1995) Platelet activation: signal transduction via membrane receptors. Thromb Haemost 74:111–116

Clemetson KJ, Lüscher EF (1988) Membrane glycoprotein abnormalities in pathological platelets. Biochim Biophys Acta 947:53–73

Coller BS, Ward P, Ceruso M, Scudder LE, Springer K, Kutok J, Prestwich GD (1992) Thrombin receptor activating peptides: importance of the N-terminal serine and its ionization state as judged by pH dependence, nuclear magnetic resonance spectrosopy, and cleavage by aminopeptidase M. Biochemistry 31:11713–11720

Connolly TM, Condra C, Feng DM, Cook JJ, Stranieri MT, Reilly CF, Nutt RF, Gould RJ (1994) Species variability in platelet and other cellular responsiveness to thrombin receptor-derived peptides. Thromb Haemost 72:627–633

Cook JJ, Sitko GR, Bednar B, Condra C, Mellott MJ, Feng M-D, Nutt RF, Shafer JA, Gould RJ, Connolly TM (1995) An antibody against the exosite of the cloned thrombin receptor inhibits experimental arterial thrombosis in the African green monkey. Circulation 91:2961–2971

Dash D, Aepfelbacher M, Siess W (1995a) The association of $pp125^{FAK}$, $pp60^{Src}$, CDC42Hs and Rap1b with the cytoskeleton of aggregated platelets is a reversible process regulated by calcium. FEBS Lett 363:231–234

Dash D, Aepfelbacher M, Siess W (1995b) Integrin $\alpha_{IIb}\beta_3$-mediated translocation of CDC42Hs to the cytoskeleton in stimulated human platelets. J Biol Chem 270:17321–17326

De Marco L, Mazzucato M, Masotti A, Fenton JW II, Ruggeri ZM (1991) Function of glycoprotein Ib$\alpha$ in platelet activation induced by $\alpha$-thrombin. J Biol Chem 266:23776–23783

De Marco L, Mazzucato M, Masotti A, Ruggeri ZM (1994) Localization and characterization of an $\alpha$-thrombin-binding site on platelet glycoprotein II$\alpha$. J Biol Chem 269:6478–6484

Dong J, Li CQ, López JA (1994) Tyrosine sulfation of the glycoprotein Ib-IX complex: identification of sulfated residues and effect on ligand binding. Biochemistry 33:13946–13953

Fenton JW II, Bing DH (1986) Thrombin active site regions. Semin Thromb Hemost 12:200–208

Fenton JW II, Olson TA, Zabinski MP, Wilner GD (1988) Anion-binding exosite of human $\alpha$-thrombin and fibrin(ogen) recognition. Biochemistry 27:7106–7112

Gagnon AW, Manning DR, Catani L, Gewirtz A, Poncz M, Brass LF (1991) Identification of $G_{z\alpha}$ as a pertussis toxin-insensitive G protein in human platelets and megakaryocytes. Blood 78:1247–1253

Gennity JM, Siess W (1991) Thrombin inhibits the pertussis-toxin-dependent ADP-ribosylation of a novel soluble $G_i$-protein in human platelets. Biochem J 279:643–650

Gerszten RE, Chen J, Ishii M, Ishii K, Wang L, Nanevicz T, Turck CW, Vu TKH, Coughlin SR (1994) Specificity of the thrombin receptor for agonist peptide is defined by its extracellular surface. Nature 368:648–651

Glusa E, Paintz M (1994) Relaxant and contractile responses of porcine pulmonary arteries to a thrombin receptor activating peptide (TRAP). Naunyn Schmiedebergs Arch Pharmacol 349:431–436

Grandaliano G, Valente AJ, Abboud HE (1994) A novel biologic activity of thrombin: stimulation of monocyte chemotactic protein production. J Exp Med 179:1737–1741

Greco NJ, Jamieson GA (1991) High and moderate affinity pathways for $\alpha$-thrombin-induced platelet activation. Proc Soc Exp Biol Med 198:792–799

Harmon JT, Jamieson GA (1985) Thrombin binds to a high-affinity ~900000-dalton site on human platelets. Biochemistry 24:58–64

Harmon JT, Jamieson GA (1986) The glycocalicin portion of platelet glycoprotein Ib expresses both high and moderate affinity receptor sites for thrombin. A soluble radioreceptor assay for the interaction of thrombin with platelets. J Biol Chem 261:13224–13229

Harmon JT, Jamieson GA (1988) Platelet activation by thrombin in the absence of the high-affinity thrombin receptor. Biochemistry 27:2151–2157

Hein L, Ishii K, Coughlin SR, Kobilka BK (1994) Intracellular targeting and trafficking of thrombin receptors. A novel mechanism for resensitisation of a G-protein coupled receptor. J Biol Chem 269:27719–27726

Herbert JM, Lamarche I, Dol F (1992) Induction of vascular smooth muscle cell growth by selective activation of the thrombin receptor. FEBS Lett 301:155–158

Hoffman M, Church FC (1993) Response of blood leukocytes to thrombin receptor peptides. J Leukoc Biol 54:145–151
Holmsen H, Dangelmeier CA, Holmsen HK (1981) Thrombin-induced platelet responses differ in requirement for receptor occupancy. Evidence for tight coupling of occupancy and compartimentalized phosphatidic acid formation. J Biol Chem 256:9393–9396
Horvat R, Palade GE (1995) The functional thrombin receptor is associated with the plasmalemma and a large endosomal network in cultured human umbilical vein endothelial cells. J Cell Sci 108:1155–1164
Hui KY, Jakubowski JA, Wyss VL, Angleton EL (1992) Minimal sequence requirement of thrombin receptor agonist peptide. Biochem Biophys Res Commun 184:790–796
Huang R, Sorisky A, Church WR, Simons ER, Rittenhouse SE (1991) "Thrombin" receptor-directed ligand accounts for activation by thrombin of platelet phospholipase C and accumulation of 3-phosphorylated phosphoinositides. J Biol Chem 266:18435–18438
Hung DT, Vu TKH, Wheaton VI, Ishii K, Coughlin SR (1992a) Cloned platelet thrombin receptor is necessary for thrombin-induced platelet activation. J Clin Invest 89:1350–1353
Hung DT, Vu THK, Wheaton VI, Charo IF, Nelken NA, Esmon N, Esmon CT, Coughlin SR (1992b) "Mirror image" antagonists of thrombin-induced platelet activation based on thrombin receptor structure. J Clin Invest 89:444–450
Ishii K, Hein L, Kobilka B, Coughlin SR (1993) Kinetics of thrombin receptor cleavage on intact cells. Relation to signaling. J Biol Chem 268:9780–9786
Ishii K, Chen J, Ishii M, Koch WJ, Freedman NJ, Lefkowitz RJ, Coughlin SR (1994) Inhibition of thrombin receptor signaling by a G-protein coupled receptor-kinase. J Biol Chem 269:1125–1130
Jackson SP, Schoenwaelder SM, Yuan Y, Rabinowitz I, Salem HH, Mitchell CA (1994) Adhesion receptor activation of phosphatidylinositol 3-kinase. J Biol Chem 269:27093–27099
Jalink K, Moolenaar WH (1992) Thrombin receptor activation causes rapid neural cell rounding and neurite retraction independent of classic second messengers. J Cell Biol 118:411–419
Jandrot-Perrus M, Didry D, Guillin MC, Nurden AT (1988) Cross-linking of $\alpha$- and $\gamma$-thrombin to distinct binding sites on human platelets. Eur J Biochem 174:359–367
Jenkins AL, Bootman MD, Berridge MJ, Stone SR (1994) Differences in intracellular calcium signaling after activation of the thrombin receptor by thrombin and agonist peptide in osteoblast-like cells. J Biol Chem 269:17104–17110
Joseph S, MacDermot J (1993) The N-terminal thrombin receptor fragment SFLLRN, but not catalytically inactive thrombin-derived agonists, activate U937 human monocytic cells: evidence for receptor hydrolysis in thrombin-dependent signalling. Biochem J 290:571–577
Kinlough-Rathbone RL, Perry DW, Guccione MA, Rand ML, Packham MA (1993a) Degranulation of human platelets by the thrombin receptor peptide SFLLRN: comparison with degranulation by thrombin. Thromb Haemost 70:1019–1023
Kinlough-Rathbone RL, Rand ML, Packham MA (1993b) Rabbit and rat platelets do not respond to thrombin receptor peptides that activate human platelets. Blood 82:103–106
Kinlough-Rathbone RL, Perry DW, Packham MA (1995) Contrasting effects of thrombin and the thrombin receptor peptide, SFLLRN, on aggregation and release of $^{14}C$-serotonin by human platelets pretreated with chymotrypsin or serratia marcescens protease. Thromb Haemost 73:122–125
Kroll MH, Harris TS, Moake JL, Handin RI, Schafer AI (1991) Von Willebrand factor binding to platelet GPIb initiates signals for platelet activation. J Clin Invest 88:1568–1573

Kruse HJ, Mayerhofer C, Siess W, Weber PC (1995) Thrombin-receptor activating peptide sensitizes the human endothelial thrombin receptor. Am J Physiol 268:C36–C44

Lau LF, Pumiglia K, Côté YP, Feinstein MB (1994) Thrombin-receptor agonist peptides, in contrast to thrombin itself, are not full agonists for activation and signal transduction in human platelets in the absence of platelet-derived secondary mediators. Biochem J 303:391–400

Liu LW, Vu TKH, Esmon CT, Coughlin SR (1991) The region of the thrombin receptor resembling hirudin binds to thrombin and alters enzyme specificity. J Biol Chem 266:16977–16980

Marchese P, Murata M, Mazzucato M, Pradella P, De Marco L, Ware J, Ruggeri ZM (1995) Identification of three tyrosine residues of glycoprotein Ib$\alpha$ with distinct roles in von Willebrand factor and $\alpha$-thrombin binding. J Biol Chem 270:9571–9578

Mathews II, Padmanabhan KP, Ganesh V, Tulinsky A, Ishii M, Chen J, Turck CW, Choughlin SR (1994) Crystallographic structures of thrombin complexed with thrombin receptor peptides: existence of expected and novel binding modes. Biochemistry 33:3266–3279

Molino M, Blanchard N, Belmonte E, Tarver AP, Abrams C, Hoxie JA, Cerletti C, Brass LF (1995) Proteolysis of the human platelet and endothelial cell thrombin receptor by neutrophil-derived cathepsin G. J Biol Chem 270:11168–11175

Negrescu EV, Luber de Quintana K, Siess W (1995) Platelet shape change induced by thrombin receptor activation. Rapid stimulation of tyrosine phosphorylation of novel protein substrates through an integrin- and $Ca^{2+}$-independent mechanism. J Biol Chem 270:1057–1061

Nelken NA, Soifer SJ, Okeefe J, Vu TK, Charo IF, Choughlin SR (1992) Thrombin receptor expression in normal and atherosclerotic human arteries. J Clin Invest 90:1614–1621

Nieuwland R, Van Willigen G, Akkerman JWN (1994) Different pathways for control of $Na^+/H^+$ exchange via activation of the thrombin receptor. Biochem J 297:47–52

Norton KJ, Scarborough RM, Kutok JL, Escobedo MA, Nannizzi L, Coller BS (1993) Immunologic analysis of the cloned platelet thrombin receptor activation mechanism: evidence supporting receptor cleavage, release of the N-terminal peptide, and insertion of the tethered ligand into a protected environment. Blood 82:2125–2136

Offermanns S, Laugwitz KL, Spicher K, Schultz G (1994) G proteins of the $G_{12}$ family are activated via thromboxane $A_2$ and thrombin receptors in human platelets. Proc Natl Acad Sci USA 91:504–508

Rasmussen UB, Vouret-Craviari V, Jallat S, Schlesinger Y, Pagès G, Pavirani A, Lecocq JP, Pouysségur J, Van Obberghen-Schilling E (1991) cDNA cloning and expression of a hamster $\alpha$-thrombin receptor coupled to $Ca^{2+}$ mobilization. FEBS Lett 288:123–128

Rasmussen UB, Gachet C, Schlesinger Y, Hanau D, Ohlmann P, Van Obberghen-Schilling E, Pouysségur J, Cazenave JP, Pavirani A (1993) A peptide ligand of the human thrombin receptor antagonizes $\alpha$-thrombin and partially activates platelets. J Biol Chem 268:14322–14328

Ruggiero M, Lapetina EG (1986) Sustained proteolysis is required for human platelet activation by thrombin. Thromb Res 42:247–255

Scarborough RM, Naughton MA, Teng W, Hung DT, Rose J, Vu TKH, Wheaton VI, Turck CW, Coughlin SR (1992) Tethered ligand agonist peptides. Structural requirements for thrombin receptor activation reveal mechanisms of proteolytic unmasking of agonist function. J Biol Chem 267:13146–13149

Seiler SM, Goldenberg HJ, Michel IM, Hunt JT, Zavoico GB (1991) Multiple pathways of thrombin-induced platelet activation differentiated by desensitization and a thrombin exosite inhibitor. Biochem Biophys Res Commun 181:636–643

Siess W (1989) Molecular mechanisms of platelet activation. Physiol Rev 69:58–178

Tesfamariam B (1994) Distinct receptors and signaling pathways in $\alpha$-thrombin- and thrombin receptor peptide-induced vascular contractions. Circ Res 74:930–936
Vassallo RR, Kieber-Emmons T, Cichowski K, Brass LF (1992) Structure-function relationships in the activation of platelet thrombin receptors by receptor-derived peptides. J Biol Chem 267:6081–6085
Vicente V, Houghten RA, Ruggeri ZM (1990) Identification of a site in the $\alpha$ chain of platelet glycoprotein Ib that participates in von Willebrand factor binding. J Biol Chem 265:274–280
Vouret-Craviari V, Van Obberghen-Schilling E, Rasmussen UB, Pavirani A, Lecocq JP, Pouysségur J (1992) Synthetic $\alpha$-thrombin receptor peptides activate G protein-coupled signaling pathways but are unable to induce mitogenesis. Mol Biol Cell 3:95–102
Vouret-Craviari V, Auberger P, Pouysségur J, Van Obberghen-Schilling E (1995a) Distinct mechanisms regulate 5-$HT_2$ and thrombin receptor desensitization. J Biol Chem 270:4813–4821
Vouret-Craviari V, Grall D, Chambard JC, Rasmussen UB, Pouysségur J, Van Obberghen-Schilling E (1995b) Post-translational and activation-dependent modifications of the G protein-coupled thrombin receptor. J Biol Chem 270:8367–8372
Vu TKH, Hung DT, Wheaton VI, Coughlin SR (1991a) Molecular cloning of a functional thrombin receptor reveals a novel proteolytic mechanism of receptor activation. Cell 64:1057–1068
Vu TKH, Wheaton VI, Hung DT, Charo I, Coughlin SR (1991b) Domains specifying thrombin-receptor interaction. Nature 353:674–677
Wicki AN, Clemetson KJ (1985) Structure and function of platelet membrane glycoproteins Ib and V. Effects of leukocyte elastase and other proteases on platelet responses to von Willebrand factor and thrombin. Eur J Biochem 153:1–11
Wilcox JN, Rodriguez J, Subramanian R, Ollerenshaw J, Zhong Ch, Hayzer DJ, Horaist Ch, Hanson SR, Lumsden A, Salam TA, Kelley AB, Harker LA, Runge M (1994) Characterization of thrombin receptor expression during vascular lesion formation. Circ Res 75:1029–1038
Wilhelm B, Siess W (1993) Activation of the cloned platelet thrombin receptor decreases the pertussis-toxin-dependent ADP-ribosylation of the membrane and soluble inhibitory guanine-nucleotide-binding-$\alpha$-proteins. Eur J Biochem 216:81–88
Woolkalis MJ, DeMelfi TM, Blanchard N, Hoxie JA, Brass LF (1995) Regulation of thrombin receptors on human umbilical vein endothelial cells. J Biol Chem 270:9868–9875
Zhong C, Hayzer DJ, Corson MA, Runge MS (1992) Molecular cloning of the rat vascular smooth muscle thrombin receptor. J Biol Chem 267:16975–16979

CHAPTER 6

# Platelet ADP/Purinergic Receptors

C. GACHET and J.-P. CAZENAVE

## A. Introduction

Platelet aggregation by adenosine 5′-diphosphate (ADP) plays a key role in the development and extension of arterial thrombosis. ADP has also been implicated in vascular smooth muscle cell proliferation leading to restenosis at sites of vascular injury (CROWLEY et al. 1994). ADP-removing systems infused to animals species with platelet storage pool disease, for example Fawn-Hooded rats lacking ADP in their platelet dense granules, have proven valuable tools to assess the role of ADP in vivo (BORN 1985; MAFFRAND et al. 1988). Specific inhibitors of the ADP activation pathway such as the antiaggregatory thienopyridine compounds ticlopidine and clopidogrel (SCHRÖR 1993) markedly prolong bleeding time and are used clinically as antithrombotic drugs (VERRY et al. 1994). Furthermore, a rare congenital bleeding disorder with impairment of ADP-induced platelet aggregation (CATTANEO et al. 1992; NURDEN et al. 1995) strikingly resembles the acquired thrombopathy resulting from ticlopidine or clopidogrel intake (DI MINNO et al. 1985; GACHET et al. 1995). Obviously, a better knowledge of the ADP platelet activation pathway is of major importance for the understanding of the physiology of primary hemostasis and for the development of new antithrombotic strategies.

ADP was the first low-molecular-weight platelet aggregating agent to be identified. Following observations that a small molecule derived from red cells stimulated platelet adhesion to glass (HELLEM 1960), the same compound was found to induce platelet aggregation (OLLGAARD 1961) and was finally identified as ADP (GAARDNER et al. 1961). ADP plays a major role in platelet activation. Contained at very high concentrations in the platelet dense granules (REIMERS 1985), it is released by exocytosis when platelets are stimulated by other aggregating agents such as thrombin or collagen and thus contributes to and reinforces platelet aggregation. In addition, low concentrations of ADP potentiate or amplify the effects of all other agents, even weak agonists such as adrenaline (LANZA et al. 1988) or serotonin (MACFARLANE 1987). At present, despite more than 30 years of investigations aimed at assessing the central role of ADP in hemostasis, the platelet ADP receptor remains basically unknown.

# B. Platelet Responses to ADP

## I. Platelet Functions

Addition of ADP to washed human platelets results in (1) shape change, (2) exposure of the fibrinogen binding site on the $\alpha_{IIb}\beta_3$ integrin and fibrinogen binding, (3) reversible aggregation in the presence of fibrinogen and physiological concentrations of $Ca^{2+}$, and (4) desensitization.

### 1. Shape Change

Platelet shape change has been extensively studied using transmission electron microscopy, scanning electron microscopy, light transmission, and other direct or indirect morphological techniques. The normally discoid circulating platelets become spherical and throw out from their surface projections of varying shape, which then develop into multiple pseudopodia (O'BRIEN and HEYWOOD 1966). Since these changes also occur in the presence of EDTA in concentrations sufficient to completely prevent aggregation, shape change is clearly not dependent on extracellular calcium. The disc-to-sphere transformation has a half-life of about 6s, and the pseudopodia develop from the spheroidal form also with a half-life of about 6s (DERANLEAU et al. 1982). Shape change involves an enormous increase in the surface area of platelets, with the surface-connected canalicular system providing the reservoir of excess membrane (FROJMOVIC and MILTON 1982). The changes in external shape of blood platelets are accompanied by considerable internal rearrangements (WHITE 1968). There is swelling of the peripheral cytoplasm that accompanies the extrusion of long pseudopods, and there is a centripetal movement of the granules that occurs with constriction of the marginal bundle of microtubules to a more central position where it fits like a collar around the more centrally located granules. Microtubules are necessary for platelet shape change and aggregation, as has been demonstrated using colchicine and vinca alkaloids, compounds acting specifically on microtubules (MENCHE et al. 1980). A transient drop in metabolic ATP occurs during shape change, indicating that metabolic energy is consumed (MACFARLANE 1987).

### 2. Binding of Fibrinogen

The platelet fibrinogen receptor is the membrane glycoprotein (GP) IIb-IIa complex (PHILLIPS et al. 1988). In patients with Glanzmann's thrombasthenia, an inherited hemorrhagic disorder in which the GPIIb-IIIa complex is absent, reduced, or abnormal (BRAY 1994), platelets are unable to bind fibrinogen when activated by ADP or any other aggregating agent and consequently do not aggregate. GPIIb and GPIIIa are the two most prominent glycoproteins in platelet membranes. These molecules form a $Ca^{2+}$-dependent heterodimer, a member of the integrin family termed $\alpha_{IIb}\beta_3$, which serves as an activation-dependent receptor for the adhesive proteins fibrinogen, fibronectin, and von

Willebrand factor. How ADP and other agonists alter the GPIIb-IIIa complex, enabling it to bind fibrinogen, is basically unknown. The intracellular tail of GPIIIa seems to be the site of action of unidentified transduction proteins (SHATTIL 1995). Soluble fibrinogen does not interact with resting platelets, although it has been found to be internalized into the $\alpha$-granules by receptor-mediated endocytosis in a GPIIb-IIIa-dependent manner (HANDAGAMA et al. 1993). Upon stimulation with ADP, specific binding sites become available at the cell surface. Binding is time dependent, divalent-cation dependent, and saturable with respect to the concentrations of ADP and fibrinogen (MARGUERIE et al. 1979), while the single class of binding sites shows an average density of 38000 ± 10000 per platelet at equilibrium and affinity constants in the micromolar range. Extracellular cations are not required for receptor induction but are necessary for the interaction of fibrinogen with its binding site. Depending on the assay conditions used, it is possible to measure fibrinogen binding to ADP-stimulated intact platelets under physiological conditions (HARFENIST et al. 1980). When equilibrium is not reached, ADP-stimulated fibrinogen binding is not more than about 1000 molecules per platelet in the presence of 0.1 mg/ml fibrinogen, a concentration that is not saturating but is sufficient to support reversible ADP-induced platelet aggregation. This accords well with the reversible ADP-induced platelet aggregation observed at 37°C in the presence of physiological $Ca^{2+}$ concentrations and in the absence of the release reaction (MUSTARD et al. 1978).

## 3. ADP-Induced Platelet Aggregation

Two types of aggregation response can be distinguished in vitro: a primary aggregation that is reversible and does not involve the release reaction (exocytic secretion of granule contents) and a secondary aggregation that is irreversible and associated with the release reaction. Primary reversible aggregation is induced by low concentrations of platelet stimuli in the presence of extracellular $Ca^{2+}$ in the suspending medium, with platelet aggregation and deaggregation being associated with fibrinogen binding to and dissociating from its receptor. Secondary irreversible aggregation observed at high agonist concentrations is mainly a consequence of the release reaction. The formation of large, irreversible platelet aggregates is mediated by endoperoxides and thromboxane $A_2$ ($TXA_2$) formed from phospholipid-derived arachidonic acid, by ADP, serotonin, and $Ca^{2+}$ secreted from dense granules, and by adhesive proteins secreted from $\alpha$-granules (SIESS 1989). This secretion from $\alpha$-granules provides high, localized concentrations of fibrinogen, fibronectin, thrombospondin, and von Willebrand factor at the platelet surface. The release reaction is inhibited by aspirin which irreversibly blocks cyclooxygenase, an enzyme responsible for the transformation of arachidonic acid into prostaglandin endoperoxides and $TXA_2$ (ZUCKER and PETERSON 1968; HAMBERG et al. 1974).

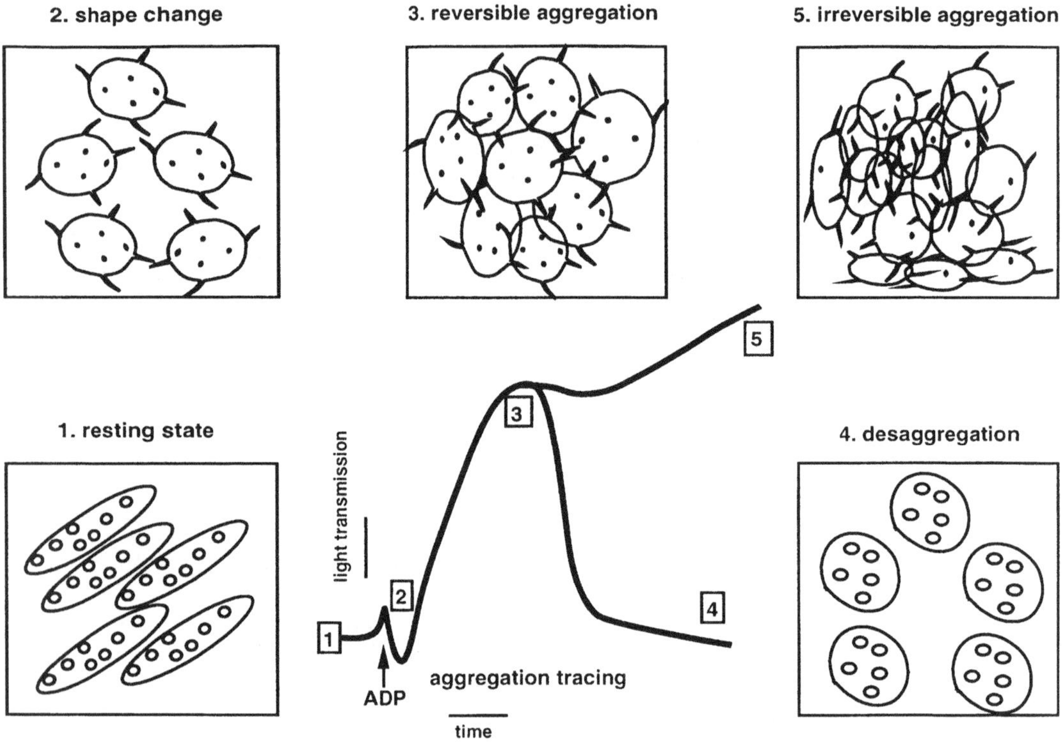


**Fig. 1.** Two types of aggregation response can be distinguished in vitro: a primary aggregation that is reversible and does not involve the release reaction and a secondary aggregation that is irreversible and associated with the release reaction. Primary reversible aggregation is induced by low concentrations of platelet stimuli in the presence of extracellular $Ca^{2+}$ in the suspending medium, with platelet aggregation and deaggregation being associated with fibrinogen binding to and dissociating from its receptor. Secondary irreversible aggregation observed at high agonist concentrations is mainly a consequence of the release reaction. The formation of large, irreversible platelet aggregates is mediated by endoperoxides and thromboxane $A_2$ ($TXA_2$) formed from phospholipid-derived arachidonic acid. When fibrinogen and free $Ca^{2+}$ are present and the platelet suspension is stirred, aggregation starts 3–5 s after addition of ADP (2), is half maximal 10–30 s later, and reaches a maximum (3) and then decreases within 2–3 min (4). During the so-called primary phase of aggregation, secretion does not occur and $TXA_2$ is not formed. In contrast, when ADP is added to a suspension of citrated platelet-rich plasma, after the first wave of primary aggregation a second wave occurs that is related to the secretion of granule contents and $TXA_2$ formation. This second wave of aggregation is not reversible (5)

When fibrinogen and free $Ca^{2+}$ are present and the platelet suspension is stirred, aggregation starts 3–5 s after the addition of ADP, is half maximal 10–30 s later, and reaches a maximum and then decreases within 2–3 min. During the so-called primary phase of aggregation, secretion does not occur and $TXA_2$ is not formed. In contrast, when ADP is added to a suspension of citrated platelet-rich plasma, after the first wave of primary aggregation a second wave occurs that is related to the secretion of granule contents and $TXA_2$ formation. This second wave of aggregation is not reversible. Factors responsible for the ADP-induced release reaction of human platelets have been studied extensively (MUSTARD et al. 1975), and it is current knowledge that low concentrations of $Ca^{2+}$ ($\mu$M range) support release from platelets stimulated with ADP in a stirred suspension, inducing extensive secondary aggregation (PACKHAM et al. 1987, 1989). These effects of $Ca^{2+}$ concentration on $TXA_2$ synthesis are not well understood. However, ADP-induced aggregation in citrated platelet-rich plasma or in artificial media containing no $Ca^{2+}$ is artefactual and does not correspond to the physiological response of platelets to ADP (Fig. 1).

#### 4. Desensitization

Desensitization of platelets to ADP is very rapid, and trace amounts of ADP are sufficient to render platelets refractory to this agonist. Since platelets that have become refractory to a new challenge with ADP display enhanced responsiveness to many other agonists (HALLAM and SCRUTTON 1987), care must be taken during platelet isolation to avoid ADP-induced desensitization by using ADP scavengers such as apyrase (ARDLIE et al. 1971; MUSTARD et al. 1972). The mechanism of platelet desensitization to ADP is not known but will probably be elucidated when the molecular structure of the receptor has been identified.

### II. Signal Transduction

At the intracellular level, platelet activation following ADP binding to its receptor leads to a transient rise in free cytoplasmic $Ca^{2+}$, resulting from both $Ca^{2+}$ influx and mobilization of internal stores. ADP also inhibits stimulated adenylyl cyclase, but this is not the cause of platelet aggregation (MACFARLANE 1987). It has been shown that ADP stimulates the binding of [$^{35}$S]GTP$\gamma$S to platelet membranes in man and in rat (GACHET et al. 1992a,b), suggesting that the ADP receptor is coupled to G proteins, further identified as $G_{i2}$ (OHLMANN et al. 1995).

#### 1. Changes in Cytosolic Free Calcium Concentration

Although most aggregating agents stimulate phospholipase C (PLC) and the subsequent production of inositol (1,4,5)-trisphosphate ($IP_3$) which is responsible for the mobilization of intracellular $Ca^{2+}$ stores, it is not clear whether ADP stimulates PLC or not (HEERMSKERK and SAGE 1994). In some studies, ADP has been found to produce rapid but very transient increases in $IP_3$ (RAHA et al. 1993; HEERMSKERK et al. 1993). However, other reports have found ADP to be unable to activate PLC or the production of $IP_3$ (FISHER et al. 1985; VICKERS et al. 1990; PACKHAM et al. 1993), and the exact mechanism by which this agonist stimulates the release of intracellular $Ca^{2+}$ stores in platelets thus remains to be established. ADP also causes an extremely rapid influx of

$Ca^{2+}$ from the extracellular medium (SAGE and RINK 1986), possibly through nonselective receptor-operated cation channels (MAHAUT-SMITH et al. 1992; GEIGER and WALTER 1993). It has recently been suggested that this $Ca^{2+}$ entry could involve $P_{2X1}$ purinoceptors (MACKENZIE et al. 1996).

### 2. Changes in Inositol Phospholipids

The most important unresolved question concerns the capability of the platelet ADP receptor to activate PLC. Studies on ADP-induced changes in inositol phospholipids are inconsistent, probably on account of experimental differences concerning the concentration of $Ca^{2+}$ in the medium, the presence or absence of aspirin, the presence of fibrinogen, and the concentration of ADP itself. Three groups have reported no change in the [$^{32}$P] labelling of phosphatidylinositol 4,5-bisphosphate ($PtdInsP_2$) for up to 20s after ADP stimulation (DANIEL et al. 1986; FISHER et al. 1985; MACINTYRE et al. 1985). Although two groups (DANIEL et al. 1986; LLOYD et al. 1973a,b) observed an increase in the labelling of phosphatidic acid (PA), three other groups (FISHER et al. 1985; MACINTYRE et al. 1985; MCNICOL et al. 1989) did not detect any such change. More recent studies (VICKERS et al. 1990; DUNCAN et al. 1993) showed no increase in $IP_3$ or diacylglycerol (DAG) upon stimulation of human platelets with ADP. Moreover, the changes observed in $PtdInsP_2$ and PtdInsP appeared to be due to ADP-stimulated binding of fibrinogen and/or to the aggregation process rather than to activation of PLC (VICKERS 1993). Lastly, using DAG analogues, a DAG kinase inhibitor, staurosporine, and okadaic acid in functional studies, PACKHAM et al. (1993) demonstrated primary ADP-induced aggregation to be independent of PLC or protein kinase C (PKC) activation.

### 3. Inhibition of Adenylyl Cyclase

Increasing the concentration of cyclic AMP (cAMP) within platelets results in inhibition of most of their responses, especially aggregation in response to weak stimuli. The most potent activator of adenylyl cyclase is prostacyclin ($PGI_2$), which produces an approximately fivefold increase in cAMP levels in intact cells, as do $PGE_1$ and $PGD_2$. Adenosine is a much weaker stimulator of adenylyl cyclase, while other agonists such as $\alpha_2$-adrenergic agents, thrombin, and ADP reduce the effects of prostaglandins on cAMP accumulation in intact cells. The ability of ADP to inhibit the accumulation of cAMP in intact platelets exposed to $PGE_1$ was first reported by COLE et al. (1971) and later investigated in more detail by HASLAM (1973). ADP was found to inhibit $PGE_1$ stimulation of cAMP formation in intact cells by about 90%. This effect was not due to an acceleration of the action of phosphodiesterases of cyclic nucleotides since it was not modified by phosphodiesterase inhibitors (HASLAM 1973). ADP has also been reported to inhibit adenylyl cyclase in lysed platelet preparations. COOPER and RODBELL (1979) found ADP to reduce the synthesis of cAMP in the presence of $PGE_1$ and GTP by about 25% in a dose-dependent manner, and this could be observed even in the absence of added GTP, suggesting a mechanism different from that of adrenaline which acts through the regulatory GTP-binding protein $Gi_2$ (SIMONDS et al. 1989). MELLWIG and JAKOBS (1980) found substantial inhibition of cAMP synthesis by ADP when GTP was added to the reaction mixture with or without $PGE_1$ or cholera toxin. This effect was not competitive if adenylyl 5′-imidodiphosphate (AMP-P(NH)P) was used as the substrate for adenylyl cyclase instead of ATP which competes with ADP at the receptor level; it was not abolished by adenosine deaminase and was not apparent when the enzyme was activated by the nonhydrolyzable analogue of GTP, GTP$\gamma$S. Although this action of ADP on adenylyl cyclase is insufficient to cause platelet aggregation since other inhibitors of cAMP accumulation do not induce aggregation and platelets exposed to phosphodiesterase inhibitors are able to aggregate after addition of ADP (HASLAM 1973), it nevertheless provides an interesting model to investigate the stimulus-response coupling of the ADP pathway.

## 4. Involvement of G Proteins

Results concerning the effect of ADP on adenylyl cyclase in membrane preparations (COOPER and RODBELL 1979; MELLWIG and JAKOBS 1980) were not in agreement with the requirements for GTP in the expression of its inhibitory activity, thus leaving unresolved the question of the involvement of G proteins in the ADP signal transduction pathway. By measuring the binding of [$^{35}$S]GTP$\gamma$S to platelet membranes stimulated with ADP or 2-methyl-thio-ADP (2MeSADP, a potent ADP receptor agonist), we found a dose-dependent increase of [$^{35}$S]GTP$\gamma$S binding that was competitively and specifically inhibited by ATP or its analogue (Sp)-ATP$\alpha$S (GACHET et al. 1992b). This suggested that the platelet ADP receptor might belong to the seven transmembrane domain G-protein-coupled receptor family (STRADER et al. 1994). In order to identify activated G proteins, we used an approach combining photolabelling of receptor-activated G proteins with 4-azidoanilido-[$\alpha$-$^{32}$P]GTP and immunoprecipitation of the G-protein $\alpha$-subunits with subtype-specific antibodies (LAUGWITZ et al. 1994). Stimulation of human platelet membranes with ADP resulted in an increase in 4-azidoanilido-[$\alpha$-$^{32}$P]GTP incorporation into the immunoprecipitates of $G_{\alpha i}$ but not of $G_{\alpha q}$ proteins, whereas stimulation with the thromboxane analogue U46619 resulted in an increase in 4-azidoanilido-[$\alpha$-$^{32}$P]GTP incorporation into the immunoprecipitates of $G_{\alpha q}$ but not of $G_{\alpha i}$ proteins while thrombin activated both G proteins. This effect of ADP was concentration-dependent and inhibited by the platelet ADP receptor antagonist ATP. Using specific antisera against subtypes of $G_i$ proteins, we found that ADP stimulated labelling of the $G_{\alpha i2}$ immunoprecipitate but not of the $G_{\alpha i3}$ precipitate. These data suggest that ADP inhibits cAMP formation by activation of $Gi_2$ proteins and provide additional evidence in support of the hypothesis that human platelet ADP receptors do not activate PLC through $G_q$ (OHLMANN et al. 1995).

# C. Platelet ADP Receptors

## I. Binding Studies

On the basis of agonist selectivity and signalling properties (BURNSTOCK 1978; DUBYAC and EL MOATASSIM 1993), the platelet receptor for ADP has been classified as a $P_{2T}$ receptor of the $P_2$ purinoceptor family. Its main characteristic is that ADP is the natural agonist and ATP a competitive antagonist. Although 2MeSADP has been described as a potent and selective agonist of the platelet ADP receptor (MACFARLANE et al. 1983), the biochemical structure of this receptor remains unknown. Studies of ADP binding to intact platelets or platelet membrane preparations are complicated by technical problems including metabolization of ADP at the surface of platelets, uptake of the adenosine formed, and high levels of unspecific binding to cytoskeletal proteins, while use of fixed platelets has the disadvantage that chemical changes in the sites could affect their accessibility to ligands. In general, only one class of binding site has been detected despite some studies suggesting the possible existence of two. Depending on experimental conditions, the number of sites per cell has been variously reported as 500 to 100000, and calculated dissociation constants range from 0.8 $nM$ to 3 $\mu M$ (reviewed in GACHET and CAZENAVE 1991). Recent studies have all used 2MeSADP as a selective platelet ADP receptor ligand and are more or less in agreement with the binding data (MILLS et al. 1992; SAVI et al. 1994a; NURDEN et al. 1995; GACHET et al. 1995).

In our studies using [$^{33}$P]2MeSADP, we found that intact washed human platelets possessed approximately 600 high-affinity binding sites per platelet ($K_d = 5.4\,nM$) and rat platelets about 1200 sites per platelet. This binding was competitively displaced by all ADP analogues tested, the rank order of potency being 2MeSADP > ADP$\beta$S > Sp-ATP$\alpha$S > ADP ≈ ATP > ATP$\gamma$S. As expected for a $P_{2T}$ receptor (DUBYAC and EL MOATASSIM 1993), 2MeSADP had the highest affinity while ADP and ATP were equivalent. Sp-ATP$\alpha$S has already been shown to have a higher affinity than ADP in binding experiments using fixed platelets (AGARWAL et al. 1989). ADP$\beta$S behaves as a partial agonist in terms of platelet aggregation (CUSACK and HOURANI 1981), and we found the same binding affinity ($K_i = 0.89\,\mu M$) as others (AGARWAL et al. 1989, 0.7 $\mu M$) in a fixed platelet system. The fact that we measured lower affinity for ADP itself ($K_i = 2.27\,\mu M$) may have been due to ectonucleotidase activity. Guanidine, uridine, and inosine nucleotides were without significant effect. Similar results were obtained using platelet membranes, although the affinity of [$^{33}$P]2MeSADP for its receptor in membrane preparations was lower ($K_d \approx 15\,nM$), possibly due to alterations in the structure of the binding sites arising from impaired anchorage of the receptor to membrane or cytoskeletal components. The specific binding of [$^{33}$P]2MeSADP was nevertheless displaced by ADP analogues in a manner comparable for intact platelets.

Using intact rat platelets, we found that the thienopyridine clopidogrel, which is known to specifically inhibit platelet aggregation induced by ADP or 2MeSADP, led to a dose-dependent reduction in the number of [$^{33}$P]2MeSADP binding sites, up to 70% of control values at the highest doses given to the animals (25–100mg/kg), while ADP-induced platelet aggregation was completely blocked (GACHET et al. 1995). These results suggest a non-competitive mechanism of binding inhibition and are in agreement with other recent studies (MILLS et al. 1992; SAVI et al. 1994a). Moreover, as also reported by others (SAVI et al. 1994b), rat platelets possess a "clopidogrel-insensitive" binding site representing 30% of the 2MeSADP binding sites in control platelets. These authors measured different affinities of ADP for this second binding site, suggesting that rat and possibly human platelets have two ADP binding sites, and the implications of these results will be discussed in the "two-receptor model" below.

## II. ADP-Binding Proteins

Several ADP-binding proteins have been proposed as putative ADP receptors, for instance a 28-kDa protein (PEARCE et al. 1978), a 61-kDa protein (ADLER and HANDIN 1979), and the 100-kDa protein called aggregin (COLMAN 1990). Recently, a 43-kDa protein has been shown to incorporate a photoaffinity analogue of ADP (CRISTALLI and MILLS 1993) that competitively inhibits the binding of 2MeSADP. This protein plausibly represents the receptor that mediates inhibition of adenylyl cyclase and platelet aggregation since its molecular weight lies in the range of those of other G-protein-coupled receptors. Nevertheless, this receptor has not yet been definitively identified by biochemical or molecular biology techniques.

## III. The P2 Purinoceptor Family

Extracellular adenine nucleotides interact with $P_2$ purinergic receptors to regulate a broad range of physiological processes. $P_2$ purinoceptors are widely distributed in many different cell types and tissues including endothelial cells, smooth muscle cells, epithelial cells, mastocytes, neurons, and blood cells. The platelet receptor for ADP has been classified as a $P_{2T}$ receptor of the $P_2$ purinoceptor family, which was established by BURNSTOCK in 1985 on the basis of the relative potencies and signalling properties of various purine agonists (DUBYAC and EL MOATASSIM 1993). The five subtypes of this family originally described ($P_{2X}$, $P_{2U}$, $P_{2Y}$, $P_{2T}$, $P_{2Z}$) have been reclassified into $P_{2X}$ (ligand-gated ion channels; BRAKE et al. 1994; VALERA 1994) and $P_{2Y}$ (G-protein-coupled) purinoceptor families (WILLIAMS 1995). Seven subtypes of $P_{2Y}$ and six subtypes of $P_{2X}$ have been reported to date, cloned, sequenced, and expressed from many tissues and species.

The $P_{2T}$ receptor may be distinguished from other purinergic receptors in that ADP is its natural agonist whereas ATP is a competitive antagonist. Furthermore, in contrast to other $P_2$ purinoceptors that are expressed in many tissues, the $P_{2T}$ receptor seems to be specific to platelets and has never been described in other cells, except in a few studies in which this receptor was found to be expressed by megakaryoblastic cell lines such as K562 (MURGO and SISTARE 1992), Dami leukemia cells (MURGO et al. 1994), human erythroleukemia (HEL) cells (SHI et al. 1995), and Meg-01 cells (HECHLER et al. 1994). The receptor-operated $Ca^{2+}$-influx stimulatory effects and adenylyl cyclase inhibitory effects of ADP on platelets are suggestive of the existence of an as yet unidentified type of purinoceptor or of two different receptors coupled to two different second messenger pathways. Although the cloning of a cDNA for a $P_{2T}$-like purinoceptor has been reported (BARNARD et al. 1994), the sequence of this receptor has not yet been published. Data have also been reported concerning the possible expression of unique $P_{2X}$ receptors on the platelet membrane that are responsible for cAMP accumulation through activation of $G_s$ by ATP and consequent inhibition of platelet aggregation induced by other agonists (SOSLAU et al. 1993). Apart from the fact that many of the effects of ATP described by these authors may be explained by its antagonistic action on released ADP that could be present in their platelet suspensions, the molecular structures of the recently cloned cDNAs for $P_{2X}$ receptors are not indicative of receptors able to activate G proteins. Nevertheless, platelet responses to $\alpha,\beta$-methyleneadenosine 5′-triphosphate ($\alpha,\beta$meATP), a selective $P_{2X1}$ receptor agonist, have been reported recently (MACKENZIE et al. 1996). However, we were unable to detect any binding of $[^3H]$ $\alpha,\beta$meATP to intact platelets (unpublished observation).

## IV. Current Hypothesis of a Two-Receptor Model

The first "two-platelet ADP receptor model" arose from studies using the affinity analogue probe 5′-p-fluorosulfonylbenzoyl adenosine (FSBA) which inhibits platelet aggregation without inhibiting the effect of ADP on adenylyl cyclase (BENNETT et al. 1978). In addition, the drug p-mercuribenzene sulfonate (pCMBS) inhibits the action of ADP on adenylyl cyclase but not its effects on shape change and aggregation, while 2-azido-ADP and 2MeSADP are considerably more potent as inhibitors of cAMP accumulation than as aggregating agents (MACFARLANE 1987). FSBA has been reported to be a weak platelet agonist since it induces a dose-dependent cytoplasmic $Ca^{2+}$ rise. The same authors also reported that FSBA does not inhibit ADP-induced $Ca^{2+}$ movements or its effect on adenylyl cyclase, whereas pCMBS abolishes both these responses but not shape change, suggesting that ADP-induced $Ca^{2+}$ mobilization and inhibition of adenylyl cyclase are mediated by platelet binding sites distinct from those mediating shape change (RAO and KOWALSKA 1987). In order to label the ADP receptor by affinity, FSBA has been used as a stable ADP analogue (COLMAN 1990). When platelets are exposed to FSBA at 37°C for 30 min, a 100-kDa protein is covalently labelled, and this labelling parallels the inhibition of ADP-induced shape change, fibrinogen receptor exposure, and aggregation. However, the inhibition of adenylyl cyclase by ADP is not affected by FSBA. The 100-kDa protein called aggregin has thus been proposed as the platelet receptor mediating ADP-induced shape change, fibrinogen receptor exposure, and aggregation.

Evidence against this first model of two receptors, one responsible for shape change and aggregation and the other responsible for inhibition of adenylyl cyclase, was recently reviewed (HOURANI and HALL 1994) and is based on pharmacological studies with $P_{2T}$ receptor antagonists (CUSACK and HOURANI 1982; HALL and HOURANI 1993, 1994) and on investigations using thienopyridines as pharmacological tools. Studies of the mechanism of action of the antiaggregatory thienopyridine compounds have now changed our understanding of the ADP-induced platelet activation/aggregation pathway. The thienopyridines ticlopidine and clopidogrel, which are antithrombotic drugs (SCHRÖR 1993), are potent and specific inhibitors of ADP-induced platelet aggregation. These compounds also inhibit aggregation in response to low concentrations of other agonists by blocking the amplification of platelet activation by released ADP. The site of action of thienopyridines has recently been identified. Although ticlopidine inhibits the binding of fibrinogen to $\alpha_{IIb}\beta_3$, this is the consequence of a more proximal modification since we have shown that ticlopidine and PCR 4099 (a racemic form of clopidogrel) neither quantitatively nor qualitatively modify $\alpha_{IIb}\beta_3$ integrin (GACHET et al. 1990a). It has also been demonstrated that ticlopidine and clopidogrel selectively neutralize ADP inhibition of $PGE_1$-activated adenylyl cyclase in intact human, rat, and rabbit platelets without affecting the action of adrenaline (human and rabbit platelets; GACHET et al. 1990b; DEFREYN et al. 1991). Moreover, clopidogrel has been shown to selectively block the effect of ADP as an inducer of G-protein activation in rat platelet membranes (GACHET et al. 1992b). In contrast, several effects of ADP on platelets, including in particular shape change and $Ca^{2+}$ movements, are not modified by these compounds.

The question is still unresolved as to whether one or two receptors for ADP exist on the platelet membrane, one of which might be altered by thienopyridines. It has been proposed that these substances might act on the coupling mechanism between an ADP receptor and unidentified G proteins linked to this receptor (GACHET et al. 1992b). Using intact rat platelets, we demonstrated that the thienopyridine clopidogrel led to a dose-dependent reduction in the number of [$^{33}$P]2MeSADP binding sites, up to 70% of control values at the highest doses given to the animals (25–100 mg/kg) (GACHET et al. 1995). These results are in agreement with other recent studies (MILLS et al. 1992; SAVI et al. 1994a). However, this does not explain why, under conditions in which ADP-induced aggregation and inhibition of adenylyl cyclase are completely blocked by thienopyridine treatment, low concentrations of ADP or 2MeSADP promote platelet shape change and elevation of intracellular $Ca^{2+}$ concentrations (GACHET et al. 1990b). Such findings point to the existence of a second platelet ADP receptor, insensitive to clopidogrel and responsible for shape change and $Ca^{2+}$ entry. This receptor possibly represents 30% of the ADP receptors on the platelet plasma membrane (SAVI et al. 1994b; GACHET et al. 1995). In further support of this hypothesis, when we measured the binding

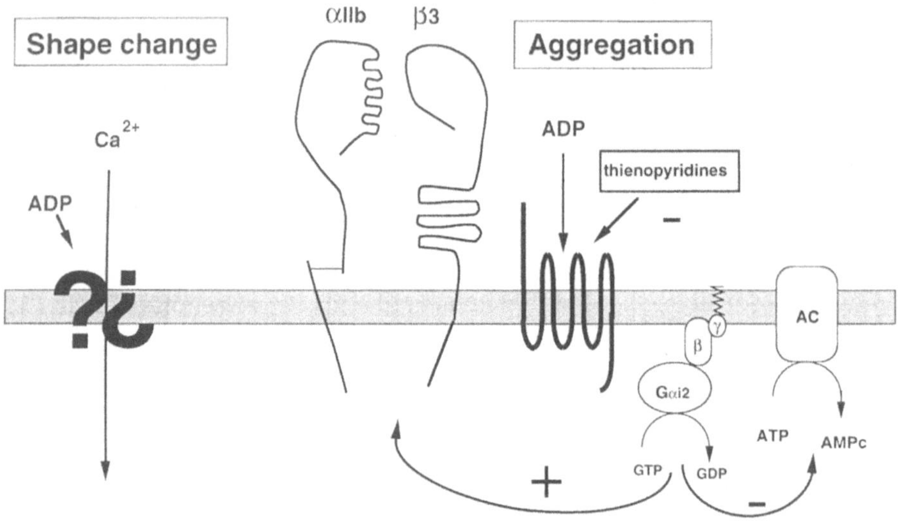


**Fig. 2.** Hypothetical model for ADP-induced platelet activation. Thienopyridine compounds completely and selectively block ADP-induced platelet aggregation, G-protein activation, and inhibition of adenylyl cyclase (*AC*), but do not affect shape change or $Ca^{2+}$ influx. Binding studies using the nonhydrolyzable [$^{33}$P]2MeSADP ligand showed that clopidogrel treatment reduces the number of sites by 70% on rat platelets (from 1200 to 450) and leaves the residual binding sites resistant to treatment. Moreover, patients with congenital impairment of ADP-induced platelet aggregation but normal shape change display the same residual proportion of [$^{33}$P]2MeSADP binding sites. Altogether, these data strongly suggest the presence of two ADP receptors, one responsible for shape change and rapid $Ca^{2+}$ influx (which are independent effects) and the other a seven transmembrane domain $G_{i2}$-protein-coupled receptor responsible for $Ca^{2+}$ mobilization from internal stores, inhibition of adenyly cyclase, exposure of the fibrinogen binding site on $\alpha_{IIb}\beta_3$ integrin, and platelet aggregation

of [$^{33}$P]2MeSADP to the platelets of a patient with a congenital deficiency of ADP-induced platelet aggregation thought to be related to a receptor defect (CATTANEO et al. 1992), we observed a reduction of 70% in the number of binding sites as compared to control platelets, without modification of binding affinity. Strikingly, the clinical profile and platelet functions of this patient are the same as when thienopyridine compounds are administered to humans or animals. The main feature is a strong and selective inhibition of aggregation induced by ADP despite conserved shape change. At the intracellular level, it has been shown (CATTANEO et al. 1992) that whereas ADP-induced responses are blocked in the platelets of this patient as after thienopyridine treatment, the intracellular $Ca^{2+}$ elevation resulting from $Ca^{2+}$ influx is not modified either in these platelets or by thienopyridine therapy. In summary, as we have proposed (GACHET et al. 1995), these data suggest the existence of two ADP receptors on platelets. One receptor responsible for platelet aggregation and inhibition of adenylyl cyclase, coupled to G proteins and sensitive to thienopyridine inhibition, might represent the so-called $P_{2T}$ receptor, while a second receptor insensitive to thienopyridine treatment might be responsible for shape change and rapid $Ca^{2+}$ influx (Fig. 2). Whether this second receptor is of the $P_{2X}$ type requires further experiments.

### V. Future Perspectives

MURGO et al. (1994) have described $P_{2U}$ and $P_{2T}$ purinoceptors on the Dami human megakaryocytic leukemia cell line and a $P_{2T}$ receptor on K562 erythroleukemia cells (MURGO and SISTARE 1992) on the basis of calcium measurements. Our results (HECHLER et al. 1994) for intracellular calcium changes in Meg-01 cells are similar to their findings, although in our hands Dami and K562 cells did not bind [$^{33}$P]2MeSADP, possibly due to variations in cell lines between laboratories. SHI et al. (1995) have reported that HEL cells express a purinoceptor that displays characteristics of the platelet $P_{2T}$ receptor and probably also express a $P_{2U}$ purinoceptor. SAVI et al. (1994c) have characterized specific binding sites for [$^{3}$H]2MeSADP on Dami and Meg-01 cells. Since simultaneous expression of $P_{2T}$ and $P_{2U}$ purinoceptors has been observed on several cell lines with potential for megakaryocytic differentiation, namely, Dami, HEL, and Meg-01 cells, this purinoceptor coexpression would appear to be a general feature of such cell lines. The presence of a $P_{2U}$ nucleotide receptor on cells with megakaryocytic features is an original finding, as $P_{2U}$ has never been shown to be expressed on platelets before and the relevance of this receptor on platelet precursors is as yet obscure. Conversely, the key role of the platelet ADP receptor in primary hemostasis has long been established, and its expression on platelet precursors is not surprising. In addition, since megakaryocytes are the progenitor cells of platelets, megakaryoblastic cell lines would be expected to express the platelet-type $P_{2T}$ purinoceptor. These cells are thus of great relevance for study of the platelet $P_{2T}$ receptor, as such nucleated cell lines could provide the necessary material to clone this as yet poorly characterized platelet purinoceptor. The proposed model of two platelet ADP receptors will require further investigation, in which cloning of the platelet $P_2$ receptors would appear to be a most promising approach.

## D. Conclusions

The central role of ADP in hemostasis and thrombosis is presently well established. Antiaggregatory thienopyridines irreversibly blocking the interaction of ADP with its platelet receptor responsible for aggregation have proven valuable tools to enlarge our understanding of this pathway. These compounds and other $P_{2T}$ antagonists currently under investigation have been shown to be active in animal models of arterial thrombosis (MAFFRAND et al. 1988; KIM et al. 1992; HUMPHRIES et al. 1995), and some have already been evaluated in man or are used clinically. Identification of the platelet $P_{2T}$ receptor(s) will provide an exciting impulse to basic and pharmacological research in this domain.

*Acknowledgements.* We thank J.N. Mulvihill for reviewing the English of the manuscript.

*Note Added in Proof.* While the manuscript was handled by the editors, new data have been generated in our laboratory pertaining to the field of platelet purinoceptors. In particular, we have established a stable Jurkat cell line expressing a human recombinant $P2Y_2$ receptor with a pharmacological profile matching that of the elusive P2T receptor. We have shown that this receptor is present in platelets together with a functional $P2X_1$ receptor and is also present in megakaryoblastic cell lines. These data would fit with a model (reviewed in GACHET et al., Platelets, 1997) of at least two receptors, one, the $P2X_1$ receptor responsible for rapid $Ca^{2+}$ entry and one, the $P2Y_1$ receptor responsible for aggregation and inhibition of adenylyl cyclase.

## References

Adler JR, Handin RI (1979) Solubilization and characterization of a platelet membrane ADP-binding protein. J Biol Chem 254:3866–3872

Agarwal AK, Tandon NN, Greco NJ, Cusack NJ, Jamieson GA (1989) Evaluation of the binding to fixed platelets of agonists and antagonists of ADP-induced platelet aggregation. Thromb Haemost 62:1103–1106

Ardlie NG, Perry DW, Packham MA, Mustard JF (1971) Influence of apyrase on stability of suspension of washed rabbit platelets. Proc Soc Exp Biol Med 136:1021–1023

Barnard EA, Burnstock G, Webb TE (1991) G-protein-coupled-receptors for ATP and other nucleotides: a new receptor family. Trends Pharmacol Sci 15:67–70

Bennett JS, Colman RF, Colman RW (1978) Identification of adenine nucleotide binding proteins in human platelet membranes by affinity labelling with 5′-p-fluorosulfonylbenzoyl adenosine. J Biol Chem 253:7346–7354

Born GVR (1985) Adenosine diphosphate as a mediator of platelet aggregation in vivo: an editorial point of view. Circulation 72:741–742

Brake AJ, Wagenbach MJ, Julius D (1994) New structural motif for ligand-gated ion channels defined by an ionotropic ATP receptor. Nature 371:519–523

Bray PF (1994) Inherited diseases of platelet glycoproteins: considerations for rapid molecular characterization. Thromb Haemost 72:492–502

Burnstock G (1978) A basis for distinguishing two types of purinergic receptors. In: Bolis L, Straub RN (eds) Cell membrane receptors for drugs and hormones: a multidisciplinary approach. Raven, New York, pp 107–118

Cattaneo M, Lecchi A, Randi AM, McGregor JL, Mannuci PM (1992) Identification of a new congenital defect of platelet function characterized by severe impairment of platelet responses to adenosine diphosphate. Blood 80:2787–2796

Cole B, Robinson A, Hartmann RC (1971) Studies on the role of cyclic AMP in platelet functions. Ann N Y Acad Sci 185:477–487

Colman RW (1990) Aggregin: a platelet ADP receptor that mediates activation. Faseb J 4:1425–11435

Cooper DMF, Rodbell M (1979) ADP is a potent inhibitor of human platelet plasma membrane adenylate cyclase. Nature 282:517–518

Cristalli G, Mills DC (1993) Identification of a receptor for ADP on blood platelets by photoaffinity labelling. Biochem J 291:875–881

Crowley ST, Dempsey EC, Horwitz KD, Horwitz LD (1994) Platelet-induced vascular smooth muscle cell proliferation is modulated by the growth amplification factors serotonin and adenosine diphosphate. Circulation 90:1908–1918

Cusack NJ, Hourani SMO (1981) Partial agonist behavior of adenosine 5′-O-(2-thiodiphosphate) on human platelets. Br J Pharmacol 73:4056–4508
Cusack NJ, Hourani SMO (1982) Adenosine 5′-diphosphate antagonists and human platelets: no evidence that aggregation and inhibition of activated adenylate cyclase are mediated by different receptors. Br J Pharmacol 76:221–227
Daniel JL, Dangelmaier CA, Selak M, Smith JB (1986) ADP stimulates $IP_3$ formation in human platelets. FEBS Lett 206:299–303
Defreyn G, Gachet C, Savi P, Driot F, Cazenave J-P, Maffrand J-P (1991) Ticlopidine and clopidogrel (SR 25990C) selectively neutralize ADP inhibition of $PGE_1$-activated platelet adenylate cyclase in rats and rabbits. Thromb Haemost 65:186–190
Deranleau DA, Dubler D, Rothen C, Lüscher EF (1982) Transient kinetics of the rapid shape change of unstirred human blood platelets stimulated with ADP. Proc Natl Acad Sci USA 79:7297–7301
Di Minno G, Cerbonne AM, Mattioli PL, Turco S, Iovine C, Mancini M (1985) Functionally thrombasthenic state in normal platelets following the administration of ticlopidine. J Clin Invest 75:328–338
Dubyac GR, El Moatassim C (1993) Signal transduction via P2-purinergic receptors for extracellular ATP and other nucleotides. Am J Physiol 265:C577–C606
Duncan EM, Tunbridge L, Lloyd JV (1993) An increase in phosphatidic acid in the absence of changes in diacylglycerol in human platelets stimulated with ADP. Int J Biochem 25:23–27
Fisher GS, Bakshian S, Baldassare JJ (1985) Activation of human platelets by ADP causes a rapid rise in cytosolic free calcium without hydrolysis of phosphatidylinositol-4,5-bisphosphate. Biochem Biophys Res Commun 129:858–864
Frojmovic MM, Milton JG (1982) Human platelet size, shape and related functions in health and disease. Physil Rev 62:186–261
Gaardner A, Jonsen J, Laland S, Hellem A, Owen PA (1961) Adenosine diphosphate in red cells as a factor in the adhesiveness of human blood platelets. Nature 192:531–532
Gachet C, Cazenave J-P (1991) ADP induced blood platelet activation: a review. Nouv Rev Fr Hematol 33:347–358
Gachet C, Cazenave J-P, Ohlmann P, Bouloux C, Defreyn G, Driot F, Maffrand J-P (1990a) The thienopyridine ticlopidine selectively prevents the inhibition effects of ADP but not of adrenaline on cAMP levels raised by stimulation of the adenylate cyclase of human platelets by $PGE_1$. Biochem Pharmacol 40:2683–2687
Gachet C, Stierlé A, Cazenave J-P, Ohlmann P, Lanza F, Bouloux C, Maffrand J-P (1990b) The thienopyridine PCR 4099 selectively inhibits ADP-induced platelet aggregation and fibrinogen binding without modifying the membrane glycoprotein IIb-IIIa complex in rat and in man. Biochem Pharmacol 40:229–238
Gachet C, Cazenave J-P, Ohlmann P, Hilf G, Wieland T, Jakobs KH (1992a) ADP receptor-induced activation of guanine-nucleotide-binding proteins in human platelet membranes. Eur J Biochem 207:259–263
Gachet C, Savi P, Ohlmann P, Maffrand JP, Jakobs KH, Cazenave J-P (1992b) ADP receptor-induced activation of guanine nucleotide binding proteins in rat platelet membranes. An effect selectively blocked by the thienopyridine clopidogrel. Thromb Haemost 68:79–83
Gachet C, Cattaneo M, Ohlmann P, Lecchi A, Hechler B, Chevalier J, Cassel D, Mannucci P, Cazenave J-P (1995) Purinoceptors on blood platelets: further phamacological and clinical evidence to suggest the presence of two ADP receptors. Br J Haematol 91:434–444
Geiger J, Walter U (1993) Properties and regulation of human platelet cation channels In: Siemen D, Hescheler J (eds) Nonselective cation channels: pharmacology, physiology and biophysics. Birkhäuser, Basel, pp 281–288
Hall DA, Hourani SMO (1993) Effects of analogues of adenine nucleotides on increases in intracellular calcium mediated by $P_{2T}$-purinoceptors on human blood platelets. Br J Pharmacol 108:728–733

Hall DA, Hourani SMO (1994) Effects of suramin on increases in cytosolic calcium and on inhibition of adenylate cyclase induced by adenosine 5′-diphosphate in human platelets. Biochem Pharmacol 47:1013–1018
Hallam TJ, Scrutton MC (1987) Desensitization. In: Holmsen H (ed) Platelets responses and metabolism, vol 2. CRC, Boca Raton, pp 105–118
Hamberg M, Svensson J, Samuelson B (1974) Mechanism of the antiaggregating effect of aspirin on human platelets. Lancet 2:223–224
Handagama PJ, Scarborough PM, Schuman MA, Bainton DF (1993) Endocytosis of fibrinogen into megakaryocytes and platelet $\alpha$ granules is mediated by $\alpha_{IIb}\beta_3$ (glycoprotein IIb-IIIa). Blood 82:135–138
Harfenist EJ, Packham MA, Mustard JF (1980) Reversibility of the association of fibrinogen with rabbit platelets exposed to ADP. Blood 56:189–198
Haslam RJ (1973) Interactions of the pharmacological receptors of blood platelets with adenylate cyclase. Ser Haematol 6:333–350
Hechler B, Hanau D, Cazenave J-P, Gachet C (1994) Presence of functional $P_{2T}$ and $P_{2U}$ purinoceptors on the human megakaryoblastic cell line, Meg-01. Characterization by functional and binding studies. Nouv Rev Fr Hematol 37:231–240
Heermskerk JWM, Sage SO (1994) Calcium signalling in platelets and other cells. Platelets 5:295–316
Heermskerk JWM, Vis P, Feijge MAH, Hoyland J, Mason WT, Sage SO (1993) Roles of phospholipase C and $Ca^{2+}$-ATPase in calcium responses of single, fibrinogen-bound platelets. J Biol Chem 268:356–363
Hellem AJ (1960) The adhesiveness of human blood platelets in vitro. Scand J Clin Lab Invest 12:1–117
Hourani SMO, Hall DA (1994) Receptors for ADP on human blood platelets. Trends Pharmacol Sci 15:103–108
Humphries RG, Robertson MJ, Leff P (1995) A novel series of $P_{2T}$ purinoceptor antagonists: definition of the role of ADP in arterial thrombosis. Trends Pharmacol Sci 16:179–181
Kim BK, Zamecnik P, Taylor G, Guo MJ, Blackburn GM (1992) Antithrombotic effect of b,b′-monochloromethylene diadenosine 5′,′″-P(1), P(4)-tetraphosphate. Proc Natl Acad Sci USA 89:11056–11058
Lanza F, Beretz A, Stierlé A, Hanau D, Kubina M, Cazenave J-P (1988) Epinephrine potentiates human platelet activation but is not an aggregating agent. Am J Physiol 255:1276–1288
Laugwitz K-L, Spicher K, Schultz G, Offermans S (1994) Identification of receptor-activated G-proteins: selective immunoprecipitation of photolabeled G-protein $\alpha$-subunits. Methods Enzymol 237:283–294
Lloyd JV, Nishizawa EE, Joist JH, Mustard JF (1973a) Effect of ADP-induced aggregation on $^{32}PO_4$ incorporation into phosphatidic acid and the phosphoinositides in rabbit platelets. Br J Haematol 24:589–604
Lloyd JV, Nishizawa EE, Mustard JF (1973b) Effect of ADP-induced shape change on incorporation of $^{32}P$ into platelet phosphatidic acid and mono, di and triphosphatidylinositol. Br J Haematol 25:77–99
MacFarlane DE (1987) Agonists and receptors: adenosine diphosphate. In: Holmsen H (ed) Platelet responses and metabolism, vol 2. CRC, Boca Raton, pp 19–36
MacFarlane DE, Srivastava PC, Mills DC (1983) 2-Methylthioadenosine [$\beta$-$^{32}$P]diphosphate: an agonist and radioligand for the receptor that inhibits the accumulation of cyclic AMP in intact blood platelets. J Clin Invest 71:420–428
MacIntyre DE, Pollock WK, Shaw AM, Bushfield M, McMillan LJ, McNicol A (1985) Agonist-induced inositol phospholipid metabolism and $Ca^{2+}$ flux in human platelet activation. Adv Exp Med Biol 192:127–144
MacKenzie AB, Mahaut-Smith MP, Sage SO (1996) Activation of receptor-operated channels via $P_{2X1}$ not $P_{2T}$ purinoceptors in human platelets. J Biol Chem 271:2879–2881

Maffrand J-P, Bernat A, Delebassée D, Defreyn G, Cazenave J-P, Gordon JL (1988) ADP plays a key role in thrombogenesis in rats. Thromb Haemost 59:225–230
Mahaut-Smith MP, Sage SO, Rink TJ (1992) Rapid ADP-evoked currents in human platelets recorded with the nystatin permeaboilized patch technique. J Biol Chem 267:3060–3065
Marguerie GA, Plow EF, Edgington TS (1979) Human platelets possess an inducible and saturable receptor specific for fibrinogen. J Biol Chem 254:5357–5363
McNicol A, Gerrard JM, MacIntyre DE (1989) Evidence for two mechanisms of thrombin-induced platelet activation: one proteolytic, one receptor-mediated. Biochem Cell Biol 67:332–336
Mellwig KP, Jakobs KH (1980) Inhibition of adenylate cyclase by ADP. Thromb Res 18:7–17
Menche D, Israel A, Karpatkin K (1980) Platelets and microtubules. Effect of colchicine and $D_2O$ on platelet aggregation and release induced by calcium ionophore A23187. J Clin Invest 66:284–291
Mills DC, Puri R, Minniti C, Grana G, Freedmann MD, Colman RF, Colman RW (1992) Clopidogrel inhibits the binding of ADP analogues to the receptor mediating inhibition of platelet adenylate cyclase. Arteriosclerosis Thromb 12:430–436
Murgo AJ, Sistare FD (1992) K562 leukemia cells express $P_{2T}$ adenosine diphosphate purinergic receptors. J Pharmacol Exp Ther 261:580–585
Murgo AJ, Contrera JG, Sistare FD (1994) Evidence for separate calcium signaling $P_{2T}$ and $P_{2U}$ purinoceptors in human megakaryocytic Dami cells. Blood 83:1258–1267
Mustard JF, Perry DW, Ardlie NG, Packham MA (1972) Preparation of suspensions of washed platelets from humans. Br J Haematol 22:193–204
Mustard JF, Perry DW, Kinlough-Rathbone RL, Packham MA (1975) Factors responsible for ADP-induced release reaction of human platelets. Am J Physiol 228:1857–1865
Mustard JF, Packham MA, Kinlough-Rathbone RL, Perry DW, Regoeczi E (1978) Fibrinogen and ADP-induced platelet aggregation. Blood 52:453–466
Nurden P, Savi P, Heilmann E, Bihour C, Herbert JM, Maffrand J-P, Nurden A (1995) An inherited bleeding disorder linked to a defective interaction between ADP and its receptor on platelets. J Clin Invest 95:1612–1622
O'Brien JR, Heywood JB (1966) Effects of aggregating agents and their inhibitors on the mean platelet shape. J Clin Pathol 19:148–153
Ohlmann P, Laugwitz K-L, Spicher K, Nurnberg K, Schultz G, Cazenave J-P, Gachet C (1995) The human platelet ADP receptor activates $G_{i2}$ proteins. Biochem J 312:775–779
Ollgaard E (1961) Macroscopic studies of platelet aggregation: nature of an aggregating factor in red blood cells and platelets. Thromb Diath Haemorrh 6:86–97
Packham MA, Kinlough-Rathbone RL, Mustard JF (1987) Thromboxane $A_2$ causes feedback amplification involving extensive thromboxane $A_2$ formation upon close contact of human platelets in media with a low concentration of ionized calciul. Blood 70:647–651
Packham MA, Bryant NL, Guccione MA, Kinlough-Rathbone RL, Mustard JF (1989) Effect of concentration of $Ca^{2+}$ in the suspending medium on the responses of human and rabbit platelets to aggregating agents. Thromb Haemost 62:968–976
Packham MA, Livne AA, Ruben DH, Rand ML (1993) Activation of phospholipase C and protein kinase C has little involvement in ADP-induced primary aggregation of human platelets: effect of diacylglycerols, the diacylglycerol kinase inhibitor R59022, staurosporine and okadaic acid. Biochem J 290:849–856
Pearce PH, Wright JM, Egan CM, Scrutton MC (1978) Interaction of human blood platelets with the 2′,3′-dialdehyde and 2′,3′-dialcohol derivatives of adenosine 5′-diphosphate and adenosine 5′-triphosphate. Eur J Biochem 88:543–554
Phillips DR, Charo IF, Parise LV, Fitzgerald LA (1988) The platelet membrane glycoprotein IIb-IIIa complex. Blood 71:831–843

Raha S, Jones GD, Gear AR (1993) Sub-second oscillation of inositol 1,4,5-trisphophate and inositol 1,3,4,5-tetrakisphosphate during platelet activation by ADP and thrombin: lack of correlation with calcium kinetics. Biochem J 292:643–646

Rao AK, Kowalska MA (1987) ADP-induced shape change and mobilization of cytoplasmic ionized calcium are mediated by distinct binding sites on platelets: 5′-p-fluorosulfonylbenzoyladenosine is a weak platelet agonist. Blood 70:751–756

Reimers HJ (1985) Adenine nucleotides in blood platelets. In: Longenecker GL (ed) The platelets: physiology and pharmacology. Academic, Orlando, p 85

Sage SO, Rink TJ (1986) Kinetic differences between thrombin-induced and ADP-induced calcium influx and release from internal stores in fura-2-loaded human platelets. Biochem Biophys Res Commun 136:1124–1129

Savi P, Laplace MCL, Herbert JM (1994a) Evidence for the existence of two different ADP-binding sites on rat platelets. Thromb Res 76:157–169

Savi P, Laplace MCL, Maffrand J-P, Herbert JM (1994b) Binding of [$^3$H]-2 methylthio-ADP to rat platelets. Effect of clopidogrel and ticlopidine. J Pharmacol Exp Ther 269:772–777

Savi P, Troussard A, Herbert JM (1994c) Characterization of specific binding sites for [$^3$H]2MeSADP on megakaryoblastic cell lines in culture. Biochem Pharmacol 48:83–86

Schrör K (1993) The basic pharmacology of ticlopidine and clopidogrel. Platelets 4:252–261

Shattil SJ (1995) Function and regulation of the $\beta 3$ integrins in hemostasis and vascular biology. Thromb Haemost 74:149–155

Shi XP, Yin KC, Gardell SJ (1995) Human erythroleukemic (HEL) cells express a platelet $P_{2T}$-like ADP receptor. Thromb Res 77:235–247

Siess W (1989) Molecular mechanisms of platelet activation. Physiol Rev 69:58–178

Simonds WF, Goldsmith PK, Codina J, Unson CG, Spiegel AM (1989) $G_{i2}$ mediates $\alpha_2$-adrenergic inhibition of adenylyl cyclase in platelet membranes: in situ identification with $G_\alpha$ C-terminal antibodies. Proc Natl Acad Sci USA 86:7809–7813

Soslau G, Brodsky I, Parker J (1993) Occupancy of P2 purinoceptors with unique properties modulates the function of human platelets. Biochim Biophys Acta 1177:199–207

Strader CD, Fong TM, Tota MR, Underwood D, Dixon RAF (1994) Structure and function of G protein coupled receptors. Annu Rev Biochem 63:101–132

Takeuchi K, Ogura M, Saito H, Satok M, Takeuchi M (1991) Production of platelet-like particles by a human megakaryoblastic leukemia cell line Meg-01. Exp Cell Res 193:223–226

Valera S, Hussy N, Evans RJ, Adami N, North RA, Surprenant A, Buell G (1994) A new class of lignad-gated ion channel defined by $P_{2X}$ receptor for extracellular ATP. Nature 371:516–519

Verry M, Panak E, Cazenave J-P (1994) Antiplatelet therapy in the prevention of ischaemic stroke. Nouv Rev Fr Hematol 36:213–228

Vickers JD (1993) ADP-stimulated fibrinogen binding is necessary for some of the inositol phospholipid changes found in ADP-stimulated platelets. Eur J Biochem 216:231–237

Vickers JD, Kinlough-Rathbone RL, Packham MA, Mustard JF (1990) Inositol phospholipid metabolism in human platelets stimulated by ADP. Eur J Biochem 193:521–528

White JG (1968) Fine structural alterations induced in platelets by adenosine diphosphate. Blood 31:604–622

Williams M (1995) Purinoceptor nomenclature: challenges for the future. In: Belardinelli L, Pelleg A (eds) Adenosine and adenine nucleotides: from molecular biology to integrative physiology. Kluwer, Dordrecht, p 39

Zucker MB, Peterson J (1968) Inhibition of adenosine diphosphate-induced secondary aggregation and other platelet functions by acetylsalicylic acid ingestion. Proc Soc Exp Biol Med 127:547–551

CHAPTER 7

# Platelet Prostaglandin Receptors

F. Ushikubi, M. Hirata, and S. Narumiya

## A. Introduction

Prostanoids, consisting of prostaglandins (PGs) and thromboxane (TX), are oxygenated metabolites of arachidonic acids. They are produced by a variety of cells in response to stimuli and are released from the cells immediately after synthesis. They exert a variety of actions in the body that are mediated via cell surface receptors specific for each prostanoid. The prostanoid receptors specific for TX, PGI, PGE, PGF, and PGD are called the TP, IP, EP, FP, and DP receptors, respectively. There are four subtypes of the EP receptor called the $EP_1$, $EP_2$, $EP_3$ and $EP_4$ receptors (Coleman et al. 1990, 1994). Recent molecular biological studies have confirmed the presence of these receptors and revealed that these eight types and subtypes of prostanoid receptors belong to the family of G-protein-coupled rhodopsin-type receptors with seven transmembrane domains. Moreover, there are several amino acid sequence motifs conserved by the prostanoid receptors but not shared by other rhodopsin-type receptors, suggesting that they constitute a new subfamily within this receptor superfamily (for a review see Ushikubi et al. 1995).

Prostanoids play important roles in the regulation of platelet function. The major metabolite of arachidonic acid in platelets is $TXA_2$, which is produced when platelets are stimulated and functions as an ultimate activator of platelets. For example, the second-wave aggregation and secretion induced by ADP and epinephrine are completely dependent on the production of $TXA_2$. In contrast, $PGI_2$ and $PGD_2$ inhibit platelet activation. $PGI_2$ is produced mainly by vascular endothelial and smooth muscle cells, and its action as an antiplatelet agent is considered important for the homeostasis of the circulatory system. $PGD_2$ is formed by a variety of cells including mast cells and cells in pia mater. Within the circulatory system it is formed from $PGH_2$ released by activated platelets via the action of albumin and works as a negative feedback regulator of platelet activation. Finally, $PGE_2$ has effects potentiating the actions of other platelet stimulants at nanomolar concentrations. However, the physiological significance of this action is still unclear. We review here recent studies on the receptors mediating these prostanoid actions in platelets.

# B. Thromboxane $A_2$ Receptor

## I. Structure and Ligand Binding Specificity

Previous pharmacological and biochemical studies identified the thromboxane $A_2$ (TP) receptor as the specific binding site for $TXA_2$ agonists/antagonists in various tissues including the platelets (for a review see HALUSHKA et al. 1995) and indicated that this receptor couples to G proteins. However, the exact nature of the receptor remained unknown until the human TP receptor was purified from blood platelets (USHIKUBI et al. 1989) and its cDNA cloned from megakaryocytic leukemia cell and placenta libraries (HIRATA et al. 1991). These studies revealed that the receptor was a protein consisting of 343 amino acids and a G-protein-coupled, rhodopsin-type receptor with seven transmembrane domains (Fig. 1). When expressed in COS cells, the cloned receptor showed the ligand binding specificity identical to that of the TP receptor in platelets (Table 1). U-46619, a $PGH_2$ analogue, bound to this receptor with high affinity, indicating that $PGH_2$ shares the receptor with $TXA_2$.

The TP receptor has several features of a rhodopsin-type receptor (Fig. 1). Firstly, there are two asparagine residues in the amino-terminal extracellular portion, and these sites are glycosylated (HIRATA et al. 1991). The difference in molecular mass of the purified protein of 57 kDa from its calculated molecular weight of 37 kDa can be accounted for by glycosylation of the receptor (MAIS

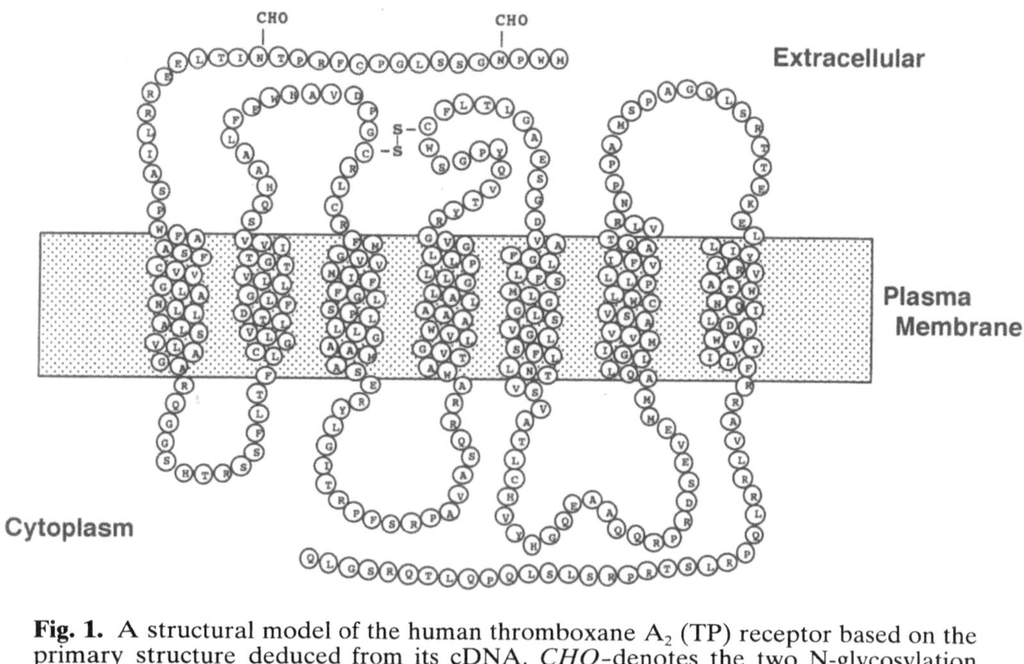


**Fig. 1.** A structural model of the human thromboxane $A_2$ (TP) receptor based on the primary structure deduced from its cDNA. *CHO*-denotes the two N-glycosylation sites. *-S-S-* indicates a putative disulfide bond

**Table 1.** Ligand binding properties of the cloned mouse prostanoid receptors

| Receptor | $K_d$ (*nM*) Ligand | Rank order of binding affinity[a] |
|---|---|---|
| TP | 1.2 [$^3$H]S-145 | S-145>ONO-3708>$STA_2$>U-46619>>$PGD_2$>$PGE_2$, $PGF_{2\alpha}$ |
| IP | 4.5 [$^3$H]iloprost | cicaprost>iloprost>$PGE_1$>carbacyclin>$PGD_2$, $PGE_2$, $STA_2$>$PGF_{2\alpha}$ |
| DP | 40 [$^3$H]$PGD_2$ | $GD_2$>BW245C>BWA868C>$STA_2$>$PGE_2$>iloprost>$PGF_{2\alpha}$ |
| $EP_3$ | 3 [$^3$H]$PGE_2$ | $PGE_2$=$PGE_1$>>iloprost>$PGD_2$>$PGF_{2\alpha}$ M&B-28767, GR-63799X>sulprostone>>butaprost, SC-19220=0 |

$K_d$, dissociation constant; TP, thromboxane $A_2$ receptor; IP, prostaglandin I receptor; DP, prostaglandin D receptor; $EP_3$, prostaglandin E receptor.
[a] The affinities of the various ligands for the receptors were determined in displacement experiments. The upper line shows the rank order of the "standard" prostanoids and the lower line that of various analogues in the $EP_3$ receptor.

et al. 1992). Secondly, the TP receptor conserves two cysteine residues, one in the first and the other in the second extracellular loop. These two residues are presumed to form a disulfide bond that is important for the maintenance of the receptor structure. Thirdly, an aspartic acid residue in the second transmembrane domain is also conserved. This residue is presumed to play an important role in transduction processes (FRASER et al. 1989).

In contrast, there are several regions conserved specifically among the prostanoid receptors which are localized in transmembrane domains and in the second extracellular loop (Fig. 2). Because these highly conserved regions are shared by all types of prostanoid receptors and even by those from different species, they may constitute the binding domains for common structures of prostanoid molecules. For example, the arginine residue in the seventh transmembrane domain was proposed responsible for the binding of the carboxyl group of prostanoid molecules by analogy to the retinal binding site $Lys^{296}$ of rhodopsin (HIRATA et al. 1991; NARUMIYA et al. 1993). In accordance with this hypothesis, FUNK et al. (1993) have shown that a point mutation at this arginine residue in the human TP receptor results in the loss of ligand binding activity.

In addition to the human receptor, the mouse and rat TP receptors have also been cloned (NAMBA et al. 1992; KITANAKA et al. 1995). They consist of 341 amino acids, and the homology between the two receptors is as high as 93%. Their sequence homologies to the human receptor are 76% and 72%, respectively. Species differences in platelet sensitivity to various TX analogues are well known. For example, rabbit platelets differ from human, cat, and canine platelets in their response to two agonist, $CTA_2$ and $PTA_2$ (BURKE et al. 1983). The potency of ONO-11120, an antagonist of the TP receptor, to rabbit platelets is two orders of magnitude lower than in human platelets (NARUMIYA et al. 1986). These species differences may reflect differences in the structures

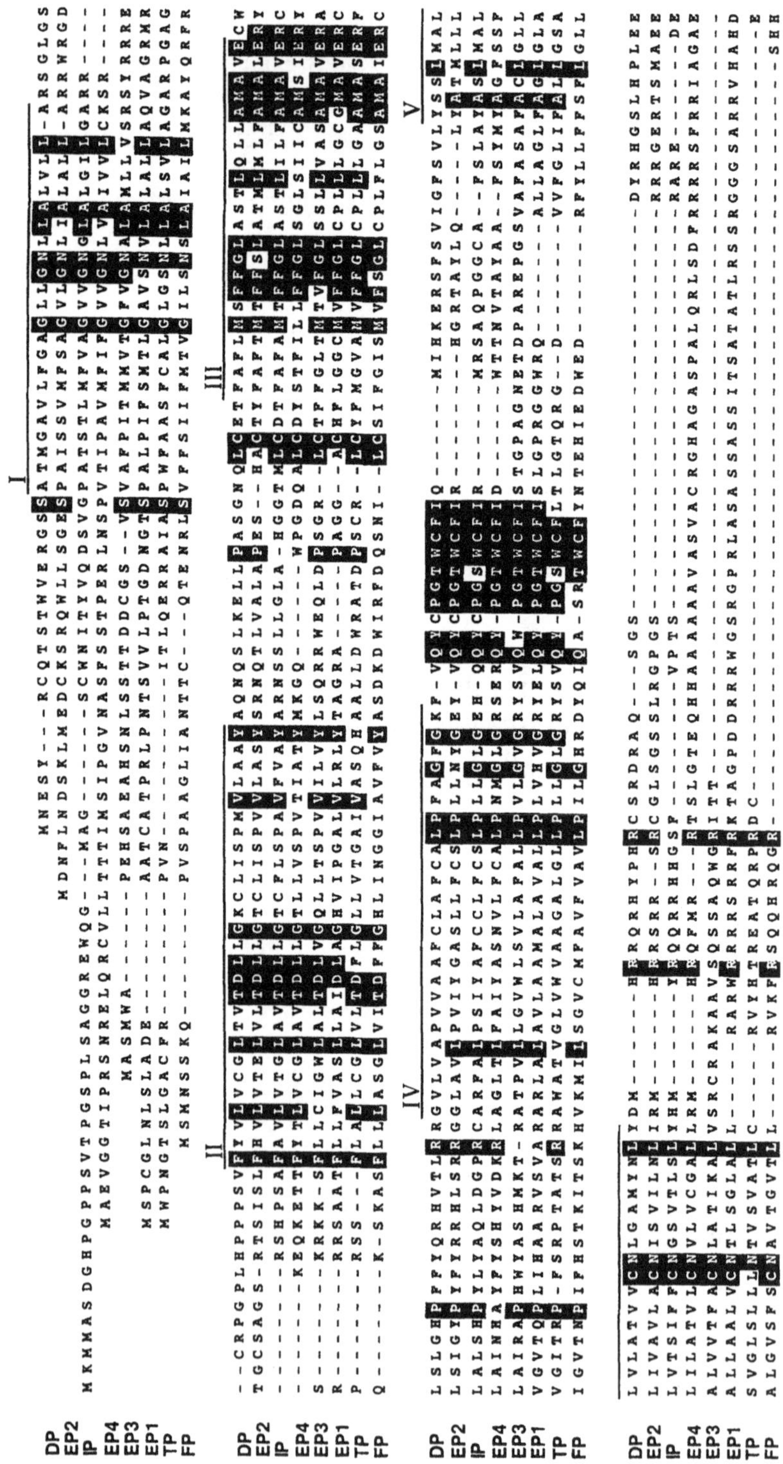

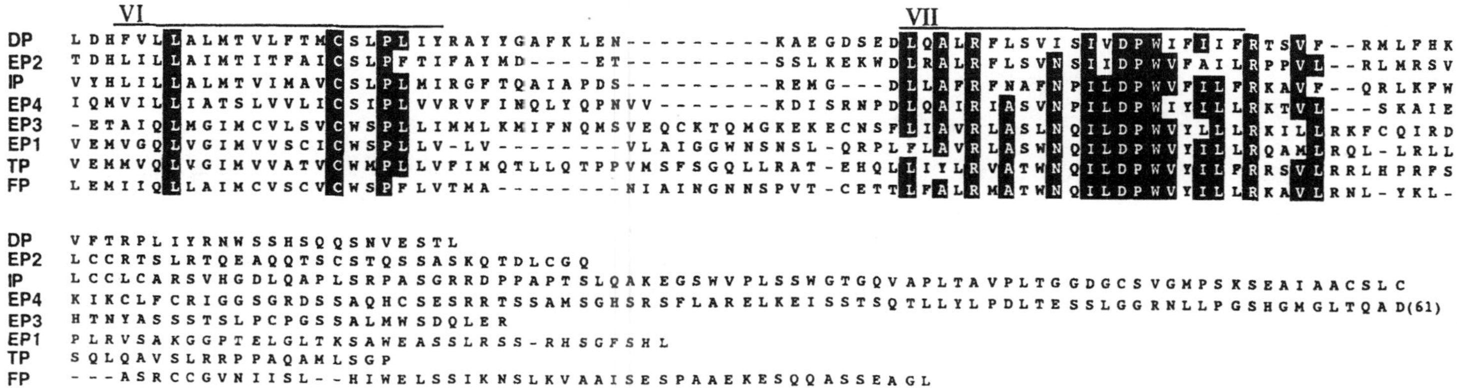


**Fig. 2.** The amino acid sequence alignment of the mouse prostanoid receptors. The amino acid sequences of the mouse PGD receptor (*DP*), PGI receptor (*IP*), four subtypes of the PGE receptors (*$EP_1$*, *$EP_2$*, *$EP_3$* and *$EP_4$*), the $TXA_2$ receptor (*TP*), and the PGF receptor (*FP*) are aligned to show optimal homology. The approximate positions of the putative transmembrane regions are indicated by *horizontal lines* above the appropriate sequences. Amino acids which are highly conserved are *highlighted*. PGD, prostaglandin D; PGI, prostaglandin I; PGE, prostaglandin E; $TXA_2$, thromboxane $A_2$

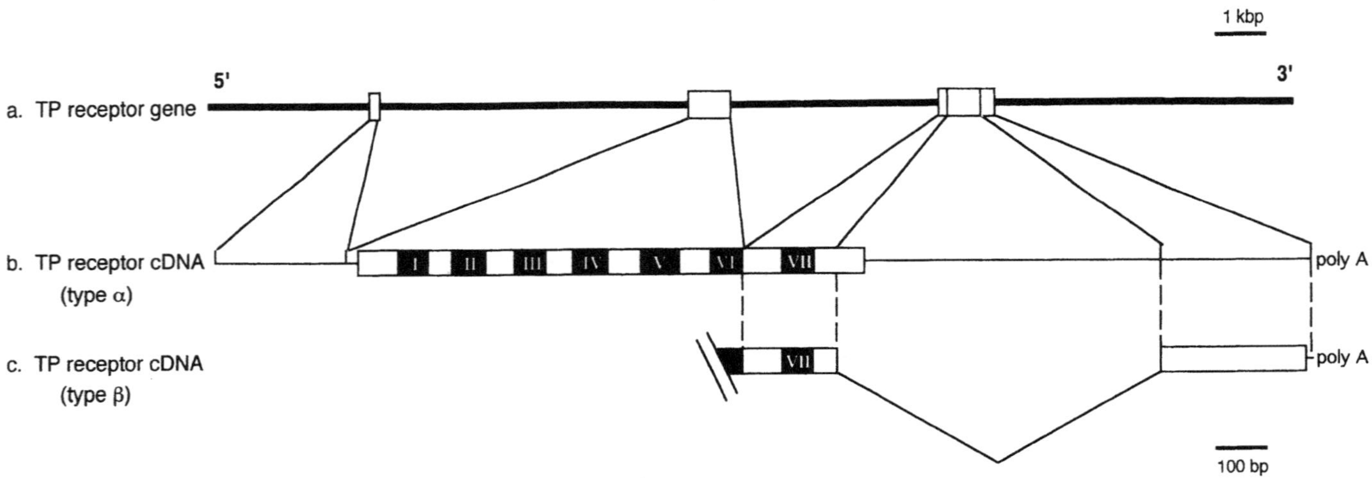


**Fig. 3a–c.** A schematic map of the human thromboxane $A_2$ (*TP*) receptor gene and cDNA. **a** Diagram of the human *TP* receptor gene. Exons are represented by *boxes*. **b, c** The structure of the two splicing variants of human TP receptor. The noncoding regions are represented by *horizontal lines*, the regions encoding the nonmembrane-spanning domains are represented by *open boxes*, and the regions encoding the transmembrane domains are represented by *closed boxes* and are *numbered*. Polyadenylation sites (*poly A*) are indicated

of the receptors, despite the high sequence homology among receptors from various species.

The structure of the human TP receptor gene has been reported. It contains three exons separated by two introns in the 5′-noncoding region and at the end of the sixth transmembrane domain (NÜSING et al. 1993; Fig. 3). This exon-intron relationship appears to be conserved in other types of prostanoid receptors and across various species, such as in the mouse DP receptor (M. HIRATA et al. 1994), the mouse $EP_1$ receptor (BATSHAKE et al. 1995), and the human $EP_3$ receptor (REGAN et al. 1994). There are additional exons in the carboxyl-terminal tails of some of the receptors, and alternative splicing of these exons creates several isoforms that differ from each other only in their carboxyl-terminal cytoplasmic tails. Recently, such a spliced variant of the TP receptor has been cloned from human umbilical vein endothelial cells (RAYCHOWDHURY et al. 1994; Fig. 3). Both isoforms of the TP receptor exist in platelets, but their signal transduction seems to be different (HIRATA et al. 1996; see below).

## II. Signal Transduction

When platelets are activated by $TXA_2$, serial responses such as shape change, aggregation, and secretion take place. These responses are thought to be induced by the TP receptor via coupling G proteins. However, the TP receptor has multiple signal transduction pathways. For example, it was reported that the TP receptor couples to G proteins and induces the activation of phospholipase (PL) C (BRASS et al. 1987). $TXA_2$ stimulates $PLA_2$ in platelets, and this action is reported to be mediated by a G protein distinct from that required for PLC activation (FUSE and TAI 1987). $TXA_2$ also induces $Ca^{2+}$ influx, which is regulated by G proteins in some situations (HALUSHKA et al. 1989; BRASS et al. 1993). There are conflicting reports concerning the inhibition of platelet adenylate cyclase by $TXA_2$ (for a review see HALUSHKA et al. 1989). The major question is how this multipicity of signalling is created. To explain this, the multiplicity of the TP receptor itself has been suggested. Several pharmacological studies suggested the heterogeneity of the TP receptor not only among various tissues of different species (HALUSHKA et al. 1989; OGLETREE and ALLEN 1992), but also within a single cell type, the blood platelet (DORN 1989; TAKAHARA et al. 1990). However, in spite of extensive efforts, there have been no reports on the presence of the second type of the TP receptor or its gene. Then again, as described above, there are at least two isoforms of the TP receptor created by alternative splicing of the single gene product, which may partly explain the above findings. Multiplicity of signalling could also be explained by the multiplicity of G proteins coupling to the receptor. In fact, several species of G proteins have been reported to participate in signalling via the platelet TP receptor. These include $G_q$ (SHENKER et al. 1991), $G_q$ and an 85 kDa unidentified G protein (KNEZEVIC et al. 1993), $G_q$ and $G_{i2}$ (USHIKUBI et al. 1994), and $G_{12}$ and $G_{13}$ (OFFERMANNS et al. 1994). Thus, the multiplicity of sigal transduction pathways activated via the TP receptor could be derived from both the receptor isoforms and the multiple coupling G proteins. In this respect, our analysis of the signalling of two TP receptor isoforms presents a model. We found that both of the receptor isoforms equally activate phospholipase C, but they either negatively or positively regulate the activity of adenylate cyclase; one isoform receptor couples to $G_q$ and $G_i$ and the other to $G_q$ and $G_s$ (HIRATA et al. 1996). Interestingly, the recently identified TP receptor mutant in human platelets with a point mutation at $Arg^{60}$ (T. HIRATA et al. 1994) cannot induce aggregation but can induce platelets' shape change and activation of phospholipase $A_2$ (FUSE et al. 1993). The $\alpha$ isoform of this mutant receptor cannot couple to $G_q$ and $G_s$, but its $\beta$ isoform can couple only to $G_i$. These results may suggest the possibility of the pathway mediated by $G_i$ and utilized by one species of isoform being involved in shape change and phospholipase $A_2$ activation (HIRATA et al. 1996).

A $TXA_2$ agonist has been shown to induce tyrosine phosphorylation of several proteins (NAKASHIMA et al. 1990) and activation of $p72^{syk}$ (MAEDA et al. 1993) in platelets. This signalling may participate in the mitogenic action of $TXA_2$ in other types of cells (HANASAKI et al. 1990; NIGAM et al. 1993; RUIZ et al. 1992). The precise mechanism and the species of G proteins involved, however, are not yet known.

## III. Regulation

The function of the receptor is regulated by activation and suppression of its gene expression and by the homologous and heterologous desensitization of the expressed receptor. There are several response elements in the promoter region of the TP receptor gene. They include a phorbol ester response element, a glucocorticoid response element, and three acute-phase response elements (NÜSING et al. 1993). In accordance with this results, phorbol ester induced the TP receptor expression in human erythroleukemia (HEL) cells, a megakaryocyte-like cell line (NAKAJIMA et al. 1989; KINSELLA et al. 1994), and in CHRF-288 megakaryoblastic cells (DORN et al. 1994). However, glucocorticoid and inducers of the acute-phase response, such as IL-1, IL-6, lipopolysaccharide, C-reactive protein, and tumor necrosis factor, could not induce this gene expression in HEL cells (KINSELLA et al. 1994). In contrast, testosterone induced gene expression for the TP receptor in HEL cells (MATSUDA et al. 1993), and increased human platelet thromboxane $A_2$ receptor density and aggregation response (AJAYI et al. 1995). Testosterone has been implicated as a significant risk factor of thrombotic diseases in young male athletes who abuse anabolic steroids.

The function of the TP receptor is also regulated by homologous and heterologous desensitization (MURRAY et al. 1990; OKWU et al. 1992; SPURNEY et al. 1994). Homologous desensitization involves initial uncoupling of the receptor from G protein(s), followed by a loss of the receptor from the plasma membrane (MURRAY and FITZGERALD 1989). Receptor loss by the mechanism of internalization was observed for the TP receptor in CHRF-288 megakaryoblastic cells and K562 cells (DORN 1991, 1992). Phosphorylation of the receptor by the receptor kinase may participate in the homologous desensitization of the receptor, as noted in other rhodopsin-type receptors (HAUSDORFF et al. 1990). In fact, agonist-induced phosphorylation of the TP receptor was reported in human platelets (OKWU et al. 1994). In contrast, heterologous desensitization occurs when the platelets are stimulated by agents that activate protein kinases C and A, and may involve the phosphorylation of the receptor by these kinases. The kinases were reported to phosphorylate the TP receptor in its carboxyl-terminal cytoplasmic portion (KINSELLA et al. 1994).

## C. Prostacyclin Receptor

### I. Structure and Ligand Binding Specificity

The prostacyclin receptor (IP receptor) has been identified pharmacologically by using stable prostacyclin ($PGI_2$) analogues in platelets and other tissues (reviewed in HALUSHKA et al. 1989). Among these analogues, iloprost and carbacyclin show significant agonistic activity on the $EP_1$ receptor. In contrast, cicaprost shows little cross-reactivity on the $EP_1$ receptor and is considered a specific IP ligand (reviewed in COLEMAN et al. 1990). Using radiolabeled compounds of these analogues, the IP receptor has been identified biochemically in platelets (TSAI et al. 1989; ITO et al. 1992), arterial preparations (TOWN et al. 1982), and several cell lines, including a neuroblastoma-glioma hybrid NCB-20 (BLAIR et al. 1981) and a mastocytoma P-815 (HASHIMOTO et al. 1990). The IP receptor cDNA was first isolated from P-815 cells (NAMBA et al. 1994). The cDNAs for human PGI receptor were isolated from a megakaryoblastic cell line and various human tissues including lung, thymus, and intestine (KATSUYAMA et al. 1994; NAKAGAWA et al. 1994; BOIE et al. 1994). Isolation of the rat clone (SASAKI et al. 1994) has also been reported.

The mouse IP receptor cDNA encodes a polypeptide of 417 amino acids with a calculated molecular mass of 45 kDa (NAMBA et al. 1994). The human IP receptor differs from the mouse receptor most notably at the translation initiation site: Its N terminus is 30 amino acids shorter than that of the mouse, and its initiating methionine corresponds to the second methionine (Met31) of the mouse receptor (Fig. 3). From this methionine through to the C-terminal tail, the human, mouse, and rat IP receptors are very similar in their primary structures. In this region, the mouse and the rat clones have 98% sequence identity while the human clone is 79% identical to the mouse and rat clones.

Studies of the cloned IP receptors expressed by transfection indicate that the mouse, human, and rat IP receptors have similar ligand specificities. To summarize, the binding affinities of the ligands tested were in the following order: iloprost = cicaprost > carbacyclin ≈ $PGE_1$ > $PGE_2$, $STA_2$ ≥ $PGD_2$. In human studies, Boie et al. (1994) reported that the equilibrium binding profile of [$^{3}$H]iloprost fitted best with a two-site model while Katsuyama et al. (1994) and Nakagawa et al. (1994) reported the presence of a single site. In lung and neural cells, two populations of iloprost binding sites with different affinities have been demonstrated (MacDermot et al. 1981; Keen et al. 1991) that became indistinguishable in the presence of a GTP analogue (Keen et al. 1991). Therefore, the discrepancies among ligand binding characteristics observed in expressed human IP receptors may have arisen from differences of the cell lines used for expression and reflect the different states of IP receptor-G protein coupling in each cell line.

## II. Signal Transduction

$PGI_2$ has been shown to raise cAMP in platelets. This inhibits $Ca^{2+}$ mobilization and aggregation induced by ADP and TP agonists. IP agonists have been known to induce differentiation of preadipocytes in vitro (Vassaux et al. 1992). The induction of the adipocytic phenotype was mediated by adenylate cyclase activation and $[Ca^{2+}]_i$ increase. When CHO cells expressing the mouse IP receptor cDNA were examined, iloprost induced a significant rise in cAMP levels. This response was observed starting from 10 *pM* of iloprost. In addition, iloprost induced inositol phospholipid breakdown in transformed CHO cells (Namba et al. 1994). However, this rise in inositol phospholipid breakdown required concentrations of the agonist 1000-fold higher than those required for adenylate cyclase activation. This inositol phospholipid response was insensitive to either cholera toxin or pertussis toxin treatment. Similar phospholipase C activation was observed with the human IP receptor (Katsuyama et al. 1994). These observations suggest that IP receptors can couple to a G protein that activates phospholipase C, probably Gq. It has not been reported whether IP receptors can activate phospholipase C in platelets. Such activation may not be easily observed since adenylate cyclase activation at lower IP agonist concentrations results in inhibition of phospholipase C activation in platelets.

## III. Regulation

IP receptors undergo homologous desensitization accompanied by reduction in receptor number (Blair et al. 1982) and Gs protein level (Edwards et al. 1987). The desensitized platelets show increased adhesiveness to endothelial cells (Darius et al. 1995). Thus, the vascular endothelium affects not only the aggregation but also the adhesion of platelets to the endothelium through $PGI_2$ production. These observations are particularly relevant in clinical settings where administration of IP agonists may be indicated as an antithrombotic therapy.

The gene for the human IP receptor has been isolated (Ogawa et al. 1995). The overall structure of the gene resembles that of the TP receptor gene: It consists of three exons; the second and third exons bear the coding sequence; and an intron divides the coding sequence at the end of the sixth transmembrane domain. The promoter region of the human IP receptor lacks a TATA box. In this region, several regulatory elements are present: SP-1, AP-2, APRE, and GRE. These elements are also present in the upstream region of the human TP receptor gene. Promoter activity of this region has yet to be tested.

## D. Prostaglandin D Receptor

### I. Structure

The prostaglandin D (DP) receptor has been identified in human platelets (WHITTLE et al. 1978; TOWN et al. 1983), basophils (VIRGOLINI et al. 1992), and neutrophils (ROSSI and O'FLAHERTY 1989) and has been shown to activate adenylate cyclase. In rats, $PGD_2$ shows little anti-aggregative activity in platelets (TOWN et al. 1986). To date, mouse and human DP receptors have been cloned (M. HIRATA et al. 1994; BOIE et al. 1994). The mouse DP receptor is a polypeptide of 357 amino acids with a calculated molecular mass of 40kDa. The mouse and human DP receptors show 75% identity in their primary structures (Fig. 4). Like other prostanoid receptor genes, the DP receptor is encoded by two exons that are separated by an intron at the end of the sixth transmembrane domain (HIRATA et al. 1994; BOIE et al. 1995).

### II. Ligand Binding Specificity and Signal Transduction

Both mouse and human DP receptors showed the presence of a single class of binding sites in expression studies. The mouse DP receptor showed a dissociation constant $K_d$ of 40 *nM* for $[^3H]PGD_2$ (HIRATA et al. 1994), comparable to the figure previously reported for human platelets (COOPER and AHERN 1979). In contrast, the cloned human DP receptor showed a higher affinity with $K_d$ at 1.5 *nM* (BOIE et al. 1994) when expressed in HEK 293 cells. This difference in affinity may be explained by the fact that the $K_d$ of the human DP receptor increased 20-fold in the presence of GTPγS (BOIE et al. 1994). Therefore, high $K_d$ values observed for mouse recombinant and human platelet DP receptors may reflect a state of receptor-G protein coupling that is different from HEK293 cells. Human DP receptors show similar binding affinities to the DP specific ligands BW245C and BWA868C as $PGD_2$ (BOIE et al. 1995), whereas their affinities for the mouse receptor were tenfold lower than that of $PGD_2$ (M. HIRATA et al. 1994). $PGD_2$ was able to activate adenylate cyclase in CHO cells expressing the mouse DP receptor. Although BW245C showed an affinity about tenfold lower than that of $PGD_2$ in the binding assay, it was tenfold more potent than $PGD_2$ in stimulating adenylate cyclase. The human receptor also showed a better cAMP response to BW245C, which showed eightfold higher potency than $PGD_2$. These results suggest that BW245C is capable of activating

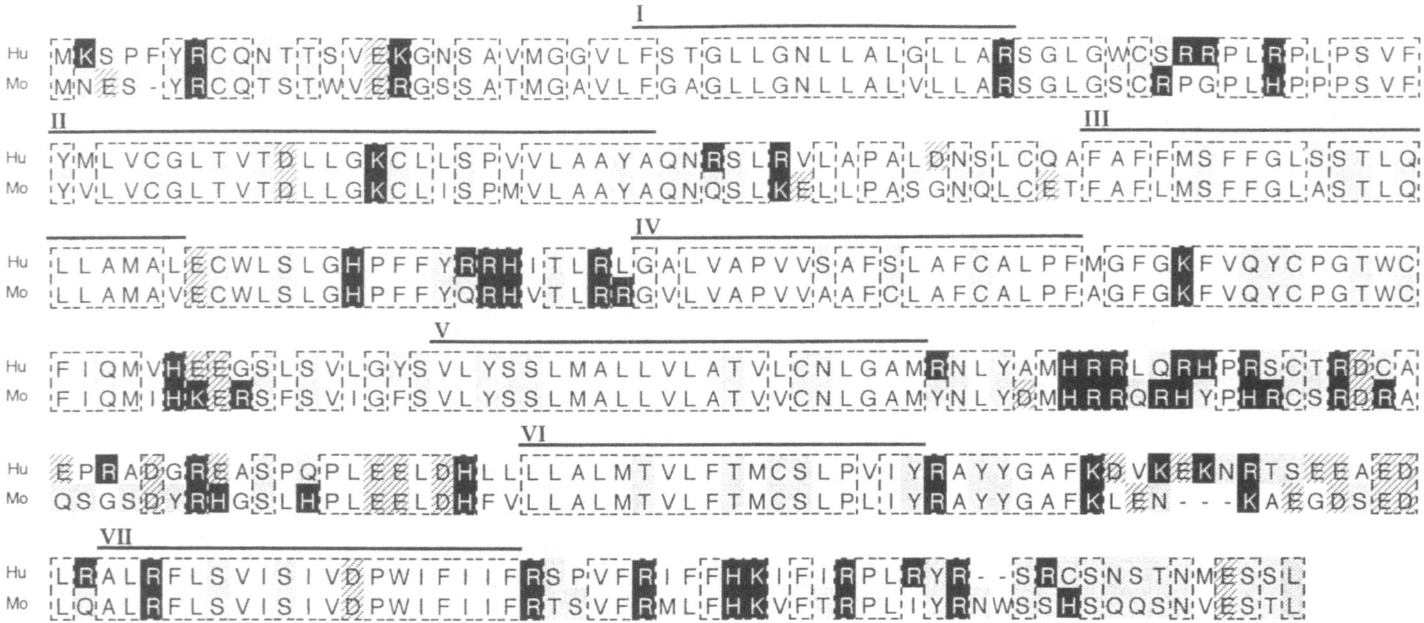


**Fig. 4.** Sequence alignment of human (*Hu*) and mouse (*Mo*) prostaglandin D (DP) receptors. Human and mouse DP receptors are aligned to show maximal sequence homology. The amino acids identical in both clones are *boxed*. Hydrophilic residue are *shadowed*, basic residues are *highlighted*, and acidic residues are indicated by *diagonal hatchings*. The seven transmembrane domains are indicated by *Roman numerals*

adenylate cyclase by binding to a small fraction of DP receptors. Such efficient agonistic activity may be accompanied by a high G-protein turnover rate. Human DP receptors were able to raise $[Ca^{2+}]_i$ in HEK 293 cells, which was abolished by chelating the extracellular $Ca^{2+}$. However, inositol phospholipid breakdown upon agonist stimulation was not observed in either mouse or human receptors. $PGD_2$ was shown to induce a rise in $[Ca^{2+}]_i$ without affecting inositol phospholipids in bovine adrenal nonchromaffin cells (OKUDA-ASHITAKA et al. 1993). In the bovine coronary artery, $PGD_2$ evokes endothelium-dependent relaxation by generating NO (BRAUN and SCHRÖR 1992). Such activation of NO synthase may result from the rise in $[Ca^{2+}]_i$ evoked by $PGD_2$.

## E. Prostaglandin E Receptor Subtype; the $EP_3$ Receptor

$PGE_2$ has a biphasic effect on platelet response, potentiating ADP-and collagen-induced aggregation at low concentrations and inhibiting aggregation at higher concentrations (GRAY and HEPTINSTALL 1985). The studies of ASHBY (1988) suggested that $PGE_2$ inhibits adenylate cyclase through $G_i$ in human platelets. These findings taken together indicate that $PGE_2$ acts on a specific receptor on the platelet surface to enhance aggregation by inhibiting adenylate cyclase, and the variability of this effect is due to an opposing action mediated by the IP receptor, with which $PGE_2$ cross-reacts. A specific binding site for $[^3H]$-$PGE_2$ was demonstrated on human platelets (EGGERMAN et al. 1986). In this study, two classes of specific $PGE_2$ binding sites were shown, one with a $K_d$ of $0.5\,nM$ and a $B_{max}$ of $1\,fmol/10^8$ platelets and the other with a $K_d$ of $20\,nM$ and a $B_{max}$ of $10\,fmol/10^8$ platelets. MATTHEWS and JONES (1993) quantitatively compared the potentiating effects of various PGE analogues on aggregation respose induced by ADP and U-46619 in human platelets (Fig. 5). The rank order of potentiation is: sulprostone = 16, 16-dimethyl $PGE_2$ > M&B 28767 > misoprostol > GR 63778X = 17-phenyl-$\omega$-trinor $PGE_2$, and this potentiation was blocked weakly by AH 6809, the $EP_1$ receptor antagonist. These results are consistent with the involvement of the $EP_3$ receptor in this action.

The human, bovine, rabbit, rat, and mouse $EP_3$ receptors have been cloned (SUGIMOTO et al. 1992; NAMBA et al. 1994; TAKEUCHI et al. 1993; YANG et al. 1994; REGAN et al. 1994; BREYER et al. 1994; ADAM et al. 1994). They have seven transmembrane domains and the homologies among them ranges from 84% to 97%. These receptors have several isoforms derived from alternative splicing that differ from each other only in the carboxyl tails as seen in the TP receptor isoforms. Because the carboxyl tail of one isoform has no homology to that of the other isoforms and because the carboxyl tails of some $EP_3$ receptor isoforms are conserved from one species to another, up to seven $EP_3$ receptor isoforms are likely to be present in any given species (REGAN et al. 1994). Since these isoforms share all seven transmembrane domains and extracellular loops, they have identical ligand binding properties (Table 1). In

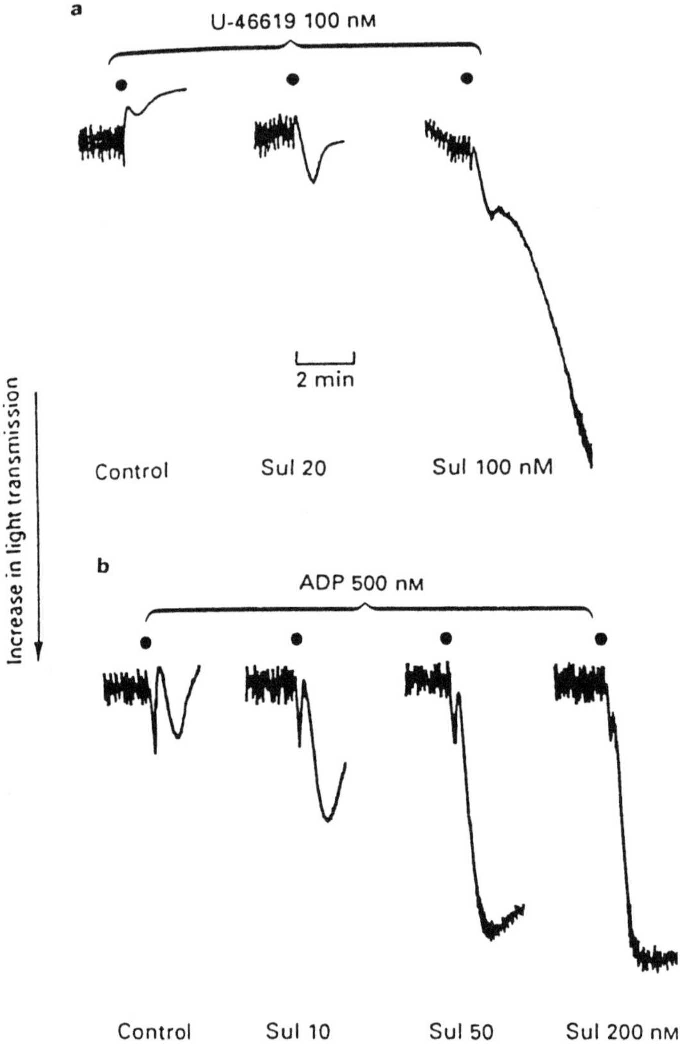


**Fig. 5a,b.** Potentiating effects of sulprostone (*Sul*) on U-46619-induced (**a**) and ADP-induced (**b**) platelet aggregation in human PRP. U-46619/ADP were added at (o) and sulprostone was added 2 min before each dose of aggregating agent. (From MATTHEWS and JONES 1993)

contrast, each isoform of the bovine $EP_3$ receptor couples to different G protein(s) and induces different signalling pathway(s) (NAMBA et al. 1994), and those of the mouse $EP_3$ receptor differ in their efficacy of G-protein coupling (SUGIMOTO et al. 1993) and in their sensitivity to agonist-induced desensitization (NEGISHI et al. 1993).

Although there are no reports on cloning the platelet $EP_3$ receptor, one study cloned it from human erythroleukemia (HEL) cells which express several platelet markers. KUNAPULI et al. (1994) screened a $\lambda$gt11 HEL cell cDNA library using a PCR-generated probe produced by using primers based on the sequence of the mouse $EP_3$ receptor. The cloned receptor exhibited the ligand binding properties of the $EP_3$ receptor, and its sequence corresponded to the $\alpha$-type isoform of the mouse $EP_3$ receptor. $PGE_2$ inhibited forskolin-stimulated cyclic AMP formation in COS-1 cells expressing this receptor. This result suggests that platelets may express the $\alpha$-type $EP_3$ receptor isoforms that are coupled to inhibition of adenylate cyclase. However, it is not clear whether other isoforms of the $EP_3$ receptor are present on platelets.

## References


Adam M, Boie Y, Rushmore TH, Müller G, Bastien L, McKee KT, Metters KM, Abramovitz M (1994) Cloning and expression of the human $EP_3$ prostanoid receptor. FEBS Lett 338:170–174

Ajayi AA, Mathur R, Halushka PV (1995) Testosterone increases human platelet thromboxane $A_2$ receptor density and aggregation responses. Circulation 91:2742–2747

Armstrong RA, Lawrence RA, Jones RJ, Wilson NH, Collier A (1989) Functional and ligand binding studies suggest heterogeneity of platelet prostacyclin receptors. Br J Pharmacol 97:657–668

Ashby B (1988) Cyclic AMP turnover in response to prostaglandins in intact platelets: evidence for separate stimulatory and inhibitory prostaglandin receptors. Second Messengers Phosphoproteins 12:45–47

Batshake B, Nilsson C, Sundelin J (1995) Molecular characterization of the mouse prostanoid $EP_1$ receptor gene. Eur J Biochem 231:809–814

Blair IA, Cresp TM, MacDermot J (1981) Divalent cations increase [$^3$H] prostacyclin binding to membranes of neuronal somatic hybrid cells. Br J Pharmacol 73:691–695

Blair IA, Leigh PJ, MacDermot J (1982) Desensitization of prostacyclin receptors in a neuronal hybrid cell line. Br J Pharmacol 77:121–127

Boie Y, Rushmore TH, Darmongoodwin A, Grygorczyk R, Slipetz DM, Metters KM, Abramovitz M (1994) Cloning and expression of a cDNA for the human prostanoid IP receptor. J Biol Chem 269:12173–12178

Brass LF, Shaller CC, Belmonte EJ (1987) Inositol 1,4,5-triphosphate-induced granule secretion in platelets: evidence that the activation of phospholipase C mediated by platelet thromboxane receptors involves a guanine nucleotide binding protein-dependent mechanism distinct from that of thrombin. J Clin Invest 79:1269–1275

Brass LF, Hoxie JA, Manning DR (1993) Signalling through G proteins and G protein coupled receptors during platelet activation. Thromb Haemost 63:220–223

Braun M, Schrör K (1992) Prostaglandin $D_2$ relaxes bovine coronary arteries by endothelium-dependent nitric oxide-mediated cGMP formation. Circ Res 71:1305–1313

Breyer RM, Emeson RB, Tarng JL, Breyer MD, Davis LS, Abromson RM, Ferrenbach SM (1994) Alternative splicing generates multiple isoforms of a rabbit prostaglandin $E_2$ receptor. J Biol Chem 269:6163–6169

Burke SE, Lefer AM, Nicolaou KC, Smith GM, Smith JB (1983) Responsiveness of platelets and coronary arteries from different species to synthetic thromboxane and prostaglandin endoperoxide analogues. Br J Pharmacol 78:287–292

Coleman RA, Kennedy I, Humphrey PPA, Bunce K, Lumley P (1990) Prostanoids and their receptors. In: Emmett JC (ed) Membranes and receptors. Pergamon, Oxford, pp 643–714 (Comprehensive medicinal chemistry, vol 3)

Coleman RA, Grix SP, Head SA, Louttit JB, Mallett A, Sheldrick RLG (1994) A novel inhibitory receptor in piglet saphenous vein. Prostaglandins 47:151–168

Cooper B, Ahern D (1979) Characterization of the platelet prostaglandin $D_2$ receptor. Loss of prostaglandin $D_2$ receptors in platelets of patients with myeloproliferative disorders. J Clin Invest 64:586–590

Corsini A, Folco GC, Fumagalli R, Nicosia S, Noe' MA, Oliva D (1987) (5Z)-Carbacyclin discriminates between prostacyclin receptors coupled to adenylate cyclase in vascular smooth muscles and platelets. Br J Pharmacol 90:255–261

Darius H, Veit K, Binz C, Fisch A, Meyer J (1995) Diminished inhibition of adhesion molecule expression in prostacyclin receptor desensitized human platelets. Agents Actions Suppl 45:77–83

Dorn GW (1989) Distinct platelet thromoboxane $A_2$/prostaglandin $H_2$ receptor subtypes: a radioligand binding study of human platelets. J Clin Invest 84:1883–1891

Dorn GW (1991) Mechanism for homologous downregulation of thromboxane $A_2$ receptors in cultured human chronic myelogenous leulemia (K562) cells. J Pharmacol EXP Ther 259:228–234

Dorn GW (1992) Regulation of response to thromboxane $A_2$ in CHRF-288 megakaryocytic cells. Am J Physiol 262:C991–C999

Dorn GW, Davis MG, D'Angelo DD (1994) Gene expression during phorbol ester-induced differentiation of cultured human megakaryoblastic cells. Am J Physiol 266:C1231–C1239

Edwards RJ, MacDermot J, Wilkens AJ (1982) Prostacyclin analogues reduce ADP-ribosylation of the a-subunit of the regulator Gs protein and diminish adenosine (A2) responsiveness of platelets. Br J Pharmacol 90:501–510

Edwards RJ, MacDermot J, Wilkens AJ (1987) Prostacyclin analogues reduce ADP-ribosylation of the $\alpha$-subunit of the regulator Gs protein and diminish adenosine ($A_2$) responsiveness of platelets. Br J Pharmacol 90:501–510

Eggerman TL, Andersen NH, Robertson RP (1986) Separate receptors for prostacyclin and prostaglandin $E_2$ on human gel-filtered platelets. J Pharmacol Exp Ther 236:568–573

Fraser CM, Wang CD, Robinson DA, Gocayne JD, Venter JC (1989) Site-directed mutagenesis of m1 muscarinic acetylcholine receptors: conserved aspartic acids play important roles in receptor function. Mol Pharmacol 36:840–847

Funk CD, Furci L, Moran N, FitzGerald GA (1993) Point mutation in the seventh hydrophobic domain of the human thromboxane $A_2$ receptor allows discrimination between agonist and antagonist binding sites. Mol Pharmacol 44:934–939

Fuse I, Tai HH (1987) Stimulations of arachidonate release and inositol-1,4,5-triphosphate formation are mediated by distinct G proteins in human platelets. Biochem Biophys Res Commun 146:659–665

Fuse I, Mito M, Hattori A, Higuchi W, Shibata A, Ushikubi F, Okuma M, Yahata K (1993) Defective signal transduction induced by thromboxane $A_2$ in a patient with mild bleeding disorder: impaired phospholipase C activation despite normal phospholipase $A_2$ activation. Blood 81:994–1000

Giles H, Leff P, Bolofo ML, Kelly MG, Robertson AD (1989) Classification of prostaglandin DP-receptors in platelets and vasculature using BWA868C, a novel, selective and potent competitive antagonist. Br J Pharmacol 96:291–300

Gorman RR, Bunting S, Miller OV (1977) Modulation of human platelet adenylate cyclase by prostacyclin (PGX). Prostaglandins 13:377–388

Gray S, Heptinstall S (1985) The effects of $PGE_2$ and CL 115,347, an antihypertensive $PGE_2$ analogue, on human blood platelet behavior and vascular contractility. Eur J Pharmacol 114:129–137

Halushka PV, Mais DE, Mayeux PR, Morinelli TA (1989) Thromboxane, prostaglandin and leukotriene receptors. Annu Rev Pharmacol Toxicol 10:213–239

Halushka PV, Allan CJ, Davis-Bruno KL (1995) Thromboxane $A_2$ receptors. J Lipid Med Cell Signal 12:361–378

Hanasaki K, Nakano T, Arita H (1990) Receptor-mediated mitogenic effect of thromboxane $A_2$ in vascular smooth muscle cells. Biochem Pharmacol 40:2535–2542

Hashimoto H, Negishi M, Ichikawa A (1990) Identification of a prostacyclin receptor coupled to the adenylate cyclase system via a stimulatory GTP-binding protein in mouse mastocytoma P-815 cells. Prostaglandins 40:491–505

Hausdorff WP, Caron MG, Lefkowitz RJ (1990) Turning off the signal: desensitization of beta-adrenergic receptor function. Faseb J 4:2881–2889

Hirata M, Hayashi Y, Ushikubi F, Yokota Y, Kageyama R, Nakanishi S, Narumiya S (1991) Cloning and expression of cDNA for a human thromboxane $A_2$ receptor. Nature 349:617–620

Hirata M, Kakizuka A, Aizawa A, Ushikubi F, Narumiya S (1994) Molecular characterization of a mouse prostaglandin D receptor and functional expression of the cloned gene. Proc Natl Acad Sci USA 91:11192–11196

Hirata T, Kakizuka A, Ushikubi F, Fuse I, Okuma M, Narumiya S (1994) $Arg^{60}$ to Leu mutation of the human thromboxane $A^2$ receptor in a dominantly inherited bleeding disorder. J Clin Invest 94:1662–1667

Hirata T, Ushikubi F, Kakizuka A, Okuma M, Narumiya S (1996) Two thromboxane $A_2$ receptor isoforms in human platelets: opposite coupling to adenylyl cyclase with different sensitivity to $Arg^{60}$ to Leu mutation. J Clin Invest 97:949–956

Ito S, Hashimoto H, Negishi M, Suzuki M, Koyano H, Noyori R, Ichikawa A (1992) Identification of the prostacyclin receptor by use of [15-$^3H_1$]19-(3-azidophenyl)-20-norisocarbacyclin, an irreversible specific photoaffinity probe. J Biol Chem 267:20326–20330

Katsuyama M, Sugimoto Y, Namba T, Irie A, Negishi M, Narumiya S, Ichikawa A (1994) Cloning and expression of a cDNA for the human prostacyclin receptor. FEBS Lett 344:74–78

Keen M, Kelly E, MacDermot J (1991) Guanine nucleotide sensitivity of [$^3$H] iloprost binding to prostacyclin receptors. Eur J Pharmacol 207:111–117

Kinsella BT, O'Mahony DJ, FitzGerald GA (1994) Phosphorylation and regulated expression of the human thromboxane $A_2$ receptor. J Biol Chem 269:29914–29919

Kitanaka J, Hashimoto H, Sugimoto Y, Sawada M, Negishi M, Suzumura A, Marunouchi T, Ichikawa A, Baba A (1995) cDNA cloning of a thromboxane $A_2$ receptor from rat astrocytes. Biochim Biophys Acta 1265:220–223

Knezevic I, Borg C, Le Breton GC (1993) Identification of $G_q$ as one of the G proteins which copurify with human platelet thromboxane $A_2$/prostaglandin $H_2$ receptors. J Biol Chem 268:26011–26017

Kunapuli SP, Fen Mao G, Bastepe M, Liu-Chen LY, Li S, Cheung PP, DeRiel JK, Ashby B (1994) Clonging and expression of a prostaglandin E receptor $EP_3$ subtype from human erythroleukemia cells. Biochem J 298:263–267

Lester HA, Steer ML, Levizki A (1982) Prostaglandin-stimulated GTP hydrolysis associated with activation of adenylate cyclase in human platelet membranes. Proc Natl Acad Sci USA 79:719–723

MacDermot J, Barnes PJ, Waddell KA, Dollery CT, Blair IA (1981) Prostacyclin binding to guinea pig pulmonary receptors. Eur J Pharmacol 75:127–130

Maeda H, Taniguchi T, Inazu T, Yang C, Nakagawara G, Yamamura H (1993) Protein-tyrosine kinase $p72^{syk}$ is activated by thromboxane $A_2$ mimetic U44069 in platelets. Biochem Biophys Res Commun 197:62–67

Mais DE, True TA, Martinelli MJ (1992) Characterization of the human thromboxane $A_2$/prostaglandin $H_2$ receptor; evidence for N-glycosylation. Eur J Pharmacol 227:267–274

Matsuda K, Mathur R, Duzic E, Halushka PV (1993) Androgen regulation of thromboxane $A_2$/prostaglandin $H_2$ receptor expression in human erythroleukemia cells. Am J Physiol 265:E928–E934

Matthews JS, Jones RL (1993) Potentiation of aggregation and inhibition of adenylate cyclase in human platelets by prostaglandin E analogues. Br J Pharmacol 108:363–369

Murray R, FitzGerald GA (1989) Regulation of thromboxane receptor activation in human platelets. Proc Natl Acad Sci USA 86:124–128

Murray R, Shipp E, FitzGerald GA (1990) Prostaglandin endoperoxide/thromboxane $A_2$ receptor desensitization; cross-talk with adenylate cyclase in human platelets. J Biol Chem 265:21670–21675
Nakagawa O, Tanaka I, Usui T, Harada M, Sasaki Y, Itoh H, Yoshimasa T, Namba T, Narumiya S, Nakao K (1994) Molecular cloning of human prostacyclin receptor cDNA and its gene expression in cardiovascular system. Circulation 90:1643–1647
Nakajima M, Yamamoto M, Ushikubi F, Okuma M, Fujiwara M, Narumiya S (1989) Expession of thromboxane $A_2$ receptor in cultured human erythroleukemia cells and its induction by 12-O-teradecanoylphorbol-13-acetate. Biochem Biophys Res Commun 158:958–965
Nakashima S, Koike T, Nozawa Y (1990) Genistein, a protein tyrosine kinase inhibitor, inhibits thromboxane $A_2$-mediated human platelet responses. Mol Pharmacol 39:475–480
Namba T, Sugimoto Y, Hirata M, Hayashi Y, Honda A, Watabe A, Negishi M, Ichikawa A, Narumiya S (1992) Mouse thromboxane $A_2$ receptor cDNA cloning, expression and northern blot analysis. Biochem Biophys Res Commun 184:1197–1203
Namba T, Oida H, Sugimoto Y, Kakizuka A, Negishi M, Ichikawa A, Narumiya S (1994) cDNA cloning of a mouse prostacyclin receptor. J Biol Chem 269:9986–9992
Narumiya S, Toda N (1985) Different responsiveness of prostaglandin $D_2$-sensitive systems to prostaglandin $D_2$ and its analogues. Br J Pharmacol 85:367–375
Narumiya S, Okuma M, Ushikubi F (1986) Binding of a radioiodinated 13-azapinane thromboxane antagonist to platelets: correlation with antiaggregatory activity in different species. Br J Pharmacol 88:323–331
Negishi M, Sugimoto Y, Irie A, Narumiya S, Ichikawa A (1993) Two isoforms of prostaglandin E receptor $EP_3$ subtype. J Biol Chem 268:9517–9521
Ney P, Schrör K (1991) $PGD_2$ and its mimetic ZK110.841 are potent inhibitors of receptor-mediated activation of human neutrophils. Eicosanoids 4:21–28
Nigam S, Eskafi S, Roscher A, Weitzel H (1993) Thromboxane $A_2$ analogue U46619 enhances tumor cell proliferation in HeLa cells via specific receptors which are apparently distinct from $TXA_2$ receptors on human platelets. FEBS Lett 316:99–102
Nüsing RM, Hirata M, Kakizuka A, Eki T, Ozawa K, Narumiya S (1993) Characterization and chromosomal mapping of the human thromboxane $A_2$ receptor gene. J Biol Chem 268:25253–25259
Offermanns S, Laugwitz KL, Spicher K, Schultz G (1994) G proteins of the $G_{12}$ family are activated via thromboxane $A_2$ and thrombin receptors in human platelets. Proc Natl Acad Sci USA 91:504–508
Ogawa Y, Tanaka I, Inoue M, Yoshitake Y, Isse N, Nakagawa O, Usui T, Itoh H, Yoshimasa T, Narumiya S, Nakao K (1995) Structural organization and chromosomal assignment of the human prostacyclin receptor gene. Genomics 27:142–148
Ogletree ML, Allen GT (1992) Interspecies differences in thromboxane receptors: studies with thromboxane receptor antagonists in rat and guinea-pig smooth muscles. J Pharmacol Exp Ther 260:789–794
Okuda-Ashitaka E, Sakamoto K, Giles H, Ito S, Hayaishi O (1993) Cyclic-AMP-dependent $Ca^{2+}$ influx elicited by prostaglandin $D_2$ in freshly isolated non-chromaffin cells from bovine adrenal medulla. Biochim Biophys Acta 1176:148–154
Okwu AK, Ullian ME, Halushka PV (1992) Homologous desensitization of human platelet thromboxane $A_2$/prostaglandin $H_2$ receptors. J Pharmacol Exp Ther 262:238–245
Okwu AK, Mais DE, Halushka, PV (1994) Agonist-induced phosphorylation of human platelet $TXA_2/PGH_2$ receptors. Biochim Biophys Acta 1221:83–88

Raychowdhury MK, Yukawa M, Collins LJ, McGrail SH, Kent KC, Ware JA (1994) Alternative splicing produces a divergent cytoplasmic tail in the human endothelial thromboxane $A_2$ receptor. J Biol Chem 269:19256–19261
Regan JW, Bailey TJ, Donello JE, Pierce KL, Pepperl DJ, Zhang D, Kedzie KM, Fairbairn CE, Bogardus AM, Woodward DF, Gil DW (1994) Molecular cloning and expression of human $EP_3$ receptors; evidence of three variants with differing carboxyl termini. Br J Pharmacol 112:377–385
Rossi AG, O'Flaherty JT (1989) Prostaglandin binding sites in human polymorphonuclear neutrophils. Prostaglandins 37:641–653
Ruiz P, Rey L, Spurney R, Coffman T, Viciana A (1992) Thromboxane augmentation of alloreactive T cell function. Transplantation 54:498–505
Sasaki Y, Usui T, Tanaka I, Nakagawa O, Sando T, Takahashi T, Namba T, Narumiya S, Nakao K (1994) Cloning and expression of a cDNA for rat prostacyclin receptor. Biochim Biophys Acta 1224:601–605
Senior J, Sangha R, Baxter GS, Marshall K, Clayton JK (1992) In vitro characterization of prostanoid FP-, DP-, IP-, and TP-receptors in the non-pregnant human myometrium. Br J Pharmacol 107:215–221
Shenker A, Goldsmith P, Unson C, Spiegel A (1991) The G protein coupled to the thromboxane a2 receptor in human platelets is a member of the novel $G_q$ family. J Biol Chem 266:9309–9313
Spurney RF, Middleton JP, Raymond JR, Coffman TM (1994) Modulation of thromboxane receptor activation in rat glomerular mesangial cells. Am J Physiol 267:F467–F478
Sugimoto Y, Namba T, Honda A, Hayashi Y, Negishi M, Ichikawa A, Narumiya S (1992) Cloning and expression of a cDNA for mouse prostaglandin E receptor $EP_3$ subtype. J Biol Chem 267:6463–6466
Sugimoto Y, Negishi M, Hayashi Y, Namba T, Honda A, Watabe A, Hirata M, Narumiya S, Ichikawa A (1993) Two isoforms of the $EP_3$ receptor with different carboxyl-terminal domains: identical ligand binding properties and different coupling properties with $G_i$ proteins. J Biol Chem 268:2712–2718
Takahara K, Murray R, FitzGerald GA, Fitzgerald DJ (1990) The response to thromboxane $A_2$ analogues in human platelets: discrimination of two binding sites linked to distinct effector systems. J Biol Chem 265:6836–6844
Takeuchi K, Abe T, Takahashi N, Abe K (1993) Molecular cloning and intrarenal localization of rat prostaglandin $E_2$ receptor $EP_3$ subtype. Biochem Biophys Res Commun 194:885–891
Thierauch K-H, Stürzebecher CS, Schillinger E, Rehwinkel H, Radüchel B, Skuballa W, Vorbrüggen H (1988) Stable 9$\alpha$- or 11$\alpha$-halogen-15-cyclohexyl-prostaglandins with high affinity on the $PGD_2$-receptor. Prostaglandins 35:855–868
Town M-H, Schillinger E, Speckenbach A, Prior G (1982) Identification and characterization of a prostacyclin-like receptor in bovine coronary arteries with a specific and stable prostacyclin analogue, iloprost, as a radioactive ligand. Prostaglandins 24:61–72
Town M-H, Casals-Stenzel J, Schillinger E (1983) Pharmacological and cardiovascular properties of a hydantoin derivative, BW245C, with high affinity and selectivity for $PGD_2$ receptors. Prostaglandins 25:13–30
Tsai A-H, Hsu M-J, Vijjeswarapu H, Wu KK (1989) Solubilization of prostacyclin membrane receptors from human platelets. J Biol Chem 264:61–67
Ushikubi F, Nakajima M, Hirata M, Okuma M, Fujiwara M, Narumiya S (1989) Purification of the thromboxane $A_2$/prostaglandin $H_2$ receptor from human blood platelets. J Biol Chem 264:16496–16501
Ushikubi F, Nakamura K, Narumiya S (1994) Functional reconstitution of platelet thromboxane $A_2$ receptor with $G_q$ and $G_{i2}$ in phospholipid vesicles. Mol Pharmacol 46:808–816
Ushikubi F, Hirata M, Narumiya S (1995) Molecular biology of prostanoid receptors; an overview. J Lipid Med Cell Signal 12:343–359

Vassaux G, Gaillard D, Ailhaud G, Négrel R (1992) Prostacyclin is a specific effector of adipose cell differentiation. J Biol Chem 267:11092–11097
Virgolini I, Li S, Sillaber C, Majdic O, Sinzinger H, Lechner K, Bettelheim P, Valent P (1992) Characterization of prostaglandin (PG)-binding sites expressed on human basophils. J Biol Chem 267:12700–12708
Whittle BJR, Moncada S, Vane JR (1978) Comparison of the effects of prostacyclin ($PGI_2$), prostaglandin $E_1$ and $D_2$ on platelet aggregation in different species. Prostaglandins 16:373–388
Yang J, Xia M, Goetzl EJ, An S (1994) Cloning and expression of the $EP_3$-subtype of human receptors for prostaglandin $E_2$. Biochem Biophys Res Commun 198:999–1006

CHAPTER 8

# Platelet Adhesion and Aggregation Receptors

K.J. CLEMETSON and J. POLGÁR

## A. Introduction

Platelets stop bleeding from damaged blood vessels and initiate repair processes. They contain many important components for these functions, of which surface glycoproteins are critical for two processes, adhesion and aggregation. In platelets, adhesion refers to the attachment of platelets to subendothelium or to other cells, while platelet–platelet "adhesion" is called *aggregation* to differentiate these processes clearly. Primary adhesion is the binding of resting platelets to subendothelium and secondary adhesion the binding of activated (via partial primary adhesion or temporary association with a thrombus) platelets to subendothelium. Primary adhesion of resting platelets involves several different stages. On initial contact the glycoprotein (GP) Ib-V-IX complex (CD42) binds to von Willebrand factor associated with collagen on the subendothelium surface. The multiple attachments formed stop the platelet, hold it on the surface and start the activation process involving other receptors. The collagen receptor, GPIa-IIa, is an important secondary receptor for platelet adhesion and activation and is critical for inducing the spreading process involving GPIIb-IIIa to assure the intimate contact of the spread platelet with the surface. Other secondary receptors also help to strengthen the links to the surface and act as a reserve in pathological situations. These include fibronectin, laminin and vitronectin receptors. GPIIb-IIIa is particularly important in aggregation and in spreading. Aggregation involves conformational changes in GPIIb-IIIa exposing a fibrinogen binding site, followed by activation of additional signalling steps inducing GPIIb-IIIa association with the cytoskeleton, leading to clustering which is necessary for aggregation of platelets held together by fibrinogen bridging. After the initial bridging, additional glycoprotein receptors such as for thrombospondin probably strengthen and stabilise the interaction between the platelets. Finally, the platelet has a further group of receptors involved in binding other cells that may play a role in the repair process such as neutrophils and monocytes. One of these, PECAM, is expressed on resting platelets while P-selectin is in the $\alpha$-granules and is only expressed on activated platelets that have undergone release. In this chapter the platelet surface receptors involved in adhesion and aggregation will be described and their structure/function relationships discussed.

**Table 1.** Platelet glycoprotein nomenclature

| Platelet nomenclature | Integrin | CD |
|---|---|---|
| GPIa | $\alpha_1$ | CDw49b |
| Ib | | CD42b/c |
| Ic | $\alpha_5$ | CDw49e |
| Ic′ | $\alpha_6$ | CDw49f |
| IIa | $\beta_1$ | CD29 |
| IIb | $^{\alpha}$IIb | CD41 |
| IIIa | $\beta_3$ | CD61 |
| IIIb (IV) | | CD36 |
| V | | CD42d |
| VI | | |
| VII | | |
| VIII | | |
| IX | | CD42a |

GP, glycoprotein; CD, cluster of differentiation.

# B. Platelet Surface Glycoproteins

Platelet glycoprotein nomenclature was originally based on migration in various gel electrophoresis systems and detection by several surface-labelling methods. More general systems of nomenclature have been developed based on the integrin ($\alpha/\beta$) and cluster of differentiation (CD) systems, so that the recent tendency has been towards rationalization of nomenclature, and corresponding names are indicated in Table 1.

# C. The Glycoprotein Ib-V-IX Complex

## I. Introduction

The glycoprotein (GP) Ib-V-IX (CD42a, b, c and d) complex consists of four chains each coded by separate genes present on different chromosomes. GPIb (CD42b and c) contains GPIb$\alpha$ (150 kDa; CD42b; gene on chromosome 17p12-ter; WENGER et al. 1989) and GPIb$\beta$ (27 kDa; CD42c; chromosome 22; BENNETT 1990) linked by a disulphide bond (PHILLIPS and POH AGIN 1977) while GPIX (22 kDa; CD42a; gene on chromosome 3; HICKEY et al. 1990) is strongly non-covalently associated in a 1:1 ratio (DU et al. 1987) and GPV (CD42d; 82 kDa) weakly non-covalently associated with the complex in a 1:2 ratio (GPV: GPIb; MODDERMAN et al. 1992). There are about 25000 copies of GPIb-IX per platelet (BERNDT et al. 1985). Current knowledge about the structure of the GPIb-V-IX complex and some of its associated proteins is shown in Fig. 1.

## 1. Glycoprotein Ibα

GPIbα consists of several distinct domains (LÓPEZ et al. 1987; TITANI et al. 1987). The N-terminal region consists of a loop formed by a disulphide bond followed by a leucine-rich domain consisting of six and a half repeats of a 24-amino-acid, leucine-rich sequence very similar to that found in an increasing number of proteins (for a review see KOBE and DEISENHOFER 1994, 1995). All of the GPIb complex proteins contain this motif, the function of which is still unknown. Recently, the X-ray crystal structure of porcine ribonuclease inhibi-

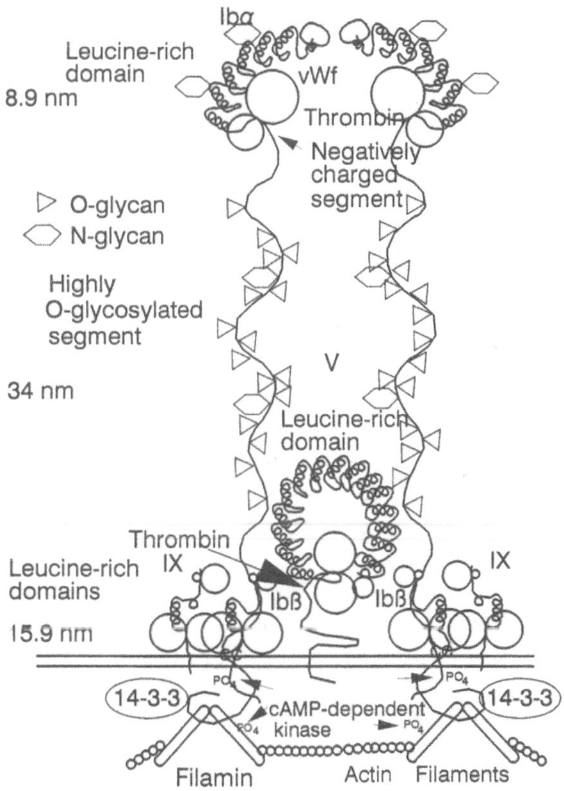


**Fig. 1.** Diagram of the platelet glycoprotein (GP) Ib-V-IX complex showing the probable sandwich arrangement with one GPV between two GPIb-IX molecules. The von Willebrand factor (*vWf*) and thrombin binding sites lie in the outer leucine-rich domains/double loop region which is held out from the platelet surface by the highly glycosylated domain. The other subunits, GPIbβ and GPIX, are close to the membrane. GPV contains a thrombin cleavage site (indicated by an *arrow*). The cytoplasmic domains are linked both to the membrane associated cytoskeleton via filamin but also to a signal transduction cascade probably via the 14-3-3 protein associated with the complex. The cytoplasmic domain of GPIbβ contains a serine residue that can be phosphorylated by cAMP-dependent kinase regulating actin polymerisation

tor, which contains 16 such repeats, was determined, and they were shown to form a novel $\alpha/\beta$ protein fold arranged in a horse-shoe shaped structure (KOBE and DEISENHOFER 1993). This domain also contains a single free cysteine residue. The low reactivity of the thiol group may be due to it being buried in the middle of the leucine-rich sequences. Following this region, two disulphide bonds form an overlapping double loop (HESS et al. 1991). This domain is important for the binding sites of the molecule and will be dealt with in more detail later. Just below the double loop comes a sequence rich in negatively charged amino acids that then switches abruptly to positively charged and is then followed by a domain with five repeats, nine amino acids long, rich in threonine and serine residues that are O-glycosylated. The structure of this region strongly resembles that of the mucins with a high density of short O-linked oligosaccharides. Before reaching the membrane there is an unglycosylated region and just above the membrane there are two cysteines. It is still not known which of these forms the link to GPIb$\beta$. The transmembrane region (29 amino acids) is followed by a 96-amino-acid cytoplasmic domain that can associate with actin-binding protein (ANDREWS and FOX 1991) and hence with the membrane-associated cytoskeleton. There are four putative N-glycosylation sites (TSUJI and OSAWA 1987; KORREL et al. 1988) and many O-glycosylation sites (KORREL et al. 1984, 1985; TSUJI et al. 1983).

### 2. Glycoprotein Ib$\beta$

Although much smaller than the $\alpha$-chain, the $\beta$-chain has certain similarities to it (LÓPEZ et al. 1988). The N-terminal region contains two disulphide loops followed by a single 24-amino-acid leucine-rich repeat, then comes a further two disulphide loops, the single cysteine just above the membrane forming the link to the $\alpha$-chain, the transmembrane region and a 34-amino-acid cytoplasmic domain. Just below the membrane lies a cysteine that can be palmitylated (MUSZBEK and LAPOSATA 1989), and in the middle of the cytoplasmic domain lies serine 166 that can be phosphorylated (WYLER et al. 1986) by cAMP dependent kinase (WARDELL et al. 1989). It is not yet clear whether this domain is also involved in the association with actin-binding protein which can also be phosphorylated by cAMP dependent kinase (COX et al. 1984). GPIb$\beta$ has a single N-glycosylation site with a lactosamine biantennary oligosaccharide within the leucine-rich domain (WICKI and CLEMETSON 1987). There is no O-glycosylation.

### 3. Glycoprotein IX

The structure of GPIX, overall, closely resembles that of GPIb$\beta$ (HICKEY et al. 1989, 1990). The N-terminal region contains two disulphide loops followed by a single 24-amino-acid leucine-rich repeat; then come two further disulphide loops, a short sequence before the transmembrane region and a six-amino-acid cytoplasmic domain. Just within the membrane from the cytoplasmic surface lies a cysteine that can be palmitylated (MUSZBEK and LAPOSATA 1989). As GPIb$\beta$ it has a single N-glycosylation site with a lactosamine biantennary oligosaccharide within the leucine-rich domain (WICKI and CLEMETSON 1987). There is also no O-glycosylation.

### 4. Glycoprotein V

GPV has recently been cloned by two groups (HICKEY et al. 1993; LANZA et al. 1993). It shares many general structural features with the other members of the complex, in particular GPIb$\alpha$. Thus, from the N terminus, it also contains two disulphide loops followed by 15 leucine-rich repeats, then two disulphide loops followed by the thrombin cleavage site (but no hirudin-like anionic site). After that comes a sequence containing one N-glycosylation site and two O-glycosylation sites but no mucin-like repeats. Then comes the transmembrane region and a 16 amino acid cytoplasmic domain with no phosphorylation sites. Overall there are eight N-glycosylation sites, with six of these in the leucine-rich repeat region. There are no palmitylation sites. The thrombin cleavage site does not appear to have a direct or indirect role in platelet activation by thrombin (BIENZ et al. 1986), unlike the binding site on GPIb$\alpha$. A possible role in modulating GPIb-V-IX function was suggested by the inhibitory effect of alloantibodies from a Bernard-Soulier syndrome patient on ristocetin-induced aggregation of normal platelets (DROUIN et al. 1989).

### 5. Polymorphism Within GPIb$\alpha$

No polymorphisms have yet been reported for the other members of the complex but several are known within GPIb$\alpha$. One of these, thought to be the Sib$^a$-Sib$^b$ system (ISHIDA et al. 1991) responsible for alloantibody induction, is a Thr145 (89%) /Met145 (11%) (MURATA et al. 1992) polymorphism. Another, seemingly with no immunological consequences, involves the duplication or tripling of a 13-amino-acid sequence between Ser399 and Thr411 (LÓPEZ et al. 1992; MOROI et al. 1984).

## II. Function of the Glycoprotein Ib-V-IX Complex

### 1. Bleeding Disorders

Much of what we know about the function of the GPIb-V-IX complex comes from studies of inherited bleeding disorders in which expression of this complex on the platelets is defective.

#### *a) Bernard-Soulier Syndrome*

In the Bernard-Soulier syndrome (BSS) GPIb-V-IX is either absent, severely depleted or non-functional (GEORGE et al. 1984; CLEMETSON and LÜSCHER 1988). Indeed, studies of this disorder provided most of the early evidence for the role of the GPIb complex (NURDEN and CAEN 1975; JENKINS et al. 1976) and for the association of GPIb with GPIX and GPV (CLEMETSON et al. 1982; BERNDT et al. 1983). Similarly, such studies provided the first clues about the physiological role of this complex. A particularly important observation was that BSS platelets adhere poorly, if at all, to subendothelium at all shear rates (WEISS et al. 1974, 1978), emphasising the importance of the GPIb-V-IX complex in this primary phase of haemostasis. BSS is a rare, autosomal, recessive genetic disorder. The long bleeding time, thrombocytopenia and morphologically abnormal, unusually large ("giant") platelets are characteristic. Resting BSS platelets are incapable of interacting with von Willebrand factor (vWf) (BITHELL et al. 1972; HOWARD et al. 1973) and therefore show dramatically decreased adhesion to subendothelium of the damaged vascular wall. Aggregation to other agonists except thrombin is normal (BITHELL et al. 1972). In the classic form of BSS all four chains are virtually completely absent (i.e. less than 1% can be detected). In vitro they do not aggregate to either ristocetin or botrocetin in the presence of vWf (ZUCKER et al. 1977), nor directly to asialo vWf (DE MARCO and SHAPIRO 1981) or animal vWf such as bovine or porcine.

Over the years a number of variable BSS cases have been described (DE MARCO et al. 1990; DROUIN et al. 1988; POULSEN and TAANING 1990; WARE et al. 1993; ZWIERZINA et al. 1983; WRIGHT et al. 1993). Although these patients have symptoms similar to those of the classic cases, they show a wide variety of molecular difference. Several cases of BSS were shown to have mutations in or near the leucine-rich repeats of GPIb$\alpha$ or GPIX, indicating that these domains are critical for the complex function (MILLER et al. 1992; WARE et al. 1993).

### *b) Platelet-Type von Willebrand's Disease*

In platelet-type, or pseudo, von Willebrand's disease, platelets show an abnormal affinity for normal von Willebrand factor and as a result aggregate and are removed from the circulation (MILLER et al. 1983). Such patients therefore develop thrombocytopenia and, consequently, bleeding problems. The molecular defect in platelet-type von Willebrand disease has been established in two families. In one family there is a G-to-T point mutation in the gene for GPIb$\alpha$, causing a valine-for-glycine substitution at position 233 of the protein sequence (MILLER et al. 1991). In the other family an A-to-G mutation leads to a valine-for-methionine substitution in position 239 (RUSSELL and ROTH 1993). The same mutation in position 239 has been described in two additional families (TAKAHASHI et al. 1995). These close mutations lie in a 40-amino-acid, disulphide-bridged loop thought to form part of the vWf binding site (HESS et al. 1991). The equivalent disorder caused by mutations in vWf is Type IIB von Willebrand's disease (DE MARCO et al. 1985). Such mutations have been localised to the region of vWf identified as the GPIb binding site (COONEY et al. 1991; RANDI et al. 1991). Thus, changes in the conformation of either the vWf binding site on GPIb or the GPIb binding site on vWf can lead to binding of one to the other, pointing to a role for such changes in the physiological functioning of the GPIb/vWf axis. Although most of the mutations found on the GPIb binding site of vWf induced increased affinity for GPIb, mutations resulting in decreased affinity for GPIb are also known (HILBERT et al. 1995).

### 2. GPIb-vWf Binding

The vWf-binding domain was localised to the N-terminal 45-kDa region of GPIb by antibodies that block vWf-related platelet function (ALI-BRIGGS et al. 1981; COLLER et al. 1983; RUAN et al. 1981) and by examining vWf-binding to platelets treated with proteases (BROWER et al. 1985; WICKI and CLEMETSON 1985). Further, peptides derived from the GPIb sequence have been used as competitive inhibitors of the platelet/vWf interaction in attempts to localise the binding site more precisely (VICENTE et al. 1990; KATAGIRI et al. 1990). Since vWf does not normally interact directly with GPIb it was necessary to induce binding using either ristocetin, botrocetin or asialo vWf (discussed below). The sequence Ser251-Tyr279 inhibits ristocetin/vWf-induced platelet agglutination but also botrocetin-induced platelet agglutination at higher concentrations. The other study reported that Asp235-Lys262 and Asp249-Asp274 inhibit ristocetin/vWf-induced platelet aggregation. Lastly, it was shown by site directed mutagenesis that sulphated tyrosine residues at positions 276, 278 and 279 are essential for both ristocetin- and bothrocetin-mediated vWf binding to GPIb (MARCHESE et al. 1995).

### 3. Non-physiological Activators of the GPIb/vWf Axis

Since vWf in plasma does not normally interact with GPIb and it has so far been difficult to assemble a simple in vitro system that accurately reflects the way in which vWf is activated in the subendothelium, various non-physiological methods have been used to induce vWf/GPIb interactions. The simplest of these has been to use either animal vWf (generally bovine or porcine; COOPER et al. 1979; KIRBY 1982) or human vWf which has been treated with neuraminidase to remove sialic acid (DE MARCO and SHAPIRO 1981; GRALNICK et al. 1985). Presumably, all of these vWf's agglutinate human platelets because they are already in a conformation that can bind GPIb. Since we do not yet have X-ray crystallographic data on the appropriate fragments of vWf and GPIb, the interactions involved remain unclear. Other reagents capable of inducing interactions between normal human vWf and platelet GPIb are ristocetin and botrocetin (HOWARD et al. 1984; GIRMA et al. 1990). Ristocetin has been known for a number of years (HOWARD and FIRKIN 1971) and has been useful for diagnosis of bleeding disorders related to the GPIb/vWf axis (ZUCKER et al. 1977). How this glycopeptide antibiotic induces the vWf/GPIb interaction is still controversial, but it is thought to bind to both vWf and GPIb (SIXMA et al. 1991). Recent results show that two ristocetin dimers bridge the two proteins, and charge neutralisation as well as interactions with proline residues in the vicinity of the binding sites on GPIb$\alpha$ and the A1 loop of vWf take part in the binding (HOYLAERTS et al. 1995).

Botrocetin from *Bothrox jararaca* (READ et al. 1978; ANDREWS et al. 1989) interacts with von Willebrand factor and induces it to bind to GPIb on platelets. Two-chain botrocetin belongs to the C-type ($Ca^{2+}$ dependent) lectin family; however, the effect on vWf was not inhibited by EDTA or a variety of sugars (USAMI et al. 1993). Human vWf that has been treated with neuraminidase to remove sialic acid (asialo vWf) also binds spontaneously to platelet GPIb. Although this has been presented as a charge effect (sialic acid carries most of the negative charge of GPIb and on the platelet surface), the fact that further treatment to remove galactose residues makes the vWf again unresponsive would rather argue for an important role of oligosaccharides in the conformational changes in vWf. Recently, some other snake peptides have been discovered that affect the vWf/GPIb axis. Alboaggregin-B from *Trimeresurus albolabris* has been shown to induce platelet agglutination to vWf by binding to GPIb (PENG et al. 1991) whereas echicetin from *Echis carinatus* (PENG et al. 1993), agkicetin from *Agkistrodon acutus* (CHEN and TSAI 1995), tokaracetin from *Trimeresurus tokarensis* (KAWASAKI et al. 1995), flavocetin-A and -B from *Trimeresurus flavoviridis* (TANIUCHI et al. 1995) and jararaca GPIb-BP from *Bothrops jararaca* (KAWASAKI et al. 1996) block platelet agglutination by attachment to the same site. Although alboaggregin-B and botrocetin have very similar sequences (YOSHIDA et al. 1993; USAMI et al. 1996) they have apparently quite dissimilar mechanisms, providing an interesting example of evolutionary adaptation.

Consideration of all these various examples allows the construction of a model for the physiological process. Although it is not yet possible to test all the parameters involved, there is some evidence for most of them. Thus, the presence of vWf is necessary on the subendothelium, and it must be associated with specific components (e.g. collagen) in order to undergo a conformational change so that it can bind to GPIb on the platelets. The results obtained from the studies of both genetic disorders and non-physiological reagents imply that it is possible to enhance the vWf/GPIb interaction from both sides. If the problem of primary haemostasis is considered, it is clear that platelets need to adhere to the subendothelium under a wide range of conditions, and changes in both vWf and in GPIb may provide the necessary flexibility to handle this. At high shear rates in particular it is important that the platelets touching the subendothelium are stopped rapidly and maintained in intimate contact with the surface. Probably an important factor is that many binding sites are involved simultaneously so that the multimeric form of vWf, the large number of GPIb molecules on the platelet surface and their distribution may be important. Having stopped the platelet and brought it into contact with the subendothelium, it is necessary to activate it and make it spread. This activation process may be started by the interaction of GPIb and vWf, together with the forces exerted on the platelet by shear.

### III. Expression of GPIb-V-IX

Since very little is known about how the various chains of the complex interact with each other and with other molecules (such as in the cytoskeleton), it is of considerable interest to be able to express these in various combinations on non-megakaryocytic cells. Two studies have appeared, one reporting that expression of all three (Ib$\alpha$, Ib$\beta$ and IX) chains is necessary for an efficient overall expression of any (López et al. 1992) while the other found that Ib$\alpha$ by itself, although extensively degraded intracellularly, could be expressed in intact form in small amounts (Meyer et al. 1993). It is to be expected that the ability to reconstitute parts of the GPIb-V-IX system will enable the role of the various constituents to be analysed better.

## D. GPIIb-IIIa

GPIIb-IIIa are major glycoproteins on the platelet surface with about 50000 copies/platelet and constituting about 1%–2% of the total platelet protein. GPIIb-IIIa can act as a receptor for a number of adhesive proteins, fibrinogen (Marguerie et al. 1979), fibronectin (Parise and Phillips 1986; Plow et al.

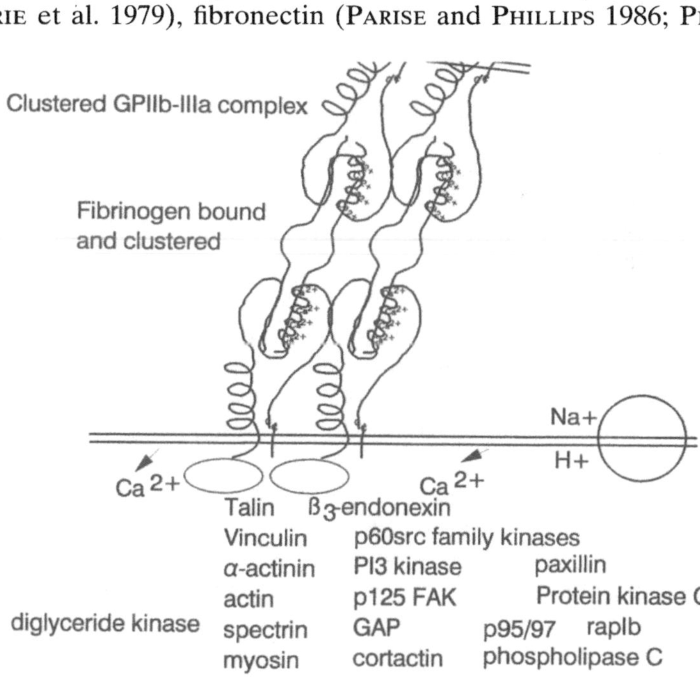


**Fig. 2.** Diagram of the glycoprotein complex GPIIb-IIIa showing the structure of the two subunits and postulated interactions with fibrinogen leading to clustering and associations with cytoplasmic and cytoskeletal components. No hierarchical organisation is shown because the order of engagement of the various components is still not established

1985), von Willebrand factor (Plow et al. 1985), and possibly thrombospondin (Karczewski et al. 1989) and vitronectin (Mohri and Ohkubo 1991). GPIIb-IIIa is normally cryptic in resting platelets and becomes activated and able to bind fibrinogen in activated platelets. A separate pool of GPIIb-IIIa is present on the membranes of platelet α-granules which fuse to the plasma membranes during exocytosis. Binding of fibrinogen induces changes in GPIIb-IIIa including formation of calcium channels, linkage to cytoskeleton and activation of tyrosine kinases. Changes in fibrinogen, or simply clustering, induced by binding to GPIIb-IIIa or to surfaces allow it to interact with GPIIb-IIIa on unactivated platelets, activating them in turn. Figure 2 shows a model of how GPIIb-IIIa interacts with fibrinogen in inducing clustering and lists some of the molecules known to associate with the cluster GPIIb-IIIa or activated during platelet aggregation via GPIIb-IIIa. These are all complex processes and are incompletely understood. Many studies use monoclonal antibodies that recognise particular epitopes in the GPIIb-IIIa complex. These may be epitopes that are only exposed on GPIIb-IIIa after activation, only exposed after binding of fibrinogen (Ligand Induced Binding Sites, LIBS; Du et al. 1991) or only exposed on fibrinogen after binding to GPIIb-IIIa (Receptor Induced Binding Sites, RIBS; Zamarron et al. 1991), representing particular functional states. Studies on the kinetics of these changes provide vital information on how platelets react. With the development of anti-platelet reagents, such as snake peptides that act by binding to the RGD binding site and blocking fibrinogen binding, these techniques can be used to follow the changes in conformation in GPIIb-IIIa. Normally, if fibrinogen or other adhesive proteins are absent, i.e. in washed platelets, the activation of GPIIb-IIIa is transient. This is also seen in the reversible aggregation that occurs with ADP-activated platelets where there is no release. The mechanism of activation of GPIIb-IIIa is still not known but investigation of the effects of various reagents on the kinetics of "opening" by measurement of fibrinogen or specific antibody binding may cast light on factors influencing this process.

Glanzmann's thrombasthenia is a bleeding disorder caused by the absence, deficiency or malfunction of GIIb-IIIa (George et al. 1990), and its study has provided much of the information about the function of this complex. In the classic disorder, deletions or mutations in either chain prevent or reduce expression of the complex (Newman et al. 1991). In some of the rarer patients with more or less normal amounts of GPIIb-IIIa but otherwise typical symptoms, mutations have been found affecting the association of the complex (Nurden et al. 1987), fibrinogen binding (Lanza et al. 1992) or signal transduction by the complex (Chen et al. 1992). The use of conformation-specific antibodies and flow cytometry has often simplified detection of the functional step that is affected.

The synthesis of GPIIb-IIIa has been studied by several groups in both megakaryocytes and megakaryocytic cell lines. Separate precursors for GPIIb and IIIa were detected using [$^{35}$S]methionine labelling and immunoprecipitation. The large and small chains of GPIIb are made as a common precursor.

Pulse-chase experiments showed that early assembly of the precusor forms of GPIIb and IIIa is necessary for the maturation of GPIIb and the transport of the complex to the cell surface (DUPERRAY et al. 1991). This provides a molecular basis for Glanzmann's thrombasthenia by explaining how defects in either GPIIb or IIIa result in the lack of expression of the other protein as well. GPIIb is one of the rare proteins that is completely specific for megakaryocytes and platelets. GPIIIa is also expressed on a wide range of other cells such as endothelial cells and monocytes in the vitronectin receptor (see below). Studies on the promoter region of the GPIIb gene have started to cast some light on the mechanisms of gene control within the megakaryocyte/erythrocyte lineage and its temporal regulation (MARTIN et al. 1993).

While the major role of the integrin GPIIb-IIIa ($\alpha_{IIb}\beta_3$) is clearly in platelet-platelet aggregation, there is considerable evidence that it also plays an important role in adhesion. Thus, Glanzmann's thrombasthenia platelets which are deficient in GPIIb-IIIa show reduced adhesion to the subendothelium, though not as dramatic as in BSS in which GPIb is missing. Studies with various antibodies point to two effects here. One is that platelets that have adhered and have been wrenched from the subendothelium by shear are activated and GPIIb-IIIa can bind to vWf-coated exposed subendothelium downstream of the initial adhesion site (RUGGERI et al. 1983). The second is that GPIIb-IIIa is apparently necessary for the spreading of the adhering platelets on the subendothelium and Glanzmann's thrombasthenia platelets show a more reduced contact area than normal (LAWRENCE and GRALNICK 1987). This role can be explained by activated GPIIb-IIIa binding to vWf and other adhesive proteins on the vessel wall (SAVAGE et al. 1992) and thus increasing the association between the platelet cytoskeleton and the subendothelium, but also by the important function of GPIIb-IIIa in signal transduction via tyrosine kinases and phosphatases and in activation of pathways involved in cytoskeletal rearrangement (FERRELL and MARTIN 1989; GOLDEN et al. 1990; KIEFFER et al. 1992; HAIMOVICH et al. 1993). In the absence of GPIIb-IIIa or if it is blocked, these cytoskeletal changes are prevented or severely reduced, thus preventing spreading from occurring efficiently. Recently, studies on recombinant GPIIb-IIIa expressed in CHO cells showed that an area of IIIa containing Tyr747 and Ile757 is essential for GPIIb-IIIa-mediated spreading (YLÄNNE et al. 1995). Recent studies have also shown that unactivated GPIIb-IIIa is also the receptor for surface-bound fibrinogen (ZAMARRON et al. 1991). Whether this may also mediate interactions between platelets and subendothelium, appears unlikely but cannot be completely excluded. Several polymorphisms in GPIIb and IIIa have been reported that are clinically important because of possible antibody production after transfusion that could lead to problems with later transfusions. On GPIIIa these include $PL^{A1}$ and $PL^{A2}$ localised to Leu33/Pro33 (NEWMAN et al. 1989) and $Pen^a$ and $Pen^b$ localised to Arg143/Gln143 (WANG et al. 1992), and on GPIIb they include $Bak^a$ and $Bak^b$ localised to Ile843/Ser843 (LYMAN et al. 1990).

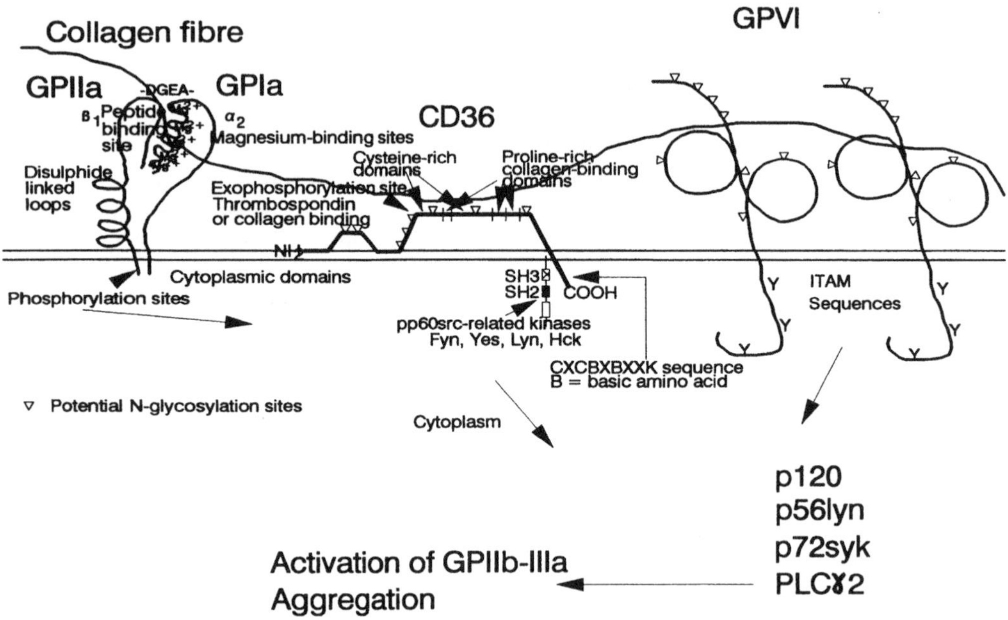


**Fig. 3.** Structures of the various postulated collagen receptors. Those shown for GPIa-IIa and CD36 are based on the determined sequences while the one for GPVI is still hypothetical. Normally it is thought that all three receptors are necessary for an adequate response to collagen; however, CD36 is absent in Naka- individuals without any haemostatic problems or an obvious effect on collagen responsiveness of platelets. This may imply that CD36 has an amplifying role on the signal via the other receptors that can be compensated for by other mechanisms. Unlike most other receptor mechanisms in platelets, that of collagen appears the most dependent upon tyrosine kinases/phosphatases in signal transduction and may well involve clustering of the receptors, bringing kinases into contact with activating sequences. *GP*, glycoprotein; *CD*, cluster of differentiation; *PLC*, phospholipase C

## E. GPIa-IIa

This integrin is present in much smaller amounts than GPIIb-IIIa (2000 copies) on the platelet surface, a fact probably related to its role in adhesion and activation instead of aggregation. There is sound evidence for GPIa-IIa as a collagen receptor (SANTORO et al. 1988) based on studies with patients whose platelets have defects in this integrin complex (NIEUWENHUIS et al. 1985; KEHREL et al. 1988) and also on the use of specific antibodies (STAATZ et al. 1989; COLLER et al. 1989). GPIa-IIa seems to be involved in the adhesion of platelets to various types (type I through VIII) of collagens (SAELMAN et al. 1994). In platelet adhesion, when GPIa-IIa is not functional, the platelets remain poorly associated with the subendothelium with only a few points of attachment and are not activated or spread, probably mainly reflecting the adhesion through the GPIb/vWf axis (NIEUWENHUIS et al. 1986). Studies on collagen receptors are complicated by the fact that other platelet glycoproteins have also been implicated, including CD36 (discussed in this chapter) and GPVI (Ryo et al. 1992). GPIa-IIa binds collagen constitutively (Coller et al. 1989) that activates platelets (tyrosine phosphorylation of $P125^{FAK}$), giving clear evidence for outside-in signalling. Figure 3 shows a model of all three postulated platelet collagen receptors, indicating how they may interact via different sites on collagen to induce signal transduction leading to GPIIb-IIIa activation and platelet aggregation.

## F. GPIc-IIa

GPIc-IIa is the fibronectin-binding integrin on platelets (Giancotti et al. 1987) that is present in fairly low amounts (1000 copies per platelet). The $\alpha$-chain is cleaved post-translationally into a small and large subunit as with GPIIb. The detailed mechanisms involved in the activation induced via this integrin have not yet been studied. With both GPIc-IIa and GPIc'-IIa (see below), RGD sequences are thought to be involved in the recognition site.

## G. GPIc′-IIa

GPIc′-IIa is one of the laminin receptors on platelets (Sonnenberg et al. 1988), and the other is a 67-kDa non-integrin (Tandon et al. 1991). It is also present in fairly low amounts (1000 copies per platelet). No detailed studies have yet been carried out to investigate the mechanisms of activation via this integrin. The role of fibronectin receptors (GPIc-IIa = $\alpha_5\beta_1$) and laminin receptors (GPIc′-IIa = $\alpha_6\beta_1$ and p67) in adhesion remains uncertain (Piotrowicz et al. 1988; Parmentier et al. 1991) despite some indication that they may play a supportive role. Thus, one patient who was reported to have a GPIIa deficiency (unpublished) leading to a lack of all integrins of this class on the platelets had more severe bleeding problems than those patients with simply GPIa-IIa deficiencies in whom the other integrins are normal.

## H. CD36 (GPIIIb or GPIV)

CD36 is a single chain molecule of 100kDa that is found in the membranes of a wide variety of cells including platelets, monocytes, epithelial and endothelial cells (Oquendo et al. 1989). CD36 is a highly compact and hydrophobic molecule, and it is resistant to proteolytic degradation in the membrane. Part of this protease resistance undoubtedly comes from the fact that the extracellular domains are both N- and O-glycosylated and contain six cysteines participating in three disulphide bonds. CD36 is thought to act both as a thrombospondin and a collagen receptor on platelets, though the situation is probably more complex because both thrombospondin and collagen appear to have several different receptors (see Fig. 3). Perhaps the reason why platelets deficient in CD36 have the same capacity as normal platelets to bind thrombospondin (KEHREL et al. 1991; MCKEOWN et al. 1994) and show normal collagen-induced platelet aggregation (YAMAMOTO et al. 1992) and normal signal transduction (DANIEL et al. 1994) is that the lack of CD36 is compensated for by alternative pathways (see Fig. 3). Even so, CD36 may be particularly important because it is able to transduce signals, and several anti-CD36 monoclonal antibodies are able to activate platelets and monocytes by cross-linking CD36 and the Fc receptor (Fc$\gamma$RII in platelets). It remains unclear whether conformational changes occur during platelet activation, making it an activatable receptor like GPIIb-IIIa, and whether it associates with other receptors such as GPIIb-IIIa upon activation. Recent studies using synthetic peptides have localised two domains, 93–110 and 139–155, involved in the receptor function for thrombospondin (LEUNG et al. 1992). These results indicated that the 139–155 sequence induced a conformational change in thrombospondin (RIBS) allowing it to bind the 93–110 sequence, indicating that not only CD36 but also thrombospondin is affected by the binding. CD36 is also a receptor for oxidised LDL (ENDEMANN et al. 1993; NICHOLSON 1995) and for malaria-infected erythrocytes (OCKENHOUSE et al. 1989) which bind via the circumsporozite protein sequestrin (OCKENHOUSE et al. 1991).

## I. P-selectin (CD62, GMP-140, PADGEM)

P-selectin is a granule membrane glycoprotein with a molecular mass of 140 kDa (hence GMP-140). PADGEM is derived from Platelet Activation Dependent Granule External Membrane protein (HSU-LIN et al. 1984). Out of a total of 789 amino acids, starting from the N terminus the structure contains a 120-amino-acid lectin-like domain, a 40-amino-acid epidermal growth factor-like domain, nine 62-amino-acid repeats similar to complement-binding protein, a transmembrane domain and a 35-amino-acid cytoplasmic domain. P-selectin is the platelet member of a family that includes the endothelial cell (E-) and leukocyte (L-) representatives (JOHNSTON et al. 1989). An alternatively spliced message for P-selectin contains no transmembrane domain and would be predicted to code for a soluble form (USHIYAMA et al. 1993). In resting platelets, P-selectin is found in the membrane of the $\alpha$-granules and in endothelial cells in equivalent structures, the Weibel-Palade bodies (MCEVER et al. 1989; BONFANTI et al. 1989). After platelet stimulation with agonists such as thrombin, the release reaction from the granules occurs and P-selectin, together with other granule membrane constituents, is transferred via membrane fusion to the plasma membrane. While resting platelets express 1000 or less P-selectin molecules, the amount rises to about 10000 when they are activated. This exposed P-selectin can then bind to carbohydrate structures of the sialyl Lewis$^x$ and sialyl Lewis$^a$ class on glycolipids or glycoproteins of neutrophils and myeloid cells (ERBE et al. 1993; LARSEN et al. 1990) and may be involved in elimination of activated platelets from the circulation on one hand or in binding neutrophils to a platelet thrombus on the other. Leukocyte “rolling”, in which leukocytes are held in loose contact with the endothelial surface but are nevertheless moved along by the blood flow, can also occur via P-selectin on a platelet layer (BUTTRUM et al. 1993). P-selectin is rapidly phosphorylated on serine, threonine and tyrosine when platelets are activated, but phosphothreonine and phosphotyrosine are rapidly dephosphorylated, leaving only phosphoserine after 5 min (CROVELLO et al. 1993). P-selectin and $pp60^{c\text{-}src}$ might be closely associated in the membranes of platelets and $pp60^{c\text{-}src}$ might be the kinase responsible for phosphorylation of P-selectin on tyrosine (MODDERMAN et al. 1994). The function of this phosphorylation is still unknown. P-selectin is acylated with palmitic acid and stearic acid at cysteine 766 through a thioester linkage (FUJIMOTO et al. 1993).

## J. PECAM-1 (CD31)

PECAM-1 (Platelet Endothelial Cell Adhesion Molecule; 130 kDa) is a platelet adhesion receptor but is also found on endothelial cells, neutrophils and monocytes (NEWMAN et al. 1990). It is a highly glycosylated molecule with 40% carbohydrates and shows similarities to both the Fc portion of IgG and to carcinoembryonic antigen. Six Ig-like domains are present. The C-terminal region contains both a transmembrane domain and a 118 amino acid long serine- and threonine-rich cytoplasmic domain. Phosphorylation of this domain seems to be important for regulating the activity of the molecule. PECAM-1 is rapidly phosphorylated on serine residues after platelet activation and becomes associated with the platelet cytoskeleton (NEWMAN et al. 1992; ZEHNDER et al. 1992). The role of PECAM-1 in platelet function is still obscure. It seems certain that it is not involved directly in platelet aggregation, implying that it has an adhesive function either to subendothelial components or, more likely, to other cells. Recent results have demonstrated that possible ligands for PECAM in heterotypic adhesion may be cell surface glycosaminoglycans (DE LISSER et al. 1993), and a consensus binding sequence, LKREKN, present in the second immunoglobulin-like homology domain was shown to be involved.

## K. Inhibition of Platelet Adhesion/Aggregation as a Prophylactic Tool or for Treatment of Acute Thrombotic Events

Platelets contain a large number of membrane glycoproteins involved in various adhesive and other recognition interactions. Many of these could be interesting targets for pharmacological manipulation. Already, a number of inhibitors of GPIIb/IIIa, the $\alpha_{IIb}\beta_3$ integrin, are in clinical testing and show promise for treatment of acute and chronic diseases involving platelet activation. However, it is worth developing alternative strategies to inhibition of aggregation, and the complex process of adhesion presents an attractive target. Clearly, the GPIb/vWf axis is the best understood (if still only partially) of these mechanisms. Methods based upon peptides from the GPIb binding region of vWf or from the vWf binding region of GPIb could be used as a first approach to this in the same way that RGD-containing peptides or snake venom peptides were used as the first approach to blocking aggregation. The availability of snake venom peptides capable of blocking the GPIb-vWf interaction in the nanomolar range (see above) may also provide a useful starting point. Just as anti-GPIIb-IIIa/fibrinogen drugs simulate the situation in Glanzmann's thrombasthenia, anti-GPIb/vWf drugs should simulate, at least partially, BSS. Some examples have already been reported in which the GPIb-binding domain of vWf in recombinant form was demonstrated to inhibit platelet adhesion to the extracellular matrix (Dardik et al. 1993; Prior et al. 1993), and a recombinant fragment of GPIb$\alpha$ has been shown to inhibit vWf binding to GPIb and also to collagen. The $IC_{50}$ was, however, $4\mu M$, which is nevertheless high compared to RGDS ($IC_{50} = 0.1\mu M$), by no means the most efficient of this class of substances. vWf binding to GPIb-V-IX can mediate activation and translocation of tyrosine and lipid kinases independently of other agonists (Jackson et al. 1994; Razdan et al. 1994) so that the GPIb-V-IX complex is an excellent target for developing selective inhibitors acting on platelet adhesion and/or activation and aggregation. Another attractive target might be the collagen receptor of which even less is known. Here, it would be possible to differentiate between adhesion and more general activation through specific inhibitors, allowing pacification of a wound without thrombus development. This might offer the sought-after prevention of pathological thrombosis with minimal effects on physiological haemostasis which is the pharmacologist's dream. The next decade will see a rapid expansion of our knowledge of structure/functional relationships throughout this area which should allow the development of inhibitors with potentially new applications in various situations.

*Acknowledgements.* Support for the work described here, carried out at the Theodor Kocher Institute, from the Swiss National Science Foundation Grants 31-32416.91 and 31-042336.94, by a grant from Hoffmann-La Roche Ltd. and by the supply of buffy coats from the Central Laboratory of the Swiss Red Cross Blood Transfusion Service is gratefully acknowledged.

## References

Ali-Briggs EF, Jenkins CSP, Clemetson KJ (1981) Antibodies against platelet membrane glycoproteins: crossed immunoelectrophoresis studies with antibodies that inhibit ristocetin-induced platelet aggregation. Br J Haematol 48:305–318

Andrews RK, Fox JEB (1991) Interaction of purified actin-binding protein with the platelet membrane glycoprotein Ib-IX complex. J Biol Chem 266:7144–7147

Andrews RK, Booth WJ, Gorman JJ, Castaldi PA, Berndt MC (1989) Purification of botrocetin from *Bothrops jararaca* venom. Analysis of the botrocetin-mediated interaction between von Willebrand factor and the human platelet membrane glycoprotein Ib-IX complex. Biochemistry 28:8317–8326

Bennett JS (1990) The molecular biology of platelet membrane proteins. Semin Hematol 27:186–204

Berndt MC, Gregory C, Chong BH, Zola H, Castaldi PA (1983) Additional glycoprotein defects in Bernard-Soulier's syndrome: confirmation of genetic basis by parental analysis. Blood 62:800–807
Berndt MC, Gregory C, Kabral A, Zola H, Fournier D, Castaldi PA (1985) Purification and preliminary characterization of the glycoprotein Ib complex in the human platelet membrane. Eur J Biochem 151:637–649
Bienz D, Schnippering W, Clemetson KJ (1986) Glycoprotein V is not the thrombin-trigger on human blood platelets. Blood 68:720–725
Bithell TC, Parekh SJ, Strong RR (1972) Platelet-function studies in the Bernard-Soulier syndrome. Ann N Y Acad Sci 201:145–160
Block KL, Ravid K, Phung QH, Poncz M (1994) Characterization of regulatory elements in the 5′-flanking region of the rat GPIIb gene by studies in a primary rat marrow culture system. Blood 84:3385–3393
Bonfanti R, Furie BC, Furie B, Wagner DD (1989) PADGEM (GMP140) is a component of Weibel-Palade bodies of human endothelial cells. Blood 73:1109–1112
Brower MS, Levin RI, Gary K (1985) Human neutrophil elastase modulates platelet function by limited proteolysis of membrane glycoproteins. J Clin Invest 75:657–666
Buttrum SM, Hatton R, Nash GB (1993) Selectin-mediated rolling of neutrophils on immobilized platelets. Blood 82:1165–1174
Chen YL, Tsai IH (1995) Functional and sequence characterization of agkicetin, a new glycoprotein Ib agonist isolated from *Agkistrodon Acutus* venom. Biochem Biophys Res Commun 210:472–477
Chen YP, Djaffar I, Pidard D, Steiner B, Cieutat AM, Caen JP, Rosa JP (1992) Ser-752 → Pro mutation in the cytoplasmic domain of integrin $\beta_3$ subunit and defective activation of platelet integrin $\alpha_{IIb}\beta_3$ (glycoprotein IIb-IIIa) in a variant of Glanzmann thrombasthenia. Proc Natl Acad Sci USA 89:10169–10173
Clemetson KJ, Lüscher EF (1988) Membrane glycoprotein abnormalities in pathological platelets. Biochim Biophys Acta 947:53–73
Clemetson KJ, McGregor JL, James E, Dechavanne M, Lüscher EF (1982) Characterization of the platelet membrane glycoprotein abnormalities in Bernard-Soulier syndrome and comparison with normal surface labelling techniques and high resolution two-dimensional gel electrophoresis. J Clin Invest 70:304–311
Coller BS, Peerschke EI, Scudder LE, Sullivan CA (1983) Studies with a murine monoclonal antibody that abolishes ristocetin-induced binding of von Willebrand factor to platelets: additional evidence in support of GPIb as a platelet receptor for von Willebrand factor. Blood 61:99–110
Coller BS, Beer JH, Scudder LE, Steinberg MH (1989) Collagen-platelet interactions: evidence for a direct interaction of collagen with platelet GPIa/IIa and an indirect interaction with platelet GPIIb/IIIa mediated by adhesive proteins. Blood 74:182–192
Cooney KA, Nichols WC, Bruck ME, Bahou WF, Shapiro AD, Bowie EJW, Gralnick HR, Ginsburg D (1991) The molecular defect in type IIB von Willebrand disease. Identification of four potential missense mutations within the putative GpIb binding domain. J Clin Invest 87:1227–1233
Cooper HA, Clemetson KJ, Lüscher EF (1979) Human platelet membrane receptor for bovine von Willebrand factor (platelet aggregating factor): an integral membrane glycoprotein. Proc Natl Acad Sci USA 76:1069–1073
Cox AC, Carroll RC, White JG, Ray GHR (1984) Recycling of platelet phosphorylation and cytoskeletal assembly. J Cell Biol 98:8–15
Crovello CS, Furie BC, Furie B (1993) Rapid phosphorylation and selective dephospholylation of P-selectin accompanies platelet activation. J Biol Chem 268:14590–14593
Daniel JL, Dangelmaier C, Strouse R, Smith JB (1994) Collagen induces normal signal transduction in platelets deficient in CD36 (platelet glycoprotein IV). Thromb Haemost 71:353–356

Dardik R, Ruggeri ZM, Savion N, Gitel S, Martinowitz U, Chu V, Varon D (1993) Platelet aggregation on extracellular matrix: effect of a recombinant GPIb-binding fragment of von Willebrand factor. Thromb Haemost 70:522–526

De Lisser HM, Yan HC, Newman PJ, Müller WA, Buck CA, Albelda SM (1993) Platelet/endothelial cell adhesion molecule-1 (CD31)-mediated cellular aggregation involves cell surface glycosaminoglycans. J Biol Chem 268:16037–16046

De Marco L, Shapiro SS (1981) Properties of human asialo-factor VIII. A ristocetin-independent platelet aggregating agent. J Clin Invest 68:321–328

De Marco L, Girolami A, Zimmerman TS, Ruggeri ZM (1985) Interaction of purified type IIB von Willebrand factor with the platelet membrane glycoprotein Ib induces fibrinogen binding to the glycoprotein IIb/IIIa complex and initiates aggregation. Proc Natl Acad Sci USA 82:7424–7428

De Marco L, Mazzucato M, Fabris F, De Roia D, Coser P, Girolami A, Vicente V, Ruggeri ZM (1990) Variant Bernard-Soulier syndrome type Bolzano. A congenital bleeding disorder due to a structural and functional abnormality of the platelet glycoprotein Ib-IX complex. J Clin Invest 86:25–31

Drouin J, McGregor JL, Parmentier S, Izaguirre CA, Clemetson KJ (1988) Residual amounts of glycoprotein Ib concomitant with near-absence of glycoprotein IX in platelets from Bernard-Soulier patients. Blood 72:1086–1088

Drouin J, McGregor JL, Clemetson KJ, Hashemi S (1989) An alloantibody to platelet glycoprotein V in a patient with Bernard-Soulier syndrome inhibits ristocetin-induced platelet aggregation (Abstr). Blood 74:31

Du X, Beutler L, Ruan CH, Castaldi PA, Berndt MC (1987) Glycoprotein Ib and glycoprotein IX are fully complexed in the intact platelet membrane. Blood 69:1524–1527

Du X, Plow EF, Frelinger AL, O'Toole TE, Loftus JC, Ginsberg MH (1991) Ligands "activate" integrin $\alpha_{IIb}\beta_3$ (platelet GPIIb-IIIa). Cell 65:409–416

Duperray A, Berthier R, Marguerie G (1991) Biosynthesis and processing of platelet glycoproteins in megakaryocytes. In: Harris JR (ed) Blood cell biochemistry, vol 2. Plenum, New York, pp 37–58

Endemann G, Stanton LW, Madden KS, Bryant CM, White RT, Protter AA (1993) CD36 is a receptor for oxidized low density lipoprotein. J Biol Chem 268:11811–11816

Erbe DV, Watson R, Presta LG, Wolitzky BA, Foxall C, Brandley BK, Lasky LA (1993) P- and E-selectin use common sites for carbohydrate ligand recognition and cell adhesion. J Cell Biol 120:1227–1235

Ferrell JE Jr, Martin GS (1989) Tyrosine-specific protein phosphorylation is regulated by glycoprotein IIb-IIIa in platelets. Proc Natl Acad Sci USA 86:2234–2238

Fujimoto T, Strou E, Whatley RE, Prescott SM, Muszbek L, Laposata M, McEver RP (1993) P-selectin is acylated with palmitic acid and stearic acid at cysteine 766 through a thioester linkage. J Biol Chem 268:11394–11400

George JN, Nurden AT, Phillips DR (1984) Molecular defects in interactions of platelets with the vessel wall. N Engl J Med 311:1084–1098

George JN, Caen JP, Nurden AT (1990) Glanzmann's thrombasthenia: the spectrum of clinical disease. Blood 75:1383–1395

Giancotti FG, Languino LR, Zanetti A, Peri G, Tarone G, Dejana E (1987) Platelets express a membrane protein complex immunologically related to the fibroblast fibronectin receptor and distinct from GPIIb/IIIa. Blood 69:1535–1538

Girma JP, Takahashi Y, Yoshioka A, Diaz J, Meyer D (1990) Ristocetin and botrocetin involve two distinct domains of von Willebrand factor for binding to platelet membrane glycoprotein Ib. Thromb Haemost 64:326–332

Golden A, Brugge JS, Shattil SJ (1990) Role of platelet membrane glycoprotein IIb-IIIa in agonist-induced tyrosine phosphorylation of platelet proteins. J Cell Biol 111:3117–3127

Gralnick HR, Williams SB, Coller BS (1985) Asialo von Willebrand factor interactions with platelets: interdependence of glycoproteins Ib and IIb/IIIa for binding and aggregation. J Clin Invest 75:19–25
Haimovich B, Lipfert L, Brugge JS, Shattil SJ (1993) Tyrosine phosphorylation and cytoskeletal reorganization in platelets are triggered by interaction of integrin receptors with their immobilized ligands. J Biol Chem 268:15868–15877
Hess D, Schaller J, Rickli EE, Clemetson KJ (1991) Determination of the disulphide bonds in human platelet glycocalicin. Eur J Biochem 199:389–393
Hickey MJ, Williams SA, Roth GJ (1989) Human platelet glycoprotein IX: an adhesive prototype of leucine-rich glycoproteins with flank-center-flank structures. Proc Natl Acad Sci USA 86:6773–6777
Hickey MJ, Deaven LL, Roth GJ (1990) Human platelet glycoprotein IX: characterization of cDNA and localization of the gene to chromosome 3. FEBS Lett 274:189–192
Hickey MJ, Yagi M, Hagen FS, Roth GJ (1993) Human platelet glycoprotein V: Structure-function relationships within the Ib-V-IX system of adhesive leucine-rich glycoproteins. Proc Natl Acad Sci USA 90:8327–8331
Hilbert L, Gaucher C, Mazurier C (1995) Identification of two mutations (Arg611Cys and Arg611His) in the A1 loop of von Willebrand factor (vWF) responsible for type 2 von Willebrand disease with decreased platelet-dependent function of vWF. Blood 86:1010–1018
Howard MA, Firkin BG (1971) Ristocetin – a new tool in the investigation of platelet aggregation. Thromb Diath Haemorrh 26:362–369
Howard MA, Hutton RA, Hardisty RM (1973) Hereditary giant platelet syndrome: a disorder of a new aspect of platelet function. Br Med J 2:586–588
Howard MA, Perkin J, Salem HH, Firkin BG (1984) The agglutination of human platelets by botrocetin: evidence that botrocetin and ristocetin act at different sites on the factor VIII molecule and platelet membrane. Br J Haematol 57:25–35
Hoylaerts MF, Nuyts K, Peerlinck K, Deckmyn H, Vermylen J (1995) Promotion of binding of von Willebrand factor to platelet glycoprotein Ib by dimers of ristocetin. Biochem J 306:453–463
Hsu-Lin SC, Berman CL, Furie BC, August D, Furie B (1984) A platelet membrane protein expressed during platelet activation and secretion. J Biol Chem 259:9121–9126
Ishida F, Saji H, Maruya E, Furihata K (1991) Human platelet-specific antigen, Sib$^a$, is associated with the molecular weight polymorphism of glycoprotein Ib$\alpha$. Blood 78:1722–1729
Jackson SP, Schoenwaelder SM, Yuan Y, Rabinowitz I, Salem HH, Mitchell CA (1994) Adhesion receptor activation of phosphatidylinositol 3-kinase. Von Willebrand factor stimulates the cytoskeletal association and activation of phosphatidylinositol 3-kinase and pp60$^{c\text{-}src}$ in human platelets. J Biol Chem 269:27093–27099
Jenkins CSP, Phillips DR, Clemetson KJ, Meyer D, Larrieu MJ, Lüscher EF (1976) Platelet membrane glycoproteins implicated in ristocetin-induced aggregation: studies of the proteins on platelets from patients with Bernard-Soulier syndrome and von Willebrand's disease. J Clin Invest 57:112–124
Johnston GI, Cook RG, McEver RP (1989) Cloning of GMP-140, a granule membrane protein of platelets and endothelium: sequence similarity of proteins involved in cell adhesion and inflammation. Cell 56:1033–1044
Karczewski J, Knudsen KA, Smith L, Murphy A, Rothman VL, Tuszvnski GP (1989) The interaction of thrombospondin with platelet glycoprotein GPIIb-IIIa. J Biol Chem 264:21322–21326
Katagiri Y, Hayashi Y, Yamamoto K, Tanoue K, Kosaki G, Yamazaki H (1990) Localization of von Willebrand factor and thrombin-interactive domains on human platelet glycoprotein Ib. Thromb Haemost 63:122–126
Kawasaki T, Taniuchi Y, Hisamichi N, Fujimura Y, Suzuki M, Titani K, Sakai Y, Kaku S, Satoh N, Takenaka T, Handa M, Sawai Y (1995) Tokaracetin, a new platelet

antagonist that binds to platelet glycoprotein Ib and inhibits von Willebrand factor-dependent shear-induced platelet aggregation. Biochem J 308:947–953

Kawasaki T, Fujimura Y, Usami Y, Suzuki M, Miura S, Sakurai Y, Makita K, Taniuchi Y, Hirani K, Titani K (1996) Complete amino acid sequence and identification of the platelet glycoprotein Ib-binding site of jararaca GPIb-BP, a snake venom protein isolated from *Bothrops jararaca*. J Biol Chem 271:10635–10639

Kehrel B, Balleisen L, Kokott R, Mesters R, Stenzinger W, Clemetson KJ, van de Loo J (1988) Deficiency of intact thrombospondin and membrane glycoprotein Ia in platelets with defective collagen-induced aggregation and spontaneous loss of disorder. Blood 71:1074–1078

Kehrel B, Kronenberg A, Schwippert B, Niesing-Bresch D, Niehues U, Tschöpe D, van der Loo J, Clemetson KJ (1991) Thrombospondin binds normally to glycoprotein IIIb deficient platelets. Biochem Biophys Res Commun 179:985–991

Kieffer N, Guichard J, Breton-Gorius J (1992) Dynamic redistribution of major platelet surface receptors after contact-induced platelet activation and spreading: an immunoelectron microscopy study. Am J Pathol 140:57–73

Kirby EP (1982) The agglutination of human platelets by bovine factor VIII: R. J Lab Clin Med 100:963–976

Kobe B, Deisenhofer J (1993) Crystal structure of porcine ribonuclease inhibitor, a protein with leucine-rich repeats. Nature 366:751–756

Kobe B, Deisenhofer J (1994) The leucine-rich repeat: a versatile binding motif. Trends Biochem Sci 19:415–421

Kobe B, Deisenhofer J (1995) A structural basis for the interactions between leucine rich-repeats and protein ligands. Nature 374:183–186

Korrel SAM, Clemetson KJ, van Halbeek H, Kamerling JP, Sixma JJ, Vliegenthart JFG (1984) Structural studies on the O-linked carbohydrate chains of human platelet glycocalicin. Eur J Biochem 140:571–576

Korrel SAM, Clemetson KJ, van Halbeek H, Kamerling JP, Sixma JJ, Vliegenthart JFG (1985) The structure of a fucose-containing O-glycosidic carbohydrate chain of human platelet glycocalicin. Glycoconj J 2:229–234

Korrel SAM, Clemetson KJ, van Halbeek H, Kamerling JP, Sixma JJ, Vliegenthart JFG (1988) Identification of a tetrasialylated monofucosylated *N*-linked carbohydrate chain in human platelet glycocalicin. FEBS Lett 228:321–326

Lanza F, Stierlé A, Foournier D, Morales M, André G, Nurden AT, Cazenave JP (1992) A new variant of Glanzmann's thrombasthenia (Strasbourg I). Platelets with functionally defective glycoprotein IIb-IIIa complexes and a glycoprotein IIIa $^{214}$Arg $\rightarrow$ $^{214}$Trp mutation. J Clin Invest 89:1995–2004

Lanza F, Morales M, de La Salle C, Cazenave JP, Clemetson KJ, Shimomura T, Phillips DR (1993) Cloning and characterization of the gene encoding the human platelet glycoprotein (GP) V. A member of the leucine-rich glycoprotein (LRG) family cleaved during thrombin-induced platelet activation. J Biol Chem 268:20801–20807

Larsen E, Palabrica T, Sajer S, Gilbert GE, Wagner DD, Furie BC, Furie B (1990) PADGEM-dependent adhesion of platelets to monocytes and neutrophils is mediated by a lineage-specific carbohydrate, LNF III (CD15). Cell 63:467–474

Lawrence JB, Gralnick HR (1987) Monoclonal antibodies to the glycoprotein IIb-IIIa epitopes involved in adhesive protein binding: effects on platelet spreading and ultrastructure on human arterial subendothelium. J Lab Clin Med 109:495–503

Leung LLK, Li WX, McGregor JL, Albrecht G, Howard RJ (1992) CD36 peptides enhance or inhibit CD36-thrombospondin binding. A two-step process of ligand-receptor interaction. J Biol Chem 267:18244–18250

López JA, Chung DW, Fujikawa K, Hagen FS, Papayannopoulou T, Roth GJ (1987) Cloning of the alpha chain of human glycoprotein Ib: a transmembrane protein with homology to leucine-rich alpha2-glycoprotein. Proc Natl Acad Sci USA 84:5615–5619

López JA, Chung DW, Fujikawa K, Hagen FS, Davie EW, Roth GJ (1988) The $\alpha$ and $\beta$ chains of human platelet glycoprotein Ib are both transmembrane proteins containing a leucine-rich amino acid sequence. Proc Natl Acad Sci USA 85:2135–2139

López JA, Ludwig EH, McCarthy BJ (1992) Polymorphism of human glycoprotein Ib$\alpha$ results from a variable number of tandem repeats of a 13-amino acid sequence in the mucin-like macroglycopeptide region. Structure/function implications. J Biol Chem 267:10055–10061

Lyman S, Aster RH, Visentin GP, Newman PJ (1990) Polymorphism of human platelet membrane glycoprotein IIb associated with the $Bak^{\alpha}/Bak^{\beta}$ alloantigen system. Blood 75:2343–2348

Marchese P, Murata M, Mazzucato M, Pradella P, De Marco L, Ware J, Ruggeri ZM (1995) Identification of three tyrosine residues of glycoprotein Ib$\alpha$ with distinct roles in von Willebrand factor and $\alpha$-thrombin binding. J Biol Chem 270:9571–9578

Marguerie GA, Plow EF, Edgington TS (1979) Human platelets possess an inducible and saturable receptor for fibrinogen. J Biol Chem 245:5357–5363

Martin F, Prandini MH, Thevenon D, Marguerie G, Uzan G (1993) The transcription factor GATA-1 regulates the promoter activity of the platelet glycoprotein IIb gene. J Biol Chem 268:21606–21612

McEver RP, Beckstead JH, Moore KL, Marshall-Carlson L, Bainton DF (1989) GMP-140, a platelet alpha-granule membrane protein, is also synthesized by vascular endothelial cells and is localized in Weibel-Palade bodies. J Clin Invest 84:92–99

McKeown L, Vail M, Williams S, Kramer W, Hansmann K, Gralnick H (1994) Platelet adhesion to collagen in individuals lacking glycoprotein IV. Blood 83:2866–2871

Meyer S, Kresbach G, Häring P, Schumpp-Vornach B, Clemetson KJ, Hadváry P, Steiner B (1993) Expression and characterization of functionally active fragments of the platelet GPIb-IX complex in mammalian cells. Incorporation of GPIb$\alpha$ into the cell surface membrane. J Biol Chem 268:20555–20562

Miller JL, Kupinski JM, Castella A, Ruggeri ZM (1983) Von Willebrand factor binds to platelets and induces aggregation in platelet-type but not type IIB von Willebrand disease. J Clin Invest 72:1532–1542

Miller JL, Cunningham D, Lyle VA, Finch CN (1991) Mutation in the gene encoding the $\alpha$ chain of platelet glycoprotein Ib in platelet-type von Willebrand disease. Proc Natl Acad Sci USA 88:4761–4765

Miller JL, Lyle VA, Cunningham D (1992) Mutation of leucine-57 to phenylalanine in a platelet glycoprotein Ib$\alpha$ leucine tandem repeat occurring in patients with an autosomal dominant variant of Bernard-Soulier disease. Blood 79:439–446

Modderman PW, Admiraal LG, Sonnenberg A, von dem Borne AEGK (1992) Glycoproteins V and Ib-IX form a noncovalent complex in the platelet membrane. J Biol Chem 267:364–369

Modderman PW, von dem Borne EG, Sonnenberg A (1994) Tyrosine phosphorylation of P-selectin in intact platelets and in a disulphide-linked complex with immunoprecipitated $pp60^{c\text{-}src}$. Biochem J 299:613–621

Mohri H, Ohkubo T (1991) How vitronectin binds to activated glycoprotein IIb-IIIa complex and its function in platelet aggregation. Am J Clin Pathol 96:605–609

Moroi M, Jung SM, Yoshida N (1984) Genetic polymorphism of platelet glycoprotein Ib. Blood 64:622–629

Murata M, Furihata K, Ishida F, Russell SR, Ware J, Ruggeri ZM (1992) Genetic and structural characterization of an amino acid dimorphism in glycoprotein Ib$\alpha$ involved in platelet transfusion refractoriness. Blood 79:3086–3090

Muszbek L, Laposata M (1989) Glycoprotein Ib and glycoprotein IX in human platelets are acylated with palmitic acid through thioester linkages. J Biol Chem 264:9716–9719

Newman PJ, Derbes RS, Aster RH (1989) The human platelet alloantigens, PlA1 and PlA2, are associated with a leucine33/proline33 amino acid polymorphism in mem-

brane glycoprotein IIIa, and are distinguishable by DNA typing. J Clin Invest 83:1778–1781
Newman PJ, Berndt MC, Gorski J, White GC II, Lyman S, Paddock C, Müller WA (1990) PECAM-1 (CD31) cloning and relation to adhesion molecules of the immunoglobulin gene superfamily. Science 247:1219–1222
Newman PJ, Seligsohn U, Lyman S, Coller BS (1991) The molecular genetic basis of Glanzmann thrombasthenia in the Iraqi-Jewish and Arab populations in Israel. Proc Natl Acad Sci USA 88:3160–3164
Newman PJ, Hillery CA, Albrecht R, Parise LV, Berndt MC, Mazurov AV, Dunlop LC, Zhang J, Rittenhouse SE (1992) Activation-dependent changes in human platelet PECAM-1: phosphorylation, cytoskeletal association, and surface membrane redistribution. J Cell Biol 119:239–246
Nicholson AC, Frieda S, Pearce A, Silverstein RL (1995) Oxidised LDL binds to CD36 on human monocyte-derived macrophages and transfected cell lines: evidence implicating the lipid moiety of the lipoprotein as the binding site. Arteriosclerosis Thromb 15:269–275
Nieuwenhuis HK, Akkerman JWN, Houdijk WPM, Sixma JJ (1985) Human blood platelets showing no response to collagen fail to express surface glycoprotein Ia. Nature 318:470–472
Nieuwenhuis HK, Sakariassen KS, Houdijk WPM, Nievelstein PFEM, Sixma JJ (1986) Deficiency of platelet membrane glycoprotein Ia associated with a decreased platelet adhesion to subendothelium: a defect in platelet spreading. Blood 68:692–696
Nurden AT, Caen JP (1975) Specific roles for platelet surface glycoproteins in platelet function. Nature 255:720–722
Nurden A, Rosa JP, Fournier D, Legrand C, Didry D (1987) A variant of Glanzmann's thrombasthenia with abnormal glycoprotein IIb-IIIa complexes in the platelet membrane. J Clin Invest 79:962–969
Ockenhouse CF, Tandon NN, Magowan C, Jamieson GA, Chulay JD (1989) Identification of a platelet membrane glycoprotein as a *falciparum* malaria sequestration receptor. Science 243:1469–1471
Ockenhouse CF, Klotz FW, Tandon NN, Jamieson GA (1991) Sequestrin, a CD36 recognition protein on *Plasmodium falciparum* malaria-infected erythrocytes identified by anti-idiotype antibodies. Proc Natl Acad Sci USA 88:3175–3179
Oquendo P, Hundt E, Lawler J, Seed B (1989) CD36 directly mediates cytoadherence of *Plasmodium falciparum* parasitized erythrocytes. Cell 58:95–101
Parise LV, Phillips DR (1986) Fibronectin-binding properties of the purified platelet glycoprotein IIb-IIIa complex. J Biol Chem 261:14011–14017
Parmentier S, Catimel B, McGregor L, Leung LLK, McGregor JL (1991) Role of glycoprotein IIa ($\beta_1$ subunit of very late activation antigens) in platelet functions. Blood 78:2021–2026
Peng M, Lu W, Kirby EP (1991) Alboaggregin-B: a new platelet agonist that binds to platelet membrane glycoprotein Ib. Biochemistry 30:11529–11536
Peng M, Lu W, Beviglia L, Niewiarowski S, Kirby EP (1993) Echicetin: a snake venom protein that inhibits binding of von Willebrand factor and alboaggregins to platelet glycoprotein Ib. Blood 81:2321–2328
Phillips DR, Poh Agin P (1977) Platelet plasma membrane glycoproteins. Evidence for the presence of nonequivalent disulfide bonds using nonreduced-reduced two-dimensional gel electrophoresis. J Biol Chem 252:2121–2126
Piotrowicz RS, Orchekowski RP, Nugent DJ, Yamada KY, Kunicki TJ (1988) Glycoprotein Ic-IIa functions as an activation-independent fibronectin receptor on human platelets. J Cell Biol 106:1359–1364
Plow EF, Pierschbacher MD, Ruoslahti E, Marguerie GA (1985) The effect of Arg-Gly-Asp-containing peptides on fibrinogen and the von Willebrand factor binding to platelets. Proc Natl Acad Sci USA 82:8057–8061

Poulsen L, Taaning E (1990) Variation in surface platelet glycoprotein Ib expression in Bernard-Soulier syndrome. Haemostasis 20:155–161

Prior CP, Chu V, Cambou B, Dent JA, Ebert B, Gore R, Holt J, Irish T, Lee T, Mitschelen J, McClintock RA, Searfoss G, Ricca GA, Tarr C, Weber D, Ware JL, Ruggeri ZM, Hrinda M (1993) Optimization of a recombinant von Willebrand factor fragment as an antagonist of the platelet glycoprotein Ib receptor. Biotechnology 11:709–713

Randi AM, Rabinowitz I, Mancuso DJ, Mannucci PM, Sadler JE (1991) Molecular basis of von Willebrand disease type IIB. Candidate mutations cluster in one disulfide loop between proposed platelet glycoprotein Ib binding sequences. J Clin Invest 87:1220–1226

Razdan K, Hellums JD, Kroll MH (1994) Shear-stress-induced von Willebrand factor binding to platelets causes the activation of tyrosine kinase(s). Biochem J 302:681–686

Read MS, Shermer RW, Brinkhous KM (1978) Venom coagglutinin: an activator of platelet aggregation dependent on von Willebrand factor. Proc Natl Acad Sci USA 75:4514–4518

Ruan C, Tobelem G, McMichael AJ, Drouet L, Legrand Y, Degos L, Kieffer N, Lee H, Caen JP (1981) Monoclonal antibody to human platelet glycoprotein I. II. Effects on human platelet function. Br J Haematol 49:511–519

Ruggeri ZM, De Marco L, Gatti L, Bader R, Montgomery RR (1983) Platelets have more than one binding site for von Willebrand factor. J Clin Invest 72:1–12

Russell SD, Roth GJ (1993) Pseudo-von Willebrand disease: a mutation in the platelet glycoprotein Ib$\alpha$ gene associated with a hyperactive surface receptor. Blood 81:1787–1791

Ryo R, Yoshida A, Sugano W, Yasunaga M, Nakayama K, Saigo K, Adachi M, Yamaguchi N, Okuma M (1992) Deficiency of P62, a putative collagen receptor, in platelets from a patient with defective collagen-induced platelet aggregation. Am J Hematol 39:25–31

Saelman EUM, Nieuwenhuis HK, Hese KM, de Groot PG, Heijnen HFG, Sage EH, Williams S, McKeown L, Gralnick HR, Sixma JJ (1994) Platelet adhesion to collagen types I through VIII under conditions of stasis and flow is mediated by GPIa/IIa ($\alpha_2\beta_1$-integrin). Blood 83:1244–1250

Santoro SA, Rajpara SM, Staatz MD, Woods VL Jr (1988) Isolation and characterization of a platelet surface collagen binding complex related to VLA-2. Biochem Biophys Res Commun 153:217–223

Savage B, Shattil SJ, Ruggeri ZM (1992) Modulation of platelet function through adhesion receptors. A dual role for glycoprotein IIb-IIIa (integrin $\alpha_{IIb}\beta_3$) mediated by fibrinogen and glycoprotein Ib-von Willebrand factor. J Biol Chem 267:11300–11306

Sixma JJ, Schiphorst ME, Verweij CL, Pannekoek H (1991) Effect of deletion of the A1 domain of von Willebrand factor on its binding to heparin, collagen and platelets in the presence of ristocetin. Eur J Biochem 196:369–375

Sonnenberg A, Modderman PW, Hogervorst F (1988) Laminin receptor on platelets is the integrin VLA-6. Nature 336:487–489

Staatz WD, Rajpara SM, Wayner EA, Carter WG, Santoro SA (1989) The membrane glycoprotein Ia-IIa (VLA-2) complex mediates the $Mg^{++}$-dependent adhesion of platelets to collagen. J Cell Biol 108:1917–1924

Takahashi H, Murata M, Moriki T, Anbo H, Furukawa T, Nikkuni K, Shibata A, Handa M, Kawai Y, Watanabe K, Ikeda Y (1995) Substitution of Val for Met at residue 239 of platelet glycoprotein Ib$\alpha$ in Japanese patients with platelet-type von Willebrand disease. Blood 85:727–733

Tandon NN, Holland EA, Kralisz U, Kleinman HK, Robey FA, Jamieson GA (1991) Interaction of human platelets with laminin and identification of the 67kDa laminin receptor on platelets. Biochem J 274:535–542

Taniuchi Y, Kawasaki T, Fujimura Y, Suzuki M, Titani K, Sakai Y, Kaku S, Hisamichi N, Satoh N, Takenaka T, Handa M, Sawai Y (1995) Flavocetin-A and -B, two high molecular mass glycoprotein Ib binding proteins with high affinity purified from *Trimeresurus flavoviridis* venom inhibit platelet aggregation at high shear stress. Biochim Biophys Acta 1244:331–338

Titani K, Takio K, Handa M, Ruggeri ZM (1987) Amino acid sequence of the von Willebrand factor-binding domain of platelet membrane glycoprotein Ib. Proc Natl Acad Sci USA 84:5610–5614

Tsuji T, Osawa T (1987) The carbohydrate moiety of human platelet glycocalicin: the structures of the major Asn-linked sugar chains. J Biochem 101:241–249

Tsuji T, Tsunehisa S, Watanabe Y, Yamamoto K, Tohyama H, Osawa T (1983) The carbohydrate moiety of human platelet glycocalicin. J Biol Chem 258:6335–6339

Usami Y, Fujimura Y, Suzuki M, Ozeki Y, Nishio K, Fukui H, Titani K (1993) Primary structure of two-chain botrocetin, a von Willebrand factor modulator purified from the venom of *Bothrops jararaca*. Proc Natl Acad Sci USA 90:928–932

Usami Y, Suzuki M, Yoshida E, Sakurai Y, Hirano K, Kawasaki T, Fujimura Y, Titani K (1996) Primary structure of alboaggregin-B purified from the venom of *Trimeresurus albolabris*. Biochem Biophys Res Commun 219:727–733

Ushiyama S, Laue TM, Moore KL, Erickson HP, McEver RP (1993) Structural and functional characterization of monomeric soluble P-selectin and comparison with membrane P-selectin. J Biol Chem 268:15229–15237

Vicente V, Houghten RA, Ruggeri ZM (1990) Identification of a site in the $\alpha$ chain of platelet glycoprotein Ib that participates in von Willebrand factor binding. J Biol Chem 265:274–280

Wang R, Furihata K, McFarland JG, Friedman K, Aster RH, Newman PJ (1992) An amino acid polymorphism within the RGD binding domain of platelet membrane glycoprotein IIIa is responsible for the formation of the $Pen^a/Pen^b$ alloantigen system. J Clin Invest 90:2038–2043

Wardell MR, Reynolds CC, Berndt MC, Wallace RW, Fox JE (1989) Platelet glycoprotein Ib beta is phosphorylated on serine 166 by cyclic AMP-dependent protein kinase. J Biol Chem 264:15656–15661

Ware J, Russell SR, Marchese P, Murata M, Mazzucato M, De Marco L, Ruggeri ZM (1993) Point mutation in a leucine-rich repeat of platelet glycoprotein Ib$\alpha$ resulting in the Bernard-Soulier syndrome. J Clin Invest 92:1213–1220

Weiss HJ, Tschopp TB, Baumgartner HR, Sussman II, Johnson MM, Egan JJ (1974) Decreased adhesion of giant (Bernard-Soulier platelets to subendothelium: further implications on the role of the von Willebrand factor in hemostasis. Am J Med 57:920–925

Weiss HJ, Turitto VT, Baumgartner HR (1978) Effect of shear rate on platelet interaction with subendothelium in citrated and native blood. I. Shear rate-dependent decrease of adhesion in von Willebrand's disease and the Bernard-Soulier syndrome. J Lab Clin Med 92:750–765

Wenger RH, Wicki AN, Kieffer N, Adolph S, Hameister H, Clemetson KJ (1989) The 5′ flanking region and chromosomal localization of the gene encoding human platelet membrane glycoprotein Ib$\alpha$. Gene 85:519–524

Wicki AN, Clemetson KJ (1985) Structure and function of platelet membrane glycoproteins Ib and V: effects of leukocyte elastase and other proteases on platelet response to von Willebrand factor and thrombin. Eur J Biochem 153:1–11

Wicki AN, Clemetson KJ (1987) The glycoprotein Ib complex of human blood platelets. Eur J Biochem 163:43–50

Wright SD, Michaelides K, Johnson DJD, West NC, Tuddenham EGD (1993) Two different missense mutations in the glycoprotein IX gene in a family with Bernard-Soulier syndrome: affected individuals are compound heterozygotes. Blood 81:2339–2347

Wyler B, Bienz D, Clemetson KJ, Lüscher EF (1986) Glycoprotein Ib$\beta$ is the only phosphorylated major membrane glycoprotein in human platelets. Biochem J 234:373–379
Yamamoto N, Akamatsu N, Yamazaki H, Tanoue K (1992) Normal aggregations of glycoprotein IV (CD36)-deficient platelets from seven healthy Japanese donors. Br J Haematol 81:86–92
Ylänne J, Huuskonen J, O'Toole TE, Ginsberg MH, Virtanen I, Gahmberg CG (1995) Mutation of the cytoplasmic domain of the integrin $\beta_3$ subunit. Differential effects on cell spreading, recruitment to adhesion plaques, endocytosis, and phagocytosis. J Biol Chem 270:9550–9557
Yoshida E, Fujimura Y, Miura S, Sugimoto M, Fukui H, Narita N, Usami Y, Suzuki M, Titani K (1993) Alboaggregin-B and botrocetin, two snake venom proteins with highly homologous amino acid sequences but totally distinct functions on von Willebrand factor binding to platelets. Biochem Biophys Res Commun 191:1386–1392
Zamarron C, Ginsberg MH, Plow EF (1991) A receptor-induced binding site in fibrinogen elicited by its interaction with platelet membrane glycoprotein IIb-IIIa. J Biol Chem 266:16193–16199
Zehnder JL, Hirai K, Shatsky M, McGregor JL, Levitt LJ, Leung LLK (1992) The cell adhesion molecule CD31 is phosphorylated after cell activation. Down-regulation of CD31 in activated T lymphocytes. J Biol Chem 267:5243–5249
Zucker MB, Kim SJ, McPherso J, Grant RA (1977) Binding of factor VIII to platelets in the presence of ristocetin. Br J Haematol 35:535–549
Zwierzina WD, Schmalzl F, Kunz F, Dworzak E, Linker H, Geissler D (1983) Studies in a case of Bernard-Soulier Syndrome. Acta Haematol (Basel) 69:195–203

CHAPTER 9
# Platelet G Proteins and Adenylyl and Guanylyl Cyclases

D. KOESLING and B. NÜRNBERG

## A. Introduction

Platelets are important constituents of hemostasis under normal and pathophysiological circumstances (MARCUS and SAFIER 1993). Following injury, platelet aggregation helps prevent excessive blood loss and also plays an essential role in atherosclerosis and thrombosis, where it is implicated in the pathogenesis of life-threatening diseases, e.g., myocardial infarction and stroke. The hallmark of platelet function is aggregation that is regulated by an array of activating and inhibiting hormones and factors released by surrounding cells or from platelets themselves. Activators of platelets include enzymes such as the protease thrombin, lipid mediators such as platelet activating factor (PAF) and thromboxane $A_2$ ($TXA_2$), nucleotides (ADP), and various neurotransmitters, e.g., catecholamines (via $\alpha_2$-adrenoceptors) and serotonin. In contrast, platelet aggregation is inhibited by adenosine, prostaglandins, e.g., prostacyclin ($PGI_2$), or catecholamines (via $\beta$-adrenoceptors). All these ligands bind to cell surface receptors exhibiting a high degree of structural homology regardless of whether or not these receptors elicit stimulating or inhibiting signals (HOURANI and CUSACK 1991).

The receptors belong to the superfamily of G-protein-coupled receptors sharing characteristic features such as seven hydrophobic domains presumably spanning the thrombocytic plasma membrane, potential N-glycosylation sites in the extracellular N-terminus, and GIP-sensitivity of agonist binding. (WATSON and ARKINSTALL 1994). G-protein-coupled transmembrane signaling requires the sequential and reversible assembly of a protein complex composed of three elements: (a) heptahelical (or serpentine) G-protein-coupled receptor (GUDERMANN et al. 1995), (b) transducer (i.e., heterotrimeric G protein; NÜRNBERG et al. 1995), and (c) effector (i.e., enzyme, ion channel, or transporter; CLAPHAM and NEER 1993; DICKEY and BIRNBAUMER 1993). (A fourth element called regulators of G-protein signalling (RGS) has been suggested to involved in desensitization mechanisms (ROUSH 1996).) The intermolecular functional coupling of heptahelical receptors to G proteins allows a tremendous amplification of the hormonal signal on the one hand since one ligand-bound receptor activates many G proteins within a short time, and on the other hand it produces branch points enabling divergent signal transduction routes (HILLE 1992; BIRNBAUMER 1992; BOURNE and NICOLL 1993;

Offermanns and Schultz 1994). This is in contrast to transmembrane signals carried by integral membrane proteins consisting of intramolecularly linked receptor-effector domains.

Notably, it is not the receptor but the receptor-coupled G protein that determines the effector pathway responsible for either platelet activation or inhibition. For example, stimulation of receptors functionally coupling to the G protein $G_q$, e.g., the $TXA_2$ receptor, activates platelets by inducing a PI-response whereas $G_s$-coupled receptors such as the prostacyclin ($PGI_2$) receptor stimulate adenylyl cyclase activity, thus raising intracellular cAMP concentration which inhibits platelet aggregation. Apart from cAMP, another cyclic nucleotide, cGMP, is an important inhibitory second messenger. The enzyme responsible for the formation of cGMP in platelets, the soluble guanylyl cyclase, functions as the receptor for nitric oxide produced in a calcium-dependent process in the neighboring endothelial cells (Koesling et al. 1993). This short overview will focus on the structures and functions of heterotrimeric G proteins and adenylyl and guanylyl cyclases present in human platelets. Monomeric GTPases and G-protein-coupled receptors and effectors with the exception of adenylyl cyclases are reviewed elsewhere in this volume.

# B. G Proteins

## I. General Considerations

G proteins are membrane-associated heterotrimers attached to the cytoplasmic surface and also located on intracellular membranes and vesicles. In addition, they are presumably involved in membrane trafficking and vesicular transport mechanisms of the cell (Hepler and Gilman 1992; Nürnberg and Ahnert-Hilger 1996). They belong to the superfamily of GTP-binding proteins or GTPases (EC 3.6.1.-) and are extremely conserved throughout the animal kingdom, underlining their pivotal role in cellular regulation (Rodbell et al. 1971; Bourne et al. 1990, 1991; Wilkie et al. 1992; Bourne and Nicoll 1993). G proteins are composed of three different subunits termed $\alpha$, $\beta$, and $\gamma$, with molecular masses of approximately 39–52 kDa, 35–39 kDa, and 6–8 kDa, respectively.

### 1. Activation of G Proteins

G proteins undergo a cycle of activated and inactivated states, allowing reversible and specific transmission of hormonal signals (Fig. 1; Gilman 1987). Activation of the G protein is initiated by its interaction with distinct cytoplasmic segments of a ligand-activated heptahelical receptor (Gudermann et al. 1995), resulting in the release of bound GDP. Most likely, this first event is the rate-limiting step of the G-protein activation reaction followed by high affinity binding of cytosolic GTP. $Mg^{2+}$ is very tightly associated with GTP and $G\alpha$ in

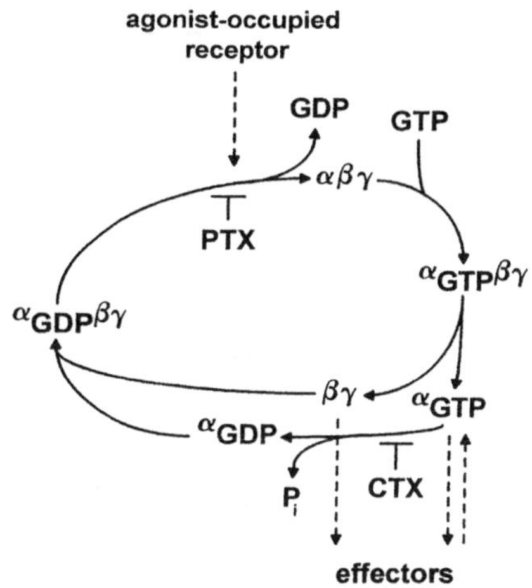


**Fig. 1.** Activation/inactivation cycle of G proteins. $\alpha$, $\beta$, and $\gamma$ represent the G$\alpha$, G$\beta$, and G$\gamma$ subunits of the G protein. Pertussis toxin (*PTX*) functionally uncouples the G protein from the receptor whereas cholera toxin (*CTX*) abolishes the GTPase activity of the G protein

this complex and is required for activation of the G$\alpha$ subunit and for subsequent hydrolytic activity (Higashijima et al. 1987). Upon G-protein activation, the G$\alpha$ subunit dissociates from the G$\beta\gamma$ complex. Both the activated G$\alpha$ subunit and the G$\beta\gamma$ complex modulate effector proteins until hydrolysis of the GTP bound to the G$\alpha$ subunit terminates signaling (intrinsic GTPase activity). The GTPase activity of the G$\alpha$ subunit is modulated by some effectors (e.g., cGMP-dependent phosphodiesterase, phospholipase C-$\beta$, L-type calcium channel), thereby functioning as GAPs (like the GTPase-activating proteins of the ras family). Following GTP hydrolysis, the inactive GDP-bound G$\alpha$ subunit dissociates from the effector and reassociates with the G$\beta\gamma$ complex to form a G$\alpha\beta\gamma$ heterotrimer. This inactive heterotrimer becomes available for subsequent activation cycles. Interestingly, the receptor-stimulated GDP/GTP-exchange reaction varies among different G proteins: $G_i$ proteins are fast nucleotide exchangers, $G_s$, $G_z$, and G proteins of the $G_q$ subfamily show lower rate constants, and for $G_{12}$ and $G_{13}$ minute guanine nucleotide exchange rates have been reported (Fields et al. 1994; Laugwitz et al. 1994). As pointed out above, it is assumed that the $\alpha$-helical domain of the $\alpha$ subunit plays a crucial role in controlling the kinetics of GDP release as it functions like a lid protecting the guanine nucleotide binding pocket.

Some bacterial toxins are of extreme value in identifying G-protein-coupled pathways. Cholera toxin (CT) and pertussis toxin (PT) catalyze ADP-ribosylation of various G$\alpha$ subunits (Gierschik and Jakobs 1992; Nürnberg 1997). CT modifies an arginine residue located in the nucleotide-binding region of the G$\alpha$ subunit (see below). G$\alpha_s$ and G$\alpha_{olf}$ are modified by CT independently of receptor ligand whereas G$\alpha_t$ and G$\alpha_{i/o}$ are modified in dependence on receptor ligand. A functional consequence of CT-catalyzed ADP-ribosylation is the abolishment of endogenous GTPase activity, resulting in continuous activation of the $G\alpha$ subunit. In contrast, PT catalyzes the transfer of an ADP moiety from $NAD^+$ to a cysteine residue of the $\alpha$ subunit of $G_i/G_o$ and $G_t$. The amino acid motif required for PT-mediated ADP-ribosylation is similar to the consensus site of isoprenylation of $G\gamma$ subunits except for the C-terminal aromatic amino acid. Since the modified cysteine is located four residues upstream of the carboxy terminus, PT-catalyzed ADP-ribosylation prevents interaction of the receptor with the G protein representing a functional receptor-G-protein uncoupling. Nevertheless, PT-modified G proteins are capable of interacting with effectors though the time required to observe maximal effect by GTP$\gamma$S can be markedly increased (JAKOBS et al. 1984; AHNERT-HILGER et al., submitted).

## 2. Diversity of G Proteins

Currently, the heterotrimer is classified by the nature of its $G\alpha$ subunit (Table 1). On the basis of their amino acid similarity, G-protein $\alpha$ subunits are classified in four families: $G_s$, $G_i$, $G_q$, and $G_{12}$ (SIMON et al. 1991). Nucleotide sequence analysis shows that the genes of two subfamilies, i.e., $G_i$ and $G_q$, are closely related, suggesting that the members of these subfamilies segregated as

**Table 1.** Heterotrimeric regulatory guanine nucleotide binding proteins (G proteins)

| | Occurrence | Effector | PT/CT |
|---|---|---|---|
| $G_s$ subfamily | | | |
| $G_{s(s,l)}$ | Ubiquitous | Adenylyl cyclases ↑ | CT |
| $G_{olf}$ | Olfactory epithelium | Adenylyl cyclases ↑ | CT |
| $G_i$ subfamily | | | |
| $Transducin_{T,C}$ | Retinal rods and cones | CGMP PDE ↑ | PT, CT |
| Gustducin | Taste cells | ? | PT, CT |
| $G_{i1}$ | Mainly neuronal cells | Adenylyl cyclase ↓ | PT, CT |
| $G_{i2}$ | Ubiquitous | Adenylyl cyclase ↓ | PT, CT |
| $G_{i3}$ | Mainly nonneuronal cells | Adenylyl cyclase ↓ | PT, CT |
| $G_{o1,2}$ | Neuronal, endocrine cells | Calcium channels ↓ | PT, CT |
| $G_z$ | Neuronal cells, platelets | ? | PT, CT |
| $G_q$ subfamily | | | |
| $G_q$ | Ubiquitous | Phospholipase C-$\beta$ ↑ | – |
| $G_{11}$ | Nonhematopoietic cells | Phospholipase C-$\beta$ ↑ | – |
| $G_{14}$ | Many cells | Phospholipase C-$\beta$ ↑ | – |
| $G_{15,16}$ | Hematopoietic cells | Phospholipase C-$\beta$ ↑ | – |
| $G_{12}$ subfamily | | | |
| $G_{12}$ | Ubiquitous | ? | – |
| $G_{13}$ | Ubiquitous | ? | – |

G proteins are named according to their $\alpha$ subunits. Some typical effector functions are listed. ↑, ↓ indicate increased and decreased effector activity, respectively. PT and CT indicate whether the $\alpha$ subunits are substrates of pertussis or cholera toxins. For more details, see text.

**Table 2.** G protein $\beta$ and $\gamma$ subunits

| Subunit | Occurrence | | |
|---|---|---|---|
| $\beta_1$ | Ubiquitous | | |
| $\beta_2$ | Ubiquitous | | |
| $\beta_3$ | Many cells | | |
| $\beta_4$ | Many cells | | |
| $\beta_5$ | Neuronal cells | | |
| $\beta_6$ | ? | | |
| $\gamma_1$ | Retinal rods | Subfamily: | I: $\gamma_1$, $\gamma_9$, $\gamma_{11}$ |
| $\gamma_2$ | Many cells | | II: $\gamma_5$, $\gamma_{10}$ |
| $\gamma_3$ | Many cells | | III: $\gamma_2$, $\gamma_3$, $\gamma_4$, $\gamma_7$, $\gamma_8$, $\gamma_{12}$ |
| $\gamma_4$ | Many cells | | |
| $\gamma_5$ | Many cells | | |
| $\gamma_7$ | Many cells | | |
| $\gamma_8$ | Olfactory cilia, brain | | |
| $\gamma_9$ | Retinal cones | | |
| $\gamma_{10}$ | Many cells | | |
| $\gamma_{11}$ | ? | | |
| $\gamma_{12}$ | ? | | |

pairs of closely linked genes whereas members of the $G_s$ and $G_{12}$ subfamilies segregated as unlinked genes during evolution (WILKIE et al. 1992). Whereas $G_{i1}$, $G_{i2}$, and $G_{i3}$ are highly homologous proteins (85%–95% similarity of amino acid residues), the $G_{12}$ proteins share only 67% similar residues. Currently, 23 different $\alpha$ subunits including several splice variants of $G_s$, $G_i$, and $G_o$ are known which are the products of 17 different genes (see Table 1; SIMON et al. 1991). Meanwhile, the number of cloned and sequenced $G\beta$ and $G\gamma$ subunits has reached a significant amount (Table 2), comprising 5 G-protein $G\beta$ ($G\beta_{1-5}$) and 11 $G\gamma$ ($G\gamma_{1-5,7-12}$) subunits (HEPLER and GILMAN 1992; INIGUEZ-LLUHI et al. 1993; RAY et al. 1995). Among $G\beta$ subunits, $G\beta_5$ exhibits the greatest difference in amino acid sequence with 53% identity to other $G\beta$ subunits whereas the deduced amino acid sequences of $G\beta_1$ to $G\beta_4$ display a high degree of homology with at least 79% identity (WATSON et al. 1994). In contrast, $G\gamma$ subunits demonstrate a higher structural diversity, e.g., the bovine retinal $G\gamma_1$ subunit and the $G\gamma_2$ subunit are only 36% identical (SPRING and NEER 1994). The differences in primary structures argue for an important role of $G\gamma$ (in addition to $G\alpha$ subunits) in the specificity of receptor-G-protein interactions (CLAPHAM and NEER 1993; INIGUEZ-LLUHI et al. 1993; KISSELEV et al. 1995). This assumption is supported by a study showing that isoprenylated C-terminal peptides derived from the $G\gamma_1$ amino acid sequence are sufficient for stabilizing the active form of the rhodopsin photoreceptor and for uncoupling the rhodopsin-$G_t$ interaction in dependence on concentration (KISSELEV et al. 1994). Moreover, $G\beta$ subunits contain essential structural attributes necessary for G-protein-subunit and G-protein-protein interactions. They belong to the family of WD-repeat proteins (NEER et al. 1994) whose common feature is a conserved core of the repeating unit (n = 4–8) of 36 to 46 amino acids usually ending with a Trp-Asp (WD). WD-repeat proteins regulate different cellular functions, such as cell division, cell fate determination, gene transcription, transmembrane signaling, mRNA modification, and vesicle fusion. Furthermore, $G\beta$ subunits contain coiled-coil forming structures interacting with other subunits at their N-terminal domain ($Met^{1}$ – $Ser^{31}$ of $G\beta_1$; LUPAS et al. 1991, 1992; GARRITSEN et al. 1993). Coiled coils are stabilized protein structures resulting from the interaction of two or more right-handed $\alpha$-helices that wind around each other in a left-handed supercoil (SIMONDS et al. 1993).

### 3. Structure of G Proteins

Profound work on several crystallized G proteins by two independent groups now provides impressive images on the three-dimensional structures of the subunits in their inactive and active states (for a review see WITTINGHOFER 1994; RENS-DOMIANO and HAMM 1995; COLEMAN and SPRANG 1996; NEER and SMITH 1996). Interestingly, only the $G\alpha$ subunit and not $G\beta\gamma$ undergo major structural changes upon activation. The conformation of the GDP-liganded $G\alpha$ in a $G\alpha\beta\gamma$ heterotrimer is different from GTP$\gamma$S- or GDP-liganded $G\alpha$ alone. Whereas the GTP$\gamma$S-bound molecule exhibits its active conformation, the structure of the GDP-bound $G\alpha$ may be artificial partly due to differences in crystal structures and crystal-crystal contacts (BOURNE 1995: MIXON et al. 1995). The $G\alpha$ subunit consists of two major parts, i.e., a GTPase domain and an $\alpha$-helical domain that are connected by two flexible linker loops. Between these two domains lies a deep cleft in which the guanine nucleotide is tightly bound. The structure of the GTPase domain is the same as that of other members of the GTP-binding class of proteins despite a low level of amino acid identity (16%–18%), e.g., the monomeric GTPase p21ras and the bacterial protein synthesis elongation factor Tu (EF-TU; NOEL et al. 1993; LAMBRIGHT et al. 1994, 1996; COLEMAN et al. 1994; MIXON et al. 1995; WALL et al. 1995). The helical domain has no known function, but there is evidence that it might act as an internal "GTPase-activating-protein". Whereas most amino acids of $G\alpha$ remain in the same conformation regardless of whether GDP or GTP is bound, some of the residues show flexibility depending on the nucleotide binding state. Three segments, i.e., switch I, II, and III regions, and the amino and carboxy terminus change their conformation upon guanine nucleotide exchange. Of the amino acids interacting with the guanine nucleotide, $glutamine^{204}$ and $arginine^{178}$ ($Gln^{200}$ and $Arg^{174}$ in $G\alpha_t$) are mobile residues that do not bind nucleotides in the basic state but interact strongly with substrates (GTP and bound water) during hydrolysis of GTP. $Arg^{178}$ stabilizes the charge developing in the transition state while $Gln^{204}$ orients and polarizes the attacking water in GTP hydrolysis. Both residues are present on comparable loci found in all heterotrimeric $G\alpha$ subunits. They were previously predicted as a prerequisite for GTPase activity of various G proteins on the basis of site-directed mutagenesis experiments (GOODY 1994). It is this arginine residue that is sensitive to cholera toxin (see above). Interestingly, the monomeric GTPase p21lras does not have an arginine residue in a position analogous to Arg[178] (in G$\alpha_{i1}$), thus explaining the enzyme's weak hydrolytic activity and its inability to bind the G-protein-activating $AlF_4^-$ complex.

G$\beta\gamma$ behaves as a functional monomer dissociating only under denaturing conditions. The G$\beta\gamma$ complex binds and stabilizes only the GDP-bound form of G$\alpha$. It does not undergo structural changes upon activation of the heterotrimer as the G$\alpha$ subunit does. The G$\beta$ subunit folds into a highly symmetric $\beta$ propeller formed by amino acid sequences belonging to the WD-repeat motif (see above; SONDEK et al. 1996). Each of the seven propeller blades consists of a small four-stranded twisted $\beta$ sheet, the innermost $\beta$ strand being nearly parallel to the axis of a central tunnel. The conserved GH-X-WD cores of the WD-repeat motif contribute predominantly to the inner three strands of each blade, which can be viewed as connectors linking the conserved cores, forming the loop connecting the two outer strands and the outer most strand. Hence, each WD repeat does not form one blade structure, but rather overlaps two $\beta$ blades. WD repeats 2 and 6 are the most divergent among the seven. The amino terminal domain including some 47 amino acids forms an extended polypeptide chain that girds the top of the propeller. The interface between G$\beta$ and the serpentine G$\gamma$ which makes very few contacts with itself is extensive. The parallel $\alpha$-helical coiled coil formed by the amino termini of the two subunits is a notable feature that had been predicted from sequence analysis (see above). Most of the G$\gamma$ subunit is stretched along the side and bottom of the G$\beta$ subunit containing blades 5, 6, 7, and 1. The loops and turns on this surface must retain the ability to bind G$\gamma$ and to discriminate among different G$\gamma$'s. The docking of G$\alpha$ to G$\beta\gamma$ involves extensive contacts: binding of the G$\alpha$ amino terminal $\alpha$-helix to the side of the G$\beta$ propeller parallel to its central tunnel and binding of the catalytic domanin of G$\alpha$ to the top surface of the $\beta$ propeller.

Cationic cytoplasmic loops of the heptahelical receptor probably have free access to a large surface of G$\beta$ that is remarkably negative and includes blades 6 and 7. The cytoplasmic loops may also contact much of G$\gamma$ and the mouth of the tunnel through the propeller. The receptor can probably interact with the amino terminus of G$\alpha$, displacing it from the surface of G$\beta\gamma$ and thus promoting subunit dissociation. Furthermore, interaction of receptor with the carboxy terminus of G$\alpha$ is clearly important. In the heterotrimeric conformation, subunits are not able to interact with effectors. Since G$\alpha$ blocks interaction of G$\beta\gamma$ with all its known effectors without inducing a conformational change in G$\beta\gamma$, it is likely that G$\alpha$ sterically interferes with binding of many effectors. While G$\alpha$- and effector-binding sites may overlap, they are surely not identical. Research employing the oligonucleotide antisense technology revealed a surprisingly high specific G-protein-subunit composition transducing signals from receptors to effectors (for a review see KALKBRENNER et al. 1996). However, the observed specificity of heterotrimer subunit composition appears to be restricted to the integrity of the cell and is lost in isolated cell membranes (NÜRNBERG et al. 1994).

## 4. Co- and Posttranslational Modifications of Platelet G Proteins

Most G-protein $\alpha$ subunits are modified by co- and posttranslational events (YAMANE and FUNG 1993; CASEY 1994). These modifications are essential for membrane association and increase the stability of the heterotrimer complex. During elongation of nascent $G\alpha_{i/o}$ proteins, the initial methionine is cleaved, followed by myristoyl amidation of the glycine residue. In addition to myristate (C-14 carbon acid), various unsaturated C-14 fatty acids and the C-12 lauryl acid are linked to the glycine residue of the retinal $G\alpha$ of transducin (CASEY 1994). Myristoylation appears to be irreversible. This protein modification enhances lipophilicity of the $G\alpha$ subunit on the one hand, and on the other hand it contributes to the specificity of protein-protein interactions. Simultaneously, myristoylation enhances the potency of the $G\alpha$ subunit to activate effectors.

Recently, palmitoylation of all $G\alpha$ subunits except transducin was reported as a posttranslational, reversible modification occurring at a cysteine residue near the N-terminus via a labile thioester bond and possibly at other sites (DEGTYAREV et al. 1993; LINDER et al. 1993; PARENTI et al. 1993; VEIT et al. 1994). After cell fractionation, nonmyristoylated $G\alpha$ subunits carrying a palmitate (C-16 carbon acid) accumulated in the particulate fraction whereas the protein devoid of fatty acid modification was found in the cytosol. However, it remains questionable whether the only function of palmitoyl acylation is anchoring the $G\alpha$ subunit in the membrane. For instance, the degree of palmitoylation of $G\alpha$ subunits is regulated in dependence on receptor (MUMBY et al. 1994). Furthermore, nonpalmitoylated chimeras of $G\alpha_s/G\alpha_t$ and $G\alpha_q/G\alpha_t$ are unable to stimulate their respective effectors, i.e., adenylyl cyclases and phospholipases C (WEDEGAERTNER et al. 1993). Interestingly, arachidonate-acylated $G\alpha$ subunits ($G\alpha_i$, $G\alpha_q$, $G\alpha_z$, $G\alpha_{13}$) were detected in platelets (HALLAK et al. 1994).

Reports on phosphorylation of G proteins are scarce. In the case of the PT-insensitive G protein $G_z$, selective phosphorylation of a serine residue near the N-terminus by protein kinase C is well established (LOUNSBURY et al. 1991, 1993). Although it was shown that the affinity of phosphorylated $G\alpha_z$ deceases its affinity to $G\beta\gamma$, the functional consequences are unkown (FIELDS and CASEY 1995). Other $G\alpha$ subunits, e.g, $G\alpha_s$, also possess consensus sites recognized by protein kinases, allowing in vitro phosphorylation under certain experimental conditions (HAUSDORF et al. 1992). $G_s$ appears to be susceptible to phosphorylation at residues $tyr^{37}$ and $tyr^{377}$ which lie near the site of $G\beta\gamma$ binding in the amino terminus and at the extreme carboxy terminus known to participate in receptor coupling (MOYERS et al. 1995). On the basis of experimental evidence, it is speculated that tyrosine phosphorylation of $G_s$ is a consequence of cross talk between receptor tyrosine kinase signaling and G-protein-coupled receptor pathways interrupting activation of adenylyl cyclases by receptors (LIEBMANN et al. 1996). In neuroblastoma/glioma cells, activation of protein kinase C stimulated phosphorylation of $G_{i2}$ with kinetics similar to its effect on adenylyl cyclase (STRASSHEIM and MALBON 1994). Stable transfection of CHO cells with cGMP protein kinase I$\alpha$ (cGMP-pk) suppressed the thrombin-induced increase in inositol 1,4,5-trisphosphate and cytosolic calcium. This regulatory step was found to be PT-sensitive. Further studies showed that cGMP-pk phosphorylated a PT-sensitive G protein in these cells and that purified cGMP-pk was able to phosphorylate reconstituted $G_i$ proteins but not $G_o$ (PFEIFER et al. 1995). Whether these events are coincidental remains to be clarified.

G-protein $\beta\gamma$ dimers exhibit higher lipophilicity than G$\alpha$ subunits, which correlates with isoprenylation of G$\gamma$ subunits (YAMANE and FUNG 1993; CASEY 1994). The signal sequence causing this modification is a CAAX-box motif at the C-terminus common of all G$\gamma$ subunits and some monomeric GTPase. This motif triggers a number of posttranslational events initiated by prenylation of the cysteine residue through a thioether bond followed by endoproteolytic cleavage of the three carboxy terminal amino acids (AAX) and methylation at the new C-terminus, which is the prenylated cysteine residue. Efforts are currently being undertaken to elucidate the mechanism of G$\beta\gamma$ assembly and functional consequences of the various posttranslational modifications. First results suggest that prenylation is not required for G$\beta\gamma$ assembly which occurs prior to the proteolytic processing of G$\gamma$ (HIGGINS and CASEY 1994). Interestingly, both isoprenylation and carboxymethylation of G$\gamma_1$ appear to contribute to the efficiency of membrane association, subunit interaction between G$\alpha$ and G$\beta\gamma$, and functional coupling of the G protein with receptors or effectors (CASEY 1994; FUKADA et al. 1994; DIETRICH et al. 1994). Some G$\gamma$, e.g., the retina-specific G$\gamma_1$ subunit, are modified by a C-15 sesquiterpene (farnesyl moiety) whereas other G$\gamma$ isoforms are modified by a C-20 diterpene (geranylgeranyl group). This difference in isoprenylation of G$\gamma$ affects lipophilicity of the G$\beta\gamma$ dimer, i.e., retinal G$\beta_1$G$\gamma_1$ dimers are soluble without detergents in contrast to all other known G$\beta\gamma$ dimers. Furthermore, the type of isoprenylation has an impact on all functions of G$\beta\gamma$ dimers studied so far. In general, C-20-isoprenylated G$\beta\gamma$ complexes appear to exhibit higher potency in modulating effectors than C-15-isoprenylated G$\beta_1\gamma_1$.

## II. G Proteins Expressed in Platelets

Platelets express G proteins of all four subfamilies (see Table 1). Isoforms were identified by employing specific antisera and by G-protein labeling with bacterial toxins and guanine nucleotide analogues. These studies revealed that platelets contain at least the short form of the $G_s$ protein which is responsible for stimulation of adenylyl cyclases (BRASS et al. 1993). Of the $G_i$ proteins responsible for inhibition of adenylyl cyclase activity and functionally significant release of G$\beta\gamma$ subunits, certainly $G_{i2}$ and $G_{i3}$ are expressed. It is thought that $G_{i2}$ makes up two-thirds of the total $G_i$ pool in platelets (SIMONDS et al. 1989). Conflicting results exist concerning the presence of $G_{i1}$ in platelets

(WILLIAMS et al. 1990; OHLMANN et al. 1995). The absence of $G_{i1}$ in platelets is more likely since purification of G proteins from human platelet membranes yielded only $G\alpha_{i2}$ and $G\alpha_{i3}$ (KÖRNER et al. 1995). Platelets exhibit a high concentration of the pertussis-toxin-insensitive $G_z$ (or $G_x$) belonging to the $G_i$-subfamily. Purified from brain (CASEY et al. 1990) or human platelets (NÜRNBERG et al., unpublished work), it coelutes with an unkown 26-kDa GTP binding protein. The physiological relevance of this observation remains cryptic. $G_z$ exhibits extremely slow basal GTPase activity similar to monomeric GTPases. Though inhibiting various adenylyl cyclase isoforms, $G_z$-activating receptors or $G_z$-regulated effectors are currently unknown (KOZASA and GILMAN 1995). Proteins of the $G_q$ subfamily activate phospholipases C-$\beta$ in a pertussis-toxin-insensitive manner. The major PLC-stimulating G protein found in platelets is $G_q$ whereas $G_{11}$ is not detectable in platelets or other hematopoietic cells (MILLIGAN et al. 1993). Little is known about other $G_q$-subfamily members. Both member of the $G_{12}$ subfamily are found in platelets, with $G_{13}$ present at higher concentrations (SPICHER et al. 1994). There are scarce reports on the expression pattern of $G\beta\gamma$ subunits in platelets. Membranes and purified platelet $G\beta\gamma$ complexes show immunoreactivity towards $G\beta_1$- and $G\beta_2$-specific antibodies. Both isoforms appear to be expressed at similar concentrations (NÜRNBERG et al. 1996). Of the various $G\gamma$ subunits isolated from platelets, we identified $G\gamma_2$ by immunoblot analysis and proved the abscence of $G\gamma_3$ (SPICHER et al., unpublished work). Recently, transient high energy phosphorylation of $G\beta$ subunits was demonstrated to occur in cell membranes of leucocytes, placental tissue, brain, and liver but only marginally in platetelet membranes (WIELAND et al. 1993; NÜRNBERG et al. 1996). The reason for this is probably a missing cofactor. This phosphorylation reaction is assumed to be an alternative route involved in activation of G proteins (KALDENBERG-STASCH et al. 1994).

## III. G Proteins as Modulators of Platelet Activation

Platelet activation initiated by G-protein-dependent receptors is subdivided into activating pathways elicited by PT-insensitive G proteins of the $G_q$ subfamily and by $G\beta\gamma$ subunits released from $G_i$ proteins in a PT-sensitive manner (Table 3). As found so far, proteins of the $G_{12}$ subfamily are activated only by $G_q$-coupled receptors. In contrast, inhibition of platelet activation is accomplished by receptors activating CT-sensitive members of the $G_s$ subfamily.

### 1. Platelet-Activating G Proteins

$G_q$ is ubiquitously expressed and exhibits low basal GTPase activity and GTP binding (SIMON et al. 1991; STERNWEIS and SMRCKA 1992; LAUGWITZ et al. 1994). $G_q$ stimulates phospholipase C-$\beta$ isoforms (PLC-$\beta$), so receptor-activated $G_q$ proteins trigger the PT-insensitive phosphoinositide breakdown (PI response;

**Table 3.** G-protein coupling to platelet heptahelical receptors

| Receptors activating platelets | Coupling G proteins |
|---|---|
| ADP receptor ($P_{2T}$-R) | $G_{i2}$ |
| $\alpha$-Adrenergic receptor ($\alpha_{2A}$-R) | $G_{i2}$ |
| Platelet activating factor receptor (PAF-R) | $G_q$ and $G_i$ (?) proteins[a] |
| Serotonin receptor (5-$HT_{2A}$) | $G_q$ proteins[a] |
| Thrombin receptor | $G_{i2}$, $G_{i3}$, $G_q$, $G_{12}$, $G_{13}$ |
| Thromboxane receptors ($TXA_{A2s,1}$-R) | $G_q$, $G_{12}$, $G_{13}$ |
| Vasopressin receptor ($V_{1A}$-R) | $G_q$ proteins[a] |
| **Receptors inhibiting platelets** | |
| Adenosin receptor ($A_{2A}$-R) | $G_s$[b] |
| $\beta$-Adrenergic receptor ($\beta_2$-R) | $G_s$[b] |
| Prostacyclin receptor ($PGI_2$-R) | $G_s$[b] |

[a] No data available identifying the exact isoforms activated by the respective receptors in platelets. $G_{11}$ is not expressed in platelets.
[b] No functional differences between the four Gs splice variants are known.

DENNIS et al. 1991) leading to mobilization of $Ca^{2+}$ from intracellular stores, $Ca^{2+}$ influx, and protein kinase C activation by diacylglycerols (NISHIZUKA 1995). Utilizing photoaffinity labeling of receptor-activated G proteins, it was shown that activation of $G_q$ is induced by $TXA_2$- and thrombin receptors in human platelet membranes (OFFERMANNS et al. 1994). The specific coupling of $TXA_2$ receptors to $G_q$ was also confirmed by reconstitution of an extracted recombinant $TXA_2$ receptor with purified G proteins into phospholipid vesicles (HARHAMMER et al. 1996). In a systematic study using recombinant proteins and peptides corresponding to the sequence of $G\alpha_q$, Arkinstall and coworkers were able to map sites of $G\alpha_q$ responsible for interaction with PLC-$\beta_1$ (ARKINSTALL et al. 1995). They found two domains located in the C-terminal third of the ras-like domain of $G\alpha_q$: amino acids 251–265 and 306–309 corresponding to helices $\alpha 3$ and $\alpha 4$, respectively. Both helices are adjacent to one another and are located at the molecular surface of $G\alpha_q$. These segments are part of a continuous surface including regions that change conformation upon nucleotide exchange and $G\alpha$ activation, i.e., switch regions II and III. In particular, $\alpha 3$-helix, which together with the $\beta 4$ segment forms the switch III region, was recognized as an important effector region of the $G\alpha$ subunit. Very recently evidence arised that $G_q$ proteins may regulate calcium hemostasis of nonexcitable cells not only by PLC-$\beta$ activation but also by an additional mechanism through direct activation of nonselective cation channels leading to influx of calcium into the cell (OBUKHOV et al. 1996). Whether this mechanism occurs in platelets remains to be elucidated.

While PLC-$\beta 1$ is much more sensitive to $G\alpha_q$ proteins, PLC-$\beta 2$ and PLC-$\beta 3$ isoenzymes are also stimulated by $G\beta\gamma$ complexes (CAMPS et al. 1992). Though $G\beta\gamma$ complexes are released from any activated receptor, e.g., from $G_q$-coupled receptors and also in a PT-sensitive manner from $G_i$-coupled receptors, it is currently assumed that physiologically significant $G\beta\gamma$ concentrations activating PLC-$\beta$ were only contributed only by $G_i$ proteins following stimulation of respective receptors. Of the platelet activating G-protein-coupled receptors, some like the $TXA_2$ receptor stimulate PLC by a PT-insensitive mechanism. In contrast, thrombin as well as other agonists also elicit PI response through receptors coupling to $G_i$ proteins (see Table 3). In fact, the thrombin receptor couples to both $G_q$ and $G_i$ proteins, thereby activating two pathways leading to PLC stimulation of which the PT-sensitive one covers more than 90% of the response (Brass et al. 1990). Following receptor activation of $G_i$ proteins, $G\beta\gamma$ complexes are released, mediating the stimulatory signal to PLC whereas the $G\alpha_i$ monomers inhibit adenylyl cyclase activity, hence suppressing platelet inhibition (see below). Specific receptor-G-protein coupling was observed. The $\alpha_{2A}$ adrenergic- and the ADP receptor were shown to couple to $G_{i2}$ proteins but not $G_{i3}$ in human platelets, thereby eliciting a PT-sensitive PI response and inhibiting cAMP formation (Simonds et al. 1989; Ohlmann et al. 1995).

Currently, a hot spot of research appears to be elucidating the functional linkage between growth factor receptor (GFR)- and G-protein-coupled receptor (GPCR) pathways. GFR and GPCR receptor pathways are assumed to be regulated by intracellular signal cross talk and to use identical downstream effector routes, e.g., the MAP kinase pathways (Herskowitz 1995; Karin 1995). In this context, $G_i$ proteins are thought to play a crucial role in mediating cross talk signals through both their $\alpha$ and $\beta\gamma$ subunits (Daub et al. 1996; Moxham and Malbon 1996; Van Biesen et al. 1996). Very recently, evidence was presented suggesting another $G_i$-modulated enzyme present in platelets, i.e., phosphoinositide 3-kinase (PI3K; Thomason et al. 1994). Indeed, a novel PI3K$\gamma$ cloned from human blood cells was shown to be activated by both $G\alpha_i$ and $G\beta\gamma$ subunits whereas previous PI3K isoforms are insensitive to G proteins but are regulated by different mechanisms including tyrosine kinases (Stoyanov et al. 1995). Thus, a remarkable similarity of the PI3K family to properties of the PLC family appears to be immanent. Stimulation of PI3K activity by $G_i$-coupling receptors is of particular interest since PI3K activity is assumed to modulate various cellular responses including mitogenesis, inhibition of apoptosis, intracellular protein trafficking, and cytoskeleton rearrangement (for a recent review see Divecha and Irvine 1995; Carpenter and Cantley 1996).

Though very much effort has been made to elucidate the physiological roles of the members of the $G_{12}$ subfamily, not much is known so far. They have been proposed as protooncogenes exhibiting the most remarkable transforming potential of all heterotrimeric G proteins (Chan et al. 1993; Jiang et al. 1993; Xu et al. 1993, 1994). $G_{12}$ and $G_{13}$ are expressed ubiquitously at very low abundance and are also present in platelets (Spicher et al. 1994). Their $G\alpha$ subunits share less than 50% amino acid identity with other G protein subfamilies. Both G proteins appear to be activated relatively slowly by $TXA_2$ und thrombin receptors in human platelets (Offermanns et al. 1994). In addition, there are reports that $G_{13}$ mediates the bradykinin-induced slow inhibition of calcium currents in a neuroblastoma/glioma cell line and that $G_{12}$ and $G_{13}$ stimulate an $Na^+/H^+$ exchanger via different pathways (VOYNO-YASENETSKAYA et al. 1994; DHANASEKARAN et al. 1994; WILK-BLASZCZAK et al. 1994). There is some evidence that $G_{12}$ proteins may modulate the minor GTPases of the Rho family. Expression of GTPase-deficient mutants of $G_{12}$ and $G_{13}$ induce stress fiber formation in fibroblasts and activate c-jun kinases (BUHL et al. 1995; VARA PRASAD et al. 1995). One may therefore speculate that shape changes occurring during platelet aggregation are induced by $G_{12/13}$-effects on the cytoskeleton following activation of $TXA_2$- and thrombin receptors.

### 2. Platelet-Inhibiting G Proteins

Agonists that stimulate adenylyl cyclases and increase cAMP levels, e.g., $PGI_2$ or adenosine, inhibit platelet function (see Table 3). These effects are mediated by $G_s$. The stimulatory G protein ($G_s$) directly activates all known adenylyl cyclases (TAUSSIG and GILMAN 1995; IYENGAR 1993). In resting cells the cAMP concentration is approximately 1 pmol/$10^8$ cells. $PGI_2$ is the most potent physiological stimulator of adenylyl cyclases in platelets which, like forskolin, increases cAMP concentration tenfold in the presence of inhibitors of cAMP phosphodiesterase (BRASS et al. 1990). A number of $G_s$-coupled receptors in other cells, e.g., the $D_1$ dopamine receptor, secretin receptors, glycoprotein hormone receptors, and $H_2$ histamine receptor, are shown to additionally couple to G proteins activating PLC, i.e., $G_q$ and $G_i$ proteins (the latter one via $G\beta\gamma$ complexes; GUDERMANN et al. 1995). Interestingly, phospholipase C activation of these $G_s$-coupled receptors requires higher receptor density and higher agonist concentrations than the stimulation of adenylyl cyclase by the same agonist, and the effect is rather weak even under optimal conditions. Phospholipase-C-stimulating ability of primarily $G_s$-coupled receptors has been implicated as the mechanism of dual signaling phenomena observed in response to some agonists in various tissues, mainly because no corresponding $G_q$-coupled receptors have been identified for them. Physiological situations, however, meeting the requirements of receptor densities and agonist concentrations high enough to activate PLC are imaginable for only very few of the ligands mentioned above.

## C. Adenylyl Cyclases

Inhibition of platelets is acomplished by raising the intracellular cAMP concentration through activation of $PGI_2$, adenosine, and, at the weakest, $\beta$-adrenergic receptors. Adenylyl cyclases catalyze the formation of cAMP from ATP. They are integral membrane glycoproteins, with native proteins exhibiting molecular masses between 110 and 180 kDa (TANG and GILMAN 1992; IYENGAR 1993; TAUSSIG and GILMAN 1995). Eight different isoforms have been identified to date whose primary structures predict proteins with two intensely hydrophobic domains, each hypothesized to contain six transmembrane helices and two 40-kDa cytosolic domains. The cytosolic domains contain sequences that are similar to each other, to the corresponding regions of related adenylyl cyclases, and to the catalytic domains of membrane-bound and soluble guanylyl cyclases. In analogy to the guanylyl cyclases, both cytosolic domains of the adenylyl cyclases are required for catalytic activity (see also Sect. D).

All eight known isoforms are activated by both forskolin, a naturally occurring diterpene (LAURENZA et al. 1989), and the GTP-bound $\alpha$ subunit of the stimulatory G protein $G_s$ (Table 4). All are inhibited by certain adenosine analogues termed P-site inhibitors, e.g. 2′-deoxy-3′-AMP. In addition, all of the isoforms of adenylyl cyclase are further regulated in type-specific patterns by other signals, i.e., other G protein subunits and calcium. For example, adenylyl cyclase type I is inhibited by $G\beta\gamma$ whereas types II and IV are activated. However, stimulatory effects of $G\beta\gamma$ alone are barely detectable, but the magnitude of stimulation is substantial in the presence of $G\alpha_s$. Since stimulation of type II enzyme requires higher concentrations of $G\beta\gamma$ than of $G\alpha_s$, it is assumed that the source of $G\beta\gamma$ is $G_i$ or $G_o$. The $\alpha$ subunits of the PT-sensitive $G_{i/o}$ proteins also affect activity of adenylyl cyclases. All three $G\alpha_i$ isoforms strongly inhibit $G\alpha_s$- or forskolin-activated type V and VI isoforms (TAUSSIG et al. 1993). In addition, type I isoform is also sensitive to $G\alpha_i$ and additionally to $G\alpha_o$. However, inhibition is largely confined to activity observed in the presence of calmodulin or forskolin but not after stimulation with $G\alpha_s$. Types I and V are also inhibited by the PT-insensitive member of the $G_i$-subfamily, $G_z$, that is present in relatively reasonable quantities in platelets (KOZASA and GILMAN 1995). Some adenylyl cyclase isoforms are stimulated by $Ca^{2+}$/calmodulin at nanomolar concentrations, i.e., types I, VIII, and to a lesser extent type III. Other adenylyl cyclases, i.e., types II, IV, V, VI, are insensitive to calmodulin. Inhibition of cAMP formation in intact cells has been shown to follow elevation of $Ca^{2+}$ concentrations. In some cases this has been correlated with expression of type V or type VI isoforms.

Platelet adenylyl cyclases are not $Ca^{2+}$/calmodulin-dependent, not affected by $G\beta\gamma$ complexes, but stimulated by $G_s$ and forskolin and strongly inhibited by $G_i$ proteins (BRASS et al. 1990). Hence, it is likely that platelets express adenylyl cyclase V and VI isoforms. Some platelet activators such as thrombin, epinephrine, or ADP induce the release of $G\alpha_i$ subunits which inhibit cAMP formation whereas others, i.e., thromboxane, do not.[1] These data suggest

[1] A recent report demonstrates the selective coupling of a partially purified $TXA_2$ receptor from platelets with purified $G_q$ and $G_{i2}$ proteins (USHIKUBI et al. 1994). This is in contrast to findings employing native platelet membranes (OFFERMANNS et al. 1994), two different $TXA_2$ receptors overexpressed in the baculovirus/Sf9 expression system (LEOPOLDT and NÜRNBERG, unpublished work), and reconstitition of the $TXA_2$ receptor with purified G proteins (HARHAMMER et al. 1996).

**Table 4.** Regulation of mammalian adenylyl cyclases

| Type[a] | Effect of G proteins | | | | $Ca^{2+}$/calmodulin | Forskolin |
|---|---|---|---|---|---|---|
| | $G\alpha_s$ | $G\alpha_i$ | $G\alpha_z$ | $G\beta\gamma$ | | |
| I | ↑ | ↓ | ↓ | ↓ | ↑ | ↑ |
| II | ↑ | – | – | ↑ | – | ↑ |
| III | ↑ | – | – | – | ↑ | ↑ |
| IV | ↑ | – | – | ↑ | – | ↑ |
| V | ↑ | ↓ | ↓ | – | – | ↑ |
| VI | ↑ | ↓ | – | – | – | ↑ |
| VIII | ↑ | – | – | ↓ | ↑ | ↑ |

↑, stimulation; ↓, inhibition; –, no effect
[a] Regulation of adenylyl cyclase VII is probably similar to that of II and IV.

that a decrease in platelet cAMP levels is not a prerequisite for platelet activation.

Most of the cellular effects of cAMP are mediated by protein kinases. cAMP-dependent protein kinases exist as inactive tetramers consisting of two catalytic subunits and two regulatory subunits, which bind cAMP and then release a free, catalytically active subunit. Several different forms of regulatory and catalytic subunits have been identified. While the importance of the ubiquitous cAMP-dependent protein kinases for the regulation of diverse cellular functions (e.g., metabolism, motility, ion channels, membrane transport, gene expression) is widely accepted, the functional consequences of the heterogeneity within the cAMP kinases are not clear. The biochemical and molecular properties of the cAMP-dependent kinases have been reviewed by Taylor et al. (1990), Meinicke et al. (1990), and McKnight (1991).

# D. Guanylyl Cyclases

## I. Membrane-Bound Guanylyl Cyclases

In contrast to the adenylyl cyclases, which are restricted to the plasma membrane, guanylyl cyclases exist in membrane-bound and soluble forms. To date seven different membrane-bound guanylyl cyclases have been identified. All isoforms belong to the group of receptor-linked enzymes as they contain an N-terminal extracellular receptor-binding domain, one membrane-spanning region and a C-terminal intracellular catalytic domain (Garbers and Lowe 1994; Garbers et al. 1994). All membrane-bound guanylyl cyclases serve the same function, i.e., formation of cGMP, but they are activated by different peptide hormones. GC-A is stimulated by the natriuretic peptides ANP and BNP (Schoenfeld et al. 1995), which are primarily cardiac hormones regulating salt and water balance as well as blood pressure in mammals. Cyclic GMP formed by GC-A is important for the maintenance of smooth muscle tone; mice whose

GC-A gene has been disrupted exhibit chronically elevated blood pressure (LOPEZ et al. 1995). GC-B displays the highest affinity for the C-type natriuretic peptide whereas GC-C is the target of the peptide hormone guanylin (CURRIE et al. 1992; SCHULZ et al. 1992). GC-C and guanylin occur mainly in the intestine, and since GC-C is also activated by the heat-stable enterotoxins of *Escherichia coli* known to cause diarrhea, GC-C and Guanylin are probably physiologically involved in the regulation of water and salt balance in the intestine. The other isoforms of membrane-bound guanylyl cyclase are restricted to sensory tissues as GC-D is expressed only in olfactory neurons (FÜLLE et al. 1995) and GC-E, GC-F, and ret-GC are found only in the retina (YANG et al. 1995). It is unknown whether physiological ligands exist for these isoforms.

In accordance with activation by various peptide hormones, the primary structures of the ligand-binding domains of the membrane-bound guanylyl cyclases reveal only minor (GC-A to GC-B) or no homologies (GC-A to GC-C or GC-ret or GC-D), but all isoforms share a highly conserved intracellular domain. The C-terminal part of about 300 amino acids within this intracellular domain is also conserved in the subunits of the soluble enzyme and found in the cytosolic domains of the adenylyl cyclases. This conserved catalytic domain is coupled to different ligand-binding domains in the membrane-bound guanylyl cyclases and is part of the subunits of the soluble form of the enzyme. In this respect, the guanylyl cyclases appear to be similar to the families of protein tyrosine kinases and protein tyrosine phosphatases whose members also contain conserved intracellular catalytic regions linked to different extracellular ligand-binding domains and exist in additional soluble forms (KOESLING et al. 1991). In addition to the similarity in the overall structure, the N-terminal regions of the intracellular polypeptide chain of the membrane-bound guanylyl cyclases reveal homologies to the catalytic domain of the protein tyrosine kinases. These domains are probably involved in ATP binding, which is required for the peptide-induced stimulation of the membrane-bound guanylyl cyclases (CHINKERS et al. 1991). Unlike other single transmembrane receptors, membrane-bound guanylyl cyclases appear to oligomerize in a ligand-independent fashion and to exist as dimers or higher ordered structures (CHINKERS and WILSON 1992; LOWE 1992).

## II. Soluble Guanylyl Cyclases

In contrast to the homomeric form of the membrane-bound enzyme, soluble guanylyl cyclase consists of two different subunits, $\alpha$ and $\beta$. Four different subunits ($\alpha_1$, $\alpha_2$, $\beta_1$, and $\beta_2$) have been identified to date (KOESLING et al. 1993). The $\alpha_1\beta_1$ heterodimer corresponds to the enzyme purified from lung (HUMBERT et al. 1990), and the $\alpha_2$ and $\beta_2$ subunits have been identified by homology screening (YUEN et al. 1990; HARTENECK et al. 1991). The so-called $\alpha_3$ and $\beta_3$ subunits (GIUILI et al. 1992) represent the human variants of the $\alpha_1$ and $\beta_1$ subunits rather than new isoforms of the subunits, and shifts in the reading frame probably account for the differences between the $\alpha_3$ and $\alpha_1$ subunits. All guanylyl cyclase subunits contain the C-terminal catalytic domain conserved in all cyclases. While the C-terminal catalytic domains have been shown to be sufficient for cGMP formation, the regulatory properties of the enzyme have been attributed to the N-terminal region of the subunits (WEDEL et al. 1995). Expression experiments show that and $\alpha$ and a $\beta$ subunit are required for the formation of a catalytically active enzyme (HARTENECK et al. 1990; BUECHLER et al. 1991) as only the coexpression of $\alpha_1$ and $\beta_1$ or $\alpha_2$ and $\beta_1$ yields enzymes capable of forming cGMP. The $\alpha_2\beta_1$ heterodimer exhibits three- to ninefold lower basal and stimulated activities in all expression systems tested (BEHRENDS et al. 1995). Although the $\alpha_2$ subunit is able to form a catalytically active heterodimer with the $\beta_1$ subunit, it is unknown whether the $\alpha_2$ subunit physiologically associates with the $\beta_1$ subunit or with an unknown partner. PCR studies performed in our laboratory indicate an almost ubiquitous distribution of $\alpha_1$, $\beta_1$, and $\alpha_2$, suggesting the existence of $\alpha_1\beta_1$ and $\alpha_2\beta_1$ heterodimers. Relatively high guanylyl cyclase activity has been found in platelets, lung, and brain. Recently, a splice variant of the $\alpha_2$ subunit was identified containing an additional exon within the catalytic domain (BEHRENDS et al. 1995). The additional exon encodes 31 amino acids and shows homologies to the respective region in the catalytic domain of the adenylyl cyclase. The splice variant occurs in some but not all tissues expressing the $\alpha_2$ subunit. Although catalytically inactive in coexpression experiments, the splice variant is able to dimerize with the $\beta_1$ subunits and to compete with the $\alpha_2$ subunit for dimerization with the $\beta_1$ subunit. Possibly, the splice variant of the $\alpha_2$ subunits represents a dominant negative subunit that can down-regulate guanylyl cyclase activity in certain situations by inhibiting the formation of catalytically active heterodimers. Like the $\alpha_2$ subunit, the $\beta_2$ subunit has not been detected on the protein level, and the recombinant subunit does not form an active enzyme with any of the other subunits (unpublished results). It is possible that the $\beta_2$ subunit needs another yet unidentified subunit to yield a cGMP-forming heterodimer.

### 1. Regulation of Soluble Guanylyl Cyclases

Soluble guanylyl cyclase contains heme as a prosthetic group, and the heme moiety mediates the stimulation by NO which is as much as 250-fold for the purified enzyme (HUMBERT et al. 1990). The heme moiety has been shown to exist in a pentacoordinate $Fe^{2+}$ state with a histidine as the axial ligand (STONE and MARLETTA 1994). NO binds to the sixth coordination position of the heme iron, thereby leading to a breakage of the histidine to iron bond with subsequent conformational changes (STONE et al. 1995). Histidine 105 of the $\beta_1$ subunit is a likely candidate for the axial heme-coordinating residue as substitution with phenylalanine yields an NO-insensitive heme-deficient enzyme (WEDEL et al. 1994). Although stimulation of the enzyme by NO had been identified in the late 1970's, the physiological significance of NO-induced activation of soluble guanylyl cyclase did not become clear until the endothelium-derived relaxing factor (EDRF) was identified as NO (Palmer et al. 1987). EDRF is synthesized in endothelial cells in response to vasodilatory agonists such as acetylcholine, histamine, and bradykinin and causes vasodilation via stimulation of soluble guanylyl cyclase in smooth muscle cells (Ignarro et al. 1987). EDRF can also stimulate soluble guanylyl cyclase in platelets which leads to inhibition of platelet aggregation. The formation of NO is not restricted to endothelial cells and has instead been demonstrated in a variety of other tissues, and three isoforms of NO-forming enzymes (NO synthases) have already been identified (for a review see Nathan and Xie 1994). Apart from its role in the mediation of endothelial relaxation, NO is considered to act as a central and peripheral neuronal messenger and has an important function in the immune response of macrophages (for a review see Schmidt and Walter 1994). In the latter case, large amounts of NO formed by the inducible NO synthase of macrophages presumably have a direct cytotoxic effect and do not exert their action through stimulation of soluble guanylyl cyclase. However, soluble guanylyl cyclase appears to be the most important effector molecule for NO acting as an intra- and intercellular signal molecule at low (*nM*) concentrations.

The release of NO is not restricted to physiological sources but can also originate from exogenous drugs such as nitrovasodilators (Murad et al. 1978; Böhme et al. 1981) such as nitroglycerin, isosorbide dinitrate, and isosorbide mononitrate which are commonly used for the treatment of angina pectoris. These substances spontaneously release very little NO and have to undergo biotransformation at their site of action. Two biotransformation systems, glutathione-S-transferase and cytochrome P450, have been suggested to be involved in the metabolic activation of organic nitrates (for a review see Bennett et al. 1994). Furthermore, tolerance of the vasodilator effects of organic nitrates is associated with impairment of this metabolic activation process rather then being due to desensitization of soluble guanylyl cyclase. In contrast to these substances, sodium nitroprusside and sydnomimines liberate NO spontaneously.

Recently, another normally noxious gas, CO, which has been known to stimulate soluble guanylyl cyclase severalfold (Brüne and Ulrich 1988; Brüne et al. 1990), has been suggested to act as a physiological activator, mainly because heme oxygenase-2, a CO-forming enzyme, colocalizes with soluble guanylyl cyclase throughout the brain (Verma et al. 1993). Heme oxygenase releases CO in the process of degradation of heme to biliverdin, yet this appears to be a rather costly way of producing a signal molecule. Of course, the possibility of other CO-generating systems and the conceivable existence of other, more CO-sensitive isoforms of soluble guanylyl cyclase have to be taken into account.

There has been a lot of speculation about regulatory sulfhydryl groups acting as redox sensors of the enzyme (Waldman and Murad 1987). The effects of thiol modifiying substances are minute and have only been investigated using crude enzyme preparations (Goldberg et al. 1978; Haddox et al.

1978; BRANDWEIN et al. 1981; TSAI et al. 1981; WU et al. 1992). Unless thiol modifying effects are confirmed using better characterized enzyme preparations, the thiol switch of the enzyme will remain obscure (FRIEBE et al., submitted).

Recently, the quinoxalin derivative 1H-[1,2,4]oxadiazolo[4,3-a]-quinoxalin-1-one (ODQ) has been described as a potent and selective inhibitor of soluble guanylyl cyclase, providing an important tool to discriminate between cGMP-dependent and -independent NO signaling (GARTHWAITE et al. 1995). Investigations with purified guanylyl cyclase revealed that the drug binds to the enzyme in an NO-competitive manner thereby oxidizing the prosthetic heme group to the ferric form (SCHRAMMEL et al., in press).

## 2. Biosynthesis of Nitric Oxide

Nitric oxide is synthesized enzymatically from the amino acid L-arginine by different isoforms of NO synthases (NOS; NATHAN and XIE 1994). The isoforms have been purified, characterized, cloned, and sequenced. Two of these NOS isoenzymes are constitutively expressed in neurons (nNOS) and vascular endothelial cells (eNOS), while the expression of a third isoform (iNOS) is induced by various cytokines in macrophages and a number of other nucleated mammalian cells, including hepatocytes, vascular smooth muscle, and glial cells. Constitutive NOSs are $Ca^{2+}$/calmodulin-dependent enzymes and are physiologically activated by hormones or neurotransmitters that enhance the intracellular $Ca^{2+}$ concentration. The $Ca^{2+}$-independent iNOS has been shown to contain calmodulin as a tightly bound subunit even under nominally $Ca^{2+}$-free conditions, suggesting that all NOS isoforms are dependent on calmodulin.

The different NOS isoforms catalyze the same reaction and exhibit similar biochemical properties (MARLETTA 1993; MAYER and WERNER 1995). They all contain heme, FMN, FAD, and $H_4$ biopterin as cofactors and catalyze a two-step redox process under the consumption of NADPH and molecular oxygen. L-arginine is first hydroxylated to the intermediate $N^G$-hydroxy-L-arginine and then immediately undergoes oxidative cleavage to yield NO and L-citrulline. The oxygen atoms incorporated during each of the two reaction steps are derived from molecular oxygen. The heme group is involved in the activation of molecular oxygen, and the required reducing equivalents are derived from NADPH and shuttled in a $Ca^{++}$-dependent manner via the flavins to the heme. Thus, NOSs resemble the two-component cytochrome P-450 hydroxylating systems, in which an FAD- and FMN-containing cytochrome P-450 reductase shuttles NADPH-derived electrons to the prosthetic heme group of an associated cytochrome P-450. Moreover, the C-terminal half of the NOS shows pronounced sequence similarities to the cytochrome P-450 reductase (BREDT et al. 1991), and the bidomain structure of NOS was recently confirmed by separation of the reductase and oxygenase domains subsequent to tryptic cleavage of the rat brain enzyme (SHETA et al. 1994).

## III. cGMP Receptor Proteins

The cellular effects of cGMP are mediated by three different receptor proteins (for a review see SCHMIDT et al. 1993; LINCOLN and CORNWELL 1993), the best characterized cGMP receptor protein being the cGMP-dependent protein kinase. The two other classes of receptor proteins for cGMP are cGMP-regulated ion channels and cGMP-binding cAMP phosphodiesterases. Each receptor protein has been shown to mediate some of the physiological actions of cGMP in specific cell types. It is also possible that a specific cell type may express more than one type of cGMP receptor protein. The capacity of GMP to interact with each of the different receptor proteins raises the possibility that cGMP may increase intracellular protein phosphorylation by activating cGMP kinase, may decrease phosphorylation in the cell by activation of cAMP hydrolysis, or may regulate monovalent cation fluxes. The different effector proteins have to be taken into account when looking at the different effects caused by an increase of cGMP in a given system. In smooth muscle cells and platelets, for example, cGMP is known to lead to a decrease in the intracellular $Ca^{2+}$ concentration, an effect considered to be mediated mainly by the cGMP-regulated kinase, whereas the activation of the rod photoreceptor cGMP channel results in an increase in intracellular calcium. Since the next chapter in this volume deals with cGMP-regulated phosphodiesterases, these enzymes will not be discussed here.

### 1. cGMP-Dependent Protein Kinases

Two major types of cGMP-dependent protein kinases, the soluble type I form and the membrane-bound type II form, have been recognized, both belonging to the group of serine/threonine kinases (Fig. 2; for review see WALTER 1989; HOFMANN et al. 1992). Type I kinase is found in a variety of cell types, with comparatively high concentrations in smooth muscle, Purkinje cells of the cerebellum and platelets, whereas type II kinase shows a more limited distribution and is expressed primarily in the intestinal epithelium and brain. Type I kinase is a homodimer, and two isoforms, called $\alpha$ and $\beta$, have been shown to exist. The $\alpha$ and $\beta$ isoforms are most likely products of alternative splicing as they differ only in their 89 and 104 N-terminal amino acids, respectively. Since the $\alpha$ form exhibits a tenfold higher affinity for cGMP than the $\beta$ enzyme, the N-terminal region probably accounts for differences in protein conformation affecting cGMP binding. The overall structure of the type I kinase can be subdivided into the amino terminal domain containing a leucin zipper-like motif mediating dimerisation of the subunits, a regulatory cGMP binding domain containing two binding sites for cGMP which exhibit positive cooperativeness, and catalytic domain which is highly conserved to the catalytic domain of the cAMP-dependent protein kinase. Apart from mediating dimerization, the amino terminal domain also has an autoinhibitory effect on kinase activity in the absence of cGMP, as the removal of 77 N-terminal amino acids yields a constitutively active, cGMP-independent kinase. In addition, the N-

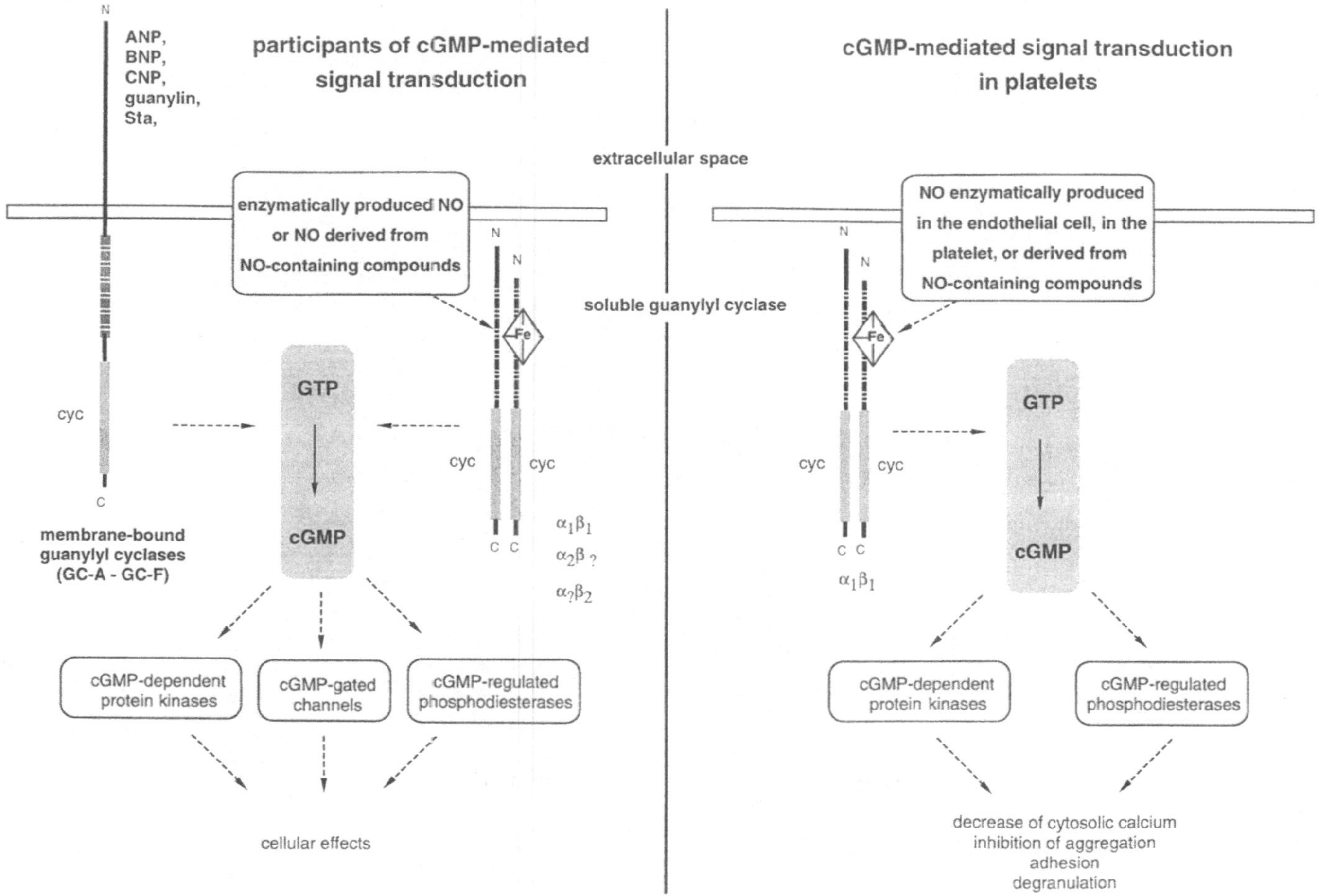


**Fig. 2.** cGMP-signaling. Shown on the left is a schematic representation of the two cGMP forming enzymes, membrane-bound and soluble guanylyl cyclase, their activators, and the three effector proteins mediating cGMP effects; on the right, the members of the cGMP signal transduction chain present in platelets are shown. *ANP*, A-type natriuretic peptide; *BNP*, B-type natriuretic peptide; *CNP*, C-type natriuretic peptide; *Sta*, heat-stable enterotoxin of *E. coli*; *cyc*, catalytic domain

terminal region contains an autophosphorylation site. The presence of cAMP greatly enhances autophosphorylation, which subsequently reduces the concentration of cAMP required for activation about tenfold without major effects on cGMP-induced activation. It is presently unknown whether activation of the cGMP-regulated protein kinase by cAMP and autophosphorylation are of physiological significance.

In contrast to the dimeric structure of the type I kinase, type II kinase is a monomer that has recently been cloned and sequenced (Jarchau et al. 1994). The derived amino acid structure shows the characteristic domain organization of the type I kinase with a tandemly duplicated cGMP-binding domain followed by a typical protein kinase domain. Although no regions were identified in the amino acid sequence that account for membrane localization, type II kinase expressed in HEK 293 cells was found in the particulate subcellular fraction. Recently, it was shown that type II kinase phosphorylates the cystic fibrosis transmembrane conductor regulator (CFTR), thereby leading to an increase in $Cl^-$ secretion in the intestine (French et al. 1995).

By activation of cGMP kinase type I, cyclic GMP decreases $Ca^{2+}$ concentration in a number of cells including smooth muscle cells and platelets by a variety of mechanisms that are not well understood (reviewed in Hofmann et al. 1992; Schmidt et al. 1993; Lincoln and Cornwell 1993). cGMP-dependent kinase has been reported to inhibit $Ca^{2+}$ entry through plasma membrane channels or to stimulate $Ca^{2+}$ efflux across the membrane by activation of a $Na^+/Ca^{2+}$ exchanger or phosphorylation of a 240-kDa protein regulator of $Ca^{2+}$-ATPase. An increase in $Ca^{2+}$ reuptake into intracellular $Ca^{2+}$ stores by phosphorylation of phospholamban, a regulator of the endoplasmic reticulum $Ca^{2+}$-ATPase, may also contribute to the reduction of cytoplasmic $Ca^{2+}$. Cyclic GMP-dependent protein kinase has also been reported to inhibit agonist-induced $IP_3$ formation and subsequent $Ca^{2+}$ release from the endoplasmic reticulum. Cyclic GMP kinase I$\alpha$ transfected in CHO cells inhibited thrombin-stimulated $IP_3$ synthesis and the subsequent rise in $Ca^{2+}$. Concomitantly, phosphorylation of Pertussis toxin-sensitive $G_i$ proteins occurred (see also sect. B.I.4.) (Pfeifer et al. 1995). In addition, the $IP_3$ receptor has been shown to be phosphorylated by the cGMP-dependent protein kinase, and this phosphorylation has been suggested to be involved in the reduction of $Ca^{2+}$. Other targets of the cGMP kinases have been reviewed (Lincoln and Cornwell 1993); relevant substrates concerning the cGMP-mediated effects in platelets will be discussed below.

### 2. cGMP-Gated Channels

In the vision system of vertebrates, cGMP plays a key role by direct regulation of a $Ca^{2+}$-permeable cation channels (for a review see Yau 1994; Biel et al. 1995), which are kept open in the dark state by high concentrations of cGMP, acting as a channel ligand, and which close during visual excitation as a result of increased cGMP hydrolysis triggered by illumination. By molecular cloning of the rod cGMP-channel it was shown that the topology of this channel is similar to that of several other monovalent ion channels such as $K^+$ and $Na^+$ channels and that it contains six membrane-spanning regions (KAUPP et al. 1989). The carboxy terminal domain contains one cGMP binding site revealing significant homology to the cGMP binding site of the cGMP-dependent protein kinases. Despite the homology, the affinity of the ion channel for cGMP is rather low, and the concentrations of cGMP required for activation are one power higher than for the low-affinity binding site of the cGMP-dependent protein kinases.

Another nucleotide-regulated channel has been found in the olfactory epithelium (LUDWIG et al. 1990). The primary stucture of this channel shares 60% identical amino acids with the rod channel. In contrast to the rod channnel, which exhibits a 30- to 40-fold higher affinity for cGMP than for cAMP, the olfactory channel is only two- to fourfold more sensitive to cGMP than cAMP. The physiological regulation of this olfactory channel appears to be closely linked to the activation of adenylyl cyclase rather than guanylyl cyclase since odorant receptors have been identified which act via G proteins to stimulate adenylyl cyclase or lead to a breakdown of phosphoinositides (BUCK and AXEL 1991). The cGMP-regulated channel has also been identified in the aorta (BIEL et al. 1993), showing that the existence of cGMP-gated channels is not limited to sensory tissue.

This notion is supported by the identification of the cone photoreceptor, cGMP-activated channel which is also expressed in kidney and testis (BIEL et al. 1994). The amino acid sequence of these channels is quite homologous to the rod and olfactory cGMP-regulated proteins.

Recently, modulatory subunits of the cGMP-gated channels have been identified by homology screening (CHEN et al. 1993). Although these proteins show homologies and reveal overall structures similar to the channel proteins, they are not able to form functional channels when expressed alone. However, when coexpressed with the respective cGMP-sensitive channel, they modulate channel activity and produce several characteristics that had been observed with the native channels but had not been found in the recombinant channels. These modulatory subunits may also be involved in the interaction of the cGMP-activated channels with $Ca^{2+}$/calmodulin (CHEN et al. 1994). cGMP-activated channels show a decrease in affinity for cyclic nucleotides after the influx of $Ca^{2+}$ into the cell, and this is the key regulatory component of olfactory and light adaptation. Calmodulin, by binding either to the modulatory subunit or to the channel protein, has been shown to mediate the decrease in the affinity for cyclic nucleotides.

# E. Physiological Role of Cyclic Nucleotides in Platelets

The NO-cGMP and the prostaglandin-cAMP signaling systems play important roles in platelet function as cGMP and cAMP have been shown to inhibit agonist-induced aggregation, adhesion, and degranulation of these cells (Waldmann and Walter 1989). Activation of platelets by various agonists involves the mobilization of cytosolic $Ca^{2+}$ from intracellular and extracellular sources, and cGMP and cAMP inhibit the agonist-induced elevation of $Ca^{2+}$ and subsequent reorganization of the cytoskeleton. There is considerable evidence that the inhibitory effects of cAMP or cGMP are mediated in human platelets by cAMP- and cGMP-dependent protein kinases, respectively, and their physiological substrates. The role of cyclic nucleotide-dependent protein kinases in platelets has been extensively reviewed in Halbbrügge and Walter (1993) and Butt and Walter (1995). In addition to the activation of the cGMP-dependent protein kinase, cGMP also affects a cGMP-dependent cAMP phosphodiesterase in platelets (see subsequent chapter of this volume). The cGMP-mediated inhibition of the cGMP-inhibited phosphodiesterase type III and the resulting increase in cAMP concentration probably represent the molecular basis for the observed synergism between the cAMP and cGMP pathways in human platelets (Horstrup et al. 1994; Nolte et al. 1994). The enzyme responsible for cGMP formation in platelets is the soluble guanylyl cyclase as the membrane-bound enzyme has not yet been detected. Information about the isoforms of the adenylyl cyclases present in platelets is provided in Sect. D.

## I. cGMP-Formation in Platelets

As outlined above, soluble guanylyl cyclase is stimulated by endogenous NO or by NO derived from NO-containing drugs. Although the matter has not been extensively investigated, there have been reports that platelets contain the constitutively expressed endothelial NOS as well as the inducible forms (Radomski et al. 1990; Muruganandam and Mutus 1994; Mehta et al. 1995). As the activation of platelets involves the elevation of the intracellular $Ca^{2+}$ level, the existence of the $Ca^{2+}$-dependent NOS suggests that, parallel to activation of platelets, a cGMP-mediated feedback loop is stimulated. The NO forming activity in the platelets appears to be rather low, but NO produced by NOS in the endothelial cells is sufficient for causing stimulation of soluble guanylyl cyclase in platelets and a subsequent rise of intracellular cGMP concentration, as has been shown by coincubation of endothelial cells and platelets (Nolte et al. 1991). Platelets do not contain the enzyme system catalyzing the NO release from nitrovasodilators, especially from organic nitrates. Again, the concentration of NO being released by the vascular smooth muscle containing the drug-metabolizing enzyme is probably high enough to lead to stimulation of platelet soluble-guanylyl cyclase.

Recently, YC-1 [3-5′-hydroxymethyl-2′-furyl)-1-benzylindazole] has been shown to inhibit platelet aggregation and to lead a marked increase in cGMP levels in human platelets [Wu et al. 1995]. In addition, the substance increased soluble guanylyl cyclase activity in a supernatant fraction from human platelets. The activation was NO-independent as hemoglobin did not prevent the

YC-1 effect. Further experiments with the purified enzyme have to prove whether YC-1 is able to stimulate soluble guanylyl cyclase by an NO-independent mechanism.

The presence of high amounts of soluble guanylyl cyclase in platelets has been known for a long time. Stimulated activity of the enzyme in cytosolic fractions amounts to 5 nmol/mg min. In intact platelets, sodium nitroprusside which liberates NO spontaneously caused about a tenfold elevation of cGMP levels (0.4 $\mu$M to 3.8 $\mu$M; EIGENTHALER et al. 1992). Despite the relatively high amount of enzyme, several attempts to purify soluble guanylyl cyclase from platelets have so far failed. Within the course of the purification procedure, soluble guanylyl cyclase was found to tightly and irreversibly bind to actin, and this presumably unspecific interaction prevented isolation of the enzyme. As peptide antibodies against the C-terminal sequence of soluble guanylyl cyclase were unsuccessful in immunohistochemical studies, investigations of a possible association of soluble guanylyl cyclase with the cytoskeleton have not been possible. With respect to the different isoforms of soluble guanylyl cyclase, the $\alpha_1\beta_1$ heterodimer has been identified in platelets with antibodies in Western blot analysis (GUTHMANN et al. 1992). As at the time antibodies against the $\alpha_2$ and $\beta_2$ subunits were not available, the existence of an $\alpha_2$ or $\beta_2$ containing heterodimer is still unkown. PCR studies in HEL cells showed the existence of the $\alpha_2$ subunit, suggesting the occurrence of this subunit in platelets as well.

## II. cGMP- and cAMP-Dependent Protein Kinases in Platelets

The effects of cGMP in platelets are mediated mainly by cGMP kinase type I since a specific activator of cGMP kinase (-pCPT-cGMP) inhibits the rise in $Ca^{2+}$ caused by activation of normal platelets, but is unable to do so in activated cGMP-dependent protein kinase deficient platelets from patients with chronic myelognous leukemia (EIGENTHALER et al. 1993). The concentration of cGMP kinase present in human platelets is the highest measured in any specific cell type and almost reaches the concentration of structural proteins (EIGENTHALER et al. 1992). In contrast to other tissues, more than 80% of platelet cGMP-dependent protein kinase are associated with the particulate fraction of platelet homogenates and have the characteristics of a peripheral membrane protein.

Similar to the level of cGMP kinase, the concentration of cAMP-dependent protein kinase in platelets is several times higher than that estimated for other cells and tissues. Like most mammalian cells, human platelets also possess at least two types of cAMP-dependent protein kinase. A recent detailed analysis of the isoform expression and subcellular distribution of cAMP-dependent protein kinase is not available.

Comparison of cyclic nucleotide levels and cyclic nucleotide binding sites on the kinases reveals that the cAMP concentration is close to the concentration of cAMP binding sites (EIGENTHALER et al. 1992). Therefore, a small increase in the cAMP level is sufficient for activating the majority of cAMP protein kinases. In contrast, the cGMP concentration in unstimulated platelets is less than one-tenth the concentration of cGMP binding sites on the cGMP kinases, and indeed the rate of protein phosphorylation mediated by cGMP protein kinase in intact platelets appears to be proportional to the cGMP level.

## III. Substrates for cAMP- and cGMP-Dependent Protein Kinases in Platelets

cAMP- and cGMP-dependent protein kinases have similar but not identical substrate specificities; the cAMP-dependent protein kinase is usually more effective than the cGMP-dependent protein kinase in in vitro experiments. To date, only a few substrates for the cAMP- and cGMP-dependent kinases have been identified and characterized. Myosin light chain kinase has been shown to be phosphorylated by the cAMP-dependent kinase in intact cells (HATHAWAY et al. 1981). This is associated with a decreased affinity of this enzyme for calmodulin and therefore inhibition of myosin light chain activity. This effect certainly contributes to the cAMP-mediated inhibition of myosin light chain kinase. Although the enzyme is also phosphorylated by the cGMP-dependent protein kinase in vitro, this phosphorylation does not alter enzyme activity and has not been observed in intact cells.

Phosphorylation of a cGMP-inhibited cAMP phosphodiesterase by the cAMP-dependent protein kinase increases phosphodiesterase activity, which probably represents a negative feedback mechanism of cellular cAMP levels (MACPHEE et al. 1988).

A 22-kDa protein phosphorylated by the cAMP-dependent protein kinase had originally been proposed to be involved in the regulation of the $Ca^{2+}$ transport into intracellular membrane vesicles or at plasma membranes. However, this substrate was later identified as a small GTP-binding protein, Rap-1b. Phosphorylation of Rap-1b is associated with translocation from the membrane to the cytosol, and, interestingly, activation of human platelets results in a quantitative association of Rap-1b with the cytoskeleton (UEDA et al. 1992).

In addition to Rap-1b, other platelet cAMP-dependent protein kinase substrates have been shown to be cytoskeletal proteins. A 24-kDa protein was identified as the $\beta$ subunit of glycoprotein Ib. Phosphorylation of glycoprotein Ib$\beta$ is thought to be important for the inhibition of collagen-induced actin polymerization during platelet activation (FOX and BERNDT 1989). Phosphorylation of caldesmon, a 240-kDa actin-binding protein, may be involved in the inhibiton of cytoskeleton reorganization during platelet activation (HETTASCH and SELLERS 1991).

Another major substrate for both cAMP- and cGMP-dependent protein kinases has been identified. This protein of about 46 kDa, which has been designated vasodilator-stimulated phosphoprotein (VASP), contains three distinct phosphorylation sites and is stoichiometrically phosphorylated in vitro and in platelets. VASP phosphorylation in intact platelets correlates well with inhibition of phospholipase C activation, $Ca^{2+}$ mobilization, and platelet activation (HALBBRÜGGE et al. 1990). In addition to platelets, VASP has been observed in a wide variety of other cells and tissues and has been shown to be an actin-binding protein associated with stress fibers and focal contact sites or adhesion plaques (REINHARD et al. 1992). VASP has been purified, cloned, and sequenced (HAFFNER et al. 1995). Within the amino acid sequence, a central prolin-rich domain contains a GPPPPP motif as a single copy and as a threefold tandem repeat as well as the three phosphorylation sites for the cyclic nucleotide-dependent protein kinases.

## IV. Cellular Responses Leading to Platelet Inhibition

Despite the identification of several substrates of the cAMP and cGMP protein kinases, the molecular mechanism of the inhibitory effect of both protein kinases on platelet activation has not been completely elucidated. Certainly, cAMP and cGMP cause inhibition at several sites in the activation cascade involving phopholipase C, $Ca^{2+}$ mobilization, myosin light chain kinase, and cytoskeletal reorganization (HALBBRÜGGE and WALTER 1993; BUTT and WALTER 1995).

Cyclic nucleotide-elevating vasodilators inhibit the phosphatidylinositol cycle at an early step of the activation cascade, probably at the level of phospholipase C activation. However, direct regulation of phospholipase C by protein phosphorylation has not been demonstrated in platelets. Recently, cAMP- and cGMP-dependent protein kinases were shown to inhibit agonist-induced $Ca^{2+}$-mobilisation from intracellular stores as well as secondary, store-related $Ca^{2+}$ influxes whereas the ADP receptor-operated cation channel remained uneffected (GEIGER et al. 1992). The $IP_3$ receptor has been shown to be phosphorylated by the cAMP- and cGMP-dependent protein kinases (SUPPATTAPONE et al. 1988; JOSEPH and RYAN 1993; KOMALAVILAS and LINCOLN 1994), but this has not been demonstrated in platelets, and the functional consequences of the phosphorylation are controversal. On the other hand, recently the inhibition and phosporylation of the $IP_3$ receptor by cyclic nucleotide-elevating vasodilators in platelets has been reported (CAVALLINI et al. 1996). In addition to the inhibition of $Ca^{2+}$ mobilization and $Ca^{2+}$ influx, activation of $Ca^{2+}$ removal from the cytoplasm by intracellular or plasma membrane-bound $Ca^{2+}$-pumps ($Ca^{2+}$-ATPases) could be a possible mechanism of cyclic nucleotide action. At this time, the existence of both types of $Ca^{2+}$-ATPases in platelets and the regulation by the cAMP- and cGMP-dependent protein kinase system remains unclear (BUTT and WALTER 1995).

The activation of platelets is also associated with reorganization of the cytoskeleton. In human platelets, PG-$E_1$ has been shown to completely inhibit collagen-induced actin polymerization. In platelets from patients with

Bernard-Soulier syndrome, which is due to a lack of the glycoprotein Ib$\beta$ complex, PG-$E_1$ did not prevent actin polymerization (FOX and BERNDT 1989). It is therefore conceivable that cAMP protein kinase mediated phosphorylation of the $\beta$ subunit of glycoprotein Ib$\beta$ complex causes the inhibiton of collagen-induced acting polymerization.

Phosphorylation of VASP by cAMP and cGMP kinases correlates well with the inhibition of platelet activation at an early step in the activation cascade, presumably at the level of phospholipase C activation (HALBBRÜGGE et al. 1990; GEIGER et al. 1992). Other findings demonstrated that VASP phosphorylation in human cells other than platelets does not correlate with inhibition of agonist-evoked calcium mobilization from intracellular stores, indicating that phosphorylation of VASP does not lead to or parallel inhibition of calcium transients in all cell types (MEINICKE et al. 1994). Recently, VASP was shown to act as a ligand of profilins (REINHARD et al. 1995). Profilins are small monomeric actin-, phosphoinositide-, and poly-L-prolin-binding proteins that have been implicated in actin polymerization and the cross talk between the growth factor/phosphoinositide pathways and the actin cytoskeleton (SOHN and GOLDSCHMIDT-CLERMONT 1994). Possibly, VASP phosphorylation alters the interaction of profilins with other ligands, which in turn affects the profilin-actin interaction as well as the phospholipase C pathway. In this context, it is interesting that cAMP-and cGMP- stimulated phosphorylation of VASP correlates well with the inhibitions of fibrinogen receptor activation (HORSTRUP et al. 1994). The fibrinogen receptor (also known as glycoprotein IIb/IIa or integrin $\alpha_{IIb}\beta_3$) is activated in human platelets by agonists such as thrombin, ADP, and thomboxane $A_2$, with activation perhaps mediated by protein kinase C or G proteins. Binding of soluble fibrinogen to the activated fibrinogen receptor leads to stimulation of a focal adhesion kinase, c-src, and other nonreceptor tyrosine kinases that phosphorylate a number of cytoskeletal proteins prior to adhesion. Of course, additional studies are necessary to establish that a VASP-profilin interaction may account for the close relation observed between VASP phosphorylation and concomitant phospholipase C and fibrinogen receptor inhibiton. However, VASP represents a possible link between signal transduction of cyclic nucleotides and actin filament formation.

*Acknowledgements.* We like to thank Günter Schultz for his support and our colleagues for helpful discussion. The authors' own research reported herein was supported by Deutsche Forschungsgemeinschaft, Sonderforschungsbereich 366, and Fonds der Chemischen Industrie. Many important references could not be included because of space limitations but are indicated in the reviews cited.

## References

Ahnert-Hilger G, Nürnberg B, Exner T et al. The heterotrimeric G-protein $G\alpha_{o2}$ inhibits catecholamine uptake into secretory granules (submitted)

Arkinstall S, Chabert C, Maundrell K, Peitsch M (1995) Mapping regions of $G\alpha_q$ interacting with $PLC\beta_1$ using multiple overlapping synthetic peptides. FEBS Lett 364:45–50
Behrends S, Harteneck C, Schultz G, Koesling D (1995) A variant of the $\alpha_2$ subunit of soluble guanylyl cyclase contains an insert homologous to a region within adenylyl cyclases and functions as a dominant negative protein. J Biol Chem 270:21109–21113
Bennett BM, McDonald BJ, Nigam R, Simon WC (1994) Biotransformation of organic nitrates and vascular smooth muscle cell function. Trends Pharmacol Sci 15:245–249
Biel M, Altenhofer W, Hullin R, Ludwig J, Freichel M, Flockerzi V, Dascal N, Kaupp UB (1993) Primary structure and functional expression of a cyclic nucleotide-gated channel from rabbit aorta. FEBS Lett 329:134–138
Biel M, Zong X, Distler M, Bosse E, Klugbauer N, Murakami M, Flockerzi V, Hofmann F (1994) A new member of the cyclic nucleotide-gated channel family expressed in testis, kidney, and heart. Proc Natl Acad Sci USA 91:3505–3509
Biel M, Zong X, Hofmann F (1995) Molecular diversity of cyclic nucleotide-gated cation channels. Naunyn Schmiedebergs Arch Pharmacol 353:1–10
Birnbaumer L (1992) Receptor-to-effector signaling through G proteins: roles for $\beta\gamma$ dimers as well as $\alpha$ subunits. Cell 71:1069–1072
Böhme E, Grossmann G, Herz J, Mülsch A, Spies C, Schultz G (1981) Regulation of cGMP formation by soluble guanylyl cyclase stimulation by NO-containing compounds. Adv Cyclic Nucleot Prot Phosphorylat Res 17:259–266
Bourne HR (1995) Trimeric G proteins: surprise witness tells a tale. Science 270:933–934
Bourne HR, Nicoll R (1993) Molecular machines integrate coincident synaptic signals. Neuron [Suppl] 10:65–75
Bourne HR, Sanders DA, McCormick F (1990) The GTPase superfamily: a conserved switch for diverse cell functions. Nature 348:125–132
Bourne HR, Sanders DA, McCormick F (1991) The GTPase superfamily: a conserved switch for diverse cell functions. Nature 349:117–127
Brandwein HJ, Lewicki JA, Murad F (1981) Reversible inactivation of guanylate cyclase by mixed disulfide formation. J Biol Chem 256:2958–2962
Brass LF, Manning DR, Shattil SJ (1990) GTP-binding proteins and platelet activation. Prog Hemost Thromb 127–175
Brass LF, Hoxie JA, Kieber-Emmons T, Manning DR, Poncz M, Woolkalis M (1993) Agonist receptors and G proteins as mediators of platelet activation. In: Authi KS (ed) Mechanisms of platelet activation and control. Plenum, New York, pp 17–36
Bredt DS, Hwang PM, Glatt CE, Lowenstein C, Reed RR, Snyder SH (1991) Cloned and expressed nitric oxide synthase structurally resembles cytochrome P-450 reductase. Nature 351:714–718
Brüne B, Ullrich V (1988) Inhibition of platelet aggregation by carbon monoxide is mediated by activation of guanylate cyclase. Mol Pharmacol 32:497–504
Brüne B, Schmidt KU, Ullrich V (1990) Activation of soluble guanylate cyclase by carbon monoxide and inhibition by superoxide anion. Eur J Biochem 192:683–686
Buck L, Axel R (1991) A novel multigene family may encode odorant receptors: molecular basis for odorant recognition. Cell 65:175–187
Buechler WA, Nakane M, Murad F (1991) Expression of soluble guanylate cyclase activity requires both enzyme subunits. Biochem Biophys Res Commun 174:351–357
Buhl AM, Johnson NL, Dhanesakaran N, Johnson GL (1995) $G\alpha_{12}$ and $G\alpha_{13}$ stimulate Rho-dependant stress fiber formation and focal adeshion assembly. J Biol Chem 270:24631–24634
Butt E, Walter U (1995) Signal transduction by cyclic nucleotide-dependent protein kinases in platelets. In: Lapetina E (ed) The platelet. Advances in molecular and cell biology (in press)

Camps M, Carozzi A, Schnabel P, Scheer A, Parker PJ, Gierschik P (1992) Osienzyme-selective stimulation of phospholipase C-$\beta 2$ by G protein $\beta\gamma$-subunits. Nature 360:684–686
Carpenter CL, Cantley LC (1996) Phosphoinositide kinases. Curr Opin Cell Biol 8:153–158
Casey PJ (1994) Lipid modifications of G proteins. Curr Opin Cell Biol 6:219–225
Casey PJ, Fong HKW, Simon MI, Gilman AG (1990) $G_z$, a guanine nucleotide-binding protein with unique biochemical properties. J Biol Chem 265:2383–2390
Cavallini L, Coassin L, Borean A, Alexandre A (1996) Prostacyclin und sodium nitroprusside inhibit the activity of the platelet inositol 1,4,5-trisphosphate receptor and promote its phosphorylation. J Biol Chem 271:5545–5551
Chan AML, Fleming TP, McGovern ES, Chedid M, Miki T, Aaronson SA (1993) Expression cDNA cloning of a transforming gene encoding the wild-type $G\alpha_{12}$ gene product. Mol Cell Biol 13:762–768
Chen TY, Peng YW, Dhallan RS, Ahamed B, Reed RR, Yau KW (1993) A new subunit of the cyclic nucleotide-gated cation channel in retinal rods. Nature 362:764–767
Chen TY, Illing M, Hsu YT, Yau KW, Molday RS (1994) Subunit 2 (or $\beta$) of retinal rod cGMP-gated cation channel is a component of the 240kDa channel associated protein and mediates $Ca^{2+}$-calmodulin modulation. Proc Natl Acad Sci USA 91:11757–11761
Chinkers M, Wilson EM (1992) Ligand-independent oligomerization of natriuretic peptide receptor/guanylyl cyclase expressed in a baculovirus system. J Biol Chem 267:18589–18597
Chinkers M, Singh S, Garbers DL (1991) Adenine nucleotides are required for activation of rat atrial natriuretic peptide receptor/guanylyl cyclase expressed in a baculovirus system. J Biol Chem 266:4088–4093
Clapham DE, Neer EJ (1993) New roles for G protein $\beta\gamma$-dimers in transmembrane signalling. Nature 365:403–406
Coleman DE, Sprang SR (1996) How G proteins work: a continuing story. Trends Biol Sci 21:41–44
Coleman DE, Berghuis AM, Lee E, Linder ME, Gilman AG, Spring SR (1994) Structures of active conformations of $G_{i\alpha 1}$ and the mechanism of GTP hydrolysis. Science 265:1405–1412
Currie MG, Fok KF, Kato J, Moore RJ, Hamra FK (1992) Guanylin: an endogenous activator of intestinal guanylate cyclase. Proc Natl Acad Sci USA 89:947–951
Daub H, Weiss, FU, Wallasch C, Ullrich A (1996) Role of transactivation of the EGF receptor in signalling by G-protein-coupled receptors. Nature 379:557–560
Degtyarev MY, Spiegel AM, Jones TL (1993) Increased palmitoylation of the $G_s$ protein $\alpha$ subunit after activation by the $\beta$-adrenergic receptor or cholera toxin. J Biol Chem 268:23769–23772
Dennis EA, Rhee SG, Billah MM, Hannun YA (1991) Role of phospholipases in generating lipid second messengers in signal transduction. FASEB J 5:2068–2077
Dhanasekaran N, Vara Prasad MVVS, Wadsworth SJ, Dermott JM, van Rossum G (1994) protein kinase C-dependent and -independent activation of $Na^+/H^+$ exchanger by $G\alpha_{12}$ class of G proteins. J Biol Chem 269:11802–11806
Dickey BF, Birnbaumer L (eds) (1993) GTPases. Springer, Berlin Heidelberg New York (Handbook of experimental pharmacology, vol 108)
Dietrich A, Meister M, Brazil D, Camps M, Gierschik P (1994) Stimulation of phospholipase C-$\beta 2$ by recombinant guanine-nucleotide-binding protein $\beta\gamma$-dimers produced in a baculovirus/insect cell expression system. Requirement of $\gamma$-subunit isoprenylation for stimulation of phospholipase C. Eur J Biochem 219:171–178
Divecha N, Irvine RF (1995) Phospholipid signaling. Cell 80:269–278
Eigenthaler M, Nolte C, Halbbrügge M, Walter U (1992) Concentration and regulation of cyclic nucleotides, cyclic nucleotide-dependent protein kinases and one of the major substrates in human platelets. Eur J Biochem 205:471–481

Eigenthaler M, Ullrich H, Geiger J, Horstrup K, Hönig-Líedl P, Wiebecke K, Walter U (1993) Defective nitrovasodilatorstimulated protein phosphorylation and calcium regulation in cGMP-dependent protein kinase deficient human platelets of chronic myelocytic leukemia. J Biol Chem 268:13526–13531
Fields TA, Casey PJ (1995) Phosphorylation of $G_{z\alpha}$ by protein kinase C blocks interaction with the $\beta\gamma$ complex. J Biol Chem 270:23119–23125
Fields TA, Linder ME, Casey PJ (1994) Subtype-specific binding of azidoanilido-GTP by purified G protein $\alpha$ subunits. Biochemistry 33:6877–6883
Fox JEB, Berndt MC (1989) Cyclic AMP-dependent phosphorylation of glycoprotein 1 b inhibits collagen-induced polymerization of actin in platelets. J Biol Chem 264:8701–8707
French PJ, Bijman J, Edixhoven M, Vaandrager AB, Scholte BJ, Lohmann SM, Nairn AC, de Jonge HR (1995) Isotype specific activation of cyctic fibrosis transmembrane conductance regulator-chlorid channels by cGMP-dependent protein kinase II. J Biol Chem 270:26626–26631
Fukada Y, Matsuda T, Kokame K, Takao T, Shimonishi Y, Akino T, Yoshizawa T (1994) Effects of carboxyl methylation of photoreceptor G protein $\gamma$-subunit in visual transduction. J Biol Chem 269:5163–5170
Friebe A, Wedel B, Harteneck C, Foerster J, Schultz G, Koesling D Functions of conserved cysteines of soluble guanylyl cyclase. Biochemistry (in press)
Fülle HJ, Vassar R, Foster DC, Yang RB, Axel R, Garbers DL (1995) A receptor guanylyl cyclase expressed specifically in olfactory sensory neurons. Proc Natl Acad Sci USA 92:3571–3575
Garbers DL, Lowe DG (1994) Guanylyl cyclase receptors. J Biol Chem 269:30741–30744
Garbers DL, Koesling D, Schultz G (1994) Guanylyl cyclase receptors. Mol Biol Cell 5:1–5
Garthwaite J, Southam E, Boulton CL, Nielsen EB, Schmidt K, Mayer B (1995) Potent and selective inhibition of nitric oxide-sensitive guanylyl cyclase by 1H-[1,2,4]oxadiazolo[4,3-a]quinoxalin-1-one. Mol Pharmacol 48:184–188
Garritsen A, van Galen PJM, Simonds WF (1993) The N-terminal coiled-coil domain of $\beta$ is essential for $\gamma$ association: a model for G protein $\beta\gamma$ subunit interaction. Proc Natl Acad Sci USA 90:7706–7710
Geiger J, Nolte C, Butt E, Sage SO, Walter U (1992) Role of cAMP and cGMP-dependent protein kinase in nitrovasodilator inhibition of agonist-evoked calcium elevation in human platelets. Proc Natl Acad Sci USA 89:1031–1035
Gierschik P, Jakobs KH (1992) ADP-ribosylation of signal-transducing guanine nucleotide binding proteins by cholera and pertussis toxin. In: Herken H, Hucho F (eds) Selective neurotoxicity. Springer, Berlin Heidelberg New York, pp 807–839 (Handbook of experimental pharmacology, vol 102)
Gilman AG (1987) G proteins: transducers of receptor-generated signals. Annu Rev Biochem 56:615–649
Giuili G, Scholl U, Bulle F, Guellaen G (1992) Molecular cloning of the cDNAs coding for the two subunits of soluble guanylyl cyclase from human brain. FEBS Lett 304:83–88
Goldberg ND, Graff G, Haddox MK, Stephenson JH, Glass DB, Moser ME (1978) Redox modulation of splenic cell guanylate cyclase activity: activation by hydrophilic and hydrophobic oxidants represented by ascorbic and dehydroascorbic acids, fatty acids hydoperoxides, and prostaglandin endoperoxides. Adv Cycl Nucl Res 9:101–130
Goody RS (1994) How G proteins turn off. Nature 372:220–221
Gudermann T, Nürnberg B, Schultz G (1995) Receptors and G proteins as primary components of transmembrane signal transduction, part 1: G-protein coupled receptors: structure and function. J Mol Med 73:51–63
Guthmann F, Mayer B, Koesling D, Kukovetz WR, Böhme E (1992) Characterization of soluble platelet guanylyl cyclase with peptide antibodies. Naunyn Schmiedebergs Arch Pharmacol 346:537–541

Haddox MK, Stephenson JH, Moser ME, Goldberg ND (1978) Oxidative-reductive modulatoin of guinea pig splenic cell guanylate cyclase activity. J Biol Chem 253:3143–3152

Haffner C, Jarchau T, Reinhard M, Hoppe J, Lohmann S, Walter U (1995) Molecular cloning, structural analysis and functional expression of the proline-rich focal adhesion and microfilament-associated protein VASP. EMBO J 14:19–27

Halbbrügge M, Walter U (1993) The regulation of platelet functions by protein kinases. In: Huang C-K, Sha'afi RI (eds) Protein kinases in blood cell function. CRC, Boca Raton, pp 245–298

Halbrügge M, Friedrich C, Eigenthaler M, Schanzenbächer P, Walter U (1990) Stoichiometric and reversible phosphorylation of a 46-kDa protein in human platelets in response to cGMP- and cAMP-elevating vasodilators. J Biol Chem 265:3088–3093

Hallak H, Muszbek L, Laposata M, Belmonte E, Brass LF, Manning DR (1994) Covalent binding of arachidonate to G protein $\alpha$ subunits of human platelets. J Biol Chem 269:4713–4716

Harhammer R, Nürnberg B, Harteneck C, Leopoldt D, Exner T, Schultz G (1996) Distinct biochemical properties of the native members of the $G_{12}$-protein subfamily characterization of purified $G\alpha_{12}$ from rat brain. Biochem J 319:165–171

Harteneck C, Koesling D, Söling A, Schultz G, Böhme E (1990) Expression of soluble guanylate cyclase: catalytic activity requires two subunits. FEBS Lett 272:221–223

Harteneck C, Wedel B, Koesling D, Malkewitz J, Böhme E, Schultz G (1991) Molecular cloning and expression of a new $\alpha$-subunit of soluble guanylyl cyclase. FEBS Lett 292:217–222

Hathaway DR, Eaton CR, Adelstein RS (1981) Regulation of human platelet myosin light chain kinase by the catalytic subunit of cyclic AMP-dependent protein kinase. Nature 291:252–254

Hausdorf WP, Pitcher JA, Luttrell DK, Linder ME, Kurose H, Parsons SJ, Caron MG, Lefkowitz RJ (1992) Tyrosine phosphorylation of G protein $\alpha$ subunits by $pp60^{c\text{-}src}$. Proc Natl Acad Sci USA 89:5720–5724

Hepler JR, Gilman AG (1992) G proteins. Trends Biochem Sci 17:383–387

Herskowitz I (1995) MAP kinase pathway in yeast: for mating and more. Cell 80:187–197

Hettasch JM, Sellers J (1991) Caldesmon phosphorylation in intact human platelets by cAMP-dependent protein kinase and protein kinase C. J Biol Chem 266:11876–11881

Higashijima T, Ferguson KM, Sternweis PC, Smigel MD, Gilman AG (1987) Effects of $Mg^{2+}$ and the $\beta\gamma$-subunit complex on the interactions of guanine nucleotides with G proteins. J Biol Chem 262:762–766

Higgins JB, Casey PJ (1994) In vitro processing of recombinant G protein $\gamma$ subunits. J Biol Chem 269:9067–9073

Hille B (1992) G protein-coupled mechanisms and nervous signaling. Neuron 9:187–195

Hofmann F, Dostmann W, Keilbach A, Lanfgraf W, Ruth P (1992) Structure and physiological role of cGMP-dependent kinase. Biochim Biophys Acta 1135:51–60

Horstrup K, Jablonka B, Hönig-Liedl P, Just M, Kocksiek K, Walter U (1992) Phosphorylation of the focal adhesion protein VASP at serin157 in intact human platelets correltaes with fibrinogen receptor inhibition. Eur J Biochem 255:21–27

Horstrup K, Jablonka B, Honig-Liedl P, Just M, Kochsiek K, Walter U (1994) Phosphorylation of focal adhesion vasodilatator-stimulated phosphorprotein at Ser157 in intact human platelets correlates with fibrinogen receptor inhibition. Eur J Biochem 225:21–27

Hourani SMO, Cusack NJ (1991) Pharmacological receptors on blood platelets. Pharmacol Rev 43:243–298

Humbert P, Niroomand F, Fischer G, Mayer B, Koesling D, Hinsch KH, Gausepohl H, Frank R, Schultz G, Böhme E (1990) Purification of soluble guanylate cyclase from bovine lung by a new immunoaffinity chromatographic method. Eur J Biochem 190:273–278

Iniguez-Lluhi J, Kleuss C, Gilman AG (1993) The importance of G protein $\beta\gamma$ subunits. Trends Cell Biol 3:230–236

Ignarro LJ, Buga GM, Wood KS, Byrns RE, Chandhuri G (1987) Endothelium-derived relaxing factor produced and released from artery and vein is nitric oxide. Proc Natl Acad Sci USA 84:9265–9269

Iyengar R (1993) Molecular and functional diversity of mammalian $G_s$-stimulated adenylyl cyclases. FASEB J 7:768–775

Jakobs KH, Aktories K, Schultz G (1984) Mechanism of pertussi toxin action on the adenylate system: inhibition of the turn-on reaction of the inhibitory regulatory site. Eur J Biochem 140:177–181

Jarchau T, Häusler C, Markert T, Pöhler D, Vandekerckhove J, DeJonge HR, Lohmann SM, Walter U (1994) Cloning, expression, and in situ localisation of rat intestinal cGMP-dependent protein kinase II. Proc Natl Acad Sci USA 91:9426–9430

Jiang H, Wu D, Simon MI (1993) The transforming activity of activated $G\alpha_{12}$. FEBS Lett 330:319–322

Joseph SK, Ryan SV (1993) Phosphorylation of the inositol trisphosphate receptor in isolated rat hepatocytes. J Biol Chem 268:23059–23065

Kaldenberg-Stasch S, Baden M, Fessler B, Jakobs KH, Wieland T (1994) Receptor-stimulated guanine nucleoside triphosphate binding to G proteines: nucleotide exchange and $\beta$-subunit-mediated phosphotransfer reactions. Eur J Biochem 221:25–33

Kalkbrenner F, Dippel E, Wittig B, Schultz G (1996) Specificity of the receptor-G protein interaction: using antisense techniques to identify the functions of G protein-subunits. Biochim Biophys Acta (in press)

Karin M (1995) The regulation of AP-1 activity by mitogen-activated protein kinases. J Biol Chem 270:16483–16486

Kaupp UB, Niidome T, Tanbe T, Terada S, Bonigk W, Stuhmer W, Cook NJ, Kanagawa K, Matsuo H, Hirose T, Miyata T, Numa S (1989) Primary structure and functional expression form complementary DNA of the rod photoreceptor cyclic GMP-gated channel. Nature 342:762–766

Kisselev OG, Ermolaeva MV, Gautam N (1994) A farnesylated domain in the G protein $\gamma$ subunit is a specific determinant of receptor coupling. J Biol Chem 269:21399–21402

Kisselev OG, Pronin A, Ermolaeva MV, Gautam N (1995) Receptor-G protein coupling is established by a potential conformational switch in the $\beta\gamma$ complex. Proc Natl Acad Sci USA 90:9102–9106

Koesling D, Böhme E, Schultz G (1991) Guanylyl cyclases, a growing family of signal-transducing enzymes. FASEB J 5:2785–2791

Koesling D, Böhme E, Schultz G (1993) Guanylyl cyclases as effectors of hormone and neurotransmitter receptors. New Compreh Biochem 24:325–336

Komalavilas P, Lincoln T (1994) Phosphorylation of the inositol 1,4,5-triphosphate receptor by cyclic GMP-dependent protein kinase. J Biol Chem 269:9520–9526

Körner C, Nürnberg B, Uhde M, Braulke T (1995) Mannose 6-phosphate/insulin-like growth factor II receptor interaction with G-proteins, analysis of mutant cytoplasmatic receptor domains. J Biol Chem 270:287–295

Kozasa T, Gilman AG (1995) Purification of recombinant G proteins from Sf9 cells by hexahistidine tagging of associated subunits. Characterization of $\alpha_{12}$ and inhibition of adenylyl cyclase by $\alpha_z$. J Biol Chem 270:1734–1741

Lambbright DG, Noel JP, Hamm HE, Sigler PB (1994) Structural determinants for activation of a heterotrimeric G protein. Nature 369:621–628

Lambbright DG, Sondek J, Bohm A, Skiba NP, Hamm HE, Sigler PB (1996) The 2.0 Å crystal structure of a heterotrimeric G protein. Nature 379:311–319

Laugwitz KL, Spicher K, Schultz G, Offermanns S (1994) Identification of receptor-activated G proteins: selective immunoprecipitation of photolabled G protein $\alpha$-sunbunits. Methods Enzymol 237:283–294
Laurenza A, McHugh Sutkowski E, Seamon KB (1989) Forskolin: a specific stimulator of adenylyl cyclase or a diterpene with multiple sites of action? Trends Pharmacol Sci 10:442–447
Liebmann C, Graness A, Kovalenko M, Adomeit A, Nürnberg B, Wetzker R, Boehmer FD (1996) Tyrosine phosphorylation of Gs$\alpha$ and inhibition of bradykinin-induced activation of the cyclic AMP pathway in A431 cells via epidermal growth factor inhibits. J Biol Chem 271:31098–31105
Lincoln TM, Cornwell TL (1993) Intracellular cyclic GMP receptor proteins. FASEB J 7:328–338
Linder ME, Middleton P, Hepler JR, Taussig R, Gilman AG, Mumby SM (1993) Lipid modifications of G proteins: $\alpha$ subunits are palmitoylated. Proc Natl Acad Sci USA 90:3675–3679
Lopez MJ, Wong SKF, Kishimoto I, Dubois S, Mach V, Friesen J, Garbers DL, Beuve A (1995) Salt resistant hypertension in mice lacking the guanylyl cyclase-A receptor for atrial natriuretic peptide. Nature 378:65–68
Lounsbury KM, Casey PJ, Brass LF, Manning DR (1991) Phosphorylation of $G_z$ in human platelets selectivity and site of modification. J Biol Chem 266:22051–22056
Lounsbury KM, Schlegel B, Poncz M, Brass LF, Manning DR (1993) Analysis of $G_{z\alpha}$ by site-directed mutagenesis. J Biol Chem 268:22051–22056
Lowe DG (1992) Human atrial naturiuretic peptide receptor-A guanylyl cyclase is self-assoziated prior to hormone binding. Biochemistry 31:10421–10425
Ludwig J, Margalit T, Eismann E, Lancet D, Kaupp UB (1990) Primary structure of cAMP-gated channel from bovine olfactory epithelium. FEBS Lett 270:24–29
Lupas A, van Dyke M, Stock J (1991) Predicting coiled coils from protein sequences. Science 252:1162–1164
Lupas AN, Lupas JM, Stock JB (1992) Do G protein subunits associate via a three-stranded coiled coil? FEBS Lett 314:105–108
Macphee CH, Reifsnyder DH, Moore TA, Lerea KM, Beavo JA (1988) Phosphorylation results in activation of a cAMP phosphodiesterase in human platelets. J Biol Chem 263:10353–10358
Marcus AJ, Safier LB (1993) Thromboregulation: multicellular modulation of platelet reactivity in hemostasis and thrombosis. FASEB J 7:516–522
Marletta MA (1993) Nitric oxide synthase structure and mechanism. J Biol Chem 268:12231–12234
Mayer B, Werner ER (1995) In search of function for tetrahydropioperin in the biosynthesis of nitric oxide. Naunyn Schmiedebergs Arch Pharmacol 351:453–463
McKnight GS (1991) Cyclic AMP second messenger systems. Curr Opin Cell Biol 3:213–217
Mehta JL, Chen LY, Kone BC, Mehta P, Turner P (1995) Identification of constitutive and inducible forms of nitric oxide synthase in human platelets. J Lab Clin Med 125:370–377
Meinicke M, Buechler W, Fischer L, Lohmann S, Walter U (1990) cAMP-dependent protein kinase: subunit diversity and functional role in gene expression. In: Jeserich G, Althaus HH, Waehnelt TV (eds) Cellular and molecular biology of myelination. Springer, Berlin Heidelberg New York (NATO ASI series H, vol 43)
Meinicke M, Geiger J, Butt E, Sandberg M, Jahnsen T, Chakraborty T, Walter U, Jarchau T, Lohmann S (1994) Human cGMP-dependent protein kinase I$\beta$ overexpression increases phosphorylation of an endogenous focal contact associated protein VASP without altering the thrombin-evoked calcium response. Mol Pharmacol 46:283–290
Milligan G, Mullaney I, McCallum JF (1993) Distribution and relative levels of expression of the phosphoinositidase-C-linked G proteins $G_q\alpha$ and $G_{11}\alpha$: abscence of

$G_{11}\alpha$ in human platelets and haematopoietically derived cell lines. Biochim Biophys Acta 1179:208–212

Mixon MB, Lee E, Coleman DE, Berghuis AM, Gilman AG, Sprang SR (1995) Tertiary and quaternary structural changes in $G_{i\alpha 1}$ induced by GTP hydrolysis. Science 270:954–960

Moxham CM, Malbon CC (1996) Insulin action impaired by deficiency of the G-protein subunit $G_{i\alpha 2}$. Nature 379:840–844

Moyers JS, Linder ME, Shannon JD, Parsons SJ (1995) Identification of the in vitro phosphorylation sites on $G_s\alpha$ mediated by $pp60^{c\text{-}src}$. Biochem J 305:411–417

Mumby SM, Kleuss C, Gilman AG (1994) Receptor regulation of G protein palmitoylation. Proc Natl Acad Sci USA 91:2800–2804

Murad F, Mittal CK, Arnold WP, Katsuki S, Kimura H (1978) Guanylate cyclase: activation by azide, nitrocompounds, nitric oxide, and hydroxyl radical and inhibition by hemoglobin and myoglobin. Adv Cyclic Nucleotide Res 9:145–158

Muruganandam A, Mutus B (1994) Isolation of nitric oxide synthase from human platelets. Biochim Biophys Acta 1200:1–6

Nathan C, Xie Q (1994) Nitric oxide synthases: roles, tolls, and controls. Cell 78:915–918

Neer EJ, Smith TF (1996) G protein heterodimers: new structures propel new questions. Cell 84:175–178

Neer EJ, Schmidt CJ, Nambudripad R, Smith TF (1994) The ancient regulatory-protein family of WD-repeat proteins. Nature 371:297–300

Nishizuka Y (1995) Protein kinase C and lipid signaling for sustained cellular responses. FASEB J 9:484–496

Noel JP, Hamm HE, Sigler PB (1993) The 2.2 Å crystal structure of transducin-$\alpha$ complexed with GTP$\gamma$S. Nature 366:654–663

Nolte C, Eigenthaler M, Schanzenbächer P, Walter U (1991) Endothelial cell-dependent phosphorylation of a platelet protein mediated by cAMP- and cGMP-elevating factors. J Biol Chem 266:14808–14812

Nolte C, Eigenthaler M, Horstrup K, Honig-Liedl P, Walter U (1994) Synergistic phosphorylation of the focal adhesion-associated vasodilatator-stimulated phosphoprotein in intact human platelets in response to cGMP- and cAMP-elevating platelets inhibitors. Biochem Pharmacol 48:1569–1575

Nürnberg B (1997) Pertussis toxin as a cell biology tool. In: Aktories K (ed) Bacterial toxins. Chapman and Hall, Weinheim (in press)

Nürnberg B, Ahnert-Hilger G (1996) Potential roles of heterotrimeric G proteins of the endomembrane system. FEBS Lett 389:61–65

Nürnberg B, Degtiar VE, Harhammer R, Uhde M, Hescheler J, Schultz G (1994) Hormone-induced $G_o$-subtype-specific inhibition of calcium currents. Naunyn Schmiedebergs Arch Pharmacol 349:R13

Nürnberg B, Gudermann T, Schultz G (1995) Receptors and G proteins as primary components of transmembrane signal transduction, part 1: G proteins: structure and function. J Mol Med 73:123–132, corrections: 73:379

Nürnberg B, Harhammer R, Exner T, Schulze RA, Wieland T (1996) Species- and tissue-dependent diversity of G-protein $\beta$-phosphorylation. Evidence for a cofactor. Biochem J 318:717–722

Obukhov A, Harteneck C, Zobel A, Harhammer R, Kalkbrenner F, Leopoldt D, Lückhoff A, Nürnberg B, Schultz G (1996) Activation of the drosophila cation channel trpl by $\alpha$-subunits of the $G_q$-protein subfamily. EMBO J 15:5833–5838

Offermanns S, Schultz G (1994) Complex information processing by the transmembrane signaling system involving G proteins. Naunyn Schmiedebergs Arch Pharmacol 350:329–338

Offermanns S, Laugwitz KL, Spicher K, Schultz G (1994) G proteins of the $G_{12}$ family are activated via thromboxane $A_2$ and thrombin receptors in human platelets. Proc Natl Acad Sci USA 91:504–508

Ohlmann P, Laugwitz KL, Nürnberg B, Spicher K, Schultz G, Cazenave JP, Gachet C (1995) The human platelet ADP-receptor activates $Gi_2$ G proteins. Biochem J 312:775–779

Palmer RMJ, Ferrige AG, Moncada S (1987) Nitric oxide release accounts for the biological activity of endothelium-derived relaxing factor. Nature 327:524–526

Parenti M, Vigano MA, Newman CMH, Milligan G, Magee AI (1993) A novel N-terminal motif for palmitoylation of G protein $\alpha$ subunits. Biochem J 291:349–353

Pfeifer A, Nürnberg B, Kamm S, Uhde M, Schultz G, Ruth P, Hofmann F (1995) Cyclic GMP-dependent protein kinase blocks pertussis toxin-sensitive hormone receptor signaling pathways in chinese hamster ovary cells. J Biol Chem 270:9052–9059

Radomski MW, Palmer RMJ, Moncada S (1990) An L-arginine/nitric oxide pathway present in human platelets regulates aggregation. Proc Natl Acad Sci USA 87:5193–5197

Ray K, Kunsch C, Bonner, Robishaw J (1995) Isolation of cDNA clones encoding eight different human G protein $\gamma$ subunits, including three novel forms designated $\gamma 4$, $\gamma 10$, and $\gamma 11$ subunits. J Biol Chem 270:21765–21771

Reinhard M, Halbrügge M, Scheer U, Wiegand C, Jockusch B, Walter U (1992) The 46/50kDa phosphoprotein VASP purified from human platelets is a novel protein associated with actin filaments and focal contacts. EMBO J 11:2063–2070

Reinhard M, Giehl K, Abel K, Haffner C, Jarchau T, Hoppe V, Jockusch B, Walter U (1995) The proline-rich focal adhesion and microfilament protein VASP is a ligand for profilinins. EMBO J 14:1583–1589

Rens-Domiano S, Hamm HE (1995) Structural and functional relationships of heterotrimeric G-proteins. FASEB J 9:1059–1066

Rodbell M, Krans HMJ, Pohl SL, Birnbaumer L (1971) The glucogon-sensitive adenyl cyclase system in plasma membranes of rat liver. IV. Binding of glucagon: effect of guanyl nucleotides. J Biol Chem 246:1872–1876

Roush W (1996) Regulating G protein signaling. Science 271:1056–1057

Schmidt HHHW, Walter U (1994) NO at work. Cell 78:919–925

Schmidt HHHW, Lohmann SM, Walter U (1993) The nitric oxide and cGMP signal-transduction system: regulation and mechanism of action. Biochim Biophys Acta 1178:153–175

Schoenfeld JR, Sehl P, Quan C, Burnier JP, Lowe DG (1995) Agonist selectivity for three species of natriuretic peptide receptor A. Mol Pharmacol 47:172–180

Schrammel A, Behrends S, Schmidt K, Koesling D, Mayer B (1996) Characterisation of 1H-[1,2,4]oxadiazolo[4,3-a]quinoxalin-1-one (ODQ) as a heme site inhibitor of nitric oxide-sensitive guanylyl cyclase. Mol Pharmacol (in press)

Schulz S, Chrisman TD, Garbers DL (1992) Cloning and expression of guanylin. J Biol Chem 267:16019–16021

Sheta EA, McMillian K, Masters BSS (1994) Evidence for a bidomain structure of constitutive cerebellar nitric oxide synthase. J Biol Chem 269:15147–15153

Simon MI, Strathmann MP, Gautam N (1991) Diversity of G proteins in signal transduction. Science 252:802–808

Simonds WF, Goldsmith PK, Codina J, Unson CG, Spiegel AM (1989) $G_{i2}$ mediates $\alpha_2$-adrenergic inhibition of adenylate cyclase in platelet membranes: in situ identification with $G\alpha$ C-terminal antibodies. Proc Natl Acad Sci USA 86:7809–7813

Simonds WF, Manji HK, Garritsen A (1993) G proteins and $\beta$ARK: a new twist for the coiled coil. Trends Biochem Sci 18:315–317

Sohn RH, Goldschmidt-Clermont PJ (1994) Profilin: at the crossroads of signal transduction and the actin cytoskeleton. Bioessays 16:465–472

Sondek J, Lambbright DG, Noel JP, Hamm HE, Sigler PB (1994) GTPase mechanism of G protein from the 1.7 Å crystal structure of transducin $\alpha \cdot GDP \cdot AlF_4^-$. Nature 372:276–279

Sondek J, Bohm A, Lambbright DG, Hamm HE, Sigler PB (1996) Crystal structure of a G protein $\beta\gamma$ dimer at 2.1 Å resolution. Nature 379:369–374

Spicher K, Kalkbrenner F, Zobel A, Harhammer R, Nürnberg B, Söling A, Schultz G (1994) $G_{12}$ and $G_{13}$ $\alpha$-subunits are immunochemically detectable in membranes of most tissues of various mammalian species. Biochem Biophys Res Commun 198:906–914

Spring DJ, Neer EJ (1994) A 14-amino acid region of the G protein $\gamma$ subunit is sufficient to confer selectivity of $\gamma$ binding to the $\beta$ subunit. J Biol Chem 269:22882–22886

Sternweis PC, Smrcka AV (1992) Regulation of phospholipase C by G proteins. Trends Biochem Sci 17:502–506

Stone JR, Marletta MA (1994) Soluble guanylyl cyclase from bovine lung: activation with nitric oxide and carbon monoxide and spectral characterisation of the ferrous and ferric states. Biochemistry 33:5636–5640

Stone JR, Sands RH, Dunham R, Marletta MA (1995) Elektron paramagnetic resonance spectral evidence for the formation of a pentacoordinate nitrosyl-heme complex on soluble guanylyl cyclase. Biochem Biophys Res Commun 207:572–577

Stoyanov B, Volinia S, Hanck T, Rubio I, Loubtchenkov M, Malek D, Stoyanova S, Vanhaesebroeck B, Dhand R, Nürnberg B, Gierschik P, Seedorf K, Hsuan JJ, Waterfield MD, Wetzker R (1995) Cloning and characterization of a G protein-activated human phosphatidylinositol-3 kinase. Science 269:690–693

Strassheim D, Malbon CC (1994) Phosphorylation of $G\alpha_{i2}$ attenuates inhibitory adenylyl cyclase in neuroblastoma/glioma hybrid (NG-108-15) cells. J Biol Chem 269:14307–14313

Suppattapone S, Danoff SK, Theibert A, Joseph S, Steiner J, Snyder SH (1988) Cyclic AMP-dependent phosphorylation of a brain inositol trisphosphate receptor decreases its release of calcium. Proc Natl Acad Sci USA 85:8747–8750

Tang WJ, Gilman AG (1992) Adenylyl cyclases. Cell 70:869–972

Taussig R, Gilman AG (1995) Mammalian membrane-bound adenylyl cyclases. J Biol Chem 270:1–4

Taussig R, Iñiguez-Lluhi J, Gilman AG (1993) Inhibition of adenylyl cyclases by $G_{i\alpha}$. Science 261:218–221

Taylor SS, Buechler JA, Yonemoto W (1990) cAMP-dependent protein kinase: framework for a diverse family of regulatory enzymes. Annu Rev Biochem 59:971–1005

Thomason PA, James SR, Casey PJ, Downes CP (1994) A G protein $\beta\gamma$-subunit-responsive phosphoinositide 3-kinase activity in human platelet cytosol. J Biol Chem 269:16525–16528

Tsai SC, Adamik R, Manganiello VC, Vaughan M (1981) Reversible inactivation of soluble liver guanylate cyclase by disulfides. Biochem Biophys Res Commun 100:637–643

Ueda M, Oho C, Takisawa H, Ogihara S (1992) Interaction of the low-molecular-mass, guanine-nucleotide-binding protein with the actin-binding protein and its modulation by the cAMP dependent protein kinase in bovine platelets. Eur J Biochem 203:347–352

Ushikubi F, Nakamura H-I, Narumiya S (1994) Functional reconstitution of platelet thromboxane $A_2$ receptors with Gq and $Gi_2$ in phospholipid vesicles. Mol Pharmacol 46:808–816

Van Biesen T, Hawes BE, Raymond BE, Luttrell LM, Koch WJ, Lefkowitz RJ (1996) Go-protein $\alpha$-subunits activate mitogen-activated protein kinase via a novel protein kinase C-dependant mechanism. J Biol Chem 271:1266–1269

Vara Prasad MVVS, Dermott JM, Heasley LE, Johnson GL, Dhanasekaran N (1995) Activation of Jun kinase/stress-activated protein kinase by GTPase-deficient mutants of $G\alpha_{12}$ and $G\alpha_{13}$. J Biol Chem 270:18655–18659

Veit M, Nürnberg B, Spicher K, Harteneck C, Ponimaskin E, Schultz G, Schmidt MFG (1994) The $\alpha$-subunits of $G_{12}$ and $G_{13}$ are palmitoylated, but not amidically myristoylated. FEBS Lett 339:160–164

Verma A, Hirsch DJ, Glatt CE, Ronnett GV, Snyder SH (1993) Carbon monoxide: a putative neuronal messenger. Science 259:381–384

Voyno-Yasenetskaya T, Conklin BR, Gilbert RL, Hooley R, Bourne HR, Barber DL (1994) $G\alpha_{13}$ stimulates Na-H exchange. J Biol Chem 269:4721–4724
Waldman SA, Murad F (1987) Cyclic GMP synthesis and function. Pharmacol Rev 39:163–196
Waldmann R, Walter U (1989) Cyclic nucleotide elevating vasodilatators inhibit platelet aggregation at an early step of the activation cascade. Eur J Pharmacol 159:317–320
Wall MA, Coleman DE, Lee E, Iñiguez-Lluhi J, Posner BA, Gilman AG, Sprang SR (1995) The structure of the G protein heterotrimer $G_{i\alpha1}\beta_1\gamma_2$. Cell 83:1047–1058
Walter U (1989) Physiological role of cGMP and cGMP-dependent protein kinase in the cardiovascular system. Rev Physiol Biochem Pharmacol 113:42–87
Watson AJ, Katz A, Simon MI (1994) A fifth member of the mammalian G protein $\beta$-subunit family. J Biol Chem 269:22150–22156
Watson S, Arkinstall S (1994) The G protein linked receptor facts book. Academic, New York
Wedegaertner PB, Chu DH, Wilson PT, Levis MJ, Bourne HR (1993) Palmitoylation is required for signaling functions and membrane attachment of $G_q\alpha$ and $G_s\alpha$. J Biol Chem 268:25001–25008
Wedel B, Humbert P, Harteneck C, Foerster J, Malkewitz J, Böhme E, Schultz G, Koesling D (1994) Mutation of His-105 of the $\beta_1$-subunit yields a nitric oxide-insensitive form of soluble guanylyl cyclase. Proc Natl Acad Sci USA 91:2592–2596
Wedel B, Harteneck C, Foerster J, Friebe A, Schultz G, Koesling D (1995) Functional domains of soluble guanylyl cyclase. J Biol Chem 270:24871–24875
Wieland T, Nürnberg B, Ulibarri I, Kaldenberg-Stasch S, Schultz G, Jakobs KH (1993) Guanine nucleotide specific high energy phosphate transfer by G protein $\beta$-subunits. J Biol Chem 268:18111–18118
Wilk-Blaszczak MA, Singer WD, Gutowski S, Sternweis PC, Berladetti F (1994) The G protein $G_{13}$ mediates bradykinin inhibition of voltage-dependent calcium current. Neuron 13:1215–1224
Wilkie TM, Gilbert DJ, Olsen AS, Chen XN, Amatruda TT, Korenberg JR, Trask BJ, de Jong P, Reed RR, Simon MI, Jenkins NA, Copeland NG (1992) Evolution of the mammalian G protein $\alpha$ subunit multigene family. Nat Genet 1:85–91
Williams AG, Woolkalis MJ, Poncz M, Manning DR, Gewirtz AM, Brass LF (1990) Identification of the pertussis toxin-sensitive G proteins in platelets, megakaryocytes, and human erythroleukemia cells. Blood 76:721–730
Wittinghofer A (1994) The structure of transducin $G\alpha_t$: more to view than just ras. Cell 76:201–204
Wu XB, Brune B, von Appen F, Ullrich V (1992) Reversible activation of soluble guanylate cyclase by oxidizing agents. Arch Biochem Biophys 294:75–82
Wu CC, Ko FN, Kuo SC, Lee FY, Teng CM (1995) YC-1 inhibited human platelet aggregation through NO-independent activation of soluble guanylyl cyclase. Br J Pharmacol 116:1973–1978
Xu N, Bradley L, Ambdukar I, Gutkind JS (1993) A mutant $\alpha$ subunit of $G_{12}$ potentiates the eicosanoid pathway and is highly oncogenic in NIH 3T3 cells. Proc Natl Acad Sci USA 90:6741–6745
Xu N, Voyno-Yasenetskaya T, Gutkind JS (1994) Potent transforming activity of the $G_{13}$ $\alpha$ subunit defines a novel family of oncogenes. Biochem Biophys Res Commun 201:603–609
Yamane HK, Fung BKK (1993) Covalent modifications of G proteins. Annu Rev Pharmacol Toxicol 32:201–241
Yang RB, Foster DC, Garbers DL, Fülle HJ (1995) Two membrane forms of guanylyl cyclase found in the eye. Proc Natl Acad Sci USA 92:602–606
Yau KW (1994) Cyclic nucleotide-gated channels: an expanding new family of ion channels. Proc Natl Acad Sci USA 91:3481–3483
Yuen PST, Potter LR, Garbers DL (1990) A new form of guanylyl cyclase is preferentially expressed in rat kidney. Biochemistry 29:10872–10878

CHAPTER 10

# Platelet Phosphodiesterases

E. BUTT and U. WALTER

## A. Introduction

Platelets play a central role in thrombosis and hemostasis. In human platelets, an increase in intracellular levels of cAMP and cGMP is associated with the inhibition of agonist-evoked platelet responses including change of cell shape, adhesion, aggregation, and release of granule content (for a review see SIESS 1989). Cyclic nucleotide phosphodiesterases (PDEs) catalyze the hydrolysis of 3′-5′-cyclic nucleotides to the corresponding nucleoside 5′-monophosphates and thereby play a crucial part in the regulation of cyclic nucleotide concentrations (BEAVO and REIFSNYDER 1990; BELTMAN et al. 1993). Currently, seven different but homologous PDE families are recognized, and most of these families contain multiple isoforms (BEAVO 1995). In human platelets three distinct forms of cyclic nucleotide PDEs exist that differ in their kinetic and physical characteristics, substrate selectivities (cAMP or cGMP) as well as their regulation by various natural and pharmacological agents. These three PDEs are cGMP-stimulated phosphodiesterase type II (cGS-PDE) with a $K_m$ of 35 $\mu M$ for cGMP and 50 $\mu M$ for cAMP (GRANT et al. 1990), the cGMP-inhibited phosphodiesterase type III (cGI-PDE) with the lowest $K_m$ (0.02 $\mu M$) for cGMP and a $K_m$ of 0.2 $\mu M$ for cAMP (GRANT et al. 1992), and the cGMP-specific PDE type V. PDE V hydrolyzes cGMP ($K_m$ = 5 $\mu M$) at a rate approximately 100 times faster than the hydrolytic rate for cAMP (HAMET et al. 1984; Table 1). Most (85%) of the total platelet phosphodiesterase activity is present in the cytosolic fraction (MACPHEE et al. 1986).

## B. Catalytic and Regulatory Properties of Human Platelet Phosphodiesterases

### I. cGMP-Stimulated Phosphodiesterase

The cGMP-stimulated phosphodiesterase from human platelets is a homodimer with a molecular mass of 206 kDa. With cAMP as substrate, enzyme activity exhibits positive cooperativeness with a Hill coefficient of approximately 2 (STROOP and BEAVO 1991). Binding of low "physiological"

**Table 1.** Kinetic properties of phosphodiesterase isozymes present in human platelets

| Family | Isozyme | $K_m$ ($\mu M$) | | $V_{max}$ ($\mu$mol/min/mg) | |
|---|---|---|---|---|---|
| | | cAMP | cGMP | cAMP | cGMP |
| II | cGMP-Stimulated[a,b] | 50 | 35 | 120 | 120 |
| III | cGMP-Inhibited[b] | 0.2 | 0.02 | 3.0 | 0.3 |
| V | cGMP-Specific[b,c] | 150 | 5 | <0.2 | 3–10 |

[a] Grant et al. (1992).
[b] Manganiello et al. (1990).
[c] Francis et al. (1992).

concentrations of cGMP (0.1–1.0 $\mu M$) to the allosteric site induces an apparent conformational change resulting in enhanced cAMP hydrolyzing activity at nonsaturating cAMP concentrations without effects on $V_{max}$ (120 $\mu$mol/min/mg). At higher concentrations of cGMP (beyond 10–20 $\mu M$), cAMP hydrolysis is not stimulated further but actually inhibited. This is due to the competition between cAMP and cGMP at the catalytic site (Wada et al. 1987). Very little is known about the regulation or the physiological role of cGS-PDE in intact human platelets. However, it is interesting to note that two distinct cAMP-hydrolyzing PDEs (the cGMP-inhibited and the cGMP-stimulated PDE) are affected by cGMP albeit in opposite ways. With its high $V_{max}$ and $K_m$ (see Fig. 1), the cGS-PDE may have a role in hydrolyzing elevated levels of cAMP, returning cAMP to basal levels. Moreover, cGS-PDE is one of several important components responsible for the extensive "cross talk" between cAMP and cGMP pathways.

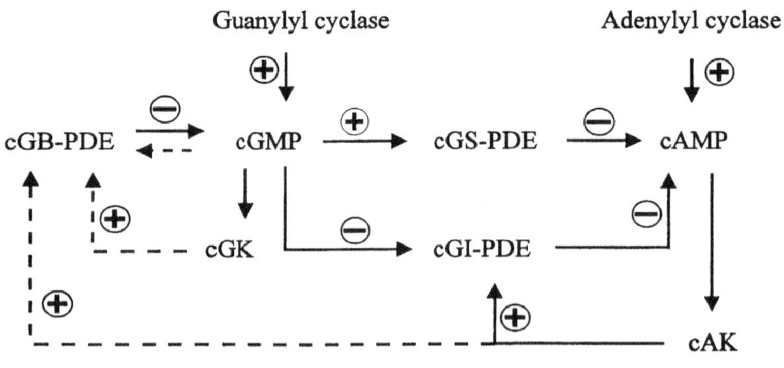


**Fig. 1.** Model summarizing the regulation and complex "feedback" mechanisms that control cAMP and cGMP levels in human platelets. *Dashed lines*, functional consequences not established; *cGI-PDE*, cGMP-inhibited phosphodiesterase; *cGS-PDE*, cGMP-stimulated phosphodiesterase; *cGB-PDE*, cGMP-binding phosphodiesterase; *cAK*, cAMP-dependent protein kinase; *cGK*, cGMP-dependent protein kinase

## II. cGMP-Inhibited Phosphodiesterase

The cGMP-inhibited phosphodiesterase was first purified from human platelets by MACPHEE and coworkers (MACPHEE et al. 1986) and appears to be similar to the isozymes from bovine aorta smooth muscle and bovine platelets (RASCON et al. 1992). More than 80% of the total low-$K_m$ platelet cAMP phosphodiesterase activity is provided by type III PDE. The native enzyme has an apparent molecular mass of approximately 210kDa. The enzyme hydrolyzes cAMP and cGMP according to normal Michaelis-Menten kinetics ($K_m$ values of 0.2 and 0.02 $\mu M$, respectively). However, the $V_{max}$ for cGMP is very low compared to that of cAMP. Since cGMP binds tightly to the enzyme but is hydrolyzed poorly, it is an important inhibitor of the cAMP hydrolytic activity of this PDE. In fact, cGMP appears to be the most potent physiological competitive inhibitor of cAMP hydrolysis ($IC_{50} = K_m$; GRANT et al. 1992; DEGERMAN et al. 1994). The protein is localized in the cytosolic fraction of lysed platelets (MACPHEE et al. 1986), whereas in adipocytes and hepatocytes, most of the cGI-PDE is associated with the particular fraction (MANGANIELLO et al. 1990). In vitro, human platelet cGI-PDE is phosphorylated by the cAMP-dependent protein kinase (cAK) with a maximal incorporation of 0.4–1.8mol $^{32}$P/mol enzyme resulting in a 50% increase in activity. Stimulation of intact washed platelets with agents that increase cAMP concentration (e.g., forskolin, prostaglandins) results in phosphorylation and activation of cGMP-inhibited phosphodiesterase. This effect is mediated by cAK (MACPHEE et al. 1988; GRANT et al. 1988) although the phosphorylation site(s) and stochiometry have not yet been reported. Such activation following direct or receptor-mediated stimulation of adenylyl cyclase (AC) serves as a mechanism for attenuating the amplitude and duration of the cAMP signal (see Fig. 1). Due to the very low level of cGMP in resting platelets (0.2 $\mu M$; EIGENTHALER et al. 1992) it is possible that cGI-PDE controls the basal level of cAMP under these conditions.

## III. cGMP-Specific Phosphodiesterase

A cGMP-binding cGMP-specific phosphodiesterase has been described in human platelets (WEISHAAR et al. 1986). However, much of the characterization of this PDE has been carried out with the lung isozyme (THOMAS et al. 1990a,b). The PDE is a homodimer with a molecular mass of 190kDa. Each monomer is composed of two functional domains, a cGMP binding site and a cGMP catalytic site with half-maximal cGMP binding and cGMP hydrolytic activities at 0.2 $\mu M$ and 5 $\mu M$ ($K_m$), respectively (FRANCIS et al. 1992, 1994). The cGB-binding site of PDE V is distinct from the allosteric site of the cGMP-stimulated PDE type II (STROOP and BEAVO 1991). Binding of cGMP to the noncatalytic site of the cGMP-binding PDE is most likely due to a conformational change after occupation of the catalytic site. Recently, CORBIN and coworkers identified two putative zinc binding motifs, both located in the catalytic domain of cGB-PDE (FRANCIS et al. 1994). In addition, the enzyme exhibits a site that can be phosphorylated by the cGMP-dependent protein kinase (cGK) and, at a lower rate, by cAK (THOMAS et al. 1990b; MCALLISTER-LUCAS et al. 1993). Binding of cGMP to the allosteric site of cGB-PDE is required for it to become a suitable substrate for cGK. However, the effect of this phosphorylation is unknown. It is nevertheless interesting to note that the activity of type V PDE in guinea pig lung is increased by cAK phosphorylation (BURNS et al. 1992).

# C. Synergistic Inhibition of Platelet Function by Cyclic Nucleotide Elevating Agents

Compounds such as prostaglandin $E_1$ (PG-$E_1$) and prostacyclin inhibit platelet function by stimulating adenylyl cyclase and thereby increasing cAMP levels. In contrast, nitrovasodilators [e.g., sodium nitroprusside (SNP), 3-morpholinosydnonimine hydrochloride (SIN-1), and the endothelium-derived relaxing factor (EDRF)] are believed to inhibit platelet aggregation by stimulating soluble guanylyl cyclase, thus increasing the concentration of cGMP. The intracellular effects of cAMP in human platelets are mediated by the cAMP-dependent protein kinase (cAK) while cGMP exerts its effects by activation of cGMP-dependent protein kinase (cGK) and regulation of phosphodiesterases (for a review see HALBRÜGGE and WALTER 1993). Both nitrovasodilator- and prostaglandin-regulated second messenger pathways have independent effects but may also act synergistically as shown by LEVIN et al. (1982) and others (MACDONALD et al. 1988; LIDBURY et al. 1989). This synergism appears to be due to a significant nitrovasodilator-induced increase in platelet cAMP in addition to the well-established cGMP increase (MAURICE and HASLAM 1990; ANDERSSON and VINGE 1991). The elevated cAMP concentration is caused by an inhibition of the cGMP-inhibited phosphodiesterase type III secondary to an NO-mediated rise in cGMP.

Recently, our laboratory analyzed the synergistic effects of these platelet inhibitors beyond the analysis of cyclic nucleotide concentrations at the level of protein phosphorylation. In human platelets, a 46-kDa protein termed vasodilator-stimulated phosphoprotein (VASP) is phosphorylated by cAK and cGK (HALBRÜGGE et al. 1990). Threshold concentrations of either SNP or PG-$E_1$ alone had only moderate effects on VASP phosphorylation while the combination of both vasodilators caused supra-additive protein phosphorylation (see Fig. 2). In contrast, the simultaneous activation of cAK and cGK by selective membrane-permeable activators resulted in only additive effects on VASP phosphorylation. These data also demonstrated that low intracellular levels of cGMP effectively inhibit cGMP-inhibited phosphodiesterase despite the high levels of cGK known to be present in human platelets (NOLTE et al. 1994).

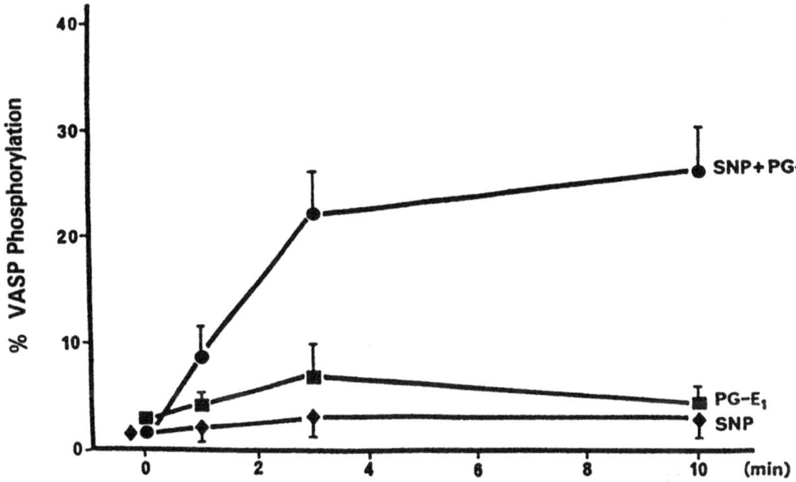


**Fig. 2.** Synergistic effects of cAMP and cGMP elevating agents on protein phosphorylation in intact human platelets. The time course of the effects of SNP (0.1 $\mu M$), PG-$E_1$ (0.005 $\mu M$), or the combination of SNP (0.1 $\mu M$) and PG-$E_1$ (0.005 $\mu M$) on VASP phosphorylation in intact human platelets are shown. *SNP*, sodium nitroprusside; *PG-$E_1$*, prostaglandin $E_1$; *VASP*, vasodilator-stimulated phosphoprotein. (Modified from NOLTE et al. 1994)

## D. Regulation of Platelet Phosphodiesterases by Insulin

Human platelets contain insulin receptors that are phosphorylated at the $\beta$-subunit in a dose-dependent manner in response to insulin stimulation (FALCON et al. 1988). Recently, LOPEZ-APARICIO and coworkers demonstrated that the incubation of human platelets with insulin results in serine phosphorylation and activation of cGMP-inhibited phosphodiesterase. The resulting decrease in cAMP levels and cAK activity renders the platelets more sensitive towards aggregating stimuli and less sensitive towards anti-aggregating agents (LOPEZ-APARICIO et al. 1992). This is of relevance to the observation that patients with diabetes mellitus exhibit an increased sensitivity towards aggregating drugs. A model demonstrating the insulin effect on human platelets is presented in Fig. 3. The insulin-stimulated serine kinase itself appears to be activated through insulin-dependent serine/threonine phosphorylation (LOPEZ-APARICIO et al. 1993). A similar pathway is present in rat adipocytes where insulin-activation of cGI-PDE is a key step in the antilipolytic action of insulin (MANGANIELLO et al. 1990).

However, insulin has also been reported to decrease platelet sensitivity to aggregating agents by increasing the platelet concentration of cGMP, perhaps due to an insulin-receptor-mediated platelet guanylyl cyclase activation (TROVATI et al. 1994).

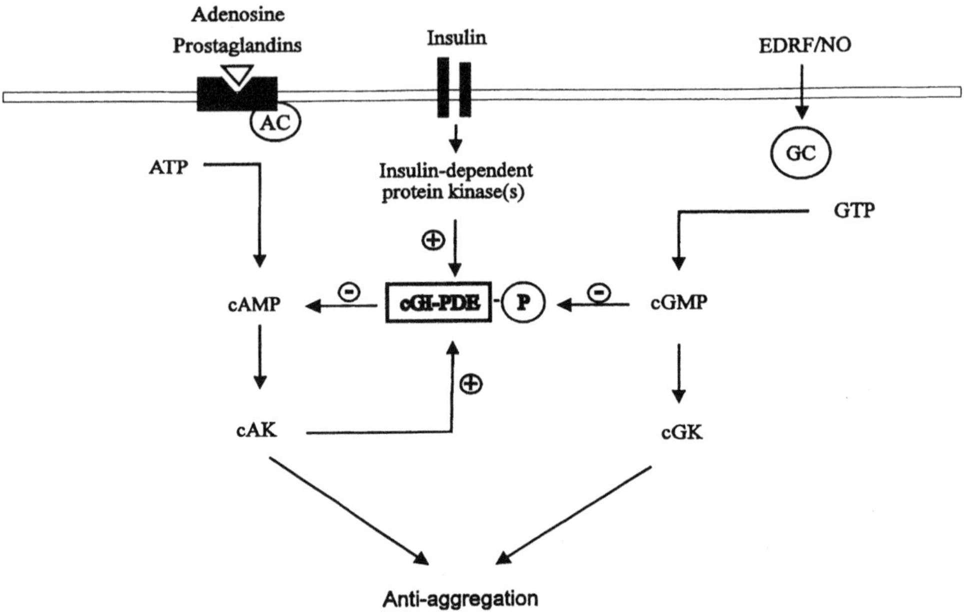


**Fig. 3.** Model summarizing the role of cGMP, cAMP, and the cGMP-inhibited phosphodiesterase in platelet function. *AC*, adenylyl cyclase; *GC*, guanylyl cyclase; *cGI-PDE*, cGMP-inhibited phosphodiesterase; *cAK*, cAMP-dependent protein kinase; *cGK*, cGMP-dependent protein kinase; *EDRF*, endothelium-derived relaxing factor

# E. Effect of Phosphodiesterase Inhibitors on Platelet Function

## I. General Aspects

Many studies have been carried out to identify drugs that selectively inhibit cyclic nucleotide phosphodiesterase activities. Some of the first agents identified, theophylline and 3-isobutyl-1-methyl-xanthin (IBMX) were found to be relatively unselective and had additional modes of action including adenosine receptor antagonism. Currently, attempts are being made to synthesize more specific inhibitors of the major PDE isozymes that are responsible for regulation of cAMP and cGMP concentrations. Possible targets are not only platelets but also airway smooth muscle cells and inflammatory cells (BEAVO and REIFSNYDER 1990; TORPHY and UNDEN 1991; NICHOLSON et al. 1991).

Most of the more recently developed selective PDE inhibitors contain a heterocyclic structure with a $\pi$- or lone-pair electron group and an electron-rich moiety. Recent studies suggested that the heterocyclic amide function probably simulates the electrophilic center in the natural phosphate moiety present in cAMP and cGMP (ERHARDT 1992). The selectivity of the phosphodiesterase is given by the additional substituents simulating the different binding profiles of cAMP and cGMP to the catalytic site (e.g., syn- versus anticonformation, hydrogen donor and/or acceptor function; BELTMAN et al. 1995; BUTT et al. 1995).

## II. Specificity

Since 80% of the total low-$K_m$ cAMP activity in platelets is due to cGI-PDE, selective inhibitors of this PDE may exhibit useful anti-aggregating properties. Cilostazol is one such compound used as an antithrombotic drug with a potential to prevent platelet aggregation. This agent selectively inhibited cGI-PDE from human platelets with an $IC_{50}$ of $0.2\,\mu M$ (TANI et al. 1990). Other drugs including milrinone, OPC 3911 (a cilostamide analog), and siguazodan showed similar results. These compounds raised the platelet cAMP level via cGI-PDE inhibition and thus activated the cAK, causing intracellular protein phosphorylation as well as inhibition of $Ca^{2+}$ mobilization (ALVAREZ et al. 1986; LANZA et al. 1987; LINDGREN et al. 1990; MURRAY et al. 1990, 1992). In addition, the anti-aggregating effects of adenylyl cyclase activators were enhanced by cGI-PDE inhibitors (LINDGREN et al. 1990). These studies support the concept that cGMP-inhibited PDE is an important modulator of platelet function.

Cilostamide was also reported to selectively inhibit PDE III (Table 2; HIDAKA et al. 1979) since it potentiated the inhibition of platelet aggregation induced by PG-$E_1$. However, when used in antithrombotic studies, cilostamide inhibited platelet aggregation induced by a thromboxane analog without significant elevation of cAMP concentration. Recent results suggested that cilostamide may have inhibitory effects on thromboxane-induced platelet activation through additional mechanisms that do not involve cGI-PDE inhibition and cAK (NISHIKAWA et al. 1992).

Dypyridamol and Zaprinast have been studied and characterized as selective inhibitors of PDE V activity (Table 2). Clinically, dypyridamol has been used as an antithrombotic agent. In experiments with subthreshold concentrations of NO, dipyridamole potentiated the NO effect presumably via inhibition of the cGMP-specific phosphodiesterase (BULT et al. 1991). However, only part of its efficacy appears to be due to cGB-PDE inhibition (MCELROY and PHILIP 1975). Stimulation of adenylyl cyclase and prostacyclin formation (DE LA CRUZ et al. 1994) as well as suppression of adenosine uptake into erythrocytes (GRESELE et al. 1986) by dipyridamole may be important effects of this drug.

Unto this day, the cGS-PDE inhibitor ENHA (erythro-9-[2-hydroxy-3-nonyl]adenosine), initially referred to as MEP 1 (PODZUWEIT and MÜLLER 1993), has not been evaluated with human platelets. The drug has been tested only in cardiac myocytes where it was found to selectively inhibit PDE II in a noncompetitive manner with respect to cGMP activation of the enzyme (MERY et al. 1993, 1995).

This short review of the results obtained with phosphodiesterase inhibitors is not meant to be complete. Here, it should be emphasized that these

**Table 2.** Selective inhibitors of specific phosphodiesterase isozymes present in human platelets

| | Phosphodiesterase isozyme family $IC_{50}$ ($\mu M$) II | III | V |
|---|---|---|---|
| *PDE II inhibitor* | | | |
| ENHA[a] | 0.8 | >100 | n.d. |
| *PDE III inhibitor* | | | |
| Milrinone (WIN 47203)[b] | 180 | 0.3 | 110 |
| SK&F 95654[c] | >100 | 0.7 | >100 |
| Siguazodan (SK&F 94836)[d] | | 0.84 | |
| Cilostamide[e] | 15 | 0.005 | 5.5 |
| Anagrelide (BL4162A)[f] | | 0.05 | |
| *PDE V inhibitor* | | | |
| Dipyridamole[g] | 8 | 40 | 0.9 |
| Zaprinast (M&B 22948)[g,h] | 78 | 225 | 0.8 |
| *Nonspecific inhibitor* | | | |
| IBMX[b] | 14 | 2 | 8 |
| Theophylline[g] | 210 | 70 | 820 |
| Pentoxifylline (Trental)[i] | | 45–150[j] | |

[a] PODZUWEIT et al. (1992).
[b] HARRISON et al. (1986).
[c] MURRAY et al. (1992).
[d] MURRAY et al. (1990).
[e] HIDAKA and ENDO (1984).
[f] GILLESPIE E (1988).
[g] WEISHAAR et al. (1986).
[h] GILLESPIE and BEAVO (1989).
[i] BEAVO JA (1995).
[j] Refers to range of $IC_{50}$ values seen with different isozymes.

drugs (especially when used in vivo) may have effects and side effects not related to PDE inhibition. Also, in the case of the in vivo side effects of PDE inhibitors, the distribution of the different phosphodiesterases in various tissues and cell types has to be considered. For example, PDE III and V are expressed in vascular smooth muscle cells and have a critical physiological role in modulating vascular and cardiac function (COLUCCI 1991; LINDGREN et al. 1990; ECKLY and LUGNIER 1994). Such "side activities" appear to be a major limitation to the use of these compounds as antithrombotic drugs.

## F. Concluding Remarks

The recent explosion of our knowledge of diversity and regulation of PDE isoforms has led to an extensive reevaluation of PDEs as possible therapeutic targets (BEAVO 1995; SHETH and COLMAN 1995). Clearly, a detailed analysis of the cell-type-specific expression of the PDE isoforms and the possible functional consequences associated with differing expression are the focus of current efforts of many investigators. In addition, a reevaluation of old PDE inhibitors for new clinical applications as well as the development of second and third generation PDE inhibitors appear to be very promising in the light of PDE diversity. The proven efficacy of antiplatelet drugs such as aspirin and ticlopidin in various cardiovascular diseases, the continuous existing need for better and safer antiplatelet drugs, and the established function of PDE inhibitors as antiplatelet agents suggest that the development and clinical evaluation of more specific PDE inhibitors ideally only affecting platelet function may be possible and of considerable pharmacological and medical interest.

*Acknowledgements.* The authors thank J.A. Beavo (Seattle) for discussions and support. The authors are supported by the Deutsche Forschungsgemeinschaft (SFB 176/TP21).

## References

Alvarez R, Banerjee GL, Bruno JJ, Jones GL, Littschwager K, Strosberg AM, Venuti MC (1986) A potent and selective inhibitor of cyclic AMP phosphodiesterase with potential cardiotonic and antithrombotic properties. Mol Pharmacol 29:554–560

Andersson TLG, Vinge E (1991) Interaction between isoprenaline, sodium nitroprusside, and isozyme selective phosphodiesterase inhibitors on ADP-induced aggregation and cyclic nucleotide levels in human platelets. J Cardiovasc Pharmacol 18:237–242

Beavo JA (1995) Cyclic nucleotide phosphodiesterases: functional implicaions of multiple isoforms. Physiol Rev 75:725–748

Beavo JA, Reifsnyder DH (1990) Primary sequence of cyclic nucleotide phosphodiesterase isozymes and the design of selective inhibitors. Trends Pharmacol Sci 11:150–155

Beltman J, Sonnenburg WK, Beavo JA (1993) The role of protein phosphorylation in the regulation of cyclic nucleotide phosphodiesterases. Mol Cel Biochem 127/128:239–253

Beltman J, Becker DE, Butt E, Jensen GS, Rybalkin SD, Jastorff B, Beavo JA (1995) Characterization of cyclic nucleotide phosphodiesterases with cyclic GMP analogs: topology of the catalytic site. Mol Pharmacol 47:330–339

Bult H, Fret HRL, Jordaens FH, Herman AG (1991) Dipyridamole potentiates platelet inhibition by nitric oxide. Thromb Haemost 66:343–349

Burns F, Dodger IW, Pyne NJ (1992) The catalytic subunit of protein kinase A triggers activation of the type V cyclic GMP-specific phosphodiesterase from guinea-pig lung. Biochem J 283:487–491

Butt E, Beltman J, Becker DE, Jensen GS, Rybalkin SD, Jastorff B, Beavo JA (1995) Characterization of cyclic nucleotide phosphodiesterases with cyclic AMP analogs: topology of the catalytic site and comparison with other cyclic AMP-binding proteins. Mol Pharmacol 47:340–347

Colucci WS (1991) Cardiovascular effects of milrinone. Am Heart J 121:1945–1947

De la Cruz JP, Ortega G, de la Cuesta FS (1994) Differential effects of the pyrimido-pyrimidine derivatives, dipyridamole and mopidamol, on platelet and vascular cyclooxygenase activity. Biochem Pharmacol 47:209–215

Degerman E, Moos M, Rascon A, Vasta V, Meacci E, Smith CJ, Lindgren S, Andersson KE, Belfrage P, Manganiello V (1994) Single-step affinity purification,

partial structure and properties of human platelet cGMP inhibited cAMP phosphodiesterase. Biochem Biophys Acta 1205:189–198
Eckly AE, Lugnier C (1994) Role of phosphodiesterases III and IV in the modulation of vascular cyclic AMP content by the NO/cyclic GMP pathway. Br J Pharmacol 113:445–450
Eigenthaler M, Nolte C, Halbrügge M, Walter U (1992) Concentrations and regulation of cyclic nucleotides, cyclic nucleotide-dependent protein kinases and one of their major substrates in human platelets. Eur J Biochem 205:471–481
Erhardt PW (1992) Second-generation phosphodiesterase inhibitors: structure activity relationship and receptor models. In: Beavo JA, Housley MD (eds) Isozymes of cyclic nucleotide phosphodiesterases. Wiley, New York, p 317
Falcon C, Pflieger G, Deckmyn H, Vermylen J (1988) The platelet insulin receptor: detection, partial characterization, and search for a function. Biochem Biophys Res Commun 157:1190–1196
Francis SH, Thomas MK, Corbin JD (1992) Cyclic GMP-binding cyclic GMP-specific phosphodiesterase from lung. In: Beavo JA, Housley MD (eds) Isoenzymes of cyclic nucleotide phosphodiesterases. Wiley, New York, p 117
Francis SH, Colbran JL, McAllister-Lucas LM, Corbin JD (1994) Zinc interactions and conserved motifs of the cGMP-binding cGMP-specific phosphodiesterase suggest that it is a zinc hydrolase. J Biol Chem 269:22477–22480
Gillespie E (1988) Anagrelide: a potent and selective inhibitor of platelet cyclic AMP phosphodiesterase enzyme activity. Biochem Pharmacol 37:2866–2868
Gillespie E, Beavo JA (1989) Mol Pharmacol 36:773
Grant PG, Mannarino AF, Colman RW (1988) cAMP-mediated phosphorylation of the low-$K_m$ cAMP phosphodiesterase markedly stimulates its catalytic activity. Proc Natl Acad Sci USA 85:9071–9075
Grant PG, Mannarino AF, Colman RW (1990) Purification and characterization of a cyclic GMP-stimulated cyclic nucleotide phosphodiesterase from the cytosol of human platelets. Tromb Res 59:105–119
Grant PG, DeCamp DL, Baily JM, Colman RF, Colman RW (1992) Low-$K_m$ cyclic AMP phoshodiesterase from human platelets: stimulation of activity by phosphorylation of the enzyme and affinity labeling of the active site. Second Messengers Phosphoprotein Res 25:73–85
Gresele P, Arnout J, Deckmyn H, Vermylen J (1986) Mechanism of the antiplatelet action of dipyridamole in whole blood: modulation of adenosine concentration and activity. Tromb Haemost 55:12–18
Halbrügge M, Walter U (1993) The regulation of platelet function by protein kinases. In: Huang CK, Sha'afi RI (eds) Protein kinases in blood cell function. CRC, London, p 245
Halbrügge M, Friedrich C, Eigenthaler M, Schanzenbächer P, Walter U (1990) Stoichiometric and reversible phosphorylation of a 46-kDa protein in human platelets in response to cGMP and cAMP-elevating vasodilators. J Biol Chem 265:3088–3093
Hamet P, Coquil JF, Bousseau-Lafortune S, Franks DJ, Tremblay J (1994) Cyclic GMP binding and phosphodiesterase: implication for platelet function. Adv Cyclic Nucleotide Protein Phosphorylation Res 16:119–136
Harrison SA, Reifsnyder DH, Gallis B, Cadd GG, Beavo JA (1986) Isolation and characterization of bovine cardiac muscle cGMP-inhibited phosphodiesterase: a receptor for new cardiotonic drugs. Mol Pharmacol 29:506–514
Hidaka H, Endo T (1984) Selective inhibitors of three forms of cyclic nucleotide phosphodiesterase – basic and potential clinical applications. Adv Cyclic Nucleotide Protein Phosphorylation Res 16:245–259
Hidaka H, Hayashi H, Kohri H, Kimura Y, Hosokawa T, Igawa T, Saitoh Y (1979) Selective inhibitor of platelet cyclic adenosine monophosphate phosphodiesterase, cilostamide, inhibits platelet aggregation. J Pharmacol Exp Ther 211:26–30

Lanza F, Beretz A, Stierle A, Corre G, Cazenave JP (1987) Cyclic nucleotide phosphodiesterase inhibitors prevent aggregation of human platelets by raising cyclic AMP and reducing cytoplasmic free calcium mobilization. Thromb Res 45:477–484

Levin RI, Weksler BB, Jaffe EA (1982) The interaction of sodium nitroprusside with human endothelial cells and platelets: nitroprusside and prostacyclin synergistically inhibit platelet function. Circulation 66:1299–1307

Lidbury PS, Antunes RMJ, de Nucci G, Vane JR (1989) Interactions of iloprost and sodium nitroprusside on vascular smooth muscle and platelet aggregation. Br J Pharmacol 98:1275–1280

Lindgren SHS, Andersson TLG, Vinge E, Andersson KE (1990) Effect of isozyme-selective phosphodiesterase inhibitors on rat aorta and human platelets: smooth muscle tone, platelet aggregation and cAMP levels. Acta Physiol Scand 140:209–219

Lopez-Aparicio P, Rascon A, Manganiello VC, Andersson KE, Belfrage P, Degerman E (1992) Insulin-induced phosphorylation and activation of the cGMP-inhibited phosphodiesterase in human platelets. Biochem Biophys Res Commun 186:517–523

Lopez-Aparicio P, Belfrage P, Manganiello VC, Kono T, Degerman E (1993) Stimulation by insulin of a serine kinase in human platelets that phosphorylates and activates the cGMP-inhibited cAMP phosphodiesterase. Biochem Biophys Res Commun 193:1137–1144

MacDonald PS, Read MA, Dusting GJ (1988) Synergistic inhibition of platelet aggregation by endothelium-derived relaxing factor and prostacyclin. Thromb Res 49:437–449

MacPhee CH, Harrison SA, Beavo JA (1986) Immunological identification of the major platelet low-$K_m$ cAMP phosphodiesterase: probable target for anti-thrombotic agents. Proc Natl Acad Sci USA 83:6660–6663

MacPhee CH, Reifsnyder DH, Moore TA, Lerea KM, Beavo JA (1988) Phosphorylation results in activation of a cAMP phosphodiesterase in human platelets. J Biol Chem 263:10353–10358

Manganiello VC, Smith CJ, Degerman E, Belfrage P (1990) Cyclic GMP-inhibited cyclic nucleotide phosphodiesterase. In: Beavo JA, Housley MD (eds) Isoenzymes of cyclic nucleotide phosphodiesterases. Wiley, New York, p 87

Maurice DH, Haslam RJ (1990) Molecular basis of the synergistic inhibition of platelet function by nitrovasodilators and activators of adenylate cyclase: inhibition of cAMP breakdown by cyclic GMP. Mol Pharmacol 37:671–681

McAllister-Lucas LM, Sonnenburg WK, Kadlecek A, Seger D, Trong HL, Colbran JL, Thomas MK, Walsh KA, Francis SH, Corbin JD, Beavo JA (1993) The structure of bovine lung cGMP-binding, cGMP-specific phosphodiesterase deduced from cDNA clone. J Biol Chem 268:22863–22873

McElroy FA, Philip RB (1975) Relative potencies of dipyridamole and related agents as inhibitors of cyclic nucleotide phosphodiesterase: possible explanation of mechanism of inhibition of platelet function. Life Sci 17:1479–1490

Mery PF, Fischmeister R, Podzuweit T, Müller A (1993) Cyclic GMP-mediated inhibition of Ca current in frog ventricular myocytes is reversed by MEP 1, a selective inhibitor of the cGMP-stimulated phosphodiesterase. J Physiol (Lond) 459:421

Mery PF, Pavoine C, Pecker F, Fischmeister R (1995) Erythro-9-(2-hydroxy-3-nonyl)adenine inhibits cyclic GMP-stimulated phosphodiesterase in isolated cardiac myocytes. Mol Pharmacol 48:121

Murray KJ, England PJ, Hallam TJ, Maguire J, Moores K, Reeves ML, Simpson AWM, Rink TJ (1990) The effects of siguazodan, a selective phosphodiesterase inhibitor, on platelet function. Br J Pharmacol 99:612–616

Murray KJ, Eden RJ, Dolan JS, Grimsditch DC, Stutchbury CA, Patel B, Knowles A, Worby A, Lynham JA, Coates WJ (1992) The effect of SK&F 95654, a novel

phosphodiesterase inhibitor, on cardiovascular, respiratory and platelet function. Br J Pharmacol 107:463–470
Nicholson CD, Challiss RA, Shahid M (1991) Differential modulation of tissue function and therapeutic potential of selective inhibitors of cyclic nucleotide phosphodiesterase isoenzymes. Trends Pharmacol Sci 12:19–27
Nishikawa M, Komada F, Morita K, Deguchi K, Shirikawa S (1992) Inhibition of platelet aggregation by the cAMP phosphodiesterase inhibitor, cilostamide, may not be associated with activation of cAMP-dependent protein kinase. Cell Signal 4:453–463
Nolte C, Eigenthaler M, Horstrup K, Hönig-Liedl P, Walter U (1994) Synergistic phosphorylation of the focal adhesion-associated vasodilator-stimulated phosphoprotein in intact human platelets in response to cGMP- and cAMP-elevating platelet inhibitors. Biochem Pharmacol 48:1569–1575
Podzuweit T, Mueller A, Opie LH (1993) Anti-arrhytmic effects of selective inhibition of myocardial phosphodiesterase II. Lancet 341:760
Rascon A, Lindgren S, Stavenow L, Belfrage P, Andersson KE, Manganiello VC, Degerman E (1992) Purification and properties of the cGMP-inhibited cAMP phosphodiesterase from bovine aortic smooth muscle. Biochim Biophys Acta 1134:149–156
Robichon A (1991) A new cGMP phosphodiesterase isolated from bovine platelets is substrate for cAMP and cGMP-dependent protein kinases: evidence for a key role in the process of platelet activation. J Cell Biochem 47:147–157
Sheth SB, Colman RW (1995) Regulatory and catalytic domains of platelet cAMP phosphodiesterases: targets for drug design. Semin Hematol 32:110–119
Siess W (1989) Molecular mechanism of platelet activation. Physiol Rev 69:58–177
Stroop SD, Beavo JA (1991) Structure and function studies of the cGMP-stimulated phosphodiesterase. J Biol Chem 266:23802–23809
Tani T, Sakurai K, Kimura Y, Ishikawa T, Hidaka H (1990) Pharmacological manipulation of tissue cyclic AMP by inhibitors: effect of phosphodiesterase inhibitors on the function of platelets and vascular endothelial cells. Adv Second Messenger Phosphoprotein Res 25:215–227
Thomas MK, Francis SH, Corbin JD (1990a) Characterization of a purified bovine lung cGMP-binding cGMP phosphodiesterase. J Biol Chem 265:14964–14970
Thomas MK, Francis SH, Corbin JD (1990b) Substrate- and kinase-directed regulation of phosphorylation of a cGMP-binding phosphodiesterase by cGMP. J Biol Chem 265:14971–14978
Torphy TJ, Unden BJ (1991) Phosphodiesterase inhibitors: new opportunities for the treatment of asthma. Thorax 46:512–523
Trovati M, Massucco P, Mattiello L, Mularoni E, Cavalot F, Anfossi G (1994) Insulin increases guanosine-3′,5′-cyclic monophosphate in human platelets. A mechanism involved in the insulin anti-aggregating effect. Diabetes 43:1015–1019
Wada H, Manganiello VC, Osborne JC (1987) Analysis of the kinetics of cyclic AMP hydrolysis by the cyclic GMP-stimulated cyclic nucleotide phosphodiesterase. J Biol Chem 262:13938–13945
Weishaar RE, Burrows SD, Koylarz DC, Quade MM, Evans DB (1986) Multiple molecular forms of cyclic nucleotide phosphodiesterase in cardiac and smooth muscle and platelets: characterization and effects of various reference phosphodiesterase inhibitors and cardiotonic agents. Biochem Pharmacol 35:786–800

CHAPTER 11

# Platelet Phospholipases C and D

S. NAKASHIMA, Y. BANNO, and Y. NOZAWA

## A. Introduction

In response to many aggregative substances or vessel wall injury, platelets rapidly undergo a series of responses: adhesion, shape change, aggregation, and secretion, to finally form a hemostatic plug. During this diverse process, dynamic biochemical and morphological changes take place, and $Ca^{2+}$ is thought to play a crucial role in these platelet responses. Phosphoinositide hydrolysis by phosphoinositide-specific phospholipase C (PI-PLC) triggers initial mobilization of $Ca^{2+}$ from intracellular storage sites. Upon exposure to various stimuli, phosphatidylinositol 4,5-bisphosphate ($PIP_2$) is degraded by PLC, yielding two second messengers, inositol 1,4,5-trisphosphate ($IP_3$) and 1,2-diacylglycerol (DG). Moreover, it has recently begun to become clear that phospholipase D (PLD), which hydrolyzes phosphatidylcholine (PC) to phosphatidic acid (PA) and choline, plays an important role in the formation of second messengers such as PA and DG in many types of cells. Thus, membrane phospholipids function as precursors for second messengers and phospholipases catalyzing phospholipid hydrolysis: PLC, PLD, phospholipase $A_2$, and sphigomyelinase, which are collectively called *signal transducing phospholipases*. This chapter will summarize our current understanding of the function and regulatory mechanisms of platelet PLC and PLD.

## B. Receptor-Mediated Activation of Phosphoinositide-Specific Phospholipase C

### I. Activation of Phosphoinositide-Specific Phospholipase C in Platelets

The implication of receptor-mediated activation of phospholipid metabolism in cellular responses was suggested by HOKIN and HOKIN in the early 1950s (reviewed by HOKIN 1985). In $^{32}P$-labeled platelets, FIRKIN and WILLIAMS (1961) first demonstrated the labeling of PA and phosphoinositides. In these early stages, labeling of these lipids was thought to be coupled with secretory responses (so-called stimulus-secretion coupling). However, the labeling of these lipids turned out to be the consequence of hydrolysis of phosphoinositide by PI-PLC. In platelets, RITTENHOUSE-SIMMONS (1979) demonstrated the possible degradation of phosphoinositide by PI-PLC by measuring the formation of DG. However, the major DG source was believed to be phosphatidylinositol (PI). In the so-called PI turnover, degradation of PI was thought to be a primary event and a product, DG was subsequently phosphorylated to PA followed by resynthesis of PI. Later, the hydrolysis of $PIP_2$ by PI-PLC turned out to occur more promptly.

Although $PIP_2$ is a quantitatively minor membrane phospholipid, activation of PI-PLC results in the generation of two second messengers, $IP_3$ and DG. $IP_3$ causes increase in intracellular $Ca^{2+}$ concentration by mobilizing $Ca^{2+}$ from intracellular storage sites (probably the dense tubular system in platelets), and DG induces activation of protein kinase C (PKC). $Ca^{2+}$ ionophores or activators of PKC, such as oleoyl-acetylglycerol (OAG) and phorbol myristate acetate (PMA), cause platelet activation. Addition of both agents synergistically stimulates platelet responses, suggesting that $Ca^{2+}$ and PKC coordinately regulate platelet functions.

A wide variety of platelet agonists including thrombin, collagen, thromboxane $A_2$ ($TXA_2$), vasopressin, platelet-activating factor (PAF), and serotonin stimulates $PIP_2$ hydrolysis. However, adrenaline had no effect on $PIP_2$ metabolism. There are controversial results concerning the effect of ADP on PI-PLC activation.

## II. Formation of Second Messengers

### 1. Inositol 1,4,5-trisphosphate

Usually, the formation of $IP_3$ in response to the agonist is rapid and transient. For example, when human platelets were stimulated with 1 U/ml of thrombin, the amount of $IP_3$ increased promptly and reached a peak within 2 s (Fig. 1). 1 U/ml of thrombin produced nearly 10 pmol $IP_3$ per $10^9$ platelets. Based on an intraplatelet volume of 10 $\mu$l per $10^9$ cells, the intraplatelet concentration of $IP_3$ increases from 0.1 $\mu M$ to 1 $\mu M$.

$IP_3$ mobilizes $Ca^{2+}$ from permeabilized platelets and platelet membrane vesicles. The $ED_{50}$ value of $IP_3$ is nearly 1 $\mu M$, which corresponds to that formed in thrombin-stimulated platelets. $IP_3$ acts on the receptor that composes an $IP_3$-gated $Ca^{2+}$ channel. Three forms of $IP_3$ receptor subtypes have been cloned. These subtypes have structural similarities and consist of three functionally different domains: an N-terminal $IP_3$-binding domain, a regulatory domain, and a C-terminal transmembrane domain which forms channels (MONKAWA et al. 1995). Structural and functional analysis revealed that receptors form homo- and heterotetramers, as reported in the voltage-gated $K^+$ channels and the cyclic nucleotide-gated ion channels.

In addition to $Ca^{2+}$ mobilization, $IP_3$ causes $TXA_2$ formation, secretory response, and aggregation in permeabilized platelets. With the exception of $Ca^{2+}$ mobilization, these responses are however inhibited by pretreatment of platelets with the cyclooxygenase inhibitors indomethacin and aspirin or with

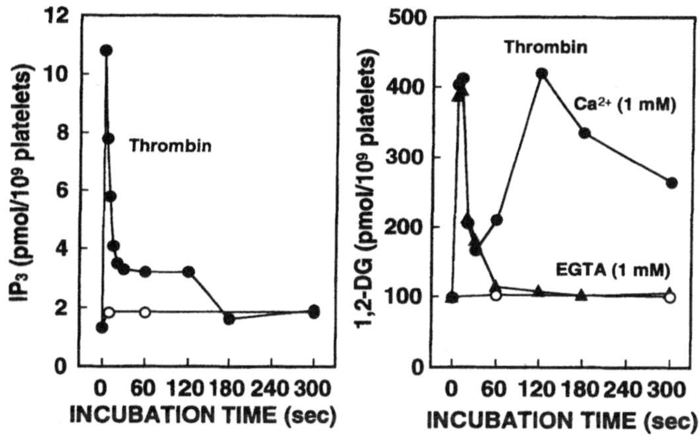


**Fig. 1.** $IP_3$ and diacylglycerol formation in thrombin-stimulated human platelets. Gel-filtered human platelets were stimulated with 1 U/ml of thrombin for indicated periods of time. *DG*, diacylglycerol; *$IP_3$*, inositol 1,4,5-trisphosphate; *EGTA*, ethylene glycol-bis($\beta$-aminoethyl ether) N,N,N′,N′-tetraacetic acid

$TXA_2$ receptor antagonists (AUTHI et al. 1986). These results indicate that $IP_3$-induced $Ca^{2+}$ mobilization does not directly elicit secretion and aggregation and that these responses in permeabilized platelets are dependent on the formation of $TXA_2$.

1,4,5-$IP_3$ is rapidly dephosphorylated to 1,4-$IP_2$ by $IP_3$ 5-phosphatase or further phosphorylated to 1,3,4,5-$IP_4$ by a specific 3-kinase. 1,3,4,5-$IP_4$ is then degraded by $IP_4$ 5-phosphatase to 1,3,4-$IP_3$ which has no $Ca^{2+}$ mobilizing activity. Inositol phosphates are finally dephosphorylated to inositol by sequential actions of phosphatases, and inositol is reutilized for PI synthesis. Besides the $Ca^{2+}$ mobilizing activity of 1,4,5-$IP_3$, several actions have been proposed for the other inositol phosphates. In addition to 1,4,5-$IP_3$, the formation of 1,2-cyclic 4,5-$IP_3$ has been reported in thrombin-stimulated platelets. 1,2-cyclic 4,5-$IP_3$ has a $Ca^{2+}$ mobilizing potency similar to that of 1,4,5-$IP_3$ in permeabilized platelets. In *Xenopus* oocytes or fibroblasts, microinjection of 1,3,4,5-$IP_4$ results in an increase in intracellular $Ca^{2+}$ concentration in the presence but not in the absence of extracellular $Ca^{2+}$, suggesting that $IP_4$ functions as a regulator of $Ca^{2+}$ influx. Several proteins that specifically bind $IP_4$ have recently been isolated, such as synaptotagmin II and Ras GTPase activating protein (rasGAP). Thus, it is reasonable to infer that $IP_4$ plays an important role in signal transduction, although definitive evidence has not been obtained in platelets.

## 2. 1,2-Diacylglycerol

Diacylglycerol (DG) functions as an intracellular second messenger that activates PKC. PKC comprises a family of related enzymes that are differentially expressed in various types of cells. Immunoblot analyses have revealed that platelets contain $\alpha$, $\beta$, $\delta$, $\varsigma$, $\theta$ and $\eta'$ (a 95-kDa protein recognized by anti-$\eta$

antibody) isozymes. In response to stimuli such as thrombin, PKC isozymes with the exception of the $\delta$ subtype translocate from the cytosol to the membrane. Thrombin and collagen show biphasic DG production in platelets (NAKASHIMA et al. 1991; WERNER and HANNUN 1991). A kinetic relationship has been demonstrated between the second sustained DG formation and PKC translocation (BALDASSARE et al. 1992). Moreover, in the absence of extracellular $Ca^{2+}$, the second sustained DG formation is abolished when PKC translocation is largely impaired. WERNER and HANNUN (1991) proposed on the basis of temporal analysis that the second sustained DG accumulation plays a key role in the induction of irreversible platelet aggregation.

In addition to PKC activation, DG can cause changes in the physical properties of the membrane (membrane perturbation; NOZAWA et al. 1991). The activation of PI-PLC and phospholipase $A_2$ in an in vitro assay system can be explained by its membrane perturbing effects. A membrane-fusogenic property of DG is also suggested. Furthermore, DG is metabolized by DG kinase to another potential second messenger, PA, and to glycerol by the sequential action of DG and monoacylglycerol (MG) lipases.

### 3. Phosphatidic Acid and Lysophosphatidic Acid

Exogenously added PA causes aggregation and augments thrombin-induced platelet activation in human platelets (KROLL et al. 1989). Thus, PA is regarded as a potential second messenger. However, platelet activation by PA is inhibited by the pretreatment of platelets with aspirin or indomethacin, indicating that the effects of PA are mediated by cyclooxygenase products.

Addition of lysophosphatidic acid (LPA) to platelet suspension induces shape change, secretion of granule constituents, and aggregation. LPA is produced by and secreted from activated platelets. Thus, LPA is considered to function as a secondary agonist. LPA is also known to have growth factor-like activity and actually stimulate proliferation of fibroblasts. LPA released from activated platelets may participate in the wound-healing process. In fibroblasts LPA stimulates the cascade of signaling events: hydrolysis of $PIP_2$ by PI-PLC, activation of PKC, activation of mitogen-activated protein (MAP) kinase, arachidonic acid release, and inhibition of adenylate cyclase (MOOLENAAR 1995). A putative LPA receptor (38–40kDa) is reported to be present on the plasma membrane. It is of great importance to note that platelets are chief suppliers of LPA, a multifunctional mediator.

## III. Modulation of Phosphoinositide-Specific Phospholipase C

Although PKC causes platelet aggregation and secretory response, pretreatment of platelets with phorbol esters or OAG results in partial inhibition of subsequent responses to agonists, such as $PIP_2$ hydrolysis, $Ca^{2+}$ mobilization, and secretory responses. Inhibitors of PKC reversed these inhibitory effects. These observations suggest the possibility that PKC may exert a feedback inhibition on PI-PLC activation, an initial biochemical event leading to platelet activation.

Platelet responses are attenuated by cyclic nucleotides, cyclic AMP (cAMP), and cyclic GMP (cGMP). Although several inhibitory mechanisms have been proposed such as a decrease in thrombin binding to its receptor and an activation of a $Ca^{2+}$ pump that decreases intraplatelet $Ca^{2+}$ concentrations, both cAMP and cGMP inhibit agonist-induced PI-PLC activation. For example, pretreatment of platelets with membrane-permeable cAMP analogs such as dibutyryl cAMP or with agents that activate adenylate cyclase such as prostaglandin $I_2$ ($PGI_2$) and $E_1$ ($PGE_1$) results in the inhibition of agonist-induced $PIP_2$ hydrolysis. Platelets also contain guanylate cyclase which produces cGMP from GTP. Unsaturated fatty acids and EDRF (endothel-derived relaxation factor), which has recently been identified as NO, activate guanylate cyclase. Increase in platelet cGMP content by membrane-permeable cGMP analogs such as 8-bromo-cGMP or treatment with NO results in inhibition of $PIP_2$ hydrolysis, aggregation, and secretory response.

Pretreatment of platelets with inhibitors of protein tyrosine kinase (PTK) such as ST-638, genistein, tyrphostin AG-213, and erbstatin results in agonist-induced platelet aggregation and secretion. Platelets contain the $\gamma 2$ subtype of PI-PLC (PLC-$\gamma 2$) which is tyrosine-phosphorylated by agonist stimulation.

## C. Regulation of Phosphoinositide-Specific Phospholipase C

### I. Multiplicity of Phosphoinositide-Specific Phospholipase C

Phosphoinositide-specific phospholipase C (PLC) has been documented to be an important enzyme that generates two second messengers, $IP_3$ and DG, in the receptor-stimulated signal transduction in mammalian cells. These two messengers participate in elevating cytosolic calcium concentration and activating protein kinase(s) C. Since the pioneering work in human platelets by RITTENHOUSE (1983) demonstrating a single PLC capable of hydrolyzing phosphoinositides (PIP > $PIP_2$ > PI), it has well been established that platelets contain multiple isoforms (60–400 kDa) of PLC (NOZAWA et al. 1993). Several distinct PLC enzymes have been purified from a variety of mammalian cells and fall into at least three types based on their sequence homology: PLC-$\beta$ (130–154 kDa), PLC-$\gamma$ (145–146 kDa), and PLC-$\delta$ (85–88 kDa). Within each type there are subtypes: $\beta 1$, $\beta 2$, $\beta 3$, $\beta 4$; $\gamma 1$, $\gamma 2$; $\delta 1$, $\delta 2$, $\delta 3$, $\delta 4$. In addition, there are several reports concerning purification of PLCs with molecular masses of 60–70 kDa from a variety of tissues and cells. Some of them are presumed to be derived from proteolytic truncation of a PLC isoform. Little sequence homology is present among the three types except for domains X (170 amino acids) and Y (260 amino acids). The domains, called *src* homology regions 2 and 3 (SH2 and SH3), are present in the structure of the PLC-$\gamma$ type.

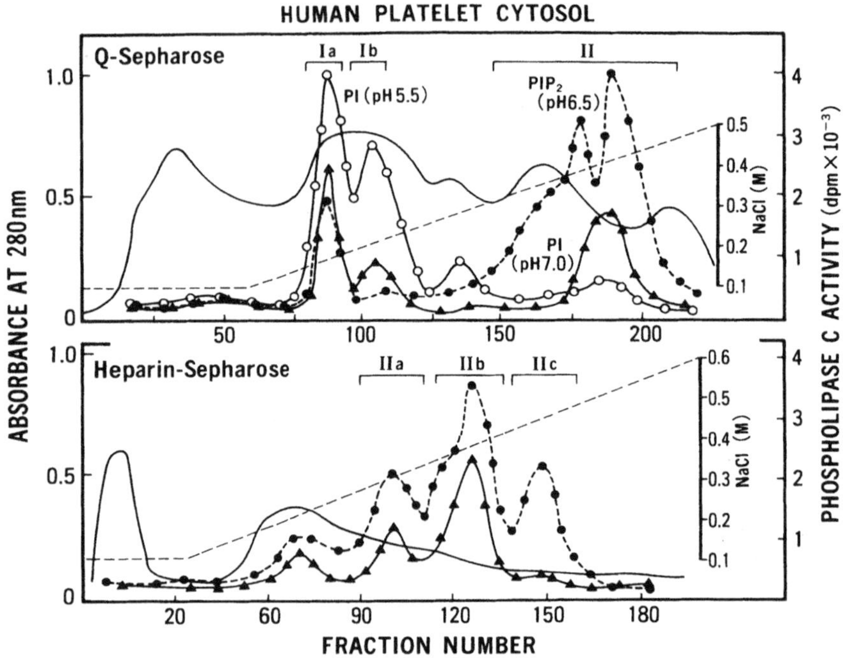


**Fig. 2.** Q-Sepharose and heparin-Sepharose column chromatographies of human platelet cytosolic PLC isoforms. The human platelet cytosolic fraction was applied onto a Q-Sepharose column and the peak fractions indicated by *bars* were separately pooled (*upper panel*). The peak II fraction was applied onto heparin-Sepharose column and eluted with NaCl buffer (*lower panel*). NaCl concentration (*dashed line*); absorbance at 280 nm (*solid line*); phospholipase C (*PLC*) activity for phosphatidylinositol (*PI*) at pH 5.5 (○); *PLC* activity for *PI* at pH 7.0 (▲); *PLC* activity for phosphatidylinositol 4,5-bisphosphate ($PIP_2$) at pH 6.5 (●)

The PLC-$\delta$ type exhibits a high level of the $Ca^{2+}$-binding loop in the EF-hand domain. Furthermore, recent studies using the crystal structure of PLC$\delta$1 have revealed the presence of a pleckstrin homology (PH) domain (Ferguson et al. 1995) and a C-terminal $\beta$-sandwich domain (C2 domain; Essen et al. 1996).

Various active fractions of PLCs from the human platelet cytosol and membrane fractions have been separated and identified by using specific antibodies for individual PLC isoforms (Banno et al. 1992). When the human platelet cytosol is subjected to Fast Q-Sepharose and heparin-Sepharose column chromatographies, several PLC activity peaks are resolved (Fig. 2). On the basis of cross-reactivity with antibodies raised against brain PLC isoforms, the four PLC fractions were identified as PLC-$\gamma$1, -$\gamma$2, -$\beta$, -$\delta$, and two unidentified PLCs. The former four isoforms are also reported by Baldassare et al. (1993). In the whole lysate of human platelets, PLC-$\beta$2 and PLC-$\beta$3 are detected by Western blotting, but PLC-$\beta$1 is hardly detectable and PLC-$\beta$4 is not. The antibody for PLC-$\beta$3 shows a positive reaction with three different bands (155 kDa, 140 kDa, 100 kDa; Banno et al. 1995). The 140-kDa type of

PLC-$\beta$3 is detected in human platelet cytosol, while 155-kDa and 100-kDa types are mainly present in the membrane fractions. The 140-kDa and 100-kDa PLC-$\beta$3 appear to be truncated from 155-kDa PLC-$\beta$3 or generated by alternative splicing. By Western blotting, PLC-$\gamma$2 is more densely stained than PLC-$\gamma$1, suggesting a higher content of the former in human platelets. The PLC-$\delta$ type has not yet been defined in human platelets.

## II. G-Protein-Mediated Phospholipase C-$\beta$ Activation

Several lines of evidence indicate that G proteins are involved in receptor-coupled phosphoinositide hydrolysis in platelets. From the experiments using permeabilized platelets, HASLAM and DAVIDSON (1980) first provided evidence for possible involvement of G proteins in PLC activation. In human platelets, thrombin, PAF, vasopressin, and $TXA_2$ activate PLC via G protein(s) (BRASS et al. 1993). Evidence for different types of G proteins coupled to PLC activation in human platelets is scant. Under conditions in which thrombin-induced phosphoinositide hydrolysis is inhibited by pertussis toxin (PT), the ability of a $TXA_2$ analog, U46619, to activate PLC is unimpaired. This raises the possibility that members of the Gi family can mediate PLC activation by thrombin and that PT-insensitive G proteins couple to the $TXA_2$ receptor. These PT-insensitive G proteins include one or more members of the Gq family, Gq$\alpha$ and G11$\alpha$, to regulate PLC in $TXA_2$-stimulated platelets (SHENKER et al. 1991). An antibody directed against Gq$\alpha$ inhibits $TXA_2$-dependent phosphoinositide hydrolysis by PLC-$\beta$ in a reconstitution system, suggesting the presence of the signal pathway of $TXA_2$-receptor-coupled Gq$\alpha$-PLC-$\beta$ (BALDASSARE et al. 1993).

In contrast, the identity of a G protein coupled to the thrombin receptor leading to PLC activation in platelets is controversial. PT has been shown to inhibit thrombin-induced phosphoinositide hydrolysis in permeabilized platelets and membrane fractions, indicating involvement of Gi family (BRASS et al. 1993). However, it has also been reported that a PT-insensitive Gq is coupled to the thrombin receptor (BENKA et al. 1995). The thrombin receptor is cloned and known to be a member of the seven transmembrane span receptor family and activated via a novel proteolytic mechanism (VU et al. 1991). In fibroblasts overexpressing the thrombin receptor, phosphoinositide hydrolysis is largely insensitive to PT, whereas the adenylate cyclase response is completely blocked by the same treatment, suggesting that the thrombin receptor can activate both PLC and adenylate cyclase via at least two distinct G proteins, most likely Gq and Gi2. In addition to the Gq family, PT-insensitive G proteins G12 and G13 are also observed to be activated via $TXA_2$ and thrombin receptors in human platelets (OFFERMANNS et al. 1994).

Reconstitution studies with purified PLC and G proteins show that the PLC-$\beta$, but not PLC-$\gamma$ and PLC-$\delta$, is regulated by a G protein via two distinct mechanisms, one through an $\alpha$ subunit of the Gq family insensitive to PT, and the other through the $\beta\gamma$ subunits. The $\alpha$q activates PLC-$\beta$ isoforms in the order of PLC-$\beta$1 > PLC-$\beta$3 > PLC-$\beta$2, whereas $\beta\gamma$ subunits stimulate PLC-$\beta$ in the order of PLC-$\beta$3 > PLC-$\beta$2 > PLC-$\beta$1. The PLC-$\beta$ isoforms are obviously not effectors of $\alpha$12 and $\alpha$13 since PLC-$\beta$1 is regulated by $\alpha$ subunits of the Gq family but not by G12 or G13, whereas PLC$\beta$2 is primarily regulated by $\beta\gamma$ subunits. The PLC-$\beta$3 appears to be regulated by both $\beta\gamma$ and $\alpha$q. Human platelets contain mainly PLC-$\beta$2 and PLC-$\beta$3. Therefore, in thrombin- and $TXA_2$-stimulated human platelets, the activation of PLC-$\beta$3 occurs via $\alpha$q/11 and that of PLC-$\beta$2, -$\beta$3 via $\beta\gamma$ subunits.

Despite abundant studies, the identity of PT-sensitive G protein(s) that activate(s) PLC in response to thrombin is still unclear. Human platelets contain large amounts of the Gi family; Gi2 is a major component, but Gi1 and Gi3 are much less abundant (NOZAWA et al. 1993). There has been growing evidence that the $\beta\gamma$ subunit of Gi plays a role in activating PLC-$\beta$2 and

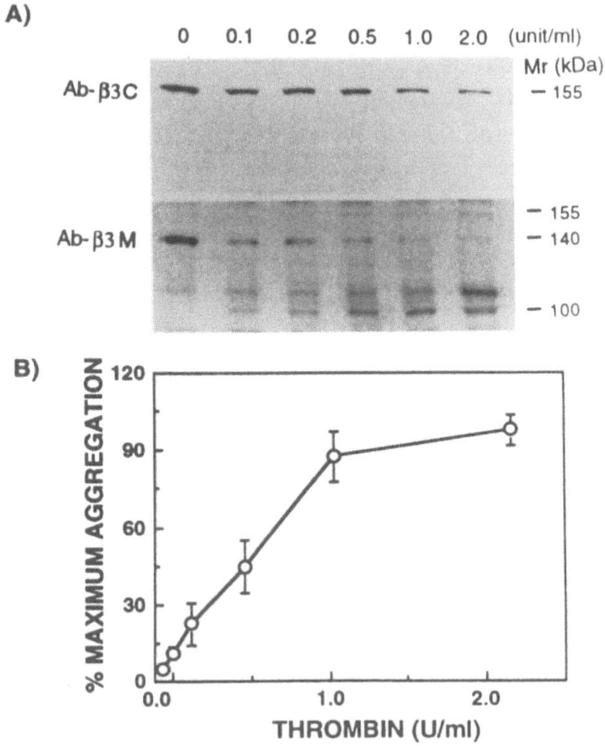


**Fig. 3A,B.** Cleavage of thrombin-induced phospholipase-C-$\beta_3$ (PLC-$\beta$3) and platelet aggregation. **A** Platelets in modified Hepes-Tyrode buffer supplemented with 1 *mM* $CaCl_2$ were treated with various concentrations of thrombin for 10 min with stirring. Total platelet lysates were resolved by SDS-PAGE and immunoblotted with Anti-PLC-$\beta$3 antibodies. *Ab-β3C*, antibody to amino acid residues 1202-1217 of PLC-$\beta$3; *Ab-β3M*, antibody to 550-561 of PLC-$\beta$3. **B** Platelet aggregations for 5 min in A were measured in an aggregometer and expressed as the percentage of maximum aggregation

PLC-$\beta$3. A truncated PLC-$\beta$3 form (100kDa) produced by calpain was activated to a greater extent by $\beta\gamma$subunits than was the intact 155kDa enzyme in vitro (BANNO et al. 1994), indicating that the limited proteolysis by calpain, the most abundant neutral protease in platelets, renders the 155kDa enzyme feasible to interact with $\beta\gamma$ subunits without altering its catalyzing activity. When examined with a specific antibody (Fig. 3), stimulation by thrombin and collagen of human platelets was observed to evoke cleavage of PLC-$\beta$3 (155kDa and 140kDa) to a 100kDa truncated form via calpain action (BANNO et al. 1995). Furthermore, the integrin $\alpha_{IIb}\beta_3$-ligand interaction induces translocation to the actin-rich cytoskeleton of PLC-$\beta$3 (155kDa and 140kDa) and various signal molecules (BANNO et al. 1996). It has been shown that the activation of calpain is induced as a consequence of binding of adhesive ligand to integrin $\alpha_{IIb}\beta_3$. These findings lead us to propose that PLC-$\beta$3 (155kDa and 140kDa) is cleavaged after translocation to the cytoskeleton to PLC-$\beta$3 (100kDa) by calpain activation and that its activation by the $\beta\gamma$ subunit leads to the second phase of DG accumulation. Thus, the activation of relocated PLC-$\beta$3 may play a role in late platelet-to-platelet aggregation.

Small $M_r$ G proteins (21–28kDa) are also thought to be involved in regulation of PI-PLC in human platelets. PLC-$\gamma$1 is associated with RasGAP in the platelet cytosol (LAPETINA and FARRELL 1993). Upon thrombin stimulation, RasGAP interacts with Rap1B in platelet membranes, allowing the interaction of PLC-$\gamma$1 with phosphoinositides. Membrane-bound Rap1B is phosphorylated by cAMP-dependent protein kinase and transferred from membrane to cytosol, thereby resulting in PLC inhibition.

## III. Tyrosine Kinase-Mediated Phospholipase C-$\gamma$ Activation

Platelets contain high levels of nonreceptor PTK activity, largely due to the proto-oncogene products pp60$^{c\text{-}src}$ and p72$^{syk}$ and also to a lesser PTK such as *yes*, *fyn*, *lyn*, and *hck* products. Tyrosine phosphorylation of various proteins occurs in platelets stimulated with various agonists (CLARK et al. 1994). Several studies have implicated tyrosine phosphorylation as a major regulator of the activation of PLC in platelets. In other cells, tyrosine phosphorylation by the receptor or PTK, e.g., PDGF and EGF receptors, is well known to regulate the activity of PLC-$\gamma$isoforms. Moreover, tyrosine phosphorylation of PLC-$\gamma$1 or PLC-$\gamma$2 has been demonstrated by nonreceptor PTKs such as *src*, *fyn*, and *lck* products in some hematopoietic cells upon stimulation of the antigen receptors. In human platelets, PLC-$\gamma$2, but not PLC-$\gamma$1, was first reported to be tyrosine-phosphorylated by stimulation with thrombin (TATE and RITTENHOUSE 1993). Similar results are obtained in Fc$\gamma$RII and collagen-stimulated platelets (BLAKE et al. 1994). Furthermore, the collagen-induced phosphorylation of PLC-$\gamma$2 is inhibited by staurosporine but not by the protein kinase C inhibitor, suggesting that collagen-dependent platelet responses occur as a result of tyrosine phosphorylation-dependent activation of PLC-$\gamma$2, but not as a G protein-coupled receptor activation (DANIEL et al. 1994).

Activation of platelets by collagen and wheat-germ agglutinin stimulated tyrosine phosphorylation of $pp72^{syk}$ (OHTA et al. 1992; BLAKE et al. 1994). These observations lead us to presume that $pp72^{syk}$ may participate in PLC-$\gamma$2 tyrosine phosphorylation in platelets stimulated by collagen and wheat-germ agglutinin.

In vitro studies with purified PLC-$\gamma$1 indicate that the tyrosine-phosphorylated enzyme does not increase its catalytic activity. Accordingly, evidence has not yet been presented to indicate direct PLC-$\gamma$2 activation by PTKs in stimulated platelets. An alternative explanation is the role of the cytoskeleton. Unphosphorylated PLC-$\gamma$1 is selectively inhibited in the presence of the actin-binding proteins (profilin, gelsolin) that have a high binding affinity for $PIP_2$, whereas only tyrosine-phosphorylated PLC-$\gamma$1 can catalyze the hydrolysis of profilin-bound $PIP_2$ in fibroblasts (GOLDSCHMIDT-CLERMONT and JANMEY 1991). Association of PLC-$\gamma$1 with actin via its SH3 domain has been demonstrated in hepatocytes, where EGF-induced translocation of PLC-$\gamma$1 to the cytoskeleton is well correlated with its tyrosine phosphorlation and increased catalytic activity (YANG et al. 1994). Actin is a major component of the cytoskeleton system in the platelet, and the actin-binding protein is involved in the association of GPIa, a collagen receptor with actin filaments. In human platelets, PLC-$\gamma$is physically associated with an actin/gelsolin complex (BANNO et al. 1992). Furthermore, other signaling enzymes such as PI3-kinase, PKC, and Src also translocate to the cytoskeleton in activated human platelets (ZHANG et al. 1992). These observations indicate that the phosphoinositide metabolism would occur at cytokeleton anchoring points to the membranes.

## D. Roles and Regulation of Phospholipase D in Platelet Activation

### I. Activation of Phospholipase D in Platelets

Several studies have demonstrated the biphasic production of DG in agonist-stimulated cells. In most cases, the first phase of DG increase is rapid and transient, which is concurrent with $IP_3$ formation. In contrast, the second sustained DG accumulation is reported to occur in the absence of $IP_3$ elevation, suggesting that DG comes form sources other than $PIP_2$. Quantitative and temporal discrepancy between $IP_3$ and DG formation prompted investigators to examine the source of secondary DG. In many types of cells, PC appears to take a major part in the production of DG, although some PI hydrolysis takes place, and in some cells the major source is phosphatidylethanolamine (PE; EXTON 1994). PC can be hydrolyzed by phospholipase D to PA and choline and/or by the putative PC-specific PLC to DG and phosphocholine. Recent studies indicate that the major mechanism is activation of PLD, yielding PA. DG arises due to the subsequent action of PA phosphohydrolase (PAP) on PA. PC hydrolysis by PC-PLC and/or PLD can be monitored by measuring the production of [$^3$H]phosphocholine and [$^3$H]choline in [$^3$H]choline-labeled cells. However, the relative contribution of PLD and PC-PLC pathways in producing second messengers is not clear due to the rapid interconversions between choline and phosphocholine, and between DG and PA.

PLD activity can be measured by the transphosphatidylation reaction, in which the phosphatidyl group of phospholipids is transferred to a primary alcohol such as ethanol or butanol. Since this reaction is only catalyzed by PLD, the formation of a metabolically stable phosphatidylalcohol is an unequivocal indicator of PLD activity. In order to preferentially label PC, radioactive saturated fatty acids such as [$^3$H]myristic acid and [$^3$H]palmitic acid are used for measurement of phosphatidylalcohol formation. In neutrophils and platelets, radioactive 1-alkyl-2-lyso-PC (lysoPAF) is often used for PC labeling.

In the presence of ethanol, thrombin induces phosphatidylethanol formation (RUBIN 1988), indicating the activation of PLD. From the analysis of [$^3$H]choline formation in [$^3$H]choline-labeled platelets, it appears that thrombin and collagen activate PLD whereas ADP, adrenaline, and platelet-activating factor (PAF) have no effect (PETTY and SCRUTTON 1993). Similar findings are obtained for thrombin, collagen, the $Ca^{2+}$ ionophore A23187, and $TXA_2$ that stimulate the formation of phosphatidylethanol (MARTINSON et al. 1995). In thrombin- or collagen-stimulated platelets, PLD activation occurs concurrently with aggregation (PETTY and SCRUTTON 1993; CHIANG 1994). When aggregation is prevented by chelation of extracellular $Ca^{2+}$ or without stirring, PLD activity is significantly reduced, suggesting that PLD activation is caused by aggregation. Interestingly, high density lipoprotein ($HDL_3$) has been reported to activate PLD through binding to its receptor glycoprotein (GP) IIb/IIIa complex (NOFER et al. 1995). In contrast, thrombin stimulates PLD activity in the presence of RGDS peptides or in platelets from Glanzmann's thromboasthenia patient lacking GP IIb/IIIa (MARTINSON et al. 1995), suggesting that PLD activation occurs in the absence of GPIIb/IIIa activation. Moreover, PA, a product of PLD, enhances the binding of GP IIb/IIIa to fibrinogen and accelerates agonist-induced platelet aggregation (CHIANG 1994; MARTINSON et al. 1995). Thus, it is proposed that PLD may play an important role in platelet aggregation. In contrast, HASLAM's group (COORSSON and HASLAM 1993) has proposed the functional coupling between PLD activation and secretory response, as reported in neutrophils, mast cells, and pancreatic islet cells. Their proposal is based on the finding that in permeabilized platelets, GTP$\gamma$S-induced PLD activation is concurrent with secretory response in a $Ca^{2+}$-independent manner. Further information is necessary to define the role of PLD in platelet functions.

In human platelets, thrombin stimulates a biphasic accumulation of DG, with an early phase reaching a peak within 10s and a late sustained phase peaking at 2min (Fig. 1; NAKASHIMA et al. 1991). The time course of primary DG production corresponds well with that of $IP_3$ formation, which is rapid and transient. The second sustained phase of DG accumulation is observed after $IP_3$ returns to basal level, indicating that DG is produced from phospholipid(s) other than $PIP_2$. The mass analysis of phospholipids shows the decrease in PC and PI upon thrombin stimulation. However, the PLD pathway plays a minor role (at most 13% of total) in PA formation in thrombin-stimulated human platelets (HUANG et al. 1991). Moreover, PA produced by the PLD pathway is hardly metabolized to DG. Accordingly, analysis of phosphatidylbutanol formation reveals that human platelets exhibits only 2% of the PLD activity of human neutrophils. There is also evidence that in an in vitro assay system, human platelets show much lower PLD activity than the rat brain or HL60 cells (BANNO et al., unpublished data). We have examined the source of the second sustained phase of DG (NAKASHIMA et al. 1991). Although PC and PI significantly decrease upon thrombin stimulation, our results suggest that the majority of PC decrease may not be mediated by PLD but rather due to hydrolysis by phospholipase $A_2$. It is also suggested that the major part of the second sustained phase of DG may come from PI (NAKASHIMA et al. 1991). The second sustained phase of DG accumulation was extracellular $Ca^{2+}$-dependent (Fig. 1). As described above, platelet PLC-$\beta$3 is cleaved by calpain, a $Ca^{2+}$-dependent protease. Thus, it is tempting to speculate that the truncated form of PLC-$\beta$3 may play a role in the sustained DG formation from PI.

## II. Regulation of Phospholipase D Activity

The mechanism underlying receptor-mediated PLD activation is not well understood. However, several factors have been proposed including calcium, protein kinase C, protein tyrosine kinase, heterotrimer GTP-binding proteins, and small molecular GTP-binding proteins in many types of cells (Fig. 4). In several cell types, PLD activation requires an increase in intracellular $Ca^{2+}$ concentration because depletion of extracellular $Ca^{2+}$ reduces phosphatidylalcohol formation and the $Ca^{2+}$ ionophore A23187 potentiates PLD activity. PKC also plays a role in PLD activation by agonists since phorbol esters and synthetic DG analogs promote phosphatidylalcohol formation. Furthermore, down-regulation of PKC by long-term exposure of the cells to PMA attenuates PLD activation. Thus, PLD activation occurs downstream of PI-PLC in several types of cells. In fact, activation of PI-PIC-$\gamma$ is essential for the activation of PLD in PDGF- or EGF-stimulated fibroblasts. In contrast, in some types of cells PLD activity is independent of PLC activation. For example, activation of rat mast cell PLD in response to stem cell factor (SCF) occurs without production of $IP_3$ (KOIKE et al. 1993). Activation of PLD by receptors with intrinsic PTK or associated with nonreceptor PTK as well as by those receptors coupled with GTP-binding protein is inhibited by PTK inhibitors such as genistein. Thus, PLD can be activated via protein tyrosine phosphorylation, although the detailed mechanism has not yet been defined. The involvement of GTP-binding proteins in PLD activation was suggested by the studies using permeabilized cells and isolated membranes. Addition of GTP$\gamma$S

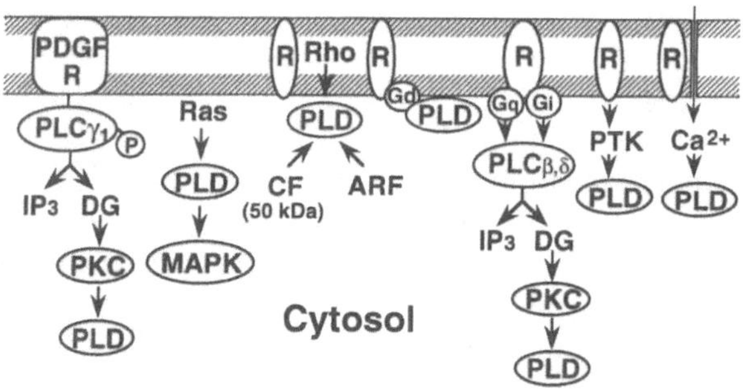


**Fig. 4.** Various mechanisms involving regulation of phospholipase D. In platelets, protein kinase C (*PKC*), $Ca^{2+}$ and protein tyrosine kinase (*PTK*) play roles in phospholipase D (*PLD*) activation. In addition, the involvement of GTP-binding protein(s) was also suggested by the data obtained in GTPγS-stimulated permeabilized platelets. However, the protein involved has not yet been identified. *MAPK*, MAP kinase; *CF*, cytosolic factor; *PLC*, phospholipase C

to these cells resulted in PC hydrolysis by PLD. In neutrophils and HL60 cells, activation of PLD by GTPγS requires a low molecular weight GTP-binding protein, ADP-ribosylation factor (ARF). In addition to ARF, other GTP-binding proteins such as Rho family proteins, Ral, and Ras activate membrane-bound PLD in several types of cells. The evidence for the involvement of Rho family proteins comes from the findings that RhoGDI, the Rho family-specific GDP dissociation inhibitor, inhibits PLD activation by GTPγS in in vitro assay systems of neutrophils, HL60 cells, and hepatocytes. We have recently shown the synergistic activation of membrane-bound PLD activity in HL60 cells by RhoA and PKCα (Ohguchi et al. 1996). The possible involvement of Ral was demonstrated in v-Src transformed fibroblasts. In *Xenopus* oocytes, injected Ras causes PLD activation which leads to activation of MAP kinase. Thus, the activation mechanisms of PLD may be dependent on types of cells or agonists.

In agonist-stimulated platelets, PLD seems to be activated following PI-PLC activation because $Ca^{2+}$ ionophore A23187 or PMA stimulates PLD activity (Huang et al. 1991; Van der Meulen and Haslam 1990; Chiang 1994). Prevention of agonist-induced PLD activation by inhibitors of PKC indicates involvement of PKC. However, HDL3 causes PLD activation through its receptor GP IIb/IIIa in the absence of phosphoinositide turnover (Nofer et al. 1995). The activation of the GP IIb/IIIa complex is not essential for but potentiates PLD activation by thrombin and collagen (Petty and Scrutton 1993; Chiang 1994). In permeabilized rabbit platelets, GTPγS or GTP stimulate PLD, suggesting the involvement of GTP-binding protein (Van der Meulen and Haslam 1990; Coorsson and Haslam 1993). Other nucleoside triphosphates, such as ATPγS, ITP, XTP, UTP, and CTP, also stimulate PLD activity in rabbit platelet membranes (Fan et al. 1994). These effects of nucleoside triphosphates are explained by the formation of GTPγS or GTP by nucleoside diphosphate kinase in platelet membranes. GTP$\gamma$S-stimulated PLD activity is observed in the absence of $Ca^{2+}$, but is enhanced by $Ca^{2+}$ or pretreatment of platelets with PMA before permeabilization. Therefore, full activation of platelet PLD requires PKC, $Ca^{2+}$, and GTP-binding protein. A possible implication of PTK in platelet PLD activation is also suggested. Pretreatment of platelets with PTK inhibitors, such as ST638, ST271, and tyrphostin A25, results in inhibition of phosphatidylethanol formation in thrombin-stimulated platelets and in GTP$\gamma$S-stimulated permeabilized platelets (MARTINSON et al. 1994).

## E. Remarks

Upon activation, platelets undergo a network of biochemical events that lead to the formation of haemostatic plug. There is no doubt that phosphoinositide metabolism plays an important part during these processes. Due to extensive investigations during the past decade, the function and regulatory mechanism of PLC are well characterized. However, those of PLD remain unclear. To this end, purification to homogeneity and molecular cloning of PLD are required. Considering the recent progress in signal transduction research, the framework covering the molecular mechanism of PLC and PLD and also their roles in platelet functions is sure to be constructed.

*Acknowledgements.* This work was supported by grants from the Ministry of Culture, Science and Education of Japan.

## References

Authi KS, Evenden BJ, Crawford N (1986) Metabolic and functional consequences of introducing inositol 1,4,5-trisphosphate into saponin-permeabilized human platelets. Biochem J 233:709–718

Baldassare JJ, Henderson PA, Burns D, Loomis C, Fisher GJ (1992) Translocation of protein kinase C isozymes in thrombin-stimulated human platelets. J Biol Chem 267:15585–15590

Baldassare JJ, Tarver AP, Henderson PA, Mackin WM, Sahagan, B, Fisher GJ (1993) Reconstitution of thromboxane $A_2$ receptor-stimulated phosphoinositide hydrolysis in isolated platelet membranes: involvement of phosphoinositide-specific phospholipase C-$\beta$ and GTP-binding protein Gq. Biochem J 291:235–240

Banno Y, Nakashima T, Kumada T, Ebisawa K, Nonomura Y, Nozawa Y (1992) Effects of gelsolin on human platelet cytosolic phosphoinositide-phospholipase C isozymes. J Biol Chem 267:6488–6494

Banno Y, Asano T, Nozawa Y (1994) Proteolytic modification of membrane-associated phospholipase C-$\beta$ by $\mu$-calpain enhances its activation by G-protein $\beta\gamma$ subunits in human platelets. FEBS Lett 340:185–188

Banno Y, Nakashima S, Hachiya T, Nozawa Y (1995) Endogenous cleavage of phospholipase C-$\beta$3 by agonist-induced activation of calpain in human platelets. J Biol Chem 270:4318–4324

Banno Y, Nakashima S, Ohzawa M, Nozawa Y (1996) Differential translocation of phospholipase C isozymes to integrin-mediated cytoskeletal complexes in thrombin-stimulated human platelets. J Biol Chem 271:14989–14994

Benka ML, Lee M, Wang G-R, Buckman SA, Burlacu A, Cole L, DePina A, Dias P, Granger A, Grant B, Amanda H-L, Karki S, Mann S, Marcu O, Nussenzweig A, Piepenhagen P, Raje M, Roegiers F, Rybak S, Salic A, Smith-Hall J, Waters J, Yamamoto N, Yanowitz J, Yeow K, Busa WB, Mendelsohn ME (1995) The thrombin receptor in human platelets is coupled to a GTP binding protein of the Gq$\alpha$ family. FEBS Lett 363:49–52

Blake RA, Schieven GL, Watson SP (1994) Collagen stimulates tyrosine phosphorylation of phospholipase C-$\gamma$2 but not phospholipase C-$\gamma$1 in human platelets. FEBS Lett 353:212–216

Brass LF, Hoxie JA, Kieber-Emmons T, Manning DR, Poncz M, Woolkalis M (1993) Agonist receptors and G proteins as mediators of platelet activation. In: Authi KS, Watson SP, Kakkar VV (eds) Mechanisms of platelet activation and control. Plenum, New York, p 17

Chiang TM (1994) Activation of phospholipase D in human platelets by collagen and thrombin and its relation ship to platelet aggregation. Biochim Biophys Acta 1224:147–155

Clark EA, Shattil SJ, Brugge JS (1994) Regulation of protein tyrosine kinases in platelets. Trends Biochem Sci 19:464–469

Coorsson JR, Haslam RJ (1993) GTP$\gamma$S and phorbol ester act synergistically to stimulate both $Ca^{2+}$-independent secretion and phospholipase D acivity in permeabilized human platelets. FEBS Lett 316:170–174

Daniel JL, Dangelmaier C, Smith JB (1994) Evidence for a role for tyrosine phosphorylation of phospholipase C$\gamma$2 in collagen-induced platelet cytosolic calcium mobilization. Biochem J 302:617–622

Essen LO, Perisic O, Cheung R, Katan M, Williams RL (1996) Crystal structure of a mammalian phosphoinositide-specific phospholipase C$\delta$. Nature 380:595–602

Exton JH (1994) Phosphatidylcholine breakdown and signal transduction. Biochim Biophys Acta 1212:26–42

Fan X-T, Sherwood JL, Haslam RJ (1994) Stimulation of phospholipase D in rabbit platelet membranes by nucleoside triphosphates and by phosphocreatine: roles of membrane-bound GDP, nucleoside diphosphate kinase and creatine kinase. Biochem J 299:701–709

Ferguson KM, Lemmon MA, Schlessinger J, Sigler PB (1995) Structure of the high affinity complex of inositol trisphosphate with a phospholipase C pleckstrin homology domain. Cell 83:1037–1046

Firkin BG, Williams WJ (1961) The incorporation of radioactive phosphorus into the phospholipids of human leukemic leukocytes and platelets. J Clin Invest 40:423 432

Goldschmidt-Clermont DJ, Janmey PA (1991) Profilin, a weak cap for actin and ras. Cell 66:419–421

Haslam RJ, Davidson MM (1984) Receptor-induced diacylglycerol formation in permeabilized platelets: possible role for a GTP-binding protein. J Recept Res 4:605–629

Hokin LW (1985) Receptors and phosphoinositide-generated second messengers. Annu Rev Biochem 54:205–235

Huang R, Kucera GL, Rittenhouse SE (1991) Elevated cytosolic $Ca^{2+}$ activates phospholipase D in human platelets. J Biol Chem 266:1652–1655

Koike T, Hirai K, Morita Y, Nozawa Y (1993) Stem cell factor-induced signal transduction in rat mast cells. J Immunol 151:359–366

Kroll MH, Zavoico GB, Schafer AI (1989) Second messenger function of phosphatidic acid in platelet activation. J Cell Physiol 139:558–564

Lapetina EG, Farrell FX (1993) Rap1B and platelet function. In: Authi KS, Watson SP, Kakkar VV (eds) Mechanisms of platelet activation and control. Plenum, New York, p 49

Martinson EA, Scheible S, Presek P (1994) Inhibition of phospholipase D of human platelets by protein tyrosine kinase inhibitors. Cell Mol Biol 40:627–634

Martinson EA, Scheible S, Greinacher A, Presek P (1995) Platelet phospholipase D is activated by protein kinase C via an integrin -independent mechanism. Biochem J 310:623–628

Monkawa T, Miyawaki A, Sugiyama T, Yoneshima H, Yamamoto-Hino M, Furuichi T, Saruta T, Hasegawa M, Mikoshiba K (1995) Heterotrimeric complex formation of inositol 1,4,5-trisphosphate receptor subunits. J Biol Chem 270:14700–14704

Moolenaar WH (1995) Lysophosphatidic acid, a multifunctional phospholipid messenger. J Biol Chem 270:12949–12952

Nakashima S, Suganuma A, Matsui A, Nozawa Y (1991) Thrombin induces a biphasic 1,2-diacylglycerol production in human platelets. Biochem J 275:355–361

Nofer JR, Walter M, Kehrel B, Seedorf U, Assmann G (1995) $HDL_3$ activates phospholipase D in normal but not in glycoprotein IIb/IIIa-deficient platelets. Biochem Biophys Res Commun 207:148–154

Nozawa Y, Nakashima S, Nagata K (1991) Phospholipid-mediated signaling in receptor activation of human platelets. Biochim Biophys Acta 1082:219–238

Nozawa Y, Banno Y, Nagata K (1993) Regulation of phosphoinositide-specific phospholipase C activity in human platelets. In: Authi KS, Watson SP, Kakkar VV (eds) Mechanisms of platelet activation and control. Plenum, New York, p 37

Offermanns S, Laugwitz K-L, Spicher K, Schultz G (1994) G proteins of the G12 family are activated via thromboxane $A_2$ and thrombin receptors in human platelets. Proc Natl Acad Sci USA 91:504–508

Ohguchi K, Banno Y, Nakashima S, Nozawa Y (1996) Regulation of membrane-bound phospholipase D by protein kinase C in HL60 cells: synergistic action of small GTP-binding protein RhoA. J Biol Chem 271:4366–4373

Ohta S, Taniguchi T, Asahi M, Kato Y, Nakagawara G, Yamamura H (1992) Protein-tyrosine kinase $p72^{syk}$ is activated by wheat germ agglutinin in platelets. Biochem Biophys Res Commun 185:1128–1132

Petty AC, Scrutton MC (1993) Release of choline metabolites from human platelets: evidence for activation of phospholipase D and of phosphatidylcholine-specific phospholipase C. Platelets 4:232–239

Randall RW, Bonser RW, Thompson NT, Garland LG (1990) A novel and sensitive assay for phospholipase D in intact cells. FEBS Lett 264:87–90

Rittenhouse SE (1983) Human platelets contain phospholipase C that hydrolyzes polyphosphoinositides. Proc Natl Acad Sci USA 80:5417–5420

Rittenhouse-Simmons S (1979) Production of diglyceride from phosphatidylinositol in activated human platelets. J Clin Invest 63:580–587

Rubin R (1988) Phosphatidylethanol formation in human platelets. Biochem Biophys Res Commun 156:1090–1096

Shenker A, Goldsmith P, Unson CG, Spiegel AM (1991) The G protein coupled to the thromboxane $A_2$ receptor in human platelets is a member of the novel Gq family. J Biol Chem 266:9309–9313

Tate BF, Rittenhouse SE (1993) Thrombin activation of human platelets causes tyrosine phosphorylation of PLC-$\gamma$2. Biochim Biophys Acta 1178:281–285

Van der Meulen J, Haslam RJ (1990) Phorbol ester treatment of intact rabbit platelets greatly enhances both the basal and guanosine 5′[-thio]triphosphate-stimulated phospholipase D activities of isolated platelet membranes. Biochem J 271:693–700

Vu T-KH, Hung DH, Wheaton VI, Coughlin SR (1991) Molecular cloning of a functional thrombin receptor reveals a novel proteolytic mechanism of receptor activation. Cell 64:1057–1068

Werner MH, Hannun YA (1991) Delayed accumulation of diacylglycerol in platelets as a mechanism for regulation of onset of agregation and secretion. Blood 78:435–444

Yang LJ, Rhee SG, Williamson (1994) Epidermal growth factor-induced activation and translocation of phospholipase C-$\gamma$1 to the cytoskeleton in rat hepatocytes. J Biol Chem 269:7156–7162

Zhang J, Fry MJ, Waterfield MD, Jaken S, Liao L, Fox LEB, Rittenhouse SE (1992) Activated phosphoinositide 3-kinase associates with membrane skeleton in thrombin-exposed platelets. J Biol Chem 267:4686–4692

CHAPTER 12

# Protein Kinase C and Its Interactions with Other Serine-Threonine Kinases

J.A. WARE and J.D. CHANG

## A. Introduction and Definition

Despite the diversity of platelet agonists, ranging from soluble molecules that bind to receptors on the platelet surface to shear stress (KROLL et al. 1993) and various artificial surfaces (WARE et al. 1991), the changes in platelet function that they induce are stereotypical. These include cytoskeletal reorganization (shape change), conversion of the $\alpha$IIb-$\beta$3 integrin complex to a ligand-binding form, release of contents from storage granules, and clot retraction. Moreover, the pathways leading to expression of these diverse functions involve the generation of second messengers in a pattern that is somewhat similar among the agonists (WARE and COLLER 1994); these mediators in turn interact with intracellular effectors, of which the most widely studied are protein kinases, enzymes that transfer phosphoryl groups from ATP to an acceptor polypeptide substrate. Two general categories of protein kinases have been identified: those that mediate the phosphorylation of proteins on tyrosine residues and those that phosphorylate substrates on serine or threonine residues. There are many families of serine-threonine kinases, only some of which have been identified in platelets (Table 1); the first "second-messenger kinase" to be identified and characterized was cyclic AMP-dependent protein kinase (termed protein kinase A), followed by $Ca^{2+}$-calmodulin-dependent protein kinase (which was initially labeled B), and $Ca^{2+}$- and phospholipid-dependent protein kinase, or protein kinase C (PKC), were discovered subsequently. Although many tyrosine and serine-threonine kinases have been linked to the activation of platelets and other cells, the evidence linking PKC activation to platelet function is comparatively strong, and thus the pharmacology and physiology of PKC in platelets has been studied extensively.

PKC was discovered in 1977 by Nishizuka and colleagues, and, as implied above, was initially characterized on the basis of its biochemical properties; its unique requirement for both phospholipid and calcium distinguished it from both PKA and PKB. Two additional discoveries led to the explosion of interest in PKC. First, the fact that it is the major intracellular receptor for a family of tumor-promoting phorbol esters was recognized (CASTAGNA et al. 1982), not only raising the possibility that PKC participates in some neoplastic processes, but also providing a useful tool with which to dissect the consequences of its

**Table 1.** Protein serine/threonine kinases in platelets

| | |
|---|---|
| I. | Regulated by ligands (second messengers): |
| | A. Cyclic nucleotides: cAMP- and cGMP-dependent protein kinases (PKA and PKG) |
| | B. $Ca^{2+}$ calmodulin: myosin light chain kinases (MLCK), $Ca^{2+}$/calmodulin kinases I, II, III, and IV (CaM kinases I–IV) |
| | C. Diacylglycerol, unsaturated fatty acids, $Ca^{2+}$: protein kinase C (PKC) |
| II. | Regulated by phosphorylation: Phosphorylase kinase, MAP kinase, S6 kinases |
| III. | Regulated by changes in protein substrate: G-protein-receptor kinases |
| IV. | Regulation by unknown mechanisms: E.g., casein kinases I and II |

activation. Second, the finding that exposure of platelets and other cells to a wide variety of agonists stimulates rapid turnover of inositol phospholipids to produce 1,2-diacylglycerol (DG), which can activate PKC at levels of $Ca^{2+}$ found in platelet cytoplasm, suggested that PKC activation occurs shortly after stimulation of platelets. Since then, many mediators in addition to phosphoinositide-derived DG have been found to activate PKC (BELL and BURNS 1991; NISHIZUKA 1995), and there is now a growing appreciation of the multiplicity of pathways that can lead to PKC activation in agonist-stimulated platelets.

# B. Molecular Structure and Heterogeneity

The biochemical pathways that lead to PKC activation, as well as PKC itself, appear to be present in all mammalian cells. However, the fact that PKC activity actually results from multiple gene products, only some of which are expressed in platelets and which have distinct mechanisms of biochemical regulation, may explain the specificity of responses both among cell types and within an individual cell (DEKKER and PARKER 1994). The PKC isoenzymes that have been identified from various tissues and cells have been divided into three subfamilies: (a) the "conventional" or calcium-dependent PKCs first identified; (b) the "novel" or diacylglycerol-dependent, calcium-independent PKCs; and (c) the "atypical" PKCs, whose members, like those of the other two families, require phospholipid for maximal activation, but require neither DG nor $Ca^{2+}$ and do not bind or respond to phorbol ester (NISHIZUKA 1988; see Fig. 1). Members of the cPKC subfamily include alpha ($\alpha$), beta ($\beta$) I/II, and gamma ($\gamma$); the two $\beta$ isoenzymes differ by an alternatively spliced short segment in the carboxyl terminus. Novel PKCs identified thus far include delta ($\delta$), epsilon ($\varepsilon$) (NISHIZUKA 1988), eta ($\eta$) (OSADA et al. 1990; BACHER et al. 1991), and theta ($\Theta$) (OSADA et al. 1992; BAIER et al. 1993; CHANG et al. 1993), while the atypical PKC isoenzymes include zeta ($\zeta$) (NISHIZUKA 1988), iota/

lambda ($\iota/\lambda$) (SELBIE et al. 1993; AKIMOTO et al. 1994) (which appear to be the same gene product but unfortunately received different names nearly simultaneously), and mu ($\mu$) (JOHANNES et al. 1994). PKC $\mu$ was classified as atypical because of its failure to bind phorbol ester; however, this latter finding has been disputed by other investigators. Because of several unique features, including an N-terminal peptide leader sequence and a wider separation between the areas of cysteine repeats, some have argued that PKC $\mu$ should be considered as representative of a separate family of protein kinases, termed PKD (VALVERDE et al. 1994). All PKC isoenzymes have certain molecular features in common; they are composed of a single polypeptide chain that can be divided into an N-terminal regulatory domain and a C-terminal catalytic or kinase domain that binds ATP and phosphorylates substrates. Certain regions

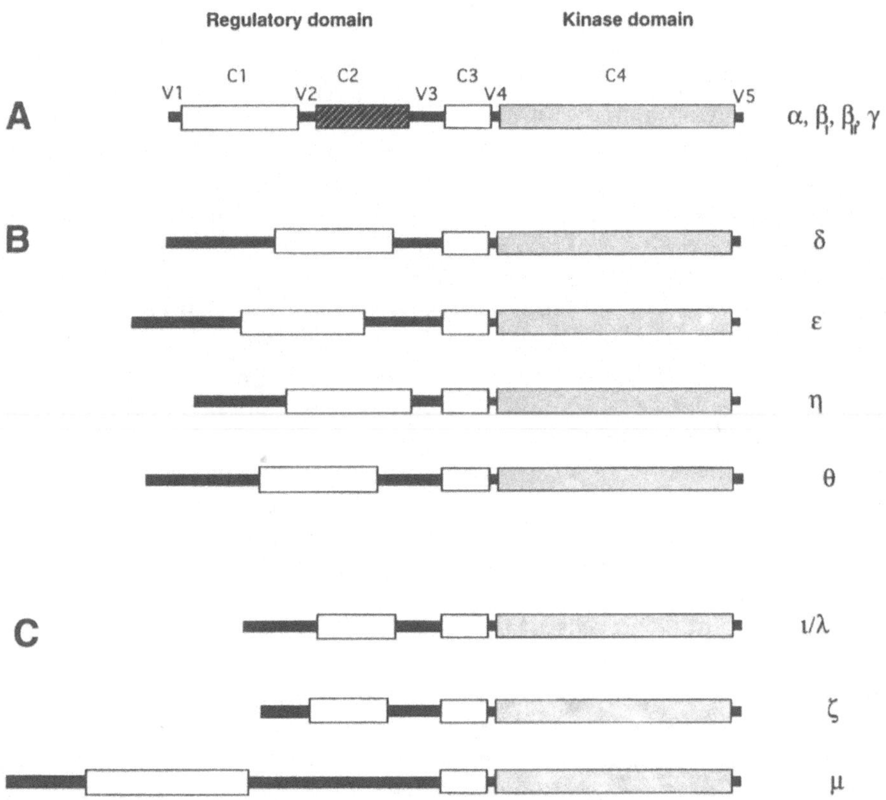


**Fig. 1A–C.** Structure of protein kinase C gene (PKC) family. Each PKC isoenzyme is composed of a single polypeptide chain that contains both regulatory and kinase domains. The PKC isoenzymes are composed of a series of conserved ($C_1$–$C_4$) and variable regions ($V_1$–$V_5$), including cystein-rich motifs contained in the $C_1$ region, a calcium-binding motif in the $C_2$ region, an ATP-binding sequence in the $C_3$ region, the substrate binding site within the $C_4$ region, and a pseudosubstrate domain in the $V_1$ region. The PKC isoenzymes are grouped according to the subfamily: **A** cPKC; **B** nPKC; **C** aPKC. (See text)

(C) within these two domains are highly conserved among all isoenzymes, and are separated by variable (V) regions that bear little resemblance to each other or among the various isoenzymes. Within the regulatory domain, areas of cysteine repeats form a zinc-finger motif in the first conserved region (C1) and represent the interaction site of effector molecules phorbol ester and DG; two of these repeat regions are required for such binding to occur. In the cPKC subfamily, a second conserved region is also found which appears to confer $Ca^{2+}$ responsiveness to the members of this subfamily (NISHIZUKA 1992). Also contained in the regulatory domain, immediately N-terminal to the cysteine repeats, is a sequence that resembles a consensus substrate site for PKC, except for the substitution of alanine for the phosphorylatable serine. This "pseudosubstrate" region functions to engage the substrate binding site of the catalytic domain, and thus prevents interaction of the catalytic domain with its true substrates in the absence of allosteric activators such as DG which induce an unfolding of the polypeptide that exposes the active site (KEMP and PEARSON 1991). The peptide sequence of this pseudosubstrate is similar among PKC-$\alpha$, -$\beta$, and -$\gamma$, but differs from that of PKC-$\eta$ and -$\varepsilon$, and from that in PKC-$\delta$ and -$\Theta$, which are similar to each other. This difference in pseudosubstrate sequence is at least one factor that is responsible for the specificity in substrate recognition by the various isoenzymes (DEKKER et al. 1993). Oligopeptides derived from these pseudosubstrate areas, such as RFARKGALRQKNVEHEVKN, which corresponds to residues 19–36 in PKC-$\alpha$, are potent competitive substrate antagonists of purified PKC and have been useful inhibitory tools in experiments in which they are added to platelets that have been rendered permeable (see e.g. KING et al. 1991). Expression of the isoenzymes varies considerably among tissues, and also in the subcellular fractions within individual cells. Some isoenzymes, such as PKC-$\delta$, -$\zeta$, and -$\alpha$, are expressed in nearly every cell type, while others are expressed in only a few tissues; for instance, PKC-$\gamma$ is expressed only in the central nervous system, and PKC-$\Theta$ is expressed chiefly in skeletal muscle, hematopoietic cells, and platelets (CHANG et al. 1993). In many cells, PKC-$\delta$ is bound to the surface membrane, while other isoenzymes are found in the cytoplasm or nuclear membrane of the same cells. Human blood platelets contain multiple isoenzymes of PKC; only two of these isoenzymes (PKC $\alpha$ and $\beta$) have been purified from platelet lysate. The remainder of the isoenzymes have been identified by use of isoenzyme-specific reagents, such as antibodies and cDNA probes (CRABOS et al. 1992; GRABAREK et al. 1992; BALDASSARE et al. 1992). Both antibodies and PCR-based analysis of reverse-transcribed RNA have demonstrated the presence of PKC-$\delta$ and PKC-$\zeta$ as well as PKC-$\alpha$ and -$\beta$ (GRABAREK et al. 1992). The newest member of the novel PKC family, PKC-$\Theta$, was identified at about the same time by three different groups in mouse brain, hematopoietic cells, and human platelets (OSADA et al. 1992; BAIER et al. 1993; CHANG et al. 1993). In addition to the variation of PKC isoenzymes among mature cell types, it is clear that many PKC isoenzymes are differentially expressed during development. For example, although at least five ($\alpha$, $\beta$, $\delta$, $\zeta$, and Θ) isoenzymes of PKC are present in human platelets, the megakaryoblastic precursor cell line HEL does not express either protein or RNA corresponding to PKC-$\alpha$, in contrast to the case with the other five isoenzymes (GRABAREK et al. 1992). Furthermore, the human erythroleukemic line K562 which express PKC-$\alpha$, -$\beta$, and -$\zeta$ can be induced to differentiate along megakaryocytic lines, which is associated with an increase in PKC-$\alpha$ and -$\zeta$, but a decrease in PKC-$\beta$. Differentiation also resulted from forced overexpression of PKC-$\alpha$ or was caused by inhibition of PKC-$\beta$ by antisense methodology (MURRAY et al. 1993). These results suggest that these members of the cPKC isoenzyme subfamily may mediate distinct functions in the differentiation and maturation of megakaryocytic cells.

## C. PKC Activation by Lipid Mediators in Platelets

Membrane phospholipids, in particular phosphatidylserine, are required for the activation of all forms of PKC identified thus far; however, these lipids are not likely to serve as the physiologic regulators of PKC activity, which is more likely to be a function of one or more lipid mediators generated transiently following stimulation. The best-known mechanism for activation of PKC in platelets and in other cells remains the generation of DG via hydrolysis of membrane phospholipids by a phospholipase (phosphoinositidase) C, which is activated by receptor-mediated agonists. Most physiologic evidence also suggests that DG is the principal mediator for PKC in activated platelets; measurements of DG mass in stimulated platelets demonstrate a close temporal correlation between its formation, PKC activation, and platelet aggregation and release (WERNER et al. 1992). In the past several years, however, other mechanisms for generating DG have been discovered, as have been pathways by which several other PKC-activating lipid metabolites can be generated as a result of signal-induced hydrolysis of various membrane phospholipids (BELL and BURNS 1991; NISHIZUKA 1995).

Of the three groups of phospholipase C, most interest in platelets has been focused on PLC $\beta$; this group of phospholipases, comprised of three members, is activated by the $\alpha$-subunit of the Gq protein, which is associated with many G-protein-linked receptors, including many that mediate responses to platelet agonists (MANNING and BRASS 1991). Hydrolysis of phosphatidylinositol by this mechanism produces the most rapid rise in DG which disappears rapidly because it is converted into phosphatidic acid by diacylglycerol kinase (NISHIZUKA 1995).

A slower rise in DG results from hydrolysis of phosphatidylcholine (PC); such hydrolysis can result, at least theoretically, from the action of either a PLC or a phospholipase D (PLD) specific for PC (Fig. 2). Most evidence favors the latter enzyme as the source for slower, more sustained PKC activation (BISHOP et al. 1992). The immediate product of PLD's action on PLC is phosphatidic acid (PA), which can be converted to DG by a PA phosphatase;

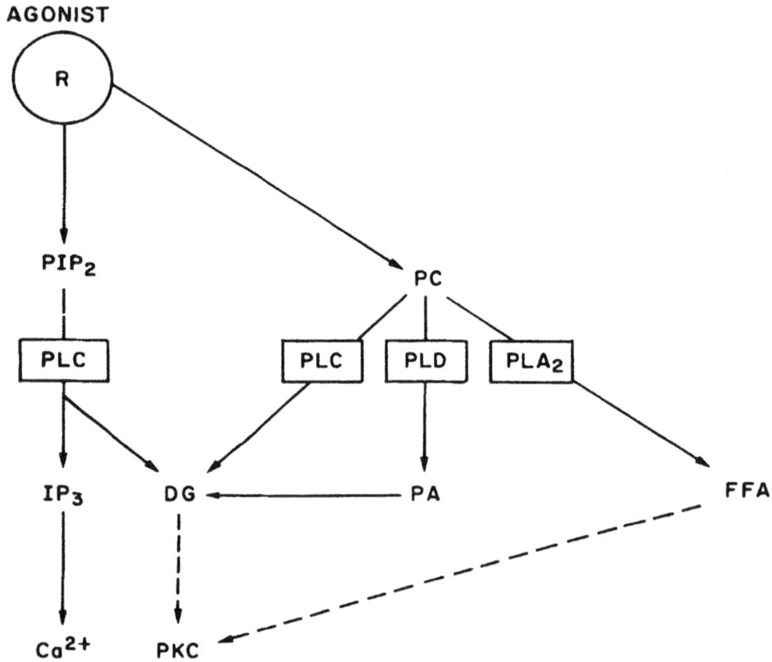


**Fig. 2.** Schematic diagram of phospholipase action. Receptor (*R*)-binding by agonists can activate several phospholipases, including phospholipase C (*PLC*), phospholipase D (*PLD*), and phospholipase $A_2$ ($PLA_2$), which act on components of the surface membrane (e.g., phosphatidylinositol 4,5-bisphosphate and phosphocholine). The products of *PLC*'s action on phosphatidylinositol 4,5-bisphosphate ($PIP_2$) include inositol trisphosphate ($IP_3$) which mobilizes $Ca^{2+}$ and diacylglycerol (*DG*). This latter mediator can also be produced directly or indirectly via phosphatidic acid (*PA*) by *PLC* or *PLD*; action on phosphocholine (*PC*) produces free fatty acids (*FFA*) that can activate protein kinase C (*PKC*) as can diacylglycerol. The *PLC* for which $PIP_2$ is a substrate probably differs from the *PLC* that hydrolyzes *PC*

whether this latter action is the sole mechanism by which PKC is activated by PLD or, alternatively, whether phosphatidic acid itself is an important direct activator of PKC, is not known. It has been estimated that only 10%–20% of total diacylglycerol mass in activated platelets is derived from PLD (HASLAM and COORSSEN 1993). In platelets, PLD appears to require $Ca^{2+}$ mobilization before significant hydrolysis of phosphatidylcholine can occur (HUANG et al. 1991), and prior activation of PKC appears to enhance this hydrolysis markedly (HASLAM and COURSSEN 1993).

Another class of PKC-activating effector molecules found in platelets in addition to DG includes of arachidonic acid, its derivatives, and other cis-unsaturated fatty acids (NISHIZUKA 1995). The biogenesis of these products, as well as of lyso-PC, results from the action of phospholipase $A_2$ on PC. PKC can be directly activated by cis-unsaturated fatty acids, including oleic, linolenic, linoleic, docosahexaenoic, and arachidonate acids (NISHIKAWA et al. 1988; KHAN et al. 1991) that greatly enhance the activation of PKC by DG and phorbol esters, even when cytoplasmic $Ca^{2+}$ is at levels characteristic of those found in unstimulated platelets. Activation by arachidonate appears to activate some, but not all PKC isoenzymes (SHEARMAN et al. 1989). The $\delta$-isoenzyme of PKC, which is present in platelets, is slightly activated by some cis-unsaturated fatty acids in the absence of diacylglycerol, but its full activation by diacylglycerol is markedly inhibited by these fatty acids (suggesting that cis-unsaturated fatty acids are a partial agonist for PKC-$\delta$), in contrast to the case of other isoenzymes such as PKC-$\varepsilon$ (NISHIZUKA 1995). Oleate, on the other hand, stimulates PKC derived from platelets, in particular the novel or non-$Ca^{2+}$-dependent isoenzymes, and potentiates PKC stimulation by DG (KHAN et al. 1993). The mechanism by which free fatty acids activate PKC is not clear, but it appears to interact with the enzyme at a location that differs from that of DG, which binds to the cysteine-rich repeats in the regulatory domain (YOSHIDA et al. 1992). The release of serotonin from platelets by DG is potentiated by cis-unsaturated fatty acids, as is the phosphorylation of the 47-kDa substrate pleckstrin; these fatty acids themselves are inactive, with the exception of arachidonate (NISHIKAWA et al. 1988). Lyso-PC can also potentiate PKC activation by DG in cell-free assays, although not to the extent of free fatty acids (NISHIZUKA 1995). It is not known whether such an effect occurs in intact cells under physiologic conditions.

Lipid products of the lipoxygenase pathway can also activate PKC, at least in biochemical assays. Some, but not all of the isoenzymes of PKC are activated by 12-hydroxyeicosatetraenoic acid (GRABAREK and WARE 1993), which accumulates in micromolar concentrations in platelets stimulated by a variety of agonists (WARE and COLLER 1994). Lipoxin A can also activate PKC (SHEARMAN et al. 1989); it is not formed by platelets alone, but, in the company of neutrophils, lipoxin can be produced, and may thus activate PKC in neighboring platelets (ROMANO and SERHAN 1992).

The phosphatidylinositol-3 (PI-3) kinase pathway has attracted much interest because of the accumulation of inositide phosphorylation products in various cells following addition of physiologic agonists such as thrombin (NOLAN and LAPETINA 1990; KUCERA and RITTENHOUSE 1990), however, suggest that it has roles other than those related to transducing signals from the cell surface to the nucleus. Initial studies of the interaction between these two enzymes revealed that PKC activation was essential for the products of PI-3 kinase to accumulate (see below). Recent studies, however, have revealed that PI metabolites phosphorylated at position 3 can activate individual isoenzymes of PKC, in particular members of the novel and atypical families (TOKER et al. 1994; NAKANISHI and EXTON 1992). Whether these metabolites are an important source of PKC activation in platelets is unclear; PI-3 kinase inhibitors block receptor-mediated phosphorylation of pleckstrin and release of 5-hydroxytryptamine (YATOMI et al. 1992), but the specificity of these inhibitors is not established. Whether the PI-3 kinase metabolites have other roles in the transduction of signals in activated platelets is not yet known.

Finally, it should be pointed out that this discussion has focused largely on second messengers that have been found to accumulate in the cytoplasm of platelets following their stimulation with agonists that for the most part interact with G-protein-linked receptors. In the physiologic and pathologic settings, platelets become activated not only by these agonists, but also by artificial surfaces, shear stress, and the interaction of adhesive proteins with their integrin receptors (as a result of "outside-in" signaling; CLARK and BRUGGE 1995). In all of these situations, increased activation of platelet protein kinase C has been documented, but it is not clear in any of the cases that this activation results from the antecedent mobilization of the same lipid mediators that follow receptor-mediated activation (KROLL et al. 1993).

## D. PKC Substrates, Binding Proteins, and Translocation

In general, much more is known about the regulators of PKC activation than about its substrates and the mechanism by which it mediates its effects. The major PKC substrate identified in platelets is a 47-kDa protein named pleckstrin which is expressed in many cells of hematopoietic lineage (GAILANI et al. 1990). This protein is highly phosphorylated by PKC and serves as a useful marker for PKC activation in intact cells. Pleckstrin phosphorylation has been correlated both with release of platelet granule contents and with fibrinogen binding and aggregation (SAITOH et al. 1989). Its function, however, is not yet clear. Recent recognition of a conserved motif in pleckstrin (the pleckstrin homology domain) that appears to mediate protein-protein interactions in other proteins (HARLAN et al. 1994) raises the possibility that a similar function will be found for pleckstrin itself, but no published evidence for such a role as yet exists. A PKC substrate that, unlike pleckstrin, is present in most cell types has been named myristoylated alanine-rich C kinase substrate, or MARCKS (BLACKSHEAR 1993). In other cells, MARCKS may participate in the regulation of cytoskeletal reorganization, but whether a similar role is seen in platelets is not known. PKC phosphorylates several other intracellular proteins found in platelets; both myosin heavy chain and light chain are phosphorylated in characteristic locations reflective of consensus domains for PKC substrates (CONTI et al. 1991; HIGASHIHARA et al. 1991). A phosphomonoesterase specific for the 5′ position of 1,4,5-inositol trisphosphate is also phosphorylated by PKC (CONNOLLY et al. 1986), raising the possibility that removal of this phosphate, and thus termination of $IP_3$-mediated $Ca^{2+}$ mobilization, can explain part of the inhibitory effect of PKC on $Ca^{2+}$ mobilization (HAYNES 1993). A partial list of other platelet proteins that are phosphorylated by PKC includes Gz, a G protein expressed at high levels in platelets (LOUNSBURY et al. 1991); mitogen-activated protein kinase S6 (SAMIEI et al. 1993); caldesmon (HETTASCH and SELLERS 1991); coagulation factor V (RAND et al. 1994); the $\alpha$-granule membrane component P-selectin (FUJIMOTO and MCEVER 1993); and the $\beta$3-subunit of the platelet's integrin receptor for fibrinogen and other adhesive proteins (HILLERY et al. 1991). Although in some cases a role for these substrates in platelet physiology has been established, the importance of phosphorylation by PKC to that function awaits further assessment.

Related to both mechanism of activation and the identification of substrates is the intracellular movement of individual PKC isoenzymes or overall enzymatic activity following stimulation, a phenomenon termed translocation. Exposure of platelets to thrombin or phorbol esters induces the net translocation of PKC-$\alpha$, -$\beta$, and -$\zeta$ in platelets, but the net distribution of PKC-$\delta$ is not changed (BALDASSARE et al. 1992). The mechanism for translocation is not clear, but appears to require $Ca^{2+}$ mobilization. Furthermore, the significance of this movement to PKC activation or agonist-induced functional response is not known; although the movement of individual isoenzymes allows their contact with membrane phospholipids that are presumably required for PKC activation, not all agonists induce translocation, including some direct PKC activators that clearly induce phosphorylation of PKC substrates in intact cells (GRABAREK and WARE 1993). Moreover, translocation is demonstrated much more readily with phorbol ester stimulation of platelets than with physiologic agonists, even those that induce greater phosphorylation of PKC substrates and cellular function. Thus, although intracellular movement of PKC may deliver individual isoenzymes to their relevant substrates and may therefore be an event of physiologic significance, it should not be regarded as a sensitive marker of enzymatic activation.

# E. Functional Roles for PKC in Platelets

Most of the evidence implicating PKC in platelet function is derived from experiments utilizing phorbol esters, bryostatins, or other direct PKC activators and from experiments in which chemical inhibitors of PKC are employed together with physiologic agonists. It should be recognized that the physiologic relevance of the former system can be questioned and that the inhibitors have limited specificity for PKC itself with unknown effects on individual isoenzymes (WILKINSON and HALLAM 1994). Nevertheless, although individual findings based on these experiments should be interpreted with caution, there are certain trends that emerge from multiple lines of investigation using several different reagents and thus suggest several critical roles for PKC in platelet regulation (Table 2). Perhaps the strongest link is between PKC activation and platelet aggregation; addition of phorbol esters causes both primary and secondary aggregation, apparently without shape change (CHANG and WARE 1995). Inhibition of PKC by any of several chemical inhibitors prevents aggregation and fibrinogen binding induced by most, if not all, platelet agonists, and increased PKC activity and phosphorylation of pleckstrin is closely correlated

**Table 2.** PKC and platelet function

I. Agonist-induced responses mediated via PKC activation:
   A. Phosphorylation of pleckstrin and other intracellular proteins
   B. Mobilization of intracellular stores of $Ca^{2+}$ or inhibition thereof (depending upon experimental conditions)
   C. Expression of the fibrinogen receptor function of integrin $\alpha_{IIb}\beta_3$ (mediating secondary irreversible aggregation* in the presence of fibrinogen)
   D. Secretory exocytosis (*release*) of granule contents
   E. Outside-in transmembrane signaling (including protein tyrosine phosphorylation, cytoskeletal reorganization, formation of focal adhesion plaques, cell spreading, and retraction of fibrin clots)
   F. Agonist-receptor desensitization involving negative feedback regulation of receptor-mediated polyphosphoinositide turnover

II. Platelet responses not mediated via PKC activation
   A. Shape change
   B. Earliest phase of (primary) aggregation*
   C. Ristocetin-induced platelet agglutination

*The initial phase of platelet activation (including shape change, adhesion, and the earliest events in aggregation) occurring prior to formation of the hemostatic plug is though to be independent of, or to precede, PKC activation. Activation of PKC appears to commit the platelet to full activation (as induced by the strong agonist thrombin), irreversible aggregation, and secretory exocytosis of granule contents.

with fibrinogen binding and an active state of the $\alpha$IIb-$\beta$3 integrin complex (VAN WILLIGAN and AKKERMAN 1992). Prevention of PKC activation by addition of a peptide modeled on the pseudosubstrate region that would be expected to be a potent and selective inhibitor of cPKCs also blocks binding of a monoclonal antibody that recognizes only the activated conformation of $\alpha$IIb-$\beta$3 to the surface of permeabilized, activated platelets (SHATTIL et al. 1992). Thus, PKC appears to play an important role in the "inside-out" signaling process that leads to fibrinogen binding to $\alpha$IIb-$\beta$3 and subsequent platelet aggregation. PKC is also likely to participate in "outside-in" signaling which is induced only after $\alpha$IIb-$\beta$3 has bound fibrinogen and/or after platelet aggregation. Outside-in transmembrane signaling leads to attachment of the cytoplasmic domain of integrin complexes to the platelet cytoskeleton and assembly of a number of structural and signaling proteins, including PKC and several tyrosine-phosphorylated molecules in complexes near the sites of platelet-platelet contact (CLARK and BRUGGE 1995). Many tyrosine-phosphorylated proteins, including the focal adhesion kinase FAK125 and src, either become phosphorylated on tyrosine residues or become associated with focal adhesions and/or the cytoskeleton in a PKC-dependent manner. One possible physiologic correlate for these "late" events is clot retraction.

In addition to its effects on platelet aggregation, PKC activation has been reported to enhance the ability of $Ca^{2+}$ mobilization to cause release of platelet granule contents (WALKER and WATSON 1993), although phorbol ester alone causes minimal release of granule contents from unstimulated platelets. PKC activation has been linked to thromboxane $A_2$ synthesis (FONT et al. 1992) and is required for histamine generation which appears to be a prerequisite for thrombin- and collagen-induced release of granule contents (GERRARD et al. 1993).

# F. Relation of PKC Activation to Other Platelet Serine-Threonine Kinases and Their Effectors

The interactions of PKC with other pathways of platelet signal transduction have received much attention. In some cases, PKC activation potentiates the effect of other mediators and their target kinases; in others, this interaction is inhibited. Cyclic AMP and cyclic GMP are intracellular mediators generated by activation of their respective cyclases, whose regulation is discussed elsewhere in this volume. Their effects are mediated, at least in part, by activation of cyclic-AMP-dependent kinase (PKA) and by cyclic-GMP-dependent kinase (PKG), two serine-threonine kinases found in platelets. The net effect of increasing levels of cAMP and cGMP in the platelet cytosol is to inhibit activation by functional agonists (WALTER et al. 1993); the weight of evidence suggests that neither of these pathways can generate "positive" mediators. The mechanism by which activation of PKA or PKG might mediate inhibitory effects is unknown, although at least one of their substrates has been identified (BUTT et al. 1994). Generation of either cGMP or cAMP prevents platelet activation at a very proximal step, perhaps at the level of the interaction between G proteins and phospholipase C $\beta$ since both $Ca^{2+}$ mobilization and PKC activation are inhibited. The effect goes beyond PLC inhibition, however, because elevation of cGMP prevents aggregation induced by agonists such as PMA and $Ca^{2+}$ ionophore, which bypass PLC (NGUYEN et al. 1991). Thus, it is likely that inhibition induced by these cyclic nucleotides and their kinases occurs at multiple levels. Furthermore, elevation of cyclic GMP and $Ca^{2+}$ appears to induce PKC activation and pleckstrin phosphorylation by an unknown mechanism (NGUYEN et al. 1991). Conversely, elevation of intracellular $Ca^{2+}$ and PKC activation appear to promote efflux of cGMP from activated platelets (WU et al. 1993).

PKC activation has a similarly complicated relation to $Ca^{2+}$ mobilization and its target kinases which include myosin light chain kinase and $Ca^{2+}$/calmodulin-dependent protein kinase. As noted in a previous section, $Ca^{2+}$ mobilization and PKC activation exert powerful synergistic effects on platelet release and aggregation. If PKC is activated by phorbol ester, the action of phospholipase C is inhibited, preventing $Ca^{2+}$ mobilization (BISHOP et al. 1992). PKC can activate enzymes that both enhance and prevent the formation of inositol trisphosphate (CONNOLLY et al. 1986; OBERDISSE et al. 1990), but the net effect appears to favor $Ca^{2+}$ sequestration (HAYNES 1993), although PMA can mobilize $Ca^{2+}$ in platelets under some circumstances (WARE et al. 1989). Preliminary data reveal that PKC activation can limit influx of $Ca^{2+}$ via receptor-operated $Ca^{2+}$ channels in platelets stimulated by thrombin; selective inhibition of PKC-$\beta$ promotes thrombin-induced influx (XU and WARE 1995). Myosin light chain is phosphorylated by both myosin light chain kinase and PKC at different sites and with different effects on conformation (HIGASHIHARA et al. 1991), but the functional implications of this are not known.

Finally, it appears that the products of PI-3 kinase that accumulate rapidly following thrombin stimulation of platelets do so by a mechanism requiring PKC (KING et al. 1991). As noted above, these metabolites also selectively activate members of the PKC family, and thus the details of their interaction are likely to prove complicated.

## G. Abnormalities of Platelet PKC in Disease

Platelets from patients with various diseases have been shown to have abnormalities in PKC activity and phosphorylation of substrates; some of these abnormalities reflect a general disturbance in platelet intracellular metabolism, while others appear to be a selective abnormality in PKC. In platelets from individuals with hypertension, both increased activity and pleckstrin phosphorylation have been noted (HALLER et al. 1992), which together with an elevation in cytoplasmic $Ca^{2+}$ may contribute to the increased platelet reactivity and tendency toward arterial thromboembolism seen in these patients. Platelets are among the few cells besides neurons to accumulate and discharge serotonin and thus are commonly used as models for certain aspects of neuronal tissue behavior in pharmacologic studies. Therefore, it is intriguing that abnormalities in platelet PKC have been detected in such diverse neurologic conditions as mania, schizophrenia, and Alzheimer's disease; additionally, a selective abnormality in PKC-$\beta$ involving its translocation from cytosol to membrane in platelets has been correlated with normal aging (WARE and COLLER 1994). The pathogenesis and significance of these associations are unknown.

## H. Summary and Future Areas of Research

In recent years, work from many laboratories has established that individual isoenzymes of PKC are differentially regulated and expressed, phosphorylate diverse substrates, and mediate selective functions in many cell types. Although the special characteristics of platelets make assignment of such functions much more difficult, the available evidence suggests that individual isoenzymes, of which there are representatives of all the subfamilies in platelets, are likely to have similar effects in these cells. Thus, selective functions of platelets may be targets for manipulation by pharmacologic agents. Attractive areas for future research include elucidation of the indirect mechanism by which $\alpha$IIb-$\beta$3 is activated by PKC, functional assessment of known PKC substrates and identification of novel pharmacologic targets for this enzyme, definition of the interaction between PKC activation and integrin-induced signaling, particularly in reference to tyrosine kinases, phosphatases, and their substrates, and definitive assignment of discrete platelet functions to individual PKC isoenzymes. Thus, despite the abundance of biochemical evidence implicating PKC and other protein serine-threonine kinases in the agonist-induced expression of a variety of cellular responses, there is a noteworthy lack of information concerning the specific mechanisms whereby the activity of specific effector molecules is modulated. In addition, the identity of the target (intermediary or effector) molecules is often unknown. The time is presently at hand for these unsolved problems to be addressed in novel animal models using current technologies involving transgene expression and ablative gene targeting by homologous recombination.

## References

Akimoto K, Mizuno K, Osada S, Hirai S, Tanuma S, Suzuki K, Ohno S (1994) A new member of the third class in the protein kinase C family, PKC lambda, expressed dominantly in an undifferentiated moust embryonal carcinoma cell line and also in many tissues and cells. J Biol Chem 269:12677–12683

Bacher N, Zisman Y, Berent E, Livneh E (1991) Isolation and characterization of PKC-L, a new member of the protein kinase C-related gene family specifically expressed in lung skin and heart. Mol Cell Biol 11:126–133

Baier G, Telford D, Giampa L, Coggeshall KM, Baier-Bitterlich G, Isakov N, Altman A (1993) Molecular cloning and characterization of PKC theta, a novel member of the protein kinase C (PKC) gene family expressed predominantly in hematopoietic cells. J Biol Chem 268:4997–5004

Baldassare JJ, Henderson PA, Burns D, Loomis C, Fisher GJ (1992) Translocation of protein kinase C isozymes in thrombin-stimulated human platelets. Correlation with 1,2-diacylglycerol levels. J Biol Chem 267:15585–15590

Bell RM, Burns DJ (1991) Lipid activation of protein kinase C. J Biol Chem 266:4661–4664

Bishop WR, Pachter JA, Pai JK (1992) Regulation of phospholipid hydrolysis and second messenger formation by protein kinase C. Adv Enzyme Regul 32:177–192

Blackshear PJ (1993) The MARCKS family of cellular protein kinase C substrates. J Biol Chem 268:1501–1504

Butt E, Abel K, Krieger M, Palm D, Hoppe V, Hoppe J, Walter U (1994) cAMP- and cGMP-dependent protein kinase phosphorylation sites of the focal adhesion vasodilator-stimulated phosphoprotein (VASP) in vitro and in intact human platelets. J Biol Chem 269:14509–14517

Castagna M, Takai Y, Kaibuchi K, Sano K, Kikkawa U, Nishizuka Y (1982) Direct activation of calcium-activated, phospholipid-dependent protein kinase by tumor-promoting phorbol esters. J Biol Chem 257:7847–7851

Chang JD, Xu Y, Raychowdhury MK, Ware JA (1993) Molecular cloning and expression of a cDNA encoding a novel isoenzyme of protein kinase C (nPKC). A new member of the nPKC family expressed in skeletal muscle, megakaryoblastic cells, and platelets. J Biol Chem 268:14208–14214

Clark EA, Brugge JS (1995) Integrins and signal transduction pathways. Science 268:233–238

Connolly TM, Lawing WJ Jr, Majerus PW (1986) Protein kinase C phosphorylates human platelet inositol trisphosphate 5′-phosphomonoesterase, increasing the phosphatase activity. Cell 46:951–958

Conti MA, Sellers JR, Adelstein RS, Elzinga M (1991) Identification of the serine residue phosphorylated by protein kinase C in vertebrate nonmuscle myosin heavy chains. Biochemistry 30:966–970

Crabos M, Fabbro D, Stabel S, Erne P (1992) Effect of tumour-promoting phorbol ester, thrombin and vasopression on translocation of three distinct protein kinase C isoforms in human platelets and regulation by calcium. Biochem J 288:891–896

Dekker LV, Parker PJ (1994) Protein kinase C – a question of specificity. Trends Biochem Sci 19:73–77

Dekker LV, McIntyre P, Parker PJ (1993) Altered substrate selectivity of PKC-eta pseudosubstrate site mutants. FEBS Lett 329:129–133

Font J, Azula FJ, Marino A, Nieva N, Trueba M, Macarulla JM (1992) Intracellular $Ca^{2+}$ mobilization and not calcium influx promotes phorbol ester-stimulated thromboxane $A_2$ synthesis in human platelets. Prostaglandins 43:383–395

Fujimoto T, McEver RP (1993) The cytoplasmic domain of P-selectin is phosphorylated on serine and threonine residues. Blood 82:1758–1766

Gailani D, Fisher TC, Mills DC, Macfarlane DE (1990) P47 phosphoprotein of blood platelets (pleckstrin) is a major target for phorbol ester-induced protein phosphorylation in intact platelets, granulocytes, lymphocytes, monocytes and cultured leukaemic cells: absence of P47 in haematopoietic cells. Br J Haematol 74:192–202

Gerrard JM, McNicol A, Sazena SP (1993) Protein kinase C, membrane fusion and platelet granule secretion. Biochem Soc Trans 21:289–293

Grabarek J, Ware JA (1993) Protein kinase C activation without membrane contact in platelets stimulated by bryostatin. J Biol Chem 268:5543–5549

Grabarek J, Raychowdhury MK, Ravid K, Kent KC, Newman PJ, Ware JA (1992) Identification and functional characterization of protein kinase C isozymes in platelets and HEL cells. J Biol Chem 267:10011–10017

Haller H, Lindschau C, Quass P, Distler A (1992) Protein phosphorylation and intracellular free calcium in platelets of patients with essential hypertension. Am J Hypertens 5:117–124

Harlan JE, Hajduk PJ, Yoon HS, Fesik SW (1994) Pleckstrin homology domains bind to phosphatidylinositol-4,5-bisphosphate. Nature 371:168–170

Haslam RJ, Coorssen JR (1993) Evidence that activation of phospholipase D can mediate secretion from permeabilized platelets. Adv Exp Med Biol 344:149–164

Haynes DH (1993) Effects of cyclic nucleotides and protein kinases on platelet calcium homeostasis and mobilization. Platelets 4:231–242

Hettasch JM, Sellers JR (1991) Caldesmon phosphorylation in intact human platelets by cAMP-dependent protein kinase and protein kinase C. J Biol Chem 266:11876–11881

Higashihara M, Takahata K, Kurokawa K (1991) Effect of phosphorylation of myosin light chain by myosin light chain kinase and protein kinase C on conformation change and ATPase activities of human platelet myosin. Blood 78:3224–3231

Hillery CA, Smyth SS, Parise LV (1991) Phosphorylation of human platelet glycoprotein IIIa (GPIIIa). Dissociation from fibrinogen receptor activation and phosphorylation of GPIIIa in vitro. J Biol Chem 266:14663–14669

Huang RS, Kucera GL, Rittenhouse SE (1991) Elevated cytosolic $Ca^{2+}$ activates phospholipase D in human platelets. J Biol Chem 266:1652–1655

Johannes FJ, Prestle J, Eis S, Oberhagemann P, Pfizenmaier K (1994) PKC $\mu$ is a novel, atypical member of the protein kinase C family. J Biol Chem 269:6140–6148

Kemp BE, Pearson RB (1991) Intrasteric regulation of protein kinases and phosphatases. Biochim Biophys Acta 1094:67–76

Khan W, el Touny S, Hannun YA (1991) Arachidonic and cis-unsaturated fatty acids induce selective platelet substrate phosphorylation through activation of cytosolic protein kinase C. FEBS Lett 292:98–102

Khan WA, Blobe G, Halpern A, Taylor W, Wetsel WC, Burns D, Loomis C, Hannun YA (1993) Selective regulation of protein kinase C isoenzymes by oleic acid in human platelets. J Biol Chem 268:5063–5068

King WG, Kucera GL, Sorisky A, Zhang J, Rittenhouse SE (1991) Protein kinase C regulates the stimulated accumulation of 3-phosphorylated phosphoinositides in platelets. Biochem J 278:475–480

Kroll MH, Hellums JD, Guo Z, Durante W, Razdan K, Hrbolich JK, Schafer AI (1993) Protein kinase C is activated in platelets subjected to pathological shear stress. J Biol Chem 268:3520–3524

Kucera GL, Rittenhouse SE (1990) Human platelets form 3-phosphorylated phosphoinositides in response to alpha-thrombin, U46619, or GTP gamma S. J Biol Chem 265:5345–5348

Lounsbury KM, Casey PJ, Brass LF, Manning DR (1991) Phosphorylation of Gz in human platelets. Selectivity and site of modification. J Biol Chem 266:22051–22056

Manning DR, Brass LF (1991) The role of GTP-binding proteins in platelet activation. Thromb Haemost 66:393–399

Murray NR, Baumgardner GP, Burns DJ, Fields AP (1993) Protein kinase C isotypes in human erythroleukemia (K562) cell proliferation and differentiation: evidence that betaII protein kinase C is required for proliferation. J Biol Chem 268:15847–15853

Nakanishi H, Exton JH (1992) Purification and characterization of the zeta isoform of protein kinase C from bovine kidney. J Biol Chem 267:16347–16354

Nguyen BL, Saitoh M, Ware JA (1991) Interaction of nitric oxide and cGMP with signal transduction in activated platelets. Am J Physiol 261:H1043–H1052

Nishikawa M, Hidaka H, Shirakawa S (1988) Possible involvement of direct stimulation of protein kinase C by unsaturated fatty acids in platelet activation. Biochem Pharmacol 37:3079–3089

Nishizuka Y (1988) The molecular heterogeneity of protein kinase C and its implications for cellular regulation. Nature 334:661–665

Nishizuka Y (1995) Protein kinase C and lipid signaling for sustained cellular responses. FASEB J (in press)

Nolan RD, Lapetina EG (1990) Thrombin stimulates the production of a novel polyphosphoinositide in human platelets. J Biol Chem 265:2441–2445

Oberdisse E, Nolan RD, Lapetina EG (1990) Thrombin and phorbol ester stimulate inositol 1,3,4,5-tetrakisphosphate 3-phosphomonoesterase in human platelets. J Biol Chem 265:726–730

Osada S, Mizuno K, Saido TC, Akita Y, Suzuki K, Kuroki T, Ohno S (1990) A phorbol ester receptor/protein kinase, nPKC eta, a new member of the protein kinase C family predominantly expressed in lung and skin. J Biol Chem 265:22434–22440

Osada S, Mizuno K, Saido TC, Suzuki K, Kuroki T, Ohno S (1992) A new member of the protein kinase C family, nPKC theta, predominantly expressed in skeletal muscle. Mol Cell Biol 12:3930–3938

Rand MD, Kalafatis M, Mann KG (1994) Platelet coagulation factor Va: the major secretory platelet phosphoprotein. Blood 83:2180–2190

Romano M, Serhan CN (1992) Lipoxin generation by permeabilized human platelets. Biochemistry 31:8269–8277

Saitoh M, Salzman EW, Smith M, Ware JA (1989) Activation of protein kinase C in platelets by epinephrine and A23187: correlation with fibrinogen binding. Blood 74:2001–2006

Samiei M, Sanghera JS, Pelech SL (1993) Activation of myelin basic protein and S6 peptide kinases in phorbol ester- and PAF-treated sheep platelets. Biochim Biophys Acta 1176:287–298

Selbie LA, Schmitz-Pfeiffer C, Sheng Y, Biden TJ (1993) Molecular cloning and characterization of PKC iota, an atypical isoform of protein kinase C derived from insulin-secreting cells. J Biol Chem 268:24296–24302

Shattil SJ, Cunningham M, Wiedmer T, Zhao J, Sims PJ, Brass LF (1992) Regulation of glycoprotein IIb-IIIa receptor function studied with platelets permeabilized by the pore-forming complement proteins C5b-9. J Biol Chem 267:18424–18431

Shearman MS, Naor Z, Sekiguchi K, Kishimoto A, Nishizuka Y (1989) Selective activation of the gamma-subspecies of protein kinase C from bovine cerebellum by arachidonic acid and its lipoxygenase metabolites. FEBS Lett 243:177–182

Toker A, Meyer M, Reddy KK, Falck JR, Aneja R, Aneja S, Parra A, Burns DJ, Ballas LM, Cantley LC (1994) Activation of protein kinase C family members by the novel polyphosphoinositides PtdIns-3,4-P2 and PtdIns-3,4,5-P3. J Biol Chem 269:32358–32367

Valverde AM, Sinnett-Smith J, Van Lint J, Rozengurt E (1994) Molecular cloning and characterization of protein kinase D: a target for diacylglycerol and phorbol esters with a distinctive catalytic domain. Proc Natl Acad Sci USA 91:8572–8576

van Willigen G, Akkerman JW (1992) Regulation of glycoprotein IIB/IIIa exposure on platelets stimulated with alpha-thrombin. Blood 79:82–90

Walker TR, Watson SP (1993) Synergy between $Ca^{2+}$ and protein kinase C is the major factor in determining the level of secretion from human platelets. Biochem J 289:277–282

Walter U, Eigenthaler M, Geiger J, Reinhard M (1993) Role of cyclic nucleotide-dependent protein kinases and their common substrate VASP in the regulation of human platelets. Adv Exp Med Biol 344:237–249

Ware JA, Coller BS (1994) Platelet morphology, biochemistry and function. In: Beutler E, Lichtman MA, Coller BS, Kipps TJ (eds) Williams' hematology. McGraw Hill, New York, pp 1161–1201

Ware JA, Saitoh M, Smith M, Johnson PC, Salzman EW (1989) Response of aequorin-loaded platelets to activators of protein kinase C. Am J Physiol 256:C35–C43

Ware JA, Kang J, DeCenzo MT, Smith M, Watkins SC, Slayter HS, Saitoh M (1991) Platelet activation by a synthetic hydrophobic polymer. Blood 78:1713–1720

Werner MH, Beilawska AE, Hannun YA (1992) Quantitative analysis of diacylglycerol second messengers in human platelets: correlation with aggregation and secretion. Mol Pharmacol 41:382–386

Wilkinson SE, Hallam TJ (1994) Protein kinase C: is its pivotal role in cellular activation overstated? Trends Pharmacol Sci 15:53–57

Wu XB, Brune B, Von Appen F, Ullrich V (1993) Efflux of cyclic GMP from activated human platelets. Mol Pharmacol 43:564–568

Xu Y, Ware JA (1994) Inhibition of protein kinase C $\beta$ expression by antisense cDNA enhances thrombin-stimulated influx in megakaryocytic cells. Circulation [Suppl] (Abstract) 90:I–397

Xu Y, Ware JA (1995) Selective inhibition of thrombin recepter-mediated $Ca^{2+}$ entry by protein kinase C beta. J Biol Chem 270:23887–23890

Yatomi Y, Hazeki O, Kume S, Ui M (1992) Suppression by wortmannin of platelet responses to stimuli due to inhibition of pleckstrin phosphorylation. Biochem J 285:745–751

Yoshida K, Asaoka Y, Nishizuka Y (1992) Platelet activation by simultaneous actions of diacylglycerol and unsaturated fatty acids. Proc Natl Acad Sci USA 89:6443–6446

CHAPTER 13
# Platelet Protein Tyrosine Kinases

P. PRESEK and E.A. MARTINSON

## A. Background

The first protein tyrosine kinase (PTK) to be identified, detected by Hunter and Sefton in 1979, was the transforming protein of the Rous sarcoma virus (HUNTER and SEFTON 1980). This 60-kDa phosphoprotein is encoded by the viral *src* oncogene (*v-src*) and is denoted pp60$^{v\text{-}src}$. Retroviral oncogenes simply represent oncogenic forms of cellular genes, which are termed protooncogenes (BISHOP 1983). Although the first PTKs were identified as transforming proteins, the great majority of cellular genes encoding PTKs were discovered by screening cDNA libraries for genetic sequence homology. Meanwhile, it has become clear that most of the cellular PTKs fall into one of two classes. The first class consists of receptor PTKs that mediate signalling by many growth factors and polypeptide hormones such as insulin and neurotrophins (FANTL et al. 1993; VAN DER GEER et al. 1994). The members of the second class are defined by their intracellular localization and lack of extracellular receptor and transmembrane domains. These enzymes are termed intracellular – or more commonly – nonreceptor PTKs.

Receptor PTKs are critically involved in a variety of cellular responses including proliferation, differentiation, and cell survival. Despite their structural and functional diversity, they all share the classical principle of membrane receptor signalling: the ligand, typically a growth factor, binds to an extracellular high affinity receptor domain that initiates signal transmission into the cell, thereby activating the cytoplasmic PTK catalytic domain (Fig. 1A). Combinations of genetic and biochemical approaches have revealed that autophosphorylation of defined tyrosine residues in the cytoplasmic domain of most of the different receptor PTKs mediates subsequent signal transduction through a common set of signalling molecules. Most of these molecules belong to a class of proteins that contain distinct domains that physically interact with specific motifs or posttranslationally modified sites of other cellular proteins.

On the other hand, activation mechanisms of the nonreceptor PTKs and their role in cellular signalling differ from those of the receptor PTKs due to their lack of an intrinsic extracellular receptor domain. Compared with the receptor PTKs, which are directly activated by agonist binding, nonreceptor PTKs are indirectly activated as a consequence of extracellular stimulation by

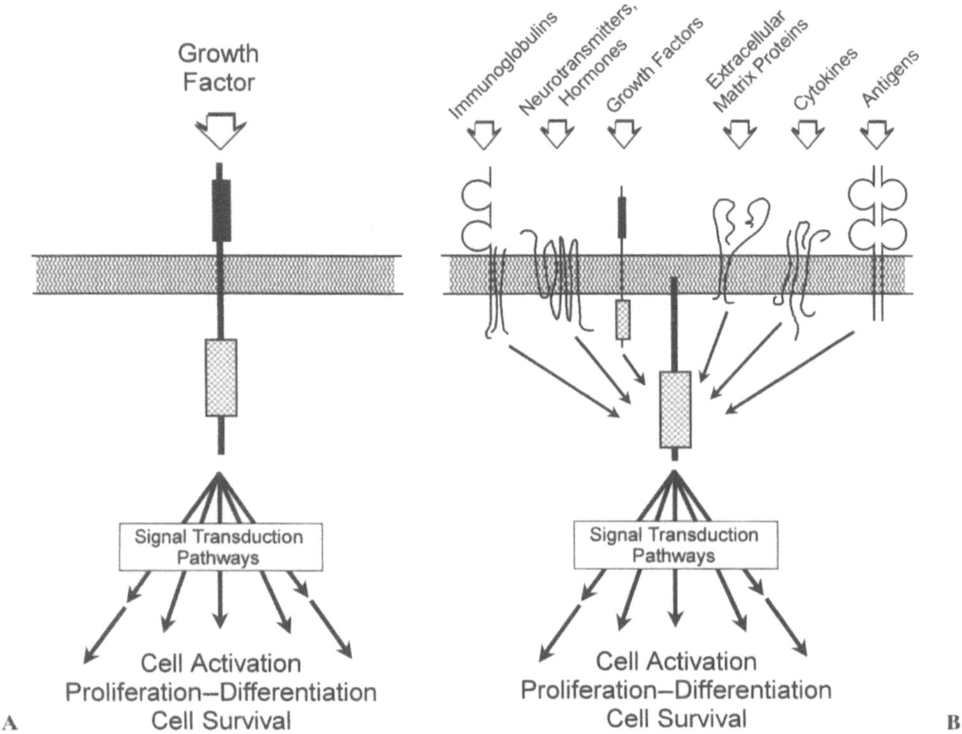


**Fig. 1A,B.** Generalized signal transduction mechanisms by protein tyrosine kinases (*PTKs*). **A** Signalling by receptor PTKs, showing transmission of a growth factor-induced signal across the cell membrane to the catalytic domain (*hatched box*), resulting in the activation of various singal transduction pathways leading to cellular responses. **B** Nonreceptor PTK signalling, showing the relaying of signals from a variety of cell-surface receptors through a nonreceptor PTK. This PTK may be associated with the membrane, as in the case of the Src family kinases, or with membrane-bound proteins

a broad variety of biologically active agents (Fig. 1B). Many of the nonreceptor PTKs associate directly with cell-surface proteins, and a number of them participate in the signal transduction pathways of different cell-surface receptors, including receptor PTKs. For example, it has been established that a number of cytokine receptors, most of which do not have intrinsic PTK activity, are capable of recruiting and/or activating a variety of nonreceptor PTKs, thereby initiating downstream signalling events (TANIGUCHI 1995). Another example concerns T cell activation: interaction of the T cell receptor with an antigenic peptide presented by the major histocompatibility complex leads to the assembly of multimeric signalling complexes that include various intracellular PTKs, which in turn transmit the activation signal to intracellular effectors (MUSTELIN and BURN 1993; DEFRANCO 1995). In general, upon cell stimulation, one or more of the nonreceptor PTKs may become activated, leading to mitogenesis or differentiation but also to more acute responses such as changes in cell shape or motility. These responses are cell type-specific; thus, not all responses are elicited by the same stimulus in different cell types, being dependent on the presence of various downstream effector mechanisms.

Several different groups of nonreceptor PTKs have been identified including Src, Itk, JAK, Fes/Fps, Syk/ZAP-70, Abl, Csk, FAK, and more recently PYK (Bolen 1993; Lev et al. 1995). All of these groups appear to represent distinct families of protein tyrosine kinases, ranging from the Src family with nine members (Erpel and Courtneidge 1995) to families with three presently characterized members, the focal adhesion kinase (FAK) family and the proline rich tyrosine kinase (PYK) family (Sasaki et al. 1995; Avraham et al. 1995; Lev et al. 1995). Analysis of nonreceptor PTK primary structure has revealed three major common domains, termed Src homology (SH) domains. These are referred to as SH1, SH2, and SH3, beginning from the C terminus of the Src molecule. SH1 represents the catalytic domain, which is common to all PTKs and has the greatest sequence homology. The presence of SH2 and SH3 domains (50–100 amino acids in length) varies from none (FAK and the JAK family kinases), to both (Src, Abl and Itk families), to the SH3 domain alone (Fes/Fps family), and to two SH3 domains (Syk/ZAP-70 family). Members of the JAK family contain two catalytic domains in series. SH2 and SH3 domains are found in many proteins that control a number of signalling pathways, especially those involving tyrosine kinases (Pawson and Schlessinger 1993). SH2-containing proteins bind to phosphorylated tyrosine residues within distinct sequence motifs on activated receptor PTKs or nonreceptor PTKs and other cytoplasmic proteins. SH3 domains bind to proline-rich motifs of approximately ten amino acids found on a variety of proteins, notably components of the cytoskeleton. Thus, it has become clear that SH2 and SH3 domains function as building blocks in controlling the localization of proteins within cells, thereby specifying cascades of protein-protein interactions that are parts of cellular signalling pathways (Pawson 1995).

Most of the nonreceptor PTKs are expressed ubiquitously, whereas some, e.g., the Src family members Lck, Hck, Fgr, and Blk, are mainly expressed in certain hematopoietic cells (Bolen 1993). Blood platelets contain high amounts of different nonreceptor PTKs, among which Src (see below, Sect. B) is probably the most abundant. Platelets respond to a broad variety of physiologically relevant extracellular stimuli with a rapid and marked increase in tyrosine phosphorylation of numerous individual proteins, as first described by Ferrell and Martin in 1988. Furthermore, upon platelet activation, several PTKs have been consistently shown to become activated. PTK inhibitors are able to block both protein tyrosine phosphorylation and certain platelet functions, and inhibition of protein tyrosine phosphatases increases protein tyrosine phosphorylation and induces platelet activation (see below, Sect. C). These observations led to the hypothesis that PTKs play a critical role in platelet physiology. Blood platelets additionally provide an excellent model system in which to study the intracellular signalling pathways associated with PTKs. Although platelets do not allow genetic manipulations, platelets are available from individuals with inborn platelet disorders that can be used to investigate the importance of certain proteins in protein tyrosine phosphorylation and other responses (see below, Sect. C, Part II).

In resting platelets, only a small number of proteins in the range of 50–62 and 120–130 kDa are detectably phosphorylated on tyrosine residues. These include several unidentified proteins as well as the 60-kDa PTK Src, which is the most prominent. Treatment of platelets with different aggregation-eliciting agents increases tyrosine phosphorylation of multiple proteins ranging from 30 kDa to over 200 kDa (FERRELL and MARTIN 1988; NAKAMURA and YAMAMURA 1989; FERRELL and MARTIN 1989; GOLDEN and BRUGGE 1989; INAZU et al. 1991; NAKASHIMA et al. 1991; DHAR et al. 1990; MURPHY et al. 1993; M.M. HUANG et al. 1992). Tyrosine phosphorylation of these proteins has been reported to occur in at least three temporal "waves" (FERRELL and MARTIN 1988), implying that different PTKs and protein tyrosine phophatases might become sequentially activated and/or inactivated during platelet activation. Additionally, substrate availability could be changed by protein and PTK translocation between various intracellular compartments. Upon thrombin stimulation, the first set of proteins is tyrosine-phosphorylated very rapidly and reaches a maximum within 15–20 s. Further individual proteins become tyrosine-phosphorylated within 1–3 min and 3–5 min after thrombin addition, respectively (FERRELL and MARTIN 1988). There appear to be discrepancies among the various reports concerning tyrosine-phosphorylated proteins in both resting and stimulated platelets and the precise time course of phosphorylation. This might be due to different platelet isolation procedures, some of which may cause partial preactivation, or to different antiphosphotyrosine antibodies used, as has been demonstrated for the polyclonal antibodies 903 and 3231 (GOLDEN and BRUGGE 1989). Whether differences in tyrosine phosphorylation of individual proteins observed with various compounds (e.g., thromboxane $A_2$ receptor agonists vs. thrombin) represent distinct intracellular responses and are not just due to methodological differences remains to be verified. Isolation and identification of individual proteins should resolve these discrepancies.

# B. Protein Tyrosine Kinases in Platelets

## I. Src Family Kinases

Five members of the Src family of nonreceptor PTKs have been detected in platelets: Src itself, Fyn, Lyn, Yes, and Hck (Fig. 2). This family is characterized by a tyrosine kinase catalytic (SH1) region, an SH2 and an SH3 domain, and a myristylated glycine residue at the amino terminus (Fig. 3). An

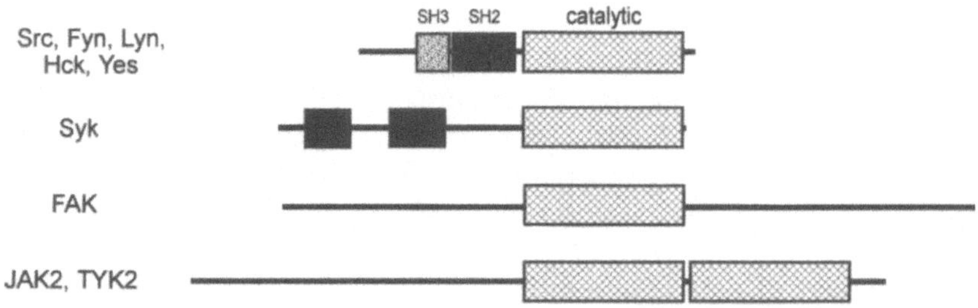


**Fig. 2.** Protein tyrosine kinases (PTKs) of platelets. Src family and other PTKs found in platelets are shown, with various domains and approximate relative size indicated

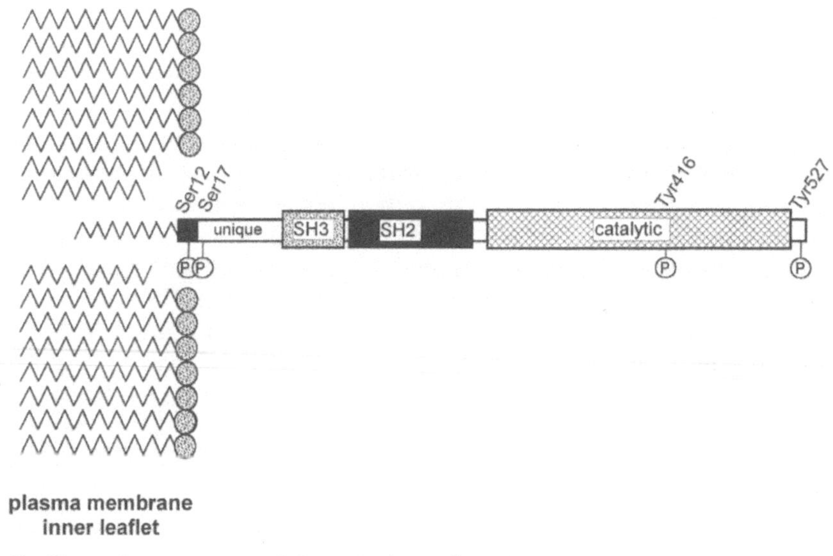


**Fig. 3.** Domain structure of Src. Various domains of Src are depicted, as well as principal sites of phosphorylation (*P*)

autophosphorylation site is located within the catalytic domain, and another tyrosine residue that is phosphorylated is situated at the carboxy terminus. The Src family PTKs differ mainly in an approximately 70-residue sequence near the amino terminus described as the "unique" region, the significance of which is not yet understood. This sequence, however, has been useful in the production of family member-specific antibodies. There is a plethora of information concerning Src in platelets, in contrast to the other four Src family members found in this cell type; therefore, we will describe the structure and regulation of this enzyme in greatest detail.

## 1. Src

The 60-kDa phosphoprotein Src, often denoted pp60$^{c\text{-}src}$ (Fig. 3), is found in the highest concentration in platelets compared to any other tissue or cell type, comprising 0.2%–0.4% of total platelet protein (GOLDEN et al. 1986). It is tethered to cellular membranes by a myristylated glycine residue at the amino terminus (RESH 1993, 1994) and is therefore found in the membrane fraction, mostly the surface plasma membrane (FERRELL et al. 1990; ANAND et al. 1993), of unstimulated platelets. In addition, it has recently been demonstrated that N-terminal basic residues, specifically lysine and arginine, are important to the membrane binding of Src family PTKs (SILVERMAN and RESH 1992; SIGAL et al. 1994; RESH 1994). This is thought to occur through electrostatic interactions with negatively charged phospholipids residing on the inner leaflet of the plasma membrane.

Under resting conditions, Src is phosphorylated at Tyr527 near the carboxy terminus, a state that is inhibitory to the kinase activity of the enzyme and correlates with its low basal activity (COOPER et al. 1986; KMIECIK and SHALLOWAY 1987; ERPEL and COURTNEIDGE 1995). The amino acid sequence numbers used here, which are most commonly used in the literature, refer to the product of the chicken *c-src* gene, where Src was first characterized. The human sequence numbering varies by only several residues (ANDERSON et al. 1985; TANAKA et al. 1987). In fibroblasts, Tyr527 is nearly stoichiometrically (100%) phosphorylated (COOPER et al. 1986). Src is phosphorylated at this position by Csk (for c-terminal Src kinase), a PTK present in other cell types but not as yet identified in platelets (OKADA et al. 1991). On the other hand, phosphorylation of Src at Tyr416 within the catalytic domain correlates with increased activity (CARTWRIGHT et al. 1986; KMIECIK et al. 1988; REUTER et al. 1990).

When platelets are activated by an agonist such as thrombin, the activity of Src increases (WONG et al. 1992; LIEBENHOFF et al. 1993; CLARK and BRUGGE 1993). This increase is rapid, occurring within 5–15 s of thrombin addition (LIEBENHOFF et al. 1993; CLARK and BRUGGE 1993). Phosphorylation of Src at Tyr527 concomitantly decreases, at least on a subfraction of Src molecules. At the same time, Src translocates to a cytoskeletal fraction (GRONDIN et al. 1991; ODA et al. 1992a; HORVATH et al. 1992; PUMIGLIA and FEINSTEIN 1993; CLARK and BRUGGE 1993) experimentally defined as being insoluble in the nonionic detergent Triton-X100 (FOX et al. 1992). Cytoskeletally localized Src is phosphorylated at Tyr416 (CLARK and BRUGGE 1993). Table 1 shows the phosphorylation, activation, and translocation status of Src associated with various stages of platelet activation.

On the basis of data from a number of different cell types and in vitro experiments, a plausible model for Src activation has emerged (MATSUDA et al. 1990; ROUSSEL et al. 1991; CANTLEY et al. 1991; COOPER and HOWELL 1993). In the inactive state, the molecule is folded so that phosphorylated Tyr527 is bound intramolecularly to the SH2 domain. Such an intramolecular SH2-

**Table 1.** Stages of platelet activation and associated PTKs

| Platelet Activation Stage | PTKs Phosphorylated (P), Activated (A), or Translocated (T) | | |
|---|---|---|---|
| Resting | Src (P)<br>Fyn (P) | | |
| Shape Change | | Syk (A, T) | |
| Early aggregation | Src (**P**,**A**,*T*)<br>Fyn (**P**,**A**,*T*)<br>Yes (P,*T*) | Syk (**P**,**A**,*A*) | |
| Aggregation | | Syk (*T*), JAK2 (P) | FAK (*P*,*A*,T) |
| Adhesion | Src (A)? | Syk (A)? | FAK (P) |
| Secretion | ? | | |

Phosphorylation, activation, and translocation of various PTKs to the cytoskeleton have been observed to coincide with events during platelet activation (see text and literature on the subject). Aggregation has been separated into early aggregation (5–30 s) and aggregation (>30 s). Events known to occur independently of aggregation are shown in **bold** typeface, while events that are aggregation-dependent are shown in *italics*.

phosphotyrosine interaction has been directly demonstrated for the Crk II adaptor protein (Rosen et al. 1995). The SH3 domain of Src also participates in an intramolecular interaction, although this has not yet been well defined (Seidel-Dugan et al. 1992; Superti-Furga et al. 1993; Erpel et al. 1995). Dephosphorylation of Tyr527 by an unknown protein tyrosine phosphatase results in the unfolding of Src, freeing the catalytic domain and allowing autophosphorylation on Tyr416. This model explains why the viral homologue of Src, v-Src, which lacks the C-terminal Tyr527, is constitutively active (and therefore transforming). Also in accordance with this model, overexpression of protein tyrosine phosphatase-$\alpha$ in fibroblasts leads to persistent activation of Src accompanied by dephosphorylation of Tyr527 (Zheng et al. 1992), and hypophosphorylation of the C-terminal tyrosine in melanocytes appears to be due to an elevated protein tyrosine phosphatase activity (O'Connor et al. 1995). In addition, it might be that a protein interferes with the phosphorylation of Tyr527 by a Csk-like regulatory kinase, as is thought to be the case with middle T antigen in polyoma virus-tranformed cells (Courtneidge 1985; Cantley et al. 1991).

In an alternative model for Src activation, a molecule containing a high-affinity binding motif, e.g., a phosphorylated tyrosine residue or a proline-rich sequence, may compete with the C terminus for binding to the SH2 or SH3 domains, respectively, and force Src to unfold, exposing phosphorylated Tyr527 to the action of a protein tyrosine phosphatase (Erpel and Courtneidge 1995). It has been shown in competition experiments in vitro that phosphotyrosine-containing peptides can activate Src, presumably by displacing phosphorylated Tyr527 from its binding site on the SH2 domain

(LIU et al. 1993). This mechanism is believed to account for the activation of Src by the PDGF receptor, which associates with Src by virtue of Src's SH2 domain (MORI et al. 1993; ERPEL and COURTNEIDGE 1995). Thus, it is not yet clear whether the putative unfolding of the kinase, which results in activation, is due to dephosphorylation of Tyr527, or whether dephosphorylation occurs because the enzyme is forced to unfold due to the association with other proteins. It may be that dephosphorylation is not even necessary for the in vivo activation mechanism since a significant decrease in Tyr527 phosphate is not usually detected (LIEBENHOFF et al. 1993; CLARK and BRUGGE 1993).

Regardless of the precise mechanism of activation of Src, the SH2 and SH3 domains of the molecule become available to interact with other proteins, for instance in the cytoskeleton, some of which might be Src substrates. The translocation of Src to the cytoskeleton might be facilitated by dissociation from the plasma membrane. During platelet activation Src becomes phosphorylated at Ser 12 (FERRELL and MARTIN 1988; LIEBENHOFF et al. 1993), a residue in the membrane-binding amino terminal sequence (RESH 1994). Phosphorylation of Src in this region could decrease its hydrophobicity, as has been shown in fibroblasts (WALKER et al. 1993), and an increase in negative charge from a phosphate in this domain would be expected to repel negatively charged phospholipids such that the enzyme becomes less tightly associated with the membrane. Ser 12 phosphorylation might also facilitate the unfolding of the enzyme and thus its activation. Src is phosphorylated at Ser 12 by protein kinase C (GOULD et al. 1985), another kinase whose activity rapidly increases during platelet activation (SANO et al. 1983; CRABOS et al. 1992). A decrease in the $K_m$ of Src during platelet activation has been observed, possibly signifying enhanced recognition of cytoskeletal protein substrates (LIEBENHOFF et al. 1993). Along these lines, it has been demonstrated that Src exists as two kinetically distinct forms with different $K_m$ values (BUDDE 1993). Phosphorylation of Src at Ser 17 (Fig. 3), which may occur via the cyclic AMP-dependent protein kinase, does not seem to have consequences for the regulation of Src activity that has thus far been detected (COOPER 1990).

## 2. Other Src Family Members

The other Src family PTKs that have been identified in platelets (Fig. 2) and their corresponding apparent molecular masses are: Fyn, 59 or 60 kDa (HORAK et al. 1990); Lyn, two proteins of 54 and 58 kDa (M.M. HUANG et al. 1991); Yes, 62 kDa (ZHAO et al. 1990); and Hck, 61 kDa (M.M. HUANG et al. 1991). These proteins are expressed in platelets at concentrations five to ten times lower than that of Src. The two forms of Lyn are due to alternate splicing of mRNA (YI et al. 1991). Little data is available that distinguishes between characteristics and regulation of these kinases, especially with regard to their presence in platelets. In fact, introduction of a *src* null mutation in mice produced no abnormalities in platelets (SORIANO et al. 1991), indicating a possible functional redundancy between Src family members. Several distinct features of these kinases that have been identified are discussed below.

Besides having a myristate moiety on the N-terminal glycine, a characteristic feature of the Src family, Fyn has been shown to be palmitylated on N-terminal cysteine residues (SHENOY-SCARIA et al. 1993; ALLAND et al. 1994; RESH 1994). This palmitylation has been shown to regulate the association of Fyn with glycosyl-phosphatidylinositol-anchored proteins in lymphocytic cell lines (SHENOY-SCARIA et al. 1993) and to control the distribution of Fyn between membrane and soluble fractions (ALLAND et al. 1994). Whether palmitylation of Fyn occurs in platelets, however, and what importance it might have for platelet function is unknown. Src itself lacks N-terminal cysteine residues and is not palmitylated (ALLAND et al. 1994), but Lyn, Yes, and Hck each contain one cysteine in this region that represents possible (but unproven) sites of modification.

It has been demonstrated that Fyn, Lyn, and Yes associate with glycoprotein (Gp) IV, a receptor for thrombospondin and collagen, in platelet lysates (M.M. HUANG et al. 1991). Src was not detected in anti-Gp IV immunoprecipitates, indicating that these other three Src family kinases possess some features that are distinct from those of Src. It is not known whether Fyn, Lyn, and Yes are specifically involved in signalling through Gp IV, but Fyn becomes activated and translocates to the cytoskeleton upon binding of a GP IV-specific antibody (C. Theiling and P. Presek, unpublished observations). In platelets, these three kinases were also detected complexed with $p21^{ras}$GAP, a GTPase-activating protein, whereas Src was absent (CICHOWSKI et al. 1992).

## II. Syk

This 72-kDa PTK was originally cloned from the particulate fraction of a porcine spleen homogenate and was therefore given the name Syk for *s*pleen t*y*rosine *k*inase (TANIGUCHI et al. 1991). It is localized largely in lymphoid cells or cells of hematopoietic lineage (ZIONCHECK et al. 1988; TANIGUCHI et al. 1991; YAMADA et al. 1991; OHTA et al. 1992). The deduced sequence of 628 amino acids contains two SH2 domains and a kinase catalytic domain (Fig. 2; TANIGUCHI et al. 1991). No SH3 sequence was detected. Additionally, Syk contains no N-terminal Met-Gly sequence for myristylation and is therefore not constitutively localized to the plasma membrane like Src family PTKs. The structure of Syk most closely resembles ZAP-70, the nonreceptor PTK involved in signal transduction from antigen receptors in lymphocytes (CHAN et al. 1991; WEISS and LITTMAN 1994).

Syk was subsequently detected in platelets (OHTA et al. 1992; TANIGUCHI et al. 1993) and found to represent 0.1%–0.2% of total platelet protein (OHTA et al. 1992). The kinase becomes phosphorylated on tyrosine residues following stimulation of platelets with agonists such as thrombin, collagen, a thromboxane $A_2$ mimetic, or ADP (MAEDA et al. 1993; CLARK et al. 1994b; FUJII et al. 1994). For thrombin, this occurs within 10s of stimulation and is sustained for at least 10 min, coinciding with early tyrosine phosphorylation events and with phosphorylation of Src but occurring before phosphorylation of FAK, the focal adhesion kinase (see below and Table 1; CLARK et al. 1994b). Unlike the case of this latter PTK, phosphorylation of Syk is not dependent on platelet aggregation, as it occurs in the absence of stirring. Phosphorylation is partly dependent on fibrinogen binding, however, because it is partially inhibited by antibodies to Gp IIb-IIIa that antagonize fibrinogen binding, and stimulated by antibodies that induce fibrinogen binding.

The kinase activity of Syk increases rapidly upon platelet activation, both for exogenous protein substrates as well as autologously (OHTA et al. 1992; TANIGUCHI et al. 1993; MAEDA et al. 1993; CLARK et al. 1994b). This activation occurs within 5 s of thrombin treatment, reaches a peak at 10 s, and decreases to basal levels by 60 s (TANIGUCHI et al. 1993). The deactivation process appears to be regulated by intracellular $Ca^{2+}$ since depletion of platelet $Ca^{2+}$ results in sustained activity. Activation of platelet Syk may be due to phosphorylation of the enzyme on tyrosine since activation and tyrosine phosphorylation coincide in both platelets and other cell types (KIENER et al. 1993; YAMADA et al. 1993; CLARK et al. 1994b; FUJII et al. 1994). Syk phosphorylation and activation might be initiated by Src or Src family PTKs that are also activated at early time points during platelet activation (CLARK et al. 1994b); this appears to be the case with the antigen receptor in B cells (KUROSAKI et al. 1994). Clustering or ligation of integrins on the platelet surface by Gp IIb-IIIa-specific antibody Fab fragments (CLARK et al. 1994b) or the lectin wheat germ agglutinin (OHTA et al. 1992) leads to Syk activation. Syk is the only platelet PTK known to be activated under these conditions and may be responsible for integrin-dependent tyrosine phosphorylation that occurs during glycoprotein clustering and aggregation (CLARK et al. 1994a). The kinase also becomes phosphorylated and activated when platelets are treated with collagen (YANAGA et al. 1995), which probably binds to a glycoprotein- or integrin-like receptor. For example, treatment with antibodies to Gp VI, whose deficiency results in impairment of collagen-mediated platelet aggregation, leads to activation of both Syk and Src (ICHINOHE et al. 1995). Also, collagen-mediated activation of Syk, as well as Src, is insensitive to treatments that elevate the concentration of platelet cAMP, which is not the case for thrombin-mediated activation of these kinases (ICHINOHE et al. 1995).

Under resting conditions, Syk is found distributed between cytoskeletal and Triton X100-soluble fractions (TOHYAMA et al. 1994). During platelet activation Syk translocates to a cytoskeletal fraction at very early times beginning with shape change (TOHYAMA et al. 1994; YANAGI et al. 1994; MAEDA et al. 1995). This coincides with activation and phosphorylation of the enzyme, which might contribute to the translocation mechanism. In addition, a redistribution of Syk from a cytosolic to a membrane fraction upon platelet activation has been observed (YANAGI et al. 1994). These translocations of Syk between various cellular fractions are likely to be due to its association with specific signalling molecules. Along these lines, it was recently demonstrated that Syk associates with the platelet Fc$\gamma$ receptor upon receptor cross-linking on the platelet surface (CHACKO et al. 1994; YANAGA et al. 1995). Immune receptors, including the Fc$\gamma$ receptor, contain an intracellular recognition sequence denoted ARAM (for antigen receptor activation motif) that becomes phosphorylated on tyrosine, probably by Src family kinases, upon receptor activation, allowing the association of molecules containing SH2 domains (WEISS and LITTMAN 1994). This appears to be the case for the association of Syk with the Fc$\gamma$ receptor in platelets, which is tyrosine-phosphorylated upon ligand binding (HUANG et al. 1992). The thrombin receptor and other G-protein-coupled receptors do not contain an ARAM, however, so there must be an intermediate signalling molecule or molecules involved in the translocation and activation of Syk by thrombin, or these receptors might utilize another signal transduction mechanism. This was recently demonstrated by the finding that an inhibitor of protein kinase C, Ro 31-8220, and a chelator of intracellular $Ca^{2+}$, BAPTA-AM, together inhibit phosphorylation of Syk in response to thrombin (YANAGA et al. 1995).

## III. JAK Family Kinases

Four kinases of the JAK family, JAK1, JAK2, JAK3 and the related kinase TYK2, were recently identified by immuno (Western) blotting in platelets (RODRIGUEZ-LINARES and WATSON 1994; EZUMI et al. 1995a). The JAK family (for Janus kinase or, for some, just another kinase) represents a group of nonreceptor PTKs that have been shown to be involved in signalling from cytokine and growth factor receptors (IHLE et al. 1994). The JAK family kinases have two tyrosine kinase domains (Fig. 2), but the N-terminal one is not expected to have catalytic activity due to the lack of several crucial residues. Neither SH2 nor SH3 domains have been detected. JAK2 becomes tyrosine-phosphorylated upon platelet activation with thrombin (RODRIGUEZ-LINARES and WATSON 1994). Phosphorylation of JAK2 is slow in comparison to phosphorylation of Src or Syk (Table 1), occurring after 30–60s, and is sustained for up to 10min. Thrombin-induced JAK2 phosphorylation may be downstream of phosphoinositide hydrolysis by phospholipase C (PLC) and ensuing signalling pathways, since $Ca^{2+}$ and protein kinase C appear to play a role. On the other hand, stimulation of platelets with thrombopoietin, which has no effect on PLC activity, also induces time- and dose-dependent tyrosine phosphorylation of JAK2 and of TYK2 but not of JAK1 and TYK1 (MIYAKAWA et al. 1995; EZUMI et al. 1995a; MIYAKAWA et al. 1996; RODRIGUEZ-LINARES and WATSON 1996).

## IV. Focal Adhesion Kinase (FAK)

FAK was originally identified in fibroblasts as a cytosolic PTK that colocalizes with integrins to focal adhesion plaques (SCHALLER et al. 1992; GUAN and SHALLOWAY 1992; HANKS et al. 1992; SCHALLER and PARSONS 1994), the intracellular sites where cells make focal contacts with the extracellular matrix. This 125-kDa protein (Fig. 2), which has been observed in many tissues, consists of a PTK catalytic domain bounded by an amino terminus shown to bind to the $\beta$ subunit of integrins and a carboxy terminus containing a focal adhesion targeting domain (SCHALLER and PARSONS 1994). FAK contains no SH2 or SH3 binding motifs, but is itself phosphorylated on a number of tyrosine residues that could induce its association with other SH2-containing molecules. In fibroblasts, it has been demonstrated that Src or Src family kinases mediate phosphorylation of FAK at some of these tyrosines upon cellular adhesion, including two adjacent tyrosine residues within the catalytic domain that appear to regulate activity (CALALB et al. 1995). Autophosphorylation also occurs, and the sequence surrounding the autophosphorylated tyrosine residue resembles a consensus sequence for the binding of the SH2 domains of Src family kinases. Indeed, it has been shown that Src and Fyn stably associate with FAK via their SH2 domains (COBB et al. 1994; SCHALLER et al. 1994; XING et al. 1994).

Platelets contain two members of the FAK family, FAK itself and a newly identified member, RAFTK. This latter kinase has only been identified in a platelet cDNA library and nothing is yet known about its regulation in platelets (AVRAHAM et al. 1995). FAK becomes phosphorylated on tyrosine upon stimulation by thrombin and collagen for 30–60 s and displays increased kinase activity (LIPFERT et al. 1992). Thus, FAK phosphorylation occurs significantly later than Src or Syk phosphorylation (Table 1), which implies that FAK might play a role in a later phase of platelet activation. In addition, phosphorylation of FAK is strictly dependent on Gp IIb-IIIa since it is not observed in platelets from Glanzmann's thrombasthenia patients, who lack the Gp IIb-IIIa complex, or when an antibody to Gp IIb-IIIa that blocks fibrinogen binding is present. FAK phosphorylation is also not initiated by fibrinogen binding alone since it does not occur in unstirred platelet suspensions treated with thrombin (LIPFERT et al. 1992). FAK does become phosphorylated, however, when platelets are induced to spread and adhere to immobilized fibrinogen, and it was shown that the release of ADP is essential for this process (HAIMOVICH et al. 1993). Protein kinase C also appears to play a crucial role since bisindoylmaleimide (GF 109203X), an inhibitor of protein kinase C in platelets, reduces tyrosine phosphorylation of FAK and two other platelet proteins during adhesion to fibrinogen (HAIMOVICH et al. 1996). Stimulation of platelets with anti-Gp IIb-IIIa Fab fragments, which enhance fibrinogen binding, is not sufficient to elicit phosphorylation of FAK (M.M. HUANG et al. 1993); a second signalling pathway involving protein kinase C and $Ca^{2+}$ must also be activated (SHATTIL et al. 1994b). This is in contrast to Syk, which can be activated by anti-Gp IIb-IIIa Fab fragments alone, as mentioned above. Furthermore, FAK does not become phosphorylated or activated in the presence of cytochalasin D, which prevents actin polymerization during platelet activation (LIPFERT et al. 1992; HAIMOVICH et al. 1993; SHATTIL et al. 1994b). Taken together, these findings suggest that phosphorylation and activation of FAK are closely tied to events accompanying aggregation and the engagement of integrins and that cross talk from a pathway involving protein kinase C and $Ca^{2+}$ originating from G-protein-linked receptors is crucial to FAK activation (Clark et al. 1994a).

Coinciding with FAK activation and phosphorylation, a fraction of total platelet FAK translocates to the cytoskeleton during platelet activation (Dash et al. 1995). In fibroblasts, FAK has been shown to associate via its C-terminal focal adhesion targeting sequence with the cytoskeletal protein paxillin, which becomes phosphorylated on tyrosine and localizes to cellular adhesions during the spreading of cells on a martix or stimulation with various growth factors (Burridge et al. 1992; Turner 1994; Schaller and Parsons 1994). Tyrosine-phosphorylated paxillin is then available to bind SH2-domain-containing proteins; it has also been shown to bind the SH3 domain of Src (Weng et al. 1993) and therefore may be involved in the unfolding and activation of Src (Schaller and Parsons 1994). Paxillin has not yet been identified in platelets, but a similar phosphotyrosine-containing protein could carry out a comparable function at the sites of integrin clustering or adhesion in platelets.

## V. Receptor Protein Tyrosine Kinases

In human platelets two different receptor PTKs have been described: the PDGF (platelet-derived growth factor) $\alpha$-receptor (Vassbotn et al. 1994) and the stem cell factor (*c-kit*) receptor (Grabarek et al. 1994). PDGF, an $\alpha$-granule component, is released by exocytosis (secretion) from activated platelets (Siess 1989). Ligand binding to the PDGF receptor induces receptor dimerisation and autophosphorylation, which correlates with activation of the receptor kinase domain (van der Geer and Hunter 1994). Binding analysis showed about 3000 binding sites per platelet for PDGF with an apparent $K_D$ value of $1.2 \times 10^{-8}$ M (Bryckaert et al. 1989). The presence of PDGF receptors suggests an autocrine activation mechanism and a function for PDGF in the nonproliferating and differentiated platelet. Consistent with this hypothesis, stimulation of platelets with low concentrations of thrombin increases tyrosine phosphorylation of the PDGF receptor by two- to fivefold. The increase is completely inhibited by the presence of neutralizing monoclonal anti-PDGF antibodies in the incubation medium (Vassbotn et al. 1994). Thus, the effect of thrombin on PDGF receptor phosphorylation might be due to released endogenous PDGF during platelet activation rather than to thrombin-induced intracellular transactivation of the PDGF receptor, as has been reported for the epidermal growth factor receptor in fibroblasts (Daub et al. 1996). PDGF seems not to influence collagen-induced hydrolysis of phosphatidylinositol 4,5-biphosphate ($PIP_2$) and phosphatidylinositol 4-phosphate but inhibits the subsequent replenishment of $PIP_2$ (Bryckaert et al. 1989). PDGF also totally inhibits collagen-induced phosphatidic acid formation.

Stem cell factor is a hematopoietic growth factor that binds to the cell-surface protein product of the *c-kit* protooncogene (Broxmeyer et al. 1991). c-

Kit is a member of the receptor PTK superfamily and has sequence homology to receptors for other polypeptide growth factors like PDGF and CSF. The presence of *c-kit* mRNA in platelets was recently demonstrated by polymerase chain reaction and its protein product was detected by radioligand binding and immunoblot analysis (GRABAREK et al. 1994). It was found to be constitutively phosphorylated on tyrosine. Further work will be required to confirm this finding and investigate the importance of stem cell factor and PDGF in platelet physiology.

## C. Potential Roles of Protein Tyrosine Kinases in Platelet Functions

Although PTKs have been shown to be involved in mitogenesis and gene regulation in several cell types, their role in the terminally differentiated, anucleate platelet is less clear. Inhibitors of PTKs (the naturally occurring compounds herbimycin A, genistein, and erbstatin, and a number of synthetic tyrphostins and cinnamide derivatives) are able to block both protein tyrosine phosphorylation and various platelet functions such as shape change, adhesion, secretion, and aggregation (DHAR et al. 1990; SALARI et al. 1990; ASAHI et al. 1992; RENDU et al. 1992; GUINEBAULT et al. 1993; FUJII et al. 1994; FURMAN et al. 1994; NEGRESCU et al. 1995). Furthermore, inhibition of protein tyrosine phosphatases by peroxyvanadate increases protein tyrosine phosphorylation and leads to platelet activation (INAZU et al. 1990; PUMIGLIA et al. 1992; BLAKE et al. 1993). Several reports, however, have demonstrated little or no effect of genistein on aggregation and/or secretion (OZAKI et al. 1993; NAKASHIMA et al. 1991; MURPHY et al. 1993). These platelet events also seem to be unaffected by a 24-h treatment with herbimycin A (SCHOENWAELDER et al. 1994), while a short 5-min treatment was capable of reducing platelet aggregation (FUJII et al. 1994). These discrepant data may reflect the fact that the PTK inhibitors have been defined by their ability to inhibit enzyme activity of isolated PTKs in vitro. Consequently, the specificity and mode of action of a particular PTK inhibitor should be rigorously investigated in an intact cell, where factors such as cell permeability and the presence of a myriad of other kinases controlling other responses come into play. Genistein, for example, has been demonstrated by some authors to prevent platelet activation by competition for thromboxane $A_2$ binding sites rather than via inhibition of PTKs (NAKASHIMA et al. 1991; MCNICOL 1993). If this is so, then the observed inhibition of tyrosine phosphorylation and platelet aggregation in the presence of genistein would simply be a consequence of a disrupted eicosanoid signalling pathway. A more carefully designed experiment would include inhibitors of eicosanoid metabolism such as indomethacin or acetylsalicylic acid to rule out effects due to this pathway. Thus, although care must be taken in interpreting data from inhibitor studies, results from a number of different laboratories provide enough evidence to postulate an important role for PTKs in platelet physiology.

## I. Second Messenger Pathways

### 1. Phospholipases

Phospholipases (PL) C, D, and $A_2$ activities increase during platelet activation (see Chap. 11 by NAKASHIMA et al., this volume). In a number of studies, PTK inhibitors have been used to investigate the role of PTKs in phospholipase activation in platelets. Thus, genistein was found to inhibit PLC-mediated phosphoinositide hydrolysis stimulated by platelet-activating factor (DHAR et al. 1990; Murphy et al. 1993), implicating tyrosine phosphorylation in the mechanism of activation of PLC. The inhibitory effect of genistein on PLC was confirmed in the presence of acetylsalicylic acid, which rules out possible effects due to inhibition of thromboxane $A_2$ binding (MURPHY et al. 1993). Furthermore, other PTK inhibitors such as erbstatin and tyrphostins have also been demonstrated to inhibit PLC responses of platelets (SALARI et al. 1990; RENDU et al. 1992; WATSON et al. 1995), and a tyrphostin was found to inhibit the recovery of PLC$\gamma$1 in anti-phosphotyrosine immunoprecipitates (GUINEBAULT et al. 1993). Consistent with these data, PLC$\gamma$1 translocates to the platelet cytoskeleton (GUINEBAULT et al. 1994) where Src and other PTKs become localized during platelet activation. In another approach, it was shown that an anti-Src antibody inhibits platelet-activating factor-stimulated PLC activity in electropermeabilized platelets (DHAR and SHUKLA 1994). These data point either to a direct phosphorylation of PLC$\gamma$1 on tyrosine during platelet activation or association of the phospholipase with other tyrosine-phosphorylated signalling molecules. The latter case may be more likely, as it was recently demonstrated that PLC$\gamma$1 is not tyrosine-phosphorylated in response to thrombin treatment (TORTI and LAPETINA 1992). Along these lines, it has recently been shown that Syk associates with the SH2 domains of PLC$\gamma$1, and that PLC$\gamma$1 and Syk can be co-immunoprecipitated from activated B lymphocytes (SILLMAN and MONROE 1995). On the other hand, an entirely different mechanism involving $Ca^{2+}$ has been proposed to explain the effects of PTK inhibitors on PLC activity (WATSON et al. 1995): the PTK inhibitor tyrphostin ST271 affects PLC activity indirectly via its inhibitory effects on the agonist-induced rise in intracellular $Ca^{2+}$ (see below).

Further evidence for the involvement of PTKs in PLC activation in platelets comes from studies using the protein tyrosine phosphatase inhibitor vanadate. Vanadate must first be oxidized by $H_2O_2$ to form peroxyvanadate before it can gain entry to intact cells. Peroxyvanadate causes a net increase in phosphorylation of platelet proteins on tyrosine and activation of PLC as well as platelet aggregation and secretion (INAZU et al. 1990; PUMIGLIA et al. 1992;

BLAKE et al. 1993). In addition, PLC$\gamma$1 itself becomes phosphorylated on tyrosine in response to peroxyvanadate (BLAKE et al. 1993). These events could be reversed by a tyrphostin PTK inhibitor or a nonselective protein kinase inhibitor, but not by a specific inhibitor of protein kinase C (PUMIGLIA et al. 1992; BLAKE et al. 1993), suggesting that a PTK mediates the activation and phosphorylation of PLC. The effect of peroxyvanadate on PLC, however, might be more indirect; this compound also increases intraplatelet [$Ca^{2+}$] (PUMIGLIA et al. 1992), which appears to play an important role in the PLC response (WATSON et al. 1995). This does not exclude, though, the potential involvement of PTKs in the activation of PLC.

Recently, it has been demonstrated that treatment of platelets with collagen or cross-linking of antibodies bound to the Fc$\gamma$II receptor leads to tyrosine phosphorylation of PLC$\gamma$2 (DANIEL et al. 1994; BLAKE et al. 1994a,b). Thrombin was found to have no effect (BLAKE et al. 1994b), although this differs from results of an earlier study (TATE and RITTENHOUSE 1993). A third report demonstrated a very slight phosphorylation of PLC$\gamma$2 by thrombin treatment compared to that produced by collagen (DANIEL et al. 1994). Thrombin and a number of other platelet agonists bind to receptors that have seven transmembrane-spanning domains and activate a G protein, which then stimulates PLC. Receptors for collagen and IgG molecules, however, are not thought to activate a G protein; thus, activation of PTKs may be a signal transduction mechanism utilized by these receptors to activate PLC$\gamma$2. As described above, the Fc$\gamma$II receptor has an ARAM sequence that becomes phosphorylated on tyrosine, and Syk has been shown to associate with the activated receptor in platelets (CHACKO et al. 1994). It has been speculated that Syk phosphorylates PLC$\gamma$2 on tyrosine following activation of the Fc$\gamma$II receptor or the receptor for collagen (BLAKE et al. 1994b). The $\gamma$ isoforms of PLC have SH2 domains that bind to phosphorylated tyrosine residues; this is how PLC$\gamma$1 binds to and becomes activated by the EGF receptor in fibroblastoid cell lines (NISHIBE et al. 1990). PLC$\gamma$2 could associate with Syk, or with the ARAM of a receptor, via a comparable interaction.

PLD is activated in platelets by various agonists (RUBIN 1988; R. HUANG et al. 1991; PETTY and SCRUTTON 1993; MARTINSON et al. 1995), resulting principally in the hydrolysis of phosphatidylcholine and the production of choline and the putative second messenger phosphatidic acid. It was recently shown that this response is dependent on the activation of protein kinase C (MARTINSON et al. 1995) and that PLD can be partially inhibited by a number of PTK inhibitors both in intact platelets and in permeabilized platelets stimulated with the G protein agonist GTP$\gamma$S (MARTINSON et al. 1994). Furthermore, peroxyvanadate also activates PLD. Thus, the activation of PLD in platelets appears to consist of a series of steps involving various kinases, including PTKs, and G proteins or small GTP-binding proteins.

The activation of $PLA_2$ has also been linked to tyrosine phosphorylation. Genistein and an erbstatin analogue inhibit arachidonate release from platelets stimulated with thrombin or collagen (HARGREAVES et al. 1994); this inhibition was shown to be independent of these inhibitors' effect on $Ca^{2+}$ influx (see below). Furthermore, vanadate has been shown to enhance arachidonate release from permeabilized platelets (McNicol et al. 1993). Along these lines, it has been demonstrated that a cytosolic $PLA_2$ in human platelets is phosphorylated upon thrombin treatment (Kramer et al. 1993). A possible basis for these observations lies in the ability of thrombin to stimulate the tyrosine phosphorylation and activation of mitogen-activated protein kinases (MAP kinases) in platelets (Nakashima et al. 1994; Papkoff et al. 1994). MAP kinases are serine/threonine protein kinases that are activated by dual phosphorylation on tyrosine and threonine, and they have been shown to phosphorylate and activate $PLA_2$ (Lin et al. 1993). A new form of MAP kinase, known as p38 MAP kinase, has recently been identified in human platelets and has been shown to be activated and tyrosine-phosphorylated in response to thrombin before other MAP kinases (Kramer et al. 1995). The kinase that phosphorylates MAP kinase, a dual specificity kinase denoted MAP kinase kinase or MEK (Her et al. 1993), has not been characterized in platelets. On the other hand, a MAP kinase-independent pathway for activation of $PLA_2$ in platelets has recently been described for thrombin and collagen (Börsch-Haubold et al. 1995).

## 2. Phosphoinositide 3-Kinase

Phosphoinositides (PI) are phosphorylated at position 3 on the inositol moiety by PI 3-kinase to produce phosphatidylinositol 3,4-bisphosphate and phosphatidylinositol 3,4,5-trisphosphate. The function of these phospholipids in cellular regulation is not yet understood, but PI 3-kinase is rapidly activated in a number of cell types by a variety of receptors, implying that it is involved in signal transduction (Kapeller and Cantley 1994; Fry 1994; Otsu 1993). Formation of 3-phosphorylated species of PI has been detected in platelets stimulated by various agonists (Kucera and Rittenhouse 1990; Nolan and Lapetina 1990; Sultan et al. 1990). Upon platelet activation, a fraction of platelet PI 3-kinase associates with the cytoskeleton where Src is also detected; this translocation correlates with activation of the lipid kinase (Vulliet et al. 1989). It has been shown that PI 3-kinase rapidly associates with Src and Fyn following exposure of platelets to thrombin (Gutkind et al. 1990) and that PI 3-kinase can be immunoprecipitated with antiphosphotyrosine antibodies, suggesting that tyrosine phosphorylation is involved (Guinebault et al. 1993). Syk and FAK also reportedly associate with PI 3-kinase in platelets (Yanagi et al. 1994; Guinebault et al. 1995). The regulatory subunit of PI 3-kinase, p85, contains an SH3 and two SH2 domains. Either or both of these sequences might be responsible for the holoenzyme's association with various other signalling molecules. Direct tyrosine phosphorylation of p85 has not been observed in platelets in vivo (Guinebault et al. 1995; Yatomi et al. 1994), although tyrphostin treatment of platelets reduces its presence in the cytoskeleton, suggesting it may interact with tyrosine-phosphorylated proteins

(GUINEBAULT et al. 1995). Use of a specific inhibitor of PI 3-kinase has shown that the enzyme is involved in the prolonged activation of Gp IIb-IIIa occurring in irreversible aggregation (KOVACSOVICS et al. 1995).

## 3. GTP-Binding Proteins

GTP-binding proteins are critical transducers of receptor-initiated and intracellular signals. They fall into two broad categories: either heterotrimeric G proteins that couple cell-surface receptors to intracellular effectors, or small GTP-binding proteins that are involved in many intracellular processes such as cytoskeletal rearrangements and secretory pathways. A number of these proteins have been identified in platelets (BRASS and MANNING 1991; NEMOTO et al. 1992; TORTI and LAPETINA 1994). There have been several reports linking PTKs to pathways involving G proteins or GTP-binding proteins in platelets. Epinephrine was shown to induce the association of Src with the $\alpha$ subunit of the G protein $G_i$ that becomes activated during thrombin stimulation (TORTI et al. 1992). Also, as mentioned above, thrombin treatment causes Fyn, Lyn, and Yes to associate with the GTPase-activating protein $p21^{ras}$GAP and results in its tyrosine phosphorylation (CICHOWSKI et al. 1992; TORTI and LAPETINA 1992). $p21^{ras}$GAP, which contains one SH3 and two SH2 domains, is found in a "signalling complex" with the small, membrane-bound GTP-binding protein rap 1B and PLC$\gamma$1 in thrombin-stimulated platelets (TORTI and LAPETINA 1992). It has been suggested that this association is responsible for bringing the phospholipase to the membrane where it comes into contact with its substrate phospholipids (PETERSON and LAPETINA 1994).

## 4. Calcium

There is a good deal of evidence to suggest that platelet PTKs are involved in the rise in intracellular $Ca^{2+}$ observed during platelet activation. PTK inhibitors such as genistein, tyrphostins, herbimycin A, and an erbstatin analogue have been shown to prevent or reduce the $Ca^{2+}$ response to various agonists (ASAHI et al. 1992; MURPHY et al. 1993; SARGEANT et al. 1993a,b; OZAKI et al. 1993; FUJII et al. 1994; POOLE and WATSON 1995; WATSON et al. 1995). These effects are thought to be due either to inhibition of $Ca^{2+}$ release from intracellular stores, perhaps via inibition of PLC (MURPHY et al. 1993; POOLE and WATSON 1995), and/or to inhibition of $Ca^{2+}$ influx (MURPHY et al. 1993; SARGEANT et al. 1993a,b). Evidence against a role of protein tyrosine phosphorylation in promoting $Ca^{2+}$ influx has been presented for the $Ca^{2+}$ response to an anti-CD9 monoclonal antibody, which induces platelet activation in a Fc$\gamma$II receptor-dependent fashion (KURODA et al. 1995).

On the other hand, intracellular $Ca^{2+}$ seems to be linked causally to protein tyrosine phosphorylation, as the $Ca^{2+}$ ionophore A23 187 induces tyrosine phosphorylation of various proteins in platelets (FERRELL and MARTIN 1988; TAKAYAMA et al. 1991; VOSTAL et al. 1991), one of which has been identified as vinculin (VOSTAL and SHULMAN 1993). Consistent with these observations, A23187 also activates the PTKs Src and Syk in platelets (LIEBENHOFF et al.

1993; WANG et al. 1994), although intracellular $Ca^{2+}$ also appears to be involved in the rapid down-regulation of Syk activity following thrombin stimulation, and a rise in intracellular $Ca^{2+}$, per se, is not required (TANIGUCHI et al. 1993). Chelation of extracellular $Ca^{2+}$ reduces thrombin-stimulated tyrosine phosphorylation of a subset of proteins (FALET and RENDU 1994); however, whether this is due to the accompanying blockade of aggregation and signalling through Gp IIb-IIIa that occurs under these conditions is unclear (see below under Aggregation). Experiments using thapsigargin, an inhibitor of the $Ca^{2+}$-ATPase linked to intracellular $Ca^{2+}$ stores, and the intracellular $Ca^{2+}$ chelator BAPTA have demonstrated that cytosolic and stored $Ca^{2+}$ control protein tyrosine phosphorylation antagonistically, with the former promoting and the latter inhibiting tyrosine phosphorylation, respectively (VOSTAL et al. 1991; SARGEANT et al. 1994). Stored $Ca^{2+}$ may therefore activate a protein tyrosine phosphatase, decreasing tyrosine phosphorylation of proteins that may be involved in regulating platelet $Ca^{2+}$ homeostasis in the putative mechansim of store-regulated $Ca^{2+}$ entry (JENNER et al. 1994; HEEMSKERK and SAGE 1994; see Chap. 15 by AUTHI, this volume). Finally, recent evidence suggests that the $Ca^{2+}$-dependent neutral protease calpain, which becomes activated in platelets upon thrombin treatment (FOX et al. 1993b), may be involved in the regulation protein tyrosine phosphorylation observed during platelet activation. Thus, it was shown that protein tyrosine phosphatase 1B becomes activated and redistributes from a membrane to a cytosolic location as a result of calpain action (FRANGIONI et al. 1993) and that an inhibitor of calpain, calpeptin, enhances protein tyrosine phosphorylation in response to thrombin (ARIYOSHI et al. 1995).

### 5. Cyclic Nucleotides

In accordance with their inhibitory effects on platelet function, cyclic AMP and cyclic GMP also inhibit protein tyrosine phosphorylation in platelets (PUMIGLIA et al. 1990; ODA et al. 1992b). The mechanism by which this occurs presumably involves the cyclic AMP-dependent protein kinase, a serine/threonine kinase. Although this enzyme has been shown to phosphorylate Src on Ser 17 (Fig. 3), this has not been shown to have an effect on Src activity, and treatment of platelets with cyclic AMP-elevating agents does not appear to enhance phosphorylation at this site (COOPER 1990; LIEBENHOFF et al. 1993; LIEBENHOFF and PRESEK 1993). Therefore, it may be that cyclic nucleotides inhibit protein tyrosine phosphorylation via their ability to modulate other second messenger pathways, such as those linked to intracellular $Ca^{2+}$.

## II. Platelet Responses

### 1. Shape Change

Shape change is the earliest response observed after treatment of platelets with physiological agonists. ADP-induced shape change, which is accompanied by elevation of intracellular [$Ca^{2+}$], coincides with the rapid tyrosine phosphorylation of 40-, 62-, and 130-kDa proteins (ODA et al. 1992b). With the use of YFLLRNP, a heptapeptide ligand derived from the sequence of the cloned thrombin receptor, shape change can be induced without $Ca^{2+}$ mobilization or protein kinase C activation (RASMUSSEN et al. 1993). YFLLRNP-induced shape change is accompanied by a very rapid, calcium-independent increase in the tyrosine phosphorylation of a set of unidentified proteins of 62, 68, and 130 kDa (NEGRESCU et al. 1995) that is similar to that observed with ADP. All three phosphoproteins are found mainly in the cytoskeletal fraction. The 130-kDa protein is not identical to the cytoskeletal protein vinculin or the PTK FAK. Although the PTK(s) responsible for phosphorylation of these proteins is presently unknown, Syk might be a likely candidate. Syk becomes activated very early during platelet activation through a calcium-independent mechanism (TANIGUCHI et al. 1993) and Syk activation is correlated with shape change following platelet stimulation by the thromboxane $A_2$ analogue, STA2 (MAEDA et al. 1995). Compounds that increase platelet cyclic AMP and cyclic GMP levels are able to prevent ADP- and YFLLRNP-induced shape change and protein tyrosine phosphorylation (ODA et al. 1992b; NEGRESCU et al. 1995), but it is not clear from these experiments whether inhibition of shape change is caused by inhibition of tyrosine phosphorylation or vice versa. Since preincubation of platelets with PTK inhibitors, however, inhibits shape change, Syk activation, and tyrosine phosphorylation, there is sound evidence that Syk and the tyrosine phosphorylated p62, p68, and p130 might play an essential role in the induction of shape change (NEGRESCU et al. 1995; MAEDA et al. 1995).

## 2. Aggregation

The sequential signalling events that lead to platelet aggregation can be separated experimentally into several steps, each of them linked to different events in PTK activation and protein tyrosine phosphorylation. Upon thrombin stimulation, tyrosine phosphorylation of a number of proteins is observed within the first 20 s. Several sets of proteins become tyrosine phosphorylated independent of fibrinogen binding to its receptor Gp IIb-IIIa and subsequent aggregation. This has been demonstrated with platelets where fibrinogen binding has been competitively prevented by the use of the tetrapeptide RGDS or with platelets from individuals having Glanzmann's thrombasthenia (FERRELL and MARTIN 1989; GOLDEN et al. 1990; FOX et al. 1993a; M.M. HUANG et al. 1993; EZUMI et al. 1995b). Intracellular, agonist-induced signalling cascades that are independent of activated Gp IIb-IIIa, fibrinogen binding, and/or aggregation are able to mediate tyrosine phosphorylation of these sets of proteins, although participation of secretion-dependent events cannot be excluded. Under normal conditions, agonist-induced Gp IIb-IIIa activation is followed by fibrinogen binding. Fibrinogen binding to and oligomerization of Gp IIb-IIIa in the absence of aggregation can be experimentally mimicked by platelet treatment with an anti-GP IIIa antibody Fab fragment, which converts the integrin into a high-affinity binding state. Using this approach with unstirred platelets to avoid cell-cell contact and aggregation, tyrosine phosphorylation of several proteins between 50–68 kDa and 130–140 kDa occurs (M.M. HUANG et al. 1993). Under these conditions, secretion from neither dense nor alpha-granules is detectable, indicating that no event further downstream of fibrinogen binding is necessary for this "outside-in" signalling pathway. The aforementioned studies with thrombasthenic platelets support this point of view; thrombin-induced protein tyrosine phosphorylation of a number of proteins, including the sets of 50–68 kDa and 130–140 kDa proteins, is diminished in comparison with normal platelets. Nonreceptor PTKs such as Src, Fyn, and Syk that become rapidly activated independent of aggregation (see Table 1) might be responsible for phosphorylating these proteins. Src has been found to be activated in thrombin-stimulated thrombasthenic platelets within seconds (CLARK and BRUGGE 1993), and also in platelets preincubated with RGDS (P. Presek, unpublished observation). Fyn activation has also been observed in platelets when fibrinogen binding was precluded by RGDS (C. Theiling and P. Presek, unpublished results). In a more detailed analysis, it has been shown that Syk becomes activated and tyrosine-phosphorylated under conditions in which only fibrinogen binding to platelets and cross-linking of Gp IIb-IIIa are allowed, indicating that this outside-in signalling pathway is indeed able to mediate PTK activation (CLARK et al. 1994b). Full activation of Syk, however, seems to require Gp IIb-IIIa engagement since blockade of fibrinogen binding reduces tyrosine phosphorylation of Syk. Therefore, it has been suggested that Syk plays a role in both early, integrin-independent tyrosine phosphorylation events and those dependent upon subsequent integrin engagement (Table 1; FUJII et al. 1994; CLARK et al. 1994b). These latter events might be important for intracellular signalling cascades induced and maintained by aggregation.

Proteins that become rapidly tyrosine-phosphorylated upon platelet stimulation by thrombin that have been identified are p21$^{ras}$GAP (CICHOWSKI et al. 1992), vinculin (VOSTAL and SHULMAN 1993), PLC$\gamma$2 (TATE and RITTENHOUSE 1993), MAP kinase (NAKASHIMA et al. 1994; PAPKOFF et al. 1994), and cortactin (p80/85), an F-actin-binding protein originally found enriched in the cell cortex (membrane ruffles and lamellipodia) of fibroblasts. Only tyrosine phosphorylation of the latter two proteins, however, has been proven to occur independently of GpIIb-IIIa during platelet aggregation (PAPKOFF et al. 1994; FOX et al. 1993a). Aggregation induces tyrosine phosphorylation of additional proteins, including the PTK FAK. FAK phosphorylation and activation do not occur in thrombasthenic platelets or if fibrinogen binding is prevented by the use of a monoclonal antibody directed against Gp IIb-IIIa (Table 1; LIPFERT et al. 1992). The aggregation-dependent activation of FAK may indicate functions for this PTK in late aggregation events such as the formation of large aggregate structures or cellular retraction of fibrin polymers (clot retraction), as has also been suggested for Src (PUMIGLIA and FEINSTEIN

1993; SCHOENWAELDER et al. 1994). During platelet activation by thrombin and other agonists, Syk and the Src family members Src, Yes, and Fyn (Fig. 4) associate with the Triton X-100-insoluble, experimentally defined cytoskeletal fraction in an aggregation-dependent manner (GRONDIN et al. 1991; HORVATH et al. 1992; ODA et al. 1992a; CLARK and BRUGGE 1993; LIEBENHOFF and PRESEK 1993; FOX et al. 1993a; PUMIGLIA and FEINSTEIN 1993; CLARK et al. 1994b). FAK also associates with the cytoskeleton upon aggregation, but the importance of fibrinogen binding in this response has not yet been established (Table 1; DASH et al. 1995). In parallel with cytoskeletal association of several PTKs, various proteins in the cytoskeletal compartment become tyrosine-phosphorylated (PUMIGLIA and FEINSTEIN 1993; FOX et al. 1993a; ALTMULLER and PRESEK 1995). Tyrosine phosphorylation, however, is not necessarily linked to cytoskeletal translocation. Under conditions where aggregation is unaffected, inhibition of tyrosine phosphorylation of cortactin does not affect its translocation (OZAWA et al. 1995). Tyrosine phosphorylation of several cytoskeletal proteins also occurs in vitro by endogeneous PTKs that are present in the cytoskeletal fraction after thrombin stimulation (PUMIGLIA and FEINSTEIN 1993). Tyrosine phosphorylation of these proteins might be important for platelet events that occur subsequent to PTK translocation. The exact significance of cytoskeletal protein phosphorylation for aggregation, secretion, and/or later events, however, has not been established. Along these lines, there is some evidence that these proteins might be involved in the formation of cytoskeletal complexes

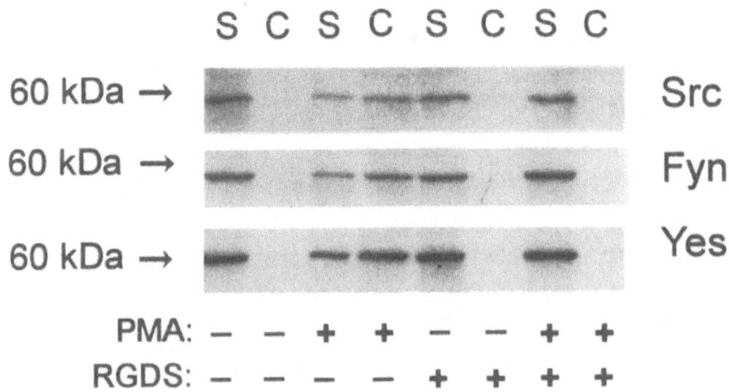


**Fig. 4.** Translocation of Src family kinases to a cytoskeletal fraction in human platelets. Washed human platelets (see MARTINSON et al. 1994) were treated with 500 $\mu M$ of the tetrapeptide *RGDS* for 20 min with stirring at 37°. Phorbol 12-myristate, 13-acetate (*PMA*; 300 $nM$), an activator of protein kinase C that causes platelet aggregation, was then added for 3 min, and Triton X-100-soluble (*S*) and -insoluble (*C*, containing cytoskeleton) fractions were prepared as described (ALTMULLER and PRESEK 1995). Proteins were separated by SDS-polyacrylamide gel electrophoresis, blotted onto nitrocellulose membranes, and probed with antibodies specific for Src, Fyn, or Yes. Bands were visualized by an $^{125}$I-labelled second abtibody and autoradiography. The position of a 60-kDa molecular mass marker protein is indicated. (C. Theiling and P. Presek, unpublished observations)

analogous to those detected at focal adhesion sites in adherent cells (Fox et al. 1993a; CLARK et al. 1994a).

Stimulation of platelets with PDGF and stem cell factor, both ligands for receptor PTKs, seems only to modulate platelet responses to other agonists. While PDGF takes part in a negative feedback control mechanism shown for thrombin- or collagen-induced aggregation and secretion (BRYCKAERT et al. 1986, 1989; VASSBOTN et al. 1994), stem cell factor seems to be involved in a signalling cascade that enhances aggregation in response to epinephrine or ADP (GRABAREK et al. 1994). Neither ligand has an effect on its own on aggregation and secretion. A role for priming platelet responses was also observed for thrombopoietin. Thrombopoietin enhances agonist-induced secondary aggregation and dense granule secretion and induces tyrosine phosphorylation of several proteins including Shc and the nonreceptor PTKs JAK2 and TYK2 (MIYAKAWA et al. 1995; EZUMI et al. 1995; MIYAKAWA et al. 1996; RODRIGUEZ-LINARES and WATSON 1996). The members of the JAK family are known to mediate some effects of cytokines or hematopoetic growth factors by recruitment and tyrosine phosphorylation of Stat proteins (signal transducers and activatiors of transcription; IHLE et al. 1994). Consistent with these findings, Stat3 and Stat5 have been identified to become tyrosine phosphorylated in platelets in response to thrombopoietin (EZUMI et al. 1995a; MIYAKAWA et al. 1996). What role Stats might play in platelets, which lack a nucleus, is unknown.

## 3. Adhesion

Tyrosine phosphorylation of various proteins has also been observed when platelets are allowed to adhere to and spread on fibrinogen and collagen matrices. Adhesion of platelets to fibrinogen is associated with phosphorylation of several proteins, most prominently FAK and two unidentified proteins of 101 and 105 kDa (HAIMOVICH et al. 1993). Moreover, platelet spreading and tyrosine phosphorylation are also dependent on cytoskeletal reorganization and on coordinated signalling through secondarily released agonists such as ADP, since both platelet spreading and inducible tyrosine phosphorylation are inhibited under conditions that block these secondary events. Gp Ia-IIa mediates adhesion of platelets to fibrillar collagen. Although Gp Ia-IIa is necessary for adhesion and platelet responses to collagen, other collagen receptors such as Gp IV may additionally contribute to collagen-dependent platelet tyrosine phosphorylation, since in platelets genetically lacking Gp Ia-IIa, no change in tyrosine phosphorylation has been observed (SHATTIL et al. 1994a). One early response during adhesion of platelets to collagen is tyrosine phosphorylation of PLC$\gamma$2 (DANIEL et al. 1994; BLAKE et al. 1994b). Platelet adhesion to collagen via Gp Ia-IIa has been reported to cause rapid tyrosine phosphorylation of FAK (HAIMOVICH et al. 1993; POLANOWSKA-GRABOWSKA et al. 1993). Besides FAK, other proteins that seem to differ from those observed during adhesion to fibrinogen become tyrosine-phosphorylated during adhesion to collagen. These phosphorylation events, as well as adhesion, cannot be prevented by blocking Gp IIb-IIIa. Tyrosine phosphorylation during shear stress-induced platelet aggregation, however, is regulated by Gp IIb-IIIa and Gp Ib-IX, the receptor for von Willebrand factor, and requires intact cytoskeleton and endogenous ADP (ODA et al. 1995). Under arterial flow conditions, PTK inhibitors have little or no effect on tyrosine phosphorylation or adhesion to collagen (POLANOWSKA-GRABOWSKA et al. 1993). The PTK inhibitor herbimycin A, however, was able to reduce collagen-evoked Syk activation (FUJII et al. 1994). Syk may contribute to $Ca^{2+}$ mobilization mediated by collagen, as has been suggested for tyrosine phosphorylation of PLC$\gamma$2 (DANIEL et al. 1994). These findings support the idea that activation of PTKs such as FAK, Syk, and probably Src (Table 1) causes platelet adhesion to the extracellular matrix via tyrosine phosphorylation of important signalling proteins.

## References

Alland L, Peseckis SM, Atherton RE, Berthiaume L, Resh MD (1994) Dual myristylation and palmitylation of Src family member p59$^{fyn}$ affects subellular localization. J Biol Chem 269:16701–16705
Altmüller A, Presek P (1995) Rapid protein tyrosine phosphorylation in the cytoskeleton of stimulated human platelets. Biochim Biophys Acta 1265:61–66
Anand R, Wilkinson JM, Kellie S (1993) Localisation of pp60$^{c\text{-}src}$ to the surface membrane of human platelets. Oncogene 8:3013–3020
Anderson SK, Gibbs CP, Tanaka A, Kung HJ, Fujita DJ (1985) Human cellular src gene: nucleotide sequence and derived amino acid sequence of the region coding for the carboxy-terminal two-thirds of pp60$^{c\text{-}src}$. Mol Cell Biol 5:1122–1129
Ariyoshi H, Oda A, Salzman EW (1995) Participation of calpain in protein-tyrosine phosphorylation and dephosphorylation in human blood platelets. Arterioscler Thromb Vasc Biol 15:511–514
Asahi M, Yanagi S, Ohta S, Inazu T, Sakai K, Takeuchi F, Taniguchi T, Yamamura H (1992) Thrombin-induced human platelet aggregation is inhibited by protein-tyrosine kinase inhibitors, ST638 and genistein. FEBS Lett 309:10–14
Avraham S, London R, Fu Y, Ota S, Hiregowdara D, Li J, Jiang S, Pasztor LM, White RA, Groopman JE, Avraham H (1995) Identification and characterization of a novel related adhesion focal tyrosine kinase (RAFTK) from megakaryocytes and brain. J Biol Chem 270:27742–27751
Bishop JM (1983) Cellular oncogenes and retroviruses. Annu Rev Biochem 52:301–354
Blake RA, Walker TR, Watson SP (1993) Activation of human platelets by peroxovanadate is associated with tyrosine phosphorylation of phospholipase C-gamma and formation of inositol phosphates. Biochem J 290:471–475
Blake RA, Asselin J, Walker T, Watson SP (1994a) Fc gamma receptor II stimulated formation of inositol phosphates in human platelets is blocked by tyrosine kinase inhibitors and associated with tyrosine phosphorylation of the receptor. FEBS Lett 342:15–18
Blake RA, Schieven GL, Watson SP (1994b) Collagen stimulates tyrosine phosphorylation of phospholipase C-gamma 2 but not phospholipase C-gamma 1 in human platelets. FEBS Lett 353:212–216
Börsch-Haubold AG, Kramer RM, Watson SP (1995) Cytosolic phospholipase A2 is phosphorylated in collagen- and thrombin-stimulated human platelets independent of protein kinase C and mitogen-activated protein kinase. J Biol Chem 270:25885–25892

Bolen JB (1993) Nonreceptor tyrosine protein kinases. Oncogene 8:2025–2031
Brass LF, Manning DR (1991) G proteins and low-molecular-weight GTP-binding proteins in platelets. Trends Cardiovasc Med 1:92–98
Broxmeyer HE, Maze R, Miyazawa K, Carow C, Hendrie PC, Cooper S, Hangoc G, Vadhan-Raj S, Lu L (1991) The kit receptor and its ligand, steel factor, as regulators of hemopoiesis. Cancer Cells 3:480–487
Budde RJA (1993) Evidence for kinetically distinct forms of pp60$^{c\text{-}src}$ with different Km values for their protein substrate. J Biol Chem 268:24868–24872
Burridge K, Turner CE, Romer LH (1992) Tyrosine phosphorylation of paxillin and pp125$^{FAK}$ accompanies cell adhesion to extracellular matrix: a role in cytoskeletal assembly. J Cell Biol 119:893–903
Byckaert, MC, Rendu F, Tobelem G, Caen J (1986) PDGF modifies phosphoinositide metabolism and inhibits aggregation and release in human platelets. Biochem Biophys Res Commun 135:52–57
Byckaert MC, Rendu F, Tombelem G, Wasteson A (1989) Collagen-induced binding to human platelets of platelet-derived growth factor leading to inhibition of p43 and p20 phosphorylation. J Biol Chem 264:4336–4341
Calalb MB, Polte TR, Hanks SK (1995) Tyrosine phosphorylation of focal adhesion kinase at sites in the catalytic domain regulates kinase activity: a role for Src family kinases. Mol Cell Biol 15:954–963
Cantley LC, Auger KR, Carpenter C, Duckworth B, Graziani A, Kapeller R, Soltoff S (1991) Oncogenes and signal transduction. Cell 64:281–302
Cartwright CA, Kaplan PL, Cooper JA, Hunter T, Eckhart W (1986) Altered sites of tyrosine phosphorylation in pp60$^{c\text{-}src}$ associated with polyomavirus middle tumor antigen. Mol Cell Biol 6:1562–1570
Chacko GW, Duchemin AM, Coggeshall KM, Osborne JM, Brandt JT, Anderson CL (1994) Clustering of the platelet Fc gamma receptor induces noncovalent association with the tyrosine kinase p72$^{syk}$. J Biol Chem 269:32435–32440
Chan AC, Irving BA, Fraser JD, Weiss A (1991) The zeta chain is associated with a tyrosine kinase and upon T-cell antigen receptor stimulation associates with ZAP-70, a 70-kDa tyrosine phosphoprotein. Proc Natl Acad Sci USA 88:9166–9170
Cichowski K, McCormick F, Brugge JS (1992) p21$^{ras}$GAP association with Fyn, Lyn, and Yes in thrombin-activated platelets. J Biol Chem 267:5025–5028
Clark EA, Brugge JS (1993) Redistribution of activated pp60$^{c\text{-}src}$ to integrin-dependent cytoskeletal complexes in thrombin-stimulated platelets. Mol Cell Biol 13:1863–1871
Clark EA, Shattil SJ, Brugge JS (1994a) Regulation of protein tyrosine kinases in platelets. Trends Biochem Sci 19:464–469
Clark EA, Shattil SJ, Ginsberg MH, Bolen J, Brugge JS (1994b) Regulation of the protein tyrosine kinase pp72$^{syk}$ by platelet agonists and the integrin alpha IIb beta 3. J Biol Chem 269:28859–28864
Cobb BS, Schaller MD, Leu TH, Parsons JT (1994) Stable association of pp60$^{src}$ and pp59$^{fyn}$ with the focal adhesion-associated protein tyrosine kinase, pp125$^{FAK}$. Mol Cell Biol 14:147–155
Cooper JA (1990) The src family of protein tyrosine kinases. In: Kemp BE (ed) Peptides and Protein Phosphorylation. CRC Press, Inc., Boca Raton, p 85
Cooper JA, Howell B (1993) The when and how of Src regulation. Cell 73:1051–1054
Cooper JA, Gould KL, Cartwright CA, Hunter T (1986) Tyr527 is phosphorylated in pp60$^{c\text{-}src}$: implications for regulation. Science 231:1431–1434
Courtneidge SA (1985) Activation of the pp60$^{c\text{-}src}$ kinase by middle T antigen binding or by dephosphorylation. EMBO J 4:1471–1477
Crabos M, Fabbro D, Stabel S, Erne P (1992) Effect of tumour-promoting phorbol ester, thrombin and vasopressin on translocation of three distinct protein kinase C isoforms in human platelets and regulation by calcium. Biochem J 288:891–896

Daniel JL, Dangelmaier C, Smith JB (1994) Evidence for a role for tyrosine phosphorylation of phospholipase C gamma 2 in collagen-induced platelet cytosolic calcium mobilization. Biochem J 302:617–622

Dash D, Aepfelbacher M, Siess W (1995) The association of pp125$^{FAK}$, pp60$^{src}$, CDC42Hs and Rap 1B with the cytoskeleton of aggregated platelets is a reversible process regulated by calcium. FEBS Lett 363:231–234

DeFranco AL (1995) Transmembrane signaling by antigen receptors of B and T lymphocytes. Curr Opin Cell Biol 7:163–175

Dhar A, Shukla SD (1994) Electrotransjection of pp60$^{v\text{-}src}$ monoclonal antibody inhibits activation of phospholipase C in platelets. A new mechanism for platelet-activating factor responses. J Biol Chem 269:9123–9127

Dhar A, Paul AK, Shukla SD (1990) Platelet-activating factor stimulation of tyrosine kinase and its relationship to phospholipase C in rabbit platelets: studies with genistein and monoclonal antibody to phosphotyrosine. Mol Pharmacol 37:519–525

Daub H, Weiss FU, Wallasch C, Ullrich A (1996) Role of transactivation of the EGF receptor in signalling by G-protein-coupled receptors. Nature 379:557–560

Erpel T, Courtneidge SA (1995) Src family protein tyrosine kinases and cellular signal transduction pathways. Curr Opin Cell Biol 7:176–182

Erpel T, Superti-Furga G, Courtneidge SA (1995) Mutational analysis of the Src SH3 domain: the same residues of the ligand binding surface are important for intra- and intermolecular interactions. EMBO J 14:963–975

Ezumi Y, Takayama H, Okuma M (1995a) Thrombopoietin, c-mpl ligand, induces tyrosine phosphorylation of Tyk2, Jak2, and Stat3, and enhances agonist-induced aggragation in platelets in vitro. FEBS Lett 374:48–52

Ezumi Y, Takayama H, Okuma M (1995b) Differential regulation of protein-tyrosine phosphatases by integrin alphaIIb-beta3 through cytoskeletal reorganization and tyrosine phosphorylation in human platelets. J Biol Chem 270:11927–11934

Falet H, Rendu F (1994) Calcium mobilisation controls tyrosine protein phosphorylation independently of the activation of protein kinase C in human platelets. FEBS Lett 345:87–91

Fantl WJ, Johnson DE, Williams LT (1993) Signalling by receptor tyrosine kinases. Annu Rev Biochem 62:453–481

Ferrell JE Jr, Martin GS (1988) Platelet tyrosine-specific protein phosphorylation is regulated by thrombin. Mol Cell Biol 8:3603–3610

Ferrell JE Jr, Martin GS (1989) Tyrosine-specific protein phosphorylation is regulated by glycoprotein IIb-IIIa in platelets. Proc Natl Acad Sci USA 86:2234–2238

Ferrell JE Jr, Noble JA, Martin GS, Jacques YV, Bainton DF (1990) Intracellular localization of pp60$^{c\text{-}src}$ in human platelets. Oncogene 5:1033–1036

Fox JE, Reynolds CC, Boyles JK (1992) Studying the platelet cytoskeleton in Triton X-100 lysates. Methods Enzymol 215:42–58

Fox JE, Lipfert L, Clark EA, Reynolds CC, Austin CD, Brugge JS (1993a) On the role of the platelet membrane skeleton in mediating signal transduction. Association of GP IIb-IIIa, pp60$^{c\text{-}src}$, pp62$^{c\text{-}yes}$, and the p21$^{ras}$ GTPase-activating protein with the membrane skeleton. J Biol Chem 268:25973–25984

Fox JE, Taylor RG, Taffarel M, Boyles JK, Goll DE (1993b) Evidence that activation of platelet calpain is induced as a consequence of binding of adhesive ligand to the integrin, glycoprotein IIb-IIIa. J Cell Biol 120:1501–1507

Frangioni JV, Oda A, Smith M, Salzman EW, Neel BG (1993) Calpain-catalyzed cleavage and subcellular relocation of protein phosphotyrosine phosphatase 1B (PTP-1B) in human platelets. EMBO J 12:4843–4856

Fry MJ (1994) Structure, regulation and function of phosphoinositide 3-kinases. Biochim Biophys Acta 1226:237–268

Fujii C, Yanagi S, Sada K, Nagai K, Taniguchi T, Yamamura H (1994) Involvement of protein-tyrosine kinase p72$^{syk}$ in collagen-induced signal transduction in platelets. Eur J Biochem 226:243–248

Furman MI, Grigoryev D, Bray PF, Dise KR, Goldschmidt-Clermont PJ (1994) Platelet tyrosine kinases and fibrinogen receptor activation. Circ Res 75:172–180
Golden A, Brugge JS (1989) Thrombin treatment induces rapid changes in tyrosine phosphorylation in platelets. Proc Natl Acad Sci USA 86:901–905
Golden A, Nemeth SP, Brugge JS (1986) Blood platelets express high levels of the pp60$^{c-src}$-specific tyrosine kinase activity. Proc Natl Acad Sci USA 83:852–856
Golden A, Brugge JS, Shattil SJ (1990) Role of platelet membrane glycoprotein IIb-IIIa in agonist-induced tyrosine phosphorylation of platelet proteins. J Cell Biol 111:3117–3127
Gould KL, Woodgett JR, Cooper JA, Buss JE, Shalloway D, Hunter T (1985) Protein kinase C phosphorylates pp60$^{src}$ at a novel site. Cell 42:849–857
Grabarek J, Groopman JE, Lyles YR, Jiang S, Bennett L, Zsebo K, Avraham H (1994) Human kit ligand (stem cell factor) modulates platelet activation in vitro. J Biol Chem 269:21718–21724
Grondin P, Plantavid M, Sultan C, Breton M, Mauco G, Chap H (1991) Interaction of pp60$^{c-src}$, phospholipase C, inositol-lipid, and diacyglycerol kinases with the cytoskeletons of thrombin-stimulated platelets. J Biol Chem 266:15705–15709
Guan JL, Shalloway D (1992) Regulation of focal adhesion-associated protein tyrosine kinase by both cellular adhesion and oncogenic transformation. Nature 358:690–692
Guinebault C, Payrastre B, Sultan C, Mauco G, Breton M, Levy-Toledano S, Plantavid M, Chap H (1993) Tyrosine kinases and phosphoinositide metabolism in thrombin-stimulated human platelets. Biochem J 292:851–856
Guinebault C, Payrastre B, Mauco G, Breton M, Plantavid M, Chap H (1994) Rapid and transient translocation of PLC-gamma 1 to the cytoskeleton of thrombin-stimulated platelets. Evidence for a role of tyrosine kinases. Cell Mol Biol Noisy le grand 40:687–693
Guinebault C, Payrastre B, Racaudsultan C, Mazarguil H, Breton M, Mauco G, Plantavid M, Chap H (1995) Integrin-dependent translocation of phosphoinositide 3-kinase to the cytoskeleton of thrombin-activated platelets involves specific interactions of p85 alpha with actin filaments and focal adhesion kinase. J Cell Biol 129:831–842
Gutkind JS, Lacal PM, Robbins KC (1990) Thrombin-dependent association of phosphatidylinositol-3 kinase with p60$^{c-src}$ and p59$^{fyn}$ in human platelets. Mol Cell Biol 10:3806–3809
Haimovich B, Lipfert L, Brugge JS, Shattil SJ (1993) Tyrosine phosphorylation and cytoskeletal reorganization in platelets are triggered by interaction of integrin receptors with their immobilized ligands. J Biol Chem 268:15868–15877
Haimovich B, Kaneshiki N, Ji P (1996) Protein kinase C regulates tyrosine phosphorylation of pp125FAK in platelets adherent to fibrinogen. Blood 87:152–161
Hanks SK, Calalb MB, Harper MC, Patel SK (1992) Focal adhesion protein-tyrosine kinase phosphorylated in response to cell attachment to fibronectin. Proc Natl Acad Sci USA 89:8487–8491
Hargreaves PG, Licking EF, Sargeant P, Sage SO, Barnes MJ, Farndale RW (1994) The tyrosine kinase inhibitors, genistein and methyl 2,5-dihydroxycinnamate, inhibit the release of ($^{3}$H)arachidonate from human platelets stimulated by thrombin or collagen. Thromb Haemost 72:634–642
Heemskerk JWM, Sage SO (1994) Calcium signalling in platelets and other cells. Platelets 5:295–316
Her JH, Lakhani S, Zu K, Vila J, Dent P, Sturgill TW, Weber MJ (1993) Dual phosphorylation and autophosphorylation in mitogen-activated protein (MAP) kinase activation. Biochem J 296:25–31
Horak ID, Corcoran ML, Thompson PA, Wahl LM, Bolen JB (1990) Expression of p60$^{fyn}$ in human platelets. Oncogene 5:597–602
Horvath AR, Muszbek L, Kellie S (1992) Translocation of pp60$^{c-src}$ to the cytoskeleton during platelet aggregation. EMBO J 11:855–861

Huang MM, Bolen JB, Barnwell JW, Shattil SJ, Brugge JS (1991) Membrane glycoprotein IV (CD36) is physically associated with the Fyn, Lyn, and Yes protein-tyrosine kinases in human platelets. Proc Natl Acad Sci USA 88:7844–7848

Huang MM, Indik Z, Brass LF, Hoxie JA, Schreiber AD, Brugge JS (1992) Activation of Fc gamma RII induces tyrosine phosphorylation of multiple proteins including Fc gamma RII. J Biol Chem 267:5467–5473

Huang MM, Lipfert L, Cunningham M, Brugge JS, Ginsberg MH, Shattil SJ (1993) Adhesive ligand binding to integrin alpha IIb beta 3 stimulates tyrosine phosphorylation of novel protein substrates before phosphorylation of $pp125^{FAK}$. J Cell Biol 122:473–483

Huang R, Kucera GL, Rittenhouse SE (1991) Elevated cytosolic $Ca^{2+}$ activates phospholipase D in human platelets. J Biol Chem 266:1652–1655

Hunter T, Sefton BM (1980) Transforming gene product of Rous sarcoma virus phosphorylates tyrosine. Proc Natl Acad Sci USA 77:1311–1315

Ichinohe T, Takayama H, Ezumi Y, Yanagi S, Yamamura H, Okuma M (1995) Cyclic AMP-insensitive activation of c-src and syk protein-tyrosine kinases through platelet membrane glycoprotein VI. J Biol Chem 270:28029–28036

Ihle JN, Witthuhn BA, Quelle FW, Yamamoto K, Thierfelder WE, Kreider B, Silvennoinen O (1994) Signaling by the cytokine receptor superfamily: JAKs and STATs. Trends Biochem Sci 19:222–227

Inazu T, Taniguchi T, Yanagi S, Yamamura H (1990) Protein-tyrosine phosphorylation and aggregation of intact human platelets by vanadate with $H_2O_2$. Biochem Biophys Res Commun 170:259–263

Inazu T, Taniguchi T, Ohta S, Miyabo S, Yamamura H (1991) The lectin wheat germ agglutinin induces rapid protein-tyrosine phosphorylation in human platelets. Biochem Biophys Res Commun 174:1154–1158

Jenner S, Farndale RW, Sage SO (1994) The effect of calcium-store depletion and refilling with various bivalent cations on tyrosine phosphorylation and $Mn^{2+}$ entry in fura-2-loaded human platelets. Biochem J 303:337–339

Kapeller R, Cantley LC (1994) Phosphatidylinositol 3-kinase. Bioessays 16:565–576

Kiener PA, Rankin BM, Burkhardt AL, Schieven GL, Gilliland LK, Rowley RB, Bolen JB, Ledbetter JA (1993) Cross-linking of Fc gamma receptor I (Fc gamma RI) and receptor II (Fc gamma RII) on monocytic cells activates a signal transduction pathway common to both Fc receptors that involves the stimulation of p 72 Syk protein tyrosine kinase. J Biol Chem 268:24442–24448

Kmiecik TE, Shalloway D (1987) Activation and suppression of $pp60^{c\text{-}src}$ transforming ability by mutation of its primary sites of tyrosine phosphorylation. Cell 49:65–73

Kmiecik TE, Johnson PJ, Shalloway D (1988) Regulation by the autophosphorylation site in overexpressed $pp60^{c\text{-}src}$. Mol Cell Biol 8:4541–4546

Kovacsovics TJ, Bachelot C, Toker A, Vlahos CJ, Duckworth B, Cantley LC, Hartwig JH (1995) Phosphoinositide 3-kinase inhibition spares actin assembly in activating platelets but reverses platelet aggregation. J Biol Chem 270:11358–11366

Kramer RM, Roberts EF, Manetta JV, Hyslop PA, Jakubowski JA (1993) Thrombin-induced phosphorylation and activation of $Ca^{2+}$-sensitive cytosolic phospholipase A2 in human platelets. J Biol Chem 268:26796–26804

Kramer RM, Roberts EF, Strifler BA, Johnstone EM (1995) Thrombin induces activation of p 38 MAP kinase in human platelets. J Biol Chem 270:27395–27398

Kucera GL, Rittenhouse SE (1990) Human platelets form 3-phosphorylated phosphoinositides in response to alpha-thrombin, U46619, or GTPgammaS. J Biol Chem 265:5345–5348

Kuroda K, Ozaki Y, Qi R, Asazuma N, Yatomi Y, Satoh K, Nomura S, Suzuki M, Kume S (1995) $Fc_{\gamma}II$ receptor-mediated platelet activation induced by anti-CD9 monoclonal antibody opens $Ca^{2+}$ channels which are distinct from those associated with $Ca^{2+}$ store depletion. J Immunol 155:4427–4436

Kurosaki T, Takata M, Yamanashi Y, Inazu T, Taniguchi T, Yamamoto T, Yamamura H (1994) Syk activation by the Src-family tyrosine kinase in the B cell receptor signaling. J Exp Med 179:1725–1729

Lev S, Moreno H, Martinez R, Canoll P, Peles E, Musacchio JM, Plowman GD, Rudy B, Schlessinger J (1995) Protein tyrosine kinase PYK 2 involved in $Ca^{2+}$-induced regulation of ion channel and MAP kinase functions. Nature 376:737–745

Liebenhoff U, Presek P (1993) Modulation of the protein tyrosine kinase $pp60^{c\text{-}src}$ upon platelet activation: a protein kinase C inhibitor blocks translocation. In: Preissner KT, Rosenblatt S, Kost C, Wegerhoff J, Mosher DF (eds) Biology of Vitronectins and their Receptors. Elsevier, Amsterdam, p 171

Liebenhoff U, Brockmeier D, Presek P (1993) Substrate affinity of the protein tyrosine kinase $pp60^{c\text{-}src}$ is increased on thrombin stimulation of human platelets. Biochem J 295:41–48

Lin LL, Wartmann M, Lin AY, Knopf JL, Seth A, Davis RJ (1993) cPLA2 is phosphorylated and activated by MAP kinase. Cell 72:269–278

Lipfert L, Haimovich B, Schaller MD, Cobb BS, Parsons JT, Brugge JS (1992) Integrin-dependent phosphorylation and activation of the protein tyrosine kinase $pp125^{FAK}$ in platelets. J Cell Biol 119:905–912

Liu X, Brodeur SR, Gish G, Zhou S, Cantley LC, Laudano AP, Pawson T (1993) Regulation of c-Src tyrosine kinase activity by the Src SH2 domain. Oncogene 8:1119–1126

Maeda H, Taniguchi T, Inazu T, Yang C, Nakagawara G, Yamamura H (1993) Protein-tyrosine kinase $p72^{syk}$ is activated by thromboxane A2 mimetic U44069 in platelets. Biochem Biophys Res Commun 197:62–67

Maeda H, Inazu T, Nagai K, Maruyama S, Nakagawara G, Yamanura H (1995) Possible involvement of protein tyrosine kinases such as p72syk in the disc-sphere change response of porcine platelets. J Biochem (Tokyo) 117:1201–1208

Martinson EA, Scheible S, Presek P (1994) Inhibition of phospholipase D of human platelets by protein tyrosine kinase inhibitors. Cell Mol Biol 40:627–634

Martinson EA, Scheible S, Greinacher A, Presek P (1995) Platelet phospholipase D is activated by protein kinase C via an integrin alphaIIb-beta3-independent mechanism. Biochem J 310:623–628

Matsuda M, Mayer BJ, Fukui Y, Hanafusa H (1990) Binding of transforming protein, $P47^{gag\text{-}crk}$, to a broad range of phosphotyrosine-containing proteins. Science 248:1537–1539

McNicol A (1993) The effects of genistein on platelet function are due to thromboxane receptor antagonism rather than inhibition of tyrosine kinase. Prostaglandins Leukot Essent Fatty Acids 48:379–384

McNicol A, Robertson C, Gerrard JM (1993) Vanadate activates platelets by enhancing arachidonic acid release. Blood 81:2329–2338

Mori S, Ronnstrand L, Yokote K, Engstrom A, Courtneidge SA, Claesson-Welsh L, Heldin CH (1993) Identification of two juxtamembrane autophosphorylation sites in the PDGF beta-receptor; involvement in the interaction with Src family tyrosine kinases. EMBO J 12:2257–2264

Murphy CT, Kellie S, Westwick J (1993) Tyrosine-kinase activity in rabbit platelets stimulated with platelet-activating factor. The effect of inhibiting tyrosine kinase with genistein on platelet-signal-molecule elevation and functional responses. Eur J Biochem 216:639–651

Mustelin T, Burn P (1993) Regulation of src family tyrosine kinases in lymphocytes. Trends Biochem Sci 18:215–220

Miyakawa Y, Oda A, Druker BJ, Kato T, Miyazaki H, Handa M, Ikeda Y (1995) Recombinant thrombopoietin induces rapid protein tyrosine phosphorylation of janus kinase 2 and Shc in human blood platelets. Blood 86:23–27

Miyakawa Y, Oda A, Druker BJ, Miyazaki H, Handa M, Ohashi H, Ikeda Y (1996) Thrombopoietin induces tyrosine phosphorylation of Stat3 and Stat5 in human blood platelets. Blood 87:439–446

Nakamura S, Yamamura H (1989) Thrombin and collagen induce rapid phosphorylation of a common set of cellular proteins on tyrosine in human platelets. J Biol Chem 264:7089–7091

Nakashima S, Koike T, Nozawa Y (1991) Genistein, a protein tyrosine kinase inhibitor, inhibits thromboxane A2-mediated human platelet responses. Mol Pharmacol 39:475–480

Nakashima S, Chatani Y, Nakamura M, Miyoshi N, Kohno M, Nozawa Y (1994) Tyrosine phosphorylation and activation of mitogen-activated protein kinases by thrombin in human platelets: possible involvement in late arachidonic acid release. Biochem Biophys Res Commun 198:497–503

Negrescu EV, de-Quintana KL, Siess W (1995) Platelet shape change induced by thrombin receptor activation. Rapid stimulation of tyrosine phosphorylation of novel protein substrates through an integrin- and $Ca^{2+}$-independent mechanism. J Biol Chem 270:1057–1061

Nemoto Y, Namba T, Teru-uchi T, Ushikubi F, Morii N, Narumiya S (1992) A rho gene product in human blood platelets. I. Identification of the platelet substrate for botulinum C3 ADP-ribosyltransferase as rhoA protein. J Biol Chem 267:20916–20920

Nishibe S, Wahl MI, Hernandez-Sotomayor SM, Tonks NK, Rhee SG, Carpenter G (1990) Increase of the catalytic activity of phospholipase C-gamma 1 by tyrosine phosphorylation. Science 250:1253–1256

Nolan RD, Lapetina EG (1990) Thrombin stimulates the production of a novel polyphosphoinositide in human platelets. J Biol Chem 265:2441–2445

O'Connor TJ, Bjorge JD, Cheng H-C, Wang JH, Fujita DJ (1995) Mechanism of c-Src activation in human melanocytes: elevated level of protein tyrosine phosphatase activity directed against the carboxy-terminal regulatory tyrosine. Cell Growth Differ 6:123–130

Oda A, Druker BJ, Smith M, Salzman EW (1992a) Association of $pp60^{src}$ with Triton X-100-insoluble residue in human blood platelets requires platelet aggregation and actin polymerization [published erratum appears in J Biol Chem 1993 Mar 5;268(7):5339]. J Biol Chem 267:20075–20081

Oda A, Druker BJ, Smith M, Salzman EW (1992b) Inhibition by sodium nitroprusside or PGE1 of tyrosine phosphorylation induced in platelets by thrombin or ADP. Am J Physiol 262:C701–C707

Oda A, Yokoyama K, Murata M, Tokuhira M, Nakamura K, Handa M, Watanabe K, Ikeda Y (1995) Protein tyrosine phsophorylation in human platelets during shear stress-induced platelet aggragation (SIPA) is regulated by glycoprotein (GP) Ib/IX as well as GP IIb/IIIa and requires intact cytoskeleton and endogenous ADP. Thromb Haemostas 74:736–742

Ohta S, Taniguchi T, Asahi M, Kato Y, Nakagawara G, Yamamura H (1992) Protein-tyrosine kinase $p72^{syk}$ is activated by wheat germ agglutinin in platelets. Biochem Biophys Res Commun 185:1128–1132

Okada M, Nada S, Yamanashi Y, Yamamoto T, Nakagawa H (1991) CSK: a protein-tyrosine kinase involved in regulation of src family kinases. J Biol Chem 266:24249–24252

Otsu M (1993) Recent progress in the studies of phosphatidylinositol 3-kinase. Seikagaku 65:1299–1316

Ozaki Y, Yatomi Y, Jinnai Y, Kume S (1993) Effects of genistein, a tyrosine kinase inhibitor, on platelet functions. Genistein attenuates thrombin-induced $Ca^{2+}$ mobilization in human platelets by affecting polyphosphoinositide turnover. Biochem Pharmacol 46:395–403

Ozawa K, Kashiwada K, Takahashi M, Sobue K (1995) Translocation of cortactin (p80/85) to the actin-based cytoskeleton during thrombin receptor-mediated platelet activation. Exp Cell Res 221:197–204

Papkoff J, Chen RH, Blenis J, Forsman J (1994) p42 mitogen-activated protein kinase and p90 ribosomal S6 kinase are selectively phosphorylated and activated

during thrombin-induced platelet activation and aggregation. Mol Cell Biol 14:463–472
Pawson T (1995) Protein modules and signalling networks. Nature 373:573–580
Pawson T, Schlessinger J (1993) SH2 and SH3 domains. Current Biology 3:434–442
Peterson SN, Lapetina EG (1994) Platelet activation and inhibition. Novel signal transduction mechanisms. Ann NY Acad Sci 714:53–63
Petty AC, Scrutton MC (1993) Release of choline metabolites from human platelets: evidence for activation of phospholipase D and of phosphatidylcholine-specific phospholipase C. Platelets 4:23–29
Polanowska-Grabowska R, Geanacopoulos M, Gear AR (1993) Platelet adhesion to collagen via the alpha 2 beta 1 integrin under arterial flow conditions causes rapid tyrosine phosphorylation of $pp125^{FAK}$. Biochem J 296:543–547
Poole AW, Watson SP (1995) Regulation of cytosolic calcium by collagen in single human platelets. Br J Pharmacol 115:101–106
Pumiglia KM, Feinstein MB (1993) Thrombin and thrombin receptor agonist peptide induce tyrosine phosphorylation and tyrosine kinases in the platelet cytoskeleton. Translocation of $pp60^{c\text{-}src}$ and integrin alpha IIb beta 3 (glycoprotein IIb/IIIa) is not required for aggregation, but is dependent on formation of large aggregate structures. Biochem J 294:253–260
Pumiglia KM, Huang CK, Feinstein MB (1990) Elevation of cAMP, but not cGMP, inhibits thrombin-stimulated tyrosine phosphorylation in human platelets. Biochem Biophys Res Commun 171:738–745
Pumiglia KM, Lau LF, Huang CK, Burroughs S, Feinstein MB (1992) Activation of signal transduction in platelets by the tyrosine phosphatase inhibitor pervanadate (vanadyl hydroperoxide). Biochem J 286:441–449
Rasmussen UB, Gachet C, Schlesinger Y, Hanau D, Ohlmann P, Van-Obberghen-Schilling E, Pouyssegur J, Cazenave JP, Pavirani A (1993) A peptide ligand of the human thrombin receptor antagonizes alpha-thrombin and partially activates platelets. J Biol Chem 268:14322–14328
Rendu F, Eldor A, Grelac F, Bachelot C, Gazit A, Gilon C, Levy-Toledano S, Levitzki A (1992) Inhibition of platelet activation by tyrosine kinase inhibitors. Biochem Pharmacol 44:881–888
Resh MD (1993) Interaction of tyrosine kinase oncoproteins with cellular membranes. Biochim Biophys Acta 1155:307–322
Resh MD (1994) Myristylation and palmitylation of Src family members: the fats of the matter. Cell 76:411–413
Reuter C, Findik D, Presek P (1990) Characterization of purified $pp60^{c\text{-}src}$ protein tyrosine kinase from human platelets. Eur J Biochem 190:343–350
Rodriguez-Linares B, Watson SP (1994) Phosphorylation of JAK2 in thrombin-stimulated human platelets. FEBS Lett 352:335–338
Rodriguez-Linares B, Watson SP (1996) Thrombopoeitin potentiates activation of human platelets in association with Jak2 and Tyk2 phosphorylaltion. FEBS Lett 316:93–98
Rosen MK, Yamazaki T, Gish GD, Kay CM, Pawson T, Kay LE (1995) Direct demonstration of an intramolecular SH2-phosphotyrosine interaction in the Crk protein. Nature 374:477–479
Roussel RR, Brodeur SR, Shalloway D, Laudano AP (1991) Selective binding of activated $pp60^{c\text{-}src}$ by an immobilized synthetic phosphopeptide modeled on the carboxyl terminus of $pp60^{c\text{-}src}$. Proc Natl Acad Sci USA 88:10696–10700
Rubin R (1988) Phosphatidylethanol formation in human platelets: evidence for thrombin-induced activation of phospholipase D. Biochem Biophys Res Commun 156:1090–1096
Salari H, Duronio V, Howard SL, Demos M, Jones K, Reany A, Hudson AT, Pelech SL (1990) Erbstatin blocks platelet activating factor-induced protein-tyrosine phosphorylation, polyphosphoinositide hydrolysis, protein kinase C activation, serotonin secretion and aggregation of rabbit platelets. FEBS Lett 263:104–108

Sano K, Takai Y, Yamanishi J, Nishizuka Y (1983) A role of calcium-activated phospholipid-dependent protein kinase in human platelet activation. Comparison of thrombin and collagen actions. J Biol Chem 258:2010–2013
Sargeant P, Farndale RW, Sage SO (1993a) ADP- and thapsigargin-evoked $Ca^{2+}$ entry and protein-tyrosine phosphorylation are inhibited by the tyrosine kinase inhibitors genistein and methyl-2,5-dihydroxycinnamate in fura-2-loaded human platelets. J Biol Chem 268:18151–18156
Sargeant P, Farndale RW, Sage SO (1993b) The tyrosine kinase inhibitors methyl 2,5-dihydroxycinnamate and genistein reduce thrombin-evoked tyrosine phosphorylation and $Ca^{2+}$ entry in human platelets. FEBS Lett 315:242–246
Sargeant P, Farndale RW, Sage SO (1994) Calcium store depletion in dimethyl BAPTA-loaded human platelets increases protein tyrosine phosphorylation in the absence of a rise in cytosolic calcium. Exp Physiol 79:269–272
Sasaki H, Nagura K, Ishino M, Tobioka H, Kotani K, Sasaki T (1995) Cloning and characterization of cell adhesion kinase beta, a novel protein-tyrosine kinase of the focal adhesion kinase subfamily. J Biol Chem 270:21206–21219
Schaller MD, Parsons JT (1994) Focal adhesion kinase and associated proteins. Curr Opin Cell Biol 6:705–710
Schaller MD, Borgman CA, Cobb BS, Vines RR, Reynolds AB, Parsons JT (1992) $pp125^{FAK}$ a structurally distinctive protein-tyrosine kinase associated with focal adhesions. Proc Natl Acad Sci USA 89:5192–5196
Schaller MD, Hildebrand JD, Shannon JD, Fox JW, Vines RR, Parsons JT (1994) Autophosphorylation of the focal adhesion kinase, $pp125^{FAK}$, directs SH2-dependent binding of $pp60^{src}$. Mol Cell Biol 14:1680–1688
Schoenwaelder SM, Jackson SP, Yuan Y, Teasdale MS, Salem HH, Mitchell CA (1994) Tyrosine kinases regulate the cytoskeletal attachment of integrin alpha IIb beta 3 (platelet glycoprotein IIb/IIIa) and the cellular retraction of fibrin polymers. J Biol Chem 269:32479–32487
Seidel-Dugan C, Meyer BE, Thomas SM, Brugge JS (1992) Effects of SH2 and SH3 deletions on the functional activities of wild-type and transforming variants of c-Src. Mol Cell Biol 12:1835–1845
Shattil SJ, Ginsberg MH, Brugge JS (1994a) Adhesive signaling in platelets. Curr Opin Cell Biol 6:695–704
Shattil SJ, Haimovich B, Cunningham M, Lipfert L, Parsons JT, Ginsberg MH, Brugge JS (1994b) Tyrosine phosphorylation of $pp125^{FAK}$ in platelets requires coordinated signaling through integrin and agonist receptors. J Biol Chem 269:14738–14745
Shenoy-Scaria AM, Gauen LKT, Kwong J, Shaw AS, Lublin DM (1993) Palmitylation of an amino-terminal cysteine motif of protein tyrosine kinases $p56^{lck}$ and $p59^{fyn}$ mediates interaction with glycosyl-phosphatidylinositol-anchored proteins. Mol Cell Biol 13:6385–6392
Sigal CT, Zhou W, Buser CA, McLaughlin S, Resh MD (1994) Amino-terminal basic residues of Src mediate membrane binding through electrostatic interaction with acidic phospholipids. Proc Natl Acad Sci USA 91:12253–12257
Sillman AL, Monroe JG (1995) Association of $p72^{syk}$ with the src homology (SH2) domains of PLC gamma 1 in B lymphocytes. J Biol Chem 270:11806–11811
Silverman L, Resh MD (1992) Lysine residues form an integral component of a novel NH2-terminal membrane targeting motif for myristylated $pp60^{v\text{-}src}$. J Cell Biol 119:415–425
Soriano P, Montgomery C, Geske R, Bradley A (1991) Targeted disruption of the c-src proto-oncogene leads to osteopetrosis in mice. Cell 64:693–702
Sultan C, Breton M, Mauco G, Grondin P, Plantavid M, Chap H (1990) The novel inositol lipid phosphatidylinositol 3,4-bisphosphate is produced by human blood platelets upon thrombin stimulation. Biochem J 269:831–834
Superti-Furga G, Fumagalli S, Koegl M, Courtneidge SA, Draetta G (1993) Csk inhibition of c-Src activity requires both the SH2 and SH3 domains of Src. EMBO J 12:2625–2634

Takayama H, Nakamura T, Yanagi S, Taniguchi T, Nakamura S, Yamamura H (1991) Ionophore A23187-induced protein-tyrosine phosphorylation of human platelets: possible synergism between $Ca^{2+}$ mobilization and protein kinase C activation. Biochem Biophys Res Commun 174:922–927

Tanaka A, Gibbs CP, Arthur RR, Anderson SK, Kung HJ, Fujita DJ (1987) DNA sequence encoding the amino-terminal region of the human c-src protein: implications of sequence divergence among src-type kinase oncogenes. Mol Cell Biol 7:1978–1983

Taniguchi T (1995) Cytokine signaling through nonreceptor protein tyrosine kinases. Science 268:251–255

Taniguchi T, Kobayashi T, Kondo J, Takahashi K, Nakamura H, Suzuki J, Nagai K, Yamada T, Nakamura S, Yamamura H (1991) Molecular cloning of a porcine gene syk that encodes a 72-kDa protein-tyrosine kinase showing high susceptibility to proteolysis. J Biol Chem 266:15790–15796

Taniguchi T, Kitagawa H, Yasue S, Yanagi S, Sakai K, Asahi M, Ohta S, Takeuchi F, Nakamura S, Yamamura H (1993) Protein-tyrosine kinase $p72^{syk}$ is activated by thrombin and is negatively regulated through $Ca^{2+}$ mobilization in platelets. J Biol Chem 268:2277–2279

Tate BF, Rittenhouse SE (1993) Thrombin activation of human platelets causes tyrosine phosphorylation of PLC-gamma 2. Biochim Biophys Acta 1178:281–285

Tohyama Y, Yanagi S, Sada K, Yamamura H (1994) Translocation of $p72^{syk}$ to the cytoskeleton in thrombin-stimulated platelets. J Biol Chem 269:32796–32799

Torti M, Lapetina EG (1992) Role of rap 1B and $p21^{ras}$ GTPase-activating protein in the regulation of phospholipase C-gamma 1 in human platelets. Proc Natl Acad Sci USA 89:7796–7800

Torti M, Lapetina EG (1994) Structure and function of rap proteins in human platelets. Thromb Haemost 71:533–543

Torti M, Crouch MF, Lapetina EG (1992) Epinephrine induces association of $pp60^{src}$ with Gi alpha in human platelets. Biochem Biophys Res Commun 186:440–447

Turner CE (1994) Paxillin: a cytoskeletal target for tyrosine kinases. Bioessays 16:47–52

van-der-Geer P, Hunter T, Lindberg RA (1994) Receptor protein-tyrosine kinases and their signal transduction pathways. Annu Rev Cell Biol 10:251–337

Vassbotn F, Havnen OK, Heldin C-H, Holmsen H (1994) Negative feedback regulation of human platelets via autocrine activation of the platelet-derived growth factor $\alpha$-receptor. J Biol Chem 269:13874–13879

Vostal JG, Shulman NR (1993) Vinculin is a major platelet protein that undergoes $Ca^{2+}$-dependent tyrosine phosphorylation. Biochem J 294:675–680

Vostal JG, Jackson WL, Shulman NR (1991) Cytosolic and stored calcium antagonistically control tyrosine phosphorylation of specific platelet proteins. J Biol Chem 266:16911–16916

Vulliet PR, Hall FL, Mitchell JP, Hardie DG (1989) Identification of a novel proline-directed serine/threonine protein kinase in rat pheochromocytoma. J Biol Chem 264:16292–16298

Walker F, deBlaquiere J, Burgess AW (1993) Translocation of $pp60^{c\text{-}src}$ from the plasma membrane to the cytosol after stimulation by platelet-derived growth factor. J Biol Chem 268:19552–19558

Wang X, Sada K, Yanagi S, Yang C, Rezaul K, Yamamura H (1994) Intracellular calcium dependent activation of $p72^{syk}$ in platelets. J Biochem Tokyo 116:858–861

Watson SP, Poole A, Asselin J (1995) Ethylene glycol bis(beta-aminoethyl ether)-N,N,N′,N′-tetraacetic acid (EGTA) and the tyrphostin ST271 inhibit phosphlipase C in human platelets by preventing $Ca^{2+}$ entry. Mol Pharmacol 47:823–830

Weiss A, Littman DR (1994) Signal transduction by lymphocyte antigen receptors. Cell 76:263–274

Weng Z, Taylor JA, Turner CE, Brugge JS, Seidel-Dugan C (1993) Detection of Src homology 3-binding proteins, including paxillin, in normal and v-Src-transformed Balb/c 3T3 cells. J Biol Chem 268:14956–14963

Wong S, Reynolds AB, Papkoff J (1992) Platelet activation leads to increased c-src kinase activity and association of c-src with an 85-kDa tyrosine phosphoprotein. Oncogene 7:2407–2415

Wu H, Parsons JT (1993) Cortactin, an 80/85-kilodalton $pp60^{src}$ substrate, is a filamentous actin-binding protein enriched in the cell cortex. J Cell Biol 120:1417–1426

Xing Z, Chen HC, Nowlen JK, Taylor SJ, Shalloway D, Guan JL (1994) Direct interaction of v-Src with the focal adhesion kinase mediated by the Src SH2 domain. Mol Biol Cell 5:413–421

Yamada T, Taniguchi T, Nagai K, Saitoh H, Yamamura H (1991) The lectin wheat germ agglutinin stimulates a protein-tyrosine kinase activity of $p72^{syk}$ in porcine splenocytes. Biochem Biophys Res Commun 180:1325–1329

Yamada T, Taniguchi T, Yang C, Yasue S, Saito H, Yamamura H (1993) Association with B-cell-antigen receptor with protein-tyrosine kinase $p72^{syk}$ and activation by engagement of membrane IgM. Eur J Biochem 213:455–459

Yanaga F, Poole A, Asselin J, Blake R, Schieven GL, Clark EA, Law C-L, Watson SP (1995) Syk interacts with tyrosine-phosphorylated proteins in human platelets activated by collagen and cross-linking of the $Fc_{\gamma}$-IIa receptor. Biochem J 311:471–478

Yanagi S, Sada K, Tohyama Y, Tsubokawa M, Nagai K, Yonezawa K, Yamamura H (1994) Translocation, activation and association of protein-tyrosine kinase $p72^{syk}$ with phosphatidylinositol 3-kinase are early events during platelet activation. Eur J Biochem 224:329–333

Yatomi Y, Ozaki Y, Satoh K, Kume S (1994) Synthesis of phosphatidylinositol 3,4-bisphosphate is regulated by protein-tyrosine phosphorylation but the p85 alpha subunit of phosphatidylinositol 3-kinase may not be a target for tyrosine kinases in thrombin-stimulated human platelets. Biochim Biophys Acta 1212:337–344

Yi TL, Bolen JB, Ihle JN (1991) Hematopoietic cells express two forms of lyn kinase differing by 21 amino acids in the amino terminus. Mol Cell Biol 11:2391–2398

Zhao YH, Krueger JG, Sudol M (1990) Expression of cellular-yes protein in mammalian tissues. Oncogene 5:1629–1635

Zheng XM, Wang Y, Pallen CJ (1992) Cell transformation and activation of $pp60^{c\text{-}src}$ by overexpression of a protein tyrosine phosphatase. Nature 359:336–339

Zioncheck TF, Harrison ML, Isaacson CC, Geahlen RL (1988) Generation of an active protein-tyrosine kinase from lymphocytes by proteolysis. J Biol Chem 263:19195–19202

CHAPTER 14
# Protein Phosphatases in Platelet Function

S.P. WATSON

## A. Introduction

Reversible phosphorylation of proteins regulates a great number of functional responses within cells, including many of those associated with platelet activation. The importance of phosphorylation is emphasised by estimates that approximately one third of cellular proteins contain covalently bound phosphate and that between 2% and 3% of all eukaryotic genes encode for proteins involved in the regulation of protein phosphorylation (HUBBARD and COHEN 1993; HUNTER 1995). There may be as many as 2000 genes in the human genome encoding proteins involved in the regulation of protein dephosphorylation, of which fewer that 10% have been cloned (COHEN 1993; HUNTER 1995).

There are believed to be similar numbers of protein serine/threonine phosphatases (PS/TP) and protein tyrosine phosphatases (PTP) within a cell (HUNTER 1995). A large family of dual specificity protein phosphatases that share little homology to PS/TPs and PTPases has also recently been identified. The concept that protein phosphatases act simply to reverse the action of protein kinases is highly misleading. The activities of many protein phosphatases are dynamically controlled through a combination of regulatory subunits, protein phosphorylation, subcellular distribution and endogenous inhibitors (HUBBARD and COHEN 1993; INAGAKI et al. 1994; SHENOLIKAR 1994; HUNTER 1995). There are also major differences between the intrinsic activities of the PS/TP and PTP families. The intrinsic activity and intracellular levels of most PS/TPs parallels that of intracellular serine/threonine kinases (COHEN 1992). In contrast, intrinsic PTP activity is often two to three orders of magnitude greater than that of protein tyrosine kinase activity (SUN and TONKS 1994), accounting for the low level of protein tyrosine phosphorylation in most cells, including platelets, and for the transient nature of responses.

Progress in determining the role of individual protein phosphatases in the cell has been slow. This is in part because of the lack of specific inhibitors and the much greater attention paid to the role of protein kinases in the belief that they are responsible for initiating cellular responses, although this is not always the case. Table 1 summarizes the major types of inhibitors that have been used in the study of protein phosphatases and their specificity. There are no specific inhibitors of PTPs available and the few non-specific agents that have

**Table 1.** Inhibitors of protein phosphatases

| Inhibitor | Target | Comments |
|---|---|---|
| Protein serine/threonine phosphatase inhibitors | | |
| Calyculin | PP2A ≈ PP1 | Phosphorylated polyketide from marine sponge; cell-permeable |
| Inhibitors 1 and 2 | PP1 | Protein; cell-impermeable; activated by protein kinase A |
| Microcystin | PP2A ≈ PP1; forms a covalent interaction PP1 and PP2A | Cyclic peptide from blue-green algae; does not enter all cells |
| Okadaic acid | PP2A >> PP1; inhibits PP2A selectively at $1\,nM$ in dilute assay but much higher concentrations required in intact cell | Polyether carboxylic acid from marine sponge; cell-permeable |
| Cyclosporin | PP2B; forms a complex with cyclophilin in cell | Cyclic endecapeptide from fungi; immunosuppressant |
| FK506 | PP2B; binds to FKBP in cell | Macrocyclic lactone from *Streptomyces*; immunosuppressant |
| Protein tyrosine phosphatase inhibitors | | |
| Vanadate | Non-selective | Less potent than peroxovanadate in intact cells |
| Peroxovanadate | Non-selective | Formed from $H_2O_2$ and vanadate; cell-permeable; inhibits many other enzymes |
| Phenylarsine oxide | Non-selective; effective over $1–10\,\mu M$ | Interacts with free sulfhydryl groups in the active site of many tyrosine phosphatases; inhibits many other enzymes |

Vanadate and peroxovanadate also inhibit the dual specificity phosphatases MKP1 and PAC1. PP2B and PP2C require $Ca^{2+}$ and $Mg^{2+}$, respectively, and can therefore be inhibited by appropriate chelating reagents.

been described, e.g. derivatives of vanadate and phenylarsine oxide (PAO), induce a wide spectrum of additional effects (see below). There are, however, several specific inhibitors of PS/TPs derived from mammalian cells, e.g. inhibitor 1 (Table 1), and from other organisms, e.g. the sea sponge toxins okadaic acid and calyculin A and the blue-green algae toxin microcystin (Table 1). Okadaic acid and calyculin A are frequently used to establish whether a cellular response is regulated by reversible serine/threonine protein phosphorylation because their targets dephosphorylate a vast number of proteins within eukaryotic cells (MACKINTOSH and MACKINTOSH 1994).

Relatively few studies have characterised the subtypes of PS/TPs and PTPs in platelets. This is in part because of the limitation in the application of molecular biology procedures in a cell that lacks a nucleus. Since it is now realised, however, that the role of protein phosphatases is far more complex than originally thought and that even at rest there is a high level of PTPase activity, there is an urgent need for a full characterisation of these enzymes in platelets. The majority of studies that have investigated the role of protein phosphatases in platelets have been based on the use of membrane-permeable inhibitors, and this will form much of the discussion in this chapter.

## B. Protein Serine/Threonine Phosphatases in Platelets

### I. Classification of Serine/Threonine Protein Phosphatases

The PS/TPs are divided into four main groups based on the original scheme of Ingebritsen and Cohen (Ingebritsen and Cohen 1983a,b; Ingebritsen and Cohen 1983). PP1 is inhibited by the thermostable proteins inhibitor-1 and inhibitor-2 whereas PP2s are insensitive to the two protein inhibitors. PP1 and PP2A do not require divalent cations, share approximately 50% amino acid sequence identity in their catalytic domains and have broad and overlapping specificities; cDNA studies have demonstrated that their catalytic subunits are highly conserved through evolution (Cohen and Cohen 1989). PP2B (also known as calcineurin) is a $Ca^{2+}$/calmodulin-dependent enzyme, structurally related to PP1 and PP2A, which exhibits high activity towards relatively few substrates. PP2B is inhibited by the two immunosuppressant drugs cyclosporin and FK506 which are commonly used in transplant surgery. Cyclosporin and FK506 exert their clinical effects through binding to intracellular proteins, the immunophilins; the drug-protein complex is a potent inhibitor of PP2B in T-cells, leading to inhibition of interleukin-2 release and T-cell proliferation (Schreiber 1992). The fourth member of this class, PP2C, does not contain the regions that have been predicted to be essential for the catalytic activity of the other three major PS/TPs and requires $Mg^{2+}$ and to a limited extent $Mn^{2+}$ for activity. PP1, PP2A and PP2C also dephosphorylate histidine residues (Kim et al. 1993) although the importance of histidine phosphorylation in mammalian cells is not established.

Isoforms of the four main PS/TPs have since been identified by molecular cloning and include PP1$\alpha$, PP1$\beta$ and PP1$\gamma$ isoforms and PP2$\alpha$ and PP2$\beta$ (Cohen 1993). In addition, several novel protein phosphatases have been identified from their cDNAs including new species in *Drosophila melanogaster* and *Saccharomyces cerevisiae*. One mammalian PS/TP that has been added to the original classification of Ingebritsen and Cohen (1983a,b) is PP4 (initially known as PPX; da Cruz e Silva et al. 1988). It shares 65% amino acid identity to PP2A, is potently inhibited by okadaic acid but is unaffected by inhibitor-1 or -2 (Cohen et al. 1990; Cohen 1993). PP2A and PP4 have overlapping but distinct substrate specificities. A further PS/TP that is also related to PP1 and PP2A but is inhibited by intermediate concentrations of okadaic acid, PP5, has been identified (HUNTER 1995). Estimates of the number of PS/TPs and their catalytic subunits in the human genome based on the number of enzymes identified in *Drosophila* suggest that there may be as many as 900 genes encoding proteins involved in the regulation of serine and threonine dephosphorylation (COHEN 1993).

## II. Inhibitors of Protein Serine/Threonine Phosphatases

Several specific and cell-permeable inhibitors of PP1 and PP2A have been described of which the best charaterised are okadaic acid and calyculin A from the sea sponges *Halichondria okada* and *Discodermia calyx*, respectively. Okadaic acid is a polyether fatty acid that is a major component of diarrhetic seafood poisoning. Under appropriate assay conditions, okadaic acid inhibits PP2A completely at 1 *nM* but is approximately three and 100 times less potent against PP4 and PP1, respectively. PP2B is far less sensitive to okadaic acid and PP2C is insensitive. Calyculin A inhibits PP2A with similar potency to okadaic acid but also inhibits PP1 over the same concentration range (Table 2). The conservation of the toxin sensitivities of PP1 and PP2A throughout the animal world has contributed to their widespread use. The effect of these two inhibitors in intact cells is not as powerful as their affinities may suggest because of relatively high intracellular concentrations of phosphatases (HIGASHIHARA et al. 1992; COHEN et al. 1989). Thus, if the intracellular concentration of the phosphatase is 1 $\mu M$, then at least 1 $\mu M$ of inhibitor will be required to block its function even if the affinity of the enzyme-inhibitor interaction is in the nanomolar range. Moreover, this problem cannot be addressed in cell-free systems by dilution of cellular extracts because most PS/TPs have the capacity to dephosphorylate a wide spectrum of proteins under such conditions. Other, less specific inhibitors include NaF which is a general inhibitor of PS/TPs, with the exception of PP2C, but has many other actions that limit its use to prevention of dephosphorylation in extraction buffers. Similarly, orthovanadate inhibits PP1, PP2A and PP2C but is more widely used as a tyrosine phosphatase inhibitor (see below). The chelation of intracellular cations represents another mechanism of phosphatase inhibition.

## III. The Presence of Protein Serine/Threonine Phosphatases in Platelets

The existence of PS/TPs in platelets is illustrated by the transient phosphorylation of proteins such as pleckstrin and myosin light chain which occurs following activation. Platelets contain all four major classes of PS/TPs, namely PP1, PP2A, PP2B and PP2C (WALKER and WATSON 1992; TALLANT and WALLACE 1985; for a review see SAKON et al. 1994). LEREA (1991) reported that

**Table 2.** Ki values of protein serine/threonine phosphatase inhibitors

| | PP1 | PP2A | PP4 | Reference |
|---|---|---|---|---|
| Calyculin A | 0.2 $nM$ | 0.2 $nM$ | – | ISHIHARA et al. 1989 |
| Okadaic acid | 20 $nM$ | 0.2 $nM$ | 0.7 $nM$ | BIALOJAN and TAKAI 1988; COHEN 1993 |

80% of PP1 and PP2A phosphatase activity is soluble using [$^{32}$P]phosphorylase a as substrate. The soluble activity is inhibited markedly (>80%) by protein inhibitor 2 but only partially (~20%) by 2 $nM$ okadaic acid, demonstrating that it is due to the presence of PP1; the remaining activity is inhibited by 1 $\mu M$ okadaic acid, indicating the presence of PP2A. ERDODI et al. (1992) purified three forms of bovine platelet phosphatase activity using [$^{32}$P]phosphorylase a as substrate yielding proteins of 36, 37 and 41 KDa. Two of these correspond to PP1 and PP2A phosphatases, but the third was distinct in that it was inhibited by an intermediate concentration of okadaic acid but not by protein inhibitor 2. It is possible that it may correspond to the PP2A-like phosphatase PP4 described above. NISHIKAWA et al. (1994) characterised PP1 and PP2A activity in human platelets by measurement of myosin light chains phosphatase activity in the absence of divalent cations. NISHIKAWA et al. (1994) also found that approximately 80% of the activity is cytosolic although, in contrast to the results of LEREA (1991), they estimated that 31% of the phosphatase activity was due to PP1 and 58% to PP2A. The PP1 fraction was resolved further by Western blotting into three isozymes (NISHIKAWA et al. 1994). WALKER and WATSON (1992) measured PP1 and PP2A levels through [$^{32}$P]phosphorylase a hydrolysis and found comparable levels of PP1 and PP2A evenly distributed in the cytosol and particulate fractions. These authors also detected PP2C activity by measurement of [$^{32}$P]casein hydrolysis in the presence of okadaic acid and absence of $Ca^{2+}$. Approximately 66% of PP2C activity is present in the cytosolic fraction. A role for PP1 and PP2A in platelets has been demonstrated by the potent effects of okadaic acid and calyculin A on intact cells (see below). TALLANT and WALLACE (1985) reported the partial purification of the $Ca^{2+}$/calmodulin-dependent phosphatase PP2B (calcineurin) and demonstrated that it is converted to an active, calmodulin-independent form by the $Ca^{2+}$-dependent protease calpain I (TALLANT et al. 1988). SAKON et al. (1990) reported the partial purification of five PS/TPs in human platelets, several of which are activated by divalent cations, but all display activity in the presence of EDTA and exhibit broad, overlapping substrate specificity.

## IV. Effect of Okadaic Acid and Calyculin A in Platelets

Several groups have characterised the cations of the protein phosphatase inhibitors okadaic acid and calyculin A in platelets and described inhibition of agonist-induced platelet activation. On the face of it this may appear surprising in view of the pivotal role of protein phosphorylation in the onset of functional responses and must reflect the existence of potent inhibitory pathways.

Karaki et al. (1989) were the first group to describe almost complete inhibition of thrombin-induced aggregation and $Ca^{2+}$ mobilisation in rabbit platelets by 1 $\mu M$ okadaic acid with a tenfold lower concentration of okadaic acid having little effect. The action of okadaic acid was subsequently characterised in greater detail in human platelets by others. Walker and Watson (1992) and Higashihara et al. (1992) reported that 1 $\mu M$ okadaic acid induced a steady increase in phosphorylation of wide spectrum of proteins over a time course of 20–60 min. Consistent with this, the inhibitory effect of okadaic acid on thrombin-induced platelet activation was time-dependent with inhibition observed as early as 60 s and continuing to increase up to 60 min (Walker and Watson 1992 and Higashihara et al. 1992). Other groups have also described marked increases in phosphorylation of a wide spectrum of proteins in the presence of okadaic acid in platelets, in particular of an uncharacterised protein of 50 kDa (Lerea 1991; Murata et al. 1992; Nishikawa et al. 1994). Phosphorylation of the 50 kDa protein correlates with inhibition of platelet activation by okadaic acid and is potentiated markedly in prostacyclin-stimulated cells (Lerea 1991; Murata et al. 1992), providing evidence that it may be the same protein as the major substrate for cyclic nucleotide-dependent kinases in platelets, namely the 50 kDa protein VASP (Walter et al. 1993). The increase in phosphorylation of VASP induced by okadaic acid is not mediated by an increase in the level of cAMP (Lerea 1991). Calyculin A induces an increase in phosphorylation of a wide spectrum of proteins in human (Nishikawa et al. 1994) and rabbit platelets (Murphy and Westwick 1994), including the 50 kDa protein.

The functional effects of okadaic acid and calyculin A on platelets are dependent on the concentration of inhibitor. There is general agreement that okadaic acid (0.3–10 $\mu M$) and calyculin A (10–300 $nM$) induce parallel inhibition of aggregation, dense granule secretion and mobilisation of $Ca^{2+}$ in human and rabbit platelets stimulated by thrombin, platelet activating factor (PAF), collagen, adrenaline and ADP (experiments performed in the absence of cyclooxygenase inhibitors; Lerea 1991; Walker and Watson 1992; Higashihara et al. 1992; Murata et al. 1992; Chiang 1993; Murphy and Westwick 1994; Nishikawa et al. 1994). The inhibitory effect of both compounds is more pronounced against lower concentrations of stimuli and can almost be fully overcome by high concentrations of strong agonists, e.g. thrombin. Calyculin A has no inhibitory effect on ionomycin-induced dense granule secretion, suggesting that it does not interfere with the secretory process itself (Murphy and Westwick 1994).

The major mechanism underlying the actions of okadiac acid and calyculin A described above is through inhibition of phosphoinositide metabolism (Walker and Watson 1992; Murphy and Westwick 1994). It is well established that phospholipase C (PLC) in platelets is inhibited by protein kinase C

(WATSON and LAPETINA 1995) and protein kinase A (WATSON et al. 1984)-dependent pathways and that these appear to be mediated by distinct mechanisms (WALKER and WATSON 1992). Phosphorylation of the 50kDa protein VASP is closely linked with the inhibitory effects of cyclic nucleotides in platelets and may underlie one component of the action of okadaic acid and calyculin A. A second component is through potentiation of a protein kinase C-dependent inhibitory pathway. The protein kinase C inhibitor, Ro 31-7549, reverses the effect of calyculin A on PLC in rabbit platelets (MURPHY and WESTWICK 1994), and the inhibitory effect of okadaic acid on formation of inositol phosphates in human platelets is partially additive with that of protein kinases A and C (WALKER and WATSON 1992).

It is well established that thrombin stimulation induces phosphorylation of pleckstrin (47kDa) and myosin light chains (20kDa) in platelets. In the presence of okadaic acid or calyculin A, thrombin-induced phosphorylation of pleckstrin in human platelets is markedly reduced in parallel with inhibition of PLC (HIGASHIHARA et al. 1992; NISHIKAWA et al. 1994). In contrast, phosphorylation of myosin light chains is potentiated despite inhibition of PLC, suggesting they are substrates for PP1 and/or PP2A (HIGASHIHARA et al. 1992; NISHIKAWA et al. 1994). Interpretation of the role of PP1 and PP2A in the dephosphorylation of pleckstrin is complicated because of parallel inhibition of PLC. AHARONOVITZ et al. (1995) have presented convincing evidence that pleckstrin is a substrate for PP1 and/or PP2A in platelets using the general inhibitor of protein kinases, staurosporine. In control cells, phosphorylation of pleckstrin induced by the membrane-permeable diacylglycerol, dihexanoylglycerol, decays to basal levels within several minutes following addition of staurosporine. No decrease is observed, however, in the presence of 1 $\mu M$ okadaic acid, suggesting that it is mediated by PP1 and/or PP2A. The explanation why agonist-induced phosphorylation of myosin light chains is potentiated whereas that of pleckstrin is inhibited by okadaic acid is presumably related to the difference in kinetics of phosphorylation and dephosphorylation of the two substrates.

In view of the wide substrate specificity of PP1 and PP2A, it can be anticipated that okadaic acid and calyculin A will have additional functional effects in platelets unrelated to inhibition of phosphoinositide metabolism. MURATA et al. (1992, 1993) described inhibition of $Ca^{2+}$ entry in thrombin-stimulated platelets by concentrations of calyculin A and okadaic acid that do not inhibit $Ca^{2+}$ elevation and dense granule secretion in the absence of extracellular $Ca^{2+}$ (the absence of an inhibitory effect under the later conditions is evidence, albeit indirect, that phosphoinositide metabolism is unimpaired). The concentration of calyculin A required to inhibit entry of extracellular $Ca^{2+}$ ($\leq 20\,nM$) has been reported by others to have little effect on phosphoinositide metabolism (MURPHY and WESTWICK 1994). Paradoxical, the concentration range for okadaic acid-induced inhibition of $Ca^{2+}$ entry (0.2–2 $\mu M$) is similar to the described for inhibition of phosphoinositide metabolism in platelets. Direct comparisons between studies from separate research groups, however, made only on the basis of inhibitor concentrations may be misleading because of differences in experimental design, e.g. incubation time, cell number, purity of material, etc. The mechanism underlying the inhibitory action of calyculin A and okadaic acid on $Ca^{2+}$ entry is not known.

The same concentrations of calyculin A (20 *nM*) and okadaic acid (2 *μM*) that inhibit $Ca^{2+}$ entry induce dramatic changes in morphology of resting platelets revealed by confocal and electron microscopy (YANO et al. 1994, 1995). Both inhibitors disrupt the microtubule network and induce redistribution of microtubules into pseudopod-like processes (YANO et al. 1995). These changes correlate with an increase in phosphorylation of a 90 kDa protein in okadaic acid-stimulated cells which co-precipitates with tubulin, suggesting that this protein plays a major role in the regulation of microtubule organisation. Okadaic acid and calyculin A also induce constriction of actin filaments without an increase in filamentous (F)-actin but do not alter the level of myosin light chain phosphorylation. In thrombin-activated platelets, okadaic acid and calyculin A induce a more prominent change in shape accompanied by generation of extremely long pseudopods containing an array of microtubules and actin filaments and formation of a dense mass of actin filaments in the centre of the cells; polymerisation of actin, however, is inhibited (YANO et al. 1995). These results suggest that okadaic acid and calyculin A induce platelet morphological changes through disruption of both the actin filament and microtubule networks. It is possible that reorganisation of the cytoskeletal network by okadaic acid and calyculin A may underlie the inhibition of $Ca^{2+}$ entry in thrombin-stimulated platelets described above (MURATA et al. 1992, 1993) and the ability of the two phosphatase inhibitors to more than double thrombin stimulation of microparticle formation (YANO et al. 1994).

Calyculin A and okadaic acid inhibit PP2A at similar concentrations, but calyculin A is approximately two orders of magnitude more potent than okadaic acid against PP1. The PS/TP that underlies the majority of the inhibitory actions described above would appear to be of the PP1 class based on the higher potency of calyculin A relative to okadaic acid. It should be emphasised, however, that relatively few studies have made a direct comparison of the actions of calyculin A and okadaic acid and that there exists a wide variability in the reported concentration dependency of the action of okadaic acid, possibly because of variations in quality of material. The high concentrations of the two agents required to inhibit activation in these experiments, relative to those the produce complete inhibition of PP1 and PP2A in controlled assays (Table 1), reflect the high levels of these two protein phosphatases within platelets.

## V. Dephosphorylation of Cofilin in Platelets

HASLAM and LYNHAM (1977) and IMAOKA et al. (1983) described the dephosphorylation of a protein of 18 kDa in platelets activated by collagen or thrombin. They recently identified this protein as cofilin through sequencing of peptide fragments and use of specific antibodies (DAVIDSON and HASLAM 1994). Dephosphorylation of cofilin is a relatively slow event beginning after dense granule secretion is nearly complete and correlating with the late stage of aggregation. Cofilin is a pH sensitive actin-depolymerising protein that binds to both G- and F-actin, and there is evidence that its F-actin-depolymerising activity is stimulated by dephosphorylation (reviewed in DAVIDSON and HASLAM 1994). DAVIDSON and HASLAM (1994) have questioned, however, whether dephosphorylation of cofilin contributes to the cytoskeletal remodeling that occurs during platelet aggregation because the level of newly dephosphorylated cofilin is no more than 1.5% of the total platelet actin and F-actin levels *increase* during platelet activation.

There appear to be at least two forms of PS/TPs underlying dephosphorylation of cofilin. Cofilin dephosphorylation is stimulated in intact and permeabilised platelets by the $Ca^{2+}$ ionophore A23187 or $10\,\mu M$ $Ca^{2+}$, respectively, demonstrating involvement of PP2B. In addition, cofilin dephosphorylation can be stimulated in permeabilised cells by GTP-$\gamma$-S in the absence of $Ca^{2+}$ elevation and through an okadaic acid-insensitive mechanism. This suggests that cofilin dephosphorylation can occur through the action of PP2C or through an unidentified PS/TP. The $Ca^{2+}$- and GTP-sensitive pathways may act together to regulate cofilin dephosphorylation in intact cells (DAVIDSON and HASLAM 1994).

### VI. Protein Serine/Threonine Phosphatases in Platelets: Concluding Remarks

Our understanding of the role PS/TPs in general, and the function of individual PS/TPs in platelet function is at a rudimentary stage. Studies with calyculin A and okadaic acid have provided evidence for roles of PP1 and/or PP2A in feedback inhibition of phosphoinositide metabolism, inhibition of $Ca^{2+}$ entry and cytoskeletal reorganisation. In all of these cases, however, the role of the two PS/TPs is to reverse the increase in phosphorylation of proteins such as myosin light chains and pleckstrin that occurs during platelet activation or to potentiate feedback pathways of protein kinases A and C. There is no evidence that PS/TPs have a more direct role in the onset of platelet responses through dephosphorylation of proteins that are phosphorylated under basal conditions.

## C. Protein Tyrosine Phosphatases

Tyrosine phosphorylation of proteins was first described by ECKHART et al. (1979) with the first tyrosine kinase identified as the src protein of Rous Sarcoma virus (COLLETT et al. 1980). Many growth factor receptors and oncogenes were subsequently identified as tyrosine kinases, fueling the initial interest in the role of tyrosine phosphorylation in growth and differentiation (ULLRICH and SCHLESSINGER 1990). Tyrosine kinase activity in platelets and other nonproliferating cells was first described by TUY et al. (1983), although it was unclear at this stage whether it was simply a remnant resulting from its role in growth and differentiation in the precursor cell. It is now established, however, that tyrosine phosphorylation is involved in many pathways unrelated to gene regulation including cytoskeletal remodelling and activation of PLC$\gamma$ isoforms.

Tyrosine phosphorylation is estimated to account for less than 0.1% of total protein phosphorylation and requires sensitive techniques for its detection. Tyrosine phosphorylation of platelet proteins was first detected by measurement of alkali-resistant protein phosphorylation (TUY et al. 1983). This approach has several limitations including selective loss of proteins and incomplete hydrolysis of phosphoserine and phosphothreonine residues. The development of antiphosphotyrosine antibodies has led to a rapid expansion in the study of protein tyrosine phosphorylation. It should be stressed, however, that various antiphosphotyrosine antibodies differ considerably in affinities for tyrosine-phosphorylated proteins. As a consequence, caution is required in comparing patterns of tyrosine phosphorylation between different studies.

## I. Classification of Protein Tyrosine Phosphatases

The first major PTPase activity to be identified, PTP1B, was purified as a cytosolic protein from human placenta extracts by Tonks and colleagues in 1988. Determination of its amino acid sequence revealed a striking level of sequence homology to the tandem intracellular domains of the leukocyte common antigen, CD45, found exclusively on haematopoietic cells (for reviews see TONKS et al. 1992; BRADY-KALNAY and TONKS 1994). Within less than four years, the use of molecular biology techniques allowed more than 30 PTPases encoded by distinct genes to be identified (TONKS et al. 1992). This work established PTPs as structurally distinct from PS/TPs and divided them into membrane-bound and cytosolic forms. The unique feature that defines the PTPase family is the sequence [I/V]H*C*XAGXXR[S/T]G contained within a catalytic domain of approximately 240 amino acids. The cysteinyl residue is critical for activity as the catalytic mechanism proceeds via the generation of a cysteine-phosphate intermediate (GUAN and DIXON 1990). In contrast to PS/TPs, no PTP regulatory subunits or endogenous protein inhibitors have been identified (HUNTER 1995).

The majority of transmembrane PTPases are characterised by the presence of two PTPase domains, although the activity of the first of these is several orders of magnitude greater than that of the second. In contrast to the similarity in structure of their intracellular segments, the transmembrane PTPases exhibit wide structural diversity in their extracellular domains that has led to their subdivision into five separate classes (TONKS et al. 1992). The possibility that these molecules represent cell surface receptors whose activities are regulated by an appropriate ligand has been investigated by many groups but with only limited success. PTP$\mu$ has been shown to mediate cell-cell aggregation through a homophilic binding mechanism, suggesting that its ligand may be itself expressed on an opposing cell (BRADY-KALNAY et al. 1993). This interaction, however, does not result in a change in phosphatase activity, and its significance may be to restrict distribution within the cell. Membrane bound PTPases exhibit considerable basal activity in the order of 10–1000 times higher than that of tyrosine kinases (TONKS et al. 1992). It is therefore attractive to speculate that interaction with a ligand may negatively regulate PTPase activity or sequester the enzyme in a limited region of the membrane, restricting its range of substrates.

The cytosolic PTPases have a single tyrosine phosphatase domain that is usually present with other motifs involved in targeting the enzyme to subcellular locations. For example, the related PTPases PTP1C and PTP1D have tandem SH2 domains enabling interaction with phosphotyrosine residues. Basal activity of cytosolic PTPases is also 10–1000 times greater than that of cellular tyrosine kinases.

## II. Distribution of Protein Tyrosine Phosphatases in Platelets

Platelets contain remarkably high levels of PTPases that can be fully inhibited by vanadate, molybdate and a variety of sulfhydryl reagents including $HgCl_2$ (LEREA et al. 1989; SMILOWITZ et al. 1991). Approximately 80% of the PTPase activity is particulate and a Triton-solubilized extract elutes from a DEAE column as a broad, heterogeneous peak suggesting multiple enzyme species (SMILOWITZ et al. 1991). DAWICKI and STEINER (1993) developed a novel assay for the detection of PTPase activity in platelets based on the separation of platelet protein by SDS-PAGE followed by blotting to polyvinylidene difluoride membranes surface-labelled with [$^{32}$P-Tyr]($Glu_4$,Tyr)$_n$. PTPase activity was renatured through a series of washing steps and identified by the appearance of clear bands on the autoradiograph. This led to the identification of 53 and 50kDa PTPases in platelet particulate and cytosolic fractions, respectively, although the activity was not characterised in further detail. The 53kDa PTPase may be PTP1B as this is known to be expressed at high levels in platelets, where it makes up approximately 0.2% of detergent soluble protein (FRANGIONI et al. 1993). LI et al. (1994a and 1995) measured PTPase activity in platelets using [$^{32}$P]phosphotyrosyl poly(Glu-Tyr) as substrate and detected substantial activity distributed between cytosol (60%) and particulate fractions (40%). PTPase activity in the cytoskeletal fraction was low, being less than 1.5% of cellular activity, but increased by eightfold in platelets stimulated with 1u/ml thrombin for 2min (LI et al. 1994a). The increase in PTPase activity in the cytoskeleton occurs in parallel with the redistribution of several signalling proteins and is dependent on actin polymerisation as this is inhibited in cells pretreated with cytochalasin D (LI et al. 1994b). GU et al.

(1991) have also provided evidence for a novel PTPase, PTPase MEG, that is associated with the cytoskeleton in megakaryocytyes. They isolated a cDNA from the megakaryocytic cell line, MEG-01, encoding a PTPase with an amino-terminal region that is 45% identical to the amino terminus of the cytoskeletal protein human erythrocyte protein 4.1. PTPase MEG is also closely related to the cytoskeletal elements ezrin and talin and may have an important role in the function of the cytoskeleton.

## III. Regulation of Tyrosine Phosphorylation by Irreversible Platelet Aggregation

FERRELL and MARTIN (1988) measured thrombin-induced tyrosine phosphorylation using an affinity-purified antiphosphotyrosine antiserum raised against polymerised phosphotyrosine and divided the response into three distinct phases based on the kinetics of phosphorylation. There is a rapid phosphorylation of four proteins of 27, 34, 68 and 70kDa which peaks at 5–20s. Phosphorylation of the two low molecular weight proteins is transient while that of the 68 and 70kDa bands is variable. A second group exhibits peak phosphorylation between 60–180s and includes proteins of 60, 105, 115 and 130kDa. A third group exhibits delayed phosphorylation peaking at 3–5min and includes three proteins of 100,108 and 126kDa. Phosphorylation of this group of proteins is inhibited by conditions that prevent irreversible aggregation, demonstrating that it is regulated by binding of fibrinogen to its receptor, glycoprotein IIb-IIIa (FERRELL and MARTIN 1989). Tyrosine phosporylation of the majority of bands is maintained for several minutes and the time courses of subsequent decreases vary considerably from experiment to experiment (FERRELL and MARTIN 1988).

TAKAYAMA et al. (1993) described the transient tyrosine phosphorylation of a 115kDa protein and 75 and 115kDa proteins in thrombin- and thromboxane-stimulated platelets, respectively. Phosphorylation of both proteins is maintained in thrombasthenic patients or in the presence of RGDS, an inhibitor of fibrinogen binding, demonstrating regulation of PS/TP activity through binding of fibrinogen to glycoprotein IIb-IIIa. LUBER and SIESS (1994) also described fibrinogen receptor-dependent tyrosine dephosphorylation during activation of platelets stimulated by thrombin or a thrombin-receptor-activating peptide. They observed a decrease in tyrosine phosphorylation of a 38kDa doublet within 7.5s of stimulation and dephosphorylation of a 140kDa doublet within 15–30s which occurred after a transient increase. Dephosphorylation of both proteins was abolished in the presence of RGDS and almost completely inhibited in the presence of EGTA, which is known to disrupt the fibrinogen receptor. Both groups measured tyrosine phosphorylation through the use of the antiphosphotyrosine monoclonal antibody PY20, and it is possible that the 115kDa protein described by TAKAYAMA et al. (1993) is the same as the 140kDa protein of LUBER and SIESS (1994) in view of the poor resolution of proteins in this region of the gel. The 38kDa doublet was present in low levels in platelets and would only have been seen in long exposures of the gel in the experiments of TAKAYAMA et al. (1993). FRANGIONI et al. (1993) also described aggregation-dependent changes in thrombin-induced tyrosine phosphorylation using the monoclonal antibody 4G10. In particular, they identified four proteins of 75, 112, 125 and 140kDa which exhibit reduced phosphorylation following irreversible aggregation.

In an attempt to identify PTPases that contribute to the increase in activity associated with the cytoskeleton and irreversible aggregation, LI et al. (1994a) measured the level of the PTPase PTP1C (also known as SH-PTP1, HCP or SHP) by immunodetection. PTP1C is expressed predominantly in haematopoietic cells (PLUTZKY et al. 1992) and contains two SH2 domains providing possible sites for interaction with the cytoskeleton. PTP1C is believed to have a key role in haematopoiesis in view of the profound disturbance in bone marrow-derived cells in *motheaten* mice which have a mutation leading to lack of expression of this PTPase (KOZLOWSKI et al. 1993). LI et al. (1994a) determined the distribution of PTP1C with a polyclonal antibody and observed that it is present in higher levels in the membrane than in the cytosol fraction, but is almost undetectable in the cytoskeleton. Following stimulation with thrombin, PTP1C undergoes translocation to the cytoskeleton over a time course similar to that of the increase in PTPase activity in the cytoskeleton, peaking at approximately 60s. Translocation of PTP1C is inhibited in the presence of RGDS, which prevents the binding of fibrinogen to its receptor, and cytochalasin D, which prevents actin polymerisation. In a subsequent study, LI et al. (1995) described phosphorylation of PTP1C on both serine and tyrosine residues in platelets stimulated by thrombin and observed that this is associated with a 60% increase in its catalytic activity. Phosphorylation occurs as early as 10s after stimulation by thrombin and prior to translocation to the cytoskeleton. It is independent of $Ca^{2+}$, ADP and thromboxane formation, is unaltered in the presence of the fibrinogen receptor antagonist RGDS and is weakly inhibited by the selective protein kinase C inhibitor GF109203X. The function of phosphorylation of PTP1C is not known. The PTPase P58 also undergoes translocation to the cytoskeleton in activated platelets (LI et al. 1994b).

The PTPase PTP1B, which constitutes approximately 0.2% of detergent soluble protein, also undergoes redistribution in platelets following aggregation (FRANGIONI et al. 1993). The full length form of PTP1B has a molecular weight of 50kDa and contains a C-terminal targeting sequence of 35 amino acids which directs its subcellular distribution to the endoplasmic reticulum. Following irreversible platelet aggregation induced by agonists, e.g. thrombin, thromboxanes and ADP, second messengers, e.g. $Ca^{2+}$ ionophore, or by direct activation of the fibrinogen receptor, GPIIb-IIIa, PTP1B redistributes to the cytosolic fraction as a 42kDa metabolite (FRANGIONI et al. 1993). Cleavage of PTP1B occurs in the C-terminal targeting sequence and appears to be mediated by the $Ca^{2+}$-dependent protease, calpain I, as it is inhibited by the selective inhibitor calpeptin (FRANGIONI et al. 1993). PTP1B cleavage brings about a twofold increase in its specific activity. Cleavage is inhibited by conditions that prevent irreversible aggregation such as the absence of stirring or presence of RGDS (FRANGIONI et al. 1993). It is estimated that between 40% and 100% of PTP1B is converted to the 42 kDa form following irreversible aggregation (FRANGIONI et al. 1993). Cleavage of PTP1B is not accompanied by a general metabolism of proteins as Coomassie stains of activated platelets are almost identical to controls, and the PTPase T cell phosphatase, which is closely related to PTP1B, remains unaltered following A23187 addition.

Irreversible aggregation of platelets by the $Ca^{2+}$ ionophore A23187 leads to a global loss of tyrosine-phosphorylated proteins in the particulate fraction, including four prominent bands of 60, 62, 64 and 74 kDa, and the appearance of two proteins in the cytosol of 52 and 54 kDa (FRANGIONI et al. 1993). Dephosphorylation induced by A23187 occurs in parallel with the cleavage of PTP1B and is inhibited completely by calpeptin (FRANGIONI et al. 1993). The marked level of protein dephosphorylation induced by PTP1B relative to the small increase in its activity suggests that increased access to substrates has a major role in determining the intracellular action of PTP1B. These data strongly suggest that *under appropriate conditions* cleavage of PTP1B results in near global dephosphorylation of particulate proteins. PTP1B may also be responsible for the reduced phosphorylation of the four bands of tyrosine-phosphorylated proteins of 75, 112, 125 and 140 kDa observed in thrombin-stimulated platelets under conditions that permit irreversible aggregation. The reduction in phosphorylation of these four proteins occurs in parallel with cleavage of PTP1B and is inhibited by conditions that prevent aggregation, namely the absence of stirring or presence of RGDS. Confirmation that dephosphorylation of this group of proteins is mediated by PTP1B through the use of calpeptin has not been performed, and it is possible that it may result from the action of other PTPases such a PTP1C and PTPase p58. There is also evidence for association of other PTPases with the cytoskeleton.

Irreversible aggregation is associated with clot retraction, and it is possible that redistribution of PTPases may play a role in this response. Alternatively, redistribution may be a remnant of an activity that is important in other cells; for example, cleavage of PTP1B may represent and endogenous pathway of mammalian cell death (FRANGIONI et al. 1993). In order to establish the significance of regulation of PTPases by fibrinogen, it will be essential to identify the protein that undergo dephosphorylation. One protein dephosphorylated in thrombin-stimulated platelets before translocation to the cytoskeleton is the tyrosine kinase src (CLARK and BRUGGE 1993). Src is constitutively phosphorylated in its inactive state on tyrosine 527, permitting an intermolecular association with its SH2 domain and inhibition of its kinase activity. Dephosphorylation of tyrosine 527 leads to activation of src and autophosphorylation at tyrosine 416. The time course of dephosphorylation and translocation of src is similar to that of phoshorylation and translocation or PTP1C, suggesting the these events may be linked (LI et al. 1995).

## IV. Effect of Vanadate and Its Derivatives in Platelets

Vanadate ions were first described as potent inhibitors of PTPases by SAWRUP et al. (1982), with complete inhibition observed at concentrations between 10–100 $\mu M$. Over the same concentration range, vanadate induces activation of permeabilised platelets although it has little effect in intact cells presumably due to poor membrane permeability. It has been known for a long time, however, that vanadate ions combine with hydrogen peroxide to form peroxovanadate (also known as pervanadate of vanadyl hydroperoxide) and that this complex induces potent effects in intact cells, for example acting as an insulinomimetic (FANTUS et al. 1989). Concentrations of vanadate or hydrogen peroxide required to form peroxovanadate have little effect on their own.

INAZU et al. (1990) reported an increase in tyrosine phosphorylation of four proteins of 38, 53, 76 and 80 kDa in peroxovanadate-stimulated cells measured with their own antiphosphotyrosine polyclonal antibodies. The onset of tyrosine phosphorylation occurred after a lag of approximately 60 s and preceded aggregation; maximal aggregation was seen with 100 $\mu M$ vanadate and 1 $mM$ $H_2O_2$. A similar concentration response relationship was observed by PUMIGLIA et al. (1992) who noted that the time course and magnitude of activation were inversely related to cell number, possibly because of the presence of intracellular reducing activity. These authors observed an increase in tyrosine phosphorylation of 27 protein bands over a 15 min period in peroxovanadate-stimulated cells measured using a polyclonal antiphosphotyrosine antibody. The increase in tyrosine phosphorylation was estimated to be 29-fold greater than that induced by thrombin, was maintained for the least 20 min and was associated with metabolism of phosphoinositides, elevation of intracellular $Ca^{2+}$, phosphorylation of pleckstrin, shape change and secretion from dense and $\alpha$-granules. The authors emphasised that this pattern of response was similar to that observed in the activation of platelets by thrombin, suggesting that peroxovanadate may induce activation through increased phosphoinositide metabolism. The tyrphostin RG 50864 (an analogue of erbstatin) inhibited completely the increase in tyrosine phosphorylation and onset of functional responses induced by peroxovanadate, although an inactive structural analogue, tyrphostin 1, had no effect. In contrast, RG 50864 did not inhibit dense granule secretion induced by thrombin or the phorbol ester PMA despite full inhibition of tyrosine phosphorylation. The inability of RG 50864 to inhibit platelet activation by thrombin suggests that the two stimuli regulate PLC through distinct mechanisms. Consistent with this, PUMIGLIA et al. (1992) demonstrated that GDP-$\beta$-S inhibited completely thrombin-induced activation of permeabilised platelets but had no effect on the response to vanadate.

BLAKE et al. (1993) also observed a marked increase in tyrosine phosphorylation using the antiphosphotyrosine antibody PY20 following stimulation with peroxovanadate including proteins of 36–40, 54, 60, 65, 70, 85, 98, 120 and 140 kDa. Phosphorylation was detected within 5 s of stimulation and increased up to 5 min, the longest time period investigated. The more rapid onset of phosphorylation in this study may reflect the much higher concentrations of vanadate (400 $\mu M$) and hydrogen peroxide (4 $mM$) used and the fact that the two reagents were added directly to the platelets in order to generate the peroxovanadate complex. BLAKE et al. (1993) provided direct evidence of activation of PLC by peroxovanadate through measurement of inositol phosphates. Activation of PLC by peroxovanadate is inhibited by the nonselective inhibitor of tyrosine and serine/threonine kinases, staurosporine, and is associated with tyrosine phosphorylation of PLC$\gamma$1 (BLAKE et al. 1993). Formation of inositol phosphates by collagen is also inhibited completely by staurosporine, providing evidence for a functional role of tyrosine phosphorylation in agonist-stimulated platelets (BLAKE et al. 1993).

Several groups have investigated the action of vanadate in permeablised cells. LEREA et al. (1989) demonstrated that a combination of 100 $\mu M$ sodium vanadate and 10 $\mu M$ ammonium molybdate inhibited platelet PTPase activity by more than 97% and increased tyrosine phosphorylation of an unidentified 50 kDa protein in electropermeabilised platelets. A higher concentration of the two phosphatase inhibitors, 250 $\mu M$ vanadate and 25 $\mu M$ molybdate, induced phosphorylation of a further unidentified protein of 38 kDa. Verification of tyrosine phosphorylation of the 50 kDa protein was based on alkali resistance and was achieved through the use of an antiphosphotyrosine antiserum that had been raised against phosphotyramine-KLH conjugates. Vanadate and molybdate also stimulated secretion from dense and $\alpha$-granules (measured by release of 5-hydroxytryptamine and PDGF, respectively), formation of inositol phosphates and phosphorylation of pleckstrin. These responses were not inhibited in the presence of aspirin (LEREA et al. 1989). MCNICOL et al. (1993) observed concentration-dependent aggregation of saponin-permeabilised platelets in the presence of vanadate (7.5–100 $\mu M$). Activation was associated with stimulation of PLC, phosphorylation of pleckstrin and release of arachidonic acid. In contrast to the results obtained by LEREA et al. (1989) in electropermeabilised platelets, MCNICOL et al. (1993) described complete inhibition of all responses, with the exception of arachidonic acid liberation, in the presence of cyclooxygenase inhibitors, demonstrating that activation is mediated through formation of thromboxanes. Using the antiphosphotyrosine monoclonal antibody 4G10, MCNICOL et al. (1993) observed tyrosine phosphorylation of two proteins of 80 and 90 kDa and several low molecular weight proteins of 26, 29, 32, 40 and 42 kDa; phosphorylation of the low molecular weight proteins was inhibited in the presence of the cyclooxygenase/lipoxygenase inhibitor BW755C. WATSON et al. (1993) also described vanadate-induced activation of saponin-permeabilised platelets and its inhibition by the cyclooxygenase inhibitor indomethacin. These authors observed transient tyrosine phosphorylation of limited number of unidentified proteins, in particular of 66 and 70 kDa, using the monoclonal antibody PY20 and noted that this was maintained the presence of indomethacin. The requirement for arachidonate conversion to thromboxanes in saponin-permeabilised cells is reminiscent of activation of platelets by a combination of inositol 1,4,5-trisphosphate and protein kinase C activation (WATSON et al. 1986).

All of these studies are in agreement that vanadate and peroxovanadate induce platelet activation in association with increased tyrosine phosphorylation and activation of PLC. The explanation for the small degree of variability in results between individual studies may reflect differences between intact and semi-permeabilised cells, differences in the action of the two vanadate species or differences in the methods used to detect phosphotyrosine residues. For example, FANTUS et al. (1989) have described differences in the action of vanadate and peroxovanadate on inhibition of PTPase activity. Vanadate ions are also known to induce several other actions including serving as a GTP mimic and inhibition of $Ca^{2+}$-ATPase, although PUMIGLIA et al. (1992) have provided evidence that neither mechanism has a role in vanadate-induced platelet activation. Inhibition of PTPases may also indirectly stimulate tyrosine kinases by permitting uncontrolled autophosphorylation, and it is possible that this has different functional effects in intact and semi-permeabilised platelets.

Studies with vanadate and its derivative are important from an historical perspective in that they provided some of the first evidence for a role of tyrosine phosphorylation in platelet activation. The observed increase in tyrosine phosphorylation of PLC$\gamma$1 in platelets stimulated by peroxovanadate, however, appears to have no physiological counterpart in platelets activated by physiological agonists such as thrombin and collagen (TORTI and LAPETINA 1992; BLAKE et al. 1994b). Phosphorylation of PLC$\gamma$1 by peroxovanadate may be the consequence of the remarkable increase in tyrosine phosphorylation that occurs in platelets activated by this powerful stimulus which is estimated to be more than 29 times greater than that induced by thrombin (PUMIGLIA et al. 1992). In contrast, BLAKE et al. (1994b) and DANIEL et al. (1994) have described tyrosine phosphorylation of PLC$\gamma$2 in platelets by collagen, Fc$\gamma$RIIA receptor crosslinking and wheat germ agglutinin. Consistent with this, activation of platelets by these three stimuli is inhibited by staurosporine or the tyrphostin ST271, demonstrating that tyrosine phosphorylation of PLC$\gamma$2 is the molecular basis of platelet activation by stimuli that induce receptor clustering (POOLE et al. 1993; BLAKE et al. 1994a; POOLE and WATSON 1995; unpublished observations).

FEINSTEIN et al. (1993) carried out a detailed investigation of the role of tyrosine phosphorylation in platelet activation by the G protein receptor stimulus thrombin based on comparisons with activation of platelets by peroxovanadate. FEINSTEIN et al. (1993) concluded that tyrosine phosphorylation was not necessary for the activation of phospholipases C and $A_2$ or functional responses such as shape change, aggregation and secretion induced by thrombin. They were unable to rule out a faciliatory role of tyrosine phosphorylation in the regulation of these responses as the tyrphostin RG50864 has a weak inhibitory effect against low concentrations of thrombin. This conclusion should be treated with caution as the tyrphostin RG50864 has other inhibitory actions independent of tyrosine phosphorylation including inhibition of oxidative metabolism (YOUNG et al. 1993). The absence of a role of tyrosine phosphorylation in the regulation of PLC is consistent with the observation that thrombin does not induce significant tyrosine phosphorylation of PLC$\gamma$ isoforms in platelets (DANIEL et al. 1994; BLAKE et al. 1994b).

## V. Effect of Phenylarsine Oxide in Platelets

Phenylarsine oxide (PAO) is a trivalent arsenical compound that induces a wide spectrum of effects in biologic systems through covalent modification of thiol groups. In particular, PAO is a potent inhibitor of PTPases by virtue of the presence of a cysteine residue in their active site. PAO stimulates tyrosine phosphorylation of multiple proteins in platelets with a threshold concentration at approximately 1 $\mu M$ and peaking at between 10–100 $\mu M$ (OETKEN et al. 1992; YANAGA and WATSON, unpublished). DAWICKI and STEINER (1993) reported that PAO is inactive against platelet-membrane PTPase activity in concentrations less than 1 $\mu M$, suggesting that the increase in tyrosine phosphorylation may be due to inhibition of cytosolic PTPases. The increase in protein tyrosine phosphorylation induced by PAO is considerably lower in magnitude than that induced by peroxovanadate and is associated with inhibition of activation rather than stimulation of functional responses (SUGATANI et al. 1987; OETKEN et al. 1992; GREENWALT and TANDON 1994; YANAGA et al. 1995a).

PAO inhibits platelet activation by at least two distinct mechanisms and its action is therefore dependent on the molecular basis of platelet activation by the stimulus under study. PAO completely inhibits the action of stimuli that induce receptor clustering such as collagen and cross-linking of Fc$\gamma$RIIA with half maximal and maximal inhibition at approximately 0.2 and 1 $\mu M$, respectively (GREENWALT and TANDON 1994; YANAGA et al. 1995a). Importantly, inhibition occurs at concentrations that have little effect on total cellular protein tyrosine phosphorylation. The inhibitory effect of PAO against collagen and Fc$\gamma$RIIA-crosslinking is reversed by the membrane-permeable disulfhydryl reagent 2,3-dimercaptopropanol, which removes PAO from proteins to form a stable cyclic adduct, but not by the impermeable analogue dimercaptopropanesulphonic acid (GREENWALT and TANDON 1994; YANAGA and WATSON, unpublished). Preincubation of PAO with collagen or the Fc$\gamma$RII-specific antibody IV.3, which was used to activate the receptor in the studies of YANAGA et al. (1995a), did not inhibit the ability of either agent to stimulate platelet activation (GREENWALT and TANDON 1994; YANAGA and

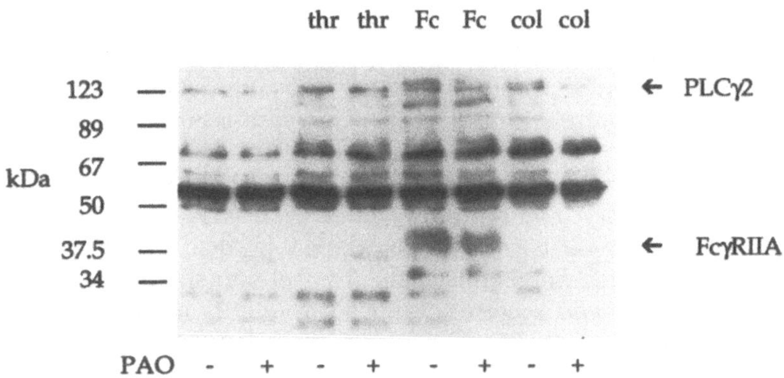


**Fig. 1.** Effect of phenylarsine oxide on protein tyrosine phosphorylation. Platelets were stimulated with thrombin (1 u/ml; 60 s), FcγRIIA crosslinking [using mAb IV.3 (1 μg/ml) and $F(ab')_2$ (30 μg/ml) as described by YANAGA et al. 1995b] or collagen (100 μg/ml; 90 s) in the absence or presence of phenylarsine oxide (*PAO*; 1 μ*M*) and measured for tyrosine phosphorylation using the monoclonal antibody 4G10 as described (YANAGA et al. 1995a). Platelets were preincubated with PAO for 60 s before agonis stimulation. The position of phospholipase C γ2 and FcγRIIA are shown. Substantial tyrosine phosphorylation of phospholipase C γ2 is induced by collagen and FcγRIIA crosslinking but not by thrombin and is inhibited selectively by PAO; in contrast, tyrosine phosphorylation of ohter proteins including FcγRIIA is not inhibited (for further details see YANAGA et al. 1995a)

WATSON, unpublished). These results demonstrate that the inhibitory effect of PAO is mediated through an intracellular site of action. The molecular basis of the action of PAO against collagen and FcγRIIA is through inhibition of tyrosine phosphorylation of PLCγ2, with tyrosine phosphorylation of other proteins including the tyrosine kinase syk being relatively unaltered (YANAGA et al. 1995a). This is illustrated in the antiphosphotyrosine blot in Fig. 1. Collagen and FcγRIIA crosslinking induce tyrosine phosphorylation of multiple proteins, including a 140 kDa protein whose phosphorylation is inhibited in the presence of PAO. This band has been identified as PLCγ2 through immunoprecipitation using specific antibodies (YANAGA et al. 1995a). The mechanism underlying inhibition of PLCγ2 tyrosine phosphorylation by PAO is not known, although it is interesting to note that PAO also inhibits activation of PLCγ isoforms in other cells including T lymphocytes and fibroblasts (YANAGA et al. 1995a; FLETCHER et al. 1993), suggesting that it may be a general inhibitor of phosphorylation of PLCγ isoforms.

PAO also inhibits aggregation and secretion induced by G protein receptor stimuli including thrombin and PAF in human and rabbit platelets but at a concentration approximately tenfold greater than that at which it inhibits activation by collagen and FcγRIIA crosslinking (SUGATANI et al. 1987; GREENWALT and TANDON 1994; YANAGA and WATSON, unpulished). PAO (10 μ*M*) does not inhibit the onset of shape change induced by PAF of thrombin, in contrast to the complete inhibition of all functional responses induced by collagen and FcγRIIA crosslinking (SUGATANI et al. 1987; GREENWALT and TANDON 1994; YANAGA and WATSON, unpublished), and potentiates thrombin-induced formation of inositol phosphates (YANAGA et al. 1995a). The latter observation is consistent with the fact that activation of platelets by thrombin is mediated through PLCβ isoforms with insignificant tyrosine phosphorylation of PLCγ isoforms, even in the presence of PAO (BLAKE et al. 1994b; YANAGA et al. 1995a). The inhibitory effect of PAO against thrombin and PAF is reversed by the membrane-permeable disulfhydryl reagent 2,3-dimercaptopropanol, demonstrating and intracellular site of action. PAO also inhibits aggregation and secretion induced by ADP, adrenaline and arachidonic acid (which induces activation via metabolism to thromboxanes and endoperoxides) although shape change and elevation of $Ca^{2+}$ are not inhibited (GREENWALT and TANDON 1994). The inhibitory effect of PAO against ADP and adrenaline occurs at an approximately tenfold lower concentration than that required to inhibit these responses in platelets stimulated by thrombin and PAF, and this may reflect an essential role of phospholipase $A_2$ in stimulation of aggregation and secretion by "weak agonists". A relatively low concentration of PAO (1 $\mu M$) also induces significant inhibition of phospholipase $A_2$ in platelets stimulated by PAF and the ionophore A23187 (SUGATANI et al. 1987). The inhibitory effect of PAO is not mediated through elevation of cAMP or metabolism of ATP (SUGATANI et al. 1987). There is little evidence to support a role for inhibition of PTPase activity in the action of PAO as several of its inhibitory actions occur at concentrations below those that increase cellular protein tyrosine phosphorylation.

## VI. Possible Role of Protein Tyrosine Phosphatases in $Ca^{2+}$ Entry

A role for PTPase activity in the control of $Ca^{2+}$ entry following depletion of intracellular $Ca^{2+}$ stores has been proposed by VOSTAL et al. (1991). These authors noted that $Ca^{2+}$ store depletion leads to tyrosine phosphorylation of a 130 kDa protein and that subsequent filling of the store leads to its dephosphorylation. The 130 kDa protein was identified as vinculin in a later study (VOSTAL and SHULMAN 1993). In a highly speculative model, VOSTAL et al. (1991) proposed that the activity of an uncharacterised PTPase is regulated by the degree of filling of the $Ca^{2+}$ store and that inhibition of this PTPase has a role in regulating $Ca^{2+}$ entry. Direct evidence in support of this model has not emerged although SARGEANT et al. (1994) reported that the increase in protein tyrosine phosphorylation that occurs on store depletion is not inhibited by chelation of intracellular $Ca^{2+}$, an observation consistent with the model.

## VII. Regulation of Protein Tyrosine Phosphatases by Serine/Threonine Phosphorylation

As discussed above, the high level of PTPase activity relative to that of tyrosine kinases suggests that this family of enzymes should be subject to tight control. One potential mechanism of regulation is through phosphorylation on serine and threonine residues. Several PTPases are known to be phosphoproteins in vivo, including PTP1B which undergoes rapid phosphorylation on serine in platelets following stimulation by thrombin, possibly through the action of protein kinase C (FRANGIONI et al. 1993). Phosphorylation of PTP1B is associated with only modest changes in phosphatase activity (SUN and TONKS 1994). Recently, the cytosolic PTPase, PTP-PEST, was shown to be phosphorylated in vivo in HeLa cells on serine 39 and 435 by protein kinases A and C, resulting in a decrease in activity (GARTON and TONKS 1994). These examples illustrate the potential for cross-talk between serine/threonine and tyrosine-regulated pathways and may contribute to the increase in tyrosine phosphorylation in platelets challenged with G protein agonists such as thrombin.

### VIII. Protein Tyrosine Phosphatases in Platelets: Concluding Remarks

Although it is now more than twelve years since the presence of tyrosine kinases in platelets was first reported, it is only recently that the functional significance of this group of enzymes is beginning to emerge. In contrast, almost nothing is known about the *physiological* role of PTPases. The much greater activity of PTPases relative to tyrosine kinases in platelets suggests that this family of enzymes may participate directly in pathways involved in platelet activation. For example, it can be speculated from the action of PAO that a PTPase may play a role in the activation of PLC$\gamma$2 by stimuli that induce receptor clustering. The molecular events underlying tyrosine phosphorylation of PLC$\gamma$2 are not characterised although strong evidence for an essential role of the tyrosine kinase syk in this pathway has been provided (YANAGA et al. 1995b). Current models of signalling by immune receptors, including Fc$\gamma$RIIA, place a member of the src family of tyrosine kinases upstream of syk or the equivalent enzyme in T cells, zap-70 (RAVETCH 1994; WEISS and LITTMAN 1994). Members of the src family of tyrosine kinases are activated by dephosphorylation (see Sect. C.III.). In B and T cell receptor signalling there is overwhelming evidence to support a role for the PTPase CD45 upstream of activation of a src-related kinase (RAVETCH 1994; WEISS and LITTMAN 1994). Whether a related PTPase (platelets do not express CD45) performs a similar function in platelets activated by Fc$\gamma$RIIA crosslinking, and possibly by collagen, is not known and is currently under investigation in my laboratory.

## D. Dual Specificity Protein Phosphatases

The large family of dual specificity phosphatases have the ability to dephosphorylate tyrosine and serine/threonine residues. The first member of this family was described in vaccinia virus by GUAN et al. (1991). Dual specificity phosphatases possess the PTPase signature motif but otherwise have little sequence identity with other protein phosphatases. They have a narrow substrate specificity. Examples include MKP1 and PAC1 which dephosphorylate mitogen-activated protein kinase requiring phosphorylation on tyrosine and threonine residues for activation (Sun et al. 1993; Ward et al. 1994). There are no selective inhibitors of dual specificity protein phosphatase, and their presence in platelets has not been investigated. The transient activation of mitogen-activated protein kinase in thrombin-stimulated platelets (Börsch-Haubold et al. 1995) may reflect the presence of a dual specificity phosphatase.

# E. Concluding Remarks

Reversible protein phosphorylation is regulated by changes in activity of kinases and phosphatases or by protein redistribution. The much greater activity of PTPases relative to tyrosine kinases suggests that PTPases may play a direct role in platelet activation. It is apparent, however, from this short review that relatively little can be concluded in regard to the functional role of protein phosphatases through the use of non-specific inhibitors or in studies measuring total protein phosphorylation rather that phosphorylation of individual proteins. A complete description of the roles of PS/TPs and PTPases in platelets will require development of inhibitors with increased selectivity, a full characterisation of the complement of enzymes within the cell and elucidation of the mechanisms underlying functional responses such as secretion and aggregation. This review was completed in August 1995.

*Acknowledgement.* It is a pleasure to acknowledge the contribution of members of my laboratory over the last few years to the work described in this article, including Judith Asselin, Robert Blake, Trevor Walker and Fumi Yanaga. I am grateful to Angelika Börsch-Haubold, Jon Gibbins and Anne Robinson for critical reading of the manuscript. Work in the author's laboratory is supported by the Wellcome Trust and British Heart Foundation. SPW is a Royal Society Research Fellow.

## References

Aharonovitz O, Livne AA, Granot Y (1995) 42 kDa protein as a substrate for protein phosphatase(s) in intact human blood platelets. Platelets 6:17–23

Bialojan C, Takai A (1988) Inhibitory effect of marine-sponge toxin, okadaic acid, on protein phosphatases. Specificity and kinetics. Biochem J 256:283–290

Blake R, Walker T, Watson SP (1993) Activation of human platelets by peroxovanadate is associated with tyrosine phosphorylation of phospholipase C $\gamma$ and formation of inositol phosphates. Biochem J 290:471–475

Blake R, Asselin J, Walker TW, Watson SP (1994a) Fc$\gamma$ receptor II stimulated formation of inositol phosphates in human platelets is blocked by tyrosine kinase inhibitors and associated with tyrosine phosphorylation of the receptor. FEBS Lett 342:15–18

Blake RA, Schieven GL, Watson SP (1994b) Collagen stimulates tyrosine phosphorylation of phospholipase C-$\gamma$2 but not phospholipase C-$\gamma$1 in human platelets. FEBS Lett 353:212–216

Börsch-Haubold AG, Kramer RM, Watson SP (1995) Cytosolic phospholipase A2 is phosphorylated in collagen- and thrombin-stimulated human platelets independent of protein kinase C and mitogen activated protein kinase. J Biol Chem 270:25885–25892

Brady-Kalnay SM, Tonks NK (1994) Protein tyrosine phosphatases: from structure to function. Trends Cell Biol 4:73–76

Brady-Kalnay SM, Flint AJ, Tonks NK (1993) Homophilic binding of PTP mu, a receptor type tyrosine phosphatase, can mediate cell-cell aggregation. J Cell Biol 122:961–972

Chiang TM (1993) The role of protein phosphatases 1 and 2A collagen-platelet interaction. Arch Biochem Biophys 302:55–63

Clark EA, Brugge JS (1993) Redistribution of activated pp60c-src to integrin-dependent cytoskeletal complexes in thrombin-stimulated platelets. Mol Cell Biol 13:1863–1871

Clark EA, Shattil SJ, Brugge JS (1994) Regulation of protein tyrosine kinases in platelets. Trends Biol Sci 19:464–469

Cohen P (1992) Signal integration at the level of protein kinases, protein phosphatases and their substrates. Trends Biochem Sci 17:408–413

Cohen P (1993) Important roles for novel protein phosphatases dephosphorylating serine and threonine residues. Biochem Soc Trans 21:884–888

Cohen P, Cohen P (1989) Protein phosphatases come of age. J Biol Chem 264:21435–21438

Cohen P, Klump, Schelling DL (1989) An improved procedure for identifying and quantitating protein phosphatases in mammalian tissues. FEBS Lett 250:596–600

Cohen P, Brewis ND, Hughes V, Mann DJ (1990) Protein serine/threonine phosphatases: an expanding family. FEBS Lett 268:355–359

Collett MS, Purchio AF, Erikson RL (1980) Avian sarcoma virus-transforming protein $pp60^{src}$ shows protein kinase activity specific for tyrosine. Nature 285:167–169

da Cruz e Silva OB, da Cruz e Silva EF, Cohen P (1988) Identification of a novel protein phosphatase catalytic subunit by cDNA cloning. FEBS Lett 242:106–110

Daniel JL, Dangelmaier C, Smith JB (1994) Evidence for a role for tyrosine phosphorylation of phospholipase $C\gamma2$ in collagen-induced platelet cytosolic calcium mobilization. Biochem J 302:617–622

Davidson MML, Haslam RJ (1994) Dephosphorylation of cofilin in stimulated platelets: roles for a GTP-binding protein. Biochem J 301:41–47

Dawicki DD, Steiner M (1993) Identification of a protein-tyrosine phosphatase from human platelet membranes by an immobilon-based solid phase assay. Anal Biochem 213:245–255

Eckhart W, Hutchinson MA, Hunter T (1979) An activity phosphorylating tyrosine in polyoma T antigen immunoprecipitates. Cell 18:925–933

Erdodi F, Csortos C, Sparks L, Muranyi A, Gergely P (1992) Purification and characterization of three distinct types of protein phosphatase catalytic subunits in bovine platelets. Arch Biochem Biophys 298:682–687

Fantus IG, Kadota S, Deragon G, Foster B, Posner BI (1989) Pervanadate [peroxide(s) of vanadate] mimics insulin action in rat adipocytes via activation of the insulin receptor tyrosine kinase. Biochemistry 28:8864–8871

Feinstein MB, Pumiglia K, Lau L-F (1993) Tyrosine phosphorylation in platelets: its regulation and possible role in platelet functions. In: Authi KS, Watson SP, Kakkar VV (eds) Mechanisms of platelet activation and control. Adv Exp Med Biol 344. Plenum, New York, pp 129–149

Ferrell JE, Martin GS (1988) Platelet tyrosine-Specific protein phosphorylation is regulated by thrombin. Mol Cell Biol 8:3603–3610

Ferrell JE, Martin GS (1989) Tyrosine-specific protein phosphorylation is regulated by glycoprotein IIb-IIIa in platelets. Proc Natl Acad Sci USA 86:2234–2238

Fletcher MC, Samelson LE, June CH (1993) Complex effects of phenylarsine oxide in T cells: induction of tyrosine phosphorylation and calcium mobilization independent of CD45 expression. J Biol Chem 268:23697–23703

Frangioni JV, Oda A, Smith M, Salzman EW, Neel BG (1993) Calpain-catalyzed cleavage and subcellular relocation of protein phosphotyrosine phosphatase 1B (PTP-1B) in human platelets. EMBO J 12:4843–4856

Garton AJ, Tonks NK (1994) PTP-PEST: a protein tyrosine phosphatase regulated by serine phosphorylation. EMBO J 13:3763–3771

Greenwalt DE, Tandon NN (1994) Platelet shape change and $Ca^{2+}$ mobilization induced by collagen, but not thrombin or ADP, are inhibited by phenylarsine oxide. Br J Haematol 88:830–838

Gu M, York JD, Warshawsky I, Majerus PW (1991) Idenfication, cloning, and expression of a cytosolic megakaryocyte protein tyrosine phosphatase with sequence homology to cytoskeletal 4.1. Proc Natl Acad Sci USA 88:5867–5871

Guan K, Dixon JE (1990) Protein tyrosine phosphatase activity of an essential virulence determinant in *Yersinia*. Science 249:553–556

Guan K, Broyles SS, Dixon JE (1991) A Tyr/Ser protein phosphatase encoded by Vaccinia Virus. Nature 350:359–362

Haslam RJ, Lynham JA (1977) Relationship between phosphorylation of blood platelet proteins and secretion of platelet granule constituents. I. Effects of different aggregating agents. Biochem Biophys Res Commun 77:714–722

Higashihara M, Takahata K, Kurokawa K, Ikeba M (1992) The inhibitory effects of okadaic acid on platelet function. FEBS Lett 307:206–210

Hubbard MJ, Cohen P (1993) On target with a new mechanism for the regulation of protein phosphorylation. Trends Biochem Sci 18:172–177

Hunter T (1995) Protein kinases and phosphatases: the Yin and Yang of protein phosphorylation and signalling. Cell 80:225–236

Imaoka T, Lynham JA, Haslam RJ (1983) Purification and characterization of the 47000-dalton protein phosphorylated during degranulation of human platelets. J Biol Chem 258:11404–11414

Inagaki N, Ito M, Nakaneo T, Inagaki M (1994) Spatiotemporal distribution of protein kinase and phosphatase activities. Trends Biochem Sci 19:448–453

Inazu T, Taniguchi T, Yanagi S, Yamamura H (1990) Protein tyrosine phosphorylation and aggregation of intact human platelets by vanadate with $H_2O_2$. Biochem Biophys Res Commun 170:259–263

Ingebritsen TS, Cohen P (1983a) The protein phosphatases involved in cellular regulation: 1. classification and substrate specificities. Eur J Biochem 132:255–261

Ingebritsen TS, Cohen P (1983b) Protein phosphatases: properties and role in cellular regulation. Science 221:331–338

Ishihara H, Martin BL, Brautigan DL (1989) Calyculin A and okadaic acid: inhibitors of protein phosphatase activity. Biochem Biophys Res Commun 159:871–877

Karaki H, Mitsui M, Nagase H (1989) Inhibitory effects of a toxin okadaic acid, isolated from the black sponge on smooth muscle and platelets. Br J Pharmacol 98:590–596

Kim Y, Huang J, Cohen P, Matthews HR (1993) Protein phosphatases 1, 2A and 2C are protein histidine phosphatases. J Biol Chem 268:18513–18518

Kozlowski M, Mlinaric-Rascan I, Feng G-S, Shen R, Pawson T, Siminovitch KA (1993) Expression and catalytic activity of the tyrosine phosphatase PTP1C is severely impaired in motheaten and viable motheaten mice. J Exp Med 178:2157–2163

Lerea KM (1991) Thrombin-induced effects are selectively inhibited following treatment of intact human platelets with okadaic acid. Biochemistry 30:6819–6824

Lerea KM, Tonks NK, Krebs EG, Fischer EH, Glomset JA (1989) Vanadate and molybdate increase tyrosine phosphorylation in a 50-kilodalton protein and stimulate secretion in electropermeabilized platelets. Biochemistry 28:9826–9292

Li RY, Gaits F, Ragab A, Ragab-Thomas JM, Chap H (1994a) Translocation of an SH2-containing protein tyrosine phosphatase (SH-PTP1) to the cytoskeleton of thrombin-activated platelets. FEBS Lett 343:89–93

Li RY, Ragab A, Gaits F, Ragab-Thomas JM, Chap H (1994b) Thrombin-induced redistribution of protein-tyrosine-phosphatases to the cytoskeletal complexes in human platelets. Cell Mol Biol (Noisy-le-grand) 40:665–675
Li RY, Gaits F, Ragab A, Ragab-Thomas JM, Chap H (1995) Tyrosine phosphorylation of an $SH_2$-containing protein tyrosine phosphatase is coupled to platelet thrombin receptor via a pertussis toxin-sensitive heterotrimeric G-protein. EMBO J 14:2519–2526
Luber K, Siess W (1994) Integrin-dependent protein dephosphorylation on tyrosine induced by activation of the thrombin receptor in human platelets. Cell Signal 3:279–284
MacKintosh C, MacKintosh RW (1994) Inhibitors of protein kinases and phosphatases. Trends Biochem Sci 19:444–448
McNicol A, Robertson C, Gerrard JM (1993) Vanadate activates platelets by enhancing arachidonic acid release. Blood 81:2329–2338
Murata K-H, Sakon M, Kambayashi J-I, Yukawa M, Yano Y, Fujitani K, Kawasaki T, Shiba E, Mori T (1992) The possible involvement of protein phosphatase 1 in thrombin-induced $Ca^{2+}$ influx of human platelets. J Cell Biochem 51:442–445
Murata K-H, Sakon M, Kambayashi J-I, Yukawa M, Ariyoshi H, Shiba E, Kawasaki T, Kang J, Mori T (1993) The effects of okadaic acid and calyculin A on thrombin induced platelet reaction. Biochem Int 26:327–334
Murphy CT, Westwick J (1994) Role of type 1 and type 2A phosphatases in signal transduction of platelet-activating-factor-stimulated rabbit platelets. Biochem J 301:531–537
Nishikawa M, Toyoda H, Saito M, Mortia K, Tawara I, Deguchi K, Kuno T, Shima H, Nagao M, Shirakawa S (1994) Calyculin and okadaic acid inhibit human platelet aggregation by blocking protein phosphatases types 1 and 2A. Cell Signal 6:59–71
Oetken C, von Willebrand M, Autero M, Ruutu T, Andersson L, Mustelin T (1992) Phenylarsine oxide augments tyrosine phosphorylation in hematopoietic cells. Eur J Haematol 49:208–214
Plutzky J, Neel BG, Rosenberg RD (1992) Isolation of a src homology 2-containing tyrosine phosphatase. Proc Natl Acad Sci USA 89:1123–1127
Poole AW, Watson SP (1995) Collagen Stimulates mobilisation of $Ca^{2+}$ in single platelets through a tyrosine kinase dependent pathway. Br J Pharmacol 115:101–106
Poole AW, Blake R, Asselin J, Watson SP (1993) Tyrosine kinase inhibitors block the increase in $Ca^{2+}$ that occurs on the adhesion of single platelets to collagen. Br J Pharmacol 110:41P
Pumiglia KM, Lau L-F, Huang C-K, Burroughs S, Feinstein MB (1992) Activation of signal transduction in platelets by the tyrosine phosphatase inhibitor pervanadate (vanadyl hydroperoxide). Biochem J 286:441–449
Ravetch JV (1994) Fc receptors: rubor redox. Cell 78:553–560
Sakon M, Kambayashi J-I, Kajiwara Y, Uemura Y, Shiba E, Kawasaki T, Mori T (1990) Platelet protein phosphatases and their endogenous substrates. Biochem Int 22:149–161
Sakon M, Kambayashi J-I, Murata KH (1994) The involvement of protein phosphatases in platelet activation. Platelets 5:130–134
Sargeant P, Farndale RW, Sage SO (1994) Calcium store depletion in dimethyl BAPTA-loaded human platelets increases protein tyrosine phosphorylation in the absence of a rise in cytosolic calcium. Exp Physiol 79:269–272
Sawrup G, Cohen S, Garbers DL (1982) Inhibition of phosphotyrosyl protein phosphatase activity by vanadate. Biochem Biophys Res Commun 107:1104–1109
Schreiber SI (1992) Immunophilin-sensitive protein phosphatase action in cell signalling pathways. Cell 70:365–368
Shenolikar S (1994) Protein serine/threonine phosphatases: new avenues for cell regulation. Annu Rev Cell Biol 10:55–86
Smilowitz HM, Aramli L, Xu D, Epstein PM (1991) Phosphotyrosine phosphatase activity in human platelets. Life Sci 49:29–37

Sugatani J, Steinhelper ME, Saito K, Olson MS, Hanahan DJ (1987) Potential involvement of vicinal sulfhydryls in stimulus-induced rabbit platelet activation. J Biol Chem 262:16995–17001
Sun H, Tonks NK (1994) The coordinated action of protein tyrosine phosphatases and kinases in cell signaling. Trends Biochem Sci 19:480–485
Sun H, Charles CH, Lau LF, Tonks NK (1993) MKP-1 (3CH134), an immediate early gene product, is a dual specificity phosphatase that dephosphorylates MAP kinase in vivo. Cell 75:487–493
Takayama H, Ezumi Y, Ichinohe T, Okuma M (1993) Involvement of GPIIb-IIIa on human platelets in phosphotyrosine-specific dephosphorylation. Biochem Biophys Res Commun 194:472–477
Tallant EA, Wallace RW (1985) Characterisation of a calmodulin-dependent protein phosphatase from human platelets. J Biol Chem 260:7744–7751
Tallant EA, Brumley LM, Wallace RW (1988) Activation of a calmodulin-dependent phosphatase by a $Ca^{2+}$-dependent protease. Biochemistry 27:2205–2211
Tonks NK, Yang Q, Gebbink MFBG, Franza BR, Hill DE, Sun H, Brady-Kalnay S (1992) Protein tyrosine phosphates: the problems of a growing family. Cold Spring Harb Symp Quant Biol 17:87–94
Torti M, Lapetina EG (1992) Role of rap1B and p21ras GTPase-activating protein in the regulation of phospholipase C-gammal in human platelets. Proc Natl Acad Sci USA 89:2277–2279
Tuy FPD, Henry J, Rosenfield C, Kahn A (1983) High tyrosine kinase activity in normal nonproliferating cells. Nature 305:435–438
Ullrich A, Schlessinger J (1990) Signal transduction by receptors with tyrosine kinase activity. Cell 61:203–212
Vostal JG, Shulman NR (1993) Vinculin is major platelet protein that undergoes $Ca^{2+}$-dependent tyrosine phosphorylation. Biochem J 294:675–680
Vostal JG, Jackson WL, Shulman NR (1991) Cytosolic and stored calcium control tyrosine phosphorylation of specific platelet proteins. J Biol Chem 266:16911–16916
Walker T, Watson SP (1992) Okadaic acid inhibits activation of phospholipase C in human platelets by mimicking the actions of protein kinases A and C. Br J Pharmacol 105:627–632
Walter U, Eigenthaler M, Geiger J, Reinhar M (1993) Role of cyclic nucleotide-dependent protein kinases and their common substrate VASP in the regulation of human platelets In: Authi KS, Watson SP, Kakkar VV (eds) Mechanisms of platelet activation and control. Adv Exp Med Biol 344. Plenum, New York, pp 237–251
Ward Y, Gupta S, Jensen P, Wartmenn M, Davis R, Kelly K (1994) Control of MAP kinase activation by the mitogen-induced threonine/tyrosine phosphatase PAC1. Nature 367:651–654
Watson SP, Lapetina EG (1985) 1,2-Diacylglycerol and phorbol ester inhibit agonist-induced production of inositol phosphates in human platelets; possible implications for negative feedback regulation of inositol phospholipid hydrolysis. Proc Natl Acad Sci USA 82:2623–2626
Watson SP, McConnell RT, Lapetina EG (1984) The rapid formation of inositol phosphates in human platelets by thrombin is inhibited by prostacyclin. J Biol Chem 259:13199–13203
Watson SP, Ruggiero M, Abrahams SL, Lapetina EG (1986) Inositol 1,4,5-trisphosphate induces aggregation and release of 5-hydroxytryptamine from saponin-permeabilised human platelets. J Biol Chem 261:5368–5372
Watson SP, Blake RA, Lane T, Walker TR (1993) The use of inhibitors of protein kinases and protein phosphatases to investigate the role of protein tyrosine phosphorylation in platelet activation. In: Authi KS, Watson SP, Kakkar VV (eds) Mechanisms of platelet activation and control. Adv Exp Med Biol 344. Plenum, New York, pp 105–118

Weiss A, Littman DR (1994) Signal transduction by lymphocyte antigen receptors. Cell 76:263–274

Yanaga F, Asselin J, Schieven G, Watson SP (1995a) Phenylarsine oxide inhibits tyrosine phosphorylation of phospholipase Cγ2 in human platelets and phospholipase Cγ1 in NIH-3T3 fibroblasts. FEBS Lett 368:377–380

Yanaga F, Poole A, Asselin J, Blake R, Schieven G, Clark EA, Law C-L, Watson SP (1995b) Syk interacts with tyrosine phosphorylated proteins in human platelets activated by collagen and crosslinking of the Fcγ-IIA receptor. Biochem J 311:471–478

Yano Y, Kambayashi J, Shiba E, Sakon M, Oiki E, Fukuda K, Kawasaki T, Mori T (1994) The role of protein phosphorylation and cytoskeletal reorganisation in microparticle formation from the platelet plasma membrane. Biochem J 299:303–308

Yano Y, Sakon M, Kambayashi J, Kawasaki T, Senda T, Tanak K, Yamada F, Shibata N (1995) Cytoskeletal reorganisation of human platelets induced by the protein phosphatase 1/2A inhibitors okadaic acid and calyculin. Biochem J 307:439–449

Young SW, Poole RC, Hudson AT, Halestrap AP, Denton RM, Tavare JM (1993) Effects of tyrosine kinase inhibitors of protein-kinase independent systems. FEBS Lett 316:278–282

CHAPTER 15

# $Ca^{2+}$ Homeostasis in Human Platelets

K.S. AUTHI

## A. Introduction

### I. $Ca^{2+}$ Homeostasis

$Ca^{2+}$ is an agent of universal significance that plays a key role in triggering many platelet functions. The understanding of its regulation is paramount to our appreciation of the mechanisms of platelet activation. The complete activation by all of the platelet agonists involves elevation of the cytosolic levels of $Ca^{2+}$. The rise in cytosolic $Ca^{2+}$ levels can vary from basal levels of approximately 100 *nM* to micromolar levels depending upon the agonist (RINK and HALLAM 1984). Furthermore, direct elevation of $Ca^{2+}$ levels by ionophores such as A23187 which does not involve receptor occupancy results in complete activation of platelets.

The maintenance of low cytosolic $Ca^{2+}$ concentrations necessary to keep platelets in a resting state and to reestablish a resting state after activation is achieved by a combination of $Ca^{2+}$ + $Mg^{2+}$ ATPases (referred to as $Ca^{2+}$ ATPases) and a $Na^+/Ca^{2+}$ exchanger. Recently there has been considerable interest in the identification and localisation of $Ca^{2+}$ ATPases present in human platelets. Most cells contain two types of $Ca^{2+}$ ATPases: the plasma membrane $Ca^{2+}$ ATPases (PMCA types 1–4; CARAFOLI 1994) and those present on intracellular stores known commonly as sarco-endoplasmic reticulum $Ca^{2+}$ ATPases (SERCA; BURK et al. 1989). The PMCAs are known to be stimulated by calmodulin and have a larger molecular size (approximately 130–140 kDa on SDS-PAGE) than the SERCAs which migrate as approximately 100–110 kDa peptides. Two types of $Na^+/Ca^{2+}$ exchangers have been identified of which the platelet has been suggested to contain the retinal type (KIMURA et al. 1993).

When platelets are activated by agonists, $Ca^{2+}$ elevation in the cytosol occurs via the release of $Ca^{2+}$ from intracellular stores and influx from the outside medium. The relationship between surface receptor occupancy by agonists, the resulting phosphoinositide metabolism and $Ca^{2+}$ elevation is well established (Fig. 1). Occupancy of surface receptors by agonists such as thrombin and thromboxane leads to the activation of the heterotrimeric G proteins releasing the GTP-bound $\alpha$ subunit and $\beta\gamma$ subunits. In vitro both the GTP-$\alpha$ subunit (from Gq) and $\beta\gamma$ subunits (derived from Gi) are able to

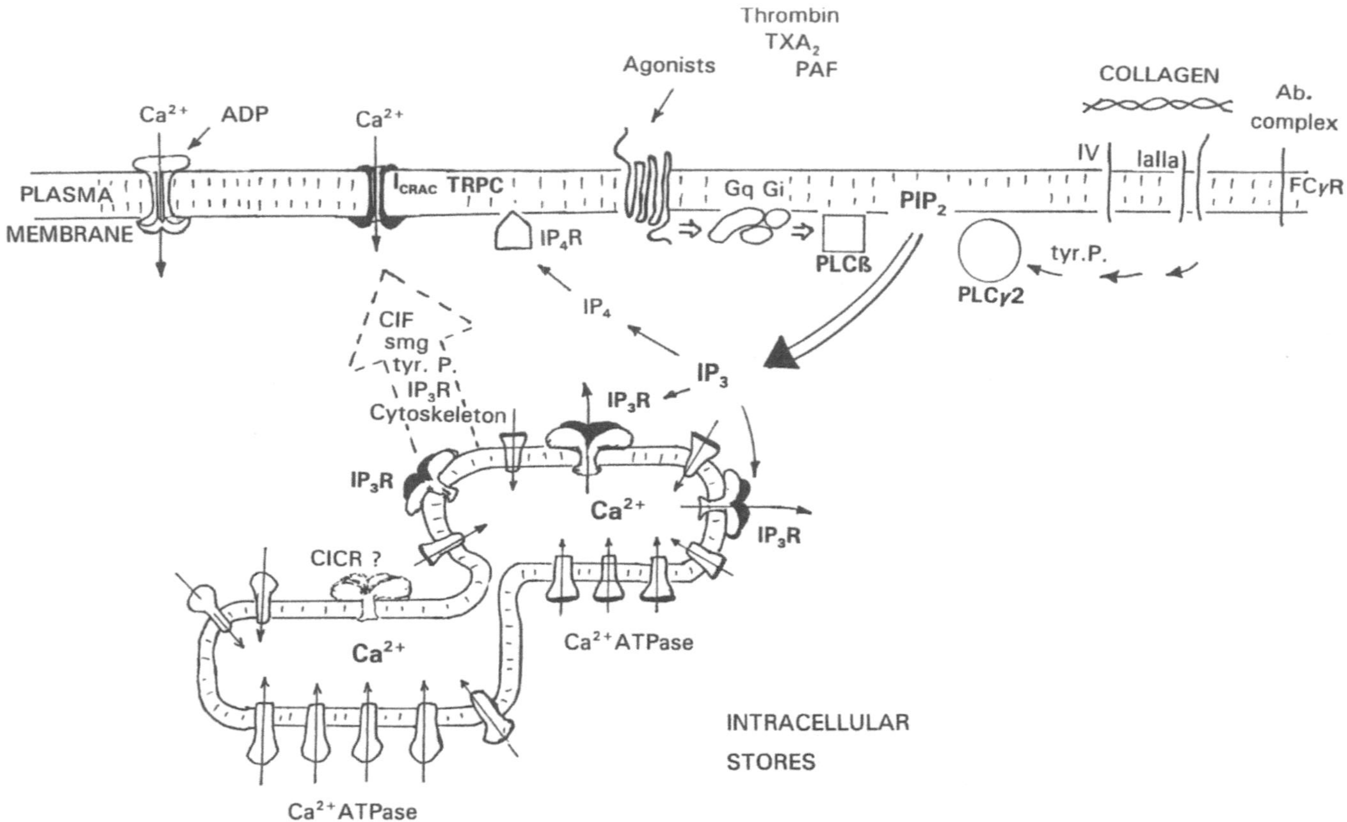


**Fig. 1.** Agonist receptors and $Ca^{2+}$ mobilisation in human platelets. For full explanation see text. In(1,4,5)$P_3$ ($IP_3$) can be generated from $PIP_2$ after activation of either *PLCβ* via G protein linked receptors or *PLCγ2* via collagen or *FCγR* activation. $IP_3$ releases $Ca^{2+}$ from intracellular stores via the $IP_3R$. $Ca^{2+}$ influx arises predominantly as a consequence of store depletion with communication between the stores and the plasma membrane $Ca^{2+}$ channel ($I_{CRAC}$, *TRPC*) not established. Proposed links include *CIF*, small G proteins (*smg*), tyrosine phosphorylation and cytoskeletal linkages. In(1,3,4,5)$P_4$ ($IP_4$) may act on an $IP_4R$ whose relationship to the $Ca^{2+}$ channel is not known. Alternatively $Ca^{2+}$ influx may occur (with *ADP*) after action on a receptor operated channel (*ROC*). At least two types of *$Ca^{2+}$ ATPases* are known to be present on intracellular stores, SERCA 2b and SERCA 3

stimulate phospholipase C of the $\beta$ class, resulting in the hydrolysis of phosphatidylinositol (4,5) bisphosphate ($PIP_2$) yielding inositol 1,4,5-trisphosphate ($In(1,4,5)P_3$) and diacylglycerol (DAG). This relationship is well established for platelet agonists such as thrombin, thromboxane $A_2$, PAF, and vasopressin where the surface receptors for these agonists have been cloned and are known to belong to the family of G protein linked receptors. Information for the surface receptors for ADP that elevates cytosolic $Ca^{2+}$ but is a weak effecter for $In(1,4,5)P_3$ formation is still awaited with much interest. Phosphoinositide-derived second messengers stimulated by collagen have recently been linked to tyrosine phosphorylation of phospholipase $C\gamma2$ (DANIEL et al. 1994; BLAKE et al. 1994).

$In(1,4,5)P_3$ causes $Ca^{2+}$ release from intracellular stores via its intracellular receptor and this role is well established. However, on the basis of studies on isolated membranes and permeabilised cells, as least two types of rapidly releasable intracellular $Ca^{2+}$ stores have been suggested, one of them sensitive to $In(1,4,5)P_3$. The mechanisms associated with $Ca^{2+}$ release from the $In(1,4,5)P_3$ insensitive stores is unknown. It is possible that linkage between the intracellular pools occurs upon agonist stimulation, allowing maximal release to occur via the $In(1,4,5)P_3$ receptor ($IP_3R$). This linkage may exhibit GTP dependence. Additionally, initial $Ca^{2+}$ release mediated by the $IP_3R$ may propagate $Ca^{2+}$ release from the $In(1,4,5)P_3$ insensitive pool via a $Ca^{2+}$ induced $Ca^{2+}$ release mechanism (CICR), analogous to the process occurring via the ryanodine receptor in cardiac muscle. However, to date, the presence of ryanodine receptors in platelets has not been reported and CICR may occur via the $IP_3R$.

Platelet activation is also associated with $Ca^{2+}$ influx. The mechanisms associated with $Ca^{2+}$ influx are more controversial and complex. Principally, $Ca^{2+}$ influx can occur via three main mechanisms. (i) $Ca^{2+}$ influx is regulated by the filling state of the intracellular stores; also known as store regulated or capacitative $Ca^{2+}$ entry mechanism. This is currently the most widely accepted mechanism. However, the link between stores and $Ca^{2+}$ entry are not well understood with many routes proposed. (ii) $Ca^{2+}$ influx may occur via second mesenger operated $Ca^{2+}$ channels such as the action of $In(1,4,5)P_3$ or $In(1,3,4,5)P_4$ with receptors in or associated with the plasma membranes. (iii) The agonist–receptor interaction may directly gate a receptor operated $Ca^{2+}$ channel (ROC) such as has been suggested for the platelet agonist ADP.

## B. Maintenance of Low Cytosolic $Ca^{2+}$ Concentrations

It is accepted that platelets maintain cytosolic concentrations of $Ca^{2+}$ at approximately 50–120 $nM$ under resting conditions. As the external $Ca^{2+}$ concentration is at millimolar levels (and this may also be reached in the intracellular storage organelles such as the endoplasmic reticulum, ER), the platelet maintains a 10000:1 gradient across its plasma and intracellular membranes. These gradients are maintained by the combined actions of the $Ca^{2+}$ ATPases and a $Na^{+}/Ca^{2+}$ exchange activity.

The $Ca^{2+}$ ATPases interact with $Ca^{2+}$ with high affinity ($K_m < 0.5\,\mu M$), making them active at low $Ca^{2+}$ concentrations, and constitute the most important mechanism for maintaining $Ca^{2+}$ levels low. As stated earlier, two families of $Ca^{2+}$ ATPases are known: PMCAs that extrude $Ca^{2+}$ to the extracellular medium and the SERCAs that translocate the cation into intracellular stores. Platelets contain high levels of two known isoforms of SERCA pumps and only low levels of PMCAs whose isoform identities are yet to be described. Both types of $Ca^{2+}$ ATPases belong to the P-class of $Ca^{2+}$ pumps (characterised by the formation of a phospho-enzyme complex on aspartyl residue during the reaction mechanism) and are inhibited by vanadate, though with differing affinities. The $Ca^{2+}$ ATPase inhibitor thapsigargin (Tg) inhibits all SERCA pumps tested so far with no effect on PMCAs (LYTTON et al. 1991). The PMCAs are stimulated by calmodulin and have a larger molecular size (124–136kDa) than the SERCAs (approximately 100kDa). The two families of $Ca^{2+}$ ATPases also differ in other modes of regulation, namely that the PMCAs are direct targets for protein kinases [such as cAMP-dependent protein kinase (cAMP-PK)], and the two types of $Ca^{2+}$ ATPases differ in the stoichiometry of $Ca^{2+}$ translocation with ATP hydrolysis, this being 1:1 ($Ca^{2+}$ transported:ATP hydrolysed) for the PMCA and 2:1 for SERCA.

## I. Platelet SERCAs

Presently three genes are known that code for the intracellular membrane $Ca^{2+}$ ATPase of muscle and non-muscle tissues, SERCA 1, 2 and 3 genes (BRANDL et al. 1986; BURK et al. 1989). The transcription of the SERCA 1 and SERCA 2 genes are subject to tissue and differentiation stage-dependent alternative processing. SERCA 1 (a + b forms) codes for the $Ca^{2+}$ ATPases present in "fast" skeletal muscle, SERCA 2 (also a + b), of which SERCA 2a, is expressed in heart and "slow" skeletal muscle, and SERCA 2b is ubiquitously expressed and has been assigned a "house keeping" role (LYTTON et al. 1992). The SERCA 3 pump has a tissue specific distribution as detected by Northern blot hybridization analysis, being highly expressed in large intestine, small intestine, spleen and lung with low levels present in liver, pancreas and kidney (BURK et al. 1989).

Platelets express the SERCA 2b and SERCA 3 pumps (ENOUF et al. 1992; PAPP et al. 1992; WUYTACK et al. 1994; BOBE et al. 1994; BOKKALA et al. 1995), and though no quantitative estimation of these two isoforms has been made, information from phosphoprotein analysis suggests the two to be present in similar amounts (PAPP et al. 1992). Amino acid sequence alignments of SERCA 2b and SERCA 3 pumps indicate approximately 75% homology. The two pumps can be resolved in SDS-PAGE systems with SERCA 2b and SERCA 3 migrating as 100 and 97kDa proteins, respectively, in acid gels

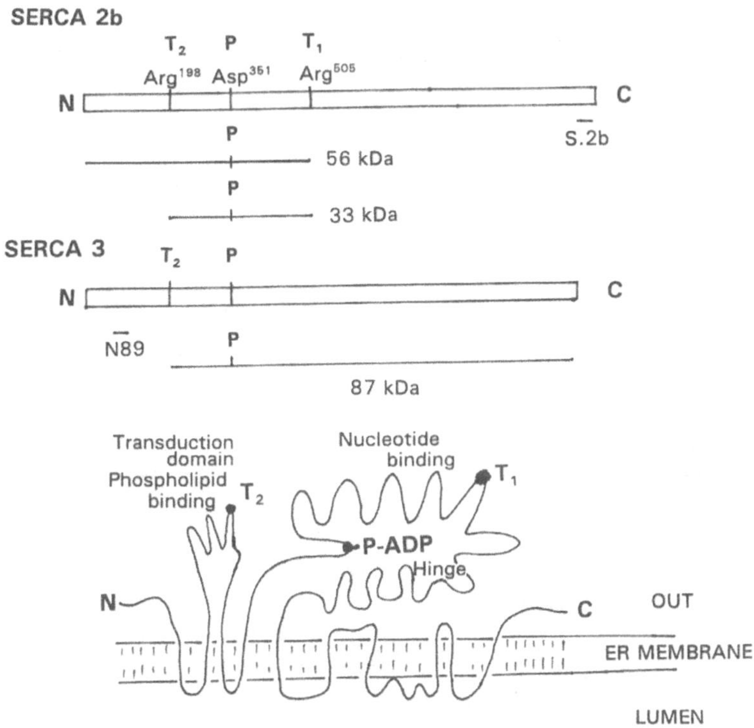


A

B

**Fig. 2A,B.** $Ca^{2+}$ ATPase isoforms in human platelets and proposed membrane topology of SERCA $Ca^{2+}$ pumps. **A** Scheme of *SERCA 2b* and *SERCA 3* proteins is shown with tryptic cleavage sites $T_1$ and $T_2$ and resulting fragments. *P* reflects the phosphorylation site and indicated epitopes for *N89* (S.3) and *S.2b* specific antisera. (Adapted from WUYTACK et al. 1994) **B** Model of membrane topology of SERCA $Ca^{2+}$ pumps. Indicated are trypsin cleavage sites, phosphorylation site, nucleotide binding domain, hinge domain, phospholipid binding domain and transduction domain. (Adapted from TOYOFUKU et al. 1993)

(PAPP et al. 1991, 1992) and as 115 and 100kDa proteins, respectively, in Laemmli gels (WUYTACK et al. 1994). Characterisation of the two isoforms in platelets have employed extensive biochemical, immunological and molecular biological techniques (for a review see AUTHI 1993). These have included usage of tryptic fragmentation pattern analysis, formation of phosphoenzyme complexes, $Ca^{2+}$ ATPase inhibitors, specific antibody recognition and amplification of mRNA transcripts using reverse transcriptase-polymerase chain reaction (RT-PCR). Two trypsin cleavage sites have been suggested in SERCA 2 at $Arg^{505}$ ($T_1$) and $Arg^{198}$ ($T_2$), explaining the characteristic fragmentation pattern seen in earlier studies (Fig. 2A). SERCA 2b shows trypsin digestion products of 55 and 35kDa (Fig. 2), both fragments contain the $Asp^{351}$ that is phosphorylated in a $Ca^{2+}$-dependent manner during $Ca^{2+}$ transport and is potently inhibited by Tg (PAPP et al. 1991, 1992). The SERCA 3 pump does not contain $Arg^{505}$, $T_1$ is missing, but $Arg^{198}$ is present with trypsin cleavage resulting in the formation of an 80kDa fragment containing the phosphorylation site, whose formation is increased in the presence of $La^{3+}$ (PAPP et al. 1991) and inhibited relatively more potently by 2,5-di-(t-butyl)-1,4-benzohydroquinone (tBuBHQ) compared with SERCA 2b (PAPP et al. 1992). A range of polyclonal and monoclonal antibodies specific for SERCA 2 and SERCA 3 pumps are available that have confirmed identification and their use with highly purified human platelet plasma and intracellular membranes reveals the location of SERCA pumps to be on intracellular membranes (BOKKALA et al. 1995).

Recently, the presence of another isoform of SERCA distinct from SERCA 2b or SERCA 3 in human platelets has been suggested on the basis of distinct mild tryptic digestion patterns (at very low trypsin to membrane ratios) giving rise to peptides of molecular size 73, 68 and 40kDa, recognised by a monoclonal antibody, PL/IM 430 (KOVACS et al. 1994). This antibody was previously raised to highly purified intracellular membranes prepared by high voltage free flow electrophoresis and has been shown to inhibit $Ca^{2+}$ sequestration into permeabilised platelets and intracellular membranes by more than 50% (HACK et al. 1988a,b). This unique antibody recognises an as yet unknown epitope, leading to inhibition of $Ca^{2+}$ sequestration without any apparent effect on ATP hydrolysis. However, recent studies by BOKKALA et al. (1995) suggest that PL/IM 430 recognises SERCA 3. PL/IM 430-recognised degradation fragments of 70, 74 and 40kDa are often seen in intracellular membrane fractions prepared by free flow electophoresis, probably due to trypsin-like proteases activated during the relatively lengthy protocol of this procedure compared to other methods that yield less pure fractions. To a lesser extent, a SERCA 3 specific antiserum (N89) raised to a peptide sequence corresponding to residues 29–39 near the N terminus of the rat SERCA 3 pump also recognised similar degradation products. More importantly, PL/IM 430 immunoprecipitated $Ca^{2+}$ ATPase showed immunereactivity with N89 antiserum but not with a SERCA 2b specific antiserum (BOKKALA et al. 1995). This strongly suggests that PL/IM 430 recognises a SERCA 3 $Ca^{2+}$ ATPase or that it co-immuneprecipitated a SERCA 3 $Ca^{2+}$ ATPase. It is possible that sub-isoforms of SERCA 3 (i.e. a or b) as for SERCA 1 and SERCA 2 exist with different antibody affinities, but further studies are required for clarification. The 80kDa fragment of SERCA 3 arising from trypsin cleavage is not recognised by PL/IM 430.

## II. Membrane Organisation, Properties and Regulation of Platelet SERCAs

The amino acid sequences deduced from cloning studies of all three SERCA pumps have similar hydropathy profiles indicating the same transmembrane organisation. There are ten predicted transmembrane domains (Fig. 2B) with two main cytoplasmic globular domains. One lies between the predicted transmembrane domains 4 and 5 containing the phosphorylation site, nucleotide binding sites and a hinge region which allows the former two domains to be in close proximity (TOYOFUKU et al. 1992). The other is located between the predicted transmembrane domains 2 and 3 containing the transduction domain that couples ATP hydrolysis to $Ca^{2+}$ transport and a phospholipid-responsive domain. Interestingly, cryo-electron microscopic studies of the sarcoplasmic reticulum $Ca^{2+}$ ATPase suggest that the globular domains arrange into the shape of a bird's head (TOYOSHIMA et al. 1993). All of the P-class pumps contain the phosphate-intermediate-forming Asp in a conserved sequence CSDKTGD.

Important for regulation, SERCA 1 and SERCA 2 $Ca^{2+}$ pumps, but not SERCA 3, contain two regions known to interact with the $Ca^{2+}$ ATPase regulatory protein phospholamban. These include a stretch of amino acids from residues 370–402 (JAMES et al. 1989; TOYOFUKU et al. 1993) and the region 467–743 determined in experiments expressing chimeric $Ca^{2+}$ ATPases and phospholamban in COS-1 cells (TOYOFUKU et al. 1993). Phospholamban is a membrane protein that is highly expressed in the heart. Binding of phospholamban to SERCA 1 and 2 pumps leads to an inhibition of the pump activity by an apparent reduction in $Ca^{2+}$ affinity. Inhibition of $Ca^{2+}$ ATPase activity by phospholamban is relieved if phospholamban is phosphorylated by cAMP-PK, $Ca^{2+}$/calmodulin dependent protein kinase (WEGENER et al. 1989) or protein kinase C (MOVSESIAN et al. 1984). The expressed SERCA 1 and 2 $Ca^{2+}$ ATPases show a higher affinity for $Ca^{2+}$ (~135 and 128 $nM$, respectively) than SERCA 3 (~460 $nM$; BURK et al. 1989; TOYOFUKU et al. 1993). Co-expression of phospholamban with the $Ca^{2+}$ pumps followed by stimulation of cAMP-PK alters the $Ca^{2+}$ affinities of the SERCA 1 and 2 pumps but not that of SERCA 3 (TOYOFUKU et al. 1993).

Platelets contain SERCA 2b that could therefore be regulated by cAMP-PK, $Ca^{2+}$/calmodulin kinase or PKC. Indeed, a number of studies on intact platelets have indicated that elevation of cAMP leads to the increase of $Ca^{2+}$ sequestration into intracellular stores (TAO et al. 1992; BRUNE and ULLRICH 1992). However, the molecular mechanism underlying cAMP mediated stimulation of $Ca^{2+}$ sequestration is highly controversial. Phospholamban is not expressed in platelets (ADUNYAH et al. 1988), but the small G protein rap 1B ($M_r$ = 22 kDa) has been implicated as having an analogue role to phospholamban (KÄSER-GLANZMANN et al. 1977; ENOUF et al. 1985, 1987; ADUNYAH and DEAN 1987; HETTASCH and LE BRETON 1987; FISHER and WHITE 1987; COURVAZIER et al. 1992). It should be stated that additional proposed roles for rap proteins include stimulation of phospholipase $C\gamma 1$ (TORTI and LAPETINA 1992), cytoskeletal reorganisation (FISCHER et al. 1990) and interactions with the mitogen-activated protein (MAP) kinase pathway (BURGERING and BOSS 1995). An upregulation of both rap 1B and the 97 kDa $Ca^{2+}$ ATPase recognised by PL/IM 430 was noted in megakaryocytic cell lines (though no other cAMP-PK substrates were examined) and in platelets of hypertensive rats (MAGNIER et al. 1994). But a corresponding down-regulation of the 97 kDa $Ca^{2+}$ ATPase was not found when rap 1B was reported to occur at decreased levels in a patient with conjestive heart failure and in a patient with severe cardiomyopathy (MAGNIER et al. 1995). In these patients the decreased levels of rap 1B were associated with a decreased level of phosphorylation of rap 1B and a decreased stimulated rate of $Ca^{2+}$ transport, with the authors suggesting that rap 1B was involved with regulation of the 97 kDa $Ca^{2+}$ ATPase. However, correlation of rap 1B phosphorylation by cAMP-PK with stimulation of $Ca^{2+}$ ATPase activities has not been found by other investigators (WHITE et al. 1989; O'ROURKE et al. 1989). Stimulation of $Ca^{2+}$ uptake was identified to occur with the presence of phosphate ions in preparations of the catalytic subunit of cAMP-PK. A partially purified preparation of the 22 kDa protein was found not to affect $Ca^{2+}$ ATPase activities (FISHER and WHITE 1989) and, another complication, that rap 1B is localised to the plasma membrane (WHITE et al. 1993; EL-DAHER et al. 1996) whereas the SERCA class $Ca^{2+}$ ATPases are found on intracellular membranes. However, these observations do not preclude the possibility that other unidentified regulatory proteins that are targets for cAMP-PK may be present on intracellular membranes. Indeed, we have observed intracellular membranes to contain proteins of $M_r$ 16, 22, 27 and 30 kDa which are targets for cAMP-PK and may represent candidate regulatory proteins for $Ca^{2+}$ ATPases in platelets (EL-DAHER et al. 1996). If rap 1B were conclusively shown to regulate the 97 kDa (SERCA 3) $Ca^{2+}$ ATPase, it would represent a unique mechanism as it would have to be released from the plasma membrane and translocate to intracellular membranes containing SERCA 3. Figure 3 depicts possible modes of regulation of platelet $Ca^{2+}$ ATPases by protein phosphorylation.

Relatively little is definitively known about the regulation of the SERCA 3 pump distinct from its stimulation by $Ca^{2+}$ with the expressed rat protein showing a significantly lower affinity towards $Ca^{2+}$ than its SERCA 2 counterpart (BURK et al. 1989). Its deduced amino acid sequence from cloning studies indicates the presence of several serine residues that could be targets for cAMP-PK but particularly $Ser^{17}$ in the sequence Arg-Arg-Phe-Ser-Val that is unique to SERCA 3. However, this phosphorylation has not been determined at the protein level. WUYTACK et al. (1994) reported that while human and rat SERCA 2b were identical, SERCA 3 showed diversity between species. Human and rat SERCA 3 migrate with different electrophoretic mobilities on Laemmli-type electrophoresis systems (human 103 kDa band vs rat 100 kDa band). With antibodies, an N-terminal-specific antiserum (N89) recognised both human and rat SERCA 3, a C- terminal-specific antiserum (C90) only recognised the rat protein, and PL/IM 430 recognises the protein in human but not rat platelets. Additionally, PCR amplification of a limited part of the SERCA 3 sequence obtained from human lymphoblastoid Jurkat cells when compared to the rat sequence showed only 86% identity. These observations suggest that the platelet SERCA 3 $Ca^{2+}$ pump exhibits unique biochemical characteristics that require more studies for full evaluation.

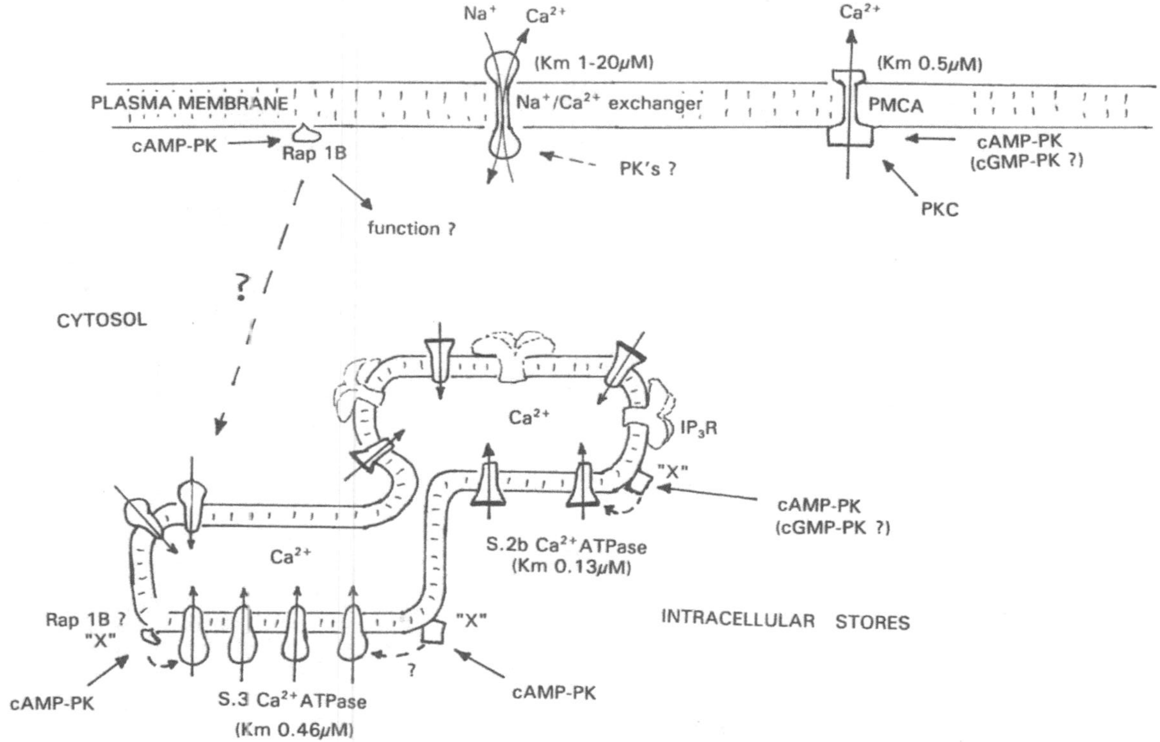


**Fig. 3.** Possible modes of regulation of $Ca^{2+}$ ATPases by protein kinases in human platelets. For full details see text. *PMCAs* are targets for *cAMP-PK* and *PKC*. Possible phosphorylation of the $Na^{+}/Ca^{2+}$ exchanger has not been demonstrated. The *S.2b $Ca^{2+}$ ATPase* can be stimulated by a phospholamban-type mechanism; however, the proteins phosphorylated by *cAMP-PK* on intracellular stores are yet to be identified. Regulation of *S.3 $Ca^{2+}$ ATPase* by a mechanism involving phoshorylation (perhaps involving *Rap 1b*?) remains to be clarified

## III. Organellar Distribution of Platelet SERCA Pumps

Many studies on isolated platelet membranes, permeabilised and intact cells have identified at least two intracellular $Ca^{2+}$ stores based on $IP_3$ sensitivity, use of PL/IM 430 inhibition of $Ca^{2+}$ sequestration, and the use of agonists and $Ca^{2+}$ ATPase inhibitors. However, we are unable to definitively suggest that different SERCAs serve distinct $Ca^{2+}$ pools. There have appeared limitations in every technique used to study intracellular $Ca^{2+}$ pools with these contributing to the discrepancies observed. In isolated platelet membranes, $Ca^{2+}$ transport into the vesicles is largely attributed to activation of SERCA 2b and SERCA 3 pumps and a proportion (usually about 30%) of the total $Ca^{2+}$ taken up is In(1,4,5)$P_3$-releasable (O'ROURKE et al. 1985; AUTHI and CRAWFORD 1985; ADUNYAH and DEAN 1985). In saponin-permeabilised platelets a larger proportion (at least 50%, usually 60%–70%) of the SERCA mediated $Ca^{2+}$ store is $IP_3$-sensitive (HACK et al. 1988b; AUTHI, unpublished observations), indicating that membrane preparations result in either a loss of a fraction of the In(1,4,5)$P_3$-sensitive pool either due to an alteration in the SERCA pumping mechanism, or a proportion of the In(1,4,5)$P_3$-sensitive pool may have functionally compromised $IP_3$Rs. Currently, the only specific tool available to inhibit $Ca^{2+}$ uptake mediated by any SERCA pump is the antibody PL/IM 430. In platelet intracellular membranes a maximum of 80% and in saponin-permeablised platelets a maximum 65% inhibition has been observed (HACK et al. 1988a,b). In the latter studies the kinetics of In(1,4,5)$P_3$-mediated $Ca^{2+}$ release were similar in the absence or presence of PL/IM 430, where in the latter approximately half of the $Ca^{2+}$ uptake was inhibited (HACK et al. 1988b). If $Ca^{2+}$ transport is carried out in the presence of PL/IM 430, In(1,4,5)$P_3$ added at equilibrium readily depleted the remaining stored $Ca^{2+}$. Additionally, co-incubation of PL/IM 430 with In(1,4,5)$P_3$ at the start of the incubations results in near total inhibition of uptake into the stores (AUTHI 1993). These results indicate that the PL/IM 430-recognised $Ca^{2+}$ ATPase predominantly transports $Ca^{2+}$ into an In(1,4,5)$P_3$ insensitive $Ca^{2+}$ store. Recently PAPP et al. (1993) have suggested that the 97kDa $Ca^{2+}$ ATPase recognised by PL/IM 430 serves the In(1,4,5)$P_3$ sensitive pool on the basis of experiments using platelet mixed membranes where the extent of In(1,4,5)$P_3$-induced $Ca^{2+}$ release was inhibited from 35% to 15% in the presence of the antibody. These contrasting results should be taken with caution when comparing data obtained with permeabilised platelets as a considerable proportion of the In(1,4,5)$P_3$ sensitive pool has been compromised, and indeed in some experiments PAPP et al. only report a maximal 25% release of $Ca^{2+}$ by In(1,4,5)$P_3$. Taken together it is likely that different SERCA pumps do not serve totally distinct In(1,4,5)$P_3$ sensitive and insensitive pools, but there probably occurs considerable overlap, also with other factors such as cytoplasmic localisation that may contribute to functional heterogeneity.

In saponin permeabilised platelets the $Ca^{2+}$ ATPase inhibitors Tg and tBuBHQ, at concentrations that totally inhibit the $Ca^{2+}$ pumps, release more

$Ca^{2+}$ than In(1,4,5)$P_3$ from a pool that totally overlaps the In(1,4,5)$P_3$ sensitive pool (AUTHI et al. 1993). In this system co-addition of In(1,4,5)$P_3$ with either inhibitor releases the same quantity of $Ca^{2+}$ as with inhibitor alone. This information is important in resolving the pools of $Ca^{2+}$ released by Tg and tBuBHQ and makes these inhibitors useful to deplete the In(1,4,5)$P_3$ sensitive pool in intact platelets. Additionally, when defining $Ca^{2+}$ pools it is important that experiments are carried out in platelets with inhibited cyclooxygenase as many $Ca^{2+}$ elevating agents (e.g. Tg) potently induce arachidonic acid release with resulting thromboxane formation, and the $Ca^{2+}$ elevation measured may be due to the action of released thromboxane. In indomethacin-treated Fura 2 loaded platelets both Tg and tBuBHQ elevate $[Ca^{2+}]_i$ to approximately similar extents which in the presence of extracellular EGTA reflects $Ca^{2+}$ release from intracellular stores. From experiments carried out in saponin permeabilised platelets this would reflect $Ca^{2+}$ release predominantly from an In(1,4,5)$P_3$ sensitive pool (AUTHI et al. 1993).

Many studies on intact cells have reported that Tg and tBuBHQ release approximately half of the $Ca^{2+}$ stores utilised by agonists such as thrombin, but there is debate about the nature and definition of these pools (BRUNE and ULLRICH 1990, 1991, 1992; AUTHI et al. 1993). Addition of Tg leads to the depletion of its respective pool and a new steady state of $Ca^{2+}$ is reached in the cytosol. Subsequent addition of thrombin results in further release of $Ca^{2+}$ which must then reflect release of $Ca^{2+}$ from an In(1,4,5)$P_3$ insensitive pool as the In(1,4,5)$P_3$ sensitive pool has already been depleted with Tg. The mechanism of this action by agonists (i.e. formation of another messenger or redistribution of the cation between pools) is not understood. BRUNE and ULLRICH (1991, 1992) have additionally reported that Tg releases $Ca^{2+}$ after a thrombin induced $Ca^{2+}$ release and reuptake cycle and suggest that the $Ca^{2+}$ ATPase inhibitor and agonist [In(1,4,5)$P_3$] pools are distinct. Further, agonists potientiate a subsequent Tg induced rise in $Ca^{2+}$. This agonist-mediated redistribution of $Ca^{2+}$ into pools that are Tg sensitive is dependent upon receptor occupancy, suggesting intracellular events downstream of receptor occupancy (BRUNE et al. 1994) perhaps involving a GTP dependent mechanism (MULLANEY et al. 1988). An apparent acceleration of $Ca^{2+}$ distribution into Tg sensitive pools was also effected if either $PGI_2$ or sodium nitroprusside were added after the peak of $Ca^{2+}$ elevation by thrombin, with the subsequent rise of $Ca^{2+}$ induced by Tg being insensitive to the presence of cyclic nucleotides (BRUNE and ULLRICH 1992). In intact platelets it has not been possible to determine if cyclic nucleotides can stimulate $Ca^{2+}$ uptake into the In(1,4,5)$P_3$ sensitive pool; however, such a demonstration has been made in isolated membranes (ENOUF et al. 1987), confirming considerable overlap of Tg and In(1,4,5)$P_3$ sensitive pools (see also discussion of $Ca^{2+}$ pools in single platelet $Ca^{2+}$ responses, below). The Tg sensitive pool may be further subdivided into a fraction that is highly sensitive to Tg (released by $2\,nM$) with the remaining fraction released by either maximal concentrations of Tg or tBuBHQ (CAVALLINI et al. 1995). Both fractions when depleted induce $Ca^{2+}$ entry.

## IV. Plasma Membrane Calcium ATPases

A great deal is known about the properties of PMCAs, mostly from studies in other cells and from the expressed proteins (for a detailed review see CARAFOLI 1992, 1994). Presently four genes are known that code the proteins with further heterogeneity arising due to alternative splicing of the transcripts of all four genes. The products of genes 1 and 4 have been assigned "house keeping" roles and their transcripts have been found in most tissues; however, those of genes 2 and 3 are restricted and found in a tissue-specific manner (STAUFFER et al. 1993). Brain contains mRNA from all four genes. The presence of these enzymes at the protein level has recently been confirmed using isoform specific antibodies (STAUFFER et al. 1995).

PMCAs are known to interact with $Ca^{2+}$ with high affinity ($K_m = 0.5\,\mu M$) but are expressed in much lower amounts than their SERCA counterparts and thus have a lower capacity. In the red cell plasma membrane where it has been well studied, it represents less than 0.1% of the total membrane protein. In platelet mixed membranes SERCA represents approximately 4%–5% of the total protein (DEAN and SULLIVAN 1982), which is expected to enrich approximately threefold in purified intracellular membranes. Even so, the PMCA is still considered to represent the most important mechanism for $Ca^{2+}$ efflux for most cells except for specialised tissues such as the heart and neurones where the $Na^+/Ca^{2+}$ exchanger is thought to be as important. The primary structure has been deduced from cloning studies (SCHULL and GREEB 1988; VERMA et al. 1988), and the predicted topology in membranes is similar to that of SERCA except for an additional cyoplasmic domain arising due to an extended C terminus containing the calmodulin binding domain (CBD) and phosphorylation sites for protein kinases that are involved in regulation. PMCAs can be regulated by multiple mechanisms with most involving interactions with the C terminal region. In the resting state the pump is thought to be auto-inhibited with the C terminal CBD region folding over the main body of the enzyme covering the active site. Synthetic peptides corresponding to the CBD region have been shown to inhibit the pump and have been used to map two sites of interaction near the active site (FALCHETTO et al. 1991). Addition of calmodulin or phosphorylation of the pumps by cAMP-PK or PKC results in removal of the C terminal CBD from the active site, resulting in activation of the pump (Fig. 4A). Both treatments lead to stimulation of the $V_{max}$ and a decrease of $K_m$ ($Ca^{2+}$) for PMCA. Calmodulin exhibits a high affinity interaction with PMCA ($K_d = 1\,nM$), a property that has been successfully used to affinity-purify PMCA from a number of tissues (e.g. NIGGLI et al. 1979). This mode of regulation of PMCA by the CBD region has similarities to SERCA regulation by phospholamban. Other mechanisms of regulation suggested for PMCA include: limited proteolysis particularly by calpain

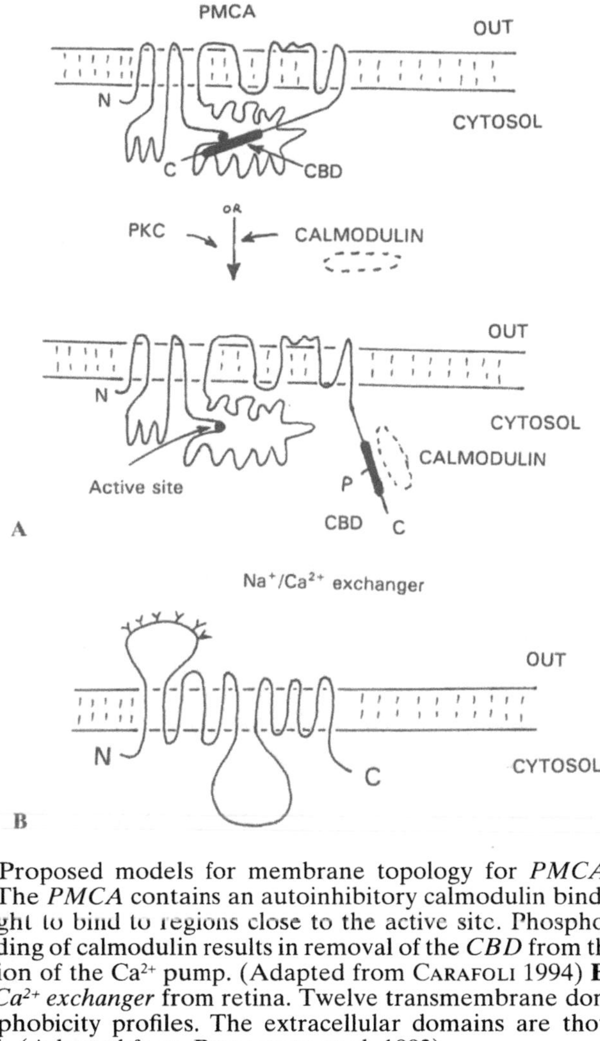


**Fig. 4A,B.** Proposed models for membrane topology for *PMCA* and *$Na^+/Ca^{2+}$ exchanger*. **A** The *PMCA* contains an autoinhibitory calmodulin binding domain (*CBD*) that is thought to bind to regions close to the active site. Phosphorylation (*p*) of the *CBD* or binding of calmodulin results in removal of the *CBD* from the active site region with activation of the $Ca^{2+}$ pump. (Adapted from CARAFOLI 1994) **B** Membrane topology of *$Na^+/Ca^{2+}$ exchanger* from retina. Twelve transmembrane domains are proposed from hydrophobicity profiles. The extracellular domains are thought to be heavily glycosylated. (Adapted from REILANDER et al. 1992)

acting near the C terminus (releasing the CBD region; AU 1986; JAMES et al. 1988), dimerisation again through the CBD region (VORHERR et al. 1991) and activation of the pump by acidic phsopholipids (which is readily demonstrated in the C terminus truncated $Ca^{2+}$ pump), a property that is also shared by SERCAs.

## V. Platelet PMCAs

Unlike the SERCAs, we know relatively little about the properties of platelet PMCAs. Their presence was suggested from studies of intact cells labelled with fluorescent indicators examining rates of $Ca^{2+}$ efflux after activation

(Pollock et al. 1987; Yoshida and Nachmias 1987; Johansson and Haynes 1988; Johansson et al. 1992). Only recently has identification at the protein level been made using antibodies, and its enzymatic insensitivity to Tg demonstrated (Cheng et al. 1994; Bokkala et al. 1995; Dean and Quinton 1995). That such a characterisation has only recently been reported is evidence to its low level of expression and to the fact that this is an activity easily lost during membrane preparation.

Using intact cells, Pollock et al. (1987) reported that thrombin and phorbol esters stimulated $Ca^{2+}$ efflux under conditions in which uptake into intracellular organelles had been short circuited by ionomycin treatment. Addition of either thrombin or phorbol ester to the peak of $Ca^{2+}$ elevation caused by ionomycin resulted in a faster rate of return towards basal $Ca^{2+}$ levels than in controls. This effect was suggested to occur via a PKC mediated mechanism acting on a PMCA. A similar approach using Quin II overloaded platelets measuring rates of cystolic $Ca^{2+}$ decrease after elevation was used to demonstrate that PMCAs were important for $Ca^{2+}$ efflux in resting platelets due to their high affinity for $Ca^{2+}$ and that elevated cAMP and cGMP stimulated $Ca^{2+}$ extrusion by an apparent increase of the $V_{max}$ without affecting the $K_m$ (Johansson and Haynes 1992; Johansson et al. 1992). Therefore, a mechanism for $Ca^{2+}$ extrusion at the plasma membrane could be demonstrated that was stimulated by PKC and cyclic nucleotide elevation.

A number of studies have attempted to correlated activities ascribed to PMCA with blood pressure in patients with hypertension in whom increased levels of cytosolic $Ca^{2+}$ are seen in resting platelets. Resink et al. (1985) reported increased $Ca^{2+}$ ATPase activity in membranes that was calmodulin stimulated and inhibited by orthovanadate and suggested that a higher PMCA may reflect a mechanism to protect the cell against $Ca^{2+}$ overload. However, other investigators have reported opposite findings with decreased PMCA activities in hypertensive patients (Takaya et al. 1990; Dean et al. 1994). In the latter study Dean et al. (1994), using Tg to exclude SERCA activity, reported that a decrease of PMCA activity correlated with diastolic blood pressure and suggested that a higher cytoplasmic $Ca^{2+}$ concentration may be the direct cause of a decreased PMCA in these patients. A further complication has been presented by Zidek et al. (1992) who have reported the presence of a humoral factor in the plasma of hypertensive individuals that leads to inhibition of PMCA. The nature of this factor remains to be determined.

Immunodetection of low levels of PMCA has recently been reported in platelet membranes purified using different techniques (Cheng et al. 1994; Bokkala et al. 1995; Dean and Quinton 1995). For this purpose the antibody 5F10 (or MA3-914) has been used extensively, and with platelet mixed and purified plasma membranes (but not intracellular membranes), immune recognition on Western blots of two bands at 150 and 135 kDa have been observed (Bokkala et al. 1995). Whether these reflect two isoforms for the PMCAs present in platelets (arising from splicing events or post-translational changes) or whether the smaller molecular-size species is a product of the larger, is not known, but both peptides were also detected in red-cell membranes. Recently the red cell membrane has been shown to contain PMCA 1 and 4 (STAUFFER et al. 1995), suggesting the platelet may also have a similar distribution. Additionally, the $Ca^{2+}$ ATPase activity in purified platelet plasma membranes has been demonstrated to be Tg insensitive, confirming its PMCA identity. This demonstration should now allow its better characterisation and also the determination of its relative contribution to the removal of $Ca^{2+}$ from the platelet cytosol.

## VI. $Na^+/Ca^{2+}$ Exchanger

Platelets contain a $Na^+/Ca^{2+}$ exchanger whose relevance in $Ca^{2+}$ efflux is generally thought to be of less importance than the PMCA. Its reduced popularity has been attributed to its lower affinity for $Ca^{2+}$ ($K_m = 1$–$20\,\mu M$) compared to PMCA; however, its large capacity may render it very important if it is located close to areas of high $Ca^{2+}$ concentration. Its activity has been easily demonstrated in both intact platelets and isolated membrane preparations (e.g. BRASS 1984; RENGASAMY et al. 1987; JOHANSSON and HAYNES 1988). BRASS (1984) suggested that the $Na^+/Ca^{2+}$ exchange mechanism was an important regulator of the cystolic pool of $Ca^{2+}$ using $^{45}Ca$ labelling experiments. Unstimulated platelets resuspended in $Na^+$-substituted medium showed an increased rate of $Ca^{2+}$ flux across the plasma membrane, an increase in the size of the intracellular storage pools with decreased efflux of $Ca^{2+}$. Efflux of $Ca^{2+}$ mediated by the $Na^+/Ca^{2+}$ exchanger has been resolved in rhod-2 labelled platelets with $K_m = 2.3$–$6.7\,\mu M$ with a $V_{max}$ approximately two times that of the extrusion pump (JOHANSSON and HAYNES 1988). Studies of the exchanger on isolated membranes have confirmed the characteristics observed with intact cells and localized the activity to plasma membranes (RENGASAMY et al. 1987; PATEL and AUTHI, unpublished observations).

Recently, two types of $Na^+/Ca^{2+}$ exchangers have been classified from cloning and mechanistic studies examining the stoichiometry of $Ca^{2+}$ exchange. Most tissues contain the cardiac type exchanger which exchanges three $Na^+$ for one $Ca^{2+}$. Cloning studies suggest the protein to contain 970 amino acids with a molecular size of 108 kDa (NICOLL et al. 1990) that is observed to migrate as both a 120 and 160 kDa-size species on SDS-PAGE. The second type of exchanger (retinal type) which has a very limited tissue distribution has been well studied in rod photoreceptors. It transports $K^+$ as well as $Na^+$ and $Ca^{2+}$ with a stoichiometry of four $Na^+$ for one $Ca^{2+}$ and one $K^+$. Cloning studies suggest the size of the exchanger protein to be 130 kDa (1149 amino acids) but the functionally expressed protein migrates as a 210 kDa band on SDS-PAGE (REILANDER et al. 1992). The purified exchanger protein from rod outer segments also migrates on SDS-PAGE as a 220 kDa protein, a discrepancy that may arise due to heavy glycosylation of the protein in addition to an acidic domain present in the molecule at residues 950–994, giving rise to anomalous migration characteristics. The retinal type $Na^+/Ca^{2+}$ exchanger has only a very limited sequence homology to the cardiac type exchanger in addition to the mechanistic differences, but the hydropathy profiles are similar with both proteins thought to have a similar membrane topology (Fig. 4B). Recently, the $Na^+/Ca^{2+}$ exchanger in platelets has been suggested to be similar to the retinal type protein transporting $K^+$ and $Ca^{2+}$ in exchange for $Na^+$ (KIMURA et al. 1993). The possibility that the $Na^+/Ca^{2+}$ exchange mechanism may be regulated by protein phosphorylation has been suggested from studies in aortic smooth muscle cells (expressing the cardiac type exchanger) where PMA via stimulation of PKC caused a twofold increase of $Na^+/Ca^{2+}$ exchange activity (VIGNE et al. 1988). Only limited studies have been published examining the platelet exchanger with JOHANSSON et al. (1992) reporting no increase in the activity of $Na^+/Ca^{2+}$ exchange after elevation of intracellular cAMP.

## C. Mechanisms of $Ca^{2+}$ Elevation in Platelets

### I. $Ca^{2+}$ Release from Intracellular Stores

#### 1. Inositol (1,4,5) Trisphosphate

The role of In(1,4,5)$P_3$ in causing $Ca^{2+}$ release from platelet intracellular stores is well established and documented. Its formation by agonists from $PIP_2$ and metabolism via phosphorylation to In(1,3,4,5)$P_4$ and dephosphorylation to In(1,3,4)$P_3$ and subsequently to inositol have been well described (DANIEL et al. 1989; KING et al. 1990; for a review see DANIEL 1990). However, its role in $Ca^{2+}$ influx and the relationship of the intracellular pool of $Ca^{2+}$ sensitive to In(1,4,5)$P_3$ to other pools is still poorly defined. Recent studies have focussed on the properties of the $IP_3$Rs, characteristics of the $Ca^{2+}$ release mechanism and its relationship to $Ca^{2+}$ entry mechanisms.

#### 2. General Properties of $IP_3$Rs

Recent studies of the characterisation of $IP_3$Rs have revealed considerable information. Presently the primary structures of three different receptor types in the $IP_3$R gene family ($IP_3$ R 1, 2 and 3 genes, located on different chromosomes) have been determined with a further possible fourth that remains to be studied (FURUICHI et al. 1989; MIGNERY et al. 1990; SUDHOF et al. 1991; ROSS et al. 1992). At the protein level approximately 60%–70% sequence homology exists between the different types. There is further heterogeneity of the type I $IP_3$R arising due to alternate splicing events at two regions (SI and SII) contributing to additional 15 and 40 amino acids, respectively (NAKAGAWA et al. 1991; DANOFF et al. 1991).

The $IP_3$Rs are approximately 300 kDa in size, migrating on SDS-PAGE as 230–250 kDa bands. Four subunits constitute a $Ca^{2+}$ channel. There is considerable interest as to whether the $Ca^{2+}$ channels comprise of homo- or heterotetramers. Immuno-affinity purified $IP_3$R 1 can form tetramers that induce $Ca^{2+}$ translocation in liposomes. Additionally, immunoprecipitation of $IP_3Rs$ using isoform-specific antibodies from tissues that contain multiple $IP_3$ R isoforms not only isolates the type of receptor that the antibody recognises but can also co-immunoprecipitate other types, suggesting the existence of heterotetramers (MONKAWA et al. 1995).

Functionally, the $IP_3R$ protein can be divided up into three parts, an $In(1,4,5)P_3$ binding domain at the N terminus, a channel portion at the C terminus, and a regulatory domain coupling the two terminal domains. Since the $In(1,4,5)P_3$ binding domain is at opposite ends to the channel domain, ligand binding would induce considerable conformational changes that lead to opening of the $Ca^{2+}$ channel. The regulatory domain has concensus sites for phosphorylation by cAMP-PK, cGMP-dependent protein kinase (cGMP-PK), calmodulin binding (type 1 receptor), $Ca^{2+}$ and ATP binding that modulate receptor function. $IP_3Rs$ type I and type II contain putative sites for phosphorylation by tyrosine kinases (HARNICK et al. 1995). The $In(1,4,5)P_3$ binding domain was found to reside in the N terminal 650 amino acids (MIYAWAKI et al. 1991), and recently the binding site has been localised to residues 476–501 using a photoaffinity analogue of $In(1,4,5)P_3$ (MOUREY et al. 1993). Overexpression of the ligand binding domains of the different $IP_3Rs$ or of the whole proteins reveals distinct binding affinities for $In(1,4,5)P_3$ in the order $IP_3R$ type II > $IP_3R$ type I > $IP_3R$ type III (NEWTON et al. 1994; MARANTO 1994; MAEDA et al. 1990). The channel domain is currently thought to contain six membrane spanning domains (M1 to M6) that exhibit high sequence homology with the channel domains of ryanodine receptors, the $Ca^{2+}$ channels of skeletal and cardiac muscle. Regions between the putative M5 and M6 domains contain two N-linked glycosylation sites (Asn 2475 and Asn 2503) and a pore forming region similar to pore regions of other cation channels (both voltage gated and cyclic nucleotide gated; MICHIKAWA et al. 1994), a characteristic of the superfamily of ion channels. Studies from isolated membranes rich in $IP_3Rs$ and with the purified receptors reconstituted into liposomes have demonstrated the receptors to have single channel conductance (of 10 pS–26 pS) co-activated by $Ca^{2+}$ and $In(1,4,5)P_3$ with four opening states that may be modulated by ATP (EHRLICH and WATRAS 1988; FERRIS and SNYDER 1992; BEZPROZVANNY et al. 1991; MAEDA et al. 1991; BEZPROZVANNY and EHRLICH 1993).

## 3. Platelet $IP_3Rs$

Recent studies suggest platelets to contain more than one isoform of the $IP_3R$. Using an antibody to the mouse type I $IP_3R$ recognising a C terminal peptide, BOURGUIGNON et al. (1993a) demonstrated saponin-permeabilised platelets contained an $IP_3R$ which with immunohistochemical staining was closely localized to the periphery of the cell and associated with the cytoskeleton. Additionally, cytochalasin D and colchicine (microfilament and microtubule inhibitors, respectively) inhibited by 50% both $In(1,4,5)P_3$ binding and $Ca^{2+}$

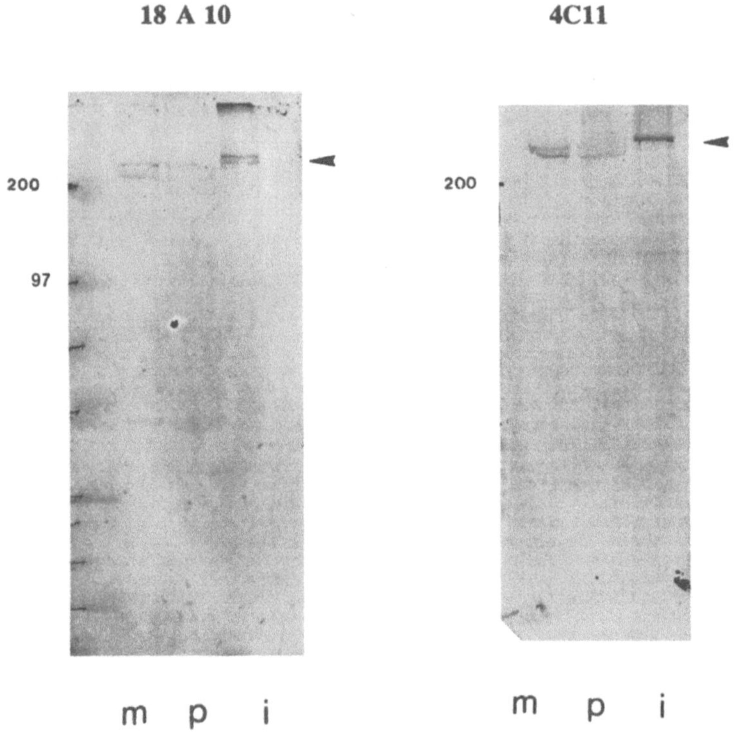


**Fig. 5.** Distribution of $IP_3Rs$ in platelet plasma and intracellular membranes prepared using free flow electrophoresis. Platelet mixed membranes (*m*), plasma membranes (*p*) and intracellular membranes (*i*) are subjected to SDS-PAGE followed by Western blotting with either *18A10* antibody recognising $IP_3R$ 1 or *4C11* antibody recognising both $IP_3R$ 1 and $IP_3R$ 3. *200* reflects molecular size marker. For full details see text and El-Daher et al. 1996

release. These studies are supported by the finding that a small proportion of $IP_3Rs$ is associated with the cytoskeletal protein ankyrin that links membrane proteins to the actin-associated microfilaments (Joseph and Samanta 1993). Thus, the $IP_3R$ may exhibit contact points for components of the cytoskeleton that serve to anchor the In(1,4,5)$P_3$ sensitive organelle at specific regions in the cytosol. An antibody raised to a sequence similar for all isoforms in the N-terminal region was used to detect the $IP_3R$ in intracellular membranes isolated using glycerol density gradients (Dean and Quinton 1995), with no staining seen in the plasma membranes. In our study using plasma and intracellular membranes prepared using free flow electrophoresis (Fig. 5), the antibody 18A10 (known to recognise type 1 $IP_3R$) detected a doublet band at approximately 250 kDa with intracellular membranes, and there was no recognition with plasma membranes, but the antibody 4C11 which also detects the protein with intracellular membranes additionally recognised low levels of a 230 kDa protein on plasma membranes that was not a contaminant from intracellular membranes (El-Daher et al. 1996). As the antibody 4C11 is thought not to be specific for the type I $IP_3R$ but also recognises type III $IP_3R$, this suggests that platelets additionally contain the type III receptor. We have recently confirmed this suggestion using the antibodies CT1 and CT3 that recognise the type I and type III $IP_3$ receptors, respectively. CT1 only detects the protein in intracellular membranes, and CT3 only detects the protein in plasma membranes (PATEL et al., to be published). This suggests that in platelets there is a distinct localisation and by implication a differentiation of function for $IP_3$ receptor isoforms with the type I receptor involved in $Ca^{2+}$ release from intracellular stores and the type III receptor involved in $Ca^{2+}$ influx.

The type I $IP_3R$ has been recently purified from human platelets and has been suggested to have distinct properties to the same from rat brain (O'ROURKE et al. 1995). Firstly, it binds very poorly to concanavalin A and other lectins, and secondly, it was not significantly phosphorylated by cAMP-PK. The latter has important implications in "cross-talk regulation" by other second messengers and adds to the controversy concerning the mechanism of action of cAMP on $Ca^{2+}$ homeostasis. Increased cAMP is inhibitory to $Ca^{2+}$ elevation in platelets with suggested stimulatory effects at the level of $Ca^{2+}$ ATPases (see earlier) and inhibitory ones at the level of $IP_3Rs$. All three human $IP_3R$ types contain concensus phosphorylation sites for cAMP-PK at different serines in the primary structure (type I–1589, 1755; type II–1687; type III–934, 1133; FURUICHI et al. 1994; YAMAMOTO-HINO et al. 1994); however, so far only the type I $IP_3R$ has been demonstrated to be phosphorylated by cAMP-PK and cGMP-PK (SUPATTAPONE et al. 1988; YAMAMOTO et al. 1989; FERRIS et al. 1991; KOMALAVILAS and LINCOLN 1994). Phosphorylation of the purified receptor from rat brain leads to a decrease of $In(1,4,5)P_3$ induced $Ca^{2+}$ release (SUPATTAPONE et al. 1988); however, immuno-purified type I $IP_3R$ (using specific anti-type I antibodies) showed an increase of $Ca^{2+}$ flux induced by $In(1,4,5)P_3$ after cAMP-PK phosphorylation (NAKADE et al. 1994), an effect also seen with the $IP_3R$ from rat liver (JOSEPH and RYAN 1993). In platelets the effects on $In(1,4,5)P_3$-induced $Ca^{2+}$ release are even more controversial with cAMP-PK reportedly shown to stimulate $In(1,4,5)P_3$-induced $Ca^{2+}$ release (ENOUF et al. 1987), to cause modest (30%) inhibition (QUINTON and DEAN 1992), an inhibitory effect of cAMP independent of cAMP-PK (TOHMATSU et al. 1989), or to cause no effect at all (O'ROURKE et al. 1989). Recently, QUINTON et al. (1996) have demonstrated phosphorylation of the type I $IP_3R$ in platelet membranes by both endogenous kinases and cAMP-PK. An adequate explanation of these contrasting effects is not currently available, except perhaps that different preparation techniques may yield this large protein in varying altered conformation states that may restrict the ability of kinases to phosphorylate the protein. With the studies of O'ROURKE et al. (1995), it is possible that the phosphorylation site has an altered conformation and a comparison with that from rat brain may not be applicable. This is due to the latter containing the larger spliced version (type Ia) that has been shown to be phosphorylated at two sites by cAMP-PK compared to the type I receptor from peripheral tissues that is the shorter version (type Ib) apparently phosphorylated at only one of these sites (DANOFF et al. 1991). With platelet intracellular membranes, we have observed phosphorylation of a 250kDa protein consistent with detection of the protein band with either 18A10 or 4C11 antibodies (EL-DAHER et al. 1996).

### 4. In(1,4,5)$P_3$-Induced $Ca^{2+}$ Release from Platelet Intracellular Stores

The characteristics of In(1,4,5)$P_3$ mediated $Ca^{2+}$ release from platelet intracellular stores are similar to those in other cells, i.e. it is rapid (O'ROURKE et al. 1985; AUTHI and CRAWFORD 1985; ADUNYAH and DEAN 1985), exhibits "quantal" phenomenon (all or nothing emptying of intracellular stores) with the extent, not rate, dependent on the concentration of In(1,4,5)$P_3$ applied. The quantal release phenomenom may represent a property of the protein as the purified $IP_3R$ reconstituted into liposomes also shows quantal release (FERRIS et al. 1992). The release event is also potentiated by agents that interact with sulfhydryl groups (e.g. thimerosal), sensitizing the $Ca^{2+}$ channel (ADUNYAH and DEAN 1986). Recently, both thimerosal and mercury ($HgCl_2$) have been shown to induce $Ca^{2+}$ elevation and potent expression of procoagulant activity in platelets (HECKER et al. 1989; GOODWIN et al. 1995). The $Ca^{2+}$ release event has been shown to be coupled to hydrolysis of In(1,4,5)$P_3$ which may contribute to the quantal process (EBERHARD and ERNE 1993). In(1,4,5)$P_3$ induced $Ca^{2+}$ release shows sensitivity to cytoplasmic $Ca^{2+}$ in a biphasic manner (AUTHI and CRAWFORD 1985) similar to observations with ion channel studies (BEZPROZVANNY et al. 1991). A large number of studies in other cells has demonstrated that luminal $Ca^{2+}$ regulates the sensitivity of stores to In(1,4,5)$P_3$ with high concentrations promoting $Ca^{2+}$ release and low concentrations inhibitory (e.g. MISSIAEN et al. 1992). This property is also true for platelets as cells with overloaded intracellular stores (obtained by incubating platelets in a $Na^+$ free medium) show a greater thrombin induced $Ca^{2+}$ mobilisation than control cells (KIMURA et al. 1994). In membrane preparations In(1,4,5)$P_3$ induced $Ca^{2+}$ release is also significantly inhibited by agents that block $Ca^{2+}$-activated $K^+$ channels (apamin), suggesting that a counterflow of $K^+$ ions occurs in tandem with $Ca^{2+}$ release induced by In(1,4,5)$P_3$, maintaining electroneutrality (O'ROURKE et al. 1994). In the latter study the authors identified a 63kDa protein that may serve to regulate $K^+$ permeability. In saponin permeabilized platelets the $Ca^{2+}$ release induced by In(1,4,5)$P_3$ has been shown to be tightly linked to activation of phospholipase $A_2$, synthesis of thromboxane $B_2$ leading to the stimulation of aggregation, and secretion of dense granule constituents (AUTHI et al. 1986, 1987; WATSON et al. 1986).

## II. Nature of $Ca^{2+}$ Release from Intracellular Stores

Measurements of cytosolic $Ca^{2+}$ variation in resuspended cells labelled with fluorescent indicators such as Fura 2 indicate that agonist addition results in a rapid elevation of $[Ca^{2+}]_i$. Measurement of $Ca^{2+}$ elevation in single cells reveals the $Ca^{2+}$ signal to exhibit complex spatial and temporal aspects (reviewed by BERRIDGE 1993). A repetitive pattern of $Ca^{2+}$ spikes is observed whose frequency is sensitive to agonist concentration and the level of external $Ca^{2+}$. There are many different oscillatory patterns that vary from cell to cell and also within cells, responding to different agonists. Generally, $Ca^{2+}$ oscillations occur either as sinusoidal, where fluctuation occurs on top of an elevated level of calcium, or transient, where the spikes originate from near base-line levels of cystolic $Ca^{2+}$. Examples of cell types exhibiting $Ca^{2+}$ oscillations include: parotid and pancreatic acinar cells for the former type and hepatocytes, endothelial cells and rat chromaffin cells for the latter type. A number of models have been proposed to explain $Ca^{2+}$ oscillations (RINK and JACOB 1989; TSIEN and TSIEN 1990; MEYER and STRYER 1991; BERRIDGE 1991); only two will be mentioned in this review as many of these represent variations of similar themes. MEYER and STRYER (1991) suggested with a surface-receptor controlled model that the onset of each peak results from $In(1,4,5)P_3$ formation with the released $Ca^{2+}$ stimulating the activation of phospholipase C via positive feedback system. Termination of the signals is established by sequestration of the cation into intracellular stores. Thus, fluctuations of $In(1,4,5)P_3$ represent an important component in this model and have been demonstrated to occur in a number of cell-types, e.g. fibroblasts (HAROOTUNIAN et al. 1991). Alternatively, BERRIDGE (1991) proposed a two pool model containing $In(1,4,5)P_3$ sensitive and insensitive pools. The $In(1,4,5)P_3$ sensitive pool is situated close to the plasma membrane from which the initial $Ca^{2+}$ release occurs after receptor occupancy. The released $Ca^{2+}$ is taken up by an $IP_3$ insensitive pool, and as the store overloads itself, $Ca^{2+}$ release occurs via a CICR mechanism, leading to a cycle of uptake and release events. CICR is suggested to occur in a manner similar to that which occurs in cardiac muscle where the CICR channels or ryanodine receptors (RYR) have been well characterised. The RYR, so called because they bind the plant alkaloid ryanodine (FLIESCHER and INUI 1989), represent the second type of intracellular $Ca^{2+}$ release channel. Currently, three genes are known to encode the proteins: RYR 1, RYR 2 and RYR 3 genes. At the protein level RYR 1 and RYR 2 have been well studied and are expressed predominantly in skeletal muscle (RYR 1), cardiac muscle and brain (RYR 2). Most of the information for RYR 3 is still at the level of the cDNA, but transcripts have been detected in a variety of tissues (HAKAMATA et al. 1994). Thus, RYRs may have a widespread distribution. Like the $IP_3Rs$ the RYRs are large proteins (500 kDa) with four subunits making a channel, and the proteins contain regions of high sequence homology with the $IP_3Rs$ particularly near the C terminal channel regions. Recently, cyclic ADP ribose (cADPr), a metabolite of $NAD^+$, has been found to cause $Ca^{2+}$ release from a number of invertebrate and vertebrate tissues via an $In(1,4,5)P_3$ independent mechanism (GALIONE et al. 1991; KOSHIYAMA et al. 1991). The $Ca^{2+}$ releasing action of cADPr is potent and stimulated by calmodulin (LEE et al. 1994). It has been suggested to interact with the RYR from cardiac (RYR 2) but not skeletal muscle (Meszaros et al. 1993), though this has not been confirmed by other workers (Fruen et al. 1994; Sitsapesan et al. 1994), and photoaffinity labeling with an analogue of cADPr binds to proteins of a size smaller than the RYRs (Walseth et al. 1993). While many details concerning the action of cADPr remain to be worked out (including whether intracellular concentrations of cADPr can be regulated by agonists, and does it merely serve as a cofactor with $Ca^{2+}$ for the CICR), the possibility that it may represent the mechanism for $Ca^{2+}$ release from $In(1,4,5)P_3$ insensitive stores is important.

The $Ca^{2+}$ signals have also been observed to be spatially organised as waves that originate at a particular locus and migrate through the cell as a regenerative wave, sometimes even crossing between cells. That $In(1,4,5)P_3$ is the important initiating stimulus for these complex $Ca^{2+}$ signals is shown by the fact that both heparin (which inhibits $In(1,4,5)P_3$ binding) and the $IP_3R$ antibody 18A10 (which inhibits $In(1,4,5)P_3$ induced $Ca^{2+}$ release) block the propagation of $Ca^{2+}$ signals (Miyazaki et al. 1992a,b).

## III. $Ca^{2+}$ Oscillations in Platelets

Fewer studies have been carried out measuring $Ca^{2+}$ signals on single platelets than other cells, principally because of the size of the cell limiting the resolution that can be obtained and probably also because the process of immobilisation of platelets to surfaces will induce partial activation. However, all of these studies suggest that platelets behave in a similar manner as other cells. Firstly, the studies suggest that platelets immobilised to adherent surfaces such as poly-ethyleneimine (Nishio et al. 1991), cell-Tak (Tsunoda et al. 1991), fibronectin (Ozaki et al. 1992a) or fibrinogen (Heeemskerk et al. 1992, 1993) respond to ADP, serotonin or low concentrations of thrombin with transient $Ca^{2+}$ oscillations. The oscillations appear jagged with three-dimensional graphics showing multiple hot spots throughout the cytoplasm (Nishio et al. 1991). The oscillations were found to be stimulated in the presence of an inhibitor of PKC and inhibited in the presence of activators of PKC (Nishio et al. 1991) or agents that elevate cyclic nucleotides (Heemskerk et al. 1993). A role for $Ca^{2+}$ influx in $Ca^{2+}$ oscillations, which may serve to replenish intracellular stores, is not totally clear as it was reported to be essential for $Ca^{2+}$ oscillations (Nishio et al. 1991; Ozaki et al. 1992a) or to have only the effect of increasing the amplitude of the spikes (Heemskerk et al. 1992, 1993). An important experiment was that a single platelet can respond to sequential additions of two different agonists. The underlying mechanism for $Ca^{2+}$ oscillations may employ components of both the receptor controlled and two pooled models. Clearly, $Ca^{2+}$ oscillations are observed under conditions in which PLC activity is moderate (low thrombin concentrations, ADP, or serotonin stimulation) and $In(1,4,5)P_3$ levels may fluctuate. Additionally, there is evidence for oscillations of $In(1,4,5)P_3$ [and $In(1,3,4,5)P_4$] formation when platelets are stimulated with ADP, thrombin or shear forces under quenched flow conditions (RAHA et al. 1993) which provide support for the receptor controlled model. Resequestration by intracellular $Ca^{2+}$ pumps is necessary for oscillations as firstly addition of Tg after ADP abolishes oscillations induced by ADP (resulting in sustained $Ca^{2+}$ elevation) and Tg itself induces sustained elevations with little evidence of oscillations (HEEMSKERK et al. 1993). Thus, both fluctuations in PLC activity and functional $Ca^{2+}$ ATPases support $Ca^{2+}$ oscillations. It has been noticed that Tg induced $Ca^{2+}$ elevation is increased if the platelets have been pre-activated by agonists, iononycin or thimerosal which sensitises $Ca^{2+}$ channels (BRUNE and ULLRICH 1991; HEEMSKERK et al. 1993). HEEMSKERK et al. (1993) have suggested that $Ca^{2+}$ released by In(1,4,5)$P_3$ resequesters into a Tg sensitive pool which now enlarged may represent the major contributor for CICR in platelets. The identity of this pool requires further definition as it is responsive to a second addition of another agonist and $Ca^{2+}$ release from it via a CICR mechanism may be mediated by a putative RYR or the $IP_3R$ on which $Ca^{2+}$ is a co-stimulator. The effects of cADPr on platelet systems is still to be reported. Another agent, sphingosine-1-phosphate which has been shown to cause $Ca^{2+}$ elevation in a number of mammalian cells was reported to cause $Ca^{2+}$ mobilisation in human platelets leading to shape change and aggregation without causing secretion (YATOMI et al. 1995). It is apparently stored in platelets and released upon agonist stimulation. The pool of $Ca^{2+}$ mobilised with respect to In(1,4,5)$P_3$ or Tg sensitivity was not determined.

## IV. $Ca^{2+}$ Influx Mechanisms

As suggested in the introduction, $Ca^{2+}$ influx mechanisms may be classified into three main types (store regulated, second messenger operated or receptor operated). There is evidence in platelet studies for support of all three types; however, we are still far from fully understanding the details of any one. Most platelet agonists are believed to cause influx resulting as a consequence of intracellular events, i.e. generation of second messengers, with a measureable delay detected between agonist addition and initiation of $Ca^{2+}$ influx as measured using the stopped-flow technique; however, this is not the case with ADP where $Ca^{2+}$ influx occurs without a measurable delay and therefore may operate a ROC-like channel (SAGE and RINK 1987). There has been a considerable effort in trying to determine the intracellular events that regulate $Ca^{2+}$ influx.

## V. Capacitative or Store-Regulated $Ca^{2+}$ Entry

PUTNEY (1986) proposed the capacitance model for $Ca^{2+}$ entry in which the depletion of the intracellular stores by agonists [via In(1,4,5)$P_3$] somehow signalled the opening of the $Ca^{2+}$ channel in the plasma membrane so that the store may be replenished. It is a decrease of $Ca^{2+}$ concentration in the lumen of the intracellular store that signals the opening of the plasma membrane $Ca^{2+}$

channel (PUTNEY 1990). This mode of $Ca^{2+}$ entry has received particular support from studies using Tg (and tBuBHQ) which by depleting intracellular stores in the absence of second messenger generation, induce $Ca^{2+}$ entry (TAKEMURA et al. 1989). There is evidence for more than one type of $Ca^{2+}$ entry channel in non-excitable tissues (FASOLATO et al. 1994). A store regulated $Ca^{2+}$ current referred to as ICRAC ($Ca^{2+}$ release activated calcium current) has been reported in the plasma membranes of mast cells using patch clamping that has very low unitary conductance (20 fS) and high selectivity for $Ca^{2+}$ (HOTH and PENNER 1992; ZWEIFACH and LEWIS 1993). This channel has not been identified, but the transient receptor potential (*trp*) gene product in *Drosophila* photoreceptors which has a higher conductance level and less selectivity than ICRAC has been suggested to be a functional homologue (MONTELL and RUBIN 1989; MINKE and SELINGER 1991; HARDIE and MINKE 1993; SELINGER et al. 1993). Indeed, when expressed is Sf9 cells, it is sensitive to store depletion (VACA et al. 1994), and another trp-like (*trpl*) gene product expressed in Sf9 cells was found to be responsive to agonists of G-protein-coupled receptors (HU and SCHILLING 1995). Most of the plasma membrane $Ca^{2+}$ currents identified in non-excitable cells have higher conductance and a lower selectivity for $Ca^{2+}$ than ICRAC (FASOLATO et al. 1994). Recently, homologues referred to as TRP channel related proteins (TRPC 1 and TRPC 3) have been described in human tissues whose characterisation should provide valuable insights into the identity of ICRAC (WES et al. 1995). The TRPC 1 cDNA codes for an 810 amino acid protein that resembles an ion-channel structure containing six putative transmembrane domains, does not contain residues that confer voltage gating properties and has regions similar to those found in ankyrin (ankyrin repeats) and dystrophin that may mediate protein-protein interactions (WES et al. 1995). Thus, there may exist a family of non-voltage gated plasma membrane $Ca^{2+}$ channels.

The link between store depletion and the plasma membrane $Ca^{2+}$ channel is not known. There has been a proliferation of the number of possible routes and mechanisms that provide this link and include diffusable factors and protein coupling mechanisms. RANDRIAMAMPITA and TSIEN (1993) and PAREKH et al. (1993) proposed that depletion of intracellular stores causes the release of a soluble factor termed $Ca^{2+}$ influx factor (CIF) (RANDRIAMAMPITA and TSIEN 1993) which may regulate the opening of the plasma membrane $Ca^{2+}$ channel. The structure of CIF has not been clarified but is postulated to be a low molecular weight substance containing a phosphate group and in its original description could diffuse across membranes as its action was apparent on intact cells. The action of CIF may also involve a phosphorylation mechanism as protein phosphatase inhibitors such as okadaic acid affected the responsiveness of cells to CIF, and the phosphate group of CIF required as phosphatase treatment abolishes activity (RANDRIAMAMPITA and TSIEN 1993, 1995; PAREKH et al. 1993). While the experiments of the above authors have been reproduced by other workers (DAVIES and HALLETT 1995; THOMAS and HANLEY 1995), Putney's group (GILON et al. 1995) have suggested that there are impurities in the acid extracts of CIF that induce $Ca^{2+}$ responsiveness by acting at surface receptors generating $In(1,4,5)P_3$ and whose activity could be blocked by intracellular injection of heparin. KIM et al. (1995) who also reported impurities have been able to further purify by hptlc a substance ($M_r$ = 600 Da) that reportedly only induced $Ca^{2+}$ entry when applied intracellularly to *Xenopus* ocytes. This preparation has been reported to activate a non-selective cation conductance that appears unrelated to the properties of ICRAC, casting questions as to whether it may represent the link between the stores and ICRAC (CLAPHAM 1995).

The role of protein phosphatases (PP) in $Ca^{2+}$ influx has been recently examined in detail in human platelets (KOIKE et al. 1994). Calyculin and tautomycin which inhibit PP 1 and 2A with equal potency suppressed $Ca^{2+}$ influx induced by thrombin without an effect on $Ca^{2+}$ mobilisation from intracellular stores. Okadaic acid, which is more potent at inhibition of PP 2A than PP 1, suppressed both $Ca^{2+}$ mobilisation and influx, suggesting a more prominant role for PP 1 in $Ca^{2+}$ influx. In experiments that effectively separated $Ca^{2+}$ mobilisation from $Ca^{2+}$ influx, thrombin is added to platelets in the absence of external $Ca^{2+}$ to release intracellular stores, followed by $Ca^{2+}$ addition after 6 min that results in $Ca^{2+}$ influx. Under these conditions all three inhibitors added 3 min prior to $Ca^{2+}$ suppressed influx. However, with thrombin mediated $Mn^{2+}$ entry and Tg mediated $Ca^{2+}$ entry all three inhibitors were less effective. There results suggest that store depletion may induce two types of inward $Ca^{2+}$ currents: one with a higher selectivity for $Ca^{2+}$ than $Mn^{2+}$ and sensitive to PP inhibition, and the other with a lower selectivity to $Ca^{2+}$ (allowing $Mn^{2+}$ to enter avidly) that is less sensitive to PP inhibition. With respect to other cations, thrombin or ionomycin treatment led to the entry of $Ca^{2+}$, $Sr^{2+}$ or $Ba^{2+}$ (OZAKI et al. 1992b). These cations can also fill the stores that have been depleted, yet only $Ca^{2+}$ or $Sr^{2+}$, but not $Ba^{2+}$, inhibited the entry of $Ca^{2+}$ as measured by $Mn^{2+}$ entry (OZAKI et al. 1992b). Thus, there is selectivity with the $Ca^{2+}$ sensing mechanism in the store lumen that regulates $Ca^{2+}$ entry. In other studies thrombin has been suggested to open two $Ca^{2+}$ channels on the basis of experiments with wortmannin (HASHIMOTO et al. 1992). At micromolar concentrations wortmannin inhibits myosin light chain kinase in addition to inhibition of phosphatidylinositol-phosphate 3-kinase (PI-3-K). Pre-incubation of platelets with wortmannin inhibited the plateau of the $Ca^{2+}$ signal induced by thrombin without an effect on the initial $Ca^{2+}$ mobilisation from intracellular stores or initial $Ca^{2+}$ entry. Wortmannin may inhibit the opening of a late cation channel coinciding with activation of myosin light chain kinase activation (HASHIMOTO et al. 1992).

Capacitative $Ca^{2+}$ entry has also been reported to include a step involving GTP hydrolysis (non-hydrolysable analogues of GTP, e.g. GTP[S], are inhibitory) and probably involves a small G protein (BIRD and PUTNEY 1993; FASOLATO et al. 1993; JACONI et al. 1993); however, the identity of the G protein remains to be clarified. It is of interest that the newly identified $In(1,3,4,5)P_4$ receptor is a member of the GTPase activating protein family (see below).

A number of protein candidates have been suggested that may provide a protein coupling link between stores and the $Ca^{2+}$ channel in the plasma membrane. These have included cytochome P-450, a 130kDa protein (vinculin) and the $IP_3R$ itself. Cytochrome P-450 was suggested on the basis that inhibitors of cytochrome P-450 such an econazole cause inhibition of both $Ca^{2+}$ and $Mn^{2+}$ entry (ALONSO et al. 1991; ALVAREZ et al. 1991). However, this finding was not fully supported by studies from SARGEANT et al. (1992) who reported that cytochome P-450 inhibitors, though effective against Tg induced cation entry, were relatively ineffective at inhibiting agonist mediated cation entry, and studies using the stopped flow techniques showed no effect on the initial phase of the $Ca^{2+}$ signal. A further report suggested that econazole-inhibited Tg induced $Ca^{2+}$ influx by a mechanism other than cytochome P-450 inhibition (VOSTAL and FRATANTONI 1993).

A 130kDa protein in human platelets has been implicated to signal information from the $Ca^{2+}$ stores to the plasma membrane (VOSTAL et al. 1992). $Ca^{2+}$ release from the stores led to the tyrosine phosphorylation of the 130kDa protein and $Ca^{2+}$ entry. Once the stores were full again, a tyrosine phosphatase activity (possibly in the stores?) dephosphorylates this protein, triggering an end to $Ca^{2+}$ influx. This protein was later identified as vinculin (VOSTAL and SHULMAN 1993) which is present in the platelet cytosol in abundance and is known to bind both talin and actin. Since these studies, tyrosine kinase inhibitors have been found to inhibit $Ca^{2+}$ entry mediated by both agonists and $Ca^{2+}$ ATPase inhibitors in platelets and other cells with the inhibitors shown not to have an affect on PLC activation or store depletion (SARGEANT et al. 1993a,b; JENNER et al. 1994; LEE et al. 1993). Furthermore, refilling of the stores with either $Ca^{2+}$ or $Sr^{2+}$ (but not with $Ba^{2+}$) leads to the tyrosine dephosphorylation of proteins and an inhibition of $Ca^{2+}$ entry (OZAKI et al. 1992b; JENNER et al. 1994). Platelets in particular are rich in tyrosine kinases, but at present there appears no information concerning the identity of the tyrosine kinase involved in $Ca^{2+}$ influx or the tyrosine phosphatase suggested to terminate the signal.

The $IP_3R$ has also been suggested to be involved in protein coupling mechanisms between the stores and a $Ca^{2+}$ channel in the plasma membrane initially thought to be, or associated with, a putative $IP_4R$ in the plasma membrane (IRVINE 1990). The proposal was largely based on known interactions between the ryanodine receptor and the dihydropyridine receptor in skeletal muscle. The $IP_3R$ serves a similar function as the ryanodine receptor in mediating $Ca^{2+}$ release, has considerable sequence homology (particularly in the channel regions), and in some cell systems $In(1,4,5)P_3$ and $In(1,3,4,5)P_4$ synergised to promote $Ca^{2+}$ entry (MORRIS et al. 1987). Support for this model has been further provided from studies of purified platelet plasma and intracellular membranes where a high-affinity receptor for $In(1,3,4,5)P_4$ has been localised to plasma membranes with the predominant localisation for $IP_3Rs$ on intracellular membranes (AUTHI 1992; CULLEN et al. 1994). The $In(1,3,4,5)P_4$ receptor has recently been purified (CULLEN et al. 1995a), cloned

(CULLEN et al. 1995b) and identified as a member of the GAP1 family (see below). A protein interaction between the $IP_3R$ and $IP_4R$ is still to be demonstrated.

There is increasing evidence that components of the cytoskeleton may play an important role in linking In(1,4,5)$P_3$ sensitive pools to the plasma membranes. In earlier studies a number of investigators found In(1,4,5)$P_3$ binding sites on membranes that co-migrated with plasma-membrane enriched fractions (DUNLOP and LARKINS 1988; GUILLEMETE et al. 1988). Later it was found that treatments that caused disruptions in the cytoskeleton separated In(1,4,5)$P_3$ binding activity from fractions enriched in plasma membranes and led to the proposal that the cytoskeleton was involved in linking the two structures (ROSSIER et al. 1991; SHARP et al. 1992; LIEVREMONT et al. 1994). As indicated above, the $IP_3Rs$ from rat brain and T-lymphoma cells have been reported to bind ankyrin (JOSEPH and SAMANTA 1993; BOURGUIGNON et al. 1993, 1995a), and the *trp*, *trpl* and human TRPC 1 products were also postulated to contain ankyrin binding motifs (WES et al. 1995) that may provide a link between the $Ca^{2+}$ channel in the plasma membrane to the $IP_3R$. With the T-lymphoma cells the binding of ankyrin to the $IP_3R$ has been reported to inhibit $Ca^{2+}$ release and the binding of In(1,4,5)$P_3$ to its receptor, with the protein interaction taking place on an 11 amino-acid stretch near the C terminal region at a sequence that is similar for all $IP_3Rs$, and reported to lie between the putative M5 and M6 transmembrane domains (BOURGUIGNON et al. 1995a). Interestingly, this suggests that this region of the $IP_3R$ should be on the cytoplasmic side of the organelle and not the luminal side as earlier thought (MICHIKAWA et al. 1994) and that the In(1,4,5)$P_3$ binding region which resides in the N terminal part of the protein is probably either spatially close to the C-terminal "channel" part of the molecule, or ankyrin binding induces conformational changes that lead to inhibition of In(1,4,5)$P_3$ binding. Furthermore, ankyrin has been shown to bind to the RYR in T-lymphoma cells via the same stretch of amino acid residues with also a loss of ryanodine binding (BOURGUIGNON et al. 1995b). This may suggest that an interaction of the $IP_3R$ with a plasma membrane protein via ankyrin binding motifs probably occurs in the absence of $Ca^{2+}$ release occurring through the $IP_3R$, effectively freeing the protein to promote $Ca^{2+}$ influx through conformational changes.

## VI. Possible Direct Involvement of Second Messengers in $Ca^{2+}$ Entry

A number of studies since 1987 have suggested that In(1,4,5)$P_3$ or In(1,3,4,5)$P_4$ may be directly involved in $Ca^{2+}$ entry. While some of these studies could be effectively assimilated into one of the proposed routes for $Ca^{2+}$ entry via a capacitative mechanism, this is not the case for all the reported effects. KUNO and GARDNER (1987) first suggested that In(1,4,5,)$P_3$ could open $Ca^{2+}$ channels with a unitary conductance of 5 pS, carrying both $Ca^{2+}$ and $Ba^{2+}$ in excised plasma membrane patches of lymphocytes. With platelet plasma membranes prepared using Percoll gradients, RENGASAMY and FEINBERG (1988) showed that In(1,4,5)$P_3$ could release $Ca^{2+}$ after loading the membranes using a $Na^+$/$Ca^{2+}$ exchange mechanism. More recently $IP_3$Rs from a number of tissues have been identified as having carbohydrate residues such as sialic acid and N-acetylglucosamine that are more prominantly associated with plasma membrane proteins (KHAN et al. 1992a,b), and in these studies In(1,4,5)$P_3$ binding activities associated with plasma membrane enriched fractions showed a reduced specificity for In(1,4,5)$P_3$ compared to $IP_3$Rs of microsomal fractions. Whether this refected differential distribution of $IP_3$R isoforms was not determined. $IP_3$Rs on surface invaginated portions of the plasma membranes (caveolae) of endothelial cells have been localised by immunochemical staining using the antibody 4C11 (FUJIMOTO et al. 1992). This antibody (4C11) also detects low levels of an $IP_3$R (possibly type III) in platelet plasma membranes purified by free flow electrophoresis (EL-DAHER et al. 1996). However, in all of these studies there has been no definitive demonstration of a plasma membrane location for $IP_3$Rs, and it is probable that they are present on organelles that are tightly associated with the plasma membrane and may be important for the protein coupling $Ca^{2+}$ entry pathway described in the previous section. Such a protein coupling mechanism would be fast and may explain why with thrombin and the thromboxane mimetic, $Ca^{2+}$ influx is recorded before $Ca^{2+}$ release in human platelets with the stopped flow technique (SAGE et al. 1989).

A specific function for In(1,3,4,5)$P_4$ related to $Ca^{2+}$ entry, though suggested soon after it was discovered (MORRIS et al. 1987), was met with considerable doubt because most cells show good capacitative $Ca^{2+}$ entry phenomenon, and phosphorylation of In(1,4,5)$P_3$ to In(1,3,4,5)$P_4$ was not necessary for $Ca^{2+}$ influx in some cells (BIRD et al. 1991). Recently, however, a number of studies have re-awakened interest in mechanisms of $Ca^{2+}$ entry involving In(1,3,4,5)$P_4$. LUCKHOFF and CLAPHAM (1992) reported that In(1,3,4,5)$P_4$ with $Ca^{2+}$ opened channels with a conductance of 2 pS in endothelial cells that were permeable to $Ca^{2+}$ and $Mn^{2+}$, and this effect was not seen with In(1,4,5)$P_3$. Intracellular application of In(1,3,4,5)$P_4$ [not In(1,4,5)$P_3$] has been reported to cause $Ca^{2+}$ influx in Ras-transformed and control fibroblasts with a greater effect on Ras-transformed cells, suggesting also a role for Ras in $Ca^{2+}$ influx (HASHII et al. 1993, 1994). Co-addition of In(1,3,4,5)$P_4$ with $Ca^{2+}$ has been shown to cause the activation of $K^+$ channels in plasma membranes of smooth muscle cells (MOLLEMAN et al. 1991). Support for a function for In(1,3,4,5)$P_4$ has been strenthened by reports of specific binding sites for In(1,3,4,5)$P_4$ and the purification of In(1,3,4,5)$P_4$ binding proteins of Mr 182, 174, 104, 84, and 42 kDa (THEIBERT et al. 1991, 1992; DONI and REISER 1991; REISER et al. 1991; CULLEN and IRVINE 1992; CULLEN et al. 1994, 1995a; for a review see IRVINE and CULLEN 1993). In addition, In(1,3,4,5)$P_4$ binding proteins such as synaptotagmin II have been suggested to have important modulatory functions in vesicle transport and neurotransmitter release and can interact with plasma membrane proteins such as neurexin, syntaxin and N-type $Ca^{2+}$ channels (e.g. PERIN et al. 1990, 1991; LEVEQUE et al. 1992). With synaptotagmin the inositol polyphosphate binding property has been elucidated to be contained in the C2B domain that contains a large number of charged lysine residues (FUKUDA et al. 1994). Interestingly, only the C2B domains of synaptotagmin I and II when expressed as GST-fusion proteins exhibited inositol polyphosphate binding activities and not those present in raphilin 3A or the C2A domains present in synaptotagmin I, II, raphilin 3A or PKC$\alpha$. Additionally, the C2B domains of synaptotagmin I and II, but not of raphilin 3A, bound phospholipids in a $Ca^{2+}$ independent manner, suggesting selectivity in the functions of C2B domains in different proteins. The cloned In(1,3,4,5)$P_4$ binding protein (GAP1$^{IP4BP}$) contains regions identified as C2A, C2B, RasGAP and pleckstrin homology domains (CULLEN et al. 1995b) with In(1,3,4,5)$P_4$ specifically activating GTPase activity towards Ras. The cloned protein is thought to be identical to the purified protein localised to platelet plasma membranes. Platelets contain low levels of Ras (KLINZ et al. 1992) and high levels of Rap proteins, particularly Rap1B, for which many functions have been proposed (see above), including a role in $Ca^{2+}$ transport (COURVAZIER et al. 1992), although in that study affects of Rap1B were suggested to be correlated to inhibition of $Ca^{2+}$ ATPases (see above).

We have recently carried out a study examining $Ca^{2+}$ fluxes using platelet plasma membranes prepared by free flow electrophoresis, into which $Ca^{2+}$ was loaded by a $Na^+/Ca^{2+}$ exchange mechanism (PATEL et al., to be published). In(1,3,4,5)$P_4$ added at equilibrium rapidly released up to a maximum of 40% trapped $Ca^{2+}$, under conditions in which ionomycin released greater than 85%. This indicates an involvement of In(1,3,4,5)$P_4$ with $Ca^{2+}$ entry in human platelets probably mediated by the In(1,3,4,5)$P_4$ binding protein. Interestingly, addition of In(1,4,5)$P_3$ or In(1,3,4)$P_3$ also induced $Ca^{2+}$ release again to a maximum of 40%; In(1,4)$P_2$ or $IP_6$ tested up to 100 $\mu M$ were not effective. $Ca^{2+}$ flux mediated by $IP_3$Rs is probably mediated by the type III $IP_3$R localised to the platelet plasma membrane and detected with the antibodies 4C11 and CT3. We have been able to distinguish the $Ca^{2+}$ releasing activities of In(1,3,4,5)$P_4$ from those of In(1,4,5)$P_3$ with the use of the catalytic subunit of cAMP-PK (cAMP-PK CAT). In the presence of cAMP-PK CAT under conditions in which phosphorylation of a number of cAMP-PK substrates is observed the $Ca^{2+}$ releasing function of In(1,4,5)$P_3$ is inhibited but that of In(1,3,4,5)$P_4$ is not affected. This suggests that the $Ca^{2+}$ release activity of In(1,3,4,5)$P_4$ occurs through a mechanism distinct from the $IP_3$R and probably involves the In(1,3,4,5)$P_4$ binding protein (PATEL et al., to be published). The involvement of Ras related G proteins in $Ca^{2+}$ flux using this membrane preparation is being investigated. Additionally, whether the type III $IP_3$R localised to his membrane is a plasma membrane protein or is present on a membrane tightly associated with the plasma membrane through protein interactions is important to clarify. Figure 6 summarises the main pathways associated with $Ca^{2+}$ influx in human platelets.

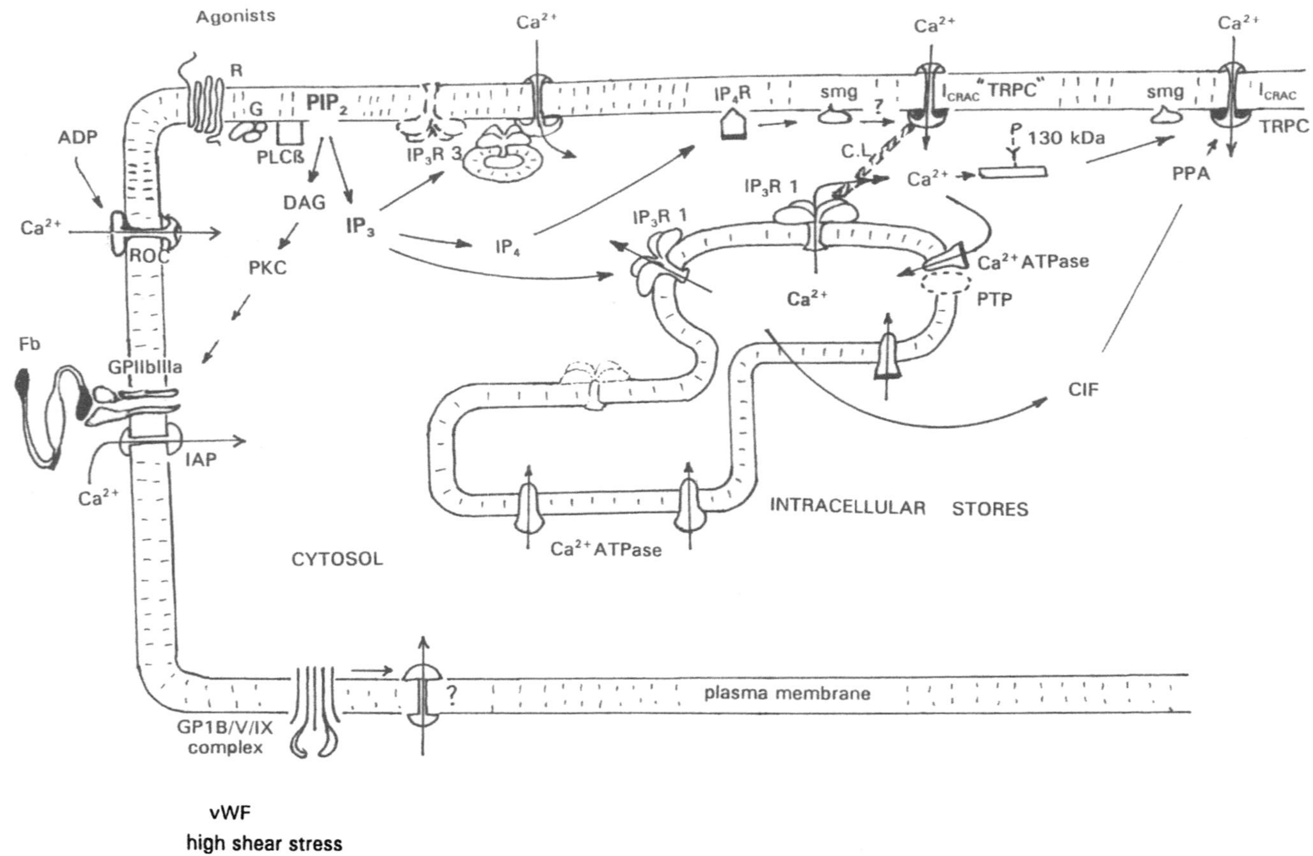


**Fig. 6.** Summary of the complexities associated with $Ca^{2+}$ influx mechanisms in platelets. *smg*, small G protein; *PTP*, protein tyrosine phosphatase; *PPA*, protein phosphatase; *C.L.*, cytoskeletal link (probably involving ankyrin); *vWF*, von Willibrand Factor; *Fb*, fibrinogen; *IAP*, integrin associated protein. For details see text

## VII. Integrin-Associated Plasma Membrane $Ca^{2+}$ Flux

The glycoprotein IIbIIIa (GPIIbIIIa) complex ($\alpha$IIb$\beta$3 integrin) is a major platelet surface receptor involved in platelet aggregation and adhesion reactions. The GPIIbIIIa is a $Ca^{2+}$ dependent complex that binds fibrinogen in activated but not resting platelets. This integrin is an important bidirectional signalling molecule with intracellular events able to regulate the affinity of integrin interaction with adhesive glycoproteins, and in turn the binding of glycoproteins induces the activation of non-receptor tyrosine kinases and the recruitment of signalling molecules into the cytoskeleton (for a review see SHATTIL 1995). A number of studies have suggested that GPIIbIIIa is involved with $Ca^{2+}$ influx. BRASS (1985) reported that the maximum rate of $Ca^{2+}$ exchange in thrombasthenic platelets lacking functional GPIIbIIIa was only half that for normal platelets. Purified GPIIbIIIa when reconstituted into liposomes was found to allow $Ca^{2+}$ transport (RYBAK and RENZULLI 1989; FUJIMOTO et al. 1991a). Further, channel activities (10 pS conductance with selectivity for $Ba^{2+}$ over $Na^+$ and inhibition by $Ni^{2+}$) were detected in membranes incorporated into bilayers from thrombin stimulated platelets (ZSCHAUER et al. 1988) that were inhibited when the GPIIbIIIa complex was dissociated or impaired in membranes prepared from thrombasthenic platelets (FUJIMOTO et al. 1991b). Survival of the channels during the fractionation procedure indicates that covalent modification may have occurred. Platelets labelled with the fluorescent protein aequorin and stimulated with either thrombin or collagen showed deceased $Ca^{2+}$ influx when in the presence of the GPIIbIIIa antibody TM83 or GRGDSP (YAMAGUICHI et al. 1987). Though these studies suggest the involvement of GPIIbIIIa with $Ca^{2+}$ influx it is unlikely that the complex itself is a $Ca^{2+}$ channel. A large number of monoclonal antibodies recognising GPIIbIIIa are available that either lead to activation of platelets (via Fc receptors) or inhibition of aggregation, but most do not directly affect $Ca^{2+}$ influx (e.g. HORNBY et al. 1991), and $Ca^{2+}$ influx is normal when measured in thrombasthenic platelets stimulated by ADP (POWLING and HARDISTY 1985). But, $Ca^{2+}$ influx may occur via a channel closely associated with the GPIIbIIIa complex. In support of this, a 50 kDa integrin associated protein (IAP) has been identified in many cells including platelets that is closely associated with the GPIIIa ($\beta_3$ subunit; BROWN et al. 1990). Cloning and primary structure of IAP suggests that it has five putative transmembrane domains, and an antibody to IAP was found to inhibit $Ca^{2+}$ influx mediated by fibronectin attachment to endothelial cells but not that mediated by histamine, suggesting a specific integrin-mediated $Ca^{2+}$ influx mechanism separate from that mediated by surface agonists (SCHWARTZ et al. 1993). There is further evidence that when platelets are exposed to high shear stress, von Willebrand Factor binding to GP1B may induce $Ca^{2+}$ influx that is not affected by cAMP elevation (CHOW et al. 1992; IKEDA et al. 1993). Thus, platelets may exhibit $Ca^{2+}$ influx pathways regulated by the high levels of integrins present on their surface.

## VIII. $Ca^{2+}$ Influx Via Receptor-Operated $Ca^{2+}$ Channel (ROC)

$Ca^{2+}$ entry pathways stimulated by second messengers acting predominantly through a capacitative entry pathway provide an explanation of the $Ca^{2+}$ influx seen with most platelet agonists such as thrombin, thromboxane and PAF. However, there is evidence from rapid time measurements using the stopped-flow technique that ADP is able to induce a $Ca^{2+}$ influx signal before release from intracellular stores and without a measurable delay after agonist addition (SAGE and RINK 1985; SAGE et al. 1989, 1990). These studies indicated that $Ca^{2+}$ influx by ADP was biphasic with the fist phase probably linked to the opening of a ROC-like channel and the second linked to store depletion. Elevation of cAMP or cGMP inhibits the second phase but not the first phase of $Ca^{2+}$ entry (SAGE et al. 1990; GEIGER et al. 1992). Both entry pathways were also permeable to $Mn^{2+}$, $Ba^{2+}$ and $Na^{+}$ (MAHAUT-SMITH et al. 1990). However, it should be noted that in another approach using a continuous and quenched flow technique with not quite so rapid time measurements, JONES and GEAR (1988) reported that ADP did give a similar $Ca^{2+}$ release pattern to thrombin with an initial release occurring from an intracellular store. While this controversy over the initial increase by ADP may reflect differences in the techniques used, the possibility that ADP may directly gate a ROC has been further supported in experiments using platelets attached to a patch clamp. MAHAUT-SMITH et al. (1990) were able to record channel activity permeable to $Ca^{2+}$, $Ba^{2+}$ and $Na^{+}$ (i.e. non-selective) when ADP but not thrombin was included in the patch pipette. Additionally, using the nyastatin permeabilisation technique, again ADP but not thrombin induced inward cation currents (single channel conductance of 15 pS, non-selective permeability), suggesting an action of ADP at a ROC-like channel (MAHAUT-SMITH et al. 1992). The non-selectivity of this ROC-like channel is supported by earlier observations that ADP also evoked entry of $Na^{2+}$, as well as $Ca^{2+}$ and $Mn^{2+}$ (SAGE et al. 1991). Recently, the rapid $Ca^{2+}$ entry induced by ADP (and ATP) has been suggested to be mediated by a $P_{2\times1}$ purinoceptor (MACKENZIE et al. 1996).

## IX. Cyclic Nucleotide Effects on $Ca^{2+}$ Influx

The cyclic nucleotides cAMP and cGMP affect platelet activation at many stages of the activation pathway predominantly via the activation of cAMP-PK or cGMP-PK, leading to the phosphorylation of a number of protein substrates (WALDMAN et al. 1986, 1987). Both cAMP-PK and cGMP-PK have been localised to be predominantly associated with the plasma membrane of human platelets (EL-DAHER et al. 1996). With respect to $Ca^{2+}$ elevation mechanisms, there is a primary effect at inhibition of In(1,4,5)$P_3$ formation (WATSON et al. 1984), in addition to stimulation of $Ca^{2+}$ efflux, and sequestration, and inhibition of $Ca^{2+}$ mobilisation (see above). The direct activation of either cAMP-PK or cGMP-PK using analogues of cAMP or cGMP on intact platelets has also been shown to inhibit Tg stimulated $Ca^{2+}$ release and resulting influx

(GEIGER et al. 1994). There have been a number of studies suggesting little direct effect of cAMP or cGMP elevation on $Ca^{2+}$ influx as for example the initial phase of $Ca^{2+}$ entry mediated by ADP acting at a ROC-like channel (GEIGER et al. 1992, 1994) or on $Ca^{2+}$ influx resulting after stores have been depleted (DONI et al. 1994; HEEMSKERK et al. 1994; GEIGER et al. 1994). Thus, whichever route turns out to be the prominent mechanism for store-depletion induced $Ca^{2+}$ entry (protein coupling, diffusable messengers etc.), there would appear to be few elements directly sensitive to elevation of cyclic nucleotides.

*Acknowledgements.* I would like to thank my colleagues Drs. S.S. El-Daher, Y. Patel and S. Bokkala for many discussion. The author's work has been supported by grants from the British Heart Foundation, the Wellcome Trust and the Thrombosis Research Trust.

## References

Adunyah SE, Dean WL (1985) Inositol trisphosphate induced $Ca^{2+}$ release from human platelet membranes. Biochem Biophys Res Commun 128:1274–1280
Adunyah SE, Dean WL (1986) Effects of sulfhydryl reagents and other inhibitors on the $Ca^{2+}$ transport and inositol trisphosphate-induced $Ca^{2+}$ release from human platelets. J Biol Chem 261:13071–13075
Adunyah SE, Dean WL (1987) Regulation of platelet membrane $Ca^{2+}$ transport by cAMP- and calmodulin dependent phosphorylation. Biochim Biophys Acta 930:401–409
Adunyah SE, Jones LR, Dean WL (1988) Structural and functional comparison of a 22kDa protein from internal human platelet membranes with cardiac phospholamban. Biochim Biophys Acta 941:63–74
Alonso MT, Alvarez J, Montero M, Sanchez A, Garcia-Sancho J (1991) Agonist induced $Ca^{2+}$ influx into human platelets is secondary to emptying of intracellular $Ca^{2+}$ stores. Biochem J 280:783–789
Alvarez J, Montero M, Garcia-Sancho J (1991) Cytochrome P-450 may link intracellular $Ca^{2+}$ stores with plasma membrane $Ca^{2+}$ influx. Biochem J 274:193–197
Au KS (1986) Activation of erythrocyte membrane $Ca^{2+}$-ATPase by calpain. Biochem Biophys Acta 905:273 278
Authi KS (1992) Localisation of the $IP_3$ binding site on platelet membranes isolated by high voltage free flow electrophoresis. FEBS Lett 298:173–176
Authi KS (1993) $Ca^{2+}$ homeostasis and intracellular pools in human platelets. In: Authi KS, Watson SP, Kakkar VV (eds) Mechanisms of platelet activation and control. Plenum, London, pp 83–104
Authi KS, Crawford N (1985) Inositol (1,4,5) trisphosphate induced release of sequestered calcium from highly purified human platelet intracellular membranes. Biochem J 230:247–253
Authi KS, Evenden BJ, Crawford N (1986) Metabolic and functional consequences of introducing Inositol (1,4,5) trisphosphate into saponin permeabilised platelets. Biochem J 233:707–718
Authi KS, Hornby EJ, Evenden BJ, Crawford N (1987) Inositol (1,4,5) trisphosphate induced rapid formation of thromboxane $B_2$ in saponin permeabilised human platelets. Mechanism of action. FEBS Lett 213:95–101
Authi KS, Bokkala S, Patel Y, Kakkar VV, Munkonge FM (1993) $Ca^{2+}$ Release from platelet intracellular stores by thapsigargin and 2,5 di-(t-butyl) 1,4-benzohydroquinone. Relationship to $Ca^{2+}$ pools and relevance in platelet activation. Biochem J 294:119–126
Berridge MJ (1991) Cytoplasmic calcium oscillations. A two pool model. Cell Calcium 12:63–72

Berridge MJ (1993) Inositol trisphosphate and calcium signalling. Nature 361:315–325
Bezprozvanny I, Ehrlich BE (1993) ATP modulates the function of inositol trisphosphate at two sites. Neuron 10:1175–1184
Bezprozvanny I, Watras J, Ehrlich BE (1991) Bell shaped calcium response curves of Ins (1,4,5)$P_3$ – and calcium-gated channels from endoplasmic reticulum of cerebellum. Nature 351:751–754
Bird GSJ, Putney JW (1993) Inhibition of thapsigargin-induced calcium entry by microinjected guanine nucleotide analogues. Evidence for the involvement of a small G-protein in capacitative calcium entry. J Biol Chem 268:21486–21488
Bird GSJ, Rossier MF, Hughes AR, Shears SB, Armstrong DL, Putney JW Jr (1991) Activation of $Ca^{2+}$ entry into acinar cells by a non-phosphorylatable inositol trisphosphate. Nature 352:162–165
Blake RA, Schieven G-L, Watson SP (1994) Collagen stimulates tyrosine phosphorylation of phospholipase C$\gamma$2 but not phospholipase C$\gamma$1 in human platelets. FEBS Lett 353:212–216
Bobe R, Bredoux R, Wuytack F, Quarck R, Kovacs T, Papp B, Corvasier E, Magnier C, Enouf J (1994) The rat platelet 97-kDa $Ca^{2+}$ ATPase isoform is the sarco/endoplasmic reticulum $Ca^{2+}$ ATPase 3 protein. J Biol Chem 269:1417–1424
Bokkala S, El-Daher SS, Kakkar VV, Wuytack F, Authi KS (1995) Localization and identification of $Ca^{2+}$ ATPases in highly purified human platelet plasma and intracellular membranes. Evidence that the monoclonal antibody PL/IM 430 recognises the SERCA 3 $Ca^{2+}$ ATPase in human platelets. Biochem J 306:837–842
Bourguignon LYW, Jin H (1995) Identification of the ankyrin binding domain of the mouse T-lymphoma cell inositol 1,4,5-trisphosphate ($IP_3$) receptor and its role in regulation of $IP_3$ mediated internal $Ca^{2+}$ release. J Biol Chem 270:7257–7260
Bourguignon LYW, Iida N, Jin H (1993a) The involvement of the cytoskeleton in regulating $IP_3$ receptor mediated internal $Ca^{2+}$ release in human blood platelets. Cell Biol Int Rep 17:751–758
Bourguignon LYW, Jin H, Iida N, Brandt NR, Zhang SH (1993b) The involvement of ankyrin in the regulation of inositol 1,4,5,-trisphosphate receptor mediated internal $Ca^{2+}$ release from $Ca^{2+}$ storage vesicles. J Biol Chem 268:7290–7297
Bourguignon LYW, Chu A, Jin H, Brandt NR (1995) Ryanodine receptor-ankyrin interaction regulates internal $Ca^{2+}$ release in mouse T-lymphoma cells. J Biol Chem 270:17917–17922
Brandl C, Green N, Korczak B, MacLennan DH (1986) Two $Ca^{2+}$ ATPase genes; homologies and mechanistic implications of deduced amino acid sequences. Cell 44:597–607
Brass LF (1984) The effect of $Na^+$ on $Ca^{2+}$ homeostasis in unstimulated platelets. J Biol Chem 259:12571–12575
Brass LF (1985) $Ca^{2+}$ transport across the platelet plasma membrane. A role for membrane glycoproteins IIb and IIIa. J Biol Chem 260:2231–2236
Brown E, Hooper L, Ho T, Gresham H (1990) Integrin associated protein: a 50kDa plasma membrane antigen physically and functionally associated with integrins. J Cell Biol 111:2785–2794
Brune B, Ullrich V (1990) $Ca^{2+}$ mobilisation in human platelets by receptor agonist and $Ca^{2+}$ ATPase inhibitors. FEBS Lett 284:1–4
Brune B, Ullrich V (1991) Different $Ca^{2+}$ pools in human platelets and their role in thromboxane $A_2$ formation. J Biol Chem 266:19232–19237
Brune B, Ullrich V (1992) Cyclic nucleotides and intracellular $Ca^{2+}$ homeostasis in human platelets. Eur J Biochem 207:607–613
Brune B, Appen FV, Ullrich V (1994) Receptor occupancy regulates $Ca^{2+}$ entry and intracellular $Ca^{2+}$ redistribution in activated platelets. Biochem J 304:993–999
Burgering BMT, Bos JL (1995) Regulation of Ras-mediated signalling; more than one way to skin a cat. Trends Biochem Sci 20:18–21
Burk SE, Lytton J, MacLennan DH, Schull GE (1989) cDNA cloning, functional expression and mRNA tissue distribution of a third organellar $Ca^{2+}$ pump. J Biol Chem 264:18561–18568

Carafoli E (1992) The calcium pump of the plasma membrane. J Biol Chem 267:2115–2118

Carafoli E (1994) Biogenesis; plasma membrane calcium ATPase; 15 years of work on the purified enzyme. FASEB J 8:993–1002

Cavallini L, Coassin M, Alexandre A (1995) Two classes of agonist sensitive $Ca^{2+}$ stores in platelets as identified by their differential sensitivity to 2,5-di(tert-butyl)-1,4-benzohydroquinone and thapsigargin. Biochem J 310:449–452

Cheng HY, Magoesi M, Cooper RS, Penniston JT, Burke JL (1994) Expression of plasma membrane $Ca^{2+}$ pump epitopes in human platelets and megakaryoblasts. Cell Physiol Biochem 4:31–43

Chow TW, Hellums JD, Moake JL, Kroll MH (1992) Shear stress-induced von Willebrand factor binding to platelet glycoprotein Ib initiates calcium influx associated with aggregation. Blood 80:113–120

Clapham D (1995) Replenishing the stores. Nature 375:634–635

Courvazier E, Enouf J, Papp B, Gunzburg J de, Tavitian A, Levy-Toledano S (1992) Evidence for a role of rap 1 protein in the regulation of human platelet $Ca^{2+}$ fluxes. Biochem J 281:325–331

Cullen PJ, Irvine RF (1992) Inositol 1,3,4,5-tetrakisphosphate binding sites in neuronal and non-neuronal tissues. Biochem J 288:149–154

Cullen PJ, Patel Y, Kakkar VV, Irvine RF, Authi KS (1994) Specific binding sites for inositol 1,3,4,5-tetrakisphosphate are located predominantly in the plasma membrane of human platelets. Biochem J 298:739–742

Cullen PJ, Dawson AP, Irvine RF (1995a) Purification and characterisation of an In(1,3,4,5)$P_4$ binding protein from pig platelets; possible identification of a non-neuronal In(1,3,4,5)$P_4$ receptor. Biochem J 305:139–143

Cullen PJ, Hsuan JJ, Truong O, Letcher AJ, Jackson TR, Dawson AP, Irvine RF (1995b) Identification of a specific Ins (1,3,4,5)$P_4$ binding protein as a member of the GAP 1 family. Nature 376:527–530

Daniel JL (1990) Inositol phosphate metabolism and platelet activation. Platelets 1:117–126

Daniel JL, Dangelmair CA, Smith JB (1989) Calcium modulates the generation of inositol 1,3,4,-trisphosphate in human platelets by the activation of inositol 1,4,5-triphosphate 3-kinase. Biochem J 253:789–794

Daniel JL, Dangelmaier C, Smith JB (1994) Evidence for a role for tyrosine phosphorylation of phospholipase Cγ2 in collagen-induced platelet cytosolic calcium mobilisation. Biochem J 302:617–622

Danoff SK, Ferris CD, Donata C, Fisher CA, Munemitsu S, Ullrich A, Snyder SH, Ross CA (1991) Inositol 1,4,5-triphosphate receptors; distinct neuronal and non-neurnal forms by alternative splicing differ in phosphorylation. Proc Natl Acad Sci USA 88:2951–2955

Davies EV, Hallet MB (1995) A soluble cellular factor directly stimulates $Ca^{2+}$ entry in neutrophils. Biochem Biophys Res Commun 206:348–354

Dean WL, Quinton JM (1995) Distribution of plasma membrane $Ca^{2+}$ ATPase and inositol 1,4,5 triphosphate receptor in human platelets membranes. Cell Calcium 17:65–70

Dean WL, Sullivan DM (1982) Structural and functional properties of a $Ca^{2+}$ ATPase from human platelets. J Biol Chem 257:14390–14394

Dean WL, Pope JE, Brier ME, Aronoff GR (1994) Platelet calcium transport in hypertension. Hypertension 23:31–37

Doni F, Reiser G (1991) Purification of a high affinity inositol 1,3,4,5-tetrakisphosphate receptor from brain. Biochem J 275:453–457

Doni MG, Cavallini L, Alexandre A (1994) $Ca^{2+}$ influx in platelets activated by thrombin and by depletion of the stores. Effects of cyclic nucleotides. Biochem J 303:599–605

Dunlop ME, Larkins RG (1988) GTP- and inositol 1,4,5-trisphosphate induced release of $Ca^{2+}$ from a membrane store co-localised with pancreatic-islet-cell plasma membrane. Biochem J 253:583–586

Eberhard M, Erne P (1993) Inositol 1,4,5-trisphosphate-induced calcium release in permeabilized platelets is coupled to hydrolysis of inositol 1,4,5-trisphosphate to inositol 1,4-bisphosphate. Biochem Biophys Res Commun 195:19–24
Ehrlich B, Watras J (1988) Inositol 1,4,5-trisphosphate activates a channel from smooth muscle sarcoplasmic reticulum. Nature 336:583–586
El-Daher SS, Eigenthaler M, Walter U, Furuichi T, Miyawaki A, Mikoshiba K, Kakkar VV, Authi KS (1996) Distribution and activation of cAMP- and cGMP- dependent protein kinases in highly purified human platelet plasma and intracellular membranes. Thromb Haemost 76
Enouf J, Bredoux R, Boucheix C, Mirshahi M, Soria C, Levy-Toledano S (1985) Possible involvement of two proteins (phosphoprotein and CD9 (p24)) in regulation of platelet calcium fluxes. FEBS Lett 183:398–402
Enouf J, Giraud F, Bredoux R, Bourdeau N, Levy-Toledano S (1987) Possible role of a cAMP dependent phosphorylation in the calcium release mediated by inositol 1,4,5-trisphosphate in human platelet membrane vesicles. Biochim Biophys Acta 928:76–82
Enouf J, Bredoux R, Papp B, Djaffer I, Lompre AM, Keiffer N, Gayet O, Clemetson K, Wuytack F, Rosa J-P (1992) Human platelets express the SECRCA 2b isoform of the $Ca^{2+}$ transport ATPase. Biochem J 286:135–140
Falchetto R, Vorherr T, Brunner J, Carafoli E (1991) The plasma membrane $Ca^{2+}$ pump contains a site that interacts with its calmodulin binding domain. J Biol Chem 266:2930–2936
Falchetto R, Vorherr T, Carafoli E (1992) The calmodulin binding site of the plasma membrane $Ca^{2+}$ pump interacts with the transduction domain of the enzyme. Protein Sci 1:1613–1621
Fasolato C, Hoth M, Penner R (1993) A GTP-dependent step in the activation mechanism of capacitative calcium influx. J Biol Chem 268:20737–20740
Fasolato C, Innocenti B, Pozzan T (1994) Receptor activated $Ca^{2+}$ influx; how many mechanisms for how many channels. Trends Pharmacol Sci 15:77–83
Ferris CD, Snyder SO (1992) Inositol 1,4,5,-trisphosphate associated calcium channels. Annu Rev Physiol 54:469–488
Ferris CD, Cameron AM, Bredt DS, Huganir RL, Snyder SH (1991) Inositol 1,4,5-trisphosphate receptor is phosphorylated by cyclic AMP-dependent protein kinase at serine 1755 and 1589. Biochem Biophys Res Commun 175:192–198
Ferris CD, Cameron AM, Huganir RL, Snyder SH (1992) Quantal calcium release by purified reconstituted inositol 1,4,5-trisphosphate receptors. Nature 356:350–352
Fischer TH, White GC II (1987) Partial purification and characterization of thrombolamban, a 22 kDa cAMP dependent protein kinase substrate in platelets. Biochem Biophys Res Commun 149:700–706
Fischer TH, White GC II (1989) cAMP-dependent protein kinase substrates in platelets. Evidence that thrombolamban, a 22-kDa substrate and the $Ca^{2+}$ ATPase are not associated proteins. Biochem Biophys Res Commun 159:644–650
Fischer TH, Gatling MN, Lacal JC, White GC II (1990) Rap 1B, a cAMP-dependent protein kinase substrate associates with the cytoskeleton. J Biol Chem 265:19405–19408
Fliescher S, Inui M (1989) Biochemistry and biophysics of excitation-contraction coupling. Annu Rev Biophys Chem 18:333–364
Fruen BR, Mickelson SR, Shomer NH, Velez P, Louis CF (1994) cADP ribose does not affect cardiac or skeletal muscle ryanodine receptors. FEBS Lett 352:123–126
Fujimoto T, Fujimura K, Kuramoto A (1991a) Functional $Ca^{2+}$ channel produced by purified platelet membrane glycoprotein IIb-IIIa complex incorporated into planar phospholipid bilayer. Thromb Haemost 66:598–603
Fujimoto T, Fujimura K, Kuramoto A (1991b) Electrophysiological evidence that glycoprotein IIb-IIIa complex is involved in calcium channel activation on human platelet plasma membrane. J Biol Chem 266:16370–16375

Fujimoto T, Nakade S, Miyawake A, Mikoshiba K, Ogawa K (1992) Localisation of inositol 1,4,5-trisphosphate receptor-like protein in plasmalemmal caveolae. J Cell Biol 119:1507–1513

Fukuda M, Amga J, Niiobe M, Aimoto S, Mikoshiba K (1994) Inositol 1,3,4,5-tetrakisphosphate binding to C2B domain of $IP_4BP$/synoptotagmin II. J Biol Chem 269:29206–29211

Furuichi T, Yoshikawa S, Miyawaki A, Wada K, Maeda N, Mikoshiba K (1989) Primary structure and functional expression of the inositol 1,4,5-trisphosphate binding protein P400. Nature 342:32–38

Furuichi T, Kohda K, Miyawaki A, Mikoshiba K (1994) Intracellular channels. Curr Opin Neurobiol 4:294–303

Galione A, Lea HC, Busa WB (1991) $Ca^{2+}$ induced $Ca^{2+}$ release in sea urchin egg homogenates; modulation by cADP ribose. Science 253:1143–1146

Geiger J, Nolte C, Butt E, Sage SO, Walter U (1992) Role of cGMP and cGMP-dependent protein kinase in nitrovasodilator inhibition of agonist-evoked calcium elevation in human platelets. Proc Natl Acad Sci USA 89:1031–1035

Geiger J, Nolte C, Walter U (1994) Regulation of calcium mobilization and entry in human platelets by endothelium-derived factors. Am J Physiol 267 (Cell Physiol 36):C236–C244

Gilon P, Bird GSJ, Bian X, Yakel BL, Putney JW (1995) The calcium mobilizing actions of a Jurkat cell extract on mammaliam cells and Xenopus laevis oocytes. J Biol Chem 270:8050–8055

Goodwin CA, Wheeler-Jones CPD, Namiranian S, Bokkala S, Kakkar VV, Authi KS, Scully MF (1995) Increased expression of procoagulant activity on the surface of human platelets exposed to heavy-metal compounds. Biochem J 308:15–21

Guillemette G, Balla T, Baukal AJ, Catt KJ (1988) Characterisation of inositol 1,4,5-trisphosphate receptors and $Ca^{2+}$ mobilisation in a hepatic plasma membrane fraction. J Biol Chem 263:4541–4548

Hack N, Wilkinson JM, Crawford N (1988a) A monoclonal antibody (PL/IM 430) to platelet intracellular membranes which inhibits the uptake of $Ca^{2+}$ without affecting the $Ca^{2+}Mg^{2+}$ ATPase. Biochem J 250:355–361

Hack N, Authi KS, Crawford N (1988b) Introduction of antibody (PL/IM 430) to a 100 kDa protein into permeabilised platelets inhibits intracellular sequestration of calcium. Biosci Rep 8:379–388

Hakamata Y, Nishmura S, Nakai J, Nakashima Y, Kita Y, Imoto K (1994) Involvement of the brain-type of ryanodine receptor in T-cell proliferation. FEBS Lett 252:206–210

Hardie RC, Minke B (1993) Novel $Ca^{2+}$ channels underlying transduction in *Drosophila* photoreceptors; implications for phosphoinositide mediated $Ca^{2+}$ mobilization. Trends Neurosci 16:371–376

Harnick DJ, Jayaraman YM, Mulieri P, Go LO, Marks AR (1995) The human type 1 inositol 1,4,5-trisphosphate receptor from T lymphocytes. Structure, localization and tyrosine phosphorylation. J Biol Chem 270:2833–2840

Harootunian AT, Kao JP, Paranjape S, Tsien RY (1991) Generation of calcium oscillations in fibroblasts by positive feedback between calcium and $IP_3$. Science 251:75–78

Hashii M, Nozawa Y, Higashida H (1993) Bradykinin-induced cytosolic $Ca^{2+}$ oscillation and inositol tetrakis-phosphate induced $Ca^{2+}$ influx in voltage-clamped ras-transformed NIH/3T3 fibroblasts. J Biol Chem 268:19403–19410

Hashii M, Hirata M, Ozaki S, Nozawa Y, Higashida H (1994) $Ca^{2+}$ influx gated by inositol 3,4,5,6-tetrakisphosphate in NIH/3T3 fibroblasts. Biochem Biophys Res Commun 200:1300–1306

Hashimoto Y, Ogihara A, Nakanishi S, Matsuda Y, Kurokawa K, Nomomura Y (1992) Two thrombin associated $Ca^{2+}$ channels in human platelets. J Biol Chem 267:17078–17081

Hecker M, Brune B, Decker K, Ullrich V (1989) The sulfhydryl reagent thimerosal elicits human platelet aggregation by mobilisation of intracellular calcium and secondary prostaglandin endoperoxide formation. Biochem Biophys Res Commun 159:961–968

Heemskerk JWM, Hoyland J, Mason WT, Sage SO (1992) Spiking in cytosolic calcium concentration in single fibrinogen-bound fura-2-loaded human platelets. Biochem J 283:379–383

Heemskerk JWM, Vis P, Feijge MAH, Hoyland J, Mason WT, Sage SO (1993) Roles of phospholipase C and $Ca^{2+}$ ATPase in $Ca^{2+}$ responses of single, finbrinogen-bound platelets. J Biol Chem 268:356–363

Heemskerk JWM, Feijge MAH, Sage SO, Walter U (1994) Indirect regulation of $Ca^{2+}$ entry be cAMP-and cGMP-dependent protein kinases and phospholipase C in rat platelets. Eur J Biochem 223:543–551

Hettasch JM, LeBreton GC (1987) Modulation of $Ca^{2+}$ fluxes in isolated platelet vesicles; effects of cAMP dependent protein kinase and protein kinase inhibitor on $Ca^{2+}$ sequestration and release. Biochim Biophys Acta 931:41–58

Hornby EJ, Brown S, Wilkinson JM, Mattock C, Authi KS (1991) Activation of human platelets by exposure to a monoclonal antibody, PM 6/248, to glycoprotein IIb–IIIa. Br J Haematol 79:277–285

Hoth M, Penner R (1992) Depletion of intracellular calcium stores activates a calcium current in mast cells. Nature 355:353–356

Hu Y, Shilling WP (1995) Receptor mediated activation of recombinant trpl expressed in Sf8 insect cells. Biochem J 305:605–611

Ikeda Y, Handa M, Kamata T et al. (1993) Transmembrane calcium influx associated with von Willebrand factor binding to GP Ib in the initiation of shear-induced platelets aggregation. Thromb Haemost 69:496–502

Irvine RF (1990) "Quantal" calcium release and the control of $Ca^{2+}$ entry by inositol phosphates – a possible mechanism. FEBS Lett 263:5–9

Irvine RF, Cullen PJ (1993) Will the real inositol 1,3,4,5-tetrakisphosphate receptor please stand up. Curr Biol 3:540–543

Jaconi MEE, Lew DP, Monod A, Krause K-H (1993) The regulation of store-dependent $Ca^{2+}$ influx in HL-60 granulocytes involves GTP-sensitive elements. J Biol Chem 268:26075–26078

James P, Maeda M, Fischer R, Verma AK, Krebs J, Penniston JT, Carafoli E (1988) Identification and primary structure of a calmodulin binding domain of the $Ca^{2+}$ pumps of human erythrocytes. J Biol Chem 263:2905–2910

James P, Inei M, Tada M, Chiesi M, Carafoli E (1989) Nature and site of phospholamban regulation of the $Ca^{2+}$ pump of sarcoplasmic reticulum. Nature 342:90–92

Jenner S, Farndale RW, Sage SO (1994) The effect of calcium store depletion and refilling with various bivalent cations on tyrosine phosphorylation and $Mn^{2+}$ entry in fura-2 loaded human platelets. Biochem J 303:337–339

Johansson JS, Haynes DH (1988) Deliberate quin 2 overload as a method for in situ characterization of active calcium extrusion systems and cytoplasmic calcium binding. Application to the human platelet. J Membr Biol 104:147–163

Johansson JS, Haynes DH (1992) Cyclic GMP increases the rate of the $Ca^{2+}$ extrusion pump in intact platelets but has no direct effect on the dense tubular $Ca^{2+}$ accumulation system. Biochim Biophys Acta 1105:40–50

Johansson JS, Nied LE, Haynes DH (1992) Cyclic AMP stimulates the $Ca^{2+}$ ATPase mediated $Ca^{2+}$ extrusion from human platelets. Biochim Biophys Acta 1105:19–28

Jones GD, Gear ALR (1988) Subsecond $Ca^{2+}$ dynamics in ADP and thrombin stimulated platelets; a continuous flow approach using Indo-1. Blood 71:1539–1543

Joseph SK, Ryan SV (1993) Phosphorylation of the inositol trisphosphate receptor in isolated rat hepatocytes. J Biol Chem 268:24509–23065

Joseph SK, Samanta S (1993) Detergent solubility of the inositol trisphosphate receptor in rat brain membranes. J Biol Chem 268:6477–6486
Käser-Glansmann R, Jakabova M, George JN, Luscher EF (1977) Stimulation of $Ca^{2+}$ uptake in platelet membrane vesicles by adenosine 3′5′-cyclic monophosphate and protein kinase. Biochim Biophys Acta 466:429–440
Khan AA, Steiner JP, Klein MA, Schneider MF, Snyder SH (1992a) $IP_3$ receptor; localisation to plasma membrane of T cells and co-capping with the T cell receptor. Science 257:815–818
Khan AA, Steiner JP, Snyder SH (1992b) Plasma membrane inositol 1,4,5-trisphosphate receptor of lymphocytes; selective enrichment of sialic acid and unique binding specificity. Proc Natl Aca Sci USA 89:2849–2853
Kim HY, Thomas D, Hanley MR (1995) Chromatographic resolution of an intracellular calcium influx factor from thapsigargin activated Jurkat cells. J Biol Chem 270:9706–9708
Kimura M, Aviv A, Reeves JP (1993) $K^+$-dependent $Na^+/Ca^{2+}$ exchange in human platelets. J Biol Chem 268:6874–6877
Kimura M, Cho JH, Reeves JB, Aviv A (1994) Inhibition of $Ca^{2+}$ entry by $Ca^{2+}$ overloading of intracellular $Ca^{2+}$ stores in human platelets. J Physiol 479:1–10
King WG, Downes CP, Prestwich DG, Rittenhouse SE (1990) $Ca^{2+}$ stimulated and protein kinase C inhibitable accumulation of inositol 1,3,4,6-tetrakisphosphate in human platelets. Biochem J 270:125–131
Klinz F-J, Seifert R, Schwaner I, Gausepohl H, Frank R, Schultz G (1992) Generation of specific antibodies against the rap1A, rap1B and rap2 small GTP-binding proteins. Analysis of rap and ras proteins in membranes from mammalian cells. Eur J Biochem 207:207–213
Koike Y, Ozaki R, Satoh RQA, Kurota K, Yatomi Y, Kume S (1994) Phosphatase inhibitors suppress $Ca^{2+}$ influx induced by receptor mediated intracellular $Ca^{2+}$ store depletion in human platelets. Cell Calcium 15:381–390
Komalavilas P, Lincoln TM (1994) Phosphorylation of the inositol 1,4,5-triphosphate receptor by cyclic GMP-dependent protein kinase. J Biol Chem 269:8701–8707
Koshiyama H, Lee H-C, Tashjian A-H Jr (1991) Novel mechanisms of intracellular $Ca^{2+}$ release in pituitary cells. J Biol Chem 266:16985–16988
Kovacs T, Corvazier E, Papp B, Magnier C, Bredoux R, Enyedi A, Sarkadi B, Enouf J (1994) Controlled proteolysis of $Ca^{2+}$ ATPases in human platelet and non-muscle cell membrane vesicles. Evidence for a multi-sarco endoplasmic reticulum $Ca^{2+}$ ATPase system. J Biol Chem 269:6177–6184
Kuno M, Gardner P (1987) Ion channels activated by inositol 1,4,5-trisphosphate in plasma membrane of human T-lymphocytes. Nature 355:356–358
Lee HC, Aarhus R, Graeff R, Gurnack ME, Walseth TF (1994) Cyclic ADP ribose activation of the ryanodine receptor is mediated by calmodulin. Nature 370:307–309
Lee KM, Toscas K, Villereal ML (1993) Inhibition of bradykinin-and thapsigargin-induced $Ca^{2+}$ entry by tyrosine kinase inhibitors. J Biol Chem 268:9945–9948
Leveque C, Hoshino T, David P, Shoji-Kasai Y, Leys K, Omori A, Lang B, El Far O, Sato K, Martin-Montot N, Newsom-Davies J, Takahashi M, Seagar MJ (1992) The synaptic vesicle protein synaptotagmin associates with calcium channels and is a putative Lambert-Eaton myasthenic syndrome antigen. Proc Natl Acad Sci USA 89:3625–3629
Lievremont J-P, Hill A-M, Hilly M, Mauger J-P (1994) The inositol 1,4,5-trisphosphate receptor is localized in special sub-regions of the endoplasmic reticulum in rat liver. Biochem J 300:419–427
Luckhoff P, Clapham DE (1992) Inositol 1,3,4,5-tetrakisphosphate activates an endothelial $Ca^{2+}$ permeable channel. Nature 326:301–304
Lytton J, Westlin M, Hanley MR (1991) Thapsigargin inhibits the sarcoplasmic or endoplasmic reticulum $Ca^{2+}$ ATPase family of $Ca^{2+}$ pumps. J Biol Chem 266:17067–17071

Lytton J, Westlin M, Burt SE, Schull GE, MacLennan DH (1992) Functional comparison between isoforms of the sarcoplasmic or endoplasmic reticulum family of calcium pumps. J Biol Chem 267:14483–14489

MacKenzie AB, Mahaut-Smith MP, Sage SO (1996) Activation of receptor-operated cation channels via $P_{2\times1}$ not $P_{2T}$ purinoceptors in human platelets. J Biol Chem 271:2879–2881

Maeda N, Niinobe M, Mikoshiba K (1990) A cerebellar purkinje cell marker P400 protein is an inositol 1,4,5-trisphosphate ($InP_3$) receptor protein. Purification and characterisation of $InP_3$ receptor complex. EMBO J 9:61–67

Maeda N, Kawasaki T, Nakade S, Yokota N, Taguchi T, Kasai M, Mikoshiba K (1991) Structural and functional characterization of inositol 1,4,5-triphosphate receptor channel from mouse cerebellum. J Biol Chem 266:1109–1116

Magnier C, Bredoux R, Kovacs T, Quarck R, Papp B, Corvazier E, Gunzburg J, Enouf J (1994) Correlated expression of the 97kDa sarcoendoplasmic reticulum $Ca^{2+}$-ATPase and rap 1B in platelets and various cell lines. Biochem J 297:343–350

Magnier C, Courvazier E, Aumont M-C, Le Jemetral TH, Enouf J (1995) Relationship between Rap 1 protein phopshorylation and regulation of $Ca^{2+}$ transport in platelets; a new approach. Biochem J 310:469–475

Mahaut-Smith MP, Sage SO, Rink TJ (1990) Receptor activated single channels in intact human platelets. J Biol Chem 265:10479–10483

Mahaut-Smith MP, Sage SO, Rink TJ (1992) Rapid ADP-evoked currents in human platelets recorded with the nystatin permeabilized patch technique. J Biol Chem 267:3060–3065

Maranto AR (1994) Primary structures, ligand binding and location of the human type 3 inositol 1,4,5-trisphosphate receptor expressed in intestinal epithelium. J Biol Chem 269:1222–1230

Meszaros LG, Bak J, Chu A (1993) Cyclic ADP ribose as an endogenous regulator of the non-skeletal type ryanodine receptor $Ca^{2+}$ channel. Nature 364:76–79

Meyer T, Stryer L (1991) $Ca^{2+}$ spiking. Annu Rev Biophys Chem 20:153–174

Michikawa T, Hamanaka H, Otsu H, Yamamoto A, Miyawaki A, Furuichi T, Tashiro Y, Mikoshiba K (1994) Transmembrane topology and sites of N-glycosylation of inositol 1,4,5-trisphosphate receptor. J Biol Chem 269:9184–9189

Mignery GA, Newton CL, Archer BT III, Sudhof TC (1990) Structure and expression of the rat inositol 1,4,5-trisphosphate receptor. J Biol Chem 265:12679–12685

Minke B, Selinger Z (1991) Inositol lipid pathway in fly photoreceptors: excitation, calcium mobilization and retinal degeneration. In: Osbourne NA, Chader GJ (eds) Prog Retinal Res 11. Pergamon, Oxford, pp 99–124

Missiaen L, De Smedt H, Droogmans G, Casteels R (1992) $Ca^{2+}$ release induced by inositol 1,4,5-trisphosphate is a steady-state phenomenon controlled by luminal $Ca^{2+}$ in permeabilized cell. Nature 357:599–601

Miyazaki SI, Yuzaki M, Nakade K, Shirakawa H, Nakanishi S, Nakade S, Mikoshiba K (1992a) Block of $Ca^{2+}$ wave and $Ca^{2+}$ oscillation by antibody to the inositol 1,4,5-trisphosphate receptor in fertilized hamster eggs. Science 257:251–255

Miyazaki SI, Shirakawa H, Nakade K, Honda Y, Yuzaki M, Nakade S, Mikoshiba K (1992b) Antibody to the inositol trisphosphate receptor blocks thimerosal-enhanced $Ca^{2+}$-induced $Ca^{2+}$ release and $Ca^{2+}$ oscillations in hamster eggs. FEBS Lett 309:180–184

Miyawaki A, Furuichi T, Ruou Y, Yoshikawa S, Nakagawa T, Saitoh T, Mikoshiba K (1991) Structure function relationships of the mouse inositol 1,4,5-trisphosphate receptor. Proc Natl Aca Sci USA 88:4911–4915

Molleman A, Hoiting B, Duin M, van den Akker J, Nelemans A, Hertog AD (1991) Potassium channels regulated by inositol (1,3,4,5) tetrakisphosphate and internal calcium in DDT, MF-2 smooth muscle cells. J Biol Chem 266:5758–5663

Monkawa T, Miyawaki A, Sugiyama T, Yoneshima H, Yamamoto-Hino M, Furuichi T, Saruta T, Hasegawa M, Mitoshiba K (1995) Heterotetramic complex formation of inositol 1,4,5-triphosphate receptor subunits. J Biol Chem 270:14700–14704

Montell C, Rubin GM (1989) Molecular characterization of the drosophilia trp locus: a putative integral membrane protein required for phototransduction. Neuron 2:1313–1323

Morris AP, Gallacher DV, Irvine RF, Peterson OH (1987) Synergism of inositol trisphosphate and tetrakisphosphate in activating $Ca^{2+}$ dependent $K^{+}$ channels. Nature 330:653–655

Mourey RJ, Estavez VA, Maracek JF, Barrow RK, Prestwick GD, Snyder SH (1993) Inositol 1,4,5-trisphosphate receptors; labelling the inositol 1,4,5-trisphosphate binding site with photoaffinity ligands. Biochemistry 32:1719–1726

Movsesian MA, Nishikawa M, Adelstein RS (1984) Phosphorylation of phospholamban by calcium activated, phospholipid dependent protein kinase. Stimulation of cardiac sarcoplasmic reticulum calcium uptake. J Biol Chem 259:8029–8032

Mullaney JM, Yu M, Ghosh TK, Gill DL (1988) $Ca^{2+}$ entry into the inositol 1,4,5-trisphosphate releasable calcium pool is mediated by a GTP regulatory mechanism. Proc Natl Acad Sci USA 85:2499–2503

Nakagawa T, Okano H, Furuichi T, Aruga J, Mikoshiba K (1991) The subtypes of the mouse inositol 1,4,5-trisphosphate receptor are expressed in a tissue specific manner. Proc Natl Acad Sci USA 88:6244–6248

Nakade S, Phee SK, Hamanaka H, Mikoshiba K (1994) Cyclic AMP-dependent phosphorylation of an immuno affinity-purified homotetrameric inositol 1,4,5-trisphosphate receptor (type 1) increases $Ca^{2+}$ flux in reconstituted lipid vesicles. J Biol Chem 269:6735–6742

Newton CL, Mignery GA, Sudhof T (1994) Co-expression in vertebrate tissues and cell lines of multiple inositol 1,4,5-trisphosphate ($InsP_3$) receptors with distinct affinities for $InsP_3$. J Biol Chem 269:28613–28619

Nicoll DA, Longoni S, Philipson KD (1990) Molecular cloning and functional expression of the cardiac sarcolemmal $Na^{+}$ $Ca^{2+}$ exchanger. Science 250:562–565

Niggli V, Pennsiton JT, Carafoli E (1979) Purification of the $Ca^{2+}$ + $Mg^{2+}$ ATPase from human erythrocyte membranes using a calmodulin affinity column. J Biol Chem 254:9955–9958

Nishio H, Ikegami Y, Segawa T (1991) Fluorescence digital image analysis of serotonin-induced calcium oscillations in single blood platelets. Cell Calcium 12:177–184

O'Rourke FA, Halenda SP, Zavioco GB, Feinstein MB (1985) Inositol 1,4,5-trisphosphate release $Ca^{2+}$ from a $Ca^{2+}$ transporting membrane vesicle fraction derived from human platelets. J Biol Chem 260:956–962

O'Rourke FA, Zavioco GB, Feinstein MB (1989) Release of $Ca^{2+}$ by inositol 1,4,5-trisphosphate in platelet membrane vesicles is not dependent on cAMP dependent protein kinase. Biochem J 257:715–721

O'Rourke F, Soons K, Flammenhauft R, Watras J, Bio-Larue C, Mathews E, Feinstein MB (1994) $Ca^{2+}$ release by inositol 1,4,5,-trisphosphate is blocked by the $K^{+}$ channel blockers apamin and tetrapentyl ammonium ion, and a monoclonal antibody to a 63 kDa membrane protein; reversal of blockage by $K^{+}$ ionophores nigericin and valinomycin and purification of the 63-kDa antibody-binding protein. Biochem J 300:673–383

O'Rourke F, Matthews E, Feinstein MB (1995) Purification and characterisation of the human type 1 In(1,4,5)$P_3$ receptor from platelets and comparison with receptor subtypes in other normal and transformed blood cells. Biochem J 312:499–503

Ozaki Y, Yatomi Y, Wakasugi S, Shirasawa Y, Saito H, Kume S (1992a) Thrombin-induced calcium oscillation in human platelets and Meg-01, a megakaryoblastic leukemia cell line. Biochem Biophys Res Commun 183:864–871

Ozaki Y, Yatomi Y, Kume S (1992b) Evaluation of platelet calcium ion mobilisation by the use of varient divalent ions. Cell Calcium 13:19–27

Papp B, Enyedi A, Kovacs T, Sarkadi B, Wuytack F, Thastrup O, Gardos G, Bredoux R, Levy-Toledano S, Enouf J (1991) Demonstration of two forms of the $Ca^{2+}$

pumps by Thapsigargin inhibition and radioimmunoblotting in platelet membrane vesicles. J Biol Chem 266:14593–14596
Papp B, Enyedi A, Paszty K, Kovacs T, Sarkadi B, Gardos G, Magnier C, Wuytack F, Enouf J (1992) Simultaneous presence of two distinct endoplasmic reticulum type calcium isoforms in human cells. Biochem J 288:297–302
Papp B, Paszty K, Kovacs T, Sarkadi B, Gardos G, Enouf J, Enyedi A (1993) Characterisation of the inositol trisphosphate-sensitive and insensitive calcium store by selective inhibition of the endoplasmic reticulum type calcium pump isoforms in isolated platelet membranes. Cell Calcium 14:531–538
Parekh AB, Terlaw IT, Struhmer W (1993) Depletion of $InP_3$ stores activates a $Ca^{2+}$ and $K^+$ current by means of a phosphatase and a diffusable messenger. Nature 264:814–818
Perin MS, Fried VA, Mignery GA, Jahn R, Sudhof TC (1990) Phospholipid binding by a synaptic vesicle protein homologous to the regulatory region of protein kinase C. Nature 345:260–263
Perin MS, Johnston PA, Ozcelik T, Jahn R, Franke V, Sudhof TC (1991) Structural and functional conservation of synaptotagmin (p65) in *Drosophila* and humans. J Biol Chem 266:615–622
Pollock WK, Sage SO, Rink TJ (1987) Stimulation of $Ca^{2+}$ efflux from Fura 2 loaded platelets activated by thrombin or phorbol myristate acetate. FEBS Lett 210:132–136
Powling MJ, Hardisty RM (1985) Glycoprotein IIbIIIa complex and $Ca^{2+}$ influx into stimulated platelets. Blood 66:731–734
Putney JW Jr (1986) A model for receptor regulated calcium entry. Cell Calcium 7:1–12
Putney JW Jr (1990) Capacitative calcium entry revisited. Cell Calcium 11:611–624
Putney JW Jr, Bird GSJ (1993) The signal for capacitative calcium entry. Cell 75:149–201
Quinton TM, Dean WL (1992) Cyclic AMP dependent phosphorylation of the inositol 1,4,5-trisphosphate receptor inhibits $Ca^{2+}$ release from platelet membranes. Biochem Biophys Res Commun 184:893–899
Quinton TM, Brown KD, Dean WL (1996) Inositol 1,4,5-trisphosphate-mediated $Ca^{2+}$ release from platelet internal membranes is regulated by differential phosphorylation. Biochemistry 35:6865–6871
Raha S, Jones GD, Gear ARL (1993) Sub-second oscillations of inositol 1,4,5-trisphosphate and inositol 1,3,4,5-tetrakisphosphate during platelet activation by ADP and thrombin; lack of correlation with calcium kinetics. Biochem J 292:643–646
Randriamampita C, Tsien RY (1993) Emptying of intracellular $Ca^{2+}$ stored releases a novel small messenger that stimulates $Ca^{2+}$ influx. Nature 364:809–814
Randriamampita C, Tsien RY (1995) Degradation of a calcium influx factor (CIF) can be blocked by phosphatase inhibitors or chelation of $Ca^{2+}$. J Biol Chem 270:29–32
Reilander H, Achilles A, Friedel V, Maul G, Lottspeich F, Cook NJ (1992) Primary structure and functional expression of the $Na^+/Ca^{2+}$ $K^+$ exchanger from bovine rod photoreceptors. EMBO J 11:1689–1695
Reiser G, Schafer R, Donie F, Hulser E, Nehls-Sahabandu M, Mayr GW (1991) A high affinity inositol 1,3,4,5-tetrakisphosphate receptor protein from brain is specifically labelled by a newly synthesised photoaffinity analogue, N-(4-azidosalicyl)aminoethanol(1)-1-phospho-D-myo-inositol 3,4,5-trisphosphate. Biochem J 280:533–539
Rengasamy A, Feinberg H (1988) Inositol 1,4,5-trisphosphate induced $Ca^{2+}$ release from platelet plasma membrane vesicles. Biochem Biophys Res Commun 150: 1021–1026
Rengasamy A, Soura S, Feinberg H (1987) Platelet $Ca^{2+}$ homeostasis; $Na/Ca^{2+}$ exchange in plasma membrane vesicles. Thromb Haemost 57:337–340
Resink TJ, Tkachuk V-A, Erne P, Buhler FR (1985) Platelet membrane calmodulin calcium-adenosine triphosphatase. Hypertension 8:159–166

Rink TJ, Hallam TJ (1984) What turns platelets on. TIBS 9:215–219
Rink TJ, Jacob R (1989) Calcium oscillations in non-excitable cells. Trends Neurosci 12:43–46
Ross CA, Danoff SK, Schell MJ, Snyder SH, Ullrich A (1992) Three additional inositol 1,4,5-trisphosphate receptors; molecular cloning and differential localisation in brain and peripheral tissues. Proc Natl Acad Sci USA 89:4265–4269
Rossier MG, Bird GSJ, Putney JW Jr (1991) Subcellular distribution of the calcium storing inositol 1,4,5-trisphosphate-sensitive organelle in rat liver. Possible linkage to the plasma membrane through the actin microfilaments. Biochem J 274:643–650
Ryback MEM, Renzulli LA (1989) Ligand inhibition of the platelet glycoprotein IIb-IIIa complex function as a calcium channel in liposomes. J Biol Chem 264:14617–14720
Sage SO, Rink TJ (1985) Kinetic differences between thrombin induced and ADP-induced $Ca^{2+}$ influx and release from internal stores in Fura 2 loaded human platelets. Biochem Biophys Res Commun 136:1124–1129
Sage SO, Rink TJ (1987) The kinetics of changes in intracellular $Ca^{2+}$ concentration in Fura 2-loaded human platelets. J Biol Chem 262:16364–16369
Sage SO, Merrit JE, Hallam TJ, Rink TJ (1989) Receptor mediated $Ca^{2+}$ entry in Fura 2 loaded human platelets stimulated with ADP and thrombin. Dual wavelength studies with $Mn^{2+}$. Biochem J 258:923–926
Sage SO, Reast R, Rink TJ (1990) ADP evokes biphasic $Ca^{2+}$ influx in Fura 2 loaded human platelets. Evidence for $Ca^{2+}$ entry regulated by the intracellular $Ca^{2+}$ store. Biochem J 265:675–680
Sage SO, Rink TJ, Mahaut-Smith MP (1991) Resting and ADP evoked changes in cytosolic free sodium concentration in human platelets loaded with the indicator SBFI. J Physiol 441:559–573
Sargeant P, Clarkson WD, Sage SO, Heemskerk JWM (1992) Calcium influx evoked by $Ca^{2+}$ store depletion in human platelets is more susceptible to cytochome P-450 inhibitors than receptor-mediated calcium entry. Cell Calcium 13:553–564
Sargeant P, Farndale RW, Sage SO (1993a) The tyrosine kinase inhibitors methyl 2,5-dihydroxycinnamate and genistein reduce thrombin-evoked tyrosine phosphorylation and entry in human platelets. FEBS Lett 315:242–246
Sargeant P, Farndale RW, Sage SO (1993b) ADP-and thapsigargin-evoked $Ca^{2+}$ entry and protein-tyrosine phosphorylation are inhibited by the tyrosine kinase inhibitors geinstein and methyl 2,5-dihydroxycinnamate in Fura-2-loaded human platelets. J Biol Chem 268:18151–18156
Schull GE, Greeb J (1988) Molecular cloning of two isoforms of the plasma membrane $Ca^{2+}$ transporting ATPase from rat brain. Structural and functional domains exhibit similarity to $Na^{+}$, $K^{+}$ and other cation transport ATPases. J Biol Chem 263:8646–8657
Schwartz MA, Brown EJ, Fazeli B (1993) A 50-kDa integrin associated protein is required for integrin regulated $Ca^{2+}$ entry in endothelial cells. J Biol Chem 268:19931–19934
Selinger Z, Doza YN, Minke B (1993) Mechanisms and genetics of photoreceptors desensitization in *Drosophila* flies. Biochim Biophys Acta 1179:283–299
Sharp AH, Snyder SH, Nigram AK (1992) Inositol 1,4,5-trisphosphate receptors. Localization in epithelial tissue. J Biol Chem 267:7444–7449
Shattil S (1995) Functions and regulation of the $\beta3$ integrins in hemostasis and vascular biology. Thromb Haemost 74:149–155
Sitsapesan R, McGarry SJ, Williams AJ (1994) Cyclic ADP ribose competes with ATP for the adenine nucleotide binding site on the cardiac ryanodine receptor $Ca^{2+}$ release channel. Circ Res 75:596–600
Stauffer TP, Hilfiker H, Carafoli E, Strehler EE (1993) Quantitative analysis of alternative splicing options for human plasma membrane calcium pumps genes. J Biol Chem 268:25993–26003

Stauffer TP, Guerin D, Carafoli E (1995) Tissue distribution of the four gene products of the plasma membrane $Ca^{2+}$ pump. A study using specific antibodies. J Biol Chem 270:12184–12190

Sudhof TC, Newton CL, Archer BI III, Ushkaryov YA, Mignery GA (1991) Structure of a novel $IP_3$ receptor. EMBO J 10:3199

Supattapone S, Danoff SK, Theibert A, Joseph S, Steiner J, Snyder SH (1988) Cyclic AMP-dependent phosphorylation of a brain inositol trisphosphate receptor decreases its release of calcium. Proc Natl Acad Sci USA 85:8747–8750

Takaya J, Lasker N, Bamforth R, Butkin M, Byrd LH, Aviv A (1990) Kinetics of $Ca^{2+}$ ATPase activation in platelet membranes of essential hypertensives and normotensives. Am J Physiol 258:C988–C994

Takemura H, Hughes AR, Thastrup O, Putney JW (1989) Activation of calcium entry by the tumor promoter thapsigargin in parotid acinar cells. Evidence that an intracellular calcium pool and not an inositol phosphate regulates calcium fluxes at the plasma membrane. J Biol Chem 264:12266–12271

Tao J, Johansson JS, Haynes DH (1992) Stimulation of dense tubular $Ca^{2+}$ uptake in human platelets by cAMP. Biochim Biophys Acta 1105:29–39

Theibert AB, Estevez VA, Ferris CD, Danoff SK, Barrow RK, Prestwich GD, Snyder SH (1991) Inostiol 1,3,4,5-tetrakisphosphate and inositol hexakisphosphate receptor proteins. Isolation and characterisation from rat brain. Proc Natl Acad Sci USA 88:3165–3169

Theibert AB, Estevez VA, Mourey RJ, Maracek JF, Barrow RK, Prestwich GD, Snyder SH (1992) Photoaffinity labelling and characterisation of inositol 1,3,4,5-tetrakisphosphate and inositol hexakisphosphate binding proteins. J Biol Chem 267:9071–9079

Thomas D, Hanley MR (1995) Evaluation of calcium influx factors from stimulated Jurkat T-lymphocytes by microinjection into *Xenopus* oocytes. J Biol Chem 270:6429–6432

Tohmatsu T, Nishida A, Nagao S, Nakashima S, Nozawa Y (1989) Inhibitory action of cAMP on inositol 1,4,5-trisphosphate induced $Ca^{2+}$ release in saponin permeabilised platelets. Biochim Biophys Acta 1013:190–193

Torti M, Lapetina EG (1992) Role of rap1B and $p21^{ras}$ GTPase-activating protein in the regulation of phospholipase C-$\gamma 1$ in human platelets. Proc Natl Acad Sci USA 89:7796–7800

Toyofuku T, Kurzydlowski K, Lytton J, MacLennan DH (1992) The nucleotide binding/hinge domain plays a crucial role in determining isoform-specific $Ca^{2+}$ dependence of organellar $Ca^{2+}$ ATPases. J Biol Chem 267:14490–14496

Toyofuku T, Kurzydlowski K, Tada M, MacLennan DH (1993) Identification of regions in the $Ca^{2+}$ ATPase of sarcoplasmic reticulum that affect functional association with phospholamban. J Biol Chem 268:2809–2815

Toyoshima C, Sasabe H, Stokes DL (1993) Three dimensional cryo-electron microscopy of the calcium ion pump in sarcoplasmic reticulum membrane. Nature 362:469–471

Tsien RW, Tsien RY (1990) Calcium channels, stores and oscillations. Annu Rev Cell Biol 6:715–760

Tsunoda Y, Matsuno K, Tashiro Y (1988) Spatial distribution and temporal change of cytoplasmic free calcium in human platelets. Biochem Biophys Res Commun 156:1152–1159

Tsunoda Y, Matsuno K, Tashiro Y (1991) Cytosolic acidification leads to $Ca^{2+}$ mobilization from intracellular stores in single populational parietal cells and platelets. Exp Cell Res 193:356–363

Vaca L, Simkins WG, Hu Y, Kunze DL, Schilling WP (1994) Activation of recombinant trp by thapsigargin in Sf9 insect cells. Am J Physiol 267:C1501–C1505

Verma AK, Filoteo AG, Stanford DR, Wieben E-D, Penniston JT, Strehler EE, Fischer R, Heim R, Vogel G, Matheus S, Strehler-Page MA, James P, Vorherr T, Krebs J, Carafoli E (1988) Complete primary structure of a human plasma membrane $Ca^{2+}$ pump. J Biol Chem 263:14152–14159

Vigne P, Breittmayer J-P, Duval D, Frelin C, Lazdunski M (1988) The $Na^{+}/Ca^{2+}$ antiporter in aortic smooth muscle cells. Characterization and demonstration of an activation by phorbol esters. J Biol Chem 263:8078–8083

Vorherr T, Kessler T, Hofmann F, Carafoli E (1991) The calmodulin binding domain mediates the self-association of the plasma membrane $Ca^{2+}$ pump. J Biol Chem 266:22–27

Vostal JG, Fratantoni JC (1993) Econazole inhibits thapsigargin induced platelet calcium influx by mechanisms other than cytochome P-450 inhibition. Biochem J 295:525–529

Vostal JG, Shulman NR (1993) Vinculin is the major platelet protein that undergoes $Ca^{2+}$ dependent tyrosine phosphorylation. Biochem J 294:675–680

Vostal JG, Jackson WL, Shulman NR (1992) Cytosolic and stored calcium antagonistically control tyrosine phosphorylation of specific platelet proteins. J Biol Chem 266:16911–16916

Waldman R, Bauer S, Gobel C, Hofman F, Jacobs KH, Walter U (1986) Demonstration of cGMP dependent protein kinase and cGMP dependent phosphorylation in cell free extracts of platelets. Eur J Biochem 158:203–210

Waldman R, Nieberding M, Walter U (1987) Vasodilator stimulated protein phosphorylation in platelets is mediated by cAMP- and cGMP-dependent protein kinases. Eur J Biol 167:441–448

Walseth TF, Aarhus R, Kerr JA, Lee H-C (1993) Identification of cyclic ADP ribose binding proteins by photoaffinity labelling. J Biol Chem 268:26686–26691

Watson SP, McConnel RT, Lapetina EG (1984) The rapid formation of inositol phosphates in human platelets by thrombin is inhibited by prostacyclin. J Biol Chem 259:13199–13203

Watson SP, Ruggeiro M, Abrahams SL, Lapetina EG (1986) Inositol trisphosphate induces aggregation and release of 5-hydroxytryptamine from saponin permeabilised human platelets. J Biol Chem 261:5368–5372

Wegener AD, Lindemann JP, Simmerman HKB, Jones LR (1989) Phospholamban phosphorylation in intact ventricles. Phosphorylation of serine 16 and threonine 12 in response to beta-adrenergic stimulation. J Biol Chem 264:11468–11474

Wes PD, Chevesich J, Jeromin A, Rosenberg C, Stetten G, Montell C (1995) TRCP 1, a human homolog of a *Drosophila* store operated channel. Proc Natl Acad Sci USA 92:9652–9656

White GC II, Barton DW, White TE, Fisher TH (1989) Cyclic AMP-dependent protein kinase does not increase calcium transport in platelet microsomes. Thromb Res 56:575–581

White GC II, Crawford N, Fischer TH (1993) Cytoskeletal interactions of Rap 1B in platelets. In: Authi KS, Watson SP, Kakkar VV (eds) Mechanisms of platelet activation and control. Plenum, London, pp 187–194

Wuytack F, Papp B, Verboomen H, Raeymaekers L, Dode L, Bobe R, Enouf J, Bokkala S, Authi KS, Casteels R (1994) A sarco/endoplasmic reticulum $Ca^{2+}$ ATPase 3-type $Ca^{2+}$ pump is expressed in platelets in lymphoid cells and in mast cells. J Biol Chem 269:1410–1416

Yamaguichi A, Yamamoto N, Kitagawa H, Tanoue K, Yamazaki H (1987) $Ca^{2+}$ influx mediated through the GPIIbIIIa complex during platelet activation. FEBS Lett 225:228–232

Yamamoto H, Maeda N, Niinobe M, Miyamoto E, Mikoshiba K (1989) Phosphorylation of P400 protein by cyclic AMP-dependent protein kinase and $Ca^{2+}$ calmodulin-dependent protein kinase II. J Neurochem 53:917–923

Yamamoto-Hino M, Sugiyama T, Hikichi K, Mattei MG, Hasegawa K, Seking S, Sakurada K, Miyawaki A, Furuichi T, Hasegawa M, Mikoshiba K (1994) Cloning and characterisation of human type 2 and type 3 inositol 1,4,5 trisphosphate receptors. Receptors Channels 2:9–22

Yatomi Y, Ruan F, Hakomori S-I, Igarashi Y (1995) Sphingosine-1-phosphate: a platelet activating sphingilipid released from agonist stimulated human platelets. Blood 86:193–202

Yoshida K, Nachmias VT (1987) Phorbol ester stimulates calcium sequestration in saponised human platelets. J Biol Chem 262:16048–16054
Zidek W, Rustmeyer T, Schluter W, Karas M, Kisters K, Graefe V (1992) Isolation of an ultrafiltrable $Ca^{2+}$ ATPase inhibitor from the plasma of ureamic patients. Clin Sci 82:659–665
Zschauer A, van Breemen C, Buhler FR, Nelson MT (1988) Calcium channels in thrombin activated human platelet membrane. Nature 334:703
Zweifach A, Lewis RS (1993) Mitogen-regulated $Ca^{2+}$ current of T lymphocytes is activated by depletion of intracellular stores. Proc Natl Acad Sci USA 90:6295–6294

CHAPTER 16

# Regulation of Platelet Function by Nitric Oxide and Other Nitrogen- and Oxygen-Derived Reactive Species

E. SALAS, H. MISZTA-LANE, and M.W. RADOMSKI

## A. Introduction

The biosynthesis of nitric oxide (NO) in the vascular system provides a simple but efficient regulatory mechanism to support hemostasis and prevent thrombosis. In this chapter, we review evidence for the importance of NO as hemostatic molecule. We also describe the generation and platelet effects of other nitrogen- and oxygen-derived reactive species and their interactions with NO. In addition, we review how the changes in the generation, actions, and metabolism of reactive species may affect the pathogenesis of vascular diseases. Finally, we present the pharmacological approaches to NO substitution therapy that can modulate the clinical course of vascular pathologies.

## B. A Historical Perspective

The current awareness of the biological significance of NO as a regulator of platelet function has evolved from the pioneering findings of FURCHGOTT and ZAWADZKI (1980). They demonstrated that the vascular endothelium, when stimulated with acetylcholine, released an unstable substance (endothelium-derived relaxing factor, EDRF) that caused vasorelaxation of rat aorta rings. The search for the chemical identity of this factor led to the proposal that the free radical NO gas is EDRF (FURCHGOTT 1988; IGNARRO et al. 1987) and to the description of NO biosynthesis, i.e., the L-arginine–to–NO pathway (PALMER et al. 1987, 1988). This pathway has been shown to be of crucial importance for the regulation of homeostasis in cardiovascular, central and peripheral nervous respiratory, gastrointestinal, genitourinary, endocrine, and immune systems (RADOMSKI 1995).

In the vascular system, the abluminal release of NO by endothelial cells results in the vasodilator tone, increased blood flow, and decreased resistance and increased conductivity of the vessel wall (for review see RADOMSKI 1995). In this chapter, we focus on the generation, release, and actions of NO in the lumen of arteries and veins, aimed at preservation of blood fluidity. It is noteworthy that the conductile and fluid properties of the vessel wall are mutually dependent in maintaining vascular homeostasis.

# C. Nitric Oxide Synthase Isoenzymes

## I. General Characteristics

At least three separate genes encode the nitric oxide synthase (NOS) enzyme family of proteins (E.C.1.14.13.39). These are isoenzyme I or nNOS (first identified in neurones), II or iNOS (cytokine-induced), and III or eNOS (first identified in endothelial cells; for review see KNOWLES 1994). Studies on the chromosomal localization of human NOS have assigned eNOS to chromosome 7, nNOS to chromosome 12 and iNOS to chromosome 17. The isoforms of NOS belong to the family of the cytochrome $P_{450}$ reductase enzymes. The native NOS appears to be a homodimer and is localized both in the soluble and particulate fractions of the cell; its molecular weight ranges from 130 to 160 kDa per subunit. Consensus binding sites for flavin adenine dinucleotide (FAD), flavin mononucleotide (FMN), NADPH, and calmodulin are conserved in all isoforms. Binding sites for L-arginine, $H_4$ biopterin, heme and molecular oxygen have been also postulated (Fig. 1). Some NOS (e.g., nNOS and eNOS) are constitutively active, i.e., their activation does not require new enzyme protein synthesis but usually depends on increased intracellular $Ca^{2+}$ levels to stimulate the binding of calmodulin to its site (for literature see SESSA 1994). Other NOS (e.g., iNOS) are inducible, i.e., their activation requires new protein synthesis. Endotoxin and cytokines often induce the expression of iNOS. A distinct feature of iNOS is also its independence of $Ca^{2+}$ levels; this may be due to the fact that it contains calmodulin as a tightly bound subunit at low levels of $Ca^{2+}$ found in resting cells. The iNOS, although functionally $Ca^{2+}$ independent, may be regulated at transcriptional, translational, and posttranslational levels (for references, see XIE and NATHAN 1994).

## II. Platelet Nitric Oxide Synthase

In 1990 we found that platelets generate NO via a NOS-dependent pathway (RADOMSKI et al. 1990a,b). The rationale for this work was based on the

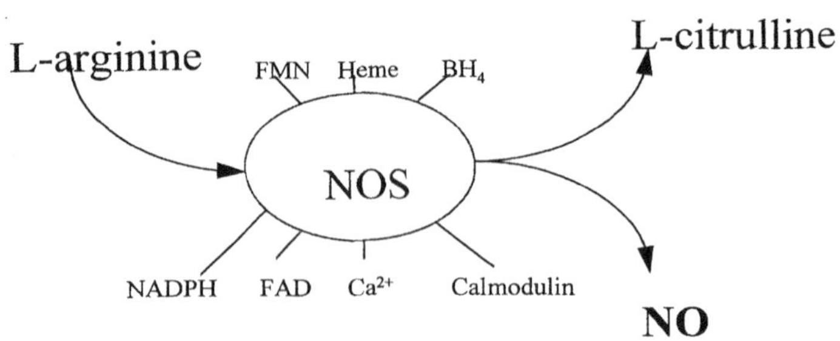


**Fig. 1.** Regulation and reaction of nitric oxide synthase (*NOS*). The enzyme converts *L-arginine* to *NO* and *L-citrulline*, a coproduct of this reaction. *NOS* is a heme-containing enzyme that requires flavin nucleotides (*FMN* and *FAD*), tetrahydrobiopterin (*$BH_4$*), *$Ca^{2+}$-calmodulin* and *NADPH* for its activity

observations that the activity of soluble guanylyl cyclase (GC-S) and the levels of cyclic GMP (cGMP) were elevated during platelet aggregation by aggregating agents such as collagen and arachidonic acid. These observations prompted the hypothesis that cGMP may be one of the mediators of platelet activation that acts by opposing the platelet-inhibitory activity of cAMP (GOLDBERG et al. 1975). However, in the early 1980s the work by Ignarro and colleagues clearly demonstrated that the biological actions of cGMP inhibited platelet aggregation (MELLION et al. 1981).

In order to resolve this controversy, we hypothesized that aggregated platelets express NOS and that NO acts by an autocrine regulatory mechanism that downregulates aggregation via cGMP-dependent mechanism (RADOMSKI et al. 1990a,b). This hypothesis was supported by some earlier work performed before the discovery of the L-arginine–to–NO pathway which demonstrated that L-arginine, the substrate for NOS, inhibited platelet aggregation both ex vivo (CAREN and CORBO 1973) and in vitro (HOUSTON et al. 1983). Using platelet aggregometry we confirmed that L-arginine was an inhibitor of platelet aggregation and found that inhibition of NOS potentiated aggregation induced by different aggregating agents in vitro (RADOMSKI et al. 1990b). We also found that cytosolic fractions obtained from human platelet homogenates constitutively expressed $Ca^{2+}$-dependent NOS activity which was functionally linked to the activity of GC-S (RADOMSKI et al. 1990a). The presence of NOS was subsequently confirmed in human platelets (PRONAI et al. 1991; NORIS et al. 1993; CADWGAN and BENJAMIN 1993; BODZENTA-LUKASZYK et al. 1994; MURUGANANDAM and MUTUS 1994; NAKATSUKA and OSAWA 1994; LAUNAY et al. 1994; MEHTA et al. 1995; ZHOU et al. 1995; SASE and MICHEL 1995; DELACRETAZ et al. 1995; PIGAZZI et al. 1995; LANTOINE et al. 1995; ADAMS et al. 1995; CHEN and MEHTA 1996) and described in rabbit (GOLINO et al. 1992), porcine (BERKELS et al. 1994), canine (YAO et al. 1992), and rat platelets (THOM et al. 1994).

Recently, both the message and protein of platelet NOS were characterized. SASE and MICHEL (1995) synthesised cDNA from platelet RNA and found it to be identical to eNOS DNA. MEHTA and colleagues (MEHTA et al. 1995; CHEN and MEHTA 1996) found expression of eNOS mRNA but not nNOS mRNA in platelets. Using Western blot analysis and a monoclonal antibody against eNOS they also detected the expression of a protein with a molecular weight of 140–150kDa. The purification of NOS from platelets showed that the native protein is dimeric $Ca^{2+}$, calmodulin, NADPH, FAD, $H_4$ biopterin, and L-arginine dependent, and that its activity is sensitive to stimulation with ADP and to inhibition by inhibitors of NOS (Table 1). Interestingly, the reported molecular weights of purified subunits of NOS are 80 and 130kDa (MURUGANANDAM and MUTUS 1994; MEHTA et al. 1995). It is tempting to speculate that the 80-kDa protein is a post-transcriptionally modified eNOS or a novel NOS isoform. All of these data clearly indicate that platelets constitutively express NOS with a biochemical profile similar to nNOS and eNOS.

**Table 1.** Characteristics of nitric oxide synthase and the soluble guanylyl cyclase in platelets

| | NOS | GC-S |
|---|---|---|
| Protein | Homodimer | Heterodimer $\alpha_1\beta_1$ |
| | 80–130 kDa | 73 and 70 kDa |
| Cofactors | Heme? | Heme present |
| | $Ca^{2+}$ | $Ca^{2+}$ |
| | Calmodulin | $Mg^{2+}$ |
| | NADPH | |
| | $H_4$biopterin | |
| Substrate | L-arginine | GTP |
| Products | NO | cGMP |
| | L-citrulline | |
| Inhibitors of enzyme activity | Arginine analogs | ODQ |
| | Methylene blue | LY83583 |
| | | Methylene blue |

See text for abbreviations.

The expression of iNOS in cells requires de novo protein synthesis. Platelets are anucleate and acquire most of their proteins by transfer from parent cells, i.e., megakaryocytes (Lelchuk et al. 1990). Therefore, we suggested that most of platelet iNOS activity is likely to be derived from megakaryocytes that can express both constitutive and inducible NOS (Lelchuk et al. 1992). However, recent work has established the presence of platelet-specific mRNAs in platelets and showed that they do synthesize proteins (Djaffar et al. 1991). In addition, the expression of iNOS as shown by Western analysis has been recently demonstrated in cytokine-stimulated as well as resting platelets (Chen and Mehta 1996). Thus, the platelet iNOS may be both of platelet and megakaryocyte origin.

Interestingly, it appears that platelets themselves can modulate the expression of iNOS. Indeed, platelet-derived growth factor (PDGF), released from platelet granules by collagen, has been shown to downregulate interleukin-1$\beta$ (IL-1$\beta$)-induced expression of iNOS in rat vascular smooth muscle cells (Durante et al. 1994). In contrast, the exposure of cells to platelet surface membranes could promote the expression of iNOS since activated platelets have membrane-bound IL-1$\beta$ capable of cytokine induction in endothelial cells (Hawrylewicz et al. 1991).

## III. The Endothelial NOS

Endothelial cells also express a constitutive isoform of NOS (eNOS). The eNOS genes from human umbilical vein and bovine aorta endothelial cells have been cloned (Marsden et al. 1992; Sessa et al. 1992). This is a $Ca^{2+}$-, NADPH-, flavin-, and biopterin-dependent enzyme. The N-terminal myristoylation of eNOS is a unique feature among the NOS family, which permits this isoform to be compartmentalized and associated with the particulate fraction of the endothelium (SESSA 1994).

The exposure of endothelial cells to endotoxin and cytokines results in the expression of iNOS activity (RADOMSKI et al. 1990c). The iNOS gene in endothelial cells has not yet been cloned; however, Northern and Western blots analyses have confirmed the cytokine-induced expression of both mRNA and protein for iNOS in rat vascular endothelial cells (KANNO et al. 1994). Its biochemical characterization showed that this isoform, similarly to that from macrophages, requires NADPH, $H_4$ biopterin, and L-arginine and is $Ca^{2+}$-independent (RADOMSKI et al. 1990c; GROSS et al. 1991). The expression of endothelial iNOS is inhibited by transforming growth factor (TGF-$\beta$), inhibitors of protein synthesis, and glucocorticoids (RADOMSKI et al. 1990c; KANNO et al. 1994). TGF-$\beta$ and glucocorticoids may block iNOS gene expression at the transcriptional level.

## IV. Nitric Oxide Synthase Reaction

The isoforms of NOS utilize the guanido nitrogen atom of L-arginine and incorporate molecular oxygen to generate NO and L-citrulline. NOS catalyzes the hydroxylation of L-arginine to, $N^{\omega}$-hydroxy-L-arginine, and intermediate in the pathway. $N^{\omega}$-hydroxy-L-arginine is then oxidized to yield NO and L-citrulline. Some arginine analogs, including $N^{G}$-monomethyl-L-arginine (L-NMMA), inhibit both parts of the NOS reaction (for refs. KNOWLES 1994; RADOMSKI 1995).

## V. Metabolic Fate of NO

Both physicochemical properties and the microenvironment determine the metabolic fate of NO. In principle, there are three redox states of NO: free radical (NO·), nitrosonium ($NO^+$), and nitroxyl ($NO^-$). NO· is a paramagnetic and diffusible molecule. The diffusivity of NO· closely resembles that of oxygen (TAHA et al. 1992) and at a given temperature is lower in proteins and membranes than in water. NO· is likely to accumulate in lipids since its partitioning is greater in hydrocarbons than in water. The presence of endogenous higher oxides of oxygen, nitrite, and nitrate, suggests that the reaction of NO· with molecular oxygen may be one of the major determinants of its metabolism. However, the experiments using recently developed porphyrinic electrochemical microsensors indicate that molecular oxygen plays a minor role in the direct oxidation process of NO· in biological systems (for literature see BECKMAN and TSAI 1994). Indeed, it appears that NO· reacts with oxygen to form nitrite only at high, nonphysiological concentrations. In contrast, NO· reacts at a near diffusion-limited rate with superoxide to generate peroxynitrite ($ONOO^-$; for literature see BECKMAN and TSAI 1994). In blood NO· reacts with red cell hemoglobin to form methemoglobin and nitrate (WENNMALM et al. 1992).

$NO^+$ may be involved in the reactions of *S-nitrosylation* to form *S*-nitrosothiols. There is some evidence that *S-nitrosylation* of endogenous thiols such as glutathione and albumin results in the formation of *S*-nitrosothiols and prolongation of the biological half-life of NO. *S-Nitrosylation* of reactive thiols can also modify the activity of receptor and enzyme proteins (LIPTON et al. 1993). Interestingly, $ONOO^-$ can also result in *S-nitrosylation* and the generation of *S*-nitrosothiols (MORO et al. 1994).

Nitroxyl has been proposed to be a physiologically relevant form of NO (MURPHY and SIES 1991). However it is possible that most biological actions of $NO^-$ (if proven to be generated in vivo) are indirect and result from its conversion to NO· (FUKUTO et al. 1992).

## VI. Interactions of NO with Biomolecules

Nitric oxide has an affinity to heme in heme proteins including GC-S (CRAVEN and DERUBERTIS 1978). Recently, the sequestration of NO gas by subcellular fractions of vascular smooth muscle and platelets was examined using the chemiluminescence-headspace gas technique (LIU et al. 1993). This study demonstrated that NO is sequestered preferentially by subcellular fractions of these cells that contain GC-S activity and that the sequestration of NO in these fractions stimulates the catalytic activity of GC-S. The reaction leads to the conversion of magnesium guanosine 5′-triphosphate to guanosine 3′,5′-monophosphate (cGMP). Although the evidence for the inhibitory action of cGMP on platelet responses is overwhelming, the levels of this inhibition are still under investigation (Fig. 2). Elevated concentrations of cGMP activate cGMP-dependent protein kinase and the phosphorylation of various target proteins (WALTER 1989). Among the proteins which are phosphorylated in response to cGMP, the best characterized is a 46/50-kDa vasodilator-stimulated phosphoprotein (VASP, HAFFNER et al. 1995). In adhering platelets VASP is associated with actin filaments and focal contact areas, i.e., transmembrane junctions between microfilaments and the extracellular matrix (REINHARD et al. 1992). In particular, the association of VASP with the platelet cytoskeleton may be of importance for its inhibitory effect on the fibrinogen receptor (LORSTRUP et al. 1994).

cGMP-induced protein phosphorylation may be also involved in the uptake of serotonin by platelets (LAUNAY et al. 1994). cGMP decreases basal and stimulated concentrations of intracellular $Ca^{2+}$ (NAKASHIMA et al. 1986; JOHANSSON and HAYNES 1992). A number of $Ca^{2+}$ handling systems have been identified in platelets (see the chapter by K. Aunthi, this volume) including receptor-operated channels, passive leak, $Ca^{2+}$-ATPase extrusion pump, the $Na^+/Ca^{2+}$ exchanger, $Ca^{2+}$-accumulating ATPase pump of the dense tubular membrane (an intraplatelet membrane $Ca^{2+}$ store), and passive leakage and receptor-operated $Ca^{2+}$ channels in the dense tubular membrane. In principle, all of these processes could be affected by cGMP. It has been shown that

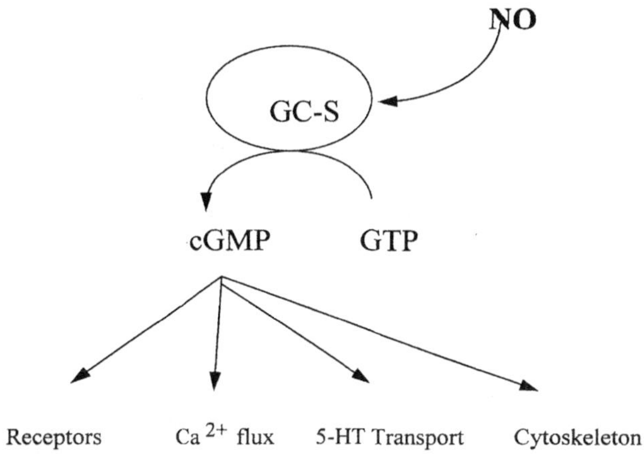


**Fig. 2.** Effects of *cGMP* on platelet function. Cyclic GMP (*cGMP*) is synthesized from GTP following stimulation of the soluble guanylyl cyclase (*CS-S*) by NO. *5-HT*, 5-hydroxytryptamine

cGMP increases the activity of the $Ca^{2+}$-ATPase extrusion pump and leakage across the plasma membrane (JOHANSSON and HAYNES 1992). In addition, cGMP causes inhibition of $Ca^{2+}$ mobilization from intraplatelet stores including the dense tubular membrane (NAKASHIMA et al. 1986). In contrast to cAMP, cGMP does not stimulate the dense tubular $Ca^{2+}$ pump. Consequently, cGMP does not result in increased dense tubular sequestration of $Ca^{2+}$, whereas cAMP does. Thus, cGMP may be "a better $Ca^{2+}$ antagonist" (JOHANSSON and HAYNES 1992). Under some conditions cGMP, by inhibiting cAMP phosphodiesterase, may delay the hydrolysis of cAMP and enhance the biological effects of the latter nucleotide (MAURICE and HASLAM 1990). However, the physiological and pharmacological relevance of this "cross-talk" between cAMP and cGMP pathways is unclear (RADOMSKI et al. 1992).

Metabolism of membrane phospholipids may also be a target for the action of cGMP. Indeed, the inhibition of both phospholipase C and $A_2$ has been implicated in the mechanism of its action on platelets (NAKASHIMA et al. 1986; SANE et al. 1989). Finally, cGMP downregulates the function of some platelet receptors including the fibrinogen receptor IIb/IIIa and protein kinase C-induced expression of P-selectin (SALAS et al. 1994; MUROHARA et al. 1995; MENDELSOHN et al. 1990). The biological actions of cGMP are terminated by cGMP phosphodiesterase and by its efflux from platelets, and may also depend on the activity of protein phosphatases (WALTER 1989; WU et al. 1993).

There is also an on-going dispute as to the contribution of cGMP-independent effects in the regulation of platelet function by NO. In addition to GC-S, NO may target the following: arachidonic acid cyclooxygenase and 12-lipoxygenase enzymes, some enzymes of mitochondrial respiratory chain, glyceraldehyde-3-phosphate dehydrogenase (GAP-DH), and intracellular thi-

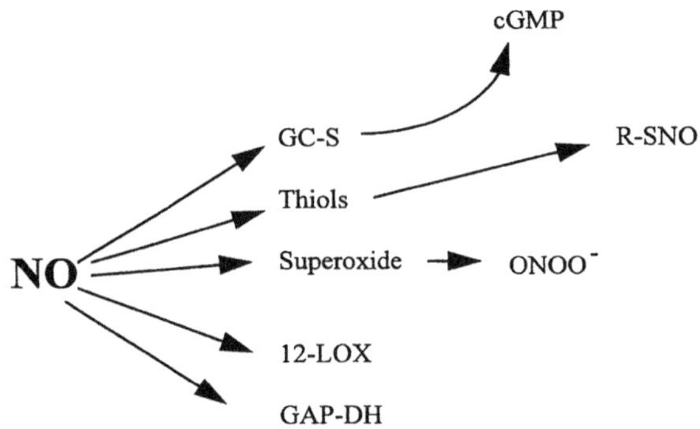


**Fig. 3.** Some of affinity targets of nitric oxide (*NO*) in platelets. In addition to *GC-S*, *NO* can interact with *thiols* resulting in the generation of *S*-nitrosothiols (*R-SNO*) and *superoxide* forming peroxynitrite (*ONOO⁻*), 12-lipoxygenase (*12-LOX*) and glyceraldehyde-3-phosphate dehydrogenase (*GAP-DH*)

ols such as glutathione (Fig. 3). Recent studies have demonstrated that NO is unlikely to interact with heme-containing cyclooxygenase and may selectively inhibit the activity of 12-lipoxygenase, an enzyme that does not contain heme (NAKATSUKA and OSAWA 1994). Nitric-oxide-induced inhibition of mitochondrial enzymes and GAP-DH could interfere with oxidative and glycolytic pathways of ATP formation (BRUNE and LAPETINA 1989); the significance of these effects for platelet function remains to be established. Interestingly, GAP-DH activation may be required for the transport of NO from some *S*-nitrosothiols to platelets (McDONALD et al. 1993).

Studies attempting to investigate the contributions of cGMP-dependent and -independent actions to the platelet-inhibitory activity of NO have been hampered by the lack of a potent and selective inhibitor of GC-S. We have recently characterized 1*H*-[1,2,4]oxadiazolo[4,3-a]quinoxalin-1-one (ODQ) and found it to be a potent and selective inhibitor of GC-S (MORO et al. 1996). We have shown that the antiaggregatory effect of NO in vitro could by completely reversed by ODQ, and is thus cGMP dependent.

The activity of GC-C can be increased by hydrogen peroxide (AMBROSIO et al. 1994), arachidonic acid hydroperoxides (BRUNE and ULLRICH 1991), and carbon monoxide (BRUNE and ULLRICH 1987), as well as by NO. The significance of these stimuli for the regulation of the soluble guanylyl cyclase in vivo remains to be determined. Among other possible reactants of NO are albumin, hemoglobin, and thiol-containing cellular receptors and ion channels. *S-Nitrosylation* of albumin leads to the generation of *S*-nitrosoalbumin which could serve as a biological depot for NO (STAMLER et al. 1992). In contrast, red cell hemoglobin is believed to be an NO scavenger that decreases the amount of biologically active NO (RADOMSKI et al. 1987a). The interactions of NO with platelet receptors and ion channels remain to be investigated.

## VII. Nitric Oxide – Physiological Regulator of Platelet Function

Early studies using indirect (activation of GC-S) and direct (electrochemical measurement of NO) assays failed to detect the generation of NO by resting platelets (RADOMSKI et al. 1990a; MALINSKI et al. 1993a). Recently, ZHOU and colleagues (1995) have found that platelets isolated from human blood, in the absence of aggregating agents, can generate NO. This release was partially inhibited by agents known to inhibit platelet activation, suggesting that platelets could have been activated during isolation procedures. However, if proven to occur in vivo, the basal production of NO in platelets could generate as much as 20 nmol NO per litre of blood (based on $1–3 \times 10^{11}$ platelets in 1 l of blood).

The platelet NOS becomes activated during platelet adhesion to collagen (POLANOWSKA-GRABOWSKA and GEAR 1994) and aggregation induced by collagen, ADP, and arachidonic acid (RADOMSKI et al. 1990a,b, MALINSKI et al. 1993a; ZHOU et al. 1995; LANTOINE et al. 1995). Nitric oxide generated during these reactions downregulates the extent of platelet activation. Since this enzyme is strictly $Ca^{2+}$-dependent and since platelet aggregation is associated with an increase in intraplatelet levels of $Ca^{2+}$, it is possible that this cation controls the activation of the platelet NOS. However, recent studies using direct electrochemical measurements of NO released from human platelets have shown that NO is released during collagen- but not thrombin-induced aggregation (MALINSKI et al. 1993a; LANTOINE et al. 1995). Since both agonists increase intraplatelet $Ca^{2+}$ levels it is possible that factors other than $Ca^{2+}$ regulate NOS in its platelet microenvironment, although the purified protein is $Ca^{2+}$-dependent.

The amounts of NO available for regulation of platelet function are supplemented by its generation in vascular endothelial cells. Studies with inhibitors of NOS suggest that eNOS generates NO constantly to provide the vasodilator tone (REES et al. 1989). The physiological stimuli for generation of NO by the endothelium are not yet fully understood, but flow and shear stress were found to stimulate the synthesis of NO via activation of the potassium $K_{Ca}$ channel (COOKE et al. 1991).

In 1987 it was shown that cultured and fresh endothelial cells, when stimulated with bradykinin, released NO in quantities sufficient to inhibit platelet adhesion (RADOMSKI et al. 1987b,c; SNEDDON and VANE 1988). Moreover, the coronary and pulmonary vasculature generated NO to inhibit platelet adhesion under constant flow conditions (VENTURINI et al. 1989; POHL and BUSSE 1989).

Platelet aggregation in vitro induced by a variety of agonists and by shear stress is inhibited by NO released from fresh or cultured endothelial cells (RADOMSKI et al. 1987a,d; FURLONG et al. 1987; BUSSE et al. 1987; MACDONALD et al. 1988; ALHEID et al. 1989; HOUSTON et al. 1990; BROEKMAN et al. 1991). This NO also causes disaggregation of preformed platelet aggregates (RADOMSKI et al. 1987d).

Animal studies have demonstrated that basal NO release or release stimulated by cholinergic stimuli and substance P result in inhibition of platelet aggregation induced by some aggregating agents or endothelial injury and increase bleeding time (ROSENBLUM et al. 1987; BHARDWAJ et al. 1988; HOGAN et al. 1988; HUMPHRIES et al. 1990; HERBACZYNSKA-CEDRO et al. 1991; MAY et al. 1991; GOLINO et al. 1992; YAO et al. 1992; HOUSTON and BUCHANAN 1994). In addition, there is luminal release of NO from human vasculature causing increases in intraplatelet cGMP levels (ANDREWS et al. 1994a). Finally, the administration of the NOS inhibitor L-NMMA to healthy volunteers increased platelet aggregation and granule release (BODZENTA-LUKASZYK et al. 1994), whereas L-arginine, the substrate for NO synthesis, led to the inhibition of platelet activation (CAREN and CORBO 1973; ADAMS et al. 1995). Thus, a concerted action of cellular NOS regulates platelet activation, causing inhibition of adhesion and aggregation and induction of disaggregation. Both the vasodilator (HOUSTON and BUCHANAN 1994) and platelet-inhibitory (YAO et al. 1992; GOLINO et al. 1992; BODZENTA-LUKASZYK et al. 1994; ADAMS et al. 1995) components contribute to the hemostatic action of NO.

The contribution of NO released from neutrophils (MCCALL et al. 1989) to the regulation of platelet function in vivo remains to be established. However, it is of interest that the NO–cGMP system also inhibits the adhesion and chemotaxis of stimulated neutrophils (KUBES et al. 1991; MOILANEN et al. 1993).

The synthesis and release of a single inhibitor is unlikely to account for regulation of platelet aggregation. We have shown that NO and prostacyclin act in synergism as inhibitors of platelet aggregation and inducers of disaggregation (RADOMSKI et al. 1987d). In addition, the synergistic induction of platelet disaggregation has been demonstrated by combining glyceryl trinitrate (an NO donor), prostaglandin $E_1$, and tissue plasminogen activator, which act via cGMP, cAMP, and plasmin-dependent mechanisms, respectively (STAMLER et al. 1989). Furthermore, the inhibition of NO generation in vivo abolishes the antithrombotic activity of aspirin (which inhibits the generation of prothrombotic eicosanoids) and ketanserin (which antagonizes the actions of serotonin) (GOLINO et al. 1992). Thus, it is likely that platelet aggregation in vivo is regulated by the synergistic action of inhibitors of platelet function.

## VIII. Role of NO in the Pathogenesis of Vascular Disorders

The vasodilator and platelet-regulatory functions of endothelium are impaired during the course of essential hypertension, diabetes, and atherosclerotic coronary artery disease (DEBELDER and RADOMSKI 1994). However, the pathomechanism of these changes remains unclear. Since oxidative modification of low-density lipoproteins (LDL) plays a key role in atherogenesis, a number of studies (reviewed by RADOMSKI and SALAS 1995a) have examined the effects of native and oxidized LDL on NO-mediated vascular functions. In most of these studies lipoproteins decreased the bioactivity of NO. Several hypotheses have been proposed to explain these effects of LDL, including inhibition of NOS activity and direct inactivation of NO (LUSCHER et al. 1993; FLAVAHAN 1992; COOKE and TSAO 1992 and references therein). The decreased bioactivity of NO in atherosclerosis could also result from changes in the metabolism of this molecule and generation of $ONOO^-$ from superoxide and inducible NO (BECKMAN et al. 1994). These data indicate that the generation of secondary oxidants such as $ONOO^-$ may play an important role in the pathogenesis of atherosclerosis (RADOMSKI and SALAS 1995a). In addition, LDL may inhibit arginine uptake into platelets and thereby decrease NOS activity and promote thrombosis (CHEN and MEHTA 1994a), an effect which is reversible by the administration of L-arginine in the diet (TSAO et al. 1994). In contrast to LDL, high-density lipoproteins was found to decrease platelet function by increasing NOS activity in platelets (CHEN and MEHTA 1994b).

An impaired NO generation or action may also underlie the pathomechanism of vasospastic and thrombotic changes in essential hypertension (CADWGAN and BENJAMIN 1993; CALVER et al. 1992), diabetes (AMADO et al. 1993), and coronary artery disease (KOMAMURA et al. 1994). Indeed, the endogenous NO inhibited microthromboembolism in the ischemic heart, protected myocardium against intracoronary thrombosis, and decreased platelet deposition due to carotid endarterectomy (KOMAMURA et al. 1994; OLSEN et al. 1996).

Vasoconstriction and increased platelet activation are characteristic for preeclampsia, a severe disease that may complicate normal pregnancy. DELACRETAZ et al. (1995) have shown that intraplatelet NOS activity is impaired in preeclampsia and this may contribute to the pathogenesis of this disease.

The expression of iNOS has been also implicated in the pathogenesis of septicemia and septic shock. The iNOS is induced following the invasion of gram-negative bacteria and exposure of cells to endotoxin (LPS) and cytokines (for literature see DEBELDER and RADOMSKI 1994; RADOMSKI 1995). LPS- and cytokine-mediated expression of iNOS is likely to have complex repercussions for vascular hemostasis. On the one hand, inducible NO acts to promote hemostasis and inhibit thrombosis since the inhibition of its generation by NOS inhibitors greatly potentiated cytokine-stimulated platelet adhesion to cultured human endothelial cells (RADOMSKI et al. 1993), precipitated renal glomerular thrombosis (SCHULTZ and RAIJ 1992), and exacerbated sepsis-induced renal hypoperfusion (SPAIN et al. 1994). On the other hand, the exposure of endothelial cells to cytokine-induced NO may result in cell toxicity and destruction (PALMER et al. 1992), and it has been reported that the inhibition of NOS may be beneficial in the treatment of septic shock (KILBOURN et al. 1990; WRIGHT et al. 1992). A partial explanation for this discrepancy may be that the currently available inhibitors of iNOS are not selective and inhibit the activities of other NOS isoenzymes. Indeed, an intact generation of constitutive NO may be important to maintain the integrity of the microvasculature during sepsis (RADOMSKI and SALAS 1995b). Clinical investigations are in progress to address the efficacy and safety of NOS inhibition treatment in this condition.

Bleeding is a well-known complication of uremia that is attributed to the suppression of platelet function by the disease process. Interestingly, L-arginine and some other platelet-inhibitory metabolites of the urea cycle are accumulated in uremia (HOROWITZ et al. 1970) and this is associated with an increase in tumor necrosis factor alpha levels (NORIS et al. 1993). It has been shown that platelets obtained from uremic patients generate more NO than controls; therefore, an increased expression and/or activity of NOS may play a role in platelet dysfunction observed in uraemia (NORIS et al. 1993).

Platelets play a role in the pathogenesis of tumor metastasis by increasing the formation of tumor-cell–platelet aggregates, thus facilitating cancer cell arrest in the microvasculature. We have demontrated that tumor-cell-induced platelet aggregation in vitro is modulated by the ability of tumor cells to generate NO, and this correlated with their propensity for metastasis (RADOMSKI et al. 1991). Indeed, human colon carcinoma cells isolated from metastases exhibited lower NO activity than cells isolated from the primary tumor. Moreover, the expression of iNOS by murine melanoma cells inversely correlated with their ability to form metastases in vivo (DONG et al. 1994). These data suggest that differential synthesis of NO may distinguish between cells of low and high metastatic potential. Interestingly, NOS has been found in some human gynecological malignancies, and the highest NOS activities were detected in poorly differentiated tumors (THOMSON et al. 1994). Thus, further work is needed to unravel the biological significance of NO generation by cancer cells.

## IX. Pharmacological Regulation of Platelet Function by NO Gas and NO Donor Drugs

Nitric oxide gas is a potent inhibitor of platelet activation in vitro; however, its antiplatelet activity is limited by short chemical (6–30 s) and biological (1–4 min) half-lives (RADOMSKI et al. 1987a). Interestingly, inhalation of NO gas by healthy volunteers and experimental animals resulted in a longer-lasting (30 min) inhibition of platelet hemostasis as inferred from the prolongation of bleeding time (HOGMAN et al. 1993). This phenomenon may be explained either by accumulation and subsequent slow release of NO from the lipid part of the cell membrane (MALINSKI et al. 1993b) or formation of endogenous NO donors. Moreover, chronic inhalation of NO gas has been shown to inhibit neointima formation and restenosis following balloon-induced arterial injury (LEE et al. 1996). Finally, NO inhibits platelet activation brought about by contact with artificial surfaces including extracorporeal circuits (SLY et al. 1995). Thus, administration of NO gas may represent a new pharmacological strategy for noninvasive inhibition of thrombosis and prevention of blood cell activation in the extracorporeal circulation.

The pharmacological activity of NO donors is mostly due to the release of NO (FEELISCH and NOACK 1987). The main advantage of these compounds is their half-lives which are longer than that of NO gas. The NO donors can be classified accordingly to their ability to release NO in vitro. The first group is comprised of the drugs that require metabolic transformation to release NO.

Organic nitrates are the most prominent members of this class. In vitro, these compounds are poor spontaneous releasers of NO and require the presence of a thiol cofactor (e.g., *N*-acetylcysteine) for acceleration of this liberation (FEELISCH 1991). However, in vivo, the release of NO from organic nitrates is greatly enhanced by thiols and enzyme(s) which still remain to be identified. These enzymes are present in the vascular tissues but not in platelets (GERZER et al. 1988). Indeed, organic nitrates are weak inhibitors of aggregation of isolated platelets in vitro (GERZER et al. 1988). However, following the stimulation of platelet aggregation by ADP in vitro, nitroglycerin has been shown to exert a potent disaggregating effect which may be due to a temporary elevation in the sensitivity of GC-S to NO (CHIRKOV et al. 1991).

Nitrate-induced inhibition of platelet aggregation in vitro can be greatly potentiated in the presence of thiols or cultured vascular cells (LOSCALZO 1985; FEELISCH 1991; BENJAMIN et al. 1991). This indicates that the conversion of organic nitrates by the vascular tissue in vivo can result in the release of sufficient amounts of NO for inhibition of platelet function. Indeed, in healthy volunteers, oral and intravenous administrations of glyceryl trinitrate (nitroglycerin) and isosorbide mononitrates resulted in inhibition of platelet aggregation ex vivo (DECATERINA et al. 1984; DRUMMER et al. 1991; WALLEN et al. 1993; CHIRKOV et al. 1993; KARLBERG et al. 1993). In addition, glyceryl trinitrate and isosorbide dinitrate inhibited experimental thrombosis and reocclusion after thrombolysis in dogs and rats (PLOTKINE et al. 1991; WERNS et al. 1994).

The effectiveness of organic nitrates as antithrombotics increases with the extent of vascular injury. In normal pigs, the deposition of platelets on arterial segments following injury from balloon angioplasty is inhibited by intravenous infusion of nitroglycerin when arterial injury is deep (extending through the internal elastic lamina) rather than mild (deendothelialization only; LAM et al. 1988). Furthermore, short- and long-term administration of nitroglycerin and isosorbide dinitrate to patients suffering from coronary artery disease and acute myocardial infarction resulted in a significant inhibition of platelet adhesion and aggregation (GEBALSKA 1990; DIODATI et al. 1990; SINZINGER et al. 1992). Of particular interest are the interactions of organic nitrates with other inhibitors of platelet function. Organic nitrates act synergistically with prostacyclin or aspirin as inhibitors of platelet aggregation (KARLBERG et al. 1993; DECATERINA et al. 1988). Also isosorbide dinitrate was found to potentiate the antiplatelet activity of prostaglandin $E_1$ in patients with peripheral vascular disease (SINZINGER et al. 1990).

The second group of NO donors is composed of compounds which release NO without the need for metabolic activation. Sodium nitroprusside, molsidomine, and SIN-1 are clinically used members of this group. Because of its powerful vasodilator action sodium nitroprusside is often used to treat vascular emergencies associated with hypertensive crisis. Since this compound shows some antiplatelet activity both in vitro and in vivo (Levin et al. 1982; Hines and Barash 1989) its acute clinical effects may also be mediated, in part, through inhibition of platelet function.

Molsidomine and its active metabolite SIN-1 inhibit experimental thrombosis and platelet aggregation in healthy volunteers and in patients suffering from acute myocardial infarction (Wautier et al. 1989). Interestingly, SIN-1 generates superoxide and $ONOO^-$ in addition to NO (Hogg et al. 1993). Since $ONOO^-$ causes platelet aggregation and counteracts the platelet inhibitory activity of NO (Moro et al. 1994), the formation of this radical may offset the anti-platelet activity of NO released from SIN-1.

Despite almost 140 years of use, organic nitrates remain one of the safest and most widely applied cardiovascular drugs. However, since organic nitrates under some conditions are known to induce tachyphylaxis there have been many attempts to synthesize tachyphylaxis-free NO donors. One of the more promising groups of drugs are cysteine-containing nitrates. The incorporation of a cellular thiol cysteine into the structure of organic nitrate resulted in a high effectiveness of these compounds as inhibitors of platelet and leukocyte functions both in vitro and in vivo (Lefer et al. 1993).

Another challenging problem is tissue selectivity. The platelet-inhibitory actions of organic nitrates cannot be separated from their effects on the vascular wall. The concept of platelet-selective NO donors has arisen from our experiments with *S*-nitrosoglutathione (GSNO; Radomski et al. 1992). *S*-Nitrosoglutathione is a tripeptide *S*-nitrosothiol that is formed by *S-nitrosylation* of glutathione, the most abundant intracellular thiol. We have found that the intravenous administration of GSNO into a conscious rat inhibits platelet aggregation at doses that have only small effects on the blood pressure (Radomski et al. 1992). Moreover, similar platelet/vascular differentiation is detected following inta-arterial administration of GSNO into the circulation of the human forearm (DeBelder et al. 1994). Finally, we have infused GSNO into patients undergoing balloon angioplasty and found that this NO donor effectively protected platelets from activation at the site of angioplastic injury without altering blood pressure (Langford et al. 1994). Interestingly, the exposure of human neutrophils to NO led to depletion of glutathione stores, activation of the hexose monophosphate shunt, synthesis of endogenous GSNO, and inhibition of superoxide generation by neutrophils. Exposure to synthetic GSNO resulted in similar effects (Clancy et al. 1994). These observations show that GSNO is a potent regulator of platelet and neutrophil functions and it may be a prototype for the development of blood-cell-selective NO donors.

What is the position of organic nitrates among "classic" inhibitors of platelet function? Acetlylsalicylic acid (aspirin) is by far the most widely used antiplatelet drug in clinical practice, and its benefits in terms of decreasing mortality due to reinfarction have been unequivocally demonstrated (ISIS-2 1988; ISIS-3 1992; for review see PATRONO 1989 and FERNANDEZ-ORTIZ et al. 1994), while those of organic nitrates have not yet been established. A meta-analysis found a significant reduction in mortality when intravenous glyceryl trinitrate or nitroprusside were used during the acute course of myocardial infarction (YUSUF et al. 1988). Moreover, when combined with *N*-acetylcysteine, glyceryl trinitrate substantially reduced myocardial infarction in unstable angina, an effect compatible with an antiplatelet effect of glyceryl trinitrate (HOROWITZ et al. 1988).

Surprisingly, GISSI III (1994) and ISIS (1993) studies failed to show a clinically beneficial effect of organic nitrates on mortality after myocardial infarction. However, further analysis of GISSI III suggests that the apparent additive effect of glyceryl trinitrate and lisinopril could be attributed to antiplatelet effects of this NO donor (ANDREWS et al. 1994b). In addition, it is possible that nitrates may act by reducing the infarct size in small rather than large infarcts so that the neutral results of GISSI-3 and ISIS-4 may be explained by the heterogeneity of the effects (MORRIS et al. 1994). Interestingly, aspirin, a cyclooxygenase inhibitor, blocks only thromboxane-mediated platelet aggregation (for literature see PATRONO 1989) leaving the remaining pathways of adhesion and aggregation unopposed. In contrast, NO inhibits the activation cascade of mediators, generated by all known pathways of platelet aggregation (JENSEN and HOLMSEN 1995), and some pathways of platelet adhesion to subendothelium (SHAHBAZI et al. 1994).

Inhibition of platelet adhesion may be of particular importance to decrease platelet accumulation due to ischemia–reperfusion injury following organ ischemia, and drugs such as NO donors that increase cGMP levels may afford effective antiplatelet action (CHINTALA et al. 1994). Thus there is a clear need for further clinical studies to determine the role and the effectiveness of traditional organic nitrates and new NO donors in the treatment of vascular thrombotic and ischemic disorders.

## D. Other Oxygen- and Nitrogen-Derived Reactive Species

Early studies reported that oxidants may affect platelet aggregation (CANOSO et al. 1974; CLARK and KLEBANOFF 1980; DELPRINCIPE et al. 1985; HANDING et al. 1977; SALVEMINI et al. 1989). However, the results obtained were largely diverse. Several investigators described a platelet-inhibitory effect of hydrogen peroxide ($H_2O_2$) (CANOSO et al. 1974; HOLMSEN and ROBKIN 1977; LEVINE et al. 1976), whereas other reports suggested that this oxidant enhances aggregation (CLARK and KLEBANOFF 1980; DELPRINCIPE et al. 1985; PATSCHEKE 1979). Interestingly, low concentrations of $H_2O_2$ may enhance the platelet-inhibitory effect of NO (NASEEM and BRUCKDORFER 1995) while increased generation of peroxide counteracts platelet inhibition by NO and may lead to arterial thrombosis (FREEDMAN et al. 1996). Thus, the net effect of $H_2O_2$ on platelets may depend on the presence of NO in the microenvironment of activated platelets.

Superoxide, another reactive oxidant may also affect platelet function (HANDING et al. 1977; SALVEMINI et al. 1989). Indeed, superoxide has been shown to increase both platelet adhesion and aggregation when thrombin but not collagen was used as the aggregating agent (SALVEMINI et al. 1989). Interestingly, some actions of superoxide may be indirect and result from its reaction with NO to form $ONOO^-$ (BECKMAN and TSAI 1994). Recently, we found that $ONOO^-$ stimulated the aggregation of human platelets and reversed inhibition of aggregation by NO and prostacyclin (MORO et al. 1994). These actions of $ONOO^-$ are due to activation of some platelet receptors including IIb/IIIa (MORO et al. 1994). The proaggregatory effect of $ONOO^-$ can be attenuated in the presence of thiols (MORO et al. 1994) and glucose (MORO et al. 1995), which convert this oxidant to NO donors. Thus, the platelet microenvironment is likely to determine the net outcome (stimulation or inhibition of aggregation) of interactions between platelets and reactive species.

## I. Mechanisms of Action of Oxygen-Derived Radicals on Platelets

### 1. Thromboxane Generation

Cyclooxygenase, the key enzyme of prostaglandin metabolism, catalyzes two different reactions: oxygenation of arachidonic acid to prostaglandin $G_2$ and its subsequent conversion to prostaglandin $H_2$. Early studies suggested that the former reaction requires peroxides while the latter leads to the formation of reactive oxygen species, which in turn may irreversibly inactivate the enzyme (HEMLER et al. 1978; HEMLER and LANDS 1980; KRKREJA et al. 1986; LAMBEIR et al. 1985). Recent studies showed that both exogenous and endogenous $H_2O_2$ generated during platelet aggregation stimulate platelet cyclooxygenase and thromboxane formation (AMBROSIO et al. 1994). In addition, $H_2O_2$ markedly shortened the lag phase that precedes aggregation by arachidonic acid and triggered an irreversible aggregation of platelets previously exposed to low amounts of arachidonic acid or collagen, but not thrombin, ADP, or A23187 (AMBROSIO et al. 1994). Moreover, catalase (AMBROSIO et al. 1994) and aspirin (PATRICIO et al. 1991) completely prevented aggregation and markedly reduced thromboxane generation when arachidonate was used as an aggregating agent. These observations suggest that $H_2O_2$ results in platelet aggregation only in the presence of agonists acting through the arachidonic acid pathway (PATRICIO et al. 1991).

### 2. Cyclic Nucleotides

The activity of the adenylyl cyclase system is not affected by $H_2O_2$ (AMBROSIO et al. 1994). In contrast, more than a tenfold increase in cGMP levels was observed in platelets following exposure to $H_2O_2$, suggesting that $H_2O_2$ stimulates guanylyl cyclase (Bruke and Wolin 1987; Wolin et al. 1990). Interestingly, other peroxides including 12-hydroperoxyeicosatetraenoic acid are known to stimulate GC-S (Brune and Ullrich 1991). The mechanism of this stimulation may depend on the oxidative modification of the functional groups (i.e., sulfydryls) of GC-S.

### 3. Stimulation of Platelet Serotonin Transport

Uptake of serotonin by platelets occurs readily, and serotonin sequestered in the dense granules in the platelets is protected from metabolism in the platelet. The active transport of serotonin by platelets plays an important role in maintaining the circulating concentration of free serotonin below the levels required to activate receptors on the platelet plasma membrane and vascular smooth muscle cells (Vanhoutte 1990).

Menadione (2-methyl-1,4-naphthoquinone) is a compound that can undergo redox cycling to generate superoxide anion and hydrogen peroxide. It has been shown that menadione stimulates the active transport of serotonin into platelets most likely through the generation of $H_2O_2$ (Bosin and Schaltenbrad 1991).

### 4. Activation of Platelet Receptors

We have shown that peroxynitrite causes aggregation of washed platelets through the activation of the IIb/IIIa receptor (Moro et al. 1994). In addition, the exposure of platelets to $ONOO^-$ results in the translocation of P-selectin from ∝-granules to the open canalicular system and platelet surface membrane. This activation is associated with an increase in intraplatelet $Ca^{2+}$ levels (Brown et al. submitted).

### 5. Formation of Nitric Oxide Donors

Oxidant-induced platelet activation is attenuated by antioxidant mechanisms acting in the platelet microenvironment. Indeed, we have demonstrated that both platelet and plasma thiols react with $ONOO^-$ and convert this oxidant to *S*-nitrosothiols, i.e., NO donors (Moro et al. 1994). Some endogenous aldehydes including glucose detoxify $ONOO^-$ by converting it to glucose nitrates (Moro et al. 1995). This protective mechanism can be, however, exhausted (Moro et al. 1995), and this may render platelets vulnerable to the oxidant attack.

## E. Conclusions

Nitric oxide generated by endothelial cells, platelets, and leukocytes has proved to be an important physiological and regulatory mediator modulating vessel wall hemostasis and preventing thrombosis. The metabolism of NO to $ONOO^-$ may explain some of the cellular toxicity associated with the formation of excessive amounts of NO. This may play an important role in the pathomechanism of vasospastic and thrombotic disorders. Understanding the regulatory and damaging properties of NO, other nitrogen- and oxygen-derived reactive species, and their interactions with other hemostatic factors may help in improving current antithrombotic therapy.

*Acknowledgements.* This work was supported by an establishment award from the Alberta Heritage Foundation for Medical Research and a grant from Eli Lilly Pharmaceuticals. MWR is a Scholar of Alberta Heritage Foundation for Medical Research. ES is an Eli Lilly and Alberta Heritage Foundation for Medical Research Post-Doctoral Fellow.

## References

Adams MR, Forsyth CJ, Jessup W, Robinson J, Celermajer DS (1995) Oral L-arginine inhibits platelet aggregation but does not enhance endothelium-dependent dilation in healthy young men. J Am Coll Cardiol 26:1054–1061

Alheid U, Reichwehr I, Forstermann U (1989) Human endothelial cells inhibit platelet aggregation by separately stimulating platelet cyclic AMP and cyclic GMP. Eur J Pharmacol 164:103–110

Amado JA, Salas E, Botana MA, Poveda JJ, Berrazueta JR (1993) Low levels of intraplatelet cGMP in IDDM. Diabetes Care 16:809–811

Ambrosio G, Golino P, Pascucci IN et al. (1994) Modulation of platelet function by reactive oxygen metabolites. Am J Physiol 267:H308–H318

Andrews NP, Dakak N, Schenke WH, Quyyumi AA (1994a) Platelet–endothelium interactions in humans: changes in platelet cyclic guanosine monophosphate content in patients with endothelial dysfunction. Circulation 90:I-397

Andrews R, May JA, Vickers J, Heptinstall S (1994b) Inhibition of platelet aggregation by transdermal glyceryl trinitrate. Br Heart J 72:575–579

Beckman J, Tsai JH (1994) Reactions and diffusion of nitric oxide and peroxynitrite. The Biochemist 16:8–10

Beckman JS, Yao ZY, Anderson P et al. (1994) Extensive nitration of protein tyrosines observed in human atherosclerosis detected by immunocytochemistry. Biol Chem Hoppe Seyler 375:81–87

Benjamin N, Dutton JAE, Ritter JM (1991) Human vascular smooth muscle cells inhibit platelet aggregation when incubated with glyceryl trinitrate: evidence for generation of nitric oxide. Br J Pharmacol 102:847–850

Berkels R, Klaus W, Boller M, Rosen R (1994) The calcium modulator nifedipine exerts its antiaggregatory property via a nitric oxide mediated process. Thromb Haemost 72:309–312

Bhardwaj R, Page CP, May GR, Moore PK (1988) Endothelium-derived relaxing factor inhibits platelet aggregation in human whole blood in vitro and in the rat in vivo. Eur J Pharmacol 157:83–91

Bodzenta-Lukaszyk A, Gabryelewicz A, Lukaszyk A et al. (1994) Nitric oxide synthase inhibition and platelet function. Thromb Res 75:667–672

Bosin TR, Schaltenbrad SL (1991) Stimulation of platelet serotonin transport by substituted 1,4-naphthoquinone-induced oxidant stress. Biochem Pharmacol 41:967–974

Broekman MJ, Eiroa AM, Marcus AJ (1991) Inhibition of human platelet reactivity by endothelium-derived relaxing factor from human umbilical vein endothelial cells in suspension: blockade of aggregation and secretion by an aspirin-insensitive mechanism. Blood 78:1033–1040

Brown AS, Moro MA, Darley-Usmar V et al. Actions of peroxynitrite on human platelets. Circulation (submitted)

Bruke TM, Wolin MS (1987) Hydrogen peroxide elicits pulmonary arterial relaxation and guanylate cyclase activation. Am J Physiol 252:H721–H732
Brune B, Lapetina EG (1989) Activation of a cytosolic ADP-ribosyltransferase by nitric oxide-generating agents. J Biol Chem 264:8455–8458
Brune B, Ullrich V (1987) Inhibition of platelet aggregation by carbon monoxide is mediated by activation of guanylate cyclase. Mol Pharmacol 32:497–504
Brune B, Ullrich V (1991) 12-Hydroperoxyeicosatetraenoic acid inhibits main platelet function by activation of soluble guanylate cyclase. Mol Pharmacol 39:671–678
Busse R, Luckhoff A, Bassenge E (1987) Endothelium-derived relaxing factor inhibits platelet activation. Naunyn Schmiedebergs Arch Pharmacol 336:566–571
Cadwgan TM, Benjamin N (1993) Evidence for altered platelet nitric oxide synthesis in essential hypertension. J Hypertens 11:417–420
Calver A, Collier J, Moncada S, Vallance P (1992) Effect of local infusion of $N^G$-monomethyl-L-arginine in patients with hypertension. The nitric oxide dilator mechanism appears abnormal. J Hypertens 10:1025–1031
Canoso RT, Rodvien R, Scoon K, Levine PH (1974) Hydrogen peroxide and platelet function. Blood 43:645–656
Caren R, Corbo L (1973) Response of plasma lipids and platelet aggregation to intravenous arginine. Proc Soc Exp Biol Med 143:1067–1071
Chen LY, Mehta JL (1994a) High density lipoprotein antagonizes the stimulatory effect of low density liporotein on platelet function by affecting L-arginine-nitric oxide pathway. Circulation 90(4):I-30
Chen LY, Mehta JL (1994b) Inhibitory effect of high-density lipoprotein on platelet function is mediated by increase in nitric oxide synthase activity in platelets. Life Sci 23:1815–1821
Chen LY, Mehta JL (1996) Further evidence of the presence of constitutive and inducible nitric oxide synthase isoforms in human platelets. J Cardiovasc Pharmacol 27:154–158
Chintala MS, Bernardino V, Chiu PJ (1994) Cyclic GMP but not cyclic AMP prevents renal platelet accumulation after ischemia–reperfusion in anesthetized rats. J Pharmacol Exp Ther 271:1203–1208
Chirkov YY, Belushkina NN, Tyshchuk IS, Severina IS, Horowitz JD (1991) Increase in reactivity of human platelet guanylyl cyclase during aggregation potentiates the disaggregating capacity of sodium nitroprusside. Clin Exp Pharmacol Physiol 18:517–524
Chirkov YY, Naujalis JI, Sage RE, Horowitz JD (1993) Anitplatelet effects of nitroglycerin in healthy subjects and in patients with stable angina pectoris. J Cardiovasc Pharmacol 21:384–389
Clancy RM, Levartovsky D, Leszczynska-Piziak J, Yegudin J, Abramson SB (1994) Nitric oxide reacts with intracellular glutathione and activates the hexose monophosphate shunt in human neutrophils: evidence for S-nitrosoglutathione as a bioactive intermediary. Proc Natl Acad Sci USA 91:3680–3684
Clark RA, Klebanoff SJ (1980) Neutrophil–platelet interaction mediated by myeloperoxidase and hydrogen peroxide. J Immunol 124:399–405
Cooke JP, Tsao P (1992) Cellular mechanisms of atherogenesis and the effects of nitric oxide. Curr Opin Cardiol 7:799–804
Cooke JP, Rossitch E Jr, Andon NA, Loscalzo J, Dzau VJ (1991) Flow activates an endothelial potassium channel to release an endogenous nitrovasodilator. J Clin Invest 88:1663–1671
Craven PA, DeRubertis FR (1978) Restoration of the reponsiveness of purified guanylate cyclase to nitrosoguanidine, nitric oxide, and related activators by heme and heme proteins. Evidence for involvement of the paramagnetic nitrosyl–heme complex in enzyme activation. J Biol Chem 253:8433–8443
DeBelder AJ, Radomski MW (1994) Nitric oxide in the clinical arena. J Hypertens 12:617–624

DeBelder AJ, MacAllister R, Radomski MW, Moncada S, Vallance PJ (1994) Effects of S-nitrosoglutathione in the human forearm circulation. Evidence for selective inhibition of platelet activation. Cardiovasc Res 28:691–694
DeCaterina T, Giannessi D, Crea F et al. (1984) Inhibition of platelet function by injectable isosorbide dinitrate. Am J Cardiol 53:1683–1687
DeCaterina R, Giannessi, Bernini W, Mazzone A (1988) Organic nitrates: direct antiplatelet effects and synergism with prostacyclin. Antiplatelet effects of organic nitrates. Thromb Haemost 59:207–211
Delacretaz E, De Quay N, Waeber B et al. (1995) Differential nitric oxide synthase activity in human platelets during normal pregnancy and pre-eclampsia. Clin Sci 88:607–610
DelPrincipe DA, Menichelli A, De Matteis W et al. (1985) Hydrogen peroxide has a role in the aggregation of human platelets. FEBS Lett 185:142–146
Diodati J, Theroux P, Latour JG et al. (1990) Effects of nitroglycerin at therapeutic doses on platelet aggregation in unstable angina pectoris and acute myocardial infarction. Am J Cardiol 66:683–688
Djaffar I, Vilette D, Bray PF, Rosa JP (1991) Quantitative isolation of RNA from human platelets. Thromb Res 62:127–135
Dong Z, Staroselsky AH, Qi X, Xie K, Fiedler I (1994) Inverse correlation between expression of inducible nitric oxide synthase activity and production of metastasis in K-1735 murine melanoma cells. Cancer Res 54:789–793
Drummer C, Valta-Seufzer U, Karrenbrock B, Heim JM, Gerzer R (1991) Comparison of antiplatelet properties of molsidomine, isosorbide-5-mononitrate and placebo in healthy volunteers. Eur Heart J 12:541–549
Durante W, Schini VB, Kroll MH et al. (1994) Platelets inhibit the induction of nitric oxide synthesis by interleukin-1b in vascular smooth muscle cells. Blood 83:1831–1838
Feelisch M (1991) The action and metabolism of organic nitrates and their similarity with endothelium-derived relaxing factor (EDRF). In: Moncada S, Higgs EA, Berrazueta JR (eds) Clinical relevance of nitric oxide in the cardiovascular system. Edicomplet, Madrid, pp 29–43
Feelisch M, Noack EA (1987) Correlation between nitric oxide formation during degradation of organic nitrates and activation of guanylyl cyclase. Eur J Pharmacol 139:19–30
Fernandez-Ortiz A, Jang IK, Fuster A (1994) Antiplatelet and antithrombin therapy. Coron Artery Dis 5:297–305
Flavahan NA (1992) Atherosclerosis or lipoprotein-induced endothelial dysfunction. Circulation 85:1927–1938
Freedman JE, Loscalzo J, Benoit SE et al. (1996) Decreasd platelet inhibition by nitric oxide in two brothers with a history of arterial thrombosis. J Clin Invest 97:979–987
Fukuto JM, Chiang K, Hszieh R, Wong P, Chaudhuri G (1992) The pharmacological activity of nitroxyl: a potent vasodilator with activity similar to nitric oxide and/or endothelium-derived relaxing factor. J Pharmacol Exp Ther 263:546–551
Furchgott RF (1988) Studies on relaxation of rabbit aorta by sodium nitrite: the basis for the proposal that the acid-activatable inhibitory factor from bovine retractor penis is inorganic nitrite and that endothelium-derived relaxing factor is nitric oxide. In: Vanhoutte PM (ed) Vascular smooth muscle, peptides, autonomic nerves, and endothelium. Raven, New York, pp 401–414
Furchgott RF, Zawadzki JV (1980) The obligatory role of endothelial cells in the relaxation of arterial smooth muscle by acetylcholine. Nature 288:373–376
Furlong B, Henderson AH, Lewis MJ, Smith JA (1987) Endothelium-derived relaxing factor inhibits in vitro platelet aggregation. Br J Pharmacol 90:687–692
Gebalska J (1990) Platelet adhesion and aggregation in relation to clinical course of acute myocardial infarction (in Polish). M.D. thesis, Warsaw
Gerzer R, Karrenbrock B, Siess W, Heim JM (1988) Direct comparison of the effect of nitroprusside, SIN-1 and various nitrates on platelet aggregation and soluble guanylyl cyclase activity. Thromb Res 52:11–21

GISSI III study group (1994) GISSI-3: effects of lisinopril and trandsdermal glyceryl trinitrate singly and together on 6-week mortality and ventricular function after acute myocardial infarction. Lancet 343:1115–1122

Goldberg ND, Haddox MK, Nicol SE et al. (1975) Biological regulation through opposing influences of cyclic GMP and cyclic AMP: the yin yang hypothesis. Adv Cyclic Nucleotide Res 5:307–330

Golino P, Capelli-Bigazzi M, Ambrosio G et al. (1992) Endothelium-derived relaxing factor modulates platelet aggregation in an in vivo model of recurrent platelet activation. Circ Res 71:1447–1456

Gross SS, Jaffe E, Levi R, Kilbourn RG (1991) Cytokine-activated endothelial cells express an isotype of nitric oxide synthase which is tetrahydrobiopterin-dependent, calmodulin-independent and inhibited by arginine analogs with a rank-order of potency characteristic of activated macrophages. Biochem Biophys Res Commun 178:823–829

Haffner C, Jarchau T, Reinhard M et al. (1995) Molecular cloning, structural analysis and functional expression of the proline-rich focal adhesion and microfilament-associated protein VASP. EMBO J 14:19–27

Handing RI, Karabin R, Boxer GJ (1977) Enhancement of platelet function by superoxide anion. J Clin Invest 59:959–965

Hawrylewicz CM, Howells GL, Felmann M (1991) Platelet-derived interleukin induces human endothelial adhesion molecule expression and cytokine production. J Exp Med 174:785–790

Hemler ME, Lands WEM (1980) Evidence for a peroxide-initiated free radical mechanism of prostaglandin biosynthesis. J Biol Chem 255:6253–6261

Hemler ME, Graff G, Lands WEM (1978) Accelerative autoactivation of prostaglandin biosynthesis by $PGG_2$. Biochem Biophys Res Commun 85:1325–1331

Herbaczynska-Cedro K, Lembowicz K, Pytel B (1991) $N^G$-monomethyl-L-arginine increases platelet deposition on damaged endothelium in vivo. A scanning electron microscopy study. Thromb Res 64:1–9

Hines R, Barash PG (1989) Infusion of sodium nitroprusside induces platelet dysfunction in vitro. Anesthesiology 71:805–806

Hogan JC, Lewis MJ, Henderson AH (1988) In vivo EDRF activity influences platelet function. Br J Pharmacol 94:1020–1022

Hogg N, Darley-Usmar VM, Wilson MT, Moncada S (1993) Oxidation of alpha-tocopherol in human low density lipoprotein by the simultaneous generation of superoxide and nitric oxide. FEBS Lett 326:199–203

Hogman M, Frostell C, Arnberg H, Hedenstierna G (1993) Bleeding time prolongation and NO inhalation. Lancet 341:1664–1665

Holmsen H, Robkin L (1977) Hydrogen peroxide lowers ATP levels in platelets without altering adenylate energy charge and platelet function. J Biol Chem 252:1752–1757

Horowitz HI, Stein IM, Cohen BD, White JG (1970) Further studies on the platelet-inhibitory effect of guanidinosuccinic acid and its role in uremic bleeding. Am J Med 49:336–345

Horowitz JD, Henry CA, Syrjanen ML et al. (1988) Combined use of nitroglycerin and N-acetylcysteine in the management of unstable angina pectoris. Circulation 77:787–794

Horstrup K, Jablonka B, Hönig-Liedl P et al. (1994) Phosphorylation of focal adhesion vasodilator-stimulated phosphoprotein at Ser157 in intact human platelets correlates with fibrinogen receptor inhibition. Eur J Biochem 225:21–27

Houston DS, Buchanan MR (1994) Influence of endothelium-derived relaxing factor on platelet function and hemostasis in vivo. Thromb Res 74:25–37

Houston DS, Gerrard JM, Mc Crea J, Glover S, Butler AM (1983) The influence of amines on various platelet responses. Biochim Biophys Acta 734:267–273

Houston DS, Robinson P, Gerrard JM (1990) Inhibition of intravascular platelet aggregation by endothelium-derived relaxing factor: reversal by red blood cells. Blood 76:953–958

Humphries RG, Tomlinson W, O'Connor SE, Leff P (1990) Inhibition of collagen- and ADP-induced platelet aggregation by substance P in vivo: involvement of endothelium-derived relaxing factor J Cardiovasc Pharmacol 16:292–297
Ignarro LJ, Buga GM, Wood KS, Byrns RE, Chaudhuri G (1987) Endothelium-derived relaxing factor produced and released from artery and vein is nitric oxide. Proc Natl Acad Sci USA 84:9625–9629
ISIS-2 (1988) (Second International Study of Infarct Survival) Collaborative Group: randomized trial of intravenous streptokinase, oral aspirin, both or neither among 17,187 cases of suspected acute myocardial infarction: ISIS-2. Lancet 2:349–360
ISIS-3 (1992) a randomized comparison of streptokinase vs tissue plasminogen activator vs anistreplase and of aspirin plus heparin vs aspirin alone among 41,299 cases of suspected acute myocardial infarction: Lancet 339:753–770
ISIS collaborative group, Oxford U.K. (1993) ISIS-4: randomixed study of oral isosorbide mononitrate in over 50,000 patients with suspected acute myocardial infarction. Circulation 88:I-394
Jensen BO, Holmsen H (1995) Nitric oxide (NO)-platelet interactions: inhibition is independent of the prostanoid and ADP pathways. Platelets 6:83–90
Johansson JS, Haynes DH (1992) Cyclic GMP increases the rate of the calcium extrusion pump in intact platelets but has no direct effect on the dense tubular calcium accumulation system. Biochim Biophys Acta 1105:40–50
Kanno K, Hirata Y, Taihei I, Iwashina M, Marumo F (1994) Regulation of inducible nitric oxide synthase gene by interleukin-1b in rat vascular endothelial cells. Am J Physiol 267:H2318–H2324
Karlberg KE, Ahlner J, Henriksson P, Torfgard K, Sylven C (1993) Effects of nitroglycerin on platelet aggregation beyond the effects of acetylsalicylic acid in healthy subjects. Am J Cardiol 71:361–364
Kilbourn RG, Gross SS, Adams J et al. (1990) $N^{G}$-methyl-L-arginine inhibits tumor necrosis factor-induced hypotension: implications for the involvement of nitric oxide. Proc Natl Acad Sci USA 87:3029–3032
Knowles RG (1994) Nitric oxide synthases. Biochemist 16:3–7
Komamura K, Node K, Kosaka H, Inoue M (1994) Endogenous nitric oxide inhibits microthromboembolism in the ischemic heart. Circulation 90:I-345
Krkreja RK, Kontons HA, Hess ML, Ellis EF (1986) PGH synthase and lipoxygenase generate superoxide in the presence of NADH or NADPH. Circ Res 59:612–619
Kubes P, Suzuki M, Granger DN (1991) Nitric oxide: an endogenous modulator of leukocyte adhesion. Proc Natl Acad Sci USA 88:4651–4655
Lam JYT, Chesebro JH, Fuster V (1988) Platelets, vasoconstriction, and nitroglycerin during arterial wall injury. A new antithrombotic role for an old drug. Circulation 78:712–716
Lambeir AM, Markey CM, Dunford HB, Marnett LJ (1985) Spectral properties of the higher oxidation states of prostaglandin H synthase. J Biol Chem 260:14894–14896
Langford EJ, Brown AS, Wainwright RJ et al. (1994) Inhibition of platelet activity by S-nitrosoglutathione during coronary angioplasty. Lancet 344:1458–1460
Lantoine F, Brunet A, Bedioui F, Devynck J, Devynck MA (1995) Direct measurement of nitric oxide production in platelets: relationship with cytosolic $Ca^{2+}$ concentration. Biochem Biophys Res Commun 215:842–848
Launay JM, Bondoux D, Oset-Gasque MJ et al. (1994) Increase of human platelet serotonin uptake by atypical histamine receptors. Am J Physiol 266:R526–R536
Lee JS, Adrie C, Jacob HJ et al. (1996) Chronic inhalation of nitric oxide inhibits neointimal formation after balloon-induced arterial injury. Circ Res 78:337–342
Lefer DJ, Nakanishi K, Vinten-Johansen J (1993) Endothelial and myocardial cell protection by a cysteine-containing nitric oxide donor after myocardial ischemia and reperfusion. J Cardiovasc Pharmacol 22[Suppl 7]:S34–S43
Lelchuk R, Carrier M, Hancock V, Martin JF (1990) The relationship between megakaryocyte nuclear DNA content and gene expression. Int J Cell clon 8:277–282

Lelchuk R, Radomski MW, Martin JF, Moncada S (1992) Constitutive and inducible nitric oxide synthases in human megakaryoblastic cells. J Pharmacol Exp Ther 262:1220–1224

Levin RL, Weksler BB, Jaffe EA (1982) The interaction of sodium nitriprusside with human endothelial cells and platelets: nitroprusside and prostacyclin synergistincally inhibit platelet function. Circulation 66:1299–1307

Levine PH, Weinger RS, Simon J, Scoon KL, Krinsky NI (1976) Leukocyte–platelet interaction. Release of hydrogen peroxide by granulocytes as a modulator of platelet reactions. J Clin Invest 57:955–963

Lipton SA, Choi YB, Pan ZH et al. (1993) A redox-based mechanism for the neuroprotective and neurodestructive effects of nitric oxide and related nitrosocompounds. Nature 364:626–632

Liu Z, Nakatsu K, Brien JF et al. (1993) Selective sequestration of nitric oxide by subcellular components of vascular smooth muscle and platelets: relationship to nitric oxide stimulation of the soluble guanylyl cyclase. Can J Physiol Pharmacol 71:938–945

Loscalzo J (1985) N-Acetylcysteine potentiates inhibition of platelet aggregation by nitroglycerin. J Clin Invest 76:703–708

Luscher TF, Tanner FC, Tschudi MR, Noll G (1993) Endothelial dysfunction in coronary artery disease. Annu Rev Med 44:395–418

Macdonald PS, Read MA, Dusting GJ (1988) Synergistic inhibition of platelet aggregation by endothelium-derived relaxing factor and prostacyclin. Thromb Res 49:437–449

Malinski T, Radomski MW, Taha Z, Moncada S (1993a) Direct electrochemical measurement of nitric oxide released from human platelets. Biochem Biophys Res Commun 194:960–965

Malinski T, Taha Z, Grunfeld S et al. (1993b) Diffusion of nitric oxide in the aorta wall monitored in situ by porphyrinic microsensors. Biochem Biophys Res Commun 193:1076–1082

Marsden PA, Shappert KT, Chen HS et al. (1992) Molecular cloning and characterization of human endothelial nitric oxide synthase. FEBS Lett 307:287–293

Maurice DH, Haslam RJ (1990) Molecular basis of the synergistic inhibition of platelet function by nitrovasodilators and activators of adenylate cyclase: inhibition of cyclic AMP breakdown by cyclic GMP. Mol Pharmacol 37:671–681

May GR, Crook P, Moore PK, Page CP (1991) The role of nitric oxide as an endogenous regulator of platelet and neutrophil activation within the pulmonary circulation. Br J Pharmacol 102:759 763

McCall TB, Boughton-Smith NK, Palmer RMJ, Whittle BJR, Moncada S (1989) Synthesi of nitric oxide from L-arginine by neutrophils. Biochem J 261:293–296

McDonald B, Reep B, Lapetina EG, Molina y Vedia L (1993) Glyceraldehyde-3-phosphate dehydrogenase is required for the transport of nitric oxide in platelets. Proc Natl Acad Sci USA 90:11122–11126

Mehta JL, Chen LY, Mehta P (1995) Identification of constitutive and inducible forms of nitric oxide synthase in human platelets. J Lab Clin Med 25:753–760

Mellion BT, Ignarro LJ, Ohlstein EH et al. (1981) Evidence for the inhibitory role of guanosine 3'-5'-monophosphate in ADP-induced human platelet aggregation in the presence of nitric oxide and related nitrovasodilators. Blood 57:946–955

Mendelsohn ME, O'Neill S, George D, Loscalzo J (1990) Inhibition of fibrinogen binding to human platelets by S-nitroso-N-acetylcysteine. J Biol Chem 265:19028–19034

Moilanen E, Vuorinen P, Metsa-Ketela T, Vapaatalo H (1993) Inhibition by nitric oxide donors of human polymorphonuclear leucocyte functions. Br J Pharmacol 109:852–858

Moro MA, Darley-Usmar VM, Goodwin DA et al. (1994) Paradoxical fate and biological action of peroxynitrite on human platelets. Proc Natl Acad Sci USA 91:6702–6706

Moro MA, Darley-Usmar VM, Lizasoain I et al. (1995) The formation of nitric oxide donors from peroxynitrite. Br J Pharmacol 116:1999–2004
Moro MA, Russell RJ, Cellek S et al. (1996) cGMP mediates the vascular and platelet actions of nitric oxide Confirmation using an inhibitor of soluble guanylyl cyclase. Proc Natl Acad Sci USA 93:1480–1485
Morris JL, Zaman AG, Smyllie JH, Cowan JC (1994) The effect of intravenous nitrate on infarct size; evidence of no benefit in small but not large infarcts. Br Heart J 71:77
Murohara T, Parkinson SJ, Waldman SA, Lefer AM (1995) Inhibition of nitric oxide biosynthesis provides P-selectin expression in platelets. Role of protein kinase C. Art Thromb Vasc Biol 15:2068–2075
Murphy ME, Sies H (1991) Reversible conversion of nitroxyl anion to nitric oxide by superoxide dismutase. Proc Natl Acad Sci USA 88:10860–10864
Muruganandam A, Mutus B (1994) Isolation of nitric oxide synthase from human platelets. Biochim Biophys Acta 1200:1–6
Nakashima S, Tohmatsu T, Hattori H, Okano Y, Nozawa Y (1986) Inhibitory action of cyclic GMP on secretion, phosphoinositide hydrolysis and calcium mobilization in thrombin-stimulated human platelets. Biochem Biophys Res Commun 135:1099–1104
Nakatsuka M, Osawa Y (1994) Selective inhibition of the 12-lipoxygenase pathway of arachidonic acid metabolism by L-arginine or sodium nitroprusside in intact human platelets. Biochem Biophys Res Commun 200:1630–1634
Naseem KM, Bruckdorfer KR (1995) Hydrogen peroxide at low concentrations strongly enhances the inhibitory effect of nitric oxide on platelets. Biochem J 310:149–153
Noris M, Benigni A, Boccardo P et al. (1993) Enhanced nitric oxide synthesis in uremia: implications for platelet dysfunction and dialysis hypotension. Kidney Int 44:445–450
Olsen SB, Ayala B, Tang DB et al. (1994) Enhancement of platelet deposition by cross-linked hemoglobin in a rat carotid endarterectomy model. Circulation 90:I-345
Olsen SB, Tang DB, Jackson MR et al. (1996) Enhancement of platelet deposition by cross-linked hemoglobin in a rat carotid endarterectomy model. Circulation 93(2):327–332
Palmer RMJ, Ferrige AG, Moncada S (1987) Nitric oxide release accounts for the biological activity of endothelium-derived relaxing factor. Nature 327:524–526
Palmer RMJ, Ashton DS, Moncada S (1988) Vascular endothelial cells synthesize nitric oxide from L-arginine. Nature 333:664–666
Palmer RMJ, Bridge L, Foxwell NA, Moncada S (1992) The role of nitric oxide in endothelial cell damage and its inhibition by glucocorticoids. Br J Pharmacol 105:11–12
Patricio D, Iuliano L, Ghiselli A, Alessandri C, Violi F (1991) Hydrogen peroxide as trigger of platelet aggregation. Haemostasis 21:169–174
Patrono C (1989) Aspirin and human platelets: from clinical trials to acetylation of cyclooxygenase and back. Trends Pharmacol Sci 10:453–458
Patscheke H (1979) Correlation of activation and aggregation of platelets: discrimination between anti-activating and anti-aggregating agents. Haemostasis 8:654–681
Pigazzi A, Fabian A, Johnson J, Upchurch GR, Loscalzo J (1995) Identification of nitric oxide synthase in human megakaryocytes and platelets. Circulation [Suppl]92:I365
Plotkine M, Allix M, Guillou J, Boulu R (1991) Oral administration of isosorbide dinitrate inhibits arterial thrombosis in rats. Eur J Pharmacol 201:115–116
Pohl U, Busse R (1989) EDRF increases cyclic GMP in platelets during passage through the coronary vascular bed. Circ Res 65:1798–1803
Polanowska-Grabowska R, Gear ARL (1994) Role of cyclic nucleotides in rapid platelet adhesion to collagen. Blood 83:2508–2515

Pronai L, Ichimori K, Nozaki H et al. (1991) Investigation of the existence and biological role of L-arginine/nitric oxide pathway in human platelets by spin-trapping/EPR studies. Eur J Biochem 202:923–930
Radomski MW (1995) Nitric oxide – biological mediator, modulator and effector molecule. Ann Med 27:321–330
Radomski MW, Salas E (1995a) Nitric oxide – biological mediator, modulator and factor of injury: its role in the pathogenesis of atherosclerosis. Atherosclerosis 118[Suppl]:S69–S80
Radomski MW, Salas E (1995b) Platelet regulation and damage in vascular thrombotic and septic disorders. In: Fink MP, Payen D (eds) Role of nitric oxide in sepsis and ARDS. Springer, Berlin Heidelberg New York, pp 138–154
Radomski MW, Palmer RMJ, Moncada S (1987a) Comparative pharmacology of endothelium-derived relaxing factor, nitric oxide and prostacyclin in platelets. Br J Pharmacol 92:181–187
Radomski MW, Palmer RMJ, Moncada S (1987b) Endogenous nitric oxide inhibits human platelet adhesion to vascular endothelium. Lancet 2:1057–1058
Radomski MW, Palmer RMJ, Moncada S (1987c) The role of nitric oxide and cGMP in platelet adhesion to vascular endothelium. Biochem Biophys Res Commun 148:1482–1489
Radomski MW, Palmer RMJ, Moncada S (1987d) The anti-aggregating properties of vascular endothelium: interactions between prostacyclin and nitric oxide. Br J Pharmacol 92:639–646
Radomski MW, Palmer RMJ, Moncada S (1990a) An L-arginine/nitric oxide pathway present in human platelets regulates aggregation. Proc Nat Acad Sci USA 87:5193–5197
Radomski MW, Palmer RMJ, Moncada S (1990b) Characterization of the L-arginine: nitric oxide pathway in human platelets. Br J Pharmacol 101:325–328
Radomski MW, Palmer RMJ, Moncada S (1990c) Glucocorticoids inhibit the expression of an inducible but not the constitutive, nitric oxide synthase in vascular endothelial cells. Proc Natl Acad Sci USA 87:10043–10047
Radomski MW, Jenkins DC, Holmes L, Moncada S (1991) Human colorectal adenocarcinoma cells: differential nitric oxide synthesis determines their ability to aggregate platelets. Cancer Res 51:6073–6078
Radomski MW, Rees DD, Dutra A, Moncada S (1992) S-Nitrosoglutathione inhibits platelet activation in vitro and in vivo. Br J Pharmacol 107:745–749
Radomski MW, Vallance P, Whitley G, Foxwell N, Moncada S (1993) Platelet adhesion to human vascular endothelium is modulated by constitutive and cytokine induced nitric oxide. Cardiovasc Res 27:1380–1382
Rees DD, Palmer RMJ, Moncada S (1989) Role of endothelium-derived nitric oxide in the regulation of blood pressure. Proc Natl Acad Sci USA 86:3375–3378
Reinhard M, Halbrügge M, Scheer U et al. (1992) The 46/50kDa phosphoprotein VASP purified from human platelets is a novel protein associated with actin filaments and focal contacts. EMBO J 11:2063–2070
Rosenblum WI, Nelson GH, Povlishock JT (1987) Laser-induced endothelial damage inhibits endothelium-dependent relaxation in the cerebral microcirculation of the mouse. Circ Res 60:169–176
Salas E, Moro MA, Askew S et al. (1994) Comparative pharmacology of analogues of S-nitroso-N-acetyl-DL-penicillamine in platelets. Br J Pharmacol 112:1071–1076
Salvemini D, Nucci G, Sneddon JM, Vane JR (1989) Superoxide anions enhance platelet adhesion and aggregation. Br J Pharmacol 97:1145–1150
Sane DC, Bielawska A, Greenberg CS, Hannun YA (1989) Cyclic GMP analogs inhibit gamma thrombin-induced arachidonic acid release in human platelets. Biochem Biophys Res Commun 165:708–714
Sase K, Michel T (1995) Expression of constitutive endothelial nitric oxide synthase in human blood platelets. Life Sci 57:2049–2055

Schultz PJ, Raij L (1992) Endogenously synthesized nitric oxide prevents endotoxin-induced glomerular thrombosis. J Clin Invest 90:1718–1725
Sessa WC (1994) The nitric oxide synthase family of proteins. J Vasc Res 31:131–143
Sessa WV, Harrison JK, Barber CM et al. (1992) Molecular cloning and expression of cDNA encoding endothelial cell nitric oxide synthase. J Biol Chem 267:15274–15276
Shahbazi T, Jones N, Radomski MW, Moro MA, Gingell D (1994) Nitric oxide donors inhibit platelet spreading on surfaces coated with fibrinogen but not fibronectin. Thromb Res 75:631–642
Sinzinger H, Fitscha P, O'Grady J et al. (1990) Synergistic effect of prostaglandin $E_1$ and isosorbide dinitrate in peripheral vascular disease. Lancet 335:627–628
Sinzinger H, Virgolini, I, O'Grady J, Raushha F, Fitscha P (1992) Modification of platelet function by isosorbide dinitrate in patients with coronary artery disease. Thromb Res 65:323–335
Sly MK, Prager MD, Eberhart RC, Jessen ME, Kulkarni PV (1995) Inhibition of surface-induced platelet activation by nitric oxide. ASAIO J 41:M394–M398
Sneddon JM, Vane JR (1988) Endothelium-derived relaxing factor reduces platelet adhesion to bovine endothelial cells. Proc Natl Acad Sci USA 85:2800–2804
Spain DA, Wilson MA, Garrison RN (1994) Nitric oxide synthase inhibition exacerbates sepsis-induced renal hypoperfusion. Surgery 116:322–331
Stamler JS, Vaughan DE, Loscalzo J (1989) Synergistic disaggregation of platelets by tissue-type plasminogen activator, prostaglandin $E_1$ and glyceryl trinitrate. Circ Res 65:796–804
Stamler JS, Simon DJ, Osborne JA et al. (1992) S-Nitrosylation of proteins with nitric oxide: synthesis and characterization of biologically active compounds. Proc Natl Acad Sci USA 89:444–448
Taha Z, Kiechle F, Malinski T (1992) Oxidation of nitric oxide by oxygen in biological systems monitored by porphyrinic microsensor. Biochem Biophys Res Commun 188:734–739
Thom SR, Ohnishi T, Ischiropoulos H (1994) Nitric oxide released by platelets inhibits neutrophil $B_2$ integrin function following acute carbon monoxide poisoning. Toxicol Pharmacol 128:105–110
Thomson L, Lawton FG, Knowles RG et al. (1994) Nitric oxide synthase activity in human gynecological cancer. Cancer Res 54:1352–1354
Tsao PS, Theilmeier G, Singer AH, Leung LLK, Cooke JP (1994) L-Arginine attenuates platelet reactivity in hypercholesterolemic rabbits. Arterioscler Thromb 14:1529–1533
Vanhoutte P (1990) Vascular effects of serotonin and ischemia. J Cardiovas Pharmacol 16[Suppl 3]:S15–S19
Venturini CM, Del Vecchio PJ, Kaplan JE (1989) Thrombin-induced platelet adhesion to endothelium is modified by endothelial derived relaxing factor (EDRF). Biochem Biophys Res Commun 159:349–354
Wallen NH, Larsson PT, Broijersen A, Andersson A, Hjemdahl P (1993) Effects of oral dose of isosorbide dinitrate on platelet function and fibrinolysis in healthy volunteers. Br J clin Pharmacol 35:143–151
Walter U (1989) Physiological role of cGMP and cGMP-dependent protein kinase in the cardiovascular system. Rev Physiol Biochem Pharmacol 113:41–88
Wautier JL, Weill D, Kadeva H, Maclouf J, Soria C (1989) Modulation of platelet function by SIN-1A. J Cardiovasc Pharmacol 14:S111–S114
Wennmalm A, Benthin G, Petersson AS (1992) Dependence of the metabolism of nitric oxide (NO) in healthy human whole blood on the oxygenation of its red cell haemoglobin. Br J Pharmacol 106:507–508
Werns SW, Rote WE, Davis JH, Guevara T, Lucchesi BR (1994) Nitroglycerin inhibits experimental thrombosis and reocclusion after thrombolysis. Am Heart J 127:727–737

Wolin MS, Rodenburg JM, Messina EJ, Kaley G (1990) Similarities in the pharmacological modulation of reactive hyperemia and vasodilation to hydrogen peroxide in rat skeletal muscle arterioles: effects of probes for endothelium-derived mediators. J Pharmacol Exp Ther 253:508–512
Wright CE, Rees DD, Moncada S (1992) Protective and pathological role of nitric oxide in endotoxin shock. Cardiovasc Res 26:48–57
Wu XB, Brune B, von Appen F, Ullrich V (1993) Efflux of cyclic GMP from activated platelets. Mol Pharmacol 43:564–568
Xie Q, Nathan C (1994) The high-output nitric oxide pathway: role and regulation. J Leukoc Biol 56:576–582
Yao SK, Ober JC, Krishnaswami A et al. (1992) Endogenous nitric oxide protects against platelet aggregation and cyclic flow variations in stenosed and endothelium-injured arteries. Circulation 86:1302–1309
Yusuf S, MacMahon S, Collins R, Peto R (1988) Effect of intravenous nitrates on mortality in acute myocardial infarction: an overview of the randomised trials. Lancet i:1088–1092
Zhou Q, Hellermann GR, Solomonson LP (1995) Nitric oxide release from resting platelets. Thromb Res 77:87–96

CHAPTER 17

# Platelet 5-Hydroxytryptamine Transporters

G. RUDNICK

## A. Introduction

Blood platelets contain two distinct transport systems for serotonin (5-hydroxytryptamine, 5-HT). The first of these systems moves 5-HT from the plasma into the cytoplasm, and the second transports cytoplasmic 5-HT into the storage organelle, or dense granule. Both systems move 5-HT uphill, against a concentration gradient, and are therefore coupled to the input of metabolic energy. In each case this energy comes from hydrolysis of cytoplasmic ATP, which is used to generate transmembrane ion gradients.

The components responsible for 5-HT transport into and storage within platelets are also found in other cells, including neurons, chromaffin cells, mast cells, basophils, and placenta, where they are also utilized for 5-HT transport and storage. Recent isolation of cDNA clones for these proteins has allowed the examination of various tissues where 5-HT is accumulated and stored in secretory granules. These studies have established the existence of mRNA encoding the serotonin transporter (SERT) and the vesicular monoamine transporter (VMAT) in all such tissues examined. Knowledge of the mechanisms for 5-HT transport and storage in platelets, therefore, sheds light on 5-HT transport in many other cell types.

At the plasma membrane the $Na^+$, $K^+$-ATPase directly creates $Na^+$, (out > in) and $K^+$ (in > out) gradients and, indirectly, through generation of a transmembrane electrical potential ($\Delta\Psi$, inside negative), creates a $Cl^-$ gradient (out > in). The transmembrane gradients of $Na^+$, $K^+$, and $Cl^-$ serve, in turn, as driving forces for 5-HT transport. SERT is also a component of the plasma membrane, and it couples the inward flux of $Na^+$ and $K^+$ as well as the outward flux of $K^+$ to the entry of 5-HT.

A similar situation is found in the dense granule membrane, where two membrane components are responsible for 5-HT accumulation. The first component is an $H^+$-pumping ATPase that appears to be related to similar enzymes in other secretory granules, endosomes, lysosomes, and coated vesicles. This ATPase generates a transmembrane pH difference ($\Delta$pH, acid inside) and $\Delta\Psi$ (positive inside). The $\Delta$pH and $\Delta\Psi$ are then utilized by the second component, VMAT, which exchanges cytoplasmic 5-HT for one or more intragranular hydrogen ions.

## B. The Plasma Membrane 5-Hydroxytryptamine Transporter (SERT)

The platelet 5-HT transporter (SERT) resides in the platelet plasma membrane and is a member of a large family of neurotransmitter transporters. The $Na^+$- and $Cl^-$-coupled neurotransmitter transporters are a group of integral membrane proteins encoded by a closely related family of recently cloned cDNAs (GUASTELLA et al. 1990; NELSON et al. 1990; BLAKELY et al. 1991; GIROS et al. 1991, 1992; HOFFMAN et al. 1991; KILTY et al. 1991; MAYSER, et al. 1991; PACHOLCZYK et al. 1991; SHIMADA et al. 1991; USDIN et al. 1991; BORDEN et al. 1992; CLARK et al. 1992; FREMEAU et al. 1992; Q.-R. LIU et al. 1992a,b; SMITH et al. 1992a,b; YAMAUCHI et al. 1992; GUIMBAL and KILIMANN 1993; LESCH et al. 1993a,b). These carrier proteins couple the transmembrane movement of $Na^+$, $Cl^-$, and in some systems $K^+$ to the reuptake of neurotransmitters released into the synaptic cleft (KANNER and SCHULDINER 1987; Fig. 1). In the nervous system, they function to regulate neurotransmitter activity by removing extracellular transmitter. Inhibitors that interfere with this regulation include antidepressant drugs and stimulants such as the amphetamines and cocaine. The platelet plasma membrane 5-HT transporter is a member of this family and is identical in its primary amino acid sequence to the brain 5-HT transporter (LESCH et al. 1993b). In humans and rats, a single gene apparently encodes the 5-HT transporter expressed in all tissues. The platelet plasma membrane 5-HT transporter is most closely related to transporters for the catecholamines norepinephrine (NE; PACHOLCZYK et al. 1991) and dopamine (DA; KILTY et al. 1991; SHIMADA et al. 1991; USDIN et al. 1991; GIROS et al. 1992), and these three stand out as a distinct subfamily. These biogenic amine transporters are all inhibited by cocaine, and share other structural and mechanistic properties.

The best studied of the three plasma membrane biogenic amine transporters is the 5-HT transporter. Much of the available data on 5-HT transport

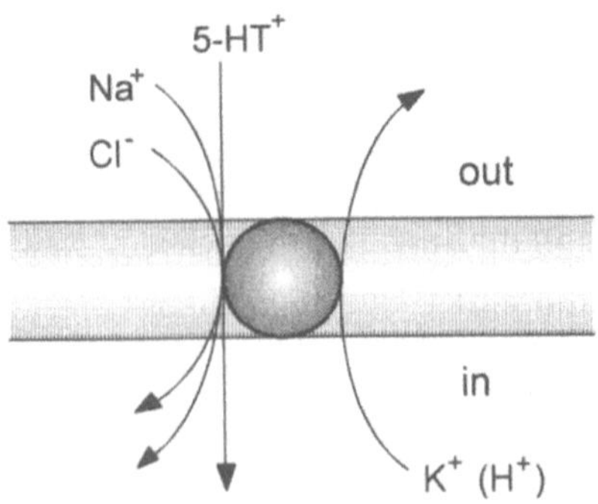


**Fig. 1.** Schematic mechanism for 5-hydroxytryptamine (*5-HT*) transport. SERT, the serotonin transporter, couples the inward flux of $Na^+$ and $Cl^-$ to influx of 5-HT. In a separate step of the same cycle, a $K^+$ ion is transported out of the cell

comes from studies using platelets. Although transport studies in intact platelets can yield valuable information, their usefulness is limited by our inability to control the cytoplasmic and intragranular ion composition. To circumvent these difficulties, we have used preparations of vesicles derived from plasma membrane and dense granule membrane as model systems (RUDNICK 1986). The availability of membrane vesicles from platelets (BARBER and JAMIESON 1970), rat basophilic leukemia cells (KANNER and BENDAHAN 1985), and placenta (BALKOVETZ et al. 1989) has allowed a detailed understanding of the 5-HT transporter's mechanism.

To demonstrate 5-HT transport in plasma membrane vesicles, ion gradients must be imposed across the vesicle membrane. This is accomplished by first incubating a dilute vesicle suspension in potassium phosphate buffer, collecting the vesicles by centrifugation, and then resuspending them in the same buffer. This process equilibrates the vesicle interior with $K^+$. The equilibrated vesicle suspension is diluted into iso-osmotic NaCl containing [$^3$H]5-HT and incubated at 25°–37°C. After a predetermined time, part or all of the reaction mixture is diluted with ice-cold buffer and filtered on a nitrocellulose filter that is subsequently washed and counted for vesicular 5-HT.

## I. Ionic Requirements

The platelet plasma membrane 5-HT transporter, like other neurotransmitter transporters in this gene family, requires both $Na^+$ and $Cl^-$ in the external medium for neurotransmitter influx. Early studies by SNEDDON (1969) and LINGJAERDE (1969) on the 5-HT transporter of intact platelets led to the suggestion that both $Na^+$ and $Cl^-$ are cotransported with 5-HT. Studies in both intact platelets and in platelet plasma membrane vesicles demonstrated that while $Cl^-$ could be replaced by $Br^-$, and to a lesser extent by $SCN^-$ or $NO_2^-$, $Na^+$ could not be replaced by other cations (LINGJAERDE 1969; SNEDDON 1969; RUDNICK 1977; NELSON and RUDNICK 1982).

### 1. Coupling to $Na^+$

Evidence that $Na^+$ and $Cl^-$ actually were cotransported with 5-HT came from studies using platelet plasma membrane vesicles (RUDNICK 1986). When a $Na^+$ concentration gradient (out > in) was imposed across the vesicle membrane in the absence of other driving forces, this gradient was sufficient to drive 5-HT accumulation (RUDNICK 1977). Coupling between $Na^+$- and 5-HT transport followed from the fact that $Na^+$ could drive transport only if its own gradient was dissipated. Thus, $Na^+$ influx must have accompanied 5-HT influx. $Na^+$-coupled 5-HT transport into membrane vesicles was insensitive to inhibitors of other $Na^+$ transport processes such as ouabain and furosemide, supporting the hypothesis that $Na^+$- and 5-HT fluxes are coupled directly by the transporter (RUDNICK 1977; NELSON and RUDNICK 1981). Many of these results have been reproduced in membrane vesicle systems from cultured rat basophilic leukemia cells (KANNER and BENDAHAN 1985), mouse brain synaptosomes (O'REILLY and REITH 1988), and human placenta (BALKOVETZ et al. 1989).

### 2. Coupling to $Cl^-$

The argument that $Cl^-$ is cotransported with 5-HT is somewhat less direct, as it has been difficult to demonstrate 5-HT accumulation with only the $Cl^-$ gradient as a driving force. However, the transmembrane $Cl^-$ gradient influences 5-HT accumulation when a $Na^+$ gradient provides the driving force. Thus, raising internal $Cl^-$ decreases the $Cl^-$ gradient and inhibits 5-HT uptake. External $Cl^-$ is required for 5-HT uptake, and $Cl^-$ can be replaced only by $Br^-$ and, to a lesser extent, by $SCN^-$, $NO_3$-, and $NO_2$- (NELSON and RUDNICK 1982). In contrast, 5-HT efflux requires internal but not external $Cl^-$ (NELSON and RUDNICK 1982). One alternative explanation for stimulation of transport by $Cl^-$ (on the same membrane face as 5-HT) is that $Cl^-$ might be required to electrically compensate for rheogenic (charge moving) 5-HT transport. This possibility was ruled out by the observation that a valinomycin-mediated $K^+$ diffusion potential (interior negative) was unable to eliminate the external $Cl^-$ requirement for 5-HT influx (NELSON and RUDNICK 1982).

### 3. Coupling to $K^+$

Perhaps the greatest surprise during the early studies of 5-HT transport was the discovery that $K^+$ efflux was directly coupled to 5-HT influx. Previously, $K^+$ countertransport had been invoked for other transport systems to explain stimulation by internal $K^+$ (CRANE et al. 1965; EDDY 1968; EDDY et al. 1970). In those other cases, however, it became clear that a membrane potential generated by $K^+$ diffusion was responsible for driving rheogenic transport processes. Stimulation of 5-HT transport into platelet plasma membrane vesicles by intravesicular $K^+$ also was attributed initially to a $K^+$ diffusion potential since internal $K^+$ stimulates transport but is not absolutely required (RUDNICK 1977). Subsequent measurements, however, showed that $K^+$ stimulated transport even if the membrane potential was close to zero (RUDNICK and NELSON 1978). When valinomycin was added to increase the $K^+$ conductance of the membrane, a $K^+$ diffusion potential (inside negative) was generated. Transport was essentially the same whether or not a diffusion potential was imposed in addition to the $K^+$ concentration gradient (NELSON and RUDNICK 1979). Thus, the $K^+$ gradient did not seem to act indirectly through the membrane potential but, rather, directly by exchanging with 5-HT.

## II. Reversal of Transport

The 5-HT transporter is quite capable of catalyzing efflux as well as influx. Efflux is stimulated by internal $Na^+$ and $Cl^-$ and by external $K^+$ and is inhibited by the tricyclic antidepressant imipramine, which also inhibits 5-HT influx into intact platelets, membrane vesicles, and synaptosomes. This property renders ambiguous the question of whether the transport assay measures transport in the physiological direction. From the experimental design of the 5-HT transport assay, the directionality of 5-HT transport is determined by artificially imposed ion gradients, and not by the orientation of the vesicles. The preparation of plasma membranes is vesicular, as evidenced by their ability to retain 5-HT, and at least 70% of the vesicles in the preparation have the same orientation as the intact platelet from studies of glycoprotein accessibility to hydrolytic enzymes (Barber and Jamieson 1970). Nevertheless, it is still difficult to state with certainty that 5-HT is transported only into right-side-out vesicles.

Amphetamines represent a class of stimulants that increase extracellular levels of 5-HT and other biogenic amines. Their mechanism differs from simple inhibitors like cocaine, although it also involves biogenic amine transporters. Amphetamine derivatives are apparently substrates for biogenic amine transporters and lead to transmitter release by a process of transporter-mediated exchange (Fischer and Cho 1979; Rudnick and Wall 1992a). Both catecholamine- and 5-HT transporters are affected by amphetamines.

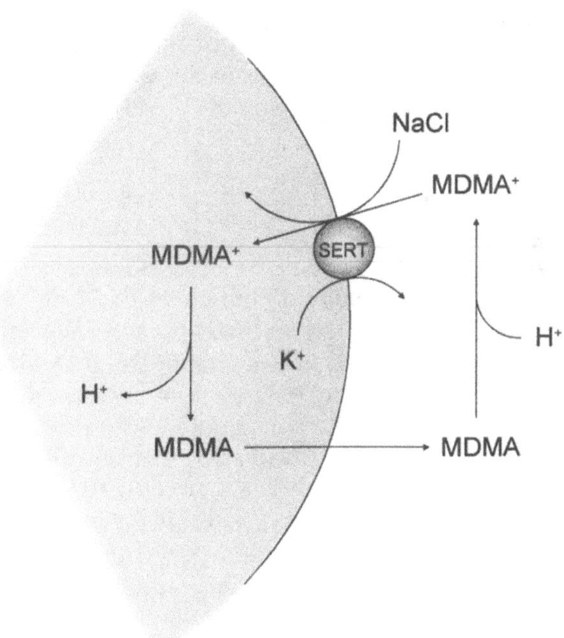


**Fig. 2.** Interaction of 3,4-methylenedioxymethamphetamine (*MDMA*) with the 5-hydroxytryptamine (5-HT) transporter. MDMA is a substrate for the serotonin transporter *SERT* and, like 5-HT, is transported into cells together with $Na^+$ and $Cl^-$ and in exchange for $K^+$. Since it is membrane permeant in its neutral form, MDMA deprotonates intracellularly and leaves the cell, at which time it can reprotonate and serve again as a substrate for SERT. This futile transport cycle may lead to dissipation of cellular $Na^+$, $Cl^-$, and $K^+$ gradients and acidification of the cell interior

In particular, compounds such as *p*-chloroamphetamine and 3,4-methylenedioxymethamphetamine (MDMA, also known as "ecstasy") preferentially release 5-HT and also cause degeneration of serotonergic nerve endings (MAMOUNAS et al. 1991).

This process of exchange stimulated by amphetamines results from two properties of amphetamine and its derivatives. These compounds are substrates for biogenic amine transporters, and they also are highly permeant across lipid membranes. As substrates, they are taken up into cells expressing the transporters, and as permeant solutes, they rapidly diffuse out of the cell without requiring participation of the transporter. The result is that an amphetamine derivative will cycle between the cytoplasm and the cell exterior in a process that allows $Na^+$ and $Cl^-$ to enter the cell and $K^+$ to leave each time the protonated amphetamine enters. Additionally, an $H^+$ ion will remain inside the cell if the amphetamine leaves as the more permeant neutral form (Fig. 2). This dissipation of ion gradients and internal acidification may possibly be related to the toxicity of amphetamines in vivo. The one-way utilization of the transporter (only for influx) leads to an increase in the availability of inward-facing transporter binding sites for efflux of cytoplasmic 5-HT. This, together with dissipation of the $Na^+$, $Cl^-$, and $K^+$ gradients, results in net 5-HT efflux. In addition, the ability of amphetamine derivatives to act as weak base ionophores at the dense granule membrane leads to leakage of granular 5-HT into the cytoplasm (SULZER and RAYPORT 1990; RUDNICK and WALL 1992b).

## III. Mechanism

It is interesting to consider how the platelet 5-HT transporter, with a molecular weight of 74 Kda, is able to couple the fluxes of 5-HT, $Na^+$, $Cl^-$, and $K^+$ in a stoichiometric manner. The problem faced by a coupled transporter is more complicated than that faced by an ion channel since a channel can function merely by allowing its substrate ions to flow across the lipid bilayer. Such uncoupled flux will dissipate the ion gradients and will not utilize them to concentrate another substrate. However, the structural similarities between transporters and ion channels may give a clue to the mechanism of transport (Fig. 3). Just as an ion channel may have a central aqueous cavity surrounded by amphipathic membrane-spanning helices, a transporter may have a central binding site that accommodates $Na^+$, $Cl^-$, and substrate. The difference in mechanism between a transporter and a channel may be that while a channel assumes open (conducting) and closed (nonconducting) states, a transporter also can assume two states that differ only in the accessibility of the central binding site. In each of these states, the site is exposed to only one face of the membrane, and the act of substrate translocation represents a conformational change to the state in which the binding site is exposed on the opposite face (Fig. 3). Thus, the transporter may behave like a channel with a gate at each face of the membrane, but only one gate is usually open at any point in time.

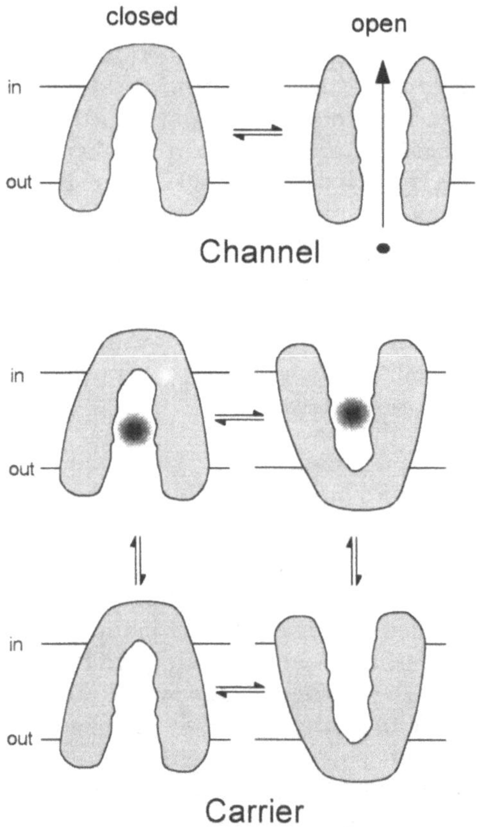


**Fig. 3.** A comparison between channel and carrier mechanisms. Channels operate by interconverting between open (*top right*) and closed (*top left*) states. A similar interconversion in a carrier is between states facing the cytoplasm (*middle* and *lower right*) and facing the cell exterior (*middle* and *lower left*)

Channel activity has been observed recently for the serotonin transporter (MAGER et al. 1994), and this activity may result from times when both gates are open, indicating a basic mechanistic and structural similarity between transporters and channels.

For this mechanism to lead to cotransport of ions with substrate molecules, the transporter must obey a set of rules governing the conformational transition between its two states (Fig. 3). For cotransport of $Na^+$, $Cl^-$, and 5-HT, the rule would allow a conformational change only when the binding site was occupied with $Na^+$, $Cl^-$, and substrate or when the site was completely empty. To account for $K^+$ countertransport with 5-HT, the conformational change could occur when the binding site contained either $K^+$ or $Na^+$, $Cl^-$, and 5-HT. This simple model of a binding site exposed alternately to one side of the membrane or the other can explain most carrier-mediated transport.

While the simple model described above was able to explain much of the transport phenomena observed with the 5-HT transporter, it did not predict the channel properties of the transporter observed by MAGER et al. (1994). It is difficult to understand if the conductive states of the 5-HT transporter are physiologically significant or if they represent unproductive side reactions of the transporter separate from its catalytic cycle. If one imagines the permeability pathway of the protein to consist of a central binding site for 5-HT, $Na^+$, $Cl^-$, and $K^+$ separated by gates from the internal and external media, then the rules for efficient coupling would predict that both gates should not be open simultaneously. However, the observed ion channel activity could result if these rules were not always obeyed, and there was a significant probabibility that one of the intermediates in the transport cycle was in equilibrium with a form in which both gates were open.

## IV. Purification

As a strategy to identify biogenic amine transporters, a number of groups have attempted to purify the 5-HT transporter to homogeneity. The protein has been solubilized in an active form using digitonin (TALVENHEIMO and RUDNICK 1980). Using this solubilized preparation, two groups achieved significant purification of the 5-HT transporter from platelet and brain using affinity resins based on citalopram, a high affinity ligand for the transporter (BIESSEN et al. 1990; GRAHAM et al. 1992). In neither case was the protein purified to homogeneity or reconstituted in liposomes to recover transport activity. LAUNAY et al. (1992) have apparently purified the transporter to homogeneity using affinity columns based on 5-HT itself or 6-fluorotryptamine, but also did not reconstitute transport activity with the purified protein. Reconstitution of the transporter has, however, been demonstrated with a urea-cholate extract of placental membranes (RAMAMOORTHY et al. 1992). As these studies were in progress, molecular cloning of other $Na^+$-dependent transporter cDNAs was providing an independent approach for studying their structure.

## V. Cloning

The identification of cDNAs for $Na^+$-dependent biogenic amine transporters has contributed greatly to our understanding of the molecular structure and function of these important proteins. The first cDNA encoding a biogenic amine carrier was identified by PACHOLCZYK et al. (1991) using an expression cloning strategy in COS cells. A norepinephrine transporter cDNA (NET) was isolated from an SK-N-SH human neuroblastoma library on the basis of its ability to direct the transport of $^{125}I$-labeled *m*-iodobenzylguanidine, a norepinephrine analog. Comparison of the predicted amino acid sequence for NET with that for the $Na^+$-dependent $\gamma$-aminobutyric acid (GABA) transporter GAT-1 previously identified by GUASTELLA et al. (1990) revealed that these transporters are members of a multigene family. Given the significant degree

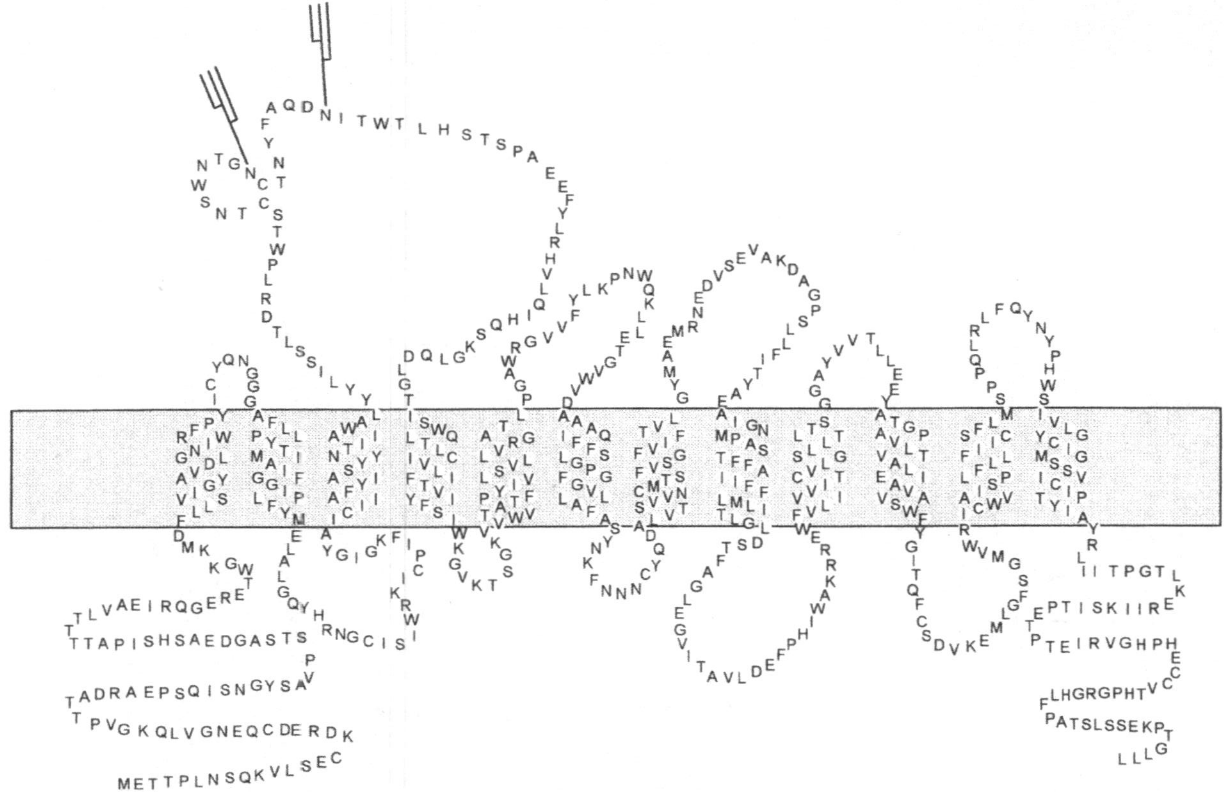

rSERT

**Fig. 4.** The 5-hydroxytryptamine (5-HT) amino acid sequence. Using hydropathy analysis, transmembrane domains were identified. The predicted structure has both N- and C-termini in the cytoplasm and the large, glycosylated loop on the cell surface

of homology between the two transporters, several groups designed degenerate oligonucleotide primers and have used them in homology-based polymerase chain reaction strategies for identifying additional members of the family. Several members of this transporter gene family have been identified, including carriers for DA (Giros et al. 1991, 1992; Kilty et al. 1991; Shimada et al. 1991; Usdin et al. 1991) and 5-HT (Blakely et al. 1991; Hoffman et al. 1991; Lesch et al. 1993a,b). Hydropathy analyses of these transporter sequences are virtually superimposable and predict proteins with 12 transmembrane domains having both their amino and carboxyl termini on the cytoplasmic side of the plasma membrane (Fig. 4). There is a significant degree of identity across this family of transporters (25%) despite their very different substrate specificities. Homology is highest in the proposed transmembrane domains (TM) and particularly sparse in the extracellular loop connecting TM3 and TM4. This large extracellular loop is the location of one to four potential N-linked glycosylation sites in each transporter sequence. In each mammalian species examined, only one 5-HT transporter gene has been found, with minor differences between species (Blakely et al. 1991; Lesch et al. 1993b). This contrasts markedly with 5-HT receptors, of which there are many types and subtypes (Fuller 1992).

## VI. Regulation

Much of the research to date on the regulation of 5-HT transport has examined the effects of agents that activate several components of second messenger systems on 5-HT transport in platelets and other cells (Myers and Pitt 1988; Myers et al. 1989; Alexi and Azmitia 1991; Cool et al. 1991; Anderson and Horne 1992; King et al. 1992; Ramamoorthy et al. 1993, 1995; Jayanthi et al. 1994; Miller and Hoffman 1994). 5-HT transport in the JAR human placental choriocarcinoma cell line (JAR cells) is enhanced by incubation of the cells in cholera toxin (Cool et al. 1991; Ramamoorthy et al. 1993). Cholera toxin treatment resulted in an increase in the $V_{max}$ and a decrease in the $K_M$ for 5-HT transport. The effects of cholera toxin could be mimicked by isobutylmethylxanthine (IBMX), dibutyryl cAMP, and forskolin and antagonized by the protein kinase inhibitor N-(2-aminoethyl)-5-isoquinolone sulfonamide. A second study compared the effects of cholera toxin on 5-HT transport in JAR cells and PC12 rat adrenal pheochromocytoma cells (PC12 cells; King et al. 1992). Incubation of cells in cholera toxin resulted in differential regulation of 5-HT transport in JAR and PC12 cells. Cholera toxin stimulated 5-HT transport in JAR cells, yet potently inhibited transport in PC12 cells. Once again the effects of cholera toxin were mimicked by forskolin and IBMX in both cell lines. Assuming that the 5-HT transporters in these two cell lines are the same, these findings suggest that different cellular components may play a role in determining how transport is regulated in different cell types.

In vivo and in vitro experiments have shown that activation of the inositol trisphosphate ($IP_3$) cascade by agents such as phorbol myristate acetate (PMA) can alter 5-HT transport (MYERS and PITT 1988; MYERS et al. 1989). PMA treatment of a perfused rabbit lung preparation resulted in a significant decrease in 5-HT removal (MYERS and PITT 1988). Although the decrease in 5-HT uptake observed in these studies may have been the result of any number of harmful effects that PMA may have on cells, more recent studies have shown that these effects are likely to have resulted from activation of the $IP_3$ cascade and protein kinase C (PKC; MYERS et al. 1989; ANDERSON and HORNE 1992). Blood platelet 5-HT transport was inhibited by treatment with PMA and mezerein, a phorbol ester analog (ANDERSON and HORNE 1992). Treatment with these phorbol esters had little effect on the $K_M$, but caused a consistent reduction in $V_{max}$ values. The reduction in $V_{max}$ was blocked by staurosporine, a PKC inhibitor, and was not associated with changes in transporter number or ion gradients. In JAR cells, staurosporine stimulated 5-HT transport by a cAMP-independent pathway (RAMAMOORTHY et al. 1995). These results suggest that the transporter activity is regulated by activation of PKC. In a separate study, PMA treatment of bovine artery endothelial cells resulted in a concentration-dependent decrease in 5-HT transport that could be mimicked by the PKC activators phorbol-12,13-dibutyrate and mezerein, but not with the inactive isomer phorbol-12,13-didecanoate. Cells pretreated with the PKC inhibitor staurosporine were resistant to the effects of PMA. Of particular interest is the finding that PMA treatment did not increase total PKC activity in these cells but caused a translocation of active PKC from the cytosol to the plasma membrane. PMA treatment may have affected 5-HT uptake by translocating activated PKC to the membrane. There it may have acted on the transporter or other membrane proteins such as the $Na^+$, $K^+$-ATPase which regulate ion gradients required for transport.

More recent evidence from both RBL cells and platelets has implicated phosphorylation by cGMP-dependent protein kinases in the direct regulation of 5-HT transporters (LAUNAY et al. 1994; MILLER and HOFFMAN 1994). In both cell types, transport rate increased following treatment of the cells with an agent that stimulates the production of nitric oxide (NO). In platelets, histamine acting at a novel receptor was reported to be the stimulus for NO generation (LAUNAY et al. 1994). In RBL cells, an adenosine receptor in the plasma membrane was found to be coupled to synthesis of NO (MILLER and HOFFMAN 1994). In both cases, NO is thought to activate guanylate cyclase, leading to phosphorylation of the transporter by cGMP-dependent protein kinase (MONCADA et al. 1989).

Although it is clear from the literature that $Na^+$-dependent transport is regulated by a number of agents, the mechanisms by which this regulation occurs remain to be elucidated. Regulation might be the result of direct phosphorylation of transport proteins by various kinases. Consensus sequences for phosphorylation sites recognized by various kinases have been identified in the 5-HT transporter amino acid sequence (BLAKELY et al. 1991). A canonical site for phosphorylation by cAMP-dependent protein kinase was identified in the 5-HT transporter sequence located in the amino terminus near the initiation codon (BLAKELY et al. 1991).

# C. Dense Granule Membrane Vesicles

The platelet lysate from which plasma membrane vesicles are isolated also contains vesicles derived from dense granules. The two independent 5-HT transport systems are both active in the crude lysate, but by choosing the appropriate conditions, it is possible to measure either plasma membrane or granule membrane transport without interference from the other system. In low-concentration $Na^+$ medium the plasma membrane 5-HT transporter is inactive, while transport into granule membrane vesicles is driven by external ATP. The plasma membrane transporter is measured by imposing $Na^+$, $Cl^-$, and $K^+$ gradients as described above in the absence of ATP. The granule amine transporter is driven by ATP-dependent $H^+$ pumping and does not require $Na^+$. Thus, transport into granule-derived membrane vesicles is assayed in low-concentration $Na^+$ medium containing ATP. The specificity of the approach is demonstrated by the fact that ATP-dependent 5-HT transport in low-concentration $Na^+$ is inhibited by reserpine and not by imipramine, while imipramine but not reserpine blocks $Na^+$-gradient-driven 5-HT accumulation in the absence of ATP.

## I. Driving Forces

The platelet dense granule amine transporter couples 5-HT transport to the transmembrane $\Delta pH$ (acid inside) and $\Delta\Psi$ (positive inside) generated by the ATP-driven $H^+$ pump; one molecule of cytoplasmic 5-HT is exchanged for one or more intragranular protons. Experimentally, transport of 5-HT into isolated dense granule membrane vesicles can be driven either by supplying ATP to the $H^+$-pumping ATPase or by artificially imposing an electrochemical $H^+$ potential ($\Delta\mu_{H^+}$). In the absence of ATP, a $\Delta\mu_{H^+}$ (inside acid and positive) is imposed by first preloading vesicles with $K^+$-containing, low-pH medium and then transferring them to $K^+$-free, higher pH medium containing nigericin which exchanges $H^+$ and $K^+$. This manipulation generates a $\Delta pH$ (interior acid) across the vesicle membrane that is maintained by nigericin-catalyzed $H^+$ influx driven by the $K^+$ gradient. Efflux of the permeant anions $Cl^-$ and $SCN^-$ generates a $\Delta\Psi$ (interior positive) that also drives transport.

Accumulation of biogenic amines within secretory organelles is similar in a variety of cell types storing NE, DA, 5-HT, Epi, and histamine. In addition to platelet dense granules, organelles that contain this system include synaptic vesicles, adrenal chromaffin granules, and mast cell and basophil secretory granules (JOHNSON and SCARPA 1978; MARON et al. 1979; JOHNSON et al. 1980;

FISHKES and RUDNICK 1982; KANNER and BENDAHAN 1985). In the absence of ATP, imposition of a transmembrane pH difference (ΔpH) provides a driving force for biogenic amine accumulation (JOHNSON and SCARPA 1978; PHILLIPS 1978; SCHULDINER et al. 1978). A consequence of ATP-driven $H^+$ pumping into these organelles is the development of a transmembrane electrical potential difference (ΔΨ; BASHFORD et al. 1975). This membrane potential is also a driving force for amine accumulation (CASEY et al. 1977; HOLZ 1978; JOHNSON et al. 1979; NJUS and RADDA 1979; KANNER et al. 1980).

## II. Purification

Biogenic amine accumulation into chromaffin granules has been used as a model for 5-HT storage within platelet dense granules and other monoaminergic secretory organelles. In these granules, biogenic amine accumulation is blocked by reserpine and tetrabenezine, two inhibitors that have been used to identify the vesicular amine transporter. Henry and coworkers (ISAMBERT et al. 1989) used [$^{125}$I]iodoazidoketanserin to label a polypeptide of approximately 80 kDa. Schuldiner and coworkers took advantage of the slow dissociation of reserpine from the transporter (RUDNICK et al. 1990) to isolate the transporter in an active form from detergent-solubilized bovine chromaffin granule membranes (STERN-BACH et al. 1990). The purified preparation consisted of an 80 kDa glycoprotein that could reconstitute tetrabenezine-sensitive $H^+$/5-HT exchange in proteoliposomes.

These two inhibitors, reserpine and tetrabenezine, apparently have different mechanisms for inhibiting 5-HT transport into granules. Reserpine binding is competitive with 5-HT (SCHERMAN and HENRY 1984) and binds with very high affinity (DARCHEN et al. 1989). However, binding and the consequent inhibition of transport are slow (RUDNICK et al. 1990). Reserpine dissociation is so slow as to be difficult to detect except under conditions in which the protein is likely to be denatured (RUDNICK et al. 1990). The rate of reserpine binding is dependent on the $\Delta\mu_{H^+}$ generated by the ATP-dependent $H^+$ pump (WEAVER and DEUPREE 1982; SCHERMAN and HENRY 1984; RUDNICK et al. 1990). In contrast, tetrabenezine binding to the transporter is not stimulated by the $\Delta\mu_{H^+}$ (SCHERMAN et al. 1983; HENRY and SCHERMAN 1989). Moreover, reserpine has little effect on tetrabenezine binding, event at concentrations that inhibit transport completely. Transport substrates, such as 5-HT, inhibit tetrabenezine binding only at concentrations 100 times greater than their $K_M$ values (SCHERMAN et al. 1983; DARCHEN et al. 1989; HENRY and SCHERMAN 1989). These results suggest that tetrabenezine binds at a site distinct from that of reserpine and exerts an allosteric inhibition of transport.

## III. Cloning

Two expression cloning strategies led to the isolation of cDNAs encoding vesicular amine transporters. Edwards and coworkers (Y. LIU et al. 1992b) were investigating the mechanism by which PC-12 cells resist the toxic action of the neurotoxin $MPP^+$ (DANIELS and REINHARD 1988) which was known to be accumulated by chromaffin granules (DANIELS and REINHARD 1988). Transfection of CHO fibroblasts, normally sensitive to $MPP^+$, with cDNA from PC-12 cells led to the appearance of $MPP^+$-resistant transfectants that concentrated DA into vesicular structures. When the PC-12 cDNA responsible for this property was isolated and sequenced (Y. LIU et al. 1992a), it was found to predict peptide sequences highly homologous to those of peptides derived from the purified bovine chromaffin granule amine transporter (STERN-BACH et al. 1990).

Independently, Hoffman and coworkers (ERICKSON et al. 1992) transfected CV-1 cells with cDNA prepared from rat basophilic leukemia (RBL) cell mRNA. RBL cells take up 5-HT using the plasma membrane 5-HT transporter and store it in secretory granules containing a vesicular amine transporter. Some of the cDNA clones that conferred increased 5-HT uptake on CV-1 cells encoded a vesicular transporter that functioned to accumulate 5-HT in acidic organelles. The sequence of the RBL vesicular amine transporter is highly homologous to the vesicular transporter cloned from PC-12 cells, and almost identical to a vesicular amine transporter cloned from rat brain (Y. LIU et al. 1992a). Despite the similarity among cloned vesicular transporters, there is essentially no homology between vesicular and plasma membrane transporters for biogenic amines. Limited homology, however, was found between vesicular amine transporters and bacterial drug resistance proteins (ERICKSON et al. 1992; Y. LIU et al. 1992b).

## References

Alexi T, Azmitia E (1991) Ethanol stimulates [3H]5-HT high-affinity uptake by rat forebrain synaptosomes; role of 5-HT receptors and voltage channel blockers. Brain Res 544:243–247

Anderson G, Horne W (1992) Activators of protein kinase-C decrease serotonin transport in human platelets. Biochim Biophys Acta 1137:331–337

Balkovetz D, Tirruppathi C, Leibach F, Mahesh V, Ganapathy V (1989) Evidence for an imipramine-sensitive serotonin transporter in human placental brush-border membranes. J Biol Chem 264:2195–2198

Barber AJ, Jamieson GA (1970) Isolation and characterization of plasma membranes from human blood platelets. J Biol Chem 245:6357–6365

Bashford C, Radda G, Ritchie G (1975) Energy-linked activities of the chromaffin granule membrane. FEBS Lett 50:21–24

Biessen E, Horn A, Robillard G (1990) Partial purification of the 5-hydroxytryptophan-reuptake system from human blood platelets using a citalopram-derived affinity resin. Biochemistry 29:3349–3354

Blakely R, Berson H, Fremeau R, Caron M, Peek M, Prince H, Bradely C (1991) Cloning and expression of a functional serotonin transporter from rat brain. Nature 354:66–70

Borden L, Smith K, Hartig P, Branchek T, Weinshank R (1992) Molecular heterogeneity of the gamma-aminobutyric acid (GABA) transport system – cloning of 2 novel high affinity GABA transporters from rat brain. J Biol Chem 267:21098–21104

Casey RP, Njus D, Radda GK, Sehr PA (1977) Active proton uptake by chromaffin granules: observation by amine distribution and phosphorus-31 nuclear magnetic resonance techniques. Biochemistry 16:972–977

Clark J, Deutch A, Gallipoli P, Amara S (1992) Functional expression and CNS distribution of a beta-alanine-sensitive neuronal GABA transporter. Neuron 9:337–348

Cool DR, Leibach FH, Bhalla VK, Mahesh VB, Ganapathy V (1991) Expression and cyclic AMP-dependent regulation of a high affinity serotonin transporter in the human placental choriocarcinoma cell line (JAR). J Biol Chem 266:15750–15757

Crane RK, Forstner G, Eichholz A (1965) Studies on the mechanism of the intestinal absorption of sugars. Biochim Biophys Acta 109:467–477

Daniels AJ, Reinhard JF (1988) Energy driven uptake of the neurotoxin 1-Methyl-4-phenylpyridinium into chromaffin granules via the cathecholamine transporter. J Biol Chem 263:5034–5036

Darchen F, Scherman D, Henry J-P (1989) Reserpine binding to chromaffin granules suggests the existence of two conformations of the monoamine transporter. Biochemistry 28:1692–1697

Eddy AA (1968) A net gain of sodium ions and a net loss of potassium ions accompanying the uptake of glycine by mouse ascites-tumour cells in the presence of sodium cyanide. Biochem J 108:195–206

Eddy AA, Indge KJ, Backen K, Nowacki JA (1970) Interactions between potassium ions and glycine transport in the yeast saccharomyces carlsbergensis. Biochem J 120:845–852

Erickson J, Eiden L, Hoffman B (1992) Expression cloning of a reserpine-sensitive vesicular monoamine transporter. Proc Natl Acad Sci USA 89:10993–10997

Fischer JF, Cho AK (1979) Chemical release of dopamine from striatal homogenates: evidence for an exchange diffusion model. J Pharmacol Exp Ther 208:203–209

Fishkes H, Rudnick G (1982) Bioenergetics of serotonin transport by membrane vesicles derived from platelet dense granules. J Biol Chem 57:5671–5677

Fremeau R, Caron M, Blakely R (1992) Molecular cloning and expression of a high affinity L-proline transporter expressed in putative glutamatergic pathways of rat brain. Neuron 8:915–926

Fuller R (1992) Basic advances in serotonin pharmacology. J Clin Psychopharmacol 53:36–45

Giros B, Elmstikawy S, Bertrand L, Caron MG (1991) Cloning and functional characterization of a cocaine-sensitive dopamine transporter. FEBS Lett 295:149–154

Giros B, Elmestikawy S, Godinot N, Zheng K, Han H, Yang-Feng T, Caron M (1992) Cloning, pharmacological characterization, and chromosome assignment of the human dopamine transporter. Mol Pharmacol 42:383–390

Graham D, Esnaud H, Langer SZ (1992) Partial purification and characterization of the sodium-ion-coupled 5-hydroxytryptamine transporter of rat cerebral cortex. Biochem J 286:801–805

Guastella J, Nelson N, Nelson H, Czyzyk L, Keynan S, Miedel MC, Davidson N, Lester AH, Kanner BI (1990) Cloning and expression of a rat brain GABA transporter. Science 249:1303–1306

Guimbal C, Kilimann M (1993) A $Na^{+}$-dependent creatine transporter in rabbit brain, muscle, heart, and kidney – cDNA cloning and functional expression. J Biol Chem 268:8418–8421

Henry J-P, Scherman D (1989) Radioligands of the vesicular monoamine transporter and their use as markers of monoamine storage vesicles. Biochem Pharmacol 38:2395–2404

Hoffman BJ, Mezey E, Brownstein MJ (1991) Cloning of a serotonin transporter affected by antidepressants. Science 254:579–580

Holz RW (1978) Evidence that catecholamine transport into chromaffin vesicles is coupled to vesicle membrane potential. Proc Natl Acad Sci USA 75:5190–5194

Isambert M-F, Gasnier B, Laduron PM, Henry J-P (1989) Photoaffinity labeling of the monoamine transporter of bovine chromaffin granules and other monoamine storage vesicles using 7-azido-8-[125I]iodoketanserin. Biochemistry 28:2265–2270
Jayanthi LD, Ramamoorthy S, Mahesh VB, Leibach FH, Ganapathy V (1994) Calmodulin-dependent regulation of the catalytic function of the human serotonin transporter in placental choriocarcinoma cells. J Biol Chem 269:14424–14429
Johnson RG, Scarpa A (1978) The internal pH of isolated serotonin containing granules of pig platelets. J Biol Chem 253:7061–7068
Johnson RG, Pfister D, Carty SE, Scarpa A (1979) Biological amine transport in chromaffin ghosts: coupling to the transmembrane proton and potential gradients. J Biol Chem 254:10963–10972
Johnson RG, Carty SE, Fingerhood BJ, Scarpa A (1980) The internal pH of mast cell granules. FEBS Lett 120:75–79
Kanner BI, Bendahan A (1985) Transport of 5-hydroxytryptamine in membrane vesicles from rat basophillic leukemia cells. Biochim Biophys Acta 816:403–410
Kanner BI, Schuldiner S (1987) Mechanism of transport and storage of neurotransmitters. CRC Crit Rev Biochem 22:1–38
Kanner BI, Sharon I, Maron R, Schuldiner S (1980) Electrogenic transport of biogenic amines in chromaffin granule membrane vesicles. FEBS Lett 111:83–86
Kilty J, Lorang D, Amara S (1991) Cloning and expression of a cocaine-sensitive rat dopamine transporter. Science 254:578–579
King SC, Tiller AA, Chang AS-S, Lam DM-K (1992) Differential regulation of the imipramine-sensitive serotonin transporter by cAMP in human JAr choriocarcinoma cells, rat PC12 pheochromocytoma cells, and C33-14-B1 transgenic mouse fibroblast cells. Biochem Biophys Res Commun 183:487–491
Launay J, Geoffroy C, Mutel V, Buckle M, Cesura A, Alouf J, DaPrada M (1992) One-step purification of the serotonin transporter located at the human platelet plasma membrane. J Biol Chem 267:11344–11351
Launay J, Bondoux D, Osetgasque M, Emami S, Mutel V, Haimart M, Gespach C (1994) Increase of human platelet serotonin uptake by atypical histamine receptors. Am J Physiol 266:R526–R536
Lesch K, Wolozin B, Estler H, Murphy D, Riederer P (1993a) Isolation of a cDNA encoding the human brain serotonin transporter. J Neural Transm Gen Sect 91:67–72
Lesch K, Wolozin B, Murphy D, Riederer P (1993b) Primary structure of the human platelet serotonin uptake site – identity with the brain serotonin transporter. J Neurochem 60:2319–2322
Lingjaerde O (1969) Uptake of serotonin in blood platelets: dependence on sodium and chloride, and inhibition by choline. FEBS Lett 3:103–106
Liu Q-R, Lopez-Corcuera B, Nelson H, Mandiyan S, Nelson N (1992a) Cloning and expression of a cDNA encoding the transporter of taurine and beta-alanine in mouse brain. Proc Natl Acad Sci USA 89:12145–12149
Liu Q-R, Nelson H, Mandiyan S, Lopez-Corcuera B, Nelson N (1992b) Cloning and expression of a glycine transporter from mouse brain. FEBS Lett 305:110–114
Liu Y, Peter D, Roghani A, Schuldiner S, Prive G, Eisenberg D, Brecha N, Edwards R (1992a) A cDNA that supresses $MPP^+$ toxicity encodes a vesicular amine transporter. Cell 70:539–551
Liu Y, Roghani A, Edwards R (1992b) Gene transfer of a reserpine-sensitive mechanism of resistance to N-methyl-4-phenylpyridinium. Proc Natl Acad Sci USA 89:9074–9078
Mager S, Min C, Henry DJ, Chavkin C, Hoffman BJ, Davidson N, Lester H (1994) Conducting states of a mammalian serotonin transporter. Neuron 12:845–859
Mamounas L, Mullen C, Ohearn E, Molliver M (1991) Dual serotoninergic projections to forebrain in the rat – morphologically distinct 5-HT axon terminals exhibit differential vulnerability to neurotoxic amphetamine derivatives. J Comp Neurol 314:558–586

Maron R, Kanner BI, Schuldiner S (1979) The role of a transmembrane pH gradient in 5-hydroxytryptamine uptake by synaptic vesicles from rat brain. FEBS Lett 98:237–240
Mayser W, Betz H, Schloss P (1991) Isolation of cDNAs encoding a novel member of the neurotransmitter transporter gene family. FEBS Lett 295:203–206
Miller KJ, Hoffman BJ (1994) Adenosine $A_3$ receptors regulate serotonin transport via nitric oxide and cGMP. J Biol Chem 269:27351–27356
Moncada S, Palmer RMJ, Higgs EA (1989) Biosynthesis of nitric oxide from L-arginine. A pathway for the regulation of cell function and communication. Biochem Pharmacol 38:1709–1715
Myers CL, Pitt BR (1988) Selective effect of phorbol ester on serotonin removal and ACE activity in rabbit lungs. J Appl Physiol 65:377–384
Myers CL, Lazo JS, Pitt BR (1989) Translocation of protein kinase C is associated with inhibition of 5-HT uptake by cultured endothelial cells. Am J Physiol 257:L253–L258
Nelson H, Mandiyan S, Nelson N (1990) Cloning of the human brain GABA transporter. FEBS Lett 269:181–184
Nelson P, Rudnick G (1979) Coupling between platelet 5-hydroxytryptamine and potassium transport. J Biol Chem 254:10084–10089
Nelson P, Rudnick G (1981) Anion-dependent sodium ion conductance of platelet plasma membranes. Biochemistry 20:4246–4249
Nelson P, Rudnick G (1982) The role of chloride ion in platelet serotonin transport. J Biol Chem 257:6151–6155
Njus D, Radda GK (1979) A potassium ion diffusion potential causes adrenaline uptake in chromaffin-granule "ghosts". Biochemistry 180:579–585
O'Reilly CA, Reith MEA (1988) Uptake of [$^3$H]serotonin into plasma membrane vesicles from mouse cortex. J Biol Chem 263:6115–6121
Pacholczyk T, Blakely R, Amara S (1991) Expression cloning of a cocaine- and antidepressant-sensitive human noradrenaline transporter. Nature 350:350–354
Phillips JH (1978) 5-Hydroxytryptamine transport by the chromaffin granule membrane. Biochem J 170:673–679
Ramamoorthy J, Ramamoorthy S, Papapetropoulos A, Catravas J, Leibach F, Ganapathy V (1995) Cyclic AMP-independent up-regulation of the human serotonin transporter by staurosporine in choriocarcinoma cells. J Biol Chem 270:17189–17195
Ramamoorthy S, Cool D, Leibach F, Mahesh V, Ganapathy V (1992) Reconstitution of the human placental 5-hydroxytryptamine transporter in a catalytically active form after detergent solubilization. Biochem J 286:89–95
Ramamoorthy S, Cool D, Mahesh V, Leibach F, Melikian H, Blakely R, Ganapathy V (1993) Regulation of the human serotonin transporter – cholera toxin-induced stimulation of serotonin uptake in human placental choriocarcinoma cells is accompanied by increased serotonin transporter messenger RNA levels and serotonin transporter-specific ligand binding. J Biol Chem 268:21626–21631
Rudnick G (1977) Active transport of 5-hydroxytryptamine by plasma membrane vesicles isolated from human blood platelets. J Biol Chem 252:2170–2174
Rudnick G (1986) Serotonin transport in plasma and dense granule membrane vesicles. In: Holmsen H (ed) Platelet responses and metabolism. CRC, Boca Raton, pp 119–133
Rudnick G, Nelson P (1978) Platelet 5-hydroxytryptamine transport an electroneutral mechanism coupled to potassium. Biochemistry 17:4739–4742
Rudnick G, Wall SC (1992a) The molecular mechanism of ecstasy [3,4-methylenedioxymethamphetamine (MDMA)] – serotonin transporters are targets for MDMA-induced serotonin release. Proc Natl Acad Sci USA 89:1817–1821
Rudnick G, Wall SC (1992b) *p*-Chloroamphetamine induces serotonin release through serotonin transporters. Biochemistry 31:6710–6718

Rudnick G, Steiner-Mordoch SS, Fishkes H, Stern-Bach Y, Schuldiner S (1990) Energetics of reserpine binding and occlusion by the chromaffin granule biogenic amine transporter. Biochemistry 29:603–608

Scherman D, Jaudon P, Henry J (1983) Characterization of the monoamine carrier of chromaffin granule membrane by binding of [2–3H]dihydrotetrabenazine. Proc Natl Acad Sci USA 80:584–588

Scherman D, Henry J-P (1984) Reserpine binding to bovine chromaffin granule membranes. Mol Pharmacol 25:113–122

Schuldiner S, Fishkes H, Kanner BI (1978) Role of a transmembrane pH gradient in epinephrine transport by chromaffin granule membrane vesicles. Proc Natl Acad Sci USA 75:3713–3716

Shimada S, Kitayama S, Lin C, Patel A, Nanthakumar E, Gregor P, Kuhar M, Uhl G (1991) Cloning and expression of a cocaine-sensitive dopamine transporter complementary DNA. Science 254:576–578

Smith K, Borden L, Hartig P, Branchek T, Weinshank R (1992a) Cloning and expression of a glycine transporter reveal colocalization with NMDA receptors. Neuron 8:927–935

Smith K, Borden L, Wang C, Hartig P, Branchek T, Weinshank R (1992b) Cloning and expression of a high affinity taurine transporter from rat brain. Mol Pharmacol 42:563–569

Sneddon JM (1969) Sodium-dependent accumulation of 5-hydroxytryptamine by rat blood platelets. Br J Pharmacol 37:680–688

Stern-Bach Y, Greenberg-Ofrath N, Flechner I, Schuldiner S (1990) Identification and purification of a functional amine transporter from bovine chromaffin granules. J Biol Chem 265:3961–3966

Sulzer D, Rayport S (1990) Amphetamine and other psychostimulants reduce pH gradient in midbrain dopaminergic neurons and chromaffin granules: a mechanism of action. Neuron 5:797–808

Talvenheimo J, Rudnick G (1980) Solubilization of the platelet plasma membrane serotonin transporter in an active from. J Biol Chem 255:8606–8611

Usdin T, Mezey E, Chen C, Brownstein M, Hoffman B (1991) Cloning of the cocaine-sensitive bovine dopamine transporter. Proc Natl Acad Sci USA 88:11168–11171

Weaver JA, Deupree JD (1982) Conditions required from reserpine binding to the catecholamine transporter on chromaffin granule ghosts. Eur J Pharmacol 80:437–438

Yamauchi A, Uchida S, Kwon H, Preston A, Robey R, Garciaperez A, Burg M, Handler J (1992) Cloning of a $Na^+$-dependent and CL–dependent betaine transporter that is regulated by hypertonicity. J Biol Chem 267:649–652

# Section III
# Platelet-Derived Factors

CHAPTER 18

# Dense Granule Factors

M.H. FUKAMI

## A. Introduction

"Platelet dense granule" is a descriptive name originating from the observation that these granules absorb electrons even in unfixed or lightly fixed, unstained mounts on electron microscopic examination. Their contents are also highly osmiophilic, with the result that they are the most electron-dense organelles after osmium staining. Other names for these granules are "amine" or "serotonin storage organelles" and "dense bodies." The title of this chapter is somewhat misleading in that "factor" is usually a name for unknown substances or proteins with some physiological function. In the case of dense granules, which have been an object of study since the first descriptions of platelets, most of the factors have long been identified, and these consist mainly of the low-molecular-weight compounds ATP, ADP, serotonin, and divalent cations. These organelle contents are released upon agonist-induced platelet secretion and apparently act to amplify the original stimulus (see Chaps. 4, 6). They are formed in the precursor megakaryocyte cell and apparently acquire all their constituents at that time, except for serotonin which is taken up by the platelet after it is released into circulation. There exist in platelets an amine transporter in the plasma membrane and another in the dense granule membrane that are responsible for serotonin uptake (see Chap. 17).

The discovery that the release of various substances from platelets was accompanied by degranulation and not by nonspecific release of cytoplasmic contents came in the late 1950s and 1960s (for a review see DA PRADA et al. 1981). It took another decade or so to establish that platelet granules consisted of several distinct types, each of which contained a separate complement of secreted substances. Of the three granule types, $\alpha$-granules (see Chap. 19) and lysosomes (see Chap. 20) contain proteins/enzymes that are secreted, whereas the dense granules apparently contain only nucleotides, divalent cations, serotonin, and small amounts of other amines; no secreted proteins have as yet been localized to these organelles, although some membrane proteins have been characterized.

# B. Contents of Platelets

## I. Nucleotides

Thirty-six years ago, BAKER et al. (1959) isolated a crude particulate fraction from blood platelets that was found to contain serotonin and ATP. Three years later, GRETTE (1962) described the phenomenon that platelets released a variety of substances into the extracellular milieu upon thrombin treatment, among them ADP which GAARDER et al. (1961) had shown to cause platelets to aggregate. At this stage, it was not yet clear whether the release was due to lysis or secretion of granules. The need to develop a more sensitive method for measuring ADP and other nucleotides in small amounts of platelets led to the use of $[^{32}P]P_i$ incubation with platelets which resulted in labelling of ADP and ATP as well as numerous other phosphorylated metabolic intermediates (HOLMSEN 1965a). When this new method was applied to the measurement of collagen-induced release of ADP from blood platelets, no radioactive ADP was found in the plasma supernatant, although enzymatic analysis showed that 2.5–4.0 $\mu$g/ml of ADP were present in plasma (HOLMSEN 1965b). In fact, all of the radioactively labelled metabolic intermediates were retained in the platelets after collagen treatment, and only trace amounts appeared in the supernatant, indicating that the platelets were intact after release. This finding was the first indication that the released nucleotides were separate, nonexchangable entities from the cellular ones. It was assumed that the nonradioactive pool of

**Table 1.** Metabolic and storage pools of adenine nucleotides in human platelets (from UGURBIL et al. 1979)

| Sample | % Radioactivity | | | | Total ATP + ADP nmol/mg protein | | |
|---|---|---|---|---|---|---|---|
| | ATP | ADP | AMP | ATP/ADP | ATP | ADP | ATP/ADP |
| Platelets control[a] | 74.3 | 15.0 | 0.5 | 4.95 | [b]29.0 | 16.9 | 1.72 |
| Secreted nucleotides[c] | 1.7 | 0.5 | 0.7 | – | 9.7 | 12.2 | 0.80 |
| Difference[d] | | | | | 19.3 | 4.7 | 4.10 |

Platelets were labelled with $[^{14}C]$adenine.
[a] The nucleotides were separated by paper electrophoresis and the radioactivity of the spots counted.
[b] An aliquot was analyzed for ATP and ADP.
[c] Supernatants of labelled platelets treated with thrombin were analyzed for radioactive and actual ATP and ADP contents; little radioactivity was present in the supernatant although 48% of total ATP and ADP were released as measured enzymatically.
[d] The difference between the total platelet content and the secreted ATP and ADP represents the actual amounts of metabolic cytoplasmic nucleotides; the ATP:ADP ratio of the difference is similar to that of the radioactivity distribution.

**Table 2.** Major constituents of human platelet dense granules (dense granules isolated as described by FUKAMI (1992); data from the same reference and from HOLMSEN and UGURBIL 1992)

| Substance | Whole platelets | Secreted amounts[a] | Dense granules |
|---|---|---|---|
| ATP | 23.8 ± 3.5 | 9.7 ± 1.3 | 440 ± 96 |
| ADP | 14.5 ± 2.4 | 14.1 ± 1.8 | 633 ± 142 |
| GTP[b] | – | – | 63 |
| GDP[b] | – | – | 18 |
| $PP_i$ | 6.3 ± 0.05 | 6.0 ± 0.52 | 236 ± 53 |
| $P_i$ | 16.7 ± 2.5 | 4.6 ± 1.6 | 248 ± 65 |
| $Ca^{2+}$ | 125 ± 27 | 88 ± 23 | 2634 ± 471 |
| $Mg^{2+}$ | – | – | 98 ± 33 |
| Serotonin | 1.5–2.5 | 1.5–2.0 | 90–100 |

Values are given as nmol/mg protein with the exception of
[a] secreted amounts that are given as nmol/mg platelet protein and
[b] guanine nucleotides representing intragranular *mM* concentration; this value happens to be approximately the same as nmol/mg protein.

secreted nucleotides was related to the degranulation of platelets that became observable with the advent of electron microscopy and to the particulate serotonin and ATP reported by BAKER et al. (1959). Subsequent subcellular fractionation studies gradually improved over the years and established that these granules with the highest specific density (g/ml) contained the storage nucleotides (FUKAMI et al. 1978; RENDU et al. 1982). Analysis of the secreted supernatant showed that the secreted nucleotides included ADP, ATP, and lesser amounts of GTP and GDP (UGURBIL and HOLMSEN 1981). Analysis of total ATP and ADP in control platelets had shown that the ATP:ADP ratio was less than 2, considerably lower than that found in most cells (see for example Table 1). When the amount of secreted ATP and ADP representing about 50%–60% of total cellular nucleotides was subtracted from the total amount of ATP and ADP in control platelets, the ratio of nonsecretable or metabolically active ATP to ADP increased to values of about 5, which is comparable to that of most cells. The major nucleotide secreted by platelets was ADP which was present in storage granules at a ratio of about 3 ADPs to 2 ATPs; this large pool of storage ADP is the reason for the low ATP:ADP ratio in intact platelets (see Tables 1 and 2). The storage pool in the dense granules is drastically reduced or absent in platelets from individuals with certain hereditary diseases, and this can be diagnosed by determining the total platelet ATP:ADP ratio which tends to approach 5 in deficient platelets (see Chap. 24). These patients have bleeding problems because their platelets lack the ADP that the primarily activated platelets secrete and that results in recruitment of other platelets to join the aggregate and hemostatic plug.

Diadenosine tetraphosphate (FLODGAARD and KLENOW 1982) and diadenosine triphosphate (LÜTHKE and OGILVIE 1983) have been reported to be present in platelets and to be almost totally released upon thrombin aggregation. Diadenosine tetraphosphate antagonized ADP-induced platelet aggregation, whereas diadenosine triphosphate became hydrolyzed by an enzyme present in the plasma (LÜTHKE and OGILVIE 1985), producing ADP (LÜTHKE and OGILVIE 1984). Diadenosine tetraphosphate levels were low in platelets from Chediak-Higashi patients and cattle with dense granule deficiency, suggesting the the dense granules are the storage sites for these compounds (KIM et al. 1985). The reported levels of these unusual nucleotides were found to be about 1%–2% of those of the dense granule contents of ATP and ADP.

## II. Divalent Cations

Platelets had been shown to contain much higher levels of calcium than other cells, and it had been suggested that calcium was involved in platelet activation (WALLACH et al. 1958). However, it was not until 1970 that it was demonstrated that calcium is among those platelet constituents secreted upon activation (MÜRER and HOLME 1970). The time course of release was shown to be the same as that of ADP and ATP with two different agonists, and the secretable pool of calcium was found not to be exchangable upon incubation of platelets with $^{45}Ca$. Electron microprobe analysis of unstained platelet mounts confirmed that high concentrations of calcium and phosphorus were contained in the dense bodies (MARTIN et al. 1974; SKAER et al. 1974; COSTA et al. 1981). Analysis of purified human platelet dense granules confirmed the concentrations and ratios of $Ca^{2+}$ to nucleotides estimated by other methods (FUKAMI 1992).

It is difficult to describe precisely the role of extracellular calcium in platelet activation in vivo because platelets in vivo are obviously exposed to both $Ca^{2+}$ and $Mg^{2+}$ in the plasma. It seems probable that both divalent cations are buffered or chelated by plasma proteins, just as all ex vivo preparations of platelets for functional studies, even platelet-rich plasma, must first be mixed with anticoagulant. The anticoagulants most commonly used are buffered solutions of citrate, a calcium chelator; heparin and hirudin which both bind thrombin are also sometimes used as anticoagulants. EDTA has also been used for in vitro experiments. Aggregation-induced platelet activation and secretion is divalent-cation-dependent, but thrombin-induced secretion of platelets is independent of divalent cations and occurs even in the presence of EDTA. Most other cells, such as mast cells and chromaffin cells, do not undergo secretion in the absence of extracellular $Ca^{2+}$. The storage granule content of $Ca^{2+}$, however, is considerable (see Table 2), and the local concentrations of $Ca^{2+}$ immediately after secretion must be very high. Since the other constituents of dense granules act to potentiate platelet activation, one could assume that $Ca^{2+}$ probably also contributes to activation. Conversely, it could be argued that the event requiring $Ca^{2+}$ – aggregation – is already over by the time that the storage $Ca^{2+}$ is released.

The $Mg^{2+}$ content of human platelet dense granules is about 5% that of $Ca^{2+}$ (see Table 2). The ratio of these two divalent cations is one of those parameters that vary considerably across species, with human platelets at one end of the scale containing mostly $Ca^{2+}$ and pig platelets at the other end of the scale containing mostly $Mg^{2+}$ (MEYERS et al. 1982; see below for a discussion of other species differences). Indeed, the fact that pig platelets with mostly $Mg^{2+}$ apparently function as well as human platelets with mostly $Ca^{2+}$ stores indicates that neither of the divalent cations in the storage pool is specifically essential or that either cation is sufficent for the events following secretion that lead to formation of the platelet plug. Experiments with platelets from storage-pool-deficient patients suggest that $Ca^{2+}$ is not important for $\alpha$-granule and lysosome secretion which occur after dense granule secretion (RENDU et al. 1987; LAGES et al. 1988). Thrombin-induced aggregation and secretion of $\alpha$-granules and lysosomes which are present in normal amounts were depressed in these patients' platelets. Addition of ADP together with thrombin restored aggregation as well as $\alpha$-granule and lysosome secretion responses; addition of $Ca^{2+}$, serotonin, or ATP together with thrombin did not correct the depressed responses. These findings suggest that the major role of divalent cations in the dense granules may not be to become available upon secretion but rather to participate in the formation of the storage complex (see Sect. C below).

## III. Amines

Serotonin was one of the first dense granule constituents described in platelets (for a review see DA PRADA et al. 1981). Blood serotonin levels were found to be attributable solely to platelet contents if care was taken to retain platelets in the sample and no platelet activation was allowed to occur. Varying amounts of other amines are also found in dense granules in animal species, but human platelets tend to contain primarily serotonin, although small amounts of noradrenaline and normetanephrine are also found in human dense granules (DA PRADA et al. 1981). As described above, the adenine nucleotides and calcium in the dense granules are metabolically relatively inactive in that $^{32}P$ from $^{32}P$-labelled cytoplasmic ATP or $^{45}Ca^{2+}$ does not become incorporated into the storage pool. Serotonin is different in that there are transporters that carry extracellular amine into the cytoplasm and transporters that move the amine into the dense granule against a tremendous gradient difference in an energy-dependent manner. In the course of 30 min, one can easily achieve incorporation of labelled serotonin into the granules, such that induction of release results in the secretion of radioactively marked serotonin (HOLMSEN et al. 1973). In fact, the plasma membrane transporter is so avid that unless one inhibits it with a specific inhibitor, imipramine, the secreted amine is so rapidly reabsorbed into the platelet that it appears as though only partial secretion occurred (WALSH and GAGNATELLI 1974). Other amines can be incorporated into storage granules when presented in sufficiently high concentrations in the case of intact platelets or to isolated organelles. The distribution of [$^3$H]methylamine in and outside isolated dense granules has shown that the internal pH of the granules in pig platelets is about 5.5 (JOHNSON et al. 1978), and this pH gradient is part of the amine uptake mechanism (see Chap. 17). The maximum storage capacity for serotonin is not known, but individuals who have eaten food rich in serotonin before blood samples are taken have above-average serotonin levels. Storage pool deficient platelets can transport serotonin across the plasma membrane, but the serotonin is rapidly degraded to 5-hydroxyindoleacetic acid and lesser amounts of other derivatives. In storage pool deficiencies in which there seem to be empty granule sacks, the amine is not stored (WEISS et al. 1993; MCNICOL et al. 1994). The dense granules in the platelet parent cell, the megakaryocyte, contain nucleotides and $Ca^{2+}$ but not enough serotonin to be highly osmiophilic. Serotonin becomes incorporated into dense granules only in the circulating platelet, and its uptake appears to be dependent on preexisting nucleotide and divalent cation stores. Although serotonin appears to have diverse effects on many target cells in the cardiovascular and central nervous sytems, it is a weak inducer of platelet activation except in feline platelet (MEYERS et al. 1979). As described above, its addition did not correct the depressed secretion of $\alpha$-granule and lysosome contents in storage pool deficient patients as did ADP. The role of serotonin that the platelet appears to be so superbly equipped to take up and store is not yet clearly defined; the platelet may only be acting as a pickup and delivery service or pickup and removal service.

### IV. Other

Trace amounts of numerous other compounds, mostly amines, have been reported in dense granules, but the only other constituents present in significant amounts besides those listed above are inorganic pyrophosphate ($PP_i$) and perhaps inorganic phosphate ($P_i$) (SILCOX et al. 1973; FUKAMI et al. 1980). These phosphates were measured in secretion supernatants and in dense granule subcellular fractions (Table 2). Almost all of the platelet $PP_i$ is secreted, but most of the cellular $P_i$ appears to be nonsecretable. Although care was taken to avoid breakdown of the $PP_i$, the the presence of $P_i$ in the dense granule may be a product of $PP_i$ hydrolysis. Neither was secreted in the radioactive $^{32}P$ form in platelets in which metabolic ATP had been labelled by $^{32}P$-$P_i$ incubation.

## C. Storage Mechanisms

The first suggestion that some sort of ordered structure existed in the dense granules came as early as 1970 when BERNEIS et al. (1970) observed that the high concentrations of nucleotides and divalent cations in the granules were unlikely to exist in free solution as this would lead to osmotic strengths of 1000 mOsM inside the granules. Since the granules appeared to be stable and did not accumulate water or undergo lysis, the contents must exist in some higher molecular weight, complex form. Solutions of ATP, $Ca^{2+}$ or $Mg^{2+}$, and serotonin were shown to form aggregates of high apparent molecular weight

(BERNEIS et al. 1970). With the arrival of whole cell nuclear magnetic resonance (NMR) techniques, several investigators began to use NMR for the study of platelet storage granule complexes (COSTA et al. 1979; UGURBIL et al. 1979). The two platelet types selected for the initial studies were human platelets with mostly $Ca^{2+}$ and pig platelets with mostly $Mg^{2+}$. Among the other species of platelets studied by NMR were cattle and rabbit platelets that have a mixture of $Ca^{2+}$ and $Mg^{2+}$ (CARROLL et al. 1980; SCHMIDT and CARROLL 1982). Since it was known that intragranular pH was approximately 5.5 (JOHNSON et al. 1978) compared to the cytoplasmic pH of 7.4, it was expected that high resolution [$^{31}$P]NMR spectra of human platelets would show two different pools of ATP and ADP, as the chemical shift is affected by pH. Much to the investigators' surprise, only one pool of adenine nucleotides consisting mostly of an ATP/(ATP + ADP) ratio of 0.95, or mostly ATP, was observed both at 4°C and 37°C with chemical shifts corresponding to pH 7.4 (UGURBIL et al. 1979). When neutralized perchloric acid extracts of of the same platelet preparation were examined, the ATP/(ATP + ADP) ratio was 0.59. These results showed that most of the ADP in the platelets, the storage pool, was not visible in NMR spectra of intact platelets. Spectra of the extracts of thrombin-treated platelet supernatants as well as enzymatic analysis confirmed that the storage pool in intact platelets did not cause NMR signals. The interpretation of these experiments was that the storage adenine nucleotides existed in a physical state that was not mobile enough to yield resonance signals. NMR spectra of pig platelets resembled that of human platelets at 4°C, but at 37°C the storage nucleotides became visible, yielding two sets of resonances corresponding to pH 7.4 (cytoplasmic nucleotides) and pH 5.5 (granule nucleotides), respectively. $Mg^{2+}$ and ATP + ADP did not seem to form aggregates at 37°C, but when the temperature was gradually decreased to below 20°C aggregate formation increased until these nucleotides also finally became NMR-invisible. Depletion of cytoplasmic nucleotides with metabolic inhibitors left only the spectrum of the storage nucleotides. $^{31}$P-NMR of 50-fold purified dense granules confirmed that the residual spectrum in platelets depleted of metabolic nucleotides was that of the dense granules (UGURBIL et al. 1984a). Aqueous mixture of ATP, ADP, MgCl, and serotonin formed viscous gels that displayed the same type of NMR spectral interactions as the actual dense granules. Serotonin also was shown to participate in these high-molecular aggregates in that [$^{19}$F]serotonin was not seen in NMR spectra of human platelets or of pig platelets at 4°C, but was visible in pig platelets at higher temperatures (COSTA et al. 1979).

$^{1}$H-NMR measurements on isolated pig platelet dense granules revealed intermolecular nuclear Overhauser effects between protons of serotonin and nucleotides as well as between the different nucleotide protons (UGURBIL et al. 1984b). These nuclear Overhauser effects show interactions, i.e., energy transfer, between adjacent protons not on the same molecule, but on adjacent molecules. Strong interactions were measured between H-2 and H-8 of adenine on one molecule and H-1′ of ribose on another molecule. The distance between the interacting protons was calculated, and a model of the most likely molecular arrangement of stacked ATP molecules and $Mg^{2+}$ was constructed. The planar adenine rings are thought to be stacked parallel to each other, each displaced a little to the right in such a way as to form a right-handed spiral like a spiral staircase, with the rings anchored to each other by interaction with $Mg^{2+}$ ions. There appears to be some space between the adenine rings for a serotonin molecule, and nuclear Overhauser effects between H-4 of serotonin and the H-2, H-8, and H-1′ of ATP were seen. However, serotonin itself was not necessary for the aggregate because observation of nucleotide interaction in dense granules depleted of serotonin by reserpine treatment of pigs before blood sampling showed nearly the same spectra as with serotonin. If ATP and ADP in pig platelet dense granules were in solution, they would result in concentrations of 400 and 250 m*M*, respectively. Their behavior as analyzed by NMR indicates that they exist as helical aggregates of 20–30 stacked nucleotides with an effective solute concentration of 17–25 m*M*, which taken together with other dense granule constitutents would be more osmotically compatible with the cytoplasm.

Human platelet dense granule nucleotides and $Ca^{2+}$ probably exist in a more crystalline form since the complex has a rotational moment too slow to produce NMR spectra. Mixing of solutions of ATP and $Ca^{2+}$ in the same proportions found in the dense granules causes precipitates rather than the clear gelphases that are formed with ATP and $Mg^{2+}$. However, the insoluble nucleotide-$Ca^{2+}$ complex in human platelets seems to be as capable of providing absorption sites for serotonin as the nucleotide-$Mg^{2+}$ complex, and the complex appears to be instantly soluble in the extracellular medium after secretion.

## D. Species Differences

There are some differences between the dense granule contents of different animal species (see Table 3). The histamine present in whole platelets is assumed to be located mostly in the dense granules, whereas the amounts of released serotonin, $Ca^{2+}$, and $Mg^{2+}$ from gel-filtered platelets are somewhat less than the total amounts (DA PRADA et al. 1981; MEYERS et al. 1982). In fact, the value for serotonin in whole platelets did not vary by more than ± 20% from that of secreted serotonin. With the exception of the horse and guinea pig, most of the other species had more serotonin than man, with the rabbit granules topping the list with 20 times more. Pig and rabbit platelets had the highest content of histamine. Histamine was detected by proton NMR in purified pig dense granules (UGURBIL et al. 1984b), and uptake studies were performed on pig platelets (FUKAMI et al. 1984). Despite their high content of histamine, the rate of exogenous histamine accumulation was some 80-100-fold slower than that of serotonin. Only 20%–40% of the total divalent cations were secreted, except for 70% $Ca^{2+}$ from human platelets and 56% $Mg^{2+}$ from

**Table 3.** Comparison of released amines, divalent cations, and nucleotides in different animal species (from DA PRADA et al. 1981 and MEYERS et al. 1982)

| Species | Serotonin | Histamine | Secreted $Ca^{2+}$ | Secreted $Mg^{2+}$ | Secreted ATP (%) | Secreted ADP (%) |
|---|---|---|---|---|---|---|
| Cat | 9.25 | 0.001 | 9.7 | 11.4 | 7.6 (25) | 3.3 (40) |
| Cow | 2.24–22.5[a] | – | 17.6 | 7.8 | 4.9 (21) | 3.9 (34) |
| Dog | 4.47 | 0.02 | 9.2 | 17.0 | 8.4 (30) | 3.3 (42) |
| Guinea pig | 0.52 | 3.18 | – | – | – | – |
| Horse | 1.0 | – | 25.2 | 12.1 | 4.0 (18) | 4.0 (44) |
| Man | 1.67 | 1.40 | 66.3 | 2.1 | 10.9 (34) | 15.1 (84) |
| Mink | 8.2 | – | 12.0 | 13.9 | 10.9 (32) | 5.4 (43) |
| Pig | 9.07 | 14.0 | 22.6 | 44.1 | 15.2 (35) | 9.6 (73) |
| Rabbit | 38.50 | 26.23 | 15.7 | 23.6 | 27.5 (50) | 4.7 (55) |

The values are given as nmol/mg of platelet protein and represent total platelet content for serotonin and histamine. Platelets were gel-filtered with $Ca^{2+}$- and $Mg^{2+}$-free Tyrode's solution, treated with 0.5 U/ml of thrombin for 3 min, and centrifuged to obtain the supernatants. The percent of release compared to total platelet ATP or ADP is given in parentheses.
[a] There was a tenfold discrepancy between the values given in the two references.

pig platelets. Some 40% of pig platelet $Ca^{2+}$ was reported to be secreted as well. The amount of $Ca^{2+}$ found in 60-fold purified pig dense granules was 188 nmol/mg protein compared to 3906 nmol/mg of $Mg^{2+}$ (SALGANICOFF et al. 1975, and unpublished data). In most of the species investigated, the ratio of secreted ATP to ADP was between 1 and 2; in rabbits it was 5.9, and in man it was 0.72. Generally, the degree of secretion was not as high in the animal platelets as in man; the fact that there is vast experience with human platelet experimentation and much less with animal platelets may be responsible for the less than optimal secretion seen with the latter. Some of the comparisons made here may not be valid for these reasons.

## E. Membrane Proteins

Pp60$^{c\text{-}src}$ is a tyrosine kinase present in platelets in relatively high concentrations; in other cells its phosphorylation is associated with mitotic events, but its function in platelets is unknown. RENDU et al. (1989) reported that high levels of pp60$^{c\text{-}src}$ were associated with a highly purified preparation of platelet dense granules. This finding has been controversial, and it remains unclear if this protein is indeed a component of the dense granule. Other workers using immunohistological techniques found pp$^{c\text{-}src}$ to be associated mainly with plasma membrane and the surface-connected canalicular system (FERRELL et al. 1990). The presence of normal amounts of pp60$^{c\text{-}src}$ in platelets from patients with storage pool deficiency was taken as an indication that this protein is not a component of the dense granule membrane (SORISKY et al. 1992). However, the association of pp60$^{c\text{-}src}$ with plasma membranes does not necessarily exclude its association with dense granules as well. Chromaffin granule membranes have been found to be associated with as much as 37% of total cell content of pp$^{c\text{-}src}$ (PARSONS and CREUTZ 1986), so there is a precedent for this localization with a secretory granule. The number of dense granules in platelets is about seven per platelet (COSTA et al. 1974), whereas chromaffin granules constitute 12.5% of the cellular volume (PHILLIPS 1982). The association of platelet P-selectin, a granule membrane protein, with pp60$^{c\text{-}src}$ described by MODDERMAN et al. (1994) suggests that pp60$^{c\text{-}src}$ could indeed be associated with secretory granules through another protein that is a part of the granule membrane. It is also possible that pp60$^{c\text{-}src}$ is associated with the plasma membrane in resting cells and becomes translocated to granule membranes and cytoskeleton in the activation-exocytosis process. A transient increase in tyrosine phosphorylation of P-selectin observed with thrombin activation was made long-lasting with the use of the phosphatase inhibitor pervanadate (MODDERMAN et al. 1994). Immunoprecipitated P-selectin was found to be linked to pp60$^{c\text{-}src}$, apparently in a disulfide linkage, since the 205-kDa complex was intact in unreduced gels, but not in reduced gels. In chromaffin cells a similar disulfide complex between pp60$^{c\text{-}src}$ and a 38-kDa membrane protein has been described (GRANDORI and HANAFUSA 1988).

There remains the problem that P-selectin has been considered as an $\alpha$-granule membrane marker (for a review see MCEVER 1991), but recently P-selectin has been reported to be present in dense granules also (ISRAELS et al. 1992). However, RENDU et al. (1989) found no pp60$^{c\text{-}src}$ in the fraction containing $\alpha$-granules. These apparent contradictions remain to be resolved.

Another dense granule membrane protein has been identified by GERRARD et al. (1991). This 40-kDa protein was present in dense granule membranes of normal platelets, and monoclonal antibodies raised against it showed that the protein was deficient in platelets from a patient with Hermansky-Pudlak syndrome, a dense granule deficiency. Subsequent studies have shown that this 40-kDa protein, named granulophysin, reacts with a monoclonal antibody to CD63, a lysosomal membrane marker (METZELAAR et al. 1991), and is expressed on the plasma membrane surface after secretion as detected by flow cytometry (NISHIBORI et al. 1993). Immunohistochemical reactivity was demonstrated between antigranulophysin and granules of skin melanocytes, neurons, endocrine gland cells, and exocrine glands excepting mucin-producing cells and surface-lining cells (HATSKELZON et al. 1993). Western blot analysis revealed staining for granulophysin in lung, adrenal gland, liver, brain, prostate, and pituitary. The granulophysin epitope appears to be present in many tissues in association with granules and may be important for granule function. Another lysosome membrane protein, LAMP 2 (see Chap. 20), has recently been reported to be associated with dense granule membranes as well (ISRAELS et al. 1996). The specificity of these membrane proteins remains to be established.

## F. Discussion and Comments

Platelet dense granules appear to be unique among secretory granules in that they alone seem to contain exclusively small molecular compounds of less than 10000 mol. wt. in such high concentrations that they result in high specific density. Adrenal chromaffin granules are similar to platelet granules in that they also contain nucleotides and amines in high concentrations, but they differ in having much lower amounts of divalent cations, and in their place, proteins – chromogranins – apparently participate in forming the storage complex (PHILLIPS 1982). It is beginning to appear that dense granule membranes may have proteins, or epitopes thereof, in common with other secretory granules (HATSKELZON et al. 1993), perhaps as components of a general, regulated exocytosis mechanism. The differences in secretion rates and strengths of stimulus required for secretion of the three granule types implies the existence of a complex differential regulation that could be mediated by integral membrane proteins unique to a given granule type, but such unique components remain to be demonstrated. Recent electron microscopic studies of serial sections of thrombin-stimulated platelets indicate that dense and α-granules appear to form compound organelles in the course of the secretion process, which implies a common mechanism (MORGENSTERN et al. 1995). The discovery of other membrane proteins, both those specific to granule type and those common to all granule types, is surely forthcoming.

## References


Baker RV, Blaschko H, Born GVR (1959) The isolation from blood platelets of particles containing 5-hydroxytryptamine and adenosine triphosphate. J Physiol 149:55–61

Berneis KH, Da Prada M, Pletscher A (1970) Metal-dependent aggregation of nucleotides with formation of biphasic liquid systems. Biochim Biophys Acta 215:547–549

Carroll RC, Edelheit EB, Schmidt PG (1980) Phosphorus nuclear magnetic resonance of bovine platelets. Biochemistry 19:3861–3867

Costa JL, Reese TS, Murphy DL (1974) Serotonin storage in platelets: estimation of storage-packet size. Science 183:537–538

Costa JL, Dobson CM, Kirk KL, Pausen FM, Valeri CR, Vecchione JJ (1979) Studies on human platelets by $^{19}F$ and $^{31}P$ NMR. FEBS Lett 99:141–146

Costa JL, Fay DD, McGill M (1981) Electron probe microanalysis of calcium and phosphorus in dense bodies isolated from human platelets. Thromb Res 22:399–405

Da Prada M, Richards JG, Kettler R (1981) Amine storage organelles in platelets. In: Gordon JL (ed) Platelets in biology and pathology 2. North Holland Biomedical, Amsterdam, (Research monographs in cell and tissue physiology 5), pp 107–145

Ferrell JE Jr, Noble JA, Martin GS, Jacques YV, Bainton DF (1990) Intracellular localization of $pp60^{c\text{-}src}$ in human platelets. Oncogene 5:1033–1036

Flodgaard H, Klenow H (1982) Abundant amounts of diadenosine $5',5'''\text{-}p^1,p^4$-tetraphosphate present and releasable, but metabolically inactive, in human platelets. Biochem J 208:737–742

Fukami MH (1992) Isolation of dense granules from human platelets. Methods Enzymol 215:36–42
Fukami MF, Bauer JS, Stewart GJ, Salganicoff L (1978) An improved method for the isolation of dense storage granules from human platelets. J Cell Biol 77:389–399
Fukami MH, Dangelmaier CA, Bauer JS, Holmsen H (1980) Secretion, subcellular localization, and metabolic status of inorganic pyrophosphate in human platelets. A major constituent of the amine-storing granules. Biochem J 192:99–105
Fukami MH, Holmsen H, Ugurbil K (1984) Histamine uptake in pig platelets and isolated dense granules. Biochem Pharmacol 33:3869–3874
Gaarder A, Jonsen J, Laland S, Hellem A, Owren PA (1961) Adenosine diphosphate in red cells as a factor in the adhesiveness of human blood platelets. Nature 192:531–532
Gerrard JM, Lint D, Sims PJ, Wiedmer T, Fugate RD, McMillan E, Robertson C, Israels SJ (1991) Identification of a platelet dense granule membrane protein that is deficient in a patient with the Hermansky-Pudlak syndrome. Blood 77:101–112
Grandori C, Hanafusa H (1988) $p60^{c\text{-}src}$ is complexed with a cellular protein in subcellular compartments involved in exocytosis. J Cell Biol 107:2125–2135
Grette K (1962) Studies on the mechanism of thrombin-catalyzed hemostatic reaction in blood platelets. Acta Physiol Scand [Suppl] 56:1–93
Hatskelzon L, Dalal BI, Shalev A, Robertson C, Gerrard JM (1993) Wide distribution of granulophysin epitopes in granules of human tissue. Lab Invest 68:509–519
Holmsen H (1965a) Incorporation in vitro of $P^{32}$ into blood platelet acid-soluble organophosphates and their chromatographic identification. Scand J Clin Lab Invest 17:230–238
Holmsen H (1965b) Collagen-induced release of adenosine diphosphate from blood platelets incubated with radioactive phosphate in vitro. Scand J Clin Lab Invest 17:239–246
Holmsen H, Ugurbil K (1992) Nuclear magnetic resonance studies of amine and nucleotide storage mechanisms in platelet dense granules. In: Myers KM, Barnes CD (eds) The platelet amine storage granule. CRC, Boca Raton, p 56
Holmsen H, Østvold A-C, Day HJ (1973) Behaviour of endogenous and newly absorbed serotonin in the platelet release reaction. Biochem Pharmacol 22:2599–2608
Israels SJ, Gerrard JM, Jacques YV, McNicol A, Nishibori M, Bainton DF (1992) Platelet dense granule membranes contain both granulophysin and P-selectin (GMP-140). Blood 80:143–152
Israels SJ, McMillan EM, Robertson C, Singhroy S, McNicol A (1996) The lysosomal granule membrane protein, Lamp-2, is also present in platelet dense granule membranes. Thromb Haemost 75:623–629
Johnson RG, Scarpa A, Salganicoff L (1978) The internal pH of isolated serotonin containing granules of pig platelets. J Biol Chem 253:7061–7068
Kim BK, Chao FC, Leavitt R, Fauci AS, Meyers KM, Zamecnik PC (1985) Diadenosine $5',5''',\text{-}p^1,p^4$-tetraphosphate deficiency in blood platelets of the Chediak-Higashi syndrome. Blood 66:735–737
Lages B, Dangelmaier CA, Holmsen H, Weiss HJ (1988) Specific correction of impaired acid hydrolase secretion in storage pool-deficient platelets by adenosine diphosphate. J Clin Invest 81:1865–1872
Lüthke J, Ogilvie A (1983) The presence of diadenosine $5',5'''\text{-}p^1,p^3$-triphosphate ($AP_3A$) in human platelets. Biochem Biophys Res Commun 115:253–260
Lüthke J, Ogilvie A (1984) Diadenosine triphosphate ($AP_3A$) mediates human platelet aggregation by liberation of ADP. Biochem Biophys Res Commun 118:704–709
Lüthke J, Ogilvie A (1985) Catabolism of $AP_3A$ and $AP_4A$ in human plasma. Purification and characterization of a glycoprotein complex with 5′nucleotide phosphodiesterase activity. Eur J Biochem 149:119–127
Martin JH, Carson FL, Race GJ (1974) Calcium-containing platelet granules. J Cell Biol 60:775–777

McEver RP (1991) Leukocyte interactions mediated by selectins. Thromb Haemast 66:80–87

McNicol A, Israels SJ, Robertson C, Gerrard JM (1994) The empty sack syndrome: a platelet storage pool deficiency associated with empty dense granules. Br J Haematol 86:574–582

Metzelaar MJ, Winjgaard PLJ, Peters PJ, Sixma JJ, Nieuwenhuis HK, Clevers HC (1991) CD63 antigen – a novel lysosomal membrane glycoprotein, cloned by a screening procedure for intracellular antigens in eukaryotic cells. J Biol Chem 266:3239–3245

Meyers KM, Seachord CL, Holmsen H, Smith JB, Prieur DJ (1979) A dominant role thromboxane formation in secondary aggregation of platelets. Nature 282:331–333

Meyers KM, Holmsen H, Seachord CL (1982) Comparative study of platelet dense granule constituents. Am J Physiol 243:R454–R461

Modderman PW, von dem Borne AEGK, Sonnenberg A (1994) Tyrosine phosphorylation of P-selectin in intact platelets and in a disulphide-linked complex with immunoprecipitated $pp60^{c\text{-}src}$. Biochem J 299:613–621

Morgenstern E, Bastian D, Dierichs R (1995) The formation of compound granules from diffeent types of secretory organelles in human platelets (dense granules and $\alpha$-granules). A cryofixation/substitution study using serial sections. Eur J Cell Biol 68:183–190

Mürer E, Holme R (1970) A study of the release of calcium from human blood platelets and its inhibition by metabolic inhibitors, N-ethylmaleimide and aspirin. Biochim Biophys Acta 222:197–205

Nishibori M, Cham B, McNicol A, Shalev A, Jain N, Gerrard JM (1993) The protein CD63 is in platelet dense granules, is deficient in a patient with Hermansky-Pudlak syndrome, and appears identical to granulophysin. J Clin Invest 91:1775–1782

Parsons SJ, Creutz CE (1986) $p60^{c\text{-}src}$ activity detected in the chromaffin granule membrane. Biochem Biophys Res Commun 134:736–742

Phillips JH (1982) Dynamic aspects of chromaffin granule structure. Neuroscience 7:1595–1609

Rendu F, Lebret M, Nurden AT, Caen JP (1982) Initial characterization of human platelet mepacrine-labelled granules isolated using a short metrizamide gradient. Br J Haematol 52:241–251

Rendu F, Maclouf J, Launay J-M, Boinot C, Levy-Toledano S, Tanzer J, Caen J (1987) Hermansky-Pudlak platelets: further studies on release reaction and protein phosphorylations. Am J Hematol 25:165–174

Rendu R, Lebret M, Danielian S, Fagard R, Levy-Toledano S, Fischer S (1989) High $pp60^{c\text{-}src}$ level in human platelet dense bodies. Blood 73:1545–1551

Salganicoff L, Hebda PA, Yandrasitz J, Fukami MH (1975) Subcellular fractionation of pig platelets. Biochem Biophys Acta 385:394–411

Schmidt PG, Carroll RC (1982) $^{31}$P Nuclear magnetic resonance of platelet dense granule adenine nucleotides. Biochim Biophys Acta 715:240–245

Silcox DC, Jacobelli S, McCarty DJ (1973) Identification of inorganic pyrophosphate in human platelets and its release on stimulation with thrombin. J Clin Invest 52:1595

Skaer RJ, Peters PD, Emmines JP (1974) The localization of calcium and phosphorus in human platelets. J Cell Sci 15:679–692

Sorisky A, Lages B, Weiss HJ, Rittenhouse SE (1992) Human platelets deficient in dense granules contain normal amounts of $pp60^{c\text{-}src}$. Thromb Res 65:77–83

Ugurbil K, Holmsen H (1981) Nucleotide compartmentation: radioisotopic and nuclear magnetic resonance studies. In: Gordon JL (ed), Platelets in biology and pathology 2. North Holland Biomedica, Amsterdam, (Research monographs in cell and tissue physiology 5), pp 147–175

Ugurbil K, Holmsen H, Shulman RG (1979) Adenine nucleotide storage and secretion in platelets as studied by $^{31}$P nuclear magnetic resonance. Proc Natl Acad Sci USA 76:2227–2231

Ugurbil K, Fukami MH, Holmsen H (1984a) $^{31}P$ NMR studies of nucleotide storage in the dense granules of pig platelets. Biochemistry 23:409–416
Ugurbil K, Fukami MH, Holmsen H (1984b) Proton NMR studies of nucleotide and amine storage in the dense granules of pig platelets. Biochemistry 23:416–428
Wallach DFH, Surgenor DM, Steele BB (1958) Calcium-lipid complexes in human platelets. Blood 13:589–598
Walsh PN, Gagnatelli G (1974) Platelet antiheparin activity: storage sites and release mechanisms. Blood 44:157–168
Weiss HJ, Lages B, Vicic W, Tsung LY, White JG (1993) Heterogeneous abnormalities of platelet dense granule ultrastructure in 20 patients with congenital storage pool deficiency. Br J Haematol 83:282–295

CHAPTER 19

# Protein Granule Factors

T.F. DEUEL

## A. Introduction

Platelets contain anatomically distinct populations of granules that include the dense and the $\alpha$-granules. However, $\alpha$-granules may associate with a third type of storage granule compartment that is believed to be the source of acid hydrolases and thus lysosomal in character. The physical nature of this association has not been established, but through the functional activities of each, the granule compartments can be readily separated. Secretion of acid hydrolases requires higher concentrations of agonists than are required for secretion from other compartments, and different agonists may trigger secretion of the contents of $\alpha$-granule and dense granule compartments but not result in release of acid hydrolases (HOLMSEN et al. 1979, 1984; KAPLAN et al. 1979). Secretion from the granule compartments requires metabolic energy (HOLMSEN et al. 1979, 1982), an indication of their importance. Furthermore, loss of secretion has been directly correlated with disorders of platelet function (KOIKE et al. 1984; HOLMSEN et al. 1987). Much remains to be learned, however. For example, the $\alpha$-granules have different staining patterns although the protein content of the granules appears similar (SANDER et al. 1983; STENBERG et al. 1984; WENCEL-DRAKE et al. 1985). Furthermore, a number of the proteins associated with the $\alpha$-granules are retained after secretion and are found localized to plasma membranes (GOGSTAD et al. 1982). The meaning of this difference in release pattern and membrane association of some but not all of the $\alpha$-granule contents is not clear, but presumably functional consequences of these differences will be found as cell-cell interactions of platelets with themselves and with the vascular endothelium become better understood. Another curiosity of the platelet granule system is that primary lysosomes are the endpoints of endocytotic vesicles with products that are targeted for proteolytic digestion by the acid hydrolases; in platelets, these hydrolases are secreted (DANGELMAIER and HOLMSEN 1980; GOGSTAD 1980). Furthermore, the various acid hydrolases are not secreted to the same degree; as example, $\beta$-galactosidase is secreted to a lesser extent than $\beta$-hexosaminidase and secretion is not complete, in contrast to secretion from dense granule and $\alpha$-granule compartments in which almost 100% of the granule contents are secreted. The acid hydrolases are secreted more slowly and require higher concentrations of agonist for release than are required from either of the two other compartments. Whereas it is difficult to establish with certainty, these and other observations suggest that the timing and extent of release of the many different platelet proteins is very finely regulated and, in turn, this regulation may be reflected in a number of functional differences in response to platelet activation.

## B. Platelet Proteins

A number of intensely studied proteins that are associated with platelet $\alpha$-granules are believed to be platelet specific (HOLT and NIEWIAROWSKI 1985). Platelet factor 4 (PF4) is an $\alpha$-granule protein that was first identified on the basis of its ability to bind to heparin (NIEWIAROWSKI and THOMAS 1969) and as an $\alpha$-granule constituent in platelets and megakaryocytes (RYO et al. 1980a,b, 1983). Human PF4 has 70 amino acids (DEUEL et al. 1977; HERMODSON et al. 1977; WALZ et al. 1977) that form tetramers in solution and associate with a specific proteoglycan carrier molecule (BARBER et al. 1972; S.S. HUANG et al. 1982) upon release from platelets. The PF4 tetramer contains two extended $\beta$-sheets that are formed by two subunits and C-terminal lysine-rich $\alpha$-helices that are arranged as antiparallel pairs on the surface of each extended $\beta$-sheet (MAYO and CHEN 1989). Consistent with its strong negative charge, PF4 released from platelets associates with cell surfaces, including endothelial cells from which it can be released by heparin (BUSCH et al. 1980; RAO et al. 1983; RUCINSKI et al. 1986, 1987). The roles of PF4 are not clear; however, PF4 may neutralize the anticoagulant activity of the surface heparin proteoglycans in vivo or perhaps compete with antithrombin III for direct binding to these molecules (BUSCH et al. 1980; MARCUM et al. 1984). PF4 also may facilitate the attachment of cells to various extracellular matrices. It stimulates histamine release from basophils, but whether this is a physiological response is not known. It also may inhibit tumor growth in nude mice and megakaryocyte maturation (LONKY et al. 1978; BRINDLEY et al. 1983; BEYTH and CULP 1984; CAPITANIO et al. 1985; TWARDZIK and TODARO 1987; DUMENCO et al. 1988) and has been shown to reverse immunosuppression in mice. However, its most potent property may be its chemotactic activity for neutrophils and monocytes at concentrations of PF4 equal to those in human serum (DEUEL et al. 1981b; OSTERMAN et al. 1982; SENIOR et al. 1983) and its ability to inhibit blood vessel proliferation in the chicken chorioallantoic membrane. Efforts are ongoing to use PF4 in vivo (DEUEL et al. 1981b; KATZ et al. 1986; BARONE et al. 1988; GEWIRTZ et al. 1989; ZUCKER et al. 1989; HAN et al. 1990b; MAIONE et al. 1990; PARK et al. 1990). Human recombinant PF4 is being tested to reverse anticoagulant activity of heparin (COOK et al. 1992) and is planned for trials as a treatment of malignancies associated with significant angiogenesis (ZUCKER and KATZ 1991).

A fascinating but poorly understood family of proteins arises from a common precursor known as the platelet basic protein (PBP). Low affinity platelet factor 4 (LA-PF4; also known as connective tissue activating peptide III, CTAP-III), β-thromboglobulin (βTG), and βTG-F are successively smaller polypeptides that result from successive N-terminal proteolytic processing of the larger propeptide. PBP and LA-PF4 are associated with megakaryocytes and platelets, whereas βTG and βTG-F are found only in platelet releasates, suggesting that these forms are the active species of the propeptide. This family of proteins binds to heparin with lower affinity than does PF4 (Rucinski et al. 1979; Holt et al. 1986), and it is likely that the affinity of this family for the extracellular matrix and other cell surfaces is lower as well.

Although yet to be definitively established, the major activity of this family of proteolytically processed proteins may be mediated by βTG-F (neutrophil activating peptide 2, NAP2). NAP2 is a potent neutrophil chemoattractant. It mobilizes cytosolic calcium and induces exocytosis within cells in a dose dependent manner (Walz and Baggiolini 1986; Holt and Niewiarowski 1989; Walz et al. 1989). NAP1 has essentially identical properties. βTG itself is a chemoattractant for human fibroblasts (Senior et al. 1983) and also may regulate megakaryocytopoiesis (Gewirtz et al. 1989; Han et al. 1990a). LA-PF4 may be a mitogen for fibroblasts, but confirmation of the experiments that led to this conclusion is needed (Castor et al. 1983; Holt et al. 1986). β-Thromboglobulin (βTG) is the proteolytic product of larger precursor proteins, including platelet basic protein (PBP) and low-affinity platelet factor 4 (LA-PF4). LA-PF4 has been independently described as the connective tissue activating peptide III (CTAP-III), identified and named as such because of its growth promoting activities (Begg et al. 1978; Rucinski et al. 1979; Castor et al. 1983; Holt and Niewiarowski 1985; Holt et al. 1986). βTG is four (N-terminal) amino acids shorter than LA-PF4. Further cleavage of 11 N-terminal amino acids results in β-TGF-F ($NAP_2$) described above. The full significance of this very complex processing pathway is unknown. However, it appears to be important, based upon the diversity of functions that have been associated with each of the proteolytic products released from the propeptide. Furthermore, both PF4 and the β-TG pathway proteins are members of the small inducible gene (SIG) family (Kawahara and Deuel 1989; Kawahara et al. 1991a,b) whose products of the small inducible genes are cytokines that moderate important activities related to inflammation and repair, immune responses, and cell growth, including NAP/IL8 (neutrophil activating peptide), $^{IP}10$, a γ-interferon inducible gene, RANTES, a gene expressed in T and B cells that also function as a lymphokine, and KC, a gene that encodes the monocyte growth activity (MsGA; Luster et al. 1985; Richmond et al. 1988; Schall et al. 1988; Baggiolini et al. 1989; Zipfel et al. 1989; Modi et al. 1990). The 9E3 gene is expressed and was first identified in Rous sarcoma transformed cells (Sugano et al. 1987). Each product of this gene family has four conserved cysteine residues and a conserved single proline residue (Zipfel et al. 1989; Kawahara et al. 1991a), and it is this identity of cysteine that characterizes this family. Because these genes are induced and expressed in response to many stimuli such as inflammation, it seems most likely that these platelet proteins also have similar roles at sites of platelet release and contribute in an important way to local responses to blood vessel injury and thrombosis.

## C. Platelet-Derived Growth Factors

The presence of growth promoting activities in platelets arose from observations that serum but not plasma supported cell growth in culture. The platelet-derived growth factor (PDGF), platelet transforming growth factor $\beta 1$ (platelet TGF $\beta 1$), and epidermal growth factor (EGF) have associated with platelet $\alpha$-granules (ANTONIADES et al. 1979; DEUEL et al. 1981a, 1982a; J.S. HUANG et al. 1982; ASSOIAN et al. 1983; OKA and ORTH 1983; HSUAN 1989).

PDGF is the major mitogen associated with platelets. It is an approximately 30-KDa disulfide linked heterodimer of approximately 50% amino-acid identical polypeptide chains (A and B). Remarkably, homodimers of PDGF (AA and BB) are widely found in tissues other than platelets, and thus the predominant heterodimeric form of PDGF is unique to platelets and to megakaryocytes. Whereas several differences in biological activities of the PDGF isoforms have been described, both PDGF AA and BB are potent mitogens and chemoattractants (DEUEL et al. 1982b; SENIOR et al. 1983). Diversity of function of the PDGF isoforms may be in part mediated by two similar receptors that differentially recognize PDGF. PDGF AA is recognized by the PDGF $\alpha$-receptor whereas PDGF BB is recognized by the related PDGF $\beta$-receptor and the PDGF $\alpha$-receptor, suggesting thus both diversity of function and cell type recognition (DEUEL 1987).

The PDGF B-chain is structurally related to the putative transforming protein $p28^{v\text{-}sis}$ of the simian sarcoma virus (DOOLITTLE et al. 1983; WATERFIELD et al. 1983). PDGF BB also may play a role in an autocrine stimulation of tumors and in the stromal responses associated with other tumors (DEUEL and HUANG 1984a,b; HUANG et al. 1984; HELDIN and WESTERMARK 1989; ROSS 1989). Interestingly, the B-chain of PDGF requires a second genetic alteration to transform cells (KIM et al. 1994). Both PDGF A and PDGF B can cooperate with gain of function Bcl-2 to induce transformation of rat fibroblasts (KIM et al. 1994). PDGF BB and PDGF AA also induce apoptosis when exposed to serum-starved cells growth arrested in $G_0$–$G_1$ (KIM et al. 1995). PDGF also may play a significant role in the pathogenesis of atherosclerosis by stimulating proliferation of arterial smooth muscle cells and through upregulation of other cytokine genes (FRIEDMAN et al. 1977; FUSTER et al. 1978; PICK et al. 1979; ROSS 1986). The PDGF isoforms may be important in certain phases of normal growth and development, in the generation of granulation tissue during wound repair, in inflammatory reactions and fibroproliferative responses (e.g., in rheumatoid arthritis, myelofibrosis, and pulmonary fibrosis), and in many other normal and abnormal responses (DEUEL et al. 1982b, 1983, 1985, 1991;

SENIOR et al. 1983; TZENG et al. 1984, 1985; BAUER et al. 1985; PORTER et al. 1988; BERK and ALEXANDER 1989; MADTES et al. 1989; PIERCE et al. 1989; ROSS 1989).

In addition to megakaryocytes, fibroblasts, smooth muscle cells, glial cells, endothelial cells, and neurons express PDGF, and thus this platelet $\alpha$-granule protein has widespread potential for autocrine and paracrine regulation (DEUEL 1987, 1991; DEUEL et al. 1988).

TGF$\beta$, first isolated from platelets (HSUAN 1989), is expressed in a number of different cells as well. The roles of both PDGF and TGF$\beta$ growth-regulatory and chemotactic properties in platelets may be important in inflammation (MUSTOE et al. 1987; PIERCE et al. 1988). Platelets are invariably seen in inflammatory sites, and platelet "plugs" are the initial event affecting hemostasis at sites of injury. The density of platelets at these sites exceeds that in plasma, and it thus is likely that the concentrations of platelet $\alpha$-granule proteins at sites of platelet plugs and thus platelet release exceed greatly those found in serum.

## D. Thrombospondin

Thrombospondin is a high-molecular-weight platelet $\alpha$-granule glycoprotein that is released in high concentrations in stimulated platelet supernatants (BAENZIGER et al. 1971; LAWLER et al. 1978, 1982, 1983). Thrombospondin also is synthesized and secreted by cultured human endothelial cells, bovine aortic endothelial cells, macrophages, and other cells in culture (MCPHERSON et al. 1981; MOSHER et al. 1982; SCHWARTZ 1989). It contains the amino acid sequence RGD that is associated with binding to the integrin family of receptors (LAWLER and HYNES 1986). Consistent with specificity of the RGD sequences, thrombospondin binds to glycoprotein IIB/IIIA, to the vitronectin receptor, and to glycoprotein IV (LAWLER and HYNES 1989), interacts with fibrinogen (LEUNG and NACHMAN 1982; TUSZYNSKI et al. 1985), and is incorporated into fibrin clots and potentiates platelet aggregation (LEUNG and NACHMAN 1982; BALE et al. 1985). Whereas much remains to be learned about the significance of these interactions, it seems likely that thrombospondin may be important in a number of roles but may be most important in stabilizing hemostatic plugs.

## E. Proteins Similar or Identical to Plasma Proteins

One of the surprising findings that has arisen from recent experiments designed to identify and characterize platelet granule proteins is that many of the platelet associated proteins are structurally similar or identical to circulating plasma proteins. Among these proteins are von Willebrand factor (vWF), von Willebrand antigen II (vWF: AgII), fibrinogen, fibronectin, Factor V, albumin, $\alpha_2$-antiplasmin, high-molecular-weight kininogen, $\alpha_2$-macroglobulin, $\alpha_1$-antitrypsin, C1 inhibitor, plasminogen, components of the alternative complement pathway, and histidine-rich glycoprotein (KEENAN and SOLUM 1972; LOPES-VIRELLA et al. 1976; NACHMAN and HARPEL 1976; NIEWIAROWSKI 1977; BAGDASARIAN and COLMAN 1978; KOUTTS et al. 1978; ZUCKER et al. 1979; HOLT and NIEWIAROWSKI 1980, 1985; CHESNEY et al. 1981; KENNEY and DAVIS 1981; PLOW and COLLEN 1981; SCOTT and MONTGOMERY 1981; TRACY et al. 1982; LEUNG et al. 1983; SCHMAIER et al. 1983, 1985; NIEWIAROWSKI et al. 1994). The functions of these proteins are largely unknown. However, increasingly, these proteins have been found to be associated with cell surfaces where they may be important in the surface localization of coagulation or themselves serve to "tether" and thus localize and bring together coagulants and perhaps other proteins for assembly of coagulant- and related complexes. Fibrinogen stored in platelet $\alpha$-granules may arise from plasma fibrinogen and function on activated platelet surfaces as an attachment substrate in platelet aggregation. vWF:AgII is expressed in high levels in platelets whereas fibronectin is found at lower concentrations and secreted from activated platelets (PLOW et al. 1979; ZUCKER et al. 1979). The fibronectin that has been localized in platelets differs from plasma fibronectin. It is detected at the surface of activated platelets (HYNES 1990) and binds to platelet surface glycoproteins, possibly functioning as an attachment protein to localize platelets and platelet release products at sites of injury (LAHAV et al. 1982; GINSBERG et al. 1983; NIEWIAROWSKI et al. 1984; PIOTROWICZ et al. 1988).

## F. Enzymes

Although their roles in platelets are unkown, a number of acid- and other hydroslases are released by platelets (ROBERT et al. 1969; HOLMSEN and DAY 1970; EHRLICH and GORDON 1976; HIDAKA and ASANO 1976; WASTESON et al. 1976; OLDBERG et al. 1980; OOSTA et al. 1982) from compartments that are similar to lysosomal compartments. Elastase, collagenase, and a neutral protease are released from platelets as well (ROBERT et al. 1969; CHESNEY et al. 1974; JAMES et al. 1985). Collectively, the spectrum of substrates that are potential targets for these enzymes is extremely broad but definitely includes the very proteins and glycosaminoglycans found at sites of injury. These broad spectrums of substrate specificity and the release of these enzymes locally support roles in the clearance and remodeling of tissues in response to inflammation and injury. The association of thrombi and platelet aggregates with inflammation supports this role as well. Other associations with pathological responses, such as the degradation of arterial elastic tissue in atherosclerosis (ROBERT et al. 1969) with the release of heparitinase from platelets and the generation of heparin-like substance from bovine aortic endothelial cells that inhibits the growth of smooth muscle cells (CASTELLOT et al. 1982; OOSTA et al. 1982), also link the activity of these enzymes with both normal and abnormal inflammatory responses. Other associations of these enzymes with potential substrates suggest other important roles of platelets, such as a peptidase released from platelets that deaminates substance P, oxytocin, and angiotensin I and the protease inhibitor nexin-2 (PN-2), a release product of the B-amyloid precursor associated with plaques of Alzheimer's disease (VAN NOSTRAND et al. 1991).

# G. Summary and Conclusions

Platelet associated proteins remain an important and perhaps even increasingly significant focus for additional study. The proteins themselves have the potential to mediate virtually all of the responses needed to initiate and sustain the progression from injury and tissue damage to tissue remodeling and healing. The platelet proteins also offer unique opportunities to learn more about compartmentalization of functional proteins up to their release to the extracellular compartments, to analyze proteolytic processing of proteins with differential functional activities, and to define and characterize different mechanisms of exocytosis of proteins resulting in a regulated released of compartments in response to different stimuli.

The ability of a single cell to direct such important and diverse activities is truly unique and suggests the platelet is a multifunctional cell of inordinate importance in many responses to injuries of diverse origins.

## References


Antoniades HN, Scher CD, Stiles CD (1979) Purification of the human platelet-derived growth factor. Proc Natl Acad Sci USA 76:1809–1813

Assoian RK, Komoriya AK, Meyers CA, Miller DM, Sporn MB (1983) Transforming growth factor-beta in human platelets: identification of a major storage site, purification, and characterization. J Biol Chem 258:7155–7159

Baenziger NL, Brodie GN, Majerus PW (1971) A thrombin-sensitive protein of human platelet membranes. Proc Natl Acad Sci USA 68:240–244

Bagdasarian A, Colman RW (1978) Subcellular localization and purification of platelet alpha1-antitrypsin. Blood 51(1):139–156

Baggiolini M, Walz A, Kunkel SL (1989) Neutrophil-activating peptide-1/interleukin 8, a novel cytokine that activates neutrophils. J Clin Invest 84(4):1045–1049

Bale MD, Westrick LG, Mosher DF (1985) Incorporation of thrombospondin into fibrin clots. J Biol Chem 260:7502–7506

Barber AG, Käser-Glanzmann R, Jakabova M, Lüscher EF (1972) Chromatography of chondroitin sulfate proteoglycan carrier for heparin neutralizing activity (platelet factor 4) released from human blood platelets. Biochim Biophys Acta 286:312–329

Barone AD, Ghrayeb J, Hammerling U, Zucker MB, Thorbecke GJ (1988) The expression in Escherichia coli of recombinant human platelet factor 4, a protein with immunoregulatory activity. J Biol Chem 263(18):8710–8715

Bauer EA, Cooper TW, Huang JS (1985) Stimulation of in vitro human skin collagenase expression by platelet-derived growth factor. Proc Natl Acad Sci USA 82:4132–4136

Begg GS, Pepper DS, Chesterman CN, Morgan FJ (1978) Complete covalent structure of human beta-thromboglobulin. Biochemistry 17(9):1739–1744

Berk BC, Alexander RW (1989) Vasoactive effects of growth factors. Biochem Pharmacol 38:219–225

Beyth RJ, Culp LA (1984) Complementary adhesive responses of human skin fibroblasts to the cell-binding domain of fibronectin and the heparan sulfate-binding protein, platelet factor-4. Exp Cell Res 155(2):537–548
Brindley LL, Sweet JM, Goetzl EJ (1983) Stimulation of histamine release from human basophils by human platelet factor 4. J Clin Invest 72(4):1218–1223
Busch C, Dawes J, Pepper DS, Wasteson A (1980) Binding of platelet factor 4 to cultured human umbilical vein endothelial cells. Thromb Res 19(1–2):129–137
Capitanio M, Niewiarowski S, Rucinski B, Tuszynski GP, Cierniewski CS, Hershock D, Kornecki E (1985) Interaction of platelet factor 4 with human platelets. Biochim Biophys Acta 839(2):161–173
Castellot JJ Jr, Favreau LV, Karnovsky MJ, Rosenberg RD (1982) Inhibition of vascular smooth muscle cell growth by endothelial cell-derived heparin. Possible role of a platelet endoglycosidase. J Biol Chem 257(19):11256–11260
Castor CW, Miller JW, Walz DA (1983) Structural and biological characteristics of connective tissue activating peptide (CTAP-III), a major human platelet-derived growth factor. Proc Natl Acad Sci USA 80(3):765–769
Chesney CM, Harper E, Colman RW (1974) Human platelet collagenase. J Clin Invest 53(6):1647–1654
Chesney CM, Pifer D, Colman RW (1981) Subcellular localization and secretion of factor V from human platelets. Proc Natl Acad Sci USA 78(8):5180–5184
Cook JJ, Niewiarowski S, Yan Z (1992) Platelet factor IV efficiently reverses heparin anticoagulation in the rat without adverse effects of heparin-protamine complexes. Circulation 85:1102
Dangelmaier CA, Holmsen H (1980) Determination of acid hydrolases in human platelets. Anal Biochem 104:182–191
Deuel TF (1987) Polypeptide growth factors: roles in normal and abnormal cell growth. Annu Rev Cell Biol 3:443–492
Deuel TF (1991) Structural and functional diversity of the platelet-derived growth factor. Curr Opin Biotech 2:802–806
Deuel TF, Huang JS (1984a) Platelet derived growth factor: structure, function, and roles in normal and transformed cells. J Clin Invest 74:669–676
Deuel TF, Huang JS (1984b) Roles of growth factor activities in oncogenesis. Blood 64:951–958
Deuel TF, Keim PS, Farmer M, Heinrikson RL (1977) Amino acid sequence of human platelet factor 4. Proc Natl Acad Sci USA 74(6):2256–2258
Deuel TF, Huang JS, Proffitt RT, Baenziger J, Chang D, Kennedy BB (1981a) Human platelet-derived growth factor: purification and resolution into two active protein fractions. J Biol Chem 256:8896–8899
Deuel TF, Senior RM, Chang D, Griffin GL, Heinrrikson RL, Kaiser ET (1981b) Platelet factor 4 is chemotactic for neutrophils and monocytes. Proc Natl Acad Sci USA 78:4584–4587
Deuel TF, Maciag T, Witte LD (1982a) Growth factors and the arterial cell wall. In: Marchesi V, Gallo R, Majerus P (eds) Differentiation and function of hematopoietic cell surfaces. Liss, New York, pp 217–223
Deuel TF, Senior RM, Huang JS, Griffin GL (1982b) Chemotaxis of monocytes and neutrophils to platelet-derived growth factor. J Clin Invest 69:1046–1049
Deuel TF, Huang JS, Huang SS, Stroobant P, Waterfield MD (1983) Expression of a platelet derived growth factor-like protein in simian sarcoma virus transformed cells science. Science 221:1348–1350
Deuel TF, Tong BD, Huang JS (1985) Autocrine regulation of SSV-transformed cell growth by PDGF-like activity. Cold Spring Harbor Laboratory conferences on growth factors and transformation: cancer cells. Cold Spring Harbor Press
Deuel TF, Silverman NJ, Kawahara RS (1988) Platelet-derived growth factor: a multifunctional regulator of normal and abnormal cell growth. BioFactors 1:213–217
Deuel TF, Kawahara RS, Mustoe TA, Pierce GF (1991) Growth factors and wound healing: platelet-derived growth factor as a model cytokine. Annu Rev Med 42:567–584

Doolittle RF, Hunkapiller MW, Hood LE, Devare SG, Robbins KC, Aaronson SA (1983) Simian sarcoma virus onc gene, v-sis, is derived from the gene (or genes) encoding a platelet-derived growth factor. Science 221:275–277
Dumenco LL, Everson B, Culp LA, Ratnoff OD (1988) Inhibition of the activation of Hageman factor (factor XII) by platelet factor 4. J Lab Clin Med 112:394–400
Ehrlich GP, Gordon JL (1976) Proteinases in platelets. In: Gordon JL Platelets in biology and pathology. Elsevier, Amsterdam
Friedman RJ, Stemerman MB, Wenz B (1977) The effect of thrombocytopenia on experimental arteriosclerotic lesion formation in rabbits: smooth muscle cell proliferation and reendothelialization. J Clin Invest 60:1191–1196
Fuster W, Bowie EJ, Lewis JC, Fass DN, Owen CA Jr, Brown AL (1978) Resistance to arteriosclerosis in pigs with von Willebrand's disease. Spontaneous and high cholesterol diet-induced arteriosclerosis. J Clin Invest 61:722
Gewirtz AM, Calabretta B, Rucinski B, Niewiarowski S, Xu WY (1989) Inhibition of human megakaryocytopoiesis in vitro by platelet factor 4 (PF4) and a synthetic COOH-terminal PF4 peptide. J Clin Invest 83:1477–1486
Ginsberg MH, Forsyth J, Lightsey A, Chediak J, Plow EF (1983) Reduced surface expression and binding of fibronectin by thrombin-stimulated thrombasthenic platelets. J Clin Invest 71:619–624
Gogstad GO (1980) A method for the isolation of alpha-granules from human platelets. Thromb Res 20:669
Gogstad GO, Hagen I, Korsmo R, Solum NO (1982) Evidence for release of soluble, but not of membrane-integrated, proteins from human platelet alpha-granules. Biochim Biophys Acta 702:81–89
Han ZC, Bellucci S, Tenza D, Caen JP (1990a) Negative regulation of human megakaryocytopoiesis by human platelet factor 4 and beta thromboglobulin: comparative analysis in bone marrow cultures from normal individuals and patients with essential thrombocythaemia and immune thrombocytopenic purpura. Br J Haematol 74:395
Han ZC, Sensebe L, Abgrall JF, Briere J (1990b) Platelet factor 4 inhibits human megakaryocytopoiesis in vitro. Blood 75:1234–1239
Heldin CH, Westermark B (1989) Platelet-derived growth factor: three isoforms and two receptor types. [Review] Trends Genet 5:108–111
Hermodson M, Schmer G, Kurache K (1977) Isolation, crystallization, and primary amino acid sequence of human platelet factor 4. J Biol Chem 252:6276–6279
Hidaka H, Asano T (1976) Platelet cyclic 3′:5′-nucleotide phosphodiesterase released by thrombin and calcium ionophore. J Biol Chem 251:7508–7516
Holmsen H, Day HJ (1970) The selectivity of the thrombin-induced platelet release reaction: subcellular localization of released and retained constituents. J Lab Clin Med 75:840–855
Holmsen H, Robkin L, Day HJ (1979) Effects of antimycin A and 2-deoxyglucose on secretion in human platelets: differential inhibition of the secretion of acid hydrolases and adenine nucleotides. Biochem J 182:413–419
Holmsen H, Kaplan KL, Dangelmaier CA (1982) Differential energy requirements for platelet responses: a simultaneous study of aggregation, three secretory processes, arachidonate liberation, phosphatidylinositol breakdown and phosphatidate production. Biochem J 208:9–18
Holmsen H, Dangelmaier CA, Rongved S (1984) Tight coupling of thrombin-induced acid hydrolase secretion and phosphatidate synthesis to receptor occupancy in human platelets. Biochem J 222:157–167
Holmsen H, Walsh PN, Koike K (1987) Familial bleeding disorder associated with impaired platelet signal processing mechanism and dificient platelet membrane glycoproteins. Br J Haematol 67:335
Holt JC, Niewiarowski S (1980) Secretion of plasminogen by washed human platelets. Circulation 62:342
Holt JC, Niewiarowski S (1985) Biochemistry of alpha granule proteins. Semin Hematol 22:151–163

Holt JC, Niewiarowski S (1989) Platelet basic protein, low-affinity platelet factor 4, and beta-thromboglobulin: purification and identification. Methods Enzymol 169: 224–233

Holt JC, Harris ME, Holt AM, Lange E, Henschen A, Niewiarowski S (1986) Characterization of human platelet basic protein, a precursor form of low-affinity platelet factor 4 and beta-thromboglobulin. Biochem 25:1988–1996

Hsuan JJ (1989) Transforming growth factors beta. Br Med Bull 45:425–437

Huang JS, Proffitt RT, Baenziger JV, Chang D, Kennedy BB, Deuel TF (1982) Human platelet-derived growth factor: purification and initial characterization. In: Marchesi V, Gallo R, Majerus P (eds) Differentiation and function of hematopoietic cell surfaces. Liss, New York, pp 225–230

Huang JS, Huang SS, Deuel TF (1984) Transforming protein of simian sarcoma virus stimulates autocrine cell growth of SSV-transformed cells through platelet-derived growth factor cell surface receptors. Cell 39:79–87

Huang SS, Huang JS, Deuel TF (1982) Proteoglycan carrier of human platelet factor 4: isolation and characterization. J Biol Chem 257:11546–11550

Hynes RO (1990) Fibronectins. Spring-Verlag, Berlin Heidelberg New York

James HL, Wachtfogel YT, James PL, Zimmerman M, Colman RW, Cohen AB (1985) A unique elastase in human blood platelets. J Clin Invest 76:2330–2337

Kaplan KL, Broekman MJ, Chernoff A, Lesznik GR, Drillings H (1979) Platelet alpha-granule proteins – studies on release and subcellular localization. Blood 53:604–618

Katz IR, Thorbecke GJ, Bell MK, Yin JZ, Clarke D, Zucker MB (1986) Protease-induced immunoregulatory activity of platelet factor 4. Proc Natl Acad Sci USA 83(10):3491–3495

Kawahara RS, Deuel TF (1989) Platelet-derived growth factor-inducible gene JE is a member of a family of small inducible genes related to platelet factor 4. J Biol Chem 264:679–682

Kawahara RS, Deng Z-W, Deuel TF (1991a) Glucocorticoids inhibit the transcriptional induction of JE, a platelet-derived growth factor-inducible gene. J Biol Chem 266:13261–13266

Kawahara RS, Deng Z-W, Deuel TF (1991b) PDGF and the small inducible gene (SIG) family. Roles in the inflammatory response. In: Westwick J, Kunkel SL, Lindley IJD (eds) Chemotactic cytokines. Plenum, New York, pp 79–87

Keenan JP, Solum NO (1972) Quantitative studies on the release of platelet fibrinogen by thrombin. Br J Haematol 23:461–466

Kenney DM, Davis AE (1981) Association of alternative complement pathway components with human blood platelets: secretion and localization of factor D and 1H globulin. Clin Immunol Immunopathol 21:351–363

Kim H-R, Upadhyay S, Korsmeyer S, Deuel TF (1994) Platelet-derived growth factor (PDGF) B and A homodimers transform murine fibroblasts depending on the genetic background of the cell. J Biol Chem 269:30604–30608

Kim H-RC, Upadhyay S, Li G, Palmer KC, Deuel TF (1995) Platelet derived growth factor induces apoptosis in growth-arrested murine fibroblasts. Proc Natl Acad Sci USA 92:9500–9504

Koike K, Rao AK, Holmsen H, Mueller PS (1984) Platelet secretion defect in patients with attention deficit disorder and easy bruising. Blood 63:427

Koutts J, Walsh PN, Plow EF, Fenton JW, Bouma BN, Zimmerman TS (1978) Active release of human platelet factor VIII-related antigen by adenosine diphosphate, collagen, and thrombin. J Clin Invest 62:1255–1263

Lahav J, Schwartz MA, Hynes RO (1982) Analysis of platelet adhesion with a radioactive chemical crosslinking reagent: interaction of thrombospondin with fibronectin and collagen. Cell 31:253–262

Lawler J, Hynes RO (1986) The structure of human thrombospondin, an adhesive glycoprotein with multiple calcium-binding-sites and homologies with several different proteins. J Cell Biol 103:1635–1648
Lawler J, Hynes RO (1989) An integrin receptor on normal and thrombasthenic platelets that binds thrombospondin. Blood 74:2022–2027
Lawler JW, Simons ER (1983) Cooperative binding of calcium to thrombospondin by cultured human endothelial cells. J Biol Chem 258:12098–12103
Lawler JW, Slayter HS, Coligan JE (1978) Isolation and characterization of high-molecular-weight glycoprotein from human platelets. J Biol Chem 253:8609–8613
Lawler JW, Chao FC, Cohen CM (1982) Evidence for calcium-sensitive structure in platelet thrombospondin: isolation and partial characterization of thrombospondin in the presence of calcium. J Biol Chem 257:12257–12261
Leung LL, Nachman RL (1982) Complex formation of platelet thrombospondin with fibrinogen. J Clin Invest 70:542–546
Leung LL, Harpel PC, Nachman RL, Rabellino EM (1983) Histidine-rich glycoprotein is present in human platelets and is released following thrombin stimulation. Blood 62:1016–1021
Lonky SA, Marsh J, Wohl H (1978) Stimulation of human granulocyte elastase by platelet factor 4 and heparin. Biochem Biophys Res Commun 85:1113–1118
Lopes-Virella MF, Sarji K, Colwell JA (1976) Platelet lipoproteins. A comparative study with serum lipoproteins. Biochim Biophys Acta 439:339–348
Luster AD, Unkeless JC, Ravetch JV (1985) Gamma-interferon transcriptionally regulates an early-response gene containing homology to platelet proteins. Nature 315:672–676
Madtes DK, Raines EW, Ross R (1989) Modulation of local concentrations of platelet-derived growth factor. Am Rev Respir Dis 140:1118–1120
Maione TE, Gary GS, Petro J, Hunt AJ, Donner AL, Bauer SI, Carson HF (1990) Inhibition of angiogenesis by recombinant human platelet factor-4 and related peptides. Science 247:77–79
Marcum JA, McKenney JB, Rosenberg RD (1984) Acceleration of thrombin-antithrombin complex formation in rat hindquarters via heparinlike molecules bound to the endothelium. J Clin Invest 74:341–350
Mayo KH, Chen MJ (1989) Human platelet factor 4 monomer-dimer-tetramer equilibria investigated by 1H NMR spectroscopy. Biochemistry 28:9469–9478
McPherson J, Sage H, Bornstein B (1981) Isolation and characterization of a glycoprotein secreted by aortic endothelial cells in culture. J Biol Chem 256:11330–11331
Modi WS, Dean M, Seuanez HN, Mukaida N, Matsushima K (1990) Monocyte-derived neutrophil chemotactic factor (MDNCF/IL-8) resides in a gene cluster along with several other members of the platelet factor 4 gene superfamily. Hum Genet 84:185–187
Mosher DF, Doyle MJ, Jaffe EA (1982) Synthesis and secretion of thrombospondin by cultured human endothelial cells. J Cell Biol 93:343–348
Mustoe TA, Pierce GF, Thomason A, Gramates P, Sporn MB, Deuel TF (1987) Accelerated healing of incisional wounds in rats induced by transforming growth factor-$\beta$. Science 237:1333–1336
Nachman RL, Harpel PC (1976) Platelet alpha2-macroglobulin and alpha1-antitrypsin. J Biol Chem 251:4512–4521
Niewiarowski J, Cierniewski CS, Tuszynski GP (1984) Association of fibronectin with the platelet cytoskeleton. J Biol Chem 259:6181–6187
Niewiarowski S (1977) Proteins secreted by the platelet. Thromb Haemost 38:924–938
Niewiarowski S, Thomas DP (1969) Platelet factor 4 and adenosine diphosphate release during human platelet aggregation. Nature 222:1269–1270
Niewiarowski S, Holt JC, Cook JJ (1994) Biochemistry and physiology of secreted platelet proteins. In: Colman RW, Hirsch J, Marder VJ, Salzman EW (eds) Hemostasis and thrombosis: basic principles and clinical practice. Lippincott, Philadelphia, pp 546–556

Oka Y, Orth DN (1983) Human plasma epidermal growth factor beta urogastrone is associated with blood platelets. J Clin Invest 72:249–259

Oldberg A, Heldin CH, Wasteson A, Busch C, Hook M (1980) Characterization of a platelet endoglycosidase degrading heparin-like polysaccharides. Biochemistry 19:5755–5762

Oosta GM, Favreau LV, Beeler DL, Rosenberg R (1982) Purification and properties of human platelet heparitinase. J Biol Chem 257:11249–11155

Osterman DG, Griffin GL, Senior RM, Kaiser ET, Deuel TF (1982) The carboxyl-terminal tridecapeptide of platelet factor 4 is a potent chemotactic agent for monocytes. Biochem Biophys Res Commun 107:130–135

Park KS, Rifat S, Eck H, Adachi K, Surrey S, Poncz M (1990) Biologic and biochemic properties of recombinant platelet factor 4 demonstrate identity with the native protein. Blood 75:1290–1295

Pick R, Chediak J, Glick G (1979) Aspirin inhibits development of coronary atherosclerosis in cynomolgus monkeys (Macaca fascicularis) fed an atherogenic diet. J Clin Invest 63:158–162

Pierce GF, Mustoe TA, Deuel TF (1988) Transforming growth factor B induces increased directed cellular migration and tissue repair in rats. In: Barbul A, Pines E, Coldwell M, Hunt TK (eds) Growth factors and other aspects of wound healing: biological and clinical implications. Liss, New York, pp 93–102

Pierce GF, Mustoe TA, Lingelbach J, Masakowski VR, Griffin GL, Senior RM, Deuel TF (1989) Platelet-derived growth factor and transforming growth factor-$\beta$ enhance tissue repair activities by unique mechanisms. J Cell Biol 109:429–440

Piotrowicz RS, Orchekowski RP, Nugent DJ, Yamada KY, Kunicki TJ (1988) Glycoprotein Ic-IIa functions as an activation-independent fibronectin receptor on human platelets. J Cell Biol 106:1359–1364

Plow EF, Collen D (1981) The presence and release of alpha 2-antiplasmin from human platelets. Blood 58:1069–1074

Plow EF, Birdwell C, Ginsburg MH (1979) Identification and quantitation of platelet-associated fibronectin antigen. J Clin Invest 63:540–543

Porter FD, Li Y-S, Deuel TF (1988) Purification and characterization of a phosphatidylinositol 4-kinase from bovine uteri. J Biol Chem 263:8989–8995

Rao AK, Niewiarowski S, James P, Holt JC, Harris M, Elfenbein B, Bastl C (1983) Effect of heparin on the in vivo release and clearance of human platelet factor 4. Blood 61:1208–1214

Richmond A, Balentien E, Thomas HG, Flaggs G, Barton DE, Spiess J, Bordoni R, Francke U, Derynck R (1988) Molecular characterization and chromosomal mapping of melanoma growth stimulatory activity, a growth factor structurally related to beta-thromboglobulin. EMBO J 7:2025–2033

Robert BY, Legrand Y, Pignand J, Caen J, Robert L (1969) Activite elastinolytique associee aux plaquettes sanguines. Pathol Biol (Paris) 17:615–622

Ross R (1986) The pathogenesis of atherosclerosis – an update. N Engl J Med 314:488–500

Ross R (1989) Peptide regulatory factors. Platelet-derived growth factor. Lancet 1:1179

Rucinski B, Niewiarowski S, James P, Walz DA, Budzynski AZ (1979) Antiheparin proteins secreted by human platelets. Purification, characterization, and radioimmunoassay. Blood 53:47–62

Rucinski B, Knight LC, Niewiarowski S (1986) Clearance of human platelet factor 4 by liver and kidney: its alteration by heparin. Am J Physiol 251:H800–H807

Rucinski B, Stewart GJ, De Feo PA, Boden G, Niewiarowski S (1987) Uptake and processing of human platelet factor 4 by hepatocytes. Proc Soc Exp Biol Med 186:361–367

Ryo R, Proffitt RT, Deuel TF (1980a) Human platelet factor 4: subcellular localization and characteristics of release from intact platelets. Throm Res 17:629–644

Ryo R, Proffitt RT, Poger ME, O'Bear R, Deuel TF (1980b) Platelet factor 4 antigen in megakaryocytes. Throm Res 17:645–652
Ryo R, Nakeff A, Huang SS, Ginsberg M, Deuel TF (1983) New synthesis of a platelet specific protein: platelet factor 4 synthesis in a megakaryocyte enriched rabbit bone marrow culture system. J Cell Biol 96:515–520
Sander HJ, Slot JW, Bouma BN (1983) Immunocytochemical localization of fibrinogen, platelet factor 4, and beta-thromboglobulin in thin frozen sections of human blood platelets. J Clin Invest 72:1277–1287
Schall TJ, Jongstra J, Dyer BJ, Jorgensen J, Clayberger C, Davis MM, Krensky AM (1988) A human T cell-specific molecule is a member of a new gene family. J Immunol 141:1018–1025
Schmaier AH, Zuckerberg A, Silverman C, Kuchibhotla J, Tuszynski GP, Colman RW (1983) High-molecular weight kininogen. A secreted platelet protein. J Clin Invest 71:1477–1489
Schmaier AH, Smith PM, Colman RW (1985) Platelet C1-inhibitor. A secreted alpha-granule protein. J Clin Invest 75:242–250
Schwartz BS (1989) Monocyte synthesis of thrombospondin. The role of platelets. J Biol Chem 264:7512–7517
Scott JP, Montgomery RR (1981) Platelet von Willebrand's antigen II: active release by aggregating agents and a marker of platelet release reaction in vivo. Blood 58:1075–1080
Senior RM, Griffin GL, Huang JS, Walz DA, Deuel TF (1983) Chemotactic activity of platelelt alpha granule proteins for fibroblasts. J Cell Biol 96:382–385
Stenberg PE, Shuman MA, Levine SP (1984) Optimal techniques for the immunocytochemical demonstration of beta-thromboglobulin, platelet factor 4, and fibrinogen in the alpha granules of unstimulated platelets. Histochem J 16:983–1001
Sugano S, Stoeckle MY, Hanafusa H (1987) Transformation by Rous sarcoma virus induces a novel gene with homology to a mitogenic platelet protein. Cell 49:321–328
Tracy PB, Eide LL, Bowie EJ, Mann KG (1982) Radioimmunoassay of factor V in human plasma and platelets. Blood 60:59–63
Tuszynski GP, Srivastava S, Switalska HI, Holt JC, Cierniewski CS, Niewiarowski S (1985) The interaction of human platelet thrombospondin with fibrinogen. J Biol Chem 260:12240–12245
Twardzik DR, Todaro GJ (1987) Platelet-related growth regulator. U.S. Patent 4645828
Tzeng DY, Deuel TF, Huang JS, Senior RM, Boxer LA, Baehner RL (1984) Platelet-derived growth factor promotes polymorphonuclear leukocyte activiation. Blood 64:951–958
Tzeng DY, Deuel TF, Huang JS, Baehner RL (1985) Platelet-derived growth factor promotes human peripheral monocyte activation. Blood 66:179–183
Van Nostrand WE, Schmaier AH, Farrow JS, Cines DB, Cunningham DD (1991) Protease nexin-2/amyloid beta-protein precursor in blood is a platelet specific protein. Biochem Biophys Res Commun 175:15–21
Walz A, Baggiolini M (1986) A novel cleavage product of beta-thromboglobulin formed in cultures of stimulated mononuclear cells activates human neutrophils. Biochem Biophys Res Commun 159:969–975
Walz DA, Wu VY, de Lamo R, Dene H, McCoy LE (1977) Primary structure of human platelet factor 4. Thromb Res 11:893–898
Walz A, Dewald B, von Tscharner V, Baggiolini M (1989) Effects of the neutrophil-activating peptide NAP-2, platelet basic protein, connective tissue-activating peptide III and platelet factor 4 on human neutrophils. J Exp Med 170:1745–1750

Wasteson A, Hook M, Westermark B (1976) Demonstration of platelet enzyme degrading heparan sulphate. FEBS Lett 64:218–221

Waterfield MD, Scrace GT, Whittle N, Stroobant P, Johnsson A, Wasteson A, Westermark B, Heldin C-H, Huang JS, Deuel TF (1983) Platelet-derived growth factor is structurally related to the putative transforming protein p28sis of simian sarcoma virus. Nature 304:35–39

Wencel-Drake JD, Painter RG, Zimmerman TS, Ginsberg MH (1985) Ultrastructural localization of human platelet thrombospondin, fibrinogen, fibronectin and von Willebrand factor in frozen thin sections. Blood 65:929–938

Zipfel PF, Balke J, Irving SG, Kelly K, Siebenlist U (1989) Mitogenic activation of human T cells induces two closely related genes which share structural similarities with a new family of secreted factors. J Immunol 142:1582–1590

Zucker MB, Katz IR (1991) Platelet factor 4: production, structure and physiologic and immunologic action. Proc Soc Exp Biol Med 197:693–702

Zucker MB, Mosesson MW, Broekman MJ, Kaplan KL (1979) Release of platelet fibronectin (cold-insoluble globulin) from alpha granules induced by thrombin or collagen; lack of requirement for plasma fibronectin in ADP-induced platelet aggregation. Blood 54:8–12

Zucker MB, Katz IR, Thorbecke GJ, Milot DC, Holt J (1989) Immunoregulatory activity of peptides related to platelet factor 4. Proc Natl Acad Sci USA 86:7571–7574

CHAPTER 20
# Lysosomal Storage

M.H. FUKAMI and H. HOLMSEN

## A. Introduction

### I. Lysosomes in General

Lysosomes are by definition membrane-bound vesicles, i.e., intracellular organelles, that contain a variety of hydrolytic enzymes characterized as having acidic pH optima (DE DUVE 1975). The lysosomal membrane has an ATP-dependent proton pump that maintains an intragranular pH of approximately 5.0. In nucleated cells these organelles are formed and processed in the endoplasmic reticulum and Golgi apparatus. Newly formed organelles, i.e., primary lysosomes, fuse with other imported vesicles containing phagocytized macromolecules to form secondary lysosomes. The macromolecules, i.e., proteins, lipids, carbohydrates, and nucleic acid moieties, subsequently undergo hydrolysis or "intracellular digestion" in the secondary lysosomes. Both constitutive and regulated secretion of lysosomal enzymes are also observed in certain cell types. Generally, lysosomes are more variable in size and appearance because of heterogeneous contents and differences in time after fusion. Lysosomes appear to be present in all eukaryotic cells (for an historical review see BAINTON 1981).

### II. Platelet Lysosomes

Since platelets lack not only nuclei but also endoplasmic reticulum and Golgi, their lysosomes are probably preformed in the megakaryocyte (see Chap. 1). It has long been supposed that platelets have no or low activity with respect to endocytosis, phagocytosis, and autophagic functions and that most of their lysosomes are probably primary lysosomes. Since the usual lysosomal housekeeping activities such as removal of old organelles and processing of endocytic vacuoles, etc. were considered not to take place in platelets, the function of platelet lysosomal activity has been assumed to be the digestion of the products of the major activity of these specialized cells, the hemostatic plug. The acid hydrolase breakdown of the accumulated clot components, i.e., platelets, fibrin, inflammatory cells, etc., is considered to be a necessary process in wound healing. However, both coated pits and vesicles as well as clathrin have now been immuno-chemically identified in platelets, and it appears that some of the α-granule contents are acquired by endocytosis of plasma factors (BEHNKE 1992; KLINGER and KLÜTER 1995). Occasional sightings of secondary lysosomes or labels for α-granule and lysosomal markers in the same organelle might indicate that the normal processing function of lysosomes also takes place in platelets (see below). Platelets also release hydrolytic enzymes to the extracellular medium by regulated secretion. Acid hydrolase secretion is partial and always lags behind secretion of dense granules (see Chap. 18) and α-granules (Chap. 19) which both occur more rapidly and to a greater extent.

Various platelet acid hydrolase activities have been reported in numerous articles over the past 50 years. There exists an excellent comprehensive review of the early work on platelet lysosomes which we will not attempt to duplicate here (VAN OOST 1986; with 192 references). This review will focus for the most part on selected literature after 1980 that further our understanding of lysosomal function and storage and secretion mechanisms. The surface expression of lysosomal membrane proteins will also be discussed.

## B. Lysosomal Hydrolases in Platelets

Acid hydrolases that have been found in platelets include glycosidases, proteases, acid phosphatase, and aryl sulfatase; some proteases and phospholipases $A_1$ and $A_2$ which have higher pH optima are also secreted (Table 1).

### I. Glycosidases

Platelet lysates were found to contain 11 of 13 acid glycosidases for which substrates were available (Table 1). No β-cellubiosidase or α-xylosidase was detected. β-*N*-Acetylglucosaminidase and β-glucuronidase were present in the highest concentrations at a ratio of 4 to 1, respectively. α- and β-Galactosidases, β-*N*-acetylgalactosaminidase and α-mannosidase were next highest in concentration.

#### 1. Secretion

Platelet lysosomal secretion differs from dense and α-granule secretion in at least three ways. The first is that lysosomal contents are only partially secreted compared to the other granules; the second is that different proportions of the various glycosidases are secreted; and the third is that lysosomal secretion requires stronger or higher concentrations of agonist and is more sensitive to metabolic inhibitors than dense and α-granule secretion.

β-*N*-Acetylglucosaminidase and β-*N*-acetylgalactosaminidase (collectively referred to as β-hexosaminidase) are secreted in amounts of up to 50%–60% of total platelet content whereas only about 30% of β-glucuronidase or β-glucosidase are secreted with maximal thrombin stimulation (HOLMSEN and DANGELMAIER 1989). Preincubation of platelets with $NH_4Cl$ or primary amines

**Table 1.** Hydrolase activities in human platelets

| Enzyme | pH Optima | Reference |
|---|---|---|
| Glycosidases | | |
| α-Arabinosidase | 4.5 | DANGELMAIER and HOLMSEN 1980[a] |
| α-Fucosidase | 4.0 | |
| β-Fucosidase | 4.0 | |
| α-Galactosidase | 3.5 | |
| β-Galactosidase | 4.0 | |
| α-Glucosidase | 4.1; 7.8 | |
| β-Glucosidase | 5.5 | |
| β-Glucuronidase | 5.0 | |
| β-*N*-Acetylgalactosaminidase | 4.5 | |
| α-*N*-Acetylglucosaminidase | 4.0 | |
| α-Mannosidase | 5.5 | |
| Proteinases | | |
| Cathepsin D[b] | 3.5–5.0 | ERLICH and GORDON 1976 |
| Cathepsin E | 3.0 | |
| Phospholipases | | |
| Phospholipase $A_1$ | 4.8 | SMITH et al. 1973 |
| Phospholipase $A_2$s (rat) | 7.0–10.0 | HORIGOME et al. 1987 |
| ″ (human) | 8.0–10.0 | KRAMER et al. 1989 |
| Other | | |
| Acid phosphatase[c,d] | 5.5 | DANGELMAIER and HOLMSEN 1980 |
| Aryl sulfatase IIA; IIB | 4.5; 5.5 | METCALF et al. 1979<br>OLDBERG et al. 1980 |
| Heparitinase | 5.5–8.0 | OOSTA et al. 1982 |

Observations in nonhuman species are specified.
[a] Both *p*-nitrophenyl and 4-methylumbelliferyl derivatives were used as substrates.
[b] Denatured hemoglobin, proteoglycan, casein.
[c] *p*-Nitrophenyl phosphate.
[d] β-Glycerophosphate.

("lysosomotropic agents") for 90–120 min followed by thrombin stimulation resulted in doubling of the β-*N*-acetylglucosaminidase secretion to over 70% of total content without causing cell lysis (VAN OOST et al. 1985). Glucuronidase and β-galactosidase secretion were also doubled with amine preincubation, but even so, more than 50% remained in the cells. Oddly enough, α-mannosidase secretion was not affected by the amines.

Analysis of β-hexosaminidase isoforms showed that both A and B isoforms exist in platelets and that they are secreted in the same proportion as is found in whole cell lysate, indicating that neither form is preferentially released (VLADUTIU et al. 1984). Acid glycosidases can also exist as a phosphomannosylated form that is the recognition group for receptor-mediated uptake into certain cells such as fibroblasts. GLASER et al. (1985) showed that platelets contained an unusually high proportion of phosphomannosylated β-glucuronidase that was rapidly taken up by fibroblasts deficient in $\beta$-glucuronidase. VLADUTIU et al. (1984) also demonstrated the presence of the phosphomannosylated (high-uptake) form of $\beta$-hexosaminidase in platelet pellets after secretion, but found that only the low-uptake, nonphosphomannosylated enzyme was secreted. They did not quantitate the proportions of high- and low-uptake forms, but their data indicated that the amount of high-uptake form was low since the uptake activity with the secreted supernatant was comparable to that of whole cell extracts. These results suggest that $\beta$-hexosaminidase exists mainly as the nonphosphomannosylated form which is also the secretable form, while proportionately more of $\beta$-glucuronidase exists as phosphomannosylated or nonsecretable enzyme, and this could be the reason why less of the latter is secreted.

Platelets themselves did not endocytose high-uptake $\beta$-hexosaminidase over a 3-h incubation period (VLADUTIU et al. 1984). These results indicate that platelet acid glycosidases most likely originate from the parent megakaryocyte and not by endocytosis of the enzymes from plasma.

The possibility that the nonsecreted glycosidases were actually exocytosed but remained bound on the outer plasma membrane has also been explored. Assay of uncentrifuged, thrombin-treated platelet suspensions showed no higher glycosidase activities than supernatants of such platelets, indicating that all of the secreted enzyme is soluble and appears in the cell supernatant (VLADUTIU et al. 1984); the alternative is that the bound enzyme is also enzymatically inactive.

This difference in degree of secretion has led to the assumption that lysosomes are heterogeneous in content and that some of them respond to activation more readily than others. It is well established that dense granule and $\alpha$-granule secretion have a different threshold to stimulatory agonists than lysosome secretion in general. With weak agonists such as ADP or low doses of thrombin, considerable secretion of dense and $\alpha$-granules takes place with relatively little release of acid hydrolases (HOLMSEN et al. 1982; AKKERMAN et al. 1983); lysosome secretion requires a relatively high dose of a strong agonist such as thrombin. Inhibitor-induced decrease in platelet metabolic ATP leads to a greater decrease of acid hydrolase secretion than of dense or $\alpha$-granule contents. In the same manner, some of the lysosomes, i.e., those containing glycosidases that are secreted more strongly than others, could also be more responsive to agonist stimulation.

## 2. Subcellular Localization

There has been a long-standing controversy as to whether platelet lysosomes and $\alpha$-granules were distinct and separate organelles (MORGENSTERN and JANZARIK 1985). Whereas mitochondria and dense granules are morphologically clearly distinguishable, the two other platelet organelle types look similar without the use of specific staining techniques. The two most frequently used methods for direct determination of the site of intracellular components are microscopy and subcellular fractionation. The first technique requires some means for specific staining or labelling of a particular substance or protein in order to visualize it. Although antibodies against glycosidases have become available, immunocytochemical staining for these lysosomal enzymes has not yet been reported. Most of the cytochemical work has been done on acid phosphatases (see below). Subcellular fractionation studies accompanied by assay of various acid glycosidases appear to have established the existence of lysosomal organelles as entities distinct from $\alpha$-granules. Acid glycosidases and $\alpha$-granule contents were clearly enriched in different fractions (BROEKMAN et al. 1974). That is to say, $\alpha$-granules were decidedly enriched in one fraction, whereas acid glycosidases were spread over most of the gradient, which indicated that they were localized in organelles of varying size and density, possibly with different response characteristics to platelet activation. These results together with the differential secretion response to varying strengths of agonist have led to the general acceptance of $\alpha$-granules and lysosomes as individual entities (for a review see FUKAMI and SALGANICOFF 1977).

Further support for these conclusions seems to come from various storage-granule-deficient patients. Platelets from patients with gray platelet syndrome, i.e., severe depletion of $\alpha$-granule contents, have normal levels of lysosomal enzymes, usually measured as $\beta$-*N*-acetylglucosaminidase and/or $\beta$-glucuronidase. However, in a case report on one patient, six acid hydrolases were measured (SRIVASTAVA et al. 1987). Normal levels of $\beta$-glucuronidase, $\beta$-galactosidase, $\beta$-*N*-acetylglucosaminidase, and arylsulfatase were found, but $\alpha$-mannosidase and $\alpha$-fucosidase levels were only 50% of those of control donors' platelets. The two latter glycosidases are infrequently measured because their levels in platelets are much lower than those of the former. The decreased level of $\alpha$-mannosidase is of interest because it is the one glycosidase described by VAN OOST et al. (1985), the secretion of which was not increased with amine preincubation (see above); VAN OOST et al. did not measure $\alpha$-fucosidase. Although $\alpha$-mannosidase release was only 38% of total, its secretion, like that of $\alpha$-granules and dense granules, was unaffected by amine preincubation. It is intriguing that the only two laboratories that have included $\alpha$-mannosidase in their studies on acid hydrolases found its storage and secretion to be somewhat anomalous when using completely independent experimental methods.

## II. Proteinases

Proteinases that have been reported to be present in platelets include cathepsins D and E (ERLICH and GORDON 1976; Table 1). ERLICH and GORDON carefully repeated the earlier work in order to characterize platelet proteinases and found that the major acid hydrolases of this cateorgy were cathepsins D and E. Elastase levels isolated from washed platelets were equivalent to the amount that would be attributiable to the amount of contaminating leukocytes. Platelets that were carefully isolated in order to minimize leukocyte contamination contained no elastase. Similary, collagenase activity could not be demonstrated in leukocyte-free platelets. Nonspecific cathepsin is released to the same extent as $\beta$-*N*-acetylglucosaminidase and $\beta$-galactosidase with thrombin stimulation (HOLMSEN and DAY 1970).

Cathepsin D has been demonstrated in platelet granules by immunocytochemical detection with electron microscopy (SIXMA et al. 1985). Colloidal gold label was localized on only one granule in every three platelet profiles, and these granules were said to be smaller than usual $\alpha$-granules. The low frequency of labelling suggests that not every lysosome may contain all the acid hydrolases found in the cell, i.e., there may exist lysosomal heterogeneity. Double labelling attempted with an $\alpha$-granule marker was inexplicably unsuccessful. In one profile of platelets treated with thrombin prior to fixation, gold particles were seen in a section of the surface-connected tubular system. Gold staining was also seen in structures that appeared to be secondary lysosomes from their heterogeneous contents in both platelets and megakaryocytes.

## III. Phospholipases

A phospholipase $A_1$ in platelets which had a pH optimum of 4.8 and deacylated phosphatidylcholine to the 2-monoacyl compound ($^{14}C$-labelled) was described by SMITH et al. (1973), but no further reports on this enzyme have since been published. A soluble phospholipase $A_2$ and lysophospholipase were first described in rat platelets by HORIGOME et al. (1987), and phospholipase $A_2$ was later isolated from human platelets (KRAMER et al. 1989). These phospholipases were secreted upon thrombin treatment and in the case of the rat enzyme showed a time course and extent resembling that of serotonin and far in excess of that of $\beta$-*N*-acetylglucosaminidase. Although the phospholipase $A_2$ isoforms were found to have broad pH optima from 7.0 to 10.0, they were stable to acid and are included in this review on lysosomes, although they may be $\alpha$-granule constituents. This 14-kDa enzyme which is secreted from stimulated platelets is the same size as other secretable phospholipase $A_2$ isoforms, e.g., from snake venoms and pancreas, and different in size from the 85-kDa cytosolic phospholipase $A_2$ which is not secreted (KRAMER et al. 1993). It was originally thought the secreted enzyme acted on platelet plasma membrane phospholipids to generate arachidonic acid which reentered the cell, but arachidonic acid liberation was also seen in the presence of EDTA. Since the secreted enzyme is $Ca^{2+}$-dependent, arachidonate formation in platelets is attributable to the cytosolic enzyme.

## IV. Other

In this category of lysosomal enzymes fall two of the most commonly studied acid hydrolase activities, that of acid phosphatase and that of aryl sulfatase (Table 1). These two enzymes were extensively used as markers for cytochemical demonstration of lysosomes by electron microscopy (see Chap. 2) and enzyme assay of lysates and subcellular fractions. As with the other hydrolases, attenuated synthetic substrates are used for these assays instead of a native substrate. Choice of substrate has led to some confusion in the case of acid phosphatase for which two types of substrates have been used, *p*-nitrophenyl phosphate, or 4-methylumbelliferyl phosphate, and β-glycerophosphate. The aromatic substrates are for an enzyme(s) that is localized to membranes, is readily solubilized under subcellular fractionation, and appears to have nonspecific phosphatase activity. The aliphatic phosphate appears to be the substrate for the secretable lysosomal enzyme (Bentfield and Bainton 1975). The aromatic phosphatase was not secreted from platelets activated with thrombin, but β-glycerophosphatase was secreted in amounts comparable to β-glucuronidase (Holmsen and Day 1970). Aryl sulfatase activity was found to be attributable to two isoforms, IIA and IIB, with molecular weights of 160000 and 60000, respectively. The sulfatases that could inactivate the slow-reacting substance of anaphylaxis were also secreted in response to thrombin to the same degree as β-glucuronidase (Metcalfe et al. 1979).

Bentfield-Barker and Bainton (1982) identified lysosomal vesicles with lead-salt staining using β-glycerophosphatase and arylsulfatase as markers. They found heavy staining for β-glycerophosphatase and arylsulfatase in small vesicles in megakaryocytes and platelets, but staining for the two enzymes was never found in the same vesicles; there was no staining in larger vesicles assumed to be α-granules. Unlike Sixma et al. (1985), no secondary lysosomes were seen. Morgenstrn and Janzarik (1985) used cytochemical staining for β-glycerophosphatase and immunochemical-protein-A gold marking for antibodies against fibrinogen and β-thromboglobulin (α-granule markers) on ultrathin serial sections. Reconstruction of the sections showed that both β-glycerophosphatase staining and gold particles could be observed in some of the same organelles.

Heparitinase is a platelet endoglycosidase that degrades heparin and heparan sulfate to oligosaccharides (Oldberg et al. 1980). The enzyme has a requirement for a sulfamino moiety and is specific for the glucuronidic linkage in the substrate. It was further shown to have a pH optimum of 6.0 to 8.0 in a highly purified preparation and a molecular weight of 134000 (Oosta et al. 1982). These authors show that with subcellular fractionation, the distribution profile of heparitinase is identical to that of β-glucuronidase and β-hexosaminidase, but there is a considerable amount of glycosidase markers present in the fraction in which platelet factor 4 (α-granule marker) is clearly most enriched with respect to relative specific activity and overall distribution. One of the most interesting aspects of heparitinase is that it is 100% secreted with thrombin treatment (Oldberg et al. 1980). Adrenaline and ADP caused only 25% and 12% secretion, respectively, but these weaker agonists also cause lesser secretion of dense granule contents, and secretion markers for these as control were not measured as in this experiment. Since none of the other acid hydrolases in platelets is fully released, this behavior of heparitinase seems quite anomalous. Functionally, it may be appropriate for heparitinase to become readily available upon platelet activation to counteract the anticoagulant effects of heparin, so that its role is not "digestive" like that of other lysosomal enzymes, but more like that of $\alpha$-granule contents in recruiting and consolidating platelet aggregates.

## C. Lysosomal Membrane Proteins

The combination of flow cytometry and monoclonal antibody preparation techniques has made it possible to study expression of proteins on the platelet plasma membranes. A monoclonal antibody that was bound by activated platelets but not resting platelets was described by Nieuwenhuis et al. (1987). This antibody, designated 2.28, colocalized with cathepsin D in immunogold double labelling studies in megakaryocytes and endothelial cells. In resting platelets the antibody label was found inside some of the granules, but not in other granules assumed to be $\alpha$-granules. Sites on the plasma membrane and surface-connected canalicular system were gold-labelled in thrombin-treated platelets, although some thrombin-treated platelets were not labelled at all. Immunoelectrophoresis showed that the protein with which this antibody reacted had a molecular weight of 53000. Antibody 2.28 did not interfere with platelet aggregation. With flow cytometry, the percentage of responding platelets after stimulation with 5U/ml (!) of thrombin was 76% compared to 4% in resting platelets. The expression of the antigen increased from 5.5% to 25% in patients before and after cardiopulmonary bypass surgery, respectively, compared to an eightfold increase in plasma $\beta$-thromboglobulin, an $\alpha$-granule marker. The presence of the antigen or of any acid hydrolase in supernatants of activated platelets was not correlated with expression of the antigen on the plasma membrane. This antigen now has the designation CD63 or LIMP-CD63 for lysosome integral membrane protein (Metzelaar et al. 1991). However, it has been found to be cross-reactive with an antibody to granulophysin, a 40-kDa protein that is present in dense granules (see Chap. 18) and in granules in many tissues (Hatskelzon et al. 1993).

Two other lysosome-associated membrane glycoproteins, LAMP-1 and LAMP-2, which have been reported in nucleated cells have been identified in platelets (Febbraio and Silverstein 1990; Silverstein and Febbraio 1992). Monoclonal antibodies to these proteins reacted with thrombin-stimulated but not with resting platelets in $^{125}$I-labelled antibody binding studies and in flow cytometry experiments. Subcellular fractionation studies showed that LAMP-1 and LAMP-2 colocalized on a sucrose gradient with $\beta$-galactosidase. Clearly, these LAMP proteins as well as the 53-kDa protein described above become available for antibody labelling after platelet stimulation with thrombin. However, a convincing study showing the correlation of the expression of lysosomal membrane antigens on the platelet surface compared to actual acid hydrolase secretion is lacking in the literature. Febbraio and Silverstein (1990) reported (although with "data not shown") that secretion of thrombospondin, an $\alpha$-granule marker, corresponded with expression of surface thrombospondin as measured by ELISA. Since thrombospondin is both secreted and rebound, this seems to be a reasonable finding. In this paper on LAMP-1 (1990), the authors have compared expression of LAMP-1 and PADGEM, which is an $\alpha$-granule membrane marker, and secretion of [$^{14}$C]-5-hydroxytryptamine, a dense granule marker. They report the $ED_{50}$ for thrombin for the above markers to be 0.05 U/ml for LAMP-1 expression and 0.005 U/ml for PADGEM expression which seem to be reasonable. The $ED_{50}$ of 0.25 U/ml for 5-hydroxytryptamine secretion, on the other hand, was extremely high, probably because reuptake of serotonin which occurs with low thrombin was not inhibited by an appropriate inhibitor (see Chap. 17). Normally, platelets secrete very little acid hydrolase at thrombin concentrations of 0.05 U/ml, but considerable secretion of dense and $\alpha$-granules is seen at this concentration (Holmsen et al. 1982). A recent report claiming that LAMP-2 is also present in platelet dense granule membranes (Israels et al. 1996) suggests that these membrane proteins are not exclusive markers for one type of granules. The low $ED_{50}$ for LAMP-1 expression reported above could indicate that LAMP-1 is also not an exclusive marker for lysosomes, and its expression may have been due to dense granule or $\alpha$-granule membrane appearance on the platelet surface.

## D. Discussion and Comments

The main recent findings on platelet lysosomes since the review written by van Oost (1986) concern the reports of the existence of secondary lysosomes and identification of lysosomal membrane proteins. The finding of apparent secondary lysosomes is of some interest because it has bearing on occasional reports of the colocalization of $\alpha$-granule and lysosomal markers in the same vesicles. Although platelet lysosomes are generally classified as primary lysosomes because they tend to be smaller than $\alpha$-granules, these studies have often involved one or two acid hydrolase and no $\alpha$-granule markers. If the premise that lysosomes are heterogeneous is valid, then nonstaining vesicles could also be lysosomes without the acid hydrolase that is being probed. From subcellular fractionation studies, it appears that $\alpha$-granules have a distinct density and sediment in a relatively tight band on a sucrose gradient, but acid hydrolases were found distributed over almost the entire gradient with only a small peak, correlating with lysosomal heterogeneity (Broekman et al. 1974). It is also consistent with normal lysosomal function that the membrane markers LAMP-1 and LAMP-2 present in other cells such as rat liver were also found in platelets. The localization of some of the hydrolases included in this chapter on lysosomes has not been studied, in particular that of heparitinase and secretable phospholipase $A_2$. Strictly speaking, these are not acid hydrolases, although they are stable to acid; they are secreted in a manner quantitatively more similar to the pattern of dense and $\alpha$-granule secretion than to lysosomal release. This brings us to the subject of the lesser degree of lysosomal secretion compared to that of the other granules. While some cells such as mast cells secrete massively, others such as chromaffin cells secrete only a small portion of total granular contents at any given time, so that it is perhaps normal that only partial lysosomal secretion occurs. The regulatory mechanisms appear to be different for lysosomal secretion with respect to energy requirement, strength of stimulus, and receptor occupancy (HOLMSEN et al. 1981). Dense and $\alpha$-granule secretion were found to continue after a brief exposure to thrombin followed by addition of excess hirudin, but lysosomal secretion was arrested upon hirudin addition. Another possible explanation for partial lysosomal secretion is heterogeneity of platelets. Functional fractionation of platelets showed that the larger platelets are more reactive, have more sialyl residues on their surface, more glycogen, and a higher ATP/ADP ratio (HAVER and GEAR 1981). These more highly reactive platelets are most likely younger platelets. Since acid hydrolase secretion is more sensitive to metabolic inhibitors and requires higher concentrations of thrombin, the partial secretion of acid hydrolases may be due in part to nonresponding older platelets. The observation of NIEUWENHUIS et al. (1987) that some 24% of platelets treated with 5 U/ml of thrombin did not in the least externally express the 53-kDa membrane antigen is compatible with the observations of HAVER and GEAR (1981). The difference in degree of secretion of various acid glycosidases might be due to the different proportions of phosphomannosylated isoforms (see above).

In conclusion, it appears that platelet lysosomes are not so uniquely different from those of other acid hydrolase secreting cells.

## References

Akkerman JWN, Gorter G, Schrama, L, Holmsen H (1983) A novel technique for rapid determination of energy consumption in platelets. Biochem J 210:145–155

Bainton DF (1981) The discovery of lysosomes. J Cell Biol 91:66s–76s

Behnke O (1992) Degrading and non-degrading pathways in fluid-phase (non-adsorptive) endocytosis in human blood platelets. J Submicrosc Cytol Pathol 24:169–178

Bentfield ME, Bainton DF (1975) Cytochemical localization of lysosomal enzymes in rat megakaryocytes and platelets. J Clin Invest 56:1635–1640

Bentfield-Barker ME, Bainton DF (1982) Identification of primary lysosomes in human megakaryocytes and platelets. Blood 59:472–481

Broekman MJ, Westmoreland NP, Cohen P (1974) An improved method for isolating alpha granules and mitochondria from human platelets. J Cell Biol 60:507–519

Dangelmaier CA, Holmsen H (1980) Determination of acid hydrolases in human platelets. Anal Biochem 104:182–191

de Duve C (1975) Exploring cells with a centrifuge. Science 189:186–194

Erlich HP, Gordon JL (1976) Proteinases in platelets. In: Gordon JL (ed) Platelets in biology and pathology. North Holland Publishing, Amsterdam, (Research monographs in cell and tissue physiology), pp 352–372

Febbraio M, Silverstein RL (1990) Identification and characterization of LAMP-1 as an activation-dependent platelet surface glycoprotein. J Biol Chem 265:18531–18537

Fukami MH, Salganicoff L (1977) Human platelet storage organelles. Thromb Haemost 38:963–970

Glaser JH, Roozen KJ, Brot FE, Sly WS (1985) Multiple isoelectric and recognition forms of human $\beta$-glucuronidase activity. Arch Biochem Biophys 166:536–542

Hatskelzon L, Dalal BI, Shalev A, Robertson C, Gerrard JM (1993) Wide distribution of granulophysin epitopes in granules of human tissue. Lab Invest 68:509–519

Haver VM, Gear ARL (1981) Functional fractionation of platelets. J Lab Clin Med 97:187–204

Holmsen H, Dangelmaier CA (1989) Measurement of secretion of lysosomal acid glycosidases. Methods Enzymol 169:336–342

Holmsen H, Day HJ (1970) The selectivity of the thrombin-induced platelet release reaction: subcellular localization of released and retained constituents. J Lab Clin Med 75:840–855

Holmsen H, Salganicoff L, Fukami MH (1977) Platelet behaviour and biochemistry. In: Ogston D, Bennett B (eds) Haemostasis: biochemistry, physiology and pathology. Wiley, London, pp 264–265

Holmsen H, Dangelmaier CA, Holmsen HK (1981) Thrombin-induced platelet responses differ in requirements for receptor occupancy: evidence of tight coupling of occupancy and compartmentalized phosphatidic acid formation. J Biol Chem 256:9393–9395

Holmsen H, Kaplan KL, Dangelmaier CA (1982) Differential energy requirements for platelet responses. Biochem J 208:9–18

Horigome K, Hayakawa M, Inoue K, Nojima S (1987) Selective release of phospholipase $A_2$ and lysophosphotidylserine-specific lysophospholipase from rat platelets. J Biochem 101:53–61

Israels SJ, McMillan EM, Robertson C, Singhroy S, McNicol A (1996) The lysosomal granule membrane protein, Lanp-2, Lamp-2, is also present in platelet dense granule membranes. Thromb Haemost 75:623–629

Klinger MH, Klüter H (1995) Immunocytochemical colocalization of adhesive proteins with clathrin in human blood platelets: further evidence for coated vesicle-mediated transport of von Willebrand factor, fibrinogen and fibronectin. Cell Tissue Res 279:453–457

Kramer RM, Hession C, Johansen B, Hayes G, McGray P, Chow EP, Tizard R, Pepinsky RB (1989) Structure and properties of a human non-pancreatic phospholipase $A_2$. J Biol Chem 264:5768–5775

Kramer RM, Roberts EF, Manette JV, Hyslop PA, Jakubowski JA (1993) Thrombin-induced phosphorylation and activation of $Ca^{2+}$-sensitive cytosolic phospholipase $A_2$ in human platelets. J Biol Chem 268:26796–26804

Metcalfe DD, Corash LM, Kaliner M (1979) Human platelet arylsulphatases: identification and capacity to destroy SRS-A. Immunology 37:723–728

Metzelaar MJ, Winjgaard PLJ, Peters PJ, Sixma JJ, Nieuwenhuis HK, Clevers HC (1991) CD63 antigen – a novel lysosomal membrane glycoprotein cloned by a screening procedure for intracellular antigens in eukaryotic cells. J Biol Chem 266:3239–3245

Morgenstern E, Janzarik H (1985) Comparative ultractyochemical studies on the secretory organelles of avian thrombocytes and human blood platelets. Acta Histochem [Suppl] XXXI:207–210

Niewenhuis HK, van Oosterhout JJG, Rozemuller E, van Iwaarden F, Sixma JJ (1987) Studies with a monoclonal antibody against activated platelets: evidence that a secreted MW 53000 lysosome-like-granule protein is exposed on the surface of activated platelets in the circulation. Blood 70:838–845

Oldberg Å, Heldin C-H, Wasteson Å, Busch C, Hook M (1980) Characterization of a platelets endoglycosidase degrading heparin-like polysaccharides. Biochemistry 19:5755–5762

Oosta GM, Favreau LV, Beeler DL, Rosenberg RD (1982) Purification and properties of human platelet heparitinase. J Biol Chem 257:11249–11255

Silverstein RL, Febbraio M (1992) Identification of lysosome-associated membrane protein-2 as an activation-dependent platelet surface glycoprotein. Blood 80:1470–1475
Sixma JJ, van den Berg A, Hasilik A, von Figura K, Geuza HJ (1985) Immuno-electron microscopical demonstration of lysosomes in humanblood platelets and megakaryocytes using anti-cathepsin D. Blood 65:1287–1291
Smith JB, Silver MJ, Webster GR (1973) Phospholipase A1 of human blood platelets. Biochem J 131:615–618
Srivastava PC, Powling MJ, Nokes TJC, Patrick AD, Dawes J, Hardistry RM (1987) Grey platelet syndrome: studies on platelet alpha-granules, lysosomes and defective response to thrombin. Br J Haematol 65:441–446
van Oost BA (1986) In vitro platelet responses: acid hydrolase secretion. In: Holmsen H (ed) Platelet responses and metabolism, vol I, responses. CRC, Boca Raton
van Oost BA, Smith JB, Holmsen H, Vladutiu GD (1985) Lysomotropic agents selectively potentiate thrombin-induced acid hydrolase secretion from platelets. Proc Natl Acad Sci USA 82:2374–2378
Vladutiu GD, Dangelmaier CA, Amigone V, van Oost BA, Holmsen H (1984) High- and low-uptake forms of $\beta$-hexosaminidase in human platelets: selective retention of the high-uptake form during stimulation with thrombin. Biochim Biophys Acta 802:435–441

CHAPTER 21

# Thromboxane $A_2$ and Other Eicosanoids

P.V. HALUSHKA, S. PAWATE, and M.L. MARTIN

## A. Introduction

The major product of the metabolism of arachidonic acid in platelets is thromboxane $A_2$ ($TXA_2$) (HAMBERG et al. 1975); thus, the main focus of this chapter will be on it. The effects of other eicosanoids that are either synthesized by platelets or impact on platelet function will also be discussed. $TXA_2$ receptors are discussed in greater detail in Chap. 7B. We will provide a brief discussion of alterations in platelet $TXA_2$ receptors in disease states. A comprehensive review of the effects of $TXA_2$ on platelet function was provided by ARITA et al. (1989). This review will focus on discoveries made since then.

## B. Platelet Arachidonic Acid Metabolism

### I. Metabolites

The eicosanoids represent the various metabolites of 5, 8, 11, 14-eicosatetraenoic acid, arachidonic acid (AA), and other related C-20 fatty acids. The eicosanoids include the prostaglandins, thromboxanes, and leukotrienes. The two major enzymatic routes for the metabolism of AA in platelets are the fatty acid cyclooxygenase and lipoxygenase pathways. In the cyclooxygenase pathway, AA is first oxygenated twice by the enzyme fatty-acid cyclooxygenase (prostaglandin G/H synthase) to yield $PGG_2$ and then $PGH_2$. In the platelet, the prostaglandin endoperoxide $PGH_2$ is enzymatically rearranged or reduced by specific synthase enzymes to give rise to the prostanoids $PGD_2$, $PGE_2$, $PGF_{2\alpha}$, and $TXA_2$. $PGH_2$ synthesized by the platelet can be transferred to endothelial cells and metabolized to $PGI_2$ (prostacyclin). For the formation of 1-series or 3-series prostanoids such as $PGE_1$ and $PGE_3$, dihomo-$\gamma$-linolenic acid (DGLA; 20:3$\omega$6) and eicosapentanoic acid (EPA; 20:5$\omega$3), respectively, are the precursor fatty acids. In the lipoxygenase pathway, AA is metabolized by one or more lipoxygenases to give rise to leukotrienes, hydroxy fatty acids, and lipoxins. These pathways have been reviewed in references (ARITA et al. 1989; PEPLOW 1992; SMITH et al. 1991; WILLIS 1987). A minor pathway that has gained attention recently is the nonenzymatic, free-radical catalyzed peroxidation of AA to isoprostanes such as 8-Epi-$PGF_{2\alpha}$ (MORROW et al. 1992a; PRATICO et al. 1994).

In platelets, the major enzymatic products of AA are $TXA_2$ and 12-monohydroxy eicosatetraenoic acid (12-HETE), while in endothelial cells $PGI_2$ formation predominates. Pharmacologic inhibition of specific enzymes in the AA cascade can shift the route of metabolism to pathways that usually are minor in a particular cell. For example, inhibition of thromboxane synthase increases the formation of $PGE_2$, $PGF_{2\alpha}$, and $PGD_2$, as well as causing an accumulation of $PGH_2$ in activated platelets (FITZGERALD et al. 1985).

Eicosanoids are not stored in cells, but are synthesized de novo in response to cell-specific proteolytic or hormonal stimuli. Since the majority of the AA in a resting cell is esterified to specific phospholipids, eicosanoid synthesis must occur in two stages: first, liberation of AA from membrane phospholipids and second, metabolism of free AA by specific enzymes. Therefore, the following subsection will focus on the liberation of AA from membrane phospholipids with special attention to the platelet.

## II. Sources of Arachidonic Acid

AA is an essential fatty acid that is stored predominantly in the sn-2 position of membrane phospholipids. Animal cells can take up AA directly or synthesize AA from linoleic acid (MEAD and WILLIS 1987). Fatty acids are esterified into membrane phospholipids by long-chain acyl-CoA synthetases and acyltransferases. The mechanism by which AA is preferentially esterified into the sn-2 position of phospholipids became clearer with the discovery of two distinct long-chain acyl-CoA sythases, one with a broad specificity for fatty acid substrates, and another that is highly specific for AA and EPA (LAPOSATA et al. 1985; WILSON et al. 1982).

In human platelets, AA is the most prevalent fatty acid, particularly in the sn-2 position of phosphatidylcholine (PC), phosphatidylethanolamine (PE), and phosphatidylinositol (PI). PC and PE contain over 72%–74% of the total AA pool in platelets (ARITA et al. 1989; BROEKMAN et al. 1976; TAKAMURA et al. 1987). The PE fraction containing the greatest amount of AA can be subdivided into diacyl-PE and plasmenylethanolamine fractions. Plasmenylethanolamine contains 66% of the AA in the PE fraction, such that it contains 25% of the total AA in platelets, making plasmenylethanolamine the major reservoir of AA in human platelets (COHEN and DERKSEN 1969; TURINI et al. 1993). Additionally, AA is preferentially localized to certain areas. It has been reported that there is a marked difference in the phospholipid composition between the plasma membrane and the intracellular membranes. The intracellular membranes, including the dense tubular system, are enriched in AA containing PE and the plasma membrane enriched in AA containing PC and PI (LAGARDE et al. 1982). The plasma membrane is asymmetric in regard to phospholipid content. The outer leaf contains 45% of the plasma membrane phospholipids, but only 10% of the total plasma membrane AA, while the inner leaf contains 90% of the plasma membrane AA. Overall, 54% of the total platelet AA is in the inner leaflet, 6% in the outer leaflet, and 49% in the intracellular membranes (PERRET et al. 1979). This asymmetric localization of arachidonic-acid-bearing phospholipids to the inner leaf of the plasma membrane and to intracellular membranes has been proposed to be physiologically important to enable a more efficient coupling between the release of AA and its transformation by enzymes in the intracellular membranes (PERRET et al. 1979).

The determination of the AA-containing phospholipids preferentially hydrolyzed upon stimulation of eicosanoid synthesis is not totally clear. TAKAMURA et al. (1987) analyzed the absolute mass changes in the phospholipid classes containing AA in platelets that were stimulated with thrombin and collagen. They found that the AA-containing species of PC, diacyl-PE, and alkenylacyl-PE were selectively hydrolyzed by both agonists in the presence of extracellular calcium. Other studies have indicated that plasmenylethanolamine containing AA is selectively hydrolyzed upon stimulation by thrombin, calcium ionophore A23187, collagen, and the $TXA_2$ receptor agonist U46619 (KAMBAYASHI et al. 1987; KRAMER and DEYKIN 1983; RITTENHOUSE-SIMMONS et al. 1977; TURINI et al. 1993; TURINI and HOLUB 1994). The specific phospholipases involved and their phospholipid specificity are currently being investigated.

## III. Release of Arachidonic Acid

In unactivated platelets the amount of free AA and the rate of eicosanoid synthesis are very low. Upon stimulation, however, the concentration of unesterified AA rapidly increases as does the rate of eicosanoid synthesis (HABENICHT et al. 1990). Similarly, the addition of free AA to resting cells results in a rapid increase in eicosanoid synthesis. Together, these finding indicate that AA cleavage from membrane phospholipids may be a major trigger for eicosanoid synthesis (BETTAZZOLI et al. 1990; DENNIS 1987).

Several mechanisms have been proposed for agonist-stimulated AA mobilization. Activation of phospholipase (PL) $A_2$, which selectively cleaves fatty acids from the sn-2 position of membrane phospholipids, has been considered the prime enzymatic route (BARTOLI et al. 1994).

### 1. Phospholipase $A_2$

Since AA is esterified in the sn-2 position of phospholipids, activation of $PLA_2$, which selectively cleaves the ester linkage at this position, has been considered a prime candidate for agonist stimulated AA mobilization (BARTOLI et al. 1994; RIENDEAU et al. 1994; FAILI et al. 1994). Three different forms of $PLA_2$ have been discovered in platelets, a 14-kDa secretory $PLA_2$ ($sPLA_2$), an 85-kDa cytosolic $PLA_2$, and a 100-kDa cytosolic $PLA_2$ (KIM and BONVENTRE 1993; KRAMER et al. 1989; KRAMER et al. 1993; TAKAYAMA et al. 1991).

The 14-kDa $sPLA_2$s require millimolar concentrations of calcium for activity and have a broad specificity for phospholipids with different acyl chains and different polar head groups (BARTOLI et al. 1994; RIENDEAU et al. 1994). While $sPLA_2$s may play a role in eicosanoid synthesis in inflammation and other conditions, these $PLA_2$s do not appear to have a role in platelet activation or liberation of AA in platelets. MOUNIER et al. (1994) report that $sPLA_2$ is not involved in AA mobilization in platelets during platelet activation or after $sPLA_2$ has been released by platelets. Similarly, 70% depletion of platelet $sPLA_2$ or inhibition of $sPLA_2$ caused no significant change in platelet activation or eicosanoid production after stimulation with thrombin or collagen (BARTOLI et al. 1994; RIENDEAU et al. 1994).

Unlike $sPLA_2$, the 85-kDa cytosolic $PLA_2$ preferentially hydrolyzes phospholipids containing AA in the sn-2 position (KRAMER et al. 1993; TAKAYAMA et al. 1991). The 85-kDa cytosolic $PLA_2$ responds to physiologic changes in calcium concentration by translocating to cellular membranes and is activated by hormonal signalling via phosphorylation of a serine residue, most likely by mitogen-activated protein kinase (BARTOLI et al. 1994; RIENDEAU et al. 1994). These findings led to the belief that the 85-kDa $cPLA_2$ is the $PLA_2$ activated for agonist induction of AA liberation. To test this hypothesis, several groups have measured platelet AA mobilization in the presence of specific competitive inhibitors of 85-kDa $PLA_2$. BARTOLI et al. (1994) found that 85-kDa $PLA_2$ inhibitors blocked most, if not all, of the AA liberated by stimulation of human platelets with thrombin. RIENDEAU et al. (1994) similarly found that inhibition of 85-kDa $PLA_2$ in calcium-ionophore-challenged platelets blocked AA mobilization.

Another class of $PLA_2$s in platelets, the 100-kDa $PLA_2$s, have been detected in bovine platelets by an antibody to the 100-kDa pig spleen $PLA_2$. The purified 100-kDa pig spleen $PLA_2$ requires micromolar calcium concentrations for activity and is somewhat selective for AA-bearing phospholipids (KIM and BONVENTRE 1993). The mechanism for activation of the 100-kDa cytosolic $PLA_2$ and the effects of the 85-kDa $PLA_2$ inhibitors on this enzyme have yet to be determined. Therefore, its role in liberation of AA for eicosanoid synthesis in platelets has not been fully explored. If this enzyme is present in the platelets of all animals, it may represent a class of calcium-dependent cytosolic $PLA_2$s that are activated by agonist stimulation via a different route (i.e., direct G protein activation) than the 85-kDa class of $PLA_2$s. This is an area that will require further investigation.

## 2. Other Routes of Arachidonic Acid Liberation

The cleavage of phospholipids by $PLA_1$ followed by cleavage by lysophospholipase to free AA or sequential cleavage of phospholipids with PLC followed by the release of AA by diacylglycerol lipase have also been proposed as possible routes for agonist-induced liberation of AA. However, since most of the agonist-stimulated AA release in platelets was inhibited by specific inhibitors of 85-kDa $cPLA_2$, it seems unlikely that they contribute to liberation of AA (RIENDEAU et al. 1994; BARTOLI et al. 1994; FAILI et al. 1994).

# C. Eicosanoids Affecting Platelet Function

## I. Thromboxane $A_2$

### 1. Introduction

Because $TXA_2$ has a half-life of approximately 30s, most studies of its actions have used stable analogues, such as I-BOP (MORINELLI et al. 1989), U-46619 (SMITH et al. 1977), and STA2(ONO 11113; DAVIS-BRUNO and HALUSHKA 1994). $TXA_2$ is a potent stimulus for smooth muscle contraction and proliferation (HALUSHKA and MAIS 1989). In platelets, $TXA_2$ causes shape change, aggregation, secretion, and exposure of fibrinogen binding sites (ARITA et al. 1989). It also acts as an amplification system for weaker agonists.

#### *a) Thromboxane $A_2$ Receptors*

$TXA_2$ and its immediate precursor prostaglandin $H_2$ produce similar effects in platelets and are thought to act through a common receptor (HALUSHKA et al. 1987, 1989; MACINTYRE et al. 1987). This receptor, known as the TP receptor, has been cloned from human placenta (HIRATA et al. 1991), K562 cells (D'ANGELO et al. 1994), mouse lung (NAMBA et al. 1992), and endothelial cells (RAYCHOWDHURY et al. 1994; see Chap. 7B for more details). The receptor belongs to the heptahelical family of G-protein-coupled receptors. The platelet $TXA_2$ receptor is thought to couple to $G_q$, $G_{12}$, $G_{13}$, and $G_{i2}$ (SHENKER et al. 1991; KNEZEVIC et al. 1993; NAKAHATA et al. 1995; OFFERMANNS et al. 1994; USHIKUBI et al. 1994). $G_q$ activates PLC, leading ultimately to an increased intracellular calcium concentration that is thought to play a role in platelet aggregation. $G_{12}$ and/or $G_{13}$ may be involved in platelet shape change because they are associated with an increase in intracellular pH (DHANASEKARAN et al. 1994; VOYNO-YASENETSKAYA et al. 1994), an early event in platelet activation. $G_{i2}$ was shown to interact with the platelet $TXA_2$ receptor in reconstituted liposomes (USHIKUBI et al. 1994), which agrees with the observation that stimulation of the $TXA_2$ receptor inhibits platelet adenylate cyclase (AVDONIN et al. 1985; BONNE et al. 1980).

There appear to be two distinct classes of binding sites for $TXA_2$ in platelets (DORN 1989). The high affinity site appears to mediate platelet shape change, adhesion, and myosin light chain phosphorylation, and a low affinity site mediates platelet aggregation and secretion (MAIS et al. 1985). The high affinity site appears to be analogous to the vascular $TXA_2$ receptor (FURCI et al. 1991). Controversy still exists whether these two sites represent two different receptor peptides or two affinity states brought about by differential coupling to G proteins (REILLY and FITZGERALD 1993).

### b) *Signal Transduction*

#### $\alpha$) Intracellular Calcium

$TXA_2$ analogues cause an increase in intracellular free calcium concentrations (ARITA et al. 1989). Activation of phospholipase C (PLC) alone cannot account for the entire increase in calcium concentration, and a PLC-independent mechanism(s) is also operative (ARITA et al. 1989). The source of the calcium may be extracellular or one of the intracellular pools. It has been proposed that a calcium channel coupled to the $TXA_2$ receptor may be involved in influx of extracellular calcium. This channel may be coupled to the receptor through a G protein. MAYEUX et al. (1991) showed that dihydropyridine (DHP) calcium channel agonists/antagonists directly interact with $TXA_2$ receptors and competitively displace I-BOP and I-PTA-OH. The significance of these observations is unknown. Picotamide, a dual TX synthase inhibitor/TX receptor antagonist, reduced the PLC activation and increase in $Ca^{2+}$ levels in human platelets (PULCINELLI et al. 1994). In dog platelets, dysfunction of $TXA_2$ receptor-linked G protein led to impaired activation of PLC in response to U46619 (JOHNSON et al. 1993). Recently, BRÜNE et al. (1994) depleted intracellular platelet calcium using Thapsigargin (TG) and 2,5-di-(t-butyl)-1,4-benzohydroquinone (tBuBHQ) and chelated the extracellular calcium using EGTA. They showed that sustained occupancy of the receptor by U46619 was necessary to cause $Ca^{2+}$ influx from extracellular sources upon restoration of extracellular calcium. Also, U46619 potentiated TG-induced calcium transients, and this was blocked by BM13177, a $TXA_2$ receptor antagonist. They explained this by assuming the existence of two pools of intracellular calcium, one sensitive to TG and the other insensitive to TG, but sensitive to agonists and cycling of calcium between the pools. The signalling mechanism that regulates this remains unknown.

#### $\beta$) Alkalinization of Intraplatelet pH

SIFFERT et al. (1990) showed that U46619 modulates the activity of the $Na^+/H^+$ exchanger as an early event and that more than half of the increase in intracellular calcium depended upon the $Na^+/H^+$ exchanger. It is known that amiloride, which blocks the $Na^+/H^+$ exchanger, inhibits platelet aggregation (SIFFERT et al. 1986). As mentioned earlier, the $TXA_2$ receptor in platelets is coupled to $G_{12}$ and $G_{13}$, which activate the $Na^+/H^+$ exchanger. Recently in rat platelets, YOKOYAMA et al. (1994) showed that the $HCO_3^-/Cl^-$-exchanger may be involved in platelet aggregation induced by U46619.

#### $\gamma$) Protein Kinase C Activation

Activation of PLC leads to the liberation of DAG and IP3, and DAG activates protein kinase C (PKC). PKC was shown to induce the phosphorylation of a

47-kDa protein in platelets (for a review see ARITA et al. 1989). Staurosporine, an inhibitor of PKC, suppresses the aggregation of platelets. Recently, this was shown to be due also to the inhibition of myosin light chain kinase (MLCK) (TAKANO 1994). TURINI et al. (1993) showed that chelerythene and calphostin C, two inhibitors of PKC, decreased aggregation in response to U46619 and that the effect of the inhibitors correlated with the dephosphorylation of a 68-kDa protein that is extensively phosphorylated in the resting state. PKC is also thought to be involved in the exposure of fibrinogen receptors (SHATTIL and BRASS 1987).

$\delta$) Tyrosine Phosphorylation

Many agonists including thrombin and $TXA_2$ have been shown to increase tyrosine phosphorylation in activated platelets (ARITA et al. 1989). MAEDA et al. (1993) showed that $p72^{syk}$ is activated by U46619. $p125^{FAK}$ was shown to be activated by U46619 acting as a costimulus with integrin (SHATTIL et al. 1994). Integrin $\alpha$IIb-$\beta$-III (glycoprotein IIb/IIIa) is known to be involved in platelet shape change and aggregation, fibrin clot retraction, and signal transduction and is regulated by tyrosine phosphorylation (SCHOENWAELDER et al. 1994). Thus, tyrosine phosphorylation is important in platelet activation. Vanadate, which inhibits protein tyrosine phosphatases, was shown to activate platelets (McNICOL et al. 1993). An inhibitor of tyrosine kinases, genistein, inhibited platelet responses to $TXA_2$, while another inhibitor, Herbimycin A, did not (NAKASHIMA et al. 1990). The effect of genistein may be due to its binding to the $TXA_2$ receptor (McNICOL 1993).

$\varepsilon$) Myosin Light Chain Kinase

Phosphorylation of myosin light chain (MLC) is an early event preceding shape change. However, HASHIMOTO et al. (1994) showed that wortmannin, which completely inhibits MLCK, had no effect on the shape change response to ADP or U46619. Conversely, wortmannin dose-dependently inhibited ADP/U46619-induced platelet aggregation. Thus, MLC may be involved in exposure of fibrinogen receptors.

## II. Prostacyclin ($PGI_2$, Epoprostanol)

Prostacyclin ($PGI_2$; for reviews see KERINS et al. 1991; VANE and BOTTING 1995) is a metabolite of arachidonic acid produced mainly by the vascular endothelium but also by subendothelial smooth muscle. Prostacyclin inhibits platelet aggregation and can induce platelet disaggregation. It is the most potent inhibitor of platelet aggregation produced in vivo and also a smooth muscle relaxant of both vascular and bronchial smooth muscle (MONCADA and VANE 1979). Recently, RAGAZZI et al. (1995) reported on an analogue of prostacyclin, (+/–)(5E)-13,14-didehydro-$\omega$-hexanor (1-hydroxycyclohexyl)-9$\alpha$-carba-prostacyclin, that has antiaggregative activity but not smooth muscle relaxing activity. This is an exciting observation since it raises the possibility that there are subtypes of $PGI_2$ receptors.

$PGI_2$ and $PGE_1$ (a cyclooxygenase metabolite of dihomo-$\gamma$-linolenic acid, DGLA) share a common receptor (HALL and STRANGE 1984; RUCKER and SCHRÖR 1983) that activates adenylate cyclase through Gs. $PGI_2$ has also been reported to increase intracellular calcium (VASSAUX et al. 1993) at higher concentrations. NAKAGAWA et al. (1994) cloned the cDNA for the $PGI_2$ receptor from a human lung cDNA library and expressed the receptor, a 386-amino-acid peptide with seven hydrophobic segments that presumably span the cell membrane.

## III. 8-Epi-prostaglandin $F_{2\alpha}$

8-Epi-prostaglandin $F_{2\alpha}$ and other isoprostanes (MORROW et al. 1990) are produced nonenzymatically from arachidonic acid via free-radical-catalyzed peroxidation. PRATICO et al. (1994) suggested that in platelets 8-epi-prostaglandin $F_{2\alpha}$ may be formed in a cyclooxygenase-dependent manner. 8-Epi-prostaglandin $F_{2\alpha}$ has been shown to be produced in vivo (MORROW et al. 1994) and to cause vasoconstriction in renal (MORROW et al. 1990) and in pulmonary (BANERJEE et al. 1992) vascular beds. SQ 29548, a thromboxane $A_2$ receptor antagonist, blocks 8-Epi-prostaglandin-$F_{2\alpha}$-induced renal vasoconstriction (TAKABASHI et al. 1992). In platelets, 8-Epi-prostaglandin $F_{2\alpha}$ blocks platelet aggregation induced by $TXA_2$ mimetics U46619 and I-BOP (MORROW et al. 1992b). Recently, YIN et al. (1994) showed that 8-Epi-prostaglandin $F_{2\alpha}$ induced shape change via the $TXA_2$ receptor and that it had no effect on platelet aggregation induced by other agonists such as high concentrations of ADP. In contrast, it dose-dependently potentiated the platelet aggregation induced by low concentrations of ADP.

## IV. Prostaglandin $D_2$ ($PGD_2$)

$PGD_2$ is derived from $PGH_2$ by the action of $PGH_2/PGD_2$ isomerase. It is a potent inhibitor of platelet activation via increases in cAMP levels (MILLS and MACFARLANE 1977; WHITTLE et al. 1978, 1985). At high concentrations it acts as a weak agonist at the $TXA_2$ receptor (HAMID-BLOOMFIELD and WHITTLE 1986). It is also reported to interact with the $PGI_2$ receptor (SCHAFER et al. 1979; SIEGL 1982; SIEGL et al. 1979a,b).

## V. 12-HETE (12-Hydroxyeicosatetraenoic Acid)

12-HETE is a product of arachidonic acid metabolism by the 12-lipoxygenase pathway and is the dominant product of this pathway in platelets. 12-HETE potentiates thrombin-induced platelet aggregation (SEKIYA et al. 1991), presumably by inhibiting cAMP formation. SEKIYA et al. proposed that 12-HETE plays a role in the switching of the aggregation response from the primary agonists (e.g., collagen) to secondary agonists (e.g., thrombin). However, much remains to be elucidated regarding the role of 12-HETE. In contrast, FONLUPT et al. (1991) found that 12-HETE inhibits $TXA_2$ agonist-induced platelet aggregation by binding to the $TXA_2$ receptor.

# D. Alterations in Thromboxane $A_2$ Synthesis in Disease States

## I. Diabetes Mellitus

Increased platelet activation or aggregability has been reported in patients with diabetes mellitus (for a review see WINOCOUR 1994). There is some controversy about whether this is a primary or secondary event, e.g., secondary to vascular diseases. The mechanisms responsible for the increased platelet activation are uncertain. The discovery that metabolism of arachidonic acid by platelets played an important role in platelet aggregation raised the possibility that increased arachidonic acid metabolism could contribute to the enhanced platelet aggregation associated with diabetes mellitus. This notion was further supported by the observation that aspirin decreased the enhanced second phase of platelet aggregation (COLWELL et al. 1978; SAGEL et al. 1975). Early support for this notion came when HALUSHKA et al. (1977) found increased platelet prostaglandin E production in diabetic subjects.

With the discovery of $TXA_2$, it became clear that this was the platelet arachidonic metabolite that had to be measured. It was then shown that patients with insulin-dependent and non-insulin-dependent diabetes mellitus had increased platelet $TXA_2$ synthesis (BUTKUS et al. 1980; DAVI et al. 1990; HALUSHKA et al. 1981b; ZIBOH et al. 1979). MAYFIELD et al. (1985) demonstrated that with improved metabolic control via intensive insulin therapy, there was a significant decrease in ex vivo platelet arachidonic-acid-induced $TXA_2$ synthesis. They also found a significant correlation between the level of platelet $TXA_2$ synthesis and the hemoglobin $A_{1c}$ level as well as a significant negative correlation with the low-density lipoprotein cholesterol level. It has been shown by several groups that elevated plasma cholesterol levels or incubation of platelets with cholesterol can increase ex vivo platelet $TXA_2$ synthesis (DAVI et al. 1992; STUART et al. 1980; TREMOLI et al. 1979). Treatment of patients with a cholesterol lowering agent also decreased platelet $TXA_2$ synthesis (DAVI et al. 1992). Thus, this could be one of the metabolic factors that contributes to an increased synthesis of $TXA_2$ by platelets in patients with diabetes mellitus. Ex vivo platelet $TXA_2$ synthesis does not necessarily predict the level of $TXA_2$ synthesis in vivo.

In order to determine if the increased ex vivo platelet $TXA_2$ synthesis was occurring in vivo, ALESSANDRINI et al. (1988) measured the main urinary metabolites of $TXA_2$, 2,3-dinor-$TXB_2$ and 11-dehydro-$TXB_2$, in insulin-dependent diabetic subjects. They found that there was no significant difference in the urinary excretion of these metabolites in these patients compared to age-matched control subjects. The reason why they did not see the anticipated elevation in the excretion of these urinary metabolites in the diabetic subjects is not known. It is also of interest to note that many of their patients had retinopathy and they also did not show any significant increases in the urinary excretion of the metabolites. DAVI et al. (1990) studied the urinary excretion of 2,3-dinor-$TXB_2$ in patients with non-insulin-dependent diabetes mellitus and macrovascular disease. They found a significant increase in the urinary excretion of the metabolite in the patients compared to the control subjects. Of particular interest was the observation that improvement in metabolic control with intensive insulin therapy resulted in a significant decrease in the excretion of the metabolites. They concluded that the increased in vivo synthesis was coming from platelets. They also concluded that the increased synthesis of $TXA_2$ was not due solely to macrovascular disease since the increased synthesis was decreased by metabolic control and furthermore that there was no evidence of patients with macrovascular disease without diabetes having increased platelet $TXA_2$ synthesis. KATAYAMA et al. (1987) also found a significant increase in the urinary excretion of 2,3-dinor-$TXB_2$ in non-insulin-dependent diabetic patients compared to control subjects. Collectively, these studies have demonstrated that there is increased platelet $TXA_2$ synthesis in patients with diabetes mellitus. The mechanism for the increased synthesis is uncertain and is almost certainly multifactorial. It does not appear to be due to elevations in glucose concentrations (BEST et al. 1979; HALUSHKA et al. 1981a). It is most likely due to a series of metabolic alterations seen in patients with diabetes.

## II. Acute Coronary Artery Syndromes

In animal models of acute coronary artery occlusion, there is increased synthesis of $TXA_2$ (for reviews see DAVIS-BRUNO and HALUSHKA 1994; FITZGERALD and FITZGERALD 1988; MORINELLI and HALUSHKA 1991). Unstable angina pectoris is associated with intermittent intracoronary artery platelet aggregation. FITZGERALD et al. (1986) measured the excretion of 2,3-dinor-$TXB_2$ in patients with unstable angina and acute myocardial infarction. They found significant increases in the excretion of the metabolite in both groups of patients compared to control subjects. However, the levels were significantly greater in the unstable angina patients compared to the acute myocardial infarction patients. They were also able to demonstrate that the increased synthesis was of platelet origin. FOEGH et al. (1994), when measuring the two major urinary metabolites of $TXA_2$, found similar results; however, they found greater increases in patients with acute myocardial infarction than in the unstable angina patients. They did not find any gender-related differences in the excretion rates. It appears that exercise-induced increases in the urinary excretion of 2,3-dinor-$TXB_2$ in postmyocardial infarction patients may have prognostic value (RASMANIS et al. 1995). Patients with a 30% or greater increase in the excretion of 2,3-dinor-$TXB_2$ had a worse 3-year prognosis than those patients who did not have such a magnitude of increase. This data is complementary to the study showing that enhanced platelet aggregation was also a poor prognostic sign (TRIP et al. 1990).

Two other clinical conditions in which there is platelet activation and increased platelet $TXA_2$ synthesis are thrombolysis therapy and percutaneous transluminal coronary artery angioplasty. FITZGERALD et al. (1988) found markedly increased excretion of 2,3-dinor-$TXB_2$ in the patients undergoing acute thrombolysis therapy with streptokinase. The increased excretion was not due to release of $TXA_2$ trapped in the coronary circulation; rather, it was due to increased platelet $TXA_2$ synthesis stimulated by streptokinase. These studies have provided the rationale for the use of aspirin in the treatment of these cardiovascular diseases (vide infra; for a review see PATRONO 1994).

## III. Pregnancy-Induced Hypertension

Pregnancy is often associated with increased platelet aggregability (ROMERO and DUFFY 1980). Although somewhat controversial, pregnancy-induced hypertension is often associated with significantly increased platelet activation in vivo often resulting in thrombocytopenia (for a review see ZAHRADNIK et al. 1991). FITZGERALD et al. (1990) measured the urinary excretion of the two major metabolites of $TXA_2$ in patients with preeclampsia compared to normal pregnant women and found that they were significantly increased in the preeclamptic patients. In addition, they found that there was a negative correlation between the level of the metabolites and the circulating-platelet counts. They were also able to conclude that the increased synthesis was coming from platelets. This is further supported indirectly by the results reported by YLIKORKALA et al. (1986) who measured urinary $TXB_2$ excretion which derives almost exclusively from renal sources. They found that excretion of $TXB_2$ was not significantly different in preeclamptic patients compared to normal pregnant women. Platelet $TXA_2$ synthesis is also increased in normal pregnancies (FITZGERALD et al. 1987). The mechanism for the increased platelet $TXA_2$ synthesis is unknown. It may in part be due to increased turnover or stimulation by other circulating factors that are increased in patients with preeclampsia. It is also uncertain which step(s) is altered in the synthesis of $TXA_2$ that results in the enhanced production.

With regard to possible mechanisms contributing to the enhanced platelet aggregability that occurs in pregnancy and particularly preeclampsia, the synthesis of prostacyclin by the endothelium is reported to be significantly decreased in these patients compared to normal pregnant women (MINUZ et al. 1988; YLIKORKALA et al. 1986; for a review see ZAHRADNIK et al. 1991).

These studies provide the rationale for the use of aspirin in the treatment of preeclampsia (GROUP 1993; SCHIFF et al. 1989; WALLENBURG et al. 1986). These results also raise the potential for the use of $TXA_2$ synthase inhibitors in the treatment of preeclampsia (KEITH et al. 1993; VAN ASSCHE et al. 1984). The latter would be particularly appealing because in theory it would reverse the biochemical abnormalities associated with preeclampsia, i.e., decreased prostacyclin formation and increased $TXA_2$ formation.

## IV. Cerebral Ischemia

Thrombotic strokes would also be expected to be associated with increased platelet $TXA_2$ synthesis by virtue of the fact that platelets play an important role in the pathogenesis of strokes, particularly transient ischemic attacks (TIAs). FISHER and ZIPSER (1985) measured the urinary excretion of $TXB_2$ in stroke patients. They found significant elevations of the levels in patients with strokes, with the highest levels in those with the most severe strokes. When they analyzed the data based on gender, they found that the elevations occurred mainly in men and that the women did not show any significant elevations. These are potentially very interesting studies if in fact female patients do not have the same elevations of $TXA_2$ synthesis as men. The authors measured urinary $TXB_2$ which may originate from the kidneys as well as systemic sites. Thus, one cannot be sure as to the origin of the $TXB_2$ in the urine of these patients. If, however, a gender difference is discovered in the platelet synthesis of $TXA_2$ in future studies, it may help to resolve the current controversy about whether or not aspirin prophylaxis is beneficial to women (for a review see PATRONO 1994).

## V. Homocystinuria

Homocystinuria is a recessively inherited disorder associated with premature atherosclerosis and arterial and venous thrombosis. DIMINNO et al. (1993) measured the urinary excretion of 11-dehydro-$TXB_2$ in patients homozygous for homocystinuria and found it to be significantly increased. They were able to demonstrate that the platelet was the source of the increased production of $TXA_2$ in vivo. Studies of the platelets ex vivo failed to demonstrate any increased synthesis rates or aggregatory responses to aggregating agents. Treatment of the patients with the antioxidant drug probucol resulted in a significant decrease in the excretion of 11-dehydro-$TXB_2$, while it had no effect on control subjects. Since homocysteine may increase the metabolism of arachidonic acid, the elevated levels may directly be responsible for the increased synthesis of $TXA_2$ and the specificity of probucol's effect only in the patients.

## VI. Sickle Cell Disease

Intravascular thrombotic events represents a major cause of morbidity and mortality in homozygous sickle cell disease. FOULON et al. (1993) measured the urinary excretion of 11-dehydro-$TXB_2$ in patients and control subjects and found that its excretion was significantly increased in the patients in remission compared to controls. During an acute exacerbation of the disease, urinary excretion was significantly increased compared to the levels found during remission. FOULON et al. also studied the aggregation response of the platelets ex vivo and found that the response to the $TXA_2$ mimetic U46619 was decreased compared to the controls. In contrast, the response to thrombin was not significantly different. They concluded that there is increased platelet activation in vivo and that the platelet $TXA_2$ receptors may be desensitized. The increased synthesis of $TXA_2$ by platelets in vivo is almost certainly secondary to activation of the platelets by other aggregating agents.

### VII. Effects of Fish Oils

Ingestion of fish oils, which are a rich source of $\omega$-3 fatty acids, has been shown to decrease the synthesis of platelet $TXA_2$ (KNAPP et al. 1986). The mechanism for this appears to be a combination of eicosapentaenoic acid acting as a competitive inhibitor of the synthesis of $TXA_2$ and also being converted to $TXA_3$, which appears to be inactive. In addition to their effects on $TXA_2$ synthesis, eicosapentaenoic and docosahexaenoic acid decrease the responsiveness of $TXA_2$ receptors (PARENT et al. 1992; SCHEURLEN et al. 1993). PARENT et al. (1992) have demonstrated that these two fatty acids can specifically modify the binding of [$^3$H]U46619, a $TXA_2$ agonist, and [$^3$H]SQ29548, a $TXA_2$ antagonist, to platelet $TXA_2$ receptors. These studies provide further insights into the possible mechanisms of the beneficial effects of fish oils in cardiovascular diseases.

## E. Alterations in Thromboxane $A_2$ Receptors in Disease States

### I. Acute Coronary Artery Syndromes

During acute myocardial infarction, there is increased platelet aggregability in response to a variety of agents when tested ex vivo (for a review see DAVIS-BRUNO and HALUSHKA 1994). MEHTA et al. (1980) found that platelets from patients with angina pectoris were more sensitive to the aggregating effects of U46619, a $TXA_2$ mimetic, compared to controls. Collectively, these observations raised the possibilities that in patients experiencing an acute myocardial infarction or unstable angina pectoris either their platelet $TXA_2$ receptor density or affinity were increased or they had enhanced signalling mechanisms. DORN et al. (1990) found that platelet $TXA_2$ receptor density, but not affinity, was increased in patients experiencing an acute myocardial infarction compared to the appropriate controls and that during the convalescent phase their receptor densities reverted to levels not significantly different from the control group and were also significantly lower than during the acute attack period. They also demonstrated increased sensitivity of the platelets ex vivo to a $TXA_2$ mimetic in the patients compared to the control groups. MODESTI et al. (1995)

confirmed and extended these observations. They found that in vitro exposure of platelets to thrombin significantly increased $TXA_2$ receptor density. Since thrombin is activated during acute myocardial infarction, they postulated that it may be responsible for the increase in platelet $TXA_2$ receptor density. They demonstrated that incubation of platelets in vitro with thrombin lead to a significant increase in receptor density. Another possible explanation for the increase in receptor density is that platelet size is increased during the acute event and that the increased surface area could allow more receptors to be exposed, accounting for the apparent increase in receptor density. They provided some evidence against this possibility, but it has not been excluded. In spite of the fact that the mechanism for the increase in receptor density has not been fully elucidated, it is nonetheless clear that receptor density increases in patients with acute myocardial infarction and represents an important contributing biochemical event in the pathogenesis of acute myocardial infarction. It should also be noted that it is paradoxical that there should be an increase in platelet $TXA_2$ receptor density in the face of marked increases in platelet $TXA_2$ synthesis. Receptor theory would predict a desensitization and/or down-regulation of the receptor (Liel et al. 1988; Murray and Fitzgerald 1989; Okwu et al. 1992).

## II. Pregnancy-Induced Hypertension

Although somewhat controversial, pregnancy-induced hypertension (PIH) appears to be associated with increased platelet turnover and enhanced platelet aggregability (Keith et al. 1993; Stubbs et al. 1986). Liel et al. (1993) measured platelet $TXA_2$ receptor density in normal pregnant women and patients with PIH. They found a significant increase in receptor density in the patients with PIH, with the most severe cases having the greatest increases. Platelets obtained from the patients with PIH were also more sensitive to a $TXA_2$ mimetic ex vivo compared to the normal pregnant subjects. The finding of increased $TXA_2$ receptor density in the PIH patients like that seen in the patients with acute myocardial infarction is paradoxical because PIH is associated with an increase in platelet $TXA_2$ synthesis (vide supra). The mechanism of increased receptor density is currently unknown but is clearly worthy of further investigation.

## III. Diabetes Mellitus

Collier et al. (1986) reported that platelets obtained from diabetic patients with proliferative retinopathy demonstrated enhanced sensitivity ex vivo to a $TXA_2$ mimetic. This raised the possibility that there was increased $TXA_2$ receptor density or affinity in patients with diabetes mellitus. Two studies in diabetic patients have resulted in conflicting results. Modesti et al. (1991) found a decreased number of receptors in diabetics' platelets compared to controls. None of these patients had proliferative retinopathy. Jaschonek et al. (1989) did not find any significant differences in platelet $TXA_2$ receptor density among controls and diabetic patients with or without retinopathy. In support of these latter findings, MORINELLI et al. (1993) did not find any alterations in platelet $TXA_2$ receptor affinity or density in rats made diabetic with streptozotocin. Thus, the current evidence would support the notion that platelet $TXA_2$ receptor density is not increased in diabetic subjects. Thus, the enhanced platelet aggregability that occurs in patients with diabetes mellitus may be due in part to increased $TXA_2$ synthesis and perhaps alterations in receptor signalling mechanisms.

### IV. Regulation by Androgenic Steroids

In the past several years there has been a series of reports of strokes or heart attacks in young male athletes abusing anabolic steroids (for a review see ROCKHOLD 1993). It is believed that these events are due to either increased platelet aggregation or vascular spasm. There is some evidence that there may be enhanced ex vivo platelet aggregability in some of the subjects abusing steroids (FERENCHICK et al. 1992). MATSUDA et al. (1994a) found that incubation with testosterone increased human-erythroleukemia-cell (a megakaryocyte-like cell) $TXA_2$ receptors. They also found that treatment of rats with testosterone also increased both platelet and aortic $TXA_2$ receptors (MATSUDA et al. 1994b). The increase in receptor density was associated with an increased responsiveness to a $TXA_2$ mimetic. AJAYI et al. (1995) treated normal male volunteers with testosterone and found that there was a twofold increase in platelet $TXA_2$ receptor density and an increased response to a $TXA_2$ mimetic. These studies raise the possibility that increased $TXA_2$ receptor density may contribute to the cardiovascular risk associated with steroid abuse and perhaps to the gender-related differences in cardiovascular disease.

## F. Aspirin, Thromboxane Synthase Inhibitors, and Thromboxane $A_2$ Receptor Antagonists in Coronary Artery Disease

Long before it was recognized that aspirin inhibits fatty acid cyclooxygenase and that metabolism of arachidonic acid to labile aggregation-stimulating substances leads to platelet aggregation, it was proposed by L.L. Craven, a physician practicing in Glendale, California (for a review see MUELLER and SCHEIDT), that aspirin might be beneficial in primary and secondary prophylaxis against acute myocardial infarctions. In a set of prescient uncontrolled clinical studies, he proposed that aspirin prevented acute myocardial infarctions and also appeared to be beneficial for secondary prophylaxis. Numerous clinical trials have subsequently shown that aspirin clearly has a role in secondary prophylaxis (for a review see PATRONO 1994). The rationale for its beneficial actions is now clear since it has been shown that there is an increase in platelet $TXA_2$ synthesis and $TXA_2$ receptor density in patients with acute coronary artery syndromes.

It was originally predicted that TX synthase inhibitors would be good antiplatelet agents by virtue of inhibiting platelet TX synthesis and causing shunting of prostaglandin $H_2$ to prostacyclin in vascular endothelial cells. However, it was subsequently shown that TX synthase inhibitors did not block platelet aggregation because prostaglandin $H_2$ was sufficient to cause platelet aggregation (HEPTINSTALL et al. 1980; GRIMM et al. 1981). Thus, attention has turned to either $TXA_2$ receptor antagonists or the combination of a TX-synthase inhibitor/receptor antagonist in a single molecule. $TXA_2$ receptor antagonists have been shown to be effective antiplatelet agents because they block the actions of both $TXA_2$ and prostaglandin $H_2$ at the receptor (for a review see DAVIS-BRUNO and HALUSHKA 1994). Recently, a dual inhibitor, ridogrel, was found to be equieffective with aspirin as an adjunct to thrombolysis therapy (INVESTIGATORS RVAPT 1994). However, it may have been more effective than aspirin in reducing new ischemic events. Unless the receptor antagonists or combined receptor antagonist/synthesis inhibitors prove to be markedly better than aspirin in head-to-head trials, it is unlikely that they will supplant aspirin since their cost would be far greater than that of aspirin.

## References


Ajayi AAL, Mathur R, Halushka PV (1995) Testosterone increases human platelet thromboxane $A_2$ receptor density and aggregation responses. Circulation 91:2742–2747

Alessandrini P, McRae J, Feman S, Fitzgerald GA (1988) Thromboxane biosynthesis and platelet function in type 1 diabetes mellitus. New Engl J Med 319:208–212

Arita H, Nakano T, Hanasaki K (1989) Thromboxane $A_2$: its generation and role in platelet activation. Prog Lipid Res 28:273–301

Avdonin PV, Svitina-Vlitini IV, Leytin VL, Tkachuk VA (1985) Interaction of stable prostaglandin endoperoxide analogs U46619 and U44069 with human platelet membranes: coupling of receptors with high-affinity GTPase and adenylate cyclase. Thromb Res 40:101–112

Banerjee M, Kang KH, Morrow JD, Roberts LJ, Newman JH (1992) Effects of a novel prostaglandin, 8-epi-$PGF_{2\alpha}$ in rabbit lung in situ. Am J Physiol 263:H660–H663

Bartoli F, Lin H-K, Ghomashchi F, Gelb MH, Jain MK, Apitz-Castro R (1994) Tight binding inhibitors of 85-kDa phospholipase $A_2$ but not 14-kDa phospholipase $A_2$ inhibit release of free arachidonate in thrombin-stimulated human platelets. J Biol Chem 269:15625–15630

Best L, Jones PBB, Preston FE (1979) Effect of glucose on platelet thromboxane biosynthesis. Lancer II:790

Bettazzoli L, Zirrolli JA, Reidhead CT, Shahgholi M, Murphy RC (1990) Incorporation of arachidonic acid into glycerophospholipids of a murine bone marrow derived mast cell. Adv Prostaglandin Thromboxane Leukot Res 20:71–78

Bonne C, Martin B, Regnault F (1980) The cyclic AMP-lowering effect of the stable endoperoxide analog U46619 in human platelets. Thromb Res 20:701–704

Broekman MJ, Handin RI, Derksen A, Cohen P (1976) Distribution of phospholipids, fatty acids, and platelet factor 3 activity among subcellular fractions of human platelets. Blood 6:963–971

Brüne B, von Appen F, Ullrich V (1994) Receptor occupancy regulates $Ca^{2+}$ entry and intracellular $Ca^{2+}$ redistribution in activated human platelets. Biochem J 304:993–999

Butkus A, Skrinska VA, Schumacher OP (1980) Thromboxane production and platelet aggregation in diabetic subjects with clinical complications. Thromb Res 19:211–223

Cohen P, Derksen A (1969) Comparison of phospholipid and fatty acid composition of human erythrocytes and platelets. Br J Haematol 17:359–371

Collier A, Tymkewycz P, Armstrong R, Young RJ, Jones RL, Clarke BF (1986) Increased platelet thromboxane receptor sensitivity in diabetic patients with proliferative retinopathy. Diabetologia 29:471–474

Colwell JA, Halushka PV, Sarji KE, Sagel J (1978) Platelet function in diabetes mellitus. Med Clin North Am 62:753–760

D'Angelo DD, Davis MG, Ali S, Dorn GW II (1994) Cloning and pharmacologic characterization of a thromboxane $A_2$ receptor from K562 (human chronic myelogenous leukemia) cells. J Pharmacol Exp Ther 271:1034–1041

Davi G, Catalano I, Averna M, Notarbartolo A, Strano A, Ciabattoni G, Patrono C (1990) Thromboxane biosynthesis and platelet function in type II diabetes mellitus. N Engl J Med 322:1769–1774

Davi G, Averna M, Catalano I, Barbagallo C, Ganci A, Notarbartolo A, Ciabattoni G, Patrono C (1992) Increased thromboxane biosynthesis in type IIa hypercholesterolemia. Circulation 85:1792–1798

Davis-Bruno KL, Halushka PV (1994) Molecular pharmacology and therapeutic potential of thromboxane $A_2$ receptor antagonists. Adv Drug Res 25:173–202

Dennis EA (1987) Regulation of eicosanoid production: role of phospholipases and inhibitors. Biotechology 5:1294–1300

Dhanasekaran N, Vara Prasad MVVS, Wadsworth SJ, Dermott JM, van Rossum G (1994) Protein kinase C-dependent and -independent activation of $Na^+/H^+$ exchanger by $G\alpha_{12}$ class of G proteins. J Biol Chem 269:11802–11806

Di Minno G, Davi G, Margaglione M, Cirillo F, Grandone E, Ciabattoni G, Catalano I, Strisciuglio P, Andria G, Patrono C, Mancini M (1993) Abnormally high thromboxane biosynthesis in homozygous homocystinuria – evidence for platelet involvement and probucol-sensitive mechanism. J Clin Invest 92:1400–1406

Dorn GW (1989) Distinct platelet thromboxane $A_2$/prostaglandin $H_2$ receptor subtypes. J Clin Invest 84:1883–1891

Dorn GW II, Liel N, Trask JL, Mais DE, Assey ME, Halushka PV (1990) Increased platelet thromboxane $A_2$/prostaglandin $H_2$ receptors in patients with acute myocardial infarction. Circulation 81:212–218

Faili A, Emadi S, Vargaftig BB, Hatmi M (1994) Dissociation between the phospholipases C aud $A_2$ activities in stimulated platelets and their involvement in arachidonic acid liberation. Br J Haematol 88:149–155

Ferenchick G, Schwartz D, Ball M, Schwartz K (1992) Androgenic-anabolic steroid abuse and platelet aggregation: a pilot study in weight lifters. Am J Med Sci 303:78–82

Fisher M, Zipser R (1985) Increased excretion of immunoreactive thromboxane $B_2$ in cerebral ischemia. Stroke 16:10–13

Fitzgerald DJ, Fitzgerald GA (1988) Eicosanoids in myocardial ischemia and injury. In: Halushka PV, Mais DE (eds) Advances in eicosanoid research. MTP Press, Norwell, pp 128–158

Fitzgerald DJ, Roy L, Catella F, Fitzgerald GA (1986) Platelet activation in unstable coronary disease. N Engl J Med 315:983–988

Fitzgerald DJ, Mayo G, Catella F, Entman SS, Fitzgerald GA (1987) Increased thromboxane biosynthesis in normal pregnancy is mainly derived from platelets. Am J Obstet Gynecol 157:325–330

Fitzgerald DJ, Catella F, Roy L, Fitzgerald GA (1988) Marked platelet activation in vivo after intravenous streptokinase in patients with acute myocardial infarction. Circulation 77:142–150
Fitzgerald DJ, Rocki W, Murray R, Mayo G, Fitzgerald GA (1990) Thromboxane $A_2$ synthesis in pregnancy-induced hypertension. Lancet I:751–754
Fitzgerald GA, Reilly IA, Pedersen AK (1985) The biochemical pharmacology of thromboxane synthase inhibition in man. Circulation 72:1194–1201
Foegh ML, Zhao Y, Madren L, Rolnick M, Stair TO, Huang KS, Ramwell PW (1994) Urinary thromboxane $A_2$ metabolites in patients presenting in the emergency room with acute chest pain. J Intern Med 235:153–161
Fonlupt P, Croset M, Lagarde M (1991) 12-HETE inhibits the binding of $PGH_2/TXA_2$ receptor ligands in human platelets. Thromb Res 63:239–248
Foulon I, Bachir D, Galacteros F, Maclouf J (1993) Increased in vivo production of thromboxane in patients with sickle cell disease is accompanied by an impairment of platelet functions to the thromboxane $A_2$ agonist U46619. Arterioscler Thromb 13:421–426
Furci L, Fitzgerald, DF, Fitzgerald GA (1991) Heterogeneity of prostaglandin $H_2$/thromboxane $A_2$ receptors: distinct subtypes mediate vascular smooth muscle contraction and platelet aggregation. J Pharmacol Exp Ther 258:74–81
Grimm LJ, Knapp DR, Senator D, Halushka PV (1981) Inhibition of platelet thromboxane synthesis by 7-(I-imidazolyl)heptanoic acid; dissociation from inhibition of aggregation. Thromb Res 24:307–317
Group, Italian Study of Aspirin in Pregnancy Study (1993) Low-dose aspirin in prevention and treatment of intrauterine growth retardation and pregnancy-induced hypertension. Lancet 341:396–400
Habenicht AJR, Salbach P, Goerig M, Zeh W, Janssen-Timmen U, Blattner C, King WC, Glomset JA (1990) The LDL receptor pathway delivers arachidonic acid for eicosanoid formation in cells stimulated by platelet-derived growth factor. Nature 345:634–636
Hall JM, Strange PG (1984) Use of a prostacyclin analogue, [$^3$H]iloprost, for studying prostacyclin-binding sites on human platelets and neuronal hybird cells. Biosci Rep 4:491–498
Halushka PV, Mais DE (1989) Basic and clinical pharmacology of thromboxane $A_2$. Drugs Today 25:383–393
Halushka PV, Lurie D, Colwell JA (1977) Increased synthesis of prostaglandin-E-like material by platelets from patients with diabetes mellitus. N Engl J Med 297:1306–1310
Halushka PV, Mayfield R, Wohltmann HJ, Rogers RC, Goldberg AK, McCoy SA, Loadholt CB, Colwell JA (1981a) Increased platelet arachidonic acid metabolism in diabetes mellitus. Diabetes 30:44–48
Halushka PV, Rogers RC, Loadholt CB, Colwell JA (1981b) Increased platelet thromboxane synthesis in diabetes mellitus. J Lab Clin Med 97:87–96
Halushka PV, Mais DE, Saussy DL Jr (1987) Platelet and vascular smooth muscle thromboxane $A_2$/prostaglandin $H_2$ receptors. Fed Proc 46:149–153
Halushka PV, Mais DE, Mayeux PR, Morinelli T (1989) Prostaglandin, thromboxane and leukotriene receptors. Annu Rev Pharmacol Toxicol 29:213–239
Hamberg M, Svensson J, Samuelsson B (1975) Thromboxanes: a new group of biologically active compounds derived from prostaglandin endoperoxides. Proc Natl Acad Sci USA 72:2994–2998
Hamid-Bloomfield S, Whittle B (1986) Prostaglandin $D_2$ interacts at thromboxane receptor sites on guinea pig platelets. Br J Pharmacol 88:931–936
Hashimoto Y, Sasaki H, Togo M, Tsukamoto K, Horie Y, Fukata H, Watanabe T, Kurokawa K (1994) Roles of myosin light-chain kinase in platelet shape change and aggregation. Biochim Biophys Acta 1223:163–169
Heptinstall S, Bevan J, Cockbill SR, Hanley SP, Parry MJ (1980) Effects of a selective inhibitor of thromboxane synthetase on human blood platelet behaviour. Thromb Res 20:219–230

Hirata M, Hayashi Y, Ushikubi F, Yokota Y, Kageyama R, Nakanishi S, Narumiya S (1991) Cloning and expression of cDNA for a human thromboxane $A_2$ receptor. Nature 349:617–620

Investigators RVAPT (1994) Randomized trial of ridogrel, a combined thromboxane $A_2$ synthase inhibitor and thromboxane $A_2$/prostaglandin endoperoxide receptor antagonist, versus aspirin as adjunct to thrombolysis in patients with acute myocardial infarction. Circulation 89:588–595

Jaschonek K, Faul C, Weisenberger H, Krönert K, Schröder H, Renn W (1989) Platelet thromboxane $A_2$/endoperoxide ($TXA_2/PGH_2$) receptors in type I diabetes mellitus. Thromb Haemost 61:535–536

Johnson GJ, Leis LA, Dunlop PC (1993) Thromboxane-insensitive dog platelets have impaired activation of phopholipase C due to receptor-linked G protein dysfunction. J Clin Invest 92:2469–2479

Kambayashi J, Kawasaki T, Tsujinaka T, Sakon M, Oshiro T, Mori T (1987) Active metabolism of phosphatidylethanolamine plasmaloogen in stimulated platelets, analyzed by high performance liquid chromatography. Biochem Int 14:241–247

Katayama S, Inaba M, Maruno Y, Omoto A, Kawazu S, Ishll J (1987) Increased thromboxane $B_2$ excretion in diabetes mellitus. J Lab Clin Med 109:711–717

Keith JJC, Spitz B, Van Assche FA (1993) Thromboxane synthetase inhibition as a new therapy for preeclampsia: animal and human studies minireview. Prostaglandins 45:3–13

Kerins DM, Murray R, Fitzgerald GA (1991) Prostacyclin and prostaglandin $E_1$: molecular mechanisms and therapeutic utility. Prog Hemost Thromb 10:307–337

Kim DK, Bonventre JV (1993) Purification of a 100 kDa phospholipase $A_2$ from spleen, lung, and kidney: antiserum raised to pig spleen phospholipase $A_2$ recognizes a similar form in bovine lung, kidney, and platelets, and immunoprecipitates phospholipase $A_2$ activity. Biochem J 294:261–270

Knapp HR, Reilly IAG, Alessandrini P, Fitzgerald GA (1986) In vivo indexes of platelet and vascular function during fish-oil administration in patients with atherosclerosis. New Engl J Med 314:937–942

Knezevic I, Borg C, Le Breton GC (1993) Identification of Gq as one of the G-proteins which copurify with human platelet thormboxane $A_2$/prostaglandin $H_2$ receptors. J Biol Chem 268:26011–26017

Kramer RM, Deykin D (1983) Arachidonoyl transacylase in human platelets. J Biol Chem 258:13806–13811

Kramer RM, Hession C, Johansen B, Hayes G, McGray P, Chow EP, Tizard R, Pepinsky RB (1989) Structure and properties of a human non-pancreatic phospholipase $A_2$. J Biol Chem 264:5768–5775

Kramer RM, Roberts EF, Manetta JV, Hyslop PA, Jakubowski JA (1993) Thrombin-induced phosphorylation and activation of $Ca^{+2}$-sensitive cytosolic phospholipase $A_2$ in human platelets. J Biol Chem 268:26796–26804

Lagarde M, Guichardant M, Menashi S, Crawford N (1982) The phospholipid and fatty acid composition of human platelet surface and intracellular membranes isolated by high voltage free flow electrophoresis. J Biol Chem 257:3100–3104

Laposata M, Reich EL, Majerus PW (1985) Arachidonoyl-CoA synthetase. J Biol Chem 260:11016–11020

Liel N, Mais DE, Halushka PV (1988) Desensitization of the platelet thromboxane $A_2$/prostaglandin $H_2$ receptors by the mimetic U46619. J Pharmacol Exp Ther 247:1133–1138

Liel N, Nathan I, Yermiyahu T, Zolotov Z, Lieberman JR, Dvilanski A, Halushka PV (1993) Increased platelet thromboxane $A_2$/prostaglandin $H_2$ receptors in patients with pregnancy induced hypertension. Thromb Res 70:205–210

MacIntyre DE, Bushfield M, Gibson I, Hopple S, MacMillan L, McNicol A, Rossi AG (1987) Human platelet receptors and receptor mechanisms for stimulatory and inhibitory lipid mediators. Colloque Inserm 0:321–334

Maeda H, Taniguchi T, Inazu T, Yang C, Nakagawara G, Yamamura H (1993) Protein-tyrosine kinase p72$^{syk}$ is activated by thromboxane $A_2$ mimetic U44069 in platelets. Biochem Biophys Res Commun 197:62–67

Mais DE, Burch RM, Saussy DL Jr, Kochel PJ, Halushka PV (1985) Binding of a thromboxane $A_2$/prostaglandin $H_2$ receptor antagonist to washed human platelets. J Pharmacol Exp Ther 235:729–734

Matsuda K, Mathur RS, Duzic E, Halushka PV (1994a) Androgen regulation of thromboxane $A_2$/prostaglandin $H_2$ receptor expression in human erythroleukemia cells. Am J Physiol 265:E928–E934

Matsuda K, Ruff A, Morinelli TA, Mathur RS, Halushka PV (1994b) Testosterone increases thromboxane $A_2$ receptor density in rat aortas and platelets. Am J Physiol 267:H887–H893

Mayeux PR, Mais DE, Halushka PV (1991) Interactions of dihydropyridine $Ca^{2+}$ channel agonists with the human platelet thromboxane $A_2$/prostaglandin $H_2$ receptor. Eur J Pharmacol 206:15–21

Mayfield RK, Halushka PV, Wohltmann HJ (1985) Platelet function during continuous insulin infusion treatment in insulin-dependent diabetic patients. Diabetes 34:1127–1133

McNicol A (1993) The effects of genistein on platelet function are due to thromboxane receptor antagonism rather than inhibition of tyrosine kinase. Prostaglandins Leukot Essent Fatty Acids 48:379–384

McNicol A, Robertson C, Gerrard JM (1993) Vanadate activates platelets by enhancing arachidonic acid release. Blood 81:2329–2338

Mead JF, Willis AL (1987) The essential fatty acids: their derivation and role. In: Willis AL (ed) CRC Handbook of eicosanoids: prostaglandins and related lipids, vol 1A. CRC, Boca Raton, pp 85–99

Mehta J, Mehta P, Conti CR (1980) Platelet function studies in coronary heart disease. IX. Increased platelet prostaglandin generation and abnormal platelet sensitivity to prostacyclin and endoperoxide analog in angina pectoris. Am J Cardiol 46:943–947

Mills D, MacFarlane DE (1977) Prostaglandins and platelet adenylate cyclase. Spectrum, New York, pp 219–233

Minuz P, Covi G, Paluani F, Degan M, Lechi C, Corsato M, Lechi A (1988) Altered excretion of prostaglandin and thromboxane metabolites in pregnancy-induced hypertension. Hypertension 11:550–556

Modesti PA, Abbate R, Gensini GF, Colella A, Serneri GGN (1991) Platelet thromboxane $A_2$ receptors in type I diabetes. Clin Sci (Colch) 80:101–105

Modesti PA, Colella A, Cecioni I, Costoli A, Biagini D, Migliorini A, Serneri GGN (1995) Increased number of thromboxane $A_2$-prostaglandin $H_2$ platelet receptors in active unstable angina and causative role of enhanced thrombin formation. Am Heart J 129:873–879

Moncada S, Vane JR (1979) Pharmacology and endogenous roles of prostaglandin endoperoxides, thromboxane $A_2$ and prostacyclin. Pharmacol Rev 30:293–331

Morinelli TA, Halushka PV (1991) Thromboxane $A_2$/prostaglandin $H_2$ receptors. Trends Cardiovasc Med 1:157–161

Morinelli TA, Oatis JE, Okwu AK, Mais DE, Mayeux PR, Masuda A, Knapp DR, Halushka PV (1989) Characterization of an [$^{125}$I]-labelled thromboxane $A_2$/prostaglandin $H_2$ receptor agonist. J Pharmacol Exp Ther 251:557–562

Morinelli TA, Tempel GE, Jaffa AA, Silva RH, Naka M, Folger W, Halushka PV (1993) Thromboxane $A_2$/prostaglandin $H_2$ receptors in streptozotocin-induced diabetes: effects of insulin therapy in the rat. Prostaglandins 45:427–438

Morrow JD, Hill KE, Burk RF, Nammour TM, Badr KF, Roberts LJ (1990) A series of prostaglandin $F_{2\alpha}$ are produced in vivo in humans by a non-cyclooxygenase, free radical-catalyzed mechanism. Proc Natl Acad Sci USA 87:9383–9387

Morrow JD, Awad JA, Boss HJ, Blair IA, Roberts LJ (1992a) Non-cyclooxygenase-derived prostanoids (F2-isoprostanes) are formed in situ on phospholipids. Proc Natl Acad Sci USA 89:10721–10725

Morrow JD, Minton TA, Roberts LJ (1992b) The F2-isoprostane, 8-epi-prostaglandin $F_{2\alpha}$, a potent agonist of the vascular thromboxane/endoperoxide receptor, is a platelet thromboxane/endoperoxide receptor antagonist. Prostaglandins 44:155–163

Morrow JD, Minton TA, Badr KF, Roberts ILJ (1994) Evidence that the $F_2$-isoprostane, 8-epi-prostaglandin $F_{2\alpha}$, is formed in vivo. Biochim Biophys Acta 1210:244–248

Mounier C, Vargaftig BB, Franken PA, Verheij HM, Bon C, Touqui L (1994) Platelet secretory phospholipase $A_2$ fails to induce rabbit platelet activation and to release arachidonic acid in contrast with venom phospholipases $A_2$. Biochim Biophys Acta 1214:88–96

Mueller RL, Scheidt S (1994) History of drugs for thrombotic disease – discovery, development, and directions for the future. Circulation 89:432–449

Murray R, Fitzgerald GA (1989) Regulation of thromboxane receptor activation in human platelets. Proc Natl Acad Sci USA 86:124–128

Nakagawa O, Tanaka I, Usui T, Harada M, Sasaki Y, Itoh H, Yoshimasa T, Namba T, Narumiya S, Nakao K (1994) Molecular cloning of human prostacyclin receptor cDNA and its gene expression in the cardiovascular system. Circulation 90:1643–1647

Nakahata N, Miyamoto A, Ohkubo S, Ishimoto H, Sakai K, Nakanishi H, Oshika H, Ohizumi Y (1995) $G_q$/11 communicates with thromboxane $A_2$ receptors in human astrocytoma cells, rabbit astrocytes and human platelets. Res Commun Mol Pathol Pharmacol 87:243–251

Nakashima S, Koike T, Nozawa Y (1990) Genistein, a protein tyrosine kinase inhibitor, inhibits thromboxane $A_2$-mediated human platelet responses. Mol Pharmacol 39:475–480

Namba T, Sugimoto Y, Hirata M, Hayashi Y, Honda A, Watabe A, Negishi M, Ichikawa A, Narumiya S (1992) Mouse thromboxane $A_2$ receptor: cDNA, cloning, expression and Northern blot analysis. Biochem Biophys Res Commun 184:1197–1203

Offermanns S, Laugwitz KL, Spicher K, Schultz G (1994) G proteins of the $G_{12}$ family are activated via thromboxane $A_2$ and thrombin receptors in human platelets. Proc Natl Acad Sci USA 91:504–508

Okwu AK, Ullian ME, Halushka PV (1992) Homologous desensitization of human platelet thromboxane $A_2$/prostaglandin $H_2$ receptors. J Pharmacol Exp Ther 262:238–245

Parent CA, Lagarde M, Venton DL, Le Breton GC (1992) Selective modulation of the human platelet thromboxane $A_2$/prostaglandin $H_2$ receptor by eicosapentaenoic and docosahexaenoic acids in intact platelets and solubilized platelet membranes. J Biol Chem 267:6541–6547

Patrono C (1994) Aspirin as an antiplatelet drug. New Engl J Med 330:1287–1294

Peplow PV (1992) Modification to dietary intake of sodium, potassium, calcium, magnesium and trace elements can influence arachidonic acid metabolism and eicosanoid production. Prostaglandins Leukot Essent Fatty Acids 45:1–19

Perret B, Chap HJ, Douste-Blazy L (1979) Asymmetric distribution of arachidonic acid in the plasma membrane of human platelets. A determination using purified phospholipases and a rapid method for membrane isolation. Biochim Biophys Acta 556:434–446

Pratico D, Lawson JA, Fitzgerald GA (1994) Cyclooxygenase-dependent formation of the isoprostane 8-epi-prostaglandin F2 alpha. Ann N Y Acad Sci 744:139–145

Pulcinelli FM, Pignatelli P, Riondino S, Parisi S, Castiglioni C, Gazzaniga PP (1994) Effect of picotamide on the calcium mobilization and phospholipase C activation in human platelets. Thromb Res 74:453–461

Ragazzi E, Chinellato A, Lille Ü, Lopp M, Doni MG, Fassina G (1995) Pharmacological properties of MM-706, a new prostacyclin derivative. Gen Pharmacol 26:703–709

Rasmanis G, Vesterqvist O, Green K, Henriksson P (1995) Implications of the prognostic importance of exercise-induced thromboxane formation in survivors of an acute myocardial infarction. Prostaglandins 49:247–253

Raychowdhury MK, Yukawa M, Collins LJ, McGrail SH, Kent KC, Ware JA (1994) Alternative splicing produces a divergent cytoplasmic tail in the human endothelial thromboxane $A_2$ receptor. J Biol Chem 269:19256–19261

Reilly M, Fitzgerald GA (1993) Cellular activation by thromboxane $A_2$ and other eicosanoids. Eur Heart J 14:88–93

Riendeau D, Guay J, Weech PK, Laliberte F, Yergey J, Li C, Desmarais S, Perrier H, Liu S, Nicoll-Griffith D, Street IP (1994) Arachidonyl trifluoromethyl ketone, a potent inhibitor of 85-kDa phospholipase $A_2$, blocks production of arachidonate and 12-hydroxyeicosatetraenoic acid by calcium ionophore-challenged platelets. J Biol Chem 269:15619–15624

Rittenhouse-Simmons S, Russel FA, Deykin D (1977) Mobilization of arachidonic acid in human platelets kinetics and $Ca^{+2}$ dependency. Biochim Biophys Acta 488:370–380

Rockhold RW (1993) Cardiovascular toxicity of anabolic steroids. Am Rev Pharmacol Toxicol 33:497–520

Romero R, Duffy TP (1980) Platelet disorders in pregnancy. Clin Perinatol 7:327–348

Rucker W, Schrör K (1983) Evidence for high affinity prostacyclin binding sites in vascular tissue: radioligand studies with a chemically stable analogue. Biochem Pharmacol 32:2405–2410

Sagel J, Colwell JA, Crook L, Laimins M (1975) Increased platelet aggregation in early diabetes mellitus. Ann Intern Med 82:733–738

Schafer AI, Cooper B, O'Hara D, Handin RI (1979) Identification of platelet receptors for prostaglandin $I_2$ and $D_2$. J Biol Chem 254:2914–2917

Scheurlen M, Kirchner M, Clemens MR, Jaschonek K (1993) Fish oil preparations rich in docosahexaenoic acid modify platelet responsiveness to prostaglandin-endoperoxide/thromboxane $A_2$ receptor agonists. Biochem Pharmacol 46:245–249

Schiff E, Peleg E, Goldenberg M, Rosenthal T, Ruppin E, Tamarkin M, Barkai G, Ben-Baruch G, Yahal I, Blankstein J, Goldman B, Mashiach S (1989) The use of aspirin to prevent pregnancy-induced hypertension and lower the ratio of thromboxane $A_2$ to prostacyclin in relatively high risk pregnancies. N Engl J Med 321:351–356

Schoenwaelder SM, Jackson SP, Yuan Y, Teasdale MS, Salem HH, Mitchell CA (1994) Tyrosine kinases regulate the cytoskeletal attachment of integrin alpha IIb beta 3 (platelet glycoprotein IIb/IIIa) and the cellular retraction of fibrin polymers. J Biol Chem 269:32479–32487

Sekiya F, Takagi J, Usui T, Kawajiri K, Kobayashi Y, Sato F, Saito Y (1991) 12*S*-Hydroxyeicosatetraenoic acid plays a central role in the regulation of platelet activation. Biochem Biophys Res Commun 179:345–351

Shattil SJ, Brass LF (1987) Induction of the fibrinogen receptor on human platelets by intracellular mediators. J Biol Chem 262:992–1000

Shattil SJ, Haimovich B, Cunningham M, Lipfert L, Parsons JT, Ginsberg MH, Brugge JS (1994) Tyrosine phosphorylation of pp $125^{FAK}$ in platelets requires coordinated signaling through integrin and agonist receptors. J Biol Chem 269:14738–14745

Shenker A, Goldsmith P, Unson CG, Spiegel AM (1991) The G protein coupled to the thromboxane $A_2$ receptor in human platelets is a member of the novel $G_q$ family. J Biol Chem 266:9309–9313

Siegl AM (1982) Receptors for $PGI_2$ and $PGD_2$ on human platelets. Methods Enzymol 86:179–193

Siegl AM, Smith JB, Silver MJ (1979a) Selective binding site for [$^3$H]-prostacyclin on platelets. J Clin Invest 63:215–220

Siegl AM, Smith JB, Silver MJ (1979b) Specific binding sites for prostaglandin $D_2$ on human platelets. Biochem Biophys Res Commun 90:291–296

Siffert W, Gengenbach S, Scheid P (1986) Inhibition of platelet aggregation by amiloride. Thromb Res 44:235–240

Siffert W, Siffert G, Scheid P, Akkerman JWN (1990) Na+/H+ exchange modulates $Ca^{2+}$ mobilization in human platelets stimulated by ADP and the thromboxane mimetic U 46619. J Biol Chem 264:719–725

Smith JB, Sedar AW, Ingerman CM, Silver MJ (1977) Prostaglandin endoperoxides: platelet shape change, aggregation and the release reaction. Academic, New York, pp 83–95

Smith WL, Marnett LJ, DeWitt DL (1991) Prostaglandin and thromboxane biosynthesis. Pharmacol Ther 49:153–179

Stuart MJ, Gerrard JM, White JG (1980) Effect of cholesterol on production of thromboxane $B_2$ by platelets in vitro. N Engl J Med 302:6–10

Stubbs TM, Lazarchick J, Van Dorsten JP, Cox J, Loadholt CB (1986) Evidence of accelerated platelet production and consumption in nonthrombocytopenic preeclampsia. Am J Obstet Gynecol 155:263–265

Takabashi K, Nammour TM, Fukunaga M, Ebert J, Morrow JD, Robert LJ, Hoover RL, Badr KF (1992) Glomerular actions of a free radical-generated novel prostaglandin, 8-epi-prostaglandin $F_2$ $\alpha$, in the rat. Evidence for interaction with thromboxane $A_2$ receptors. J Clin Invest 90:136–141

Takamura H, Narita H, Park HJ, Tanaka K, Matsuura T, Kito M (1987) Differential hydrolysis of phospholipid molecular species during activation of human platelets with thrombin and collagen. J Biol Chem 262:2262–2269

Takano S (1994) Staurosporine inhibits STA2-induced platelet aggegation by inhibition of myosin light chain phosphorylation in rabbit washed platelets. Ann N Y Acad Sci 714:315–317

Takayama K, Kudo I, Kim DK, Nagata K, Nozawa Y, Inoue K (1991) Purification and characterization of human platelet phospholipase $A_2$ which preferentially hydrolyzes an arachidonoyl residue. FEBS Lett 282:326–330

Tremoli E, Folco G, Agradi E, Galli C (1979) Platelet thromboxanes and serum-cholesterol. Lancet I:107–108

Trip MD, Cats VM, van Capelle FJL, Vreeken J (1990) Platelet hyperreactivity and prognosis in survivors of myocardial infarction. N Engl J Med 322:1549–1554

Turini ME, Holub BJ (1994) The cleavage of plasmenylethanolamine by phospholipase $A_2$ appears to be mediated by the low affinity binding site of the $TXA_2/PGH_2$ receptor in U46619-stimulated human platelets. Biochim Biophys Acta 1213:21–26

Turini ME, Gaudette DC, Holub BJ, Kirkland JB (1993) Correlation between platelet aggregation and dephosphorylation of a 68 kDa protein revealed through the use of putative PKC inhibitors. Thromb Haemost 70:648–653

Ushikubi F, Nakamura K, Narumiya S (1994) Functional reconstitution of platelet thromboxane $A_2$ receptors with $G_q$ and $G_{i2}$ in phospholipid vesicles. Mol Pharmacol 46:808–816

Van Assche FA, Spitz B, Vermylen J, Deckmijn H (1984) Preliminary observations on treatment of pregnancy-induced hypertension with a thromboxane synthetase inhibitor. Am J Obstet Gynecol 148:216–218

Vane JR, Botting RM (1995) Pharmacodynamic profile of prostacyclin. Am J Cardiol 75:3A–10A

Vassaux G, Far DF, Gaillard D, Ailhaud G, Negrel R (1993) Inhibition of prostacyclin-induced Ca2+ mobilization by phorbol esters in Ob1771 preadipocytes. Prostaglandins 46:441–451

Voyno-Yasenetskaya T, Conklin BR, Gilbert RL, Hooley R, Bourne HR, Barber DL (1994) $G\alpha 13$ Stimulates Na-H exchange. J Biol Chem 269:4721–4724

Wallenburg HCS, Dekker GA, Makovitz JW, Rotmans P (1986) Low-dose aspirin prevents pregnancy-induced hypertension and pre-eclampsia in angiotensin-sensitive primigravidae. Lancet 1:1–3

Whittle BR, Moncada S, Vane JR (1978) Comparison of the effects of prostacyclin ($PGI_2$), prostaglandin $E_1$ and $D_2$ on platelet aggregation in different species. Prostaglandins 16:373–388
Whittle BR, Hamid S, Lidbury P, Rosam AC (1985) Specificity between the anti-aggregatory actions of prostacylin, prostaglandin $E_1$ and $D_2$ on platelets. Adv Exp Med Biol 192:109–125
Willis AL (1987) The eicosanoids: an introduction and overview. In: Willis AL (ed) CRC Handbook of eicosanoids: prostaglandins and related lipids, vol 1A. CRC, Boca Raton, pp 3–46
Wilson DB, Prescott SM, Majerus PW (1982) Discovery of an arachidonoyl coenzyme A synthetase in human platelets. J Biol Chem 257:3510–3515
Winocour PD (1994) Platelet turnover in advanced diabetes. Eur J Clin Invest 24:34–37
Yin K, Halushka PV, Yan Y-T, Wong PY-K (1994) Antiaggregatory activity of 8-epi-prostaglandin $F_{2\alpha}$ and other F-series prostanoids and their binding to thromboxane $A_2$/prostaglandin $H_2$ receptors in human platelets. J Pharmcol Exp Ther 270:1192–1196
Ylikorkala O, Pekonen F, Viinikka L (1986) Renal prostacyclin and thromboxane in normotensive and preeclamptic pregnant women and their infants. J Clin Endocrinol Metab 63:1307–1312
Yokoyama K, Kudo I, Nakamura H, Inoue K (1994) A possible role for extracellular bicarbonate in U-46619-induced rat platelet aggregation. Thromb Res 74:369–376
Zahradnik HP, Schäfer W, Wetzka B, Breckwoldt M (1991) Hypertensive disorders in pregnancy – the role of eicosanoids. Eicosanoids 4:123–136
Ziboh VA, Maruta H, Lord J, Cagle WD, Lucky W (1979) Increased biosynthesis of thromboxane $A_2$ by diabetic platelets. Eur J Clin Invest 9:223–228

CHAPTER 22

# Platelet-Activating Factor: Biosynthesis, Biodegradation, Actions

Y. DENIZOT

## A. Introduction

Some of the most potent inflammatory mediators share a lipidic origin. Upon cell activation, the action of lipases on membrane phospholipids produces free fatty acids and the phospholipid backbone. Among the former are eicosanoids (prostaglandins and leukotrienes); among the latter, platelet-activating factor (PAF; SAMUELSSON et al. 1987; WALLACE 1990; Fig. 1). Studies carried out during the past decade demonstrated that PAF induces biological responses detectable at levels as low as 10 f*M* (BRAQUET et al. 1987). In 1972, BENVENISTE et al. reported that a compound originating from sensitized basophils generated aggregation of platelets. After elucidation of the PAF structure, 1-alkyl-2-acetyl-*sn*-glycero-3-phosphocholine (Fig. 2), numerous fields of research have emerged, and the subsequent results on PAF biological activities profoundly changed the perspective on its action (BRAQUET et al. 1987; PINCKARD et al. 1988). PAF is an endogenous compound synthesized by a wide range of inflammatory cell types. However, non-inflammatory cells also produce PAF, suggesting that it may be a molecule that has been conserved through evolution. This chapter briefly reviews PAF biosynthesis and degradation in eukaryotic and prokaryotic cells, the immunoregulatory actions of PAF, and some of its actions on cells and tissue structures.

## B. Biosynthesis of PAF in Mammalian Cells

Numerous methodologic difficulties appear when assessing PAF levels in organs, fluids, or secretions (such as blood, urine, saliva, stool, gastric juice, bone marrow) as well as PAF production from isolated cells in vitro. On the one hand, PAF has a short half-life of about 5 min in whole blood because of PAF metabolization by a plasma PAF acetylhydrolase activity or by numerous cell types. On the other hand, PAF is not stored in cells but is rapidly generated from PAF precursors in response to a wide range of stimuli. Moreover, measurements of PAF can be altered by the presence of an excess of other lipids. Thus, preparative procedures such as extraction with ethanol or chloroform/methanol yield crude lipids displaying PAF activity. Purification of PAF is then performed by high performance liquid chromatography or thin layer chromatography (see Chap. 23).

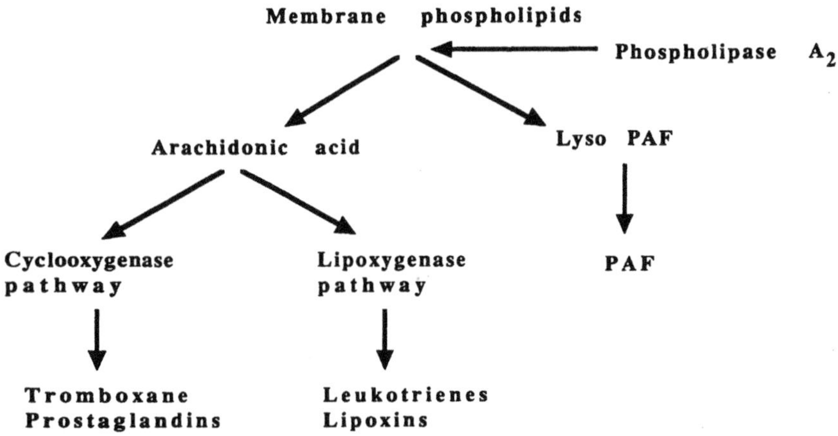


**Fig. 1.** Simplified representation of the pathways involved in eicosanoid and platelet-activating-factor (PAF) formations

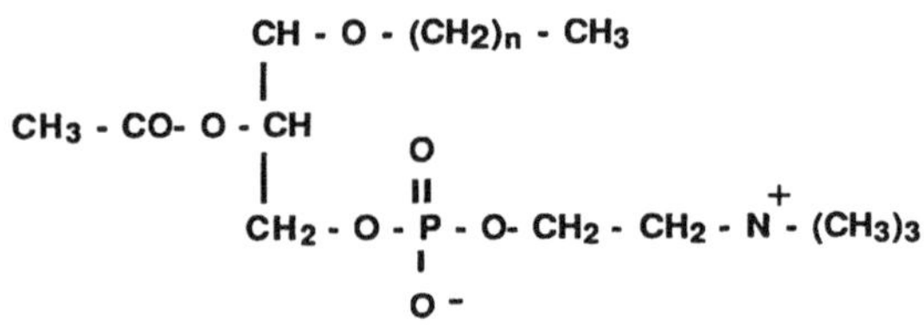

**Fig. 2.** Structure of platelet-activating factor (PAF)

Two metabolic steps are involved in PAF production: a phospholipase $A_2$ acts on choline-containing membrane alkyl-ether and produces lyso-PAF (1-*O*-alkyl-*sn*-glycero-3-phosphocholine). Following this, an acetyltransferase acetylates the lyso compound to yield the biologically active molecule. This "remodeling pathway" seems implicated in inflammatory and allergic responses. Of interest, both the acetylhydrolase and acetyltransferase activities are stimulated by calcium ionophores in numerous cell types (LEE et al. 1982; ALBERT and SNYDER 1983; NINIO et al. 1983). PAF can also be synthesized by a de novo pathway, in which a cytidine-5′-diphosphate (CDP)-choline-phosphotransferase transfers the phosphobase from CDP-choline to alkyl-acetyl-glycerol to form PAF (Fig. 3). A wide variety of stimuli modulates the de novo synthesis of PAF in several cell types (BLANK et al. 1984, 1988; HELLER et al. 1991; BAKER and CHANG 1995). This route could generate the permanent production of small amounts of PAF with some putative physiological roles (VENABLE et al. 1993). To date, reviews have recently highlighted progress on the characterization of the enzymes involved in the biosynthesis and catabolism of PAF and ether lipids (SNYDER 1994, 1995). It becomes evident that the length and the degree of unsaturation of the 1-*O*-alkyl chain of PAF influence the potency of some of its activities. For example, C16:0 PAF is the most active

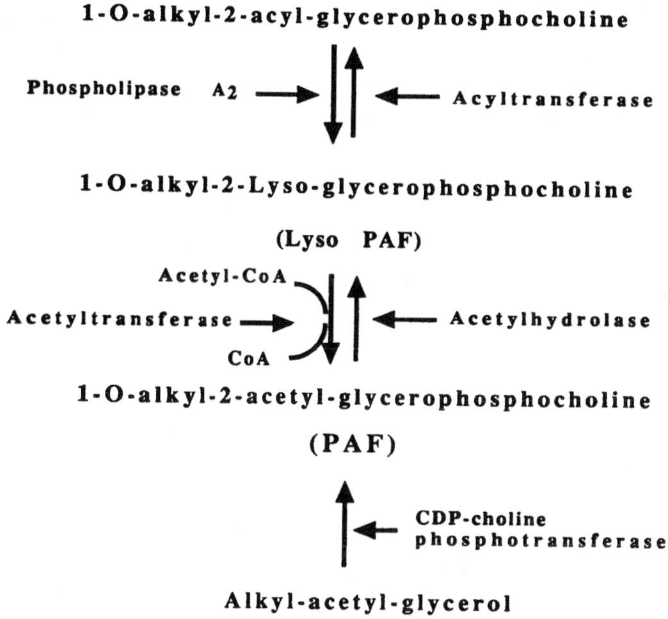


**Fig. 3.** Simplified representation of the biosynthesis and catabolism of platelet-activating factor (PAF)

species on isolated rabbit platelet aggregation, neutrophil desensitization, and $O_2^-$ production. In contrast, C18:0 and C18:1 PAF are the most active species on neutrophil chemotaxis, isolated guinea pig heart coronary flow, and neutrophil lysosomal enzyme secretion (PINCKARD et al. 1988; MCMANUS et al. 1993). The molecular heterogeneity of PAF has been widely studied in several stimulated and unstimulated cells. C16:0, C18:0, and C18:1 are the predominant PAF molecular species (MUELLER et al. 1984; PINCKARD et al. 1984; WEINTRAUB et al. 1985; ODA et al. 1985; RAMESHA and PICKETT 1987; MICHEL et al. 1988; BOSSANT et al. 1989; CLAY et al. 1984). However, at present the physiological implications of the variation in PAF composition are not known.

PAF is released from stimulated inflammatory cells such as platelets (BENVENISTE et al. 1982), neutrophils (SISSON et al. 1987; BUSSOLINO et al. 1992), monocytes (ARNOUX et al. 1982), alveolar macrophages (SUGIURA et al. 1991), eosinophils (OJIMA-UCHIYAMA et al. 1991), and vascular endothelial cells (BUSSOLINO et al. 1988). When generated by eosinophils (OJIMA-UCHIYAMA et al. 1991), neutrophils (LYNCH and HENSON 1986), and endothelial cells (HIRAFUJI et al. 1987), the newly synthesized PAF remains mainly cell-associated. By contrast, human monocytes (ELSTAD et al. 1988) and human skin fibroblasts (MICHEL et al. 1988) release half of their newly synthesized PAF. These results suggest that monocytes may be one of the major cell sources of the PAF found in the blood of patients with allergic or inflammatory disorders (GRANDEL et al. 1985; DENIZOT et al. 1995b). Among human blood cells, lymphocytes generate very low amounts of PAF (JOUVIN-MARCHE et al. 1984). A low level of acetyltransferase activity is observed in lymphocytes and may explain their inability to produce PAF. By contrast, studies report that large granular lymphocytes produce PAF through their Fc receptor activation (MALAVASI et al. 1986) and that several human cell lines of B- and T-cell origins generate PAF in response to calcium ionophore and/or phytohemagglutinin (PHA) (BUSSOLINO et al. 1984). Moreover, these stimuli induce PAF production from freshly isolated lymphoid leukemic cells (FOA et al. 1985). Finally, a $CD4^+$ T-lymphocyte clone generates PAF after stimulation with a mitogenic pair of anti-CD2 antibodies (LE GOUVELLO et al. 1992). Of importance, the capacity to produce PAF is not restricted to inflammatory cells. Thus, keratinocytes (MICHEL et al. 1990), fibroblasts (MICHEL et al. 1988), submandibular gland cells (DOHI et al. 1991), liver cells (PINZANI et al. 1994), and intestinal cells (FERRARIS et al. 1993) generate PAF.

Numerous cytokines induce PAF production in several cell types. Interleukin-1 (IL-1) stimulates PAF synthesis in monocytes (VALONE and EPSTEIN 1988) and endothelial cells (BUSSOLINO et al. 1988, 1994). TNF-$\alpha$ increases PAF production in macrophages, neutrophils, and endothelial cells (CAMUSSI et al. 1987). IL-8 stimulates PAF production from neutrophils (BUSSOLINO et al. 1992). IL-6 generates PAF from murine skin fibroblasts (PIGNOL et al. 1994). Granulocyte-macrophage colony-stimulating factor (GM-CSF) primes PAF synthesis in neutrophils (WIRTHMUELLER et al. 1989, 1990; DENICHILO et al. 1991). GM-CSF, IL-3, and IL-5 prime eosinophils to produce PAF (TRIGGIANI et al. 1992; VAN DER BRUGGEN et al. 1994). In turn, PAF generates cytokine production in several cell types. PAF enhances TNF synthesis in alveolar macrophages (DUBOIS et al. 1989), monocytes (BONAVIDA et al. 1990; ROLA-PLESZCZYNSKI 1990; RUIS et al. 1991), and B cells (SMITH et al. 1994). PAF increases IL-1 synthesis in blood monocytes (POUBELLE et al. 1991) and rat spleen monocytes (PIGNOL et al. 1987), and it stimulates IL-6 production by alveolar macrophages (THIVIERGE and ROLA-PLESZCZYNSKI 1992). In rats, the infusion of PAF alters the production of IL-1 and IL-2 from adherent splenocytes stimulated with endotoxin (PIGNOL et al. 1990), suggesting that PAF may play an immunoregulatory role in vivo. Finally, VALONE (1991) reports that PAF stimulates its own synthesis in vitro in a dose-dependent manner with a maximal effect occurring between 10 p*M* and 100 p*M*. The subsequent persistent PAF production may prolong inflammatory reactions in vivo. Similar results have been observed with human neutrophils (GOMEZ-CAMBRONERO et al. 1989).

## C. Catabolism of PAF in Mammalian Cells

The catabolism of PAF is an effective means by which cells can regulate PAF level. PAF acts through receptors on the membrane of responsive cells. PAF receptors are reported and cloned in several cell types (HONDA et al. 1991;

NAKAMURA et al. 1991). Sequence analysis reveals that the receptor belongs to the superfamily of G-protein-coupled receptors. Reviews have highlighted the progress in characterization of the molecular structure of the PAF receptor and its transduction mechanisms (IZUMI et al. 1995; SHUKLA 1992; VENABLE et al. 1993). In order to study PAF metabolism, cells are usually incubated with 1-*O*-[$^3$H]-alkyl-PAF in serum-free medium for various periods of time. After extraction and purification of metabolites, the label is found in products migrating with PAF, lyso PAF, phosphatidylcholine (alkyl-acyl-glycerophosphocholine), and neutral lipids (alkyl-acyl-glycerol). In vitro PAF is metabolized by a wide range of cells using the deacetylation-transacylation pathway. Accumulation of the deacetylated compound lyso PAF is found in human endothelial cells (BLANK et al. 1986) and rat alveolar macrophages (ROBINSON and SNYDER 1985). In contrast, in human monocytes (SALEM et al. 1990), neutrophils (TRIGGIANI et al. 1991), lymphocytes (TRAVERS et al. 1990), epithelial cells (KUMAR et al. 1987), and type II pneumonocytes (EGUCHI et al. 1994), the major PAF metabolite is a phosphatidylcholine. Human bone marrow cells metabolize PAF using the deacetylation-transacylation pathway (DENIZOT et al. 1995c). Similar to the mechanism found for human neutrophils and lymphocytes, the metabolite lyso PAF appears briefly and is quickly reacylated with a fatty acid at the *sn*-2 position of the PAF glycerol backbone. Thus, PAF is similarly metabolized by cells of the bone marrow and of the peripheral blood. Since PAF is present in human bone marrow (DENIZOT et al. 1995a,c), the catabolism of PAF by human bone marrow cells may be an important regulatory mechanism of the level of this potent inflammatory molecule.

PAF can generate biological responses detectable at levels as low as 10 *fM*. It is thus evident that regulating PAF levels is important since elevated or decreased levels of PAF might result in pathological effects. An acetylhydrolase present in blood and tissues rapidly degrades PAF by removing the acetyl moiety at the *sn*-2 position of the glycerol backbone, thus generating the inactive metabolite lyso PAF (BLANK et al. 1981; STAFFORINI et al. 1987, 1991; ANTONOPOULOS et al. 1994). A PAF acetylhydrolase-like activity is found in seminal plasma (LETENDRE et al. 1992; HOUGH and PARKS 1994), human milk (MOYA et al. 1994), and cerebrospinal fluid (HIRASHIMA et al. 1994). In human and mouse, plasma acetylhydrolase is associated with lipoprotein particles (LDL and HDL; KARABINA et al. 1994; TSAOUSSIS and VAKIRTZI-LEMONIAS 1994). While 17-$\alpha$-ethynylestradiol decreases plasma PAF acetylhydrolase, dexamethasone increases it, and testosterone has no effect (MIYAURA et al. 1991). Alcohol increases plasma acetylhydrolase in pregnant mice (SALEH et al. 1994). Taken together, these results support the putative importance of PAF in reproductive and developmental biology (HARPER 1989; see Chap. 28). Thus, elevated levels of PAF are reported in the amniotic fluid of women in labor (KOBAYASHI et al. 1994), and decreased acetylhydrolase activity is documented in the maternal plasma during the later stages of pregnancy (MAKI et al. 1988). Numerous experimental results indicate that peripheral blood monocyte-derived macrophages are a cellular source of plasma acetylhydrolase activity (STAFFORINI et al. 1990). The ability to produce enzymatic activity markedly increases when monocytes mature to macrophages (ELSTAD et al. 1989; LEE et al. 1994). Human decidual macrophages also secrete acetylhydrolase activity (NARAHARA et al. 1993). In this cell type, cigarette smoke extract, endotoxins, TNF-$\alpha$, and IL-1 inhibit the enzymatic secretion in a dose-dependent manner (NARAHARA and JOHNSTON 1993a,b). Liver cells also secrete a type of PAF acetylhydrolase activity identical to the plasma form (TARBET et al. 1991; SATOH et al. 1994).

Changes in PAF acetylhydrolase activity are associated with pathological events. Elevated levels of PAF acetylhydrolase have been documented in plasma of lizards submitted to physiological stress (LENIHAN et al. 1985), in plasma of cigarette smokers (IMAIZUMI et al. 1990), of hypertensive (SATOH et al. 1989), and of atherosclerotic patients (OSTERMANN et al. 1987). Increased PAF acetylhydrolase activity is found in patients with insulin-dependent diabetes mellitus (HOFMANN et al. 1989; however, NATHAN et al. 1992c published contradictory results), ischemic cerebrovascular disease (SATOH et al. 1988, 1992), Tangier disease (PRITCHARD et al. 1985), and chronic cholestasis (MEADE et al. 1991) as well as in cancer patients after rapid administration of 5-fluorouracil (DENIZOT et al. 1994). Decreased PAF acetylhydrolase activity is documented in serum of asthmatic children (MIWA et al. 1988). Recently, TJOELKER et al. (1995) have reported the molecular cloning and the in vitro expression of the human plasma PAF acetylhydrolase. These data are of great importance and offer the possibility of using recombinant acetylhydrolase to terminate some types of inflammatory states (BAZAN 1995). It is now recognized that acetylhydrolase activity may play a key regulatory role in the control fo PAF-mediated inflammation by decreasing PAF amounts in various biological fluids. However, it remains to be clarified whether changes in plasma (or serum) PAF acetylhydrolase activity have clinical relevance. Moreover, the assessment of acetylhydrolase activity in various diseases is usually only descriptive (elevated or decreased levels as compared to controls). HATTORI et al. (1994) have reported a very surprising and exciting result. The protein encoded by the causative gene for Miller-Dieker lissencephaly, a human brain malformation with abnormal neuronal migration, is a homologue of the 45K subunit of intracellular acetylhydrolase. This result suggests that PAF and acetylhydrolase are important in the formation of the brain cortex during differentiation and development.

## D. PAF and Bacteria

The growing importance of PAF in inflammatory processes and the similarities of its properties in many eukaryotic species suggested to investigators that PAF may be a molecule that has been conserved through evolution. Thus, the production of PAF has been investigated in bacteria, yeast cells, protozoans, and parasites.

The production of PAF has been assessed in various bacterial strains derived from *Escherichia coli* (*E. coli*) K12 grown to the late stationary phase in synthetic medium. Bacterial lipids are extracted, purified, and assessed for PAF content by aggregation of rabbit platelets. About 100 pg PAF per $10^9$ cells are found in lipids extracted from various strains of *E. coli* (THOMAS et al. 1986; DENIZOT et al. 1989b). The prokaryotic PAF exhibits the same biological (inactivation by phospholipase $A_2$ but not by lipase $A_1$, inhibition of the aggregating activity by PAF receptor antagonists) and physicochemical properties (retention time during high performance liquid chromatography) as eukaryotic PAF. However, PAF quantities are insufficient for structural analysis using mass spectrometric analysis. Since ether-linked and choline-substituted phospholipids are not reported in *E. coli* – what is the mechanism of PAF production in bacteria? The simplest interpretation is that PAF results from the bioconversion of lyso PAF present in the culture medium. Supporting this hypothesis when cultures are supplemented with C16:0 lyso PAF ($10\,\mu M$–$1\,mM$), a 10- to 240-fold increase in PAF production is observed (DENIZOT et al. 1989b). Identity of this prokaryotic PAF with eukaryotic PAF is based on mass spectrometric analysis. Of note, C18:0 lyso PAF is less efficient than C16:0 lyso PAF in enhancing PAF production by different *E. coli* strains (DENIZOT et al. 1989a). The higher efficiency of C16:0 lyso PAF may indicate that the putative bacterial acetyltransferase recognizes C16:0 ether phospholipids more specifically than C18:0 ones. Alternatively, it is possible that an unrelated transacylase more efficiently converts C16:0 lyso PAF to a whole variety of 2-*O*-acyl derivatives including the 2-*O*-acetyl compound. At present, no assay has been performed to analyse in vitro the putative bacterial acetyltransferase activity. The production of PAF in the presence of lyso PAF is reported for other bacteria such as *Salmonella typhimurium* (DENIZOT et al. 1990a) and *Helicobacter pylori* isolates collected from antral biopsy specimens of patients with gastritis and duodenal ulcer (DENIZOT et al. 1990b). Thus, bacteria are able to generate PAF in vitro and therefore contain at least the final enzymatic patways involved in PAF synthesis.

The existence of bacterial PAF alone and/or associated with endotoxins supports the hypothesis that local release of this molecule may initiate and/or enhance septic shock or gastrointestinal damage. Thus, PAF is a potent compound in vivo that induces proinflammatory events such as shock (TERASHITA et al. 1985), gastric ulceration (ROSAM et al. 1986), and ischemic bowel necrosis (GONZALEZ-CRUSSI and HSUEH 1983). Moreover, when systemically administered, endotoxins and PAF act in synergy to generate necrotizing bowel lesions. In human, a role of PAF is suspected in bacterial meningitis (ARDITI et al. 1990) and in bacterial diarrhea (DENIZOT et al. 1991a, 1992a). However, it is unclear whether bacterial PAF plays a role in initiation, development, and perpetuation of inflammatory damage during bacterial diseases or whether PAF is merely a nonspecific marker. Moreover, the ability of bacteria to produce PAF in vivo is not documented (DENIZOT and CHAUSSADE 1994), casting some doubt on the physiological significance of the acetylation of exogenous lyso PAF by bacteria. Finally, the ability of bacteria to metabolize PAF, the presence of a bacterial acetyltransferase and acetylhydrolase activity, and the influence of PAF on bacterial physiology are as yet totally unknown.

## E. PAF and Yeast Cells, Protozoans, Amoebas, and Parasites

The assessment of PAF production by yeast cells is novel. Two studies report PAF biosynthesis using strains of the *Saccharomyces* genus (NAKAYAMA et al. 1994a,b). In these experiments, lipids from yeast cells are extracted, purified, and assessed for PAF by aggregation of rabbit platelets. Yeast-cell-derived PAF is further characterized by mass spectrometry. PAF is found in the stationary phase but not during the logarithmic phase of yeast growth (NAKAYAMA et al. 1994a). PAF is detected in all the tested strains, but its level varies from one strain of *Saccharomyces cerevisiae* to another. A tenfold increase of PAF production is observed in yeast cells (i.e., 100 pg/g wet-weight cells) in response to 5-min incubation with 2-$\mu M$ calcium ionophore (NAKAYAMA et al. 1994b). Under these conditions, the newly synthesized PAF remains cell-associated. PAF quantities in yeast cells are reduced with longer incubation with calcium ionophore and are transiently accumulated in the presence of phenylmethyl-sulfonyl fluoride (PMSF), an inhibitor of PAF catabolism. These results suggest the presence of a PAF-acetylhydrolase-like activity. In yeast cells, PAF is produced by the "remodeling pathway" and is related to cell growth. At present, the metabolism of PAF by yeast cells and its role in cell physiology are unknown.

The production of PAF has also been investigated in protozoans, amoebas, and parasites. The protozoan *Tetrahymena pyriformis* has been harvested at the late log growth phase to assess the presence of PAF. Results indicate that *Tetrahymena pyriformis* contains 4 ng of PAF per $10^7$ cells. Thus, PAF appears to be a natural minor component of the protozoan lipids (LEKKA et al. 1986). The stimulation of cells with calcium ionophore has no effect on PAF formation. Unlike bacteria and yeast cells, the metabolism of PAF has been extensively studied in *Tetrahymena pyriformis*. In vivo it captures PAF and rapidly converts the molecule into phosphatidylcholine which is a major component of the protozoan phospholipids (LEKKA et al. 1989, 1990). Similar to that found in most mammalian cells, the metabolite lyso PAF appears briefly in the system and is quickly reacylated with a fatty acid at position 2 of the glycerol backbone. Thus, *Tetrahymena pyriformis* metabolizes PAF by the deacetylation-transacylation pathway. In vitro experiments provide evidence for the existence of acetylhydrolase-like activity in this protozoan (TSELEPIS et al. 1991).

PAF biosynthesis has been documented in amoebas such as *Dictyostelium discoideum* (BUSSOLINO et al. 1991). The basal PAF levels increase during cell development of *Dictyostelium discoideum*, reaching a maximum after aggregation when cells form a multicellular organism in response to starvation. Similar to most mammalian cells, PAF is found only in a cell-bound form. While the production of PAF is not reported for the gastrointestinal nematode parasite *Nippostrongylus brasiliensis*, studies indicate that it inactivates PAF by acetylhydrolase-like activity (BLACKBURN and SELKIRK 1992). These results may explain in vivo observations reporting that the gastrointestinal synthesis of PAF is significantly reduced during *Nippostrongylus brasiliensis* infection in the rat (HOGABOAM et al. 1991). They may also explain the absence of fecal PAF during human parasitic diarrhea (DENIZOT et al. 1992a).

## F. Effects of PAF on Cell Proliferation

During the past decade, numerous results have demonstrated the immunoregulatory properties of PAF. Most of them have proven that PAF interfers with the process leading to T- and B-cell activation in vitro. In most of these experiments, cell proliferation was assessed by [$^3$H]thymidine incorporation after cell stimulation. PATRIGNANI et al. (1987) showed that the addition of micromolar amounts of PAF to PHA-stimulated human blood mononuclear leukocytes did not alter cell proliferation. In contrast, ROLA-PLESZCZYNSKI et al. (1987) report that PAF ($1\,\mu M$–$10\,nM$) reduces the proliferation of PHA-stimulated blood mononuclear cells. Differences in cell culture and/or purification have been suggested to explain these contradictory results. PAF ($10\,\mu M$) increases by 50% the proliferation induced by a combination of anti-CD2 monoclonal antibodies and inhibits by 50% T-cell proliferation induced by immobilized anti-CD3 monoclonal antibodies (VIVIER et al. 1990). Thus, PAF regulates CD2- and CD3-induced T-cell proliferation in vitro differently. DULIOUST et al. (1988) report that micromolar amounts of PAF reduce PHA-stimulated proliferation of blood CD4$^+$ T cells. The inhibitory effect of PAF is due to lower expression of the high affinity IL-2 receptor on PAF-treated cells (DULIOUST et al. 1990). PAF has no effect on the spontaneous proliferation of human B lymphocytes in vitro. However, micromolar amounts of PAF diminish B-cell proliferation induced by phorbol myristate acetate (PMA) or anti-immunoglobulin M (IgM) antibodies by 40%. In contrast, nanomolar amounts of PAF stimulate PMA-mediated B-cell proliferation by 40% but not anti-IgM-induced proliferation (LEPRINCE et al. 1991). These data are in agreement with the fact that PAF receptor expression on B cells is dependent upon their activation. Thus, while PAF receptors are absent on resting B cells, they are present on B cells activated in vitro with *Staphylococcus aureus* Cowan (NGUER et al. 1992). Similarly, while PAF receptors are not found on resting T cells, they are present on membranes of T cells stimulated via the CD2 or CD3 pathways (28000 receptors/cell; binding affinity 2–9 $nM$; CALABRESSE et al. 1992). These results indicate that the in vitro activation of human T- and B cells induces membrane PAF receptors, suggesting that they could play a role in the regulation of T- and B-cell activation mechanisms, thus explaining the effect (either positive or negative) of PAF on the proliferation of stimulated lymphocytes.

While numerous studies have investigated the effects of PAF on circulating blood cells, its capacity to modulate the process of progenitor cell proliferation and maturation is much less documented. Results showed that PAF enhances in a dose-dependent manner (10–0.1 $\mu M$) the DNA synthesis in freshly isolated guinea pig bone marrow cells (Kato et al. 1988). Furthermore, guinea pig bone marrow cells treated with PAF generate compounds that affect their DNA synthesis (Kudo et al. 1991). Thus, PAF affects the proliferation of one or more classes of bone marrow cells of the monocyte-macrophage lineage through the release of cytokines. The effects of PAF on guinea pig bone marrow cells are mediated by PAF receptors (Hayashi et al. 1988). While PAF is present in human sternal (Denizot et al. 1995c) and femoral bone marrow (Denizot et al. 1995a), its role in human bone marrow cell proliferation is not documented. However, the presence of PAF in human hematopoietic organs such as bone marrow and thymus (Salem et al. 1989) suggest its putative role during human hematopoiesis. Moreover, PAF in association with another signal such as calcium ionophore stimulates apoptotic processes in an immature T-cell line (El Azzousi et al. 1993), suggesting that PAF may play a role in the control of intrathymic lymphocyte maturation. Finally, PAF enhances the differentiation and maturation of hematopoietic precursor cells from human umbilical cord blood into basophils and eosinophils (Saito et al. 1992, 1993).

# G. Some Effects of PAF on Several Cell Functions and Tissue Structures

An important effect on PAF on B-cell functions is its ability to influence the production of immunoglobulins. While nanomolar amounts of PAF enhance immunoglobulin production in several B-cell lines (Mazer et al. 1990), micromolar amounts of PAF down-regulate the IL-4-induced IgE production by human peripheral blood mononuclear cells in vitro (Deryckx et al. 1992). Thus, PAF has a regulatory role in the control of immunoglobulin isotype production in vitro. However, whether PAF regulates immunoglobulin production in vivo remains highly putative. In several B-cell lines, PAF induces important early events implicated in B-cell activation (i.e., the increase in intracellular levels of calcium, expression of the genes C-fos and C-jun, phosphoinositol hydrolysis; Mazer et al. 1991; Schulam et al. 1991). The effect of PAF on T-cell functions is much less documented. PAF activates suppressor cells (Rola-Pleszczynski et al. 1988) and modulates the expression of cell surface glycoproteins on human lymphocytes. Vivier et al. (1988) report that T cells treated for 3 days with 10-$\mu M$ PAF showed a 40% decrease in the expression of CD2 and CD3 HLA class II antigens, but not in the HLA class I antigens. However, at present the functional consequences of these effects are unclear.

Apart from its action on T- and B lymphocytes and its platelet-aggregating activity (CHIGNARD et al. 1979), PAF exhibits potent effects on a wide range of other cells. In neutrophils, PAF induces (among numerous other effects) exocytosis, migration, superoxide production and aggregation (SHAW et al. 1981), tyrosine phosphorylation (GOMEZ-CAMBRONERO et al. 1991), and elevated cytosolic free $Ca^{2+}$ concentrations (SHIBATA et al. 1994) as well as markedly enhances their responses to chemotactic peptides (GAY 1993). In monocytes, PAF stimulates cytotoxicity toward WEHI 164 cells (ROSE et al. 1990), the formation of leukotrienes (FAULER et al. 1989), superoxide anions, and hydrogen peroxides (PUSTYNNIKOV et al. 1991), and the expression of the low-affinity receptor for IgE (PAUL-EUGENE et al. 1990). In eosinophils, PAF induces activation and degranulation (EBISAWA et al. 1990; EDA et al. 1993) as well as chemotaxis in vitro and in vitro (HENOCQ and VARGAFTIG 1986; MIYAGAWA et al. 1992). PAF has no effect on human skin fibroblast proliferation (L. MICHEL, unpublished results). By contrast PAF ($50\mu M$) reduces lipoprotein degradation and alters lipid metabolism (increased synthesis of sterol or triacylglycerol and decreased synthesis of phospholipid) in cultured human lung fibroblasts (MAZIERE et al. 1988). SAWYER and ANDERSEN (1989) have reported that micromolar concentrations of PAF alter the barrier properties of the membrane lipid bilayer, showing that it is a general membrane perturbant. This action may provide an explanation for its in vitro cellular effects on extraphysiological concentrations. Finally, numerous studies report that PAF mediates cell-to-cell interactions such as platelet adhesion to endothelial cells (HIRAFUJI and SHINODA 1991), leukocyte adherence to mesenteric venular endothelium (KUBES et al. 1990), and eosinophil recruitment by transendothelial migration (MORLAND et al. 1992).

PAF acts in vitro on numerous organs and tissue structures. The effects of its in vivo administration to animals suggest that it may be an important pathophysiological regulatory compound. Numerous reviews have listed its effects on several organs such as the liver (EVANS et al. 1991), the lungs (CHUNG 1992), the heart (EVANGELOU 1994), the brain (LINDSBERG et al. 1991), the skin (DENIZOT et al. 1991b), the gastrointestinal tract (DENIZOT et al. 1992b), the retina (BAZAN et al. 1994), the kidney (CAMUSSI 1986), and the uterus (HARPER 1989). However, PAF molecular heterogeneity may cast some doubts on the pathophysiological significance of most of the experimental effects of PAF (MCMANUS et al. 1993). Most of these results are obtained in vitro with extraphysiological concentrations of PAF (i.e., micromolar amounts) or in vivo after intravenous administration of a single PAF species (i.e., C16:0 or C18:0 PAF). Moreover, results of animal studies have not been confirmed in human clinical studies. This last point is particularly evident for studies assessing the putative role of PAF in coronary heart disease, septic shock, and inflammatory gastrointestinal damage.

PAF is assumed to play a role in coronary heart and vascular diseases. Infusion of PAF to animals induces damage to microcirculation, pulmonary hypertension, and systemic hypotension (reviewed by Evangelou 1994). Intracardiac infusion of PAF in perfused guinea pig hearts induces a decrease of coronary flow and of myocardial contractility (Benveniste et al. 1983). Moreover, PAF has a direct negative inotropic effect on rat cardiomyocytes (Delbridge et al. 1994). However, a recent clinical study reports no elevated blood PAF levels in patients with myocardial infarction (Graham et al. 1994b). Variations of blood PAF levels are found in humans at the cessation of cardiopulmonary bypass and after protamine reversal of heparin (Nathan et al. 1992a,b), suggesting a putative role of PAF on cardiopulmonary bypass- or protamine-induced hematologic or hemodynamic changes. Unfortunately, the PAF receptor antagonist BN 52021 is unable to prevent hematological alterations (such as leucopenia, thrombocytopenia, and blood loss) induced by cardiopulmonary bypass or protamine reversal of heparin (Nathan et al. 1994). While a transient improvement of pulmonary vascular resistance is found at the end of cardiopulmonary bypass in the BN-52021-treated patients, the effect disappears after protamine infusion (Nathan et al. 1995). Thus, at present clinical studies provide no direct evidence to support the hypothesis of an important role of PAF in human cardiovascular disorders.

Animal studies suggest an important role of PAF in septic shock. In experimental models of shock states induced by intravenous infusion of endotoxins, the beneficial effects of PAF antagonists have been documented (Kuipers et al. 1994). PAF acetylhydrolase activity is decreased in severely ill patients with clinical sepsis (Graham et al. 1994a). However, the increase of blood PAF levels in patients with septic shock is controversial. Sörensen et al. (1994) report a significant increase of PAF in the blood of patients with septic shock (about 550 pg/ml) as compared to controls (25 pg/ml). In contrast, Graham et al. (1994b) report that plasma PAF was not different in patients with sepsis (220 pg/ml) as compared to controls (250 pg/ml). In five healthy volunteers the PAF antagonist Ro 24-4736 has been ineffective in attenuating the hemodynamic changes induced by intravenous infusion of bacterial endotoxin (Thompson et al. 1994). Recently, the PAF antagonist BN 52021 has been reported to be a putative promising treatment for patients with severe gram-negative sepsis but not with gram-positive sepsis (Dhainaut et al. 1994). However, the clinical improvements reported in this study are far less important than those observed in experimental animal studies. Thus, at present, clinical studies provide only weak direct evidence in support of the hypothesis of an important role of PAF in human sepsis.

Experimental animal studies suggest a role of PAF in the generation and/or amplification of inflammatory gastrointestinal damage. Numerous studies report that administration of PAF generates bowel necrosis (Gonzalez-Crussi and Hsueh 1983) and gastric ulceration (Rosam et al. 1986). Most studies examining the effects of PAF on the gastrointestinal tract have been evaluated in the rat after systemic administration. At present, no information is available on the effect of PAF administered through the intestinal lumen. In human, elevated levels of PAF are found in the gastric juice of patients with erosive gastritis (SOBHANI et al. 1992a) and in intestinal biopsies of patients with Crohn's disease (KALD et al. 1990; SOBHANI et al. 1992b) and ulcerative colitis (ELIAKIM et al. 1988). PAF is found in the stool of patients with pouchitis (CHAUSSADE et al. 1991), bacterial diarrhea (DENIZOT et al. 1991a), Crohn's disease (DENIZOT et al. 1992c), and ulcerative colitis (CHAUSSADE et al. 1992). However, it is still unclear whether PAF plays an important role in the inflammatory ailments of the gastrointestinal tract in humans or is merely a nonspecific marker of inflammation. These doubts are reinforced by the results of MALCHOW et al. (1994) who found no clinical benefit of a PAF antagonist in the treatment of ulcerative colitis in human, casting some doubts on the true role of PAF in inflammatory bowel disease in human.

## H. Conclusion

PAF is considered as a mediator of inflammation and allergy. However, this could be a bias providing from its historical discovery and, thus, from its subsequent name of "platelet-activating factor." As for other mediators, the true in vivo PAF actions are unknown. Numerous studies have investigated the effects of PAF on several cells in vitro. Results suggest that the various biological effects of PAF observed in vitro depend on numerous factors such as: (1) its chemical structure and the quantitative balance between the numerous PAF species, (2) PAF concentration, (3) the rate of its degradation in culture medium, (4) the presence of a second activating stimulus, (5) the state of cell activation, and (6) the presence of "accessory" cells such as fibroblasts and monocytes/macrophages.

PAF appears to be a molecule that acts on a wide variety of organs in vivo. However, most of the effects of PAF are found after intravenous infusion and/or with extraphysiological concentrations of PAF (i.e., micromolar to nanomolar amounts), casting some doubts on the ability of this inflammatory mediator to really act in vivo. It is thus difficult to compare the high amounts of PAF used in most of these studies with the f*M* amounts of PAF found in blood and organs in steady state conditions. It may thus be an objective of in vitro studies to add high amounts of PAF to cell cultures because PAF is rapidly catabolized in the culture medium by serum acetylhydrolase and/or by cells themselves. However, studies showing the effects of continuous addition or infusion of low doses of PAF (which seems to be a more relevant protocol of stimulation to obtained information on the in vivo effects of PAF) are extremely rare. In fact, at present, although a multitude of biological activities of PAF can be demonstrated in vitro, there is little evidence that PAF is implicated in some pathophysiological processes in humans.

## References

Albert DH, Snyder F (1983) Biosynthesis of 1-alkyl-2-acetyl-sn-glycero-3-phosphocholine (platelet-activating factor) from 1-alkyl-2-acyl-sn-glycero-3-phosphocholine by rat alveolar macrophages. J Biol Chem 258:97–102

Antonopoulos S, Demopoulos CA, Iatrou C, Moustakas G, Zirogiannis P (1994) Platelet-activating factor acetylhydrolase (PAF-AH) in human kidney. Int J Biochem 26:1157–1162

Arditi M, Manogue KR, Caplan M, Yogev R (1990) Cerebrospinal fluid cachectin/tumor necrosis factor-$\alpha$ and platelet-activating factor concentrations and severity of bacterial meningitis in children. J Infect Dis 162:139–147

Arnoux B, Jouvin-Marche E, Arnoux A, Benveniste J (1982) Release of PAF-acether from human blood monocytes. Agents Actions 12:713–716

Baker RR, Chang HY (1995) Fatty acyl-CoA inhibits 1-alkyl-sn-glycero-3-phosphate acetyltransferase in microsomes of immature rabbit cerebral cortex: control of the first committed step in the de novo pathway of platelet-activating factor synthesis. J Neurochem 64:364–370

Bazan NG (1995) A signal terminator. Nature 374:501–502

Bazan HEP, Tao Y, Hurst JS (1994) Platelet-activating factor antagonists and ocular inflammation. J Ocul Pharmacol 10:319–327

Benveniste J, Henson PM, Cochrane CG (1972) Leukocyte-dependent histamine release from rabbit platelets: the role of IgE, basophils and a platelet-activating factor. J Exp Med 136:1356–1377

Benveniste J, Chignard M, Le Couedic JP, Vargaftig BB (1982) Biosynthesis of platelet-activating factor (paf-acether). II. Involvement of phospholipase $A_2$ in the formation of paf-acether and lyso-paf-acether from rabbit platelets. Thromb Res 25:375–385

Benveniste J, Boullet C, Brink C, Labat C (1983) The actions of paf-acether (platelet-activating factor) on guinea-pigs isolated heart preparations. Br J Pharmacol 80:81–83

Blackburn CC, Selkirk ME (1992) Inactivation of platelet-activating factor by a putative acetylhydrolase from the gastrointestinal nematode parasite *Nippostrongylus brasiliensis*. Immunology 75:41–46

Blank ML, Lee TC, Fitzgerald V, Snyder F (1981) A specific acetylhydrolase for 1-alkyl-2-acetyl-sn-glycero-3-phosphocholine (a hypotensive and platelet-activating lipid). J Biol Chem 256:175–178

Blank ML, Lee TC, Cress EA, Malone B, Fitzgerald V, Snyder F (1984) Conversion of 1-alkyl-2-acetyl-sn-glycerols to platelet activating factor and related phospholipids by rabbit platelets. Biochem Biophys Res Commun 124:156–163

Blank ML, Spector AA, Kaduce TL, Lee TC, Snyder F (1986) Metabolism of platelet-activating factor (1-alkyl-2-acetyl-sn-glycero-3-phosphocholine) and 1-alkyl-2-acetyl-sn-glycerol by human endothelial cells. Biochim Biophys Acta 876:373–378

Blank ML, Lee YJ, Cress EA, Snyder F (1988) Stimulation of the de novo pathway for the biosynthesis of platelet-activating factor (PAF) via cytidylyltransferase activation in cells with minimal endogenous PAF production. J Biol Chem 263:5656–5661

Bonavida B, Mencia-Huerta JM, Braquet P (1990) Effects of platelet-activating factor on peripheral blood monocytes: induction and priming for TNF secretion. J Lipid Mediat 2:S65–S76

Bossant MJ, Fariotti R, De Maack F, Mahuzier G, Benveniste J, Ninio E (1989) Capillary gas chromatography and tandem mass spectrometry of paf-acether and analogs: absence of 1-O-alkyl-2-propionyl-sn-glycero-3-phosphocholine in human polymorphonuclear neutrophils. Lipids 24:121–124

Braquet P, Touqui L, Shen TY, Vargaftig BB (1987) Perspectives in platelet-activating factor research. Pharmacol Rev 39:97–145

Bussolino F, Foa R, Malavasi F, Ferrando ML, Camussi G (1984) Release of platelet-activating factor (PAF)-like material from human lymphoid cell lines. Exp Hematol 12:688–693

Bussolino F, Camussi G, Baglioni C (1988) Synthesis and release of platelet-activating factor by human vascular endothelial cells treated with tumor necrosis factor or interleukin 1$\alpha$. J Biol Chem 263:11856–11861

Bussolino F, Sordano C, Benfenati E, Bozzaro S (1991) *Dictyostelium* cells produces platelet-activating factor in response to cAMP. Eur J Biochem 196:609–615

Bussolino F, Sironi M, Bocchietto E, Mantovani A (1992) Synthesis of platelet-activating factor by polymorphonuclear neutrophils stimulated with interleukin-8. J Biol Chem 267:14598–14603

Bussolino F, Arese M, Silvestro L, Soldi R, Benfenati E, Sanavio F, Aglietta M, Bosia A, Camussi G (1994) Involvement of a serine protease in the synthesis of platelet-activating factor by endothelial cells stimulated by tumor necrosis factor-$\alpha$ or interleukin-1$\alpha$. Eur J Immunol 24:3131–3139

Calabresse C, Nguer MC, Pellegrini O, Benveniste J, Richard Y, Thomas Y (1992) Induction of high-affinity paf receptor expression during T cell activation. Eur J Immunol 22:1349–1355

Camussi G (1986) Potential role of platelet-activating factor in renal pathophysiology. Kidney Int 29:469–477

Camussi G, Bussolino F, Tetta C, Piacibello W, Aglietta M (1983) Biosynthesis and release of platelet-activating factor from human monocytes. Int Arch Allergy Appl Immunol 70:245–251

Camussi G, Bussolino F, Salvidio G, Baglioni C (1987) Tumor necrosis factor/cachectin stimulates peritoneal macrophages, polymorphonuclear neutrophils, and vascular endothelial cells to synthesize and release platelet-activating factor. J Exp Med 166:1390–1404

Chaussade S, Denizot Y, Valleur P, Nicoli J, Raibaud P, Guerre J, Hautefeuille P, Couturier D, Benveniste J (1991) Presence of PAF-acether in stool of patients with pouch ileoanal anastomosis and pouchitis. Gastroenterology 100:1509–1514

Chaussade S, Denizot Y, Colombel JF, Benveniste J, Couturier D (1992) Paf-acether in stool as a marker of intestinal inflammation. Lancet 339:739

Chignard M, Le Couedic JP, Tencé M, Vargaftig BB, Benveniste J (1979) The role of platelet-activating factor in platelet aggregation. Nature 279:799–800

Chung (1992) Platelet-activating factor in inflammation and pulmonary disorders. Clin Sci 83:127–138

Clay KL, Murphy RC, Andres JL, Lynch J, Henson PM (1984) Structure elucidation of platelet activating factor derived from human neutrophils. Biochem Biophys Res Commun 121:815–825

Dhainaut JFA,Tenaillon A, Le Tulzo Y, Schlemmer B, Solet JP, Wolff M, Holzapfel L, Zeni F, Dreyfuss D, Mira JP, De Vathaire F, Guinot P, BN 52021 Sepsis Study Group (1994) Platelet-activating factor receptor antagonist BN 52021 in the treatment of severe sepsis: a randomized, double-blind, placebo-controlled, multicenter clinical trial. Crit Care Med 22:1720–1728

Delbridge LM, Stewart AG, Goulter CM, Morgan TB, Harris PJ (1994) Platelet-activating factor and WEB-2086 directly modulate rat cardiomyocyte contractility. J Mol Cell Cardiol 26:185–193

DeNichilo MO, Stewart AG, Vadas MA, Lopez AF (1991) Granulocyte-macrophage colony-stimulating factor is a stimulant of platelet-activating factor and superoxide anion generation by human neutrophils. J Biol Chem 266:4896–4902

Denizot Y, Chaussade S (1994) Cellular origin of faecal PAF from patients with bacterial diarrhoea [letter]. Immunol Infect Dis 4:78

Denizot Y, Dassa E, Benveniste J, Thomas Y (1989a) Paf-acether production by *Escherichia coli*. Biochem Biophys Res Commun 161:939–943

Denizot Y, Dassa E, Kim HY, Bossant MJ, Salem N Jr, Thomas Y, Benveniste J (1989b) Synthesis of paf-acether from exogenous precursors by the prokaryote *Escherichia coli.* FEBS Lett 243:13–16
Denizot Y, Dassa E, Benveniste J, Thomas Y (1990a) Production of paf-acether by various bacterial strains [letter]. Gastroenterol Clin Biol 14:681–682
Denizot Y, Sobhani I, Rambaud JC, Lewin M, Thomas Y, Benveniste J (1990b) Paf-acether synthesis by *Helicobacter pylori.* Gut 31:1242–1245
Denizot Y, Chaussade S, Benveniste J, Couturier D (1991a) Presence of paf in stool of patients with infectious diarrhea [letter]. J Infect Dis 163:1168
Denizot Y, Michel L, Benveniste J, Thomas Y, Dubertret L (1991b) Paf-acether and human skin – a review. Lipids 26:1093–1094
Denizot Y, Chaussade S, De Boissieu D, Dupont C, Nathan N, Benveniste J, Couturier D (1992a) Presence of paf-acether in stool of patients with bacterial but not with viral or parasitic diarrhoea. Immunol Infect Dis 2:269–273
Denizot Y, Chaussade S, Nathan N, Benveniste J, Couturier D (1992b) Paf-acether and gastrointestinal tract. Eur J Gastroenterol Hepatol 4:871–876
Denizot Y, Chaussade S, Nathan N, Colombel JF, Bossant MJ, Cherouki N, Benveniste J, Couturier D (1992c) PAF-acether and acetylhydrolase in stool of patients with Crohn's disease. Dig Dis Sci 37:432–437
Denizot Y, Dupuis F, Roullet B, Praloran V (1994) PAF and hematopoiesis. II. Elevated levels of plasma paf acetylhydrolase after rapid infusion of 5-fluorouracil in cancer patients. Cancer Lett 85:185–188
Denizot Y, Charissoux JL, Nathan N, Praloran V (1995a) PAF and haematopoiesis. VI. Platelet-activating factor and acetylhydrolase in femoral bone marrow. J Lipid Mediat Cell Signal 12:45–47
Denizot Y, Dupuis F, Praloran V (1995b) Platelet-activating factor and human blood monocytes: an overview. Bull Inst Pasteur 93:43–51
Denizot Y, Trimoreau F, Dupuis F, Verger C, Praloran V (1995c) Presence and metabolism of platelet-activating factor in human bone marrow. Biochim Biophys Acta 1265:55–60
Deryckx S, De Waal Malefyt R, Gauchat JF, Vivier E, Thomas Y, De Vries JE (1992) Immunoregulatory functions of paf-acether. VIII. Inhibition of IL-4-induced human IgE synthesis in vitro. J Immunol 148:1465–1470
Dohi T, Morita K, Kitayama S, Tsujimoto A (1991) Calcium-dependent biosynthesis of platelet-activating factor by submandibular gland cells. Biochem J 276:175–182
Dubois C, Bissonnette E, Rola-Pleszczynski M (1989) Platelet-activating factor (PAF) enhances tumor necrosis factor production by alveolar macrophages. Prevention by PAF receptor antagonists and lipoxygenase inhibitors. J Immunol 143:964–970
Dulioust A, Vivier E, Salem P, Benveniste J, Thomas Y (1988) Immunoregulatory functions of paf-acether. I. Effect of paf-acether on $CD4^+$ cell proliferation. J Immunol 140:240–245
Dulioust A, Duprez V, Pitton C, Salem P, Hemar A, Benveniste J, Thomas Y (1990) Immunoregulatory functions of paf-acether. III. Down-regulation of $CD4^+$ T cells high-affinity IL-2 receptor expression. J Immunol 144:3123–3129
Ebisawa M, Saito H, Iikura Y (1990) Platelet-activating factor-induced activation and cytoskeletal change in cultured eosinophils. Int Arch Allergy Appl Immunol 93:93–98
Eda R, Sugiyama H, Hopp RJ, Okada C, Bewtra AK, Townley RG (1993) Inhibitory effects of formoterol on platelet-activating factor induced eosinophil chemotaxis and degranulation. Int Arch Allergy Appl Immunol 102:391–398
Eguchi H, Frenkel RA, Johnston JM (1994) Binding and metabolism of platelet-activating factor (PAF) by isolated rat type II pneumonocytes. Arch Biochem Biophys 308:426–431
El Azzouzi B, Jurgens P, Benveniste J, Thomas Y (1993) Immunoregulatory functions of paf-acether. IX. Modulation of apoptosis in an immature T cell line. Biochem Biophys Res Commun 190:320–324

Eliakim R, Karmeli F, Razin E, Rachmilewitz D (1988) Role of platelet-activating factor in ulcerative colitis. Enhanced production during active disease and inhibition by sulfasalazine and prednisolone. Gastroenterology 95:1167–1172
Elstad MR, Prescott SM, Mclntyre TM, Zimmerman GA (1988) Synthesis and release of platelet-activating factor by stimulated human mononuclear phagocytes. J Immunol 140:1618–1624
Elstad MR, Stafforini DM, McIntyre TM, Prescott SM, Zimmerman GA (1989) Platelet-activating factor acetylhydrolase increases during macrophage differentiation. A novel mechanism that regulates accumulation of platelet-activating factor. J Biol Chem 264:8467–8470
Evangelou AM (1994) Platelet-activating factor (PAF): implications for coronary heart and vascular diseases. Prostaglandins Leukot Essent Fatty Acids 50:1–28
Evans RD, Lund P, Williamson DH (1991) Platelet-activating factor and its metabolic effects. Prostaglandins Leukot Essent Fatty Acids 44:1–10
Fauler J, Sielhorst G, Frölich JC (1989) Platelet-activating factor induces the production of leukotrienes by human monocytes. Biochim Biophys Acta 1013:80–85
Ferraris L, Karmeli F, Eliakim R, Klein J, Fiocchi C, Rachmilewitz D (1993) Intestinal epithelial cells contribute to the enhanced generation of platelet-activating factor in ulcerative colitis. Gut 34:665–668
Foa R, Bussolino F, Ferrando ML, Guarini A, Tetta C, Mazzone R, Gugliotta L, Camussi G (1985) Release of platelet-activating factor in human leukemia. Cancer Res 45:4483–4485
Gay JC (1993) Mechanism and regulation of neutrophil priming by platelet-activating factor. J Cell Physiol 156:189–197
Gomez-Cambronero J, Durstin M, Molski TFP, Naccache PH, Sha'afi RI (1989) Calcium is necessary but not sufficient for the platelet-activating factor release in human neutrophils stimulated by physiological stimuli. J Biol Chem 264:21699–21704
Gomez-Cambronero J, Wang E, Johnson G, Huang CK, Sha'afi RI (1991) Platelet-activating factor induces tyrosine phosphorylation in human neutrophils. J Biol Chem 266:6240–6245
Gonzalez-Crussi F, Hsueh W (1983) Experimental model of ischemic bowel necrosis. The role of platelet-activating factor and endotoxin. Am J Pathol 112:127–135
Graham RM, Stephens CJ, Silvester W, Leong LLL, Sturm MJ, Taylor RR (1994a) Plasma degradation of platelet-activating factor in severe ill patients with clinical sepsis. Crit Care Med 22:204–212
Graham RM, Strahan ME, Norman KW, Watkins DN, Sturm MJ, Taylor RR (1994b) Platelet and plasma platelet-activating factor in sepsis and myocardial infarction. J Lipid Mediat Cell Signal 9:167–182
Grandel KE, Farr RS, Wanderer AA, Eisenstadt TC, Wasserman SI (1985) Association of platelet-activating factor with primary acquired cold urticaria. N Engl J Med 313:405–409
Harper MJK (1989) Platelet-activating factor: a paracrine factor in preimplantation stages of reproduction? Biol Reprod 40:907–913
Hattori M, Adachi H, Tsujimoto M, Aral H, Inoue K (1994) Miller-Dieker lissencephaly gene encodes a subunit of brain platelet-activating factor. Nature 370:216–218
Hayashi H, Kudo I, Kato T, Nozawa R, Nojima S, Inoue K (1988) A novel bioaction of PAF: induction of microbicidal activity in guinea pig bone marrow cells. Lipids 23:1119–1124
Heller R, Bussolino F, Ghigo D, Garbarino G, Pescarmonas G, Till U, Bosia A (1991) Stimulation of platelet-activating factor synthesis in human endothelial cells by activation of the de novo pathway. J Biol Chem 266:21358–21361
Henocq E, Vargaftig BB (1986) Accumulation of eosinophils in response to intracutaneous paf-acether and allergens in man. Lancet II:1378–1379

Hirafuji M, Shinoda H (1991) PAF-mediated platelet adhesion to endothelial cells induced by FMLP-stimulated leukocytes. J Lipid Mediat 4:347–352

Hirafuji M, Mencia-Huerta JM, Benveniste J (1987) Regulation of PAF-acether (platelet-activating factor) biosynthesis in cultured human vascular endothelial cells stimulated with thrombin. Biochim Biophys Acta 930:359–369

Hirashima Y, Endo S, Ohmori T, Kato R, Takaku A (1994) Platelet-activating factor (PAF) concentration and PAF acetylhydrolase activity in cerebrospinal fluid of patients with subarachnoid hemorrhage. J Neurosurg 80:31–36

Hofmann B, Rühling K, Spangenberg P, Ostermann G (1989) Enhanced degradation of platelet-activating factor in serum from diabetic patients. Haemostasis 19:180–184

Hogaboam CM, Befus AD, Wallace JL (1991) Intestinal platelet-activating factor synthesis during *Nippostrongylus brasiliensis* infection in the rat. J Lipid Mediat 4:211–224

Honda ZI, Nakamura M, Miki I, Minami M, Watanabe T, Seyama Y, Okado H, Toh H, Ito K, Miyamoto T, Shimizu T (1991) Cloning by fünctional expression of platelet-activating factor receptor from guinea-pig lung. Nature 349:342–346

Hough SR, Parks JE (1994) Platelet-activating factor acetylhydrolase activity in seminal plasma from the bull, stallion, rabbit, and rooster. Biol Reprod 50:912–916

Imaizumi TA, Satoh K, Yoshida H, Kawamura Y, Hiramoto M, Koyanagi M, Takamatsu S, Takamatsu M (1990) Activity of platelet-activating factor (PAF) acetylhydrolase in plasma from healthy habitual cigarette smokers. Heart Vessels 5:81–86

Izumi T, Takano T, Bito H, Nakamura M, Mutoh H, Honda ZI, Shinizu T (1995) Platelet-activating factor receptor. J Lipid Mediat Cell Signal 12:429–442

Jouvin-Marche E, Ninio E, Beaurain G, Tencé M, Niaudet P, Benveniste J (1984) Biosynthesis of paf-acether (platelet-activating factor). VII. Precursors of paf-acether and acetyltransferase activity in human leukocytes. J Immunol 133:892–898

Kald B, Olaison G, Sjödahl R, Tagesson C (1990) Novel aspect of Crohn's disease: increased content of platelet-activating factor in ileal and colonic mucosa. Digestion 46:199–204

Karabina SA, Liapikos TA, Grekas G, Goudevenos J, Tselepis AD (1994) Distribution of PAF-acetylhydrolase activity in human plasma low-density lipoprotein subfractions. Biochim Biophys Acta 1213:34–38

Kato T, Kudo I, Hayashi H, Onozaki K, Inoue K (1988) Augmentation of DNA synthesis in guinea pig bone marrow cells by platelet-activating factor (PAF). Biochim Biophys Acta 157:563–568

Kobayashi F, Sagawa N, Ihara Y, Kitagawa K, Yano J, Mori T (1994) Platelet-activating factor-acetylhydrolase activity in maternal and umbilical venous plasma obtained from normotensive and hypertensive pregnancies. Obstet Gynecol 84:360–364

Kubes P, Suzuki M, Granger DN (1990) Platelet-activating factor-induced microvascular dysfunction: role of adherent leukocytes. Am J Physiol 258:G158–G163

Kudo I, Kato T, Hayashi H, Yanoshita R, Ikizawa K, Uda H, Inoue K (1991) Guinea pig bone marrow cells treated with platelet-activating factor generate factor(s) which affect their DNA synthesis and microbicidal activity. Lipids 26:1065–1070

Kuipers B, van der Poll T, Levi M, van Deventer SJH, Ten Cate H, Imai Y, Hack CE, Ten Cate JW (1994) Platelet-activating factor antagonist TCV-309 attenuates the induction of the cytokine network in experimental endotoxemia in chimpanzees. J Immunol 152:2438–2446

Kumar R, King RJ, Martin HM, Hanahan DJ (1987) Metabolism of platelet-activating factor (alkylacetylphosphocholine) by type-II epithelial cells and fibroblasts from rat lungs. Biochim Biophys Acta 917:33–41

Lee TC, Malone B, Wasserman SI, Fitzgerald V, Snyder F (1982) Activities of enzymes that metabolize platelet-activating factor (1-alkyl-2-acetyl-glycero-3-

phosphocholine) in neutrophils and eosinophils from humans and the effect of calcium ionophore. Biochem Biophys Res Commun 105:1303–1308
Lee TC, Fitzgerald V, Chatterjee R, Malone B, Snyder F (1994) Differentiation induced increase of platelet-activating factor acetylhydrolase in HL-60 cells. J Lipid Mediat Cell Signal 9:267–283
Le Gouvello S, Vivier E, Debre P, Thomas Y, Colard O (1992) CD2 triggering stimulates the formation of platelet-activating-factor-acether from alkyl-arachidonoyl-glycerophosphocholine in a human $CD4^+$ T lymphocyte clone. J Immunol 149:1289–1293
Lekka M, Tselepis AD, Tsoukatos D (1986) 1-O-alkyl-2-acetyl-sn-glyceryl-3-phosphorylcholine (PAF) is a minor lipid component in *Tetrahymena pyriformis* cells. FEBS Lett 208:52–55
Lekka ME, Tsoukatos D, Kapoulas V (1989) Uptake of platelet-activating factor (1-O-alkyl-2-acetyl-sn-glyceryl-3-phosphorylcholine) by *Tetrahymena pyriformis*. Comp Biochem Physiol 93B:113–117
Lekka ME, Tsoukatos D, Kapoulas VM (1990) In vivo metabolism of platelet-activating factor (1-O-alkyl-2-acetyl-sn-glyceryl-3-phosphocholine) by the protozoan *Tetrahymena pyriformis*. Biochim Biophys Acta 1042:217–220
Lenihan DJ, Greenberg N, Lee TC (1985) Involvement of platelet-activating factor in physiological stress in the lizard, *Anolis carolinensis*. Comp Biochem Physiol 81C:81–86
Leprince C, Vivier E, Treton D, Galanaud P, Benveniste J, Richard Y, Thomas Y (1991) Immunoregulatory functions of paf-acether. VI. Dual effect on human B cell proliferation. Lipids 26:1204–1208
Letendre ED, Miron P, Roberts KD, Langlais J (1992) Platelet-activating factor acetylhydrolase in human seminal plasma. Fertil Steril 57:193–198
Lindsberg PJ, Hallenbeck JM, Feuerstein G (1991) Platelet-activating factor in stroke and brain injury. Ann Neurol 30:117–129
Lynch JM, Henson PM (1986) The intracellular retention of newly synthesized platelet-activating factor. J Immunol 137:2653–2661
Maki N, Hoffman DR, Johnston JM (1988) Platelet-activating factor acetylhydrolase activity in maternal, fetal, and newborn rabbit plasma during pregnancy and lactation. Proc Natl Acad Sci USA 85:728–732
Malavasi F, Tetta C, Funaro A, Bellone G, Ferrero E, Colli Franzone A, Dellabona P, Rusci R, Matera L, Camussi G, Caligaris-Cappio F (1986) Fc receptor triggering induces expression of surface activation antigens and release of platelet-activating factor in large granular lymphocytes. Proc Natl Acad Sci USA 83:2443–2447
Malchow H, Ewe K, Goebell H, Wellmann W, Leimer HG, Kempe R (1994) Failure of the specific PAF-antagonist apafant in the treatment of ulcerative colitis. Gastroenterology 106:A728 (abstract).
Mazer B, Clay KL, Renz H, Gelfand EW (1990) Platelet-activating factor enhances lg production in B lymphoblastoid cell lines. J Immunol 145:2602–2607
Mazer B, Domenico J, Sawami H, Gelfand EW (1991) Platelet-activating factor induces an increase in intracellular calcium and expression of regulatory genes in human B lymphoblastoid cells. J Immunol 146:1914–1920
Mazière JC, Mazière C, Auclair M, Mora L, Polonovski J (1988) PAF-acether decreases low density lipoprotein degradation and alters lipid metabolism in cultured human fibroblasts. FEBS Lett 236:115–118
McManus LM, Woodard DS, Deavers SI, Pinckard RN (1993) Biology of disease: PAF molecular heterogeneity: pathobiological implications. Lab Invest 69:639–650
Meade CJ, Metcalfe S, Svvennsen Jamieson N, Watson C, Calne RY, Kleber G, Neild G (1991) Serum PAF acetylhydrolase and chronic cholestasis. Lancet 338:1016–1017
Michel L, Denizot Y, Thomas Y, Jean-Louis F, Pitton C, Benveniste J, Dubertret L (1988) Biosynthesis of paf-acether by human skin fibroblasts in vitro. J Immunol 141:948–953

Michel L, Denizot Y, Thomas Y, Jean-Louis F, Heslan M, Benveniste J, Dubertret L (1990) Production of paf-acether by human epidermal cells. J Invest Dermatol 95:576–581

Miwa M, Miyake T, Yamanaka T, Sugatani J, Suzuki Y, Sakata S, Araki Y, Matsumoto M (1988) Characterization of serum platelet-activating factor (PAF) acetylhydrolase. Correlation between deficiency of serum PAF acetylhydrolase and repiratory symptoms in asthmatic children. J Clin Invest 82:1983–1991

Miyagawa H, Nabe M, Hopp RJ, Okada C, Bewtra AK, Townley G (1992) The effect of WEB 2086 on PAF-induced eosinophil chemotaxis and LTC4 production from eosinophils. Agents Actions 37:39–43

Miyaura S, Maki N, Byrd W, Johnston JM (1991) The hormonal regulation of platelet-activating factor acetylhydrolase activity in plasma. Lipids 26:1015–1020

Morland CM, Wilson SJ, Holgate ST, Roche WR (1992) Selective eosinophil leukocyte recruitment by transendothelial migration and not by leukocyte-endothelial cell adhesion. Am J Respir Cell Mol Biol 6:557–566

Moya FR, Eguchi H, Zhao B, Furukawa M, Sfeir J, Osorio M, Ogawa Y, Johnston JM (1994) Platelet-activating factor acetylhydrolase in term and preterm human milk: a preliminary report. J Pediatr Gastroenterol Nutr 19:236–239

Mueller HW, O'Flaherty JT, Wykle RL (1984) The molecular species distribution of platelet-activating factor synthesized by rabbit and human neutrophils. J Biol Chem 259:14554–14559

Nakamura M, Honda ZI, Izumi T, Sakanaka C, Mutoh H, Minami M, Bito H, Seyama Y, Matsumoto T, Noma M, Shimizu T (1991) Molecular cloning and expression of platelet-activating factor receptor from human leukocytes. J Biol Chem 266:20400–20405

Nakayama R, Kumagai H, Saito K (1994a) Evidence for production of platelet-activating factor by yeast *Saccharomyces cerevisiae* cells. Biochim Biophys Acta 1199:137–142

Nakayama R, Udagawa H, Mitsui S, Kumagai H (1994b) *Saccharomyces cerevisiae* cells produce platelet-activating factor in response to calcium ionophore A23187. Biosci Biotech Biochem 58:1115–1119

Narahara H, Johnston JM (1993a) Effects of cytokines on the secretion of platelet-activating factor-acetylhydrolase by human decidual macrophages. Am J Obstet Gynecol 169:531–537

Narahara H, Johnston JM (1993b) Smoking and preterm labor: effect of a cigarette smoke extract on the secretion of platelet-activating factor-acetylhydrolase by human decidual macrophages. Am J Obstet Gynecol 169:1321–1326

Narahara H, Nishioka Y, Johnston JM (1993) Secretion of platelet-activating factor acetylhydrolase by human decidual macrophages. J Clin Endocrinol Metab 77:1258–1262

Nathan N, Denizot Y, Feiss P, Laskar M, Arnoux B, Benveniste J (1992a) Variations of blood PAF-acether levels during coronary artery surgery. J Cardiothorac Vasc Anesth 6:692–696

Nathan N, Denizot Y, Feiss P, Cornu E, Benveniste J, Arnoux B (1992b) Variations in blood platelet-activating factor levels after protamine reversal of heparin in humans. Acta Anaesthesiol Scand 36:264–269

Nathan N, Denizot Y, Huc MC, Claverie C, Laubie B, Benveniste J, Arnoux B (1992c) Elevated levels of paf-acether in blood of patients with type 1 diabetes mellitus. Diabete Metab 18:59–62

Nathan N, Mercury P, Denizot Y, Cornu E, Laskar M, Arnoux B, Feiss P (1994) Effects of the platelet-activating factor receptor antagonist BN 52021 on hematologic variables and blood loss during and after cardiopulmonary bypass. Anesth Analg 79:205–211

Nathan N, Mercury P, Denizot Y, Cornu E, Laskar M, Arnoux B, Feiss P (1995) Effects of the PAF receptor BN 52021 on hemodynamics during and after cardiopulmonary bypass. J Cardiothor Vasc Anesth 9:647–652

Nguer CM, Pellegrini O, Galanaud P, Benveniste J, Thomas Y, Richard Y (1992) Regulation of paf-acether receptor expression in human B cells. J Immunol 149:2742–2748
Ninio E, Mencia-Huerta JM, Benveniste J (1983) Biosynthesis of platelet-activating factor (paf-acether). V. Enhancement of acetyltransferase activity in murine peritoneal cells by calcium ionophore A23187. Biochim Biophys Acta 751:298–304
Oda M, Satouchi K, Yasunaga K, Saito K (1985) Molecular species of platelet-activating factor generated by human neutrophils challenged with ionophore A23187. J Immunol 134:1090–1093
Ojima-Uchiyama A, Masuzawa Y, Sugiura T, Waku K, Fukuda T, Makino S (1991) Production of platelet-activating factor by human normodense and hypodense eosinophils. Lipids 26:1200–1203
Ostermann G, Rühling K, Zabel-Langhenning R, Winkler L, Schlag B, Till U (1987) Plasma from atherosclerotic patients exerts an increased degradation of platelet-activating factor. Thromb Res 47:279–285
Patrignani P, Valitutti S, Aiello F, Musiani P (1987) Platelet-activating factor (PAF) receptor antagonists inhibit mitogen-induced human peripheral blood T-cell proliferation. Biochem Biophys Res Commun 148:802–810
Paul-Eugene N, Dugas B, Picquot S, Lagente V, Mencia-Huerta JM, Braquet P (1990) Influence of interleukin-4 and platelet-activating factor on the Fc epsilon RII/CD23 expression on human monocytes. J Lipid Mediat 2:95–101
Pignol B, Hénane S, Mencia-Huerta JM, Rola-Pleszczynski M, Braquet P (1987) Effect of platelet-activating factor (paf-acether) and its specific receptor antagonist, BN 52021, on interleukin 1 (IL1) release and synthesis by rat spleen adherent monocytes. Prostaglandins 33:931–939
Pignol B, Hénane S, Sorlin B, Rola-Pleszczynski M, Mencia-Huerta JM, Braquet P (1990) Effect of long-term treatment with platelet-activating factor on IL-1 and IL-2 production by rat splen cells. J Immunol 145:980–984
Pignol B, Maisonnet T, Guinot P, Mencia-Huerta JM, Braquet P (1994) Effect of platelet-activating factor on in vitro and in vivo interleukin-6 production. Med Inflamm 3:281–285
Pinckard RN, Jackson EM, Hoppens C, Weintraub ST, Ludwig JC, McManus LM, Mott GE (1984) Molecular heterogeneity of platelet-activating factor produced by stimulated human polymorphonuclear leukocytes. Biochem Biophys Res Commun 122:325–332
Pinckard RN, Ludwig JC, McManus LM (1988) Platelet-activating factors. In: Gallin JI, Goldstein IM, Snyderman R (eds) Inflammation: basic principles and clinical correlates. Raven, New York, pp 139–167
Pinzani M, Carloni V, Marra F, Riccardi D, Laffi G, Gentilini P (1994) Biosynthesis of platelet-activating factor and its 1-O-acyl analogue by liver fat-storing cells. Gastroenterology 106:1301–1311
Poubelle PE, Gingras D, Demers C, Dubois C, Harbour D, Grassi J, Rola-Pleszczynski M (1991) Platelet-activating factor (PAF-acether) enhances the concomitant production of tumour necrosis factor-alpha and interleukin-1 by subsets of human monocytes. Immunology 72:181–187
Pritchard PH, Chonn A, Yeung CCH (1985) The degradation of platelet-activating factor in the plasma of a patient with familial high density lipoprotein deficiency (Tangier Disease). Blood 66:1476–1478
Pustynnikov MG, Porodenko NV, Makarova OV, Kozyukov AV, Moskaleva EY, Sokolovsky AA, Severin ES (1991) Platelet-activating factor stimulates receptor-mediated formation of reactive oxygen intermediates in human monocytes. Lipids 26:1214–1217
Ramesha CS, Pickett WC (1987) Human neutrophil platelet-activating factor: molecular heterogeneity in unstimulated and ionophore-stimulated cells. Biochim Biophys Acta 921:60–66

Robinson M, Snyder F (1985) Metabolism of platelet-activating factor by rat alveolar macrophages: lyso-PAF as an obligatory intermediate in the formation of alkylarachidonoyl glycerophosphocholine species. Biochim Biophys Acta 837:52–56

Rola-Pleszczynski M (1990) Priming of human monocytes with PAF augments their production of tumor necrosis factor. J Lipid Mediat 2:S77–S82

Rola-Pleszczynski M, Pignol B, Pouliot C, Braquet P (1987) Inhibition of human lymphocyte proliferation and interleukin 2 production by platelet-activating factor (paf-acether): reversal by a specific antagonist, BN 52021. Biochem Biophys Res Commun 142:754–760

Rola-Pleszczynski M, Pouliot C, Turcotte S, Pignol B, Braquet P, Bouvrette L (1988) Immune regulation by platelet-activating factor. I. Induction of suppressor cell activity in human monocytes and $CD8^+$ T cells and of helper cell activity in $CD4^+$ T cells. J Immunol 140:3547–3552

Rosam AC, Wallace JL, Whittle BJR (1986) Potent ulcerogenic actions of platelet-activating factor on the stomach. Nature 319:54–56

Rose JK, Debs RA, Philip R, Ruis NM, Valone FH (1990) Selective activation of human monocytes by the platelet-activating factor analog 1-O-hexadecyl-2-O-methyl-sn-glycero-3-phosphorylcholine. J Immunol 144:3513-3517

Ruis NM, Rose JK, Valone FH (1991) Tumor necrosis factor release by human monocytes stimulated with platelet-activating factor. Lipids 26:1060–1064

Saito H, Hayakawa T, Mita H, Akiyama K, Shida T (1992) PAF-induced eosinophilic and basophilic differentiation in human hematopoietic precursor cells. J Lipid Mediat 5:135–137

Saito H, Koshio T, Yanagihara Y, Akiyama K, Shida T (1993) Platelet-activating-factor-induced augmentation of production of eosinophil-lineage cells in hematopoietic precursor cells obtained from human umbilical cord blood. Int Arch Allergy Immunol 102:375–382

Saleh AA, Church MW, Johnston JM (1994) Effect of alcohol on platelet-activating factor acetylhydrolase activity in pregnant and nonpregnant mice. Alcohol Clin Exp Res 18:1009–1012

Salem P, Denizot Y, Pitton C, Dulioust A, Bossant MJ, Benveniste J, Thomas Y (1989) Presence of paf-acether in human thymus. FEBS Lett 257:49–51

Salem P, Deryckx S, Dulioust A, Vivier E, Denizot Y, Damais C, Dinarello CA, Thomas Y (1990) Immunoregulatory functions of paf-acether. IV. Enhancement of IL-1 production by muramyl dipeptide-stimulated monocytes. J Immunol 144:1338–1344

Samuelsson B, Dahlen SE, Lindgren JA, Rouzer CA, Serhan CN (1987) Leukotrienes and lipoxins: structures, biosynthesis, and biological effects. Science 237:1171–1176

Satoh K, Imaizumi TA, Kawamura Y, Yoshida H, Takamatsu S, Mizono S (1988) Activity of platelet-activating factor (PAF) acetylhydrolase in plasma from patients with ischemic cerebrovascular disease. Prostaglandins 35:685–698

Satoh K, Imaizumi TA, Kawamura Y, Yoshida H, Takamatsu S, Mizono S (1989) Increased activity of the platelet-activating factor acetylhydrolase in plasma low density lipoprotein from patients with essential hypertension. Prostaglandins 37:673–681

Satoh K, Yoshida H, Imaizumi TA, Takamatsu S, Mizuno S (1992) Platelet-activating factor acetylhydrolase in plasma lipoproteins from patients with ischemic stroke. Stroke 23:1090–1092

Satoh K, Imaizumi TA, Yoshida H, Takamatsu S (1994) High-density lipoprotein inhibits the production of platelet-activating factor acetylhydrolase by HepG2 cells. J Lab Clin Med 124:225–231

Sawyer DB, Andersen OS (1989) Platelet-activating factor is a general membrane perturbant. Biochim Biophys Acta 987:129–132

Schulam PG, Kuruvilla A, Putcha G, Mangus L, Franklin-Johnson J, Shearer WT (1991) Platelet-activating factor induces phospholipid turnover, calcium flux,

arachidonic acid liberation, eicosanoid generation, and oncogene expression in a human B cell line. J Immunol 146:1642–1648
Shaw JO, Pinckard RN, Ferrigni KS, McManus LM, Hanahan DJ (1981) Activation of human neutrophils with 1-O-hexadecyl/octadecyl-2-acetyl-sn-glyceryl-3-phosphorylcholine (platelet-activating factor). J Immunol 127:1250–1255
Shibata K, Kitayama S, Morita K, Shirakawa M, Okamoto H, Dohi T (1994) Regulation by protein kinase C of platelet-activating factor- and thapsigargin-induced calcium entry in rabbit neutrophils. Jpn J pharmacol 66:273–276
Shukla SD (1992) Platelet-activating factor receptor and signal transduction mechanisms. FASEB J 6:2296–2301
Sisson JH, Prescott SM, Mclntyre TM, Zimmerman GA (1987) Production of platelet-activating factor by stimulated human polymorphonuclear leukocytes. Correlation of synthesis with release, functional events, and leukotriene B4 metabolism. J Immunol 138:3918–3926
Smith CS, Parker L, Shearer WT (1994) Cytokine regulation by platelet-activating factor in a human B cell line. J Immunol 153:3997–4005
Snyder F (1994) Metabolic processing of PAF. Clin Rev Allergy 12:309–327
Snyder F (1995) Platelet-activating factor: the biosynthetic and catabolic enzymes. Biochem J 305:689–705
Sobhani I, Denizot Y, Vissuzaine C, Vatier J, Benveniste J, Lewin MJM, Mignon M (1992a) Significance and regulation of gastric secretion of platelet-activating factor (PAF-acether) in man. Dig Dis Sci 37:1583–1592
Sobhani I, Hochlaf S, Denizot Y, Vissuzaine C, Rene E, Benveniste J, Lewin MJM, Mignon M (1992b) Raised concentration of platelet-activating factor in colonic mucosa of Crohn's disease patients. Gut 33:1220–1225
Sörensen J, Kald B, Tagesson C, Lindahl M (1994) Platelet-activating factor and phospholipase A2 in patients with septic shock and trauma. Intensive Care Med 20:555–561
Stafforini DM, Prescott SM, Mclntyre TM (1987) Human plasma platelet-activating factor acetylhydrolase. Purification and properties. J Biol Chem 262:4223–4230
Stafforini DM, Elstad MR, McIntyre TM, Zimmerman GA, Prescott SM (1990) Human macrophages secrete platelet-activating factor acetylhydrolase. J Biol Chem 265:9682–9687
Stafforini DM, Prescott SM, Zimmerman GA, Mclntyre TM (1991) Platelet-activating factor acetylhydrolase activity in human tissues and blood cells. Lipids 26:979–985
Sugiura T, Ojima-Uchiyama A, Masuzawa Y, Fujita M, Nakagawa Y, Waku K (1991) Regulation of the biosynthesis of platelet-activating factor in alveolar macroph ages. Lipids 26:974–978
Tarbet EB, Stafforini DM, Elstad MR, Zimmerman GA, Mclntyre TM, Prescott SM (1991) Liver cells secrete the plasma form of platelet-activating factor acetylhydrolase. J Biol Chem 266:16667–16673
Terashita ZI, Imura Y, Nishikawa K, Sumida S (1985) Is platelet-activating factor (PAF) a mediator of endotoxin shock? Eur J Pharmacol 109:257–261
Thivierge M, Rola-Pleszczynski M (1992) Platelet-activating factor enhances interleukin-6 production by alveolar macrophages. J Allergy Clin Immunol 90:796–802
Thomas Y, Denizot Y, Dassa E, Boullet C, Benveniste J (1986) Synthesis of paf-acether by *E. coli* K12. CR Acad Sci III 303:699–702
Thompson WA, Coyle S, Van Zee K, Oldenburg H, Trousdale R, Rogy M, Felsen D, Moldawer L, Lowry SF (1994) The metabolic effects of platelet-activating factor antagonism in endotoxemic man. Arch Surg 129:72–79
Tjoelker LW, Wilder C, Eberhardt C, Stafforini DM, Dietsch G, Schimpf B, Hooper S, Le Trong H, Cousens LS, Zimmerman GA, Yamada Y, Mclntyre TM, Prescott SM, Gray PW (1995) Anti-inflammatory properties of a platelet-activating factor acetylhydrolase. Nature 374:549–553
Travers JB, Sprecher H, Fertel RH (1990) The metabolism of platelet-activating factor in human T-lymphocytes. Biochim Biophys Acta 1042:193–197

Triggiani M, D'Souza DM, Chilton FH (1991) Metabolism of 1-acyl-2-acetyl-sn-glycero-3-phosphocholine in the human neutrophil. J Biol Chem 266:6928–6935

Triggiani M, Schleimer RP, Tomioka K, Hubbard WC, Chilton FH (1992) Characterization of platelet-activating factor synthesized by normal and granulocyte-macrophage colony-stimulating factor-primed human eosinophils. Immunology 77:500–504

Tsaoussis V, Vakirtzi-Lemonias C (1994) The mouse plasma PAF acetylhydrolase: II. It consists of two enzymes both associated with the HDL. J Lipid Mediat Cell Signal 9:317–331

Tselepis AD, Lekka ME, Tsoukatos D (1991) A PAF-acetylhydrolase activity in *Tetrahymena pyriformis*. FEBS Lett 288:147–150

Valone FH, Epstein LB (1988) Biphasic platelet-activating factor synthesis by human monocytes stimulated with IL-1$\beta$, tumor necrosis factor, or INF-$\gamma$. J Immunol 141:3945–3950

Valone FH (1991) Synthesis of platelet-activating factor by human monocytes stimulated by platelet-activating factor. J Allergy Clin Immunol 87:715–720

Van Der Bruggen T, Kok PTM, Raaijmakers JAM, Lammers JWJ, Koenderman L (1994) Cooperation between Fc$\gamma$ receptor II and complement receptor type 3 during activation of platelet-activating factor release by cytokine-primed human eosinophils. J Immunol 153:2729–2735

Venable ME, Zimmerman GA, Mclntyre TM, Prescott SM (1993) Platelet-activating factor: a phospholipid autacoid with diverse actions. J Lipid Res 34:691–702

Vivier E, Salem P, Dulioust A, Praseuth D, Metezeau P, Benveniste J, Thomas Y (1988) Immunoregulatory functions of paf-acether. II. Decrease of CD2 and CD3 antigen expression. Eur J Immunol 18:425–430

Vivier E, Deryckx S, Wang JL, Valentin H, Peronne C, De Vries JE, Bernard A, Benveniste J, Thomas Y (1990) Immunoregulatory functions of paf-acether. VI. Inhibition of T cell activation via CD3 and potentiation of T cell activation via CD2. Int Immunol 2:545–553

Wallace JL (1990) Lipid mediators of inflammation in gastric ulcer. Am J Physiol 258:G1–G11

Weintraub ST, Ludwig JC, Mott GE, McManus LM, Lear C, Pinckard RN (1985) Fast atom bombardment-mass spectrometric identification of molecular species of platelet-activating factor produced by stimulated human polymorphonuclear leukocytes. Biochem Biophys Res Commun 129:868–876

Wirthmueller U, De Weck AL, Dahinden CA (1989) Platelet-activating factor production in human neutrophils by sequential stimulation with granulocyte-macrophage colony stimulating factor and the chemotactic factors C5a or formyl-methionyl-leucyl-phenylalanine. J Immunol 142:3213–3218

Wirthmueller U, De Weck AL, Dahinden CA (1990) Studies on the mechanism of platelet-activating factor production in GM-CSF primed neutrophils: involvement of protein synthesis and phospholipase A2 activation. Biochem Biophys Res Commun 170:556–562

CHAPTER 23
# Qualitative and Quantitative Assessment of Platelet Activating Factors

R.N. PINCKARD, D.S. WOODARD, S.T. WEINTRAUB, and L.M. MCMANUS

## A. PAF Molecular Diversity

Platelet-activating factor (PAF) often is viewed as a single molecular entity with defined biological activities and potencies. It is now clear, however, that several structurally related molecular species of PAF are produced by various cells and tissues (BRATTON et al. 1994; MCMANUS et al. 1993; PINCKARD 1989; PINCKARD et al. 1994). PAF is actually comprised of three subclasses of choline-containing, *sn*-2-acetylated phospholipids having either 1-*O*-alkyl-, 1-acyl-, or 1-*O*-alk-1′-enyl-linked carbon chains which vary in their length, degree of saturation, and/or branching [i.e., (1-*O*-alkyl/acyl/*O*-alk-1′enyl)-2-acetyl-*sn*-glycero-3-phosphocholine; Fig. 1]. While most molecular species of PAF have platelet-activating activity, there are significant variations in their respective potencies (Table 1). Therefore, these phospholipids should be designated as "PAF" provided that the appropriate subclass prefix alkyl-, acyl-, or alkenyl- be included to denote the chemical linkage at the *sn*-1 position, i.e., alkyl-PAF, acyl-PAF, and alkenyl-PAF.

An important question relative to the above is to what extent is PAF molecular diversity (patho)physiologically important? There are at least two aspects of this question that will require additional information. The first issue is whether or not individual molecular species of PAF (within and between subclasses) interact differentially with various target cells and tissues; and if so, do they differentially interact with a PAF receptor having more than one affinity state or with more than one PAF receptor subtype? In this regard, accumulating evidence indicates that there may be more than one PAF receptor affinity state or perhaps more than one PAF receptor subtype (cf. HWANG 1990, 1991, 1994; KROEGEL et al. 1989; MCMANUS et al. 1993; PINCKARD et al. 1994; see also Chap. 7). For example, there are significant differences in the rank orders of potency of several alkyl-PAF and acyl-PAF homologs for stimulating various cells and tissues in vitro and in vivo (Table 1). Even within the same cell type, i.e., the isolated human polymorphonuclear neutrophilic leukocyte or PMN, there are different rank orders of potency of alkyl-PAF and acyl-PAF homologs for stimulating different physiological responses (Table 1; PINCKARD et al. 1992). These observations support the premise that the nature and intensity of (patho)physiological responses involving PAF will be dependent on the molecular composition (i.e., types and amounts) of PAF.

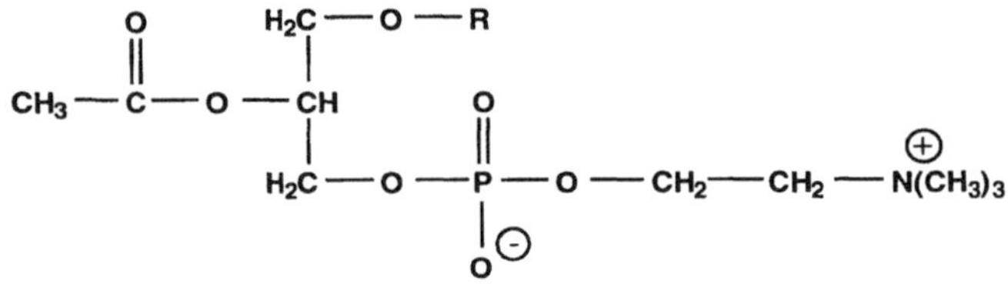

| R = | Linkage | Chain Composition | |
|---|---|---|---|
| | alkyl | 12:0 | 18:0 |
| | acyl | 14:0 | 18:1 |
| | alk-1'-enyl | 15:0 | 18:2 |
| | | 16:0 | 19:0 |
| | | 17:0 | 20:0 |
| | | $17{:}0_{BR}$ | 20:1 |

**Fig. 1.** Platelet-activating factor (PAF) molecular diversity. The *sn*-1 substituent of PAF varies in the length and degree of unsaturation of the carbon chain which is linked to the glycerol backbone by either an alkyl-, acyl-, or alk-1′-enyl bond. The number of carbons in the chain and the degree of unsaturation are indicated

**Table 1.** Relative potency of PAF molecular species

| PAF homolog | PAF relative potency | | | | | |
|---|---|---|---|---|---|---|
| | Rabbit platelet secretion | Human PMN secretion | Human PMN priming | Rabbit right ventricular hypertension | Rabbit thrombocytopenia | Rabbit neutropenia |
| 16:0-Alkyl-PAF | 0.11 ± 0.004[a] | 14.2 ± 2.3[b] | 3.4 ± 0.8[c] | 0.08[d] | 0.19[d] | 0.08[d] |
| 18:1-Alkyl-PAF | 0.15 ± 0.006 | 16.1 ± 1.1 | 5.5 ± 1.6 | 0.13 | 0.30 | 0.14 |
| 18:0-Alkyl-PAF | 0.27 ± 0.011 | 445.0 ± 105.0 | 6.3 ± 1.8 | 0.54 | <0.13 | 0.35 |
| 16:0-Acyl-PAF | 18.1 ± 0.9 | Inactive | 57.4 ± 12.3 | 43 | >150 | 34 |
| 18:1-Acyl-PAF | 53.9 ± 2.9 | Inactive | 56.1 ± 10.2 | 66 | >400 | 25 |
| 18:0-Acyl-PAF | 195.9 ± 74.9 | Inactive | 67.3 ± 8.7 | 192 | >400 | 257 |

PAF, platelet-activating factor; PMN, polymorphonuclear neutrophilic leukocyte.
[a] PAF concentration (*nM*) that stimulated 50% release of [$^3$H] serotonin from isolated rabbit platelets. Data represent the mean ± SE; n = 5–48 separate platelet preparations.
[b] PAF concentration (*nM*) that resulted in the release of 10% of the total lysozyme from isolated human PMN. Inactive = no detectable lysozyme release at a final concentration of 1 *μM*. Data represent the mean ± SE; n = 10–30 different PMN preparations.
[c] PAF concentration (*nM*) that primed isolated human PMN for 50% of maximum superoxide production after stimulation with N-formyl-methionyl-leucyl-phenylalanine (FMLP). Data represent the mean ± SE; n = 6 different PMN preparations.
[d] PAF concentration (nmole/kg) that resulted in 50% of the maximum change in right ventricular pressure, circulating platelets, or PMN in anesthetized rabbits. Each animal received a single intravenous dose of PAF. Data represent the results derived from the evaluation of 3–5 groups of animals for each PAF; n = 3–10 animals per group receiving a given dose of PAF.

Related to the above, equilibrium binding studies have indicated that there are both high and low affinity PAF receptors on the human PMN (Herbert et al. 1993; O'Flaherty et al. 1992); these receptors appear to be associated with different alkyl-PAF-induced physiological responses of the PMN (Naccache et al. 1985, 1986; Pinckard et al. 1992, 1994) and eosinophil (Kroegel et al. 1989). To date, only a single human PMN PAF receptor has been cloned (Kunz et al. 1992), and it has a high affinity. No evidence is available to support the hypothesis that there is a genetically distinct low affinity PAF receptor. Nevertheless, the cloned PMN high affinity PAF receptor can interact with more than one G protein (Ali et al. 1994). Therefore, it is possible that different G proteins could confer different conformational states on the PAF receptor, thereby altering its affinity. PAF receptor coupling to diverse G proteins would also provide an explanation as to why the PAF receptor(s) appears to be linked to separate signal transduction pathways. In any case, it is important to recognize that the interaction of individual molecular species of PAF with these receptors remains to be established.

The second issue relative to the (patho)physiological importance of PAF molecular diversity is to elucidate the molecular composition of the PAF that is synthesized by different cells and tissues and the PAF that is produced in vivo during normal as well as during various disease processes. Thus, it is imperative that the appropriate technology be employed to detect and quantitate PAF. As will be described below, certain PAF assays do not detect or are more sensitive in detecting certain molecular species of PAF. The following discussion will briefly review the various tests that have been employed to detect and quantitate PAF and possible pitfalls that might be encountered during their utilization.

Early PAF studies used available analytical technology to analyze PAF of biological origin. These techniques included PAF's chromatographic behavior relative to synthetic alkyl-PAF, the chemical and enzymatic susceptibility of PAF, and inhibition of biological activity with PAF receptor antagonists. More recent studies employing phospholipid derivatization and subsequent mass spectral (MS) analysis have significantly improved the identification and quantitation of PAF.

## B. Estimation of PAF Activity by Bioassay

### I. Platelet Bioassay

The initial characterization of PAF was based upon platelet bioassay (Benveniste et al. 1972). Indeed, the descriptive name "platelet-activating factor" reflects those original studies. PAF stimulates platelets to undergo receptor-initiated, G-protein-mediated intracellular signalling (e.g., calcium mobilization, phospholipase $A_2$ ($PLA_2$) activation, protein phosphorylation, phosphoinositide turnover) which leads to platelet aggregation, arachidonic

**Table 2.** Species-dependent platelet responsiveness to PAF

| Species | Platelet preparation | $ED_{50}$ ($nM$)[a] Aggregation | Secretion | References |
|---|---|---|---|---|
| Horse | PRP | 0.0038 | – | SUQUET and LEID 1983 |
| | Isolated | 0.0005 | – | |
| Cow | PRP | 0.036 | – | LIGGITT et al. 1984 |
| | Isolated | 0.0088 | – | |
| Guinea pig | PRP | 0.7 | 4.7 | AMBLER and WALLIS 1983 |
| | Isolated | 0.1 | – | STEWART and GRIGORIADIS 1991 |
| Rabbit | Whole blood | 23.2 | – | AMMIT and O'NEILL 1991b |
| | PRP | 2.4 | 15.0 | MCMANUS et al. 1981 |
| | Isolated | 0.06 | 0.11 | BOSSANT et al. 1990 HENSON 1990 |
| Sheep | PRP | 10 | – | MOON et al. 1990 |
| | Isolate | 0.05 | 0.05 | |
| Human | PRP | 100 | 760 | MCMANUS et al. 1981 |
| | Isolated | 0.25 | Little or none | EKHOLM et al. 1992 |
| Rat | Isolated | – | None | INARREA et al. 1984 |
| Mouse | PRP | none | – | LANARA et al. 1982 |
| | Isolated | none | – | SCODRAS et al. 1991 |

PAF, platelet-activating factor; PRP, platelet-rich plasma.
[a] $ED_{50}$ represents the concentration of PAF (usually 16:0-alkyl-PAF) required to elicit half maximal response.

acid metabolism, and the release of granule contents. Among different animal species, platelet responsiveness to alkyl-PAF is quite variable (Table 2). Of note, platelets from rats and mice are unresponsive to alkyl-PAF stimulation; platelets from these species appear to lack PAF receptors (HWANG and LAM 1986; INARREA et al. 1984).

Platelet responses to PAF are dependent upon the conditions of bioassay. For instance, platelets that have been washed free of plasma proteins are considerably more responsive to alkyl-PAF activation than are platelets in a plasma environment (i.e., platelet-rich plasma or PRP; Table 2). This difference in platelet responsiveness is because plasma contains both albumin which interferes with PAF activity and PAF acetylhydrolase which rapidly degrades PAF into an inactive lysophospholipid (see below). In isolated platelets, extracellular calcium and, for some species, exogenous fibrinogen are required for optimal responses to PAF.

### 1. Rabbit Platelet Bioassay

The majority of studies to quantitate PAF by bioassay utilized rabbit platelets. This is because PAF-induced rabbit platelet aggregation and secretion of dense granule contents is very reproducible and sensitive. A standard curve using known amounts of authentic PAF (e.g., 16:0-alkyl-PAF) is employed to estimate the equivalent biological PAF activity in unknown samples (Fig. 2). With isolated rabbit platelets, the range of sensitivity of this PAF bioassay is approximately 1–30 fmole of 16:0-alkyl-PAF. In this same bioassay, the metabolism of arachidonic acid into thromboxane $A_2$ and the release of alpha-granule constituents (platelet factor 4) can also be measured to quantify PAF (McMANUS et al. 1983). However, since both of these endpoints of platelet activation are more cumbersome to use for the rapid estimation of PAF activity, the measurement of rabbit platelet aggregation and/or secretion of [$^3$H]serotonin are preferentially and routinely employed. Of importance, PAF-induced rabbit platelet aggregation is enhanced by platelet-derived ADP (PINCKARD et al. 1979); therefore, ADP scavengers must be included in platelet aggregation assays. Neither ADP scavengers nor inhibitors of arachidonate metabolism alter PAF-induced secretion of [$^3$H]serotonin in isolated rabbit platelets.

## 2. Platelet Desensitization

After exposure to PAF, platelets (and most PAF target cells or tissues) become temporarily unresponsive or desensitized to further PAF activation. This desensitization is specific for PAF inasmuch as platelets retain complete

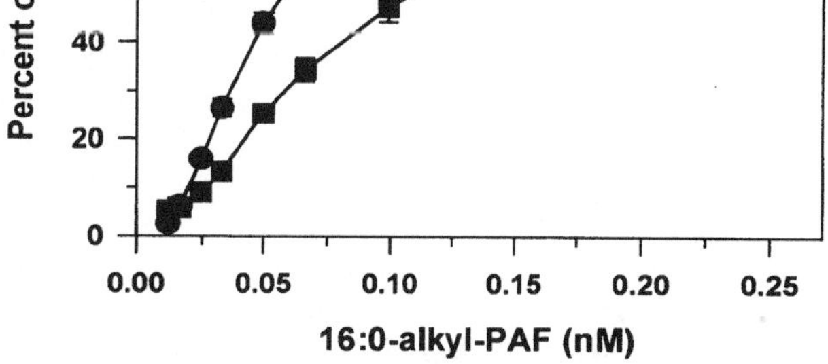


**Fig. 2.** Platelet-activating factor (PAF) bioassay: dose-response of PAF-induced rabbit platelet aggregation and secretion. Washed rabbit platelets, prelabeled with [$^3$H]serotonin (250000 platelets/$\mu$l, 1.3 *mM* $Ca^{2+}$), were stimulated with the indicated concentrations of 16:0-alkyl-PAF. Platelet aggregation was estimated (in the presence of ADP scavengers) as the maximum change in light transmittance (%). In separate assays for platelet secretion, the reaction was terminated 60 s after platelet stimulation and the extent of [$^3$H]serotonin release calculated as a percentage of total platelet [$^3$H]serotonin. PAF levels in unknown samples analyzed in parallel can be directly determined from these standard curves and expressed as fmole equivalents of 16:0-alkyl-PAF. Data represent the mean ± SE; n = 17 or 9 different platelet preparations for aggregation or secretion, respectively

responsiveness to other agonists such as thrombin (DEMOPOULOS et al. 1979). Although the precise basis for platelet desensitization to PAF is unknown, it is likely that PAF receptor availability is reduced or that some aspect of receptor-initiated intracellular signalling is attenuated. In either case, desensitization to a given molecular species of PAF confers platelet unresponsiveness to all other PAF molecular species (MCMANUS and PINCKARD, unpublished observations).

### 3. PAF Carriers in Platelet Bioassay

Important in PAF bioassay is the carrier used to solubilize phospholipids. To this end, numerous reagents have been employed including buffer alone, buffer containing 0.25% albumin, ethanol, or dimethylsulfoxide. In our experience, maximal platelet responsiveness is obtained when bovine or human serum albumin-containing aqueous solutions are utilized. It should be noted that although albumin binds to PAF (and other lipids; CLAY et al. 1990) and permits maximal platelet responsiveness, excess albumin has a detrimental effect upon PAF activity in platelet bioassay (GRIGORIADIS and STEWART 1992).

### 4. PAF Molecular Diversity in Platelet Bioassay

For the most part, biological samples containing PAF are usually not rigorously fractionated into individual molecular species of PAF prior to bioassay. This reflects inherent difficulties in accomplishing the separation of these structurally similar phospholipids. Nevertheless, this is an important consideration for unknown samples inasmuch as each molecular species of PAF has a distinct biological potency in the platelet bioassay (Table 1). Therefore, even in the best of circumstances (e.g., samples purified by normal- and/or reversed-phase high performance liquid chromatography or HPLC), several molecular species of PAF are simultaneously presented to the platelet. Little is known regarding the effects of mixtures of PAF on platelets in bioassay. However, it is anticipated that individual PAF molecular species would simultaneously compete for available PAF receptors on the platelet. This would be expected to reduce the sensitivity of the bioassay. Thus, it is likely that platelet bioassay estimation of biologically derived PAF underestimates the total (molar) PAF in a given sample. We have recently confirmed this premise in studies characterizing the PAF in normal human saliva where six molecular species of PAF were quantitated after derivatization and gas chromatographic (GC)/MS analysis (WOODARD et al. 1995; see also below). Thus, the total amounts of salivary PAF detected by platelet bioassay were only 67% of that determined by GC/MS.

## II. Other

PAF initiates receptor-mediated activation of a variety of other cells and tissues (cf. STEWART 1994; VENABLE et al. 1993). This includes the dose-dependent stimulation of PMN, eosinophils, monocytes, macrophages, vascular endothelial cells, as well as isolated heart, ileum, and lung. Although each of these PAF targets could be employed in the estimation of PAF activity in unknown samples, several factors have hampered this approach. First, the preparation of many of these PAF targets is more difficult and technically demanding than that involved in platelet isolation and bioassay. Second, the reproducibility of the results after PAF stimulation is more variable than that obtained with platelets. Third, and most importantly, the sensitivity of these other bioassays is generally reduced as compared to platelet bioassay.

## C. Estimation of PAF Activity by Radioimmunoassay (RIA)

### I. RIA Development and Use

In 1984, Nishihira and colleagues described the characterization of specific antibodies against alkyl-PAF (Nishihira et al. 1984); this antibody preparation did not react with acyl-PAF or with nonacetylated phospholipids. Based on these observations, several PAF-specific antibodies have been produced and RIAs for the estimation of PAF have been developed (Baldo et al. 1991; Karasawa et al. 1991; Smal et al. 1990; Sugatani et al. 1990; Wang and Tai 1992). The range of detection for these assays is 20 fmoles to 30 pmoles of 16:0-alkyl-PAF. Of importance, the antibodies used in these RIAs do not cross-react with other phospholipids including phosphatidylcholine, phosphatidylethanolamine, phosphatidylglycerol, phosphatidylserine, sphingomyelin, phosphatidylinositol, phosphatidic acid, cardiolipin, or lysophosphatidylcholine.

Estimation of PAF levels in biological samples has been facilitated by the commercial availability of RIA detection systems. These assays have been used to demonstrate increased levels of PAF in a variety of diverse human disease states including: (1) in the blood of asthmatic subjects (Hsieh and Ng 1993; Kurosawa et al. 1994), (2) in the plasma of patients with liver cirrhosis and disseminated intravascular coagulation (Sugatani et al. 1993), and (3) in the bronchoalveolar lavage fluid obtained from patients with adult respiratory distress syndrome (ARDS). It is anticipated that the results of these types of studies will facilitate further investigations designed to improve our understanding of the role of PAF in these and related disorders.

### II. PAF Molecular Diversity in RIA

As outlined above, most RIAs for PAF have specificities for the acetyl group at the *sn*-2 position and an alkyl linkage at the *sn*-1 position. In addition, there are differences in antibody reactivity with the various alkyl-PAF homologs in RIA which reflect variations in the degree of unsaturation and carbon chain length at the *sn*-1 position (cf. Smal et al. 1990). Thus, while the RIA for PAF is highly reproducible and sensitive and may circumvent problems inherent in PAF bioassay, RIA is limited in that it will not detect acyl- or alkenyl-PAF. In this regard, salivary PAF detected by RIA was found in significantly higher concentration than that determined by bioassay (Cooney et al. 1991; Smal and Baldo 1991); these differences were thought to reflect the presence of a PAF inhibitor in saliva that did not affect the RIA. In contrast, when used to estimate embryo-derived PAF, the PAF RIA provided comparable or somewhat reduced PAF levels as compared to the PAF bioassay (Ammit and O'Neill 1991a; Ammit et al. 1992); these differences likely reflect variations in RIA sensitivity to individual molecular species of alkyl-PAF. In unrelated studies, PAF levels estimated by RIA were found to be comparable to values derived by MS analyses (Sugatani et al. 1993). Overall, RIA quantitation of PAF appears to offer a rapid and reproducible measure of alkyl-PAF in heterogenous lipid extracts of biological origin.

## D. Estimation of PAF by Radioreceptor Assay

Based on observations that PAF receptors have high affinity (Honda et al. 1991; Kunz et al. 1992; Ye et al. 1991; see above), a competitive receptor binding assay for the detection and quantitation of PAF has been developed (Janero et al. 1988; Paulson and Nicholson 1988; Tiberghien et al. 1991). In brief, these techniques which utilize intact platelets or platelet membrane preparations (which can be stored at –20°C) are based on competition of unlabeled PAF for the specific, high-affinity binding of [$^3$H]PAF to platelet receptors. However, this assay does not take into account differences in the binding characteristics of different molecular species of PAF.

## E. Estimation of PAF Based on [$^3$H]Acetate Incorporation

Exogenously added, radiolabelled PAF precursors have been widely used to study PAF synthesis and release. There are several potential PAF precursors that can be utilized in this regard; the selection of the most appropriate PAF precursor to employ depends on which of the PAF biosynthetic pathways is operative in the cell or tissue of interest. For cells capable of de novo PAF biosynthesis (Fig. 3), radiolabelled choline would be the substrate of choice because cytidylyltransferase is the rate-limiting enzyme specific for the de novo pathway (Snyder 1994); cytidylyltransferase catalyzes the production of cytidine diphosphate (CDP)-choline. CDP-choline together with 1-*O*-alkyl-2-acetyl-*sn*-glycerol forms PAF through the action of dithiothreitol (DTT)-insensitive cholinephosphotransferase (Renooij and Snyder 1981). With respect to the remodelling (deacylation/reacetylation) pathway for PAF production (Fig. 3), two PAF precursors have been used, i.e., 1-*O*-[$^3$H]alkyl-

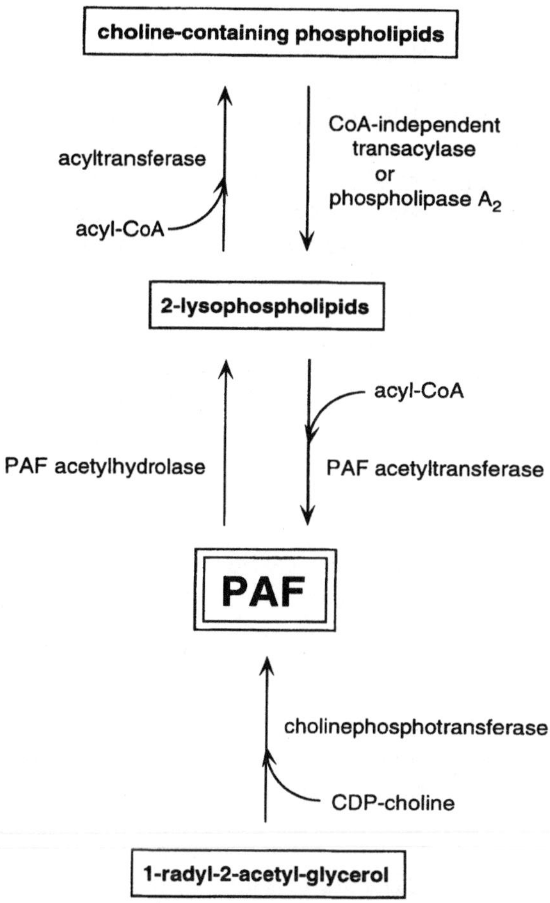


**Fig. 3.** Platelet-activating factor (PAF) metabolic pathways. PAF can be synthesized either from choline-containing phospholipids via the remodelling pathway (CoA-independent transacylase or phospholipase $A_2$ coupled with PAF acetyltransferase) or the de novo pathway (choline phosphotransferase)

*sn*-glycero-3-phosphocholine ([$^3$H]lyso-PAF) and [$^3$H]acetate. It should be pointed out that each of these substrates can also be incorporated into PAF via the de novo pathway of PAF biosynthesis (Fig. 3). However, in cells that rapidly synthesize PAF, incorporation of either [$^3$H]acetate or [$^3$H]lyso-PAF into PAF would most likely reflect the remodeling pathway for PAF biosynthesis. To date, [$^3$H]acetate has been the radiolabelled substrate of choice and when utilized appropriately, provides a very useful tool to study PAF synthesis and release.

Shortly after the structural elucidation of PAF (Demopoulos et al. 1979; Hanahan et al. 1980), radiolabelled acetate was employed to study PAF biosynthesis (Chap et al. 1981). Since that time, there have been numerous studies that have utilized either $^{14}C$- or $^{3}H$-labelled acetate to study PAF production by a variety of cells and tissues. Currently, [$^{3}H$]acetate is the preferred substrate because it is commercially available in very high specific activities, i.e., 1–10 Ci/mmol. To study PAF biosynthesis by [$^{3}H$]acetate incorporation, many investigators have employed procedures described by Mueller and coworkers (Mueller et al. 1983). In brief, [$^{3}H$]acetate (5–25 $\mu$Ci/ml or approximately 1 *nmol*, depending on the specific activity) is added to cell suspensions, cell monolayers, or tissues slices; often, albumin (either BSA or HSA) at 1–2.5 mg/ml is also added to the incubation medium. Either immediately or after 5–10 min of preincubation, the cells/tissues are then stimulated with the desired agonist for the appropriate time period; unstimulated cells/tissues incubated in the presence of an identical concentration of [$^{3}H$]acetate serve as the control. After stimulation, the lipids are extracted from the cell suspensions, tissues, or their respective supernatants. The preferred method of lipid extraction is that originally described by Bligh and Dyer (1959), often with the inclusion of 0.1%–1% glacial acetic acid (A). If bioassay, RIA, or other tests for PAF will not be conducted on these lipid extracts, unlabelled PAF (10–15 $\mu$g) often is added as carrier to improve isolation efficiency. Briefly, lipids are extracted for at least 1 h using a mixture of chloroform:methanol:water (C:M:W) at 1:2:0.8 v/v/v; thereafter, additional chloroform and water are added to obtain a ratio of 1:1:0.9 (C:M:W) that effects phase separation. The lower chloroform-rich layer, which contains the PAF, is washed several times with fresh upper phase layer (methanol/water) to remove free [$^{3}H$]acetate. Some investigators have quantitated [$^{3}H$] radioactivity in this crude lipid extract as an estimate of PAF biosynthesis; however, this is not prudent practice for several reasons. First, there is no way to know whether all of the free [$^{3}H$]acetate has been removed; moreover, it is very likely that some of the [$^{3}H$]acetate is either metabolized and/or becomes incorporated into other non-PAF, chloroform-soluble molecules. For these reasons, crude lipid extracts are subsequently purified by thin layer chromatography (TLC) on silica gel G plates employing various solvent systems including: C:M:A:W, 50:25:8:3; C:M:W, 65:35:6; C:M:W, 60:40:10. Irrespective of the solvent system used, it is imperative to cochromatograph standard amounts of authentic, radiolabelled PAF (both alkyl-PAF and acyl-PAF) in separate lanes on the same TLC plate to estimate percent recoveries for [$^{3}H$]acetate PAF in parallel.

The preceding protocol has been employed in many studies to identify and quantitate PAF. Unfortunately, erroneous conclusions have been made in some of these studies because of technical pitfalls (cf. Pinckard et al. 1994). The principal problem is that [$^{3}H$]acetate incorporates into any of the three PAF subclasses, i.e., alkyl-PAF, alkenyl-PAF, and acyl-PAF; none of the currently available TLC purification procedures (mentioned above), and few of the HPLC purification procedures, adequately separate the three PAF subclasses. Therefore, $\beta$-scintillation spectrometry of the TLC-purified lipid extract does not discriminate between PAF subclasses. Some studies have addressed this problem by quantitating [$^{3}H$] activity of the TLC-purified lipid extracts before and after their enzymatic digestion with *Rhizopus* lipase. This enzyme has phospholipase-$A_1$-like activity and hydrolyzes the *sn*-1 ester bond of diacylphospholipids. Thus, after *Rhizopus* lipase enzymatic digestion of a mixture of [$^3$H]alkyl-PAF and [$^3$H]acyl-PAF, only the [$^3$H]acyl-PAF would be hydrolyzed to 2-[$^3$H]acetyl-glycero-3-phosphocholine and the corresponding long chain fatty acids. Then, after Bligh and Dyer lipid extraction of the enzyme digest, the 2-[$^3$H]acetyl-glycero-3-phosphocholine would partition into the upper methanol/water phase leaving only alkyl-[$^3$H]PAF in the lower chloroform-rich phase. Therefore, by subtraction of the pre- and postlipase, chloroform soluble $^3$H-activity and the alkyl-PAF/acyl-PAF ratio can be determined. Unfortunately, for unknown reasons, *Rhizopus* lipase digestion has not always proven reliable. As an example, for many years, PAF produced by cultured vascular endothelial cells (EC) and detected by [$^3$H]acetate incorporation was thought to be "authentic" alkyl-PAF because of its relative resistance to lipase digestion. However, with improved analytical techniques, it now appears that EC-derived PAF consists in actuality of more than 90%–95% acyl-PAF (cf. McManus et al. 1993; Pinckard et al. 1994). From our own experience, one possible explanation for the previous misidentifications is that commercial preparations of *Rhizopus* lipase are not always active when tested with radiolabelled, synthetic acyl-PAF. That EC were finally found to synthesize acyl-PAF as opposed to alkyl-PAF was documented by both GC/MS (Clay et al. 1991) and by the derivatization and TLC separation procedures described below which differentiate [$^3$H]acetate labelled alkyl-PAF from acyl-PAF (Garcia et al. 1991; Mueller et al. 1991; Triggiani et al. 1991).

To separate alkyl-PAF from acyl-PAF by TLC, the molecules must first be derivatized (Mueller et al. 1991). This is accomplished for this method by first removing phosphocholine from the PAF by phospholipase C (PLC) enzymatic digestion. The resulting diglycerides are then acetylated by treatment with radiolabelled acetic anhydride, producing 1-*O*-alkyl-[$^3$H]2,3-diacetylglycerols and 1-acyl-[$^3$H]2,3-diacetylglycerols from alkyl-PAF and acyl-PAF, respectively. The derivatives are subsequently separated from one another by TLC on silica gel G plates utilizing a solvent system of hexane:diethyl ether:glacial acetic acid at 60:40:1 v/v/v. After development, the TLC plates are scraped at 0.5-cm increments, lipid extracted, and $^3$H radioactivity determined. Using this solvent system, 1-*O*-alkyl-[$^3$H]2,3-diacetylglycerols migrate faster than 1-acyl-[$^3$H]2,3-diacetylglycerols, allowing accurate quantitation of each PAF subclass, and hence determination of the relative amounts of [$^3$H]alkyl-PAF and [$^3$H]acyl-PAF. As with any TLC procedure, authentic standards of 1-*O*-alkyl-[$^3$H]2,3-diacetylglycerol and 1-acyl-[$^3$H]2,3-diacetylglycerol must be cochromatographed on the same TLC plate to determine the exact $R_f$ values for the 1-*O*-alkyl-[$^3$H]2,3-diacetylglycerols and 1-acyl-[$^3$H]2,3-diacetylglycerols derived from the PAF in the sample of unknown composition. It is important to note that this procedure does not provide information about the percent composition of the individual molecular species within each PAF subclass.

In summary, by carefully following all of the procedures outlined above, including the utilization of appropriate authentic standards for quality control, [$^3$H]acetate incorporation is a very useful tool to probe PAF production. Nevertheless, there are some issues that have not been fully addressed relative to the interpretation of results in these types of studies. The first issue is that [$^3$H]acetate incorporation may not represent the total mass of PAF that has been synthesized since PAF produced from endogenous stores of acetyl-CoA would not be detected; likewise, PAF that has been formed prior to cell stimulation would go undetected. In addition, PAF synthesized through the de novo pathway also might not become radiolabelled with [$^3$H]acetate because of preformed 1-radyl-2-acetyl-*sn*-glycero-3-phosphate precursor pools. Thus, utilizing the specific activity of the [$^3$H]acetate to calculate the molar amounts of PAF synthesized by a given cell or tissue must be done with some reservation. A second problem in quantitation based on [$^3$H]acetate incorporation is the variable losses of starting material that occur during the sequential processing steps, i.e., lipid extraction, TLC purification, PLC digestion, acetylation, and TLC separation. Thus, in cases in which accurate quantitation is an important issue, the use of radiolabelled $^{14}$C internal standards in the initial lipid extraction is mandated. Finally, even if all these quality control criteria are met, [$^3$H]acetate incorporation studies provide no information about the composition of the various molecular species of PAF within each PAF subclass.

## F. Analysis of PAF by Mass Spectrometry

Mass spectrometry provides a powerful way to obtain both structural and quantitative information about individual molecular species of PAF. A wide variety of mass spectrometric techniques have been developed to analyze PAF, capitalizing on the strengths of different instruments. These techniques can be generally divided into two basic groups: methods that are suitable for analysis of intact PAF and those that require derivatization; of these methods, several are predominantly utilized for structural elucidation while others are capable of quantitative analysis.

### I. Analysis of Intact PAF

In the past, a major obstacle to mass spectral analysis of intact PAF was the fact that phospholipids are both nonvolatile and thermally labile, thereby preventing direct characterization by electron impact ionization (EI) with sample introduction by either the gas chromatograph (GC) or direct insertion probe. As an early solution to this problem, desorption chemical ionization (DCI) was utilized to analyze underivatized PAF (REINHOLD and CARR 1982; THOMAS et al. 1986). While DCI analysis of PAF yields structurally informative fragmentation, the accompanying molecular ions are generally weak, and the spectra are complicated by pyrolysis of the analyte. This essentially precludes the use of DCI for quantification of PAF. Furthermore, since sample quantities in the 100-ng range are required to produce an interpretable full scan spectrum (HANAHAN and WEINTRAUB 1985), DCI has not proved particularly useful for analysis of PAF in biological samples.

The introduction of fast-atom bombardment mass spectrometry (FAB/MS) provided a much more sensitive and reliable means to analyze intact PAF than DCI. FAB mass spectra of PAF (CLAY et al. 1984; WEINTRAUB et al. 1985) are characterized by intense ions representing the protonated molecule ($[M + H]^+$) and the polar head group (m/z 184 for phosphocholine). There is also a relatively weak ion produced by loss of the 2-*O*-acetyl group. When static FAB is employed, low-nanogram amounts of PAF yield meaningful full scan spectra (WEINTRAUB et al. 1985). FAB/MS has also been utilized for quantitative analysis of PAF. By addition of a [$^2H_3$-acetyl]-alkyl-PAF internal standard, 10ng of PAF could be reproducibly quantified (CLAY et al. 1984). HAROLDSEN and GASKELL (1989) added the specificity of tandem MS to the assay by monitoring the transitions from the corresponding molecular ions to m/z 184 for endogenous PAF and m/z 185 for the [$^2H_3$-acetyl]-alkyl-PAF internal standard; they obtained a low-picogram detection limit for analysis of authentic 16:0-alkyl-PAF. Thermospray ionization MS (TSP/MS) has also been utilized for quantitative analysis of PAF (KIM and SALEM 1987; MALLET and ROLLINS 1986), but the sensitivity of on-line HPLC/TSP/MS is generally inadequate for detecting PAF in samples of biological origin.

Of the current techniques suitable for analysis of intact PAF at biologically relevant levels, electrospray ionization MS (ESI/MS) is clearly the method of choice. The ESI mass spectrum of PAF contains an intense $[M + H]^+$ (or $[M + Na]^+$, depending on sample preparation and ESI solvent) and essentially no fragmentation (SILVESTRO et al. 1993; WEINTRAUB et al. 1991a). Furthermore, ions representative of the polar head group can be generated by collision-induced dissociation. Silvestro and coworkers (SILVESTRO et al. 1993) reported detection limits of 0.3–1.5ng for analysis of alkyl-PAF homologs by HPLC/ESI/MS/MS using selected ion monitoring and multiple reaction monitoring.

## II. Analysis of PAF After Derivatization

There are many advantages to the use of GC/MS for analysis of PAF. A major factor is that detection limits for PAF are lower when electron-capture MS (EC/MS, also known as negative ion chemical ionization MS) is employed as compared to on-line HPLC/ESI/MS or FAB/MS. And, by selecting from a variety of derivatizing agents, one can choose the degree of structural information to be gained. Furthermore, many laboratories do not have routine access to an HPLC/MS instrument.

After isolation of PAF, an essential step in preparation for GC/MS analysis is removal of the polar head group. In many laboratories, this is accomplished by either phospholipase C digestion or treatment with hydrogen fluoride (HF). There are several problems that can occur during removal of the polar head group of PAF. HF is a highly reactive hazardous chemical that can cause degradation of PAF, particularly for the unsaturated molecular species. While phospholipase C treatment is much gentler, the use of this enzyme necessitates an additional purification step before derivatization; furthermore, phospholipase C treatment has the potential of adding artifacts if the enzyme preparation is not pure. After removal of the phosphocholine by either of these procedures, derivatization of the resultant diradyl-glycerol is required to produce a volatile derivative suitable for GC/MS analysis. Early studies by Saito and co-workers (ODA et al. 1985; SATOUCHI et al. 1983) employed *t*-butyldimethylsilyl (TBDMS) ethers in conjunction with analysis by GC/EI/MS with positive ion detection. TBDMS derivatives of PAF are quite stable, fractionate well by GC, and are characterized by structurally informative ions. While quantification of TBDMS ethers of PAF is routine, a major drawback to this method is that sample size requirements are substantially higher than for PAF derivatives that can be analyzed by EC/MS. For this reason, many researchers have switched to pentafluorobenzoyl (PFB) esters for GC/EC/MS analysis of PAF (RAMESHA and PICKETT 1986). The EC mass spectra of the PFB derivatives of PAF are characterized by intense molecular anions ($[M]^-$) and essentially no fragmentation. Detection limits in the subpicogram range have been reported (PICKETT and RAMESHA 1990).

Derivatization procedures have now been developed that obviate the need for removal of the polar head group prior to derivatization (SATSANGI et al. 1989; WEINTRAUB et al. 1990, 1992; WOODARD et al. 1995). In the first instance, heptafluorobutyric (HFB) anhydride was utilized for direct derivatization of PAF. HFB derivatives are extremely useful for structural elucidation of low levels of biologically derived PAF. With positive ion detection, $[M + H]^+$ ions of modest intensity are observed, providing important molecular weight information. In the negative ion mode, lower intensity fragments generated by losses of HF, ketene, and/or acetic acid are observed along with intense ions produced by a rearrangement reaction between the HFB group and any acyl moieties on the molecule. These ions provide direct evidence about the nature of the acyl substituents attached to the glycerol backbone. It is noteworthy that differentiation between alkyl-PAF and acyl-PAF can be obtained from the relative intensities of several fragments in the PAF/HFB EC mass spectrum (WEINTRAUB et al. 1991b).

New procedures have also been developed for direct formation of PFB derivatives as a way to obtain very low detection limits for quantitative analysis of PAF. Initially, PFB chloride was utilized for this purpose (WEINTRAUB et al. 1990). Subsequently, significant improvements were obtained by the use of PFB anhydride (WEINTRAUB et al. 1993b) to prepare these derivatives (WEINTRAUB et al. 1992; WOODARD et al. 1995). The optimized procedure provides recoveries of at least 65% for six molecular species of PAF, with a linear dynamic range of 0.05–500 pmole of PAF derivatized (WOODARD et al. 1995). Typically, less than 2% of the sample is required for quantitative

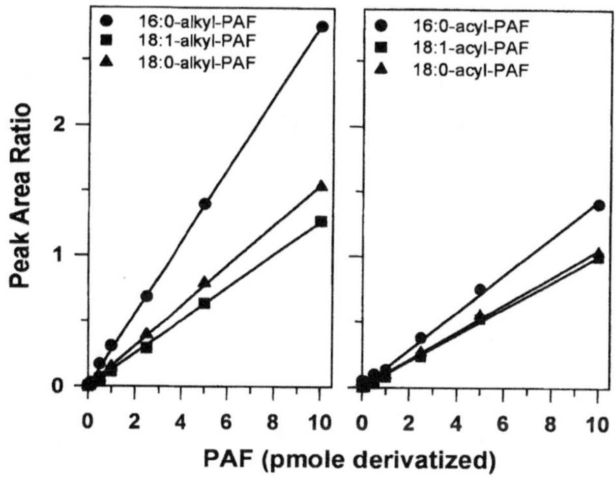


**Fig. 4.** Gas chromatography/electron capture/mass spectrometry (GC/EC/MS) analysis of individual molecular species of pentafluorobenzoyl (PFB)-derivatized platelet-activating factor (PAF). A series of mixtures containing known amounts of each alkyl- and acyl-PAF standard (0.1–10 pmole each) together with 5 pmole of internal standard ([$^2H_4$, $^{13}C_2$-acetyl]-16:0-alkyl-PAF) were derivatized with PFB anhydride and assessed by GC/EC/MS with selected ion monitoring as previously described in detail (WOODARD et al. 1995). The following ions were monitored: m/z 552, 16:0-alkyl-PAF/PFB; m/z 558, [$^2H_4$,$^{13}C_2$-acetyl]-16:0-alkyl-PAF/PFB; m/z 578, 18:1-alkyl-PAF/PFB; m/z 580, 18:0-alkyl-PAF/PFB; m/z 566, 16:0-acyl-PAF/PFB; m/z 592, 18:1-acyl-PAF/PFB; m/z 594, 18:0-acyl/PAF-PFB. The ratio of the peak area of each analyte to that of the internal standard was determined for each PAF species

analysis by GC/EC/MS with selected ion monitoring. Regression analysis of standard curves produced by direct derivatization of PAF by PFB anhydride routinely are characterized by correlation coefficients ($r^2$) of 0.9998 or better, with outstanding day-to-day reproducibility. This approach has recently been used to quantify six PAF molecular species isolated from human mixed saliva (WOODARD et al. 1995).

Although the procedures for GC/EC/MS analysis of PFB ester of PAF are very straightforward, there are two very important technical aspects that require brief comment. First, it is essential to choose an internal standard that is appropriate for PFB derivatization of PAF. We found that when [$^2H_3$-acetyl]-16:0-alkyl-PAF is employed, a significant portion of the deuterium label exchanges with hydrogen during derivatization (HAROLDSEN et al. 1991); however, since the degree of exchange is variable, the potential exists for adding an uncontrolled source error to the assay. Fortunately, both [$^2H_3$-hexadecyl]- and [$^2H_4$-hexadecyl]-alkyl-PAF are now commercially available, and the deuterium label in the alkyl chain is fully stable. Second, we found that the PFB derivatives of the six molecular species of PAF exhibit significantly different electron-capture mass spectral responses (WEINTRAUB et al. 1993a), as evidenced by the different slopes obtained for the individual standard curves (Fig. 4). Thus, for accurate quantification, a separate standard curve for each PAF molecular species of interest must be constructed in order to avoid significant errors in the quantitative result. With proper attention to these two issues, GC/EC/MS analysis of PFB derivatives can be utilized as a highly reproducible and sensitive technique for characterization of PAF molecular heterogeneity in biological samples.

## G. Concluding Comments

The identification and quantitation of PAF from a myriad of biological sources is critical to further understanding the role of PAF in normal physiology and (patho)physiological conditions. The aim of this review has been to summarize the different techniques available to the research community for the analysis of PAF. While equipment and resources may dictate the possible choices of techniques that can be utilized, every attempt should be made to insure both the appropriateness and accuracy of the chosen assay system. Therefore, for screening purposes and for estimates of relative activities, partial purification of PAF followed by quantitation by bioassay or RIA provides an acceptable approach. Additional qualitative information can be ascertained for the putative PAF sample by studying its physicochemical/enzymatic susceptibility and biological antagonism by known PAF receptor antagonists. However, mass spectral analysis currently provides the only avenue by which all of the molecular species of PAF may be identified and quantified. Notwithstanding, even the results obtained by MS may be biased by a priori decisions to look at only a limited number of the molecular species of PAF. For example, 16:0-alkyl-PAF has historically been considered as "authentic" PAF, and bioassays and RIA have been developed to detect and quantify this molecule. If only 16:0-alkyl-PAF were monitored by MS, however, at least 70% of the PAF in normal human mixed saliva (WOODARD et al. 1995), 60% of the PAF synthesized by human PMN (PINCKARD et al. 1994), and more than 95% of the PAF produced by stimulated endothelial cells (CLAY et al. 1991) would go undetected. Therefore, it is very important to carefully consider in future investigations which molecular species of PAF are to be studied. In this regard, if the (patho)physiological significance of PAF is to be fully elucidated, concerted efforts must be made to comprehensively characterize the molecular diversity of PAF in health and in disease.

*Acknowledgements.* The authors gratefully acknowledge the assisstance of Julie Friedrichs in the preparation of this manuscript. Supported in part by USPHS grants HL22555, HL46760, and DE08000 and the American Heart Association, Texas Affiliate.

### References

Ali H, Richardson RM, Tomhave ED, DuBose RA, Haribabu B, Snyderman R (1994) Regulation of stably transfected platelet activating factor receptor in RBL-2H3 cells. Role of multiple G proteins and receptor phosphorylation. J Biol Chem 269:24557–24563

Ambler J, Wallis RB (1983) A comparison between platelet aggregation and ATP secretion induced by collagen and by PAF acether and their inhibition by phenylbutazone, sulphinpyrazone and its metabolites. Thromb Res 31:577–589

Ammit AJ, O'Neill C (1991a) Comparison of a radioimmunoassay and bioassay for embryo-derived platelet-activating factor. Hum Reprod 6:872–878

Ammit AJ, O'Neill C (1991b) Rapid and selective measurement of platelet-activating factor using a quantitative bioassay of platelet aggregation. J Pharmacol Methods 26:7–21

Ammit AJ, Wells XE, O'Neill C (1992) Structural heterogeneity of platelet-activating factor produced by murine preimplantation embryos. Hum Reprod 7:865–870

Baldo BA, Smal MA, McCaskill AC (1991) A specific, sensitive and high-capacity immunoassay for PAF. Lipids 26:1136–1139

Benveniste J, Henson PM, Cochrane CG (1972) Leukocyte-dependent histamine release from rabbit platelets. The role of IgE, basophils, and a platelet activating factor. J Exp Med 136:1356–1377

Bligh EG, Dyer WJ (1959) A rapid method of total lipid extraction and purification. Can J Biochem Physiol 37:911–917

Bossant MJ, Ninio E, Delautier D, Benveniste J (1990) Bioassay of paf-acether by rabbit platelet aggregation. Methods Enzymol 187:125–130

Bratton DL, Clay KL, Henson PM (1994) Cellular sources of platelet-activating factor and related lipids. In: Cunningham FM (ed) The handbook of immunopharmacology – lipid mediators. Academic, London, pp 193–219

Chap H, Mauco G, Simon MF, Benveniste J, Douste-Blazy L (1981) Biosynthetic labelling of platelet activating factor from radioactive acetate by stimulated platelets. Nature 289:312–314

Clay KL, Murphy RC, Andres JL, Lynch L, Henson PM (1984) Structure elucidation of platelet activating factor derived from human neutrophils. Biochem Biophys Res Commun 121:815–825

Clay KL, Johnson C, Henson P (1990) Binding of platelet activating factor to albumin. Biochim Biophys Acta 1046:309–314

Clay KL, Johnson C, Worthen GS (1991) Biosynthesis of platelet activating factor and 1-*O*-acyl analogues by endothelial cells. Biochim Biophys Acta 1094:43–50

Cooney SJ, Smal MA, Baldo BA (1991) Quantitation by radioimmunoassay of PAF in human saliva. Lipids 26:1140–1143

Demopoulos CA, Pinckard RN, Hanahan DJ (1979) Platelet activating factor (PAF): evidence for 1-*O*-alkyl-2-acetyl-*sn*-glyceryl-3-phosphorylcholine as the active component (a new class of lipid chemical mediators). J Biol Chem 254:9355–9358

Ekholm J, Tatsumi Y, Nouchi T, Hanahan DJ (1992) The human platelet as a reproducible and sensitive cell for the detection and assay of platelet-activating factor. Anal Biochem 204:79–84

Garcia MC, Mueller HW, Rosenthal MD (1991) $C_{20}$ polyunsaturated fatty acids and phorbol myristate acetate enhance agonist-stimulated synthesis of 1-radyl-2-acetyl-*sn*-glycero-3-phosphocholine in vascular endothelial cells. Biochim Biophys Acta 1083:37–45

Grigoriadis G, Stewart AG (1992) Albumin inhibits platelet-activating factor (PAF)-induced responses in platelets and macrophages: implications for the biologically active form of PAF. Br J Pharmacol 107:73–77

Hanahan DJ, Weintraub ST (1985) Platelet-activating factor. In: Glick D (ed) Methods of biochemical analysis. Wiley, New York, pp 195–219

Hanahan DJ, Demopoulos CA, Liehr J, Pinckard RN (1980) Identification of platelet-activating factor isolated from rabbit basophils as acetyl glyceryl ether phosphorylcholine. J Biol Chem 255:5514–5516

Haroldsen PE, Gaskell SJ (1989) Quantitative analysis of platelet activating factor using fast atom bombardment/tandem mass spectrometry. Biomed Environ Mass Spectrom 18:439–444

Haroldsen PE, Gaskell SJ, Weintraub ST, Pinckard RN (1991) Isotopic exchange during derivatization of platelet activating factor for gas chromatography-mass spectrometry. J Lipid Res 32:723–729

Henson PM (1990) Bioassay of platelet-activating factor by release of [3H]serotonin. Methods Enzymol 187:130–134

Herbert JM, Laplace M-C, Cailleau C, Maffrand JP (1993) Effect of SR 27417 on the binding of [$^3$H]PAF to rabbit and human platelets and human polymorphonuclear leukocytes. J Lipid Mediat 7:57–58

Honda Z-I, Nakamura M, Miki I, Minami M, Watanabe T, Seyama Y, Okado H, Toh H, Ito K, Miyamoto T, Shimizu T (1991) Cloning of functional expression of platelet-activating factor receptor from guinea-pig lung. Nature 349:342–346

Hsieh KH, Ng CK (1993) Increased plasma platelet-activating factor in children with acute asthmatic attacks and decreased in vivo and in vitro production of platelet-activating factor after immunotherapy. J Allergy Clin Immunol 91:650–657

Hwang S-B (1990) Specific receptors of platelet-activating factor, receptor heterogeneity, and signal tranduction mechanisms. J Lipid Mediat 2:123–158

Hwang S-B (1991) High affinity receptor binding of platelet-activating factor in rat peritoneal polymorphonuclear leukocytes. Eur J Pharmacol 196:169–175

Hwang S-B (1994) Platelet-activating factor: receptors and receptor antagonists. In: Cunningham FM (ed) The handbook of immunopharmacology – lipid mediators. Academic, London, pp 297–360

Hwang S-B, Lam M-H (1986) Species difference in the specific receptors of platelet activating factor. Biochem Pharmacol 35:4511–4518

Inarrea P, Gomez-Cambronero J, Nieto M, Crespo MS (1984) Characteristics of the binding of platelet-activating factor to platelets of different animal species. Eur J Pharmacol 105:309–315

Janero DR, Burghardt B, Burghardt C (1988) Radioligand competitive binding methodology for the evaluation of platelet-activating factor (PAF) and PAF-receptor antagonism using intact canine platelets. J Pharmacol Methods 20:237–253

Karasawa K, Satoh N, Hongo T, Nakagawa Y, Setaka M, Nojima S (1991) Radioimmunoassay for platelet-activating factor. Lipids 26:1126–1129

Kim HY, Salem N (1987) Application of thermospray high-performance liquid chromatography/mass spectrometry for the determination of phospholipids and related compounds. Anal Chem 59:722–726

Kroegel C, Yakawa T, Westwick J, Barnes PJ (1989) Evidence for two platelet activating factor receptors on eosinophils: dissociation between PAF-induced intracellular calcium mobilization and degranulation and superoxides anion generation in eosinophils. Biochem Biophys Res Commun 162:511–521

Kunz D, Gerard NP, Gerard C (1992) The human leukocyte platelet-activating factor receptor. cDNA cloning, cell surface expression, and construction of a novel epitope-bearing analog. J Biol Chem 267:9101–9106

Kurosawa M, Yamashita T, Kurimoto F (1994) Increased levels of blood platelet-activating factor in bronchial asthmatic patients with active symptoms. Allergy 49:60–63

Lanara E, Vakirtzi-Lemonias C, Kritikou L, Demopoulos CA (1982) Response of mice and mouse platelets to acetyl glyceryl ether phosphorylcholine. Biochem Biophys Res Commun 109:1148–1156

Liggitt HD, Leid RW, Huston L (1984) Aggregation of bovine platelets by acetyl glyceryl ether phosphorylcholine (platelet activating factor). Vet Immunol Immunopathol 7:81–87

Mallet AI, Rollins K (1986) Thermospray liquid chromatography/mass spectrometry of ether phosphocholines. Biomed Environ Mass Spectrom 13:541–543

McManus LM, Hanahan DJ, Pinckard RN (1981) Human platelet stimulation by acetyl glyceryl ether phosphorylcholine (AGEPC). J Clin Invest 67:903–906
McManus LM, Fitzpatrick FA, Hanahan DJ, Pinckard RN (1983) Thromboxane B2 release following acetyl glyceryl ether phosphorylcholine (AGEPC) infusion in the rabbit. Immunopharmacology 5:197–207
McManus LM, Woodard DS, Deavers SI, Pinckard RN (1993) Biology of disease. PAF molecular heterogeneity: pathobiological implications. Lab Invest 69:639–650
Moon DG, van der Zee H, Morton KD, Krasodomski JA, Kaplan JE, Fenton JW (1990) Platelet activating factor and sheep platelets: a sensitive new bioassay. Thromb Res 57:551–564
Mueller HW, O'Flaherty JT, Wykle RL (1983) Biosynthesis of platelet activating factor in rabbit polymorphonuclear neutrophils. J Biol Chem 258:6213–6218
Mueller HW, Nollert MU, Eskin SG (1991) Synthesis of 1-acyl-2-[$^3$H]acetyl-*sn*-glycero-3-phosphocholine, a structural analog of platelet activating factor, by vascular endothelial cells. Biochem Biophys Res Commun 176:1557–1564
Naccache PH, Molski MM, Volpi M, Becker EL, Sha'afi RI (1985) Unique inhibitory profile of platelet activating factor induced calcium mobilization, polyphosphoinositide turnover and granule enzyme secretion in rabbit neutrophils towards pertussis toxin and phorbol ester. Biochem Biophys Res Commun 130:677–684
Naccache PH, Molski MM, Volpi M, Shefcyk J, Molski TF, Loew L, Becker EL, Sha'afi RI (1986) Biochemical events associated with the stimulation of rabbit neutrophils by platelet-activating factor. J Leukoc Biol 40:533–548
Nishihira J, Ishibashi T, Imai Y, Muramatsu T (1984) Mass spectrometric evidence for the presence of platelet-activating factor (1-*O*-alkyl-2-acetyl-*sn*-glycero-3-phosphocholine) in human amniotic fluid during labor. Lipids 19:907–910
O'Flaherty JT, Jacobson DP, Redman JF (1992) Regulation of platelet-activating-factor receptors and the desensitization response in polymorphonuclear neutrophils. Biochem J 288:241–248
Oda M, Satouchi K, Yasunaga K, Saito K (1985) Molecular species of platelet-activating factor generated by human neutrophils challenged with ionophore A23187. J Immunol 134:1090–1093
Paulson SK, Nicholson NS (1988) A radioreceptor binding assay for measurement of platelet-activating factor synthesis by human neutrophils. J Immunol Methods 110:209–215
Pickett WC, Ramesha CS (1990) Quantitative analysis of platelet-activating factor by gas chromatography–negative ion chemical ionization mass spectrometry. Methods Enzymol 187:142–152
Pinckard RN (1989) Platelet-activating factor (PAF): molecular heterogeneity and (patho)biology. In: Saito K, Hanahan DJ (eds) Platelet-activating factor and diseases. International Medical, Tokyo, pp 37–50
Pinckard RN, Farr RS, Hanahan DJ (1979) Physicochemical and functional identity of rabbit platelet-activating factor (PAF) released in vivo during IgE anaphylaxis with PAF released in vitro from IgE-sensitized basophils. J Immunol 123:1847–1857
Pinckard RN, Showell HJ, Castillo R, Lear C, Breslow R, McManus LM, Woodard DS, Ludwig JC (1992) Differential responsiveness of human neutrophils to the autacrine actions of 1-*O*-alkyl-homologs and 1-acyl-analogs of platelet-activating factor (PAF). J Immunol 148:3528–3535
Pinckard RN, Woodard DS, Showell HJ, Conklyn MJ, Novak MJ, McManus LM (1994) Structural and (patho)physiological diversity of PAF. Clin Rev Allergy 12:329–359
Ramesha CS, Pickett WC (1986) Measurement of sub picogram quantities of platelet activating factor (AGEPC) by gas chromatography/negative ion chemical ionization mass spectrometry. Biomed Environ Mass Spectrom 13:107–111

Reinhold VN, Carr SA (1982) Direct chemical ionization mass spectrometry with polyimide-coated wires. Anal Chem 54:499–503
Renooij W, Snyder F (1981) Biosynthesis of 1-alkyl-2-acetyl-*sn*-glycero-3-phosphocholine (platelet activating factor and a hypotensive lipid) by cholinephosphotransferase in various rat tissues. Biochim Biophys Acta 663:545–556
Satouchi K, Oda M, Yasunaga K, Saito K (1983) Applicaiton of selected ion monitoring to determination of platelet-activating factor. J Biochem 94:2067–2070
Satsangi RK, Ludwig JC, Weintraub ST, Pinckard RN (1989) A novel method for the analysis of platelet-activating factor: direct derivatization of phosphoglycerides. J Lipid Res 30:929–937
Scodras JM, Betteridge KJ, Croy BA, Johnstone JB, Rieger D (1991) Failure to detect platelet-activating factor using the splenectomized mouse bioassay. J Reprod Fertil 92:483–494
Silvestro L, Da Col R, Scappaticci E, Libertucci D, Biancone L, Camussi G (1993) Development of a high-performance liquid chromatographic-mass spectrometric technique, with an ionspray interface, for the determination of platelet-activating factor (PAF) and lyso-PAF in biological samples. J Chromatogr 647:261–269
Smal MA, Baldo BA (1991) Inhibitor(s) of platelet-activating factor (PAF) in human saliva. Lipids 26:1144–1147
Smal MA, Baldo BA, McCaskill A (1990) A specific, sensitive radioimmunoassay for platelet-activating factor (PAF). J Immunol Methods 128:183–188
Snyder F (1994) Metabolic processing of PAF. Clin Rev Allergy 12:309–328
Stewart AG (1994) Biological properties of platelet-activating factor. In: Cunningham FM (ed) The handbook of immunopharmacology – lipid mediators. Academic, London, pp 221–295
Stewart AG, Grigoriadis G (1991) Structure-activity relationships for platelet-activating factor (PAF) and analogues reveal differences between PAF receptors on platelets and macrophages. J Lipid Mediat 4:299–308
Sugatani J, Lee DY, Hughes KT, Saito K (1990) Development of a novel scintillation proximity radioimmunoassay for platelet-activating factor measurement: comparison with bioassay and GC/MS techniques. Life Sci 46:1443–1450
Sugatani J, Miwa M, Komiyama Y, Murakami T (1993) Quantitative analysis of platelet-activating factor in human plasma. Application to patients with liver cirrhosis and disseminated intravascular coagulation. J Immunol Methods 166:251–261
Suquet CM, Leid RW (1983) Aggregation of equine platelets by PAF (platelet-activating factor). Inflammation 7:197–203
Thomas MJ, Samuel M, Wykle RL, Surles JR, Piantadosi C (1986) Desorption chemical ionization mass spectrometry of 1-*O*-alkyl-2-O-acetyl-*sn*-glycero-3-phosphocholines and analogues. J Lipid Res 27:172–176
Tiberghien C, Laurent L, Junier MP, Dray F (1991) A competitive receptor binding assay for platelet-activating factor (PAF): quantification of PAF in rat brain. J Lipid Mediat 3:249–266
Triggiani M, Schleimer RP, Warner JA, Chilton FH (1991) Differential synthesis of 1-acyl-2-acetyl-*sn*-glycero-3-phosphocholine and platelet-activating factor by human inflammatory cells. J Immunol 147:660–666
Venable ME, Zimmerman GA, McIntyre TM, Prescott SM (1993) Platelet-activating factor: a phospholipid autacoid with diverse actions. J Lipid Res 34:691–702
Wang C, Tai HH (1992) A sensitive and specific radioimmunoassay for platelet-activating factor. Lipids 27:206–208
Weintraub ST, Ludwig JC, Mott GE, McManus LM, Pinckard RN (1985) Fast atom bombardment-mass spectrometric identification of molecular species of platelet-activating factor produced by stimulated human polymorphonuclear leukocytes. Biochem Biophys Res Commun 129:868–876

Weintraub ST, Lear CL, Pinckard RN (1990) Analysis of platelet-activating factor by GC/MS after direct derivatization with pentafluorobenzoyl chloride and heptafluorobutyric anhydride. J Lipid Res 31:719–725

Weintraub ST, Pinckard RN, Hail M (1991a) Electrospray ionization for analysis of platelet-activating factor. Rapid Commun Mass Spectrom 5:309–311

Weintraub ST, Pinckard RN, Heath TG, Gage DA (1991b) Novel mass spectral fragmentation of heptafluorobuyryl derivatives of acyl analogs of platelet-activating factor. J Am Soc Mass Spectrom 2:476–482

Weintraub ST, Williams RF, Pinckard RN (1992) New derivatizing agent for GC/MS analysis of PAF. Proceedings of the Kyoto '92 International Conference on Biological Mass Spectrometry, pp 362–363

Weintraub ST, Lear C, Pinckard RN (1993a) Differential electron capture mass spectral response of pentafluorobenzoyl derivatives of platelet activating factor alkyl chain homologs. Biol Mass Spectrom 22:559–564

Weintraub ST, Satsangi RK, Simmons AM, Williams RF, Pinckard RN (1993b) Synthesis of pentafluorobenzoic anhydride: a superior derivatizing agent for lipids. Anal Chem 65:2400–2402

Woodard DS, Mealey BL, Lear CS, Satsangi RK, Prihoda TJ, Weintraub ST, Pinckard RN, McManus LM (1995) Molecular heterogeneity of PAF in normal human mixed saliva: quantitative mass spectral analysis after direct derivatization of PAF with pentafluorobenzoic anhydride. Biochim Biophys Acta 1259:137–147

Ye RD, Prossnitz ER, Zou A, Cochrane CG (1991) Characterization of a human cDNA that encodes a functional receptor for platelet activating factor. Biochem Biophys Res Commun 180:105–111

CHAPTER 24

# Biosynthetic Inhibitors and Receptor Antagonists to Platelet Activating Factor

D. HANDLEY

## A. PAF: Historical Perspective and Pathology

Platelet activating factor (PAF) is now a quarter of a century old. As a mediator of inflammation, it still enjoys the focus of intensive studies ranging from preclinical pharmacology to clinical studies related to pathological involvements. Since its discovery as a soluble substance released from IgE-stimulated basophils (BENVENISTE et al. 1972) and its subsequent structural identification, PAF has been referred to as "antihypertensive polar renal medullary lipid" (APRL; BLANK et al. 1979), "acetyl glyceryl ether phosphorylcholine" (AGEPC; DEMOPOLUS et al. 1979) or by the founding name of "Paf-acether" (BENVENISTE et al. 1979). Currently, there is no accepted international nomenclature that describes the autocoid PAF family (PINCKARD et al. 1994).

Injection of PAF produces a variety of responses in mammals from rodents to man (HANDLEY 1989a, 1990b), such as thrombocytopenia, leukopenia, hypotension (HANDLEY et al. 1987b,c), acute bronchospasm (CHUNG and BARNES 1989; DENJEAN et al. 1981, 1983; HANDLEY et al. 1983, 1987c; LEWIS et al. 1984; PATTERSON et al. 1984), and acute increases in pulmonary (EVANS et al. 1987; MISAWA and TAKATA 1988; O'DONNELL et al. 1990) and vascular permeability (HANDLEY 1988) which lead to extravasation of plasma proteins and hemoconcentration (HANDLEY et al. 1984, 1985, 1987c). The hypotensive and vascular-permeability (both systemic and local) effects of PAF are dose-dependent but are independent of platelets (HANDLEY et al. 1985) or neutrophils. The vascular permability effects of PAF vary by potency with different routes of administration (HANDLEY et al. 1984) and are significantly more potent than leukotrienes (HANDLEY et al. 1986a). PAF is the most powerful bronchoconstrictor agent identified to date, over 20000-fold more potent than histamine (HEUER 1991a; PATTERSON and HARRIS 1983). Acute bronchoconstrictor effects of PAF are platelet-dependent when PAF in given parenterally (CHRISTMAN et al. 1988; LEWIS et al. 1984; COYLE et al. 1990; HANDLEY et al. 1985; VARGAFTIG et al. 1980) but are platelet-independent when PAF is given by aerosol or intratracheal administration (DESQUAND et al. 1986; VARGAFTIG et al. 1983). Species that possess platelets that are unresponsive to PAF, including rats (INARREA et al. 1984; NAMM et al. 1982) and *Cebus* primates, do not exhibit bronchoconstriction when challenged with PAF unless large doses of PAF are used (DAHLBACK et al. 1984). PAF has no oral or transdermal activity.

At the organ level, PAF has been observed to produce pulmonary artery hypertension, pulmonary edema (BURHOP et al. 1986; MOJARAD et al. 1983), circulatory coagulopathies, cardiac an renal dysfunction, anaphylaxis of the pulmonary (LELLOUCH-TUBIANA et al. 1987) and circulatory (DOEBBER et al. 1986) systems, and acute lethality (MYERS et al. 1983). The inflammatory effects of PAF are associated with endogenous release of thromboxanes, leukotrienes, and prostaglandins as well as inflammatory cell recruitment (ARNOUX et al. 1988; COLDITZ and TOPPER 1991; MCFADDEN et al. 1995). For these reasons, PAF is felt to participate in the initiation or progression of adult respiratory distress syndrome (CHRISTMAN et al. 1988), arthritis (PETTIPHER et al. 1987), asthma (BARNES and CHUNG 1987; PAGE and MORLEY 1986), cerebral-vascular ischemia (SATOH et al. 1988), cold urticaria (GRANDEL et al. 1985), colitis (WALLACE 1988), glomerulonephritis (PERICO et al. 1988), immune anaphylaxis (HALONEN et al. 1980), ischemic bowel necrosis (HSUEH et al. 1987), myocardial ischemia (STAHL et al. 1988), periodontal inflammation (GARITO et al. 1995), peritonitis (STEIL et al. 1995), reperfusion injury (IDE et al. 1995; MINOR et al. 1995), and shock (HANDLEY 1987, 1988) and ulcers (ROSAM et al. 1986). The role of PAF is less impressive or defined in cardiovascular diseases (JACKSON et al. 1986; NATHAN and DENIZOT 1995; ROBERSTON et al. 1988), pancreatitis (EMANUELLI et al. 1989; JANCER et al. 1995; ZHOU et al. 1987, 1990), acute transplant rejection (MAKOWKA et al. 1987, 1990), or atherosclerosis (HANDLEY et al. 1983; HANDLEY and SAUNDERS 1986).

The above diseases can be grouped into those mediated by immune complexes (CAMUSSI et al. 1987; HELLEWELL 1990; ITO et al. 1984; MAKOWKA et al. 1990; INARREA et al. 1985; SAUNDERS and HANDLEY 1990), by endotoxin (CHANG et al. 1987; DOEBBER et al. 1985; HANDLEY 1990b; HSUEH et al. 1987; NAGAOKA et al. 1991; NATSUME et al. 1994; WHITTLE et al. 1987), or underlying organ ischemia (LEPRAN and LEFER 1985; MONTRUCCHIO et al. 1990; STAHL et al. 1988), as each is known to induce endogenous release of PAF. Even with such an encompassing list of effects, it has been difficult to elucidate the etiology of PAF in these pathological conditions (SAUNDERS and HANDLEY 1990). Quantifying endogenous production of PAF is difficult because it may remain cell- or tissue-bound after formation, thereby necessitating quantitative extraction techniques. Measurement of PAF in body fluids such as plasma, urine, pleural, or saliva requires extraction and chromatographic separation. As PAF is cleared rapidly from the plasma with a T1/2 of 40 s (HANDLEY et al. 1987d) and the degradation product (lyso-PAF) is the same as the substrate used in its formation, conclusions regarding circulating amounts of PAF formed are difficult to establish (HANDLEY 1989a).

With this introduction, I will review the preclinical and clinical studies surrounding the development of the major series of xenobiotic PAF receptor antagonists. PAF antagonists from natural sources are less potent than xenobiotics and will not be reviewed. Compounds such as BN 52021 and BN 52063 that have been encrusted in reviews, exhibited limited potency, and have been abandoned from development will not be included. Biosynthetic inhibitors of PAF will also be detailed.

## B. Biosynthesis Inhibitors

PAF can be synthesized via two different metabolic pathways, one involving remodeling of membrane lipid constituents and the other the de novo pathway (SUMMERS and ALBERT 1995). The more critical pathway is remodeling which occurs during cellular activation and hypersensitivity reactions and, as the name implies, involves synthesis of PAF by modification of existing membrane phospholipids. The most common phospholipid used in PAF biosynthesis is alkyl lyso-GPC. The initial step is the formation of 1-*O*-alkyl-2-lyso-*sn*-glycero-3-phosphocholine (lyso-PAF) by the action of phospholipase $A_2$ ($PLA_2$). Lyso-PAF is then acetylated, with acetyl-CoA as a cofactor, by the enzyme acetyl-CoA:1-alkyl-*sn*-glycero-3-phosphochiline-acetyltransferase (EC2.3.1.67) to form PAF. The presence of an acetyl group on the 2-position of PAF enhances hydrophilicity and thus reduces the critical micellar concentration below levels seen in pathological states. Although PAF can be formed by other routes (BLANK et al. 1995), development of specific inhibitors of acetylation of lyso-PAF represents a therapeutic target for new anti-inflammatory drugs (YAMAZAKI et al. 1994).

The de novo mechanism generates PAF through a stepwise manner by the transfer of the phosphocholine group from CDP-choline to 1-*O*-alkyl-2-lyso-*sn*-glycerol. This is a major metabolite in the intermediary biosynthesis of several different ether-linked glycerolipids. This pathway may be important for physiological production of steady-state levels of PAF which are generally below the 1 ng/ml plasma level of healthy individuals (MEADE et al. 1991). In most cells and tissues, the level of PAF in the resting state is also very low.

PAF levels in most cells, fluids, or tissues are also regulated by PAF acetylhydrolyase. Most cells will produce and release PAF when given appropriate stimuli (PINCKARD et al. 1988), such as eosinophils (LEE et al. 1984), mast cells (JOLY et al. 1987; SCHLEIMER et al. 1986), pleural cells (HAYASHI et al. 1987a), polymorphonuclear leukocytes (NGUYEN et al. 1995; SISSON et al. 1987), platelets (CHIGNARD et al. 1980), macrophages (ALBERT and SNYDER 1983; RYLANDER and BEIJER 1987), glomerular mesanginal cells (WANG et al. 1988), lymphocytes (MALAVASI et al. 1986), cultured endothelial cells (McINTYRE et al. 1986; WHATELY et al. 1988), and basophils (BETZ et al. 1980; HANAHAN et al. 1980). PAF is produced by the retina (BUSSOLINO et al. 1986), heart (CAMUSSI et al. 1987), kidney (ZANGLIS and LIANOS 1987), liver (RENOOIJ and SNYDER 1981), lung (FITZGERALD et al. 1986), and spleen (RENOOIJ and

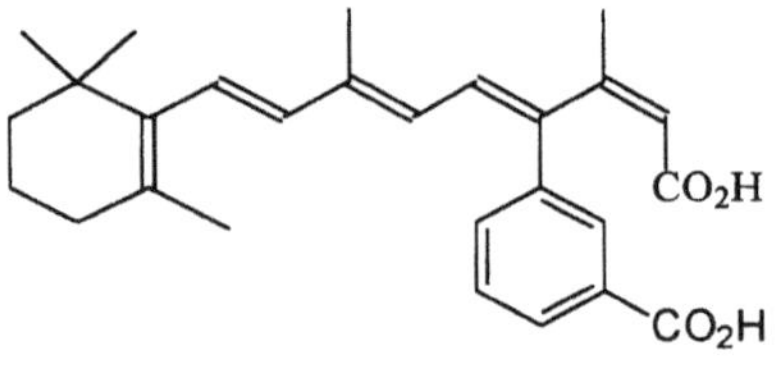


**BMS-181162**

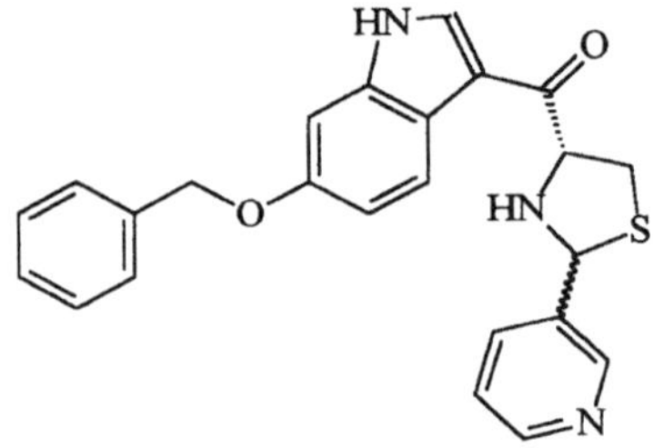


**A-79981**

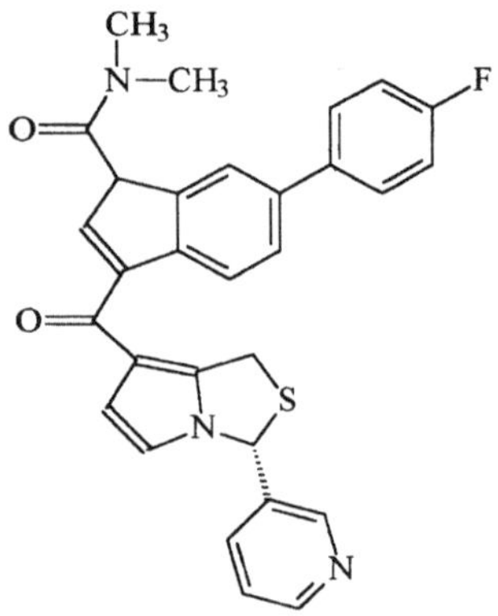


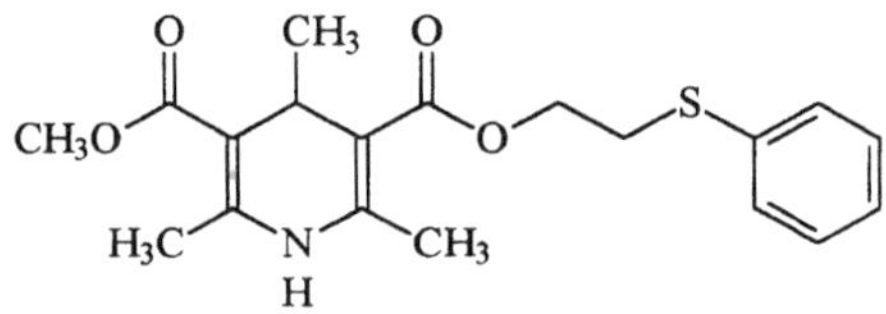


**PCA-4248**

**A-85783**

**Fig. 1.** Compound structures of BMS-181162, A-79981, A-85783, and PCA-4248

SNYDER 1981). PAF is thought to play a role in the proliferation and maturation of bone marrow lymphocytes (DENIZOT et al. 1995).

## I. Phospholipase $PLA_2$ Inhibitors

When activated, $PLA_2$ (EC 3.3.3.4) hydrolyzes the fatty acyl linkage at the *sn*-2 position to liberate arachidonic acid and other 2-lysophospholipid precursors from membrane phospholipids. However, the release of arachidonic acid from 1-alkyl-2-arachidonoyl-GLP (and other related polyenoic molecules) will yield equimolar amounts of 1-alkyl-2-lyso-GPC, a substrate which can be converted by acetyl transferase to PAF. Accordingly, inhibitors of $PLA_2$, such as steroids or antiflammins (TETTA et al. 1991), will suppress not only generation of leukotrienes and prostaglandins but also PAF. BMS-181162 (Fig. 1) specifically inhibited $PLA_2$ isolated from human polymorphonuclear leukocytes with an $IC_{50}$ of $10\,\mu M$ and blocked PAF biosynthesis in vitro and in vivo (TRAMPOSCH et al. 1993).

### II. Lyso-PAF-Transferase Inhibitors

Lyso-PAF-transferase is the rate limiting enzyme for the catalytic biosynthesis of PAF from lyso-PAF in many inflammatory cells. With the use of radioactive substrate ([$^3$H]acetyl CoA), the synthesis of radiolabeled PAF can be obtained and quantitated. Gold salts, along with other rare earth metals, have long been thought to possess medicinal and anti-inflammatory effects (HANDLEY 1989b). Several gold salts, such as gold thiosulfate and aurothioglucose, exhibited $IC_{50}$ values of 70 and 150*nM*, respectively, against lyso-PAF-transferase (RAYAMAJHI and BRUCHHAUSSEN 1989). Auranofin (Ridura), which is triethylphosphine gold complexed with carbohydrates, was only weakly effective in this system. Antiflammins, noted for inhibition of $PLA_2$, are reported to inhibit PAF acetyl transferase (TETTA et al. 1991).

## C. Preclinical and Clinical PAF Receptor Antagonists

Development of PAF antagonists was enhanced by the fact that while there is molecular heterogeneity in terms of endogenous generation (PINCKARD et al. 1994), there are only two dominant forms of PAF (C16:0 and C18:0) and these "gold standards" vary only by potency. However, the enantiomers of PAF exhibit a four-log-fold stereoselective difference in binding to platelets (HWANG and LAM 1986) and in bronchoconstriction potency (HWANG et al. 1988). There seems to be one type of receptor, proposed as a bipolarized 10–12 Å cylinder (BATT et al. 1991) of 38982 mol. wt. (KROEGEL et al. 1989). The PAF receptor has been cloned from guinea pig lung, HL-60 cells, and human neutrophils (SNYDER 1995), and there is no evidence of receptor heterogeneity or subtypes. All inflammatory actions of PAF are mediated via receptors, and there are no real differences in terms of species specificity, such that PAF antagonists developed against the human platelet receptor antagonize PAF-mediated responses in mice through (hu)man.

It was the development of specific competitive receptor antagonists to PAF which lead to relevant pharmacological approaches to understanding the involvement of PAF in disease. A variety of PAF antagonists, either structurally similar or dissimilar to PAF, have been identified from both synthetic natural sources or through synthetic chemistry. Many PAF antagonists are reported to inhibit pathological effects of exogenously administered or endogenously produced PAF and are now in early clinical trials.

None of the PAF antagonists have been approved for clinical use, although many are modestly considered specific, long-lasting, new, and highly potent. However, almost all are specifically stated as "novel" by their founders, such as A-79981 (LUO et al. 1993; SHEPPARD et al. 1994), BN 50739 (RABINOVICI et al. 1990), CV 6209 (TAKATANI et al. 1989), E 5880 (NAGAOKA et al. 1991), E6123 (SAKUMA et al. 1991), L-659,989 (HWANG et al. 1988), RP 59227 (MONDAT and CAVERO 1988; ROBAUT et al. 1988) which has remained so for many years (UNDERWOOD et al. 1992), SDZ 62-432 (HOULIHAN et al.

1995b), SDZ 64-419 (HOULIHAN et al. 1991), SM 12502 (NATSUME et al. 1994), SR 27417 (HERBERT et al. 1991), TCV-309 (STOCKMANS et al. 1991; TERASHITA et al. 1992), UK-74,505 (PONS et al. 1993), WEB 2086 (CASALS-STENZEL et al. 1987a,b; PRETOLANI et al. 1987; UKENA et al. 1989), and WEB 2347 (HEUER 1991b).

### I. Abbott Laboratories

Chemical modification of RP-59227 by replacement of the pyrrolothiazole portion with hetrocyclic moieties such as with a monocyclic thiazolidine or substituted indole leads to the generation of A-79981 (Fig. 1), which exhibits potent inhibition of PAF-induced platelet aggregation (SHEPPARD et al. 1994) and endotoxin-mediated intestinal lesion formation in the rat (LUO et al. 1993). Chemical modification to develop water-soluble pyridinium prodrugs lead to the identification of ABT-299, which is converted to A-85783 (Fig. 1), a potent PAF antagonist that is in clinical trials for treatment of sepsis (DAVIDSON et al. 1995).

### II. Alter S.A.

PCA 4248, [2-(phenylthio) ethyl-5-methoxycarbonyl 2,4,6-triethyl 1, 4-dihydropyridine-3-carboxylate is reported to inhibit PAF-, immune-complex-, and endotoxin-induced reactions in mice and rats (FERNANDEZ-GALLARDO et al. 1990). Although a dihydropyridine PAF antagonist, PCA-4248 (Fig. 1) is devoid of calcium channel blocking activity, in humans, PCA-4248 at 80mg orally showed significant ex vivo inhibition of PAF-induced platelet aggregation (ORTEGA et al. 1992) and is currently in Phase I clinical trials for sepsis.

### III. Boehringer Ingelheim KG

Initial studies showed that triazolobenzodiazepines possessed PAF receptor antagonist activity in vitro (KORNECKI et al. 1984), and it was soon verified that other anti-anxiety drugs such as etizolam (TERASAWA et al. 1987), brotizolam (CASALS-STENZEL 1987b; CUSALS-STENZEL and WEBER 1987), alprazolam (CASALS-STENZEL 1987b; CASALS-STENZEL and WEBER 1987; COX 1986), and triazolam (CASALS-STENZEL 1987b; CASALS-STENZEL and WEBER 1987; COX 1986) exhibited PAF antagonist activity in vivo (HANDLEY 1988).

PAF receptor activity could be dissociated from the CNS activity (CASALS-STENZEL 1987b; CASALS-STENZEL and WEBER 1987), and the first significant compound in this series was the thieno-triazolodiazepine hetrazepine compound WEB-2086 (apafant, Fig. 2). WEB-2086 was shown to be a potent PAF-receptor antagonist in vitro. It has been compared to L-652,731 (STEWART and DUSTING 1988) and BN52021, being over 50-fold more potent (CHUNG et al. 1987). It effectively inhibits PAF-induced responses in the guinea pig such as

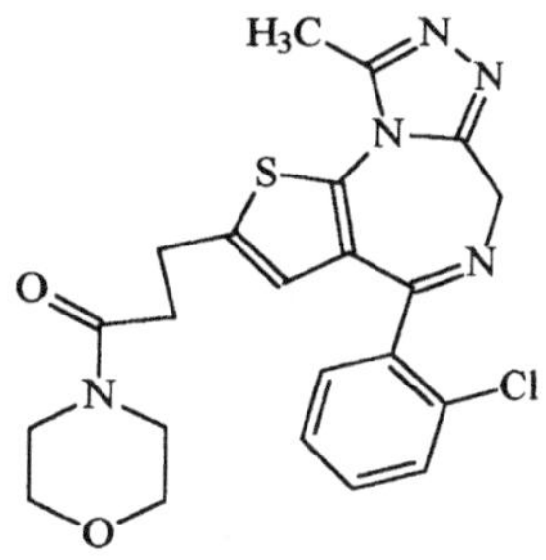

**WEB-2086**

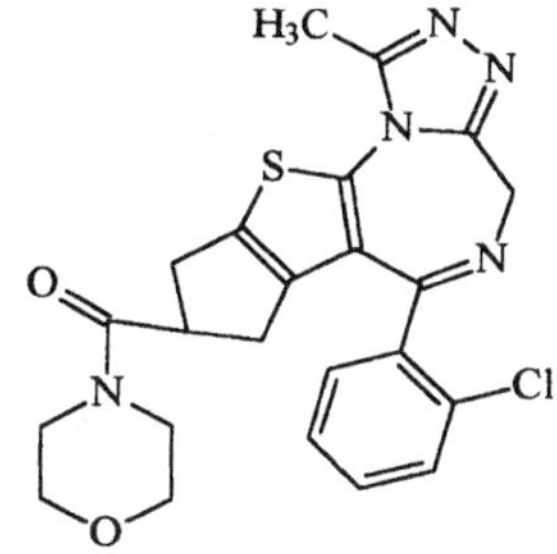

**WEB-2170**

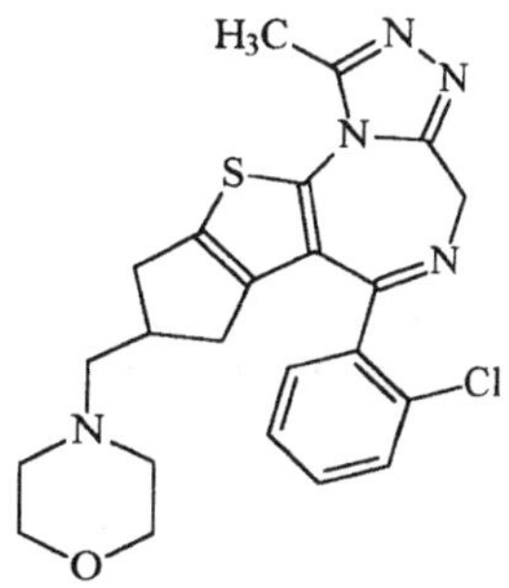

**STY-2108**

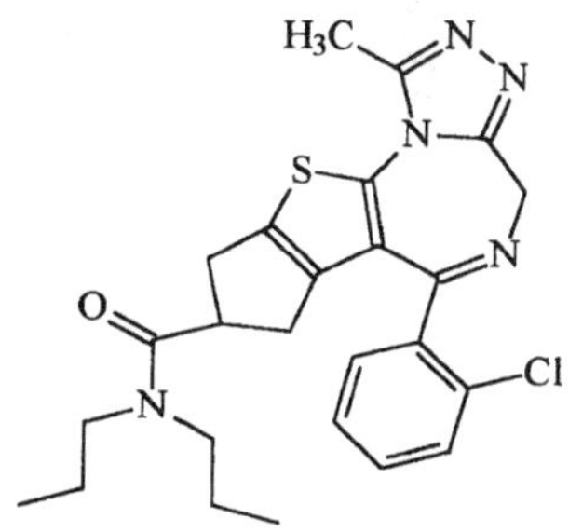

**WEB-2347**

**Fig. 2.** Compound structures of WEB-2086, WEB-2170, STY-2108, and WEB-2347

hypotension, bronchoconstriction, and lethality as well as PAF effects in rats such as hypotension and cutaneous (CASALS-STENZEL et al. 1987b; PRETOLANI et al. 1987) and bronchial (CASALS-STENZEL et al. 1987a) vascular permeability. In the guinea pig model of PAF-induced microvascular leakage, WEB-2086 was observed to be 500-fold more potent than BN 52021 (EVANS et al. 1988; HANDLEY 1988). Similarly, WEB-2086 has been demonstrated to inhibit PAF- or endotoxin-induced hypotension (HANDLEY 1988; MERLOS et al. 1991) and lethality in the rat (CASALS-STENZEL 1987c) as well as IgE-mediated anaphylaxis in the rabbit (LOHMAN and HALONEN 1990), but not anaphylaxis in the guinea pig (DESQUAND et al. 1990). WEB-2086 is active orally, parenterally, and by aerosol administration. Recent phase II studies (Germany) have examined tolerability of single or multiple oral dosing, as well as effects of aerosol administration (ADAMUS et al. 1990) on ex vivo inhibition of PAF-induced platelet aggregation (ADAMUS et al. 1989). However, administration of WEB-2086 to atopic asthmatics for 7 days orally (FREITAG et al. 1991) or by aerosol (WILKENS et al. 1991) was not protective against allergen-induced bronchoconstriction.

Recent development compounds include WEB-2170 (HEUER 1991a; bepafant, Fig. 2), STY-2108 (Fig. 2), and WEB-2347 (HEUER 1991b; Fig. 2) which have been reported with increased potency and specificity. In the case of WEB-2170, the enantiomers show a 47-fold difference with respect to potency in the human platelet aggregation test but smaller in vivo differences.

## IV. British Biotechnology Ltd.

Initial compounds were derived from benzoimidazoles and were weakly active as PAF antagonists. With the development of imidazopyridines, extremely potent inhibitors were reported, such as BB-823 (Fig. 3), which exhibits a $K_i$ of 0.015 n$M$ for the platelet receptor assay (WHITTAKER et al. 1992). This compound inhibits PAF- or endotoxin-mediated effects in vivo and displays oral activity. BB-882 (lexipafant, Fig. 3) is currently in phase III clinical trials for parenteral treatment of pancreatitis and in phase II trials for oral treatment of asthma and sepsis.

## V. Eisai Co., Ltd.

E-5880 (Fig. 3) is structurally similar to CV-6209, in which there is a 4-substituted piperazino carbamate moiety between the glycerol spacer and the lipophilic group on carbon 1, and a *O*-methyl benzoyl amide in place of acetyl

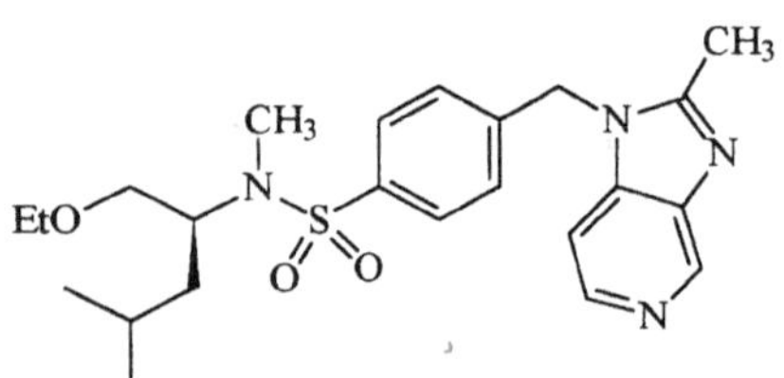

**BB-823**

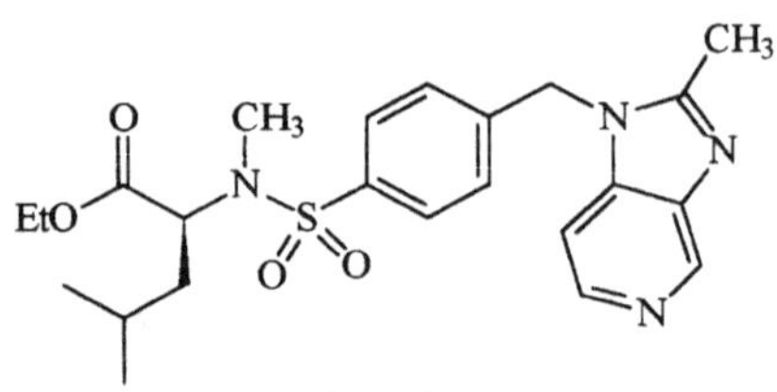

**BB-882**

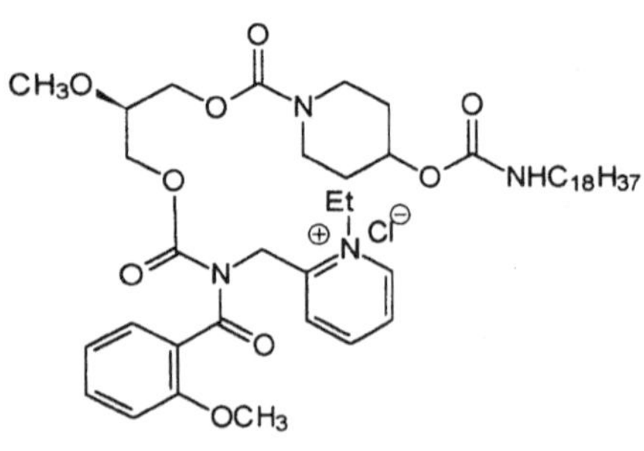

**E-5880**

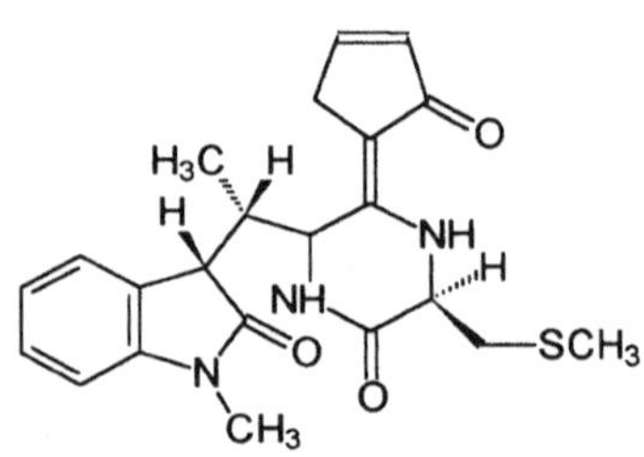

**FR-900452**

**Fig. 3.** Compound structures of BB-823, BB-882, E-5880, and FR-900452

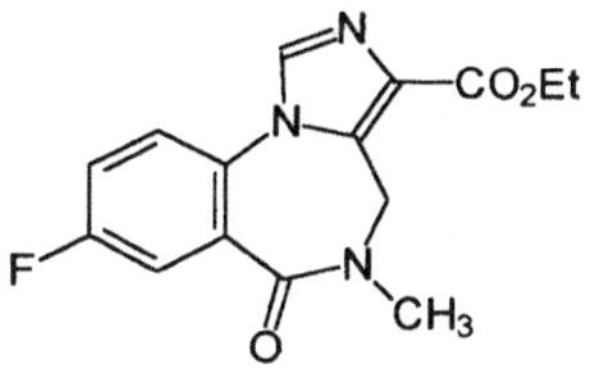

Ro 15-1788

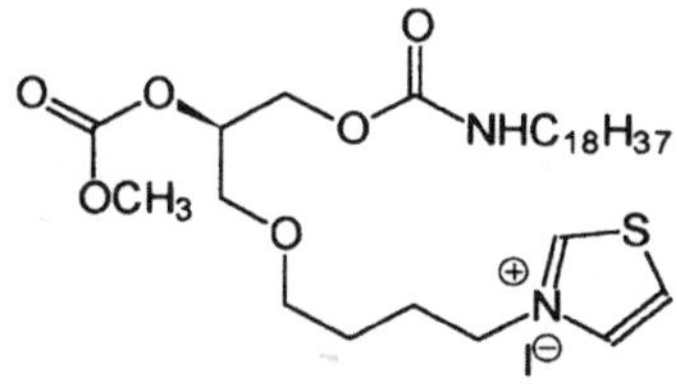

Ro 19-3704

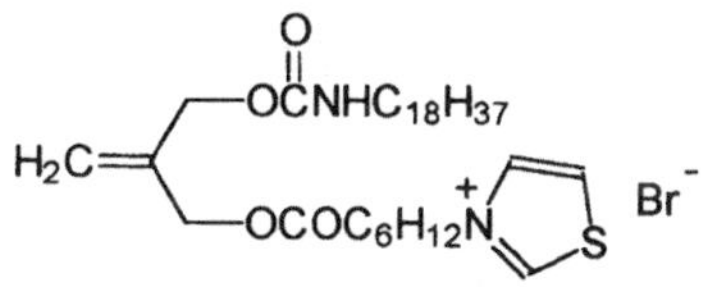

GS-1160-180

**Fig. 4.** Compound structures of Ro 15-1788, Ro 19-3704, and GS-1160-180

on carbon 3. E-5880 has similarly been shown to inhibit PAF-induced human platelet aggregation ($IC_{50} = 0.66\,nM$), protect from shock and lethality induced by PAF or endotoxin in mice and rats (NAGAOKA et al. 1991), and reduce ischemic/reperfusion injury following orthotopic liver transplantation (TAKADA et al. 1995). It is currently in Phase II clinical trials for sepsis. Recently, E-6123 has been shown to demonstrate exceptional oral potency in a number of PAF- or antigen-mediated in vivo responses (SAKUMA et al. 1991) and is reported to orally exhibit an elimination half-life of 137h (SUMMERS and ALBERT 1995).

## VI. Fujisawa Pharmaceutical Co.

FR-900452 (Fig. 3) is a cyclopentenopiperazinylindolinone that was isolated from the fermentation broth of Streptomyces phaeofaciens (OKAMOTO et al. 1986a). It exhibits moderate PAF receptor antagonist activity (OKAMOTO et al. 1986a) and inhibition of endotoxin-induced thrombocytopenia (OKAMOTO et al. 1986b).

## VII. Hoffmann-La Roche and Co., Ltd.

Ro 15-1788, ethyl 8-fluoro 5-methyl -5,6-dihydro -6-oxo-4H-imidazo (1,5-a) (1,4) benzo diazepine-3-carboxylate (Fig. 4), only slightly inhibits PAF-induced platelet aggregation (CASALS-STENZEL and WEBER 1987) and hemorrhagic shock (BITTERMAN et al. 1987). Orally or parenterally, it did not inhibit PAF-induced bronchoconstriction or hypotension in guinea pigs (PRETOLANI et al. 1987). Ro 19-3704 [3-(4((R)-2-((methoxycarbonyl) oxy-3-(octadecyl carbamoyl) oxy) propoxy) butyl) thiazolium iodide, Fig. 4] is a structural analog of PAF, in which the phosphate linker at carbon 3 of CV-3988 has been replaced by a polymethylene moiety and the methyl at carbon 2 by a methyl carbamate. It has been reported to inhibit PAF-induced human platelet aggregation with an $IC_{50}$ of 400n*M* (HADVARY and BAUMGARTNER 1985), inhibit gastrointestinal damage and endotoxin-induced hypotension in the rat (WALLACE et al. 1987), and PAF-induced platelet aggregation and allergen-induced bronchoconstriction in passively sensitized guinea pigs (LAGENTE et al. 1988).

## VIII. Leo Pharmaceutical Products Ltd.

GS-1160-180 (Fig. 4), a methylene octadecylamino carbonyl hexyl thiazolium bromide, acts as a PAF receptor antagonist and in vivo inhibits PAF-induced bronchoconstriction in the guinea pig and hypotension in the rat (GRUE-SORENSEN et al. 1988).

## IX. Merck Sharp and Dohme Research Laboratories

The first compound reported from this company was kadsurenone, an extract isolated from the Chinese herbal plant *Piper futokadsura* (SHEN et al. 1985). In vitro, kadsurenone acts as competitive receptor antagonist (HWANG et al. 1987; PONIPOM et al. 1987) and exhibits a 20 to 1 difference in the in vitro activity between the two tested enantiomers (PONIPOM et al. 1987). Kadsurenone was weakly active orally at doses of 20–50mg/kg as an inhibitor of PAF-induced

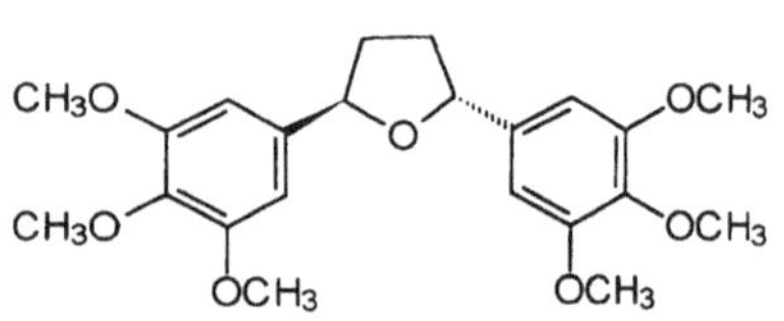


L 652,731

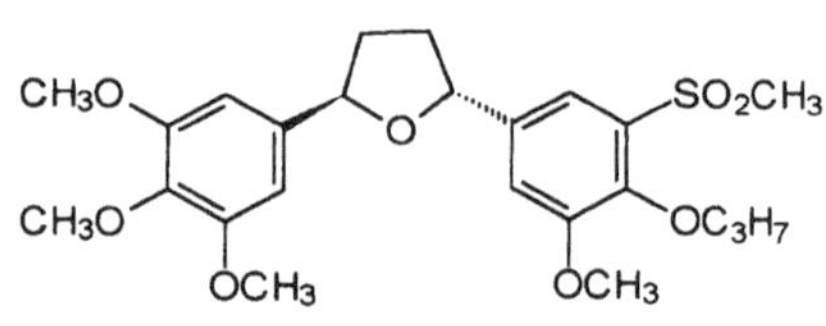


L 659,989

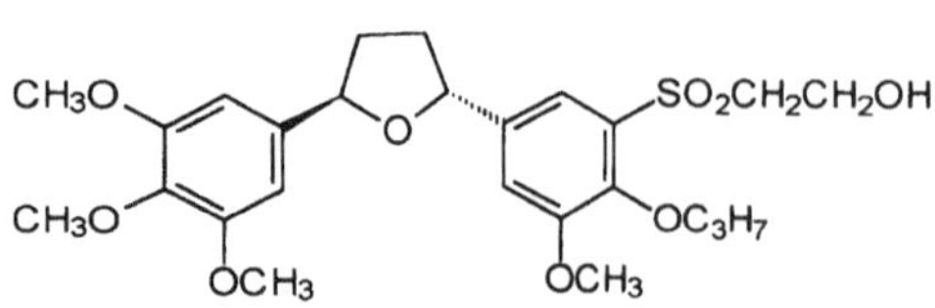


L 680,573

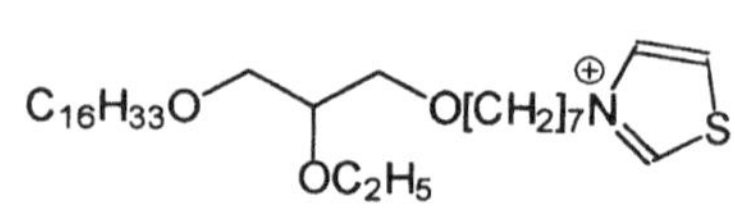


Ono 6240

**Fig. 5.** Compound structures of L 652,731, L 659,989, L 680,573, and Ono 6240

permeability effects in the rat and guinea pig (SHEN et al. 1985) but by parenteral administration effectively inhibited endotoxin-induced hypotension in the rat (DOEBBER et al. 1985). L 652,469 [14-acetoxy7b-(3′-ethylcrotonoyloxy-2-methylbutyryl-oxy)-notonipetranone] was isolated from methylene extracts of *Tussilago farfara*, but as a PAF receptor antagonist, it also exhibits weak oral activity (HWANG et al. 1987).

L 652,731 (Fig. 5), a trimethoxyphenyl tetrahydrofuran, is a very potent receptor antagonist that is orally active in several species (DOEBBER and WU 1988; HANDLEY 1988; HWANG et al. 1985, 1986; SMALLBONE et al. 1987). Well-conducted studies examined the ability of L 652,731 to inhibit PAF-induced pulmonary effects, lethality, ocular inflammation, gastric mucosal damage, altered glomerular permeability, cardiovascular effects, bronchoconstriction, and hemoconcentration in the guinea pig (see HANDLEY 1990a).

The most recent compound from the tetrahydrofuran series is L 659,989 (Fig. 5) in which the chiral enantiomers exhibit a 27-fold potency difference in vitro (HWANG et al. 1988). This compound is reported to inhibit PAF-induced bronchoconstriction in the guinea pig (HWANG et al. 1988). Recently, MK-287 (L 680,573, Fig. 5) given orally reportedly failed to inhibit acute broncho-constriction or the increase in hyperreactivity induced by allergen (BEL et al. 1991).

## X. Ono Pharmaceutical Co., Ltd.

Ono 6240 (Fig. 5) is a competitive receptor antagonist (HWANG 1987), a structural analog of PAF, and similar to CV-3988, containing a short ether at carbon 2 and a thiazolium at carbon 3. It has been shown to inhibit PAF-induced hypotension and vascular permeability effects (MIYAMOTO et al. 1985). It displays agonist properties and has been dropped from further development.

## XI. Pfizer Laboratories

Following the discovery that calcium channel blockers exhibited PAF antagonist activity, redesign of the dihydropyridines led to the development of UK-74,505 (Fig. 6) and the R-enantiomer, modipafant. In vivo, UK-74,505 was a potent PAF antagonist but weakly effective as a calcium channel blocker (PONS et al. 1993).

## XII. Rhône-Poulenc Santé Laboratories

A distinct chemical series has been developed by this company, of which RP 48,740 (Fig. 6) has been most extensively studied. This is a relatively weak PAF receptor antagonist, whose parenteral in vivo doses are often in the 3–30 mg/kg range (COEFFIER et al. 1986; HANDLEY and SAUNDERS 1986; LEFORT et al. 1988). Phase II clinical trials at oral doses of 250–1000 mg for 7 days

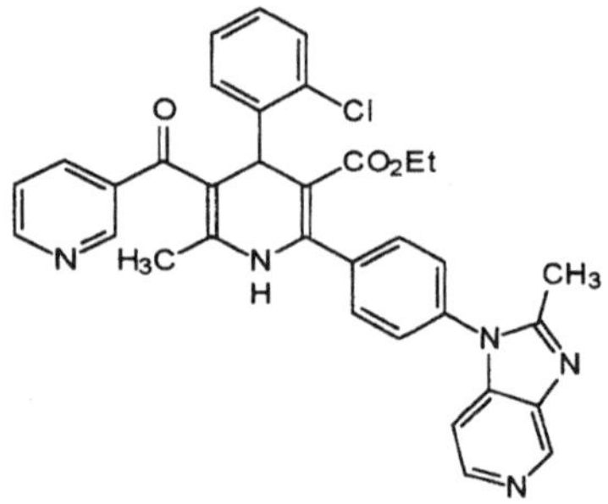


**UK-74,505**

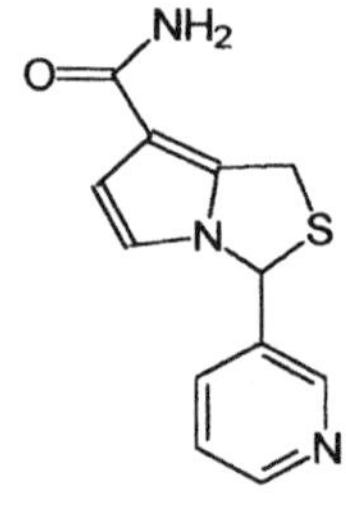


**RP 48,740**

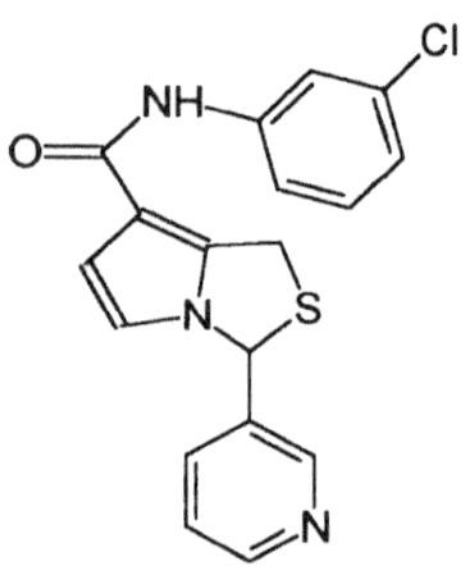


**RP 52,770**

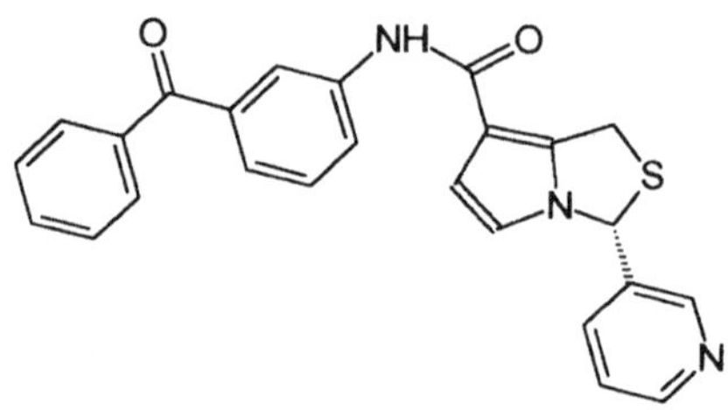


**RP 59,227**

**Fig. 6.** Compound structures of UK-74,505, RP 48,740, RP 52,770, and RP 59,227

produced no adverse effects, while ex vivo inhibition of platelet aggregation was observed (PINQUIER et al. 1991). Chemically related follow-up compounds RP 52,770 (MARQUIS et al. 1988, Fig. 6) and RP 59,227 (ROBAUT et al. 1988, Fig. 6) have recently been introduced and appear to possess much greater potency by parenteral or oral administration (UNDERWOOD et al. 1992). In the in vitro rabbit aggregation assay, the enantiomers of RP 52,770 show a 670-fold difference (ROBAUT et al. 1987), and the RP 59,227 enantiomers exhibited a 250-fold difference (MONDOT and CAVERO 1988). Named tulopafant, this compound reportedly inhibits PAF-induced bronchoconstricton in guinea pigs (ROBAUT et al. 1988) and hypotension in rats (MONDOT and CAVERO 1988).

## XIII. Sandoz Research Institute

The earliest PAF antagonist from Sandoz East Hanover was SDZ 63-072 (Fig. 7), a molecule comparable to CV-3988 but with a thiazolium group on carbon 3 and a tetrahydrofuran ring on carbon 2. This compound antagonized PAF-induced platelet aggregation (WINSLOW et al. 1987), PAF-induced $PGE_2$ synthesis by cultured glomerular mesangial cells (NEUWIRTH et al. 1987), and systemic effects of PAF, endotoxin, and immune complexes in rats (HANDLEY et al. 1986c) and primates (HANDLEY et al. 1986d). The d- and l-enantiomers show similar inhibition of PAF-induced human platelet aggregation and receptor binding (WINSLOW et al. 1987), indicating a lack of enantioselectivity. In vivo, SDZ 63-072 has been shown to effectively inhibit PAF-, endotoxin-, and immune-complex-induced hypotension (HANDLEY et al. 1986c), PAF-induced hemoconcentration in the guinea pig, dog, and primate (HANDLEY and SAUNDERS 1986; HANDLEY et al. 1987a,b), bronchoconstriction (HANDLEY et al. 1987a; PATTERSON et al. 1986), and PAF-induced dermal extravasation (HANDLEY et al. 1987a). Clinically relevant studies showed that SDZ 63-072 prevented cardiac rejection (CAMUSSI et al. 1987) and endotoxin-, PAF-, or tumor-necrosis-factor-mediated ischemic bowel necrosis (HSUEH et al. 1987). In the guinea pig, the d- and l-enantiomers of SDZ 63-072 PAF are equipotent inhibitors of induced hemoconcentration and bronchoconstriction (HANDLEY et al. 1987a), indicating a lack of enantioselectivity. SRI 63-072 was not orally active in rats, guinea pigs, or primates and was dropped from development.

SDZ 63-119 (Fig. 7) is a nonphosphorous thiazolium PAF receptor antagonist that inhibits PAF-induced aggregation (WINSLOW et al. 1987). As for SDZ 63-072, the enantiomers are equipotent in vitro as inhibitors of PAF-induced human platelet aggregation and receptor binding. In vivo, SDZ 63-119 inhibits PAF-induced hypotension in the rat, hemoconcentration in the guinea pig (HANDLEY et al. 1987a) and primate (HANDLEY et al. 1986d), and

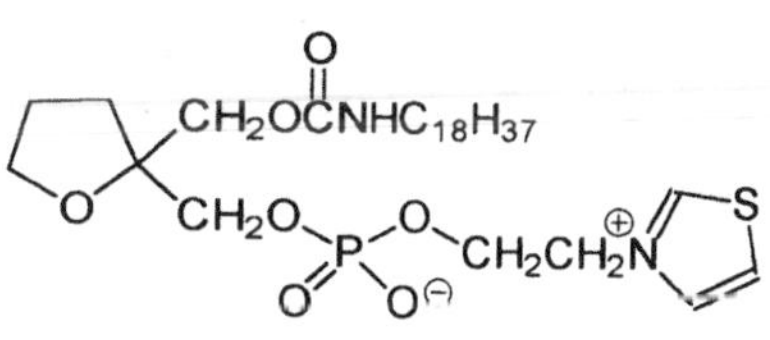

**SDZ 63-072**

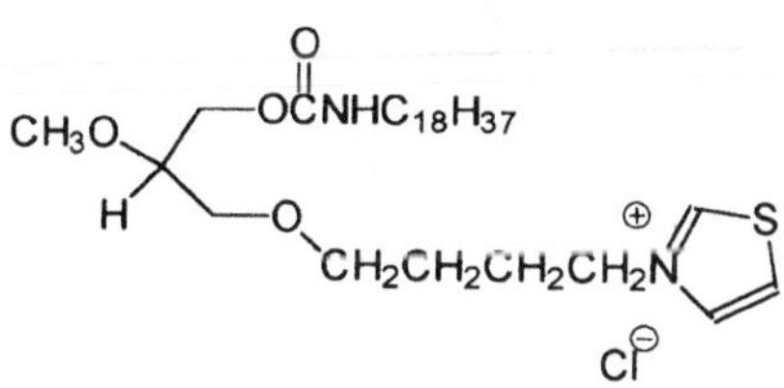

**SDZ 63-119**

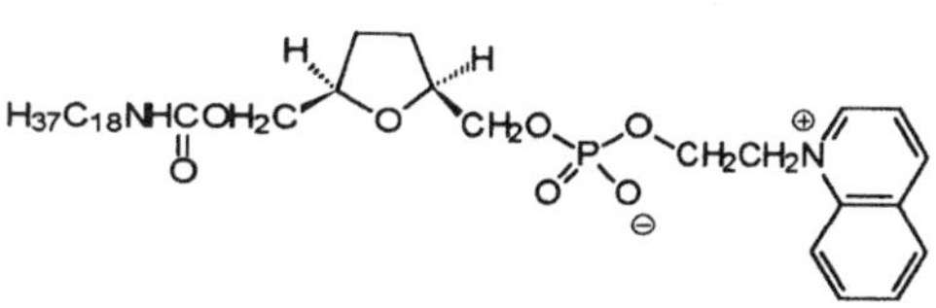

**SDZ 63-441**

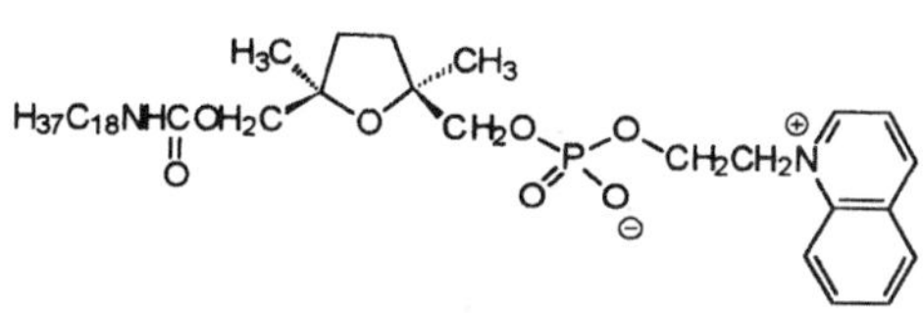

**SDZ 63-675**

**Fig. 7.** Compound structures of SDZ 63-072, SDZ 63-119, SDZ 63–441, and SDZ 63-675

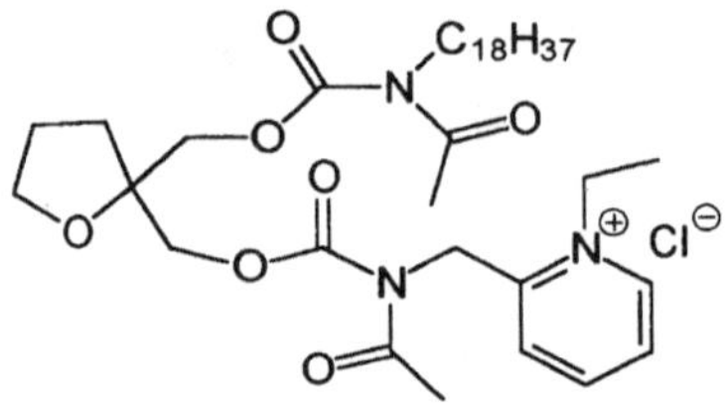

**SDZ 64-770**

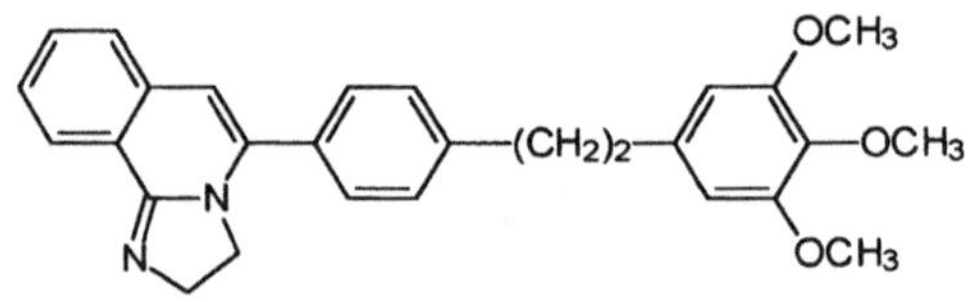

**SDZ 64-412**

**Fig. 8.** Compound structures of SDZ 64-770 and SDZ 64-412

ischemic bowel necrosis in the rat induced by PAF, endotoxin, or tumor necrosis factor (HSUEH et al. 1987). As with 63-072, this compound exhibits weak oral activity and the d- and l-forms are equipotent against PAF-induced bronchoconstriction in the guinea pig (HANDLEY et al. 1987a). SRI 63-119 was not orally active in rats, guinea pigs, or primates and was dropped from development.

In SDZ 63-441 (Fig. 7), the glycerol fragment has been replaced with a cis-2,5-bis hydromethyltetrahydrofuran unit and the thiazolium group with a quinolinium. SDZ 63-441 inhibits PAF-induced platelet aggregation (HANDLEY et al. 1986b), dose-dependently inhibits PAF-mediated vascular responses in rats, dogs, and primates (HANDLEY et al. 1987b), and is reported to be more potent than CV-3988 (ROBERSTON et al. 1987). In vivo, SDZ 63-441 inhibits airways hyperreactivity (ANDERSON et al. 1988; CHRISTMAN et al. 1987; CHUNG et al. 1986) but not acute antigen-induced bronchoconstriction in the dog or primate (PATTERSON et al. 1987, 1989; STENZEL et al. 1987). Clinically relevant studies demonstrated inhibition by SDZ 63-441 of endotoxin effects in the rat such as hypotension (HANDLEY et al. 1987b) and extravascular plasma accumulation in lungs (CHANG et al. 1987), prolongation of renal xenograft (pig-to-dog) survival (MAKOWKA et al. 1987), and inhibition of PAF-induced negative inotropic effects on human myocardium (ROBERSTON et al. 1987). These studies further substantiate the role of endogenously produced PAF as a mediator in pathological conditions involving endotoxin and sepsis. In vitro and in vivo, the chiral forms are of equal potency (HANDLEY 1988). Again, as with SRI 63-072 and SRI 63-119, there was no evidence of oral activity.

SDZ 63-500 is a cyclimmonium PAF receptor antagonist (HANDLEY et al. 1990). This compound and a number of related analogs (dealing with polar head modifications) have been shown to inhibit in vitro platelet aggregation in several species and PAF-induced responses in the rat, guinea pig, and primate (HANDLEY et al. 1986b).

SDZ 63-675 (Fig. 7) is an analog of SDZ 63-441, in which the hydrogens at carbons 2 and 5 have been replaced with methyl groups. SDZ 63-675 is slightly more potent than SDZ 63-441 and inhibits PAF-induced responses in rat, guinea pigs, dogs, and primates (HANDLEY et al. 1987c; MA and DUNHAM 1988). The (+) form of SDZ 63-675 is four fold more potent than the (–) form as an inhibitor of PAF-induced hemoconcentration in the guinea pig and in the *Cebus* primate (HANDLEY 1990a). SDZ 63-675 has been shown to attenuate a number of responses in animals, as recently reviewed (OLSON 1992).

Other developmental PAF antagonists were generated that exhibited potent inhibition of the major systemic responses to PAF, such as seen with SDZ 64-419 (HOULIHAN et al. 1991). SDZ 64-770 (Fig. 8) is structurally comparable to WEB-2086, in which the phosphothiazolium group has been replaced by an *N*-acetyl carbamate pyridinium moiety. SDZ 64-770 showed exceptional potency, long duration of activity, and dose-dependent inhibition of PAF-induced bronchoconstriction and hemoconcentration in the guinea pig, hypotension in the rat, hemoconcentration and hypotension in the dog, and hemoconcentration in the *Cebus* primate (HANDLEY and CERASOLI 1994). Such broad inhibitory effects in different species indicate an absence of receptor heterogeneity or species specificity. Furthermore, SDZ 64-770 showed potent dose-dependent reversal of endotoxin-induced hypotension in the rat with a potency that exceeded the activity of L 652,731 or WEB-2086 in the same model (HANDLEY and CERSAOLI 1994).

SDZ 64-412 (Fig. 8) is a 5-aryl imidazo (2,1-a) isoquinoline salt and is one of the most potent PAF receptor antagonists developed to date with an $IC_{50}$ of 60 n*M* that again exhibits broad inhibitory effects against PAF- or endotoxin-mediated responses (HANDLEY et al. 1988, 1992). SDZ 64-412 also demonstrates the ability to inhibit the development of PAF-induced pulmonary eosinophilia (SANJAR et al. 1990a), allergen-induced airways hyperreactivity in antigen-sensitized guinea pigs (HAVILL et al. 1990; ISHIDA et al. 1990), and lung injury following cardiopulmonary bypass (ZEHR et al. 1995).

A number of PAF antagonists, such as SRI 62-826 (HOULIHAN et al. 1994), SRI 62-834 (HOULIHAN et al. 1987), SDZ 64-412 (HOULIHAN et al. 1995a), and SRI 62-434 (HOULIHAN et al. 1995b) have demonstrated in vitro and in vivo antitumor activity.

## XIV. Sanofi Research

The first of a new series of PAF antagonists, SR-27417, i.e., [N-(2-dimethylaminoethyl)-N-(3-pyridinylmethyl) [2,4,6-triisopropyl phenyl thiazol-2-yl] amine] fumarate (Fig. 9), inhibits PAF-, endotoxin-, and antigen-mediated responses in mice (HERBETT et al. 1991) as well as thromboembolism in cats and mice (SETH et al. 1994). Treatment of mild asthmatics with SR-27417 (10 mg/d for 7 days) did not prevent early allergic responses following allergen provocation, but attenuation of late allergic responses was observed.

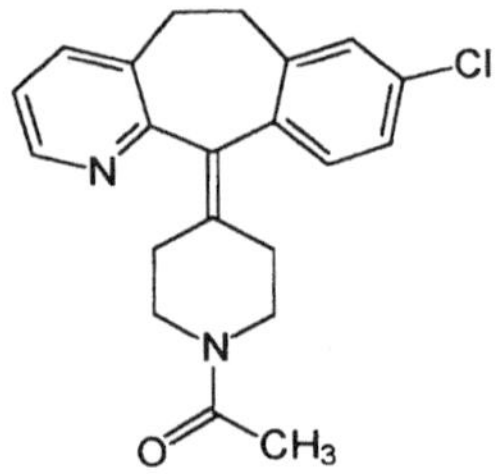

**Sch 37370**

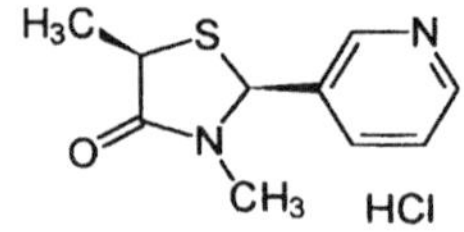



**SM-10661**

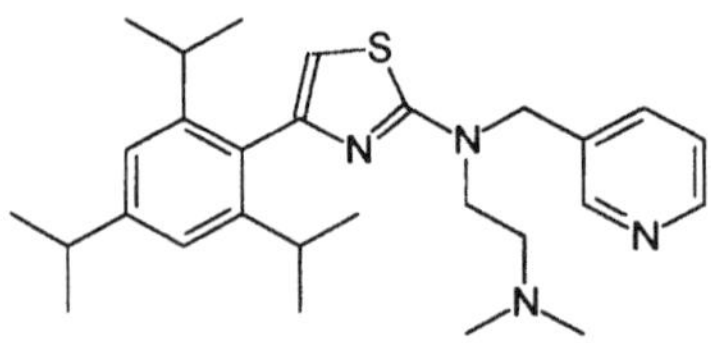

**SR-27417**

**Fig. 9.** Compound structures of Sch 37370, SM-10661, and SR-27417

## XV. Schering-Plough Research Institute

Sch 37370 (Fig. 9), i.e., [1-acetyl-4(8-chloro-5,6-dihydro 11H-benzo[5,6] cyclohepta [1,2-b] pyridine 11-ylidine] piperidine, is a dual antagonist of platelet activating factor and histamine that inhibits platelet responses to PAF and in vivo responses in the guinea pig (BILLAH et al. 1990). However, this molecule exhibited limited potency in humans (BILLAH et al. 1992).

## XVI. Solvay Pharma Laboratories

A series of pyrrolo quinoline derivatives has been generated that antagonize the inflammatory actions of PAF-, histamine-, and leukotriene-dependent phases of bronchoconstriction (PARIS et al. 1995). These compounds are orally active in guinea pigs and exhibit all three antimediator activities.

## XVII. Sumitomo Pharmaceuticals Co., Ltd.

SM-10661 (Fig. 9) is a water-soluble, cis-diastereoisomer thiazolidinone PAF antagonist that has recently been reported to exhibit moderate in vivo activity (KOMURO et al. 1991). Recently developed antagonists that are highly specific and potent include the thiazolidin series (TANABE et al. 1995), of which the one lead compound, SM-12502, has been evaluated for inhibition of endotoxin-induced hypotension and hematological effects in the rat (NATSUME et al. 1994).

## XVIII. Takeda Chemical Co., Ltd.

The first synthetic PAF receptor antagonist reported in the literature was CV-3988 (Fig. 10), a structural analog of PAF in which the quaternary moiety on carbon 3 has been replaced by a thiazolium, the ether linkage on carbon 1 replaced by a octadecyl carbamate group, and methyl ether on carbon 2 instead of an acetoxy. CV-3988 inhibits PAF-induced aggregation with an $IC_{50}$ of 7800n$M$ (TERASHITA et al. 1983) and binding of PAF on human platelets with an $IC_{50}$ of 160n$M$ (TERASHITA et al. 1985) and on liver membranes (HWANG 1987). CV-3988 has been used as a literature standard for comparison with other PAF receptor antagonists, such as Ono 6240 ((TERASHITA et al. 1987), L 652,731 (HWANG and LAM 1986; STEWART and DUSTING 1988), kadsurenone (NUNEZ et al. 1986), Ginkgolide B (TERASHITA et al. 1987), alprazolam (COX 1986), and triazolam (COX 1986).

In vivo, CV-3988 inhibited PAF-induced bronchoconstriction in the guinea pig (DANKO et al. 1988), vascular permeability changes related to hemoconcentration (HANDLEY et al. 1985, 1986a; TERASHITA et al. 1983, 1985) in the rat and guinea pig (HANDLEY 1988), but not vascular permeability changes induced by other vasoactive agents such as histamine (HANDLEY et al.

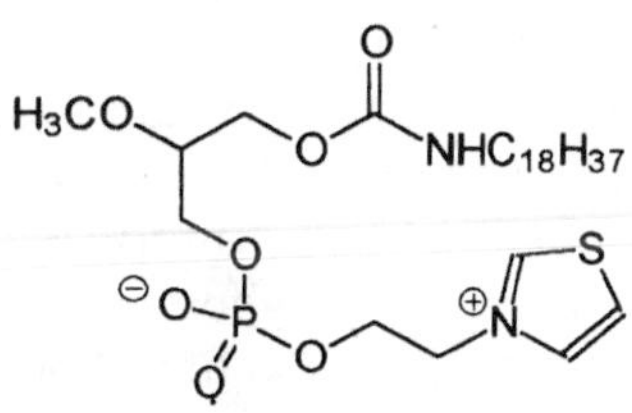


CV-3988

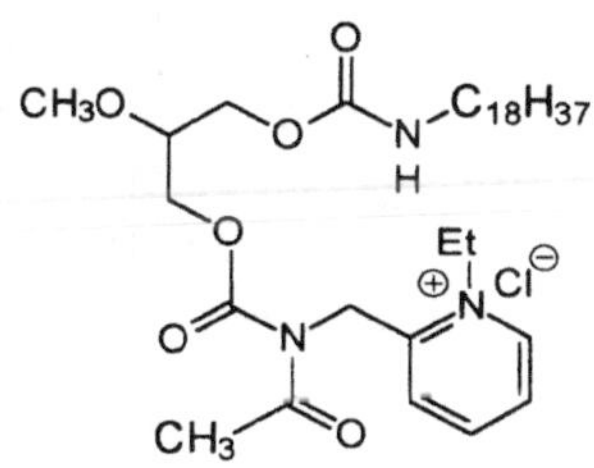


CV-6209

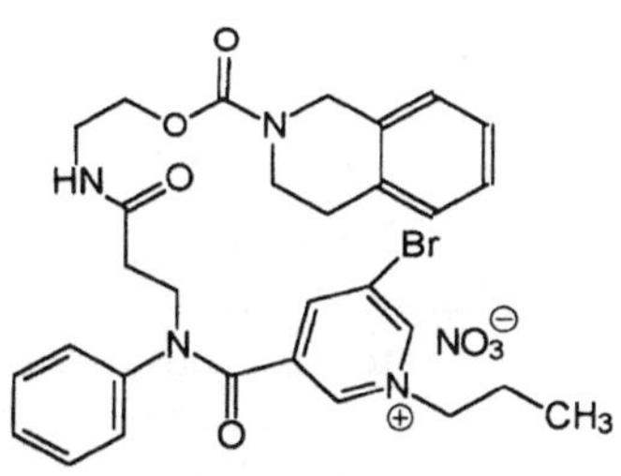


TCV-309

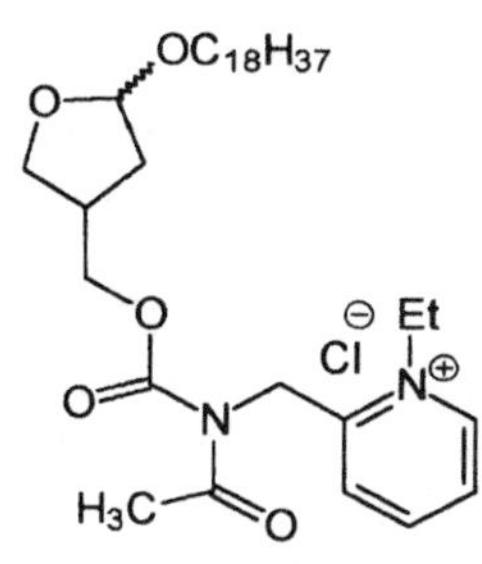


UR-11353

**Fig. 10.** Compound structures of CV-3988, CV-6209, TCV-309, and UR-11353

1986a), bradykinin (HAYASHI et al. 1987), serotonin (HAYASHI et al. 1987b), or leukorienes C4 or D4 (HANDLEY et al. 1986a). Clinically relevant studies showed that CV-3988 could prevent or reverse the effects of systematically administered endotoxin, such as shock hypotension (TERASHITA et al. 1985), increased pulmonary permeability (CHANG et al. 1987), and gastrointestinal damages (HANDLEY 1990a). As endotoxin was known to induce endogenous release of PAF (CHANG et al. 1987; DOEBBER et al. 1985; HSUEH et al. 1987; INARREA et al. 1985; WHITTLE et al. 1987), these studies strongly suggest an involvement of endogenous PAF in the pathological effects of endotoxin.

While effective in preventing lethal anaphylactic shock in sensitized mice (TERASHITA et al. 1987), CV-3988 did not inhibit acute allergen-induced bronchospasm in sensitized guinea pigs (DANKO et al. 1988). While CV-3988 is almost exclusively used parenterally as it has very poor oral activity (HANDLEY et al. 1985), it exhibits a relatively short half-life (in terms of PAF antagonism). Initial studies in man show that CV-3988 given parenterally at 2 mg/kg will inhibit ex vivo platelet responses to PAF but causes hemolysis, probably reflecting the detergent-like properties of the molecule (ARNOUT et al. 1988). It is currently being developed as a topical formulation (Chemex Inc.) for use in atopic and allergic dermatitis.

The next developmental compound was CV-6209 (Fig. 10), which has an ethyl pyridinium group on carbon 3 and whose phosphate group has been replaced by an *N*-acetyl carbamate linkage. In vitro, the enantiomers of CV-6209 at carbon 2 of the glycerol exhibit equal potency (TAKATANI et al. 1989). This compound has been shown to inhibit PAF-induced hypotension and lethality (HANDLEY 1988; MERLOS et al. 1991; TERASHITA et al. 1987) and demonstrates considerably greater potency over Ono 6240 and BN 52021, which was virtually inactive. CV-6209 has also been shown to reduce cardiac damage during myocardial ischemia (STAHL et al. 1986), suggesting that PAF is an important mediator of ischemic damage in the rat. We have found that CV-6209 possesses exceptional potency against PAF-induced responses in the rat, guinea pig, and primate (HANDLEY 1988) and a protracted duration of biological activity in vivo, measured by inhibition of PAF-induced hemoconcentration in the guinea pig (HANDLEY 1989a; HANDLEY et al. 1989). Although one of the most potent and long-acting PAF receptor antagonists developed to date, oral activity appears to be low (HANDLEY 1988), and there are as yet no reports of clinical trials with CV-6209.

The most recent compound of this series is TCV-309 (Fig. 10), i.e., 3-bromo-5 [*N*-phenyl-*N*-[2[[2-(1,2,3,4-tetrahydro 2-isoquinoly carbonyloxy) ethyl] carbomyl] ethyl]carbomyl] 1-propylpyridinium nitrate, which exhibits potent inhibition of several PAF- and endotoxin-mediated responses (TERASHITA et al. 1992; UENO et al. 1995) and prolongs lung (IDE et al. 1995) and renal (YIN et al. 1995) preservation following ischemia/reperfusion injury. A dihydroisoquinoline group has replaced the lipid side-chain on carbon 1 to reduce hemolytic side-effects. In man, while relatively potent and able to inhibit ex vivo platelet aggregation, TCV-309 is short-lived, and large doses (360 mg) are required for pharmacological activity (STOCKMANS et al. 1991).

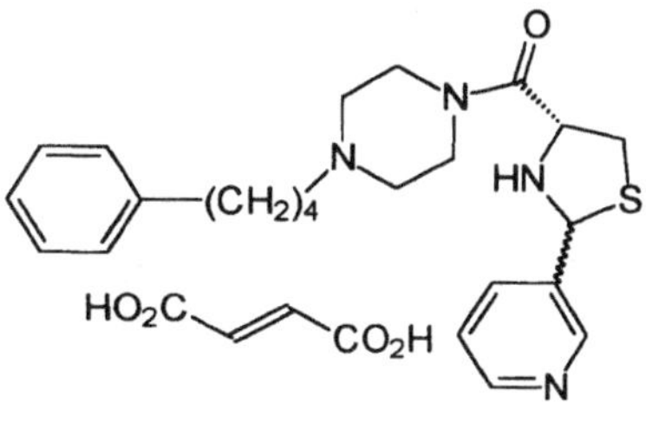


YM 461

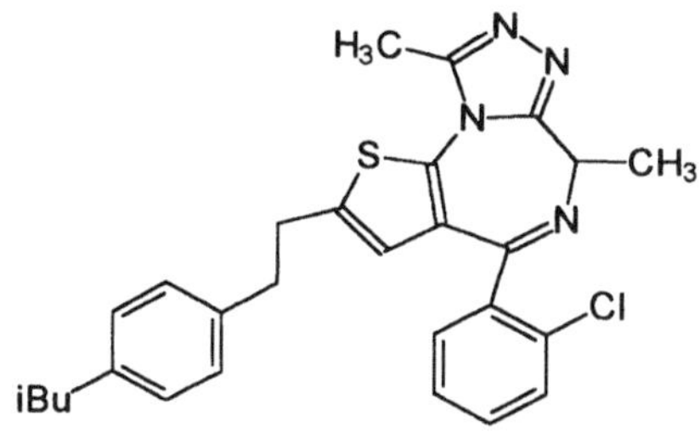


Y-24180

**Fig. 11.** Compound structures of YM 461 and Y-24180

## XIX. Uriach S.A.

UR-10324 and UR-11353 (Carceller et al. 1993; Fig. 10) are lead compounds in a 2,4-disubstituted furan series of ionic PAF antagonists. These compounds inhibit PAF-induced platelet aggregation and in vivo inhibition of PAF-induced hypotension, comparable to CV-6209 and WEB-2086 (Merlos et al. 1991). As expected, in these models BN 52021 was virtually inactive.

## XX. Yamanouchi Pharmaceutical Company

YM 461 (Fig. 11), i.e., 1-(3-phenylpropyl) 4-[2-(3-pyridyl) thiazolidin 4-ylcarbonyl] piperazine fumarate, exhibits potent and-long lasting oral activity in the sheep model of allergen-induced responses (Tomioka et al. 1989) and reasonably potent inhibition of PAF-induced platelet aggregation and PAF-induced hypotension (Merlos et al. 1991).

## XXI. Yoshitomi Pharmaceutical Industries Ltd.

Etizolam, 6-(o-chlorophenyl) 8-ethyl-1-methyl 4H-s-triazolo [3,4-c] thieno [2,3-e] [1,4] diazepine, is an antianxiety drug that inhibits PAF-induced reactions in guinea pigs and mice (Terasawa et al. 1987). A recent derivative of etizolam, Y-24180 (Fig. 11), is a potent long-acting PAF receptor antagonist (Terasawa et al. 1990) that when given orally, inhibits ex vivo PAF-induced platelet aggregation in rabbits (Takehara et al. 1990) and bronchoconstriction effects in guinea pigs (Saito et al. 1995).

# D. Conclusions

The field of PAF biosynthetic inhibitors an receptor antagonists reflects diverse compound structures as well as distinctions based on therapeutic potential, superior potency, duration of activity, and unique pharmacological actions. Clinical trials have primarily profiled therapeutic effectiveness in asthma (Handley and Cerasoli 1994) and shock. More recent compounds have been designed to antagonize multiple mediators such as histamine, leukotrienes, and PAF and include pyrrolo quinoline derivatives developed by Solvay Pharma Laboratories (PARIS et al. 1995), Sch 37370 developed by Schering-Plough (BILLAH et al. 1990), HWA-214 by Hoechst AG (ANAGNOSTOPULOS et al. 1989), hydroxyurieidyl diarytetrahydrofurane molecules (CAI et al. 1994), and UK-47,098 (COOPER et al. 1992). Based on the redundancy of inflammatory mediators, compounds that provide antagonism of several mediators should provide better therapeutic potential that single-mediator antagonists.

*Acknowledgements.* I am indebted to Ms. Ellenor Golonka for manuscript preparation, to Dr. Yun Gao for structural graphics, and to Jaimie and Megan Handley for critical manuscript review.

## References

Adamus WS, Heuer H, Meade CJ (1989) PAF-induced platelet aggregation ex vivo as a method for monitoring pharmacological activity in healthy volunteers. Methods Find Exp Clin Pharmacol 11:415–420

Adamus WS, Heuer HO, Meade CJ, Schilling JC (1990) Inhibitory effects of the new PAF acether antagonist WEB 2886 on pharmacological changes induced by PAF inhalation in human beings. Clin Pharmacol Ther 47:456–462

Albert DH, Snyder F (1983) Biosynthesis of 1-alkyl 2-acetyl sn-glycero-3-phosphocholine (platelet activating factor) from of 1-alkyl 2-acyl sn-glycero-3-phosphocholine by rat alveolar macrophages. J Biol Chem 258:97–102

Anagnostopulos H, Bartlett RR, Elben U, Stoll P (1989) Synthesis of basic substituted pyridines: a new class of antiasthmatic-antiallergic agents. Eur J Chem 24:227–232

Anderson GP, White HL, Fennessy MR (1988) Increased airways responsiveness to histamine induced by platelet activating factor in the guinea pig: possible role of lipoxygenase metabolites. Agents Actions 24:1–7

Arnout J, VanHecken A, De Lepeleire I, Miyamoto Y, Holmes I, DeSchepper P, Vermylem J (1988) Effectiveness and tolerability of CV 3988, a selective PAF antagonist, after intravenous administration in man. Br J Clin Pharmacol 25:445–454

Arnoux A, Denjean A, Page CP, Nolibe D, Morley J, Benveniste J (1988) Accumulation of platelets and eosinophils in baboon lung after Paf-acether challenge. Am Rev Respir Dis 137:855–860

Barnes PJ, Chung KF (1987) PAF closely mimics pathology of asthma. Trends Pharmacol Sci 8:285–287

Batt JP, Lamouri A, Tavet F, Heymans F, Dive G, Godfroid JJ (1991) New hypothesis on the conformation of the PAF receptor from studies on the geometry of selected platelet activating factor antagonists. J Lipid Mediat 4:343–346

Bel EH, De Smet M, Rossing TH, Timmers MC, Dijkman JH, Sterk PJ (1991) The effect of a specific PAF antagonist, MK-287, on antigen-induced early and late asthmatic reactions in man. Am Rev Respir Dis 143:A811

Benveniste J, Henson PM, Cochrane GC (1972) leukocyte dependent histamine release from rabbit platelets: the role of IgE-basophils and platelet activating factor. J Exp Med 136:1356–1368

Benveniste J, Tence M, Varenne P, Bidault J, Boullet C, Polonsky J (1979) Semi-syntheses et structure proposee du facteur activant les plaquettes (P.A.F.): PAF-aceteher, un alkyl ether analogue de la lysophatidylcholine. CR Acad Sci III 289D:1037–1041

Betz S, Lotner GZ, Henson PM (1980) Production and release of platelet activating factor (PAF): dissociation from degranulation and superoxide production in the human neutrophil. J Immunol 125:2749–2763
Billah MM, Chapman RW, Egan RW, Gilchrest H, Piwinski JJ, Sherwood J, Siegel MI, West REJ, Dreutner W (1990) Sch 37370: a potent, orally active, dual antagonist of platelet activating factor and histamine. J Pharmacol Exp Ther 252:1090–1096
Billah MM, Gilchrest H, Eckel SP, Granzow CA, Lawton PJ, Radwanskii E, Brannan MD, Affrime MB, Christopher JD, Richard W, Siegel MI (1992) Differential plasma duration of anti-platelet activating factor and antihistamine activities of oral Sch 37370 in humans. Clin Pharmacol Ther 52:151–159
Bitterman H, Lefer D, Lefer A (1987) Beneficial actions of RO 15-1788, a benzodiazepine receptor antagonist, in hemorrhagic shock. Methods Find Exp Clin Pharmacol 9:341–347
Blank ML, Synder F, Byers LW, Brooks B, Muirhead EE (1979) Antihypertensive activity of an alkyl ether analog of phosphatidylcholine. Biochem Biophys Res Commun 90:1194–1200
Blank ML, Smith ZL, Fitzgerald V, Snyder F (1995) The CoA-independent transacylase in PAF biosynthesis: tissue distribution and molecular species selectivity. Biochem Biophys Acta 1254:295–301
Burhop KE, Van Der Zee H, Bizios R, Kaplan JE, Malik AB (1986) Pulmonary vascular response to platelet activating factor in awake sheep and the role of cyclooxygenase metabolites. Am Rev Respir Dis 134:548–554
Bussolino F, Gremo F, Tetta C, Pescarmona G, Camussi G (1986) Production of platelet activating factor by chick retina. J Boil Chem 261:16502–16508
Cai X, Hussoin S, Killan D, Scannel R, Yaeger D, Eckman J, Hwang SB, Libertine-Barahan L, Qian C, Yeh C, Ip S (1994) Synthesis of substituted hydroxyurieidyl diaryltetrahydrofurans, a potent class of inhibitors of 5-lipoxygenase and platelet activating factor. In: 207th National Meeting of the American Chemical Society, MEDI 123
Camussi G, Neilsen N, Tetta C, Saunders RN, Milgrom F (1987) Release of platelet activating factor from rabbit heart perfused in vitro by sera with transplantation alloantibodies. Transplantation 44:113–118
Carceller E, Merlos M, Giral M, Almansa C, Bartrolli J, Garcia-Rafaneel J, Forn J (1993) Synthesis and structure-activity relationships of 1-acyl-4-[(methyl-3-pyridyl) cyanomethyl] piperazines as PAF antagonists. J Med Chem 36:2984–2997
Casals-Stenzel J (1987a) Effects of WEB 2086, a novel antagonist to platelet activating factor in active and passive anaphylaxis. Immunopharmacology 13:117–124
Casals-Stenzel J (1987b) Triazolodiazepines are potent antagonists of platelet activating factor (PAF) in vitro and in vivo. Arch Pharmacol 335:351–355
Casals-Stenzel J (1987c) Protective effect of WEB 2086, a novel antagonist of platelet activating factor, in endotoxin shock. Eur J Pharmacol 135:117–122
Casals-Stenzel J, Weber KH (1987) Triazolodiazepines: dissociation of their Paf (platelet activating factor) antagonistic and CNS activity. Br J Pharmacol 90:139–146
Casals-Stenzel J, Franke J, Friedrich T, Lichey J (1987a) Bronchial and vascular effects of Paf in the rate isolated lung are completely blocked by WEB 2086, a novel specific Paf antagonist. Br J Pharmacol 91:799–802
Casals-Stenzel J, Muacevic G, Weber KH (1987b) Pharmacological actions of WEB 2086, a new specific antagonist of platelet activating factor. J Pharmacol Exp Ther 241:974–981
Chang SW, Feddersen CO, Henson PM, Voelkel NF (1987) Platelet activating factor mediates hemodynamic changes and lung injury in endotoxin treated rats. J Clin Invest 79:1498–1509
Chignard M, Le Couedic JP, Vargaftig BB, Benveniste J (1980) Platelet activating factor (PAF-acether) secretion from platelets: effect of aggregating agents. Br J Haematol 46:455–464

Christman BW, Lefferts PL, Snapper JR (1987) Effect of platelet activating factor on aerosol histamine responsiveness in awake sheep. Am Rev Respir Dis 135:1267–1270

Christman BW, Lefferts PL, King GA, Snapper JR (1988) Role of circulating platelets and granulocytes in platelet activating factor (PAF)-induced pulmonary dysfunction in awake sheep. J Appl Physiol 64:2033–2041

Chung KF, Barnes PJ (1988) PAF antagonists: their potential therapeutic role in asthma. Drugs 35:93–103

Chung KF, Barnes PJ (1989) Effects of platelet activating factor on airway caliber, airway responsiveness, and circulating cells in asthmatic subjects. Thorax 44:108–115

Chung KF, Aizawa H, Becker AB, Frick O, Gold WM, Nadel JA (1986) Inhibition of antigen-induced airway hyperresponsiveness by a thromboxane synthetase inhibitor (OKY-046) in allergic dogs. Am Rev Respir Dis 134:258–261

Chung KF, Dent G, McCusker M, Guinot PH, Page CP, Barnes PJ (1987) Effect of a ginkgolide mixture (BN 52063) in antagonising skin and platelet responses to platelet activating factor in man. Lancet 1:248–251

Coeffier E, Borrel MC, Lefort J, Chignard M, Broquet C, Heymans F, Godfroid JJ, Vargaftig BB (1986) Effects of Paf-acether and structural analogues on platelet activation and bronchoconstriction in guinea pigs. Eur J Pharmacol 131:179–188

Colditz IG, Topper EK (1991) The effect of systemic treatment with platelet activating factor on the migration of eosinophils to the lung, pleural and peritoneal cavities in the guinea pig. Int Arch Allergy Appl Immunol 95:94–96

Cooper K, Fray MJ, Richardson K, Steele J (1992) The discovery of UK-74505: a potent and selective4 PAF antagonist. In: 203rd National Meeting of the American Chemical Society, MEDI 182

Coyle AJ, Spina D, Page CP (1990) PAF-induced bronchial hyperresponsiveness in the rabbit: contribution of platelets and airway smooth muscle. Br J Pharmacol 101:31–38

Cox CP (1986) Effects of CV-3988, an antagonist of platelet activating factor (PAF), on washed rabbit platelets. Thromb Res 41:211–222

Dahlback M, Bergstrand H, Sorenby L (1984) Bronchial anaphylaxis in actively sensitized Sprague Dawley rats: studies of mediators involved. Acta Pharmacol Tox 55:6–17

Danko G, Sherwood JE, Grissom B, Kreutner W, Chapman RW (1988) Effect of the PAF antagonists, CV-3988 and L-652,731 on the pulmonary and hematological responses to guinea pig anaphylaxis. Pharmacol Res Commun 20:785–798

Davidson SK, Summers JB, Albert DH, Holms JH, Heyman HR, Magocc TJ, Conway RG, Rhein DA, Carter GW (1995) N-(Acyloxyalkyl) pyridinium salts as soluble prodrugs of a potent platelet activating factor antagonist. J Med Chem 37:4423–4429

Demopoulus CA, Pinckard RN, Hanahan DJ (1979) Platelet activating factor. Evidence for 1-*O*-alkyl 2-acetyl sn-glyceryl 3-phosphorylcholine as the active component (a new class of lipid chemical mediators). J Boil Chem 254:9355–9358

Denizot Y, Trimoreau F, Dupuis F, Verger C, Praloran V (1995) PAF and haematopoiesis: III. Presence and metabolism of platelet activating factor in human bone marrow. Biochim Biophys Acta 1265:55–60

Denjean A, Arnoux B, Benveniste J, Lockhart A, Masse R (1981) Bronchoconstriction induced by intratracheal administration of platelet activating factor (PAF-acether) in baboons. Agents Actions 11:567–568

Denjean A, Arnoux B, Masse R, Lockart A, Benveniste J (1983) Acute effects of intratracheal administrtion of platelet activating factor in baboons. J Appl Physiol 55:799–804

Desquand S, Touvay C, Randon J, Lagente V, Vilain B, Parini M, Etienne A, Lefort J, Braquet P, Vargaftig BB (1986) Interference of BN 52021 (Ginkgolide B) with bronchopulmonary effects of PAF-acether in the guinea pig. Eur J Pharmacol 127:83–95

Desquand S, Lefort J, Dumarey C, Vargaftig BB (1990) The booster injection of antigen during active sensitization of guinea pig modifies the anti-anaphylactic activity of the PAF antagonist WEB 2086. Br J Pharmacol 100:217–222
Doebber T, Wu M (1988) Platelet activating factor induced cellular and pathophysiological responses in the cardiovascular system. Drug Dev Res 12:151–161
Doebber TW, Wu MS, Robbins JC, Choy BM, Chang MN, Shen TY (1985) Platelet activating factor (PAF) involvement in endotoxin-induced hypotension in rats. Studies with PAF receptor dadsurenone. Biochem Biophys Res Commun 127:799–808
Doebber TW, Wu MS, Biftu T (1986) Platelet activating factor (PAF) mediation of rat anaphylactic responses to soluble immune complexes. Studies with a PAF receptor antagonist L-752,731. J Immunol 136:4659–4668
Emanuelli G, Montrucchio G, Gaia E, Dughera L, Corvetti G, Gubetta L (1989) Experimental acute pancreatitis induced by platelet activating factor in rabbits. Am J Pathol 134:315–325
Evans TW, Chung KF, Rogers DF, Barnes PJ (1987) Effect of platelet activating factor on airway permeability: possible mechanisms. J Appl Physiol 63:479–484
Evans T, Dent G, Rogers D, Aursudkij B, Chung KF, Barnes P (1988) Effect of a Paf antagonist, WEB 2086, on airway microvascular leakage in the guinea pig and platelet aggregation in man. Br J Pharmacol 94:164–168
Fernandez-Gallardo S, Ortega M, Priego J, Casa F, Sunkel C, Sanchez Crespo M (1990) Pharmacological actions of PCA 4248, a new platelet activating factor receptor antagonist: in vivo studies. J Pharmacol Exp Ther 255:34–39
Fitzgerald MF, Moncada S, Parente L (1986) The anaphylactic release of platelet activating factor from perfused guinea pig lungs. Br J Pharmacol 88:149–153
Freitag A, Watson RM, Mastros G, Eastwood C, O'Byrne PM (1991) Effects of treatment with the oral platelet activating factor antagonist (WEB 2086) on allergen induced asthmatic responses in human subjects. Am Rev Respir Dis 143:A156
Garito ML, Prihoda TJ, McMannus LM (1995) Salivary PAF levels correlate with severity of periodontal inflammation. J Dent Res 74:1048–1056
Grandel K, Farr R, Wanderer A, Eisenstadt T, Wasserman S (1985) Association of platelet activating factor with primary acquired cold urticaria. N Engl J Med 313:405–409
Grue-Sorensen G, Nielsen I, Neilsen C (1988) Derivatives of 2-methylene propane 1,3-diol as a new class of antagonists of platelet activating factor. J Med Chem 31:1174–1178
Hadvary P, Baumgartner HR (1985) Interference of PAF antagonists with platelet aggregation and the formation of platelet thrombi. Prostaglandins 30:694–702
Hakansson L, Venge P (1990) Inhibition of neutrophil and eosinophil chemotactic responses to PAF by the PAF antagonists WEB 2086, L-652-731, and SRI 63-441. J Leukoc Biol 47:449–456
Halonen M, Palmer J, Lohman I, McManus L, Pinckard RN (1980) Respiratory and circulatory alterations induced by acetyl glyceryl ether phosphorylcholine, a mediator of IgE anaphylaxis in the rabbit. Am Rev Respir Dis 122:915–924
Hanahan DJ, Demopoulos CA, Liehr J, Pinckard RN (1980) Identification of a platelet activating factor isolated from rabbit basophils as acetylglyceryl ether phosphorylcholine. J Biol Chem 255:5514–5516
Handley DA (1987) New developments for clinical applications for PAF antagonists. J Clin Exp Med 143:415–418
Handley DA (1988) Development and therapeutic indications of PAF receptor antagonists. Drugs Future 13:137–152
Handley DA (1989a) Quantitation of in vitro and in vivo biological effects of platelet activating factor. In: Chang JY, Lewis AJ (eds) Pharmacological methods in the control of inflammation. Liss, New York, pp 23–49
Handley DA (1989b) The development and application of colloidal gold as a microscopic probe. In: Hayat MA (ed) Colloidal gold: principles, methods and applications. Academic, San Diego, vol 1, pp 1–12

Handley D (1990a) Preclinical and clinical pharmacology of platelet activating factor receptor antagonists. Med Res Rev 10:351–370

Handley DA (1990b) Platelet activating factor as a mediator of endotoxin mediated diseases. In: Handley DA, Saunders R, Houlihan W, Tomesch J (eds) Platelet activating factor in immune complex and endotoxin mediated diseases. Dekker, New York, pp 451–495

Handley DA, Cerasoli F (1994) PAF antagonists as asthma therapeutics. In: Page CP, Metzger WJ (eds) Drugs and the lung. Raven, New York, pp 467–505

Handley DA, Saunders RN (1986) Platelet activating factor and inflammation in atherosclerosis: targets for drug development. Drug Dev Res 7:361–375

Handley DA, Lee M, Saunders RN (1983) Extravasation and aortic changes accompaning acute and subacute intraperitoneal administration of PAF. In: Benveniste J, Arnout B (eds) Platelet activating factor. Elsivier, Amsterdam, pp 243–251

Handley DA, Van Valen RG, Melden MK, Saunders RN (1984) Evaluation of dose and route effects of platelet activating factor-induced extravasation in the guinea pig. Thromb Haemost 52:34–36

Handley DA, Lee ML, Saunders RN (1985) Evidence for a direct effect on vascular permeability of platelet activating factor induced hemoconcentration in the guinea pig. Thromb Haemost 54:756–759

Handley DA, Farley C, Deacon RW, Saunders RN (1986a) Evidence for distinct systemic extravasation effects of platelet activating factor, leukotrienes B4, C4, D4 and histamine in the guinea pig. Prostaglandins Leukot Med 21:269–277

Handley DA, Tomesch JC, Saunders RN (1986b) Inhibition of PAF induced responses in the rat, guinea pig, dog and primate by the receptor antagonist SRI 63-441. Thromb Haemost 56:40–44

Handley DA, Van Valen RG, Melden MK, Flury S, Lee ML, Saunders RN (1986c) Inhibition and reversal of endotoxin-, aggregated IgG- and PAF-induced hypotension in the rat by SRI 63-072, a PAF receptor antagonist. Immunopharmacology 12:11–17

Handley DA, Van Valen RG, Saunders RN (1986d) Vascular responses of platelet-activating factor in the cebus apella primate and inhibitory profiles of PAF antagonists SRI 63-072 and SRI 63-119. Immunopharmacology 11:175–182

Handley DA, Anderson RC, Saunders RN (1987a) Inhibition by SRI 63-072 and SRI 63-119 of PAF acether and immune complex effects in the guinea pig. Eur J Pharmacol 141:409–416

Handley DA, Van Valen RG, Tomesch JC, Melden MK, Jaffe JA, Ballard F, Saunders RN (1987b) Biological properties of the receptor antagonist SRI 63-441 in the PAF and endotoxin models of hypotension in the rat and dog. Immunopharmacology 13:125–132

Handley DA, Van Valen RG, Winslow CM, Tomesch JC, Saunders RN (1987c) In vitro and in vivo pharmacological effects of the PAF receptor antagonist SRI 63-675. Thromb Haemost 57:187–190

Handley D, Van Valen RG, Lee ML, Saunders RN (1987d) Cebus appela primate responses to platelet activating factor and inhibition of platelet activating factor antagonist SRI 63-072. In: Winslow CM, Lee ML (eds) New horizons in platelet activating factor research. Wiley, New York, pp 335–341

Handley DA, Van Valen RG, Melden MK, Houlihan WJ, Saunders RN (1988) Biological effects of the orally active PAD receptor antagonist SDZ 64-412. J Pharmacol Exp Ther 247:617–628

Handley DA, Houlihan W, Tomesch J, Farley C, Deacon R, Koletar J, Prashed M, Hughes J, Jaeggi C (1989) Chemistry and pharmacology of PAF antagonists. In: Sammuelsson B, Wong P, Sun F (eds) Evaluation of changes at potential metabolism sites on activity and duration of activity. Adv Prostaglandin Thromboxane Leukot Res 19:367–370

Handley DA, Lee M, Farley C, Deacon R, Melden M, Van Valen RG, Winslow CM, Saunders R (1990) Biological characterization of cyclimmonium PAF antagonists. In: Handley DA, Houlihan W, Saunders R, Tomesch J (eds) Platelet activating factor in endotoxin and immune complex mediated diseases. Dekker, New York, pp 157–175

Handley DA, DeLeo J, Havill AM (1992) Induction by aerosol allergen of sustained and nonspecific IgE-mediated airway hyperreactivity in the guinea pig. Agents Actions 37:201–203

Havill AM, Van Valen RG, Handley DA (1990) Prevention of non-specific airway hyperreactivity after allergen challenge in guinea pigs by the PAF receptor antagonist SDZ 64-412. Br J Pharmacol 99:396–400

Hayashi M, Kimura J, Yamaki K, Suwabe Y, Dozen M, Imai Y, Oh-ishi S (1987a) Detection of platelet activating factor in exudates of rats with phorbol myristate acetate induced pleurisy. Thromb Res 48:299–310

Hayashi M, Kimura J, Oh-ishi S, Tsushima S, Nomura H (1987b) Characterization of the activity of a platelet activating factor antagonist, CV-3988. Jpn J Pharmacol 44:127–134

Hellewell PG (1990) The contribution of platelet activating factor to immune complex mediated inflammation. In: Handley DA, Saunders R, Houlihan W, Tomesch J (eds) Platelet activating factor in immune complex and endotoxin mediated diseases. Dekker, New York, pp 367–386

Herbert JM, Lespy L, Maffrand JP (1991) Protective effect of SR 27417, a novel PAF antagonist, on lethal anaphylactic and endotoxin-induced shock in mice. Eur J Pharmacol 205:271–276

Heuer HO (1991a) Inhibition of active anaphylaxis in mice and guinea pigs by the new hetrazepinoic PAF antagonist bepafant (WEB 2170). Eur J Pharmacol 199:157–163

Heuer HO (1991b) WEB 2347: pharmacology of a new very potent and long acting hetrazepinic PAF antagonist and its action in repeatedly sensitized guinea pigs. J Lipid Mediat 4:39–44

Houlihan WJ, Lee ML, Nemecek GM, Handley DA, Winslow CM, Jaeggi C (1987) Antitumor activity of SRI 62-834, a cyclic ether analogue of ET-18-OCH3. Lipids 22:884–890

Houlihan WJ, Cheon SH, Handley DA, Larson DA (1991) Synthesis and pharmacology of a novel class of long-lasting PAF antagonists. J Lipid Mediat 3:91–99

Houlihan WJ, Lee ML, Munder P, Handley DA, Jaeggi SC, Winslow C (1994) Antitumor activity of a piperidine phospholipid. Drug Res 44:1348–1388

Houlihan WJ, Munder P, Handley DA, Cheon SH, Parillo VA (1995a) Antitumor activity of 5-aryl-2,3 dihydroimidazo [2,1-a] isoquinolines. J Med Chem 38:234–240

Houlihan WJ, Munder P, Handley DA, Nemecek GA (1995b) 5-(4′-piperidinomethyl phenyl) 2,3-dihydroimidazo [2,1-a] isoquinoline (SDZ 62-434). A novel antitumor agent. Drug Res 22:1143–1155

Hsueh W, Gonzalez-Crussi F, Arroyave J (1987) Platelet activating factor: an endogenous mediator of bowel necrosis endotoxemia. FASEB J 1:403–406

Hwang MN, Chang ML, Garcia QQ, Huang L, King VF (1987) L-652-469, a dual antagonist of platelet activating factor and dihydropyridines from Tussilago farfara. Eur J Pharmacol 141:269–278

Hwang SB (1987) Specific receptor sites for platelet activating factor on rat liver plasma membranes. Arch Biochem Biophys 257:339–344

Hwang SB, Lam M (1986) Species difference in the specific receptor of platelet activating factor. Biochem Pharmacol 35:4511–4518

Hwang S, Lam M, Biftu T, Beattie T, Shen T (1985) Trans 2,5 bis-(3,4,5-trimethoxyphenyl) tetrahydrofuran. J Biol Chem 260:15639–15645

Hwang SB, Lam MH, Li CL, Shen TY (1986) Release of platelet activating factor and its involvement in the first phase of carrageenin-induced rat foot edema. Eur J Pharmacol 120:33–41

Hwang SB, Lam M, Alberts A, Bugianesi R, Chabala J, Ponpipom M (1988) Biochemical and pharmacological characterization of L-659,989: an extremely potent, selctive and competitive receptor antagonist to platelet activating factor. J Pharmacol Exp Ther 246:534–541

Ide S, Kawahara K, Takahashi T, Sasaki N, Shingu H, Nagayasu T, Yamamoto S, Tagawa T, Tomita M (1995) Donor administration of PAF antagonist (TCV-309) enhances lung preservation. Transplant Proc 27:570–573

Inarrea P, Gomez-Cambronero J, Nieto M, Sanchez Crespo M (1984) Characteristics of the binding of platelet activating factor to platelets to different animal species. Eur J Pharmacol 105:309–315

Inarrea P, Gomez-Cambronero J, Pascual J, del Carmen Ponte M, Hernando L, Sanchez-Crespo M (1985) Synthesis of PAF acether and blood volume changes in Gram negative sepsis. Immunopharmacology 9:45–52

Ishida K, Thompson R, Beattie LL, Wiggs B, Schellenberg RR (1990) Inhibition of antigen-induced airway hyperresponsiveness, but not acute hypoxia nor airway eosinophilia, by an antagonist of platelet-activating factor. J Immunol 144:3907–3911

Ito S, Camussi G, Tetta C, Milgrom F, Andres G (1984) Hyperacute renal allograft rejection in the rabbit. Lab Invest 51:148–161

Jackson C, Schumacher W, Kunkel S, Driscoll E, Lucchesi B (1986) Platelet activating factor and the release of a platelet-derived coronary artery vasodilator substance in the canine. Circ Res 58:218–229

Jancar S, Abdo E, Sampietre M, Kwasniewski FH, Coelho AMM, Bonizzia A, Machado MCC (1995) Effect of PAF antagonists on cerulein-induced pancreatitis. J Lipid Mediat Cell Signal 11:41–49

Joly F, Bessou G, Benveniste J, Ninio E (1987) Ketotifen inhibits Paf-acether biosynthesis and b-hexosaminidase release in mouse mast cells stimulate with antigen. Eur J Pharmacol 144:133–139

Komuro Y, Imanishi N, Uchida M, Morooka S (1991) Biological effect of orally active platelet activating factor receptor antagonist SM-10661. Mol Pharmacol 38:378–384

Kornecki E, Ehrlich Y, Lenox R (1984) Platelet activaing factor induced aggregation of human platelets specifically inhibited by triazolobenzodiazepines. Science 226:1954–1957

Kroegel C, Yakawa T, Westwick J, Barnes PJ (1989) Evidence for two distinct platelet activating factor receptors on eosinophils: dissociation between PAF-induced intracellular calcium mobilization and superoxide anion generation in eosinophils. Biochem Biophys Res Commun 162:511–521

Lagente V, Desquand S, Hadvary P, Cirino M, Lellouch-Tubiana A, Vargaftig BB (1988) Interference of Paf antagonist Ro 19-3704 with Paf and antigen-induced bronchoconstriction in the guinea pig. Br J Pharmacol 94:27–36

Lee T, Lenihan D, Malone B, Roddy L, Wasserman S (1984) Increased biosynthesis of platelet activating factor in activated human eosinophils. J Biol Chem 259:5526–5530

Lefort J, Sedivy P, Desquand S, Randon J, Coeffier E, Floch A, Benveniste J, Vargaftig BB (1988) Pharmacological profile of 48740 R.P., a PAF acether antagonist. Eur J Pharmacol 150:257–268

Lellouch-Tubiana A, Lefort J, Simon MT, Pfister A, Vargaftig BB (1988) Eosinophil recruitment into guinea pig lungs after PAF-acether and allergen administration. Am Rev Respir Dis 137:948–954

Lepran I, Lefer AM (1985) Ischemia aggravating effects of platelet activating factor in acute myocardial ischemia. Basic Res Cardiol 80:135–141

Lewis AJ, Dervinis A, Chang J (1984) The effects of antiallergic and bronchodilator drugs on platelet activating factor (PAF-acether) bronchospasm and platelet aggregation. Agents Actions 15:636–642

Lohman IC, Halonen M (1990) Effects of the PAF antagonist WEB 2086 on PAF-induced physiological alterations and on IgE anaphylaxis in the rabbit. Am Rev Respir Dis 142:390–397

Luo G, Albert DH, Rhein DA, Conway RG, Sheppard GS, Summers JB, Carter GW (1993) A-79981: a novel PAF antagonist blocks endotoxin induced small intestinal damage in conscious rats. J Immunol 150:211A

Ma Y, Dunham E (1988) Antagonism of the vasodilator effects of a platelet activating factor precursor in anesthetized spontaneously hypertensive rats. Eur J Pharmacol 145:153–162

Makowka L, Miller C, Chapchap P, Podestra P, Pan C, Pressely D, Mazzaferro V, Esquivel C, Todo S, Banner B, Jaffe R, Saunders R, Starzel TE (1987) Prolongation of pig-to-dog renal xenograft survival by modification of the inflammatory mediator response. Annu Surg 206:482–495

Makowka L, Chapman F, Cramer D, Qian S, Sun H, Starzel TE (1990) Platelet activating factor and hyperacute rejection. Transplantation 50:359–365

Malavasi F, Tetta C, Funaro A (1986) Fc receptor triggering induced expression of surface activation antigens and release of platelet activating factor in large granulocytes. Proc Natl Acad Sci USA 83:2443–2447

Marquis O, Robaut C, Cavero I (1988) [3H] 52770 RP, a platelet activating factor receptor antagonist, and tritiated platelet activating factor label a common specific binding site in human polymorphoncuclear leukocytes. J Pharmacol Exp Ther 244:709–715

McFadden RG, Bishop MA, Caveney AN, Fraher LJ (1995) Effect of platelet activating factor (PAF) on the migration of human lymphocytes. Thorax 50:265–269

Mclntyre T, Zimmerman G, Prescott S (1986) Leukotrienes C4 and D4 stimulate human endothelial cells to synthesize platelet activating factor and bind neutrophils. Proc Natl Acad Sci USA 83:2204–2208

Meade CJ, Heuer H, Kempe R (1991) Biochemcial pharmacology of platelet activating factor (and PAF antagonists) in relation to clinical and experimental thrombocytopenia. Biochem Pharmacol 41:657–668

Merlos M, Gomez LA, Giral M, Vericat ML, Rafanell JG, Forn J (1991) Effects of PAF antagonists in mouse ear oedema induced by serveral inflammatory agents. Br J Pharmacol 104:990–994

Minor T, Yamaguchi T, Isselhard W (1995) Treatment of preservation/reperfusion injury by platelet activating factor antagonism in the rat liver graft. Transplant Proc 27:522–523

Misawa M, Takata T (1988) Effects of platelet activating factor on rat airways. Jpn J Pharmacol 48:7–13

Miyamoto T, Ohno H, Yano T, Okada T, Hamanaka N, Kawasaki A (1985) Ono-6240: a new potent antagonist of platelet activating factor. In: Hayaishi O, Yamamoto S (eds) Adv Prostaglandin Thromboxane Leukot Res 15:719–720

Mojarad M, Hamasaki Y, Said IS (1983) Platelet activating factor increases pulmonary microvascular permeability and induced pulmonary edema. Bull Eur Physiopath Respir 19:253–256

Mondot S, Cavero I (1988) Cardiovascular profile of 59227 RP, a novel potent and specific PAF receptor antagonist. Prostaglandin 35:827

Montrucchio G, Alloatti G, Mariano F, De Paulis R, Comino A, Emanuelli G, Camussi G (1990) Role of platelet activating factor in the reperfusion injury of rabbit ischemic heart. Am J Pathol 137:71–81

Myers A, Ramey E, Ramwell P (1983) Glucocorticoid protectin against PAF-acether toxicity in mice. Br J Pharmacol 79:595–598

Nagaoka J, Harada K, Kimura A, Kobayashi S, Murakami M, Yoshimura K, Yamada K, Asano O, Katayama K, Yamatsu S (1991) Inhibitory effects of the novel platelet activating factor receptor antagonist, 1-ethyl-2-[N-(2-methoxy) benzoyl-N-[(2R) 2-methoxy 3-(4-octadecylcarbamoyloxy) piperidinocarbonyl

oxypropyloxy] carbonyl] aminomethyl pyridinium chloride, in several experimentally induced shock models. Drug Res 41:719–724
Namm DH, Tadepalli AS, High JA (1982) Species specificity of the platelet response to 1-*O*-alkyl-2-acetyl sn glycero-3-phosphocholine. Thromb Res 25:341–350
Nathan N, Denizot Y (1995) PAF and human cardiovascular disorders. J Lipid Mediat Cell Signall 11:103–104
Natsume Y, Inanishi N, Koike H, Morooka S (1994) Effect of the platelet activating factor antagonist (+)-cis-3,5-dimethyl-2-(3-pyridyl) thiazolidin-4-one hydrochloride on endotoxin-induced hypotension and hematological parameters in rats. Drug Res 44:1208–1213
Neuwirth R, Singhal P, Satriano J, Braquet P, Schlondorff D (1987) Effect of platelet activating factor antagonists on cultured rat mesangial cells. J Pharmacol Exp Ther 242:409–414
Nguyen P, Petitfrere E, Potron G (1995) Mechanisms of the platelet aggregation induced by activated neutrophils and inhibitory effect of specific PAF receptor antagonist. Thromb Res 78:33–42
Nunez D, Chignard M, Korth R, Le Couedic J, Norel X, Spinnewyn B, Braquet P, Benveniste J (1986) Specific inhibition of PAF-acether induced platelet activation by BN 52021 and comparison with the PAF acether inhibitors kadsurenone and CV-3988. Eur J Pharmacol 123:197–205
O'Donnell SR, Erjefalt I, Persson GA (1990) Early and late tracheobronchial plasma exudation by platelet activating factor administered to the airway mucosal surface in guinea pigs: effects of WEB 2086 and enprofylline. J Pharmacol Exp Ther 254:65–70
Okamoto M, Yoshida K, Nishikawa N (1986a) FR-900452, a specific antagonist of platelet activating factor (PAF) produced by Streptomyces phaeofaciens. J Antibiot (Tokyo) 39:198–204
Okamoto M, Yoshida K, Nishikawa M, Kohsaka M, Aoki H (1986b) Platelet activating factor (PAF) involvement in endotoxin induced thrombocytopenia in rabbits: studies with FR-900452, a specific inhibitor of PAF. Thromb Res 42:661–671
Olson NC (1992) SRI 63-675: a platelet activating factor receptor antagonist. Cardiovasc Drug Rev 10:54–70
Ortega MP, Cillero F, Narvaiza J, Maroto ML, Smith CA, Redpath K, Priego JC (1992) Effectiveness, tolerance and plasma levels of PCA-4248, a new PAF antagonist, after administration in human healthy volunteers. In: 4th International Congress on PAF and Related Lipid Mediators, Paris, Abstr c4.2
Page CP, Morley J (1986) Evidence favouring PAF rather than leukotrienes in the pathogesesis of asthma. Pharmacol Res Commun 18:217–237
Paris D, Cottin M, Demonchaux P, Augert G, Dupassieux P, Lenoir P, Peck M, Jasserand D (1995) Synthesis, structure-activity relationships, and pharmacological evaluation of pyrrolo [3,2,1-ij] quinoline derivatives: potent histamine and platelet activating factor antagonism and 5-lipoxygenase inhibitory properties. Potential therapeutic applications in asthma. J Med Chem 38:669–685
Patterson R, Harris KE (1983) The activity of aerosolized and intracuteneous synthetic platelet activating factor (AGEPC) in rhesus monkeys with IgE mediated airway responses and normal monkeys. J Lab Clin Med 102:933–938
Patterson R, Bernstein PR, Harris KE, Krell PD (1984) Airway responses to sequential challenges with platelet actiavting factor and leukotriene D4 in rhesus monkey. J Lab Clin Med 104:340–345
Patterson R, Harris KE, Lee ML, Houlihan WJ (1986) Inhibition of rhesus monkey airway and cutaneous responses to platelet activating factor (PAF) (AGEPC) with the anti-PAF agent SRI 63-072. Int Arch Allergy Appl Immunal 81:265–272
Patterson R, Harris KE, Handley DA, Saunders RN (1987) Evaluation of the effect of a platelet activating factor (PAF) receptor antagoniston platelet activating factor and ascaris antigen-induced airway respones in rhesus monkeys. J Lab Clin Med 110:606–611

Patterson R, Harris KE, Bernstein PR, Krell RD, Handley DA, Saunders RN (1989) Effects of combined receptor antagonists of leukotriene D4 (LTD4) and platelet activating factor (PAF) on rhesus airway responses to LTD4, PAF and antigen. Int Arch Allergy Appl Immunol 88:462–470
Perico N, Delaini F, Tagliaferri M, Abbate M, Cucchi M, Bertani T, Remuzzi G (1988) Effect of platelet activating factor and its specific receptor antagonist on glomerular permeability to proteins in isolated perfused rat kidney. Lab Invest 58:163–171
Pettipher ER, Higgs GA, Henderson B (1987) PAF-acether in chronic arthritis. Agents Actions 21:98–103
Pinckard R, Ludwig JC, McManus L (1988) Platelet activating factors. In: Gallin JI, Goldstein I, Snyderman R (eds) Inflammation: basic principles and clinical correlates. Raven, New York, pp 1–46
Pinckard RN, Woodard DS, Showell HJ, Conklyn MJ, Novak MJ, McManus LM (1994) Structural and (patho)physiological diversity of PAF. Clin Rev Allergy 12:329–359
Pinquier JL, Sedivy P, Bruno R, Bompart E, Gregoire J, Strauch G, Gaillot J, Clucas A (1991) Inhibitin of ex vivo PAF-induced platelet aggregation by the PAF antagonist RP 48740: relationship to plasma concentrations in healthy volunteers. Eur J Clin Pharmacol 41:141–145
Ponipom M, Bugianesi R, Brokker S, Yue B, Hwang S, Shen T (1987) Structure-activity relationships of kadsurenone analogues. J Med Chem 30:136–142
Pons R, Rossi AG, Norman KE, Williams TJ, Nourshargh S (1993) Role of platelet activating factor (PAF) in platelet accumulation in rabbit skin: effect of the novel long-acting PAF antagonist UK-74,505. Br J Pharmacol 109:234–242
Pretolani M, Lefort J, Malanchere E, Vargaftig BB (1987) Interference by the novel PAF-acether antagonist WEB 2086 with the bronchopulmonary responses to PAF-acether and to active and passive anaphylactic shock in guinea pigs. Eur J Pharmacol 140:311–321
Rabinovici R, Yue T, Farhat M, Smith EF, Esser KM, Slivjak M, Feuerstein G (1990) Platelet activating factor (PAF) and tumor necrosis factor-a (TNFa) interactions in endotoxemic shock: studies with BN 50739, a novel PAF antagonist. J Pharmacol Exp Ther 255:256–263
Rayamajhi P, Bruchhausen Fv (1989) Inhibition in vitro of PAF biosynthesis from lyso-PAF by antirheumatic gold salts and gold compounds. Arch Pharmacol 340:R84
Renooij W, Snyder F (1981) Biosynthesis of 1-alkyl 2-acetyl sn-glycero-3-phosphocholine (platelet activating factor and a hypotensive lipid) by cholinephosphotransferase in various rat tissues. Biochim Biophys Acta 663:545–556
Robaut C, Durand G, James C, Lave D, Sedivy P, Floch A, Mondot S, Pacot D, Cavero I, Le Fur G (1987) PAF binding sites. Characterization by [3H] 52770, a pyrrolo [1,2c] thiazole derivative, in rabbit platelets. Biochem Pharmacol 36:3221–3229
Robaut C, Mondot S, Floch A, Tahraoue L, Cavero I (1988) Pharmacological profile of a novel, potent and specific PAF receptor antagonist the 59227 RP. Prostaglandins 35:838
Robertson DA, Genovese A, Levi R (1987) Negative inotropic effect of platelet activating factor on human myocardium: a pharmacological study. J Pharmacol Exp Ther 243:834–839
Robertson DA, Wang DY, Lee CO, Levi R (1988) Negative inotropic effect of platelet activating factor: association with a decrease in intracellular sodium activity. J Pharmacol Exp Ther 245:124–128
Rosam A, Wallace J, Whittle B (1986) Potent ulcerogenic actions of platelet activating factor on the stomach. Nature 319:54–56
Rylander R, Beijer L (1987) Inhalation of endotoxin stimulates alevolar macrophage production of platelet activating factor. Am Rev Respir Dis 135:83–86
Sakuma Y, Muramoto K, Harada K, Katayama S, Tsunoda H, Katayama K (1991) Inhibitory effects of a novel PAF antagonist E6123 on anaphylactic responses in

passively and actively sensitized guinea pigs and passively sensitized mice. Prostaglandins 42:541–555
Saito M, Fujimura M, Ogawa H, Matsuda T (1995) Role of thromboxane $A_2$ and platelet-activating factor in allegic bronchoconstriction in guinea pig airway in vivo. J Lipid Mediat Cell Signal 11:1–12
Sanjar S, Aoki S, Boubekeur K, Chapman I, Smith D, Kings M, Morley J (1990a) Eosinophil accumulation in pulmonary airways of guinea pigs induced by exposure to an aerosol of platelet activating factor: effect of anti-asthma drugs. Br J Pharmacol 99:267–272
Sanjar S, Aoki S, Kristersson A, Smith D, Morley J (1990b) Antigen challenge induces pulmonary eosinophil accumulation and airway hyperreactivity in sensitized guinea pigs: the effect of anti-asthma drugs. Br J Pharmacol 99:679–686
Satoh K, Imaizumi T, Kawamura Y, Yoshida H, Takamatsu S, Mizono S, Shoji B, Takamatsu M (1988) Activity of platelet-activating factor (PAF) acetylhydrolase in plasma from patients with ischemic cerebrovascular disease. Prostaglandins 35:685–698
Saunders RN, Handley DA (1987) Platelet activating factor antagonists. Annu Rev Pharmacol Toxicol 27:237–255
Saunders R, Handley DA (1990) PAF and immunopathological responses. In: Handley DA, Saunders R, Houlihan W, Tomesch J (eds) Platelet activating factor in immune complex and endotoxin mediated diseases. Dekker, New York, pp 223–246
Schleimer RP, MacGlashan DW, Peters S, Pinckard N, Adkinson N, Lichtenstein LW (1986) Characterization of inflammatory mediator release from purified human lung mast cells. Am Rev Respir Dis 133:614–617
Seth P, Kumari R, Dikshit M, Srimal RC (1994) Effect of platelet activating factor antagonists in different models of thrombosis. Thromb Res 76:503–512
Shen TY, Hwang S, Chang M, Doebber T, Lam M, Wu M, Wang X, Han G, Li R (1985) Characterizartion of a platelet activating factor receptor antagonist isolated from haifenteng (Piper futokadsura): specific inhibition if in vitro and in vivo platelet activating factor induced effects. Proc Natl Acad Sci USA 82:672–676
Sheppard GS, Pireh D, Carrera GM, Bures MG, Heyman HR, Steinman DH, Davidson H et al. (1994) 3-(2-(3-pyridinyl) thiazolid-4-oyl)-indoles, a novel series of PAF antagonists. J Med Chem 37:2011–2032
Sisson JH, Prescott SM, McIntyre TM, Zimmerman GA (1987) Production of platelet activating factor by stimulated polymorphonculear leukocytes. J Immunol 138:3918–3926
Smallbone B, Taylor NE, McDonald JW (1987) Effects of L652,731, a platelet activating factor (PAF) receptor antagonist, on PAF- and complement-induced pulmonary hypertension in sheep. J Pharmacol Exp Ther 242:1035–1040
Snyder F (1995) Platelet-activating factor and its analogs: metabolic pathways and related intracellular processes. Biochim Biophys Acta 1254:231–249
Stahl GL, Terashita ZI, Lefer AM (1986) Role of platelet activating factor on propagation of cardiac damage during myocardial ischemia. J Pharmacol Exp Ther 244:898–904
Stahl GL, Terachita Z, Lefer AM (1988) Role of platelet activating factor in propagation of cardiac damage during myocardial ischemia. J Pharmacol Exp Ther 244:898–904
Steil AA, Rodriguez MDCG, Alonso A, Sanchez Crespo M, Bosca L (1995) Platelet-activating factor: the effector of protein-rich plasma extravasation and nitric-oxide synthase induction in rat immune complex peritonitis. Br J Pharmacol 114:895–901
Stenzel H, Hummer B, Hahn H (1987) Effect of the PAF antagonist SRI 63-441 on the allergic response in wake dogs with natural asthma. Agents Actions 21:253–260
Stewart A, Dusting G (1988) Characterization of receptor for platelet activating factor on platelets, polymorphonuclear leukocytes and macrophages. Br J Pharmacol 94:1225–1233

Stockmans F, Arnout J, Depre M, DeSchepper P, Argehrn JC, Vermylen J (1991) TCV-309, a novel PAF antagonist, inhibits PAF induced human platelet aggregation ex vivo. Thromb Haemost 65: 1108–1117

Summers JB, Albert DH (1995) Platelet activating factor antagonists. Adv Pharmacol 32:67–168

Takada Y, Boudjema K, Jaeck D, Chenard MP, Wolf P, Cliqualbre J, Yamaoka Y (1995) Platelet activating factor antagonist has a protective effect on preservation/reperfusion injury of the graft in pig liver transplant. Transplant Proc 27:747–748

Takatani M, Yoshioka Y, Tasaka A, Terashita ZI, Imura Y, Nishikawa K, Tsuhima S (1989) Platelet activating factors antagonists: synthesis and structure activity studies of a novel PAF antagonist modified in the phosphorylcholine moiety. J Med Chem 32:56–64

Takehara S, Mikashima H, Muramoto Y, Terasawa M, Setoguchi M, Tahara T (1990) Pharmacological actions of Y-24180, a new specific antagonist of platelet activating factor (PAF): II. Interactions with PAF and benzodiazepine receptors. Prostaglandins 40:571–583

Tanabe Y, Yamamoto H, Murakami M, Yanagi K, Kubota Y, Okumura H, Sanemitsu Y, Suzukamo G (1995) Synthetic study of the highly potent and selective antiplatelet activating factor thiazolidin-4-one agents and related compounds. J Chem Soc Perkin Trans 1:935–947

Terasawa M, Mikashima H, Tahara T, Maruyama Y (1987) Antagonistic activity of etizolam on platelet activating factor in vivo experiments. Jpn J Pharmacol 44:381–386

Terasawa M, Aratani H, Setoguchi M, Tahara M (1990) Pharmacological actions of Y-24180: I. A potent specific antagonist of platelet activating factor. Prostaglandins 40:553–569

Terashita ZI, Tsushima S, Yoshioka Y, Nomura H, Inada Y, Nishikawa K (1983) CV-3988-a specific antagonist of platelet activating factor (PAF). Life Sci 32:1975–1982

Terashita Z, Imura Y, Nishikawa K (1985) Inhibition by CV-3988 of the binding of [3H]-platelet activating factor (PAF) to the platelet. Biochem Pharmacol 34:1491–1499

Terashita Z, Imura Y, Takatani M, Tsushima S, Nishikawa K (1987) CV-6209: a highly potent platelet activating factor antagonist in vitro and in vivo. J Pharmacol Exp Ther 242:263–268

Terashita Z, Kawamura M, Takatani M, Tsushima S, Imura Y, Nishikawa K (1992) Benefitial effects of TCV-309, a novel potent and selective platelet activating factor antagonist in endotoxin and anaphylacitic shock in rodents. J Pharmacol Exp Ther 260:748–755

Tetta C, Camussi G, Bussolino F, Herrick-Davis K, Baglioni C (1991) Inhibition of the synthesis of platelet activating factor by antiinflammatory peptides (antiflammins) without methionine. J Pharmacol Exp Ther 257:616–620

Tomioka K, Garrido R, Ahmed A, Stevenson JS, Abraham WM (1989) YM 461, PAF antagonist, blocks antigen-induced late airway responses and airway hyperresponsiveness in allergic sheep. Eur J Pharmacol 170:209–215

Tramposch KM, Chilton FH, Stanley PL, Franson RC, Havens MB, Nettleton DO, Davern LB, Sarling IM, Bonney RJ (1993) Inhibitor of phospholipase $A_2$ blocks eicosanoid and platelet activating factor biosynthesis and has topical anti-inflammatory activity. J Pharmacol Exp Ther 271:852–859

Ueno A, Ishida H, Oh-ishi S (1995) Comparative study of endotoxin-induced hypotension in kininogen-deficient rats with that in normal rats. Br J Pharmacol 114:1250–1256

Ukena D, Krogel C, Yukawa T, Sybrecht G, Barnes PJ (1989) PAF-receptors on eosinophils: identification with a novel ligand, [3H] WEB 2086. Biochem Pharmacol 38:1702–1705

Underwood SL, Lewis SA, Raeburn D (1992) RP 59227, a novel PAF receptor antagonist: effects in guinea pig models of airway hyperreactivity. Eur J Pharmacol 210:97–102

Vargaftig BB, Lefort J, Chicnard M, Benveniste J (1980) Platelet activating factor induced platelet-dependent bronchoconstriction unrelated to the formation of prostaglandin derivatives. Eur J Pharmacol 65:185–192

Vargaftig BB, Lefort J, Rotilio D (1983) Route-dependent interactions between PAF-acether and guinea pig bronchopulmonary smooth muscle: relevance of cyclooxygenase mechanisms. In: Benveniste J, Arnoux A (eds) Platelet activating factor. INSERM Symposium. Elsevier, Amsterdam, pp 307–313

Wallace JL, Steel G, Whittle B, Lagente V, Vargaftig BB (1987) Evidence for platelet activating factor as a mediator of endotoxin induced gastrointestinal damage in the rat. Gastroenterology 93:765–773

Wang J, Kester M, Dunn M (1988) The effects of endotoxin on platelet activating factor synthesis in cultured rat glomerular mesangial cells. Biochim Biophys Acta 969:217–224

Whately R, Zimmerman G, Mclntyre T, Prescott S (1988) Endothelium from diverse vascular sources synthesizes platelet activating factor. Arteriosclerosis 8:321–331

Whittaker M (1993) PAF receptor antagonists: recent advances. Curr Opin Ther Pat 3:1569–1573

Whittaker M, Beauchamp CL, Bowles SA, Cakett KS, Christodoulou MS, Galloway WA, Longstaff DS, McGuinness GP, Miller A, Timiiis DJ, Wood LM (1992) BB-823, a PAF receptor antagonist with picomolar activity. Pharmacol Comm 1:251–257

Whittle B, Boughton-Smith N, Hutcheson I, Esplugues JV, Wallace JL (1987) Increased intestinal formation of Paf in endotoxin induced damage in the rat. Br J Pharmacol 92:3–4

Wilkens JH, Wilkens H, Uffmann J, Bovers J, Fabel H, Frolich JC (1990) Effects of the PAF antagonist (BN 52063) on bronchoconstriction and platelet activation during exercise induced asthma. Br J Clin Pharmacol 29:85–91

Wilkens H, Wilkens JH, Bosse S (1991) Effects of an inhaled PAF antaongist (WEB 2086) on allergen-induced early and late asthmatic responses and increased bronchial responsiveness to methacholine. Am Rev Respir Dis 143:A812

Winslow C, Anderson R, D'Aries F, Frisch G, DeLillo A, Lee M, Saunders RN (1987) Toward understanding the mechanism of action of platelet activating factor antagonists. In: Winslow CM, Lee ML (eds) New horizons in platelet activating factor research. Wiley, New York, pp 153–168

Yamazaki R, Sugatani J, Fujii I, Kuroyangi M, Umehara K, Ueno A, Suzuki Y, Miwa M (1994) Development of a novel method for determination of acetyl-coA: 1-alkyl-sn- glycero-3-phosphocholine acetyltransferase activity and its application to screening for acetyltransferase inhibitors. Biochem Pharmacol 47:995–1006

Yin M, Kurvers HAJM, Buurman WA, Tangelder GJ, Booster MH, Daemen JHC, Kondracki S, Kootstra G (1995) Beneficial effect of platelet-activating factor antagonist TCV-309 on renal ischemia-reperfusion injury. Transplant Proc 27:774–776

Zanglis A, Lianos EA (1987) Platelet activating factor biosynthesis and degradation in the rat glomeruli. J Lab Clin Med 110:330–337

Zehr KJ, Poston RS, Lee PC, Uthoff K, Kumar P, Cho PW, Gillinov AM, Redmond JM, Wilkenstein JA, Herskowitc A, Cameron DE (1995) Platelet activating factor inhibition reduces lung injury after cardiopulmonary bypass. Ann Thorac Surg 59:328–335

Zhou W, Chao W, Levine BA, Olsen MS (1990) Evidence for platelet activating factor as a late phase mediator of chronic pancreatitis in the rat. Am J Pathol 137:1501–1508

# Section IV
# Clinical Aspects of Platelets and Their Factors

CHAPTER 25

# Platelets and the Vascular System: Atherosclerosis, Thrombosis, Myocardial Infarction

H. SINZINGER and M. RODRIGUES

## A. Introduction

More than a century ago, MALMSTEN (1859) demonstrated the strong relationship between thrombosis and atherosclerosis on coronary arteries. ROKITANSKY (1852) observed that the incorporation of a mural coagulum was a primary event in the atherogenesis of man and hypothesized that material derived from the bloodstream and "fibrin deposition" in the arterial wall caused "incrustation" and subsequent cellular "fibrotic transformation". As early as 1882, BIZZOZERO stressed the role of platelets in this particular process and in atherogenesis and reported for the first time, in a microscopic study, platelet adhesion and aggregation following an experimental mechanical injury to mesenteric arteries in animals. Similar findings on the involvement of platelets (parietal microthrombosis) in atherosclerosis were later obtained with the use of different techniques by DUGUID (1946). Since then, various studies have emphasized the incorporation of a thrombus, which is either generated by plasmatic coagulation or by platelets alone, yet mostly by combined action as a very early event in the development of human atherosclerotic lesions.

Over the past decades, the pathophysiology of acute myocardial infarction has been intensively studied, and methods of reducing the incidence of coronary artery disease, infarct size, and morbidity and mortality in patients with coronary artery disease have been extensively implemented. The growing clinical use of drugs in patients with coronary artery disease for these purposes has increased the need to directly image atherosclerotic lesions for early lesion detection and to evaluate and monitor therapeutical interventions.

The great clinical demand for early and rapid diagnosis of thromboembolic disease has originated the continuous development of a large number of thrombus-testing and -imaging techniques.

## B. Atherosclerosis

Platelet accumulation in the arterial wall is nowadays well accepted as a rather early and key event both in initiation and perpetuation of atherosclerotic lesions (ROSS and GLOMSET 1976; WEKSLER and NACHMAN 1981). Different mechanisms, such as vascular damage, the development of mural platelet thrombi as a response to injury, and the biochemical effects of intraplatelet substances that are released in response to damage may be involved.

Localization of atherosclerosis at sites where the major arteries branch or are curved suggests that hemodynamic stress causes functional changes and/or injury to the arterial wall, preferentially at sites where hemodynamic stress occurs and the endothelial cell turnover is high, and that these effects might facilitate the initiation and perpetuation of atherosclerotic lesions (Petterson et al. 1993). Endothelial damage by influencing factors (normal aging, hemodynamic factors, biochemical mechanisms, and diseases, among others) can be the initiation event in the development of atherosclerosis (Ross and Glomset 1976), and it may also be a localizing factor for atherosclerosis (Petterson et al. 1993). Although a number of studies report a significant loss of endothelial integrity accompanying atherosclerotic disease (Gown et al. 1986), it is not clear how endothelial injury might facilitate the development of atherosclerotic lesions in the intima. Intact human platelets do not adhere to a morphologically normal and functionally intact endothelium. However, it is well known that platelets in general rapidly form a thin layer and accumulate at the sites on the arterial wall where the protective endothelium has been lost. At sites of fissure (endothelium or plaque), an intraintimal thrombus which can further reseal occurs. It can develop into a mural intraluminal thrombus that can organize and/or originate an occlusive luminal thrombus, which in turn either organizes or recanalizes. All these steps lead to plaque growth.

When human endothelium is detached, the circulating platelets adhere to the damaged vessel wall and are activated by subendothelial components, notably collagen (Bills et al. 1976; D'Souza et al. 1994), microfibrils, amorphous basement-membrane-like material, and elastin (Bills et al. 1976); the platelets discharge their granule contents, such as the $\alpha$-granule proteins $\beta$-thromboglobulin, platelet factor 4, and platelet-derived growth factor (PDGF), of which the first two are known as diagnostic markers for in-vivo platelet activation. Clotting factors, associated with the platelet surface, at this time play a key role. An activation of factors XII and XI occurs (Moncada et al. 1977) at the sites of endothelial injury as well. PDGF, which causes migration and proliferation of vascular smooth muscle cells (SMC; Ross et al. 1974) and release of prostaglandins (PGs; Coughlin et al. 1980) via inducible cyclooxygenase, is of special importance for the early development of atherosclerosis. Adenosine diphosphate (ADP) causes adhesion between platelets (aggregation) and also erythrocytes as well as adhesion to those platelets already adhering to the vascular wall by a process mediated by thromboxane (TX) $A_2$ ($TXA_2$) (Walsh 1972; Packham 1994). Besides ADP, a number of substances such as thrombin, epinephrine, serotonin, vasopressin, platelet activating factor, and others induce platelet aggregation within seconds. Fibrinogen and calcium or magnesium ions are necessary for this response. The mechanisms involved in platelet aggregation include the induction of receptors for fibrinogen and the adherence of platelets to the (increased number of) receptors. In addition, one of the first events after endothelial cell disruption is the activation of the coagulation cascade and resultant generation of thrombin at the site of vessel damage, which leads to further platelet activation (with release of platelet granule contents) and aggregation as well as the production of fibrin. Aggregation of activated platelets forms the initial event in thrombus formation, followed by fibrin deposition. Thrombin catalyzes the final step of fibrin formation and, along with $TXA_2$ and ADP, is crucial for activating surrounding quiescent platelets, thereby amplifying platelet aggregation, fibrin formation, vasoconstriction, and vascular SMC proliferation (D'Souza et al. 1994; Packham 1994). A platelet aggregate thus self-perpetuates the coagulation mechanism. Glycosaminoglycans and proteoglycans of the subendothelial tissue, which are exposed directly to the bloodstream when vascular walls are injured, tend to delay plasmatic coagulation. These substances also have an inhibitory effect on platelets. Platelet-fibrin accumulations may be invaded by mononuclear elements, polymorphonuclear leukocytes, and SMC (Gorman et al. 1977), resulting in an organization and parietal incorporation of a thrombus (mural thrombus). This causes intimal thickening and thus narrowing of the vascular lumen.

There is a rich vascular supply to the human atheroma, with the necessary blood supply to maintain the intimal mass. These vessels appear to develop from the adventitia (Gown et al. 1986; Wilcox and Blumenthal 1995).

Various experimental studies suggest the existence of a self-protection phenomenon against platelet deposition, SMC proliferation, and the release of intraplatelet products via temporarily enhanced generation of significant amounts of $PGI_2$, a potent known activator of adenylate cyclase (Moncada et al. 1976). The endothelium of the arterial wall is the main source of $PGI_2$. However, after weeks there is a general decrease in $PGI_2$ production, which persists for a long period of time. $PGI_2$, together with $PGE_1$ and $PGD_2$, inhibits platelet aggregation under in vivo haemodynamic conditions and the proliferation of SMC. In this particular mechanism, PDGF might play a key role, too. Even very low doses of $PGI_2$ inhibit PDGF release from platelets. However, adhesion is blocked only at high concentrations by a stimulation of intracellular cyclic adenosine monophosphate (cAMP; Siegl et al. 1979). $PGI_2$ and $PGE_1$ seem to share receptors on the surface of platelets, whereas $PGD_2$ seems to have a different binding site. Platelets undergo rather fast desensitization at the receptor level when exposed for a longer period of time to a PG (Gorman et al. 1977) while they can be sensitized by extremely low doses of cyclooxygenase inhibitors. Cyclooxygenase rather rapidly converts arachidonic acid to the labile proaggregatory PG endoperoxides $PGH_2$ and $PGG_2$, which are finally transformed by platelets to $TXA_2$. In addition, enhanced endogenous $PGI_2$ production appears to have other beneficial effects, such as affecting the accumulation of cholesterol in the arterial wall by stimulating cholesterol ester hydrolase, thus reducing intracellular cholesterol ester levels. The number of (modified) low-density lipoprotein (LDL) receptors might be modulated, too, causing activation of platelets. Furthermore, the vascular wall, rather than the platelets, appear to induce thromboresistance at the cellular level. In vitro work demonstrated severely decreased $PGI_2$ formation in human atherosclerotic arteries (SINZINGER et al. 1979). Atherosclerosis and its risk factors (smoking, diabetes, hyperlipoproteinemia, and others) are associated with decreased platelet sensitivity to antiaggregatory PGs. Platelet activity is also decreased by nitric oxide (NO), a potent activator of guanylate cyclase. NO and $PGI_2$ act synergistically to inhibit platelet aggregation in vitro (RADOMSKI et al. 1987), a mechanism which has been proven in human in vivo as well (SINZINGER et al. 1979). NO may thus contribute to the antiaggregatory activity of the activated endothelium (PEARSON 1994).

Atherosclerotic plaques, as well as endothelial and plaque fissures, continue to undergo episodic growth over years or decades, leading to progressive atherosclerotic disease. A key problem of the disease is its silent stepwise progression for many years towards its acute manifestation which frequently ends in sudden death.

Although atherosclerosis is still the major cause of mortality in western industrialized countries, its early noninvasive diagnosis still poses many difficulties. No valid technique allowing measures of stage, extent, composition, and complications of atherosclerosis is yet available. Conventional imaging techniques are much better at defining the extent of more advanced atherosclerotic changes than at detecting earlier stages of the disease (SINZINGER and VIRGOLINI 1990) at a time when medical intervention could be by far more beneficial. The evaluation of the functional staging of the disease and of its metabolism at a rather early stage, namely, the evaluation of early changes of the vascular wall, is of great current interest and hope. Agents that interact with platelet function and the signalling pathways that are activated by various drugs are also under intensive investigation.

## C. Thrombosis

Thrombosis rarely occurs in the presence of a healthy or undamaged vascular wall because of the active involvement of endothelial cells which provide a thromboresistant surface for flowing blood (PEARSON 1994).

A significant feature common to developed atherosclerotic plaques is the presence of thrombi. Local platelet thrombosis results from the activation of platelets due to higher shear stress, pathologically altered mediators or, most commonly, loss of endothelial cells, fissure, or ulceration of the plaque. Thrombus formation is most frequently initiated by rupture of the atherosclerotic plaque or by hemorrhaging of the small plaque vessels, the vasa vasorum (WILCOX and BLUMENTHAL 1995). Aggregated platelets and the fibrin network form the essential constituents of thrombi. Evidence suggests that thrombus formation may simulate vascular lesion formation as well as SMC migration and proliferation through the action of a number of factors (mediators and cytokines) liberated from the thrombus including platelet- or macrophage-derived factors, PDGF, basic fibroblast growth factor, and thrombin (WILCOX and BLUMENTHAL 1995), among others. Mononuclear elements and macrophages may, in part, act as pseudoendothelial cells and, in part, as surface lining, consisting of macrophage accumulation on the subendothelium. Via mediation by various chemoattractants, platelets can adhere to this particular cell type, as well as to foam cells. Platelet adhesion to connective tissue material is one of the early events in the development of thrombosis (BOUNAMEAUX 1959). The most important interaction of platelets with human arterial wall seems to be mediated through collagen (MUGGLI and BAUMGARTNER 1973). The physical structure of collagen might be of greater importance than the collagen type (I–IV) itself. Von Willebrand factor is a necessary requirement for platelet adhesion. Tissue factor, a membrane bound protein that is found in the adventitia surrounding blood vessels as well as in macrophages in the necrotic core of the plaque and in foam cells, activates both the intrinsic and extrinsic pathway of coagulation. In human atherosclerotic plaques, a significant overexpression of tissue factor exists, which may explain why thrombi tend to form at sites of plaque rupture (WILCOX et al. 1989) and may play a significant role in the initiation of coagulation associated with plaque rupture (WILCOX and BLUMENTHAL 1995). The formation of mural thrombi is later followed by incorporation and cellular organization, resulting in organized thrombi of different ages and subsequent luminal stenoses. Although angiography may show "intact coronary arteries," crescendo angina, stable angina, acute infarction, and sudden death, among others, can occur in association with immediate plaque growth.

Arterial thrombi tend to develop in the presence of hemodynamic disturbances and shear stress (causing endothelial injury) and at sites of narrowing of the vessel. This might result in an increased platelet deposition and formation of platelet aggregates, leading to an occlusion of the vessel.

Thrombosis is the predominant fatal complication of human atherosclerosis. As the processes involved in the formation of hemostatic plugs and arterial thrombi are essentially the same (PACKHAM 1994), attempts to prevent arterial thrombosis are limited by the risk of interfering with the formation of hemostatic plugs and causing bleeding. Thus, rendering surfaces less thrombogenic might be a preferable procedure.

Intracardiac thrombi are frequently associated with an increased rate of early complications and mortality. However, the prognostic factors that influence the long-term prognosis of these patients are not yet well defined. A recent study (GONCALVES et al. 1995) demonstrated basal left ventricular function as being an important predictor of long-term survival in patients with intracardiac thrombus.

Carotid and coronary artery thrombosis are often diagnosed late in the course of the disease after serious or fatal complications have occurred (STRATTON and RITCHIE 1982; DEWANJEE et al. 1984).

Platelets, if deposited in excess, have been found to play an important role in failure of synthetic material, namely, in causing graft thrombosis or thromboembolic phenomena or promoting late graft failure through the stimulation of anastomotic neointimal fibrous hyperplasia (NORDESTGAARD et al. 1990).

The dissolution of thrombi that have formed in vivo is also receiving extensive study. The main approach has been the use of fibrinolytic agents (streptokinase, urokinase, tissue plasminogen activator, among others) that cause generation of the proteolytic enzyme plasmin (PACKHAM 1994). These agents have been extensively evaluated and monitored by several groups of investigators using radiolabeled platelets.

The great clinical importance of early and rapid diagnosis of thromboembolic disease and of thrombus imaging is reflected by the quite large number of different platelet function tests and imaging techniques developed over many years. However, most of them suffer from lack of sensitivity and/or specificity. A simple, specific, rapid, reliable, and inexpensive procedure is not yet available.

## D. Myocardial Infarction

Platelets play a key role in the pathogenesis of arterial occlusive disorders. Over the past two decades, numerous studies have reported platelet reactivity in patients with ischemic heart disease. It is accepted that platelets from patients with acute arterial thrombotic disorders exhibit increased in vitro aggregability and adhesiveness as well as in vivo activation, which can be detected by various activation markers notwithstanding certain methodological processing problems. How platelet hyperactivity develops in vivo remains unclear in most cases. Increased platelet activity may be a response to vascular changes, but it may also result from inherent differences in platelet reactivity among the general population or from endogenous and exogenous factors (D'SOUZA et al. 1994). Previous studies have shown that mortality following acute myocardial infarction is related to the number of risk factors present in an individual patient. A recent study (TRIP et al. 1990) indicates that spontaneous platelet aggregation is an independent risk factor for recurrent myocardial infarction and resultant mortality.

In the 1980s, the development of reperfusion strategies and particularly thrombolysis strategies dramatically reduced early and long-term mortality of patients with myocardial infarction. A total of 70%–80% of the occluded vessels are opened by thrombolysis, and the reocclusion rate may be as high as 10%–15%, which means that after thrombolysis 30%–45% of patients still have an occluded artery. However, few patients (15%–37%) with acute myocardial infarction qualify for thrombolysis. In some cases, this is because the time window for successful reperfusion has passed; in other cases, this is related to the presence of contraindications to thrombolysis (GONCALVES 1995).

Interest in the contribution of platelets to the pathological effects of myocardial ischemia has recently increased as a result of better understanding of the mechanisms underlying intracoronary thrombus formation and of the effects of thrombolytic therapy. Several potential mechanisms are implicated. Clinical and experimental evidence suggests that platelets may act in two ways: when activated, they may form aggregates that can restrict or prevent coronary perfusion, exacerbating the effects of ischemia and eventually causing myocardial infarction; alternatively, when activated, they may release various products that have myocardial electrophysiological effects (FLORES and SHERIDAN 1994) namely, by causing increase in intracellular calcium concentrations (CHIEN et al. 1990). TX seems to play an important role in mediating platelet activation during myocardial ischemia (FLORES and SHERIDAN 1994). Increased numbers of $TXA_2$-$PGH_2$ (TP) receptors were shown in patients suffering from acute myocardial infarction (DORN et al. 1990).

Additionally, in patients suffering from myocardial infarction, the number of megakaryocytes and platelets increases prior to the development of infarction and then rises during infarction and remains elevated in survivors for up to several weeks. Larger platelets appear to be more reactive and contribute to the development of infarction, probably by mechanical obstruction of microcirculation (FLORES and SHERIDAN 1994).

# E. Methodology of Radiolabeling of Platelets

There have been several attempts to develop simple and inexpensive methods allowing wide use of radiolabeled platelets in clinical and experimental applications. Several radionuclides and tracers have been examined in order to obtain high labeling efficiency (LE; which is determined by measuring cell-bound radioactivity as compared to total radioactivity) without loosing platelet viability and the other requirements for optimal radiolabeling. Monoclonal antibodies for labeling platelets specifically in vivo and thus avoiding cell separation procedures have also been explored. However, labeling of monoclonal antibodies can interfere with functional properties, leading to platelet activation (NIIYA et al. 1987; SINZINGER and VIRGOLINI 1990) and the monitoring of nonrepresentative, unphysiological kinetics. Antiplatelet antibodies may therefore only be useful if they can detect platelet properties elaborated merely during activation in fresh, i.e., active thrombi.

Most of these different radiopharmaceuticals have been successful in producing positive images. However, none of those studied so far have established significant advantages over [$^{111}$In]oxine ([$^{111}$In]8-hydroxyquinoline) serving as a gold standard introduced in 1976 (THAKUR et al. 1976).

$^{111}$In (Table 1; NAJEAN 1986; SINZINGER and VIRGOLINI 1990; DE VRIES et al. 1993) has the potential to image and quantify in vivo the distribution of platelets in man in parallel with determination of platelet kinetics and survival (THAKUR et al. 1984; TSAN 1984; LÖTTER et al. 1986). Oxine has gained wide acceptance as an $^{111}$In-chelator for platelet labeling in clinical routine owing to its simple handling technique and commercial availability.

**Table 1.** Characteristics of $^{111}In$ for platelet-labeling

| |
|---|
| Half-life – 2.83 days |
| Major energies of gamma photons – 173 keV (89%), 247 keV (94%) |
| Activity administered in gamma camera imaging |
| Kinetic studies – 30–48.6 μCi |
| Imaging – 240–300 μCi |
| Difficult storage |
| High cost |
| LE – >80% (depending on the complex) |
| No specificity of labeling |
| Elution rate (per h) – 0.27% |
| Reutilization |

**Table 2.** Parameters influencing platelet function and -labeling

| |
|---|
| Platelet density |
| Procedure |
| Collection injury |
| Anticoagulants |
| Incubation |
| Medium |
| pH |
| Speed of centrifugation |
| Type of plastic used for bags and tubes |
| Temperature |
| Time |
| Amount of complex |
| Aging (BALDINI and MEYERS 1980) |
| Plasma proteins, especially transferrin (SINZINGER and VIRGOLINI 1990) |
| Calcium ions |
| Lipids |
| Cholesterol (GRANEGGER et al. 1988) |
| Low density lipoprotein (GRANEGGER et al. 1988) |
| Additives |
| Prostaglandins (SINZINGER and FITSCHA 1987) |
| Nitric oxide (WAGNER et al. 1989) |
| Others |

Because the radiopharmaceuticals presently available label other blood cells and some plasma proteins as well (NAJEAN 1986), platelets have to be isolated prior to labeling. Only autologous platelets should be used. Although several methods have been reported for platelet isolation, in practice the platelets are isolated from anticoagulated blood by simple differential centrifugation (NAJEAN 1986; PETERS 1988; SINZINGER et al. 1984). The introduction of a closed Monovette technique (SINZINGER et al. 1984) made the methodology of platelet labeling accessible for wide clinical use. This system has the advantages of requiring only a small amount of blood (16 ml), a short ex-vivo period for platelets (less than 60 min), optimal labeling conditions (37°C, 5 min of incubation), and low costs. The in vitro manipulation of platelets for radiolabeling (including centrifugation steps, exposure to radioactive material, plasma-free medium, etc.) does not significantly change their in vivo behavior (MATHIAS and WELCH 1984). However, several variables and technical parameters influence platelet function and labeling (for a review see RODRIGUES and SINZINGER 1994; Table 2). The type of labeling solution, the anticoagulant, the temperature, time, and volume of incubation, the amount of ligand, and the concentration of the complex used have been optimized. A key factor for obtaining an optimal LE is a high platelet concentration, with at least approximately $1 \times 10^9$ (concentrated) platelets/ml necessary (SINZINGER et al. 1990b) to obtain an adequate LE and recovery (REC; i.e., the percentage of injected dose – radiolabeled platelets – remaining cell bound in the circulation for 60min).

$^{111}$In-lipophilic complexes such as [$^{111}$In]oxine have a higher LE than most other radiopharmaceuticals, and only a small amount of blood (10–15ml) is thus required to obtain the concentration of platelets necessary for radiolabeling (THAKUR et al. 1976). By a simple and quick method (SINZINGER et al. 1990b), even a peripheral platelet count of 5000 platelets/$\mu$l is enough to obtain a sufficient LE to allow in vivo kinetic studies as well as scintigraphic imaging.

Although platelet labeling is long established and a large store of information on labeling techniques exists, the studies with radiolabeled platelets are not yet widely used in clinical practice. Labeling problems, the costs of the tracer ($^{111}$In), and several other methodological factors still limit general application of platelet labeling for routine purposes.

# F. Studies with Radiolabeled Platelets

The measurement and clinical use of platelet survival time have greatly expanded during the past decades with the application of $^{51}$Cr-labeled platelets.

Clinical studies with $^{111}$In-labeled platelets began in 1978 with the publication of scintigraphic detection of in vivo thrombosis (DAVIS et al. 1978). Since then, quite a wide range of applications has been found in both clinical and experimental medicine. These applications have included the study of platelet behavior in vitro and in vivo, namely, the evaluation of kinetics, measurement

**Table 3.** Indications for kinetic studies with radiolabeled platelets

| |
|---|
| Evaluation of normal platelet kinetics (production, survival, degradation) |
| Thrombocytopenia |
| Cause(s): reduction of output, changes in biodistribution, increased sequestration of blood platelets |
| Assessment of therapeutic interventions |
| Thromboembolic diseases |
| Detection of prethrombotic states |
| Vascular disorders or endothelial abnormalities |
| Monitoring and/or evaluation of the effects of pharmacological interventions |
| Evaluation of the effects of risk factors (environmental factors, cigarette smoke, lipids, vascular injury, prosthetic materials, drugs) |
| Transfusion medicine |

of platelet survival (Table 3), in vivo distribution, sites of sequestration and accumulation of platelets (Table 4) in hematological and hemostaseological indications, and evaluation of the efficacy of therapeutic interventions (Table 5).

## I. Platelet Kinetics

Platelet kinetic studies in healthy volunteers and in patients demonstrated that platelets are consumed by three principal mechanisms: phagocytosis (in reticu-

**Table 4.** Indications for scintigraphic imaging with radiolabeled platelets

| |
|---|
| Evaluation of normal platelet biodistribution |
| Thrombosis |
| Thrombogenicity |
| Atherosclerosis |
| Aneurysms |
| Monitoring and/or evaluation of the effects of therapy |
| Pharmacological interventions |
| Surgical interventions (vascular surgery, splenectomy, transplants) |
| Synthetic material (grafts, prostheses) |

**Table 5.** Evaluation of antiplatelet therapy with radiolabeled platelets

| Vessel/Lesion | Drug | References |
|---|---|---|
| Aortocoronary bypass | ASA/DIP | DEWANJEE et al. 1984 |
| | ASA/DIP | FUSTER et al. 1979 |
| Coronary | Streptokinase | BERGMANN et al. 1983 |
| Left ventricle/thrombus | SP | STRATTON et al. 1982 |
| Carotid | Dazoxiben/ASA | RANDAL and WILDING 1982 |
| | ASA | KESSLER et al. 1985 |
| | Hydergine | STEURER et al. 1989 |
| Aorta | 13-Azaprostanoic acid | LEBRETON et al. 1984 |
| Shunt | SP | RITCHIE et al. 1982 |
| Graft/aneurysm | ASA, DIP, SP | HUANG and HARKER 1981 |
| PTFE grafts | $PGI_2$ | CALLOW et al. 1982 |
| | Ibuprofen | LOVAAS et al. 1983 |
| | ASA/DIP | NORDESTGAARD et al. 1990 |
| Graft/Dacron | SP | STRATTON and RITCHIE 1982 |
| | Ticlopidine | STRATTON et al. 1983 |
| Aortofemoral/Dacron | ASA/DIP | PUMPHREY et al. 1982 |
| Femoral | $PGI_2$ | SIZINGER and FITSCHA 1984a |
| | Taprostene | FITSCHA et al. 1985 |
| | Iloprost | FITSCHA 1986 |
| | $PGE_1$ | SINZINGER et al. FITSCHA 1987 |
| | ASA | SINZINGER et al. 1987 |
| | ISDN, $PGE_1$ | SINZINGER et al. 1990c |
| Cerebral | Dihydroergotoxine | SINZINGER et al. 1990a |

ASA, acetylsalicylic acid; DIP, dipyridamole; SP, sulfinpyrazone; PG, prostaglandin; ISDN, isosorbide dinitrate; PTFE, polytetrafluoroethylene.

loendothelial cells), thrombosis, and fragmentation of platelets into microparticles or vesicles (DEWANJEE 1994).

Assessing platelet REC, survival, and sequestration sites during the initial hours after injection of radiolabeled platelets provided key information about platelet integrity and function (NAJEAN 1986; RODRIGUES and SINZINGER 1994). REC indicates the in vivo functional integrity of platelets, and it is an important indirect measure of platelet damage during the in vitro radiolabeling procedure. Platelet survival can be estimated by measuring the rate of disappearance of radiolabeled platelets from the circulation. It reflects total platelet consumption, i.e., systemic hemostasis, and the complex in vivo interaction between platelets and the vascular surface, and it is a measure of repeated injury and platelet consumption at the level of lesions, although it is not specific. However, shortened platelet survival indicates significant in-vivo damage (HARKER 1978), wherever the location of destruction may be, and it is generally accepted as an indication of in vivo platelet activation (LÖTTER et al. 1986). Only long-lasting, continuous, and/or repeated platelet damage reduces platelet survival (SINZINGER and VIRGOLINI 1990).

The normal values for platelet survival range from 7.3 to 9.5 days (INTERNATIONAL COMMITTEE FOR STANDARDIZATION IN HAEMATOLOGY 1988). Shortened platelet survival has been reported in a variety of disorders associated with disrupted hemostasis (Table 3). Extremely short platelet survival is often accompanied by a low platelet REC which is due to an abnormal platelet population (SINZINGER and FITSCHA 1990).

Experimental work and a study of patients with femoral-artery atherosclerosis (SINZINGER and FITSCHA 1984b) revealed no correlation between platelet survival, the severity of the disease, and the extent of damaged vascular surface. Patients with ulcerated plaques and patients with acute thrombosis had a significantly higher arterial platelet uptake ratio (PUR), reflecting the activity of the lesions, than patients with older lesions, whereas there was no significant difference between ulceration and acute thrombosis. Platelet survival may not be significantly altered by even several small lesions, while a small number of large lesions may result in significant reduction or vice versa. Morphological control experiments demonstrated that local lesions may result in local disruption of hemostasis associated with enhanced platelet uptake. Even lesions not detectable by an imaging technique may, however, well result in severely shortened platelet survival, reflecting systemic hemostatic imbalance.

## II. Imaging

### 1. Atherosclerosis

Although the pathogenetic mechanisms and the pathoanatomical changes of atherosclerosis are well known, their monitoring using nuclear medicine techniques is still in its initial stage. In recent years, the early diagnosis of athero-

**Table 6.** Imaging of atherosclerosis with radiolabeled platelets

| Vessel | Conditions | Species | References |
|---|---|---|---|
| Carotid arteries | E | Dog | THAKUR et al. 1976 |
| | S | Human | DAVIS et al. 1978 |
| | S | Human | GOLDMAN et al. 1982 |
| | S | Human | ISAKA et al. 1984 |
| | S | Human | KESSLER et al. 1985 |
| | S | Human | KIMURA et al. 1982 |
| | S | Human | POWERS et al. 1982b |
| | E | Macaque | KIMURA et al. 1982 |
| Abdominal aorta | E | Macaque | POWERS et al. 1982a |
| | E | Rabbit | FINKELSTEIN et al. 1982 |
| Femoral arteries | S | Human | SINZINGER and FITSCHA 1984b |
| Coronary arteries | S | Human | EZEKOWITZ et al. 1980 |
| | E | Dog | BERGMANN et al. 1983 |
| Aneurysms | S | Human | HEYNS et al. 1982 |
| | S | Human | SINZINGER et al. 1985 |
| Endarterectomy | E | Dog | LUSBY et al. 1983 |

E, experimental; S, spontaneous.

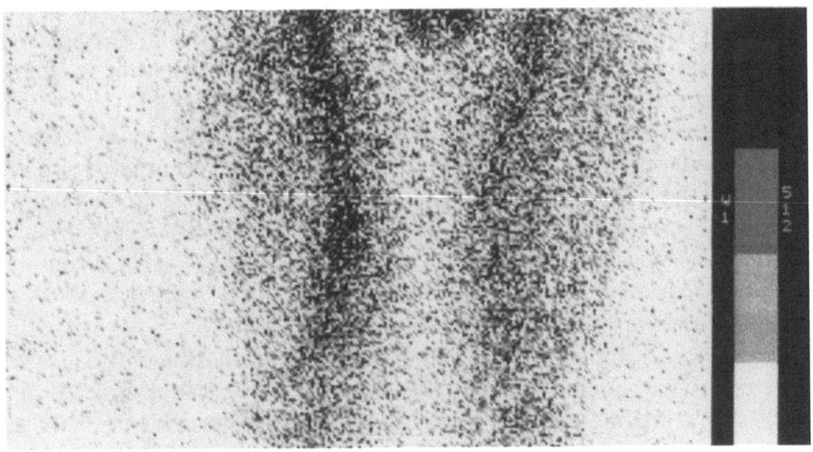

**Fig. 1.** Selective accumulation of platelets in the right femoral artery

sclerosis using these techniques has frequently been attempted. An accumulation of $^{111}$In-labeled platelets on a vascular lesion can be visualized in various parts of the vascular tree (Table 6; Fig. 1). This accumulation can be detected if the number of platelets and the duration of reactivity ("residence time") at the imaging site are sufficient. Platelet accumulation can be imaged only if the plaque to blood ratio exceeds 4.0–5.0 (DEWANJEE 1990). Continuous monitoring with portable detector systems allows permanent, around-the-clock control of local platelet kinetics.

Experimental lesions appear positive much more frequently and seem to be more extensive than human lesions. Interestingly, morphological study of human atherosclerotic lesions indicates that only part of the radioactivity seen is due to radiolabeled platelets deposited on the surface of the vessel (parietal thrombus, ulcerative lesion), while the majority of radioactivity can be found as platelet debris already incorporated into the vascular wall, mostly along with macrophages and mononuclear elements (SINZINGER et al. 1991). The pathological trapping of platelets reflects a prolonged residence time of platelets and an active atherosclerotic lesion (SINZINGER et al. 1986). This selective accumulation of radiolabeled platelets shows strong correlation to the activity of the disease at this particular site (SINZINGER and VIRGOLINI 1990), as monitored by histological examinations. No correlation was found, however, with the duration or the clinical extent of the disease (POWERS et al. 1982b; SINZINGER and VIRGOLINI 1990), as monitored by conventional ultrasonography, computerized tomography, and angiography (SINZINGER and FITSCHA 1990). Correlations with intravascular ultrasound are not yet available. Continuous platelet accumulation at a certain site in the vascular system indicates either continuous stimulation or persistent vascular injury (SINZINGER and VIRGOLINI 1990). Interestingly, an age-dependent decrease in the incidence of positive lesions in asymptomatic patients pointing to higher activity of the disease at a younger age has been found (SINZINGER and VIRGOLINI 1990; VAN RENSBERG and HEYNS 1990).

Platelet deposition and its sequelae are among the most important causes of restenosis of recanalization of peripheral and coronary atherosclerotic lesions, which occurs in as many as 30% of patients in 1 year and 66% in 2 years (MCBRIDE et al. 1988). Imaging of radiolabeled platelets provides an important diagnostic method for detection of platelet deposition and localization, and in selected cases it is helpful in indicating and monitoring of platelet-inhibitory drugs.

Active sites of atheroma with lesions identified by imaging using radiolabeled platelets are relatively stable and show minimal changes in activity over a couple of weeks unless subject to pharmacological intervention (SINZINGER and FITSCHA 1990). Increased platelet accumulation might correlate with a progression of the disease as well (SINZINGER et al. 1986). $^{111}$In-labeled platelets have thus been successfully used in experimentally induced lesions and in humans as well to assess the efficacy of therapeutical interventions on platelet deposition (Table 5).

Carotid endarterectomy sites were 100% imaged (PRICE et al. 1982). While the studies of the carotid or leg arteries (Fig. 1) show positive images in general, those for the coronary arteries do not. In coronary arteries, platelet accumulation can be imaged only after very severe intimal injury under experimental conditions. However, in humans, positive findings are extremely rare, and thus studies with radiolabeled platelets can only be used in selected cases (EZEKOWITZ et al. 1980, 1983).

Tortuous vessels, especially in the carotid region or varicose veins, hematoma, inflammation, and (rarely) tumors may cause false-positive images, requiring more detailed analysis of the images.

The contribution of resident-labeled platelets vs. circulating platelets in the vessels can be calculated in a short-term study by a sophisticated well-validated technique for obtaining information from scintigraphic images, i.e., by the dual-tracer technique and a subtraction method using a second radiotracer such as $^{99m}$Tc-labeled erythrocytes or [$^{113}$In]oxine-labeled platelets. The latter seems not only to allow good-quality subtraction but may much better reflect the haemodynamic characteristics of platelets in the circulation, allowing improved differentiation between those platelets still circulating and those already resident. However, the preparation poses many pitfalls, making it very difficult for routine purposes. Alternatively, imaging at 3 to 5 days after injection requiring a radiotracer with a long half-life, such as $^{111}$In, can be performed.

## 2. Thrombosis

None of the various techniques that have been used for early and rapid diagnosis of thrombus has gained widespread application. The gold standards for studying thrombosis are still contrast venography and angiography which, however, also have their limitations, namely, due to the use of contrast media.

Labeling platelets in order to image thrombus (Table 7) has been performed more for clinical and experimental research than as a routine diagnostic test.

Platelets accumulate only in active thrombus. A thrombus must contain at least 0.1% of the total injected dose (equivalent to a dose of 0.1 $\mu$Ci carried by roughly one million platelets; SINZINGER and VIRGOLINI 1990) to produce significant imaging enabling scintigraphic detection.

In intracardiac thrombosis, scintigraphy with radiolabeled platelets can provide useful information not available from any other test (EZEKOWITZ et al. 1980, 1984). It can identify a thrombus before it reaches sufficient size to be detected by echocardiography and may thus be used to monitor the activity of

**Table 7.** Imaging of thrombi with radiolabeled platelets

| Vessel/Lesion | References |
|---|---|
| Arteries | DAVIS et al. 1978 |
| Veins | CLARKE-PEARSON et al. 1985 |
| | DAVIS et al. 1978 |
| Left ventricle | EZEKOWITZ et al. 1980 |
| | EZEKOWITZ et al. 1984 |
| | STRATTON et al. 1982 |
| Aneurysms | FENECH et al. 1981 |
| Synthetic material (grafts, prostheses) | GOLDMAN et al. 1982 |
| | PRICE et al. 1982 |
| | YUI et al. 1982 |
| Pharmacological interventions | NORCOTT et al. 1982 |

thrombi not apparent by echocardiography . However, the problems of high blood background, small vessels, and a resultant relatively small lesion-to-background ratio are limitations for reliable imaging, preventing its widespread clinical application.

Scintigraphy with radiolabeled platelets is a useful alternative to radiological methods for demonstrating acute deep venous thrombophlebitis, being particularly useful for evaluating patients who have relative contraindications for contrast venography. It also appears to be useful for evaluating patients with chronic venous disease who develop signs and symptoms of acute thrombophlebitis (RODRIGUES and SINZINGER 1994).

In femoropopliteal bypass grafts, the measurement of platelet uptake at one week was found to be highly predictive of graft patency at one year, as validated against a clinical-end-point graft patency (GOLDMAN et al. 1983).

In studies with prosthetic grafts, it has been shown that grafts up to 10 years old still continue to accumulate platelets with persistent thrombotic activity at anastomoses (GOLDMAN et al. 1982; YUI et al. 1982).

Radiolabeled platelets thus help to provide a suitable technique for identification of the intensity and study of the kinetics of platelet deposition on thrombogenic surface. However, it was recently observed (DEWANJEE 1994) that fibrin/platelet ratio, platelet density, and size of adherent thrombus are more important for estimating tissue ischemia/necrosis in the distal organ bed. The estimation of these parameters will thus be essential for effective evaluation of platelet inhibitors, anticoagulants, and fibrinolytic drugs.

Although studies with $^{111}$In-labeled platelets have produced much useful data, techniques have been developed that are suited for the diagnosis of thrombosis, and platelet imaging has had some clinical success, no investigation using radiolabeled platelets has been widely adopted in clinical practice. The major drawbacks remain the time and technical skill required, which is not widely available for platelet labeling.

### 3. Myocardial Infarction

Only few studies with radiolabeled platelets have been performed in myocardial infarction.

In a study of patients suffering from myocardial infarction who were evaluated within 24h of the onset of chest pain, only 25% of the patients had an area of increased accumulation of radiolabeled platelets corresponding to an obstructed coronary vessel detected by angiography (SINZINGER et al. 1991). Gating of both heart and respiratory functions may be necessary to improve sensitivity.

## G. Platelet Aggregation

Enhanced and spontaneous platelet aggregation have been found in vitro in arterial thrombotic disease. Platelets can be activated in vitro before quantification of platelet aggregation (D'SOUZA et al. 1994).

Platelets encounter a number of different aggregating agents in circulation, namely, collagen, $TXA_2$, ADP, thrombin, platelet-activating factor, serotonin, epinephrine, vasopressin, and immune complexes. Other substances bind to platelets, thus playing a central role in platelet aggregation, such as fibrinogen, vitronectin, laminin, and thrombospondin. Aggregation can also be induced by other agents including arachidonic acid, some antiplatelet antibodies, viruses, bacteria, and tumor cells, among others. Receptors for some of the aggregating agents and other substances that bind to platelets have been identified (PACHKAM 1994).

Tests for platelet aggregation include measurement of platelet aggregates by the platelet-count ratio method or flow cytometry, measurement of urinary thromboxane $B_2$ metabolites, particularly 11-dehydro$TXB_2$, by radioimmunoassays, and platelet activation by the use of specific monoclonal antibodies to detect activated platelet-surface GPIIb/IIIa by flow cytometry (see Chap. 28) and shear-stress test (D'SOUZA 1994). Results of these tests are highly variable because of in vitro activation. Extreme care must be taken to avoid even minor in vitro activation while collecting or processing blood samples. However, these tests are powerful tools for studying platelet activation in vivo (D'SOUZA 1994).

The sensitivity and specificity of the above tests for risk prediction require further prospective clinical studies.

## H. Conclusions

It has been demonstrated that the deposition of platelets on the arterial wall plays an important role in the development and progression of atherosclerotic lesions. A local platelet thrombus formation bears the subsequent risk of incorporation into the vascular wall. This may result in obstruction of the vessel. Platelets have also been found to make a significant contribution to the pathophysiology of myocardial ischemia.

Further application of the studies with radiolabeled platelets should yield important information as to pathogenesis as well as to routine diagnostic tests and evaluation and monitoring of the effect of therapeutical interventions.

### References

Baldini MG, Meyers TJ (1980) Utilitá e limitazioni degli studi di sopravivenza delle piastrine. Una valutazione critica. Haematologica 65:689–716

Bergmann SR, Lerch RA, Mathias CJ, Sobel BE, Welch MJ (1983) Noninvasive detection of coronary thrombi with In-111 platelets: concise communication. J Nucl Med 24:130–135

Bills TK, Smith JB, Silver J (1976) Metabolism of (14C)arachidonic acid by human platelets. Biochim Biophys Acta 424:303–314

Bizzozero J (1982) Über einen neuen Formbestandteil des Blutes und dessen Rolle bei der Thrombose und der Blutgerinnung. Virchows Arch Pathol Anat 90:261

Bounameaux Y (1959) L'accolement des plaquettes aux fibres sousendotheliales. C R Soc Biol (Paris) 153:865–867
Callow AD, Connolly R, O'Donnel TF, Gembarowicz H (1982) Platelet arterial synthetic graft interaction and its modification. Arch Surg 117:1447–1452
Chien WW, Mohabir R, Neuman D, Leung LLK, Clusin WT (1990) Effect of platelet release products on cytosolic calcium in cardiac myocytes. Biochem Biophys Res Commun 170:1121–1127
Clarke-Pearson DL, Coleman RE, Siegel R, Synan IS, Petry N (1985) Indium-111 imaging for the detection of deep venous thrombosis and pulmonary embolism in patients without symptoms after surgery. Surgery 98:98–103
Coughlin SR, Moskowitz MA, Zetter BR, Antoniades HN, Levine L (1980) Platelet-dependent stimulation of prostacyclin synthesis by platelet derived growth factor. Nature 288:600–603
Davis HH, Siegel BA, Joist JH (1978) Scintigraphic detection of atherosclerotic lesions and venous thrombi in man by indium-111 labelled autologous platelets. Lancet I:1185–1187
De Vries RA, De Bruin M, Marx JJM, van de Wiel A (1993) Radioisotopic labels for blood cell survival studies: a review. Nucl Med Biol 20:809–817
Dewanjee MK (1990) Radiolabelled platelets in monitoring of drug efficacy in animal models. Thromb Haemorrh Dis 1:37–51
Dewanjee MK (1994) Quantitation of platelet consumption in health and disease. In Martin-Comin J (ed) Radiolabelled blood elements. Plenum, New York, pp 67–90
Dewanjee MK, Tago M, Josa M, Fuster V, Kaye MP (1984) Quantification of platelet retention in aortocoronary femoral vein bypass graft in dogs treated with dipyridamole and aspirin. Circulation 69:350–356
Dorn GW, Liel N, Trask JL, Mais DE, Assey ME, Halushka PV (1990) Increased platelet thromboxane $A_2$/prostaglandin $H_2$ receptors in patients with acute myocardial infarction. Circulation 81:212–218
D'Souza D, Wu KK, Hellums JD, Phillips MD (1994) Platelet activation and arterial thrombosis, Lancet 334:991–995
Duguid JB (1946) Thrombosis as a factor in the pathogenesis of coronary atherosclerosis. Am J Pathol Bacteriol 58:207–212
Ezekowitz MD, Smith EO, Allen EW (1980) Identification of left ventricular thrombi in humans using $^{111}$In autologous platelets. Am J Cardiol 45:463–467
Ezekowitz MD, Pope CF, Smith EO (1983) $^{111}$Indium platelet deposition at sites of percutaneous transluminal peripheral angiography. Circulation 6[Suppl 3]:144
Ezekowitz MD, Kellerman DJ, Smith EO, Streitz TM (1984) Detection of active left ventricular thrombosis during acute myocardial infarction using indium-111 platelet scintigraphy. Chest 86:35–39
Fenech A, Hussey JK, Smith FW, Dendy PP, Bennett B, Douglas AS (1981) Diagnosis of aortic aneurysm using autologous platelets labelled with indium-111. Br Med J 282:1122
Finklestein SD, Miller A, Callahan RJ, Fallon JT, Godley F, Feldman BL, Hinton RC, Roberts AB, Strauss HW, Lees RS (1982) Imaging of acute arterial injury with $^{111}$In-labelled platelets: a comparison with scanning-electron micrographs. Radiology 145:155–159
Fitscha P (1986) Prostaglandine in Pathogenese und Therapie der peripheren arteriellen Verschlußkrankeit (PVK). In: Kraupp O, Sinzinger H, Widhalm K (eds) Atherogenesis VII. Maudrich, Vienna, p 124
Fitscha P, Kaliman J, Sinzinger H (1985) Gamma-camera imaging after autologous human platelet labelling with $^{111}$In-oxine-sulphate: a key for assessing the efficacy of prostacyclin treatment in active atherosclerosis? In: Schrör K (ed) Prostaglandins and other eicosanoids in the cardiovascular system. Karger, Basel, pp 352–361
Flores NA, Sheridan DJ (1994) The pathophysiological role of platelets during myocardial ischaemia. Cardiovasc Res 28:295–302

Fuster V, Dewanjee MK, Kaye MP, Tosa M, Metke MP, Chesebro JH (1979) Noninvasive radioisotope technique for detection of platelet deposition in coronary artery bypass grafts in dogs and its reduction with platelet inhibitors. Circulation 60:1508–1512
Goldman M, Norcott HC, Hawker RJ, Hail C, Drole Z, McCollum CN (1982) Platelet accumulation on mature Dacron grafts in man. Br J Surg [Suppl] 69:380–382
Goldman M, Hail C, Dykes J, Hawker RJ, McCollum CN (1983) Does indium platelet deposition predict patency in prosthetic arterial grafts? Br J Surg 70:653–658
Goncalves L (1995) Risk stratification after myocardial infarction. Clinical evaluation before discharge. Rev Port Cardiol 14:475–482
Goncalves L, Cardoso P, Monteiro A, Isaac J, Correia NF, Lopes C, Silveira A, Providencia LA (1995) Left ventricular function is an important predictor of long-term survival in patients with intracardiac thrombus. Rev Port Cardiol 14:507
Gorman RR, Bunting S, Miller OV (1977) Modulation of human platelet adenylate cyclase by prostacyclin (PGX). Prostaglandins 13:377–388
Gown AM, Tsukada T, Ross R (1986) Human atherosclerosis. II. Immunocytochemical analysis of the cellular composition of human atherosclerotic lesions. Am J Pathol 125:191–207
Granegger S, Flores J, Widhalm K, Sinzinger H (1988) Increased low-density lipoproteins (LDL) negatively affect human platelet labelling. Folia Haematol 115:451–453
Harker LA (1978) Platelet survival time: its measurement and use. Prog Hemost Thromb 4:321–347
Heyns AP, Lötter MG, Badenhorst PN, Pieters H, Nel CJC, Minnaar PC (1982) Kinetics and fate of indium 111 oxine labelled platelets in patients with aortic aneurysms. Arch Surg 117:1170–1174
Huang TW, Harker LA (1981) $^{111}$In-platelet imaging for detection of platelet deposition in abdominal aneurysms and prosthetic arterial grafts. Am J Cardiol 47:882–889
International Committee For Standardization In Haematology (1988) Panel on diagnostic applications of radionuclides. Recommended method for indium-111 platelet survival studies. J Nucl Med 29:564–566
Isaka, Y, Kimura K, Yoneda S, Kusunoki M, Etani H, Uyama O, Tsuda Y, Abe H (1984) Platelet accumulation in carotid atherosclerotic lesions: semiquantitative analysis with $^{111}$indium platelets and $^{99m}$technetium human serum albumin. J Nucl Med 25:556–563
Kessler C, Henningsen H, Reuther R, Antalics I, Kimmig B (1985) Szintigraphie mit $^{111}$In-markierten Thrombozyten: Therapie-Kontrolle bei Acetylsalicylsäure (ASS)-behandelten Schlaganfallpatienten. Nucl Compact 16:30–31
Kimura K, Isaka Y, Etani H, Kusunoki M, Yoneda S, Nagatsuka K, Uyama O, Abe H (1982) In – 111 labeled autologous platelet scintigraphy for the detection of vascular thrombi in ischemic cerebrovascular disease. In: Raynaud C (ed) Nuclear medicine and biology. Pergamon, Paris, pp 1759–1761
Larrue J, Bricaud H, Sinzinger H (1984) Prostacyclin synthesis by proliferative aortic smooth muscle cells. Vasa 13:311–319
LeBreton GC, Lipowski JP, Feinberg H, Venton DL, Ho T, Wu KK (1984) Antagonism of thromboxane $A_2$/prostaglandin $H_2$ by 13-azaprostanoic acid prevents platelet deposition to the deendothelialized rabbit aorta in vivo. J Pharmacol Exp Ther 229:80–84
Lötter MG, Heyns AP, Badenhorst PN, Wessels P, Martin Van Zyl J, Kotze HF, Minnaar PC (1986) Evaluation of mathematic models to assess platelet kinetics. J Nucl Med 27:1192–1201
Lovaas ME, Gloviczki P, Dewanjee MK, Hollier LH, Kaye MP (1983) Inferior vena cava replacement: the role of antiplatelet therapy. J Surg Res 35:234–242
Lusby RJ, Florell LD, Endlestad BL, Price DC, Lipton MJ, Stoney RJ (1983) Vessel wall and $^{111}$In-labelled platelets response to carotid endarterectomy. Surgery 93:424–432

Mathias CJ, Welch MJ (1984) Radiolabeling of platelets. Semin Nucl Med 14:118–127
McBride W, Lange R, Hillis LD (1988) Restenosis after successful coronary angioplasty. N Engl J Med 318:1734–1737
Moncada S, Gryglewski R, Bunting S, Vane JR (1976) An enzyme isolated from arteries transforms prostaglandin endoperoxides to an unstable substance that inhibits platelet aggregation. Nature 263:663–665
Moncada S, Herman AG, Higgs EA, Vane JR (1977) Differential formation of prostacyclin ($PGI_2$ or PGX) by the layers of the arterial wall. An explanation for the antithrombotic properties of vascular endothelium. Thromb Res 12:367–374
Muggli R, Baumgartner HR (1973) Collagen-induced platelet aggregation: requirement for tropocollagen multimers. Thromb Res 3:715–728
Najean Y (1986) The choice of tracer for platelet kinetics and scintigraphic studies. Nucl Med Biol 13:159–164
Niiya K, Hodson E, Bader R, Byers-Ward V, Koziol JA, Plow EW, Ruggeri ZM (1987) Increased surface expression of the membrane glycoprotein IIb/IIIa complex induced by platelet activation. Relationship of the binding of fibrinogen and platelet aggregation. Blood 70:475–483
Norcott HC, Goldman M, Hawker RJ, Rafiqi EI, Drolc Z, McCollum CN (1982) Platelet inhibitory drugs: an in vivo method of evaluation. Thromb Haemost 48:307–310
Nordestgaard AG, Marcus CS, Wilson SE (1990) Effect of aspirin and dipyridamole on sequential graft platelet accumulation after implantation of small diameter PTFE prosthesis. Platelets 1:37–41
Packham MA (1994) Role of platelets in thrombosis and hemostasis. Can J Physiol Pharmacol 72:278–284
Pearson JD (1994) Vessel wall interactions regulating thrombosis. Br Med Bull 50:776–788
Peters AM (1988) Review of platelet labelling and kinetics. Nucl Med Commun 9:803–808
Petterson K, Björk H, Bondjers G (1993) Endothelial integrity and injury in atherogenesis. Transplant Proc 2:2054–2056
Powers WJ, Mathias CJ, Welch MJ, Sherman LA, Siegel BA, Clarkson TB (1982a) Scintigraphic detection of platelet deposition in atherosclerotic macaques: a new technique for investigation of antithrombotic drugs. Thromb Res 25:137–143
Powers WJ, Siegel BA, Davis HH, Mathias CJ, Clark HB, Welch MJ (1982b) $^{111}$Indium platelet scintigraphy in cerebrovascular disease. Neurology 32:938–943
Price DC, Lipton MJ, Lusby RJ, Engelstad BL, Stoney RJ, Prager RJ, Hartmeyer JA, Holly AS (1982) In vivo detection of thrombi with indium-111-labelled platelets. IEEE Trans Nucl Sci 29:1191–1197
Pumphrey CW, Dewanjee MK, Chesebro JH, Fuster V, Wahner HW (1982) A new method of quantifying human platelet vascular graft interaction and the effect of platelet-inhibitory therapy. In: Raynaud C (ed) Nuclear medicine and biology. Pergamon, Paris, pp 895–897
Radomski MW, Palmer RMJ, Moncada S (1987) Comparative pharmacology of endothelium-dependent relaxing factor, nitric oxide and prostacyclin in platelets. Br J Pharmacol 92:181–187
Randal MJ, Wilding RI (1982) Acute arterial thrombosis in rabbits: reduced platelet accumulation after treatment with dazoxiben hydrochloride. Br J Clin Pharmacol 15:495
Ritchie JL, Lindner A, Hamilton GW, Harker LA (1982) $^{111}$In-oxine platelet imaging in hemodialysis patients: detection of platelet deposition at vascular access sites. Nephrology 39:334–339
Rodrigues M, Sinzinger H (1994) Platelet labeling – methodology and clinical application. Thromb Res 76:399–432
Rokitansky CV (1852) Über einige der wichtigsten Krankeiten der Arterien. KK Hof- und Staatsdruckerei, Wien

Ross R, Glomset JA (1976) The pathogenesis of atherosclerosis. N Engl J Med 295:369–372
Ross R, Glomset JA, Kariya B, Harker A (1974) A platelet-dependent serum factor that stimulates the proliferation of arterial smooth muscle cells in vivo. Proc Natl Acad Sci USA 71:1207–1210
Siegl AM, Smith BJ, Silver MJ, Nicolaoŭ KC, Ahem D (1979) Selective binding site for (3H) prostacyclin on platelets. J Clin Invest 63:215–220
Sinzinger H, Fitscha P (1984a) Epoprostenol and platelet deposition in atherosclerosis. Lancet I:905–906
Sinzinger H, Fitscha P (1984b) Scintigraphic detection of femoral artery atherosclerosis with 111-indium-labelled autologous platelets. Vasa 13:350–355
Sinzinger H, Fitscha P (1987) Influence of prostaglandin $E_1$ on in vivo accumulation of radiolabelled platelets and LDL on human arteries. Vasa 17:5–10
Sinzinger H, Fitscha P (1990) Monitoring of antithrombotic activity by platelet labelling. In: O'Grady J, Linet OI (eds) Early phase drug evaluation in man. Macmillan, Houndmills, pp 337–347
Sinzinger H, Virgolini I (1990) Nuclear medicine and atherosclerosis. Eur J Nucl Med 17:160–178
Sinzinger H, Feigl W, Silberbauer K (1979) Prostacyclin generation in atherosclerotic arteries. Lancet II:479
Sinzinger H, Kolbe H, Strobl-Jäger E, Höfer R (1984) A simple and safe technique for sterile autologous platelet labelling using "Monovette" vials. Eur J Nucl Med 9:320–322
Sinzinger H, O'Grady J, Fitscha P (1985) Detection of aneurysms by gamma-camera imaging after injection of autologous labelled platelets. Lancet I:929–931
Sinzinger H, O'Grady J, Fitscha P (1986) Effect of prostaglandin $E_1$ on deposition of autologous labelled platelets onto human atherosclerotic lesions in vivo. Postgrad Med J 63:245–247
Sinzinger H, O'Grady J, Fitscha P, Kaliman J (1987) Diminished platelet residence time on active human atherosclerotic lesions in vivo – evidence for an optimal dose of aspirin? Prostaglandins Leukot Essent Fatty Acids 34:89–93
Sinzinger H, Rauscha F, Fitscha P (1990a) Dihydroergotoxin-beneficial effect on in vivo and in vitro-platelet function in patients with cerebrovascular disease. A double-blind placebo controlled study. Med Nucl 2:305–311
Sinzinger H, Virgolini I, Vinazzer H (1990b) Autologous platelet – labelling in thrombocytopenia. Thromb Res 60:223–230
Sinzinger H, Fitscha P, O'Grady J, Rauscha F, Rogatti W, Vane JR (1990c) Synergistic effect of prostaglandin $E_1$ and isosorbide dinitrate in peripheral vascular disease. Lancet I:627–628
Sinzinger H, Virgolini I, Fitscha P (1991) Platelet labelling in atherosclerosis. In: Wissler RW (ed) Atherosclerotic plaques. Plenum, New York, pp 181–188
Steurer G, Fitscha P, Ettl K, Sinzinger H (1989) Effects of hydergine on platelet deposition on "active" human carotid artery lesions and platelet function. Thromb Res 55:577–589
Stratton JR, Ritchie JL (1982) Sulfinpyrazone fails to inhibit platelet deposition on Dacron prosthetic grafts in man. Circulation 66:55
Stratton JR, Ritchie JL, Sisk EJ, McFadden KW (1982) Inhibition of indium-111 platelet deposition in left ventricular thrombi by platelet active drugs. Circulation 66:342
Stratton JR, Thiele BL, Ritchie JL (1983) Natural history of platelet deposition on Dacron aortic bifurcation grafts in the first years after implantation. Am J Cardiol 52:371–374
Thakur ML, Welch MJ, Malech HL (1976) Indium-111 labelled human platelets: studies on preparation and evaluation of in vitro and in vivo functions. Thromb Res 9:345–357

Thakur ML, Welch MJ, Malech HL (1984) Indium-111 labelled human platelets: improved method, efficacy and evaluation. J Nucl Med 22:381–385
Trip MD, Cats VM, van Capelle FJ, Vreeken J (1990) Platelet hyperreactivity and prognosis in survivors of myocardial infarction. N Engl J Med 323:1549–1554
Tran MF (1984) Kinetics and distribution of platelets in man. Am J Hematol 17:97–104
Van Rensburg EJ, Heyns AP (1990) The effect of age and hypercholesterolemia on platelet function tests. Thromb Haemorrh Dis 1:11–16
Wagner X, Granegger S, Dembinska-Kiec A, Sinzinger H (1989) Nitric oxide (NO) for radiolabelling of human platelets. In: Sinzinger H, Thakur ML (eds) Nuclear medicine research. Facultas, Vienna, p 58
Walsh PN (1972) The role of platelets in the contact phase of blood coagulation. Br J Haematol 22: 237–254
Weksler BB, Nachman RL (1981) Platelets and atherosclerosis. Am J Med 71:331
Wilcox JN, Blumenthal BF (1995) Thrombotic mechanisms in atherosclerosis: potential impact of soy proteins. J Nutr 125:631–638
Wilcox JN, Smith KM, Schwartz SM, Gordon D (1989) Localization of tissue factor in the normal vessel wall and in the atherosclerotic plaque. Proc Natl Acad Sci USA 86:2839–2843
Yui T, Uchida T, Matsuda S, Iwaya K, Umino M, Ono K, Muroi S, Owada K, Machii K, Kariyone S (1982) Detection of platelet consumption in aortic graft with 111-In labelled platelets. Eur J Nucl Med 7:77–79

CHAPTER 26

# Platelets, Vessel Wall, and the Coagulation System

R. HELLER and E.M. BEVERS

## A. Platelet–Endothelium Interaction

Platelet–endothelium interactions play a critical role in the maintenance of vascular integrity and in the regulation of thrombotic and inflammatory events. Platelets support endothelial barrier function, thereby restricting the loss of macromolecules and blood cells from the circulation. Endothelial cells in turn maintain platelet reactivity low to prevent platelet adhesion and intravascular thrombus formation. Platelet control mechanisms involve nonadhesive properties of the endothelial cell membrane, inactivation of platelet stimuli by endothelial-bound proteins, and production of prostacyclin ($PGI_2$) and endothelium-derived relaxing factor (EDRF), mediators that are well known to inhibit platelet reactivity. Importantly, the synthesis of $PGI_2$ and EDRF in endothelial cells can be amplified by compounds produced and/or released upon platelet activation (thrombin, $PGH_2$, adenine nucleotides, serotonin), representing a major feedback mechanism in thromboregulation and limiting excessive platelet aggregation. When the antithrombotic balance of the endothelium is disturbed, platelets may attach to endothelial cells through a mechanism involving adhesion receptors. Activated platelets also express interleukin-1 (IL-1) on their surface and release transforming growth factor-$\beta$ (TGF-$\beta$), which cause endothelial cells to synthesize adhesion molecules, cytokines, and plasminogen activator inhibitor type 1 (PAI-1), all of which potentiate inflammatory and thrombotic events. This section will discuss the various platelet–endothelium interactions and especially address the mechanisms underlying control of platelet reactivity by endothelial cells.

### I. Role of Platelets in Maintaining Vascular Integrity

The vascular endothelium forms an active boundary between the bloodstream and underlying tissues, allowing free exchange of water and small solutes but restricting the escape of plasma macromolecules and blood cells. Endothelial permeability and endothelial barrier function are mainly controlled by intercellular junctions. These are complex structures formed by transmembrane adhesive molecules linked to a network of cytoskeletal proteins (LUM and MALIK 1994; DEJANA et al. 1995). The mechanisms that regulate the opening and closing of endothelial junctions are still unresolved but certainly involve cytoskeletal reorganization. In general, inhibition of actin microfilament assembly increases and stabilization of the actin cytoskeleton reduces endothelial cell permeability. Inflammatory stimuli, for example, cause phosphorylation of cytoskeletal proteins via protein kinase C and/or $Ca^{2+}$-dependent myosin light-chain kinase, resulting in enhancement of vascular permeability (LUM and MALIK 1994). In contrast, agents that increase intracellular levels of cAMP or cGMP reduce endothelial permeability both under basal conditions and following stimulation by inflammatory mediators (STELZNER et al. 1989; WESTENDORP et al. 1994; DRAIJER et al. 1995a). This is attributed to the activation of cAMP- and cGMP-dependent protein kinases which have been shown to inhibit myosin light-chain kinase activation and $Ca^{2+}$ increase, respectively (SHELDON et al. 1993; DRAIJER et al. 1995a). A recent study suggested that the vasodilator-stimulated phosphoprotein (VASP), a substrate of both cAMP- and cGMP-dependent protein kinase (BUTT et al. 1994), may also be involved in the regulation of endothelial barrier function (DRAIJER et al. 1995b). Interestingly, EDRF has been shown to increase intracellular cGMP in endothelial cells and to operate as an endogenous permeability-regulatory factor (DRAIJER et al. 1995a).

Platelets have been implicated in the maintenance of the endothelial barrier function of blood vessels since it has been observed that thrombocytopenic patients have fragile and leaky vessels (GAYDOS et al. 1962). Accordingly, ultrastructural studies have demonstrated that capillaries from thrombocytopenic rabbits are thinner and have fenestrations as well as fewer luminal projections (KITCHENS and WEISS 1975). Thrombocytopenia in animals has been shown to be accompanied by increased vascular permeability measured as an increased disappearance of intravenously injected [$^{125}$I]albumin or enhanced pulmonary lymph flow (AURSNES 1974; LO et al. 1988). Platelet repletion in these animals restored the vascular alterations, suggesting a causal relationship between thrombocytopenia and disturbed vascular integrity. Additionally, it has been shown that platelets strongly reduce the permeability of capillaries for macromolecules in perfused isolated organs (GIMBRONE et al. 1969; MCDONAGH 1986).

The mechanisms underlying the stabilizing platelet effects are still unresolved. Originally it was thought that platelet support of vascular integrity is due to mechanical obstruction of intercellular gaps or junctional regions (TRANZER and BAUMGARTNER 1967) or even to incorporation of the cells into the endothelium (WOJCIK et al. 1969). More recently, however, it has been shown that platelet endothelial contact is not necessary and that soluble platelet factors are likely to mediate the stabilizing effect. In in vitro studies, intact platelets, supernatants of lysed platelets, or conditioned medium from unstimulated platelets have been found to decrease endothelial permeability to a comparable degree not further potentiated by thrombin stimulation (SHEPHARD et al. 1989; HASELTON and ALEXANDER 1992). Though controversial, several platelet-derived vasoactive compounds and a polypeptide factor have been suggested as candidate agents for the barrier-enhancing effect.

Serotonin, for example, has been shown to reduce the vascular permeability in thrombocytopenic hamsters and to induce stress fiber formation in endothelial monolayers in vitro (SHEPRO et al. 1984, 1985). Similarly, adenosine has been demonstrated to decrease endothelial permeability, and moreover, pretreatment of platelets with adenosine deaminase reduces their barrier-enhancing effect (HASELTON et al. 1993; PATY et al. 1992). A possible mechanism of adenosine-mediated effects is thought to be an accumulation of intracellular cAMP via activation of $A_2$-adenosine receptors (LEGRAND et al. 1989), leading to stabilization of actin microfilament assembly. Furthermore, a protective effect of platelets on vascular integrity may also be mediated by stimulation of endothelial EDRF synthesis with a subsequent increase in intracellular cGMP (see Sect. II.3).

Besides possible acute (most likely cytoskeleton-directed) effects of platelets or platelet-derived agents, support of endothelial proliferation by platelet factors might also contribute to the maintenance of vascular integrity. NAKAJIMA et al. (1987) demonstrated decreased mitogenic activity of sera from thrombocytopenic patients on endothelial cells in culture. This suggested that an endothelial growth-promoting effect of platelets could also be relevant in vivo. Earlier in vitro studies had already shown that coincubation of platelets with endothelial cells augmented endothelial cell growth, possibly mediated by serotonin and adenosine diphosphate (ADP; MACA et al. 1977; D'AMORE and SHEPRO 1977). MIYAZONO et al. (1987) have purified a 45-kDa protein from human platelet lysate which has been shown to stimulate growth and chemotaxis of endothelial cells in vitro and to have angiogenic activity in vivo (ISHIKAWA et al. 1989). Further studies of this platelet-derived endothelial growth factor are required to establish its possible involvement in the maintenance of vascular integrity.

## II. Control of Platelet Reactivity by Endothelial Cells

Endothelial cells lining blood vessels are normally nonthrombogenic and do not allow significant platelet adhesion. The negative surface charge of endothelium repulsing the likewise negatively charged surface of platelets is an important antiadhesive factor (SAWYER et al. 1972). Additionally, endothelial cells actively regulate their membrane properties, release platelet inhibiting factors, and inactivate platelet stimuli, thereby preventing platelet deposition and thrombus formation. These platelet control mechanisms will be discussed in detail in the following paragraphs.

### 1. 13-Hydroxy-octadecadienoic Acid (13-HODE) – An Antiadhesive Fatty Acid Metabolite

Several studies have suggested that the concentration of the monohydroxide 13-hydroxy-octadecadienoic acid (13-HODE) in endothelial cells influences adhesive properties of the cell membrane (BUCHANAN et al. 1989; HAAS et al.

1990; TLOTI et al. 1991). In unstimulated endothelial cells, 13-HODE is continuously synthesized via the lipoxygenase pathway. The precursor linoleic acid is known to be liberated from the cytosolic triacylglycerol pool in a cAMP-dependent manner. Upon cell stimulation, 13-HODE synthesis is decreased, and the lipoxygenase pathway preferentially metabolizes liberated arachidonic acid to 15-hydroxy-eicosatetraenoic acid (15-HETE). The levels of 13-HODE show good inverse correlation with platelet/endothelial cell adhesion, which led to the hypothesis that 13-HODE may prevent adhesive processes (BUCHANAN et al. 1989; HAAS et al. 1990; TLOTI et al. 1991). Accordingly, exogenously added 13-HODE is also able to reduce thrombin-induced adherence of platelets to endothelial cells (TLOTI et al. 1991) and to the extracellular matrix (AZNAR-SALATTI et al. 1991). In a recent study, 13-HODE has been shown to be colocalized with the vitronectin receptor in unstimulated endothelial cells. When cells were stimulated by IL-1, 13-HODE was no longer detectable, and the vitronectin receptor was observed on the apical surface of the endothelial cells, resulting in increased vitronectin binding and platelet adhesion. From these data, it has been suggested that 13-HODE exerts its antiadhesive action by inhibiting expression or ligand-binding properties of the vitronectin receptor (BUCHANAN et al. 1993). The role of the vitronectin receptor in mediating platelet/endothelial cell adhesion has also been confirmed in other studies (BERTOMEU et al. 1990).

### 2. Endothelium-Derived Platelet Inhibitors

Endothelial cells release at least two factors that are able to inhibit platelet activation. One of these molecules is $PGI_2$, which is formed from endothelial arachidonic acid by the sequential action of the intracellular enzymes cyclooxygenase, peroxidase, and prostacyclin synthase (GRYGLEWSKI et al. 1988). The other compound is EDRF, which has been identified as nitric oxide (NO) or a related nitrosothiol. NO is produced from the guanidino nitrogen of L-arginine by the action of the enzyme NO synthase, which is constitutively expressed in endothelial cells and regulated by intracellular calcium levels (MONCADA and HIGGS 1993). $PGI_2$ and EDRF/NO exert their effects locally instead of acting as circulating hormones. This is especially true for NO, which is rapidly inactivated by hemoglobin and can probably only affect platelets adjacent to the vessel wall (HOUSTON et al. 1990). $PGI_2$ and EDRF/NO are interlinked in several ways. Their release from endothelial cells is coupled, they have similar biological (e.g., vasodilator, platelet-suppressant, and fibrinolytic) activities, and though acting via different signaling pathways, they have some common intracellular targets (see Sect. II.4).

NO has been shown to contribute to the nonadhesive properties of the resting vascular endothelium since scavenging NO or inhibiting NO synthesis increases platelet adhesion (RADOMSKI et al. 1987a, 1993). Furthermore, NO is effective in reducing enhanced platelet adhesion to endothelial cells caused by stimulation of either platelets or endothelium by agonists such as thrombin

(RADOMSKI et al. 1987a; VENTURINI et al. 1989, 1992). In contrast, $PGI_2$ does not affect platelet adhesion under resting conditions. When endothelial cells are stimulated with thrombin, however, inhibition of $PGI_2$ production potentiates platelet adhesion, and addition of exogenous $PGI_2$ reverses the potentiating effect (HOAK et al. 1980; FRY et al. 1980; DARIUS et al. 1995). The role of $PGI_2$ as an antiadhesive agent has been confirmed by demonstrating that $PGI_2$ receptor desensitization in platelets increases the thrombin-stimulated adhesion to endothelial cells (DARIUS et al. 1995). Interestingly, it has been suggested that the antiadhesive effect of $PGI_2$ is related to its antiaggregatory action, as $PGI_2$ was not able to inhibit single platelet adhesion in perfused lungs under flow conditions, but reduced platelet aggregate adhesion to endothelial cell monolayers in vitro (VENTURINI et al. 1992). Both $PGI_2$ and EDRF are potent antiaggregatory agents in vitro as well as in vivo, and significantly, there is a clear synergism between the antiaggregatory effect of $PGI_2$ and subthreshold concentrations of NO (RADOMSKI et al. 1987b).

### 3. Platelet Effects on $PGI_2$ and EDRF/NO Synthesis

When platelets become activated, they produce and/or release compounds that augment the synthesis of $PGI_2$ and EDRF/NO. These in turn inhibit platelet adhesion and aggregation and limit or reverse the consequences of enhanced platelet reactivity. Through their effects on $PGI_2$ and EDRF release, platelets also contribute to the regulation of vascular tone as both factors are potent vasodilators (GRYGLEWSKI et al. 1988; MONCADA et al. 1992). An important amplifying mechanism of $PGI_2$ production is the transfer of platelet endoperoxides to endothelial cells (MARCUS et al. 1980; SCHAFER et al. 1984). In endothelial cells $PGI_2$ synthesis is limited due to the low conversion rate of arachidonic acid to the prostaglandin endoperoxide $PGH_2$. This rate-limiting step can be bypassed by platelet $PGH_2$ while a reciprocal transfer of endothelial endoperoxides for utilization by the platelet thromboxane synthetase does not occur (SCHAFER et al. 1984).

Another amplifying mechanism for both $PGI_2$ and EDRF synthesis is induced by several platelet factors either released from dense granules during aggregation – ADP, adenosine triphosphate (ATP), and serotonin – or produced on the surface of activated platelets (thrombin; VANHOUTTE 1988). These factors stimulate endothelial cells through receptor-dependent mechanisms which in turn activate phospholipase C and increase cytosolic free $Ca^{2+}$ concentration. $Ca^{2+}$ then initiates EDRF and $PGI_2$ production by activating both NO synthase and phospholipase $A_2$ which liberates the $PGI_2$ precursor arachidonic acid from membrane phospholipids (DE NUCCI et al. 1988). Accordingly, platelet aggregates have been shown to release EDRF from endothelial cells, an effect mainly triggered by adenine nucleotides and in part by serotonin (HOUSTON et al. 1986; FÖSTERMANN et al. 1988; SHIMOKAWA et al. 1989). In addition, platelet-derived 12-hydroxy-5,8,10-heptadecatrienoic acid

(HHT), platelet-activating factor (PAF), and platelet-derived growth factor have been shown to stimulate endothelial $PGI_2$ synthesis (SADOVITZ et al. 1987; D'HUMIERES et al. 1986; COUGHLIN et al. 1980). Interestingly, a recent study reported that unstimulated platelets cause an increase in intracellular $Ca^{2+}$ in endothelial cells, probably by mechanical interaction (IJZENDOORN et al. 1996) which may also be followed by stimulation of $PGI_2$ and EDRF synthesis.

Platelet/endothelium interactions are altered when endothelial cells become dysfunctional, e.g., when their ability to release antiaggregatory and vasoactive factors is decreased. Endothelial dysfunction develops in a variety of vascular diseases involving hypercholesterolemia, atherosclerosis, hypertension, and diabetes mellitus (LÜSCHER et al. 1995). Dysfunction has been attributed to changes in receptor expression and/or signal transduction. Furthermore, oxidized lipoproteins accumulated in atherosclerotic plaques have been shown to inactivate NO and to reduce NO synthesis (LÜSCHER et al. 1995). The local decrease of $PGI_2$ and EDRF in dysfunctional endothelium may enhance platelet adhesion and aggregation similar to the situation in which NO synthesis is pharmacologically blocked (RADOMSKI et al. 1993). In dysfunctional endothelium, aggregating platelets and platelet factors release less EDRF, and paradoxic vasoconstriction, mainly mediated by serotonin and thromboxane $A_2$, occurs instead of vasodilation (LUDMER et al. 1986; GOLINO et al. 1991). Endothelial dysfunction together with platelet hyperreactivity may also lead to platelet factor mediated endothelial retraction and injury (KISHI et al. 1989).

## 4. Mechanisms of Platelet Inhibition by $PGI_2$ and EDRF/NO

$PGI_2$ acts on platelets through a receptor-mediated activation of membrane-bound adenylate cyclase and a consecutive increase in intracellular cAMP whereas NO stimulates the guanylate cyclase of platelet cytosol and leads to a rise in intracellular cGMP. Cyclic nucleotide effects are thought to be mediated predominantly by the activation of cAMP- or cGMP-dependent protein kinases (cA-PK, cG-PK). It has been shown that these kinases exist in platelets at high concentrations and are rapidly activated upon increase in intracellular cAMP and cGMP, respectively (EIGENTHALER et al. 1992). Their functional importance has been confirmed by demonstrating that selective membrane permeant activators of cA-PK or cG-PK mimic the effects of cAMP- or cGMP-enhancing compounds on platelet activation (GEIGER et al. 1992, 1994). Moreover, decreased expression of cG-PK in platelets is associated with disturbed NO/cGMP-regulated signal transduction (EIGENTHALER et al. 1993). Besides stimulating cG-PK, cGMP can act through the cGMP-inhibited phosphodiesterase, which leads to an increase in intracellular cAMP and subsequent cA-PK activation (MAURICE and HASLAM 1990). This may represent a component of the synergistic action of cAMP- and cGMP-increasing platelet inhibitors. Another component of this synergism may lie with the target platelet substrates for the kinases and in particular with the phosphorylation of VASP (NOLTE et al. 1994). VASP is phosphorylated stoichiometrically and reversibly by both cA-PK and cG-PK, and this is closely correlated with inhibition of fibrinogen binding to platelets (HORSTRUP et al. 1994).

Several other platelet proteins have been characterized as substrates for cyclic nucleotide-regulated protein kinases, including the $\beta$-subunit of glycoprotein Ib, the low molecular weight G protein Rap1b, caldesmon, myosin light-chain kinase, the actin-binding protein filamin, and the inositol-1,4,5,-trisphosphate receptor (WALTER et al. 1993). The molecular mechanisms coupling cA-PK- and cG-PK-mediated phosphorylations to inhibition of platelet activation are not entirely understood, and effects may occur at several stages of the activation cascade, e.g., the inhibition of agonist-induced phospholipase C activation, $Ca^{2+}$ mobilization/$Ca^{2+}$ entry, fibrinogen binding, etc. Recent studies have indicated that focal contact areas might be important targets for regulation by cyclic-nucleotide-dependent protein kinases. The cA-PK- and cG-PK substrate VASP has been characterized as an actin-binding protein (REINHARD et al. 1992) and binds profilin and a zyxin-like protein which are recognized as important for cytoskeleton-membrane attachment (REINHARD 1995a,b). Profilin is additionally known to be linked to signal transduction pathways by binding phosphoinositides (GOLDSCHMIDT-CLERMONT et al. 1991). Molecular cloning and structural analysis of VASP has revealed that a central proline-rich domain is responsible for VASP–profilin interactions, whereas a C-terminal domain is important for anchoring the protein to focal adhesion sites (HAFFNER et al. 1995).

Studies in cell culture systems or intact coronary vascular beds have confirmed that the regulatory mechanisms of cyclic nucleotides mainly investigated with pharmacological compounds can also be attributed to the physiological platelet inhibitors $PGI_2$ and EDRF/NO. In coincubation experiments with platelets and endothelial cells, it has been shown that endothelial-derived $PGI_2$ and EDRF activate platelet cA-PK and cG-PK, respectively (NOLTE et al. 1991), and inhibit platelet-agonist-evoked phospholipase C activation and $Ca^{2+}$-mobilization (DURANTE et al. 1992; GEIGER et al. 1994). A short passage ($<2$ s) of platelets through an acetylcholine-stimulated coronary bed in rabbits increased both platelet cAMP and cGMP levels, and concomitantly, phosphorylation of VASP was dependent on the presence of both $PGI_2$ and EDRF (POHL and BUSSE 1989; POHL et al. 1994).

## 5. Endothelial-Bound Antiplatelet Factors

Besides releasing factors that directly inhibit platelet activation, endothelial cells control platelet reactivity by degrading or inactivating important platelet stimuli. Endothelial cells possess an ecto-ADPase which functions on the cell surface to convert ADP released from erythrocytes and activated platelets to adenosine monophosphate (AMP) and orthophosphate. The enzyme also has ATPase activity and is classified as an ATP-diphosphohydrolase or apyrase (YAGI et al. 1991; MARCUS et al. 1991; COTÉ et al. 1992). MARCUS et al. (1991) have demonstrated ADPase activity in an in vitro coincubation system in which endothelial cells inhibit platelet responsivness although $PGI_2$ and EDRF/NO effects are blocked by aspirin and hemoglobin, respectively. This study has also reported that exogenous [$^{14}$C]ADP is rapidly degraded by endothelial cells, which is accompanied by a loss of proaggregatory activity of endothelial cell supernatant. Studies with endothelial monolayers have indicated that ADPase activity is located on the luminal surface of the vessel. Interestingly, endothelial cells can also convert the ADPase product AMP to adenosine by a 5′-nucleotidase which further inhibits platelet reactivity by elevating cAMP levels.

A second endothelial factor involved in the removal of platelet stimuli at sites of injury is the integral membrane glycoprotein thrombomodulin, which is able to bind thrombin with very high affinity (ESMON 1995). Thrombin is generated on the procoagulant surface of activated platelets and is known to be not only the key enzyme of coagulation but also a potent stimulus for both platelets and endothelial cells. The complex formation with thrombomodulin changes the function of thrombin from a clot-promoting to a clot-preventing enzyme as it loses its ability to cleave fibrinogen and becomes capable of stimulating the anticoagulant protein C pathway. Thrombomodulin-bound thrombin is also no longer able to bind to platelets and to induce aggregation or secretion (ESMON et al. 1983). Accordingly, it has been shown that thrombomodulin adds to the endothelium-mediated inhibition of thrombin-induced platelet aggregation as antibodies against thrombomodulin partially reverse the effect of endothelial cells (MURATA et al. 1988). Recent studies have demonstrated that fibrinogen, thrombomodulin, and the newly discovered thrombin receptor bind to the same anion binding exosite 1 of thrombin (COUGHLIN 1994; MATHEWS et al. 1994). Subsequently, it has been suggested that thrombomodulin-mediated inhibition of thrombin-induced platelet activation may be attributed to occupation of anion binding exosite 1, which may block high-affinity interactions of thrombin with its receptor. Besides interacting with thrombomodulin, thrombin has been shown to bind to endothelial cell surface proteoglycans, an effect which might favor its inactivation by antithrombin III (Fig. 1; PREISSNER 1990).

### 6. Platelet Adhesion to Endothelial Cells

While platelet adhesion to endothelial cells is normally prevented by different platelet control mechanisms, enhanced platelet attachment to morphologically intact endothelium may occur under certain conditions. Significant platelet adhesion has been demonstrated following different treatment of endothelial

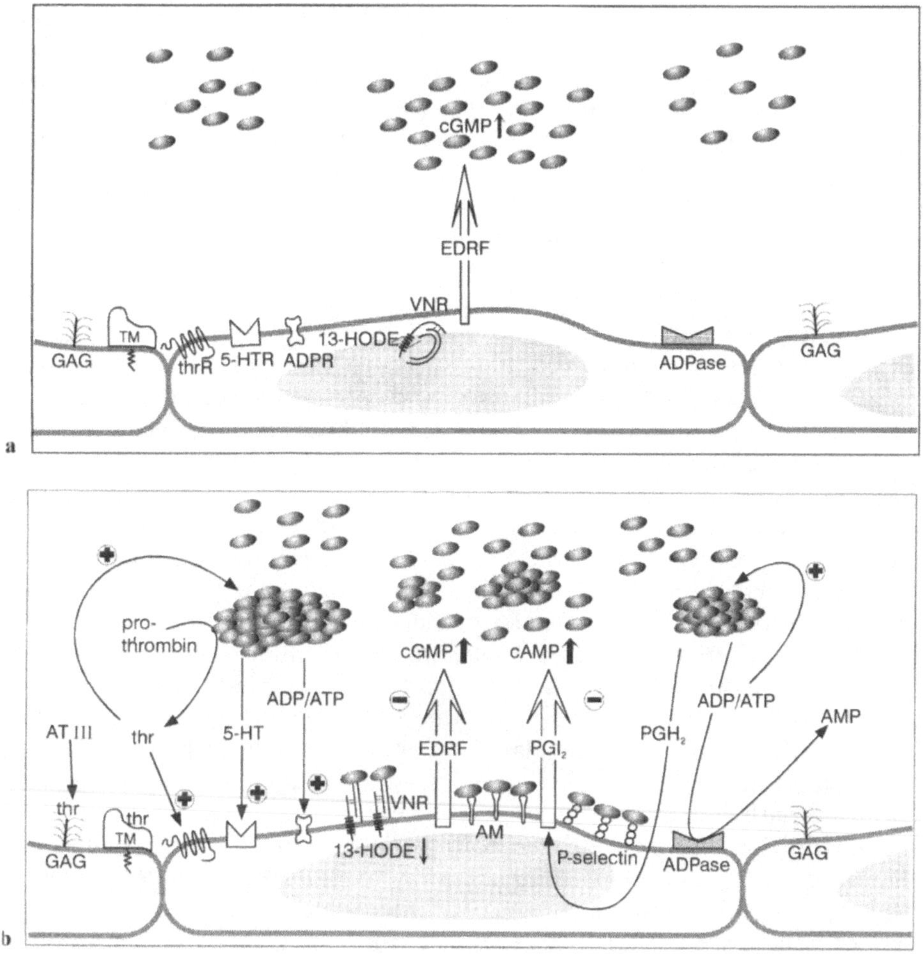


**Fig. 1a,b.** Interactions between platelets and endothelial cells. **a** Under resting conditions, endothelial cells express a nonthrombogenic surface, which is maintained by a negatively charged glycocalix and by antiadhesive molecules such as 13-hydroxy-octadecadienoic acid (*13-HODE*) and endothelium-derived relaxing factor (*EDRF*). **b** Activated platelets produce, release, or support the formation of compounds – prostaglandin $H_2$ (*$PGH_2$*), serotonin, ADP, ATP, thrombin – that augment the synthesis of EDRF and $PGI_2$ in endothelial cells. EDRF and $PGI_2$ in turn decrease platelet responsiveness to various agonists by increasing intracellular cGMP and cAMP. ADP and thrombin, which also represent important platelet stimuli, are inactivated by endothelial-bound ADPase and thrombomodulin, respectively, processes that limit platelet and endothelial cell activation. Thrombin can additionally bind to endothelial glycosaminoglycans, which supports its inactivation by antithrombin III. In stimulated endothelial cells, 13-HODE formation is inhibited, which may enhance the activation of the vitronectin receptor. When the antithrombotic balance is disturbed, platelets may also attach to endothelial cells, probably mediated by endothelial P-selectin, vitronectin receptor, and other adhesion molecules. *ADPR*, ADP receptor; *AM*, adhesion molecules; *ATIII*, antithrombin III; *GAG*, glycosaminoglycans; *5-HT*, serotonin; *5-HTR*, serotonin receptor; *thr*, thrombin; *thrR*, thrombin receptor; *TM*, thrombomodulin; *VNR*, vitronectin receptor

cells including thrombin stimulation, hypoxia/reoxygenation, virus or parasite infection, and addition of oxidants (VENTURINI et al. 1992; HEMPEL et al. 1990; CURWEN et al. 1980; TANOWITZ et al. 1990; SHATOS et al. 1991). Furthermore, plasma from patients undergoing chemotherapy or angioplasty has been shown to stimulate platelet adhesion to endothelial cells in vitro, suggesting that plasma factors, possibly cytokines, are involved in the induction of increased endothelial adhesivity (BERTOMEU et al. 1990; GAWAZ et al. 1996). Enhanced adhesion to endothelial cells may also occur with hyperreactive platelets as has been demonstrated with platelets from diabetic patients (ROSOVE et al. 1985; LUPU et al. 1993).

The molecular mediators of platelet attachment to endothelial cells are not completely specified. A possible candidate might be P-selectin, which is stored in secretory granules of platelets and endothelial cells (MCEVER 1995). Upon cellular activation by agonists such as thrombin, P-selectin is rapidly translocated to the cell surface where it is known to serve as a receptor for leukocyte adhesion. A recent study has indicated that endothelial P-selectin also mediates interactions with platelets (FRENETTE et al. 1995). Intravital microscopic observations in wild-type and P-selectin-deficient mice have demonstrated that platelets roll on endothelium in vivo and that this interaction requires endothelial but not platelet P-selectin. These results also indicate that platelets constitutively express a P-selectin ligand which may be similar to the P-selectin glycoprotein ligand-1 (PSGL-1) demonstrated in myeloid cells (FURIE and FURIE 1995). Interestingly, there is evidence that P-selectin expression in endothelial cells is regulated by endogenous NO (DAVENPECK et al. 1994), suggesting that EDRF/NO might also exert its antiadhesive action via autocrine pathways.

In addition to P-selectin, the endothelial vitronectin receptor has been suggested as mediating platelet adhesion to endothelial cells, which was confirmed by the inhibitory effect of vitronectin receptor antibodies on platelet adherence (BUCHANAN 1993; BERTOMEU et al. 1990). Glycoprotein Ib, which is not only expressed on platelets but also on endothelial cells where it can be upregulated by cytokines, may be another candidate for mediating adhesive interactions between the two cell types (KONKLE et al. 1990). Furthermore, platelet-endothelial cell adhesion molecule-1 (PECAM-1), glycoprotein IIb/IIIa, the fibronectin receptor, and fibrinogen were shown to be involved in platelet/endothelial adhesion (LUPU et al. 1993). At present, it is not clear whether endothelial-derived PAF, which is known to be associated with the endothelial cell membrane and to mediate leukocyte attachment, is also implicated in platelet adhesion. As PAF synthesis is controlled by cAMP- and cGMP-enhancing compounds, it might be another target for autocrine regulation of endothelial adhesivity by $PGI_2$ and EDRF (HELLER et al. 1991, 1992).

Besides acting directly with endothelial adhesion molecules, platelets can also adhere to leukocytes already bound to the endothelium. Platelet-leukocyte interactions are mainly mediated through P-selectin/PSGL-1 (FURIE and FURIE 1995), but other molecules such as glycoprotein IIb/IIIa, intercellular adhesion molecule-2 (ICAM-2), or $\beta_2$ integrins may also contribute (CERLETTI et al. 1995).

## III. Platelet-Mediated Inflammatory and Procoagulant Alterations of Endothelial Cells

Platelets contain a number of cytokines that cause endothelial cells to synthesize proteins, thereby aquiring new functional capacities (HAWRYLOWICZ et al. 1993). Using flow cytometry, immunohistochemistry, and electron microscopy, platelets have been shown to express IL-1 (predominantly IL-1$\beta$) which is activated upon platelet stimulation and in turn induces in endothelial cells the expression of intercellular adhesion molecule-1 (ICAM-1), interleukin-6 (IL-6), interleukin-8 (IL-8), and granulocyte/macrophage colony-stimulating factor (GM-CSF). These changes result in leukocyte attraction and adhesion to the endothelium. Platelet IL-1 activity has been shown to remain associated with the surface of intact cells or with membrane fractions and can be blocked with specific antibodies (HAWRYLOWICZ et al. 1991; KAPLANSKI et al. 1993; SEDLMAYR et al. 1995).

In contrast to IL-1, TGF-$\beta$ is stored in platelet $\alpha$-granules, released upon cell stimulation, and activated by a mechanism that probably involves plasmin (ASSOIAN et al. 1986). Platelet-derived TGF-$\beta$ has been shown to inhibit endothelial cell proliferation (MAGYAR-LEHMANN et al. 1992), to enhance the synthesis of PAI-1, and to increase the production of the potent vasoconstrictor endothelin-1 (ET-1). The latter has been demonstrated by the use of TGF-$\beta$-neutralizing antibodies that inhibited the effects of thrombin-stimulated platelet releasates or platelet lysates on PAI-1 and ET-1 formation (SLIVKA and LOSKUTOFF 1991; MURATA et al. 1995). Platelet cytokines may also be involved in the initiation of coagulation as activated platelets have been shown to stimulate tissue factor expression on endothelial cells in a platelet-number-dependent manner (JOHNSEN et al. 1983).

Platelet factor-4 (PF4), another platelet cytokine known as a heparin-binding protein, may also support inflammatory and coagulant responses. PF4 has been shown to react with heparin-like sulfated glycosaminoglycans on the endothelial surface (RYBAK et al. 1989). This may lead to a reduction of the net negative surface charge of the cells and an alteration of their reactivity with circulating blood cells. Moreover, by binding to heparin-like domains which were shown to accelerate thrombin-antithrombin complex formation (PREISSNER 1990), PF4 may prevent thrombin-inactivation on the endothelial surface.

Taken together, platelet-derived cytokines support procoagulant and proinflammatory alterations of endothelial cells as well as vasoconstriction. This may be important in the early inflammatory response as platelet-derived

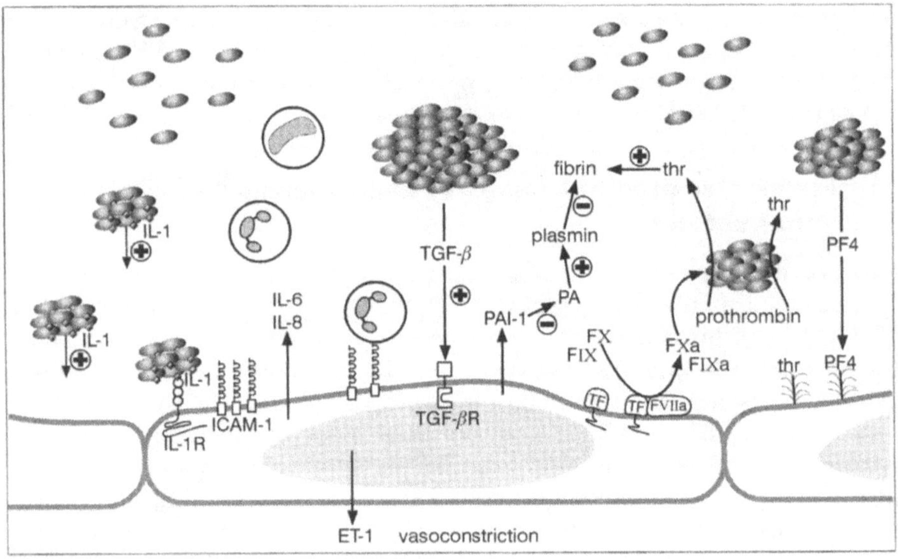


**Fig. 2.** Interactions between platelets and endothelial cells. Activated platelets express interleukin (*IL*)-1 on their surface and release transforming growth factor (*TGF*)-$\beta$ from their $\alpha$-granules, both compounds being able to cause functional reprogramming of endothelial cells. Platelet-bound IL-1 induces endothelial expression of intercellular adhesion molecule (*ICAM*)-1 and production of cytokines such as IL-6 and IL-8, while TGF-$\beta$ increases the synthesis of plasminogen activator inhibitor (*PAI*)-1 and endothelin (*ET*)-1. Platelet-derived cytokines may also lead to an induction of tissue factor. These alterations potentiate inflammatory and thrombotic events by attracting and binding leukocytes, by supporting thrombin formation, and by inhibiting fibrinolysis. Additionally, platelet factor-4 (*PF4*), which binds to heparin-like sulfated glycosaminoglycans, can interfere with thrombin inactivation by antithrombin III. *IL-1R*, IL-1 receptor; *PA*, plasminogen activator; *thr*, thrombin; *TF*, tissue factor

cytokines can mediate rapid and localized signals to the site of injury. Additionally, the cytokines may also amplify inflammatory and thrombotic events, especially when counterregulatory mechanisms (ERDF, $PGI_2$) are inhibited (Fig. 2).

## B. Platelets and Coagulation

Blood platelets are well recognized to contribute to the hemostatic process in a variety of ways. Following vascular injury, platelets adhere to subendothelial structures such as collagen or von Willebrand factor and aggregate to form the primary hemostatic plug. Most importantly, platelet activation also induces the generation of a procoagulant membrane surface on which coagulation factors can assemble and interact. This leads to a dramatic increase in the generation of thrombin, the key enzyme in hemostasis, which will further stimulate platelets and convert fibrinogen into insoluble fibrin, thereby stabilizing the platelet aggregate. The membrane surface of activated platelets also exhibits anticoagulant activity since it is known to catalyze the inactivation of coagulation factors. This represents important negative feed-back control which limits thrombin generation and prevents excessive propagation of the coagulation process. Interestingly, a part of the pro- and anticoagulant platelet activities is associated with microparticles shed from the plasma membrane upon platelet activation. Platelets also interact with the fibrinolytic system both by enhancing plasminogen activation on the platelet surface and by inhibiting its generation by the release of plasminogen activator inhibitor. By regulating pro- and anticoagulant as well as pro- and antifibrinolytic pathways, platelets play a critical role in the maintenance of vascular homeostasis. This section will discuss the role of platelets in coagulation and fibrinolysis and especially address the mechanisms involved in the maintenance of lipid asymmetry and in the formation of procoagulant membrane surfaces. As this review is focused on platelets, a selection of references was made which does not cover all literature on mechanisms in regulation of lipid asymmetry.

## I. Membrane-Dependent Reactions in Blood Coagulation

Blood coagulation at the site of vascular injury is characterized by rapid generation of thrombin, which in turn leads to increased platelet aggregation and formation of an insoluble fibrin network. Once the coagulation process is initiated, it needs both amplification and regulation in a localized area, which is achieved by the cascade arrangement of the coagulation reactions and by the formation of membrane-bound enzyme complexes. In general, four membrane-bound complexes are known: three procoagulant activities (prothrombinase and extrinsic and intrinsic tenase) that produce thrombin in a joint action and one anticoagulant complex (protein Case) that activates protein C, the major anticoagulant enzyme (Mann et al. 1990; Kalafatis et al. 1994a). In each case, a serine protease (either factor VIIa, factor IXa, factor Xa, or $\alpha$-thrombin) binds to a nonenzymatic cofactor protein (either tissue factor, factor VIIIa, factor Va, or thrombomodulin, respectively) in the presence of $Ca^{2+}$ to form an enzymatic complex on the membrane surface. The formation of extrinsic tenase (factor VIIa/tissue factor) and protein Case (thrombin/thrombomodulin) is dependent on the membrane expression of the respective cofactors which are integral membrane proteins. Both complexes function on membranes composed entirely of phosphatidylcholine, but negatively charged phospholipids can increase their activities (Esmon 1993).

In contrast, the intrinsic tenase (factor IXa/factor VIIIa) and the prothrombinase (factor Xa/factor Va) complexes are only formed on surfaces that contain anionic phospholipids (Mann et al. 1990; Kalafatis et al. 1994a), phosphatidylserine being the most important one (Rosing et al. 1988). These complexes catalyze two sequential reactions of the coagulation cascade which finally lead to the formation of thrombin. First, the tenase complex mediates the proteolytic conversion of zymogen factor X into active serine protease factor Xa. Once factor Xa is formed, it becomes assembled with factor Va to form the prothrombinase complex which converts prothrombin to thrombin. The attachment of the factors IX, X, and prothrombin to the lipid surface is accomplished via a $Ca^{2+}$-mediated bridge formation between gamma-carboxyglutamate residues of the proteins and negatively charged polar headgroups of the phospholipid molecules (CUTHFORTH et al. 1989). The binding of cofactors VIIIa and Va is $Ca^{2+}$-independent, mediated by stereoselective recognition of a phosphatidylserine moiety (GILBERT et al. 1993; COMFURIUS et al. 1994), and probably supported by hydrophobic and electrostatic interactions (KALAFATIS et al. 1994b). The stimulatory effect of phospholipids on both tenase- and prothrombinase reactions is due to a dramatic decrease in $K_m$ values of the two substrates, factor X, and prothrombin from far above to far below their respective plasma concentrations. Subsequently, both factor Xa and thrombin generation proceed at near $V_{max}$ conditions (ROSING et al. 1980; VAN DIEIJEN et al. 1981). The combined decrease in the $K_m$ of both reactions can increase the rate of thrombin formation by several orders of magnitude.

In addition to their role in the formation of procoagulant enzyme complexes, membrane surfaces are essential in the inactivation of bound coagulation cofactors. The cleavage of factor VIIIa and factor Va by activated protein C has been shown to be strongly dependent on the presence of anionic phospholipids (KALAFATIS et al. 1993). Protein C activation and the subsequent action of activated protein C represent important negative feedback control of the coagulation response.

## II. Platelet Procoagulant Activity

Activated platelets contribute to coagulation by providing a negatively charged phospholipid membrane for both the intrinsic tenase and the prothrombinase complexes (ZWAAL et al. 1992, 1996; BEVERS et al. 1993). This procoagulant activity is almost exclusively attributed to phosphatidylserine as phosphoinositides, the other anionic phospholipids, have only weak effects in coagulation and, moreover, are rapidly metabolized during stimulus-response coupling. Under resting conditions, phosphatidylserine is virtually absent in the outer leaflet of the platelet plasma membrane, and only little procoagulant activity is observed (CHAP et al. 1977). Upon platelet activation, however, normal membrane phospholipid asymmetry dissipates with subsequent exposure of phosphatidylserine in the outer leaflet of the membrane (BEVERS et al. 1983; DACHARY-PRIGENT et al. 1995). Concomitantly, the procoagulant activity of the platelet membrane increases (COMFURIUS et al. 1985; ROSING et al. 1985a), and binding sites for factor VIIIa and Va are expressed (AHMAD et al. 1989; GILBERT et al. 1991; TRACY et al. 1979; SIMS et al. 1989).

The extent of procoagulant activity has been shown to depend on the platelet-activating stimulus. Calcium ionophore, a mixture of collagen and thrombin, and the complement membrane attack complex C5b-9 are the strongest inducers of platelet procoagulant activity (Comfurius et al. 1985; Rosing et al. 1985a; Sims et al. 1989; Ahmad et al. 1989). Stimulation with collagen alone produces a moderate rise in procoagulant activity, whereas thrombin, though known to be one of the most potent platelet agonists in evoking aggregation and secretion, has a rather weak effect on procoagulant activity. Agonists such as ADP, epinephrine, and PAF have no appreciable procoagulant stimulating effect on platelets. A mild stirring of the platelet suspension is required to evoke procoagulant activity with collagen and thrombin, but is not effective with ionophore or complement. All agonists require the presence of extracellular $Ca^{2+}$ to induce the procoagulant response, strongly suggesting that an increase in intracellular $Ca^{2+}$ is essential for the formation of a procoagulant surface. Compared to platelet suspensions, adherent platelets may require different agonists for the induction of procoagulant activity. In recent studies it has been shown that adherence to von Willebrand factor, which does not promote full platelet activation, is sufficient to induce functional binding sites for the prothrombinase complex (Swords et al. 1993a,b).

The development of a procoagulant surface after platelet activation is not strictly related to other platelet responses like shape change, aggregation, or secretion. The generation of procoagulant activity, for example, reveals a distinct time course and different responsiveness to certain agonists or agonist combinations (Rosing et al. 1985a; Sims et al. 1989). Furthermore, local anesthetics and sulfhydryl reagents which inhibit aggregation and release reaction have been shown to enhance procoagulant activity (Verhallen et al. 1987; Bevers et al. 1989; Comfurius et al. 1990). Additionally, storage pool-deficient and thrombasthenic platelets which do not aggregate have a regular ability to become procoagulant (Bevers et al. 1986). This dissociation of aggregation and secretion may suggest that the induction of procoagulant activity requires higher cytoplasmatic $Ca^{2+}$ concentrations than other platelet responses (Schroit and Zwaal 1991).

It should be noted that platelets support thrombin formation not only by providing the essential negatively charged surface but also by supplying factor Va (see Sect. XI.3). Recently it has been shown that platelets also contain factor X, but the biological importance of this finding remains unknown (Holme et al. 1995).

## III. Mechanisms Involved in the Maintenance of Membrane Lipid Asymmetry

The plasma membrane of platelets and most eukaryotic cells shows differences in the phospholipid composition of the two leaflets of the bilayer (Op den Kamp 1979). While the outer leaflet is rich in choline-containing lipids, i.e., sphingomyelin and phosphatidylcholine, aminophospholipids, in particular phosphatidylserine, are preferentially located in the cytoplasmic site. This membrane lipid asymmetry is at least partially maintained by the action of an aminophospholipid translocase which specifically transports phosphatidylserine and phosphatidylethanolamine from the outer to the inner leaflet of the membrane (SCHROIT and ZWAAL 1991). This translocase activity was first characterized in erythrocytes (SEIGNEURET and DEVAUX 1984). It has been shown to be dependent on ATP and reduced sulfhydryls, and to be inhibited by elevated intracellular $Ca^{2+}$ concentrations (BITBOL et al. 1987). The observation of a stereospecific transport confirmed the protein nature of the transporter system (MARTIN and PAGANO 1987), and presently two candidates, a 32-kDa protein related to or associated with Rhesus antigen (SCHROIT et al. 1990) and the erythrocyte Mg-ATPase (MORROT et al. 1990), are suggested as being responsible for translocase activity. A concerted action of both proteins in the control of lipid asymmetry has also been proposed (SCHROIT and ZWAAL 1991).

Evidence of the presence of a specific aminophospholipid translocase was also obtained for platelets (BEVERS et al. 1989; COMFURIUS et al. 1990; TILLY et al. 1990). Thrombin stimulation of platelets has been shown to induce a four- to fivefold increase in translocase activity, which led to the proposal that this activity is important for correcting the loss of lipid asymmetry that may occur as a result of membrane fusion during the secretory event. The intracellular $Ca^{2+}$ concentration upon activation by thrombin was evidently below the level required for translocase inhibition. When platelets are stimulated with calcium ionophore, however, translocase activity is inhibited by increased levels of intracellular $Ca^{2+}$. This inhibition of translocase by $Ca^{2+}$ has been shown to be reversible as removal of intracellular $Ca^{2+}$ by EGTA produces a decrease in procoagulant activity, most likely because surface-exposed phosphatidylserine is pumped back to the inner leaflet of the membrane (BEVERS et al. 1989; COMFURIUS et al. 1990). It should be emphasized that loss of lipid asymmetry and exposure of phosphatidylserine are not merely results of inhibition of translocase activity (SCHROIT and ZWAAL 1991), but are caused by activation of a distinct mechanism, as will be described below.

In addition to the aminophospholipid translocase, a counter transport of lipids from the inner to the outer leaflet has been observed (BITBOL and DEVAUX 1988; CONNOR et al. 1992). This activity, referred to as “flopase”, was found to also be ATP dependent and inhibitable by sulfhydryl-reactive agents. However, no specificity with respect to the polar headgroup of the lipids was observed. Whether or not this outward-directed transport reflects an intrinsic property of the aminophospholipid translocase remains to be investigated.

Finally, it should be mentioned that, originally, lipid asymmetry was ascribed to specific interactions between phosphatidylserine and cytoskeletal proteins (HAEST et al. 1982). However, as reviewed by WILLIAMSON and SCHLEGEL (1994), these interactions are presently not considered to play a major role in maintenance of lipid asymmetry.

## IV. Mechanisms Involved in the Expression of Procoagulant Activity

Platelet activation can lead to a progressive loss of phospholipid asymmetry, producing increased exposure of phosphatidylserine in the outer leaflet of the membrane. This process is due to the induction of rapid bidirectional transbilayer movement (flip-flop) of all major phospholipid classes, which leads to randomization of the lipids over the bilayer membrane (BEVERS et al. 1983; SMEETS et al. 1994). The concomitant inhibition of aminophospholipid translocase activity prevents exposed phosphatidylserine from being pumped back to the inner leaflet, thus contributing to formation of a procoagulant surface. The mechanisms responsible for lipid scrambling are still unresolved, but the key event certainly is an influx of $Ca^{2+}$ ions into the cytoplasm (COMFURIUS et al. 1990). In most cases, generation of a procoagulant surface is accompanied by activation of a $Ca^{2+}$-dependent protease, calpain, with subsequent hydrolysis of cytoskeletal proteins (FOX et al. 1990, 1991; VERHALLEN et al. 1987) and shedding of small membrane vesicles from the plasma membrane (SIMS et al. 1989). This led to the proposal that either cytoskeletal degradation and/or membrane vesiculation might constitute a normal cellular mechanism for lipid scrambling. Recent evidence, however, indicates that exposure of phosphatidylserine in platelets can occur in the absence of cytoskeletal degradation and microvesicle formation (DACHARY-PRIGENT et al. 1995). Moreover, in Scott syndrome (see Sect. VI.1) calpain-mediated hydrolysis of the cytoskeletal proteins is normal, but microvesicle formation and procoagulant response are impaired (COMFURIUS et al. 1990; WIEDMER et al. 1990; BEVERS et al. 1992). Considering these data, it appears most likely that microvesicle formation requires both calpain activation and lipid scrambling. The contribution of the latter process to microvesiculation might be partially explained by the unequal rates of scrambling observed for the various lipid classes. The rate of inward transbilayer migration of sphingomyelin was found to be appreciably lower than the migration rates of the other phospholipids (WILLIAMSON et al. 1992; SMEETS et al. 1994). This may – at least transiently – cause a mass imbalance between both leaflets that results in exfoliation of the membrane and subsequent shedding of microvesicles.

Another proposed mechanism for the induction of lipid scrambling involves the formation of a membrane-perturbing complex between $Ca^{2+}$ and the metabolic resistant pool of phosphatidylinositol-4,5-bisphophate ($PIP_2$; SULPICE et al. 1994). This hypothesis, however, could not be confirmed in studies with Scott erythrocytes and with $PIP_2$-loaded lipid vesicles (BEVERS et al. 1995).

Recent studies suggested the involvement of a putative integral membrane protein, a phospholipid scramblase, in the generation of procoagulant activity (WILLIAMSON et al. 1995; ZWAAL et al. 1996). In contrast to the aminophospholipid translocase, the scramblase would be characterized by its bidirectionality, energy independence, and opposite response to the elevation

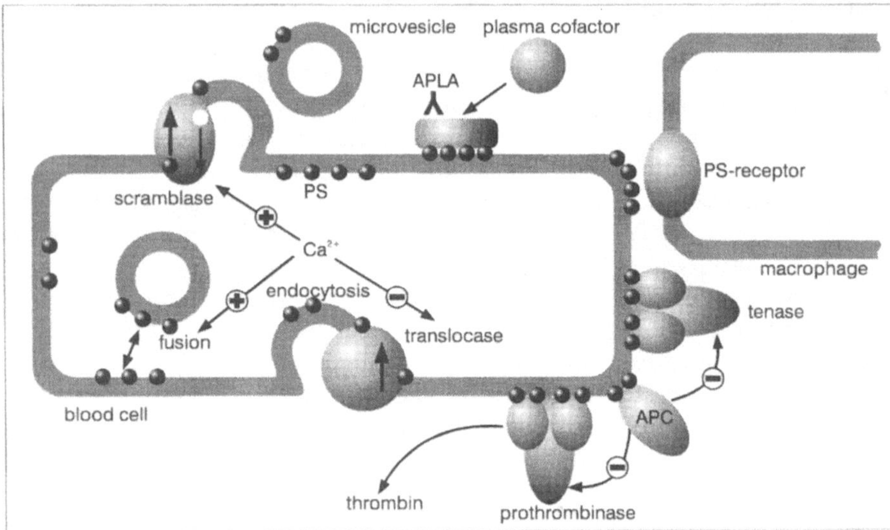


**Fig. 3.** Cellular phenomena associated with phosphatidylserine topology in blood cell membranes. In unactivated cells, membrane phospholipid asymmetry is generated by an ATP-dependent lipid pump (translocase) that specifically shuttles aminophospholipids (e.g., phosphatidylserine, *PS*) from the outer to the inner leaflet of the lipid bilayer membrane. When PS is provided from an external source, translocase activity produces mass imbalance between the two leaflets of the lipid bilayer possibly facilitating endocytosis. Influx of calcium can inhibit translocase and activates a non-ATP-dependent lipid transporter (scramblase) that produces rapid lipid flip-flop, leading to loss of membrane phospholipid asymmetry, which results in surface exposure of PS. Mass imbalance between the two membrane leaflets resulting from scramblase activity may facilitate formation of microvesicles that also expose PS. Moreover, calcium influx produces fusion of storage granules with the plasma membrane followed by exocytosis. Surface-exposed PS promotes assembly and catalysis of tenase and prothrombinase complexes, leading to dramatic acceleration of thrombin formation, while this surface also exerts negative feedback by promoting protein C (*APC*)-catalyzed inactivation of factors Va and VIIIa. In the antiphospholipid syndrome, certain plasma proteins (e.g., prothrombin, $\beta_2$-glycoprotein I) may interact with PS, leading to the exposure of cryptic epitopes towards which antiphospholipid antibodies (*APLA*) are directed. In addition, surface-exposed PS may trigger recognition by PS receptors on macrophages, leading to removal of the blood cell from the circulation. (Reprinted with permission from ZWAAL et al. 1996, copyright by CRC press, Boca Raton, Florida, 1996)

of intracellular $Ca^{2+}$ as it is switched on after interaction with $Ca^{2+}$. Initial evidence of the involvement of a protein in the randomization of lipids was obtained by showing that this process is sensitive to the sulfhydryl reagent pyridyldithioethylamine (WILLIAMSON et al. 1995). The protein nature of scramblase is further supported by the existence of an inherited bleeding disorder, Scott syndrome, characterized by impairment of scramblase activity (see Sect. VI.1). Recently it has been demonstrated that a protein fraction isolated from platelet membranes and incorporated in artificial lipid vesicles is able to enhance transbilayer movement of lipids in a $Ca^{2+}$-dependent manner (COMFURIUS et al. 1996). The actual mechanism underlying the scrambling activity and the nature of the protein or proteins that support it, however, remains to be defined (Fig. 3).

## V. Platelet Microvesicles

Platelet-derived microparticles are unilamellar vesicles with an average diameter of about 0.2 $\mu$m (ZWAAL et al. 1992). Originally they were described as "platelet dust" (WOLF 1967) possibly derived from intracellular membranes and to be externalized upon secretion (SANDBERG et al. 1985). Now it is evident that microparticles are shed from the plasma membrane of a variety of cells, including platelets. As a result, they contain the major membrane glycoproteins Ib, IIb, and IIIa as well as the membrane cytoskeletal proteins filamin, talin, and myosin heavy chain (SIMS et al. 1988, 1989; FOX et al. 1991). Moreover, coagulation factor XIII, which is located in the cytoplasm, has also been found in microvesicles (HOLME et al. 1993).

Platelet microparticles contain about 25%–30% of the procoagulant activity and factor Va binding sites formed upon cell activation. An exception is observed with platelets activated by complement C5b-9, where most of the factor Va binding sites are associated with the microvesicles (SIMS et al. 1988). Platelet microparticles have also been shown to expose high-affinity receptors for factor VIII (GILBERT et al. 1991) and to express factor Xa activity (HOLME et al. 1995).

Shedding of microvesicles may provide a mechanism for effectively increasing the accessibility of procoagulant activity around a platelet aggregate. It should be noted, however, that platelet-derived microparticles do not initiate, but only accelerate thrombin formation once coagulation has started. Accordingly, the infusion of procoagulant phospholipids in rabbits was found to be potentially thrombogenic, but only in combination with a subthreshold dose of factor Xa (GILES et al. 1982). Little information exists on the clinical significance of platelet-derived microparticles in human disorders. An enhanced occurrence of microvesicles was described in patients with autoimmune thrombocytopenia (JY et al. 1992) and in patients undergoing cardiopulmonary bypass (ABRAMS et al. 1990). A recent study reported that platelet-derived microvesicles circulated in vivo in blood from patients with activated coagulation and fibrinolysis, whereas microvesicles were rarely found in healthy individuals (HOLME et al. 1994). These observations may suggest that platelet-derived microparticles are associated with thrombotic episodes and/or reflect a hypercoagulable state, but prospective studies are needed to prove this assumption.

## VI. Platelets and Coagulation Disorders

### 1. Scott Syndrome

Scott syndrome is a rare bleeding disorder associated with a deficiency in the expression of platelet procoagulant activity while platelet aggregation and secretion responses are unaffected (WEISS et al. 1979). Compared to control cells, activated Scott platelets have a decreased ability to promote factor Xa and thrombin formation (ROSING et al. 1985b) as well as a reduced expression of membrane binding sites for factor Va and VIIIa (SIMS et al. 1989; AHMAD et al. 1989). Interestingly, platelets from Scott syndrome patients are also markedly impaired in their capacity to generate microparticles in response to all platelet activators (SIMS et al. 1989). The membrane phenomena underlying the Scott syndrome are not unique to platelets, but are also present in erythrocytes and lymphocytes (BEVERS et al. 1992; TOTI et al. 1996; KOJIMA et al. 1994) from these patients, suggesting a mutation in an early stem cell. Upon stimulation with a $Ca^{2+}$ ionophore, normal erythrocytes change shape from biconcave disks to echinocytic spheres, express procoagulant activity, and generate microvesicles. In contrast, Scott erythrocytes remain mostly discoid and show diminished expression of procoagulant surface and reduced microvesiculation (BEVERS et al. 1992). An impaired development of procoagulant activity was even seen in resealed ghosts from Scott erythrocytes, implying that the lesion resided in the plasma membrane (BEVERS et al. 1992). Scott platelets and erythrocytes are interesting models for studying mechanisms of lipid scrambling (see Sects. IV and V).

### 2. Antiphospholipid Syndrome

The antiphospholipid syndrome is characterized by the association of antiphospholipid antibodies with an increased risk of thrombosis, thrombocytopenia, or recurrent abortions (ROUBEY 1994; CHONG et al. 1995). The syndrome often develops in patients with systemic lupus erythematosus or related autoimmune disorders. Recently, evidence has been presented that antiphospholipid antibodies (for example, anticardiolipin and lupus anticoagulant immunoglobulins) recognize cryptic or neoepitopes exposed by lipid-bound proteins, i.e., $\beta_2$-glycoprotein I, prothrombin, protein C, and protein S (GALLI et al. 1990; BEVERS et al. 1991; OOSTING et al. 1993). Activated platelets and platelet-derived microvesicles as well as damaged vascular cells may participate in the generation of antiphospholipid antibodies by providing an anionic phospholipid surface with which certain plasma proteins interact in association with exposure of neoepitopes. The formation of antiphospholipid antibodies against lipid-bound proteins may then represent a normal response of the immune system towards prolonged exposure of thrombogenic surfaces rather than reflecting a disorder of the immune system with formation of thrombogenic antibodies. This would explain the apparent controversy that thrombotic complications in vivo can be associated with circulation of antibodies prolonging clotting time in vitro (ROUBEY 1994).

### 3. Factor V Quebec

Factor V Quebec describes a severe bleeding disorder characterized by an almost complete absence of functional platelet factor V while plasma factor V concentration is normal (TRACY et al. 1984). Platelet factor V is stored in platelet $\alpha$-granules and comprises about 20% of the circulating factor V (TRACY et al. 1982). Upon platelet activation, it can be released and bound to the platelet surface. The importance of platelet factor V may be attributed to its high local concentration within a platelet aggregate. Moreover, platelet factor V is stored as a partially proteolysed molecule already exhibiting significant cofactor activity which can be slightly increased by either factor Xa or thrombin (MONKOVIC and TRACY 1990). In contrast, plasma factor V circulates in its inactive form and requires full proteolytic activation. The fact that an isolated deficiency of platelet factor V is associated with a bleeding disorder points to its significant role in rapid assembly of the prothrombinase complex although platelet factor V represents the minor part of the factor V pool.

## VII. Anticoagulant Activities of Platelets and Microvesicles

Thrombin, the key enzyme of coagulation, also initiates a major anticoagulant pathway. It binds to endothelial thrombomodulin, thereby changing its catalytic properties, and is then capable of activating protein C. Activated protein C in turn inactivates the cofactors of both the tenase and the prothrombinase complex (factor VIIIa and factor Va, respectively) (ESMON 1993), a process which is markedly enhanced in the presence of anionic phospholipids (KALAFATIS et al. 1993) and protein S (WALKER 1981). While protein S, a nonenzymatic protein C cofactor, was originally suggested as enhancing the binding of activated protein C to phospholipids, it is now thought to abolish the protective action of factor Xa on factor Va inactivation (SOLYMOSS et al. 1988).

Platelets and platelet-derived microparticles are known to support the protein C pathway by providing a negatively charged surface for factor Va and VIIIa inactivation. Activated protein C has been shown to bind specifically to the platelet surface, increasing its catalytic efficiency by several orders of magnitudes (HARRIS and ESMON 1985). Accordingly, several studies reported that platelets stimulate the inactivation of factor Va by activated protein C and that platelet activation markedly accelerates this process (DAHLBÄCK et al. 1980; HARRIS and ESMON 1985; TANS et al. 1991).

A small percentage of plasma protein S is stored in platelets and can be released upon platelet activation (SCHWARZ et al. 1985). Platelet-derived protein S has been shown to be involved in factor Va inactivation on the platelet surface (HARRIS and ESMON 1985). This result, however, was not confirmed in a study by TANS et al. (1991) in which it was demonstrated that only exogenously added protein S leads to an approximately twofold increase of platelet-dependent factor Va inactivation (TANS et al. 1991).

The ability of platelets to support protein C-catalyzed inactivation of factor Va and factor VIIIa (e.g., the anticoagulant activity) and the development of procoagulant activity share many properties. Both reactions increase when platelets are activated, and the effectiveness of the various platelet agonists in generating both activities is comparable (most pronounced responses after ionophore, moderate after collagen, and low after thrombin, ADP, or epinephrine). In addition, it has been shown that microparticles shed from activated platelets possess about 25% of anticoagulant activity, which is virtually identical to that proportion observed for procoagulant activity (TANS et al. 1991). These data underscore the importance of anionic phospholipids, in particular phosphatidylserine in both pro- and anticoagulant platelet responses. Moreover, they illustrate the efficiency of the negative feedback loop exerted by activated protein C, which restricts thrombin formation at the same membrane surface on which it is generated.

## VIII. Role of Platelets in Fibrinolysis

Coagulation processes are counteracted by fibrinolysis, and the regulated interaction of coagulation and fibrinolytic systems is needed to maintain vascular homeostasis. Fibrin clots or fibrin-containing thrombi produced by thrombin can be solubilized by the serine protease plasmin which is generated from its inactive precursor plasminogen by the action of both tissue-type and urokinase-type plasminogen activators (t-PA and u-PA, respectively). Plasminogen activation is tightly controlled by PAI-1 and regulated by either fibrin surfaces or cellular surfaces where the components of the fibrinolytic system can bind and interact (PLOW et al. 1995).

Platelets have been shown to provide a cellular surface for the assembly of the plasminogen activator system (LOSCALZO 1995). Several studies have reported that plasminogen binds to resting platelets and that the binding capacity is upregulated upon platelet stimulation (MILES et al. 1986; ADELMAN et al. 1988). Glycoprotein IIb/IIIa has been shown to function as the platelet plasminogen receptor, and additional binding may occur via surface expression of platelet fibrin (MILES et al. 1986). The platelet surface also possesses a large number of specific, low-affinity binding sites for t-PA (VAUGHAN et al. 1989). Upon plasminogen and t-PA binding to platelets, plasminogen activation by t-PA is significantly enhanced and is further increased by platelet stimulation (GAO et al. 1990; OUIMET et al. 1994). Interestingly, the generated plasmin binds to platelets and augments its own generation, most likely by increasing plasminogen turnover (OUIMET et al. 1994). Surface-bound plasmin is protected from inhibition by $\alpha_2$-antiplasmin and $\alpha_2$-macroglobulin (HALL et al. 1991) and possesses increased catalytic efficiency towards its substrate (OUIMET et al. 1994).

While platelets influence fibrinolytic activity, a modulatory effect of the fibrinolytic system on platelet function is also described. Once plasmin is generated and bound to the platelet surface, it can stimulate or inhibit platelet function, with the observed effects depending on plasmin concentration, incubation time, and incubation medium (i.e., physiological buffer or plasma). In a system with washed or gel-filtered platelets, plasmin at relatively high concentrations (approximately 1 caseinolytic unit/ml) was able to elicit platelet activation through the phospholipase C pathway (SCHAFER et al. 1986). In a plasma system and at lower concentrations, plasmin led to distinct cleavage of glycoprotein IIIa, which reduced fibrinogen binding to platelets and subsequently inhibited aggregation (PASCHE et al. 1994).

The profibrinolytic effects of platelets can be counteracted by PAI-1 and by $\alpha_2$-antiplasmin, which are both secreted from activated platelets (ERICKSON et al. 1984; PLOW et al. 1981). Especially platelet-derived PAI-1 stored in $\alpha$-granules and constituting the main reservoir of PAI-1 in the blood is thought to be a relevant regulatory protein of the fibrinolytic system. Activated platelets have been shown to inhibit t-PA-mediated fibrinolysis, which was attributed to PAI-1 since antibodies against PAI-1 abolished the platelet effect (KEIJER et al. 1991) and PAI-1 deficient platelets showed only moderate antifibrinolytic activity (FAY et al. 1994). In an experimental model mimicking the formation of arterial thrombi in vivo, platelet-derived PAI-1 was retained within the thrombus. Thrombolysis resistance correlated with the local concentration of PAI-1 and was largely abolished by the inclusion of PAI-1 antibodies into the thrombi (STRINGER et al. 1994). Interestingly, it has been shown that the released PAI-1 can bind to fibrin fibers in the vicinity of platelet remnants (BRAATEN et al. 1993) and that fibrin-bound PAI-1 is fully able to form inactive complexes with t-PA (KEIJER et al. 1991).

Apart from platelet PAI-1, other mechanisms may add to the antifibrinolytic activity of platelets. Cross-linking of fibrin by platelet-derived factor XIII, increased density of fibrin in the vicinity of platelets, and platelet-mediated clot retraction, all of which render the fibrin clot more resistant to digestion by plasmin, are thought to be further fibrinolysis-limiting factors (SABOVIC et al. 1989, BRAATEN et al. 1994; KUNITADA et al. 1992; FAY et al. 1994). Taken together, these data show that platelets may have profibrinolytic as well as antifibrinolytic effects. Which effect predominates – the kinetic enhancement of plasminogen activation or its inhibition – may be determined by the state of platelet activation and the availability and affinity of the platelet surface for the components of the fibrinolytic system.

## References

Abrams CS, Ellison N, Budzynski AZ, Shattil SJ (1990) Direct detection of activated platelets and platelet-derived microparticles in humans. Blood 75:128–138

Adelman B, Rizk A, Hanners E (1988) Plasminogen interactions with platelets in plasma. Blood 72:1530–1535

Ahmad SS, Rawala Sheikh R, Ashby B, Walsh PN (1989) Platelet receptor-mediated factor X activation by factor IXa. High affinity IXa receptors induced by factor VIII are deficient on platelets in Scott syndrome. J Clin Invest 84:824–828

Assoian RK, Sporn MB (1986) Type beta transforming growth factor in human platelets: release during platelet degranulation and action on vascular smooth muscle cells. J Cell Biol 102:1217–1233

Aursnes, I (1974) Increased permeability of capillaries to protein during thrombocytopenia: an experimental study in the rabbit. Microvasc Res 7:283–295

Aznar-Salatti J, Bastida E, Haas TA, Escolar G, Ordinas A, de Groot PHG,

Buchanan MR (1991) Platelet adhesion to exposed endothelial cell extracellular matrixes is influenced by the method of preparation. Arterioscler Thromb 11:436–442
Bertomeu MC, Gallo S, Lauri D, Levine MN, Orr FW, Buchanan MR (1990) Chemotherapy enhances endothelial cell reactivity to platelets. Clin Exp Metastasis 8:511–518
Bevers EM, Comfurius P, Zwaal RFA (1983) Changes in membrane phospholipid distribution during platelet activation. Biochim Biophys Acta 736:57–66
Bevers EM, Comfurius P, Nieuwenhuis HK, Levy-Toledano S, Enouf J, Belluci S, Caen JP, Zwaal RFA (1986) Platelet prothrombin converting activity in hereditary disorders of platelet function. Br J Haematol 63:335–345
Bevers EM, Tilly RHJ, Senden JMG, Comfurius P, Zwaal RFA (1989) Exposure of endogenous phosphatidylserine at the outer surface of stimulated platelets is reversed by restoration of aminophospholipid translocase activity. Biochemistry 28:2382–2387
Bevers EM, Galli M, Barbui T, Comfurius P, Zwaal RFA (1991) Lupus anticoagulant IgG's (LA) are not directed to phospholipids only, but to a complex of lipid-bound human prothrombin. Thromb Haemost 66:629–632
Bevers EM, Wiedmer T, Comfurius P, Shattil SJ, Weiss HJ, Zwaal RFA, Sims PJ (1992) Defective $Ca^{2+}$-induced microvesiculation and deficient expression of procoagulant activity in erythrocytes from a patient with a bleeding disorder: a study of the red blood cells of Scott syndrome. Blood 79:380–388
Bevers EM, Comfurius P, Zwaal RFA (1993) Mechanisms involved in platelet procoagulant response. In: Authi KS, Watson SP, Kakkar VV (eds) Mechanisms of platelet activation and control. Plenum, New York, p 195 (Advances in experimental medicine and biology, vol 344)
Bevers EM, Wiedmer T, Comfurius P, Zhao J, Smeets EF, Schlegel RA, Schroit AJ, Weiss HJ, Williamson P, Zwaal RFA, Sims PJ (1995) The complex of phosphatidylinositol 4,5-bisphosphate and calcium ions is not responsible for $Ca^{2+}$-induced loss of phospholipid asymmetry in the human erythrocyte: A study in Scott syndrome, a disorder of calcium-induced phospholipid scrambling. Blood 86:1983–1991
Bitbol M, Fellmann P, Zachowski A, Devaux PF (1987) Ion regulation of phosphatidylserine and phosphatidylethanolamine outside-inside translocation in human erythrocytes. Biochim Biophys Acta 904:268–282
Bitbol M, Devaux PF (1988) Measurement of outward translocation of phospholipids across human erythrocyte membrane. Proc Natl Acad Sci USA 85:6783–6787
Braaten JV, Handt S, Jerome WG, Kirkpatrick J, Lewis JC, Hantgan RR (1993) Regulation of fibrinolysis by platelet-released plasminogen activator inhibitor 1: light scattering and ultrastructural examination of lysis of a model platelet-fibrin thrombus. Blood 81:1290–1299
Braaten JV, Jerome WG, Hatgan RR (1994) Uncoupling fibrin from integrin receptors hastens fibrinolysis at the platelet-fibrin interface. Blood 83:982–993
Buchanan MR, Bastida E, Escolar G, Haas TA, Weber E (1989) Vessel wall reactivity and 13-hydroxy-octadecadienoic acid synthesis. In: Samuelsson B, Wong PY-K, Sun FF (eds) Advances in prostaglandin, thromboxane and leukotriene research, vol 19. Raven, New York, p 293
Buchanan MR, Bertomeu MC, Haas TA, William Orr F, Eltringham-Smith LL (1993) Localization of 13-hydroxyoctadecadienoic acid and the vitronectin receptor in human endothelial cells and endothelial cell/platelet interactions in vitro. Blood 81:3303–3312
Butt E, Abel K, Krieger M, Palm D, Hoppe V, Hoppe J, Walter U (1994) cAMP- and cGMP-dependent protein kinase phosphorylation sites of the focal adhesion vasodilator-stimulated phosphoprotein (VASP) in vitro and in intact human platelets. J Biol Chem 269:14509–14517
Cerletti C, Evangelista V, Molino M, de Gaetano G (1995) Platelet activation by

polymorphonuclear leukocytes: role of cathepsin G and P-selectin. Thromb Haemost 74:218–223
Chap HJ, Zwaal RFA, van Deenen LLM (1977) Action of highly purified phospholipases on blood platelets. Evidence for an asymmetric distribution of phospholipids in the surface membrane. Biochim Biophys Acta 467:146–164
Chong BH, Brighton TC, Chesterman CN (1995) Antiphospholipid antibodies and platelets. Semin Thromb Hemost 21:76–84
Comfurius P, Bevers EM, Zwaal RFA (1985) The involvement of cytoskeleton in the regulation of transbilayer movement of phospholipids in human platelets. Biochim Biophys Acta 815:143–148
Comfurius P, Senden JMG, Tilly RHJ, Schroit AJ, Bevers EM, Zwaal RFA (1990) Loss of membrane phospholipid asymmetry in platelets and red cells may be associated with calcium-induced shedding of plasma membrane and inhibition of aminophospholipid translocase. Biochim Biophys Acta 1026:153–160
Comfurius P, Smeets EF, Willems GM, Bevers EM, Zwaal RFA (1994) Assembly of the prothrombinase complex on lipid vesicles depends on the stereochemical configuration of the polar headgroup of phosphatidylserine. Biochemistry 33:10319–10324
Comfurius P, Williamson P, Smeets EF, Schlegel RA, Bevers EM, Zwaal RFA (1996) Reconstitution of phospholipid scramblase activity from human blood platelets. Biochemistry 35:7631–7634
Connor J, Pak CH, Zwaal RF, Schroit AJ (1992) Bidirectional transbilayer movement of phospholipid analogs in human red blood cells. Evidence for an ATP-dependent and protein-mediated process. J Biol Chem 267:19412–19417
Coté YP, Filep JG, Battistini B, Gauvreau J, Sirois P, Beaudoin AR (1992) Characterization of ATP-diphosphohydrolase activities in the intima and media of the bovine aorta: evidence for a regulatory role in platelet activation in vitro. Biochim Biophys Acta 1139:133–142
Coughlin SR, Moskowitz MA, Zetter BR, Antonides HN, Levine L (1980) Platelet-dependent stimulation of prostacyclin synthesis by platelet-derived growth factor. Nature 288:600–604
Coughlin SR (1994) Molecular mechanisms of thrombin signaling. Sem Hematol 31:270–277
Curwen KD, Gimbrone MA, Handin RI (1980) In vitro studies of thromboresistance. The role of prostacyclin ($PGI_2$) in platelet adhesion to cultured normal and virally transformed human endothelial cells. Lab Invest 42:366–374
Cutsforth GA, Whitaker RN, Hermans J, Lentz BR (1989) A new model to describe extrinsic protein binding to phospholipid membranes of varying composition: application to human coagulation proteins. Biochemistry 28:7453–7461
D'Amore P, Shepro D (1977) Stimulation of growth and calcium influx in cultured, bovine, aortic endothelial cells by platelets and vasoactive substances. J Cell Physiol 92:177–183
D'Humieres S, Russo-Marie F, Vargaftig BB (1986) PAF-acether-induced synthesis of prostacyclin by human endothelial cells. Eur J Pharmacol 131:13–19
Dachary-Prigent J, Pasquet JM, Freyssinet JM, Nurden AT (1995) Calcium involvement in aminophospholipid exposure and microparticle formation during platelet activation: a study using $Ca^{++}$-ATPase inhibitors. Biochemistry 34:11625–11634
Dahlbäck B, Stenflo J (1980) Inhibitory effect of activated protein C on activation of prothrombin by platelet-bound factor Xa. Eur J Biochem 107:331–335
Darius H, Binz C, Veit K, Fisch A, Meyer J (1995) Platelet receptor desensitization induced by elevated prostacyclin levels causes platelet-endothelial cell adhesion. J Am Coll Cardiol 26:800–806
Davenpeck KL, Gauthier TW, Lefer AM (1994) Inhibition of endothelial-derived nitric oxide promotes P-selectin expression and actions in the rat microcirculation. Gastroenterology 107:1050–1058

De Nucci G, Gryglewski RJ, Warner TD, Vane JR (1988) Receptor-mediated release of endothelium-derived relaxing factor and prostacyclin from bovine aortic endothelial cells is coupled. Proc Natl Acad Sci USA 85:2334–2338
Dejana E, Corada M, Lampugnani MG (1995) Endothelial cell-to-cell junctions. FASEB J 9:910–918
Draijer R, Atsma DE, van der Laarse A, van Hinsbergh VWM (1995a) cGMP and nitric oxide modulate thrombin-induced endothelial permeability. Regulation via different pathways in human aortic and umbilical vein endothelial cells. Circ Res 76:199–208
Draijer R, Vaandrager AB, Nolte C, de Jonge HR, Walter U, van Hinsbergh VWM (1995b) Expression of cGMP-dependent protein kinase I and phosphorylation of its substrate, vasodilator-stimulated phosphoprotein, in human endothelial cells of different origin. Circ Res 77:897–905
Durante W, Kroll MH, Vanhoutte PM, Schafer AI (1992) Endothelium-derived relaxing factor inhibits thrombin-induced platelet aggregation by inhibiting platelet phospholipase C. Blood 79:110–116
Eigenthaler M, Nolte C, Halbrügge M, Walter U (1992) Concentration and regulation of cyclic nucleotides, cyclic-nucleotide-dependent protein kinases and one of their major substrates in human platelets. Eur J Biochem 205:471–481
Eigenthaler M, Ullrich H, Geiger J, Horstrup K, Hönig-Liedl P, Wiebecke D, Walter U (1993) Defective nitrovasodilator-stimulated protein phosphorylation and calcium regulation in cGMP-dependent protein kinase-deficient human platelets of chronic myelocytic leukemia. J Biol Chem 268:13526–13531
Erickson LA, Ginsberg MH, Loskutoff DJ (1984) Detection and partial characterisation of an inhibitor of plasminogen activator in human platelets. J Clin Invest 74:1465–1472
Esmon CT (1993) Cell mediated events that control blood coagulation and vascular injury. Annu Rev Cell Biol 9:1–26
Esmon CT (1995) Thrombomodulin as a model of molecular mechanisms that modulate protease specificity and function at the vessel surface. FASEB J 9:946–955
Esmon NL, Carroll RC, Esmon CT (1983) Thrombomodulin blocks the ability of thrombin to activate platelets. J Biol Chem 258:12238–12242
Fay WP, Eitzman DT, Shapiro AD, Madison EL, Ginsburg D (1994) Platelets inhibit fibrinolysis in vitro by both plasminogen activator inhibitor-1-dependent and -independent mechanisms. Blood 83:351–356
Förstermann U, Mügge A, Bode SM, Frölich JC (1988) Response of human coronary arteries to aggregating platelets: Importance of endothelium-derived relaxing factor and prostanoids. Circ Res 63:306–312
Fox JEB, Austin CD, Reynolds CC, Steffen PK (1991) Evidence that agonist-induced activation of calpain causes the shedding of procoagulant-containing microvesicles from the membrane of aggregating platelets. J Biol Chem 266:13289–13295
Fox JEB, Reynolds CC, Austin CD (1990) The role of calpain in stimulus-response coupling: evidence that calpain mediates agonist-induced expression of procoagulant activity in platelets. Blood 76:2510–2519
Frenette PS, Johnson RC, Hynes RO, Wagner DD (1995) Platelets roll on stimulated endothelium in vivo: an interaction mediated by endothelial P-selectin. Proc Natl Acad Sci USA 92:7450–7454
Fry GL, Czervionke RL, Hoak JC, Smith JB, Haycraft DL (1980) Platelet adherence to cultured vascular cells: influence of prostacyclin ($PGI_2$). Blood 55:271–275
Furie B, Furie BC (1995) The molecular basis of platelet and endothelial cell interaction with neutrophils and monocytes: role of P-selectin and the P-selectin ligand, PSGL-1. Thromb Haemost 74:224–227
Galli M, Comfurius P, Maassen C, Hemker HC, De Baets MH, van Breda-Vriesman PJC, Barbui T, Zwaal RFA, Bevers EM (1990) Anticardiolipin antibodies (ACA) directed not to cardiolipin but to a plasma protein cofactor. Lancet 335:1544–1547

Gao S-W, Morser J, McLean K, Shuman MA (1990) Differential effect of platelets on plasminogen activation by tissue plasminogen activator, urokinase, and streptokinase. Thromb Res 58:421–433
Gawaz M, Neumann F-J, Ott I, Schiessler A, Schömig A (1996) Platelet function in acute myocardial infarction treated with direct angioplasty. Circulation 93:229–237
Gaydos LA, Freireich EJ, Mantel N (1962) The quantitative relationship between platelet count and hemorrhage in patients with acute leukemia. New Engl J Med 266:905–909
Geiger J, Nolte C, Butt E, Sage SO, Walter U (1992) Role of cGMP and cGMP-dependent protein kinase in nitrovasodilator inhibition of agonist-evoked calcium elevation in human platelets. Proc Natl Acad Sci USA 89:1031–1035
Geiger J, Nolte C, Walter U (1994) Regulation of calcium mobilization and entry in human platelets by endothelium-derived factors. Am J Physiol 267:C236–C244
Gilbert GE, Sims PJ, Wiedmer T, Furie B, Furie BC, Shattil SJ (1991) Platelet-derived microparticles express high-affinity receptors for factor VIII. J Biol Chem 266:17261–17268
Gilbert GE, Drinkwater D (1993) Specific membrane binding of factor VIII is mediated by *O*-phospho-L-serine, a moiety of phosphatidylserine. Biochemistry 32:9577–9585
Giles AL, Nesheim ME, Hoogendoorn H, Tracy PB, Mann KG (1982) Stroma-free human platelet lysates potentiate the in vivo thrombogenicity of factor Xa by the provision of coagulant-active phospholipid. Br J Haematol 51:457–468
Gimbrone MA, Aster RH, Cotran RS, Corkery J, Jandl JH, Folkman J (1969) Preservation of vascular integrity in organs perfused in vitro with a platelet-rich medium. Nature 222:33–36
Goldschmidt-Clermont PJ, Kim JW, Machesky LM, Rhee SG, Pollard TD (1991) Regulation of phospholipase C-$\tau$1 by profilin and tyrosine phosphorylation. Science 251:1231–1233
Golino P, Piscione F, Willerson JT, Cappelli-Bigazzi M, Focaccio A, Villari B, Indolfi C, Russolillo E, Condorelli M, Chiarello M (1991) Divergent effects of serotonin on coronary-artery dimensions and blood flow in patients with coronary atherosclerosis and control patients. N Engl J Med 324:641–648
Gryglewski RJ, Botting RM, Vane JR (1988) Mediators produced by the endothelial cell. Hypertension 12:530–548
Haas TA, Bertomeu M-C, Bastida E, Buchanan MR (1990) Cyclic AMP regulation of endothelial cell triacylglycerol turnover, 13-hydroxyoctadecadienoic acid (13-HODE) synthesis and endothelial cell thrombogenecity. Biochim Biophys Acta 1051:174–178
Haest CWM (1982) Interactions between membrane skeleton proteins and the intrinsic domain of the erythrocyte membrane. Biochim Biophys Acta 694:331–352
Haffner C, Jarchau T, Reinhard M, Hoppe J, Lohmann SM, Walter U (1995) Molecular cloning, structural analysis and functional expression of the proline-rich focal adhesion and microfilament-associated protein VASP. EMBO J 14:19–27
Hall SW, Humphries JE, Gonias SL (1991) Inhibition of cell surface receptor-bound plasmin by $\alpha_2$-antiplasmin and $\alpha_2$-macroglobulin. J Biol Chem 266:12329–12336
Harris KW, Esmon CT (1985) Protein S is required for bovine platelets to support activated protein C binding and activity. J Biol Chem 260:2007–2010
Haselton FR, Alexander JS (1992) Platelets and a platelet-released factor enhance endothelial barrier. Am J Physiol 263:L670–L678
Haselton FR, Alexander JS, Mueller SN (1993) Adenosine decreases permeability of in vitro endothelial monolayers. J Appl Physiol 74:1581–1590
Hawrylowicz CM (1993) Viewpoint: a potential role for platelet derived cytokines in the inflammatory response. Platelets 4:1–10
Hawrylowicz CM, Howells GL, Feldman M (1991) Platelet derived interleukin 1 induces human endothelial adhesion molecule expression and cytokine production. J Exp Med 74:785–790

Heller R, Bussolino F, Ghigo D, Garbarino G, Pescarmona GP, Till U, Bosia A (1992) Nitrovasodilators inhibit thrombin-induced platelet-activating factor synthesis in human endothelial cells. Biochem Pharmacol 44:223–229

Heller R, Bussolino F, Ghigo D, Garbarino G, Schröder H, Pescarmona GP, Till U, Bosia A (1991) Protein kinase C and cyclic AMP modulate thrombin-induced platelet-activating factor synthesis in human endothelial cells. Biochim Biophys Acta 1093:55–64

Hempel SL, Haycraft DL, Hoak JC, Spector AA (1990) Reduced prostacyclin formation after reoxygenation of anoxic endothelium. Am J Physiol 259:C738–C745

Hoak JC, Czervionke RL, Fry GL, Smith JB (1980) Interaction of thrombin and platelets with the vascular endothelium. Federation Proc 39:2606–2609

Holme PA, Brosstad F, Solum NO (1993) The difference between platelet and plasma FXIII used to study the mechanism of platelet microvesicle formation. Thromb Haemost 70:681–686

Holme PA, Solum NO, Brosstad F, Roger M, Abdelnoor M (1994) Demonstration of platelet-derived microvesicles in blood from patients with activated coagulation and fibrinolysis using a filtration technique and Western blotting. Thromb Haemost 72:166–171

Holme PA, Brosstad F, Solum NO (1995) Platelet-derived microvesicles and activated platelets express factor Xa activity. Blood Coagul Fibrin 6:302–310

Horstrup K, Jablonka B, Hönig-Liedl P, Just M, Kochsiek K, Walter U (1994) Phosphorylation of focal adhesion vasodilator-stimulated phosphoprotein at Ser157 in intact human platelets correlates with fibrinogen receptor inhibition. Eur J Biochem 225:21–27

Houston DS, Shepherd JT, Vanhoutte PM (1986) Aggregating human platelets cause direct contraction and endothelium-dependent relaxation of isolated canine coronary arteries: role of serotonin, thromboxane $A_2$ and adenine nucleotides. J Clin Invest 78:539–544

Houston DS, Robinson P, Gerrard JM (1990) Inhibition of intravascular platelet aggregation by endothelium-derived relaxing factor: reversal by red blood cells. Blood 76:953–958

Ijzendoorn SCD, van Gool RGJ, Reutelingsperger CPM, Heemskerk JWM (1996) Unstimulated platelets evoke calcium responses in human umbilical vein endothelial cells. Biochim Biophys Acta 1311:64–70

Ishikawa F, Miyazono K, Hellman U, Drexler H, Wernstedt C, Hagiwara K, Usuki K, Takaku F, Risau W, Heldin C-H (1989) Identification of angiogenic activity and the cloning and expression of platelet-derived endothelial growth factor. Nature 338:557–562

Johnsen ULH, Lyberg T, Galdal KS, Prydz H (1983) Platelets stimulate thromboplastin synthesis in human endothelial cells. Thromb Haemost 42:69–72

Jy W, Horstman LL, Arce M, Ahn YS (1992) Clinical significance of platelet microparticles in autoimmune thrombocytopenias. J Lab Clin Med 119:334–345

Kalafatis M, Mann KG (1993) Role of the membrane in the activation of factor Va by activated protein C. J Biol Chem 268:27246–27257

Kalafatis M, Swords NA, Rand MD, Mann KG (1994a) Membrane-dependent reactions in blood coagulation: role of the vitamin K-dependent enzyme complexes. Biochim Biophys Acta 1227:113–129

Kalafatis M, Rand MD, Mann KG (1994b) Factor Va-membrane interaction is mediated by two regions located on the light chain of the cofactor. Biochemistry 33:486–493

Kaplanski G, Porat R, Aiura K, Erban JK, Gelfand JA, Dinarello CA (1993) Activated platelets induce endothelial secretion of interleukin-8 in vitro via an interleukin-1-mediated event. Blood 81:2492–2495

Keijer J, Linders M, van Zonneveld A-J, Ehrlich HJ, de Boer J-P, Pannekoek H (1991) The interaction of plasminogen activator inhibitor 1 with plasminogen activators

(tissue-type and urokinase-type) and fibrin: Localisation of interaction sites and physiologic relevance. Blood 78:401–409
Kishi Y, Numano F (1989) In vitro study of vascular endothelial injury by activated platelets and its prevention. Atherosclerosis 76:95–101
Kitchens CS, Weiss L (1975) Ultrastructural changes of endothelium associated with thrombocytopenia. Blood 46:567–578
Kojima H, Newton-Nash D, Weiss HJ, Zhao J, Sims PJ, Wiedmer T (1994) Production and characterization of transformed B-lymphocytes expressing the membrane defect of Scott syndrome. J Clin Invest 94:2237–2244
Konkle BA, Shapiro SS, Asch AS, Nachman RL (1990) Cytokine-enhanced expression of glycoprotein Ib alpha in human endothelium. J Biol Chem 265:19833–19838
Kunitada S, Fitzgerald GA, Fitzgerald DJ (1992) Inhibition of clot lysis and decreased binding of tissue-type plasminogen activator as a consequence of clot retraction. Blood 79:1420–1427
Legrand AB, Narayana TK, Ryan US (1989) Modulation of adenylate cyclase activity in cultured bovine pulmonary arterial endothelial cells: effects of adenosine and derivatives. Biochem Pharmacol 38:423–430
Lo SK, Burhop KE, Kaplan JE, Malik AB (1988) Role of platelets in maintenance of pulmonary vascular permeability to protein. Am J Physiol 254:H763–H771
Loscalzo J, Pasche B, Ouimet H, Freedman JE (1995) Platelets and plasminogen activation. Thromb Haemost 74:291–293
Ludmer PL, Selwyn AP, Shook TL, Wayne R, Mudge GH, Alexander RW, Ganz P (1986) Paradoxical vasoconstriction induced by acetylcholine in atherosclerotic coronary arteries. New Engl J Med 315:1046–1051
Lüscher TF, Wenzel RR, Noll G (1995) Local regulation of the coronary circulation in health and disease: role of nitric oxide and endothelin. Eur Heart J 16 [Suppl C]:51–58
Lum H, Malik AB (1994) Regulation of vascular endothelial barrier function. Am J Physiol 267:L223–L241
Lupu C, Manduteanu I, Calb M, Ionescu M, Simionescu N, Simionescu M (1993) Some major plasmalemma proteins of human diabetic platelets are involved in the enhanced platelet adhesion to cultured valvular endothelial cells. Platelets 4:79–84
Maca RD, Fry GL, Hoak JC, Loh P-MT (1977) The effects of intact platelets on cultured human endothelial cells. Thromb Res 11:715–727
Magyar-Lehman S, Böhlen P (1992) Purification of platelet-derived endothelial growth inhibitor and its characterization as transforming growth factor-$\beta$ type 1. Experientia 48:374–379
Mann KG, Nesheim ME, Church WR, Haley P, Krishnaswamy S (1990) Surface-dependent reactions of the vitamin K-dependent enzyme complexes. Blood 76:1–16
Marcus AJ, Weksler BB, Jaffe EA, Broekman MJ (1980) Synthesis of prostacyclin from platelet-derived endoperoxides by cultured human endothelial cells. J Clin Invest 66:979–986
Marcus AJ, Safier LB, Hajjar KA, Ullman HL, Islam N, Broekman MJ, Eiroa AM (1991) Inhibition of platelet function by an aspirin-insensitive endothelial cell ADPase. J Clin Invest 88:1690–1696
Martin OC, Pagano RE (1987) Transbilayer movement of fluorescent analogs of phosphatidylserine and phosphatidylethanolamine at the plasma membrane of cultured cells. Evidence for a protein-mediated and ATP-dependent process(es). J Biol Chem 262:5890–5898
Mathews II, Padmanabhan KP, Tulinsky A (1994) Structure of a nonadecapeptide of the fifth EGF domain of thrombomodulin complexed with thrombin. Biochemistry 33:13547–13552
Maurice DH, Haslam RJ (1990) Molecular basis of the synergistic inhibition of platelet function by nitrovasodilators and activators of adenylate cyclase: Inhibition of cyclic AMP breakdown by cyclic GMP. Mol Pharmacol 37:671–681

McDonagh PF (1986) Platelets reduce coronary microvascular permeability to macromolecules. Am J Physiol 251:H581–H587
McEver RP (1994) Role of selectins in leukocyte adhesion to platelets and endothelium. Ann NY Acad Sci 714:185–189
Miles LA, Ginsberg MH, White JG, Plow EF (1986) Plasminogen interacts with human platelets through two distinct mechanisms. J Clin Invest 77:2001–2009
Miyazono K, Okabe T, Urabe A, Takaku F, Heldin CH (1987) Purification and properties of an endothelial cell growth factor from human platelets. J Biol Chem 262:4098–4103
Moncada S, Higgs A (1993) The L-arginin-nitric oxide pathway. New Engl J Med 329:2002–2012
Moncada S, Palmer RMJ, Higgs EA (1992) Nitric oxide: physiology, pathophysiology, and pharmacology. Pharmacol Rev 43:109–142
Monkovic DD, Tracy PB (1990) Functional characterisation of human platelet-released factor V and its activation by factor Xa and thrombin. J Biol Chem 265:17132–17140
Morrot G, Zachowski A, Devaux PF (1990) Partial purification and characterisation of the human erythrocyte $Mg^{++}$-ATPase. FEBS Lett 266:29–32
Murata M, Ikeda Y, Araki Y, Murakami H, Sato K, Yamamoto M, Watanabe K, Ando Y, Igawa T, Maruyama I (1988) Inhibition by endothelial cells of platelet aggregating activity of thrombin – role of thrombomodulin. Thromb Res 50:647–656
Murata S, Matsumura Y, Takada K, Asai Y, Takaoka M, Morimoto S (1995) Role of transforming growth factor-$\beta$1 on platelet-induced enhancement of endothelin-1 production in cultured vascular endothelial cells. J Pharmacol Exp Ther 274:1524–1530
Nakajima T, Kakishita E, Nagai K (1987) Effect of serum from patients with idiopathic thrombocytopenic purpura on cultured endothelial cell growth. Am J Hematol 24:15–22
Nolte C, Eigenthaler M, Schanzenbächer P, Walter U (1991) Endothelial cell-dependent phosphorylation of a platelet protein mediated by cAMP- and cGMP-elevating agents. J Biol Chem 266:14808–14812
Nolte C, Eigenthaler M, Horstrup K, Hönig-Liedl P, Walter U (1994) Synergistic phosphorylation of the focal adhesion-associated vasodilator-stimulated phosphoprotein in intact human platelets in response to cGMP- and cAMP-elevating platelet inhibitors. Biochem Pharmacol 48:1569–1575
Oiumet H, Freedman JE, Loscalzo J (1994) Kinetics of platelet-surface plasminogen activation by tissue-type plasminogen activator. Biochemistry 33:2970–2976
Oosting JD, Derksen RHWM, Bobbink WG, Hackeng TM, Bouma BN, de Groot PG (1993) Antiphospholipid antibodies directed against a combination of phospholipids with prothrombin, protein C, or protein S: an explanation for their pathogenic mechanism? Blood 81:2618–2625
Op den Kamp JAF (1979) Lipid asymmetry in membranes. Annu Rev Biochem 48:47–71
Pasche B, Ouimet H, Francis S, Loscalzo J (1994) Structural changes in platelet glycoprotein IIb/IIIa by plasmin: determinants and functional consequences. Blood 83:404–414
Paty PSK, Sherman PF, Shephard JM, Malik AB, Kaplan JE (1992) Role of adenosine in platelet-mediated reduction in pulmonary vascular permeability. Am J Physiol 262:H771–H77
Plow EF, Collen D (1981) The presence and release of $\alpha_2$-antiplasmin from human platelets. Blood 58:1069–1074
Plow EF, Herren T, Redlitz A, Miles LA, Hoover-Plow JL (1995) The cell biology of the plasminogen system. FASEB J 9:939–945
Pohl U, Busse R (1989) EDRF increases cyclic GMP in platelets during passage through the coronary vascular bed. Circ Res 65:1798–1803

Pohl U, Nolte C, Bunse A, Eigenthaler M, Walter U (1994) Endothelium-dependent phosphorylation of vasodilator-stimulated protein in platelets during coronary passage. Am J Physiol 266:H606–H612
Preissner KT (1990) Physiological role of vessel wall related antithrombotic mechanisms: contribution of endogenous and exogenous heparin-like components to the anticoagulant potential of the endothelium. Haemostasis 20(Suppl 1):30–49
Radomski MW, Palmer RMJ, Moncada S (1987a) Endogenous nitric oxide inhibits human platelet adhesion to vascular endothelium. Lancet ii:1057–1058
Radomski MW, Palmer RMJ, Moncada S (1987b) The anti-aggregatory properties of vascular endothelium: interactions between prostacyclin and nitric oxide. Br J Pharmacol 92:639–646
Radomski MW, Vallance P, Whitley G, Foxwell N, Moncada S (1993) Platelet adhesion to human vascular endothelium is modulated by constitutive and cytokine induced nitric oxide. Cardiovasc Res 27:1380–1382
Reinhard M, Halbrügge M, Scheer U, Wiegand C, Jockusch BM, Walter U (1992) The 46/50 kDa phosphoprotein VASP purified from human platelets is a novel protein associated with actin filaments and focal adhesion contacts. EMBO J 11:2063–2070
Reinhard M, Giehl K, Abel K, Haffner C, Jarchau T, Hoppe V, Jockusch B, Walter U (1995a) The proline-rich focal adhesion and microfilament protein VASP is a ligand for profilins. EMBO J 14:1583–1589
Reinhard M, Jouvenal K, Tripier D, Walter U (1995b) Identification, purification, and characterization of a zyxin-related protein that binds the focal adhesion and microfilament protein VASP (vasodilator-stimulated phosphoprotein). Proc Natl Acad Sci USA 92:7956–7960
Rosing J, Tans G, Govers-Riemslag JWP, Zwaal RFA, Hemker HC (1980) The role of phospholipids and factor Va in the prothrombinase complex. J Biol Chem 255:274–283
Rosing J, van Rijin JLML, Bevers EM, van Dieijen G, Comfurius P, Zwaal RFA (1985a) The role of activated human platelets in prothrombin and factor X activation. Blood 65:319–322
Rosing J, Bevers EM, Comfurius P, Hemker HC, Van Dieijen G, Weiss HJ, Zwaal RFA (1985b) Impaired factor X and prothrombin activation associated with decreased phospholipid exposure in platelets from a patient with a bleeding disorder. Blood 65:557–561
Rosing J, Speijer H, Zwaal RFA (1988) Prothrombin activation on phospholipid membranes with positive electrostatic potential. Biochemistry 27:8–11
Rosove MH, Harrison JL, Frank HJL, Harwig SSL, Berliner J (1985) Plasma $\beta$-thromboglobulin is correlated with platelet adhesiveness to bovine endothelium in patients with diabetes mellitus. Thromb Res 37:251–258
Roubey RAS (1994) Autoantibodies to phospholipid-binding plasma proteins: a new view of Lupus anticoagulants and other "antiphospholipid" autoantibodies. Blood 84:2854–2867
Rybak ME, Gimbrone MA, Davies PF, Handin RI (1989) Interaction of platelet factor four with cultured vascular endothelial cells. Blood 73:1534–1539
Sabovic M, Lijnen HR, Keber D, Collen D (1989) Effect of retraction on the lysis of human clots with fibrin specific and non-specific plasminogen activators. Thromb Haemost 62:1083–1087
Sadowitz PD, Yamaja Setty BN, Stuart M (1987) The platelet cyclooxygenase metabolite 12-L-hydroxy-5,8,10-hepta-decatrienoic acid (HHT) may modulate primary hemostasis by stimulating prostacyclin production. Prostaglandins 34:749–763
Sandberg H, Bode AP, Dombrose FA, Hoechli M, Lentz BR (1985) Expression of coagulant activity in human platelets: release of membrane vesicles providing platelet factor 1 and platelet factor 3. Thromb Res 39:63–79
Sawyer PN, Srinivasan S (1972) The role of electrochemical surface properties in thrombosis at vascular interfaces: cumulative experience of studies in animals and man. Bull N Y Acad Med 48:235–256

Schafer AI, Crawford DD, Gimbrone MA (1984) Unidirectional transfer of prostaglandin endoperoxides between platelets and endothelial cells. J Clin Invest 73:1105–1112

Schafer AI, Mass AK, Ware A, Johnson PC, Rittenhouse SE, Salzman EW (1986) Platelet protein phosphorylation, elevation of cytosolic calcium, and inositol phospholipid breakdown in platelet activation induced by plasmin. J Clin Invest 78:73–79

Schroit AJ, Bloy C, Connor J, Cartron JP (1990) Involvement of Rh blood group polypeptides in the maintenance of aminophospholipid asymmetry. Biochemistry 29:10303–10306

Schroit AJ, Zwaal RFA (1991) Transbilayer movement of phospholipids in red cell and platelet membranes. Biochim Biophys Acta 1071:313–329

Schwarz HP, Heeb MJ, Wencel-Drake JD, Griffin JH (1985) Identification and quantification of protein S in human platelets. Blood 66:1452–1455

Sedlmayr P, Blaschitz A, Wilders-Truschnig M, Tiran A, Dohr G (1995) Platelets contain interleukin-1 alpha and beta which are detectable on the cell surface after activation. Scand J Immunol 42:209–214

Seigneuret M, Devaux PF (1984) ATP-dependent asymmetric distribution of spin-labeled phospholipids in the erythrocyte membrane: relation to shape changes. Proc Natl Acad Sci USA 81:3751–3755

Shatos MA, Doherty JM, Hoak JC (1991) Alterations in human vascular endothelial cell function by oxygen free radicals. Platelet adherence and prostacyclin release. Arterioscler Thromb 11:594–601

Sheldon R, Moy A, Lindsley K, Shasby S, Shasby M (1993) Role of myosin light-chain phosphorylation in endothelial cell retraction. Am J Physiol 265:L606–L612

Shephard JM, Moon DG, Sherman PF, Weston LK, Del Vecchio PJ, Minnear FL, Malik AB, Kaplan JE (1989) Platelets decrease albumin permeability of pulmonary artery endothelial cell monolayers. Microvasc Res 37:256–266

Shepro D, Sweetman HE, Hechtman HB (1984) Vasoactive agonists prevent erythrocyte extravasation in thrombocytopenic hamsters. Thromb Res 35:421–430

Shepro D, Hechtman HB (1985) Endothelial serotonin uptake and mediation of prostanoid secretion and stress fiber formation. Federation Proc 44:2616–2619

Shimokawa H, Vanhoutte PM (1989) Impaired endothelium-dependent relaxation to aggregating platelets and related vasoactive substances in porcine coronary arteries in hypercholesterolemia and atherosclerosis. Circ Res 64:900–914

Sims PJ, Faioni EM, Wiedmer T, Shattil SJ (1988) Complement proteins C5b-9 cause release of membrane vesicles from the platelet surface that are enriched in the membrane receptor for coagulation factor Va and express prothrombinase activity. J Biol Chem 263:18205–18212

Sims PJ, Wiedmer T, Esmon CT, Weiss HJ, Shattil SJ (1989) Assembly of the platelet prothrombinase complex is linked to vesiculation of the platelet plasma membrane. J Biol Chem 264:17049–17057

Slivka SR, Loskutoff DJ (1991) Platelets stimulate endothelial cells to synthesize type 1 plasminogen activator inhibitor. Evaluation of the role of transforming growth factor $\beta$. Blood 77:1013–1019

Smeets EF, Comfurius P, Bevers EM, Zwaal RFA (1994) Calcium-induced transbilayer scrambling of fluorescent phospholipid analogs in platelets and erythrocytes. Biochim Biophys Acta 1195:281–286

Solymoss S, Tuckers MM, Tracy PB (1988) Kinetics of inactivation of membrane-bound factor Va by activated protein C. Protein S modulates factor Xa protection. J Biol Chem 263:14884–14890

Stelzner TJ, Weil JV, O'Brien RF (1989) Role of cyclic adenosine monophosphate in the induction of endothelial barrier properties. J Cell Physiol 139:157–166

Stringer HAR, van Swieten P, Heijnen HFG, Sixma JJ, Pannekoek H (1994) Plasminogen activator inhibitor-1 released from activated platelets plays a key role in

thrombolysis resistance. Studies with thrombi generated in the Chandler loop. Arterioscler Thromb 14:1452–1458
Sulpice JC, Zachowski A, Devaux PF, Girand F (1994) Requirement for phosphatidylinositol 4,5-bisphosphate in the $Ca^{2+}$-induced phospholipid redistribution in the human erythrocyte membrane. J Biol Chem 269:6347–6354
Swords NA, Mann KG (1993a) The assembly of the prothrombinase complex on adherent platelets. Arterioscler Thromb 13:1602–1612
Swords NA, PB Tracy, KG Mann (1993b) Intact platelet membranes, not platelet-released microvesicles, support the procoagulant activity of adherent platelets. Arterioscler Thromb 13:1613–1622
Tanowitz HB, Burns ER, Sinha AK, Kahn NN, Morris SA, Factor SM, Hatcher VB, Bilezikian JP, Baum SG, Wittner M (1990) Enhanced platelet adherence and aggregation in Chaga's disease: A potential pathogenic mechanism for cardiomyopathy. Am J Trop Med Hyg 43:274–281
Tans G, Rosing J, Thomassen CLGD, Heeb MJ, Zwaal RFA, Griffin JH (1991) Comparison of anticoagulant and procoagulant activities of stimulated platelets and platelet-derived microparticles. Blood 77:2641–2648
Tilly RHJ, Senden JMG, Comfurius P, Bevers EM, Zwaal RFA (1990) Increased aminophospholipid translocase activity in human platelets during secretion. Biochim Biophys Acta 1029:188–190
Tloti MA, Moon DG, Weston LK, Kaplan JE (1991) Effect of 13-hydroxyoctadeca-9, 11-dienoic acid (13-HODE) on thrombin induced platelet adherence to endothelial cells in vitro. Thromb Res 62:305–317
Toti F, Satta N, Fressinaud E, Meyer D, Freyssinet JM (1996) Scott syndrome, characterized by impaired transmembrane migration of procoagulant phosphatidylserine and hemorrhagic complications, is an inherited disorder. Blood 87:1409–1415
Tracy PB, Peterson JM, Nesheim ME, McDuffie FC, Mann KG (1979) Interaction of coagulation factor V and factor Va with platelets. J Biol Chem 254:10354–10361
Tracy PB, Eide LL, Bowie EJW, Mann KG (1982) Radioimmunoassay of factor V in human plasma and platelets. Blood 60:59–63
Tracy PB, Giles AR, Mann KG, Eide LL, Hoogendoorn H, Rivard GE (1984) Factor V (Quebec): a bleeding diathesis associated with a qualitative platelet factor V deficiency. J Clin Invest 74:1221–1228
Tranzer JP, Baumgartner HR (1967) Filling gaps in the vascular endothelium with blood platelets. Nature 216:1226–1228
Van Dieijen G, Tans G, Rosing J, Hemker HC (1981) The role of phospholipid and factor VIIIa in the activation of bovine factor X. J Biol Chem 256:3433–3442
Vanhoutte PM (1988) Platelets, endothelium and blood vessel wall. Experientia 44:105–109
Vaughan DE, Mendelsohn ME, Declerck PJ, van Houtte E, Collen D, Loscalzo J (1989) Characterisation of the binding of human tissue-type plasminogen activator to platelets. J Biol Chem 264:15869–15874
Venturini CM, Fenton II JW, Minnear FL, Kaplan JE (1989) Rat platelets adhere to human thrombin-treated rat lungs under flow conditions. Thromb Haemost 62:1006–1010
Venturini CM, Weston LK, Kaplan JE (1992) Platelet cGMP, but not cAMP, inhibits thrombin-induced platelet adhesion to pulmonary vascular endothelium. Am J Physiol 263:H606–H612
Verhallen PFJ, Bevers EM, Comfurius P, Zwaal RFA (1987) Correlation between calpain-mediated cytoskeletal degradation and expression of procoagulant activity. A role for the platelet membrane skeleton in the regulation of membrane lipid asymmetry? Biochim Biophys Acta 903:206–217
Walker FJ (1981) Regulation of activated protein C by protein S. The role of phospholipid in factor Va inactivation. J Biol Chem 256:11128–11131
Walter U, Eigenthaler M, Geiger J, Reinhardt M (1993) Role of cyclic nucleotide-dependent protein kinases and their common substrate VASP in the regulation of

human platelets. In: Authi KS, Watson SP, Kakkar VV (eds) Mechanisms of platelet activation and control. Plenum, New York, p 237 (Advances in experimental medicine and biology, vol 344)
Weiss HJ, Vivic WJ, Lages BA, Rogers J (1979) Isolated deficiency of platelet procoagulant activity. Am J Med 67:206–213
Westendorp RGJ, Draijer R, Meinders AE, van Hinsbergh VWM (1994) Cyclic-cGMP-mediated decrease in permeability of human umbilical and pulmonary artery endothelial cell monolayers. J Vasc Res 31:42–51
Wiedmer T, Shattil SJ, Cunningham M, Sims PJ (1990) Role of calcium and calpain in complement-induced vesiculation of the platelet plasma membrane and in the exposure of the platelet factor Va receptor. Biochemistry 29:623–632
Williamson P, Kulick A, Zachowski A, Schlegel RA, Devaux PF (1992) $Ca^{2+}$ induces transbilayer redistribution of all major phospholipids in human erythrocytes. Biochemistry 31:6355–6360
Williamson P, Schlegel RA (1994) Back and forth: the regulation and function of transbilayer phospholipid movement in eukaryotic cells. Mol Membr Biol 11:199–216
Williamson P, Bevers EM, Smeets EF, Comfurius P, Schlegel RA, Zwaal RFA (1995) Continuous analysis of the mechanism of activated transbilayer lipid movements in platelets. Biochemistry 34:10448–10255
Wojcik JD, van Horn DL, Webber AJ, Johnson SA (1969) Mechanism whereby platelets support the endothelium. Transfusion 9:324–335
Wolf P (1967) The nature and significance of platelets products in human plasma. Br J Haematol 13:269–288
Yagi K, Shinbo M, Hashizume M, Shimba LS, Kurimura S, Miura Y (1991) ATP diphosphohydrolase is responsible for ecto-ATPase and ecto-ADPase activities in bovine aorta endothelial and smooth muscle cells. Biochem Biophys Res Commun 180:1200–1206
Zwaal RFA, Comfurius P, Bevers EM (1992) Platelet procoagulant activity and microvesicle formation. Its putative role in hemostasis and thrombosis. Biochim Biophys Acta 1180:1–8
Zwaal RFA, Comfurius P, Smeets E, Bevers EM (1996) Platelet procoagulant activity and microvesicle formation. In: Seghatchian MJ, Samama MM, Hecker SP (eds) Hypercoagulable states: fundamental aspects, acquired disorders and congenital thrombophilia. CRC Press, Boca Raton, p 29

CHAPTER 27

# Platelet Flow Cytometry – Adhesive Proteins

D. Tschoepe and B. Schwippert

## A. Introduction

Circulating platelets act as the universal interface between the vessel wall, liquid phase coagulation, and fibrinolysis, leading to physiologically balanced hemostasis at sites of vascular injury that prevents pathological thrombus formation. With regard to platelet function, the medical laboratory should answer two principle categories of diagnostic questions: (1) Is the sample platelet population reacting functionally abnormal, either hypo- or hyperreactive (cellular potential), and (2) what is the level of actual activation of the sample platelet population (cellular status)?

### I. Procedures for the Diagnosis of Platelet Disease States

Primary hemostasis consists of a cascade of functions as defined on the basis of standardized ex vivo stimulation with different agonists: adhesion, shape change, aggregation, secretion, and contraction (for a review see Philp 1981). Testing a given sample of separated platelets with conventional methods, e.g., Born's classical turbidimetric aggregation test, is effective in characterizing integral functional properties (Born 1962). Thrombelastography places platelet reactivity directly in the context of the hemostasis cascade. More recently, thrombostat implements controlled flow over a defined surface in order to measure primary hemostasis (Alshameeri and Mammen 1993). Breddin's PITT (platelet-induced thrombin generation time) test cross-links shear stress activation with the plasmatic coagulation cascade by induction of thrombin generation (Basic-Micic et al. 1992).

#### 1. Problems

There are, however, several disadvantages to these general function tests: different platelet populations in the sample tested may contribute to the overall result in different ways but cannot be discriminated; there is no possibility of identifying the ultrastructural factors contributing to aberrant functional properties; in diseases with graduated platelet dysfunction, sensitivity of these conventional techniques may be insufficient for clear-cut discrimination; and there are no generally established protocols for performing the tests and processing the results in a comparable manner.

With the exception of severe bleeding disorders or the evaluation of pharmacological interventions, it remains completely unclear under prethrombotic aspects what physiological impact can be evaluated from accelerated platelet aggregometer tracing for the single patient. Does a high functional potential (response to artificial in vitro stimulus) become operative in vivo? These problems hold true for the more advanced technologies such as the measurement of specific release proteins, prostanoid products, or indirect molecular markers of intravasal hemostasis activation (FAREED et al. 1995). Furthermore, all of these tests have been demonstrated to sensitively indicate platelet-cell interaction, thus being descriptive within a clinically apparent situation rather than being predictive for the prognosis-limiting clinical end point. From a theoretical point of view, it should be expected that a cellular signature indicating the immediate biological fate of the sample cell population could provide a general superior principle to predict clinical outcome (LOVETT 1984; VALET et al. 1993).

### 2. Requirements

Controlled clinical studies must evaluate whether and under what conditions the overall efficacy of good clinical practice (GCP) can be improved for the individual patient by precisely describing platelet status. Unfortunately, due to the problems described above, the classic methods of platelet function testing often do not provide sufficiently convenient and stable performance characteristics as a prerequisite for implementation in such clinical trials. In view of the epidemiological impact of platelets for hard clinical end points (hemorrhage and occlusive vascular events), there is an urgent need for a convenient short-term method in the clinical routine laboratory that provides those platelet parameters of best-evaluated predictability: (1) volume distribution (and count); (2) level of activation in a given platelet sample population with direct association to the intravasal conditions of the individual patient; and (3) standardized platelet reactivity to agonists.

## B. Single Platelet Flow Cytometry (SPFC)

In order to overcome the outlined problems, flow cytometry has emerged over the past years as a superior tool for platelet research, providing ultrastructural information derived directly platelet-by-platelet, similar to the entire spectrum of white cell analysis (ABRAMS et al. 1990; ABRAMS and SHATTIL 1991; ADELMAN et al. 1985, 1992; CORASH 1990; LOVETT 1984; MARTI et al. 1988; SHAFI 1986; TSCHOEPE et al. 1990a).

### I. Principle

In flow cytometric analysis, target cells are hydrodynamically focussed to a serial cell flow that crosses a laser beam of defined wave length. At the time of intercept with the laser beam, each cell generates forward light scatter dependent on cell size and right-angle light scatter dependent on cell granularity. Both parameters characterize different cell populations in the sample by providing a typical two-dimensional plot shape of the simultaneous impulse recordings. Prestaining the cells with fluorochrome dyes results in a fluorescent light emission impulse generated after laser light excitation. All of the optical signals generated by a single platelet at the time of laser intercept are simultaneously recorded by a photomultiplier matrix and triggered by the forward light scatter impulse of the respective cell. Analyzing a defined but freely adjustable number of cells, each parameter is recorded along a logarithmic intensity scale. Quantitative results can be obtained by computing the number of fluorescence-positive events and the respective mean signal intensity derived from them.

The application of flow cytometry is limited by the possibility of coupling a specific fluorescent dye to the target molecules to be analyzed. Generally, subcellular and surface-associated probes have been proposed (MURPHY 1986). A variety of direct dyes react with intraplatelet molecules or may be bound depending on functional status, e.g., thiazole-orange may be used for RNA labeling (KIENAST and SCHMITZ 1990; RINDER et al. 1993); indo-1 loading makes possible the monitoring of the calcium signal (DAVIES et al. 1990; JENNINGS et al. 1989); mepacrine is indicative of dense body loading and secretion (GAWAZ 1993); and bis-carboxyethyl carboxyfluorescein [BCECF = 2′,7′- bis-(2-carboxyethyl)-5- and -6-carboxyfluorescein] may be used for intraplatelet pH measurement. Platelet surface membrane analysis has been more extensively developed for use in clinical situations, especially with monoclonal antibodies against defined antigenic epitopes which may be either directly or indirectly coupled to fluorochrome dyes (LOVETT 1984; SHAFI 1986; SHATTIL 1992).

## II. Pitfalls and Standardization Requirements

Platelets are very sensitive to preactivation either through blood sampling or by preparation, and as a result of hydrodynamic stress during the process of analyzation. Moreover, within the peripheral blood, platelets are the smallest particles to be easily discriminated from red and white blood cells, but the light scatter properties are difficult to distinguish from the debris signal. Accordingly, the emission light intensity of stained platelets is much smaller than that of leukocytes. Using immunostaining techniques, standardization of quantitative measurements of antigenic binding sites turns out to be difficult. The handling conditions of the sample platelets vary depending on the target structure: quantitative analysis of constitutive membrane characteristics requires stabilized native platelets, whereas measurement of epitopes indicating the functional activation status requires additional fixation before and/or after addition of the staining antibodies, at any rate under clinical conditions. Another crucial issue to consider regarding the immunostaining procedure is the expression of Fc receptors on (activated) platelets (McCREE et al. 1990). Precise evaluation of (a) the sampling procedure (atraumatic standard materials), (b) the platelet-preparation and staining protocol (whole blood or washed platelets), (c) the instrument setups (type of instruments, data acquisition, and processing mode), and (d) fluorescence calibration of the fluorochrome dyes (with a titrated saturated concentration of used monoclonal antibodies) is prerequisite for any flow cytometric platelet analysis. Moreover, considering

**Table 1.** Flow sheet summary of the Düsseldorf IIIx protocol for flow cytometric analysis of indirectly immunostained activation-dependent platelet epitopes

Blood vessel puncture (arterial or undwelled venous; 1 mm-caliber needle; no stasis)

↓

Immediate stabilization (10 vol% special medium)

↓

Second step fixation (0.5% paraformaldehyde)

↓

Differential centrifugation (10 min, 250 g; 5 min, 700 g) → platelet rich plasma (PRP)

↓

Washing

↓

Blocking Fc receptors with 10 vol% rabbit serum

↓

Adjustment of platelet count (50,000/$\mu$l)

↓

Staining with monoclonal antibody panel (titrated saturated concentrations)

↓

Indirect fluorescence labeling with $F(ab)_2$-FITC fragments (titrated saturated concentrations)

↓

Washing

↓

Flow cytometric analysis (FacScan, approved instrument)

↓

List mode data processing

↓

Test result compilation:
1. Percentage of marker-positive platelets (qualitative)
2. Mean fluorescence intensity (quantitative)
3. Activation index [(1) × (2)]

FITC, fluorescein isothiocyanate.

the introductory remarks, the category of analytical question must be defined prior to choosing the adequate procedure.

## III. Assay Systems

Many aspects of assay systems have been standardized by the so-called Düsseldorf protocols (Tschoepe et al. 1990a; Table 1) which have been proposed to evaluate either constitutive membrane receptors such as glycoprotein IIB/IIIA (CD41) or activation-dependent epitopes such as P-selectin (CD62). Test results are expressed as percentage of platelets positive for the monoclonal probe and as quantitative antigen expression by the arbitrarily measured specific fluorescent light intensity within an individual patient sample (Fig. 1). Multiplication of the combined values is achieved with the activation index which proves more sensitive in discriminating deviations from the norm. In order to minimize analytical steps and preparative artifacts and to extend

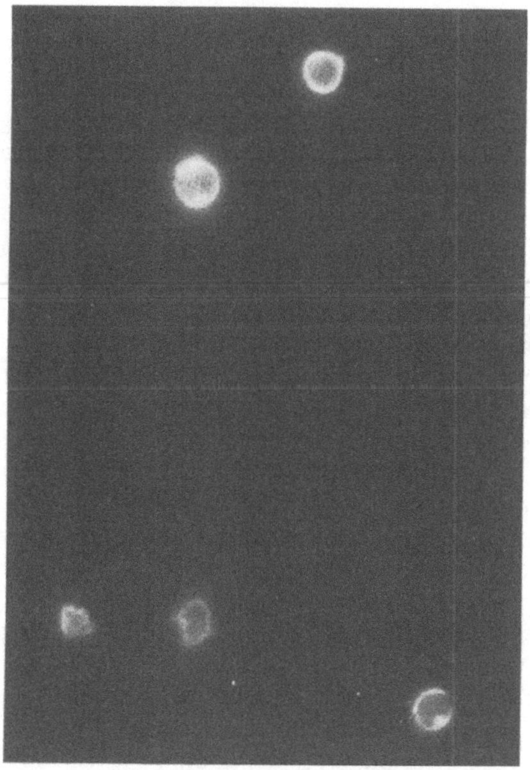

**Fig. 1.** Dye-loaded platelets. In this example, platelets were stained by a CD61 (GPIIIA, constitutive) monoclonal antibody and then labelled indirectly by $F(ab)_2$-FITC fragments directed against the first-step staining antibody. *CD*, cluster designation; *GP*, glycoprotein; *FITC*, fluorescein isothiocyanate

the technique to thrombocytopenic patients, we and others have achieved optimization of the test for direct measurement of whole blood (AULT et al. 1989; JENNINGS et al. 1986; MICHELSON 1994; TSCHOEPE et al. 1993a; WARKENTIN et al. 1990). A key modification of this advanced procedure is the requirement of parallel double staining, e.g., phycoerythrin (PE)/fluorescein isothiocyanate (FITC), where by a tagging-constitutive pan-platelet marker such as CD41 defines a platelet event and a function-dependent marker such as CD62 simultaneously characterizes the activation status (SHATTIL et al. 1987; SHATTIL 1992). Technically, this procedure requires that used probes are directly conjugated with fluorochrome dyes, which is not possible with a majority of interesting antibodies. The so-called Düsseldorf-IIIx reference protocol evaluates certain activation-dependent epitopes on washed sample platelets and has proven versatility in various applications and stability in background clinical-routine application (Table 1; for a review see TSCHOEPE et al. 1990a; TSCHOEPE and SCHWIPPERT 1993; VALET et al. 1993).

## C. Platelet Adhesion Molecules as Diagnostic Targets

The described overall platelet functions such as shear-rate dependent initial attachment, firm adhesion, and cross-bridging with subsequent aggregation are mediated by a class of proteins addressed as adhesion molecules. These act either as specific receptors or ligands. In flowing blood, platelets are marginated and thereby flow in close contact to the endothelial or subendothelial environment. In the resting state, a majority of platelet membrane proteins consists of constitutively expressed adhesive glycoproteins (GP) such as GPIB which mediates initial contact adhesion following exposure to endothelial-cell/matrix-derived ligands such as von Willebrand factor. Following activation, platelets are also capable of releasing adhesion molecules such as thrombospondin or fibrinogen which bind as ligands to the activated recognition sites of complementary glycoprotein receptors on the surface membrane such as GPIV or activated GPIIB/IIIA. This finally promotes platelet-platelet interaction (aggregation). These molecules have been individually evaluated and represent different functional platelet aspects. From a diagnostic point of view, they are attractive for indicating normal or abnormal platelet constitution and, perhaps more important, platelet activation. Availability of monoclonal antibodies directed against these structures allows the assembly of a flow cytometry test consisting of highly specific immunostaining combined with high-resolution fluorescence-activated single platelet analysis. This ultimately provides pathophysiological information for the individual patient on the molecular level of the platelets depending on the applied marker panel. Table 2 shows a recent compilation of CD-clustered platelet antigens of known functional impact. It must be mentioned that in addition to these, there exists a wide variety of better characterized platelet epitopes such as thrombospondin, exceeding this compiled cluster designation (CD) panel.

**Table 2.** Platelet adhesion molecules

| Antigen | | Structure | Inducible | kDa | Name | Functional aspect |
|---|---|---|---|---|---|---|
| CD9 | § | | | 24 | | (Signal transduction) |
| CD17 | | | | | Lactosylceramid | Glycosphingolipid/ unknown |
| CD29 | # | $\beta_1$ | | 130 | VLA $\beta$/GPIIa | Fibronectin receptor |
| CD31 | | | x | 140 | Pecam-1/GPIIa′ | Immunoglobulin superfamily |
| CD36 | | | | 90 | GPIV/GPIIIb | Thrombospondin receptor |
| CD41 | # | $\alpha_2\beta_3$ | (x) | 133 | GPIIb/IIIa complex | Fibrinogen/vWF receptor |
| CD42a | * | | | 23 | GPIX | Adherence to subendothelium/ |
| CD42b | * | | | 135/25 | GPIb$\alpha$ | vWF:Ag receptor |
| CD42c | * | | | 22 | GPIb$\beta$ | |
| CD42d | * | | | 85 | GPV | Associated with GPIb-IX complex |
| CD49b | # | $\alpha_2/\beta_1$ | | 165 | VLA-2/GPIa/IIa | Collagen receptor |
| CD49e/29 | # | $\alpha_5/\beta_1$ | | 135/25 | VLA-5 | Fibronectin receptor |
| CD49f/29 | # | $\alpha_6/\beta_1$ | | 120/25 | VLA-6/GPIc | Laminin receptor |
| CD51 | # | $\alpha_V/\beta_1$ | | 125/25 | VNR$\alpha$ | $\alpha_v$-Chain of vitronectin receptor (VNR) |
| CD51/61 | # | $\alpha_V/\beta_3$ | | 125/110 | VNR | Vitronectin receptor |
| CD61 | # | $\beta_3$ | | 110 | GPIIIa/VNR$\beta$ | $\beta$-Chain of vitronectin receptor |
| CD62p | $ | | x | 140 | P-selectin | Receptor for WBC Sialyl Lewis x |
| CD63 | | | x | 53 | GP53/"LIMP" | Lysosomal integral membrane protein |
| CD107a | | | | 110 | "LAMP1" | Lysosome-associated membrane protein 1 |
| CD107b | | | | 120 | "LAMP2" | Lysosome-associated membrane protein 2 |

Summary of CD-classified constitutive and inducible (x) antigenic platelet epitopes (prepared according to METZELAAR et al. 1989; NEWMAN 1993; ROESEN et al. 1994a).
§, platelet tetraspan protein; #, integrin; *, leucine-rich glycoprotein; $, selectin; GP, glycoprotein; LAMP, lysosome-associated membrane protein; VLA, very late activation antigen; VNR, vitronectin receptor; WBC, white blood corpuscles; vWF, v. Willebrand factor; LIMP, lysosomal integral membrane protein; CD, cluster designation.

## I. Megakaryocytic Conditioning

Platelets are cell fragments from bone-marrow-derived megakaryocytic progenitor cells. They lack a nucleus and thus do not possess all of the elements of the protein synthesizing apparatus. Constitutive platelet proteins have to be synthesized in the megakaryocytes and packed into platelet precursors, which are then shed into the peripheral circulation (SCHNEIDER and GATTERMANN 1994). Consequently beyond their functional impact, glycoproteins such as

GPIB and GPIIB/IIIA may be regarded as cell lineage markers for bone marrow progenitor (stem) cells (KANZ et al. 1988; TOMER et al. 1988). In conclusion, one has to understand that functional platelet behavior is determined by the constitutive expression pattern of these adhesive glycoproteins, among others, which reflects the regulation of the nucleated megakaryocytic progenitor compartment (HANCOCK et al. 1993; TSCHOEPE et al. 1992).

## II. Platelet Membrane Processing (Fig. 2)

Activation of platelets through a variety of agonists leads to stimulation of different intraplatelet signal pathways (JOHNSTON et al. 1987; TSCHOEPE et al. 1990c). In contrast to the ability of constitutively expressed platelet glycoproteins to interact with adhesive molecules (e.g., matrix-derived, surface-bound) in the resting state (COLLER et al. 1989; DUNSTAN 1985; STEWART et al. 1988), the very early phase of activation leads to a change in organization of the outer platelet membrane by exposing additional ligand recognition sites for soluble fibrinogen, for instance. This is achieved by a switch in the conformation of some glycoproteins such as GPIIB/IIIA following complex interaction of the biochemical signal transduction cascade with the cytosceleton (for a review see SMYTH et al. 1993). Concomitantly, intraplatelet granules fuse with the outer membrane, subsequently exposing molecules such as CD62, the P-selectin receptor for the attachment of neutrophils via the Sialyl Lewis$^X$ (CD15) ligand (DE-BRUIJNE-ADMIRAAL et al. 1992), or subsequently releasing intraplatelet proteins such as platelet factor 4, $\beta$-thromboglobulin, and thrombospondin which in turn rebound to specific surface receptors in order to further cross-link the white platelet thrombus. In parallel, microparticles are shed from the activated platelet membranes that promote liquid-phase-coagulation protease activities (Va, VIIIa), facilitating thrombin generation. Following the initial adhesion reactions, the second-wave signal amplification

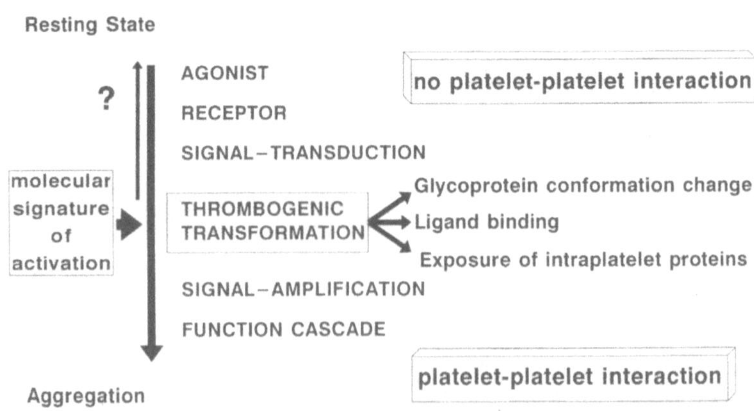


**Fig. 2.** Molecular cell processing cascade of platelet activation following agonist stimulation

system recruits platelets for depositing from the flowing blood by formation of the potent aggregating agent thromboxane $A_2$. Conversely, platelet activation is diminished under the control of prostacyclin and NO. They are derived from the endothelial cells and indirectly modulate the activation of the ligand binding site of membrane glycoproteins (by the cAMP/cGMP-proteinkinase-dependent phosphorylation status of the regulatory vasodilator-stimulated phosphoprotein (VASP); HORSTRUP et al. 1994; HELLBRÜGGE and WALTER 1993).

## III. Defining Diagnostic Epitopes

As outlined above, a diagnostic procedure must be adopted following the a priori categories of diagnostic questions to be answered: (1) reactivity (potential) and (2) activation (status). This has immediate impact on the choice of markers to be tested under individual diagnostic conditions. Alterations in the clonal megakaryocytic thrombopoiesis in bone marrow (primary) have to be assumed in case staining the sample platelets for constitutive membrane glycoproteins results in typical deficiency or overexpression patterns. Intravasal platelet activation must be assumed in case staining for function-dependent epitopes not accessible for the probe under resting conditions leads to positive results (NIEUWENHUIS et al. 1987).

### 1. The Constitutive Stain

As the intrinsic platelet membrane constituents such as glycoproteins are constitutively expressed on nearly all platelets, they are referred to as pan-platelet markers, e.g., residues of the GPIB (CD42b) or GPIIB/IIIA (CD41) molecules or subunits, e.g., GPIIIA (CD61). The functional impact of some platelet glycoproteins has been evaluated from specific defects in Bernard Soulier's syndrome and Glanzmann's thrombasthenia. Glycoprotein IB mediates the adhesion of platelets at high blood shear rates in the microcirculation by interaction with von Willebrand Factor (vWF). Glycoprotein IIB/IIIA is predominantly supposed to bear binding sites for fibrinogen and is essential for normal aggregation behavior. It was shown that both glycoproteins IB and IIB/IIIA are expressed on more than 90% of analyzed normal platelets. Using the "Düsseldorf protocols," the number of binding sites for monoclonal antibodies to GPIB/GPIIB/IIIA molecules was calculated at $39100 \times 1.3 \pm 1$ and $62700 \times 1.3 \pm 1$ (geometric mean $\times$ standard deviation factor $\pm 1$) on control platelets, respectively (TSCHOEPE et al. 1988, 1990b).

### 2. The Functional Stain (Fig. 3)

Platelet antigenicity varies in accordance to the state of activation: (1) receptor conformation change; (2) receptor ligand binding; (3) ligand conformational change; (4) translocation of intraplatelet granules; and (5) secretion of proteins that are again bound (with 1:1 stoichiometry) to specific receptors, e.g.,

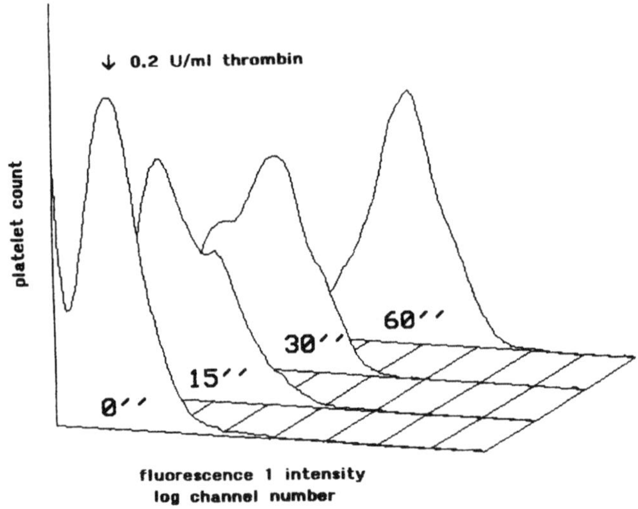


**Fig. 3.** Flow cytometric kinetics of platelet exposure of activation-dependent epitopes (e.g., CD62 = P-selectin) following agonist stimulation (e.g., 0.2 U/ml thrombin)

thrombospondin. Thereby, subcellular structures not detectable in the resting state become antigenic on single platelets that still circulate and bear the signature of very recent activation without having functionally reacted with other platelets or cells. This system represents a natural binary principle allowing classification of each platelet under investigation as activated or not. Analysis of a significant number of sample platelets subsequently allows a quantitative stroboscopic description of the activation status of the circulating platelet population at the time of blood sampling (for a review see TSCHOEPE and SCHWIPPERT 1993).

## IV. Platelet–Leukocyte Coaggregates

Initially, platelet activation leads to platelet-(sub-)endothelial, platelet-leukocyte [polymorph neutrophils (PMNs) or monocytes/macrophages (M/Ms)], or platelet-platelet adhesion. This is due to the fact that adhesion molecules regulated by biochemical signal transduction mechanisms act as ultrastructural transmitters in the sequence of events (SAVAGE et al. 1992; ROTH 1992): P-selectin (CD62) rapidly appears in the outer membrane of activated endothelial cells and platelets. It serves as receptor for oligosacharides containing sialic acid (Sialyl Lewis$^{X}$, CD15) within the monocyte/PMN membrane. When such cells adhere to locally activated endothelium (nondenuding injury), a nidus is formed for further adherence of P-selectin-positive platelets that become particularly thrombogenic by concomitant exposure of the activated receptor for soluble fibrinogen (GPIIB/IIIA). Later, when platelets become activated in response to nuding endothelial injury, exposure of

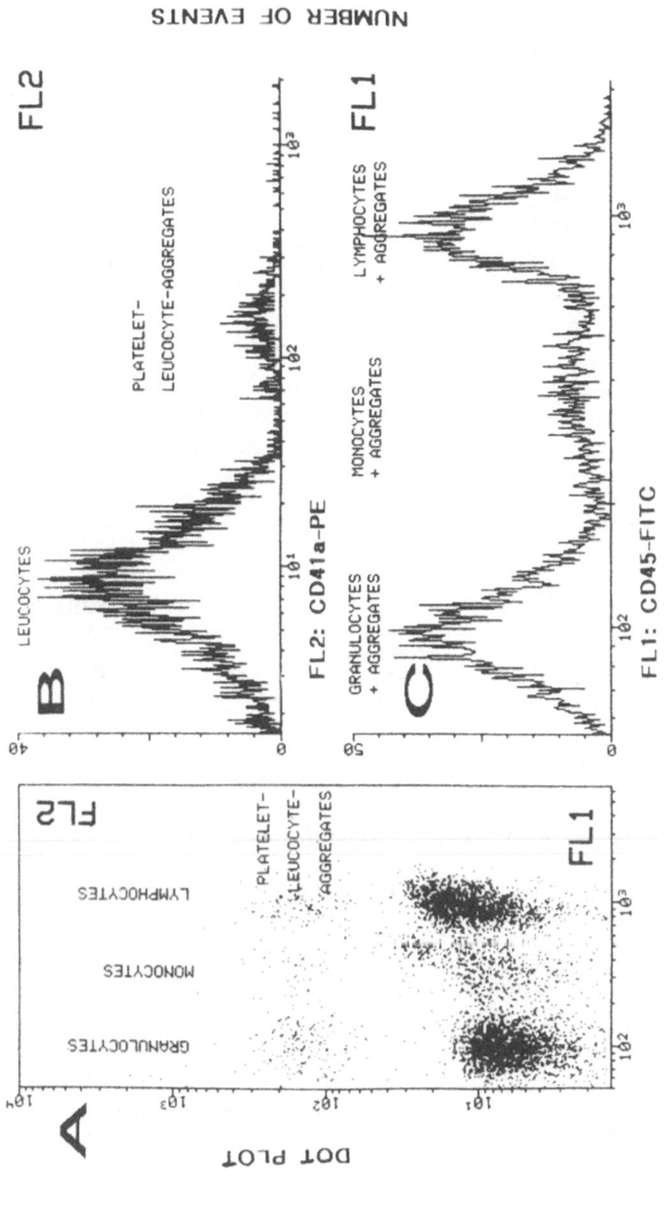


**Fig. 4. A** CD45 (pan-leukocyte marker)-PE/CD41a (pan-thrombocyte marker)-FITC double-staining flow dot plot of 5000 whole blood events for the identification of platelet-leukocyte coaggregates. **B** Histogram of CD41a-positive events; platelets include all platelet-leukocyte coaggregates. **C** Histogram of CD45-positive events. The three populations with different antigen expression intensity correspond to granulocytes (PMNs), monocytes, and lymphocytes. *CD*, cluster designation; *PE*, phycoerythrin; *FITC*, fluorescein isothiocyanate

P-selectin leads to parallel local recruitment (and activation) of white blood cells that appear to be irreversibly anchored on the platelet surface by the $\beta_2$-integrin ligand, intercellular adhesion molecule 2 (ICAM-2; DIACOVO et al. 1994; LORANT et al. 1993; MCEVER 1991; PALABRICIA et al. 1992). Presumably, this interaction also occurs in flowing blood, facilitating transcellular signalling and metabolism, e.g., induction of tissue factor transcription in circulating monocytes by P-selectin-positive platelets (CELI et al. 1994; MAUGERI et al. 1994). Cellular hemostasis must always be understood as an interdependent cell communication process in which, depending on the underlying disease, different cells than the types mentioned above act as leading pacer cells. Thus, platelet-leukocyte coaggregates interface inflammatory, atherogenetic, and thrombogenic responses and may be involved in the postocclusive no-reflow phenomenon or ischemia-reperfusion-type vascular injury (YEO et al. 1994). Therefore, attempts have been made to quantitate this platelet-leukocyte interaction by adapting the flow cytometry platelet test (RINDER et al. 1991b): the concept of pan-platelet markers, such as constitutive CD41 directly conjugated by PE, and pan-leukocyte markers, such as CD45 directly conjugated by FITC, allows double-staining of whole blood samples. Running the sample through a flow cytometer as described above, $CD45^+/CD41^+$ events are easily identified as platelet-leukocyte coaggregates. By subtyping the leukocyte population according to its CD45 expression intensity, platelet-PMN, platelet-M/M, and platelet-lymphocyte aggregates can be differentially quantitated (see Fig. 4 for example).

# D. Applications in Platelet Pathology

The central pathophysiological role of platelets has already been worked out. Since they are stem cell derived and suspended in the environment of flowing blood, thereby perfusing all organs, it is easily realized that they are involved in or respond to diseases, abnormalities, or manipulations in many clinical situations. However, from a diagnostic perspective, it must be admitted that the contribution of platelets rarely becomes substantiated in such a way that the acting physician could be guided by objective platelet parameters in his actions. Here, data is presented which demonstrate that this unsatisfactory situation can often be terminated by means of single platelet flow cytometry.

## I. Hemorrhagic Diathesis

Adequate platelet function depends on the normal expression pattern of the corresponding surface receptors. Flow cytometry is a quick and useful tool for diagnosis of disease states comprising hemorrhage associated with defective platelet glycoprotein expression, e.g.: Bernard Soulier's syndrome with GPIB (CD42b) lacking; Glanzmann's thrombasthenia with GPIIB/IIIA (CD41) lacking; leukemia with differential clonal expansion; and constitutive CD36 deficiency in Japanese subjects (Adelman et al. 1992; Ginsberg et al. 1990; Jennings et al. 1986; Kehrel et al. 1991; Nomura et al. 1993). Recently, single platelet flow cytometry has also shown that uremia leads to loss of function (fibrinogen binding) of constitutively expressed GPIIB/IIIA. Since dialysis appeared to restore these abnormalities, this finding must at least be partially viewed as a secondary phenomenon (Dechavanne et al. 1987; Gawaz and Bogner 1994; Lindahl et al. 1992a).

In diseases with thrombocytopenia, flow cytometry assays were helpful in indicating and quantifying platelet antibody binding even in the antenatal state (Ault 1988; Heim and Petersen 1988; Lazarchick and Hall 1986; Lin et al. 1990; Marshall et al. 1994; Mizumoto et al. 1991; Rosenfeld et al. 1987; Visentin et al. 1990; Worfolk and MacPherson 1991). Measurement of reticulated platelets allows an estiamte of the production activity of bone marrow megakaryocytes (Ault 1993; Kienast and Schmitz 1990). When bleeding must be clinically addressed, low-platelet-count transfusion of platelet concentrates is required. The quality of such blood products (e.g., activation-by-handling bias, therapeutic reactivity, compatible HLA cross-match status, nonplatelet contamination) can be reliably controlled by platelet activation marker testing (Fijnheer et al. 1990, Gates and MacPherson 1994; Goodall 1991; Lombardo et al. 1993; Metzelaar et al. 1993; Pedigo et al. 1993; Rinder et al. 1991c; Wang et al. 1989).

## II. Thrombotic Diathesis

Enhanced platelet glycoprotein expression is believed to contribute to prethrombotic states. In diabetes mellitus, a significant increase of 39% to $545000 \times 1.28 \pm 1$ GPIB binding sites and of 24% to $77500 \times 1.3 \pm 1$ GPIIB/IIIA binding sites per platelet could be demonstrated with the flow cytometric test and has been proposed as a molecular explanation for the well-known platelet hyperreactivity, similar to myocardial infarction patients (Giles et al. 1994; Schultheiss et al. 1994; Tschoepe et al. 1990b).

Hyperactive platelets operate by three mechanisms: (1) capillary microembolism; (2) local progression of preexisting vascular lesions by secretion of constrictive, mitogenic, and oxidative substances; and (3) trigger of the prognosis-limiting arterial thrombotic event. The predominant task of the laboratory remains to provide parameters that identify patients with activated platelets, possibly indicating the individual risk for thrombotic end points. This appears particularly important for patients who may randomly suffer from severe bleeding or thrombosis such as in myeloproliferative syndromes (Michelson 1987; Wehmeier et al. 1991). The principle of flow cytometry platelet activation marker testing has now been applied to a wide variety of disease states associated with an increased risk of arterial thrombosis (Table 3).

It was anticipated that patients with myocardial infarction show activated platelets after the event. However, the finding that patients with final stage

**Table 3.** Profile of platelet activation marker expression in patients clinically assumed to be in a prethrombotic condition

| Prethrombotic condition | CD62 | TSP | CD63 | CD41-LIBS | CD31 |
|---|---|---|---|---|---|
| Coronary heart disease* | ↑ | ↑ | (↑) | n.t. | n.t. |
| PTCA* | ↑↑ | ↑↑ | (↑) | ↑↑ | n.t. |
| Cardiopulmonary bypass | ↑ | (↑) | ↑ | ↑ | ↓ |
| Myocardial infarction* | ↑ | n.t. | ↑↑ | n.t. | n.t. |
| ARDS | ↑ | ↑ | n.t. | n.t. | n.t. |
| Generalized atherosclerosis* | (↓) | n.t. | ↑↑ | n.t. | n.t. |
| Hemodialysis/plasmapheresis | ↑ | n.t. | n.t. | ↑ | n.t. |
| Multiple organ failure/sepsis | ↑ | ↑ | ↑ | ↑ | n.t. |
| Crohn's disease* | ↑(↑) | ↑ | ↑(↑) | n.t. | n.t. |
| Type-I diabetes (recent onset)* | ↑(↑) | ↑(↑) | ↑(↑) | ↑ | n.t. |
| Type-II diabetes* | ↑ | ↑ | (↓) | ↓ | ↑↑ |
| Preeclampsia | n.t. | n.t. | ↑ | n.t. | n.t. |

Prepared according to BECKER et al. 1994; COLLINS et al. 1994; GAWAZ et al. 1994a; GEORGE et al. 1986; JANES et al. 1993; JANES and GOODALL 1994; REVERTER et al. 1994; RINDER et al. 1991a; SCHARF et al. 1992; TSCHOEPE et al. 1995, 1993c; TSCHOEPE and SCHWIPPERT 1993.
(*) indicates our own data.
PTCA, percutaneous transluminal angioplasty; ARDS, acute respiratory distress syndrome; TSP, thrombospondin; LIBS, ligand-induced binding sites; CD, cluster designation; ↑, increased expression; ↓, decreased expression of the epitope; n.t., not tested.

peripheral atherosclerosis, coronary heart disease (CHD), and simply diabetes also show a significant proportion of activated platelets in their circulation indicates a theoretical potential for predictability (BECKER et al. 1994; GALT et al. 1991; TSCHOEPE et al. 1995). The computer-based technology of single platelet flow cytometry (SPFC) allowed investigation of whether platelet activation is related to any particular subpopulation of different platelet size. As expected, there was a clear increase in binding sites for activation markers of platelet size, but this increase was more enhanced for platelets from patients with diabetes and myocardial infarction: large platelets circulate in an activated state. This finding provides functional support of the concept that large platelets are a signature of a prethrombotic state and have predictive power for the occurrence of ischemic heart attacks (SCHULTHEISS et al. 1994; TSCHOEPE et al. 1991). This has recently been confirmed for CHD patients with activated platelets who underwent percutaneous transluminal coronary angioplasty [Düsseldorf PTCA Platelet Study (DPPS); TSCHOEPE et al. 1993c]. Activated platelets as measured flow cytometrically predict early reocclusion ("acute ischemic events"). Moreover, it has to be assumed that rescue procedures for preserving arterial vessel partency lead to additional iatrogenic platelet activation, e.g., by fibrinolytic drugs or by the injection of contrast media (CHRONOS et al. 1993; EJIM et al. 1990; FITZGERALD et al. 1988; GRABOWSKI et al. 1994).

Data from which Table 3 has been compiled represent mean value comparisons including wide overlaps between the groups studied and the corresponding reference values. This illustrates the problem with patient groups assumed to be at a prethrombotic stage: some are activated, others only partially, and still others are not. Prethrombotic states are characterized by the sum of subtle changes that can hardly be discriminated and measured in detail. However, flow cytometry data give a clear numerical quantification of the activation status of the cellular hemostasis system, e.g., percentage of marker-positive platelets or activation index. This allows the use of classic definitions of clinical chemistry that ultimately allow classification of the individual patient as having or not having activated platelets. This was the prerequisite for allowing the controlled design of the Düsseldorf PTCA Platelet Study.

From Table 3 it becomes evident that platelet activation marker analysis does not provide a homogenous result with regard to the chosen marker panel. This obviously reflects the different functional aspects of the individual epitopes. The subpoulation of activated platelets itself consists of subpopulations that are differentiated by function and by the time level within the molecular processing cascade after activation stimulus of the platelets: it appears reasonable that initiation of extracorporal circulation results in a nearly uniformly increased platelet activation pattern, and the same initiation may account for the increase observed for hemodialysis or plasmapheresis (Gawaz et al. 1994a; Reverter et al. 1994; Wun et al. 1992). In contrast, it seems that in patients chronically treated by extracorporal circulation (HELP-apheresis in end-stage atherosclerosis patients), $\alpha$-degranulated P-selectin (CD62) platelets may be eliminated by specific interaction with activated leukocytes and endothelial cells of the RES system, for example (Gawaz et al. 1994b), whereas lysosomal degranulated $CD63^+$ platelets may accumulate under the chronic activation of the coagulation cascade with periprocedural thrombin generation.

## III. Inflammation

Thrombogenesis and inflammation are closely linked by the molecular cross talk of corpuscular and vessel wall cells (Fisher and Meiselman 1994). This helps to understand the as yet underestimated association of inflammatory states with prethrombotic states: the concept of adhesion molecules explains how streaming white blood cells can adhere locally to activated endothelial cells (selectin-mediated), become triggered, stick (integrin-mediated), and subsequently transmigrate or release potent damaging mediators, e.g., reactive oxygen species (Adams and Shaw 1994; Pratico et al. 1993). Sialyl $\text{Lewis}^X$ (CD15) acts as a ligand for the adherence of CD62 $(\text{P-selectin})^+$ platelets. When platelets become activated, exposure of CD62 (P-selectin) leads to parallel local recruitment (and activation) of $CD15^+$ white blood cells.

In line with this theory, patients with chronic inflammatory bowel disease such as Crohn's disease or ulcerative colitis have been shown to have elevated levels of activated platelets in their circulation (COLLINS et al. 1994). Interestingly, in this patient group platelet CD62 (P-selectin) expression in particular was most enhanced in the microcirculation (TSCHOEPE et al. 1993a). Moreover, in comparison with other investigated disease states, we also reported the highest relative eightfold increase in activated platelets in this cohort of patients (TSCHOEPE et al. 1993b). Complementary to this, using platelet flow cytometry, SCHAUFELBERGER et al. have reported increased levels of cytokine receptor (IL-1 and IL-8) positive platelets in patients with inflammatory bowel disease. They were also able to demonstrate induction of P-selectin positivity in isolated platelets preincubated with both interleukins (SCHAUFELBERGER et al. 1994). Activated platelets are part of the cellular procoagulant pathway and may act as an important factor in causing thrombotic occlusion of the functional end arteries with subsequent hypoxic damage of the affected circulation area. The detection of P-selectin $(CD62)^+$ platelets in the microcirculation indicates local platelet $\alpha$-granule release of thrombotic precursors, such as fibrinogen, adhesion ligands, thrombospondin, or von Willebrand factor, and mitogenic activity such as platelet-derived growth factor. In concert, these factors may cause structural alterations and further flow disruption in the microvessels. Accordingly, GAWAZ et al. recently showed uniformly increased levels of activated (LIBS) platelets in patients with septicemia. In contrast, patients having reached the end point of multiple organ failure showed even more increased levels of predominantly degranulated (P-selectin$^+$, thrombospondin$^+$) platelets, indicating strong degranulation presumably in the microcirculation, which is well known to be disrupted in this condition (GAWAZ et al. 1995). In this context, it is interesting to note that platelets may expose *Staphylococcus aureus* binding sites, and this phenomenon can be quantitatively measured by platelet flow cytometry (YEAMAN et al. 1992). In a more subtle model of chronic inflammation, *Helicobacter pylori* has been shown to induce platelet-leukocyte coaggregation. The inhibitory effect of a P-selectin antibody suggests a pathogenetic role in the destructive mechanisms leading to albumin leakage of rat mesentery vessels (KUROSE et al. 1994).

## E. Pharmacology

Epidemiological and pathophysiological considerations lead to the rational indication for vascular prevention therapy, e.g., with antiplatelet agents. At present, preventive therapy is usually applied following the clinical estimation of individual vascular risk rather than following the assessment of the risk-benefit ratio for an individual patient based on a laboratory stratification parameter. The assessment of platelet activation status reflecting the real-time activity of the cellular hemostasis system at the time of blood sampling has been shown for the first time to predict the acute outcome from a widespread cardiological procedure of reassuring vessel patency, PTCA (percutaneous transluminal coronary angioplasty; TSCHOEPE et al. 1993c). The consequences exceed the cardiological application setting. It must be assumed that the platelet activation marker test provides predictive information for the occurrence of (arterial) occlusive events in general. Based on this assumption, knowledge of the intravasal platelet activation status will allow direct change of the medical management of patients characterized as activated with the intention to treat them for the prevention of clinical events. Diagnostic identification of patients with an activated platelet system may help to optimize the effectiveness of the individual therapeutic approach. Drug therapy would then be stratified only to those patients at particular risk, without treating patients who will not respond adequately to antiplatelet therapy but may be affected by side effects. This amendment becomes particularly important with the increasing risk and cost level of the planned medical intervention.

Optimized pharmacological intervention does not only require stratified indication of therapy, but also customized choice of the individual drug. A surprising aspect evolving from the DPPS study was that within the acute PTCA setting, "gold-standard" anticoagulation with heparin and aspirin given acutely appeared insufficient to passivate periprocedurally activated cellular hemostasis (TSCHOEPE et al. 1993c). The second part of the study reports that monitoring the patients revealed spontaneous increases in platelet activation after PTCA as measured flow-cytometrically by the expression of $\alpha$-granule secretion markers (KOLAROV et al. 1995). This provides a rationale for prolonged and aggressive antiplatelet therapy at least in those patients identified by flow cytometric monitoring of activated platelets. Recently, fairly controlled data of the EPIC trial revealed the therapeutic potential of additional prolonged periprocedural therapy with antibody fragments directed against the fibrinogen binding domain of GPIIB/IIIA ("fibrinogen receptor blockers"; LEFKOVITS et al. 1995; CALIFF 1994; TOPOL et al. 1994). Single platelet flow cytometry appears useful in directly measuring receptor coverage by the drug, specific treatment efficacy caused by inhibition of ligand fibrinogen binding, and the overall treatment efficacy as indicated by the net peripheral platelet activation status (DUNSTAN 1985; FARADAY et al. 1994; FROJMOVIC and WONG 1991; FROJMOVIC et al. 1991; JACKSON and JENNINGS 1989; JENES et al. 1993; LINDAHL et al. 1992a).

Interestingly, the monitoring function of the activation marker test was used to demonstrate the prothrombotic effect of cocaine (KUGELMASS et al. 1993). It was useful for demonstrating the thrombasthenic potential of penicillin-type antibiotics (INGALLS and FREIMER 1992; PASTAKIA et al. 1993) as it visualized the antisecretory potential of high-dose heparin (ROHRER et al. 1992). Platelet activation marker testing has recently also been proposed for individually monitoring the effect of endurance exercise (KESTIN et al. 1993).

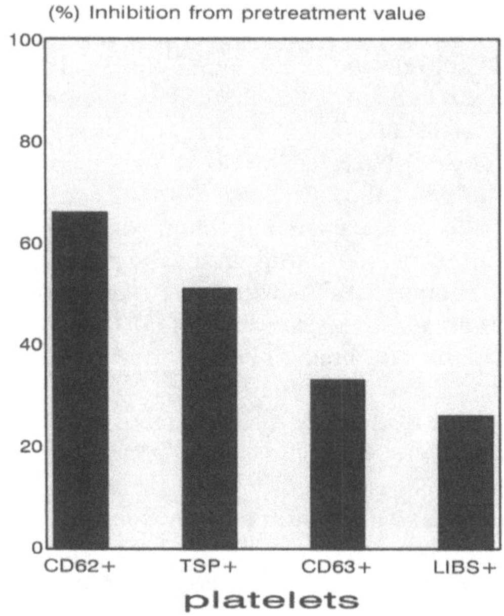


**Fig. 5.** Inhibition of platelet activation under chronic low-dose (100 mg/d) treatment with acetyl-salicylic acid; percentage of platelets of pretreatment value exposing the respective activation marker CD62, thrombospondin, CD63, or CD41-LIBS. *CD*, cluster designation; *TSP*, thrombospondin; *LIBS*, ligand-induced binding sites

This information could ultimately guide the physician to adjust the dosage and the drug combination to the individual patient. We were recently able to show that this concept is valid by monitoring chronic intake of low-dose acetylsalicylic acid resulting in a decrease in circulating activating platelets, but not in passivation of GPIIB/IIIA activation following thrombin stimulation (Fig. 5). Finally, by combining a standardized platelet isolation and stimulation model in vitro, flow cytometry will help to evaluate and understand the direct cellular effects of drugs with proposed antiplatelet potential (PHILP 1981). Using such a cellular inhibition assay of stimulated expression of platelet activation antigens, we demonstrated that acute aspirin exposure does not in any way affect adhesive platelet reactions (despite thromboxane inhibition; TSCHOEPE et al. 1994). In contrast, the strong inhibitory effect of prostacyclin mimetics upon the stimulated expression of platelet adhesive proteins suggests a potent inhibitory potential upon platelet-endothelial adhesion mechanisms exceeding the biochemically defined scope of action of these drugs (ROESEN et al. 1994b).

The arguments presented here allow the assignment of single platelet flow cytometry as a key analytical technology with the potential to switch from reactive to preventive therapy in platelet-based diseases, particularly in the field of atherothrombosis.

## References


Abrams CS, Shattil SJ (1991) Immunological detection of activated platelets in clinical disorders. Thromb Haemost 65:467–473

Abrams CS, Ellinson N, Budzynski AZ, Shattil SJ (1990) Direct detection of activated platelets and platelet-derived microparticles in humans. Blood 75:128–138

Adams DH, Shaw S (1994) Leukocyte–endothelial interactions and regulation of leukocyte migration. Lancet 343:831–836

Adelman B, Michelson AD, Handin RI, Ault KA (1985) Evaluation of platelet glycoprotein Ib by fluorescence flow cytometry. Blood 66:423–427

Adelman B, Carlson P, Handin RI (1992) Evaluation of platelet surface antigens by fluorescence flow cytometry. Methods Enzymol 215:420–428

Alshameeri R, Mammen EF (1993) Evaluation of an in vitro bleeding time device. Thrombostat 4000. Thromb Haemost 69 (Abstr):1146

Ault KA (1988) Flow cytometric measurement of platelet-associated immunoglobulin. Pathol Immunopathol Res 7:395–408

Ault KA (1993) Flow cytometric measurement of platelet function and reticulated platelets. Ann N Y Acad Sci 677:293–308

Ault KA, Rinder HM, Mitchell JG, Rinder CS, Lambrew CT, Hillmann RA (1989) Correlated measurement of platelet release and aggregation in whole blood. Cytometry 10:448–455

Basic-Micic M, Roman C, Herpel U, Kling E, Scholz B, Breddin HK (1992) Platelet induced thrombin generation time: a new sensitive global assay for platelet function and coagulation. Haemostasis 22:309–321

Becker RC, Tracy RP, Bovill EG, Mann KG, Ault K (1994) The clinical use of flow cytometry for assessing platelet activation in acute coronary syndromes. TIMI-III Thrombosis and anticoagulation group. Coron Artery Dis 5:339–345

Born GVR (1962) Aggregation of blood platelets by adenosine diphosphate and its reversal. Nature 194:927–929

Califf RM (1994) The EPIC investigators. Use of a monoclonal antibody directed against the platelet glycoprotein IIb/IIIa receptor in high-risk coronary angioplasty. N Engl J Med 7:956–961

Celi A, Pellegrini G, Lorenzet R, DeBlasi A, Ready N, Furie BC, Furie B (1994) P-Selectin induces the expression of tissue factor on monocytes. Proc Natl Acad Sci 91:8767–8771

Chronos NA, Goodall AH, Wilson DJ, Sigwart U, Buller NP (1993) Profound platelet degranulation is an important side effect of some types of contrast media used in interventional cardiology. Circulation 88:2035–2044

Coller BS, Beer JH, Scudder LE, Steinberg MH (1989) Collagen-platelet interactions: evidence for a direct interaction of collagen with platelet GPIa/IIa and an indirect interaction with platelet GPIIb/IIIa mediated by adhesive proteins. Blood 74:182–192

Collins CE, Cahill MR, Newland AC, Rampton DS (1994) Platelets circulate in an activated state in inflammatory bowel disease. Gastroenterology 106:840–845

Corash L (1990) Measurement of platelet activation by fluorescence-activated flow cytometry. Blood Cells 16:97–106

Davies TA, Weil GJ, Simons ER (1990) Simultaneous flow cytometric measurements of thrombin-induced cytosolic pH and Ca2– fluxes in human platelets. J Biol Chem 265:11522–11526

De-Bruijne-Admiraal LG, Modderman PW, Von-dem-Borne AE, Sonnenberg A (1992) P-selectin mediates CA(2–)-dependent adhesion of activated platelets to many different types of leukocytes: detection by flow cytometry. Blood 80:134–142

Dechavanne M, Ffrench M, Pages J, Ffrench P, Boukerche H, Bryon PA, McGregor JL (1987) Significant reduction in the binding of a monoclonal antibody (LYP 18) directed against the IIb/IIIa glycoprotein complex to platelets of patients having undergone extracorporal circulation. Thromb Haemost 57:106–109

Diacovo TG, deFougerolles AR, Bainton DF, Springer TA (1994) A functional integrin ligand on the surface of platelets: interacellular adhesion molecule-2. J Clin Invest 94:1243–1251
Dunstan RA (1985) Use of fluorescence flow cytometry to study the binding of various ligands to platelets. J Histochem Cytochem 33:1176–1179
Ejim OS, Powling MJ, Dandona P, Kernoff PB, Goodall AH (1990) A flow cytometric analysis of fibronectin binding to platelets from patients with peripheral vascular disease. Thromb Res 58:519–524
Faraday N, Goldschmidt-Clermont P, Dise K, Bray PF (1994) Quantitation of soluble fibrinogen binding to platelets by fluorescence-activated flow cytometry. J Lab Clin Med 123:728–740
Fareed J, Bick RL, Hoppensteadt DA, Bermes EW (1995) Molecular markers of hemostatic activation: applications in the diagnosis of thrombosis and vascular and thrombotic disorders. Clin Appl Thromb Hemost 1:87–102
Fijnheer R, Modderman PW, Veldman H, Ouwehand WH, Nieuwenhuis HK, Roos D, de Korte D (1990) Detection of platelet activation with monoclonal antibodies and flow cytometry. Changes during platelet storage. Transfusion 30:20–25
Fisher TC, Meiselman HJ (1994) Polymorphonuclear leukocytes in ischaemic vascular disease. Thromb Res 74:21–34
Fitzgerald DJ, Catella F, Roy L, Fitzgerald GA (1988) Marked platelet activation in vivo after intravenous streptokinase in patients with acute myocardial infarction. Circulation 77:142
Frojmovic M, Wong T (1991) Dynamic measurements of the platelet membrane glycoprotein IIb–IIIa receptor for fibrinogen by flow cytometry. II. Platelet size-dependent subpopulation. Biophys J 59(4):828–837
Frojmovic M, Wong T, van de Ven T (1991) Dynamic measurements of the platelet membrane glycoprotein IIb-IIIa receptor for fibrinogen by flow cytometry. I. Methodology, theory and results for two distinct activators. Biophys J 59(4):815–827
Galt SW, McDaniel MD, Ault KA, Mitchell J, Cronenwett JL (1991) Flow cytometric assessment of platelet function in patients with peripheral arterial occlusive disease. J Vasc Surg 14:747–755
Gates K, MacPherson BR (1994) Retrospective evaluation of flow cytometry as a platelet crossmatching procedure. Cytometry 18:123–128
Gawaz MP, Bogner C (1994) Changes in platelet membrane glycoproteins and platelet-leukocyte interaction during hemodialysis. Clin Invest 72:424–429
Gawaz MP, Bogner C, Gurland HJ (1993) Flow-cytometric analysis of mepacrine-labelled platelets in patients with end-stage renal failure. Haemostasis 23:284–292
Gawaz MP, Dobos G, Spath M, Schollmeyer P, Gurland HJ, Mujais SK (1994a) Impaired function of platelet membrane glycoprotein IIb-IIIa in end-stage renal disease. J Am Soc Nephrol 5:36–46
Gawaz MP, Mujais SK, Schmidt B, Gurland HJ (1994b) Platelet-leukocyte aggregation during hemodialysis. Kidney Int 46:489–495
Gawaz M, Fateh-Moghadam S, Pilz G, Gurland HJ, Werdan K (1995) Severity of multiple organ failure (MOF) but not spesis correlates with irreversible platelet degranulation. Infection 23:16–23
George JN, Pickett EB, Saucerman S, McEver RP, Kunicki TJ, Kieffer N, Newman PJ (1986) Platelet surface glycoproteins. Studies on resting and activated platelets and platelet membrane microparticles in normal subjects, and observations in patients during adult respiratory distress syndrome and cardiac surgery. J Clin Invest 78:340–348
Giles H, Smith RE, Martin JF (1994) Platelet glycoprotein IIb-IIIa and size are increased in acute myocardial infarction. Eur J Clin Invest 24:69–72
Ginsberg MH, Frelinger AL, Lam SCT, Forsyth J, McMillan R, Plow EF, Shattil SJ (1990) Analysis of platelet aggregation disorders based on flow cytometric analysis

of platelets membrane glycoprotein IIb-IIIa with conformation specific monocloncal antibodies. Blood 76:2017–2023

Goodall AH (1991) Platelet activation during preparation and storage of concentrates: detection by flow cytometry. Blood Coagul Fibrinolysis 2:377–382

Grabowski EF, Head C, Michail HA, Jang IK, Gold H, Benoit SE, Michelson AD (1994) Effects of contrast media on platelet activation using flowing whole blood aggregometry and flow cytometry of platelet membrane glycoproteins. Invest Radiol 29[Suppl 2]:198–200

Hancock V, Martin JF, Lelchuk R (1993) The relationship between human megakaryocyte nuclear DNA content and gene expression. Br J Haematol 85:692–697

Heim MC, Petersen BH (1988) Detection of platelet-associated immunoglobulin in immune thrombocytopenia by flow cytometry. Diagn Clin Immunol 5:309–313

Hellbrügge M, Walter U (1993) The regulation of platelet function by protein kinases. In: Huang CK, Shaafi RI (eds) Protein kinases in blood cell function. CRC, Boca Raton, pp 245–298

Horstrup K, Jablonka B, Just M, Kochsiek K, Walter U (1994) Phosphorylation of focal adhesion vasodilator-stimulated phosphoprotein at Ser 157 in intact human platelets correlates with fibrinogen receptor inhibition. Eur J Biochem 225:21–27

Ingalls CS, Freimer EH (1982) Detection of antibiotic-induced platelet dysfunction in whole blood using flow cytometry. J Antimicrob Chemother 29:313–321

Jackson CW, Jennings LK (1989) Heterogeneity of fibrinogen receptor expression on platelets activated in normal plasma with ADP: analysis by flow cytometry. Br J Haematol 72:407–414

Janes SL, Goodall AH (1994) Flow cytometric detection of circulating activated platelets and platelet hyper-responsiveness in pre-eclampsia and pregnancy. Clin Sci (Colch) 86:731–749

Janes SL, Wilson DJ, Chronos N, Goodall AH (1993) Evaluation of whole blood flow cytometric detection of platelet bound fibrinogen on normal subjects and patients with activated platelets. Thromb Haemost 70:659–666

Jennings LK, Ashmun RA, Wand WC, Dockter ME (1986) Analysis of human platelet glycoproteins IIb-IIIa and glanzmann's thrombasthenia in whole blood by flow cytometry. Blood 68:173–179

Jennings LK, Dockter ME, Wall CD, Fox CF, Kennedy DM (1989) Calcium mobilization in human platelets using indo-1 and flow cytometry. Blood 74(8):2674–2680

Johnston GI, Pickett EB, McEver RP, George JN (1987) Heterogeneity of platelet secretion in response to thrombin demonstrated by fluorescence flow cytometry. Blood 69:1401–1403

Kanz L, Mielke R, Fauser AA (1988) Analysis of human hemopoietic progenitor cells for the expression of glycoprotein IIIa. Exp Hematol 16:741–747

Kehrel B, Kronenberg A, Schwippert B, Niesing-Bresch D, Niehues U, Tschöpe D, van de Loo J, Clemetson KJ (1991) Thrombospondin binds normally to glycoprotein IIIb deficient platelets. Biochem Biophys Res Commun 179:985–991

Kestin AS, Ellis PA, Barnard MR, Errichetti A, Rosner BA, Michelson AD (1993) Effect of strenuous exercise on platelet activation state and reactivity. Circulation 88:1502–1511

Kienast J, Schmitz G (1990) Flow cytometric analysis of thiazole orange uptake by platelets: a diagnostic aid in the evaluation of thrombocytopenic disorders. Blood 75:116–121

Kolarov P, Tschoepe D, Nieuwenhuis HK, Gries FA, Strauer B, Schultheiß HP (1995) PTCA: periprocedural platelet activation. Part II of the Duesseldorf PTCA platelet study. Eur Heart J 17:1216–1222

Kugelmass AD, Oda A, Monahan K, Cabral C, Ware JA (1993) Activation of human platelets by cocaine. Circulation 88:876–883

Kurose I, Granger DN, Evans DJ, Evans DG, Graham DY, Miyasaka M, Anderson DC, Wolf RE, Cepinskas G, Kvietys PR (1994) Helicobacter pylori-induced mi-

crovascular protein leakage in rats: role of neutrophils, mast cells, and platelets. Gastroenterology 107:70–79
Lazarchick J, Hall SA (1986) Platelet-associated IgG assay using flow cytometric analysis. J Immunol Methods 87:257–265
Lefkovits J, Plow EF, Topol EJ (1995) Platelet glycoprotein IIb/IIIa receptors in cardiovascular medicine. Mechanisms of disease. N Engl J Med 332:1553–1559
Lin RY, Levin M, Nygren EN, Normal A, Lorenzana FG (1990) Assessment of platelet antibody by flow cytometric and ELISA techniques: a comparison study. J Lab Clin Med 116:479–486
Lindahl TL, Festin R, Larsson A (1992a) Studies of fibrinogen binding to platelets by flow cytometry: an improved method for studies of platelet activation. Thromb Haemost 68:221–225
Lindahl TL, Lundahl J, Netre C, Egberg N (1992b) Studies of the platelet fibrinogen receptor in Glanzmann patients and uremic patients. Thromb Res 67:457–466
Lombardo JF, Cusack NA, Rajagopalan C, Sangaline RJ, Ambruso DR (1993) Flow cytometric analysis of residual white blood cell concentration and platelet activation antigens in double filtered platelet concentrates. J Lab Clin Med 122:557–566
Lorant DE, Topham MK, Whatley RE, McEver RP, McIntyre TM, Prescott SM, Zimmerman GA (1993) Inflammatory roles of P-selectin. J Clin Invest 92:559–570
Lovett EJ (1984) Application of flow cytometry to diagnostic pathology. Lab Invest 50:115–140
Marshall LR, Brogden FE, Roper TS, Barr AL (1994) Antenatal platelet antibody testing by flow cytometry – results of a pilot study. Transfusion 34:961–965
Marti GE, Magruder L, Schuette WE, Gralnick HR (1988) Flow cytometric analysis of platelet surface antigens. Cytometry 9:448–455
Maugeri N, Evangelista V, Celardo A, Dell'Elba G, Martelli N, Piccardoni P, de Gaetano G, Cerletti C (1994) Polymorphonuclear leukocyte-platelet interaction: role of P-selectin in thromboxane B2 and leukotriene C4 cooperative synthesis. Thromb Haemost 72:450–456
McCrae KR, Shattil SJ, Cines DB (1990) Platelet activation induces increased Fc gamma receptor expression. J Immunol 144:3920–3927
McEver RP (1991) Selectins: novel receptors that mediate leukocyte adhesion during inflammation. Thromb Haemost 65:223–228
Metzelaar MJ, Sixma JJ, Nieuwenhuis HK (1989) Activation dependent mAB recognizing a 140kD platelet alpha-granule membrane protein, expressed after activation. In: Knapp W, Dörken B, Rieber P, Schmidt RE, Stein H, von dem Borne AEGK (eds) Leukocyte typing IV. White cell differentiation antigens. Oxford University Press, Oxford, pp 1039–1040
Metzelaar MJ, Korteweg J, Sixma JJ, Nieuwenhuis HK (1993) Comparison of platelet membrane markers for the detection of platelet activation in vitro and during platelet storage and cardiopulmonary bypass surgery. J Lab Clin Med 121:579–587
Michelson AD (1987) Flow cytometric analysis of platelet surface glycoproteins: phenotypically distinct subpopulations of platelets in children with chronic myeloid leukemia. J Lab Clin Med 110:346–354
Michelson AD (1994) Platelet activation by thrombin can be directly measured in whole blood through the use of the peptide GPRP and flow cytometry: methods and clinical applications. Blood Coagul Fibrinolysis 5:121–131
Mizumoto Y, Fujimura Y, Nishikawa K, Uchida M, Fukui H, Morii T, Narita N, Kurata Y (1991) Flow cytometric analysis of anti-platelet antibodies in patients with chronic idiopathic thrombocytopenic purpura (ITP) using acid-treated, formalin-fixed platelets. Am J Hematol 37:274–276
Murphy RF (1986) Flow cytometry in cell biology, in: Applications of fluorescence in the biomedical sciences. Liss, New York, pp 525–530
Newman PJ (1993) Recent studies on the role of CD31 (Pecam-1) in vascular cell biology. 5th international conference on human leukocyte differentiation antigens. Tissue Antigens 42:414(A)

Nieuwenhuis HK, van-Oosterhout JJ, Rozemuller E, van-Iwaarden F, Sixma JJ (1987) Studies with a monoclonal antibody against activated platelets: evidence that a secreted 53 000-molecular weight lysosome-like granule protein is exposed on the surface of activated platelets in the circulation. Blood 70(3):834–845
Nomura S, Komiyama Y, Murakami T, Funatsu A, Kokawa T, Sugo T, Matsuda M, Yasunaga K (1993) Flow cytometric analysis of surface membrane proteins on activated platelets and platelet-derived microparticles from healthy and thrombasthenic individuals. Int J Hematol 58:203–212
Palabrica T, Lobb R, Furie BC, Aronovitz M, Benjamin C, Hsu YM, Sajer SA, Furie B (1992) Leukocyte accumulation promoting fibrin deposition is mediated in vivo by P-selectin on adherent platelets. Nature 359:848–851
Pastakia KB, Terle D, Prodouz KN (1993) Penicillin inhibits agonist-induced expression of platelet surface membrane glycoproteins. Ann NY Acad Sci 677:437–439
Pedigo M, Wun T, Paglieroni T (1993) Removal by white cell-reduction filters of activated platelets expressing CD62. Transfusion 33:930–935
Philp RB (1981) In vitro tests of platelet function and platelet-inhibiting drugs. In: Philp RB (ed) Methods of testing proposed antithrombotic drugs. CRC, Boca Raton, pp 129–161
Pratico D, Iuliano L, Alessandrini C, Camastra C, Violi F (1993) Polymorphonuclear leukocyte- derived O2- species activate primed platelets in human whole blood. Am J Physiol 264:H1582–H1587
Reverter JC, Escolar G, Sanz C, Cases A, Villamor N, Nieuwenhuis HK, Lopez J, Ordinas A (1994) Platelet activation during hemodialysis measured through exposure of p-selectin: analysis by flow cytometric and ultrastructural techniques. J Lab Clin Med 124:79–85
Rinder CS, Bohnert J, Rinder HM, Mitchell J, Ault K, Hillman R (1991a) Platelet activation and aggregation during cardiopulmonary bypass. Anesthesiology 75:383–393
Rinder HM, Bonan JL, Rinder CS, Ault KA, Smith BR (1991b) Activated and unactivated platelet adhesion to monocytes and neutrophils. Blood 78:1760–1769
Rinder HM, Murphy M, Mitchell JG, Stocks J, Ault KA, Hillman RS (1991c) Progressive platelet activation with storage: evidence for shortened survival of activated platelets after transfusion. Transfusion 31:409–414
Rinder HM, Munz UJ, Ault KA, Bonan JL, Smith BR (1993) Reticulated platelets in the evaluation of thrombopoietic disorders. Arch Pathol Lab Med 117:606–610
Rohrer MJ, Kestin AS, Ellis PA, Barnard MR, Rodino L, Breckwoldt WL, Li JM, Michelason AD (1992) High-dose heparin suppresses platelet alpha granule secretion. J Vasc Surg 15:1000–1008
Rösen P, Schwippert B, Tschoepe D (1994a) Adhesive proteins in platelet-endothelial interactions. In: Schrör K, Rösen P, Tschoepe D (eds) Megakaryocytes and platelets in cardiovascular disease. Eur J Clin Invest 24[Suppl 1]:21–24
Rösen P, Schwippert B, Kaufmann L, Tschoepe D (1994b) Expression of adhesion molecules on the surface of activated platelets is diminished by PGI2- analogues and an NO (EDRF) donor: a comparison between platelets of healthy and diabetic subjects. Platelets 5:45–52
Rosenfeld CS, Nichols G, Bodensteiner DC (1987) Flow cytometric measurment of antiplatelet antibodies. Am J Clin Pathol 87(4):518–522
Roth GJ (1992) Platelets and blood vessels: the adhesion event. Immunol Today 13:100–105
Savage B, Shattil SJ, Ruggeri ZM (1992) Modulation of platelet function through adhesion receptors. J Biol Chem 267:11300–11306
Scharf RE, Tomer A, Marzec UM, Teirstein PS, Ruggeri ZM, Harker LA (1992) Activation of platelets in blood perfusing angioplasty-damaged coronary arteries. Flow cytometric detection. Arterioscler Thromb 12:1474–1487
Schaufelberger HD, Uhr MR, McGuckin C, Logan RPH, Misiewicz JJ, Gordon-Smitz EC, Beglinger C (1994) Platelets in ulcerative colitis and Crohn's disease

express functional interleukin-1 and interleukin-8 receptors. Eur J Clin Invest 24:656–663
Schneider W, Gattermann N (1994) Megakaryocytes: origin of bleeding and thrombotic disorders. In: Schrör K, Rösen P, Tschoepe D (eds) Megakaryocytes and platelets in cardiovascular disease. Eur J Clin Invest 24[Suppl 1]:16–20
Schultheiss HP, Tschoepe D, Esser J, Schwippert B, Roesen P, Nieuwenhuis HK, Schmidt-Soltau C, Strauer B (1994) Large platelets continue to circulate in an activated state after myocardial infarction. Eur J Clin Invest 24:243–247
Shafi NQ (1986) Evaluation of the platelet surface by fluorescence flow cytometry. Ann Clin Lab Sci 16:365–372
Shattil SJ (1992) Why is platelet activation useful for assessing thrombotic risk? What are the advantages and limiations of using flow cytometry to measure platelet activation. Am J Clin Nutr 56[Suppl 4]:789–790
Shattil SJ, Cunningham M, Hoxie JA (1987) Detection of activated platelets in whole blood using activation-dependent monoclonal antibodies and flow cytometry. Blood 70:307–315
Smyth SS, Joneckis CC, Parise LV (1993) Regulation of vascular integrins. Blood 81:2827–2843
Stewart MW, Etches WS, Gordon PA (1988) Analysis of vWf binding to platelets by flow cytometry. Thromb Res 50:455–460
Tomer A, Harker LA, Burstein SA (1988) Flow cytometric analysis of normal human megakaryocytes. Blood 71:1244–1252
Topol E, Califf RM, Weisman HF, Ellis SG, Tcheng JE, Worley S et al. (1994) The EPIC investigators. Randomized trial of coronary intervention with antibody against platelet IIb/IIIa integrin for reduction of clinical restenosis: results at six months. Lancet 343:881–886
Tschoepe D, Schwippert B (1993) Flow cytometric measurement of intravascular platelet activation: a new concept to assess the risk of arterial thrombosis. Gynecol Endocrinol 7:13–19
Tschoepe D, Schauseil S, Rösen P, Kaufmann L, Gries FA (1988) Demonstration of thrombocyte membrane proteins with monoclonal antibodies by a flow cytometry bioassay. Klin Wochenschr 66:117–122
Tschoepe D, Rösen P, Schwippert B, Kehrel B, Schauseil S, Esser J, Gries FA (1990a) Platelet analysis using flowcytometric procedures. Platelets 1:127–133
Tschoepe D, Roesen P, Kaufmann L, Schauseil S, Ostermann H, Gries FA (1990b) Evidence for abnormal glycoprotein receptor expression on diabetic platelets. Eur J Clin Invest 20:166–170
Tschoepe D, Spangenberg P, Esser J, Schwippert B, Rösen P, Gries FA (1990c) Flowcytometric detection of surface membrane alterations and concomitant changes in the cytoskeletal actin status of activated platelets. Cytometry 11:652–656
Tschoepe D, Rösen P, Esser J, Schwippert B, Nieuwenhuis HK, Kehrel B, Gries FA (1991) Large platelets circulate in an activated state in diabetes mellitus. Semin Thromb Hemost 17:433–438
Tschoepe D, Schwippert B, Schettler B, Kiesel U, Rothe H, Roesen P, Gries FA (1992) Increased GPIIB/IIIA expression and altered DNA-ploidy pattern in megakaryocytes of diabetic BB-rats. Eur J Clin Invest 22:591–598
Tschoepe D, Schwippert B, Schumacher B, Strauss K, Jackson A, Niederau C (1993a) Increased p-selectin (CD62) expression on platelets from capillary whole blood of patients with Crohn's disease. Gastroenterology 104:793a
Tschoepe D, Niederau C, Schwippert B, Schumacher B, Nieuwenhuis KH, Strohmeyer G (1993b) Platelets of patients with Crohn's disease circulate in an activated state. Thromb Haemost 69:796a
Tschoepe D, Schultheiss HP, Kolarov P, Schwippert B, Dannehl K, Nieuwenhuis HK, Kehrel B, Strauer B, Gries FA (1993c) Platelet membrane activation markers are predictive for increased risk of acute ischaemic events after PTCA. Circulation 88:37–42

Tschoepe D, Schwippert B, Raic I, Schmidt-Soltau C, Roesen P (1994) Influence of acetylsalicylic acid upon the intravasal platelet activation in type-I-diabetics – a pilot study. Diab Stoffw 3:391–396

Tschoepe D, Driesch E, Schwippert B, Nieuwenhuis HK, Gries FA (1995) Exposure of adhesion molecules on activated platelets in patients with newly diagnosed IDDM is not normalized by near-normoglycemia. Diabetes 44:890–894

Valet G, Tschoepe D, Gabriel H, Valet M (1993) Standardized, self learning flow cytometric list mode data classifer: classif 1 for thrombocyte and lymphocyte immune phenotyping. Ann NY Acad Sci 677:233–251

Visentin GP, Wolfmeyer K, Newman PJ, Aster RH (1990) Detection of drug-dependent, platelet-reactive antibodies by antigen-capture ELISA and flow cytometry. Transfusion 30:694–700

Wang GX, Terashita GY, Terasaki PI (1989) Platelet crossmatching for kidney transplants by flow cytometry. Transplantation 48:959–961

Warkentin TE, Powling MJ, Hardisty RM (1990) Measurement of fibrinogen binding to platelets in whole blood by flow cytometry: a micromethod for the detection of platelet activation. Br J Haematol 76:387–394

Wehmeier A, Tschoepe D, Esser J, Menzel C, Nieuwenhuis HK, Schneider W (1991) Circulating activated platelets in myeloproliferative disorders. Thromb Res 61(3):271–280

Worfolk LA, MacPherson BR (1991) The detection of platelet alloantibodies by flow cytometry. Transfusion 31:340–344

Wun T, Paglieroni T, Sazama K, Holland P (1992) Detection of plasmapheresis-induced platelet activation using monoclonal antibodies. Transfusion 32:534–540

Yeaman MR, Sullam PM, Dazin PF, Norman DC, Bayer AS (1992) Characterization of Staphylococcus aureus-platelet binding by quantitative flow cytometric analysis. J Infect Dis 166:65–73

Yeo EL, Sheppard JA, Feuerstein IA (1994) Role of P-selectin and leukocyte activation in polymorphonuclear cell adhesion to surface adherent activated platelets under physiologic shear conditions (an injury vessel wall model). Blood 83:2498–2507

CHAPTER 28
# Pathological Aspects of Platelet-Activating Factor (PAF)

F. VON BRUCHHAUSEN

## A. Introduction

Elucidation of the involvement of platelet-activating factor (PAF) in physiological and pathological processes is hampered by analytical problems concerning the low amounts of PAF to be detected, the kinetics of this mediator when given or released endogenously, and the restrictions resulting from studies with PAF receptor antagonists, the pharmacokinetics of which are not fully understood and which also reduce actions of PAF-like lipids (IMAIZUMI et al. 1995). Endogenous inhibitory factors may interfere with PAF actions (WOODARD et al. 1995). Another problem is presented by the fact that PAF, like other mediators or cytokines, acts within a framework (network), so that mutual influences occur on other mediators regarding their release and priming effects in the target, i.e., on the cellular level.

Under these limitations, PAF has proven a pathologically important agent for an outstanding number of situations and diseases which are compiled in this chapter. Reviews on this topic were edited in 1989 and recently (see BARNES et al. 1989; IMAIZUMI et al. 1995; for critisim see VARGAFTIG 1995; DENIZOT, Chap. 22). In the following review, no difference is made with respect to the molecular entity of PAF (see Chap. 23; PINCKARD et al. 1994). Sometimes, PAF-like lipids are not easily differentiated from PAF itself (IMAIZUMI et al. 1995).

## B. Involvement of PAF in Inflammatory and Allergic Responses

PAF may be involved in acute and especially in chronic inflammatory processes, either directly as a mediator or indirectly by interaction with the cytokine network, especially with tumor necrosis factor (TNF). TNF and interleukin-1 (IL-1) have a central role in inflammation (SEOW et al. 1987; LAST-BARNEY et al. 1988; AREND and DAYER 1990; HAWRYLOWICZ 1993), also affecting it indirectly in that PAF is induced by IL-1 and TNF (SUN and HSUEH 1988). The proinflammatory effect of PAF has been reviewed by SNYDER (1990), PRESCOTT et al. (1990), and PEPLOW and MIKHAILIDIS (1990). The destructive enzyme for PAF, PAF acethydrolase, exerts anti-inflammatory properties (TJÖLKER et al. 1995).

The direct role of PAF is understood through its release by several inflammatory cell types. Its precursor lyso-PAF is found in injured cornea (HURST and BAZAN 1993) and PAF itself in experimental peritoneal exudates (DAMAS et al. 1990; DAMAS and PRUNESCU 1993) which can be reduced by applying the PAF antagonist apafant (WEB 2086; DAMAS et al. 1990). For export of PAF from inflammatory cells, the contribution of a PAF releasing factor was described (MIWA et al. 1992) which is found in human serum and inflammatory exudate (OKAMOTO et al. 1993).

The release of PAF by inflammatory cells is stimulated by opsonized zymosan (SUGATANI et al. 1991a), PAF itself (SISSON et al. 1987; TESSNER et al. 1989), TNF (CAMUSSI et al. 1987; SISSON et al. 1987), interleukin-1$\beta$ (VALONE and EPSTEIN 1988), interleukin-6 (BIFFL et al. 1996), and N-formyl-methionyl-leucyl-phenylalanine (FMLP) (SISSON et al. 1987).

An early event of acute inflammatory response by several inflammatory agents (thrombin, histamine, phorbol esters) is the coexpression of P-selectin (GENG et al. 1990) and PAF (LORANT et al. 1991) within epithelial cells. PAF is a potent modulator of leukocyte adhesion to endothelial cells and emigration across the microvasculature (HUMPHREY et al. 1982a; DILLON et al. 1988a; GARCIA et al. 1988; KUBES et al. 1990a). Monocyte adhesion to endothelial cells is greatly reduced by the PAF antagonist apafant (MURPHY et al. 1994). Furthermore, PAF's action on plasma protein extravasation can be inhibited by apafant (DAMAS and PRUNESCU 1993). In mediator-mediated vascular reponses, PAF and histamine show priming effects (TOMEO et al. 1991). They both increase vascular permeability for macromolecules, PAF being an autocrine response amplifier (TOMEO et al. 1991).

Elevated blood levels of PAF have been found in bacterial infections (MAKRISTATHIS et al. 1993) and especially in sepsis of animals (INARREA et al. 1985) and of men (BUSSOLINO et al. 1987b; MEZZANO et al. 1993). In the plasma of patients with clinical sepsis, a decrease in the PAF-destroying enzyme PAF acethydrolase was found (GRAHAM et al. 1994). PAF's involvement is seen not only in gram-negative bacterial infections with endotoxin-mediated changes (TORLEY et al. 1992; see also Sect. C) but also in gram-positive streptococcal infections (MEZZANO et al. 1993). PAF was found to be elevated in peritoneal exudates of peritonitis (MONTRUCCHIO et al. 1989b). PAF antagonists also inhibited HIV expression in HIV-infected cells (WEISSMAN et al. 1993).

The indirect role of PAF in inflammatory responses is to produce other mediators: Prostaglandins and thromboxanes are produced (DE LIMA et al. 1995); PAF enhances the release of TNF-alpha and other cytokine factors (BONAVIDA and MENCIA-HUERTA 1994; MANDI et al. 1989a,b); and enhanced production of IL-1 from monocytes or macrophages by PAF has been reported (SALEM et al. 1990), particularly in a bimodal manner (PIGNOL et al. 1987a). IL-1-induced leukocyte extravasation is mediated by PAF (NOURSHAGH et al. 1995).

In rheumatoid arthritis, the involvemant of PAF has been deduced (STUFLER et al. 1992; HILLIQUIN et al. 1995a), possibly being released by polymorphonuclear leucocytes (PMNLs; HILLIQUIN et al. 1995b). Interleukins regulate production of inflammation mediators in synoviocytes, IL-6 (DECHANET et al. 1995) as well as IL-1$\beta$ and IL-6 with regard to their PAF production (GUITIERREZ et al. 1995). Moreover, PAF is found in exudates of the knee (SHARMA and MRHSIN 1990; CUSH and LIPSKY 1991; HILLIQUIN et al. 1992; ZARCO et al. 1992), in acute more than in chronic arthritis (PETTIPHER et al. 1987), and in breakdown products (PETTIPHER et al. 1987) although secretion of acethydrolase from macrophages is decreased (NARAHARA and JOHNSTON 1993). It is interesting that dexamethasone, in addition to other actions, can enhance the activity of this enzyme (MIYAURA et al. 1991).

## I. Inflammatory Cellular Responses by PAF

In the subcutaneous air pouch model induced by PAF (WONG et al. 1992), leukocyte infiltration prevails. Chemotactic properties of PAF for polymorphonuclear neutrophils (PMNs) have been observed (MARTINS et al. 1989; TAN and DAVIDSON 1995), also in the skin (CZARNETZKI and BENVENISTE 1981; ARCHER et al. 1984). For monocytes, the precursor of PAF, lyso-PAF, is chemotactically more active than PAF itself (QUINN et al. 1988). The release of PAF does not arise from mast cells (BARNES et al. 1981) but from mononuclear cells (PARENTE and FLOWER 1985), macrophages (PARENTE and FLOWER 1985), and PMNs (IMAI et al. 1991).

As far as eosinophils are concerned, PAF's actions by and on eosinophils are described in the following subsection. For their chemoattraction, PAF is important (WARDLAW et al. 1986) and more potent than interleukin-5 (WANG et al. 1989).

The accumulation of platelets in an acute inflammation process by carrageenin (VINCENT et al. 1978; ISSEKUTZ et al. 1983) is mediated by PAF and represents the basis of an early event. In many inflammatory cells (neutrophils and eosinophils), PAF elicits and potentiates superoxidanion production induced by zymosan or FMLP (MABUCHI et al. 1992; KROEGEL et al. 1992), as do macrophages (SCHMIDT et al. 1992).

As far as adhesion processes between cells in general and endothelial cells are concerned, the involvement of PAF is described in Sect. C. It should be mentioned that acceleration of wound healing by PAF has also been reported (PORRES-REYES and MUSTONE 1992).

## II. Involvement of PAF in Allergic Reactions

PAF has been claimed to regulate immune responses (BRAQUET and ROLA-PLESZCZYNSKI 1987). PAF cannot merely mimic anaphylaxis (VARGAFTIG and BRAQUET 1987), but it is also released by antigenic challenge from sensitized basophils and mast cells (BENVENISTE et al. 1972; PINCKARD et al. 1979; TENCE et al. 1980; MENCIA-HUERTA et al. 1983; BRAQUET and ROLA-PLESCZCYNSKI 1987; DUNOYER-GEINDRE et al. 1992). This correlates with enhanced activity of

PAF acetyltransferase (NINIO et al. 1987; JOLY et al. 1990). However, divergent results were obtained by the use of PAF antagonists in anaphylaxis. In passive anaphylaxis, the protective effects were greater than in active anaphylaxis, but also demonstrable by bepafant (HEUER 1991). Upon infusion of human IgG into rodents, prominent formation of PAF by mononuclear cells is induced (INARRA et al. 1983; DOEBBER et al. 1986). PAF was found in nasal fluid from sufferers of allergic rhinitis (MIADONNA et al. 1989).

It also has a role in infiltration of polymorphonuclear leukocytes (PMNs; WATANABE et al. 1990). Eosinophils are claimed to play a critical role in allergy and asthma (DE MONCHY et al. 1985; GLEICH and ADOLPHSON 1986). They are primarily involved in immune defense (SIMON and BLASER 1995), play a role in late-type inflammatory reactions, and are prominent in many chronic inflammatory diseases, particularly in allergic manifestations (SIMON and BLASER 1995).

Eosinophilia correlates with T-cell activation and production of specific cytokines (WALKER et al. 1991; SIMON et al. 1994b; ERGER and CASALE 1995) and is a consequence of inhibited eosinophilic programmed cell death (SIMON and BLASER 1995). In eosinophilia, extremely high levels of lyso-PAF acetyltransferase – the enzyme that produces PAF from its precursor – have been found, whereas normally this enzyme is not even detectable in eosinophils (LEE et al. 1982). Even more of the enzyme is found when these cells are stimulated by factors (LEE et al. 1984). PAF is not only the most potent natural chemoattractant of eosinophils (WARDLAW et al. 1986; WANG et al. 1989), but also the most potent natural activator of eosinophils (KROEGEL et al. 1992). Therefore, PAF antagonists impede eosinophil recruitment (TENG et al. 1995).

PAF receptors are expressed on eosinophils (UKENA et al. 1989) and on eosinophilic leukemia cell line EOL-1 (IZUMI et al. 1995). The number of PAF binding sites in eosinophils are approximately 200-fold higher than in human platelets (UKENA et al. 1988; UKENA et al. 1989). The actions of PAF on eosinophils are: opening of $Ca^{2+}$-activated $K^{+}$ channels (SAITO et al. 1996), degranulation (KROEGEL et al. 1989), conversion to hypodense forms (YUKAWA et al. 1989), release of eosinophil peroxidase (KROEGEL et al. 1988, 1989; UKENA et al. 1989), of other granular contents, of superoxide anion (MABUCHI et al. 1992), and of the eosinophil basic protein which is supposed to mediate allergic inflammation (SPRY 1988; LI et al. 1995) and prostanoid production (GIEMBYCZ et al. 1990; KROEGEL et al. 1990). PAF-induced pleural eosinophilia is prevented by cetirizine (MARTINS et al. 1992).

PAF inhibits human lymphocyte proliferation when stimulated and suppresses interleukin-2 production (ROLA-PLESZCZYNSKI et al. 1987; PIGNOL et al. 1987b; GEBHARDT et al. 1988; DULIOUST et al. 1990). Conversely, enhanced lymphocyte proliferation by PAF has also been found (WARD et al. 1991).

Examples of PAF's involvement in allergic diseases are: asthma bronchiale (see Sect. E); immune-complex vasculitis (WARREN et al. 1989);

microvascular immune injury (BRAQUET et al. 1989), and cardiac anaphylaxis (LEVI et al. 1984; YAACOB et al. 1995), among others.

# C. Involvement of PAF in Cardiovascular Diseases

## I. PAF in Vascular Disturbances

PAF's vascular effects have a dual character (DILLON et al. 1988b; KAMATA et al. 1989). The microvascular constrictive effect of PAF (for a review see VAN DONGEN 1991) has two basic machanisms, a direct one on vascular smooth muscle cells which have PAF receptors (HWANG et al. 1983) that involves the endothelium, and an indirect one mediated by polymorphonuclear leukocytes (PMNLs; BJÖRK and SMEDEGARD 1983). In contrast, PAF is a potent dilator of arterioles and venules (rat cremasteric muscle, see SMITH et al. 1981), apparently by direct action. This mechanism contributes to the hypotensive effect of PAF (and endotoxin, see below) as seen on mesenteric microvessels (LAGENTE et al. 1988). PAF-induced hypotension may also result from cardiodepression (BENVENISTE et al. 1983) and from an indirect PMNL-dependent effect on small vessels with subsequent reduction of peripheral resistance. Hemoconcentration due to increased vascular permeability and plasma extravasation (HANDLEY et al. 1984; PIROTZKY et al. 1984) by PAF accentuates the situation. PAF-induced vasodilation is inhibited by PAF antagonist BN 52021 (LAGENTE et al. 1988).

PAF originates from cells by action of tumor necrosis factor (TNF; BUSSOLINO et al. 1988; CAMUSSI et al. 1987), interleukin-1-alpha (BUSSOLINO et al. 1988b), and thrombin (PRESCOTT et al. 1984; CARVETH et al. 1992). Priming effects with other vascular-active mediators have been described (for histamine see TOMEO et al. 1991).

PAF can induce endothelial cell shape changes and the formation of interendothelial gaps (BUSSOLINO et al. 1987a; NORTHOVER 1989). Thus, increased vascular permeability by PAF is reported to produce these gaps between endothelial cells without involvement of platelets, neutrophils, or mast cells (SANCHEZ-CRESPO et al. 1982; SHASBY et al. 1982). PAF promotes endothelial migration and thus angiogenesis (CAMUSSI et al. 1995). In addition, leukocyte extravasation by IL-1 seems to be mediated by PAF (NOURSHAGH et al. 1995).

In contrast, PAF is a potent modulator of leukocyte adhesion to endothelial cells and of emigration across the microvasculature (HUMPHREY et al. 1982a; DILLON et al. 1988a; GARCIA et al. 1988; KUBES et al. 1990a; AKOPOV et al. 1995) This holds true for dynamic flow situations upon thrombin- and TNF-alpha-induced stimulation of PAF production by endothelial cells (MACCONI et al. 1995).

Although arterial pressure falls immediately after IV infusion of PAF in rats (ROSAM et al. 1986) and PAF is a potent antihypertensive agent (ROUBIN et al. 1983), high levels of PAF have been found only in salt-induced hypertension. Its antihypertensive action, when human atrial natriuretic peptide (hANP) is low, is attributed to its modulating role (Sakaguchi et al. 1989). In kidney-clipped rats, lyso-PAF was increased (McGowan et al. 1986). Otherwise, in hypertensive patients the PAF levels of the blood were found to be unchanged (Masugi et al. 1989). In cases of hypertension, PAF acethydrolase was increased (Blank et al. 1983; Satoh et al. 1989), yet not markedly in hypertensive pregnancies (Maki et al. 1993).

Pulmonary vasoconstriction by hypoxia is accompanied by elevated PAF levels in pulmonary alveolar fluid (Prevost et al. 1984), which suggests a down-regulating role of endogenous PAF by vasodilation (McMurthy and Morris 1986; Voelkel et al. 1986). Vasoconstriction may partially be elicited by endothelins (Oparil et al. 1995). Inhalation of PAF in normal subjects, however, did not induce any significant changes in systemic or pulmonary hemodynamics (Rodriguez-Roisin et al. 1991). Some doubts about the constrictive effect of PAF in pulmonary hypertension also resulted from experiments with apafant (McCormack et al. 1989) and SRI 63 411 (Haynes et al. 1988). Conversely, PAF induces pulmonary hypertension, which was found in experimental animals by applying PAF (Hamasaki et al. 1984; Ohar et al. 1990; Albertini and Clement 1994; Argiolas et al. 1995). This was interpreted as potentiation of TNF's action (Horvath et al. 1991; Chang 1994b) or involvement of thromboxane (Heffner et al. 1983; Pinheiro et al. 1989). From a regulatory point of view, PAF induces vascular constriction when vascular tone is low, but causes vasodilation when vascular tone is high (Barnes and Liu 1995) and in addition pulmonary arteries are relaxed and pulmonary veins contracted (Gao et al. 1995). When infused into animals, sufficient amounts of PAF transiently increase pulmonary pressure (Blank et al. 1979; Bessin et al. 1983; Pirotzky et al. 1985a), possibly indirectly (Heffner et al. 1983; Hamasaki et al. 1984; Burhop et al. 1986). Thus, on the basis of chronic infusion of PAF, and animal model of pulmonary hypertension was developed (Ohar et al. 1990). The bronchial and vascular effects of PAF could be blocked by apafant (Casals-Stenzel 1987a). PAF-induced pulmonary hypertension was reduced by inhaled nitric oxide (Albertini and Clement 1994).

In pulmonary hypertension of the neonate, involvement of PAF was assumed (Ohar et al. 1986; Soifer and Schreiber 1988). Pulmonary hypertension in ARDS (adult respiratory distress syndrome) is caused by pulmonary vascular endothelium injury worsened by hypoxia (McIntyre et al. 1994; for involvement of PAF see Sect. E).

## II. PAF in Endotoxin-Induced Shock

The shock by endotoxin is not elicited by PAF itself or its complex with lipopolysaccharide binding protein (for a review see Handley et al. 1990; Glauser et al. 1991; Vassalli 1992). Bacterial lipopolysaccharides (LPS) is known to release tumor necrosis factor (TNF; Beutler and Cerami 1988), eicosanoids (BALL et al. 1986), and PAF (TERASHITA et al. 1985; LEFER 1989; MOZES et al. 1991). TNF subsequently leads to pulmonary neutrophil sequestration and its consequences (CHANG 1994a) and PAF to hypotension, lung edema, hemoconcentration, thrombocytopenia, leukopenia, and bronchoconstriction (OLSON et al. 1990a).

The involvement of PAF in endotoxic shock was evaluated by the following observations: (1) Administration (infusion) of PAF mimics effects of endotoxin or rather the state of shock (McMANUS et al. 1980), even if its action is more venoconstrictive than arterioconstrictive (BAR-NATAN et al. 1995); (2) endotoxin induces rise of PAF in blood (CHANG et al. 1987; DOBROWSKY et al. 1991), or rather PAF is generated during the state of shock (PINCKARD et al. 1979) produced by lung and liver (ANDERSON et al. 1991; RABINOVICI et al. 1991), which is paralleled by a decrease in acethydrolase (NARAHARA and JOHNSTON 1993); endotoxin may also activate PAF receptors (NAKAMURA et al. 1992; WAGA et al. 1993); (3) PAF rises in blood before TNF rises (DOEBBER et al. 1985), with PAF antagonists impeding TNF rise and action (FLOCH et al. 1989; FERGUSON-CHANOWITZ et al. 1990; RUGGIERO et al. 1994); (4) synergistic effects of PAF and LPS at the cellular level have been reported (PIRES et al. 1994; FOUDA et al. 1995), and (5) specific PAF antagonists improve survival and provide significant protection against the state of shock (DOEBBER et al. 1985; TERASHITA et al. 1985; INNAREA et al. 1985; ETIENNE et al. 1986; ADNOT et al. 1986; HANDLEY et al. 1986; CASALS-STENZEL 1987c; YUE et al. 1990a; TANG et al. 1993) or against lung injury by endotoxin (CHANG et al. 1987, 1990; OLSON et al. 1990a,b).

Concerning the cooperation of PAF with TNF in endotoxin shock, the question arises whether TNF acts only by induction of PAF biosynthesis from hematologic and endothelial cells (CAMUSSI et al. 1987, 1989; SUN and HSUEH 1988; BUSSOLINO et al. 1988b) or inversely by its PAF-stimulated release (FLOCH et al. 1989; HAYASHI et al. 1989) from monocytes (BONAVIDA et al. 1989, 1990) or from macrophages (ROLA-PLESZCZYNSKI et al. 1988a; DUBOIS et al. 1989). It should be mentioned that other mediators in endotoxic shock such as interleukin-6 and interferon-gamma (RUGGIERO et al. 1994) are important for lethality marker function (KELLY and CROSS 1992) or as mediators of lethality (DOHERTY et al. 1992). Furthermore, leukotriene B4 is released by PAF (DUBOIS et al. 1989; CHANG et al. 1990; OLSON et al. 1990b) Interleukin-1 is also released by endotoxin (OKUSAWA et al. 1988; OHLSSON et al. 1990) which is involved in the development of tachycardia, hypotension, and tissue damage (EVERAERDT et al. 1989; OHLSSON et al. 1990). PAF is able to prime PMNs by enhancing the response to chemotactic peptides (GAY 1993). Some aspects of endotoxin also have beneficial effects in shock (URBASCHEK and URBASCHEK 1987).

The lethal outcome of endotoxic shock of pigs depends, according to MOZES et al. (1993), on whether only TNF or – for better survival – also PAF and eicosanoids are increased in blood. PAF is probably of value because it induces nitric oxide synthase (SZABO et al. 1993) and thus provokes partially favorable effects of NO in this situation (KILBOURN and GRIFFITH 1992). As far as disseminated intravascular coagulation (DIC) with formation of fibrin thrombi, loss of platelets, and activation of the fibrinolytic system are concerned, PAF levels are elevated as a consequence of endotoxin action (SAKAGUCHI et al. 1989), and the use of PAF antagonists reduces the process (IMANISHI et al. 1991; OU et al. 1994; SUMMERS and ALBERT 1995). A role of PAF is also assumed for DIC during anaphylaxis (CHOI et al. 1995). PAF's involvement in another deleterious process, multiorgan failure, is acknowledged (OU et al. 1994).

In traumatic shock, the involvement of PAF is likewise being discussed. A PAF-like substance was found in peritoneal fluid of rats with traumatic shock (STAHL et al. 1988). Using PAF antagonists (TERASHITA et al. 1988; STAHL et al. 1989), protective actions could be revealed. Thus, PAF may play an important pathophysiological role in traumatic shock. In shock of a more hemorrhagic character, PAF antagonists also demonstrate the importance of PAF in this type of shock (STAHL et al. 1988; ZINGARELLI et al. 1994). Likewise, PAF seems to be involved in anaphylactic shock (AMORIM et al. 1993). In sudden death, PAF and other mediators (endothelin) have been claimed to be involved (TERASHITA et al. 1989).

Together, these results suggest reciprocal positive interactions between PAF and TNF during endotoxemia and indicate the potential use of PAF antagonists in septic shock.

## III. Thrombovascular Diseases

At present there is little evidence for the implication of PAF in human cardiovascular diseases (NATHAN and DENIZOT 1995) despite its involvement in animal studies. In the underlying arterial thrombotic process, PAF plays a critical role in the pathogenesis of thrombosis as it is an activator of committed cells, leukocytes, platelets, and endothelial cells (HIRAFUJI and SHINODA 1991). By superfusion of the mesenteric artery with PAF, thrombosis can be induced (BOURGAIN et al. 1985, 1986), and PAF antagonists have protective effects on pulmonary microembolism (KLEE et al. 1991) and on other types of thrombosis models, especially in mice and cats (SETH et al. 1994). The same PAF antagonists had no effect in a model of venous thrombosis (SETH et al. 1994). Among the actions of PAF is the development of thrombopenia (MARTINS et al. 1987).

The involvement of PAF in coronary diseases has attracted much attention. As for the general vascular situation, PAF has coronary constrictive actions (LEVI et al. 1984; STEWARD and PIPER 1986; STAHL et al. 1987) or dilatory actions (FEUERSTEIN et al. 1984) at lower doses (SAGASCH et al. 1992). This effect is attributed to endothelium-dependent release of nitric oxide (NO) and/or lipoxygenase products. As seen with intravital videomicroscopy, the coronary vasodilator action of acetylcholine or serotonin is attenuated by

PAF (DE FILY et al. 1996). The question of whether PAF could be a potent endogenous mediator responsible for coronary vasospasm has been forwarded by SOLOVIEV and BRAQUET (1992) with the following arguments: (1) the coronary constrictive potential of PAF mentioned above; (2) coronary arterial stripes responding to PAF with tonic tension (SOLOVIEV and BRAQUET 1992), a type of contraction of phasic (transient) and tonic (sustained) character and mediated via PAF receptors on plasma membrane and intracellular receptors (SOLOVIEV and BRAQUET 1992); (3) a decrease in coronary blood flow by PAF (BARANES et al. 1986), especially following local administration (FEUERSTEIN et al. 1984; FELIX et al. 1991), together with electrocardiographic alterations induced by only 1–3 ng of PAF and with blood flow increasing following addition of PAF antagonist BN 52021 (BARANES et al. 1986); (4) hypoxia-induced coronary vasospasm due to higher rate of reaction to several agonists (noradrenaline, arachidonic acid metabolites), mediated by a postulated vasoconstrictor substance whose properties best match those of PAF, which is also (preferentially) released from endothelial cells under hypoxia and from smooth muscle cells and acts as a trigger for contractile responses; (5) a lower acethydrolase activity found in coronary artery disease (GRAHAM et al. 1992) leading to a longer half-life of PAF; (6) PAF's ability to induce down-regulation of $\beta$-adrenoreceptors and to shift the dose-response curve for isproterenol to the right; and (7) inactivation of EDRF (endothelium-derived relaxing factor; NO; VANHOUTTE et al. 1986) which would otherwise counteract. Via production of superoxide anion by PAF. Cardiopulmonary alterations by infusion of TNF-alpha and IL-1 alpha are mediated by PAF and therefore blocked by PAF antagonists (KRUSE-ELLIOTT et al. 1993). PAF has a negative inotropic effect on myocytes (MASSEY et al. 1991) in accordance with earlier findings on hearts.

Reperfusion injury following ischemia is attributed to multiple factors and processes (GRANGER and KORTHUIS 1995). Involvement of PAF in reperfusion injury has been reported for several organs. The involvement of PAF in an early event, i.e., acute circulatory collapse following reperfusion, depends on several circumstances concerning PAF (FILEP et al. 1991a): (1) increased concentration of PAF in the circulation; (2) injection of PAF into local veins to mimic several events of reperfusion; (3) pretreatment of animals with PAF antagonists to prevent circulatory collapse caused by ischemia reperfusion; (4) further actions of eicosanoids accentuating ischemia-reperfusion-induced shock in animals (FILEP et al. 1991c) via induction of PAF; and (5) pentoxifyllin ameloriating reperfusion injury and reducing PAF production (ADAMS et al. 1995).

The involvement of polymorphonuclear neutrophils (PMNs) is important for reperfusion (GRANGER and KORTHUIS 1995). Thus, reperfusion of ischemic tissue is accompanied by microvascular dysfunction following endothelial-dependent neutrophil adherence and generation of reactive oxygen species (ROS). Increased adherence of PMNs to endothelial cells is provoked by PAF

(MILHOUN et al. 1992; YOSHIDA et al. 1992; BIENVENUE and GRANGER 1993; DURAN et al. 1996), partially by expression of CD 11b/CD18 on neutrophils (HARLAN and LIU 1992; KORTHUIS et al. 1994). Generation of ROS has been reported (DROY-LEFAIX et al. 1988; KUBES et al. 1990b) as well as its reduction by PAF antagonists (DROY-LEFAIX et al. 1988; KUBES et al. 1990b). Activation of PMNs by PAF has been reported in patients and is prevented by PAF receptor antagonists in vitro (SIMINIAK et al. 1995).

## IV. Myocardial Infarction and Stroke

PAF may be involved in the pathogenesis of acute myocardial infarction. Activation of phospholipase $A_2$ (VADAS and PRUZANSKI 1986) and thus generation of lyso-PAF (LEONG et al. 1991) in damaged tissues seem to be prerequisite for the detection of PAF in the coronary sinus following damage in animals (MONTRUCCHIO et al. 1989a) or man (MONTRUCCHIO et al. 1986). PAF acethydrolase activity was found to be increased in blood (OSTERMANN et al. 1988). Furthermore, PAF antagonists were able to reduce infarct size (MARUYAMA et al. 1990) and incidence of ischemia-induced arrhythmias (WAINWRIGHT et al. 1986; KOLTAI et al. 1991; SUMMERS and ALBERT 1995). Whether any PAF antagonist can limit myocardial infarct size has been denied in the case of apafant (WEB 2086; LEONG et al. 1992). CV 3988 caused more significant amelioration of the state than PMN depletion (SAWA et al. 1994). However, another PAF antagonist, BN 50739, reduced electrophysiological response, malondialdehyde generation, and infarct size following coronary artery ligation (RANAUT and SINGH 1993), possibly by actions other than by PAF antagonism. In digoxin-induced arrhythmias, some PAF antagonists could reduce mortality rate and arrhythmia score (CAKICI et al. 1995). One source of PAF in myocardial infarction may be the accumulated polymorphonuclear leukocytes (PMNL; MULLANE et al. 1987; STRAHAN et al. 1995), as PAF is a chemoattractant of PMNLs. Myocytes and endothelial cells also contribute to PAF production (MULLANE 1987). Once myocardial ischemia has become effective, PAF seems to be an important mediator of ischemic damage (see above, this section), also in the heart (STAHL et al. 1988c; SALINAS et al. 1995).

In ischemic stroke, the brief ischemic state is followed by PAF-induced arterial constriction (KIM et al. 1993). Therefore, involvement of PAF in stroke (LINDSBERG et al. 1990b, 1991) has been assumed. As in myocardial infarction, PMNLs, as a source of PAF, play an important role in ischemic stroke (KOCHANEK and HALLENBECK 1992). PAF level in blood is increased (SATOH et al. 1992a). It was found that PAF's hydrolyzing acethydrolase activity was decreased likewise in red cell membranes as they were deformed in patients with cerebral thrombosis (YOSHIDA et al. 1992, 1993). However, PAF's activity in serum was increased (SATOH et al. 1992b). PAF antagonists are reported to ameliorete stroke (LINDSBERG et al. 1991), also by reducing infarct volume (BIELENBERG et al. 1992).

# D. Involvement of PAF in Cerebral Disturbances

Production of PAF in the brain is possible (KUMAR et al. 1988b), even under resting conditions (GORACCI 1989), though at a very low rate (BAZAN 1989). The production of PAF is dependent on acetylcholine (BUSSOLINO et al. 1988a; SOGOS et al. 1990) and is enhanced under hypoxic conditions (KUNIEVSKY and YAVIN 1994). PAF is also produced by neuronal cells, e.g., by rat cerebellar granular cells in culture (YUE et al. 1990b) synapses (CLARK et al. 1992; BAZAN et al. 1993), microglial cells (JARANOWSKA et al. 1995), and C6 glioma cells that contain phospholipase $A_2$ (QVIST et al. 1995) and the key enzyme for PAF production, acetyltransferase (VON BRUCHHAUSEN 1991). The PAF content in rat brain (decreasing with age) was measured by TOKUMURA et al. (1992) and that in the cortical area of bovine brain by THIBERGHIEN et al. (1991). Furthermore, the concentrations in the brain are reduced by the coexistence of an endogenous inhibitor (NAKAYAMA et al. 1987). PAF receptors have been found in brain (DOMINGO et al. 1988; BITO et al. 1992), glial cells (BRODY 1994), and especially in synaptic endings of the brain cortex (MARCHESELLI et al. 1990; BAZAN et al. 1993) or hippocampus (CLARK et al. 1992), as in the neural cell line N1E-115 (LALOUETTE et al. 1995). The highest expression of receptor was found in micoglia (MORI et al. 1996). In the brain, PAF is usually biosynthesized de novo.

PAF is degraded by an acethydrolase that is abundant in the brain (BLANK et al. 1981) and has been purified from bovine brain (HATTORI et al. 1993).

The following actions of PAF in brain have been stated: increase in intracellular calcium (KORNECKI and EHRLICH 1988; YUE et al. 1991; BITO et al. 1992; DISERBO et al. 1995); activation of large conductance $Ca^{2+}$ activated $K^+$ channels (DISERBO et al. 1996) as in eosinophils (SAITO et al. 1996); diminished acetylcholine release (WANG 1994); modulation of signal transduction in synapses (DOMANSKA-JANIK and ZABLOCKA 1995); diminished glutamate release at excitatory synapses by presynaptic action of PAF (JIANG et al. 1993) or increased glutamate release (CLARK et al. 1992) by mimicking some aspects of glutamate (BAZAN and RODRIGUEZ DE TURCO 1995); release of ATP from PC12 cells (KORNECKI and EHRLICH 1988); potentiation of catecholamine release (MORITA et al. 1995); release of hypothalamic hormones (DRAY et al. 1993); induction of nerve growth factor production in rat astrocytes (BRODIE 1995); and activation of astroglial cells for the generation of eicosanoids (PETRONI et al. 1993).

Since PAF enhances excitatory synaptic transmission (CLARK et al. 1992), it has the role of a second messenger in the CNS. It modulates neuronal migration (HATTORI et al. 1994) and develops the characteristics of a retrograde messenger in long-term potentiation (DEL CERRO et al. 1990; ARAI and LYNCH 1992). PAF seems to be a mediator of cross-talk circuits in the CNS between several types of cells.

Furthermore, in C6 glioma cells, PAF induces the expression of the astrocytic phenotype and the subsequent expression of a gliose-type response

(KENTROTI et al. 1991) and is involved in the expression of proto-oncogenes (SQUINTO et al. 1989).

In high concentrations however, PAF is neurotoxic (KORNECKI and EHRLICH 1988).

In higher brain functions, the involvement of PAF is scarce. By infusion into some structures, memory enhancement was deduced (IZQUIERDO et al. 1995).

Apart from its role in modulation of synaptic activity and gene expression (BAZAN and RODRIGUEZ DE TURCO 1995), PAF's involvement in some pathological states has been reported: During convulsions phospholipase $A_2$ is activated, and PAF accumulates (KUMAR et al. 1988b) along with free arachidonic acid: PAF is claimed to be the main mediator of secondary brain damage (FRERICHS et al. 1990). Thus, PAF antagonist BN 52021 is neuroprotective (PANETTA et al. 1987).

In brain trauma, the secondary autodestructive response includes phospholipid hydrolysis (FADEN and SALZMAN 1992). The amount of PAF increases (FADEN et al. 1989). Thus, PAF has been named as key mediator in neuroinjury (for a review see FRERICHS and FEUERSTEIN 1990). Reduced cerebral blood flow (KOCHANEK et al. 1988), disruption of the blood-brain barrier (KUMAR et al. 1988b), and PAF's toxic effect on neuronal cells, which is observed in culture lines (KORNECKI and EHRLICH 1988), may be significant for its pathophysiological role. Pretreatment with PAF antagonists reduced these alterations and enhanced neurological recovery (posttraumatic behavioral deficits; FADEN and TZENDZALIAN 1992). Other mediators may contribute to posttraumatic brain injury such as leukotrienes, the generation of which is additionally enhanced by PAF (FAULER et al. 1989) in monocytes or astrocytes (PETRONI et al. 1993).

Postischemic neuronal damage (for reperfusion injury of the brain see Sect. C; for a review see FEUERSTEIN et al. 1990) in which PAF is involved (SPINNEWYN et al. 1987) and in which PAF acethydrolase is increased (SATOH et al. 1988b, 1992) is reduced by several PAF antagonists (LE PONCIN LAFITTE et al. 1986; PANETTA et al. 1987; KOCHANEK et al. 1987; OBERPICHLER et al. 1990; LINDSBERG et al. 1990b; GILBOE et al. 1991). Direct measurements reveal a strong increase in concentration in gerbils (NISHIDA et al. 1996). PAF is not involved in edema production (KOCHANEK et al. 1991) nor in normal vascular regulation (KOCHANEK et al. 1991).

In subarachnoid hemorrhage, an increased PAF concentration and a decreased PAF acetylhydrolase concentration in cerebrospinal fluid were found (HIRASHIMA et al. 1994). The following cerebrospasm was prevented by a PAF antagonist (HIRASHIMA et al. 1996).

Enhancement of bacterial inflammation by PAF has been observed (TOWNSEND and SCHELD 1994) as PAF is an important mediator of bacterial meningitis (CABELLO et al. 1992). Thus, PAF acethydrolase in the brain (TOKUMURA et al. 1987) is important to prevent amplification of inflammation (BAZAN 1995). An increased PAF concentration in the cerebrospinal fluid has been reported (ARDITI et al. 1990). In contrast, PAF antagonists did not protect against the development of experimental autoimmune encephalitis (VELA et al. 1991). In HIV-related neuronal injury, PAF contributes to neurotoxicity (GELHARD et al. 1994; LIPTON et al. 1994). The Miller-Dieker syndrome might be an example of PAF involvement in neuronal migration (BAZAN 1995). The outcome of neurological score was improved in multiple sclerosis by application of a PAF antagonist (BROCHET et al. 1992).

Many of the described pathological actions of PAF in the brain may be evoked by activation of phospholipase $A_2$ (BAZAN and RODRIGUEZ DE TURCO 1995) and by accumulation of polymorphonuclear neutrophils in committed brain tissue (HALLENBECK et al. 1986; LINDSBERG et al. 1990b; KOCHANEK et al. 1991).

# E. PAF in Respiratory Diseases

## I. Bronchial Hyperresponsiveness

Bronchial hyperresponsiveness is one of the most important aspects of several bronchial diseases. It is caused by several agents (BARNES 1988), including cytokines (VAN OOSTERHOUT and NIJKAMP 1993). The role of PAF in bronchial responsiveness is primarily distinguished by the fact that normal man develops hyperresponsiveness by inhaling PAF (GATEAU et al. 1984; CUSS et al. 1986; KAYE and SMITH 1990) as do animals (HALONEN et al. 1980; VARGAFTIG et al. 1980; PATTERSON et al. 1984; MAZZONI et al. 1985; CHUNG et al. 1986; FITZGERALD et al. 1987; ANDERSON et al. 1988; CHRISTMAN et al. 1990; NIEMINEN et al. 1991a, 1994) with striking similarity to antigen-induced hyperreactivity (PAGE 1988). The effect of PAF sometimes lasts for weeks and must be indirect (BARNES 1988). Further involvement of leukotrienes seems to be important (STENTON et al. 1990a; SPENCER et al. 1991; SPENCER 1992). The effects of PAF in hyperactivity, however, were bimodal (NIEMINEN et al. 1994) and not easily reproducible in normal man (HOPP et al. 1989; LAI et al. 1990; STENTON et al. 1990; SPENCER et al. 1990, 1991; NIEMINEN et al. 1991b, 1992) as well as less pronounced than those of antigen (MARSH et al. 1985) or C 5a (IRVIN et al. 1986). IV injections of PAF into guinea pigs also induce typical signs and disturbances that go with extravascular platelet recruitment (LELLOUCH-TUBIANA et al. 1985). Furthermore, bronchial spasms by PAF were observed in humans (GATEAU et al. 1984; CUSS et al. 1986; NIEMINEN et al. 1991b, 1992). The bronchospasms by PAF in guinea pigs are reduced by capsaicin (SPINA et al. 1991), indicating the involvement of neuropeptides (PERRETTI and MANZINI 1992), and by lipoxygenase inhibitors and the slow-reacting-substance (SRS) antagonist FPL-5572 (BONNETT et al. 1983), indicating that leukotrienes are secondary mediators of histamine release (HALONEN et al. 1985). Delayed bronchospasm was seen by PAGE et al. (1985).

However, asthmatic patients do not seem to show any increase in responsiveness following PAF treatment (RUBIN et al. 1987; CHUNG and BARNES 1989). PAF primes bronchial reactivity to histamine (SAKURAI et al. 1994); the airway response to histamine is reduced by potent PAF antagonists (Morooka et al. 1992).

In particular, PAF's bronchial actions are: impairment of tracheobronchial transport in man (Nieminen et al. 1991b); induction of inflammatory reactions (Camussi et al. 1983a), chemotaxis of neutrophils (O'Flaherty et al. 1981) and eosinophils (Wardlaw et al. 1986) as well as their activation; and alterations of specific pulmonary conductance for assessment of baseline airway caliber (Nieminen et al. 1994). Eosinophil activation correlates well with allergen-induced hyperreactivity (Santing et al. 1994).

Furthermore, involvement of PAF has been observed in allergen inhalation (Page et al. 1985), administration of IgE (Kravis and Henson 1975), and ozone exposure (Kaneko et al. 1995) to macrophages or bronchial-related cell lines (Samet et al. 1992).

The source of PAF within the airways may be the inflammatory cells (neutrophils, eosinophils, platelets; Chignard et al. 1979; Clark et al. 1980; Lee et al. 1984; Henson 1987) and alveolar macrophages (Arnoux et al. 1980) as well as resident cells of the lung (Kravis and Henson 1975; Arnoux et al. 1980; Voelkel et al. 1982). Whenever PAF antagonists have been employed, substantial improvements of hyperreactivity have not always been found (Demarkarian et al. 1991; Wilkens et al. 1991; Kuitert et al. 1992), although reductions of tracheal hyperresponsiveness were reported in animals (Deeming et al. 1985; Braquet et al. 1985; Darius et al. 1986; Casals-Stenzel 1987b; Coyle et al. 1987, 1988). However, other drugs can inhibit PAF-induced hyperresponsiveness, such as cromoglycate, ketotifen, hydrocortisone, and theophylline (Page 1988).

## II. PAF in Asthma Bronchiale

Air flow obstruction in asthma bronchiale derives from bronchoconstriction, epithelial damage, mucosal edema, increased mucus secretion, and an inflammatory infiltrate rich in eosinophils (Kaliner et al. 1987; Barnes 1988). Leukotrienes (Chanarin and Johnston 1994) and PAF (Chung 1992) are involved in all of these mechanisms.

PAF induces variable isometric contractions of human bronchial smooth muscle in vitro, with rapid tachyphylaxis (Johnson et al. 1990) and potentiation by histamine. Its bronchoconstrictor action via aerosol is 100-fold more potent than that of methacholine (Cuss et al. 1986; Rubin et al. 1987) and is inhibited by PAF antagonist UK 75506 (O'Connor et al. 1991). Furthermore, PAF causes pulmonary leukocyte sequestration in normal man (Tam et al. 1992) and also increases microvascular leakage throughout the respiratory tract of guinea pigs (Evans et al. 1987; O'Donnell and Barnett 1987; Tokuyama et al. 1991) with exudation of plasma proteins across the airway epithelium (Persson et al. 1987). PAF also stimulates the secretion of mucus from explants of tracheas of a variety of species (including human; Rogers et al. 1987; Sasaki et al. 1989; Luindgren et al. 1990). Lipoxygenase metabolites seem to be involved (SASAKI et al. 1989).Inhaled PAF impairs tracheobronchial mucociliary clearance in normal subjects (NIEMINEN et al. 1991b). The pulmonary effects of PAF were inhibited by salbutamol (ROCA et al. 1995).

Endogenously generated PAF is able to induce bronchial hyperresponsiveness (see above, this section) along with recruitment and activation of inflammatory cells, such as neutrophils and eosinophils in the lung (PAGE 1990). Furthermore, intravascular platelet aggregation is involved in bronchial asthma by releasing granule proteins (PINCKARD et al. 1977; MCMANUS et al. 1979; PAGE et al. 1984), also in humans (MACCIA et al. 1977; JEFFERY et al. 1989; YAMAMOTO et al. 1993). PAF's local release by exposure of asthmatics to allergen was deduced by BEER et al. (1995).

The stringent involvement of PAF in the pathogenesis of asthma bronchiale (CHUNG 1992; KUITERT and BARNES 1996) is nevertheless far from clear. However, PAF could mimic many features of asthma within the complex and distinctive inflammatory process in the airways that is associated not only with contraction of airway smooth muscle (see BARNES et al. 1988a), in which thromboxane mediation seems possible (UHLIG et al. 1994) together with dramatic PMN sequestration (fall in circulation; CUSS et al. 1986; SMITH et al. 1988), but that is also associated with airway edema (EVANS et al. 1987) and plasma extravasation, mucus hypersecretion (GOSWAMI et al. 1987), impairment of mucociliary clearance (AURSUDKIJ et al. 1987), and bronchial hyperresponsiveness (CHUNG 1992). Additionally, PAF's interaction with eosinophils and its release of eosinoplilic basic proteins could lend further support to the basis of prolonged bronchial effects (CUSS et al. 1986; BARNES et al. 1989; BARNES and CHUNG 1988).

In blood of asthmatic patients, the PAF level was found to be increased (HSIEH and NG 1993; TSUKIOKA et al. 1993; KUROSAWA et al. 1994), especially after antigen provocation (THOMPSON et al. 1984; BEER 1984). In neutrophils of atopic asthmatic patients, higher activity of the enzyme necessary for the biosynthesis of PAF was found (MISSO et al. 1993). Conversely, in asthmatic adults (TSUKIOKA et al. 1993) and children (MIWA et al. 1988), serum PAF acethydrolase was found to be reduced. Thus, endogenous PAF may be involved (STAFFORINI et al. 1996). Locally, PAF was detected biologically but not mass spectrometrically in bronchoalveolar lavage of asthmatics (NAKAMURA et al. 1987; STENTON et al. 1990b; AVERILL et al. 1991; CREA et al. 1992).

In animal models of asthma, the use of PAF antagonists reduced eosinophil infiltration and bronchial hyperresponsiveness (LELLOUCH-TUBIANA et al. 1988; COYLE et al. 1988). In asthma patients, the use of PAF antagonists, however, was limited or intriguingly small (WILKENS et al. 1991; BEL et al. 1991; HSIEH 1991; FREITAG et al. 1993). Furthermore, treatment with the PAF antagonist WEB 2086 did not reduce steroid requirement in a double-blind randomized, placebo-controlled trial (SPENCE et al. 1994). Recently, a more potent PAF antagonist, modipafant, had no effect on clinical asthma in a placebo-controlled study (KUITERT et al. 1995). Yet it must questioned whether local concentrations of antagonists are sufficiently high, as suggested by IMAIZUMI et al. (1995).

As a special aspect, eosinophilic cell infiltration of airway mucosas (Frigas and Gleich 1986) in asthma with regard to PAF should be discussed in more detail. In this respect, bronchial asthma has striking similarity with antigen-induced eosinophil infiltration (Page 1988). PAF is known to be a priming substance of leukotriene C4 for eosinophils (Shindo et al. 1996) and a potent chemoattractant itself (Wardlaw et al. 1986) and activator of eosinophils (Wardlaw et al. 1986; Kroegel et al. 1989). Eosinophils of asthmatic patients are elevated in peripheral blood (Smith 1992), are enriched in bronchial lavage fluid (Smith 1992), have increased phospholipase $A_2$ (Mehta et al. 1990), and produce more PAF than normal (Lee et al. 1984; Schauer et al. 1992). Eosinophilia of asthma and other allergic diseases seems to be a consequence of inhibition of eosinophilic programmed cell death (Simon and Blaser 1995). Eosinophilia leads to activation of T cells, especially Th 2 cells (Corrigan et al. 1988; Walker et al. 1991; Simon et al. 1994a,b), and finally to increased production of peptidoleukotrienes (Mocbel et al. 1990), which are important mediators of asthma attack (Manning et al. 1990). PAF upregulates IgE bindung to normodense human eosinophils (Moqbel et al. 1990). Activation of eosinophils by PAF induces tracheal epithelial shedding in vitro, associated with a slowing of ciliary beat frequence, an effect inhibitable by PAF antagonists (Yukawa et al. 1989, 1990). Using PAF antagonists, eosinophilic accumulation could be reduced (Coyle et al. 1988).

Additionally, activated neutrophils (PMNs) are involved in exacerbations of asthma (Boschetto et al. 1989) and are more susceptible to chemotactic stimulation by PAF (Rabier et al. 1991). PMNs of atopic asthmatics also have higher lyso-PAF acetyltransferase activity than those of normal persons (Misso et al. 1993). The question arises if platelets per se or activated platelets in asthmatic patients contribute to the profile of asthma (Maccia et al. 1977) such that PAF could exert a role in platelet activation (Yamamoto et al. 1993), a role denied by Nieminen et al. (1994). After IV PAF injection, extravascular platelet recruitment in the lung of guinea pigs was observed by Lellouch-Tubina et al. (1985).

However, some other facts speak against a significant role of PAF in asthma: PAF doses used in many of the studies cited above were generally very high (Nieminen et al. 1994); and PAF is highly tachyphylactic at the cellular level (Chesney et al. 1985) in tissue preparations (Stimler et al. 1981) or in animals in vivo (Halonen et al. 1980; Maridonneau-Parini et al. 1985).

In conclusion, involvement of PAF in sustained hyperresponsiveness and asthma bronchiale is noticable but not high enough for PAF antagonists to be likely to have a major clinical effect.

### III. Pulmonary Edema

Infusion of PAF caused pulmonary edema secondary to concomitant induction of pulmonary hypertension (see Sect. C; Heffner et al. 1983; Hamasaki et al. 1984). In particular, an increase in lung lymph flow and the lymph-to-plasma protein concentration ratio (CHRISTMAN et al. 1988) as well as a drop in the protein reflection coefficient as a measure of pulmonary vascular permeability (BURHOP et al. 1986) occurred. Thus, an increase in pulmonary macrovascular permeability has been established (CHRISTMAN et al. 1988).

Pulmonary edema in rat induced by PAF is dependent on the generation of leukotrienes (VOELKEL et al. 1982). In connection with these and other injury factors, PAF has priming effects in animal studies of lung injury with edema formation (RABINOVICI et al. 1993b).

The mechanism of PAF's vascular action is attributed to formation of gaps between epithelial cells (BJORK and SMEDEGARD 1983; HUMPHREY et al. 1984; BOLIN et al. 1987) leading to protein leakage (TABOR et al. 1992) in adjunction with vasoconstriction (SAKAI et al. 1989; SEALE et al. 1991) and the influence of leukocytes and platelets (SIROIS et al. 1994) as well as that of superoxide dismutase (HUANG et al. 1994). In contrast, PAF also has priming effects on pulmonary arterial reactivity to other autacoids (OHAR et al. 1993) or other substances, e.g., protamine (CHEN et al. 1989).

The role of PAF in clinical pulmonary edema has been discussed. PAF seems to have no major role as strudied by PAF antagonists in adrenaline-induced animal pulmonary edema (RAO and FONTELES 1991). Conversely, pulmonary edema produced by coronary ligation in dogs was reduced completely by applying a selective PAF antagonist (TANIGUCHI et al. 1992).

## IV. Adult Respiratory Distress Syndrome (ARDS)

Involvement of PAF in ARDS is likely (HUANG et al. 1994). The processes and stages of ARDS comprise neutrophil sequestration, fibrin-platelet aggregation (HEFFNER et al. 1987), accumulation of interstitial water, and acute inflammation reponse (DEMLING 1990).

The significant role of PAF within the ARDS is evidenced by the following data: (1) PAF is found to be increased in the bronchoalveolar lavage fluid of patients with ARDS (MATSUMOTO et al. 1992) and is possibly produced by neutrophils elicited from these cells by tumor necrosis factor (TNF; MATSUMOTO et al. 1992); in lavage fluid, the activity of phospholipase $A_2$ is also increased (D.K. KIM et al. 1995); (2) exogenously administered PAF is capable of causing many of the hallmarks of lung injury seen in ARDS, such as alveolar-capillary damage, high-permeability pulmonary edema, and pulmonary vasoconstriction and bronchoconstriction (CHUNG 1992); administration of PAF together with lipopolysaccharide (LPS) produces the full profile of ARDS (RABINOVICI et al. 1993a); (3) endotoxemia (see Sect. C) is involved in ARDS (PARSON et al. 1989), its action reduced by PAF antagonists (CHANG et al. 1987, 1990; OLSON et al. 1990a,b); (4) in preterm patients and animals, a leakage of protein from pulmonary vascular space is also significant (IKEGAMI et al. 1992) and is reduced by PAF antagonists (TABOR et al. 1992); and (5) local platelet accumulation is a further item of evidence (HEFFNER et al. 1987).

### V. Lung Fibrosis

In lung fibrosis, application of a PAF antagonist actually reduced the process of fibrose development (GIRI et al. 1995).

## F. Involvement of PAF in Gastrointestinal Diseases

### I. Gastric Diseases

PAF'S involvement in gastric diseases is reviewed by WALLACE (1990) and SUGATANI et al. (1991). The content of PAF in the stomach is normally high (SUGATANI et al. 1991a); antral mucosa contains much more PAF than the corpus or antrum (SUGATANI et al. 1989). PAF content is decreased in water-immersed stress (SUGATANI et al. 1989) and ethanol-induced gastric ulcer (FUJIMURA et al. 1992) and increased in the endotoxin-damaged stomach (IBBOTSON and WALLACE 1989) as well as in damage by aspirin, indometacin, taurocholic acid, and hydrochloric acid (LIGUMSKY et al. 1990). Enhanced levels of PAF are reduced by sucralfate, but control levels are not (LIGUMSKY et al. 1990).

PAF in the stomach is preferably synthesized de novo (FERNANDEZ-GALLARDO et al. 1988), for it lacks lyso-PAF transacetylase for the remodeling pathway. Generation of PAF from gastric mucosa was two- to threefold higher in duodenal ulcer patients than in normal controls (ACKERMAN et al. 1990), returning to normal after successful treatment. It seems possible, however, that PAF and leukotrienes in ulcer may be derived from an increased number of inflammatory cells of active peptic ulcers (ACKERMAN et al. 1990). Using PF-5901, a PAF biosynthesis inhibitor, generation of PAF in damaged gastrointestinal tissue was reduced (HOGABOAM et al. 1992). In case of vascular disturbances, PAF has endothelium-dependent relaxant effects in the mesenteric and gastric arteries (CHU et al. 1988; KAMATA et al. 1989; CHIBA et al. 1990).

In the stomach, activity of the PAF-destroying enzyme acethydrolase is low (SUGATANI et al. 1989), but is increased in gastric ulcer patients (FUJIMURA et al. 1989).

PAF may be involved in the physiological regulation of gastric secretion (CUCULA et al. 1989; SOBHANI et al. 1992). By direct action of PAF on guinea pig parietal cells (NOGAMI et al. 1992) or of picomolar concentrations of PAF on isolated rabbit gastric glands, acid secretion is stimulated, which on the other hand is inhibited by omeprazole, verapamil, and BAPTA, yet not by the H2 blocking agent famotidine (SOBHANI et al. 1995). In cats, gastric PAF secretion was increased in response to histamine (SOBHANI et al. 1990), and in man, lyso-PAF increases in response to gastrin (SOBHANI et al. 1992). In a human gastric tumor cell (HGT 1) PAF production involves several mediators (SOBHANI et al. 1996a).

A role of PAF in gastric ulcers is forwarded by ROSAM et al. (1986). The arguments for the involvement of PAF are the following:

1. Infusion of PAF into rats produced extensive hemorrhagic mucosal erosions (ROSAM et al. 1986; WALLACE and WHITTLE 1986b; BINNAKA et al. 1989), especially intra-arterial infusion (WHITTLE et al. 1987b; ESPLUGUES and WHITTLE 1988). PAF administered intravenously augmented ethanol-induced gastric damage without affecting blood pressure (WALLACE and WHITTLE 1986a) so that a vascular hypotensive effect alone is not the reason for the damage.
2. Direct mucosal damage is caused by PAF. Mucosal blood flow and mucosal capillary blood flow were reduced by IV PAF (WHITTLE et al. 1986), followed by increases in the release of leukotrienes and other prostanoids (DEMBINSKA-KIEC et al. 1989) as well as subepithelial edema (WALLACE et al. 1987a). Since endotoxin inhibits pentagastrin-stimulated gastric acid secretion, its partial restoration by PAF antagonists (MARTINEZ-CUESTA et al. 1992) seems plausible in light of PAF's involvement (see Sect. C).
3. In accordance with its spasmogenic action on the stomach (LEVY 1987), PAF administered by intra-arterial infusion increased gastric intraluminal pressure (ESPLUGUES and WHITTLE 1989) which is correlated with the extent of gastric mucosal damage (WHITTLE et al. 1986). This action is also dependent on histamine and 5-hydroxytryptamine.
4. The damaging effect is apparently mediated by further mediators. Thus, PAF's effect on the stomach could be prevented by dexamethasone and free radical scavengers such as allopurinol (ETIENNE et al. 1988; WATANABE et al. 1988; YOSHIKAWA et al. 1990) or by somatostatin analogs and atropine (ETIENNE et al. 1988). The following processes seem to be involved: generation of reactive oxygen species by granulocyte activation (GAY et al. 1986), increased leukocyte adherence (SUEMATSU et al. 1989; KUBES et al. 1990a), and release of proteases (SHAW et al. 1981). The release of leukotrienes from the gastrointestinal tract by PAF infusion into rats is well documented (HSUEH et al. 1986a,b; WALLACE et al. 1988), acting on gastric mucosa (ACKERMAN et al. 1990).
5. PAF antagonists applied before damage by agents other than PAF showed prophylactic action in endotoxin-induced gastrointestinal damage (HWANG et al. 1986; WALLACE et al. 1987b; see Sect. C). PAF antagonists suppressed gastric mucosal damage by ischemia/reperfusion (BRAQUET et al. 1988; IWAI et al. 1989; BINNAKA et al. 1990; see Sect. C), hemorrhagic shock (WALLACE et al. 1989), water-immersion stress (SUGATANI et al. 1989), or dexamethasone (FILEP et al. 1991b), but not by endothelin (WALLACE et al. 1989) or nonsteroid antiinflammatory drugs (NSAID; WALLACE et al. 1990b) such as aspirin (RAINSFORD 1986). This indicates that PAF may not play a significant role in the pathophysiological process of gastric ulcers by NSAIDs.
6. The elevated gastric secretion of *Heliobacter pylori* is stimulated via PAF (SOBHANI et al. 1996b).

7. Some ulcer therapeutics such as bismuth compounds or carbenoxolone have inhibitory action on PAF-acetyltransferase in vitro (VON BRUCHHAUSEN and ROCHEL 1990).

It must be mentioned, however, that not all PAF antagonists have the same antagonistic action on gastric damage (SUGATANI et al. 1991a). Altogether, PAF's primary action is probable on the mucosal and submucosal vasculature (ACKERMAN et al. 1990). However, the proposal by SUGATANI et al. (1991a) that exogenous PAF is involved in ulcer genesis, but endogenous PAF has protective actions, has to be proven and taken into mind.

## II. Intestinal Diseases

In normal human mucosal biopsies, PAF is produced (KALD et al. 1990) as it is produced by normal rat intestinal mucosa (WHITTLE et al. 1987a; TAGESSON et al. 1988), especially by intestinal epithelial cells (by colonic more than by small intestinal cells) (KALD et al. 1992) even without special stimulation (GUSTAFSON et al. 1991). Since A 23 187 stimulates PAF production in normal intestinal epithelial cells (KALD et al. 1992), it is assumed that phospholipase $A_2$ is involved in PAF production (GUSTAFSON and TAGESSON 1989; KALD et al. 1992), the activity of which is increased in Crohn's disease (OLAISON et al. 1988). Recently, the pathway of de novo PAF biosynthesis has been found to be involved in human inflammatory bowel diseases (APPLEYARD and HILLIER 1995). A phospholipase D is involved in the metabolism of PAF in rat enterocytes (FURUKAWA et al. 1995). In the colon, PAF receptors have been evidenced (LONGO et al. 1995).

Local action of PAF on enteric mucosa is characterized by stimulation of ion transport in jejunum of rat (HANGLOW et al. 1989) and in distal colon of rat (BERN et al. 1989; BUCKLEY and HOULTS 1989) and rabbit (BERN et al. 1989; TRAVIS and JEWELL 1992) as well as by increased anion secretion (TRAVIS and JEWELL 1992; TRAVIS et al. 1995) and increased epithelial permeability (TRAVIS and JEWELL 1992), partly exerted indirectly by inhibition of short-circuit current (SCC) by PAF (by 90%; HANGLOW et al. 1989; BUCKLEY and HOULTS 1989). The response of SCC to PAF can be attributed to the anion secretion mentioned above (TRAVIS and JEWELL 1992). Furthermore, action of PAF on motor function has been described (FINDLAY et al. 1981; PONS et al. 1991, 1994). Some actions of PAF are mediated by prostaglandins and histamine (from mast cells) as based on its inhibition by indomethacin, doxantrazole, and mepyrone (TRAVIS et al. 1995).

As the most potent ulcerogenic agent, PAF – such as that of the stomach – induces a dramatic change in the mucosal vasculature (WHITTLE et al. 1986) and action on neutrophils (WALLACE et al. 1990b).

In the case of irritation of colonic mucosa, increased PAF formation has been reported by castor oil (PINTO et al. 1989, 1992), but not by senna and its active component rhein (MASOLO et al. 1992). Intestinal epithelial crypt cells are the source of inflammatory mediators (KESHV et al. 1990).

PAF was used as a model for ischemic bowel necrosis (HSUEH et al. 1987; ESPLUGUES and WHITTLE 1988; MUSEMECHE et al. 1995). It is largely mediated by oxygen radicals (CUEVA and HSUEH 1988; CAPLAN et al. 1990b) and primed by endotoxins (SUN et al. 1995). Furthermore, leukotrienes are involved (HSUEH et al. 1986b), even in neonates (CAPLAN et al. 1990b). Necrotizing effects of tumor necrosis factor (TNF) on the intestine are mediated by PAF and prevented by PAF antagonists (SUN and HSUEH 1988). PAF is also involved in necrotizing enterocolitis by hypoxia in premature infants (GONZALES-CRUSSI and HSUEH 1983; HSUEH et al. 1987; CAPLAN et al. 1990b, 1992a) which can be caused by intra-aortic administration of PAF (FURUKAWA et al. 1993). In plasma from these patients, decreased activity of PAF acethydrolase was found (CAPLAN and HSUEH 1990). It should be mentioned that hypoxia in itself provokes PAF production (in the lung; PREVOST et al. 1984) in vivo and in vitro (CAPLAN et al. 1992). The perfused gut also generates PAF, which attracts and primes neutrophils (F.J. KIM et al. 1995). In the state of necrotizing enterocolitis, blockage of nitric oxide (NO) synthesis worsens intestinal injury (MACKENDRICK et al. 1993). In this context, it is noteworthy that NO is important for the maintenance of gastrointestinal microvascular and mucosal integrity (WHITTLE et al. 1990; KUBES and GRANGER 1992; WHITTLE 1993). Therefore, blockage of NO synthesis augments intestinal damage by endotoxin at higher (HUTCHESON et al. 1990) or lower doses (LASZLO et al. 1994a).

In endotoxin-induced gastrointestinal damage, involvement of PAF (GONZALES-CRUSSI and HSUEH 1983) has been noted and is diminished by PAF antagonists (WALLACE et al. 1987a; WALLACE and WHITTLE 1986c; HSUEH et al. 1986b; WHITTLE et al. 1987a). Mucosal ulcerations following infusion of PAF pathogenetically resembled those following endotoxic shock (ROSAM et al. 1986; ESPLUGUES and WHITTLE 1988). Early release of tissue-damaging secondary mediators by lipopolysaccarides, PAF, and thromboxane $A_2$ (WALLACE et al. 1987; BOUGHTON-SMITH et al. 1989) plays a special role. PAF induces gastric hemorrhagic necrosis (ROSAM et al. 1986) and microvascular injury in the gastrointestinal tract (WALLACE et al. 1987; BOUGHTON-SMITH et al. 1992; FILEP and FOLDES-FILEP 1993) as well as hypotension. PAF may release leukotriene C4 from intestinal cells (HSUEH et al. 1986) as from other cells (VOELKEL et al. 1982; HSUEH et al. 1988; WALLACE and MACNAUGHTON 1988). Microvascular damage by PAF, however, is counterbalaced by NO synthesis (LASZLO et al. 1994).

As far as intestinal ischemia/reperfusion is concerned (see Sect. C), several agents or mediators may be involved via induced neutrophil infiltration (GRANGER 1988). Following reperfusion in dogs, increased blood levels of PAF have been observed (FILEP et al. 1989), with PAF possibly derived from endothelial cells (LEWIS et al. 1988) along with induction of cell-dependent neutrophil adhesion. In this context, it seems necessary to stress the significance of PAF as a final messenger of adhesion reactions (KUBES et al. 1990a,b; WATANABE et al. 1991).

In experimental colitis (of rats), involvement of PAF is likely, e.g., via trinitrobenzene (DE LONGO et al. 1994, 1995) or various other agents (BOUGHTON-SMITH and WHITTLE 1988). The role of mast cells has been reported (MARSHALL and BIENENSTOCK 1994).

Involvement of PAF in infective colitis is forwarded by CHAUSSADE et al. (1991). In sepsis, minor intestinal microvascular hypoperfusion by PAF is observed (BAR-NATAN et al. 1995). In the case of immune-complex enteropathies of rat, several PAF antagonists also acted partially or completely as inhibitors (BLOCH et al. 1991), so PAF may be involved.

In animal models of colitis ulcerosa by dextrane sodium sulfate, PAF and prostanoids are considered as secondary triggers (VAN DIJK et al. 1993), with TNF-alpha and interleukin-1 as primary triggers (MORTEAU et al. 1993). Production of PAF is increased two- to threefold in this type of experimental colitis (ZIJLSTRA et al. 1993), which again is reduced by prednisolone.

Pathogenesis of inflammatory bowel disease (IBD, i.e., colitis ulcerosa and Crohn's disease) is probably multifactorial (PODOLSKI 1991; FERRARIS et al. 1993). Modulation of PMN function (KAZI et al. 1995) and increased platelet activity (dysfunction; COLLLINS and RAMPTON 1995) are pathologically relevant. Systemic endotoxinemia often accompanies IBD (GARDINER et al. 1995). In colitis ulcerosa itself, intestinal cells contribute to enhanced generation of PAF (FERRARIS et al. 1993; GUIMBAUD et al. 1995; WARDLE et al. 1996), so a high content of PAF was found in stool of patients with IBD and bacterial diarrhea (CHAUSSADE et al. 1991). FERRARIS et al. (1993) showed that human intestinal mucosal cells and lamina-propria-mucosae mononuclear cells contribute to the generation of PAF in ulcerative colitis, but not in Crohn's disease, with epithelial cells especially stimulable to PAF production. This means that colonic mucosas of colitis ulcerosa show higher basal stimulated PAF production (ELIAKIM et al. 1988; KALD et al. 1990; THYSSEN et al. 1996). The colonic mucosa metabolizes PAF (APPLEYARD and HILLIER 1992). In Crohn's disease, activity of PAF acethydrolase is also increased (DENIZOT et al. 1992). Nevertheless, higher PAF concentrations in colonic mucosa were also found in Crohn's disease (KALD et al. 1990; SOBHANI et al. 1992b), as in experimental models of colitis (BOUGHTON-SMITH and WHITTLE 1988). In IBD. however, PAF is also released by stimulated granulocytes, mast cells, platelets, and endothelial cells (LYNCH et al. 1979; MUELLER et al. 1983; MCINTYRE et al. 1985; WHATLEY et al. 1989). In IBD, platelet counts are usually increased (HARRIES et al. 1991); and thrombocytosis may relate to an increase in the turnover of platelets (TALSTAD et al. 1973), enhanced production of PAF (ELIAKIM et al. 1988), or even secretion of a thrombopoietin-like hormone (CORBETT and PERRY 1983). As mentioned above, high concentrations of PAF were found in stool of patients with IBD (CHAUSSADE et al. 1991; DENIZOT et al. 1991a) such as in patients with infectious diarrhea (DENIZOT et al. 1991b) especially of bacterial origin (DENIZOT et al. 1992). Thus, PAF can be used as a stool marker for intestinal inflammation (Chaussade et al. 1992; Denizot et al. 1993).

## G. Involvement of PAF in Function and Disturbances of Reproduction

PAF apparently has a physiological role in the establishment and termination of pregnancy. It is the first hormone to be produced by mammalian embryos that can be measured in just the one-cell zygote state (Page and Abbott 1989). Under in vitro conditions, supplementation with PAF increases the implantation rate by 50% (O'Neill et al. 1989; Ando et al. 1990). Furthermore, PAF originating from type II pneumocytes is involved in the initiation of labor (Johnston et al. 1992). High maternal serum levels of the PAF-inactivating enzyme, PAF acethydrolase, decline late in gestation, so PAF may initiate uterus contractions.

In particular, the following observations on various aspects of reproduction have been made.

### I. Ovulation

Inflammation, ovulation (follicle rupture), and implantation have certain aspects in common (Espey 1980; Abisogun et al. 1989), which is demonstrated by the involvement of PAF in follicle rupture (ovulation; Abisogun et al. 1989; Li et al. 1991) and the inhibition of follicle rupture by the PAF-specific antagonist BN 52021. Instillation of PAF antagonists into the ovarian bursa of immature rats reduced the number of ova shed following injection of gonadotropins and decreased collagenolysis and vascular permeability (Abisogun et al. 1989).

Secretion of PAF by periovulatory ovine follicles has been shown (Alexander et al. 1990), and the occurrence of PAF in follicular fluid was also observed (Amiel et al. 1991; Lopez-Bernal et al. 1991).

### II. Gametocytes (Oocytes, Sperm)

The occurrence of enzymes for de novo biosynthesis of PAF was studied in oocytes by Wells and O'Neill (1994).

Mammalian sperm contains PAF (Kumar et al. 1988a; Kuzan et al. 1990; Parks et al. 1990), particularly that of man (Minhas et al. 1991). It is synthesized by the remodeling pathway (Sanwick et al. 1992; Baldi et al. 1993). The production of PAF is increased by progesterone combined with the compound A 23 187 (Baldi et al. 1993). PAF's precursor is highly concentrated in sperm (Selivonchik et al. 1980; Parks and Graham 1992; Parks and Lynch 1992). Phospholipase $A_2$ required for its production was found in several species

(MEIZEL 1984; LANGLAIS and ROBERTS 1985). PAF acethydrolase – important for the acrosome reaction – is also found in sperm as well as in seminal plasma and seminal fluid (PARKS and HOUGH 1993), though sometimes activity is not detectable in seminal plasma (BENVENISTE et al. 1988; ANGLE et al. 1991), yet in seminal vesicles (MUGURUMA et al. 1993). The destructive enzyme acethydrolase was found in very high-density lipoprotein (VHDL) fractions of human seminal plasma (LETENDRE et al. 1992; PARKS and HOUGH 1993; JARVI et al. 1993a) as well as in prostatic fluid (JARVI et al. 1993b) and vas-deferens fluid (JARVI et al. 1993b).

While there have been conflicting reports on the effect of PAF on sperm motility (RICKER et al. 1989; SILVERBERG et al. 1990; JARVI et al. 1991), PAF stimulates motility in the presence of albumin (JARVI et al. 1993a; HELLSTROM et al. 1991; JARVI et al. 1993a) and is toxic at higher concentrations (JARVI et al. 1991, 1993a).

Astonishingly, antagonists of PAF thus have spermicidal effects (HARPER 1989). In contrast, PAF is used as a cryoprotectant of human sperm (WANG et al. 1993).

In asthenozoospermic patients, PAF content of sperm was higher than in normal men (ANGLE et al. 1991).

The sperm penetration rate was increased by PAF and decreased by PAF antagonist CV 3988 (SENGOKU et al. 1993). The human spermatozoon produces PAF especially during the process of capacitation (KUZAN et al. 1990). PAF enhance both capacitation and acrosome reaction processes of spermatozoes of mice in vitro (SENGOKU et al. 1992). The capacitation is enhanced by treatment also of human spermatozoa with PAF (MINHAS 1993; LACHAPELLE et al. 1993). The acrosome reaction in human sperm was also enhanced (SENGOKU et al. 1993; KRAUTZ et al. 1994) as was in animals (RICKER et al. 1989; D'CRUZ and HAAS 1989; FUKUDA et al. 1994).

## III. Fertilization

In the fertilization process, the in-vitro presence of PAF is favorable (HARPER 1989; ROUDEBUSH et al. 1990; KUZAN et al. 1990). Thus, the in-vitro fertilization rate is enhanced by PAF (MINHAS et al. 1989; KUZAN et al. 1990; MINHAS and RICKER 1990).

## IV. Oviduct Passage and Preimplantation State

The oviduct is the site of many events normally essential to mammalian reproduction, including fertilization, gamete transport, early embryonic development, and embryo transport (YANG et al. 1992). PAF plays a role in maternal recognition of pregnancy (O'NEILL 1985) produced by the zygote (embryo). Thus, a signalling role of PAF for oviduct transport is assumed (VELASQUEZ et al. 1995). Secreted by ovaries and oviduct, an early pregnancy factor (EPF) is elicited by PAF administration (OROZCO et al. 1986; SUEOKA et al. 1988). Blastocysts take up PAF (JONES et al. 1992; KUDOLO et al. 1991).

PAF plays an important role for the preimplantation embryo as a direct autocrine growth factor (O'NEILL 1991). Embryos produce PAF (BATTYE et al. 1991), either C16:0 PAF or C18:0 PAF, or both (AMMIT et al. 1992). Reduced viability of embryos under in vitro culture conditions was attributed to reduced PAF production (RYAN et al. 1989; BATTYE et al. 1991). The use of PAF antagonists reduced the success rate of animal embryo implantation (O'NEILL et al. 1990) whereas PAF supplementation was advantageous (O'NEILL et al. 1989; KUZAN et al. 1990). PAF's action in stabilizing preimplantation embryo (O'NEILL 1992) is based on improved carbohydrate utilization, improved growth conditions, and an enhanced rate of thymidine incorporation into DNA (O'NEILL 1991).

Another action of embryo-derived PAF is to signal for extended estrous cycle (BATTYE et al. 1989), maintaining progesterone production by an antiluteolytic mechanism (SMITH and KELLY 1988). The maternal reaction to PAF seems to be early reduced vascularization (STEIN and O'NEILL 1991). Later, endometrial production of PAF peaks in the luteal phase in rabbit (ANGLE et al. 1988) and women (ALECOZAY et al. 1989).

## V. Nidation

With regard to implantation, PAF production by the embryo is more important (O'NEILL 1991) than that of the uterus. PAF antagonists instilled into the uterine lumen were able to inhibit implantation (ACKER et al. 1988, 1989). Implantation failure caused by applying PAF antagonists was also seen (ANDO et al. 1990; SPINKS et al. 1990), yet not by MILLIGAN and FINN (1990), and it was attributed to their high doses (O'NEILL 1995a). Inhibition of implantation by low doses of PAF antagonists suggested that the antagonists exert their effect by acting on the embryo (SPINKS et al. 1990).

## VI. Embryonic Development

PAF has been found in uterine tissue of rats (YASUDA et al. 1986) and rabbits (YASUDA et al. 1986) and in human endometrial cell cultures (ALECOZAY et al. 1989, 1991). The corresponding PAF receptors have been found in pregnant uteri (KUDOLO and HARPER 1989, 1990, 1992). The onset of parturition late in gestation seems to be influenced by PAF (MAKI et al. 1988). Before delivery, increased sensitivity of the myometrium to PAF was found (B.K. KIM et al. 1995). PAF acethydrolase activity of the uterus changes in the course of pregnancy (O'NEILL 1995b). Enzyme activity declines late in gestation, so PAF may initiate uterus contraction, the decline of which is strengthened by estrogens (YASUDA et al. 1996). This is in accordance with the observation that parturition is prolonged after administration of PAF antagonists (ZHU et al. 1991). In premature labor and in premature rupture, PAF level was increased in the amniotic fluid (HOFFMAN and ROMERO 1990). Apart from oxytocin, PAF also stimulates prostaglandin production in endometrial cells (KIM and FORTIER 1995), thus an indirect mechanism may exist.

## H. Involvement of PAF in Skin Diseases

PAF can be synthesized in skin (CSATO et al. 1987; CUNNINGHAM et al. 1987; MICHEL et al. 1990) The cells contributing to its production are epidermal cells (CSATO et al. 1987), including keratinocytes (CUNNINGHAM et al. 1987; BARKER et al. 1991; MICHEL and DUBERTET 1992) and fibroblasts (MICHEL et al. 1988b; MICHEL and DUBERTET 1992). Applied to skin, PAF produces inflammation and pruritus (ARCHER et al. 1984), and increased vascular permeability with protein exudation (HUMPHREY et al. 1982a; MORLEY et al. 1983; PIROTZKY et al. 1985b; MICHEL et al. 1987; ARCHER et al. 1988; FUJII et al. 1995). The dermal action of PAF is enhanced by prostaglandin E2 (MORLEY et al. 1983; MCGIVERN and BASRAN 1984) and skin response decreased in cancer patients (BURTIN et al. 1990). Together with lipopolysaccharides (LPS), PAF stimulates interleukin-1 release in keratinocytes (PIGNOL et al. 1990) and together with TNF promotes their detachment (KATAYAMA et al. 1994). In normal man, the biphasal response to PAF begins with a wheal-and-flare reaction followed by a late onset of erythema and neutrophil infiltration (ARCHER et al. 1985; MICHEL et al. 1987; MICHEL and DUBERTET 1992) which is reduced by PAF antagonists (GUINOT et al. 1986; CHUNG et al. 1987; MARKEY et al. 1990). In atopic patients, eosinophilic infiltration dominates (HENOCQ and VARGAFTIG 1986), which is also reduced by PAF antagonist BN 52063 (MARKEY et al. 1990). Vascular reaction and cell infiltration are somewhat different in animal (HUMPHREY et al. 1982b; MORLEY et al. 1983; DEWAR et al. 1984) and man (ARCHER et al. 1984, 1985; BASRAN et al. 1984). Antigen skin reaction in animals resembles that induced by PAF (FOSTER et al. 1995).

Recent findings show PAF to be a potent pruritogen in a special animal model (WOODWARD et al. 1995) and be more potent than histamine. D-PAF and lyso-PAF showed only minor action. The pruritogenic action of PAF was abolished by selective antagonists such as CV 6209, apafant, and E 6123. PAF's pruritogenic action is independent of mast cell degranulation, as formerly suggested by FJELLNER and HÄGERMARK (1985), and independent of histamine release because it is not influenced by pyrilamine (WOODWARD et al. 1995). It should be mentioned that an inhibitor of PAF biosynthesis, polidocanol, (CERINSKI-HENNIG and VON BRUCHHAUSEN 1991) was formerly (BRÜGGMANN 1988) and more recently suggested to be antipruritic. The known antipruritic action of glucocorticoids now seems understandable in light of their indirect action on phospholipase $A_2$.

These observations support the view that PAF may be an inflammatory mediator of significance in cutaneous and other inflammatory disorders (MARKEY et al. 1990) and essentially committed to inflammatory dermatoses

(PINCKARD et al. 1980; CZARNETZKY 1983; ARCHER et al. 1984, 1985; BASRAN et al. 1984; MALLET and CUNNINGHAM 1985; GRANDEL et al. 1985) that can be influenced by PAF antagonists (GUINOT et al. 1986; CHUNG et al. 1987).

In skin lesions, an increase in PAF is seen in allergic skin diseases (VALONE et al. 1987; MICHEL et al. 1988a,b), in atopic dermatitis (see below), and in psoriasis (MALLET and CUNNINGHAM 1985; see below). Cutaneous late-phase responses with eosinophilic infiltration (FADEL et al. 1987) induced in man by allergen or PAF (ARCHER et al. 1984; BASRAN et al. 1984; HENOCQ and VARGAFTIG 1986; FADEL et al. 1990) were abolished by cetirizine (FADEL et al. 1987, 1990; DE VOS et al. 1989; FADEL 1990). Eosinophilic accumulation in skin is associated especially with typical hypersensitive cutaneous reactions (HENOCQ and VARGAFTIG 1986). This accumulation is also induced by interleukin-1 and inhibited by PAF antagonists showing evidence of PAF receptor involvement (SANZ et al. 1995).

The dermal reversed-passive Arthus reaction is accompanied by platelet accumulation (HENSON and COCHRANE 1969; PONS et al. 1993) as is the local Shwartzman reaction (MOVAT and BURROWES 1985). PAF appears to play an important role in the development of the Arthus reaction. Endothelial cells, neutrophils undergoing phagocytosis, and platelets themselves are its source (CHIGNARD et al. 1980; CAMUSSI et al. 1983b; HELLEWELL and WILLIAMS 1986). Furthermore, PAF antagonists reduce edema formation associated with dermal reversed-passive Arthus reaction (DEACON et al. 1986; HELLEWELL and WILLIAMS 1986; ISSEKUTZ and SZEJDA 1986; WILLIAMS et al. 1986; WARREN et al. 1989; HELLEWELL 1990; ROSSI et al. 1992; PONS et al. 1993). In urticaria characterized by erythema, pruritus, and edema, PAF's involvement is also likely, as in cold urticaria (GRANDEL et al. 1985; HUSZ et al. 1991).

With regard to atopia (atopic dermatitis), the role of eosinophils (see Sect. B) is significant (HENOCQ and VARGAFTIG 1986; MARKEY et al. 1990). PAF is the most potent chemotactic factor of eosinophilic activity (WARDLAW et al. 1986) and eosinophils have PAF receptors (UKENA et al. 1989). The attraction to the skin is seen in allergic patients (LEIFERMAN et al. 1985; HENOCQ and VARGAFTIG 1986; BRUYNZEEL et al. 1987; MORITA et al. 1989; KAPP 1991). In atopic dermatitis, this chemotactic activity is increased (MORITA et al. 1989), which is associated with release of proteins and lipid mediators (see Sect. B). Atopic patients differ from nonatopic individuals in their response to PAF (MICHEL et al. 1988a).

PAF could be involved in other skin affections. Contact dermatitis caused by dinitrofluorobenzene is reduced by PAF antagonist BN 52063 (CSATO and CZARNETZKI 1988). PAF has been found in cold urticaria (GRANDEL et al. 1985), an inflammatory dermatosis, the symptoms of which were reduced by doxepin hydrochloride (GRANDEL et al. 1985). PAF is involved in the *Candida-albicans* killing activity of human epidermal cells as an inflammatory mediator (CSATO et al. 1990). It is detected in UV-irradiated skin (CALIGNANO et al. 1988), as in mucosa following gamma irradiation (MCMANUS et al. 1993) or in the eye following laser irradiation (VERBEY et al. 1989).

PAF has been identified in psoriatic lesions (MALLET et al. 1984; MALLET and CUNNINGHAM 1985) and is possibly produced by neutrophils (ARCHER et al. 1988; CUNNINGHAM et al. 1987). The blood levels rise in this dermatosis, going back during therapy (IZAKI et al. 1996).

In lupus erythematosus, an elevated level of PAF acethydrolase has been found as an indicator of involvement of PAF (MATSUZAKI et al. 1992) and a decreased level (TETTA et al. 1990) for systemic disease.

## I. Involvement of PAF in Diseases of Other Organs

### I. Renal Diseases

In the kidneys, PAF is preferably synthesized de novo (SCHLONDORFF and NEUWIRTH 1986; SCHLONDORFF et al. 1986; MUIRHEAD 1986), as the kidneys usually are the main source of PAF found in blood (CARAMELO et al. 1984; MCGOWAN et al. 1986; LEE et al. 1989: SNYDER 1990). Cell types that generate PAF are the renal medullary interstitial cells (MCGOWAN et al. 1986; MUIRHEAD 1986; VANDONGEN 1991), mesangial cells (ZANGLIS and LIANOS 1987), and endothelial cells as well as activated leukocytes and platelets under certain pathological conditions (SCHLONDORFF et al. 1986; SCHLONDORFF and NEUWIRTH 1986; CAMUSSI 1986; BERTANI et al. 1987; REMUZZI 1989). In contrast, activity of the destructive enzyme PAF acethydrolase is particularly high (BLANK et al. 1981); this enzyme was found to be increased in most cases of glomerulonephritis and hypertensive renal diseases (KIRSCHBAUM 1991).

PAF's involvement in renal function is likely. Thus, PAF enhances glomerular permeability for proteins (PERICO et al. 1988), and PAF antagonists blocked the noradrenaline-induced renal vascular escape in rabbits (FERREIRA et al. 1991). However, PAF's renal hemodynamic effects are poorly understood. In-vivo observations of renal vasodilatation by PAF in minor doses (HANDA et al. 1990, 1991) are contrasted by observations (at higher doses) of decreased renal blood flow and glomerular filtration rate following intrarenal PAF infusion (SCHERF et al. 1986; WANG and DUNN 1987; BADR et al. 1989; TOLINS et al. 1989). Cyclooxygenase products are involved in PAF's vasoconstrictive effect on afferent arterioles (JUNCOS et al. 1993) and nitric oxide in its possible vasodilator effect. Both effects are blocked by the PAF antagonist hexanolamino-PAF (JUNCOS et al. 1993).

PAF is involved in renin release (SCHWERTSCHLAG et al. 1987). In one-kidney, one-clip hypertension, an increase in PAF in renal effluent was seen after unclipping (MASUGI et al. 1984; MCGOWAN et al. 1986; MUIRHEAD 1986; MCGOWAN 1986; VANDONGEN 1991) that came from renal medullary cells. The subsequent decrease in peripheral resistance and drop in blood pressure are markedly attenuated by infusion of a PAF antagonist (MASUGI et al. 1984; MUIRHEAD 1986; VANDONGEN 1991). PAF antagonists can therefore substantially ameliorate the reduction in glomerular filtration rate and renal plasma flow as well as in proteinuria and histological lesions (BERTANI et al. 1987; REMUZZI 1989; FELSEN et al. 1990).

Involvement of PAF or PAF deficiency has been reported in the following pathological states: experimental nephropathies (REMUZZI 1989); glomerulonephritis (LLIANOS 1989; LLIANOS and ZANGLIS 1990; YOO et al. 1991; NORIS et al. 1993); nephrotoxic nephritis (BERTANI et al. 1987); lupus nephritis, especially in PMN recruitment (BRADIE 1995); glomerulosclerosis (CLARK and NAYLOR 1989; RUIZ-TORRES et al. 1994); endotoxin-induced acute renal insufficiency (WANG and DUNN 1987); gentamicin toxicity by supposed activation of mesangial cells (RODRIGUEZ-BARTERO et al. 1995); and ureteral obstruction (WEISMAN et al. 1990; FELSEN et al. 1990).

## II. Hepatic Disturbances

The liver is a prominent target organ of toxicity during endotoxemia (NOLAN 1981; see also Sect. C). Hepatic and sinusoidal cells (macrophages, i.e., Kupffer cells, endothelial cells, and Ito cells) release several mediators, including PAF (CHAO et al. 1989a; DECKER 1990; MUSTAFA et al. 1995). Its functional active receptors were also found on/in Kupffer cells (CHAO et al. 1989b; GANDHI and OLSON 1991).

The action of PAF on liver cells provokes an increase in intracellular pH (GARDNER et al. 1995), an increase in inositol phosphates and $Ca^{2+}$ (FISHER et al. 1989), and production of eicosanoids and superoxide (DIETER et al. 1986). Superoxide release from liver cells by PAF was markedly diminished by PAF antagonist SDZ 63–441 (BAUTISTA and SPITZER 1992). Thus, PAF has an autocrine function in Kupffer cells. It activates metabolism and glycogenolysis (SHUKLA et al. 1983; HINES et al. 1991; HINES and FISHER 1992), followed by glucose release (LAPOINTE and OLSON 1989) and hyperglycemia (LANG et al. 1988).

As in other reperfusion conditions (see Sect. C), PAF is involved in microcirculatory disturbances following global hepatic ischemia (MINOR et al. 1995). In endotoxin-induced portal hypertension, endogenous PAF plays a significant role (KITAGAWA et al. 1992). In patients with chronic cholestasis, PAF acetylhydrolase activity is found to be elevated (MEADE et al. 1991). In cirrhosis, the plasma concentration of PAF also increases (CARAMELO et al. 1987; SUGATANI et al. 1993). As in other localizations (GIRI et al. 1995), PAF antagonists may reduce fibrosis.

## III. Acute Pancreatitis

The possible involvement of PAF in acute pancreatitis is widely accepted. PAF can be produced in acinar cells of pancreatic tissue (LEONHARDT et al. 1995), especially following stimulation with cerulein (SÖLING and FEST 1986). PAF stimulates exocytosis in exocrine secretion (SÖLING et al. 1984). After IV or IP injection of PAF, vascular permeation increases in the pancreas of rats (SIROIS et al. 1988; JANCAR et al. 1988) and rabbits (EMANUELLI et al. 1989). Thus, the involvement of PAF was postulated.

Some studies based on experimental models of acute pancreatitis underline the involvement of PAF. In pancreatitis by cerulein, in which activation of trypsinogen (Steer and Meldolesi 1987), complement activation (Wisner et al. 1988), and release of oxygen radicals (Wisner et al. 1988) are involved, the use of PAF antagonists, e.g., BN 52021, led to amelioration (Fujimura et al. 1992; Alonso et al. 1994; Dabrowski et al. 1995; Hirano et al. 1995; Summers and Albert 1995), but bepafant (WEB 2701) had no effect (Jancar et al. 1995). With the latter antagonist (Zhou et al. 1993), however, effects on extravasation in pancreatitis were found.

In pancreatitis by immune complexes (Jancar et al. 1988) or taurocholate (Ais et al. 1991; Leonhardt et al. 1992), PAF is generated and released (Kald et al. 1993). Bepafant prolonged mean survival time when administered early (Leonhardt et al. 1992). In another type of pancreatitis model by cholcystokinin, involvement of PAF is likely (Alonso et al. 1994), as in lipopolysaccharide-induced pancreatitis (Emanuelli et al. 1994).

Acute pancreatitis by ischemia in rats was ameliorated with the PAF antagonist BB-882 (Formela et al. 1995).

Thus, PAF has an aggravating role in acute pancreatitis (Yotsumoto et al. 1994; Kineworth 1996). Some data also speak in favor of involvement of PAF as a late-phase mediator of chronic pancreatitis in the rat (Zhou et al. 1990).

Reperfusion damage in animals is reduced by the PAF antagonist bepafant (Leonhardt et al. 1995).

## IV. Transplant Rejection

Since PAF regulates immune processes (Braquet and Rola-Pleszczynki 1987) by inhibition of human lymphocyte proliferation when stimulated and by suppression of interleukin-2 production (Dulioust et al. 1990) as well as by induction of suppressor cells – accompanied by an increase in CD 8+ T cells and a minor decrease in CD 4+ T cells (Rola-Pleszczynski et al. 1988b) – a role in transplant rejection seems possible (Foegh et al. 1985, 1987; Braquet et al. 1989). In xenotransplantation, xenograft rejection is mediated by endothelial cell activation in which PAF is involved (Hammer 1995). In renal transplantation, the outcome of function was improved by BN 52021, possibly by reducing thromboxane production (Butterfly et al. 1995). Preservation of transplant reperfusion injury by PAF antagonists (see Sect. C) is further evidenced by Hashikura et al. (1994), Ide et al. (1995), Takada et al. (1995), and Summers and Albert (1995).

## V. Tumor Growth

In hematological malignancies of lymphoid and monocytic nature, a decreased level of PAF in blood has been reported (Denizot et al. 1995). Certain tumoricidal properties of PAF can apparently be induced by induction and modulation of macrophages (HOWARD and ERICKSON 1995) or attributed to certain PAF antagonists (HOULIHAN et al. 1995). Pulmonary metastases of a melanoma were augmented by PAF (IM et al. 1996).

# J. Conclusion

Indeed, PAF is not known to be the sole mediator in any human disease, but it may have a pivotal role in certain diseases. The most significant and clinically relevant roles of PAF are in inflammation, septic (endotoxin-induced) shock, ischemia/reperfusion, and acute pancreatitis. In other pathological conditions, involvement of PAF is of an obviously more limited relevance. In summary: Only in conditions involving immune complexes, endotoxins, or reperfusion is involvement of PAF of major concern (IMAIZUMI et al. 1995).

## References

Abisogun AO, Braquet P, Tsafiri A (1989) The involvement of platelet-activating factor in ovulation. Science 243:381–383

Acker G, Hecquet F, Etienne A, Braquet P, Mencia-Huerta JM (1988) Role of platelet-activating factor (paf) in ovo implantation in the rat: effect of the specific paf-acether antagonist BN 52021. Prostaglandins 35:233–241

Acker G, Braquet P, Mencia-Huerta JM (1989) Role of platelet-activating factor (PAF) in the initiation of the decidual reaction in the rat. J Reprod Fertil 85:623–629

Ackerman Z, Karmeli F, Ligumsky M, Rachmilewitz D (1990) Enhanced gastric and duodenal platelet-activating factor and leukotriene generation in duodenal ulcer patients. Scand J Gastroenterol 25:925–934

Adams JG, Dhar A, Shukla SD, Siver D (1995) Effect of pentoxifylline on tissue injury and platelet-activating factor production during ischemia-reperfusion injury. J Vasc Surg 21:742–749

Adnot S, Lefort J, Lagente V, Braquet P, Vargaftig BB (1986) Interference of BN 52021, a PAF-acether antagonist, with endotoxin-induced hypotension in ther guinea-pig. Pharmacol Res Commun 18[Suppl]:197–200

Argiolas F, Fabi F, del Basso P (1995) Mechanisms of pulmonary vasoconstriction and bronchoconstriction produced by PAF in the guinea-pig: role of platelets and cyclo-oxygenase metabolites. Br J Pharmacol 114:203–209

Ais G, Lopez-Farre A, Gomez-Garre DN, Novo C, Romero JM, Braquet P, Lopez-Novoa M (1991) Role of platelet-activating factor in hemodynamic derangements in an acute rodent pancreatic model. Gastroenterology 102:181–187

Akopov SE, Sercombe R, Seylaz J (1995) Leukocyte-induced endothelial dysfunction in the rabbit basilar artery: modulation by platelet-activating factor. J Lipid Mediat Cell Signal 11:267–280

Albertini M, Clement MG (1994) In pigs, inhaled nitric oxide (NO) counterbalances PAF-induced pulmonary hypertension. Prostaglandins Leukot Essent Fatty Acids 51:357–362

Alecozay AA, Casslen BG, Riehl RM, Leon FD, Harper MJK, Silva M, Nouchi TA, Hanahan DJ (1989) Platelet-activating factor (PAF) in human luteal phase endometrium. Biol Reprod 41:578–586

Alecozay AA, Harper MJK, Schenken RS, Hanahan DJ (1991) Paracrine interactions between platelet-activating factor and prostaglandins in hormonally-treated human luteal phase endometrium in vitro. J Reprod Fertil 91:301–312

Alexander BM, van Kirk EA, Murdoch WJ (1990) Secretion of platelet-activating factor by periovulatory ovine follicles. Life Sci 47:865–868

Alonso R, Montero A, Arévalo M, Garciá LJ, Sanchez-Vicente C, Rodrigez-Nodal F, Lopez-Novoa J, Calvo JC (1994) Platelet-activating factor mediates pancreatic function derangement in caerulein-induced pancreatitis in rats. Clin Sci 87:85–90

Amiel ML, Testart J, Benveniste J (1991) Platelet-activating factor-acether is a component of human follicular fluid. Fertil Steril 56:62–65

Ammit AJ, Wells XE, O'Neill C (1992) Structural heterogeneity of platelet-activating factor produced by murine preimplantation embryos. Hum Reprod 7:865–870

Amorim CZ, Cordeiro RS, Vargaftig BB (1993) Involvement of platelet-activating factor in death following anaphylactic shock in boosted and unboosted mice. Eur J Pharmacol 235:17–32

Anderson BO, Bensard DD, Harken AH (1991) The role of platelet activating factor and its antagonists in shock, sepsis and multiple organ failure. Surg Gynecol Obstet 172:415–424

Anderson GP, Whittle HL, Fennessy MR (1988) Increased airway reponsiveness to histamine induced by platelet activating factor in the guinea pig: possible role of lipoxygenase metabolites. Agents Actions 24:1–7

Ando M, Suginami H, Matsuura S (1990) Pregnancy suppression by a platelet activating factor antagonist, ONO-6240, in mice. Asia Oceania J Obstet Gynaecol 16:169–174

Angle MJ, Jones MA, McManus LM, Pinckard RN, Harper MJK (1988) Platelet-activating factor in the rabbit uterus during early pregnancy. J Reprod Fertil 83:711–722

Angle MJ, Tom RA, McClure RD (1989) The effect of platelet activating factor on sperm-egg interactions. Progr Suppl, Am Fertility Soc, Ann Meeting 1989, p S46

Angle MJ, Tom R, Khoo D, McClure RD (1991) Platelet-activating factor in sperm from fertile and subfertile men. Fertil Steril 56:314–318

Appleyard CB, Hillier K (1992) Catabolism of PAF by human colonic mucosa. Calcium dependence of the catabolizing enzymes. Biochem Pharmacol 43:2503–2509

Appleyard CB, Hillier K (1995) Biosynthesis of platelet-activating factor in normal and inflamed human colon mucosa: evidence for the involvement of the pathway of platelet-activating factor synthesis de novo in inflammatory bowel disease. Clin Sci 88:713–717

Arai A, Lynch G (1992) Antagonists of platelet-activating factor receptor block long-term potentiation in hippocampal slices. Eur J Neurosci 4:411–419

Archer CB, Page CP, Paul W, Morley J, MacDonald DM (1984) Inflammatory characteristics of platelet activating factor (PAF-acether) in human skin. Br J Dermatol 110:45–50

Archer CB, Page CP, Morley J, MacDonald DM (1985) Accumulation of inflammatory cells in response to intracutaneous platelet activating factor (PAF-acether) in man. Br J Dermatol 112:285–291

Archer CB, Cunningham FM, Greaves MW (1988) Actions of platelet-activating factor (PAF) homologues and their combinations on neutrophil chemokinesis and cutaneous inflammatory responses in man. J Invest Dermatol 91:82–85

Arditi M, Manogue KR, Caplan M, Yogev R (1990) Cerbrospinal fluid cachectin, tumor necrosis factor-alpha and platelet-activating factor concentrations and severity of bacterial meningitis in children. J Infect Dis 162:139–147

Arend WP, Dayer JM (1990) Cytokines and cytokine inhibitors or antagonists in rheumatiod arthritis. Arthritis Rheum 33:305–315

Arnoux B, Duval D, Benveniste J (1980) Release of platelet-activating factor (PAF-acether) from alveolar macrophages by the calcium ionophore A 23187 and phagocytosis. Eur J Clin Invest 10:437–441

Aursudkij B, Rogers DF, Evans TW, Alton EWFW, Chung KF, Barnes PJ (1987) Reduced tracheal mucus velocity in guinea-pig in vivo by platelet activating factor. Am Rev Respir Dis 136:A160

Averill FJ, Hubbard WC, Liu MC (1991) Detection of platelet-activating factor and lyso-PAF in bronchoalveolar lavage fluids from allergic subjects following allergen challenge. Am Rev Respir Dis 143:A811

Badr KF, de Boer DK, Takahashi K, Harris RC, Fogo A, Jacobson HR (1989) Glomerular responses to platelet-activating factor in the rat: role of thromboxane A2. Am J Physiol 256:F35–F43

Baldi E, Falsetti C, Krausz C, Cervasi G, Carloni V, Casano R, Forti G (1993) Stimulation of platelet-activating factor synthesis by progesterone and A 23187 in human spermatozoa. Biochem J 292:209–216

Ball HA, Cokk JA, Wise WC, Halushka PV (1986) Role of thromboxane, prostaglandins and leukotrienes in endotoxic and septic shock. Intensive Care Med 12:116–126

Baranes J, Hellegouarch A, Le Hagarat M, Viossat I, August M, Charbrier PE, Clostre F, Braquet P (1986) The effect of PAF-acether on the cardiovascular system and their inhibition by a new highly specific PAF-acether receptor antagonist BN 52021. Pharmacol Res Commum 18:717–737

Barker JNWN, Mitra RS, Griffiths CEM, Dixit VM, Nickoloff BJ (1991) Keratinocytes as initiators of inflammation. Lancet 337:211–214

Bar-Natan MF, Wilson MA, Spain DA, Garrison RN (1995) Platelet-activating factor and sepsis-induced small intestinal microvascular hypoperfusion. J Surg Res 58:38–45

Barnes PJ (1988) Platelet-activating factor and asthma. J Allergy Clin Immunol 81:152–158

Barnes PJ, Liew FY (1995) Nitric oxide and asthmatic inflammation. Immunol Today 16:128–130

Barnes PJ, Chung KF, Page CP (1981) Platelet-activating factor as a mediator of allergic disease. J Allergy 81:919–934

Barnes PJ, Chung KF, Page CP (1988a) Inflammatory mediators and asthma. Pharmacol Rev 40:49–84

Barnes PJ, Chung KF, Page CP (1988b) Platelet-activating factor as a mediator of allergic disease. J Allergy Clin Immunol 81:919–934

Barnes PJ, Page CP, Henson PM (eds) (1989) Platelet-activating factor and human disease. Blackwell, Oxford

Basran GS, Page CP, Paul W, Morley J (1984) Platelet activating factor: a possible mediator of the dual response to allergen. Clin Allergy 14:75–79

Battye KM, Evans G, O'Neill C (1989) A role for platelet activating factor in maintenance of the ovine corpus luteum. Austr Soc Reprod Biol, Proc 21th Ann Confer, abstr 32

Battye KM, Ammit AJ, O'Neill C, Evans G (1991) Production of platelet-activating factor by the pre-implantation sheep embryo. J Reprod Fertil 93:507–514

Bautista AB, Spitzer JJ (1992) Platelet activating factor stimulates and primes liver, Kupffer cells and neutrophils to release superoxide anion. Free Radic Res Commun 17:195–209

Bazan HEP, Reddy STK, Woodland JM, Bazan NG (1987) The accumulation of platelet-activating factor in the injured cornea may be interrelated with the synthesis of lipoxygenase products. Biochem Biophys Res Commun 149:915–920

Bazan NG (1989) Arachidonic acid in the modulation of excitable membrance function and at the onset of brain damage. Ann NY Acad Sci 559:1–16

Bazan NG (1995) A signal terminator. Nature 374:501–502

Bazan NG, Zorumski CF, Clark GD (1993) The activation of phospholipase A2 and release of arachidonic acid and other lipid mediators at the synapse: the role of platelet-activating factor. J Lipid Mediat 6:421–427

Bazan NG, Rodriguez de Turco EB (1995) Platelet-activating factor is a synapse messenger and a modulator of gene expression in the nervous system. Neurochem Int 26:435–441

Beer HJ (1984) Wirkungen des "platelet-activating factors" auf die Thrombozyten des Menschen. Doctoral Thesis, University of Zürich

Beer JH, Wütherich B, von Felten A (1995) Allergen exposure in acute asthma causes the release of platelet-activating factor (PAF) as demonstrated by the desensitization of platelets to PAF. Int Arch Allergy Immunol 106:291–296

Bel EH, Desmet M, Rossing TH, Timmers MC, Dijkman JH, Sterk PJ (1991) The effect of specific oral PAF-antagonist, MK-287, on antigen-induced early and late asthmatic reactions in man. Am Rev Respir Dis 143:A811

Benveniste J, Henson PM, Cochrane CG (1972) Leucocyte dependent histamine release from rabbit platelet: the role of IgE, basophils and platelet activating factor. J Exp Med 136:1356–1377

Benveniste J, Boullet C, Brink C, Labat C (1983) The actions of PAF-acether (platelet-activating factor) on guinea-pig isolated heart preparations. Br J Pharmacol 80:81–83

Benveniste J, Nunez D, Dureiz P, Korth R, Bidault J, Frichart J-C (1988) Preformed PAF-acether and lyso PAF-acether are bound to blood lipoproteins. FEBS Lett 226:371–376

Bern MJ, Sturbaum CW, Karayalcin SS, Berschneider HM, Wachsman JT, Powell DW (1989) Immune system control of rat and rabbit colonic electrolyte transport. Role of prostaglandins and enteric nervous system. J Clin Invest 83:1810–1820

Bertani T, Livion M, Macconi D, Morigi M, Bisogno G, Patrono C, Remuzzi G (1987) Platelet activating factor (PAF) as a mediator of injury in nephrotoxic nephritis. Kidney Int 31:1248–1256

Bessin P, Bonnet J, Apffel D, Soulard C, Desgroux L, Benveniste J (1983) Acute circulatory collapse caused by platelet-activating factor (PAF-acether) in dogs. Eur J Pharmacol 86:403–413

Beutler B, Cerami A (1988) Tumor necrosis factor, cachexia, shock and inflammation: a common mediator. Annu Rev Biochem 57:505–518

Bielenberg GW, Wagener G, Beck T (1992) Infarct reduction by the platelet activating factor antagonist apafant in rats. Stroke 23:98–103

Bienvenue K, Granger DN (1993) Molecular determinants of shear rate-dependent leukocyte adhesion in postcapillary venules. Am J Physiol 264:H1504–H1508

Bienvenue K, Russell J, Granger DN (1993) Platelet-activating factor promotes shear rate-dependent leukocyte adhesion in postcapillary venules. J Lipid Mediat 8:95–103

Biffl WL, Moore EE, Moore FA, Barnett CC, Silliman CC, Peterson VM (1996) Interleukin-6 stimulate neutrophil production of platelet-activating factor. J Leukocyte Biol 59:569–574

Binnaka T, Yamaguchi T, Kubota T (1990) Gastric hemodynamic disturbance induced by hemorrhagic shock in rats. Scand J Gastroenterol 25:555–562

Bito H, Nakamura M, Honda Z, Izumi T, Iwatsubo T, Seyama Y, Ogura A, Kudo Y, Shimizu T (1992) Platelet-activating factor (PAF) receptor in rat brain: PAF mobilizes intracellular Ca2+ in hippocampal neurons. Neuron 9:285–294

Björk J, Smedegard G (1983) Acute microvascular effects of PAF-acether, as studied by intravital microscopy. Eur J Pharmacol 96:87–94

Blank ML, Snyder F, Byers LW, Brooks B, Muirhead EE (1979) Antihypertensive activity of an alkyl ether analog of phosphatidylcholine. Biochem Biophys Res Commun 90:1194–1200

Blank ML, Lee T, Fitzgerald T, Snyder F (1981) A specific acethydrolase for 1-alkyl-2-acetyl-*sn*-glycero-3-phosphocholine (a hypertensive and platelet-activating lipid). J Biol Chem 256:175–178

Blank ML, Hall MN, Cress EA, Snyder F (1983) Inactivation of 1-alkyl-2-acetyl-sn-glycero-3-phosphocholine by a plasma acetylhydrolase: higher activities in hypertensive rats. Biochem Biophys Res Commun 113:666–671

Bloch KJ, Ng BP, Bishara SM, Bloch M (1991) Inhibition of immune complex-induced enteropathy by three different PAF receptor antagonists. Prostaglandins 41:237–249

Bolin RW, Martin TR, Albert RK (1987) Lung endothelial and epithelial permeability after platelet-activating factor. J Appl Physiol 63:1770–1774
Bonavida B, Mencia-Huerta JM (1994) Platelet-activating factor and the cytokine network in inflammatory processes. Clin Rev Allergy 12:381–395
Bonavida B, Mencia-Huerta JM, Braquet P (1989) Effect of platelet activating factor (PAF) on monocyte activation and production of tumor necrosis factor (TNF). Int Arch Allergy Appl Immunol 88:157–160
Bonavida B, Mencia-Huerta JM, Braquet P (1990) Effects of platelet activating factor on peripheral blood monocytes: induction and priming for TNF secretion. J Lipid Mediat 2:S65–S76
Bonnet J, Thibaudeau D, Bessin P (1983) Dependency of the Paf-acether induced bronchospasm on the lipoxygenase pathway in the guinea pig. Prostaglandins 26:457–466
Boschetto P, Mapp CE, Picotti G, Fabbri LM (1989) Neutrophils and asthma. Eur Respir J [Suppl 6] 2:456S–459S
Boughton-Smith NK, Whittle BJR (1988) Formation of the proinflammatory mediator, PAF-acether, in different models of colitis. Gastroenterology 94:A45
Boughton-Smith NK, Hutcheson I, Whittle BJR (1989) Relationship between PAF-acether and thromboxane A2 biosynthesis in endotoxin-induced intestinal damage in the rat. Prostaglandins 38:319–333
Boughton-Smith NK, Deakin AM, Whittle BJR (1992) Actions of nitric oxide on the acute gastrointestinal damage induced by PAF in the rat. Agents Actions (Spec issue) C4–C9
Bourgain RH, Maes L, Braquet P, Andries R, Touqui L, Braquet M (1985) The effect of 1-*O*-alkyl-2-acetyl-*sn*-glycero-3-phosphocholine (PAF-acether) on arterial wall. Prostaglandins 30:185–197
Bourgain RH, Maes L, Andries R, Braquet P (1986) Thrombus induction by endogenic PAF-acether and its inhibition by ginkgo biloba extracts in the guinea pig. Prostaglandins 32:142–144
Brady HR, Lamas S, Papayanni A, Takata S, Matsubara M, Marsden PA (1995) Lipoxygenase product formation and cell adhesion during neutrophil-glomerular endothelial cell interaction. Am J Physiol 268:F1–F12
Braquet P, Rola-Pleszczynski M (1987) Platelet activating factor and cellular immune response. Immunol Today 8:345–352
Braquet P, Etienne A, Touvay C, Bourgain R, Lefort J, Vargaftig BB (1985) Involvement of platelet-activating factor in respiratory anaphylaxis, demonstrated by Paf-acether inhibitor BN 52021. Lancet I:1501
Braquet P, Etienne A, Mencia-Huerta JM, Clostre F (1988) Effects of the specific platelet-activating factor antagonist, BN 52021, and BN 52063, on various experimental gastrointestinal ulcerations. Eur J Pharmacol 150:269–276
Braquet J, Paubert-Braquet M, Bourgain RH, Bussolino F, Hosford D (1989) PAF/cytokine auto-generated feedback networks in microvascular immune injury: consequences in shock, ischemia and graft rejection. J Lipid Res 1:75–112
Brochet B, Orogogozo JM, Dartigues JF, Henry P, Loiseau P (1992) Pilot study of ginkgolide B, a paf-acether specific inhibitor in the treatment of acute outbreaks of multiple sclerosis. Rev Neurol (Paris) 148:299–301
Brodie C (1994) Functional PAF receptor in glia cells: binding parameters and regulation of expression. Int J Dev Neurosci 12:631–640
Brodie C (1995) Platelet activating factor induces nerve growth factor production by rat astrocytes. Neurosci Lett 186:5–8
Brüggmann J (1988) Entwicklung einer validierten Rezeptur- und Defekturvorschrift für eine amästhetisierende Salben- und Gelzubereitung. Pharm Ztg 133:69–72
Bruynzeel PL, Bruynzeel-Koomen CA (1989) Skin eosinophilia in atopic dermatitis. Allerg Immunol 21:224–227
Buckley TL, Hoult JRS (1989) Platelet activating factor is a potent colonic secretagogue with actions independent of specific PAF receptors. Eur J Pharmacol 163:275–283

Burshop KE, Garcia JG, Selig WM, Lo SK, van der Zee H, Kaplan JE, Malik AB (1986) Platelet-activating factor increases lung vascular permeability to protein. J Appl Physiol 61:2210–2217
Burtin G, Noirot C, Scheinmann P, Paupe J, Benveniste J (1990) Decreased skin response to intradermal platelet-activating factor (PAF-acether) in cancer patients. Allergy Clin Immunol 86:418–419
Bussolino F, Camussi G, Aglietta M, Braquet P, Bosia A, Pescamona G, Sanavio F, D'Urso N, Marchisio PC (1987a) Human endothelial cells are target for platelet activating factor I. Platelet-activating factor induces changes in cytoskeleton structures. J Immunol 139:2439–2446
Bussolino F, Porcellini MG, Varese L, Bosia A (1987b) Intravascular release of platelet activating factor in children with sepsis. Thromb Res 48:619–623
Bussolino F, Pescarmona G, Camussi C, Cremo F (1988a) Acetylcholine and dopamine promote the production of platelet activating factor in immature cells of chick embryonic retina. J Neurochem 51:1755–1759
Bussolino F, Camussi G, Baglioni C (1988b) Synthesis and release of platelet-activating factor by human vascular endothelial cells treated with tumor necrosis factor or interleukin 1 $\alpha$. J Biol Chem 263:11856–11861
Butterfly DW, Spurney RF, Ruiz P, Pirotzky E, Braquet P, Coffman TM (1995) Role of platelet activating factor in kidney transplant rejection in the rat. Kidney Int 48:337–343
Cabello C, McIntyre DE, Forrest M, Burroughs M, Prasad S, Tuomanen E (1992) Differing roles for platelet activating factor during inflammation of the lung and subarachnoid space. J Clin Invest 90:612–618
Cakici I, Mataraci N, Ersoy S, Tunctan B, Abacioglu N, Kanzik I (1995) Effects of platelet-activating factor antagonists WEB 2086 and BN 50730 on digoxin-induced arrhythmias. Pharmacol Toxicol 76:343–347
Calignano A, Cirino G, Meli R, Persico P (1988) Isolation and identification of platelet-activating factor in UV-irradiated guinea pig skin. J Pharmacol Methods 19:89–91
Camussi G (1986) Potential role of platelet-activating factor in renal pathopsysiology. Kidney Int 29:469–477
Camussi G, Pawlowski L, Tetta C, Raffinello C, Alberton M, Brentjens J, Andres G (1983a) Acute lung inflammation induced in the rabbit by local instillation of 1-*O*-octadecyl-2-acetyl-*sn*-glyceryl-3-phosphorylcholine or of native platelet-activating factor. Am J Pathol 112:78–88
Camussi G, Aglietta M, Malavesi F, Tetta C, Piacibello W, Sanovio F, Bussolino F (1983b) The release of platelet-activating factor from human endothelial cells in culture. J Immunol 131:2397–2403
Camussi G, Bussolino F, Salvidio G, Baglioni C (1987) Tumor necrosis factor/cachectin stimulates peritoneal macrophages, polymorphonuclear neutrophils, and vascular endothelial cells to synthetisize and release platelet-activating factor. J Exp Med 166:1390–1404
Camussi G, Tetta C, Bussolino F, Baglioni C (1989) Tumor necrosis factor stimulates human neutrophils to release leukotriene B4 and platelet-activating factor. Induction of phospholipase A2 and acetyl-CoA: l-alkyl-sn-glycero-3-phosphocholine O-acetyltransferase activity and inhibition by antiproteinase. Eur J Biochem 182:661–666
Camussi G, Montrucchio G, Lupia E, de Martino A, Perona L, Arese M, Vercellone A, Toniolo A, Bussolino F (1995) Platelet-activating factor directly stimulates in vivo migration of endothelial cells and promotes in vivo angiogenesis by a heparin-dependent mechanism. J Immunol 154:6492–6501
Caplan MS, Hsueh W (1990) Necrotizing enterocolitis: role of platelet activating factor, endotoxin, and tumor necrosis factor. J Pediatr 117:S47–S51
Caplan MS, Sun X-M, Hsueh W (1990a) Hypoxia causes ischemic bowel necrosis in rats: the role of platelet-activating factor (PAF-acether). Gastroenterology 99:979–986

Caplan MS, Sun XM, Hsueh W, Hagmann R (1990b) Role of platelet activating factor and tumor necrosis factor-alpha in neonatal necrotizing enterocolitis. J Pediatr 116:960–964
Caplan MS, Kelly A, Hsueh W (1992a) Endotoxin and hypoxia-induced intestinal necrosis in rats: the role of platelet activating factor. Pediatr Res 31:428–434
Caplan MS, Adler L, Kelly A, Hsueh W (1992b) Hypoxia increases stimulus-induced PAF production and release from human umbilical vein endothelial cells. Biochim Biophys Acta 1128:205–210
Caramelo C, Fernandez-Gallardo S, Marin-Cao D, Inarrea P, Santos JC, Lopez-Novoa J, Sanchez-Crespo M (1984) Presence of platelet-activating factor in blood from humans and experimental animals. Its absence in anephric individuals. Biochem Biophys Res Commun 120:789–796
Caramelo C, Fernandez-Gallardo S, Santos JC, Inarrea P, Sanchez-Crespo M, Lopez-Novoa JM, Hernando L (1987) Increased levels of platelet-activating factor in blood from patients with cirrhosis of the liver. Eur J Clin Invest 17:7–11
Carveth HJ, Shaddy RE, Whatley RE, McIntyre TM, Prescott SM, Zimmerman GA (1992) Regulation of platelet-activating factor (PAF) synthesis and PAF-mediated neutrophil adhesion to endothelial cells activated by thrombin. Semin Thromb Hemost 18:126–134
Casals-Stenzel J (1987a) Bronchial and vascular effects of PAF in the rat isolated lung are completely blocked by WEB 2086, a novel specific PAF antagonist. Br J Pharmacol 91:799–802
Casals-Stenzel J (1987b) Effects of WEB 2086, a novel antagonist of platelet-activating factor in active and passive anaphylaxis. Immunopharmacology 13:117–124
Casals-Stenzel J (1987c) Protective effect of WEB 2086, a novel antagonist of platelet activating factor in endotoxin shock. Eur J Pharmacol 135:117–122
Cerinski-Hennig D, von Bruchhausen F (1991) Inhibition of acetylating formation of PAF (platelet-activating factor) by some polyethylene glycols and bituminosulfonates. Arch Pharmacol 344:R100
Chanarin N, Johnston SL (1994) Leukotrienes as a target in asthma therapy. Drugs 47:12–24
Chang SW (1994a) Endotoxin-induced pulmonary leukostasis in the rat: role of platelet-activating factor and tumor necrosis factor. J Lab Clin Med 123:65–72
Chang SW (1994b) TNF potentiates PAF-induced pulmonary vasoconstriction in the rat: role of neutrophils and thromboxane A2. J Appl Physiol 77:2817–2826
Chang SW, Fedderson CO, Henson PM, Voelkel NF (1987) Platelet-activating factor mediates hemodynamic changes and lung injury in endotoxin-treated rats. J Clin Invest 79:1498–1509
Chang SW, Fernyak S, Voelkel NF (1990) Beneficial effect of platelet-activating factor antagonist, WEB 2086, on endotoxin-induced lung injury. Am J Physiol 258:H153–H158
Chao W, Siafaka-Kapadai A, Olson MS, Hanahan DJ (1989a) Biosynthesis of platelet-activating factor by cultured rat Kupffer cells stimulated with calcium ionophore A 23187. Biochem J 257:823–829
Chao W, Liu H, DeBuysere M, Hanahan DJ, Olson MS (1989b) Identification of receptors for platelet-activating factor in rat Kupffer cells. J Biol Chem 264:13591–13598
Chaussade S, Denizot Y, Colombel JF, Cherouki N, Guerre J, Couturier D, Benveniste J (1991) Presence of PAF-acether (PAF) in stool of patients with bacterial diarrhea (BD), ulcerative colitis (UC) and Crohn disease (CD). Gastroenterology 100:A202
Chaussade S, Denizot Y, Colombel JF, Benveniste J, Couturier D (1992) PAF-acether in stool as marker of intestinal inflammation. Lancet 339:739
Chen CR, Voelkel NF, Chang S-W (1989) PAF potentiates protamine-induced lung edema: role of pulmonary venoconstriction. J Appl Physiol 66:1059–1068

Chesney CM, Pifer DD, Huch KM (1985) Desensitization of human platelets by platelet-activating factor (PAF). Biochem Biophys Res Commun 127:24–30

Chiba Y, Mikoda N, Kawasaki H, Ito K (1990) Endothelium-dependent relaxant action of platelet activating factor in the rat mesenteric artery. Arch Pharmacol 341:68–73

Chignard M, Le-Couedic JP, Tence M, Vargaftig BB, Benveniste J (1979) The role of platelet-activating factor in platelet aggregation. Nature 279:799–800

Chignard M, Le Couedic JP, Vargaftig BB (1980) Platelet-activating factor (PAF-acether) secretion from platelets: effect of aggregating agents. Br J Haematol 46:455–464

Choi IH, Ha TY, Lee DG, Park JS, Lee JH, Park YM, Lee HK (1995) Occurence of disseminated intravascular coagulation (DIC) in active systemic anaphylaxis: role of platelet-activating factor. Clin Exp Immunol 100:390–394

Christman BW, Lefferts PL, King GA, Snapper JR (1988) Role of circulating platelets and granulocytes in PAF-induced pulmonary dysfunction in awake sheep. J Appl Physiol 64:2033–2041

Christman BW, Lefferts PL, Snapper JR (1990) Effect of platelet-activating factor on aerosol histamine response in awake sheep. Am Rev Respir Dis 135:1267–1270

Chu K-M, Gerber JG, Nies AS (1988) Local vasodilator effect of platelet activating factor in the gastric, mesenteric and femoral arteries of the dog. J Pharmacol Exp Ther 246:996–1000

Chung KF (1986) Role of inflammation in the hyperactivity of the airways in asthma. Thorax 41:657–662

Chung KF (1992) Platelet-activating factor in inflammation and pulmonary disorders. Clin Sci 83:127–138

Chung KF, Barnes PJ (1989) Effect of inhaled platelet-activating factor on airway calibre, bronchial responsiveness and circulating cells in asthmatic subjects. Thorax 44:108–115

Chung KF, Aizawa H, Leikauf GD, Ueli IF, Evans TW, Nadel JA (1986) Airway hyperresponsiveness induced by platelet-activating factor: role of thromboxane generation. J Pharmacol Exp Ther 236:580–584

Chung KF, Dent C, McCusker M, Guinot P, Page CP, Barnes PJ (1987) Effect of a ginkgolide mixture (BN 52063) in antagonising skin and platelet responses to platelet activating factor in man. Lancet I:248–251

Clark GD, Happel LT, Zorumski CF, Bazan NG (1992) Enhancement of hippocampal excitatory synaptic transmission by platelet-activating factor. Neuron 9:1211–1216

Clark PO, Hanahan DJ, Pinckard RN (1980) Physical and chemical properties of platelet-activating factor obtained from human neutrophils and monocytes and rabbit neutrophils and basophils. Biochim Biophys Acta 628:69–75

Clark WF, Naylor CD (1989) The role of platelets in progressive glomerulosclerosis: mechanism for intraglomerular platelet activation and pathogenetic consequences. Med Hypotheses 28:51–56

Collins CE, Rampton DS (1995) Platelet dysfunction: a new dimension in inflammatory bowel disease. Gut 36:5–8

Corbett G, Perry DJ (1983) Significance of thrombocytosis. Lancet I:77

Corrigan CJ, Hartnell A, Kay AB (1988) T lymphocyte activation in acute severe asthma. Lancet I:1129–1132

Coyle A, Sjoerdsma K, Page CP, Brown L, Metzger WJ (1987) Modification of the late asthmatic response and bronchial hyperreactivity by BN 52021, a platelet-activating factor antagonist. Clin Res 35:A254

Coyle AG, Urwin SC, Page CP (1988) The effect of the selective PAF antagonist BN 52021 on PAF-and antigen-induced tracheal hyperreactivity and eosinophil accumulation. Eur J Pharmacol 148:51–58

Crea AEG, Nakhosteen JA, Lee TH (1992) Mediator concentrations in bronchoalveolar lavage fluid of patients with mild asymptomatic bronchial asthma. Eur Respir J 5:190–195
Csato M, Czarnetzki BM (1988) Effect of BN 52021, a platelet activating factor antagonist, on experimental murine contact dermatitis. Br J Dermatol 111:475–480
Csato M, Rosenback T, Moler A, Czarnetzki BM (1987) Generation of platelet aggregation activity from murine epidermal cells. J Invest Dermatol 89:440P
Csato M, Kenderessy ASZ, Judak R, Dobozy A (1990) Inflammatory mediators are involved in the Candida albicans killing activity of human epidermal cells. Arch Dermatol Res 282:348–350
Cucula M, Wallace JL, Salas A, Guarner F, Rodriguez R, Malagelada JR (1989) Central regulation of gastric secretion by platelet-activating factor in anesthetized rats. Prostaglandins 37:275–285
Cueva JP, Hsueh W (1988) Role of oxygen derived-free radicals in platelet activating factor induced bowel necrosis. Gut 29:1207–1212
Cunningham FM, Leigh I, Mallet AI (1987) The production of platelet activating factor (PAF) by human epidermal cells. Br J Pharmacol 90:117P
Cush JJ, Lipsky PE (1991) Cellular basis for rheumatoid inflammation. Clin Orthop 265:9–22
Cuss FM, Dixon CMS, Barnes PJ (1986) Effects of inhaled platelet activating factor on pulmonary function and bronchial responsiveness in man. Lancet II:189–192
Czarnetzki B (1983) Increased monozyte chemotaxis towards leukotriene B4 and platelet-activating factor in patients with inflammatory dermatoses. Clin Exp Immunol 54:486–492
Czarnetzki BM, Benveniste J (1981) Effect of synthetic PAF-acether on human neutrophil function. Agents Actions 11:549–550
Dabrowski A, Gabryelewicz A, Chyczewski L (1995) The effect of platelet activating factor antagonist (BN 52021) on acute experimental pancreatitis with reference to multiorgan oxidative stress. Int J Pancreatol 17:173–180
Damas J, Prunescu P (1993) Presence of immunoreactive platelet-activating factor in peritoneal exudate induced by zymosan. Arch Int Pharmacodyn 322:115–123
Damas J, Remade-Volon G, Nguyen TP (1990) Inhibition by WEB 2086, a PAF-acether antagonist of oedema and peritonitis by zymosan. Arch Int Pharmacodyn 306:161–169
Darius H, Lefer DJ, Smith B, Lefer AM (1986) Role of platelet-activating factor-acether in mediating guinea pig anaphylaxis. Science 232:58–60
Deacon RW, Melden MK, Saunders RN, Handley DA (1986) PAF involvement in dermal extravasation in the reverse passive Arthus reaction. Fed Proc 45:995
Dechanet J, Rissoan MC, Bauchereaut J, Miossec P (1995) Interleukin 4, but not interleukin 10, regulates the production of inflammation mediators by rheumatoid synoviocytes. Cytokine 7:176–183
Decker K (1990) Biologically active products of stimulated liver macrophages (Kupffer cells). Eur J Biochem 192:245–261
D'Cruz OJ, Haas GG (1989) Platelet-activating factor induces acrosome reaction and fertilization capacity of guinea pig spermatozoa with special reference to the possible involvement of lysophospholipids in the acrosome-reaction. Proc Serono Symp on Fertilization in Mammals, Boston, p 53 (abstr II 6)
Deeming K, Mazzoni L, Page CP, Sanjar S, Smith DL (1985) PAF antagonism in asthma. Prostoglandins 30:715
de Fily DV, Kuo L, Chilian WM (1996) PAF attenuates endothelium-dependent coronary arteriolar vasodilation. Am J Physiol 270:H2094–H2099
Del Cerro S, Arai A, Lynch G (1990) Inhibition of long-term potentiation by an antagonist of platelet-activating factor receptors. Neural Biol 54:213–217
De Lima WT, Kwasniewski FH, Sirois P, Jancar S (1995) Studies on the mechanism of PAF-induced vasopermeability in rat lungs. Prostaglandins Leukot Essent Fatty Acids 52:245–250

de Longo WE, Standeren T, Chandel B, Deshpande Y, Vernava AM, Kaminski DL (1995) Platelet-activating factor mediates trinitrobenzene induced dysmotility in the left colon. Mediat Inflamm 4:17–18
de Longo WE, Polities G, Vernava AM, Deshpande Y, Niejhoff M, Chandel B, Kulkarni A, Kaminski DL (1994) Platelet-activating factor mediates trinitrobenzene induced colitis. Prostaglandins Leukot Essent Fatty Acids 51:419–424
Demarkarian RM, Israel E, Rosenberg MA (1991) The effect of Sch-37370, a dual platelet activating factor and histamine antagonist, on the bronchoconstriction induced in asthmatics by cold, dry air isocapnic hyperventilation (ISH). Am Rev Respir Dis 143:A812
Dembinska-Kiec A, Peskar BA, Müller MK, Peskar BM (1989) The effect of platelet-activating factor on flow rate and eicosanoid release in the isolated perfused rat gastric vascular bed. Prostaglandins 37:69–91
Demling RH (1990) Current concept on the adult respiratory distress syndrome. Circ Shock 30:297–309
de Monchy JGR, Kauffman HF, Venge P, Jansen GH, Sluiter HJ, Vires K (1985) Bronchoalveolar eosinophils during allergen-induced late asthmatic reactions. Am Rev Respir Dis 131:373–376
Denizot Y, Chaussade S, Colombel JF, Benveniste J, Couturier D (1991a) Presence of PAF-acether in stools of patients suffering from inflammatory bowel disease. C R Acad Sci III 312:329–333
Denizot Y, Chaussade S, Benveniste J, Couturier D (1991b) Presence of PAF-acether in stool of patients with infectious diarrhea. J Infect Dis 163:1168
Denizot Y, Chaussade S, Nathan N, Colombel JF, Bossant MJ, Cherouki N, Benveniste J, Couturier D (1992a) PAF-acether and acetylhydrolase in stool of patients with Crohn's disease. Dig Dis Sci 37:432–437
Denizot Y, Chaussade S, de Boissieu D, Dupont C, Nathan N, Benveniste J, Coutourier D (1992b) Presence of paf-acether in stool of patients with bacterial but not with viral or parasitic diarrhea. Immunol Inf Dis 2:269–273
Denizot Y, Chaussade S, Couturier D (1993) Fecal platelet-activating factor (PAF). A marker of digestive inflammatory disorders. Presse Med 22:1704–1706
Denizot Y, Dupuis F, Trimoreau F, Praloran V, Liozon E (1995) Decreased levels of platelet-activating factor in blood of patients with lymphoid and non-lymphoid hematologic malagnancies. Blood 85:2992–2993
De Vos C, Joseph M, Leprevost C, Vorng H, Tomassini M, Carvon M, Capron A (1989) Inhibition of human eosinophil chemotaxis and of the IgE-dependent stimulation of human blood platelets by cetirizine. Int Arch Allergy Appl Immunol 88:212–215
Dewar A, Archer CB, Page CP, Paul W, Morley J, McDonald DM (1984) Cutaneous and pulmonary histopathological responses to platelet activating factor (PAF-acether) in the guinea-pig. J Pathol 144:25–34
Dieter P, Schulze-Specking A, Decker K (1986) Differential inhibition of prostaglandin and superoxide production by dexamethasone in primary cultures of rat Kupffer cells. Eur J Biochem 159:451–457
Dillon PK, Fitzpatrick MF, Ritter AB, Duran WN (1988a) The effect of platelet activating factor on leukocyte adhesion to microvascular endothelium: time course and dose-response relationships. Inflammation 12:563–573
Dillon PK, Ritter AB, Duran WN (1988b) Vasoconstrictor effects of platelet-activating factor in the hamster cheek pouch microcirculation: dose related relations and pathways of reaction. Circul Res 62:722–731
Diserbo M, Cand F, Ziade M, Verdetti J (1995) Stimulation of platelet-activating factor (PAF) receptor increases inositol phosphate production and cytosolic free Ca2+ concentrations in N1E-115 neuroblastoma cells. Cell Calcium 17:442–452
Diserbo M, Fatome H, Verdetti J (1996) Activation of large conductance Ca2+-activated K+ channels in N1E-115 neuroblastoma cells by platelet-activating factor. Biochem Biophys Res Comm 218:745–748

Dobbins DE, Buehn MJ, Dabney JM (1990) The inflammatory actions of platelet activating factor are blocked by levorotatory terbutaline. Microcirc Endothelium Lymphatics 6:437–455
Dobrowsky RT, Voyksner RD, Olson NC (1991) Effect of SRI 63–675 on hemodynamics and blood PAF levels during porcine endotoxemia. Am J Physiol 260:H1455–H1465
Doebber TW, Wu MS, Robbins JC, Choy BM, Chang MN, Shen TY (1985) Platelet-activating factor (PAF) involvement in endotoxin-induced hypotension in rats: studies with PAF-receptor antagonist Kadsurenone. Biochem Biophys Res Commun 127:799–808
Doebber TW, Wu M, Biftu T (1986) Platelet-activating factor (PAF) mediation of rat anaphylactic responses to soluble immune complexes. J Immunol 136:4659–4668
Doherty GM, Lange JR, Langstein HN, Alexander HR, Buresh CM, Norton JA (1992) Evidence for IFN-gamma as a mediator for the lethality of endotoxin and tumor necrosis factor-alpha. J Immunol 149:1666–1670
Domanska-Janik K, Zablocka B (1995) Modulation of signal transduction in rat synaptoneurosomes by platelet activating factor. Mol Chem Neuropathol 25:51–67
Domingo MT, Spinnewyn B, Chabrier PE, Braquet P (1988) Presence of specific binding sites for platelet-activating factor (PAF) in brain. Biochim Biophys Res Commun 151:730–736
Dray F, Rougeot C, Tiberghien C, Junier M-P (1993) PAF and hypothalamic secretions in the rat. NATO ASI series vol H 70. In: Massarelli R, Horrocks LA, Kanfer JN, Löffelholz K (eds) Phospholipid and signal transmission. Springer-Verlag, Berlin Heidelberg New York, pp 387–402
Droy-Lefaix MT, Drouet Y, Gerard G (1988) Involvement of platelet-activating factor in rat ischemia reperfusion gastric damage. In: Braquet P (ed) Ginkgolides-chemistry, biology, pharmacology and clinical perspectives. Prous, Barcelona, pp 563–574
Dubois C, Bissonnette E, Rola-Pleszczynski M (1989) Platelet-activating factor (PAF) enhances tumor necrosis factor production by alveolar macrophages. J Immunol 143:964–970
Dulioust A, Vivier E, Salem P, Benveniste J, Thomas Y (1988) Immunoregulatory functions of paf-acether I. Effect of paf-acether on CD4+ cell proliferation. J Immunol 140:240–245
Dulioust A, Duprez V, Pilton C (1990) Immunoregulatory function of PAF acether III. Down regulation of CD4+ T cells high affinity IL 2 receptor expression. J Immunol 144:3123–3129
Dunoyer-Geindre S, Ludi F, Perez T, Terki N, Grunowski PI, Girard JP (1992) PAF acether release on antigenic challenge. A method for the investigation of drug allergic reactions. Allergy 47:50–54
Duran WN, Milazzo VJ, Sabido F, Hobson RW (1996) Platelet-activating factor modulates leukocyte adhesion to endothelium in ischemia-reperfusion. Microvasc Res 51:108–115
Eliakim R, Karmeli F, Razin E, Rachmilewitz D (1988) Role of platelet-activating factor in ulcerative colitis. Enhanced production during active disease and inhibition by sulfasalazine and prednisolone. Gastroenterology 95:1167–1172
Emanuelli G, Montrucchio G, Gaia E, Dughera L, Corvetti G, Gubetta L (1989) Experimental acute pancreatitis induced by platelet activating factor in rabbits. Am J Pathol 134:315–326
Emanuelli G, Montrucchio G, Dughera L, Gaia E, Lupia E, Battaglia E, de Martino A, de Giuli P, Gubetta L, Camussi G (1994) Role of platelet activating factor in acute pancreatitis induced by lipopolysaccharides in rabbits. Eur J Pharmacol 261:265–272
Erger RA, Casale TB (1995) Interleukin-8 is a potent mediator of eosinophil chemotaxis through endothelium and epithelium. Am J Physiol 268:L117–L122
Espey LL (1980) Ovulation as an inflammatory reaction: a hypothesis. Biol Reprod 22:73

Esplugues JV, Whittle BJR (1988) Gastric mucosal damage induced by local intraarterial administration of PAF in the rat. Br J Pharmacol 93:222–228
Esplugues JV, Whittle BJR (1989) Mechanisms contributing to gastric motility changes induced by PAF-acether and endotoxin in rats. Am J Physiol 256:G275–G282
Etienne A, Hecquet F, Soulard C, Spinnewyn B, Clostre F, Braquet P (1986) In vivo inhibition of plasma protein leakage and Salmonella enteritidis-induced mortality in the rat by a specific paf-acether antagonist: BN 52021. Agents Actions 17:368–370
Etienne AF, Thonier F, Hecquet F, Braquet P (1988) Role of neutrophils in gastric damage induced by platelet activating factor. Arch Pharmacol 338:422–425
Evans TW, Chung KF, Rogers DF, Barnes PJ (1987) Effect of platelet-activating factor on airway vascular permeability: possible mechanisms. J Appl Physiol 63:479–484
Everaerdt B, Brouchart P, Shaw A, Fiers W (1989) Four different interleukin-1 species sensitize to the lethal action of tumor necrosis factor. Biochem Biophys Res Comm 163:378–385
Fadel R, Herpin-Richard N, Rihoux JP, Henocq E (1987) Inhibitory effect of cetirizine 2HCl on eosinophil migration. Clin Allergy 17:373–379
Fadel R, David B, Herpin-Richard N, Borgnon A, Rassemont R, Rihoux JP (1990) In vivo effects of cetirizine on cutaneous reactivity and eosinophil migration induced by platelet-activating factor (PAF-acether) in man. J Allergy Clin Immunol 86: 314–320
Faden AI, Salzman SK (1992) Pharmacological strategies in CNS trauma. Trends Pharmacol Sci 13:29–35
Faden AI, Tzendzalian PA (1992) Platelet-activating factor antagonists limit glycine changes and behavioral deficits after brain trauma. Am J Physiol 263:R909–R914
Faden AI, Tzendzallian P, Lemke M, Valone F (1989) Role of platelet-activating factor in pathophysiology of traumatic brain injury. Soc Neurosci Abstr 15:1112
Fauler JG, Sielhorst G, Frölich JC (1989) Platelet-activating factor induces the production of leukotrienes by human monocytes. Biochim Biophys Acta 1013:80–85
Felix SB, Baumann G, Niemczyk M, Ahmad Z, Hashemi T, Berdel WE (1991) Effects of platelet-activating factor on beta- and H2-receptor-mediated increase of myocardial contractile force in isolated perfused guinea pig hearts. Res Exp Med (Berl) 191:1–9
Felsen D, Loo M, Marion D, Vaughan E (1990) Involvement of platelet-activating factor and thromboxane A2 in the renal response to unilateral ureteral obstruction. J Urol 144:141–145
Ferguson-Chanowitz KM, Katocs AS, Pickett WC, Kaplan JB, Sass PM, Oronsky KL, Kerwar SS (1990) Platelet activating factor or a platelet activating factor antagonist decreases tumor necrosis factor-alpha in the plasma of mice treated with endotoxin. J Infect Dis 162:1081–1086
Fernandez-Gallardo S, Gijon MA, Garcia MC, Cano E, Sanchez-Crespo M (1988) Biosynthesis of platelet-activating factor in glandular gastric mucosa. Evidence for the involvement of the "de novo" pathway and modulation by fatty acids. Biochem J 254:707–714
Ferraris L, Karmeli F, Eliakim R, Klein J, Fiocchi C, Rachmilewitz D (1993) Intestinal epithelial cells contribute to the enhanced generation of platelet activating factor in ulcerative colitis. Gut 34:665–668
Ferreira MG, Braquet P, Fonteles MC (1991) Effects of PAF antagonists on renal vascular escape and tachyphylaxis in perfused rabbit kidney. Lipids 26:1329–1332
Feuerstein G, Boyd LM, Ezra D, Goldstein RE (1984) Effect of platelet-activating factor on coronary circulation of the domestic pig. Am J Physiol 246:H466–H471
Feuerstein G, Yue TL, Lysko PG (1990) Platelet-activating factor: a putative mediator in central nervous system injury? Stroke 21[Suppl III]:90–94
Filep J, Herman F, Braquet J, Mozes T (1989) Increased levels of platelet-activating factor in blood following intestinal ischemia in the dog. Biochem Biophys Res Comm 158:353–359

Filep JG, Foldes-Filep E (1993) Modulation by nitric oxide of platelet-activating factor-induced albumin extravasation in the conscious rat. Br J Pharmacol 110:1347–1352
Filep J, Braquet P, Mozes T (1991a) Significance of platelet-activating factor in mesenteric ischemia-reperfusion. Lipids 26:1336–1339
Filep J, Herman F, Braquet P (1991b) Platelet actvating factor may mediate dexamethasone-induced gastric damage in the rat. Lipids 26:1356–1358
Filep JG, Braquet P, Mozes T (1991c) Interactions between platelet-activating factor and prostanoids during mesenteric ischemia-reperfusion-induced shock in the anesthetized dog. Circ Shock 35:1–8
Findlay SR, Lichtenstein LM, Hanahan DJ, Pinckard RN (1981) The contraction of guinea pig ileal smooth muscle by acetyl glyceryl ether phosphorylcholine. Am J Physiol 241:C130–C133
Fisher RA, Sharma RV, Bhalla RC (1989) Platelet-activating factor increases inositol phosphate production and cytosolic free $Ca^{2+}$ concentrations in cultured rat Kupffer cells. FEBS Lett 251:22–26
Fitzgerald MF, Lees IW, Parente L, Payne AN (1987) Exposure to PAF-acether aerosol induces airway hyperresponsiveness to 5-HT in guinea pigs. Br J Pharmacol 90:112P
Fjellner B, Hägermark O (1985) Experimental pruritus evoked by platelet activating factor (PAF-acether) in human skin. Acta Derm Venereol 65:409–412
Floch A, Bousseau A, Hetier E, Floch F, Bost PE, Cavero I (1989) RP 55778, a PAF receptor antagonist, prevents and reverses LPS-induced hemoconcentration and TNF $\alpha$ release. J Lipid Mediat 1:349–360
Foegh M, Khirabadi BS, Braquet P, Ramwell PW (1985) Platelet-activating factor antagonist BN 52021 prolongs experimental cardiac allocraft survival. Prostaglardins 30:718 (Abstract)
Foegh ML, Hartmann DP, Rowles JR, Khirabadi BS, Alijani MR, Helfrich GB, Ramwell PW (1987) Leukotrienes, thromboxane, and platelet activating factor in organ transplantation. Adv Prostaglandin Thromboxane Leukot Res 6:140–146
Formela LJ, Wood LM, Whittaker M, Kingsnorth AN (1994) Amelioration of experimental acute pancreatitis with a potent platelet-activating factor antagonist. Br J Surg 81:1783–1785
Foster AP, Lees P, Cunningham FM (1995) Platelet activating factor mimics antigen-induced cutaneous inflammatory responses in sweet itch horses. Vet Immunol Immunopathol 44:115–128
Fouda SI, Molski EP, Ashour MS-E, Sha'afi RI (1995) Effect of lipopolysaccharide on mitogen-activated protein kinases and cytosolic phospholipase A2. Biochem J 308:815–822
Francescangeli E, Goracci P (1989) The de novo biosynthesis of platelet-activating factor in rat brain. Biochem Biophys Res Commun 161:107–112
Freitag A, Watson RM, Matsos G, Eastwood C, O'Byrne PM (1993) Effect of a PAF antagonist WEB-2086, on allergen induced asthmatic response. Thorax 48:594–598
Frerichs KU, Feuerstein GZ (1990) Platelet-activating factor – key mediator in neural injury? Cerebrovasc Brain Metab Rev 2:148–160
Frerichs KU, Lindsberg PJ, Hallenbeck JM, Feuerrtein GZ (1990) Platelet-activating factor and progressive brain damage following focal brain injury. J Neurosurg 73:223–233
Frigas E, Gleich GJ (1986) The eosinophil and the pathology of asthma. J Allergy Clin Immunol 77:527–537
Fujii E, Irie K, Uchida Y, Ohba K, Muraki T (1995) Role of eicosanoids but not nitric oxide in the platelet-activating factor-induced increase in vascular permeability in mouse skin. Eur J Pharmacol 273:267–272
Fujimura K, Sugatani J, Miwa M, Mizuno T, Sameshima Y, Saito K (1989) Serum platelet-activating factor acetylhydrolase activity in rats with gastric ulcers induced by water-immersion stress. Scand J Gastroenterol Suppl 162:59–62

Fujimura K, Kubota Y, Ogura M, Yamaguchi T, Binnaka T, Tani K, Kitagawa S, Mizuno T, Inoue K (1992) Role of platelet activating factor in caerulein-induced acute pancreatitis in rats: protective effects of a PAF-antagonist. J Gastroenterol Hepatol 7:199–202

Fukuda A, Roudebush WE, Thatcher SS (1994) Platelet activating factor enhances the acrosome reaction, fertilization in vitro by subzonal sperm injection and resulting embryonic development in the rabbit. Hum Reprod 9:94–99

Furukawa M, Lee EL, Johnston JM (1993) Platelet-activating factor-induced ischemic bowel necrosis: the effect of platelet-activating factor acetylthydrolase. Pediatr Res 34:237–241

Furukawa M, Muguruma K, Frenkel RA, Johnston JM (1995) Metabolic fate of platelet-activating factor in the rat enterocyte: the role of specific lysophospholipase D. Arch Biochem Biophys 319:274–280

Gao Y, Zhou H, Raj JU (1995) PAF induces relaxation of pulmonary arteries but contraction of pulmonary veins in the ferret. Am J Physiol 269:H704–H709

Garcia JGN, Azghani A, Callahan KS, Johnson AR (1988) Effect of platelet activating factor on leukocyte-endothelial cell interactions. Thromb Res 51:83–96

Gardiner KR, Halliday MI, Barclay GR, Milne L, Brown D, Stephens S, Maxwell RJ, Rowlands BJ (1995) Significance of systemic endotoxaemia in inflammatory bowel disease. Gut 36:897–901

Gardner CR, Laskin JD, Laskin DL (1995) Distinct biochemical responses of hepatic macrophages and endothelial cells to platelet-activating factor during endotoxemia. J Leukoc Biol 57:269–274

Gateau O, Arnoux B, Deriz H, Viars P, Benveniste J (1984) Acute effects of intratracheal administration of PAF-acether (platelet-activating factor) in humans. Am Rev Respir Dis 129:A3

Gay JC (1993) Mechanism and regulation of neutrophil priming by platelet-activating factor. J Cell Physiol 156:189–197

Gay JC, Beckman JK, Zaboy KA, Lukens JN (1986) Modulation of neutrophil oxidative responses to soluble stimuli by platelet-activating factor. Blood 67:931–936

Gebhardt BM, Braquet P, Bazan H, Bazan N (1988) Modulation of in vitro immune reactions by platelet activating factor and a platelet activating factor antagonist. Immunopharmacology 15:11–19

Gelhard HA, Nottet HS, Swindells S, Jett M, Dzenko KA, Genis P, White R, Wang L, Choi YB, Zhang D (1994) Platelet-activating factor: a candidate human immunodeficiency virus type 1-induced neurotoxin. J Virol 68:4628–4635

Geng JG, Bevilacqua MP, Moore KL, McIntyre TM, Prescott SM, Kim JM, Zimmerman GA, McEver RP (1990) Rapid neutrophil adhesion to activated endothelium mediated by GMP-140. Nature 343:757–760

Gandhi CR, Olson NC (1991) PAF effects on transmembrane signalling pathways in rat Kupffer cells. Lipids 26:1038–1043

Giembycz MA, Kroegel C, Barnes PJ (1990) Platelet activating factor stimulates cyclooxygenase activity in guinea pig neutrophils: concerted biosynthesis of thromboxane A2 and E series prostaglandins. J Immunol 144:3489–3497

Gilboe DD, Kinter D, Fitzpatrick JH, Emoto SE, Braquet P, Bazan NG (1991) Recovery of postischemic brain metabolism and function following treatment with a free radical scavenger and platelet-activating factor antagonists. J Neurochem 56:311–319

Giri SN, Sharma AK, Hyde DM, Wild JS (1995) An amelioration of bleomycin-induced lung fibrosis by treatment with the platelet activating factor receptor antagonist WEB 2086 in hamsters. Exp Lung Res 21:287–307

Glauser MP, Zanetti G, Baumgartner JD, Cohen J (1991) Septic shock:pathogenesis. Lancet 338:732–736

Gleich GJ, Adolphson CR (1986) The eosinophil leukocyte: structure and function. Adv Immunol 39:177–253

Gonzales-Crussi F, Hsueh W (1983) Experimental model of ischemic bowel necrosis. The role of platelet-activating factor and endotoxin. Am J Pathol 112:127–135
Goswami SK, Okashi M, Panagiotis S, Maron Z (1987) Platelet activating factor enhances mucous glycoprotein release from human airways in vitro. Am Rev Respir Dis 136:A159
Graham RM, Stephens CJ, Sturm MJ, Taylor RR (1992) Plasma platelet-activating factor degradation in patients with severe coronary artery disease. Clin Sci 82:535–541
Graham RM, Stephens CJ, Silvester W, Leong LL, Sturm MJ, Taylor RR (1994) Plasma degradation of platelet-activating factor in severely ill patients with clinical sepsis. Crit Care Med 22:204–212
Grandel KE, Farr RS, Wanderer AA, Eisenstadt TC, Wasserman SI (1985) Association of platelet-activating factor with primary acquired cold urticaria. N Engl J Med 313:405–409
Granger DN (1988) Role of xanthine oxidase and granulocytes in ischemia reperfusion injury. Am J Physiol 255:H1269–H1275
Granger DN, Korthuis RJ (1995) Physiologic mechanisms of post-ischemic tissue injury. Annu Rev Physiol 57:311–332
Guimbaud R, Izzo A, Martinolle P, Oidon N, Coutourier D, Benveniste J, Chaussade S (1995) Intraluminal excretion of PAF, lyso PAF, and acetyl-hydrolase in patients with ulcerative colitis. Digest Dis Sci 40:2635–2640
Guinot P, Braquet P, Duchier J, Cournot A (1986) Inhibitors of PAF-acether-induced wheal and flare reaction in man by a specific antagonist. Prostaglandins 32:160–163
Guitierrez S, Palacios I, Egido J, Zarco P, Miguelez R, Gonzalez E, Herrero-Beaumont G (1995) IL-1$\beta$ and IL-6 stimulate the production of platelet-activating factor (PAF) by cultured rabbit synovial cells. Clin Exp Immunol 99:364–368
Gustafson C, Tagesson C (1989) Phospholipase activation and arachidonic acid release in cultured intestinal epithelial cells (INT407). Scand J Gastroenterol 24:475–484
Gustafson C, Kald B, Sjödahl R, Tagesson C (1991) Phospholipase C from Clostridium perfringens stimulates formation and release of platelet-activating factor (PAF-acether) in cultured intestinal epithelial cells (INT-407). Scand J Gastroenterol 26:1000–1006
Hallenbeck JM, Dutka AJ, Tanishima T, Kochanek PM, Kumarov KK, Thompson CB, Obrenovitch TP, Contreras TJ (1986) Polymorphonuclear leukocyte accumulation in brain regions with low blood flow during the early postischemic period. Stroke 17:246–253
Halonen M, Palmer JD, Lohman IC, McManus LM, Pinckard RN (1980) Respiratory and circulatory alterations induced by acetyl glyceryl ether phosphorylcholine, a mediator of IgE anaphylaxis in the rabbit. Am Rev Respir Dis 122:915–924
Halonen M, Lohman IC, Dunn AM, McManus LM, Palmer JD (1985) Participation of platelet in the physiologic alterations of AGEPC response anf of IgE anaphylaxis in the rabbit. Am Rev Respir Dis 131:11–17
Hamasaki Y, Mojarad M, Saga T, Tai HH, Said SI (1984) Platelet-activating factor raises airway and vascular pressures and induces edema in lungs perfused with platelet free solutions. Am Rev Respir Dis 129:742–746
Hammer C (1995) Xenotransplantation. Dtsch Ärztebl 92:A133–A137
Handa RK, Strandhoy JW, Buckley VM (1990) Platelet-activating factor is a renal vasodilator in the anesthetized rat. Am J Physiol 258:F1504–F1509
Handa RK, Strandhoy JW, Buckley VM (1991) Vasorelaxant and C18-PAF on renal blood flow and systemic blood pressure in the anesthetized rat. Life Sci 49:747–752
Handley DA, van Valen GR, Melden MK, Saunders RN (1984) Evaluation of dose and route effects of platelet activating factor-induced extravasation in the guinea pig. Thromb Haemost 52:34–36
Handley DA, van Valen RG, Saunders RN (1986) Vascular responses of platelet-activating factor in the Cebus apella primate and inhibitory profiles of antagonists SRI 63-072 and SRI 63-119. Immunpharmacology 11:175–182

Handley DA, Houlihan WJ, Saunders RN, Tomesch JC (eds) (1990) PAF in endotoxin and immune disease. Dekker, New York, pp 63–119
Hanglow AC, Bienenstock J, Perdue MH (1989) Effects of platelet-activating factor on ion transport in isolated rat jejunum. Am J Physiol 257:G845–G850
Harlan JM, Liu DY (1992) Adhesion: its role in inflammatory disease. Freeman, New York
Harper MJK (1989) Platelet-activating factor: a paracrine factor in pre-implantation stages of reproduction? Biol Reprod 40:907–913
Harper MJK, Woodard DS, Norris CJ (1989) Spermicidal effect of antagonists of platelet-activating factor. Fertil Steril 51:890–895
Harries AD, Beeching NJ, Rogerson SJ, Nye JF (1991) The platelet count as a simple measure to distinguish inflammatory bowel disease from infective diarrhea. J Infect 22:247–250
Hashikura Y, Kawasaki S, Matsunami H, Ikegami T, Nakazawa Y, Makuuchi M (1994) Effect of platelet-activating factor on cold-preserved liver grafts. Br J Surg 81:1779–1782
Hattori M, Arai H, Inoue K (1993) Purification and characterization of bovine brain platelet-activating factor acetylhydrolase. J Biol Chem 268:18748–18753
Hattori M, Adachi H, Tsugimoto M, Arai H, Inoue K (1994) Miller-Dieker lissencephaly gene encodes a subunit of brain platelet-activating factor acetylhydrolase. Nature 370:216–218
Hawrylowicz CM (1993) Viewpoint. A potential role for platelet derived cytokines in the inflammatory response. Platelets 4:1
Hayashi H, Kudo I, Kato T, Inoue K (1989) PAF and macrophage-lineage cells. In: Saito K, Hanahan DJ (eds) Platelet-activating factor and diseases. p 51
Haynes J, Chang S, Morris KG, Voelkel NF (1988) Platelet-activating factor antagonists increase vascular reactivity in perfused rat lung. J Appl Physiol 65:1921–1928
Heffner JE, Shoemaker SA, Canham EM, Patel M, McMurthy IF, Morris HG, Repine JE (1983) Acetyl glyceryl ether phosphorylcholine-stimulated human platelets cause pulmonary hypertension and edema in isolated rabbit lung. J Clin Invest 71:351–357
Heffner JE, Sahn SA, Repine JE (1987) The role of platelets in the adult respiratory distress syndrome. Am Rev Respir Dis 135;482–492
Hellewell PG (1990) The contribution of PAF to immune-complex mediated inflammation. In: Handley DA, Saunders RN, Houlihan WJ, Tomesch JC (eds) Platelet activating factor in endotoxin and immune diseases. Dekker, New York, pp 367–386
Hellewell PG, Williams TJ (1986) A specific antagonist of platelet-activating factor suppresses oedema formation in an Arthus reaction but not oedema induced by leukocyte chemoattractants in rabbit skin. J Immunol 137:302–307
Hellstrom WJC, Wang R, Sikka SC (1991) Platelet-activating factor stimulates motion parameters of cryopreserved human sperm. Fertil Steril 56:768–770
Henocq E, Vargaftig BB (1986) Accumulation of eosinophils in response to intracutaneous PAF-acether and allergens. Lancet I:1378–1379
Henson PM (1987) Extracellular and intracellular activities of platelet activating factor. In: Winslow CM, Lee ML (eds) Horizons in platelet activating factor research. Wiley, New York
Henson PM, Cochrane CG (1969) Immunological induction of increased vascular permeability I A rabbit passive cutaneous anaphylactic reaction requiring complement, platelets, and neutrophils. J Exp Med 129:153–165
Heuer HO (1991) Inhibition of active anaphylaxis in mice and guinea pigs by the new hetrazepinoic PAF antagonist bepafant (WEB-2170). Eur J Pharmacol 199:157–163
Hilliquin P, Menkes CJ, Lavoussadi S, Benveniste J, Arnoux B (1992) Presence of paf-acether in rheumatic diseases. Ann Rheumatol 51:29–31

Hilliquin P, Houbaba H, Aissa J, Benveniste T, Menkes CJ (1995a) Correlations between PAF-acether and tumor necrosis factor in rheumatoid arthritis. Influence of parenteral corticosteroids. Scand J Rheumatol 24:169–173
Hilliquin P, Dulioust A, Gregoir C, Arnoux A, Menkes CJ (1995b) Production of PAF-acether by synovial fluid neutrophils in rheumatoid arthritis. Inflamm Res 44:313–316
Hines KL, Fisher RA (1992) Regulation of hepatic glycogenolysis and vasoconstriction during antigen-induced anaphylaxis. Am J Physiol 262:G868–G877
Hines KL, Braillon A, Fisher RA (1991) Platelet activating factor increases hepatic vascular resistance and glycogenolysis in vivo. Am J Physiol 260:G471–G480
Hirafuji M, Shinoda H (1991) Platelet-leukocyte interaction in adhesion to endothelial cells induced by platelet-activating factor in vitro. Br J Pharmacol 103:1333–1338
Hirano T, Furuyama H, Tsuchitani T (1995) Platelet activating factor (PAF) antagonist prevents impaired pancreatic blood flow in rabbits with caerulein-induced acute pancreatitis. Med Sci 23:209–212
Hirashima Y, Endo S, Ohmori T, Kato R, Takaku A (1994) Platelet-activating factor (PAF) concentration and PAF acethydrolase activity in cerebrospinal fluid of patients with subarachnoid hemorrhage. J Neurosurg 80:31–36
Hirashima Y, Endo S, Kato R, Takaku A (1996) Prevention of cerebrovasospasm following subarachnoid hemorrhage in rabbits by the platelet-activating factor antagonist, E 5880. J Neurosurg 84:826–830
Hoffman DR, Romero R, Johnston JM (1990) Detection of platelet-activating factor in amniotic fluid of complicated pregnancies. Am J Obstet Gynecol 162:525–528
Hogaboam CM, Donigi-Gale D, Shoupe TS, Wallace JL (1992) PF 5901 inhibits gastrointestinal PAF synthesis in vivo. Eur J Pharmacol 216:315–318
Hopp RJ, Bewtra AK, Agrawal DK, Townley RG (1989) Effect of PAF inhalation on nonspecific bronchial reactivity in man. Chest 96:1070–1072
Horvath CJ, Kaplan JE, Malik AB (1991) Role of platelet-activating factor in mediating tumor necrosis factor alpha-induced pulmonary vasoconstriction and plasma-lymph transport. Am Rev Respir Dis 144:1337–1341
Houlihan WJ, Munder P, Handley DA, Cheon SH, Parillo VA (1995) Antitumor activity of 5-aryl-2,3-dihydroimidazo (2,1-a) isoquinolines. J Med Chem 38:234–240
Howard AD, Erickson KL (1995) The induction and augmentation of macrophage tumoricidal responses by platelet-activating factor. Cell Immunol 164:105–112
Hsieh KH (1991) Effects of PAF antagonist, BN 52021, on the PAF-, methacholine- and allergen-induced bronchoconstriction in asthmatic children. Chest 99:877–882
Hsieh KH, Ng CK (1993) Increased plasma platelet-activating factor in children with acute asthmatic attacks and decreased in vivo and in vitro production of platelet-activating factor after immunotherapy. J Allergy Clin Immunol 91:650–657
Hsueh W, Gonzales-Crussi F, Arroyave IL (1986a) Platelet-activating factor induced ischemic bowel necrosis: an investigation of secondary mediators in its pathogenesis. Am J Pathol 122:231–239
Hsueh W, Gonzalez-Crussi F, Arroyave IL, Anderson RC, Lee ML, Houlihan WJ (1986b) Platelet activating factor-induced ischemic bowel necrosis: the effect of PAF antagonists. Eur J Pharmacol 123:79–83
Hsueh W, Gonzales-Crussi F, Arroyave IL (1986c) Release of leukotriene C4 by isolated perfused rat small intestine in response to platelet activating factor. J Clin Invest 78:108–114
Hsueh W, Gonzales-Crussi F, Arroyave JL (1987) Platelet-activating factor: is an endogenous mediator for bowel necrosis in endotoxemia. FASEB J 1:403–405
Hsueh W, Gonzalez-Crussi F, Arroyave IL (1988) Sequential release of leukotrienes and norepinephrine in rat bowel after platelet-activating factor. A mechanistic study of platelet-activating factor-induced bowel necrosis. Gastroenterology 94:1412–1418

Huang Y-CT, Nozik ES, Piantadosi CA (1994) Superoxide dismutase potentiates platelet-activating-factor-induced injury in perfused lung. Am J Physiol 266:L246–L254

Humphrey DM, Hanahan DJ, Pinckard RN (1982a) Induction of leukocytic infiltrates in rabbit skin by acetyl glyceryl ether phophorylcholine. Lab Invest 47:227–234

Humphrey DM, McManus LM, Satonchi K, Hanahan DJ, Pinckard RN (1982b) Vasoactive properties of acetyl glyceryl ether phosphorylcholine. Lab Invest 46:422–427

Humphrey DM, McManus LM, Hanahan DJ, Pinckard RN (1984) Morphologic basis of increased vascular permeability induced by acetyl glyceryl ether phosphorylcholine. Lab Invest 50:16–25

Hurst JS, Bazan HEP (1993) The platelet-activating factor precursor of the injured cornea is selectively implicated in arachidonate and eicosanoid release. Curr Eye Res 12:655–663

Husz S, Toth-Kasa I, Obal F, Jancso G (1991) A possible pathomechanism of the idiopathic cold contact urticaria. Acta Physiol Hung 77:209–215

Hutcheson IR, Whittle BJR, Boughton-Smith NK (1990) Role of nitric oxide in maintaining vascular integrity in endotoxin-induced acute intestinal damage in the rat. Br J Pharmacol 101:815–820

Hwang SB, Lee CS, Cheah MJ, Shen TY (1983) Specific receptor sites for 1-*O*-alkyl-2-acetyl-*sn*-glycero-3-phosphocholine (platelet-activating factor) on rabbit platelet and guinea pig smooth muscle membranes. Biochem 22:4756–4763

Hwang SB, Lam MH, Li CL, Shen TY (1986) Release of platelet activating factor and its involvement in the first phase of carrageenin-induced rat foot edema. Eur J Pharmacol 120:33–41

Ibbotson GC, Wallace JL (1989) Beneficial effects of prostaglandin E2 in endotoxic shock are unrelated to effects on PAF-acether synthesis. Prostaglandins 37:237–250

Ide S, Kawahara K, Takahashi T, Sasaki N, Shingu H, Nagayasu T, Yamamoto S, Tagawa T, Tomita M (1995) Donor administration of PAF antagonist (TCV-309) enhances lung preservation. Transplant Proc 27:570–573

Ikegami M, Jobe AH, Tabor BL, Rider ED, Lewis JF (1992) Lung albumin recovery in surfactant-treated preterm ventilated lambs. Am Rev Respir Dis 145:1005–1008

Im SY, Ko HM, Ko YS, Kim JW, Lee HK, Ha TY, Lee HB, Oh SJ, Bai S, Chung KC, Lee Y-B, Kang H-S, Chun SB (1996) Augmentation of tumor metastasis by platelet-activating factor. Cancer Res 56:2662–2665

Imai Y, Hayashi M, Oh-Ishi S (1991) Key role of complement and platelet-activating factor in exudate formation in zymosan-induced rat pleurisy. Jpn J Pharmacol 57:225–232

Imaizumi TA, Stafforini DM, Yamada Y, McIntyre TM, Prescott SM, Zimmerman GA (1995) Platelet-activating factor: a mediator for clinicians. J Intern Med 238:5–20

Imanishi N, Komuro Y, Morooka S (1991) Effect of a selective antagonist SM-10661 ((+/–)-cis-3,5-dimethyl-2-(3-pyridyl) thiazolidin-4-one HCl) on experimental disseminated intravascular coagulation (DIC). Lipids 26:1391–1395

Inarrea P, Gomez-Cambronero J, Pascual J, Carmel-Ponte M, Hernando L, Sanchez-Crespo M (1985) Synthesis of PAF-acether and blood volume changes in gram negative sepsis. Immunopharmacology 9:45–52

Inarrea P, Alonso F, Sanchez-Crespo M (1983) Platelet-activating factor: an effector substance of the vasopermeability changes induced by the infusion of immunoaggregates in the mouse. Immunopharmacology 6:7–14

Irvin CG, Berend N, Henson PM (1986) Airways hyperreactivity and inflammation produced by aerosolization of human C5b des arg. Am Rev Respir Dis 134:777–783

Issekutz AC, Szejda M (1986) Evidence that platelet activating factor may mediate some acute inflammatory responses. Studies with platelet activating factor antagonist CV 3988. Lab Invest 54:275–281

Issekutz AC, Ripley M, Jackson JR (1983) Role of neutrophils in the deposition of platelets during acute inflammation. Lab Invest 49:716–724

Iwai A, Itoh M, Yokoyama Y (1989) Role of PAF in ischemia-reperfusion injury in the rat stomach. Scand J Gastroenterol 24[Suppl 162]:63–66

Izaki S, Yamamoto T, Goto Y, Ishimaru S, Yudate F, Kitamura K, Matsuzaki M (1996) Platelet-activating factor and arachidonic acid metabolites in psoriatic inflammation. Brit J Dermatol 134:1060–1064

Izquierdo I, Fin C, Schmitz PK, Da Silva RC, Jerusalinsky D, Quillfeldt JA, Fereira MBG, Medina JH, Bazan NG (1995) Memory enhancement by intrahippocampal, intraamygdala, or intraentorhinal infusion of platelet-activating factor measured in an inhibitory avoidance task. Proc Natl Acad Sci 92:5047–5051

Izumi T, Kishimoto S, Takano T, Nakamura M, Miyabe Y, Nakata M, Sakanaka C, Shimizu T (1995) Expression of human platelet-activating factor receptor gene in EoL-1 cells following butyrate-induced differentiation. Biochem J 305:829–835

Jancar S, Giaccobi G, Mariano M, Mencia-Huerta JM, Sirois P, Braquet P (1988) Immune-complex pancreatitis: effects of BN 52021, a selective antagonist of platelet activating factor. Prostaglandins 35:757–770

Jancar S, Abdo EE, Sampietre SN, Kwasniewski FH, Coelho AMM, Bonizzia A, Machado MCC (1995) Effects of PAF antagonists on coerulein-induced pancreatitis. J Lipid Mediat Cell Signal 11:41–49

Jaranowska A, Bussolino F, Sogos V, Arese M, Lauro GM, Gremo F (1995) Platelet-activating factor production by human fetal microglia. Effect of lipopolysaccharides and tumor necrosis factor-alpha. Mol Chem Neuropathol 24:95–106

Jarvi K, Gagnon C, Miron P, Langlais J (1991) Platelet activating factor (PAF) metabolism by spermatozoa: role of metabolites in PAF action on sperm motility. J Cell Biol 115:836A

Jarvi K, Roberts KD, Langlais J, Gagnon C (1993a) Effect of platelet-activating factor, lyso-platelet-activating factor and lysophosphatidylcholine on sperm motion. Importance of albumin for motility stimulation. Fertil Steril 59:1266–1275

Jarvi K, Langlais J, Gagnon C, Roberts KD (1993b) Platelet-activating factor acetylhydrolase in the male reproductive tract: origin and properties. Int J Androl 16:121–127

Jeffery PK, Wardlaw AJ, Nelson FC, Collins JV, Kay AB (1989) Bronchial biopsies in asthma: an ultrastructural, quantitative study and correlation with hyperreactivity. Am Rev Respir Dis 140:1745–1753

Jiang Z-G, Yue T-L, Feuerstein G (1993) Platelet-activating factor inhibits excitatory transmission in rat neostriatum. Neurosci Lett 155:132–135

Johnson PRA, Armour CL, Black JL (1990) The action of platelet activating factor and its antagonism by WEB 2086 on human isolated airways. Eur Respir J 3:55–60

Johnston S, Spencer D, Calverley P, Winter J, Dhillon P (1992) WEB 2086, a platelet activating factor antagonist, has no steroid sparing effect in asthmatic subjects requiring inhaled corticoids. Thorax 47:871 (Abstract)

Joly F, Vilgrain J, Bossant MJ, Bessou G, Benveniste J, Ninio E (1990) Biosynthesis of paf-acether. Activators of protein kinase C stimulate cultered mast cell acetyltransferase without stimulating paf-acether synthesis. Biochem J 271:501–507

Jones MA, Kudolo GB, Harper MJ (1992) Rabbit blastocysts accumulate platelet-activating factor (PAF) and lyso-PAF in vitro. Mol Reprod Dev 32:243–250

Juncos LA, Ren Y, Arima S, Ito S (1993) Vasodilator and constrictor actions of platelet-activating factor in the isolated microperfused afferent arteriole of the rabbit kidney. Role of endothelium-derived relaxing factor/nitric oxide and cyclooxygenase products. J Clin Invest 91:1374–1379

Kald B, Olaison G, Sjodahl R, Tagesson C (1990) Novel aspect of Crohn's disease: increased content of platelet-activating factor in ileal and colonic mucosa. Digestion 46:199–204

Kald B, Gustavson C, Franzen L, Weström B, Sjödahl R, Tagesson C (1992) Platelet-activating factor (PAF-acether) formation in neonatal intestinal mucosa and in cultured intestinal epithelial cells. Eur Surg Res 24:325–332

Kald B, Kald A, Ihse I, Tagesson C (1993) Release of platelet-activating factor in acute experimental pancreatitis. Pancreas 8:440–442

Kaliner M, Eggleston PA, Mathews KP (1987) Rhinitis and asthma. JAMA 258:2851–2871

Kamata K, Mori T, Shigenobu K, Kasuya Y (1989) Endothelium-dependent vasodilator effects of platelet-activating factor on rat resistance vessels. Br J Pharmacol 98:1360–1364

Kaneko T, Ikeda H, Fu L, Nishiyama H, Okubo T (1995) Platelet-activating factor mediates the ozone-induced increase in airway microvascular leakage in guinea pigs. Eur J Pharmacol 292:251–256

Kapp A (1991) Cytokines in atopic dermatitis. In: Ruzicka T, Ring J, Przybilla B (eds) Handbook of atopic eczema. Springer-Verlag, Berlin Heidelberg New York, pp 256–262

Katayama H, Hase T, Yaoita H (1994) Detachment of cultured normal human keratinocytes by contact with TNF alpha-stimulated neutrophils in the presence of platelet-activating factor. J Invest Dermatol 103:187–190

Kaye MG, Smith LJ (1990) Effects of inhaled leukotriene D4 and platelet-activating factor on airway reactivity in normal subjects. Am Rev Respir Dis 141:993–997

Kazi N, Fields JZ, Sedghi S, Kottapalli V, Eiznhamer D, Winship D, Keshavarzian A (1995) Modulation of neutrophil function by novel colonic factors: possible role in pathophysiology of ulcerative colitis. J Lab Clin Med 126:70–80

Kelly NM, Cross AS (1992) Interleukin 6 is a better marker of lethality than tumor necrosis factor in endotoxin-treated mice. FEMS Microbiol Immunol 89:317–322

Kentroti S, Baker R, Lee K, Bruce C, Vernadakis A (1991) Platelet-activating factor increases glutamine synthetase activity in early and late passage C-6 glioma cells. J Neurosci Res 28:497–506

Keshav S, Lawson L, Chung LP, Stein M, Perry VH, Gordon S (1990) Tumor necrosis factor mRNA localized to Paneth cells of normal murine intestinal epithelium by in situ hybridization. J Exp Med 171:327–332

Kilbourn RG, Griffith OW (1992) Overproduction of nitric oxide in cytokine-mediated and septic shock. J Natl Cancer Inst 84:827–831

Kim BK, Ozaki H, Lee SM, Karaki H (1995) Increased sensitivity of rat myometrium to the contractile effect of platelet activating factor before delivery. Br J Pharmacol 115:1211–1214

Kim DK, Fukuda T, Thompson BT, Cockrill B, Hales C, Bonventre JV (1995) Bronchoalveolar lavage fluid phospholipase A2 activities are increased in human respiratory distress syndrome. Am J Physiol 269:L109–L118

Kim FJ, Moore EE, Moore FA, Biffe WL, Fontes B, Banerjee A (1995) Reperfused gut elaborates PAF that chemoattracts and primes neutrophils. J Surg Res 58:636–640

Kim JJ, Fortier MA (1995) Cell type specificity and protein kinase C dependency on the stimulation of prostaglandin E 2 and prostaglandin F-2alpha production by oxytocin and platelet-activating factor in bovine endometrial cells. Reprod Fertil 103:239–248

Kim YD, Danchak RM, Heim KF, Lees DE, Myers AK (1993) Constriction of canine coronary arteries by platelet activating factor after brief ischemia. Prostaglandins 46:269–276

Kingnorth AN (1996) Platelet-activating factor. Scand J Gasteroenterol 31[Suppl 219]:28–31

Kirschbaum B (1991) Platelet activating factor acetylhydrolase activity in the urine of patients with renal disease. Clin Chim Acta 199:139–146

Kitagawa S, Kubota Y, Yamaguchi T, Fujimura K, Binnaka T, Tani K, Ogura M, Mizuno T, Inoue K (1992) Role of endogenous platelet-activating factor (PAF) in endotoxin-induced portal hypertension in rats. J Gastroenterol Hepatol 7:481–485

Klee A, Schmid-Schönbein GW, Seiffge D (1991) Effects of platelet activating factor on rat platelets in vivo. Eur J Pharmacol 209:223–229
Kochanek PM, Hallenbeck JM (1992) Polymorphonuclear leukocytes and monocytes/macrophages in the pathogenesis of cerebral ischemia and stroke. Stroke 23:1367–1379
Kochanek PM, Dutka AJ, Kumaro KK, Hallenbeck JM (1987) Platelet activating factor receptor blockade enhances recovery after multifocal brain ischemia. Life Sci 41:2639–2644
Kochanek PM, Nemeto EM, Melick JA, Evans RW, Burkle DF (1988) Cerebrovascular and cerebrometabolic effects of intracarotid infused platelet-activating factor in rats. J Cereb Blood Flow Metab 8:546–551
Kochanek PM, Melick BS, Schoettle RJ, Magargee MJ, Evans RW, Nemoto EM (1990) Endogenous platelet activating factor does not modulate blood flow and metabolism in normal rat brain. Stroke 21:459–462
Kochanek PM, Kochanek P, Schoettle R, Uhl M, Magargee MJ, Nemoto E (1991) Platelet-activating factor antagonists do not attenuate delayed posttraumatic cerebral edema in rats. J Neurotrauma 8:19–25
Koltai M, Tosaki A, Hosford D, Esanu A, Braquet P (1991) Effect of BN 50739, a new platelet activating factor antagonist, on ischemia induced ventricular arrhythmias in isolated working rat hearts. Cardiovasc Res 24:391–397
Kornecki E, Ehrlich P (1988) Neuroregulatory and neuropathological actions of the ether-phospholipid platelet-activating factor. Science 240:1792–1794
Korthuis RJ, Anderson DC, Granger DN (1994) Role of neutrophil-endothelial cell adhesion in inflammatory disorders. J Crit Care 9:47–71
Krausz C, Gervasi G, Forti G, Baldi E (1994) Effect of platelet-activating factor on motility and acrosome reaction of human spermatozoa. Hum Reprod 9:471–476
Kravis TC, Henson PM (1975) IgE-induced release of platelet-activating factor from rabbit lung. J Immunol 115:1677–1681
Kroegel C (1992) Immunbiologie des eosinophilen Granulozyten. Pharm Unserer Zeit 21:61–70
Kroegel C, Yukawa T, Dent G, Chanez P, Chung KF, Barnes PJ (1988) Platelet-activating factor induces eosinophil peroxidase release from purified human eosinophils. Immunology 64:559–561
Kroegel C, Yukawa T, Dent G, Venge P, Chung KF, Barnes PJ (1989) Stimulation of degranulation from human eosinophils by platelet-activating factor. J Immunol 142:3518–3526
Kroegel C, Hubbard WH, Lichtenstein LM (1990) Spectrum of prostanoid generation by human blood eosinophils stimulated with platelet activating factor and calcymicin. Eur J Respir Dis 3:347S
Kroegel C, Warner JA, Giembycz MA, Matthys H, Lichtenstein LM, Barnes PJ (1992) Dual transmembrane signalling mechanisms in eosinophils: Evidence for two functionally distinct receptors for platelet-activating factor. Int Arch Allergy Immunol 99:226–229
Kruse-Elliott KT, Pino MV, Olson NC (1993) Effect of PAF receptor antagonism on cardiopulmonary alterations during coinfusion of TNF-alpha and IL-1 alpha in pigs. Am J Physiol 264:L175–L182
Kubes P, Granger DN (1992) Nitric oxide modulates microvascular permeability. Am J Physiol 262:H611–H615
Kubes P, Suzuki M, Granger DN (1991a) Platelet-activating factor-induced microvascular dysfunction: role of adherent leukocytes. Am J Physiol 258:G158–G163
Kubes P, Ibbotson G, Russell J, Wallace JL, Granger DN (1990b) Role of platelet-activating factor in ischemia/reperfusion-induced leukocyte adherence. Am J Physiol 259:G300–G305
Kudolo GB, Harper MJK (1989) Characterization of platelet-activating factor binding sites on uterine membranes from pregnant rabbits. Biol Reprod 41:587–603

Kudolo GB, Harper MJK (1990) Estimation of platelet-activating factor receptors in the endometrium of the pregnant rabbit: regulation of ligand availability and catabolism by bovine serum albumin. Biol Reprod 43:368–377

Kudolo GB, Harper MJK (1992) Pregnancy-associated remodeling of rabbit endometrial platelet-activating factor receptors. J Lipid Mediat 5:271–278

Kudolo GB, Kasamo M, Harper MJ (1991) Autoradiographic localization PAF binding sites in the rabbit endometrium during the peri-implantation period. Cell Tissue Res 265:231–241

Kuitert L, Barnes NC (1995) PAF and asthma. Time for an appraisal? Clin Exp Allerg 25:1159–1162

Kuitert LM, Hui KP, Uthayarkumar S, Burke W, Newland AC (1992) Effect of a platelet activating factor antagonist, UK 74505, on allergen induced early and late response. Am Rev Respir Dis 145:A292

Kuitert LM, Angus RM, Barnes NC, Barnes PJ, Bone MF, Chung KF, Fairfax AJ, Higenbotham TW, O'Connor BJ, Piotrowska B (1995) Effect of a novel platelet-activating factor antagonist, modipafant, in clinical asthma. Am J Resp Crit Care Med 151:1331–1335

Kumar R, Harper MJK, Hanahan DJ (1988a) Occurence of platelet activating factor in rabbit spermatozoa. Arch Biochem Biophys 260:497–502

Kumar R, Harvey SAK, Kester M, Hanahan DJ, Olson MS (1988b) Production and effects of platelet-activating factor in rat brain. Biochim Biophys Acta 963:375–383

Kunievsky B, Yavin E (1994) Production and metabolism of platelet-activating factor in the normal and ischemic fetal rat brain. J Neurochem 63:2144–2151

Kurosawa M, Yamashita T, Kurimoto F (1994) Increased levels of blood platelet-activating factor in bronchial asthmatic patients with active symptoms. Allergy 49:60–63

Kuzan FB, Geissler FT, Henderson WR (1990) Role of spermatozoal platelet-activating factor in fertilization. Prostaglandins 39:61–74

Lachapelle MH, Bouzayen R, Langlais J, Jarvi K, Bourquet J, Miron P (1993) Effect of lyso-PAF on human sperm fertilizing ability. Fertil Steril 59:863–868

Lagente V, Fortes ZB, Garcia-Leme J, Vargaftig BB (1988) PAF-acether and endotoxin display similar effects on rat mesenteric microvessels: Inhibition by specific antagonists. J Pharmacol Exp Ther 247:254–261

Lai CKW, Jenkins JR, Polosa R, Holgate ST (1990) Inhaled PAF fails to induce airway hyperresponsiveness to methacholine in normal human subjects. J Appl Physiol 68:919–926

Lalouette F, Diserbo M, Martin C, Verdetti J, Fatome M (1995) Presence of specific platelet-activating fractor binding-sites in neuroblastoma N1E-115 cells. Neurosci Lett 186:173–176

Lang CH, Dobrescu C, Hargrove C, Bagby GJ, Spitzer JJ (1988) Platelet activating factor-induced increases in glucose kinetics. Am J Physiol 254:193–200

Langlais J, Roberts KD (1985) A molecular membrane model of sperm capacitation and the acrosome reaction of mammalian spermatozoa. Gamete Res 12:183–224

Lapointe DS, Olson MS (1989) Platelet-activating factor-stimulated hepatic glycogenolysis is not mediated through cyclooxygenase-derived metabolites of arachidonic acid. J Biol Chem 264:12130–12133

Last-Barney K, Homon CA, Faanes RB, Merluzzi VJ (1988) Synergistic and overlapping activities of tumor necrosis factor-alpha and IL-1. J Immunol 141:527–530

Laszlo F, Whittle BJR, Moncada S (1994a) Time-dependent enhancement or inhibition of endotoxin-induced vascular injury in rat intestine by nitric oxide synthase inhibitors. Br J Pharmacol 111:1309–1315

Laszlo F, Whittle BJR, Moncada S (1994b) Interactions of constitutive nitric oxide with PAF and thromboxane on rat intestinal vascular integrity in acute endotoxaemia. Br J Pharmacol 113:1131–1136

Le Poncin Lafitte M, Rapin J, Rapin JR (1986) Effects of Ginkgo biloba on changes induced by quantitative cerebral micro-embolization in rats. Arch Pharmacodyn 243:236–244

Lee T-C, Malone B, Wasserman SJ, Fitzgerald V, Snyder F (1982) Activities of enzymes that metabolize platelet-activating factor (1-alkyl-2-acetyl-sn-glycero-3-phosphocholine) in neutrophils and eosinophils from humans and the effect of a calcium ionophore. Biochem Biophys Res Commun 105:1303–1308

Lee T-C, Lenihan DJ, Malone B, Roddy LL, Wasserman SJ (1984) Increased biosynthesis of platelet-activating factor in activated human eosinophils. J Biol Chem 259:5526–5530

Lee T-C, Malone B, Woodward D, Snyder F (1989) Renal necrosis and involvement of a single enzyme of the de novo pathway for the biosynthesis of platelet-activating factor in the rat kidney inner medulla. Biochem Biophys Res Commun 163:1002–1005

Lefer AM (1989) Significance of lipid mediators in shock states. Circ Shock 27:3–12

Leiferman KM, Ackerman SJ, Sampson HA, Haugen HS, Venencie PY, Gleich GJ (1985) Dermal deposition of eosinophil-granule major basic protein in atopic dermatitis. Comparison with onchocerciasis. N Engl J Med 313:282–285

Lellouch-Tubiana A, Lefort J, Pirotzky E, Vargaftig BB (1985) Ultrastructural evidence for extravascular platelet recruitment in the lung upon intravenous injection of platelet-activating factor (PAF-acether) to guinea-pigs. Br J Exp Pathol 66:345–355

Lellouch-Tubiana A, Lefort J, Simon MT, Pfister A, Vargaftig BB (1988) Eosinophil recruitment into guinea pig lungs after PAF-acether and allergen administration: modulation by prostacyclin, platelet depletion and selective antagonists. Am Rev Respir Dis 137:948–954

Leong L, Sturm M, Raylor R (1991) The lyso-precursor of platelet-activating factor (lyso-PAF) in ischaemic myocardium. J Lipid Mediat 4:277–288

Leong LL, Stephens CJ, Sturm MJ, Taylor RR (1992) Effect of WEB 2086 on myocardial infarct size and regional blood flow in the dog. Cardiovasc Res 26:126–132

Leonhardt U, Exner B, Schrock E, Ritzel U, Nebendahl K, Stöckmann F (1992) Effect of platelet-activating factor antagonist on pancreas perfusion after 24h of ischemia. Pancreas 11:160–164

Letendre ED, Miron P, Roberts KD, Langlais J (1992) Platelet-activating factor acethydrolase in human seminal plasma. Fertil Steril 57:193–198

Levi R, Burke JA, Guo ZG, Hattori Y, Hoppens CM, McManus LM, Hanahan DJ, Pinckard RN (1984) Acetyl glyceryl ether phosphorylcholine (AGEPC): putative mediator of cardiac anaphylaxis in the guinea pig. Circ Res 54:117–122

Levy JV (1987) Spasmogenic effect of platelet activating factor (PAF) on isolated rat stomach fundus strip. Biochem Biophys Res Commun 146:855–860

Lewis MS, Whatley RE, Cain P, McIntyre TM, Prescott SM, Zimmerman GA (1988) Hydrogen peroxide stimulates the synthesis of platelet-activating factor by endothelium and induces endothelial cell-dependent neutrophil adhesion. J Clin Invest 82:2045–2055

Li M-S, Sun L, Satoh T, Fisher LM, Spry CDF (1995) Human eosinophil major basic protein, a mediator of allergic inflammation, is expressed by alternative splicing from two promotors. Biochem J 305:921–927

Li X-M, Sagawa N, Ihara Y, Okagaki A, Hasegawa M, Inamori K, Itoh H, Mori T, Ban C (1991) The involvement of platelet-activating factor in thrombocytopenia and follicular rupture during gonadotropin-induced superovulation in immature rats. Endocrinol 129;3132–3138

Ligumsky M, Sestieri M, Karmeli F, Okon E, Rachmilewitz D (1990) The role of platelet activating factor in acute gastric injury and its protection by sucralfate. Aliment Pharmacol Ther 4:507–514

Lindsberg PJ, Yue T-L, Frerichs KU, Hallenbeck JM, Feuerstein G (1990a) Evidence of platelet-activating factor (PAF) as a mediator of stroke. Stroke 21:179

Lindsberg PJ, Yue T-L, Frerichs KU, Hallenback JM, Feuerstein G (1990b) Evidence of platelet-activating factor as novel mediator in experimental stroke in rabbits. Stroke 21:1452–1457

Lindsberg PJ, Hallenbeck JM, Feuerstein G (1991) Platelet-activating factor in stroke and brain injury. Ann Neurol 30:117–129
Lipton SA, Yeh M, Dreyer EB (1994) Update on current models of HIV-related neuronal injury: Platelet-activating factor, arachidonic acid and nitric oxide. Adv Neuroimmunol 4:181–188
Leonhardt U, Fayyazzi A, Seidensticker F, Stöckmann F, Söling HD, Creutzfeldt W (1992) Influence of a platelet activating factor antagonist on several pancreatitis in two experimental models. Int J Pancreatol 12:161–166
Llianos EA (1989) Lipid inflammatory mediators in glomerulonephritis. J Lab Clin Med 113:535–536
Llianos EA, Zanglis A (1990) Glomerular platelet-activating factor levels and origin in experimental glomerulonephritis. Kidney Int 37:736–740
Longo WE, Carter JD, Chandel B, Niehoff M, Standeven J, Deshpande Y, Vernava AM, Polites G, Kulkarni AD, Kaminski DL (1995) Evidence for a colonic PAF receptor. J Surg Res 58:12–18
Lopez-Bernal A, Ross C, Laird E, Newman GE, Phizaacherly PJR (1991) Platelet activating factor levels in human follicular fluid and amniotic fluid. Soc Study Fertil (abstr 61)
Lorant DE, Patel KD, McIntyre TM, McEver RP, Prescott SM, Zimmerman GA (1991) Coexpression of GMP-140 and PAF by endothelium stimulated by histamine or thrombin: a juxtacrine system for adhesion and activation of neutrophils. J Cell Biol 115:223–234
Lundgren JD, Kaliner M, Logun C, Shelhamer JH (1990) Platelet-activating factor and tracheobronchial respiratory glycoconjugate release in feline and human explants: involvement of lipoxygenase pathway. Agents Actions 30:329–337
Lynch JM, Lotner GZ, Betz SJ, Henson PM (1979) The release of platelet-activating factor by stimulated rabbit neutrophils. J Immunol 123:1219–1226
Mabuchi K, Sugiura T, Ojima-Uchiyama A, Masuzawa Y, Waku K (1992) Differential effects of platelet-activating factor on superoxide anion production in human eosinophils and neutrophils. Biochem Int 26:1105–1113
Maccia CA, Gallagher JS, Ataman G, Glueck HI, Brooks SM, Bernstein IL (1977) Platelet thrombopathy in asthmatic patients with elevated immunoglobulin. J Allergy Clin Immunol 59:101–108
Macconi D, Foppolo M, Paris S, Noris M, Aiello S, Remuzzi G, Remuzzi A (1995) PAF mediates neutrophil adhesion to thrombin or TNF-stimulated endothelial cells under shear stress. Am J Physiol 269:C42–C47
MacKendrick W, Caplan M, Hsueh W (1993) Endogenous nitric oxide protects against platelet-activating factor-induced bowel injury in the rat. Pediatr Res 34:222–228
Maki N, Hoffman DR, Johnston JM (1988) Platelet-activating factor acetylhydrolase activity in maternal, fetal, and newborn rabbit plasma during pregnancy and lactation. Proc Natl Acad Sci USA 85:728–732
Maki N, Magness RR, Miyaura S, Gant NF, Johnston JM (1993) Platelet-activating-factor acetylhydrolase activity in normotensive and hypertensive pregnancies. Am J Obstet Gynecol 168:50–54
Makristathis A, Stauffer F, Feistauer SM, Georgopulos A (1993) Bacteria induce release of platelet-activating factor (PAF) from polymophonuclear neutrophil granulocytes: possible role of PAF in pathogenesis of experimentally induced bacterial pneumonia. Infect Immun 61:1996–2002
Mallet AJ, Cunningham FM (1985) Structural identification of platelet activating factor in psoriatic scale. Biochem Biophys Res Commun 126:192–198
Mallet AJ, Cunningham FM, Daniel R (1984) Rapid isokratic high performance liquid chromatographic purification of platelet activating factor (PAF) and lyso-PAF from human skin. J Chromatogr 309:160–166
Mandi Y, Farkas G, Koltai M, Beladi I, Mencia-Huerta JM, Braquet P (1989a) The effect of the platelet-activating factor antagonist BN 52021, on human natural killer cell-mediated cytotoxicity. Immunology 67:370–374

Mandi Y, Farkas G, Koltai M, Beladi I, Braquet P (1989b) Effect of the platelet-activating factor antagonist BN 52021 on human natural killer cell cytotoxicity. Int J Allergy Appl Immunol 88:222–224
Manning PJ, Watson RM, Margolskee DJ, Williams VC, Schwarz JI, Byrne PM (1990) Inhibition of exercise-induced bronchoconstriction by MK-571 a potent leukotriene D4-receptor antagonist. N Engl J Med 323:1736–1739
Marcheselli VL, Rossowska MJ, Domingo MT, Braquet P, Bazan NG (1990) Distinct platelet-activating factor binding sites in synaptic endings and in intracellular membranes of rat cerebral cortex. J Biol Chem 265:9140–9145
Maridonneau-Parini I, Lagente V, Lefort J, Randon J, Russo-Marie F, Vargaftig BB (1985) Desensitization to PAF-induced bronchoconstriction and to activation of alveolar macrophages by repeated inhalations of PAF in the guinea pig. Biochem Biophys Res Commun 131:42–49
Markey AC, Barker JNW, Archer CB, Guinot P, Lee TH, MacDonald DM (1990) Platelet activating factor-induced clinical and histopathologic responses in atopic skin and their modification by the platelet activating factor antagonist BN 52063. J Am Acad Dermatol 23:263–268
Marsh WR, Irvin CG, Murphy KR, Behrens BL, Larsen GL (1985) Increases in airway reactivity to histamine and inflammatory cells in bronchoalveolar lavage after late asthmatic response in an animal model. Am Rev Respir Dis 131:875–879
Marshall JS, Bienenstock J (1994) The role of mast cells in inflammatory reactions of the airways, skin, and intestine. Curr Opin Immunol 6:853–859
Martinez-Cuesta MA, Barrachina MD, Pique JM, Whittle BJR, Esplugues JV (1992) The role of nitric oxide and platelet-activating factor in the inhibition by endotoxin of pentagastrin-stimulated gastric acid secretion. Eur J Pharmacol 218:351–354
Martins MA, Silva PM, Castro HC, Neto F, Cordeiro RS, Vargaftig BB (1987) Interactions between local inflammatory and systemic haematological effects of PAF-acether in the rat. Eur J Pharmacol 136:353–360
Martins MA, Silva PMR, Castro Faria Neto HC, Bozza PT, Dias PMFL, Lima MCR, Cordeiro RSB, Vargaftig BB (1989) Pharmacological modulatin of PAF-induced rat pleurisy and its role in inflammation by zymosan. Brit J Pharmacol 96:363–371
Martins MA, Pasquale CP, Silva PMR, Pires ALA, Ruffie C, Rihoux JP, Cordeiro RSB, Vargaftig BB (1992) Interference of cetirizine with late eosinophil accumulation induced by either PAF or compound 48/80. Br J Pharmacol 105:176–180
Maruyama M, Farber NE, Vercelotti GM, Jacob HS, Gross GJ (1990) Evidence for a role of platelet actvating factor in the pathogenesis of irreversible but not reversible myocardial injury after reperfusion in dogs. Am Heart J 120:510–520
Mascolo N, Autore G, Izzo AA, Biondi A, Capasso F (1992) Effects of senna and its active compounds rhein and rhein-anthrone on PAF formation by rat colon. J Pharm Pharmacol 44:693–695
Massey CV, Kohout TA, Gaa ST, Lederer WJ, Rogers TB (1991) Molecular and cellular actions of platelet-activating factor in heart cells. J Clin Invest 88:2106–2121
Masugi F, Ogihara T, Otsuka A, Saeki S, Kumahara Y (1984) Effect of l-alkyl-2-acetyl-*sn*-glycero-3-phosphorylcholine inhibitor on the reduction of one-kidney, one clip hypertension after unclipping in the rat. Life Sci 34:197–201
Masugi F, Sakaguchi K, Soeki S, Imaoka M, Ogihara T (1989) Dietary salt, blood pressure and circulating levels of l-O-hexadecyl-2-acetyl-sn-glycero-3-phosphocholine in patients with essential hypertension. J Lipid Med 1:341–348
Matsumoto K, Taki F, Kondoh Y, Taniguchi H, Takagi K (1992) Platelet-activating factor in bronchoalveolar lavage of patients with adult respiratory distress syndrome. Clin Exp Pharmacol Physiol 19:509–515
Matsuzaki M, Ishiguro Y, Tetsumoto T, Tsukada Y, Kayahara H, Tadasa K (1992) PAF acetylhydrolase activities in human systemic lupus erythematosus and lupus-prone mice. Clin Chim Acta 210:139–144

Mazzoni I, Morley J, Page CP, Sanjar S (1985) Induction of airway hyperreactivity by platelet-activating factor in the guinea pig. J Physiol Paris 365:107
McCormack DG, Barnes PJ, Evans TW (1989) Evidence against a role for PAF in hypoxic pulmonary vasoconstriction in the rat. Clin Sci 77:439–443
McGivern DV, Basran GS (1984) Synergism between platelet activating factor (PAF-acether) and prostaglandin E2 in man. Eur J Pharmacol 102:183–185
McGowan HM, Vandongen R, Codde JP, Croft KD (1986) Increased aortic PGI 2 and plasma lyso-PAF in the unclipped one-kidney hypertensive rat. Am J Physiol 251:H1361–H1364
McIntyre TM, Zimmerman GA, Satoh K, Prescott SM (1985) Cultured endothelial cells synthesize both platelet-activating factor and prostacyclin in response to histamine, bradykinin and adenosine triphosphate. J Clin Invest 76:271–280
McIntyre RC, Banerjee A, Bensard DD, Brew EC, Hahn AR, Fullerton DA (1994) Adenosine Al-receptor mechanisms antagonize $\beta$-adrenergic pulmonary vasodilation in hypoxia. Am J Physiol 267:H2179–H2185
McManus LM, Morley CA, Levine SP, Pinckard RN (1979) Platelet activating factor (PAF) induced release of platelet factor 4 (PF 4) in vitro during IgE anaphylaxis in the rabbit. J Immunol 123:2835–2841
McManus LM, Hanahan DJ, Demopoulos CA, Pinckard RN (1980) Pathobiology of the intravenous infusion of acetyl glyceryl etherphosphorylcholine (AGEPC), a synthetic platelet-activating factor (PAF), in the rabbit. J Immunol 124:2919–2924
McManus LM, Ostrom KK, Lear C, Luce EB, Gander DL, Pinckard RN, Redding SW (1993) Radiation-induced increased platelet-activating factor activity in mixed saliva. Lab Invest 68:118–124
McMurphy IF, Morris KG (1986) Platelet-activating factor causes pulmonary vasodilation in the rat. Am Rev Respir Dis 134:757–762
Meade CJ, Metcalfe S, Sovennsen R, Jamieson N, Watson C, Calne RY, Kleber G, Neild G (1991) Serum PAF acetylhydrolase and chronic cholestasis. Lancet 338:1016–1017
Mehta D, Gupta S, Gaur SN, Gangals V, Agrawal KP (1990) Increased leukocyte phospholipase A2 activity and plasma lysophosphatidylcholine levels in asthma and rhinitis and their relationship to airway sensitivity to histamine. Am Rev Respir Dis 142:157–161
Meizel S (1984) The importance of hydrolytic enzymes to an exocytotic event, the mammalian sperm acrosome reaction. Biol Rev Camb Philos Soc 59:125–157
Mencia-Huerta JM, Lewis RA, Razin E, Austen KF (1983) Antigen-initiated release of platelet-activating factor (PAF-acether) from mouse bone marrow-derived mast cells sensitized with monoclonal IgE. J Immunol 131:2958–2964
Mezzano S, Kunick M, Olavarria F, Ardilles L, Montrucchio G, Silvestro L, Biancone L, Camussi G (1993) Detection of platelet-activating factor in plasma of patients with streptococcal nephritis. Aliment Pharmacol Ther 7:357–367
Miadonna A, Tedeschi A, Arnoux B, Sala A, Zanussi C, Benveniste J (1989) Evidence of PAF-acether metabolic pathway activation in antigen challenge of upper respiratory airway. Am Rev Respir Dis 140:142–147
Michel L, Dubertet L (1992) Leukotriene B4 and platelet-activating factor in human skin. Arch Dermatol Res 284:S12–S17
Michel L, Mencia-Huerta JM, Benveniste J, Dubertet L (1987) Biologic properties of LT B4 and PAF-acether in vivo in human skin. J Invest Dermatol 88:675–681
Michel L, Denizot Y, Thomas Y, Benveniste J, Dubertet L (1988a) Release of paf-acether during allergic cutaneous reactions. Lancet II:404
Michel L, Denizot Y, Thomas J, Jean-Louis F, Pitton C, Benveniste J, Dubertet L (1988b) Biosynthesis of paf-acether by human skin fibroblasts. J Immunol 141:948–953
Michel L, Denizot Y, Thomas Y, Jean-Louis F, Heslan M, Benveniste J, Dubertet L (1990) Production of PAF-acether by human epidermal cells. J Invest Dermatol 95:576–581

Milhoun KA, Lane TA, Bloor CM (1992) Hypoxia induces endothelial cells to increase their adherence for neutrophils: role of PAF. Am J Physiol 263:H956–H962
Milligan SR, Finn CA (1990) Failure of platelet-activating factor (paf-acether) to induce decidualization in mice and failure of antagonists of paf to inhibit implantation. J Reprod Fertil 88:105–112
Minhas BS (1993) Platelet-activating factor treatment of human spermatozoa enhances fertilization potential. Am J Obstet Gynecol 168:1314–1317
Minhas BS, Ricker DD (1990) Enhancement of mouse IVF using platelet-activating factor: an examination of shorter gamete coincubation times and removal of cumulus cells. ARTA 2:75–87
Minhas BS, Kumar R, Ricker DD, Roudebush WE, Dodson MG, Fortunato SJ (1989) Effects of platelet activating factor on mouse oocyte fertilization in vitro. Am J Obstet Gynecol 161:1714–1717
Minhas BS, Kumar R, Ricker DD, Robertson JL, Dodson MG (1991) The presence of platelet-activating factor-like activity in human spermatozoa. Fertil Steril 55:372–376
Minor T, Isselhard W, Yamaguchi S (1995) Involvement of platelet activating factor in microcirculatory disturbances after global hepatic ischemia. J Surg Res 58:536–540
Mion F, Cuber JC, Minaire Y, Chayvialle JA (1994) Short term effects of indomethacin on rat small intestinal permeability. Role of eicosanoids and platelet activating factor. Gut 35:490–495
Misso NLA, Gillon RL, Taylor ML, Stewart GA, Thompson PJ (1993) Acetyl-CoA: lyso-platelet-activating factor acetyltransferase activity in neutrophils from asthmatic patients and normal subjects. Clin Sci 85:455–463
Miwa M, Miyake T, Yamanaka T, Sugatani J, Suzuki Y, Sakata S, Araki Y, Matsumoto M (1988) Characterization of serum platelet-activating factor (PAF) acetylhydrolase. Correlation between deficiency of PAF acetylhydrolase and respiratory symptoms in asthmatic children. J Clin Invest 82:1983–1991
Miwa M, Sugatani J, Ikemura T, Okamoto Y, Ino M, Saito K, Suzuki Y, Matsumoto M (1992) Release of newly synthesized platelet-activating factor (PAF) from human polymorphonuclear leukocytes under in vivo conditions. Contribution of PAF-releasing factor in serum. J Immunol 148:872–880
Miyaura S, Maki N, Byrd W, Johnston JM (1991) The hormonal regulation of platelet-activating factor acetylhydrolase activity in plasma. Lipids 26:1015–1020
Montrucchio G, Camussi G, Tetta C (1986) Intravascular release of platelet-activating factor during atrial pacing. Lancet 2:193–194
Montrucchio G, Alloatti G, Tetta C, de Luca R, Saunders RN, Emanuelli G, Camussi G (1989a) Release of platelet-activating factor from ischemic-reperfused rabbit heart. Am J Physiol 256:H1236–H1246
Montrucchio G, Mariano F, Cavalli PL, Tetta C, Viglino G, Emanuelli G, Camussi G (1989b) Platelet activating factor is produced during infectious peritonitis in CAPD patients. Kidney Int 36:1029–1036
Moqbel R, Walsh GM, Nagakura T, MacDonald AJ, Wardlaw AJ, Iiura Y, Kay AB (1990a) The effect of platelet-activating factor on IgE binding to, and IgE-dependent biological properties of human eosinophils. Immunology 70:251–257
Moqbel R, MacDonald AJ, Cromwell O, Kay AB (1990b) Release of leukotriene C4 (LTC4) from human eosinophils following adherence to IgE- and IgG-coated schistosomula of Schistosoma mansoni. Immunology 69:435–442
Mori M, Aihara M, Kume K, Hamanoue M, Kohsaka S, Shimizu T (1996) Predominant expression of platelet-activating factor receptor in the rat brain microglia. J Neurosci 16:3590–3600
Morita E, Schröder JM, Christophers E (1989) Chemotactic responsiveness of eosinophils isolated from patients with inflammatory skin diseases. J Dermatol 16:348–351
Morita K, Suemitsu T, Uchiyama Y, Miyasako T, Dohi T (1995) Platelet-activating factor mediated potentiation of stimulation-evoked catecholamine release and the

rise in intracellular free $Ca^{2+}$ concentration in adrenal chromaffin cells. J Lipid Mediat Cell Signal 11:219–230
Morley J, Page CP, Paul W (1983) Inflammatory actions of platelet activating factor (PAF-acether) in guinea pig skin. Br J Pharmacol 80:503–509
Morooka S, Uchida M, Imanishi N (1992) Platelet-activating factor (PAF) plays an important role in the immediate asthmatic response in guinea-pig by augmenting the response to histamine. Br J Pharmacol 105:756–762
Morteau O, More J, Pons L, Bueno L (1993) Platelet-activating factor and interleukin 1 are involved in colonic dysmotility in experimental colitis in rats. Gastroenterology 104:47–56
Movat HZ, Burrowes CE (1985) The local Shwartzman reaction: endotoxin-mediated inflammatory and thrombo-hemorrhagic lesions. In: Barry LJ (ed) Handbook of endotoxin 3: cellular biology of endotoxin. Elsevier, Amsterdam, pp 260–302
Mozes T, Heiligers JPC, Tak CJAM, Zijlstra FJ, Ben-Efraims S, Saxena PR, Bonta IL (1991) Platelet activating factor is one of the mediators involved in endotoxic shock in pigs. J Lipid Mediat 4:309–325
Mozes T, Ben-Efraim S, Bonta IL (1993) Lethal and non-lethal course of endotoxic shock is determined by interactions between tumor necrosis factor, platelet activating factor and eicosanoids. Pathol Biol (Paris) 40:807–812
Mueller HW, O'Flaherty JT, Wykle RL (1983) Biosynthesis of platelet activating factor in rabbit polymorphonuclear neutrophils. J Biol Chem 258:6213–6218
Muguruma K, Komatz Y, Ikeda M, Sugimoto T, Saito K (1993) Platelet-activating factor (PAF) in male reproductive organs of guinea pigs and rats. Effect of androgen on PAF in seminal vesicles. Biol Reprod 48:386–392
Muirhead E (1986) Renomedullary system of blood pressure control. Hypertension 8[Suppl I]:I38–I46
Mullane KM, Salmon JA, Kraemer R (1987) Leukocyte-derived metabolites of arachidonic acid in ischemia-induced myocardial injury. Fed Proc 46:2422–2433
Murphy JF, Bordet J-C, Wyler B, Rissoan M-C, Chomarat P, Defrance T, Miossec P, McGregor JL (1994) The vitronectin receptor ($\alpha_V\beta_3$) is implicated in cooperation with P-selectin and platelet-activating factor, in the adhesion of monocytes to activated-endothelial cells. Biochem J 304:537–542
Musemechi CA, Baker JL, Feddersen RM (1995) A model of intestinal ischemia in the neonatal rat utilizing superior mesenteric occlusion and intraluminal platelet-activating factor. J Surg Res 58:724–727
Mustafa SB, Gandhi CR, Harvey SAK, Olson MS (1995) Endothelin stimulates platelet-activating-factor synthesis by cultured rat Kupffer cells. Hepatology 21:545–553
Nakamura T, Morita Y, Kuriyama M, Ishihara K, Ito K, Miyamoto T (1987) Platelet-activating factor in late asthmatic response. Arch Allergy Appl Immunol 82:57–61
Nakamura M, Honda Z, Waga I, Matsumoto T, Noma M, Shimizu T (1992) Endotoxin transduces $Ca^{2+}$ signaling via platelet-activating factor receptors. FEBS Lett 314:125–129
Nakayama R, Yasuda K, Saito K (1987) Existence of endogenous inhibitors of platelet-activating factor (PAF) with PAF in rat uterus. J Biol Chem 262:13174–13179
Narahara H, Johnston JM (1993) Effects of endotoxins and cytokines on the secretion of platelet-activating factor acetylhydrolase by human decidual macrophages. Am J Obstet Gynecol 169:531–537
Nathan N, Denizot Y (1995) PAF and human cardiovascular disorders. J Lipid Mediat Cell Signal 11:103–104
Nieminen MM, Hill M, Irvin GG (1991a) Body temperature modulates the effect of platelet-activating factor (PAF) on airway responsiveness in the rabbit. Agents Actions 32:173–181
Nieminen MM, Moilanen E, Nyholm JE (1991b) Platelet-activating factor impairs tracheobronchial transport and increases plasma leukotriene B4 in man. Eur Respir J 4:561–570

Nieminen MM, Moilanen EK, Kokinen MO, Karvonen JI, Tuomisto L, Metsa-Ketela TJ, Vapaatalo H (1992) Inhaled budesonide fails to inhibit PAF-induced increase in plasma leukotriene B4 in man. Br J Clin Pharmacol 33:645–652
Nieminen MM, Henson PM, Irvin GG (1994) Bimodal effect of platelet-activating factor (PAF) on airway responsiveness in the rabbit. Exp Lung Res 20:559–577
Ninio E, Joly F, Hieblot C, Besson G, Mencia-Huerta JM, Benveniste J (1987) Biosynthesis of paf-acether IX. Role for a phosphorylation-dependent activation of acetyltransferase in antigen-stimulated mouse mast cells. J Immunol 139:154–160
Nishida K, Markey SP (1996) Platelet-activating factor in brain regions after transient ischemia in gerbils. Stroke 27:514–518
Nogami M, Matsunobu S, Termusa M (1990) The effect of platelet-activating factor on (14C) aminopyrine uptake by isolated guinea pig parietal cells. Biochem Biophys Res Commun 168:1047–1052
Nolan (1981) Endotoxin, reticuloendothelial function, and liver injury. Hepatology 1:458–465
Noris M, Benigni A, Boccardo P, Gotti E, Benfenati E, Aiello S, Todeschini M (1993) Urinary excretion of platelet-activating factor in patients with immune-mediated glomerulonephritis. Kidney Int 43:426–429
Northover AM (1989) The effects of TMB-8 on the shape changes of vascular endothelial cells resulting from exposure to various inflammatory agents. Agents Actions 26:367–371
Nourshagh S, Lerkin SW, Das A, Williams TJ (1995) Interleukin-1-induced leukocyte extravasation across mesenteric microvessels is mediated by platelet-activating factor. Blood 85:2553–2558
Oberpichler H, Sauer D, Roßberg C, Mennel HD, Krieglstein J (1990) PAF antagonist Ginkgolide B reduces postischemic neuronal damage in rat brain hippocampus. J Cereb Blood Flow Metab 10:133–135
O'Connor BJ, Ridge SM, Chen-Worsdell YM, Uden S, Barnes PJ, Chung KF (1991) Complete inhibition of airway and neutrophil responses to inhaled platelet-activating factor (PAF) by an oral PAF antagonist UK 74505. Am Rev Respir Dis 143:A156
O'Donnell SR, Barnett CJK (1987) Microvascular leakage to platelet activating factor in guinea-pig trachea and bronchi. Eur J Pharmacol 138:385–396
O'Flaherty JT, Wykle RL, Mueller CH, Lewis JC, Waite M, Bass DA, McCall CE, de Chatelet LR (1981) 1-O-Alkyl-sn-glyceryl-3-phosphorylcholines: a novel class of neutrophil stimulants. Am J Pathol 103:70–78
Ohar JA, Pyle JA, Hyers TM, Webster RO (1986) AGEPC-induced pulmonary vascular injury. A rabbit model of pulmonary hypertension. Am Rev Respir Dis 133:A159
Ohar JA, Pyle JA, Walter KS, Myers TM, Webster RO, Lagunoff DA (1990) A rabbit model of pulmonary hypertension induced by the synthetic platelet-activating factor acetylglyceryl ether phosphorylcholine. Am Rev Respir Dis 141:104–110
Ohar JA, Waller KS, Dahms TE (1993) Platelet-activating factor induces selective pulmonary arterial hyperreactivity in isolated perfused rabbit lungs. Am Rev Respir Dis 148:158–163
Ohlsson K, Björk P, Bergenfeldt M, Hageman R, Thompson RC (1990) Interleukin-1 receptor antagonist reduces mortality from endotoxin shock. Nature 348:550–552
Okamoto Y, Yoshida H, Ino M, Nakamura S-I, Okamoto G, Mizoguchi K, Suzuki Y, Sugatani J, Miwa M (1993) PAF-releasing factor in human serum and inflammatory exudate. J Lipid Mediat 8:151–156
Okusawa S, Gelfand JA, Ikejima T, Connolly RJ, Dinarello CA (1988) Interleukin 1 induces a shock-like state in rabbits. Synergism with tumor necrosis factor and the effect of cyclooxygenase inhibition. J Clin Invest 81:1162–1172
Olaison G, Sjödahl R, Tagesson C (1988) Increased phospholipase A2 activity of ileal mucosa in Crohn's disease. Digestion 41:136–141

Olson NC, Joyce PB, Fleisher LN (1990a) Role of platelet-activating factor and eicosanoids during endotoxin-induced lung injury in pigs. Am J Physiol 258:H1674–H1686
Olson NC, Joyce PB, Fleisher LN (1990b) Mono-hydroxyeicosatetraenoic acids during porcine endotoxemia: effect of a platelet-activating factor receptor antagonist. Lab Invest 63:221–231
O'Neill C (1985) Examination of the cause of early pregnancy-associated thrombocytopenia in mice. J Reprod Fertil 73:567–577
O'Neill C (1991) A physiological role for PAF in the stimulation of mammalian embryonic development. Trends Pharmacol Sci 12:82–84
O'Neill C (1992) Embryo-derived platelet activating factor. Reprod Fertil Dev 4:283–288
O'Neill (1995a) Platelet-activating factor-antagonists reduce implantation in mice at low doses only. Reprod Fertil Dev 7:51–58
O'Neill C (1995b) Activity of platelet-activating factor acetylhydrolase in the mouse uterus during the estrous cycle, throughout the preimplantation phase of pregnancy, and throughout the luteal phase of pseudopregnancy. Biol Reprod 52:965–971
O'Neill C, Collier M, Ammit AJ, Ryan JP, Saunders DM, Pike IL (1989) Supplementation of in-vitro fertilisation culture medium with platelet activating factor. Lancet II:769–772
O'Neill C, Wells X, Battye K (1990) Embryo-derived platelet activating factor: Interactions with the arachidonic acid cascade and the establishment and maintenance of pregnancy. Reprod Fertil Dev 2:423–441
Oparil S, Chen S-J, Meng QC, Elton TS, Yano M, Chen Y-F (1995) Endothelin-A receptor antagonist prevents acute hypoxia-induced pulmonary hypertension in the rat. Am J Physiol 268:L95–L100
Orozco C, Perkins T, Clarke FM (1986) Platelet-activating factor induces the expression of early pregnancy factor in female mice. J Reprod Fertil 78:549–555
Ostermann G, Lang A, Holz H, Rühling K, Winkler L, Till U (1988) The degradation of platelet-activating factor in serum and its discriminative value in atherosclerotic patients. Thromb Res 52:529–540
Ou MC, Kambayashi J, Kawasaki T, Uemura Y, Shinozaki K, Shiba E, Sakon M, Yukawa M, Mori T (1994) Potential etiologic role of PAF in two major septic complications: disseminated intravascular coagulation and multiple organ failure. Thromb Res 73:227–238
Page CP (1988) The role of platelet-activating factor in asthma. J Allergy Clin Immunol 81:144–149
Page CP (1990) The role of platelet activating factor in allergic respiratory disease. Br J Clin Pharmacol 30:995–1065
Page C, Abbott A (1989) PAF: new antagonists, new roles in disease and a major role in reproductive biology, Trends Pharmacol Sci 10:255–257
Page CP, Paul W, Morley J (1984) Platelets and bronchospasm. Int Arch Allergy Appl Immunol 74:347–350
Page CP, Guerreiro D, Sanjar S, Morley J (1985) Platelet activating factor (PAF-acether) may account for late-onset reactions to allergen inhalation. Agents Actions 16:30–32
Panetta T, Marcheselli VL, Braquet P, Spinnewyn B, Bazan NG (1987) Effects of platelet activating factor antagonist (BN 52021) on free fatty acids, diacylglycerols, polyphosphoinositides and blood flow in the gerbil brain: inhibition of ischemia-reperfusion-induced cerebral injury. Biochem Biophys Res Commun 149:580–587
Parente L, Flower RJ (1985) Hydrocortisone and macrocortin inhibit zymosan-induced release of lyso-PAF from rat peritoneal leucocytes. Life Sci 36:1225–1231
Parsons PE, Worthen SG, Moore EE, Tate RM, Henson PM (1989) The association of circulating endotoxin with development of the adult respiratory distress syndrome. Am Rev Respir Dis 140:294–300

Parks JE, Graham JK (1992) Effects of cryopreservation procedures on sperm membranes. Theriogenol 38:210–223
Parks JE, Lynch DV (1992) Lipid composition and thermotropic phase behavior of boar, bull, stallion, and rooster sperm membranes. Cryobiology 29:255–266
Parks JE, Hough SR (1993) Platelet-activating factor acetylhydrolase activity in bovine seminal plasma. J Androl 14:335–339
Parks JE, Hough S, Elrod C (1990) Platelet activating factor activity in the phospholipids of bovine spermatozoa. Biol Reprod 43:806–811
Patterson R, Bernstein PR, Harris KE, Krell RD (1984) Airway responses to sequential challenges with platelet-activating factor and leukotriene D4 to rhesus monkeys. J Lab Clin Med 104:340–345
Peplow PV, Mikhailidis DP (1990) Platelet-activating factor (PAF) and its relation to prostaglandins, leukotrienes and other aspects of arachidonate metabolism. Prostaglandins Leukot Essent Fatty Acids 41:71–82
Perico N, Delaini F, Tagliaferri M, Abbate M, Cucchi M, Bertani T, Remuzzi G (1988) Effect of platelet activating factor and its specific receptor antagonist on glomerular permeability to proteins in isolated perfused rat kidney. Lab Invest 58:163–171
Perretti F, Manzini S (1992) Activation of capsaicin-sensitive sensory fibres modulates PAF-induced bronchial hyperresponsiveness in anesthetized guinea pigs. Am Rev Respir Dis 148:927–931
Persson CGA, Erjefalt I, Sundler F (1987) Airway microvascular and epithelial leakage of plasma induced by PAF-acether (PAF) and capsaicin. Am Rev Respir Dis 135:A401
Petroni A, Salami M, Blasevich M, Galli C (1993) Activation of phospholipid hydrolysis and generation of eicosanoids in cultured rat astroglial cells by PAF, and modulation by n-3 fatty acids. In: Massarelli R, Horrocks LA, Kanfer JN, Löffelholz K (eds) Phospholipids and their signal transmission. Springer-Verlag, Berlin Heidelberg New York, pp 403–410
Pettipher ER, Higgs GA, Henderson B (1987) PAF-acether in chronic arthritis. Agents Actions 21:98–103
Pignol B, Henane S, Mencia-Huerta J-M, Rola-Pleszczynski M, Braquet P (1987a) Effect of platelet-activating factor (PAF-acether) and its specific receptor antagonist, BN 52021, on interleukin 1 (IL-1) release and synthesis by rat adherent monocytes. Prostaglandins 33:931–939
Pignol B, Henane S, Mencia-Huerta JM, Rola-Pleszczynski M, Braquet P (1987b) Inhibition of IL2 production and lymphocyte proliferation by platelet activating factor (PAF-acether) or analogs and reversal by the specific antagonists BN 52020 and BN 52021. Fed Proc 46:2720 (abstr)
Pignol B, Lonchampt MO, Chabrier PE, Mencia-Huerta JM, Braquet P (1990) Platelet-activating factor potentiates interleukin-1/epidermal cell-derived thymocyte-activating factor release by guinea-pig keratinocytes stimulated with lipopolysaccharide. J Lipid Mediat 2[Suppl]:S83–S91
Pinckard RN, Halonen M, Palmer JD, Butler C, Shaw JO, Henson PA (1977) Intravascular aggregation and pulmonary sequestration of platelets during IgE-induced systemic anaphylaxis in the rabbit: abrogation of lethal anaphylactic shock by platelet depletion. J Immunol 119:2185–2193
Pinckard RN, Farr RS, Hanahan DJ (1979) Physicochemical and functional identity of rabbit platelet-activating factor (PAF) released in vivo during IgE anaphyalaxis with PAF released in vitro from IgE sensitized basophils. J Immunol 123:1847–1857
Pinckard RN, Kniker WT, Lee L, Hanahan DJ, McManus LM (1980) Vasoactive effects of 1-O-alkyl-2-acetyl-sn-glyceryl-3-phosphorylcholine (AcGEPC) in human skin. J Allergy Clin Immunol 65:196
Pinckard RN, Woodard DS, Showell HJ, Conklyn MJ, Novak MJ, McManus LM (1994) Structural and (patho) physiological diversity of PAF? Clin Rev Allergy 12:329–359

Pinheiro JMB, Pitt BR, Gillis CN (1989) Roles of platelet-activating factor and thromboxane in group B streptococcus-induced pulmonary hypertension in piglets. Pediatr Res 26:420–424

Pinto A, Calignano A, Mascolo N, Autore G, Capasso F (1989) Castor oil increases intestinal formation of platelet-activating factor and acid phosphatase release in the rat. Br J Pharmacol 96:872–874

Pinto A, Autore G, Mascolo N, Sorrentino R, Biondi A, Izzo AA, Capasso F (1992) Time course of PAF formation by gastrointestinal tissue in rats after castor oil challenge. J Pharm Pharmacol 44:224–226

Pires AL, deSilva PM, Pasquale C, Castro-Faria-Neto HC, Bozza PT, Cordeiro RS, Rae GA, Braquet P, Lagente V, Martins MA (1994) Long-lasting inhibitory activity of the hetrazepinic BN 50730 on exudation and cellular alterations evoked by PAF and LPS. Br J Pharmacol 113:994–1000

Pirotzky E, Page CP, Roubin R, Pfister A, Paul W, Bonnet J, Benveniste J (1984) PAF-acether-induced plasma exudation in rat skin is independent of platelets and neutrophils. Microcirc Endothel Lymphat 1:107–122

Pirotzky E, Page C, Morley J, Bidault J, Benveniste J (1985a) Vascular permeability induced by PAF-acether (platelet-activating factor) in the isolated perfused rat kidney. Agents Actions 16:17–18

Pirotzky E, Pfister A, Benveniste J (1985b) A role for PAF-acether (platelet-activating factor) in acute skin inflammation. Br J Dermatol 113[Suppl 28]:91–94

Podolski DK (1991) Inflammatory bowel disease. N Engl J Med 325:928–937

Pons L, Droy-Lefaix MT, Braquet P, Bueno L (1991) Role of ree radicals and platelet-activating factor in the genesis of intestinal motor disturbances induced by Escherichia coli endotoxins in rat. Gastroenterology 100:946–953

Pons F, Rossi AG, Norman KE, Williams TJ, Nourshargh S (1993) Role of platelet-activating factor (PAF) in platelet accumulation in rabbit skin: effect of the novel long-acting PAF antagonist, UK-74505. Br J Pharmacol 109:234–242

Pons L, Droy-Lefaix MT, Bueno L (1994) Role of platelet-activating factor (PAF) and prostaglandins in colonic motor and secretory disturbances induced by Escherichia coli endotoxin in conscious rats. Prostaglandins 47:123–136

Porres-Reyes BH, Mustone TA (1992) Platelet-activating factor improvement in wound healing by a chemotactic factor. Surgery 111:416–423

Prescott SM, Zimmerman GA, McIntyre TM (1984) Human endothelial cells in culture produce platelet-activating factor (1-alkyl-2-acetyl-sn-glycero-3-phosphocholine) when stimulated with thrombin. Proc Natl Acad Sci USA 81:3534–3538

Prescott SM, Zimmerman GA, McIntyre TM (1990) Platelet-activating factor. J Biol Chem 265:17381–17384

Prevost MC, Cariven C, Simon MF, Chap H, Douste-Blazy L (1984) Platelet activating factor (PAF-acether) is released into rat pulmonary alveolar fluid as a consequence of hypoxia. Biochem Biophys Res Commun 119:58–63

Quinn MT, Parthasarathy S, Steinberg S (1988) Lysophosphatidylcholine: a chemotactic factor for human monocytes and its potential role in atherogenesis. Proc Natl Acad Sci USA 85:2805–2809

Qvist R, Morii H, Watanabe Y (1995) Interleukin-1$\beta$ stimulates phospholipase A2 activity in rat C6 glioma cells. Biochem Mol Biol Int 35:363–370

Rabier M, Damon M, Chanez P, Mencia-Huerta JM, Braquet P, Michel FB, Godard P (1991) Neutrophil chemotactic activity of PAF, histamine and neuromediators in bronchial asthma. J Lipid Mediat 4:265–275

Rabinovici R, Esser KM, Lysko PG, Yue T-L, Griswold DE, Hillegass LM, Bugelski PJ, Hallenbeck JM, Feuerstein G (1991) Priming by platelet-activating factor of endotoxin-induced lung injury and cardiac vascular shock. Circ Res 69:12–25

Rabinovici R, Bugelski PJ, Esser KM, Hillegass LM, Vernick J, Feuerstein G (1993a) ARDS-like lung injury produced by endotoxin in platelet-activating factor-primed rats. J Appl Physiol 74:1791–1802

Rabinovici R, Bugelski PJ, Esser KM, Hillegass LM, Griswold DE, Vernick J, Feuerstein G (1993b) Tumor necrosis factor-alpha mediates endotoxin-induced lung injury in platelet activating factor-primed rats. J Pharmacol Exp Ther 267:1550–1557

Rainsford KD (1986) Relative roles of leukotrienes and platelet activating factor in experimentally-induced gastric ulceration. Pharmacol Res Commun 18[Suppl]:209–215

Ranaut K, Singh M (1993) BN-50739: a PAF antagonist and limitations of myocardial infarct size. Methods Find Exp Clin Pharmacol 15:9–14

Rao VS, Fonteles MC (1991) Involvement of PAF-acether and eicosanoids in adrenaline-induced pulmonary edema in mice. Braz J Med Biol Res 24:319–321

Remuzzi G (1989) Eicosanoids and platelet-activating factor as possible mediators of injury in experimental nephropathies. Adv Exp Med Biol 259:221–247

Ricker DD, Minhas BS, Kumar R, Robertson JL, Dodson MG (1989) The effects of PAF on the motility of human spermatozoa. Fertil Steril 52:655–658

Roca J, Felez MA, Chung KF, Barbesa JA, Rotger M, Santos C, Rodriguez-Roisin R (1995) Salbutamol inhibits pulmonary effects of platelet activating factor in man. Am J Respir Crit Care Med 151:1740–1744

Rodriguez-Bartero A, Rodriguez-Lopez AM, Gonzalez-Sarmiento R, Lopez-Novoa JM (1995) Gentamicin activates rat mesangial cells: a role for platelet activating factor. Kidney Int 47:1346–1353

Rodriguez-Roisin RR, Roca J, Chung KF (1991) Effects of inhaled platelet-activating factor on pulmonary gas exchange in man. Am Rev Respir Dis 143:A771

Rogers DF, Alton EWFW, Aursudkij B, Boschetto P, Dewar A, Barnes PJ (1987) Effect of platelet activating factor on formation and composition of airway fluid in the guinea pig trachae. J Physiol 431:643–658

Rola-Pleszczynski M, Pignol B, Pouliot C, Braquet P (1987) Inhibition of human lymphocyte proliferation and interleukin 2 production by platelet activating factor (PAF-acether): reversal by a specific antagonist, BN 52021. Biochem Biophys Res Commun 142:754–760

Rola-Pleszczynski M, Bosse J, Bissonnette E, Dubois C (1988a) PAF-acether enhances the production of tumor necrosis factor by human and rodent lymphocytes and macrophages. Prostaglandins 35:802–806

Rola-Pleszczynski M, Pouliot C, Turcotte S, Pignol B, Braquet P, Bouvette L (1988b) Immune regulation of platelet-activating factor I. Induction of suppressor cell activity in human monocytes and CD8+ cells and of helper cell activity in CD4+ T cells. J Immunol 140:3547–3552

Rosam AC, Wallace JL, Whittle BJR (1986) Potent ulcerogenic actions of platelet-activating factor on the stomach. Nature 319:54–56

Rossi AG, Norman KE, Donigi-Gale D, Shoupe TS, Edwards R (1992) The role of complement, PAF and leukotriene B4 in a reversed passive Arthus reaction. Br J Pharmacol 107:44–49

Roubin RR, Tence M, Mencia-Huerta JM, Arnoux B, Ninio E, Benveniste J (1983) A chemically defined monokine – macrophage-derived platelet-activating factor. Lymphokines 8:249–276

Roudebush WE, Minhas BS, Ricker DD, Palmer TV, Dodson MG (1990) Platelet activating factor enhances in vitro fertilization of rabbit oozytes. Am J Obstet Gynecol 163:1670–1673

Rubin AH, Smith LJ, Patterson R (1987) The bronchoconstrictor properties of platelet-activating factor in humans. Am Rev Respir Dis 136:1145–1151

Ruggiero V, Chiapparino C, Manganello S, Pacello L, Foresta P, Martelli EA (1994) Beneficial effects of a novel platelet-activating factor receptor antagonist, ST 899, on endotoxin-induced shock in mice. Shock 2:275–280

Ruiz-Torres P, Gonzalez-Rubio M, Lucio-Cazana FJ, Ruiz-Villaespasa A, Rodriguez-Puyol M, Rodriguez-Puyol D (1994) Reactive oxygen species and platelet-activating factor synthesis in age-related glomerulosclerosis. J Lab Clin Med 124:489–495

Ryan JP, Spinks NR, O'Neill C, Ammit AJ, Wales RG (1989) Platelet-activating factor (PAF) production by mouse embryos in vitro and its effect on embryonic metabolism. J Cell Biochem 40:387–395

Sagasch VP, Zhukova AV, Braquet P (1992) Endothelium-dependent effects of platelet-activating factor in the coronary circulation. J Cardiovasc Pharmacol 20[Suppl 12]:S85–S89

Saito M, Sato R, Hisatome I, Narahashi T (1996) RANTES and platelet-activating factor open $Ca^{2+}$-activated K+ channels in eosinophils. FASEB J 10:792–798

Sakaguchi K, Masugi F, Chen YH, Inoue M, Ogihara T, Yamada K, Yamatsu I (1989) Direct evidence of elevated levels of circulating platelet-activating factor in a patient with disseminated intravascular coagulation syndrome. J Lipid Mediat 1:171–173

Sakai A, Chang S-W, Voelkel NF (1989) Importance of vasoconstriction in lipid-mediator-induced pulmonary oedema. J Appl Physiol 66:2667–2674

Sakurai T, Yamaguchi S, Iwama T, Nagai H (1994) Pharmacological studies of platelet-activating factor (PAF)-induced augmentation of response to histamine in guinea-pigs. Prostaglandins Leukot Essent Fatty Acids 51:95–99

Salem P, Deryckx S, Dulioust A, Vivier E, Denizot Y, Damais C, Dinarello CA, Thomas Y (1990) Immunoregulatory functions of paf-acether IV. Enhancement of IL-1 production by muramyl dipeptide-stimulated monocytes. J Immunol 144: 1388–1344

Salinas P, Perez MD, Fernandez-Sanpablo R, Fernandez-Gallardo S, Sanchez-Crespo M, Barrigon S (1995) Lack of platelet-activating factor release on acute myocardial ischemia in isolated ventricular septum rabbit heart. Eur J Pharmacol 293:65–70

Samet JM, Noah TL, Devlin RB, Yankaskeas JR, McKinnon K, Dailey LA, Friedman M (1992) Effect of ozone on platelet-activating factor production in phorbol-differentiated HL 60 cells, a human bronchial epithelial cell line (BEAS 56), and primary human bronchial epithelial cells. Am J Respir Cell Mol Biol 7:514–522

Sanchez-Crespo M, Alonso F, Inarrea P, Alvarez V, Egido J (1982) Vascular actions of synthetic PAF-acether (a synthetic platelet-activating factor) in the rat: evidence for a platelet independent mechanism. Immunopharmacology 4:173–185

Santing RE, Hoekstra Y, Pasman Y, Zaagsma J, Meurs H (1994) The importance of eosinophil activation for the development of allergen-induced bronchial hyperactivity in conscious, unrestrained guinea-pigs. Clin Exp Allergy 24:1157–1163

Sanwick JM, Talaat RE, Kuzan FB, Geissler FT, Chi EY, Henderson WR (1992) Human spermatozoa produce C16 PAF. Arch Biochem Biophys 295:214–216

Sanz MJ, Weg VB, Bolanowski MA, Nourshargh S (1995) IL-1 is a potent inducer of eosinophil accumulation in rat skin. Inhibition of response by a platelet-activating factor antagonist and an anti-human IL-8 antibody. J Immunol 154:1364–1373

Sasaki T, Shimura S, Ikada K, Takishima H (1989) Platelet-activating factor increases platelet-dependent glycoconjugate secretion from tracheal submucosal glands. Am J Physiol 257:L373–L378

Satoh K, Imaizumi T, Kawamura Y, Yoshida H, Takamatsu S, Mizumo S (1988) Activity of platelet-activating factor (PAF) acetyl hydrolase in plasma from patients with ischemic cerebrovascular disease. Prostaglandins 35:685–698

Satoh K, Imaizumi TA, Kawamura Y, Yoshida H, Takamatsu S, Takamatsu M (1989) Increased activity of the platelet-activating factor acetylhydrolase in plasma low density lipoprotein from patients with essential hypertension. Prostaglandins 37:673–682

Satoh K, Imaizumi T, Yoshida H, Hiramoto M, Takamatsu S (1992a) Increased levels of blood platelet-activating factor (PAF) and PAF-like lipids in patients with ischemic stroke. Acta Neurol Scand 85:122–127

Satoh K, Yoshida H, Imaizumi T, Takamatsu S, Mizuno S (1992b) Platelet-activating factor acetylhydrolase in plasma lipoproteins from patients with ischemic stroke. Stroke 23:1090–1092

Sawa Y, Schaper J, Roth M, Nagasawa K, Ballagi G, Bleese N, Schaper W (1994) Platelet-activating factor plays an important role in reperfusion injury in myocardium. Efficacy of platelet-activating factor receptor antagonist (CV-3988) as compared with leukocyte-depleted reperfusion. J Thorac Cardiovasc Surg 108:953–959

Schauer U, Koch B, Michl U, Jager R, Rieger CHL (1992) Enhanced production of platelet activating factor by peripheral granulocytes from children with asthma. Allergy 47:143–149

Scherf H, Nies AS, Schwertschlag U, Hughes M, Gerber JG (1986) Hemodynamic effects of platelet activating factor in the dog kidney in vivo. Hypertension 8:737–741

Schlondorff D, Neuwirth R (1986) Platelet-activating factor and the kidney. Am J Physiol 251:F1–F11

Schlondorff D, Goldwasser P, Neuwirth R, Satriano J, Clay K (1986) Production of platelet-activating factor in glomeruli and cultured glomerular mesangial cells. Am J Physiol 250:F1123–F1127

Schmidt J, Lindstaedt R, Szelenyi I (1992) Characterization of platelet-activating factor induced superoxide generation by guinea-pig alveolar macrophages. J Lipid Mediat 5:13–22

Schwertschlag U, Scherf H, Gerber JG, Mathias M, Nies AS (1987) Platelet activating factor induces changes on renal vascular resistance, vascular reactivity, and renin release in the isolated perfused rat kidney. Circ Res 60:534–539

Seale JP, Nourshargh S, Hellewell PG, Williams TJ (1991) Mechanism of action of platelet activating factor in the pulmonary circulation: an investigation using a novel isotopic system in rabbit isolated lung. Br J Pharmacol 104:251–257

Selivonchick DP, Schmid PC, Natarajan V, Schmid HO (1980) Structure and metabolism of phospholipids in bovine epididymal sperm. Biochim Biophys Acta 618:242–254(6)

Sengoku K, Ishikawa M, Tamate K, Shimizu T (1992) Effect of platelet activating factor on mouse sperm function. J Assist Reprod Genet 9:447–453

Sengoku K, Tamate K, Takaoka Y, Ishikawa M (1993) Effects of platelet activating factor on human sperm function invitro. Hum Reprod 8:1443–1447

Seow WK, Thong YH, Ferrante A (1987) Macrophage-neutrophil interactions: Contrasting effects of interleukin-1 and tumour necrosis factor-alpha (cachectin) on human neutrophil adherence. Immunology 62:357–361

Seth P, Kumari R, Dikshit M, Srimal RC (1994) Effect of platelet activating factor antagonists in different models of thrombosis. Thromb Res 76:503–512

Sharma JN, Mrhsin SS (1990) The role of chemical mediators in the pathogenesis of inflammation with emphasis on the kinin system. Exp Pathol 38:73–96

Shasby DM, Shasby SS, Sullivan JM, Peach MJ (1982) Role of endothelial cell cytoskeleton in control of endothelial permeability. Circ Res 51:657–661

Shaw JO, Pinckard RN, Ferrigini KS, McManus LM, Hanahan DJ (1981) Activation of human neutrophils with 1-O-hexadecyl/octadecyl-2-acetyl-sn-glyceryl-3-phosphorylcholine (platelet activating factor). J Immunol 127:1250–1253

Shindo K, Koide K, Hirai Y, Sumitomo M, Fukunura M (1996) Priming effect of platelet activating factor on leukotriene C4 from stimulated eosinophils of asthmatic patients. Thorax 51:155–158

Shukla SD, Buxton BD, Olson MS, Hanahan DJ (1983) Acetylglycerol ether phosphocholine. A potent activator of hepatic phosphoinositide metabolism and glycogenolysis. J Biol Chem 258:10212–10214

Silverberg KM, Groff TR, Riehl RM, Harper MJK, Olive DL, Schenker RS (1990) Platelet activating factor does not enhance motility of fresh or cryopreserved human sperm. 37th Annual Meeting Soc Gyn Invest (abstr 258)

Siminiak T, Egdell RM, O'Gorman DJ, Dye JF, Sheridan DJ (1995) Plasma-mediated neutrophil activation during acute myocardial infarction: role of platelet-activating factor. Clin Sci 89:171–176

Simon HU, Blaser K (1995) Inhibition of programmed eosinophil death: a key pathogenic event for eosinophilia? Immunol Today 16:53–55

Simon HU, Tsao PW, Siminovitch KA, Mills GB, Blaser K (1994a) Functional platelet-activating-factor receptors are expressed by monocytes and granulocytes but not by resting or activated T and B lymphocytes from normal individuals or patients with asthma. J Immunol 153:364–377

Simon HU, Grotzer M, Nikolaizik WH, Blaser K, Schoni MH (1994b) High altitude climate therapy reduces peripheral blood T lympocyte activation, eosinophilia, and bronchial obstructions in children with house-dust allergic asthma. Pediatr Pulmonol 17:304–311

Sirois MG, Jancar S, Braquet P, Plante GE, Sirois P (1988) PAF increases vascular permeability in selected tissues: effect of BN 52021 and L 655240. Prostaglandins 36:631–644

Sirois MG, Tavares de Lima W, de Brun Fernandez AJ, Johnson RJ, Plante GE, Sirois P (1994) Effect on rat lung vascular permeability: role of platelets and polymorphonuclear leukocytes. Br J Pharmacol 111:1111–1116

Sisson JH, Prescott SM, McIntyre TM, Zimmerman GA (1987) Production of platelet-activating factor by stimulated polymorphonuclear human leukocytes. Correlation of synthesis with release, functional events, and leukotriene B4 metabolism. J Immunol 138:3918–3926

Smith H (1992) Asthma, inflammation, eosinophils and bronchial hyperresponsiveness. Clin Exp Allergy 22:187–192

Smith KA, Prewitt RL, Byers LW, Muirhead EE (1981) Analogs of phosphatidylcholine: alpha-adrenergic antagonists from the renal medulla. Hypertension 3:460–470

Smith LJ, Rubin AHE, Patterson R (1988) Mechanism of platelet activating factor-induced bronchoconstriction in humans. Am Rev Respir Dis 137:1015–1019

Smith SK, Kelly RW (1988) Effect of platelet-activating factor on the release of PGF-2alpha and PG E2 by separated cells of human endometrium. J Reprod Fertil 82:271–276

Snyder F (1990) Platelet-activating factor and related acetylated lipids as potent biologically active cellular mediators. Am J Physiol 259:C697–C708

Sobhani I, Bado A, Denizot Y, Mignon M, Bonfils S, Lewin MJM (1990) Gastric secretion of platelet activating factor in cat. Gastroenterology 98:A128

Sobhani I, Denizot Y, Vatier J, Vissuzaine C, René E, Benveniste J, Bonfils S, Mignon M (1992a) Significance and regulation of gastric secretion of PAF in man. Dig Dis Sci 37:1583–1592

Sobhani I, Hochlaf S, Denizot Y, Vissuzaine C, René E, Benveniste J, Lewin MJM, Mignon M (1992b) Raised concentrations of platelet-activating factor in colonic mucosa of Crohn's disease. Gut 33:1220–1225

Sobhani I, Bado A, Moizo L, Laigneau J-P, Tarrade T, Braquet P, Lewin MJM (1995) Platelet-activating factor stimulates gastric acid secretion in isolated rabbit gastric glands. Am J Physiol 268:G889–G894

Sobhani I, Denizot Y, Moizo L, Laigneau JP, Bado A, Laboisse C, Benveniste J, Lewin MJM (1996a) Regulation of platelet-activating factor production in gastric epithelial cells. Eur J Clin Invest 26:53–58

Sobhani I, Bado A, Cherifi Y, Morizo L, Laigneau JP, Pospai D, Mignon M, Lewin MJ (1996b) Helicobacter pylori stimulates gastric acid secretion via platelet activating factor. J Physiol Pharmacol 47:177–186

Söling HD, Fest W (1986) Synthesis of 1-O-alkyl-2-acetyl-sn-glycero-3-phosphocholine (platelet activating factor) in exocrine glands and its control by secretagogues. J Biol Chem 261:13916–13922

Söling HD, Eibl H, Fest W (1984) Acetylcholine-like effects of 1-*O*-alkyl-2-acetyl-*sn*-glycero-3-phosphocholine ("platelet-activating factor") and its analogues in exocrine secretory glands. Eur J Biochem 144:65–72

Sogos V, Bussolino F, Pilia E, Torelli S, Gremo F (1990) Acetylcholine-induced production of PAF by human fetal brain cells in culture. J Neurosci Res 27:706–711
Soifer SJ, Schreiber MD (1988) Arachidonic acid metabolites mediate pulmonary hypertension caused by platelet activating factor in newborn lambs. Ped Res 23:525A
Soloviev AI, Braquet P (1992) Platelet-activating factor – a potent endogenous mediator responsible for coronary vasospasm. News Physiol Sci 7:166–172
Spence DPS, Johnston SL, Calverley PM, Dhillon P, Higgins C, Winning A, Winter J, Holgate ST (1994) The effect of orally active platelet-activating factor antagonist WEB 2086 in the treatment of asthma. Am J Respir Crit Care Med 149:1142–1148
Spencer DA (1992) An update on PAF. Clin Exp Allergy 22:521–524
Spencer DA, Green SE, Evans JM, Piper PJ, Costello JF (1990) Platelet-activating factor does not cause a reproducible increase in bronchial responsiveness in normal man. Clin Exp Allergy 20:525–532
Spencer DA, Evans JM, Green SE, Piper PJ, Costello JF (1991) Participation of the cysteinyl-leukotrienes in the acute bronchoconstrictor response to inhaled PAF in man. Thorax 46:441–445
Spina D, McKenniff MG, Coyle AJ, Seeds EAM, Tramontana M, Perretti F, Manzini S, Page CP (1991) Effect of capsaicin on PAF-induced bronchial hyper responsiveness and pulmonary cell accumulation in rabbit. Br J Pharmacol 103:1268–1274
Spinks NR, Ryan JP, O'Neill C (1990) Antagonists of embryo-derived platelet-activating factor act by inhibiting the ability of the mouse embryo to implant. J Reprod Fertil 88:241–248
Spinnewyn B, Blavet N, Clostre F, Bazan N, Braquet P (1987) Involvement of platelet-activating factor (PAF) in cerebral post-ischemic phase in mongolian gerbils. Prostaglandins 34:337–349
Spry CF (1988) Eosinophils. A comprehensive review, and guide to the scientific and medical literature. Oxford University Press, Oxford
Squinto SP, Block AL, Braquet P, Bazan NG (1989) Platelet-activating factor stimulates a fos/jun/AP-1 transcriptional signaling system in human neuroblastoma cells. J Neurosci Res 24:558–566
Stafforini DM, Satoh K, Atkinson DL, Tjoelker LW, Eberhardt C, Yoshida H, Imaizumi T, Takamatsu S, Zimmerman GA, McIntyre TM (1996) Platelet-activating factor acetylhydrolase deficiency. A missense mutation near the active site of an anti-inflammatory phospholipase. J Clin Invest 97:2784–2791
Stahl G, Lefer DJ, Lefer AM (1987) PAF-acether induced cardiac dysfunction in the isolated perfused guinea pig heart. Arch Pharmacol 336:459–463
Stahl GL, Craft DV, Lento PH, Lefer AM (1988a) Detection of platelet-activating factor during traumatic shock. Circ Shock 26:237–244
Stahl GL, Bitterman H, Terashita Z, Lefer AM (1988b) Salutary consequences of blockade of platelet activating factor in hemorrhagic shock. Eur J Pharmacol 149:233–240
Stahl GL, Terashita Z, Lefer AM (1988c) Role of platelet activating factor in propagation of cardiac damage during myocardial ischemia. J Pharmacol Exp Ther 244:898–904
Stahl GL, Bitterman H, Lefer AM (1989) Protective effects of a specific platelet activating factor (PAF) antagonist, WEB 2086, in traumatic shock. Thromb Res 53:327–338
Steer ML, Meldolesi J (1987) The cell biology of experimental pancreatitis. N Engl J Med 316:144–150
Stein BA, O'Neill C (1991) Reduced vascularity in the mouse fallopian tube in early pregnancy. Austral New Zeal Microcircul Society, Proc 6th Symposium (abstr 15)
Stenton SC, Ward C, Duddridge M, Harris A, Palmer JB, Hendrick DJ, Walters EH (1990a) The actions of GR 32191 B, a thromboxane receptor antagonist, on the effects of inhaled PAF on human airways. Clin Exp Allergy 20:311–317

Stenton SC, Court EN, Kingston WP, Goadby P, Kelly CA, Duddridge M, Ward C, Hendrick DJ, Walters EH (1990b) Platelet-activating factor in bronchoalveolar fluid from asthmatic subjects. Eur Respir J 3:408–413
Steward IG, Piper PG (1986) Platelet-activating factor induced vasoconstriction in rat isolated perfused hearts: contribution of cyclooxygenase and lipoxygenase arachidonic acid metabolites. Pharmacol Res Commun 18:163–172
Stimler NP, Bloor CM, Hugli TE, Wykle RL, McGall CE, O'Flaherty JT (1981) Anaphylactic actions of platelet activating factor. Am J Pathol 105:64–69
Strahan ME, Graham RM, Eccleston DS, Sturm MJ, Taylor RR (1995) Neutrophil platelet-activating-factor production and acetyltransferase activity in clinical acute myocardial infarction. Clin Exp Pharmacol Physiol 22:102–106
Stufler M, Methner A, von Bruchhausen F (1992) Auranofin (AN) but not aurothioglucose (ATG) or aurothiosulfate (ATS) inhibits PAF production of human neutrophils. Arch Pharmacol 346:R42
Suematsu M, Kurose I, Asako H, Miura S, Tsuchiya M (1989) In vivo visualization of oxyradical-dependent photoemission during endothelium-granulocyte interaction in microvascular beds treated with platelet-activating factor. J Biochem 106:355–360
Sueoka K, Dharmarajan AM, Miyazaki T, Atlas SJ, Wallach EE (1988) Platelet-activating factor-induced early pregnancy factor activity from the perfused rabbit ovary and oviduct. Am J Obstet Gynecol 159:1580–1584
Sugatani J, Fujimura K, Miwa M, Mizuno T, Sameshima Y, Saito K (1989) Occurence of platelet activating factor (PAF) in normal rat stomach and alteration of PAF level by water immersion stress. FASEB J 3:65–70
Sugatani J, Fujimura K, Mizuno T, Sameshima Y, Saito K (1991a) Review: the role of PAF in the pathogenesis of gastric ulcers. Prostaglandins Leukot Essent Fatty Acids 44:135–147
Sugatani J, Hughes KT, Miwa M (1991b) Determination of PAF in biological samples by scintillation proximity radioimmuno assay. PAF Symposium on Allergic, Respiratory and Cardiovascular Diseases, Tokyo, pp 49–50
Sugatani J, Fujimura K, Miwa M, Satouchi K, Saito K (1991c) Molecular heterogeneity of platelet-activating factor (PAF) in normal rat glandular stomach determined by gas chromatography/mass spectrometry: alteration in molecular species of PAF by water-immersion stress. Lipids 26:1347–1353
Sugatani J, Miwa M, Komiyama Y, Murakami T (1993) Quantitative analysis of platelet-activating factor in human plasma. Application to patients with liver cirrhosis and disseminated intravascular coagulation. J Immunol Methods 166:251–261
Summers JB, Albert DH (1995) Platelet activating factor antagonists. Adv Pharmacol 32:67–168
Sun J, Hsueh W (1988) Bowel necrosis induced by tumor necrosis factor in rats is mediated by platelet-activating factor. J Clin Invest 81:1328–1331
Sun XM, MacKendrick W, Tien J, Huang W, Caplan MS, Hsueh W (1995) Endogenous bacterial toxins are required for the injurious action of platelet-activating factor in rat. Gastroenterology 109:83–88
Suzuki M, Inauen W, Kvietys PR, Grisham MB, Meininger C, Schelling ME, Granger HJ, Granger DN (1989) Superoxide mediates reperfusion-induced leukocyte-endothelial cell interactions. Am J Physiol 257:H1740–H1745
Szabo C, Wu C-C, Mitchell JA, Gross SS, Thiemermann C, Vane JR (1993) Platelet-activating factor contributes to the induction of nitric oxide synthase by bacterial lipopolysaccharide. Circ Res 73:991–999
Tabor BL, Lewis JF, Ikegami M, Jobe AH (1992) Platelet-activating factor antagonists decrease lung protein leak in preterm ventilated rabbits. Am J Obstet Gynecol 167:810–814
Tagesson C, Lindahl M, Otamiri T (1988) BN52021 ameliorates mucosal damage associated with small intestinal ischaemia in rats. In: Braquet P (ed) Ginkgolides Chemistry, biology, clinical science. Prous, Barcelona, pp 553–561

Takada Y, Boudjema K, Jaeck D, Bel-Haouri M, Doghmi M, Chenard MP, Wolff P, Cinqualbrie J (1995) Effects of platelet-activating factor antagonist on preservation reperfusion injury of the graft in porcine orthotopic liver transplantation. Transplantation 59:10–15

Talstad I, Rootwelt K, Gjone E (1973) Thrombocytosis in ulcerative colitis and Crohn's disease. Scand J Gastroenterol 8:135–138

Tam FWK, Claque J, Dixon CM, Shuttle AW, Henderson BL, Peters AM, Lavender JR, Ind PW (1992) Inhaled platelet-activating factor causes pulmonary neutrophil sequestration in normal humans. Am Rev Respir Dis 146:1003–1008

Tan ND, Davidson D (1995) Comparative differences and combined effects of interleukin-8, leukotriene B-4, and platelet-activating factor on neutrophil chemotaxis of the newborn. Pediatr Res 38:11–16

Tang HM, Teshima DY, Lum BKB (1993) Effects of the PAF antagonists bepafant and L-659989 in endotoxic and septic shock. Drug Dev Res 29:216–221

Taniguchi H, Iwasaka T, Takayama Y, Sugiura T, Inada M (1992) Role of platelet-activating factor in pulmonary edema after coronary ligation in dogs. Chest 102:1245–1250

Tence M, Polansky J, Le Couedic JP, Benveniste J (1980) Release, purification and characterization of platelet activating factor (PAF). Biochimie 62:251–259

Teng CM, Lin CH, Kuo HP, Ko FN, Ishii H, Ishikawa T, Chen IS (1995) Effect of CIS 19, a novel PAF receptor antagonist, on PAF-induced eosinophil recruitment and enhancement of superoxide anion generation in guinea pigs. Arch Pharmacol 351:529–534

Terashita Z, Imura Y, Nishikawa K, Sumida S (1985) Is platelet activating factor (PAF) a mediator of endotoxin shock? Eur J Pharmacol 109:257–261

Terashita Z, Stahl GL, Lefer AM (1988) Protective effects of platelet activating factor (PAF) antagonist and its combined treatment with prostaglandin (PG) E1 in traumatic shock. J Cardiovasc Pharmacol 12:505–511

Terashita Z, Shibouta Y, Imura Y, Iwasaki K, Nishikawa K (1989) Endothelin-induced sudden death and the possible involvement of platelet activating factor (PAF). Life Sci 45:1911–1918

Tessner TG, O'Flaherty JT, Wykle RL (1989) Stimulation of platelet-activating factor synthesis by a nonmetabolizable bioactive analog of platelet-activating factor and influence of arachidonic acid metabolites. J Biol Chem 264:4794–4799

Tetta C, Bussolino F, Modena V, Montrucchio G, Segolini G, Pescarmona G, Camussi G (1990) Release of platelet-activating factor in systemic lupus erythematosus. Int Arch Allergy Appl Immunol 91:244–256

Thompson PJ, Hanson JM, Bilani H, Turner-Warwick M, Morley J (1984) Platelets, platelet-activating factor, and asthma. Am Rev Respir Dis 129:3

Thyssen E, Turk J, Bohrer A, Stenton WF (1996) Quantification of distinct molecular species of platelet activating factor in ulcerative colitis. Lipids 31[Suppl]:S255–S259

Tiberghien C, Laurent L, Junier MP, Dray F (1991) A competitive receptor binding for platelet-activating factor (PAF): Quantification of PAF in rat brain. J Lipid Mediat 3:249–266

Tjoelker LW, Wilder C, Eberhardt C, Stafforini DM, Dietsch G, Schimpf B, Hooper S, Trong HL, Cousens LS, Zimmerman GA, Yamada Y, McIntyre TM, Prescott SM, Gray PW (1995) Anti-inflammatory properties of platelet-activating factor acetylhydrolase. Nature 374:549–553

Tokumura A, Kamiyasu K, Takauchi K, Tsukatani H (1987) Evidence for existence of various homologues and analogues of platelet activating factor in a lipid extract of bovine brain. Biochem Biophys Res Commun 145:415–425

Tokumura A, Yotsumoto T, Hoshikawa T, Tanaka T, Tsukatani H (1992) Quantitative analysis of platelet-activating factor in rat brain. Life Sci 51:303–308

Tokuyama K, Lotvall J, Barnes PJ, Chung KF (1991) Airway narrowing after inhalation of platelet-activating factor in guinea-pig: Contribution of airway microvascular leakage and edema. Am Rev Respir Dis 143:1345–1349

Tolins JP, Vercelotti GM, Wilkowske M, Ha B, Jacob HS, Raij L (1989) Role of platelet activating factor in endotoxemic acute renal failure in the male rat. J Lab Clin Med 113:316–324
Tomeo AC, Egan RW, Duran WN (1991) Priming interactions between platelet activating factor and histamine in the in vivo microcirculation. FASEB J 5:2850–2855
Torley LW, Pickett WC, Carrol MI, Kohler CA, Schaub RE, Wissner A, de Joy SQ, Oronsky AL, Kerwar SS (1992) Studies on the effect of a platelet activating factor antagonist, CL 184605, in animal models of gram-negative bacterial sepsis. Antimicrob Agents Chemother 36:1971–1977
Townsend GC, Scheld WM (1994) Platelet-activating factor augments meningeal inflammation elicited by Haemophilus influenzae lipooligosaccharide in an animal model of meningitis. Infect Immun 62:3739–3744
Travis SPL, Jewell DP (1992) Regional differences in the response to platelet-activating factor in rabbit colon. Clin Sci 82:673–680
Travis SPL, Crotty B, Jewell DP (1995) Site of action of platelet-activating factor within the mucosa of rabbit distal colon. Clin Sci 88:51–57
Tsukioka K, Matsuzaki M, Nakamata M, Kayahara H (1993) Increased plasma levels of platelet-activating factor (PAF) and low serum PAF acethydrolase (PAFAH) activity in adult patients with bronchial asthma. Arerugi 42:167–171
Uhlig S, Wollin L, Wendel A (1994) Contributions of thromboxane and leukotrienes to PAF-induced impairments of lung function in the rat. J Appl Physiol 77:262–269
Ukena D, Dent G, Birke BW, Robaut C, Sybrecht GW, Barnes PJ (1988) Radioligand binding of antagonists of platelet activating factor to intact human platelets. FEBS Lett 228:285–289
Ukena D, Krögel C, Dent G, Yulawa T, Sybrecht G, Barnes PJ (1989) PAF receptors on eosinophils: identification with a novel ligand, (3H) WEB 2086. Biochem Pharmacol 38:1702–1705
Urbaschek R, Urbaschek B (1987) Tumor necrosis factor and interleukin 1 as mediators of endotoxin-induced beneficial effects. Rev Infect Dis 9:S607–S615
Vadas P, Pruzanski W (1986) Biology of disease: role of secretory phospholipase A 2 in the pathology of disease. Lab Invest 55:391–404
Valone F, Skalit M, Atkins P, Goetzl E, Zweiman B (1987) Platelet activating factor release in allergic skin sites in humans. J Allergy Clin Immunol 79:248
Valone FH, Epstein LB (1988) Biphasic platelet-activating factor (PAF) synthesis by human monocytes stimulated with interleukin 1 beta (IL-1), tumor necrosis factor (TNF) or gamma interferon (IFN). FASEB J 2:A878
van Dijk APM, Wilson JHP, Zijlstra FJ (1993) The effect of malotilate, a derivative of malotilate and a flavonoid on eicosanoid production in inflammatory bowel disease in rats. Mediat Inflamm 2:67
Vandongen R (1991) Platelet activating factor and the circulation. J Hypertens 9:771–778
Vanhoutte PM, Rubanyi GM, Miller VM, Houston DS (1986) Modulation of vascular smooth muscle contraction by the endothelium. Annu Rev Physiol 48:307–320
van Oosterhout AJ, Nijkamp FP (1993) Role of cytokines in bronchial hyperresponsiveness. Pulm Pharmacol 6:225–236
Vargaftig BB (1995) PAF. A mediator in search of a disease. Clin Rev Allergy 12:419–421
Vargaftig BB, Braquet P (1987) PAF-acether today: relevance for acute experimental anaphylaxis. Br Med Bull 43:312–335
Vargaftig BB, Lefort J, Chignard M, Benveniste J (1980) Platelet-activating factor induces a platelet-dependent bronchoconstriction unrelated to the formation of prostaglandin derivatives. Eur J Pharmacol 65:185–192
Vassalli P (1992) The pathophysiology of tumor necrosis factors. Am Rev Immunol 10:411–452
Vela L, Garcia-Merino A, Fernandez-Gallardo S, Sanchez-Crespo M, Lopez-Lozano JJ, Saus C (1991) PAF antagonists do not protect against the development of experimental autoimmune encephalomyelitis. J Neuroimmunol 33:81–86

Velasquez LA, Aguilera JG, Croxatto HB (1995) Possible role of platelet-activating factor in embryonic signaling during oviductal transport in the hamster. Biol Reprod 52:1302–1306

Verbey NL, van Delft JL, van Haeringen NJ, Braquet P (1989) Platelet-activating factor and laser trauma of the iris. Invest Ophthalmol Vis Sci 30:1101–1103

Vincent JE, Bonta IL, Zijlstra FJ (1978) Accumulation of blood platelets in carrageenin rat paw edema. Possible role in the inflammatory process. Agents Actions 8:291–295

Voelkel NF, Worthen S, Reeves JT, Henson PM, Murphy RC (1982) Non-immunological production of leukotrienes induced by platelet-activating factor. Science 218:286–288

Voelkel NF, Chang SW, Pfeffer KD, Worthen SG, McMurthy IF, Henson PM (1986) PAF antagonists: different effects on platelets, neutrophils, guinea pig ileum and PAF induced vasodilation in isolated rat lungs. Prostaglandins 32:359–372

von Bruchhausen F (1991) Lyso-PAF transacetylase and PAF-acethydrolase in C6-glioma cells. Arch Pharmacol 344:R104

von Bruchhausen F, Rochel M (1990) Bismuth(III) salts and carbenoxolone are potent inhibitors of lyso-PAF transacetylase in vitro. Arch Pharmacol 341:R70

Waga I, Nakamura M, Honda Z, Terby I, Toyoshima S, Ishiguro S, Shimizu T (1993) Two distinct signal transduction pathways for the activation of guinea-pig macrophages and neutrophils by endotoxin. Biochem Biophys Res Commun 197:465–472

Wainwright CC, Parratt JR, Bigaud M (1989) The effects of PAF antagonists on arrhythmias and platelets during acute myocardial ishaemia and reperfusion. Eur Heart J 10:235–243

Walker C, Virchow JC, Bruijnzeel PL, Blaser K (1991) T cell subsets and their soluble products regulate eosinophilia in allergic and nonallergic asthma. J Immunol 146:1829–1835

Wallace JL (1990) Lipid mediators of inflammation in gastric ulcer. Am J Physiol 258:G1–G11

Wallace JL, MacNaughton WK (1988) Gastrointestinal damage induced by platelet-activating factor: role of leukotrienes. Eur J Pharmacol 151:43–50

Wallace JL, Whittle (1986a) Picomole doses of platelet-activating factor predispose the gastric mucosa to damage by topical irritants. Prostaglandins 31:989–998

Wallace JL, Whittle BJR (1986b) Profile of gasrointestinal damage induced by platelet-activating factor. Prostaglandins 32:137–141

Wallace JL, Whittle BJR (1986c) Prevention of endotoxin-induced gastrointestinal damage by CV-3988, an antagonist of platelet-activating factor. Eur J Pharmacol 124:209–210

Wallace JL, Steel G, Whittle BJR, Lagente V, Vargaftig B (1987a) Evidence for platelet-activating factor as a mediator of endotoxin-induced gastrointestinal damage in the rat. Gastroenterology 93:765–773

Wallace JL, Steel G, Whittle BJR (1987b) Gastrointestinal plasma leakage in endotoxic shock. Inhibition by prostaglandin E2 and by platelet-activating factor antagonist. Can J Physiol Pharmacol 65:1428–1432

Wallace IL, MacNaughton WK, Guarner F, Rodriguez R, Malagelada JR (1988) Role of leukotrienes as mediators of gastric damage and hemoconcentration induced by platelet activating factor. Gastroenterology 94:A485

Wallace JL, Cirino G, DeNucci G, McKnight W, MacNaughton WK (1989) Endothelin has potent ulcerogenic and vascular constrictor actions in the stomach. Am J Physiol 256:G661–G666

Wallace JL, Hogaboam GM, McKnight GW (1990a) Platelet-activating factor mediates gastric damage induced by hemorrhagic shock. Am J Physiol 259:G140–G146

Wallace JL, Keenan CM, Granger DN (1990b) Gastric ulceration induced by nonsteroidal anti-inflammatory drugs is a neutrophil-dependent process. Am J Physiol 259:G462–G467

Wang HY, Yue TL, Feuerstein G, Friedman E (1994) Platelet-activating factor: diminished acetylcholine release from rat brain slices is mediated by a G(i) protein. J Neurochem 63:1720–1725
Wang J, Dunn MT (1987) Platelet-activating factor mediates endotoxin-induced acute renal insufficiency in rats. Am J Physiol 253:F1283–F1289
Wang JM, Rambaldi A, Biondi A, Chen ZG, Sanderson CJ, Mantovani A (1989) Recombinant human interleukin 5 a selective eosinophil attractant. Eur J Immunol 19:701–705
Wang R, Sikla SC, Veeraragavan K, Bell M, Hellström WJ (1993) Platelet activating factor and pentoxifylline as human sperm cryoprotectants. Fertil Steril 60:711–715
Ward PA, Warren JS, Varani J, Johnson KJ (1991) PAF, cytokines, toxic oxygen products and cell injury. Mol Aspects Med 12:169–174
Wardlaw AJ, Moqbel R, Cromwell O, Kay AB (1986) Platelet activating factor. A potent chemotactic and chemokinetic factor for human eosinophils. J Clin Invest 78:1701–1706
Wardle TD, Hall L, Turnberg LA (1996) Platelet activating factor: release from colonic mucosa in patients with ulcerative colitis and its effect on colonic secretion. Gut 38:355–361
Warren JS, Mandel DM, Johnson KJ, Ward PA (1989) Evidence for the role of platelet-activating factor in immune complex vasculitis in the rat. J Clin Invest 83:669–678
Watanabe K, Yano S, Kojima T (1988) Formation of gastric mucosal lesions by platelet activating factor (PAF): comparison of the effects of drugs on gastric lesions induced by hypovolumic shock and by PAF. In: Ogura Y, Kisara K (eds) Trends pharmacol res on platelet activating factor (PAF) in Japan. Ishiyaku, Tokio, pp 170–181
Watanabe M, Sugidachi A, Omata M, Hirasawa N, Mue S, Tsurufuji S, Ohuchi K (1990) Possible role for platelet-activating factor in neutrophil infiltration in allergic inflammation in rats. Int Arch Allergy Appl Immunol 92:396–400
Watanabe M, Yagi M, Omata M, Hirasawa N, Mue S, Tsurufugi S, Ohuchi K (1991) Stimulation of neutrophil adherence to vascular endothelia cells by histamine and thrombin and its inhibition by PAF antagonists and dexamethasone. Br J Pharmacol 102:239–245
Weissman D, Poli G, Bousseau A, Fauci AS (1993) A platelet-activating factor antagonist, RP 55778, inhibits cytokine-dependent induction of human immunodeficiency virus expression in chronically infected promonocytic cells. Proc Natl Acad Sci USA 90:2537–2541
Weisman SM, Felsen D, Vaughan ED (1985) Platelet-activating factor is a potent stimulus for renal prostaglandin synthesis: possible significance in unilateral ureteral obstruction. J Pharmacol Exp Ther 235:10–15
Wells X-E, O'Neill C (1994) Detection and preliminary characterization of two enzymes involved in biosynthesis of platelet-activating factor in mouse oozytes, zygotes and preimplantation embryos: dithiothreitol-insensitive cytidinediphosphocholine: 1-*O*-alkyl-2-acetyl-*sn*-glycerol choline phosphotransferase and acetyl-coenzyme A: 1-O-alkyl-2-lyso-*sn*-glycero-3-phosphocholine acetyltransferase. J Reprod Fertil 101:385–391
Whatley RE, Nelson P, Zimmerman GA, Stevens DL, Parker CJ, McIntyre TM, Prescott SM (1989) The regulation of platelet-activating factor production in endothelial cells. The role of calcium and protein kinase C. J Biol Chem 264:6325–6333
Whittle BJR, Morishita T, Ohya Y, Leung FW, Guth PH (1986) Microvascular actions of platelet-activating factor on rat mucosa and submucosa. Am J Physiol 251:G772–G778
Whittle BJR, Boughton-Smith NK, Hutcheson IR, Esplugues V, Wallace JL (1987a) Increased intestinal formation of PAF in endotoxin-induced damage in the rat. Br J Pharmacol 92:3–4

Whittle BJR, Kauffman GL, Wallace JL (1987b) Gastric vascular and mucosal damaging actions of platelet activating factor in the canine stomach. Adv Prostaglandin Thromboxane Leukot Res 17:285–292

Whittle BJR, Lopez-Belmonte J, Moncada S (1990) Regulation of gastric mucosal integrity by endogenous nitric oxide: interactions with prostanoids and sensory neuropeptides in the rat. Br J Pharmacol 99:607–611

Wilkens H, Wilkens JH, Bosse S (1991) Effects of inhaled PAF-antagonist (WEB 2086 BS) on allergen-induced early and late asthmatic responses and increased bronchial responsiveness to methacholine. Am Rev Respir Dis 143:A812

Williams TJ, Helewell PG, Jose PJ (1986) Inflammatory mechanisms in the Arthus reaction. Agents Actions 19:66–72

Wisner J, Green D, Ferrell L, Renner I (1988) Evidence for a role of oxygen free radicals on the pathogenesis of cerulein induced acute pancreatitis in rats. Gut 29:1516–1523

Wong CW, Seow WK, O'Callaghan JW, Thong YH (1992) Comparative effects of tetrandrine and berbamine on subcutaneous air pouch inflammation induced by interleukin-1, tumour necrosis factor and platelet-activating factor. Agents Actions 36:112–118

Woodard DS, Ostrom KK, McManus LM (1995) Lipid inhibitors of platelet-activating factor (PAF) in normal human plasma. J Lipid Mediat Cell Signal 12:11–28

Woodward DF, Nieves AL, Spada CS, Williams LS, Tuckett RP (1995) Characterization of a behavioral model for peripherically evoked itch suggests platelet-activating factor as a potent pruritogen. J Pharmacol Exp Ther 272:758–765

Yaacob HB, Gong NH, Shahimi MM, Piper PJ (1995) The release of leukotrienes and the vascular changes mediated by platelet activating factor and cardiac anaphylaxis. Asia Pac J Pharmacol 10:25–32

Yamamoto H, Nagata M, Tabe K, Kimura I, Kiuchi H, Sakamoto Y, Yamamoto K, Dohi Y (1993) The evidence of platelet activation in bronchial asthma. J Allergy Clin Immunol 91:79–87

Yang Y-Q, Kudolo GB, Harper MJK (1992) Binding of PAF to oviduct membranes during early pregnancy in the rabbit. J Lipid Mediat 5:77–96

Yasuda K, Satouchi K, Saito K (1986) Platelet-activating factor in normal rat uterus. Biochem Biophys Res Commun 138:1231–1236

Yasuda K, Furukawa M, Johnston JM (1996) Effects of estrogens on plasma platelet-activating factor acetylhydrolase and the timing of parturation in the rat. Biol Reprod 54:224–229

Yoo J, Schlondorff D, Neugarten J (1991) Protective effects of specific platelet-activating-factor receptor antagonists in experimental glomerulonephritis. J Pharmacol Exp Ther 256:841–844

Yoshida H, Satoh K, Imaizumi T (1992) Platelet-activating factor acetylhydrolase activity in red blood cell-stroma from patients with cerebral thrombosis. Acta Neurol Scand 86:199–203

Yoshida H, Satoh K, Takamatsu S (1993) Platelet-activating factor acetylhydrolase in red cell membranes. Does decreased activity impair erythrocyte deformability in ischemic stroke patients? Stroke 24:14–18

Yoshida N, Granger DN, Anderson DC, Koietys PR (1992) Anoxia/reoxygenation induced neutrophil adherence to cultured endothelial cells. Am J Physiol 262:H1891–H1898

Yoshikawa T, Naito Y, Ueda S, Oyamada H, Takemura T, Yoshida N, Sugino S, Kondo M (1990) Role of oxygen derived free radicals in the pathogenesis of gastric mucosal lesions in rat. J Clin Gastroenterol 12[Suppl 1]:S65–S71

Yotsumoto F, Manabe T, Kyogoku T, Hirano T, Oshio G, Yamamoto M, Yoshitomi S (1994) Platelet-activating factor involvement in the aggravation of acute pancreatitis in rabbits. Digestion 55:260–267

Yue TL, Farhat M, Rabinovici R, Pavera PY, Vogel SN, Feuerstein G (1990a) Protective effect of BN 50739, a new platelet activating factor antagonist in endotoxin treated rabbits. J Pharmacol Exp Ther 254:976–981
Yue TL, Lysko PG, Feuerstein G (1990b) Production of platelet-activating factor from rat cerebellar granule cells in culture. J Neurochem 54:1809–1811
Yue TL, Gleason MM, Gu JL, Lysko PG, Hallenbeck J, Feuerstein G (1991) Platelet-activating factor (PAF) receptor-mediated calcium mobilization and phosphoinositide turnover in neurohybrid NG 108–15 cells: studies with BN 50739, a new PAF antagonist. J Pharmacol Exp Ther 257:374–381
Yukawa T, Kroegel C, Evans P, Fukuda T, Barnes PJ (1989) Density heterogeneity of eosinophils: Induction of hypodense eosinophils by platelet activating factor. Immunology 68:140–143
Yukawa T, Read RC, Kroegel C, Rutman A, Chung KF, Wilson R, Cole PJ, Barnes PJ (1990) The effect of eosinophils and neutrophils on guinea-pig airway epithelium in vitro. Am J Respir Cell Mol Biol 2:341–354
Zanglis A, Lianos EA (1987) Platelet activating factor biosynthesis and degradation in the rat glomeruli. J Lab Clin Med 110:330–337
Zarco P, Maestre C, Herrero-Beaumont G, Gonzalez E, Garcia-Hoyo R, Navarro FJ, Braquet P, Egido J (1992) Involvement of platelet activating factor and tumour necrosis factor in the pathogenesis of joint inflammation in rabbits. Clin Exp Immunol 88:318–323
Zhou W, Chao W, Levine BA, Olson MS (1990) Evidence for platelet activating factor as a late-phase mediator of chronic pancreatitis in the rat. Am J Pathol 137:1501–1508
Zhou W, Levin BA, Olson MS (1993) Platelet-activating factor: a mediator of pancreatic inflammation during cerulein hyperstimulation. Am J Pathol 142:1504–1512
Zhu Y-P, Hoffman DR, Hwang S-B, Miyaaura S, Johnston JM (1991) Prolongation of parturation in the pregnant rat following treatment with a platelet activating factor receptor antagonist. Biol Reprod 44:39–42
Zijlstra FJ, van Dijk JP, Wilson JH (1993) Increased platelet activating factor synthesis in experimental colitis after diclofenac and 5-amino-salicylic acid. Eur J Pharmacol 249:R1–R2
Zingarelli B, Squadrito F, Bussolino F, Calapai G, Altavila D, Ioculano M, Campo GM, Canale P, Caputi AP (1994) Evidence for a role of platelet activating factor in hypovolemic shock in the rat. J Lipid Mediat Cell Signal 9:123–134

CHAPTER 29

# Therapeutic Aspects of Platelet Pharmacology

F. CATELLA-LAWSON and G.A. FITZGERALD

Pharmacological, angiographic, and biochemical studies indicate that platelets may play a key role in the initiation and evolution of atherosclerosis. They are also key to occlusive complication of the atherosclerotic plaque manifest clinically as unstable angina or myocardial infarction (LEWIS et al. 1983; CAIRNS et al. 1985; AMBROSE et al. 1986; SHERMAN et al. 1986; FITZGERALD et al. 1986; THEROUX et al. 1987).

## Aspirin

The Antiplatelet Trialists' Collaboration (1988) published an overview analysis of 25 randomized trials in 1988. This concluded that antiplatelet therapy significantly reduces the risk of cardiovascular death, nonfatal myocardial infarction (MI) and nonfatal stroke in patients with unstable angina or a past history of myocardial infarction, transient ischemic attack, or stroke by about 25%. A more recent overview by the same Collaborative Group has confirmed the efficacy of platelet inhibition in a broader spectrum of pathological conditions (Antiplatelet Trialists' Collaboration 1994).

Two clinical trials have also assessed the risk-benefit ratio of prolonged treatment with aspirin in healthy individuals at low risk of cardiovascular events. The Physicians' Health Study (Steering Committee of the Physicians' Health Study 1989) randomized more than 22000 U.S. physicians under double-blind conditions to receive either aspirin, at the dose of 325 mg every other day, or placebo. The aspirin group exhibited a 44% reduction in the risk of myocardial infarction, a non-statistically significant increased risk of stroke, and no reduction in vascular death at an average follow-up of 5 years. A similar randomized, but non blinded study was conducted among 5000 male British doctors (PETO et al. 1988). Administration of aspirin, at the daily dose of 500 mg, for approximately 6 years to this apparently healthy population did not result in a significant difference in the incidence of nonfatal myocardial infarction, stroke, or mortality from all cardiovascular causes. When the data from these two trials were combined and analyzed together, antiplatelet therapy produced a small but highly significant reduction in nonfatal myocardial infarction. No significant difference in vascular mortality was observed and a small but significant excess of hemorrhagic stroke was noted. In contrast to the results achieved in high-risk patients, it is currently believed that the benefits of long term aspirin do not clearly outweigh the risks in individuals who have not previously suffered an event. However, it should be noted that both trials were conducted among physicians who had independently decreased their risk factor profile. This was reflected by a vascular mortality rate in both trials which was approximately 15% of that expected in a comparable age and gender matched population at the time. Therefore, a larger sample size or a more prolonged period of observation would have been required to detect a reasonable reduction in this end-point due to aspirin. Several prospective studies have been initiated to test further the hypothetical utility of chronic aspirin administration in healthy individuals.

Aspirin, at a medium daily dosage of 75–325 mg, was the most widely tested antiplatelet treatment in the trials surveyed in the overview analyses by the Antiplatelet Trialist Collaborative Group (1988, 1994). These doses of aspirin, overall, were associated with a 25% reduction in vascular events. Six other antiplatelet regimens (monotherapy or combination) were tested: ticlopidine, dipyridamole, suloctidil, aspirin plus dipyridamole, and aspirin plus sulphinphyrazone. Overall, these regimens appeared to confer similar protection. However, sample size limitations precluded a precise estimate of the relative efficacy of these agents. Limited information was also provided as to the optimum dose and duration of treatment with these drugs.

The clinical benefit of aspirin and other nonsteroidal anti-inflammatory drugs (NSAIDs) derive from inhibition of the enzyme cyclooxygenase (Cox) (VANE 1971; SMITH 1992), the first step in the conversion of arachidonic acid to thromboxane (Tx) $A_2$, a potent stimulator of platelet aggregation and of vasoconstriction (HAMBERG et al. 1975; FITZGERALD 1991). NSAIDs are reversible inhibitors of Cox, while aspirin selectively acetylates the hydroxyl group of a single serine residue at position 529 of the polypepide chain inducing irreversible enzyme inhibition (ROTH et al. 1975; ROTH and MAJERUS 1975; DEWITT and SMITH 1988).

Cox, or prostaglandin-endoperoxide synthase (PGHS), exists in at least two forms (KUJUBU et al. 1991; HLA and NELSON 1992) encoded by separate genes (YOKOYAMA and TANABE 1989). Cox-1 is constitutively expressed in platelets, monocytes, and most tissues, while Cox-2 expression is less widely expressed and is readily induced by cytokines, growth factors, and tumor promoters. Inhibition of Cox-1 may account for the gastric side effects of NSAIDs, as this is the form expressed in the normal GI tract. The factors that regulate Cox-2 imply that inhibition of Cox-2 may explain the antiinflammatory activity of these compounds (MASFERRER et al. 1994) and their speculative role in the modulation of vascular responses to injury (RIMARACHIN et al. 1994).

Currently available NSAIDs inhibit both Cox-1 and Cox-2, with limited selectivity. Experiments in vitro have suggested that aspirin is a more potent inhibitor of Cox-1 than Cox-2, while other NSAIDs, such as diclofenac, acetaminophen and naproxen appear to inhibit both enzymes with equal potency (MITCHELL et al. 1944). However, in vivo antiinflammatory activity cannot be predicted on the basis of in vitro data (LANEUVILLE et al. 1994). More recently, a model of human Cox-2 expression allowing detection of ex vivo pharmacological inhibition has been described (PATRIGNANI et al. 1994). This and other strategies (MCADAM and FITZGERALD 1997) are currently being utilized to develop highly selective inhibitors of Cox-2, with the expectation that these compounds will retain antiinflammatory activity without inducing the adverse effects associated with nonselective NSAID treatment.

Future studies will also clarify whether the antiinflammatory and antithrombotic properties of aspirin are linked to different molecular mechanisms. Preliminary evidence suggests that while Cox inhibition accounts for the antithrombotic efficacy of aspirin, inhibition of the nuclear factor (NF)-kB might contribute to the antiinflammatory activity of aspirin. It has previously been difficult to explain the similar antiinflammatory potency of aspirin and salicylates, given that the latter are relatively weak reversible inhibitors of both Cox-1 and Cox-2 enzymes. It has been shown recently that both salicylates and aspirin inhibit the activation of the inducible transcription factor NF-kB, by preventing degradation of its inhibitor IkB (KOPP and GHOSH 1994). A subsequent study has expanded this observation and has demonstrated that aspirin-induced inhibition of NF-kB mobilization results in impaired monocyte adhesion to endothelial cells (WEBER et al. 1995). However, the relevance of the drug concentrations used in these in vitro experiments to the mechanism of action of aspirin and salicylate in vivo remains to be established.

Inhibition of Cox is not tissue specific and the prevention of platelet $TxA_2$ formation is accompanied by impaired vascular biosynthesis of prostacyclin ($PGI_2$), the major Cox product in endothelial cells and a potent platelet inhibitor.

Selective inhibition of platelet Cox is not possible with presently available oral preparations (FITZGERALD et al. 1993; BRADEN et al. 1991). However platelet selectivity can be achieved by taking advantage of the pharmacokinetics and pharmacodynamics of aspirin. Platelets are anucleate and therefore unable to regenerate the Cox. Thus, after exposure to aspirin, the enzyme recovers with a half-life of 5 days, equivalent to the platelet half-life. Consequently, it is possible to achieve cumulative inhibition of the enzyme in the platelet with small, repeated doses, given orally (PATRIGNANI et al. 1982). In contrast, the nucleated endothelial cell recovers enzyme activity and $PGI_2$ biosynthesis within hours.

Alternatively, circulating platelets may be inhibited by slow, local administration of aspirin, without achieving high systemic levels of the drug. For example, continuous intraduodenal administration of low-dose aspirin inhibits circulating platelets with little aspirin detected systemically. Platelets in the portal circulation are exposed to relatively high concentrations of aspirin (PEDERSEN and FITZGERALD 1984). The inhibited platelets pass into the systemic circulation, whereas the aspirin is rapidly deacetylated to salicylic acid by nonspecific esterases in plasma and in the liver. A controlled release formulation of aspirin 75 mg has been shown to inhibit platelet $TxA_2$ selectively without suppressing systemic $PGI_2$ biosynthesis (CLARKE et al. 1991; Fig. 1). Placebo-controlled studies using this formulation of aspirin are currently underway. They are focussed on the potential utility of aspirin in the prevention of pregnancy induced hypertension (PIH) and in the primary prevention of cardiovascular events in high-risk patients.

Selective inhibition of platelet Cox-1 may also be achieved by applying aspirin to the skin. Dermal aspirin exibits a very low systemic bioavailability and a recent study has shown that daily application of aspirin induces a marked inhibition of platelet Cox, as measured by serum $TxB_2$ (MCADAM et al. 1996). This occurs while preserving vascular cyclooxygenase, as assessed by the modest inhibition of basal or stimulated $PGI_2$ formation (KEIMOWITZ et al. 1993).

## Selective Thromboxane Blockade

Tx receptor (TP) antagonists and Tx synthase inhibitors have also been developed in an attempt to inhibit selectively thromboxane effects while preserving the vasodilatory and anti-platelet properties of $PGI_2$. The effect of selective blockade of $TP_s$ in the prevention of restenosis after percutaneous

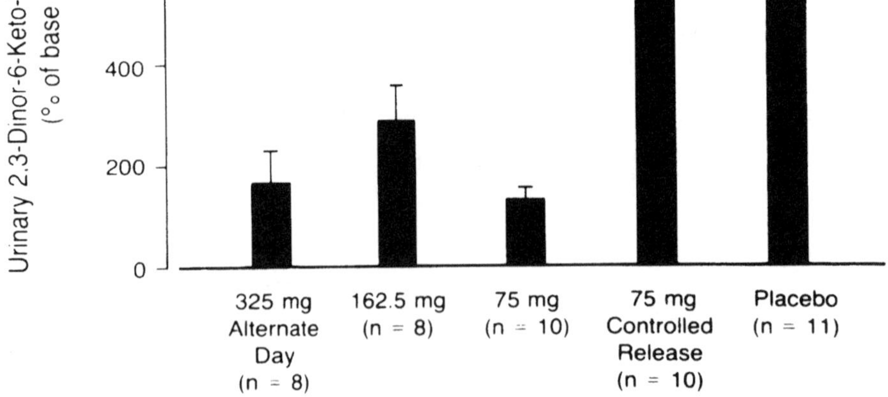


**Fig. 1.** Bradykinin-induced prostacyclin biosynthesis assessed by the urinary excretion of 2,3-dinor-6-keto-prostaglandin $F_{1\alpha}$. This index of systemic prostacyclin biosynthesis was suppressed by all the immediate-release aspirin regimens but was not inhibited after 75 mg controlled-release aspirin. (From Clarke et al. 1991, with permission)

transluminal coronary angioplasty (PTCA) has been investigated. The rationale was that lack of efficacy of aspirin in this indication is related to the inhibition of prostacyclin induced by aspirin. Aspirin alone or in combination with dypiridamole has been proven to be effective in reducing the incidence of periprocedural events in the setting of coronary angioplasty (BOURASSA et al. 1990). However, no prevention of late restenosis has been shown (SCHWARTZ et al. 1988). Long term TP antagonism with GR32191B, failed to prevent restenosis and did not favorably influence the clinical course after angioplasty (SERRUYS et al. 1991). In a subsequent study, sulotraban, another selective blocker of $TP_s$, reduced the risk of myocardial infarction at 6 months after successful PTCA. However, in contrast to aspirin, was ineffective in preventing clinical failure, the primary end-point of the study (SAVAGE et al. 1995).

$TxA_2$ synthase inhibitors have the additional advantage over TP antagonists of increasing the formation of platelet-inhibitory prostaglandins from accumulated endoperoxides: the "endoperoxide shunt" (MARCUS et al. 1980; FITZGERALD et al. 1983; SCHAFER et al. 1984; NOWAK and FITZGERALD 1989). Unfortunately, the endoperoxides also activate $TP_s$ and limit the anti-platelet effects of the synthase inhibitors (FITZGERALD et al. 1985; BERTELE' and DE GAETANO 1982). This explains the synergistic interaction of the combination of a Tx synthase inhibitor with a TP antagonist (GRESELE et al. 1987; FITZGERALD et al. 1988). Therefore, molecules which combine synthase inhibition with receptor antagonism have been developed (HOET et al. 1990). When tested in healthy volunteers. Ridogrel, a potent inhibitor of $TxA_2$ synthase and a relatively weak TP blocker, appeared to exert a stronger antiplatelet effect than aspirin (GRESELE et al. 1987; HOET et al. 1990). However, in a subsequent study performed in patients with myocardial infarction, Ridogrel was not superior to aspirin in enhancing fibrinolysis and preventing rethrombosis (The RAPT Investigators 1994). Potential explanations for the lack of superiority of Ridogrel over aspirin include the insufficient activity of Ridogrel as a TP antagonist and the reduced capacity of atherosclerotic arteries to produce prostacyclin. The need of concomitant thrombin inhibition (no heparin was administered in this study) may also explains these results. The benefits of combining a thrombin inhibitor with ridogrel has been demonstrated in a canine model of thrombosis and reperfusion (YAO et al. 1992a).

## Ticlopidine and Clopidogrel

Aspirin, other NSAIDs, $TxA_2$ synthase inhibitors, and TP antagonists are examples of pharmacological agents that affect the metabolism of arachidonic acid in platelets. Alternatively, platelet inhibition can also be achieved via antagonism of ADP-mediated platelet activation as with the thienopyridines, ticlopidine, and clopidogrel (MAFFRAND et al. 1988; HERBERT et al. 1993). The mechanism of action of these compounds has not been fully elucidated, but appears to be related to inhibition of ADP induced conformational changes of the platelet glycoprotein IIb/IIIa (DUNN et al. 1984; DI MINNO et al. 1985). This is may be consequent to the inhibition of ADP binding to its putative adenylyl cyclase- coupled receptor on platelet (MILLS et al. 1992; SAVI et al. 1994).

The efficacy of ticlopidine for the secondary prevention of stroke has been demonstrated in a placebo-controlled trial (GENT et al. 1989). However, it is controversial whether or not ticlopidine is superior to aspirin in this indication. The Ticlopidine Aspirin Stroke Study (TASS) showed that ticlopidine treatment (250mg twice per day) in patients with transient ischemic attacks is associated with a reduced incidence of stroke and death as compared with aspirin (650mg twice daily). The 3-year event rate was 17% for ticlopidine and 19% for aspirin ($p = 0.048$). However, the 95% confidence interval for the risk reduction embraced zero (–2% to 26%). The suggestion of ticlopidine's superiority was further diminished when vascular mortality, rather than total mortality was examined (HASS et al. 1989). A large number of retrospective subgroup analyses of the TASS study have also claimed the superiority of ticlopidine in the prevention of reversible events (BELLAVANCE 1993), in African-Americans (WEISBERG 1993), in diabetics (GROTTA 1993) and in patients with minor stroke (HARBISON 1992). The questionable reliability of these results is discussed in an editorial by VAN GIJN and ALGRA (1994). The putative superior efficacy of ticlopidine is balanced by an established incidence of serious adverse events (severe diarrhoea and severe neutropenia occur in 2.1% and 0.85% of the patients, respectively; GENT et al. 1989; HASS et al. 1989) and higher cost of ticlopidine, than aspirin (FITZGERALD 1990).

Ticlopidine is routinely administered after intracoronary stent implantation even though limited data are available as to the incidence of stent thrombosis in patients treated with ticlopidine and aspirin versus aspirin alone (HALL et al. 1996). A recent study has compared platelet function after stenting in patients receiving aspirin and conventional anticoagulation therapy or a ticlopidine-aspirin combination (GAWAZ et al. 1996). Surface exposure of LIBS (ligand induced binding sites), an index of the activated and ligand-occupied glycoprotein IIb/IIIa complexes (GINSBERG et al. 1990) and expression of CD62P (P-selectin or GMP-140), a marker of $\alpha$ degranulation (MCEVER 1990), were assessed before and after stenting, daily for 12 days. Increased LIBS and CD62P expression were present in patients receiving anticoagulation therapy, but not in patients receiving ticlopidine and aspirin. This suggested that only combined antiplatelet therapy may have successfully prevented platelet activation associated with the implantation of thrombogenic stent devices. A larger multicenter study is presently investigating the angiographic and clinical implications of these observations.

Clopidogrel, the successor to ticlopidine, is another thienopyridine derivative which not only inhibits ADP induced platelet aggregation, but also effects aggregation mediated by thromboxane (YAO et al. 1992b). Like ticlopidine, clopidogrel affects fibrinogen binding without modifying the platelet glycoprotein IIb/IIIa. Experimental studies have suggested that clopidogrel is more effective than aspirin as adjuvant treatment to prevent reocclusion after thrombolysis (YAO et al. 1994).

The CAPRIE study, a large multicenter study ($n$ = 19185) comparing clopidogrel and aspirin in patients at risk of ischemic events, has recently demonstrated that clopidogrel induces a very slight but significantly greater benefit than aspirin (CAPRIE Steering Committee 1996). Patients treated with clopidogrel had an annual 5.32% risk of ischemic stroke, myocardial infarction, or vascular death (primary end-point) compared with 5.83% with aspirin ($p = 0.043$). No statistically significant difference was observed in the incidence of vascular death or of death from any cause (Table 1). The very modest additional benefit induced by clopidogrel over aspirin translates into the prevention of approximately five major clinical events for each 1000 patients treated for 1 year. This marginal superiority in efficacy may be offset by the differential cost of the two treatments in many countries.

Dipyridamole is another antiplatelet agent whose mechanism of action at clinically tolerated doses is poorly understood. It appears to increase cAMP formation through phosphodiesterase inhibition. However, this effect has been reported only at plasma concentrations of the drug higher than the levels normally achieved during dypiridamole therapy. For many years, this drug was widely used despite little evidence for its clinical efficacy (FITZGERALD 1987). More recently a novel formulation has been developed which allows adminis-

**Table 1.** The CAPRIE study: Primary and secondary outcome clusters, intention to treat analysis. (From CAPRIE STEERING COMMITTEE 1996)

| | Events | Event rate per year (%) | Relative risk reduction (95% CI) | $p$ |
|---|---|---|---|---|
| **Ischemic stroke, MI, or vascular death** | | | | |
| Clopidogrel | 939/17636 | 5.32 | 8.7% (0.3–16.5) | 0.043 |
| Aspirin | 1021/17519 | 5.83 | | |
| **Ischemic stroke, MI, amputation, or vascular death** | | | | |
| Clopidogrel | 979/17594 | 5.56 | 7.6% (−0.8–15.3) | 0.076 |
| Aspirin | 1051/17482 | 6.01 | | |
| **Vascular death** | | | | |
| Clopidogrel | 350/17482 | 1.90 | 7.6% (−6.9–20.1) | 0.29 |
| Aspirin | 378/18354 | 2.06 | | |
| **Fatal stroke, MI, or death from any cause** | | | | |
| Clopidogrel | 1133/17622 | 6.43 | 7.0% (−0.9–14.2) | 0.081 |
| Aspirin | 1207/17501 | 6.90 | | |
| **Death from any cause** | | | | |
| Clopidogrel | 560/18377 | 3.05 | 2.2% (−9.9–12.9) | 0.71 |
| Aspirin | 571/18354 | 3.11 | | |

tration of doses which inhibit whole blood aggregation (MULLER et al. 1990). Recently, a large, multicenter study has shown the superiority of the combination of aspirin with this rational dose and formulation of dypiridamole as compared with aspirin alone in the prevention of cerebrovascular thrombosis (DIENER et al. 1996).

## Fibrinogen Receptor Antagonism

The efficacy of aspirin and other antiplatelet drugs, such as ticlopidine and clopidogrel, must be considered as only partial. These agents inhibit only one pathway of platelet activation, either that mediated by thromboxane or ADP. The activity of other platelet agonists, such as thrombin is unperturbed. More recently, several compounds have been designed to antagonize the binding of fibrinogen to the platelet glycoprotein (Gp) IIb/IIIa ($\alpha_{IIb}\beta_3$) complex. Fibrinogen binding to activated platelets is the final common step in platelet aggregation, regardless of the initiating event (GOGSTAD et al. 1982; PARISE and PHILLIPS 1985; PLOW and GINSBERG 1989). Gp IIb/IIIa is a member of the integrin superfamily of membrane-bound adhesion molecules. Contrasting with other integrins, Gp IIB/IIIa is unique to the platelet surface (PHILLIPS 1991) where it exists in an equilibrium between a resting state, unable to bind adhesive proteins and an active (ligand binding-competent) state (PARISE et al. 1987). Following platelet activation, the Gp IIB/IIIa complex undergoes a conformational change (inside-out signaling), leading to binding of divalent fibrinogen with consequent formation of fibrinogen bridges between platelets. This ultimately leads to platelet aggregation and thrombus formation (SHATTIL et al. 1985; GINSBERG et al. 1992; FOX 1993). The Gp IIb/IIIa–fibrinogen interaction is mediated by the peptide sequence Arg-Gly-Asp (RGD) which is located on the $\alpha$ chain of the fibrinogen molecule (DOOLITTLE et al. 1979), but also present in fibronectin, von Willebrand factor, vitronectin and thrombospondin (RUOSLAHTI and PIERSCHBACHER 1986; D'SOUZA et al. 1991). Another recognition sequence (Lys-Gln-Ala-Gly-Asp-Val) has been described at the carboxyl terminus of the $\gamma$ chain of fibrinogen (KLOCZEWIAK et al. 1984).

The first Gp IIb/IIIa antagonist, a mouse monoclonal antibody known as 7E3, was produced by Coller in 1983. Extensive pre-clinical evaluation demonstrated the great antithrombotic potential of this compound. Following promising results in phase I and II trials, the chimeric (c7E3) monoclonal antibody Fab fragment was tested in a prospective, randomized, double-blind trial of 2099 patients undergoing high-risk coronary angioplasty (the EPIC Investigators, 1994). Patients were randomized to receive either a bolus of c7E3, or a bolus followed by an infusion of c7E3, or placebo. As compared with placebo, the c7E3 bolus and infusion resulted in a 35% reduction in the incidence of death, myocardial infarction, and unplanned revascularization (The EPIC Investigators 1994). Results at six months showed that inhibition of platelet Gp IIb/IIIa was associated with a reduction in events presumed to reflect restenosis. This was assessed by a 23% reduction of ischemic events and a 26%

reduction in the need for subsequent coronary revascularization (TOPOL et al. 1994). The 7E3 antibody (ReoPro) has been approved by the Food and Drug Administration for use in patients undergoing high risk PTCA on the basis of these results.

The EPILOG study has subsequently shown that adjustments in concomitant anticoagulation (use of low-dose, weight-adjusted heparin) and vascular access management (early sheath removal) allow improvement in the safety profile without a concomitant reduction in relative efficacy (TCHENG 1996).

The efficacy of c7E3 Fab treatment was also documented in patients with refractory angina who undergo PTCA. A pilot study performed in 60 patients with dynamic ST-T changes and recurrent pain, despite intensive medical therapy, showed that effective blockade of Gp IIb/IIIa may reduce recurrent ischemia during the waiting period for PTCA and reduce the incidence of death, myocardial infarction, and urgent revascularization during the hospital stay (SIMOONS et al. 1994).

A recently completed study (CAPTURE) in 1400 patients with refractory angina has confirmed the promising results of the phase II trial. The 30 day overall incidence of death, nonfatal myocardial infarction and urgent revascularization was significantly reduced in patients receiving c7E3 Fab (0.25 mg/kg bolus injection followed by an infusion at a rate of 10 $\mu$g/min for 18–24 h) as compared to placebo patients in this randomized, double-blind, placebo controlled study (10.8% vs 16.4%) (VON DE WERF 1996).

The binding of fibrinogen to Gp IIa/IIIb may also be competitively inhibited by synthetic peptides, or by small molecules containing the RGD recognition sequence. Potential advantages of this strategy are expected to be the reversibility of the antiplatelet effect and reduced immunogenicity. Cyclic peptides containing the KGD sequence (where the normal arginine is substituted by a lysine) have also been synthesized to increase the affinity and the specificity. The addition of disulfide bonds and the consequent cyclic configuration imparts higher affinity for the integrin receptor (NICHOLS et al. 1992). The KGD sequence is responsible for the fact that the naturally occurring disintegrin, barbourin, uniquely binds to Gp IIb/IIIa and does not react with other integrins, as do the other natural peptides derived from snake venom (SCARBOROUGH et al. 1991). The potential efficacy of integrelin, a cyclic heptapeptide with a short half-life and high specificity for Gp IIa/IIIb, has been evaluated in low and high risk patients undergoing elective PTCA (TCHENG et al. 1995a). A phase III trial (TCHENG 1995b) in this indication has recently been completed. Patients were randomized to receive one of three regimens: 135 $\mu$g/kg bolus followed by a 0.75 $\mu$g/kg/min infusion, or 135 $\mu$g/kg bolus followed by a 0.5 $\mu$g/kg/min infusion, or placebo. Integrelin treatment was associated with a reduced incidence of death, myocardial infarction and unplanned revascularization during the first 24-h period. However, the incidence of major coronay ischemic events at 30 days (the primary study end-point) was 11.4% in the placebo group, 9.2% in the low-dose integrelin group ($p = 0.06$) and 9.9% ($p = 0.3$) in the group receiving high-dose integrelin.

Retrospective subgroup analyses have been performed to assess if the lack of dose-response curve could be related to the heterogeneity of the study population. For example, high risk patients may require higher doses than the low risk patients (FERGUSON et al. 1996).

The effects of integrelin as an adjunctive therapy to thrombolysis for myocardial infarction have also been evaluated (OHMAN et al. 1997). Unfortunately, the study suffered a major limitation which precluded any final conclusion. Six dose regimens of Integrilin were administered according to a sequential, open-label design. The effect of Integrilin on platelet aggregation in response to ADP 20 $\mu M$ was not dose dependent. However, the blood for the aggregation studies was improperly collected in sodium citrate, a calcium-chelating anticoagulant. This artifactually overestimated the degree of platelet inhibition induced by the drug (RESAR et al. 1996; TCHENG 1996). Thus similar inhibitory effects on platelet aggregation were noticed with four of the six regimens tested, and conclusions could not be extrapolated as to the minimum degree of platelet inhibition necessary for synergistic interaction with tissue plasminogen activator (tPA). The small number of patients evaluated at each dose level (up to 18) and the lack of homogeneity in the concomitant administration of heparin (two groups did not receive a heparin bolus, and heparin infusion was started 60 min after initiation of tPA) also hampered assessment of the dose-response relationship of Integrilin with respect to restoration of coronary blood flow. When the effects of the highest dose of Integrilin were evaluated in a double-blind, placebo-controlled, randomized study, the suggestion of Integrilin superiority over placebo in the openlabel phase was not confirmed (88% patency with Integrilin and 92% with placebo). A phase III study in unstable angina patients (PURSUIT) is currently ongoing.

More recently, synthetic non-peptide, RGD or KGD-mimetic, Gp IIb/IIIa antagonists have been produced. A phase III trial (RESTORE) with Tirofiban (MK-383), a nonpeptide tyrosine derivative antagonist of the RGD binding site, has recently been completed in more than 2000 high risk patients undergoing PTCA. Tirofiban treatment (bolus of 10 $\mu$g/kg followed by an infusion at a rate of 0.15 $\mu$g/kg/min for 36 h) was well tolerated and effective at reducing the early adverse cardiac outcomes following angioplasty. There was a 38% reduction in the composite endpoint (death, myocardial infarction, and repeat revascularization procedures due to abrupt closure or recurrent ischemia) ($p = 0.005$) two days after the procedure (a prespecified analysis). This benefit was mantained for 7 days (27% reduction in the composite endpoint; $p = 0.027$). However, 30 days after angioplasty (the primary endpoint), there was a 16% reduction in events, which was no longer statistically significant (KING et al. 1996).

Two phase III studies (PRISM and PRISM PLUS) investigating the safety and efficacy of tirofiban in patients with unstable angina, have recently been completed, and the results are expected to be presented at the next American College of Cardiology meeting (March 1997).

The reason for the apparently disparate results achieved with the monoclonal antibodies versus the synthetic molecules is presently unclear (Table 2). This could be due to the dose of the various compounds selected for phase III trials or merely reflect different end-point definitions across trials. For example, the RESTORE trial included all unplanned revascularizations, while the EPIC study included only those procedures performed on an emergency basis. When the RESTORE data were analyzed according to the EPIC definition of clinical events, Tirofiban treatment confers a 24% reduction in the composite endpoint at 30 days ($p = 0.052$) (unpublished data).

Patient heterogeneity across trials may also account for the different response rate. WEISS et al. (1996) have recently shown an association between one allele of a polymorphism in the Gp IIIa ($Pl^{A2}$) and the incidence of early coronary events. The possibility that $Pl^{A2}$ positive subjects may respond differently to Gp IIb/IIIa antagonists is presently being investigated.

It is also possible that discordant results reflect a true difference between the antibody and the synthetic compounds. In contrast to the high degree of specificity common to the small peptides and peptido-mimetics, the monoclonal antibody ReoPro (c7E3 Fab) is relatively nonspecific, in that it binds other integrins, including the vitronectin receptor (CHARO et al. 1987). Antagonism of this receptor has been associated with reduced neo-intimal proliferation in an experimental model of balloon angioplasty (CHOI et al. 1994) and has been suggested to account for the reduced restenosis observed in the EPIC trial. Also, the pharmacodynamic effects of the antibodies are long lived, as they do not dissociate from platelets. These properties may represent a virtue, resulting in long term "passivation" of ruptured atherosclerotic plaques, or a vice resulting in a prolonged risk of bleeding in the absence of an antidote.

Lamifiban (RO-44-9883) another non-peptidic GP IIb/IIIa antagonist, has been developed primarily in unstable angina. The Canadian Lamifiban study showed that this drug dose-dependently inhibits platelet aggregation and induces a significant prolongation of bleeding time only when platelet inhibition

**Table 2.** Coronary angioplasty trials: primary outcome events at 30 days

| Study (Treatment) | Active treatment (%) | Placebo (%) | $p$ | % Reduction |
|---|---|---|---|---|
| EPIC (ReoPro) | 8.3[a] | 12.8 | 0.009 | 35 |
| IMPACT II (Integrilin) | 9.9[b] | 11.4 | 0.219 | 13 |
| RESTORE (Tirofiban) | 10.3 | 12.2 | 0.160 | 16 |

[a] c7E3 bolus and infusion.
[b] Integrilin high dose.

is greater than 80%. The reported suggestion of clinical benefit in these patients must be interpreted with caution due to limited sample size (365 patients allocated to five different dose treatments; THEROUX et al. 1996). Unfortunately, despite the study being underpowered the doses for the subsequent phase III trial were selected on the basis of clinical efficacy rather than pharmacodynamic response. The PARAGON study (part A), a double-blind, placebo-controlled, randomized study, investigated two doses of lamifiban in more than 2200 patients with unstable angina and non-Q-wave myocardial infarction. The two infusion doses selected for this phase III trial were 1 and 5 μg/min, a wide dose range for a class of compounds characterized by a very steep dose-response curve. The Canadian Lamifiban study had previously shown that the low-dose (1 μg/min) induces incomplete inhibition of platelet aggregation (approximately 60% inhibition in response to ADP 10 *μM* and 45% inhibition in response to TRAP 10 *μM*) while the high-does (5 μg/min) completely inhibits platelet aggregation and prolongs bleeding time up to ten times. The incidence of major bleeding at the 4 μg/min infusion dose was more than seven times higher than placebo (5.8% vs 0.8%) in this study (THEROUX et al. 1996). Not surprisingly, the PARAGON study reported that the high dose (5 μg/min) induces an unacceptably high incidence of bleeding (Paragon Investigators 1996).

In conclusion, the clinical benefits of parenteral pharmacological blockade of platelet Gp IIb/IIIa receptor have been clearly demonstrated. Future investigations will clarify whether the apparently lower effectiveness of the peptides and nonpeptide mimetics is related to improper dose selection or to intrinsic characteristics of these compounds. Ideally, the treatment of unstable coronary syndromes in the future will involve the combination of intravenous and oral Gp IIb/IIIa antagonism. An initial high doses short term infusion would be followed by long-term therapy with orally bioavailable compounds (KEREIAKES et al. 1996). Several of these latter products are in clinical development (MULLER et al. 1995; CANNON et al. 1996; SIMPFENDORFER et al. 1996), but the risk/benefit ratio of chronic therapy with these agents has not yet been established. Given the extremely steep dose-response relationship exibited by this class of compounds for surrogate markers of efficacy (inhibition of platelet aggregation) and risk (prolongation of bleeding time), development of such a strategy will require extreme care. The optimum degree of platelet inhibition, the need for dose titration and monitoring, and the requirement for concomitant aspirin are all variables which need to be evaluated carefully. Finally, the pharmacoeconomic implications of such a strategy, rather than conventional management with aspirin and heparin, merit consideration.

## References

Ambrose JA, Winters SL, Arora RR, Eng A, Riccio A, Gorlin R, Fuster V (1986) Angiographic evolution of coronary artery morpholgy in unstable angina. J Am Coll Cardiol 7:474–478

Antiplatelet Trialists' Collaboration (1988) Secondary prevention of vascular disease by prolonged antiplatelet treatment. Brit Med J 296:320–331

Antiplatelet Trialists' Collaboration (1994) Collaborative overview of randomised trials of antiplatelet therapy-I: Prevention of death, myocardial infarction and stroke by prolonged antiplatelet therapy in various categories of patients. Brit Med J 308:81–106

Bellavance A, for the Ticlopidine Aspirin Stroke Study Group (1993) Efficacy of ticlopidine and aspirin for the prevention of reversible cerebrovascular ischemic events: the Ticlopidine Aspirin Stroke Study. Stroke 24:1452–1457

Bertele' V, Cerletti C, Schippati A, di Minno G, de Gaetano G (1981) Inhibition of thromboxane synthetase does not necessarily prevent platelet aggregation. Lancet i:1057–1058

Bertele' V, de Gaetano G (1982) Potentiation by dazoxiben, a thromboxane synthetasse inhibitor, of platelet aggregation inhibition activity of a thromboxane receptor antagonist and of prostacyclin. Eur J Pharmacol 85:331–333

Bourassa MG, Schwartz L, Lesperance J, Eastwood C, Kazim F (1990) Prevention of acute complications after percutaneous transluminal coronary angioplasty. Thrombosis Res Suppl 12:51–58

Braden GA, Knapp HR, FitzGerald GA (1991) Suppression of eicosanoid biosynthesis during coronary angioplasty by fish oil and aspirin. Circulation 84:679–685

Cairns JA, Gent M, Singer J, Finnie KJ, Froggatt GM, Holder DA, Jablonsky G, Kostuk WJ, Melendez LJ, Myers MG, Sackett DL, Sealey BJ, Tanser PH (1985) Aspirin, Sufinpyrazone, or both in unstable angina. Results of a Canadian Multicenter Trial. N Engl J Med 313:1369–1375

Cannon CP, Novotny WF, McCabe CH, Tischler MD, Borzac S, Henry TD, Feldman R, Hamilton S, Rothman JM, Braunwald E, and the TIMI Investigators (1996) Evaluation of the oral glycoprotein IIb/IIIa antagonist Ro 48–3657 in patients post acute coronary syndromes: primary results of the TIMI 12 trial. Circulation 94:3231A

CAPRIE Steering Committee (1996) A randomised, blinded, trial of clopidogrel versus aspirin in patients at risk of ischaemic events (CAPRIE). Lancet 348:1329–1339

Charo JF, Bekeart LS, Phillips DR (1987) Platelet glycoprotein IIb/IIIa-like proteins mediate endothelial cell attachment to adhesive proteins and the extracellular matrix. J Biol Chem 262:9935–9938

Choi ET, Engel L, Callow AD, Sun S, Trachtenberg J, Santoro S, Ryan US (1994) Inhibition of neointimal hyperplasia by blocking $\alpha_V\beta_3$ integrin with a small peptide antagonist $G_{pen}$ GRGDSPCA. J Vasc Surg 19:125–134

Clarke RJ, Mayo G, Price P, FitzGerald GA (1991) Suppression of thromboxane $A_2$ but not of systemic prostacyclin by controlled-release aspirin. N Engl J Med 325:1137–1141

Coller BS, Peerschke EI, Scudder LE, Sullivan CA (1983) A murine monoclonal antibody that completely blocks the binding of fibrinogen to platelets produces a thromboasthenic-like state in normal platelets and binds to glycoprotein IIb and/or IIIa. J Clin Invest 72:325–338

DeWitt DL, Smith WL (1988) Primary structure of prostaglandin G/H synthase from sheep vesicular gland determined from the complementary DNA sequence. Proc Natl Acad Sci USA 85:1412–1416 (Erratum: Proc Natl Acad Sci USA 85:5056, 1988)

Di Minno G, Cerbone AM, Mattioli PM, Turco S, Iovine C, Mancini M (1985) Functionally thromboasthenic state in normal platelets following the administration of ticlopidine. J Clin Invest 75:328–338

Diener H, Cuhla L, Forbes C, Sivenius J, Smets P, Lowenthal A (1996) European Stroke Prevention Study 2. Dipyridamole and acetylsalicylic acid in the secondary prevention of stroke. J Neurolog Sci 143:1–13

Doolittle RF, Watt KWK, Cottrell BA et al. (1979) The aminoacid sequence of the $\alpha$ chain of human fibrinogen. Nature 280:464–468
D'Souza SE, Ginsberg MH, Plow EF (1991) Arginyl-glycyl-aspartic acid (RGD): a cell adhesion motif. Trends Biochem Sci 16:246–250
Dunn FW, Soria J, Soria C, Thomaidis A, Lee H, Caen JP (1984) In vivo effect of ticlopidine on fibrinogen-platelet cofactor activity and binding of fibrinogen on platelets. Agents Actions 15:97–103
Ferguson JJ, McDonough TJ, Worley SJ et al. (1996) Clinical outcome of "high-risk" vs "elective" patients undergoing percutaneous coronary intervention: results from IMPACT II. J Am Coll Cardiol 27:180A
Fitzgerald DJ, Catella F, FitzGerald GA (1986) Platelet activation in unstable coronary disease. N Engl J Med 315:983–989
Fitzgerald DJ, Fragetta J, FitzGerald GA (1988) Prostaglandin endoperoxide modulate the response to thromboxane synthase inhibition during coronary thrombosis. J Clin Invest 82:1708–1713
FitzGerald GA, Brash AR, Oates JA, Pedersen AK (1983) Endogenous prostacyclin biosynthesis and platelet function during selective inhibition of thromboxane synthase in man. J Clin Invest 71:1336–1343
FitzGerald GA, Reilly IA, Pedersen AK (1985) The biochemical pharmacology of thromboxane synthase inhibition in man. Circulation 72:1194–1201
FitzGerald GA (1987) Modern drug therapy: dipyridamole. N Engl J Med 316:1247–1257
FitzGerald GA (1990) Ticlopidine in unstable angina. A more expensive aspirin? Circulation 82:296–298
FitzGerald GA (1991) Mechanisms of platelet activation: thromboxane $A_2$ as an amplifying signal for other agonists. Am J Cardiol 68:11B–15B
FitzGerald GA, Oates JA, Hawiger J, Maas RL, Roberts LJ, Brash AR (1993) Endogenous prostacyclin biosynthesis and platelet function during chronic aspirin administration in man. J Clin Invest 71:676–688
Fox JE (1993) The platelet cytoskeleton. Thromb Haemost 70:884–893
Funk CD, Funk LB, Kennedy ME, Pong AS, FitzGerald GA (1991) Human platelet/erithroleukemia cell prostaglandin G/H synthase: cDNA cloning, expression, mutagenesis and gene chromosomal assignment. Faseb J 5:2304–2312
Gawaz M, Neumann FJ, Ott I, May A, Schömig A (1996) Platelet activation and coronary stent implantation. Effect of antithrombotic therapy. Circulation 94:279–285
Gent M, Blakely JA, Easton JD, Ellis DJ, Hachinski VC, Harbison JW, Panak E, Roberts RS, Sicurella J, Turpie AGG, and the CATS Group (1989) The Canadian American Ticlopidine Study (CATS) in thromboembolic stroke. Lancet 6:1215–1220
Ginsberg MH, Frelinger AL, Lam SCT, Forsyth J, McMillan R, Plow EF, Shattil SJ (1990) Analysis of platelet aggregation disorders based on flow cytometric analysis of platelet membrane glycoprotein IIb/IIIa with conformation specific monoclonal antibodies. Blood 76:2017–2023
Ginsberg MH, Du X, Plow EF (1992) Inside-out integrin signalling. Curr Opin Cell Biol 4:766–771
Gogstad GO, Brosstad F, Krutnes MB, Hagen I, Solum NO (1982) Fibrinogen binding properties of the human platelet glycoprotein IIb/IIIa complex: a study using crossed-radioimmunoelectophoresis. Blood 60:663–671
Gresele P, Arnout J, Deckmyn H, Huybrechts E, Pieters G, Vermylen J (1987) Role of proaggregatory and antiaggregatory prostaglandins in hemostasis. Studies with combined thromboxane synthase inhibition and thromboxane receptor antagonism. J Clin Invest 80:1435–1445
Grotta J (1993) Is aspirin effective in preventing strokes in diabetic patients? Stroke 24:760
Hall P, Nakamura S, Maiello L, Itok A, Blengino S, Martini G, Ferraro M, Colombo A (1996) A randomized comparison of combined ticlopidine and aspirin therapy

versus aspirin therapy alone after successful intravascular ultrasound-guided stent implantation. Circulation 93:215–222

Hamberg M, Svensson J, Samuelsson B (1975) Thromboxanes: a new group of biologically active compounds derived from prostaglandin endoperoxides. Proc Natl Acad Sci USA 72:2294–2298

Harbison JW, for the Ticlopidine Aspirin Stroke Study Group (1992) Ticlopidine versus aspirin for the prevention of recurrent stroke: analysis of patients with minor stroke from the Ticlopidine Aspirin Stroke Study. Stroke 23: 1723–1727

Hla T, Neilson K (1992) Human cyclooxygenase-2 cDNA. Proc Natl Acad Sci USA 89:7384–7388

Hass WK, Easton JD, Adams HP, Pryse-Phillips W, Molony BA, Anderson S, Kamm B, and the Ticlopidine Aspirin Stroke Study Group (1989) A randomized trial comparing ticlopidine hydrocloride with aspirin for the prevention of stroke in high-risk patients. N Engl J Med 321:501–507

Herbert JM, Frehel D, Vallee E, Keffer G, Gouy D, Berger Y, Defreyn G, Maffrand JP (1993) Clopidogrel, a novel antiplatelet and antithrombotic agent. Cardiovasc Drug Rev 11:180–198

Hoet B, Falcon C, De Reys S, Arnout J, Deckmyn H, Vermylen J (1990) R68070, a combined thromboxane/endoperoxide receptor antagonist and thromboxane synthase inhibitor, inhibits human platelet activation in vitro and in vivo: a comparison with aspirin. Blood 75:646–653

Keimowitz RM, Pulvermacher G, Mayo G, Fitzgerald DJ (1993) Transdermal modification of platelet function: a dermal aspirin preparation selectively inhibits paltelet cyclooxygenase and preserves prostacyclin biosynthesis. Circulation 88:556–561

Kereiakes DJ, Runyon JP, Kleiman NS et al. (1996) Differential dose-response to oral Xemilofiban after antecedent intravenous Abciximab administration for complex coronary intervention. Circulation 94:906–910

King SB, Willerson JT, Ross AM, Herrmann HC, Lipschutz KH, and the RESTORE Investigators (1996) Time course of reduction in adverse cardiac events following angioplasty using a IIb/IIIa receptor blocker, tirofiban: The restore trial. Circulation 94:1155A

Kopp E, Ghosh S (1994) Inhibition of NF-kB by sodium salicylate and aspirin. Science 265:956–959

Kloczewiak M, Timmons S, Lukas TJ, Hawiger J (1984) Platelet receptor recognition site on human fibrinogen: synthesis and structure-function relationship of peptides corresponding to the carboxy-terminal segment of the gamma chain. Biochemistry 23:1767–1774

Kujubu DA, Fletcher BS, Varnum BC, Lim RW, Herschman HR (1991) TIS-10, a phorbol ester tumor promoter-inducible mRNA from Swiss 3T3 cells, encodes a novel prostaglandin synthase/cyclooxygenase homologue. J Biol Chem 266:12866–12872

Laneuville O, Breuer DK, Dewitt DL, Hla T, Funk CD, Smith WL (1994) Differential inhibition of human prostaglandin endoperoxide H synthases-1 and -2 by nonsteroidal anti-inflammatory drugs. J Pharmacol Exp Ther 271:927–934

Lewis HD, Davis JW, Archibald DG, Steinke WE, Smitherman TC, Doherty JE, Schnaper HW, LeWinter MM, Linares E, Pouget JM, Sabharwal SC, Chelser E, DeMots H (1983) Protective effects of aspirin against acute myocardial infarction and death in men with unstable angina. N Engl J Med 309:396–403

Maffrand JP, Defreyn G, Bernat A, Delebasse D, Tissinier AM (1988) Reviewed pharmacology of ticlopidine. Angiology 77:6–13

Marcus AJ, Weksler BB, Jaffe EA, Broekman MJ (1980) Synthesis of prostacyclin from platelet-derived endoperoxides by cultured human endothelial cells. J Clin Invest 66:979–986

Masferrer JL, Zweifel BS, Manning PT, Hauser SD, Leahy KM, Smith WG, Isakson PC, Seibert K (1994) Selective inhibition of inducible cyclooxygenase 2 in vivo is antiinflammatory and nonulcerogenic. Proc Natl Acad Sci USA 91:3228–3232

McAdam B, Keimowtz RM, Maher M, Fitzgerald DJ (1996) Transdermal modification of platelet function: an aspirin patch system results im marked suppression of platelet cyclooxygenase. J Pharmacol Exp Therap 277:559–564
McAdam BV, FitzGerald GA (1997) Enzymatic regulation of the prostaglandin response in a human model of inflammation. In: Bazan N, Vane JR (eds) New targets in inflammation. Kluwer, Lancaster
McEver RP (1990) The clinical significance of platelet membrane glycoproteins. Hematol Oncol Clin North Am 4:87–103
Mills DCB, Puri R, Hu CJ, Minniti C, Grana C, Freedman MD, Colman RF, Colman RW (1992) Clopidogrel inhibits the binding of ADP analogues to the receptor mediating inhibition of platelet adenylate cyclase. Arterioscl Thrombosis 12:430–436
Mitchell JA, AKarasereenont P, Thiemermann C, Flower RJ, Vane JR (1994) Selectivity of nonsteroidal antiinflammatory drugs as inhibitors of constitutive and inducible cyclooxygenase. Proc Natl Acad Sci USA 90:11693–11697
Muller TH, Su CA, Weisenberger H, Brickl R, Nehmiz G, Eisert WG (1990) Dipyridamole alone or combined with low-dose acetylsalicylic acid inhibits platelet aggregation in human whole blood ex vivo. Br J Clin Pharmacol 30:179–186
Muller TH, Weisenberger H, Brickl R, Kirchner M, Narjes H, Himmelsbach F, Guth B, Krause J (1995) Pharmaco-dynamics and kinetics of BIBU 52, a platelet glycoprotein (GP) IIb/IIIa antagonist and its orally active prodrug BIBU 104 in man. Thromb Haem 73:2081A
Nichols AJ, Ruffolo RR, Huffman WF, Poste G, Samanen J (1992) Development of Gp IIb/IIIa antagonists as antithrombotic drugs. Trends Pharmacol Sci 13:413–417
Nowak J, FitzGerald GA (1989) Redirection of prostaglandin endoperoxide metabolism at the platelet-vascular interface in man. J Clin Invest 83:380–385
Ohman EM, Kleiman NS, Gacioch G, Worley SJ, Navetta FI, Talley JD, Anderson HV, Ellis SG, Cohen MD, Spriggs D, Miller M, Kereiakes D, Yakubov S, Kitt MM, Sigmon KN, Califf RM, Krucoff MW, Topol EJ, for the IMPACT-AMI Investigators (1997) Combined accelerated tissue-plasminogen activator and platelet glycoprotein IIb/IIIa integrin receptor blockade with Integrilin in acute myocardial infarction: results of a randomized, placebo-controlled, dose-ranging trial. Circulation (in press)
PARAGON Investigators (1996) A randomized trial of potent platelet IIb/IIIa antagonism, heparin or both in patients with unstable angina. The PARAGON study. Circulation 94:3234A
Parise LV, Phillips DR (1985) Reconstitution of the purified platelet fibrinogen receptor: fibrinogen binding properties of the glycoprotein IIb/IIIa complex. J Biol Chem 260:10698–10707
Parise LV, Helgerson SL, Steiner B, Nannizzi L, Phillips DR (1987) Synthetic peptides derived from fibrinogen and fibronectin change the conformation of purified platelet glycoprotein IIb/IIIa. J Biol Chem 262:12597–12602
Patrignani P, Filabozzi P, Patrono C (1982) Selective cumulative inhibition of platelet thromboxane production by low-dose aspirin in healthy subjects. J Clin Invest 69:1366–1372
Patrignani P, Panara MR, Greco A, Fusco O, Natoli C, Iacobelli S, Cipollone F, Ganci A, Creminon C, Maclouf J, Patrono C (1994) Biochemical and pharmacological characterization of the cyclooxygenase activity of human blood prostaglandin endoperoxide synthases. J Pharmacol Exper Therap 271:1705–1712
Pedersen AK, FitzGerald GA (1984) Dose-related kinetics of aspirin: presystemic acetylation of platelet cyclooxygenase. N Engl J Med 311:1206–1211
Peto R, Gray R, Collins R, Wheatley K, Hennekens C, Jamrozik K, Warlow C, Hafner B, Thompson E, Norton S, Gilliland J, Doll R (1988) Randomised trial of prophylactic daily aspirin in British male doctors. Brit Med 296:313–316
Phillips DR, Charo IF, Scarborough RM (1991) Gp IIb/IIIa: the responsive integrin. Cell 65:359–362

Plow EF, Ginsberg MH (1989) GpIIb/IIIa as a prototypic adhesion receptor. Prog Hemost Thromb 9:117–156

Resar JR, Brinker JA, Gerstenblith G, Blumenthal RS, Dudek A, Coombs VJ, Goldschmidt-Clermont PJ (1996) Disparity of Integrilin inhibition of platelet aggregation and GP IIb/IIIa fibrinogen binding in angioplasty patients. Circulation 94:567A

Rimarachin JA, Jacobson JA, Szabo P, Maclouf J, Creminon C, Weksler BB (1994) Regulation of cyclooxygenase-2 expression in aortic smooth muscle cells. Arteriosclerosis Thrombosis 14:1021–1031

Roth GJ, Majerus PW (1975) The mechanism of the effect of aspirin on platelets. Acetylation of a particular fraction protein. J Clin Invest 56:624–632

Roth GJ, Stanford N, Majerus PW (1975) Acetylation of prostaglandin synthase by aspirin. Proc Natl Acad Sci USA 72:3073–3076

Ruoshlati E, Pierschbacher MD (1986) Arg-Gly-Asp: a versatile cell recognition signal. Cell 44:517–518

Savage MP, Goldberg S, Bove A et al., for the M-HEART II Study Group (1995) Effect of thromboxane $A_2$ blockade on clinical outcome and restenosis after successful coronary angioplasty. Multi-Hospital Eastern Atlantic Restenosis Trial (M-HEART II). Circulation 92:3194–3200

Savi P, Laplace MC, Maffrand JP, Herbert JM (1994) Binding of $^3$H-2-Methylthio ADP to rat platelets. Effect of clopidogrel and ticlopidine. J Pharmacol Exp Ther 269:772–777

Scarborough RM, Rose JW, Hsu MA et al. (1991) Barbourin: a GpIIb/IIIa-specific integrin antagonist from the venom of Sisturus m. barbouri. J Biol Chem 266:9359–9362

Schafer AI, Crawford DD, Gimbrone MA (1984) Unidirectional transfer of prostaglandin endoperoxides between platelets and endothelial cells. J Clin Invest 73:1105–1112

Schwartz L, Bourassa MG, Lesperance J, Aldridge HE, Kazim F, Salvatori VA, Henderson M, Bonan R, David PR (1988) Aspirin and dipyridamole in the prevention of restenosis after percutaneous transluminal coronary angioplasty. N Engl J Med 318:1714–1719

Serruys PW, Rutsch W, Heyndrickx GR, Danchin N, Mast EG, Wijns W, Rensing BJ, Vos J, Stibbe J, for the CARPORT Study Group (1991) Prevention of restenosis after percutaneous transluminal coronary angioplasty with thromboxane $A_2$-receptor blockade. A randomized, double-blind, placebo-controlled trial. Circulation 84:1568–1580

Shattil SJ, Hoxie JA, Cunningham M, Brass LF (1985) Changes in the platelet membrane glycoprotein IIb/IIIa comples during platelet activation. J Biol Chem 260:11107–11114

Sherman CT, Litvack F, Grundfest W, Lee M, Hickey A, Chaux A, Kass R, Blanche C, Matloff J, Morgenstern et al. (1986) Coronary angioscopy in patients with unstable angina pectoris. N Engl J Med 315:913–919

Simoons ML, de Boer JM, van den Brand MJBM, van Miltenburg AJ, Hoorntje JC, Heyndrickx GR, van der Wieken LR, de Bono D, Rutsch W, Schaible TF et al. (1994) Randomized trial of a GP IIb/IIIa platelet receptor blocker in refractory unstable angina. Circulation 89:596–603

Simpfenndorfer C, Kottke-Marchant Kandice, Topol E (1996) First experience with chronic platelet Gp IIb/IIIa receptor blockade: A pilot study of xemilofiban, an orally active antagonist, in unstable angina patients eligible for PTCA. J Am Coll Cardiol 27:242A

Smith WL (1992) Prostanoid biosynthesis and mechanisms of action. Am J Physiol 263:F181–F191

Steering Committee of the Physicians' Health Study Research Group (1989) Final report on the aspirin component of the ongoing physicians' health study. N Engl J Med 321:129–135

Tcheng JE, Harrington RA, Kottke-Marchant K et al. (1995a) Multicenter randomized, double-blind, placebo-controlled trial of the platelet integrin glycoprotein IIb/IIIa blocker Integrelin in elective coronary intervention. Circulation 91:2151–2157

Tcheng JE, Lincoff M, Sigmon KN et al. (1995b) Platelet glycoprotein IIb/IIIa inhibition with Integrelin during percutaneous coronary intervention: The IMPACT II Trial. Circulation 92:2595A

Tcheng JE (1996) Glycoprotein IIb/IIIa receptor inhibitors: putting the EPIC, IMPACT II, RESTORE and EPILOG trials into perspective. Am J Cardiol 78:35–40

The EPIC Investigators (1994) Use of a monoclonal antibody directed against the platelet glycoprotein IIb/IIIa receptor in high-risk coronary angioplasty. N Engl J Med 330:956–961

The RAPT Investigators (1994) Randomized trial of Ridogrel, a combined thromboxane $A_2$ synthase inhibitor and thromboxane $A_2$ / prostaglaglandin endoperoxide receptor antagonist, versus aspirin as adjunct to thrombolysis in patients with acute myocardial infarction. The Ridogrel versus aspirin patency trial (RAPT). Circulation 89:588–595

Théroux P, Latour JG, Leger-Gauthier C, De Lara J (1987) Fibrinopeptide A and platelet factor levels in unstable angina pectoris. Circulation 75:156–162

Théroux P, Kouz S, Roy L et al., on behalf of the Investigators (1996) Platelet membrane receptor glycoprotein IIb/IIIa antagonism in unstable angina. The Canadian Lamifiban Study. Circulation 94:899–905

Topol EJ, Califf RM, Weisman HF, Ellis SG, Tcheng JE, Worley S, Ivanhoe R, George BS, Fintel D, Weston M, Sigmon K, Anderson KM, Lee KL, Willerson JT, The EPIC Investigators (1994) Randomised trial of coronary intervention with antibody against platelet IIb/IIIa integrin for reduction of clinical restenosis: results at six months. Lancet 343:881–886

Van de Werf F (1996) More evidence for a beneficial effect of platelet glycoprotein IIb/IIIa blockade during coronary interventions. Latest results from the EPILOG and CAPTURE trials. Eur Heart J 17:325–326

Vane JR (1971) Inhibition of prostaglandin synthesis as a mechanism of action for aspirin-like drugs. Nature 231:232–235

van Gijn J, Algra A (1994) Ticlopidine: trials and torture. Stroke 25:1097–1098

Weber C, Erl W, Pietsch A, Weber PC (1995) Aspirin inhibits nuclear factor-kB mobilization and monocyte adhesion in stimulated human endothelial cells. Circulation 91:1914–1917

Weisberg LA (1993) The efficacy and safety of ticlopidin and aspirin in non-whites: analysis of a patient subgroup from the Ticlopidine Aspirin Stroke Study. Neurology 43:27–31

Weiss E, Bray PF, Tayback M, Schulman SP, Kickler TS, Becher LC, Weiss JL, Gerstenblith G, Goldschmidt-Clermont PJ (1996) A polymorphism of a platelet glycoprotein receptor as an inherited risk factor for coronary thrombosis. N Engl J Med 334:1090–1094

Yao SK, Ober JC, Ferguson JJ, Andersson HV, Maraganore J, Buja LM, Willerson JT (1992a) Combination of inhibition of thrombin and blockade of thromboxane $A_2$ synthetase and receptors enhances thrombolysis and delays reocclusion in canine coronary arteries. Circulation 86:1993–1999

Yao SK, Ober JC, McNatt J, Benedict CR, Rosolowsky M, Anderson HV, Cui K, Maffrand JP, Campbell WB, Buja LM, Willerson JT (1992b) ADP plays an important role in mediating platelet aggregation and cyclic flow variations in vivo in stenosed and endothelium-injured canine coronary arteries. Circ Res 70:39–48

Yao SK, Ober JC, Ferguson JJ, Maffreand JP, Anderson HV, Buja LM, Willerson JT (1994) Clopidogrel is more effective than aspirin as adjuvant treatment to prevent reocclusion after thrombolysis. Am J Physiol 267:H488–H493

Yokoyama C, Tanabe T (1989) Cloning of human gene encoding prostaglandin endoperoxide synthase and primary structure of the enzyme. Biochem Biophys Res Comm 165:888–894

# Subject Index

Kolibri 360 GmbH
Ferdinand-Jühlke-Straße 7
99095 Erfurt, Deutschland
produktsicherheit@kolibri360.de